GATE PREP SERIES

GATE

Graduate Aptitude Test in Engineering

2022

26 Years

Chapter-Wise

Solved Papers

(1996-2021)

Mechanical Engineering

G. K. PUBLICATIONS (P) LTD.

Title : GATE 2022 : Mechanical Engineering 26 Years Chapter-wise Solved Papers (1996-2021)

Language : English

Editor's Name : Vinit Garg

Typeset & Published by :

Career Launcher Infrastructure (P) Ltd.

A-45, Mohan Cooperative Industrial Area, Near Mohan Estate Metro Station, New Delhi - 110044

Marketed by :

G.K. Publications (P) Ltd.

Plot No. 9A, Sector-27A, Mathura Road, Faridabad, Haryana-121003

ISBN : **978-93-90820-92-4**

Printer's Details : Made in India, New Delhi.

For product information :

Visit ***www.gkpublications.com*** or email to ***gkp@gkpublications.com***

GATE
Mechanical Engineering

Contents

Solved Papers (Chapter-Wise)

Verbal Ability

Numerical Ability

Engineering Mathematics

Technical Section

GATE

Graduate Aptitude Test in Engineering

IIT Institutes

GATE 2021 will be conducted by
Indian Institute of Technology, Mumbai

Indian Institute of Technology, Kharagpur

GATE 2020 conducted by
Indian Institute of Technology, Delhi

Indian Institute of Technology, Chennai

Indian Institute of Technology, Guwahati

Indian Institute of Technology, Roorkee

Indian Institute of Technology, Kanpur

Preface

The Graduate Aptitude Test in Engineering (GATE) is an online exam conducted by the IITs for admissions to PG courses in IITs, IISc Bangalore, NITs and many state run universities as well as private universities. Also there are more than 37 PSUs that use GATE score for recruitments. A large number of corporates are also using GATE score as a tool to screen students for placements.

GK Publications is well known as the "publisher of choice" to students preparing for GATE and other technical examinations in the country. We published the first set of books in 1994 when GATE exam, both objective and conventional, was conducted in the paper and pencil environment, and used as a check point for entry to post graduate courses in IITs and IISCs. At that time, students had little access to technology and relied mainly on instructor led learning followed by practice with books available for these examinations.

A lot has changed since then!

Today, GATE is conducted in an online only mode with multiple choice and numerical based questions. The score is valid for three years and is used not only for post graduate courses but is also used by major PSUs for recruitment. Today's students have easy access to technology and the concept of a monologue within the classroom has changed to dialogue where students come prepared with concepts and then discuss topics. They learn a lot of things on the go with their mobile devices and practice for mock tests online.

We, as a leading publisher of GATE books, have also embraced change. Today, our books are no more guides and papers only but come with a fully supported mobile app and a web portal. The mobile app provides access to video lectures, short tests and regular updates about the exam. The web portal in additional to what is available on the app provides full length mock tests to mimic the actual exam and help you gauge your level of preparedness. The combination of practice content in print, video lectures, and short and full length tests on mobile and web makes this product a complete courseware for GATE preparation.

This book includes previous years GATE questions along with detailed solution of each question for better understanding. It will help the GATE aspirants to know an idea about the pattern of questions asked in GATE examination.

We also know that improvement is a never ending process and hence we welcome your suggestions and feedback or spelling and technical errors if any. Please write to us at gkp@gkpublications.com

We hope that our small effort will help you prepare well for the examination.

We wish you all the best!

GK Publications Pvt. Ltd.

About GATE

The Graduate Aptitude Test in Engineering (GATE) conducted by IISc and IITs has emerged as one of the bench mark tests for engineering and science aptitude in facilitating admissions for higher education (M.Tech./Ph.D.) in IITs, IISc and various other Institutes/Universities/Laboratories in India. With the standard and high quality of the GATE examination in 27 disciplines of engineering and science, Humanities and Social Sciences subjects, it identifies the candidate's understanding of a subject and aptitude and eligibility for higher studies. During the last few years, GATE score is also being used as one of the criteria for recruitment in Government Organizations such as Cabinet Secretariat, and National/State Public Sector Undertakings in India. Because of the importance of the GATE examination, the number of candidates taking up GATE exams has increased tremendously. GATE exams are conducted by the IITs and IISc as a computer based test having multiple choice questions and numerical answer type questions. The questions are mostly fundamental, concept based and thought provoking. From 2017 onwards GATE Exam is being held in Bangladesh, Ethiopia, Nepal, Singapore, Sri Lanka and United Arab Emirates. An Institute with various nationalities in its campus widens the horizons of an academic environment. A foreign student brings with him/her a great diversity, culture and wisdom to share. Many GATE qualified candidates are paid scholarships/assistantship, especially funded by Ministry of Human Resources Development, Government of India and by other Ministries. Indian Institute of Technology Bombay is the Organizing Institute for GATE 2021.

Why GATE?

Admission to Post Graduate and Doctoral Programmes

Admission to postgraduate programmes with MHRD and some other government scholarships/ assistantships in engineering colleges/institutes is open to those who qualify through GATE. GATE qualified candidates with Bachelor's degree in Engineering/Technology/Architecture or Master's degree in any branch of Science/Mathematics/Statistics/Computer Applications are eligible for admission to Master/Doctoral programmes in Engineering/Technology/Architecture as well as for Doctoral programmes in relevant branches of Science with MHRD or other government scholarships/ assistantships. Candidates with Master's degree in Engineering/Technology/Architecture may seek admission to relevant Ph.D programmes with scholarship/assistantship without appearing in the GATE examination.

Financial Assistance

A valid GATE score is essential for obtaining financial assistance during Master's programs and direct Doctoral programs in Engineering/Technology/Architecture, and Doctoral programs in relevant branches of Science in Institutes supported by the MHRD or other Government agencies. As per the directives of the MHRD, the following procedure is to be adopted for admission to the post-graduate programs (Master's and Doctoral) with MHRD scholarship/assistantship. Depending upon the norms adopted by a specific institute or department of the Institute, a candidate may be admitted directly into a course based on his/her performance in GATE only **or** based on his/her performance in GATE **and** an admission test/interview conducted by the department to which he/she has applied **and/or** the candidate's academic record. If the candidate is to be selected through test/interview for post-graduate programs, a minimum of 70% weightage will be given to the performance in GATE and the remaining 30% weightage will be given to the candidate's performance in test/interview and/or academic record, as per MHRD guidelines. The admitting institutes could however prescribe a minimum passing percentage of marks in the test/interview. Some colleges/institutes specify GATE qualification as the mandatory requirement even for admission without MHRD scholarship/assistantship.

To avail of the financial assistance (scholarship), the candidate must first secure admission to a program in these Institutes, by a procedure that could vary from institute to institute. Qualification in GATE is also a minimum requirement to apply for various fellowships awarded by many Government organizations. Candidates are advised to seek complete details of admission procedures and availability of MHRD scholarship/assistantship from the concerned admitting institution. The criteria for postgraduate admission with scholarship/assistantship could be different for different institutions. The management of the post-graduate scholarship/assistantship is also the responsibility of the admitting institution. Similarly, reservation of seats under different categories is as per the policies and norms prevailing at the admitting institution and Government of India rules. *GATE offices will usually not entertain any enquiry about admission, reservation of seats and / or award of scholarship / assistantship.*

PSU Recruitments

As many as 37 PSUs are using GATE score for recruitment. It is likely that more number of PSUs may start doing so by next year. Below is the list of PSUs:

MDL, BPCL, GAIL, NLC LTD, CEL, Indian Oil, HPCL, NBPC, NECC, BHEL, WBSEDCL, NTPC, ONGC, Oil India, Power Grid, Cabinet Secretariat, Govt. of India, BAARC, NFL, IPR, PSPCL, PSTCL, DRDO, OPGC Ltd., THDC India Ltd., BBNL, RITES, IRCON, GHECL, NHAI, KRIBHCO, Mumbai Railway Vikas Corporation Ltd. (MRVC Ltd.), National Textile Corporation, Coal India Ltd., BNPM, AAI, NALCO, EdCIL India.

Important :

1. Admissions in IITs/IISc or other Institutes for M.Tech./Ph.D. through GATE scores shall be advertised separately by the Institutes and GATE does not take the responsibility of admissions.
2. Cabinet Secretariat has decided to recruit officers for the post of Senior Field Officer (Tele) (From GATE papers of EC, CS, PH), Senior Research Officer (Crypto) (From GATE papers of EC, CS, MA), Senior Research Officer (S&T) (From GATE papers EC, CS, CY, PH, AE, BT) in the Telecommunication Cadre, Cryptographic Cadre and Science & Technology Unit respectively of Cabinet Secretariat. The details of the scheme of recruitment shall be published in National Newspaper/Employment News by the concerned authority.
3. Some PSUs in India have expressed their interest to utilize GATE scores for their recruitment purpose. The Organizations who intend to utilize GATE scores shall make separate advertisement for this purpose in Newspapers etc.

Who Can Appear for GATE?

Eligibility for GATE

Before starting the application process, the candidate must ensure that he/she meets the eligibility criteria of GATE given in Table.

Eligibility Criteria for GATE 2021

Degree/Program	Qualifying Degree/Examination	Description of Eligible Candidates	Expected Year of Completion
B.E. / B.Tech. / B.Pharm.	Bachelor's degree in Engineering / Technology (4 years after 10+2 or 3 years after B.Sc. / Diploma in Engineering / Technology)	Currently in the 3rd year or higher grade or already completed	2022
B. Arch.	Bachelor's degree of Architecture (5- year course) / Naval Architecture (4- year course) / Planning (4- year course)	Currently in the 3rd year or higher grade or already completed	2023 (for 5-year program), 2022 (for 4-year program)
B.Sc. (Research) / B.S.	Bachelor's degree in Science (Post-Diploma/4 years after 10+2)	Currently in the 3rd year or higher grade or already completed	2022
Pharm. D. (after 10+2)	6 years degree program, consisting of internship or residency training, during third year onwards	Currently in the 3rd/4th/5th/6th year or already completed	2024
M.B.B.S.	Degree holders of M.B.B.S. and those who are in the 5th/6th/7th semester or higher semester of such programme.	5th, 6th, 7th or higher semester or already completed	2022
M. Sc. / M.A. / MCA or equivalent	Master's degree in any branch of Arts/Science/Mathematics/Statistics/ Computer Applications or equivalent	Currently in the first year or higher or already Completed	2022
Int. M.E./ M.Tech. (Post-B.Sc.)	Post-B.Sc Integrated Master's degree programs in Engineering/ Technology (4-year program)	Currently in the 1st / 2nd/3rd/4th year or already completed	2024
Int. M.E./ M.Tech. or Dual Degree (after Diploma or 10+2)	Integrated Master's degree program or Dual Degree program in Engineering/Technology (5-year program)	Currently in the 3rd /4th/5th year or alreadycompleted	2023
B.Sc. / B.A. / B.Com.	Bachelor degree in any branch of Science / Arts / Commerce (3 years program)	Currently in the 3rd year or already completed	2021
Int. M.Sc. / Int. B.S. / M.S.	Integrated M.Sc. or 5-year integrated B.S.-M.S. program	Currently in the 3rd year or higher or already completed	2022
Professional Society Examinations (equivalent to B.E. / B.Tech. / B.Arch.)	B.E./B.Tech./B.Arch. equivalent examinations of Professional Societies, recognized by MoE/UPSC/AICTE (e.g. AMIE by Institution of Engineers-India, AMICE by the Institute of Civil Engineers-India and so on)	Completed Section A or equivalent of such professional courses	NA

In case a candidate has passed one of the qualifying examinations as mentioned above in 2020 or earlier, the candidate has to submit the degree certificate / provisional certificate / course completion certificate / professional certificate / membership certificate issued by the society or institute. In case, the candidate is expected to complete one of the qualifying criteria in 2021 or later as mentioned above, he/she has to submit a certificate from Principal or a copy of marks card for section A of AMIE.

Certificate From Principal

Candidates who have to submit a certificate from their college Principal have to obtain one from his/her institution beforehand and upload the same during the online submission of the application form.

Candidates With Backlogs

Candidates, who have appeared in the final semester/year exam in 2021, but with a backlog (arrears/failed subjects) in any of the papers in their qualifying degree should upload a copy of any one of the mark sheets of the final year,

OR

obtain a declaration from their Principal along with the signature and seal beforehand and upload the same during the online submission of the application form.

GATE Structure

Structure of GATE

GATE 2021 will be conducted on 27 subjects (papers). Table below shows the list of papers and paper codes for GATE 2021. A candidate is allowed to appear in ANY ONE or UP TO TWO papers of the GATE examination. However, note that the combination of TWO papers in which a candidate can appear MUST be from the pre-defined list as given in Table. Also note that for a paper running in multiple sessions, a candidate will be mapped to appear for the examination in one of the sessions ONLY.

List of GATE Papers and Corresponding Codes

GATE Paper	Code	GATE Paper	Code
Aerospace Engineering	AE	Instrumentation Engineering	IN
Agricultural Engineering	AG	Mathematics	MA
Architecture and Planning	AR	Mechanical Engineering	ME
Bio-medical Engineering	BM	Mining Engineering	MN
Biotechnology	BT	Metallurgical Engineering	MT
Civil Engineering	CE	Petroleum Engineering	PE
Chemical Engineering	CH	Physics	PH
Computer Science and Information Technology	CS	Production and Industrial Engineering	PI
Chemistry	CY	Statistics	ST
Electronics and Communication Engineering	EC	Textile Engineering and Fibre Science	TF
Electrical Engineering	EE	Engineering Sciences	XE*
Environmental Science & Engineering	ES	Humanities & Social Sciences	XH**
Ecology and Evolution	EY	Life Sciences	XL***
Geology and Geophysics	GG		

Note: *Environmental Science and Engineering (ES) and Humanities and Social Sciences (XH) are two new papers introduced in GATE 2021.*

*XE Paper Sections	Code	**XH Paper Sections	Code	***XL Paper Sections	Code
Engineering Mathematics (Compulsory) (15 marks)	A	Reasoning and Comprehension (Compulsory) (25 marks)	B1	Chemistry (Compulsory) (25 marks)	P
Any TWO optional Sections (2x35 = 70 marks)		Any ONE optional Section (60 marks)		Any TWO optional Sections (2x30 = 60 marks)	
Fluid Mechanics	B	Economics	C1	Biochemistry	Q
Materials Science	C	English	C2	Botany	R
Solid Mechanics	D	Linguistics	C3	Microbiology	S
Thermodynamics	E	Philosophy	C4	Zoology	T
Polymer Science and Engineering	F	Psychology	C5	Food Technology	U
Food Technology	G	Sociology	C6		
Atmospheric and Oceanic Sciences	H				

*XE (Engineering Sciences), **XH (Humanities & Social Sciences), ***XL (Life Sciences), papers are of general nature and will be comprised of Sections listed in the above table

Note: *Each subject/paper is of total 100 marks. General Aptitude (GA) section of 15 marks is common for all papers. Hence remaining 85 marks are for the respective subject/paper code.*

Combination of Two Papers Allowed to Appear in GATE 2021 (subject to availability of infrastructure and schedule)

Code of The First (Primary) Paper	Codes of Papers Allowed as The Second Paper
AE	XE
AG	ES
AR	CE
BM	BT / XL
BT	BM / XL
CE	AR / ES
CH	CY / PE / XE
CS	MA
CY	CH / XL
EC	IN / PH
EE	IN
ES	AG / CE
EY	XL
GG	MN / PE / PH
IN	EC / EE / PH
MA	CS / PH / ST
ME	XE
MN	GG / XE
MT	PH / XE
PE	CH / GG / XE
PH	EC / GG / IN / MA / MT / ST
PI	XE
ST	MA / PH
TF	XE
XE	AE / CH / ME / MN / MT / PE / PI / TF
XH	----
XL	BM / BT / CY / EY

General Aptitude Questions

All the papers will have a few questions that test the General Aptitude (Language and Analytical Skills), apart from the core subject of the paper.

Duration and Examination Type

All the papers of the GATE 2021 examination will be for 3 hours duration and they consist of 65 questions for a total of 100 marks. Since the examination is an ONLINE computer based test (CBT), at the end of the stipulated time (3-hours), the computer screen will automatically close the examination inhibiting any further action.

Candidates will be permitted to occupy their allotted seats 40 minutes before the scheduled start of the examination. Candidates can login and start reading the instructions 20 minutes before the start of examination. The late login time (if any) recorded by the computer system MUST NOT be beyond 30 minutes from the actual starting time of the examination. Under NO circumstances, will a candidate be permitted to login after 30 minutes from the actual examination starting time. Candidates will NOT be permitted to leave the examination hall before the end of the examination.

Pattern of Question Papers

GATE 2021 may contain questions of THREE different types in all the papers:

(i) **Multiple Choice Questions (MCQ)** carrying 1 or 2 marks each, in all the papers and sections. These questions are objective in nature, and each will have choice of four answers, out of which ONLY ONE choice is correct.

Negative Marking for Wrong Answers: For a wrong answer chosen in a MCQ, there will be negative marking. For 1-mark MCQ, 1/3 mark will be deducted for a wrong answer. Likewise, for 2-mark MCQ, 2/3 mark will be deducted for a wrong answer.

(ii) **Multiple Select Questions (MSQ)** carrying 1 or 2 marks each in all the papers and sections. These questions are objective in nature, and each will have choice of four answers, out of which ONE or MORE than ONE choice(s) is / are correct.

Note: There is NO negative marking for a wrong answer in MSQ questions. However, there is NO partial credit for choosing partially correct combinations of choices or any single wrong choice.

(iii) **Numerical Answer Type (NAT)** Questions carrying 1 or 2 marks each in most of the papers and sections. For these questions, the answer is a signed real number, which needs to be entered by the candidate using the virtual numeric keypad on the monitor (keyboard of the computer will be disabled). No choices will be shown for these types of questions. The answer can be a number such as 10 or –10 (an integer only). The answer may be in decimals as well, for example, 10.1 (one decimal) or 10.01 (two decimals) or –10.001 (three decimals). These questions will be mentioned with, up to which decimal places, the candidates need to present the answer. Also, for some NAT type problems an appropriate range will be considered while evaluating these questions so that the candidate is not unduly penalized due to the usual round-off errors. Candidates are advised to do the rounding off at the end of the calculation (not in between steps). Wherever required and possible, it is better to give NAT answer up to a maximum of three decimal places.

Note: *There is NO negative marking for a wrong answer in NAT questions. Also, there is NO partial credit in NAT questions.*

Marking Scheme – Distribution of Marks and Questions

General Aptitude (GA) Questions

In all papers, GA questions carry a total of 15 marks. The GA section includes 5 questions carrying 1-mark each (sub-total 5 marks) and 5 questions carrying 2-marks each (sub-total 10 marks).

Question Papers other than GG, XE, XH and XL

These papers would contain 25 questions carrying 1-mark each (sub-total 25 marks) and 30 questions carrying 2-marks each (sub-total 60 marks) consisting of some MCQ type questions, while the remaining may be MSQ and / or NAT questions.

GG (Geology and Geophysics) Paper

Apart from the General Aptitude (GA) section, the GG question paper consists of two parts: Part A and Part B. Part A is compulsory for all the candidates. Part B contains two sections: Section 1 (Geology) and Section 2 (Geophysics). Candidates will have to attempt questions in Part A and questions in either Section 1 or Section 2 of Part B. The choice of Section 1 OR Section 2 of Part B has to be made at the time of online registration in GOAPS. At the examination hall, candidate cannot request for change of section.

Part A consists of 25 questions carrying 1-mark each (sub-total 25 marks and some of these will be MSQ and/or numerical answer type questions while remaining questions will be MCQ type). Either section of Part B (Section 1 and Section 2) consists of 30 questions carrying 2-marks each (sub-total 60 marks and some of these will be MSQ and/or numerical answer type questions while remaining questions will be MCQ type).

XE Paper (Engineering Sciences)

A candidate appearing in the XE paper has to answer the following:

- GA – General Aptitude carrying a total of 15 marks.
- Section A – Engineering Mathematics (Compulsory): This section contains 11 questions carrying a total of 15 marks: 7 questions carrying 1-mark each (sub-total 7 marks), and 4 questions carrying 2-marks each (sub-total 8 marks). Some questions will be of numerical answer type while remaining questions will be MCQ type.
- Any two of XE Sections B to H: The choice of two sections from B to H can be made during the examination after viewing the questions. Only TWO optional sections can be answered at a time. A candidate wishing to change midway of the examination to another optional section must first choose to deselect one of the previously chosen optional sections (B to H). Each of the optional sections of the XE paper (Sections B through H) contains 22 questions carrying a total of 35 marks: 9 questions carrying 1-mark each (sub-total 9 marks) and 13 questions carrying 2-marks each (sub-total 26 marks). Some questions will be of MSQ and/or numerical answer type while remaining questions will be MCQ type.

XH Paper (Humanities and Social Sciences)

A candidate appearing in the XH paper has to answer the following:

- **GA – General Aptitude** carrying a total of 15 marks.
- **Section B1 – Reasoning and Comprehension (Compulsory):** This section contains 15 questions carrying a total of 25 marks: 5 questions carrying 1-mark each (sub-total 5 marks) and 10 questions carrying 2-marks each (sub-total 20 marks). Some questions will be of MSQ and/or numerical answer type while remaining questions will be MCQ type.
- **Any ONE of XH Sections C1 to C6:** The ONE choice of section from C1 to C6 has to be made at the time of online registration in GOAPS. At the examination hall, candidate cannot request for change of section. Each of the optional sections of the XH paper (Sections C1 through C6) contains 40 questions carrying a total of 60 marks: 20 questions carrying 1-mark each (sub-total 20 marks) and 20 questions carrying 2-marks each (sub-total 40 marks). Some questions will be of MSQ and/or numerical answer type while remaining questions will be MCQ type.

XL Paper (Life Sciences)

A candidate appearing in the XL paper has to answer the following:

- **GA – General Aptitude** carrying a total of 15 marks.
- **Section P–Chemistry (Compulsory):** This section contains 15 questions carrying a total of 25 marks: 5 questions carrying 1-mark each (sub-total 5 marks) and 10 questions carrying 2-marks each (sub-total 20 marks). Some questions will be of MSQ and/or numerical answer type while remaining questions will be MCQ type.
- **Any two of XL Sections Q to U:** The choice of two sections from Q to U can be made during the examination after viewing the questions. Only TWO optional sections can be answered at a time. A candidate wishing to change midway of the examination to another optional section must first choose to deselect one of the previously chosen optional sections (Q to U). Each of the optional sections of the XL paper (Sections Q through U) contains 20 questions carrying a total of 30 marks: 10 questions carrying 1-mark each (sub-total 10 marks) and 10 questions carrying 2-marks each (sub-total 20 marks). Some questions will be of MSQ and/or numerical answer type while remaining questions will be MCQ type.

GATE Score

After the evaluation of the answers, the actual (raw) marks obtained by a candidate will be considered for computing the GATE Score. For multi-session papers (subjects), raw marks obtained by the candidates in different sessions will be converted to Normalized marks for that particular subject. Thus, raw marks (for single session papers) or normalized marks (for multi-session papers) will be used for computing the GATE Score, based on the qualifying marks.

Calculation of Normalized Marks for Multi-Session Papers

In GATE 2021 examination, some papers may be conducted in multi-sessions. Hence, for these papers, a suitable normalization is applied to take into account any variation in the difficulty levels of the question papers across different sessions. The normalization is done based on the fundamental assumption that "in all multi-session GATE papers, the distribution of abilities of candidates is the same across all the sessions". This assumption is justified since the number of candidates appearing in multi-session papers in GATE 2021 is large and the procedure for allocation of session to candidates is random. Further, it is also ensured that for the same multi-session paper, the number of candidates allotted in each session is of the same order of magnitude.

Based on the above, and considering various normalization methods, the committee arrived at the following formula for calculating the normalized marks for the multi-session papers.

Nsormalization mark of jth candidate in the ith session $\widehat{M_{ij}}$ is given by

$$\widehat{M_{ij}} = \frac{\bar{M}_t^g - M_q^g}{\bar{M}_{ti} - M_{iq}}(M_{ij} - M_{iq}) + M_q^g$$

where

M_{ij} : is the actual marks obtained by the jth candidate in i^{th} session

$\bar{M}_t^g$: is the average marks of the top 0.1% of the candidates considering all sessions

M_q^g : is the sum of mean and standard deviation marks of the candidates in the paper considering all sessions

$\bar{M}_{ti}$: is the average marks of the top 0.1% of the candidates in the ith session

M_{iq} : is the sum of the mean marks and standard deviation of the ith session

Calculation of GATE Score for All Papers

For all papers for which there is only one session, actual marks obtained by the candidates will be used for calculating the GATE 2021 Score. For papers in multi-sessions, normalized marks will be calculated corresponding to the raw marks obtained by a candidate and the GATE 2021 Score will be calculated based on the normalized marks.

The GATE 2021 score will be computed using the formula given below.

$$\text{GATE Score} = S_q + (S_t - S_q)\frac{(M - M_q)}{(\bar{M}_t - M_q)}$$

where

M : marks obtained by the candidate (actual marks for single session papers and normalized marks for multi-session papers)

M_q : is the qualifying marks for general category candidate in the paper

$\bar{M}_t$: is the mean of marks of top 0.1% or top 10 (whichever is larger) of the candidates who appeared in the paper (in case of multi-session papers including all sessions)

S_q : 350, is the score assigned to M_q

S_t : 900, is the score assigned to $\bar{M}_t$

In the GATE 2021 the qualifying marks (M_q) for general category student in each subject will be 25 marks (out of 100) or $\mu + \sigma$, whichever is larger. Here μ is the mean and σ is the standard deviation of marks of all the candidates who appeared in the paper.

After the declaration of results, GATE Scorecards can be downloaded by the GATE qualified candidates ONLY.

The GATE 2021 Committee has the authority to decide the qualifying mark/score for each GATE paper. In case of any claim or dispute with respect to GATE 2021 examination or score, the Courts and Tribunals in Mumbai alone will have the exclusive jurisdiction to entertain and settle them.

GATE Syllabus

SECTION 1: ENGINEERING MATHEMATICS

Linear Algebra: Matrix algebra, systems of linear equations, eigenvalues and eigenvectors.

Calculus: Functions of single variable, limit, continuity and differentiability, mean value theorems, indeterminate forms; evaluation of definite and improper integrals; double and triple integrals; partial derivatives, total derivative, Taylor series (in one and two variables), maxima and minima, Fourier series; gradient, divergence and curl, vector identities, directional derivatives, line, surface and volume integrals, applications of Gauss, Stokes and Green's theorems.

Differential Equations: First order equations (linear and nonlinear); higher order linear differential equations with constant coefficients; Euler-Cauchy equation; initial and boundary value problems; Laplace transforms; solutions of heat, wave and Laplace's equations.

Complex Variables: Analytic functions; Cauchy-Riemann equations; Cauchy's integral theorem and integral formula; Taylor and Laurent series.

Probability and Statistics: Definitions of probability, sampling theorems, conditional probability; mean, median, mode and standard deviation; random variables, binomial, Poisson and normal distributions.

Numerical Methods: Numerical solutions of linear and non-linear algebraic equations; integration by trapezoidal and Simpson's rules; single and multi-step methods for differential equations.

SECTION 2: APPLIED MECHANICS AND DESIGN

Engineering Mechanics: Free-body diagrams and equilibrium; friction and its applications including rolling friction, belt-pulley, brakes, clutches, screw jack, wedge, vehicles, etc.; trusses and frames; virtual work; kinematics and dynamics of rigid bodies in plane motion; impulse and momentum (linear and angular) and energy formulations; Lagrange's equation.

Mechanics of Materials: Stress and strain, elastic constants, Poisson's ratio; Mohr's circle for plane stress and plane strain; thin cylinders; shear force and bending moment diagrams; bending and shear stresses; concept of shear centre; deflection of beams; torsion of circular shafts; Euler's theory of columns; energy methods; thermal stresses; strain gauges and rosettes; testing of materials with universal testing machine; testing of hardness and impact strength.

Theory of Machines: Displacement, velocity and acceleration analysis of plane mechanisms; dynamic analysis of linkages; cams; gears and gear trains; flywheels and governors; balancing of reciprocating and rotating masses; gyroscope.

Vibrations: Free and forced vibration of single degree of freedom systems, effect of damping; vibration isolation; resonance; critical speeds of shafts.

Machine Design: Design for static and dynamic loading; failure theories; fatigue strength and the S-N diagram; principles of the design of machine elements such as bolted, riveted and welded joints; shafts, gears, rolling and sliding contact bearings, brakes and clutches, springs.

SECTION 3: FLUID MECHANICS AND THERMAL SCIENCES

Fluid Mechanics: Fluid properties; fluid statics, forces on submerged bodies, stability of floating bodies; control-volume analysis of mass, momentum and energy; fluid acceleration; differential equations of continuity and momentum; Bernoulli's equation; dimensional analysis; viscous flow of incompressible fluids, boundary layer, elementary turbulent flow, flow through pipes, head losses in pipes, bends and fittings; basics of compressible fluid flow.

Heat-Transfer: Modes of heat transfer; one dimensional heat conduction, resistance concept and electrical analogy, heat transfer through fins; unsteady heat conduction, lumped parameter system, Heisler's charts; thermal boundary layer, dimensionless parameters in free and forced convective heat transfer, heat transfer correlations for flow over flat plates and through pipes, effect of turbulence; heat exchanger performance, LMTD and NTU methods; radiative heat transfer, Stefan- Boltzmann law, Wien's displacement law, black and grey surfaces, view factors, radiation network analysis.

Thermodynamics: Thermodynamic systems and processes; properties of pure substances, behavior of ideal and real gases; zeroth and first laws of thermodynamics, calculation of work and heat in various processes; second law of thermodynamics; thermodynamic property charts and tables, availability and irreversibility; thermodynamic relations.

Applications: Power Engineering: Air and gas compressors; vapour and gas power cycles, concepts of regeneration and reheat. I.C. Engines: Air-standard Otto, Diesel and dual cycles. Refrigeration and air-conditioning: Vapour and gas refrigeration and heat pump cycles; properties of moist air, psychrometric chart, basic psychrometric processes. Turbomachinery: Impulse and reaction principles, velocity diagrams, Pelton-wheel, Francis and Kaplan turbines; steam and gas turbines.

SECTION 4: MATERIALS, MANUFACTURING AND INDUSTRIAL ENGINEERING

Engineering Materials: Structure and properties of engineering materials, phase diagrams, heat treatment, stress-strain diagrams for engineering materials.

Casting, Forming and Joining Processes: Different types of castings, design of patterns, moulds and cores; solidification and cooling; riser and gating design. Plastic deformation and yield criteria; fundamentals of hot and cold working processes; load estimation for bulk (forging, rolling, extrusion, drawing) and sheet (shearing, deep drawing, bending) metal forming processes; principles of powder metallurgy. Principles of welding, brazing, soldering and adhesive bonding.

Machining and Machine Tool Operations: Mechanics of machining; basic machine tools; single and multi-point cutting tools, tool geometry and materials, tool life and wear; economics of machining; principles of non-traditional machining processes; principles of work holding, jigs and fixtures; abrasive machining processes; NC/CNC machines and CNC programming.

Metrology and Inspection: Limits, fits and tolerances; linear and angular measurements; comparators; interferometry; form and finish measurement; alignment and testing methods; tolerance analysis in manufacturing and assembly; concepts of coordinate-measuring machine (CMM).

Computer Integrated Manufacturing: Basic concepts of CAD/CAM and their integration tools; additive manufacturing.

Production Planning and Control: Forecasting models, aggregate production planning, scheduling, materials requirement planning; lean manufacturing.

Inventory Control: Deterministic models; safety stock inventory control systems.

Operations Research: Linear programming, simplex method, transportation, assignment, network flow models, simple queuing models, PERT and CPM.

Chapter-Wise Analysis

GATE PAPERS (Mechanical Engineering)

Subject	2011	2012	2013	2014	2015	2016	2017	2018	2019	2020	2021
1. Engineering Maths											
1 mark Questions	6	5	5	5	6	5	5	5	4	6	5
2 marks Questions	4	5	5	7	3	4	4	5	4	4	3
Total Marks	**14**	**15**	**15**	**19**	**12**	**13**	**13**	**15**	**12**	**14**	**11**
2. Engineering Mechanics											
1 mark Questions	2	1	0	0	0	2	1	1	2	1	0
2 marks Questions	2	3	1	1	4	2	1	3	2	0	2
Total Marks	**6**	**7**	**2**	**2**	**8**	**6**	**3**	**7**	**6**	**1**	**4**
3. Strength of Materials											
1 mark Questions	4	3	4	2	4	2	4	3	5	1	4
2 marks Questions	5	4	1	3	4	4	3	4	5	3	2
Total Marks	**14**	**11**	**6**	**8**	**12**	**10**	**10**	**11**	**15**	**7**	**8**
4. Theory of Machines											
1 mark Questions	0	3	2	4	2	2	2	3	3	4	3
2 marks Questions	4	1	3	3	4	3	3	4	4	3	3
Total Marks	**8**	**5**	**8**	**10**	**10**	**8**	**8**	**11**	**11**	**10**	**9**
5. Design of Machine Elements											
1 mark Questions	0	0	2	2	1	1	1	1	0	1	0
2 marks Questions	0	0	3	2	2	1	1	1	0	2	3
Total Marks	**0**	**0**	**8**	**6**	**5**	**3**	**3**	**3**	**0**	**5**	**6**
6. Fluid Mechanics											
1 mark Questions	1	1	2	2	3	3	4	3	3	2	2
2 marks Questions	2	2	2	2	2	4	3	4	2	3	2
Total Marks	**5**	**5**	**6**	**6**	**7**	**11**	**10**	**11**	**7**	**8**	**6**
7. Heat Transfer											
1 mark Questions	2	2	2	1	2	1	1	0	0	2	2
2 marks Questions	1	2	4	2	1	3	3	1	2	1	2
Total Marks	**4**	**6**	**10**	**5**	**4**	**7**	**7**	**2**	**4**	**4**	**6**

Subject	2011	2012	2013	2014	2015	2016	2017	2018	2019	2020	2021
8. Thermodynamics, IC Engines, Refrigeration & Power Plant Engineering Air-Conditioning											
1 mark Questions	2	3	2	3	2	3	2	2	3	3	4
2 marks Questions	9	5	4	5	5	3	4	3	4	5	4
Total Marks	**20**	**13**	**10**	**13**	**12**	**9**	**10**	**8**	**11**	**13**	**12**
9. Materials, Manufacturing & Industrial Engg.											
1 mark Questions	7	7	6	6	5	6	5	7	5	5	5
2 marks Questions	8	8	7	5	5	6	8	5	7	9	10
Total Marks	**23**	**23**	**20**	**16**	**15**	**18**	**21**	**17**	**19**	**23**	**25**
10. General Aptitude											
1 mark Questions	5	5	5	5	5	5	5	5	5	5	5
2 marks Questions	5	5	5	5	5	5	5	5	5	5	5
Total Marks	**15**	**15**	**15**	**15**	**15**	**15**	**15**	**15**	**15**	**15**	**15**

Verbal Ability

1 CHAPTER English Grammar

MCQ TYPE QUESTIONS

2016

1. Computers were invented for performing only high-end useful computations. However, it is no understatement that they have taken over our world today. The internet, for example, is ubiquitous. Many believe that the internet itself is an unintended consequence of the original invention. With the advent of mobile computing on our phones, a whole new dimension is now enabled. One is left wondering if all these developments are good or, more importantly, required.

Which of the statement(s) below is/are logically valid and can be inferred from the above paragraph?

(*i*) The author believes that computers are not good for us.

(*ii*) Mobile computers and the internet are both intended inventions

(*a*) (*i*) only (*b*) (*ii*) only

(*c*) both (*i*) and (*ii*) (*d*) neither (*i*) nor (*ii*)

2. All hill-stations have a lake. Ooty has two lakes.

Which of the statement(s) below is/are logically valid and can be inferred from the above sentences?

(*i*) Ooty is not a hill-station.

(*ii*) No hill-station can have more than one lake.

(*a*) (*i*) only (*b*) (*ii*) only

(*c*) both (*i*) and (*ii*) (*d*) neither (*i*) nor (*ii*)

3. Identify the correct spelling out of the given options :

(*a*) Managable (*b*) Manageable

(*c*) Mangaeble (*d*) Managible

4. Choose the statement(s) where the underlined word is used correctly :

(*i*) A prone is a dried plum.

(*ii*) He was lying prone on the floor.

(*iii*) People who eat a lot of fat are prone to heart disease.

(*a*) (*i*) and (*iii*) only (*b*) (*iii*) only

(*c*) (*i*) and (*ii*) only (*d*) (*ii*) and (*iii*) only

5. The policeman asked the victim of a theft, "What did you ?"

(*a*) loose (*b*) lose

(*c*) loss (*d*) louse

6. In a world filled with uncertainty, he was glad to have many good friends. He had always assisted them in times of need and was confident that they would reciprocate. However, the events of the last week proved him wrong.

Which of the following inference(s) is/are logically valid and can be inferred from the above passage?

(*i*) His friends were always asking him to help them.

(*ii*) He felt that when in need of help, his friends would let him down.

(*iii*) He was sure that his friends would help him when in need.

(*iv*) His friends did not help him last week.

(*a*) (*i*) and (*ii*) (*b*) (*iii*) and (*iv*)

(*c*) (*iii*) only (*d*) (*iv*) only

7. An apple costs Rs. 10. An onion costs Rs. 8.

Select the most suitable sentence with respect to grammar and usage.

(*a*) The price of an apple is greater than an onion.

(*b*) The price of an apple is more than onion.

(*c*) The price of an apple is greater than that of an onion.

(*d*) Apples are more costlier than onions.

8. He <u>turned a deaf ear</u> to my request. What does the underlined phrasal verb mean?

(*a*) ignored (*b*) appreciated

(*c*) twisted (*d*) returned

2015

9. Choose the statement where underlined word is used correctly

(*a*) The minister <u>insured</u> the victims that everything would be all right.

(*b*) He <u>ensured</u> that the company will not have to bear any loss.

(*c*) The actor got himself <u>ensured</u> against any accident.

(*d*) The teacher <u>insured</u> students of good results

10. The following question presents a sentence, part of which is underlined. Beneath the sentence you find four ways of phrasing the underline part. Following the requirements of the standard written English, select the answer that produces the most effective sentence.

Tuberculosis, together with its effects, ranks one of the leading causes of death in India.

(*a*) ranks as one of the leading causes of death

(*b*) rank as one of the leading causes of death

(*c*) has the rank of one of the leading causes of death

(*d*) are one of the leading causes of death

11. Choose the statement where underlined word is used correctly.

(*a*) When the teacher eludes to different authors, he is being elusive

(*b*) When the thief keeps eluding the police, he is being elusive

(*c*) Matters that are difficult to understand, identify or remember are allusive

(*d*) Mirages can be allusive, but a better way to express them is illusory

12. Fill in the blank with the correct idiom/phrase.

That boy from the town was a __________ in the sleepy village.

(*a*) Dog out of herd (*b*) Sheep from the heap

(*c*) Fish out of water (*d*) Bird from the flock

13. Select the appropriate option in place of underlined part of the sentence.

Increased productivity necessary reflects greater efforts made by the employees.

(*a*) Increase in productivity necessary

(*b*) Increase productivity is necessary

(*c*) Increase in productivity necessarily

(*d*) No improvement required

14. Ram and Shyam shared a secret and promised to each other that it would remain between them. Ram expressed himself in one of the following ways as given in the choices below. Identify the correct way as per standard English.

(*a*) It would remain between you and me

(*b*) It would remain between me and you

(*c*) It would remain between you and I

(*d*) It would remain with me

15. What is the adverb for the given word below? Misogynous

(*a*) Misogynousness (*b*) Misogynity

(*c*) Misogynously (*d*) Misogynous

16. In the following sentence certain parts are underlined and marked P, Q and R. One of the parts may contain certain error or may not be acceptable in standard written communication. Select the part containing an error. Choose D as your Answer: if there is no error.

The student corrected all the errors (P) that the instructor marked (Q) on the answer book (R)

(*a*) P (*b*) Q

(*c*) R (*d*) No Error

17. Choose the statement where underlined word is used correctly.

(*a*) The industrialist had a personnel jet.

(*b*) I write my experience in my personnel diary.

(*c*) All personnel are being given the day off.

(*d*) Being religious is a personnel aspect.

18. Out of the following four sentences, select the most suitable sentence with respect to grammer and usage:

(*a*) Since the report lacked needed information, it was of no use to them.

(*b*) The report was useless to them because there were no needed information in it.

(*c*) Since the report did not contain the needed information, it was not real useful to them.

(*d*) Since the report lacked needed information, it would not had been useful to them.

19. Select the alternative meaning of the underlined part of the sentence.

The chain snatchers took to their heels when the police party arrived.

(*a*) took shelter in a thick jungle

(*b*) open indiscriminate fire

(*c*) took to flight

(*d*) unconditionally surrendered

2014

20. Which of the following options is the closest in meaning to the word underlined in the sentence below?

In a democracy, everybody has the freedom to disagree with the government.

(*a*) dissent (*b*) descent

(*c*) decent (*d*) decadent

21. 'Advice' is ________.

(*a*) a verb (*b*) a noun

(*c*) an adjective (*d*) both a verb and a noun

22. Which of the options given below best completes the following sentence?

She will feel much better if she ______.

(*a*) will get some rest
(*b*) gets some rest
(*c*) will be getting some rest
(*d*) is getting some rest

23. While trying to collect (I) an envelope from under the table, (II) Mr. X fell down (III) and was losing consciousness. (IV)

Which one of the above underlined parts of the sentence is NOT appropriate?

(*a*) I (*b*) II
(*c*) III (*d*) IV

24. Which of the following options is the closest in meaning to the phrase underlined in the sentence below?

It is fascinating to see life forms cope with varied environmental conditions.

(*a*) Adopt to (*b*) Adapt to
(*c*) Adept in (*d*) Accept with

25. In a press meet on the recent scam, the minister said, "The buck stops here". What did the minister convey by the statement?

(*a*) He wants all the money
(*b*) He will return the money
(*c*) He will assume final responsibility
(*d*) He will resist all enquiries

2013

26. The professor (I) ordered to (II) the students to go (III) out of the class. (IV)

Which of the above underlined parts of the sentence is grammatically incorrect?

(*a*) I (*b*) II
(*c*) III (*d*) IV

27. Choose the grammatically INCORRECT sentence:

(*a*) He is of Asian origin.
(*b*) They belonged to Africa.
(*c*) She is an European.
(*d*) They migrated from India to Australia.

28. Choose the grammatically CORRECT sentence:

(*a*) Two and two add four.
(*b*) Two and two become four.
(*c*) Two and two are four.
(*d*) Two and two make four.

29. All engineering students (I) should learn mechanics, (II) mathematics and (III) how to do computation. (IV)

Which of the above underlined parts of the sentence is not appropriate?

(*a*) I (*b*) II
(*c*) III (*d*) IV

2012

30. Choose the most appropriate alternative from the options given below to complete the following sentence :

Suresh's dog is the one ______ was hurt in the stampede.

(*a*) that (*b*) which
(*c*) who (*d*) whom

31. Choose the grammatically **INCORRECT** sentence :

(*a*) They gave us the money back less the service charges of Three Hundred rupees.
(*b*) This country's expenditure is not less than that of Bangladesh.
(*c*) The committee initially asked for a funding of Fifty Lakh rupeesr, but later settled for a lesser sum.
(*d*) This country's expenditure on educational reforms is very less.

32. One of the parts (A, B, C, D) in the sentence given below contains an ERROR. Which one of the following is **INCORRECT** ?

I requested that he should be given the driving test today instead of tomorrow.

(*a*) requested that
(*b*) should be given
(*c*) the driving test
(*d*) instead of tomorrow

ANSWERS

MCQ Type Questions

1. (*d*)	**2.** (*d*)	**3.** (*b*)	**4.** (*d*)	**5.** (*b*)	**6.** (*b*)	**7.** (*c*)	**8.** (*a*)	**9.** (*b*)	**10.** (*a*)
11. (*b*)	**12.** (*c*)	**13.** (*c*)	**14.** (*a*)	**15.** (*c*)	**16.** (*b*)	**17.** (*c*)	**18.** (*a*)	**19.** (*c*)	**20.** (*a*)
21. (*b*)	**22.** (*b*)	**23.** (*d*)	**24.** (*b*)	**25.** (*c*)	**26.** (*b*)	**27.** (*c*)	**28.** (*d*)	**29.** (*d*)	**30.** (*a*)
31. (*d*)	**32.** (*b*)								

EXPLANATIONS

MCQ TYPE QUESTIONS

4. 'lying prone' means lie down flat. 'Prone to' means vulnerable to.

5. Lose (verb)

9. *Insured*–the person, group, or organization whose life or property is covered by an insurance policy.
Ensured–to secure or guarantee

11. Elusive: Difficult to answer.

12. From the statement, it appears that boy found it tough to adapt to a very different situation.

16. The is not required in 'Q'.

30. The correct answer is 'that'. In the given sentence 'was hurt in the stampede' determines which dog belongs to Suresh.

■■

2 CHAPTER Sentence Completion

MCQ TYPE QUESTIONS

2016

1. The man who is now Municipal Commissioner worked as ________.

(*a*) the security guard at a university

(*b*) a security guard at the university

(*c*) a security guard at university

(*d*) the security guard at the university

2. Nobody knows how the Indian cricket team is going to <u>cope with</u> the difficult and seamer-friendly wickets in Australia.

Choose the option which is closest in meaning to the underlined phrase in the above sentence.

(*a*) put up with (*b*) put in with

(*c*) put down to (*d*) put up against

3. The chairman requested the aggrieved shareholders to ________ him.

(*a*) bare with (*b*) bore with

(*c*) bear with (*d*) bare

4. The unruly crowd demanded that the accused be ________ without trial.

(*a*) hanged (*b*) hanging

(*c*) hankering (*d*) hung

5. Which of the following is CORRECT with respect to grammar and usage?

Mount Everest is ________.

(*a*) the highest peak in the world

(*b*) highest peak in the world

(*c*) one of highest peak in the world

(*d*) one of the highest peak in the world

6. Despite the new medicine's ________ in treating diabetes, it is not ________widely.

(*a*) effectiveness --- prescribed

(*b*) availability --- used

(*c*) prescription --- available

(*d*) acceptance --- proscribed

7. The students _____ the teacher on teachers' day for twenty years of dedicated teaching.

(*a*) facilitated (*b*) felicitated

(*c*) fantasized (*d*) facillitated

8. After India's cricket world cup victory in 1985, Shrotria who was playing both tennis and cricket till then, decided to concentrate only on cricket. And <u>the rest is history</u>.

What does the underlined phrase mean in this context?

(*a*) history will rest in peace

(*b*) rest is recorded in history books

(*c*) rest is well known

(*d*) rest is archaic

9. If I were you, I ________ that laptop. It's much too expensive.

(*a*) won't buy (*b*) shan't buy

(*c*) wouldn't buy (*d*) would buy

10. Choose the most appropriate set of words from the options given below to complete the following sentence.

_____ ________ is a will, ________ is a way.

(*a*) Wear, there, their (*b*) Were, their, there

(*c*) Where, there, there (*d*) Where, their, their

2015

11. Choose the most appropriate word from the options given below to complete the following sentence. The official answered ________ that the complaints of the citizen would be looked into.

(*a*) respectably (*b*) respectfully

(*c*) reputably (*d*) respectively

12. Choose the most appropriate word from the options given below to complete the following sentence.

The principal presented the chief guest with a ________, as token of appreciation.

(*a*) momento (*b*) memento

(*c*) momentum (*d*) moment

13. Choose the appropriate word/phrase, out of the four options given below, to complete the following sentence : Frogs ________.

(*a*) Croak (*b*) Roar

(*c*) Hiss (*d*) Patter

14. Extreme focus on syllabus and studying for tests has become such a dominant concern of Indian students that they close their minds to anything ______ to the requirements of the exam

(*a*) related (*b*) extraneous

(*c*) outside (*d*) useful

15. The Tamil version of _____ John Abraham-starrer Madras café _____ cleared by the Censor Board with no cuts last week but the film's distributors _____ no takers among the exhibitors for a release in Tamil Nadu _____ this Friday.

(*a*) Mr., was, found, on (*b*) a, was, found, at

(*c*) the, was, found, on (*d*) a, being, find at

16. Choose the appropriate word/phase, out of the four options given below, to complete the following sentence:

Apparent lifelessness _____ dormant life.

(*a*) harbours (*b*) lead to

(*c*) supports (*d*) affects

17. Choose the correct verb to fill in the blank below :

Let us ___________.

(*a*) Introvert (*b*) Alternate

(*c*) Atheist (*d*) Altruist

18. Choose the most appropriate word from the options given below to complete the following sentence :

If the Athlete had wanted to come first in the race, he _________ several hours everyday

(*a*) should practice (*b*) should have practiced

(*c*) practiced (*d*) should be practicing

19. Choose the appropriate word/phrase, out of the four options given below, to complete the following sentence:

Dhoni, as well as the other team members of Indian team, _____ present on the occasion.

(*a*) were (*b*) was

(*c*) has (*d*) have

20. We _______ our friend's birthday and we _____ how to make it up to him.

(*a*) completely forgot — don't just known

(*b*) forget completely — don't just know

(*c*) completely forget —just don't know

(*d*) forgot completely —just don't know

21. Which of the following options is the closest in meaning to the sentence below?

She enjoyed herself immensely at the party.

(*a*) She had a terrible time at the party

(*b*) She had a horrible time at the party

(*c*) She had a terrific time at the party

(*d*) She had a terrifying time at the party

22. Didn't you buy _____ when you went shopping?

(*a*) any paper

(*b*) much paper

(*c*) no paper

(*d*) a few paper

2014

23. Choose the most appropriate word from the options given below to complete the following sentence:

A person suffering from Alzheimer's disease _________ short-term memory loss.

(*a*) experienced (*b*) has experienced

(*c*) is experiencing (*d*) experiences

24. Choose the most appropriate word from the options given below to complete the following sentence:

_________ is the key to their happiness; they are satisfied with what they have.

(*a*) Contentment (*b*) Ambition

(*c*) Perseverance (*d*) Hunger

25. Choose the most appropriate word from the options given below to complete the following sentence:

One of his biggest _____ was his ability to forgive.

(*a*) vice (*b*) virtues

(*c*) choices (*d*) strength

26. After the discussion, Tom said to me, 'Please revert!' He expects me to _______.

(*a*) retract (*b*) get back to him

(*c*) move in reverse (*d*) retreat

27. The value of one U.S. Dollar is 65 Indian Rupees today, compared to 60 last year. The Indian Rupee has _______.

(*a*) Depressed (*b*) Depreciated

(*c*) Appreciated (*d*) Stabilized

28. Choose the most appropriate word from the options given below to complete the following sentence:

Communication and interpersonal skills are _______ important in their own ways.

(*a*) each (*b*) both

(*c*) all (*d*) either

29. Choose the most appropriate pair of words from the options given below to complete the following sentence:

She could not _______ the thought of _______ the election to her bitter rival.

(*a*) bear, loosing (*b*) bare, loosing

(*c*) bear, losing (*d*) bare, losing

30. Choose the most appropriate phrase from the options given below to complete the following sentence:

The aircraft _________ take off as soon as its flight plan was filed.

(*a*) is allowed to (*b*) will be allowed to

(*c*) was allowed to (*d*) has been allowed to

31. Choose the most appropriate word from the options given below to complete the following sentence:

Many ancient cultures attributed disease to supernatural causes. However, modern science has largely helped _____ such notions.

(*a*) impel (*b*) dispel

(*c*) propel (*d*) repel

32. If she _________ how to calibrate the instrument, she _______ done the experiment.

(*a*) knows, will have

(*b*) knew, had

(*c*) had known, could have

(*d*) should have known, would have

33. Who _______ was coming to see us this evening ?

(*a*) you said (*b*) did you say

(*c*) did you say that (*d*) had you said

34. Choose the most appropriate word from the options given below to complete the following sentence:

He could not understand the judges awarding her the first prize, because he thought that her performance was quite ___________.

(*a*) Superb (*b*) Medium

(*c*) Mediocre (*d*) Exhilarating

2013

35. Friendship, no matter how _______ it is, has its limitations.

(*a*) cordial (*b*) intimate

(*c*) secret (*d*) pleasant

36. Were you a bird, you_____ in the sky.

(*a*) would fly (*b*) shall fly

(*c*) should fly (*d*) shall have flown

37. Complete the sentence:

Dare _________ mistakes.

(*a*) commit (*b*) to commit

(*c*) committed (*d*) committing

38. The Headmaster _____ to speak to you.

Which of the following options is incorrect to complete the above sentence?

(*a*) is wanting (*b*) wants

(*c*) want (*d*) was wanting

2012

39. Choose the most appropriate alternative from the options given below to complete the following sentence :

Despite several _____ the mission succeeded in its attempt to resolve the conflict.

(*a*) attempts (*b*) setbacks

(*c*) meetings (*d*) delegations

40. Choose the most appropriate alternative from the options given below to complete the following sentence :

If the tired soldier wanted to lie down, he ___ the mattress out on the balcony.

(*a*) should take (*b*) shall take

(*c*) should have taken (*d*) will have taken

41. Choose the most appropriate word from the options given below to complete the following sentence :

Given the seriousness of the situation that he had to face, his ___ was impressive.

(*a*) beggary (*b*) nomenclature

(*c*) jealousy (*d*) nonchalance

2011

42. Choose the most appropriate word from the options given below to complete the following sentence:

If you are trying to make a strong impression on your audience, you cannot do so by being understated, tentative or ______ .

(*a*) hyperbolic (*b*) restrained

(*c*) argumentative (*d*) indifferent

43. Choose the most appropriate word(s) from the options given below to complete the following sentence.

I contemplated _____ Singapore for my vacation but decided against it.

(*a*) to visit (*b*) having to visit

(*c*) visiting (*d*) for a visit

44. Choose the most appropriate word from the options given below to complete the following sentence:

Under ethical guidelines recently adopted by the Indian Medical Association, human genes are to be manipulated only to correct diseases for which______treatments are unsatisfactory.

(*a*) similar (*b*) most

(*c*) uncommon (*d*) available

45. Choose the most appropriate word from the options given below to complete the following sentence:

It was her view that the country's problems had been_________by foreign technocrats, so that to invite them to come back would be counter-productive.

(*a*) identified

(*b*) ascertained

(*c*) exacerbated

(*d*) analysed

2010

46. Choose the most appropriate word from the options given below to complete the following sentence:

If we manage to _________ our natural resources, we would leave a better planet for our children.

(*a*) uphold (*b*) restrain

(*c*) cherish (*d*) conserve

47. Choose the most appropriate word from the options given below to complete the following sentence:

His rather casual remarks on politics _______his lack of seriousness about the subject.

(*a*) masked (*b*) belied

(*c*) betrayed (*d*) suppressed

ANSWERS

MCQ Type Questions

1. (*b*)	**2.** (*a*)	**3.** (*c*)	**4.** (*a*)	**5.** (*a*)	**6.** (*a*)	**7.** (*b*)	**8.** (*c*)	**9.** (*c*)	**10.** (*c*)
11. (*b*)	**12.** (*b*)	**13.** (*a*)	**14.** (*b*)	**15.** (*c*)	**16.** (*a*)	**17.** (*b*)	**18.** (*b*)	**19.** (*b*)	**20.** (*c*)
21. (*c*)	**22.** (*a*)	**23.** (*d*)	**24.** (*a*)	**25.** (*b*)	**26.** (*b*)	**27.** (*b*)	**28.** (*b*)	**29.** (*c*)	**30.** (*c*)
31. (*b*)	**32.** (*c*)	**33.** (*b*)	**34.** (*c*)	**35.** (*b*)	**36.** (*a*)	**37.** (*b*)	**38.** (*c*)	**39.** (*b*)	**40.** (*a*)
41. (*d*)	**42.** (*b*)	**43.** (*c*)	**44.** (*d*)	**45.** (*c*)	**46.** (*a*)	**47.** (*c*)			

EXPLANATIONS

MCQ TYPE QUESTIONS

2. From the given underlined phrase copewith means put up with.

4. 'hanged' means death by hanging 'hung' is used only with things and not with people.

6. Here 'effectiveness' is noun and 'prescribed' is verb. So these words are apt and befitting with the given word 'medicine'.

12. The principal presented the chief guest with a memento, as token of appreciation.

13. Frogs make 'croak' sound.

14. Extraneous -irrelevant or unrelated to the subject being dealt with.

15. John-Abraham starrer Madras Cafe talks about the movie not the person, so Mr. is ruled out. 'Find no takers' is not the correct phrase. At this Friday is incorrect.

So, option (*c*) is correct.

16. Apparent: looks like; dormant: hidden
Harbour: give shelter; Effect (verb): results in

18. For condition regarding something which already happened, should have practiced is the correct choice.

23. The correct word to fill in the blank is experiences. This is because the sentence is in present tense and the sense is in present indefinite.

24. Contentment is the right word which is appropriate to the meaning of the sentence. Contentment is the sign of true happiness as it gives a sense of fulfillment.

39. The word despite tells us that mission succeeded even though there were problems. Hence the correct answer is Setbacks.

42. Tone of the sentence clearly indicates a word that is similar to understated is needed for the blank. Alternatively, word should be antonym of strong (fail to make strong impression). Therefore, best choice is restrained which means controlled / reserved / timid.

43. Contemplate is a transitive verb and hence is followed by a gerund Hence the correct usage of contemplate is verb + ing form.

44. The context seeks to take a deviation only when existing / present / current / alternative treatments are unsatisfactory. So word for the blank should be a close synonym of existing / present / current / alternative. Available is the closest of all.

45. Clues: foreign technocrats did something negatively to the problems – so it is counter-productive to invite them. All other options are non-negative. Best choice is exacerbated which means aggravated or worsened.

47. Betrayed, **means** 'showed' or revealed.

3 CHAPTER

Synonyms

MCQ TYPE QUESTIONS

2016

1. The Buddha said, "Holding on to anger is like <u>grasping</u> a hot coal with the intent of throwing it at someone else; you are the one who gets burnt." Select the word below which is closest in meaning to the word underlined above.
   (a) burning (b) igniting
   (c) clutching (d) flinging

2015

2. The word similar in meaning to 'dreary' is
   (a) cheerful (b) dreamy
   (c) hard (d) dismal
3. Choose the word most similar in meaning to the given word :
   Educe
   (a) Exert (b) Educate
   (c) Extract (d) Extend
4. Choose the most suitable one word substitute for the following expression
   Connotation of a road or way
   (a) Pertinacious (b) Viaticum
   (c) Clandestine (d) Ravenous
5. Choose the word most similar in meaning to the given word:
   Awkward
   (a) Inept (b) Graceful
   (c) Suitable (d) Dreadful
6. A generic term that includes various items of clothing such as a skirt, a pair of trousers and a shirt as
   (a) fabric
   (b) textile
   (c) fibre
   (d) apparel
7. Which of the following combinations is incorrect?
   (a) Acquiescence – Submission
   (b) Wheedle – Roundabout
   (c) Flippancy – Lightness
   (d) Profligate – Extravagant

2014

8. A student is required to demonstrate a high level of <u>comprehension</u> of the subject, especially in the social sciences.
   The word closest in meaning to <u>comprehension</u> is
   (a) understanding (b) meaning
   (c) concentration (d) stability
9. While receiving the award, the scientist said, "I feel vindicated". Which of the following is closest in meaning to the word 'vindicated'?
   (a) punished (b) substantiated
   (c) appreciated (d) chastened
10. Choose the word that is opposite in meaning to the word "coherent".
    (a) sticky (b) well-connected
    (c) rambling (d) friendly
11. Match the columns:

Column-1	Column-2
1. eradicate	P. misrepresent
2. distort	Q. soak completely
3. saturate	R. use
4. utilize	S. destroy utterly

    (a) 1 : S, 2 : P, 3 : Q, 4 : R
    (b) 1 : P, 2 : Q, 3 : R, 4 : S
    (c) 1 : Q, 2 : R, 3 : S, 4 : P
    (d) 1 : S, 2 : P, 3 : R, 4 : Q

2013

12. Which of the following options is the closest in meaning to the word given below:
    Primeval
    (a) Modern (b) Historic
    (c) Primitive (d) Antique
13. Which one of the following options is the closest in meaning to the word given below?
    Nadir
    (a) Highest (b) Lowest
    (c) Medium (d) Integration
14. They were requested not to **quarrel** with others. Which one of the following options is the closest in meaning to the word **quarrel**?
    (a) make out (b) call out
    (c) dig out (d) fall out

2012

15. Which one of the following options is the closest in meaning to the word given below?

Mitigate

(*a*) Diminish (*b*) Divulge

(*c*) Dedicate (*d*) Denote

16. Which one of the following options is the closest in meaning to the word given below?

Latitude

(*a*) Eligibility (*b*) Freedom

(*c*) Coercion (*d*) Meticulousness

2011

17. Which of the following options is the closest in the meaning to the word below :

Inexplicable

(*a*) Incomprehensible (*b*) Indelible

(*c*) Inextricable (*d*) Infallible

2010

18. Which of the following options is the closest in meaning to the word below:

Circuitous

(*a*) cyclic (*b*) indirect

(*c*) confusing (*d*) crooked

ANSWERS

MCQ Type Questions

1. (*c*)	**2.** (*d*)	**3.** (*c*)	**4.** (*b*)	**5.** (*a*)	**6.** (*d*)	**7.** (*b*)	**8.** (*a*)	**9.** (*b*)	**10.** (*c*)
11. (*a*)	**12.** (*c*)	**13.** (*b*)	**14.** (*b*)	**15.** (*a*)	**16.** (*b*)	**17.** (*a*)	**18.** (*b*)		

EXPLANATIONS

MCQ TYPE QUESTIONS

1. The meaning of underlined word grasping means clutching (or holding something tightly).

2. *dreary* - depressingly dull and bleak or repetitive.

3. The word similar in meaning to Educe is Extract.

13. **Given:** A word "Nadir"

To find: Closest meaning word to 'Nadir' in the given options.

Analysis: The dictionary meaning of 'Nadir' is 'Lowest' which is potion (*b*)

Source: Google

Given: A sequence 44, 42, 40,......

To find: Maximum sum of the sequence

Analysis: To find the maximum sum, we have to ignore all the negative terms (they will only decrease the sum). Hence, we terminate the sequence at 0.

Now, the sequence is 44, 42,......,2, 0.

This sequence is an Arithmetic progression with common difference –2. Now, the problem reduces to finding the sum of this A.P.

To find the sum of A.P., we have to find the number of terms in A.P.

Let n be the number of terms

a be the first term

l be the last term

d be the common difference.

$n = ?$

$a = 44, \quad l = 0, \quad d = -2$

Now, $l = a + (n - 1)d$

15. 'Mitigate' means to make something less harmful, serious, intense etc. Hence correct answer is 'Diminis'

17. Inexplicable means not explicable; that cannot be explained, understood, or accounted for. So the best synonym here is incomprehensible.

18. **Circuitous** means round about or not direct So circuitous : indirect.

4

CHAPTER

Antonyms

MCQ TYPE QUESTIONS

2015

1. Which word is not a synonym for the word *vernacular*?

(*a*) regional

(*b*) indigeneous

(*c*) indigent

(*d*) colloquial

2011

2. Choose the word from the options given below that is most nearly opposite in meaning to the given word:

Amalgamate

(*a*) merge

(*b*) split

(*c*) collect

(*d*) separate

3. Choose the word from the options given below that is most nearly **opposite** in meaning to the given word :

Frequency

(*a*) periodicity

(*b*) rarity

(*c*) gradualness

(*d*) persistency

ANSWERS

MCQ Type Questions

1. (*c*) **2.** (*b*) **3.** (*b*)

EXPLANATIONS

MCQ TYPE QUESTIONS

1. *Vernacular* - expressed or written in the native language of a place

   Indigent - deficient in what is requisite

2. Amalgamate means combine or unite to form one organization or structure. So the best option here is split. Separate on the other hand, although a close synonym, it is too general to be the best antonym in the given question while Merge is the synonym; Collect is not related.

3. Best antonym is rarity which means shortage or scarcity.

5 CHAPTER

Reasoning Ability

MCQ TYPE QUESTIONS

2016

1. Find the odd one in the following group of words.
mock, deride, praise, jeer

(*a*) mock (*b*) deride
(*c*) praise (*d*) jeer

2. Pick the odd one from the following options.

(*a*) CADBE (*b*) JHKIL
(*c*) XVYWZ (*d*) ONPMQ

3. Pick the odd one out in the following :
13, 23, 33, 43, 53

(*a*) 23 (*b*) 33
(*c*) 43 (*d*) 53

4. R2D2 is a robot. R2D2 can repair aeroplanes. No other robot can repair aeroplanes. Which of the following can be logically inferred from the above statements?

(*a*) R2D2 is a robot which can only repair aeroplanes.
(*b*) R2D2 is the only robot which can repair aeroplanes.
(*c*) R2D2 is a robot which can repair only aeroplanes.
(*d*) Only R2D2 is a robot.

5. A poll of students appearing for masters in engineering indicated that 60% of the students believed that mechanical engineering is a profession unsuitable for women. A research study on women with masters or higher degrees in mechanical engineering found that 99% of such women were successful in their professions.
Which of the following can be logically inferred from the above paragraph?

(*a*) Many students have misconceptions regarding various engineering disciplines.
(*b*) Men with advanced degrees in mechanical engineering believe women are well suited to be mechanical engineers.
(*c*) Mechanical engineering is a profession well suited for women with masters or higher degrees in mechanical engineering.
(*d*) The number of women pursuing higher degrees in mechanical engineering is small.

6. Sourya committee had proposed the establishment of Sourya Institutes of Technology (SITs) in line with Indian Institutes of Technology (IITs) to cater to the technological and industrial needs of a developing country.
Which of the following can be logically inferred from the above sentence?
Based on the proposal,

(*i*) In the initial years, SIT students will get degrees from IIT.
(*ii*) SITs will have a distinct national objective.
(*iii*) SIT like institutions can only be established in consulation with IIT.
(*iv*) SITs will serve technological needs of a developing country.

(*a*) (*iii*) and (*iv*) only (*b*) (*i*) and (*iv*) only
(*c*) (*ii*) and (*iv*) only (*d*) (*ii*) and (*iii*) only

7. Fact : If it rains, then the field is wet.
Read the following statements:

(*i*) It rains (*ii*) The field is not wet
(*iii*) The field is wet (*iv*) It did not rain

Which one of the options given below is NOT logically possible, based on the given fact?

(*a*) If (*iii*), then (*iv*) (*b*) If (*i*), then (*iii*)
(*c*) If (*i*), then (*ii*) (*d*) If (*ii*), then (*iv*)

8. Students taking an exam are divided into two groups, **P** and **Q** such that each group has the same number of students. The performance of each of the students in a test was evaluated out of 200 marks. It was observed that the mean of group **P** was 105, while that of group **Q** was 85. The standard deviation of group **P** was 25, while that of group **Q** was 5. Assuming that the marks were distributed on a normal distribution, which of the following statements will have the highest probability of being **TRUE**?

(*a*) No student in group **Q** scored less marks than any student in group **P**.
(*b*) No student in group **P** scored less marks than any student in group **Q**.
(*c*) Most students of group **Q** scored marks in a narrower range than students in group **P**.
(*d*) The median of the marks of group **P** is 100.

9. A smart city integrates all modes of transport, uses clean energy and promotes sustainable use of resources. It also uses technology to ensure safety and security of the city, something which critics argue, will lead to a surveillance state.
Which of the following can be logically inferred from the above paragraph?
(*i*) All smart cities encourage the formation of surveillance states.
(*ii*) Surveillance is an integral part of a smart city.
(*iii*) Sustainability and surveillance go hand in hand in a smart city.
(*iv*) There is a perception that smart cities promote surveillance.
(*a*) (*i*) and (*iv*) only (*b*) (*ii*) and (*iii*) only
(*c*) (*iv*) only (*d*) (*i*) only

10. Find the missing sequence in the letter series.
B, FH, LNP, _ _ _ _.
(*a*) SUWY (*b*) TUVW
(*c*) TVXZ (*d*) TWXZ

11. Michael lives 10 km away from where I live. Ahmed lives 5 km away and Susan lives 7 km away from where I live. Arun is farther away than Ahmed but closer than Susan from where I live. From the information provided here, what is one possible distance (in km) at which I live from Arun's place?
(*a*) 3.00 (*b*) 4.99
(*c*) 6.02 (*d*) 7.01

12. Leela is older than her cousin Pavithra. Pavithra's brother Shiva is older than Leela. When Pavithra and Shiva are visiting Leela, all three like to play chess. Pavithra wins more often than Leela does. Which one of the following statements must be TRUE based on the above?
(*a*) When Shiva plays chess with Leela and Pavithra, he often loses.
(*b*) Leela is the oldest of the three.
(*c*) Shiva is a better chess player than Pavithra.
(*d*) Pavithra is the youngest of the three.

13. Based on the given statements, select the appropriate option with respect to grammar and usage.
Statements :
I. The height of Mr. X is 6 feet.
II. The height of Mr. Y is 5 feet.
(*a*) Mr. X is longer than Mr. Y.
(*b*) Mr. X is more elongated than Mr. Y
(*c*) Mr. X is taller than Mr. Y
(*d*) Mr. X is lengthier than Mr. Y

14. Social science disciplines were in existence in an amorphous form until the colonial period when they were institutionalized. In varying degrees, they were intended to further the colonial interest. In the time of globalization and the economic rise of postcolonial countries like India, conventional ways of knowledge production have become obsolete.
Which of the following can be logically inferred from the above statements?
I. Social science disciplines have become obsolete.
II. Social science disciplines had a pre-colonial origin.
III. Social science disciplines always promote colonialism.
IV. Social science must maintain disciplinary boundaries.
(*a*) II only (*b*) I and III only
(*c*) II and IV only (*d*) III and IV only

15. Two and a quarter hours back, when seen in a mirror, the reflection of a wall clock without number markings seemed to show 1 : 30. What is the actual current time shown by the clock?
(*a*) 8:15 (*b*) 11:15
(*c*) 12:15 (*d*) 12:45

16. M and N start from the same location. M travels 10 km East and then 10 km North-East. N travels 5 km South and then 4 km South-East. What is the shortest distance (in km) between M and N at the end of their travel?
(*a*) 18.60 (*b*) 22.50
(*c*) 20.61 (*d*) 25.00

17. M has a son Q and a daughter R. He has no other children. E is the mother of P and daughter-inlaw of M. How is P related to M?
(*a*) P is the son-in-law of M.
(*b*) P is the grandchild of M.
(*c*) P is the daughter-in-law of M.
(*d*) P is the grandfather of M.

18. The number that least fits this set: (324, 441, 97 and 64) is ______.
(*a*) 324
(*b*) 441
(*c*) 97
(*d*) 64

19. The overwhelming number of people infected with rabies in India has been flagged by the World Health Organization as a source of concern. It is estimated that inoculating 70% of pets and stray dogs against rabies can lead to a significant reduction in the number of people infected with rabies.

Which of the following can be logically inferred from the above sentences ?

(*a*) The number of people in India infected with rabies is high

(*b*) The number of people in other parts of the world who are infected with rabies is low

(*c*) Rabies can be eradicated in India by vaccinating 70% of stray dogs

(d) Stray dogs are the main source of rabies worldwide

20. A flat is shared by four first year undergraduate students. They agreed to allow the oldest of them to enjoy some extra space in the flat. Manu is two months older than Sravan, who is three months younger than Trideep. Pavan is one month older than Sravan. Who should occupy the extra space in the flat ?

(*a*) Manu (*b*) Sravan

(*c*) Trideep (*d*) Pavan

21. Today, we consider Ashoka as a great ruler because of the copious evidence he left behind in the form of stone carved edicts. Historians tend to correlate greatness of a king at his time with the availability of evidence today. Which of the following can be logically inferred from the above sentences?

(*a*) Emperors who do not leave significant sculpted evidence are completely forgotten.

(*b*) Ashoka produced stone carved edicts to ensure that later historians will respect him.

(*c*) Statues of kings are a reminder of their greatness.

(*d*) A king's greatness, as we know him today, is interpreted by historians.

22. Fact 1 : Humans are mammals.

Fact 2 : Some humans are engineers.

Fact 3 : Engineers build houses.

If the above statements are facts, which of the following can be logically inferred ?

I. All mammals build houses.

II. Engineers are mammals.

III. Some humans are not engineers.

(*a*) II only (*b*) III only

(*c*) I, II and III (*d*) I only

2015

23. Mr. Vivek walks 6 meters North-East, then turns and walks 6 meters South- East, both at 60 degrees to East. He further moves 2 meters South and 4 meters West. What is the straight distance in meters between the point he started from and the point he finally reached?

(*a*) $2\sqrt{2}$ (*b*) 2

(*c*) $\sqrt{2}$ (*d*) $1/\sqrt{2}$

24. There are 16 teachers who can teach Thermodynamics (TD), 11 who can teach Electrical Sciences (ES), and 5 who can teach both TD and Engineering Mechanics (EM). There are a total of 40 teachers, 6 cannot teach any of the three subjects, i.e. EM, ES or TD. 6 can teach only ES. 4 can teach all three subjects, i.e. EM, ES and TD. 4 can teach ES and TD. How many can teach both ES and EM but not TD?

(*a*) 1 (*b*) 2

(*c*) 3 (*d*) 4

25. The given question is followed by two statements: select the most appropriate option that solves the question

Capacity of a solution tank A is 70% of the capacity of tank B. How many gallons of solution are in tank A and tank B?

Statements:

I. Tank A is 80% full and tank B is 40% full

II. Tank A if full contains 14,000 gallons of solution

(*a*) Statement I alone is sufficient

(*b*) Statement II alone is sufficient

(*c*) Either statement I or II alone is sufficient

(*d*) Both the statements I and II together are sufficient

26. Operators □, ◇ and → are defined by

$$a \,\square\, b = \frac{a-b}{a+b}\,;\; a \diamond b = \frac{a+b}{a-b}\,;\; a \rightarrow b = ab.$$

Find the value $(66 \,\square\, 6) \rightarrow (66 \diamond 6)$

(*a*) −2 (*b*) −1

(*c*) 1 (*d*) 2

27. Humpty Dumpty sits on a wall every day while having lunch. The wall sometimes breaks. A person sitting on the wall falls if the wall breaks.

Which one of the statements below is logically valid and can be inferred from the above sentences?

(*a*) Humpty Dumpty always falls while having lunch

(*b*) Humpty Dumpty does not fall sometimes while having lunch

(*c*) Humpty Dumpty never falls during dinner

(*d*) When Humpty Dumpty does not sit on the wall, the wall does not break

28. If ROAD is written as URDG, then SWAN should be written as :

(*a*) VXDQ (*b*) VZDQ
(*c*) VZDP (*d*) UXDQ

29. Select the pair that best expresses a relationship similar to that expressed in the pair :

Children : Pediatrician

(*a*) Adult : Orthopaedist
(*b*) Females : Gynaecologist
(*c*) Kidney : Nephrologist
(*d*) Skin : Dermatologist

30. The head of a newly formed government desires to appoint five of the six selected members P,Q,R,S,T and U to portfolios of Home, Power, Defense, Telecom, and Finance. U does not want any portfolio if S gets one of the five. R wants either Home or Finance or no portfolio. Q says that if S gets either Power or Telecom, then she must get the other one. T insists on a portfolio if P gets one.

Which is the valid distribution of portfolios?

(*a*) P-Home, Q-Power, R-Defense, S-Telecom, T-Finance
(*b*) R-Home, S-Power, P-Defense, Q-Telecom, T-Finance
(*c*) P-Home, Q-Power, T-Defense, S-Telecom, U-Finance
(*d*) Q-Home, U-Power, T-Defense, R-Telecom, P-Finance

31. Most experts feel that in spite of possessing all the technical skills required to be a batsman of the highest order, he is unlikely to be so due to lack of requisite temperament. He was guilty of throwing away his wicket several times after working hard to lay a strong foundation. His critics pointed out that until he addressed this problem success at the highest level will continue to elude him.

Which of the statement (s) below is/are logically valid and can be inferred from the above passage?

(*i*) He was already a successful batsman at the highest level
(*ii*) He has to improve his temperament in order to become a great batsman
(*iii*) He failed to make many of his good starts count
(*iv*) Improving his technical skills will guarantee success

(*a*) (*iii*) and (*iv*) (*b*) (*ii*) and (*iii*)
(*c*) (*i*), (*ii*) and (*iii*) (*d*) (*ii*) only

32. Alexander turned his attention towards India, since he had conquered Persia.

Which one of the statements below is logically valid and can inferred from the above sentence?

(*a*) Alexander would not have turned his attention towards India had he not conquered Persia.
(*b*) Alexander was not ready to rest on his laurels, and wanted to march to India
(*c*) Alexander was completely in control of his army and could command it to move towards India
(*d*) Since Alexander's kingdom extended to Indian borders after the conquest of Persia, he was keen to move further

33. Tanya is older than Enc.

Cliff is older than Tanya.

Eric is older than Cliff.

If the first two statements are true, then the third statement is

(*a*) True (*b*) False
(*c*) Uncertain (*d*) Data insufficient

34. Given below are two statements followed by two conclusions. Assuming these statements to be true, decide which one logically follows:

Statements:

I. No manager is a leader.
II. All leaders are executive.

Conclusions:

I. No manager is a executive.
II. All executive is a manager.

(*a*) Only conclusion I follows.
(*b*) Only conclusion II follows.
(*c*) Neither conclusion I nor II follows.
(*d*) Both conclusion I and II follow.

35. Find the missing sequence in the letter series below :

A, CD, GHI, ?___, UVWXY

(*a*) LMN (*b*) MNO
(*c*) MNOP (*d*) NOPQ

36. Ms. X will be in Bagdogra from 1/5/14 to 20/5/14 and from 22/5/14 to 31/5/14. On the morning of 21/5/14, she will reach Kochi via Mumbai.

Which one of the statements below is logically valid and can be inferred from the above sentences?

(*a*) Ms. X will be in Kochi for only one day only in May
(*b*) Ms. X will be in Kochi for only one day in May
(*c*) Ms. X will be only in Kochi for one day in May
(*d*) Only Ms. X will be in Kochi for one day in May

37. Given below are two statements followed by two conclusions. Assuming these statements to be true, decide which one logically follows.

Statement:

I. All film stars are playback singers.

II. All film directors are film stars.

Conclusions:

I. All film directors are playback singers.

II. Some film stars are film directors.

(*a*) Only conclusion I follows

(*b*) Only conclusion II follows

(*c*) Neither conclusion I nor II follows

(*d*) Both conclusions I and II follow

38. Lamenting the gradual sidelining of the arts in school curricula, a group of prominent artists wrote to the Chief Minister last year, asking him to allocate more funds to support arts education in schools. However, no such increase has been announced in this year's Budget. The artists expressed their deep anguish at their request not being approved, but many of them remain optimistic about funding in the future.

Which of the statement(*s*) below is/are logically valid and can be inferred from the above statements?

i. The artists expected funding for the arts to increase this year.

ii. The Chief Minister was receptive to the idea of increasing funding for the arts.

iii. The Chief Minister is a prominent artist.

iv. Schools are giving less importance to arts education nowadays.

(*a*) *iii* and *iv* (*b*) *i* and *iv*

(*c*) *i*, *ii* and *iv* (*d*) *i* and *iii*

39. Four branches of a company are located at M, N, O, and P. M is north of N at a distance of 4 km; P is south of O at a distance of 2 km; N is southeast of O by 1 km. What is the distance between M and P in km?

(*a*) 5.34 (*b*) 6.74

(*c*) 28.5 (*d*) 45.49

40. The given statement is followed by some courses of action. Assuming the statement to be true, decide the correct option.

Statement:

There has been a significant drop in the water level in the lakes supplying water to the city.

Course of action:

I. The water supply authority should impose a partial cut in supply to tackle the situation.

II. The government should appeal to all the residents through mass media for minimal use of water.

III. The government should ban the water supply in lower areas.

(*a*) Statements I and II follow.

(*b*) Statements I and III follow.

(*c*) Statements II and III follow.

(*d*) All statements follow.

2014

41. In a group of four children, Som is younger to Riaz. Shiv is elder to Ansu. Ansu is youngest in the group. Which of the following statements is/are required to find the eldest child in the group?

Statements

1. Shiv is younger to Riaz.

2. Shiv is elder to Som.

(*a*) Statement 1 by itself determines the eldest child.

(*b*) Statement 2 by itself determines the eldest child.

(*c*) Statements 1 and 2 are both required to determine the eldest child.

(*d*) Statements 1 and 2 are not sufficient to determine the eldest child.

42. X is 1 km northeast of Y. Y is 1 km southeast of Z. W is 1 km west of Z. P is 1 km south of W. Q is 1 km east of P. What is the distance between X and Q in km?

(*a*) 1 (*b*) $\sqrt{2}$

(*c*) $\sqrt{3}$ (*d*) 2

43. Rajan was not happy that Sajan decided to do the project on his own. On observing his unhappiness, Sajan explained to Rajan that he preferred to work independently.

Which one of the statements below is logically valid and can be inferred from the above sentences?

(*a*) Rajan has decided to work only in a group.

(*b*) Rajan and Sajan were formed into a group against their wishes.

(*c*) Sajan had decided to give in to Rajan's request to work with him.

(*d*) Rajan had believed that Sajan and he would be working together.

44. Find the odd one in the following group: ALRVX, EPVZB, ITZDF, OYEIK

(*a*) ALRVX

(*b*) EPVZB

(*c*) ITZDF

(*d*) OYEIK

45. Anuj, Bhola, Chandan, Dilip, Eswar and Faisal live on different floors in a six-storeyed building (the ground floor is numbered 1, the floor above it 2, and so on). Anuj lives on an even-numbered floor. Bhola does not live on an odd numbered floor. Chandan does not live on any of the floors below Faisal's floor. Dilip does not live on floor number 2. Eswar does not live on a floor immediately above or immediately below Bhola. Faisal lives three floors above Dilip. Which of the following floor-person combinations is correct?

	Anuj	*Bhola*	*Chandan*	*Dilip*	*Eswar*	*Faisal*
(*a*)	6	2	5	1	3	4
(*b*)	2	6	5	1	3	4
(*c*)	4	2	6	3	1	5
(*d*)	2	4	6	1	3	5

46. Find the next term in the sequence: 13M, 17Q, 19S, ______

(*a*) 21W (*b*) 21V

(*c*) 23W (*d*) 23V

47. If 'KCLFTSB' stands for 'best of luck' and 'SHSWDG' stands for 'good wishes', which of the following indicates 'ace the exam'?

(*a*) MCHTX (*b*) MXHTC

(*c*) XMHCT (*d*) XMHTC

48. "India is a country of rich heritage and cultural diversity."

Which one of the following facts best supports the claim made in the above sentence?

(*a*) India is a union of 28 states and 7 union territories.

(*b*) India has a population of over 1.1 billion.

(*c*) India is home to 22 official languages and thousands of dialects.

(*d*) The Indian cricket team draws players from over ten states.

49. In which of the following options will the expression P < M be definitely true?

(*a*) M < R > P > S (*b*) M > S < P < F

(*c*) Q < M < F = P (*d*) P = A < R < M

50. Find the next term in the sequence: 7G, 11K, 13M, ______

(*a*) 15Q

(*b*) 17Q

(*c*) 15P

(*d*) 17P

51. The multi-level hierarchical pie chart shows the population of animals in a reserve forest. The correct conclusions from this information are:

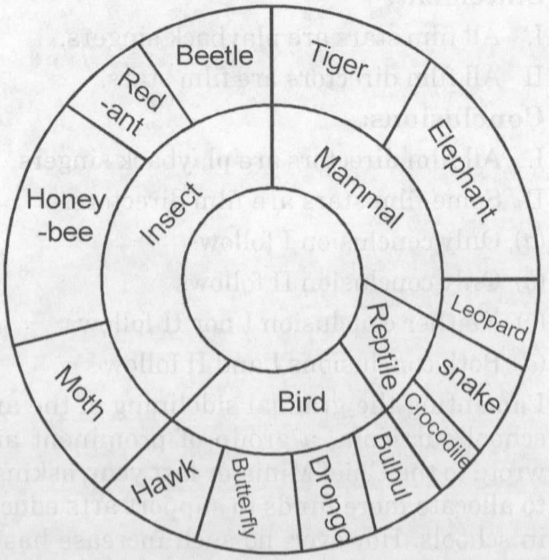

(*i*) Butterflies are birds

(*ii*) There are more tigers in this forest than red ants

(*iii*) All reptiles in this forest are either snakes or crocodiles

(*iv*) Elephants are the largest mammals in this forest

(*a*) (*i*) and (*ii*) only

(*b*) (*i*), (*ii*), (*iii*) and (*iv*)

(*c*) (*i*), (*iii*) and (*iv*) only

(*d*) (*i*), (*ii*) and (*iii*) only

52. Find the odd one in the following group

Q,W,Z,B B,H,K,M W,C,G,J M,S,V,X

(*a*) Q, W, Z, B (*b*) B, H, K, M

(*c*) W, C, G, J (*d*) M, S, V, X

53. Lights of four colors (red, blue, green, yellow) are hung on a ladder. On every step of the ladder there are two lights. If one of the lights is red, the other light on that step will always be blue. If one of the lights on a step is green, the other light on that step will always be yellow. Which of the following statements is not necessarily correct?

(*a*) The number of red lights is equal to the number of blue lights

(*b*) The number of green lights is equal to the number of yellow lights

(*c*) The sum of the red and green lights is equal to the sum of the yellow and blue lights

(*d*) The sum of the red and blue lights is equal to the sum of the green and yellow lights

54. Read the statements:

All women are entrepreneurs.

Some women are doctors.

Which of the following conclusions can be logically inferred from the above statements?

(*a*) All women are doctors

(*b*) All doctors are entrepreneurs

(*c*) All entrepreneurs are women

(*d*) Some entrepreneurs are doctors

55. Find the odd one from the following group:

(*a*) W,E,K,O (*b*) I,Q,W,A

(*c*) F,N,T,X, (*d*) N,V,B,D

56. For submitting tax returns, all resident males with annual income below ₹ 10 lakh should fill up Form P and all resident females with income below ₹ 8 lakh should fill up Form Q. All people with incomes above ₹ 10 lakh should fill up Form R, except non residents with income above ₹ 15 lakhs, who should fill up Form S. All others should fill Form T. An example of a person who should fill Form T is

(*a*) a resident male with annual income ₹ 9 lakh

(*b*) a resident female with annual income ₹ 9 lakh

(*c*) a non-resident male with annual income ₹ 16 lakh

(*d*) a non-resident female with annual income ₹ 16 lakh

57. Which number does not belong in the series below?

2, 5, 10, 17, 26, 37, 50, 64

(*a*) 17 (*b*) 37

(*c*) 64 (*d*) 26

58. A dance programme is scheduled for 10.00 a.m. Some students are participating in the programme and they need to come an hour earlier than the start of the event. These students should be accompanied by a parent. Other students and parents should come in time for the programme. The instruction you think that is appropriate for this is

(*a*) Students should come at 9.00 a.m. and parents should come at 10.00 a.m.

(*b*) Participating students should come at 9.00 a.m. accompanied by a parent, and other parents and students should come by 10.00 a.m.

(*c*) Students who are not participating should come by 10.00 a.m. and they should not bring their parents. Participating students should come at 9.00 a.m.

(*d*) Participating students should come before 9.00 a.m. Parents who accompany them should come at 9.00 a.m. All others should come at 10.00 a.m.

59. The number of people diagnosed with dengue fever (contracted from the bite of a mosquito) in north India is twice the number diagnosed last year. Municipal authorities have concluded that measures to control the mosquito population have failed in this region.

Which one of the following statements, if true, does not contradict this conclusion?

(*a*) A high proportion of the affected population has returned from neighbouring countries where dengue is prevalent

(*b*) More cases of dengue are now reported because of an increase in the Municipal Office's administrative efficiency

(*c*) Many more cases of dengue are being diagnosed this year since the introduction of a new and effective diagnostic test

(*d*) The number of people with malarial fever (also contracted from mosquito bites) has increased this year

60. Geneticists say that they are very close to confirming the genetic roots of psychiatric illnesses such as depression and schizophrenia, and consequently, that doctors will be able to eradicate these diseases through early identification and gene therapy.

On which of the following assumptions does the statement above rely?

(*a*) Strategies are now available for eliminating psychiatric illnesses

(*b*) Certain psychiatric illnesses have a genetic basis

(*c*) All human diseases can be traced back to genes and how they are expressed

(*d*) In the future, genetics will become the only relevant field for identifying psychiatric illnesses

2013

61. Select the pair that best expresses a relationship similar to that expressed in the pair:

Medicine: Health

(*a*) Science : Experiment

(*b*) Wealth : Peace

(*c*) Education : Knowledge

(*d*) Money : Happiness

62. Abhishek is elder to Savar.

Savar is younger to Anshul.

Which of the given conclusions is logically valid and is inferred from the above statements?

(*a*) Abhishek is elder to Anshul

(*b*) Anshul is elder to Abhishek

(*c*) Abhishek and Anshul are of the same age

(*d*) No conclusion follows

63. Complete the sentence:
Universalism is to particularism as diffuseness is to ____
(*a*) specificity
(*b*) neutrality
(*c*) generality
(*d*) adaptation

64. After several defeats in wars, Robert Bruce went in exile and wanted to commit suicide. Just before committing suicide, he came across a spider attempting tirelessly to have its net. Time and again, the spider failed but that did not deter it to retrain from making attempts. Such attempts by the spider made Bruce curious. Thus, Bruce started observing the near-impossible goal of the spider to have the net. Ultimately, the spider succeeded in having its net despite several failures. Such act of the spider encouraged Bruce not to commit suicide. And then, Bruce went back again and won many a battle, and the rest is history.
Which one of the following assertions is best supported by the above information?
(*a*) Failure is the pillar of success.
(*b*) Honesty is the best policy.
(*c*) Life begins and ends with adventures.
(*d*) No adversity justifies giving up hope.

65. Statement: You can always give me a ring whenever you need.
Which one of the following is the best inference from the above statement?
(*a*) Because I have a nice caller tune.
(*b*) Because I have a better telephone facility.
(*c*) Because a friend in need is a friend indeed.
(*d*) Because you need not pay towards the telephone bills when you give me a ring.

66. Statement: There were different streams of freedom movements in colonial India carried out by the moderates, liberals, radicals, socialists, and so on.
Which one of the following is the best inference from the above statement?
(*a*) The emergence of nationalism in colonial India led to our Independence.
(*b*) Nationalism in India emerged in the context of colonialism.
(*c*) Nationalism in India is homogeneous.
(*d*) Nationalism in India is heterogeneous.

67. Select the pair that best expresses a relationship similar to that expressed in the pair :
water : pipe : :
(*a*) cart : road
(*b*) electricity : wire
(*c*) sea : beach
(*d*) music: instrument

68. All professors are researchers
Some scientists are professors
Which of the given conclusions is logically valid and is inferred from the above arguments :
(*a*) All scientists are researchers
(*b*) All professors are scientists
(*c*) Some researchers are scientists
(*d*) No conclusion follows

2012

69. Wanted Temporary, Part-time persons for the post of Field Interviewer to conduct personal interviews to collect and collate economic data. Requirements: High School-pass, must be available for Day, Evening and Saturday work. Transportation paid, expenses reimbursed.
Which one of the following is the best inference from the above advertisement?
(*a*) Gender-discriminatory
(*b*) Xenophobic
(*c*) Not designed to make the post attractive
(*d*) Not gender-discriminatory

70. Given the sequence of terms, AD CG FK JP, the next term is
(*a*) OV (*b*) OW
(*c*) PV (*d*) PW

2011

71. A transporter receives the same number of orders each day. Currently, he has some pending orders (backlog) to be shipped. If he uses 7 trucks, then at the end of the 4th day he can clear all the orders. Alternatively, if he uses only 3 trucks, then all the orders are cleared at the end of the 10th day. What is the minimum number of trucks required so that there will be no pending order at the end of the 5th day?
(*a*) 4
(*b*) 5
(*c*) 6
(*d*) 7

72. Few school curricula include a unit on how to deal with bereavement and grief, and yet all students at some point in their lives suffer from losses through death and parting.
Based on the above passage which topic would not be included in a unit on bereavement?

(*a*) How to write a letter of condolence
(*b*) What emotional stages are passed through in the healing process
(*c*) What the leading causes of death are
(*d*) How to give support to a grieving friend

73. The question below consists of a pair of related words followed by four pairs of words. Select the pair that best expresses the relation in the original pair :

Gladiator : Arena

(*a*) dancer : stage
(*b*) commuter : train
(*c*) teacher : classroom
(*d*) lawyer : courtroom

74. The horse has played a little known but very important role in the field of medicine. Horses were injected with toxins of diseases until their blood built up immunities. Then a serum was made from their blood. Serums to fight with diphtheria and tetanus were developed this way.
It can be inferred from the passage, that horses were

(*a*) given immunity of diseases
(*b*) generally quite immune to diseases
(*c*) given medicines to fight toxins
(*d*) given diphtheria and tetanus serums

2010

75. The question below consists of a pair of related words followed by four pairs of words. Select the pair that best expresses the relation in the original pair.

Unemployed : Worker

(*a*) fallow : land (*b*) unaware : sleeper
(*c*) wit : jester (*d*) renovated : house

76. Hari (H), Gita (G), Irfan (I) and Saira (S) are siblings (i.e. brothers and sisters). All were born on 1st January. The age difference between any two successive siblings (that is born one after another) is less than 3 years. Given the following facts:

I. Hari's age + Gita age > Irfan's age + Saira's age
II. The age difference between Gita and Saira is 1 year. However, Gita is not the oldest and Saira is not the youngest.
III. There are no twins.

In what order were they born (oldest first)?

(*a*) HSIG
(*b*) SGHI
(*c*) IGSH
(*d*) IHSG

77. Modern warfare has changed from large scale clashes of armies to suppression of civilian populations. Chemical agents that do their work silently appear to be suited to such warfare; and regretfully, there exist people in military establishments who think that chemical agents are useful tools for their cause.
Which of the following statements best sums up the meaning of the above passage:

(*a*) Modern warfare has resulted in civil strife
(*b*) Chemical agents are useful in modern warfare
(*c*) Use of chemical agents in warfare would be undesirable
(*d*) People in military establishments like to use chemical agents in war

78. If 137 + 276 = 435 how much is 731 + 672?

(*a*) 534 (*b*) 1403
(*c*) 1623 (*d*) 1513

NUMERICAL TYPE QUESTIONS

2015

1. Fill in the missing value

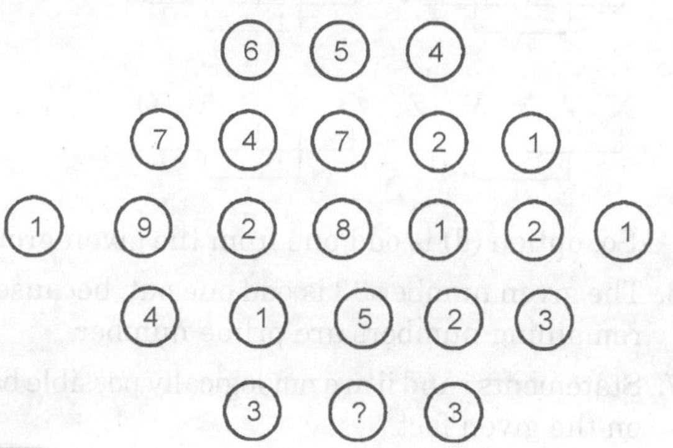

2014

2. The next term in the series 81, 54, 36, 24,.... is ______

3. Fill in the missing number in the series.
2 3 6 15 ___ 157.5 630

4. What is the next number in the series?
12 35 81 173 357 ___

ANSWERS

MCQ Type Questions

1. (*c*)	**2.** (*d*)	**3.** (*b*)	**4.** (*b*)	**5.** (*a*)	**6.** (*c*)	**7.** (*c*)	**8.** (*c*)	**9.** (*b*)	**10.** (*c*)
11. (*c*)	**12.** (*d*)	**13.** (*c*)	**14.** (*a*)	**15.** (*d*)	**16.** (*c*)	**17.** (*b*)	**18.** (*c*)	**19.** (*a*)	**20.** (*c*)
21. (*d*)	**22.** (*b*)	**23.** (*a*)	**24.** (*a*)	**25.** (*d*)	**26.** (*c*)	**27.** (*b*)	**28.** (*b*)	**29.** (*b*)	**30.** (*b*)
31. (*b*)	**32.** (*a*)	**33.** (*b*)	**34.** (*c*)	**35.** (*c*)	**36.** (*b*)	**37.** (*d*)	**38.** (*b*)	**39.** (*a*)	**40.** (*a*)
41. (*a*)	**42.** (*c*)	**43.** (*d*)	**44.** (*d*)	**45.** (*b*)	**46.** (*c*)	**47.** (*b*)	**48.** (*c*)	**49.** (*d*)	**50.** (*b*)
51. (*d*)	**52.** (*c*)	**53.** (*d*)	**54.** (*d*)	**55.** (*d*)	**56.** (*b*)	**57.** (*c*)	**58.** (*b*)	**59.** (*d*)	**60.** (*b*)
61. (*c*)	**62.** (*d*)	**63.** (*a*)	**64.** (*d*)	**65.** (*c*)	**66.** (*d*)	**67.** (*b*)	**68.** (*c*)	**69.** (*d*)	**70.** (*a*)
71. (*c*)	**72.** (*c*)	**73.** (*d*)	**74.** (*b*)	**75.** (*d*)	**76.** (*b*)	**77.** (*c*)	**78.** (*c*)		

Numerical Type Questions

1. (3) **2.** (16) **3.** (45) **4.** (725 to 725)

EXPLANATIONS

MCQ TYPE QUESTIONS

1. Given words 'mock, deride and jeer' are synonyms which means mockery. So, the given word praise is odd one.

2. C A D B E, J H K I L

(+1, +1, +1; +1, +1, +1)

X V Y W Z, O N P M Q

(+1, +1, +1; +1, +1, −1)

So, option (d) is odd one from the given group.

3. The given number 33 is odd one out, because the remaining numbers are prime number.

7. Statements i and ii are not logically possible based on the given fact.

8. Group 'P'

Mean (μ) = 105

Standard deviation (σ_1) = 25

$\Pr(\mu - \sigma \le x \le \mu + \sigma) = 0.6827$

$\therefore$ 68% within one standard deviation

$$\mu_1 - \sigma_1 = 105 - 25 = 80$$

$$\mu_1 + \sigma_1 = 105 + 25 = 130$$

$\therefore$ range = 80 to 130

Distribution of P:

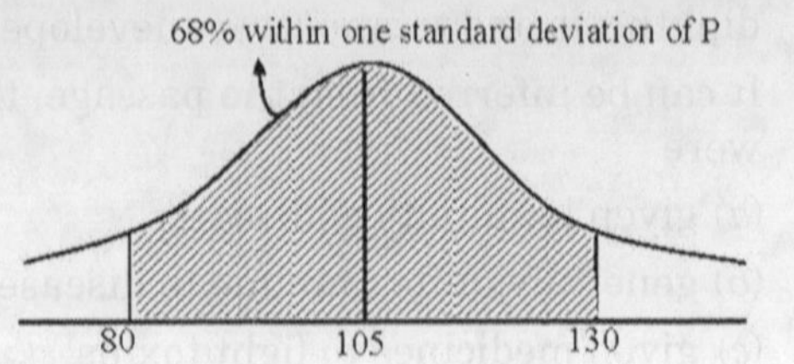

Group 'Q'

Mean (μ_2) = 85

Standard deviation (σ_2) = 5

$\Pr(\mu - \sigma \le x \le \mu + \sigma) = 0.6827$

$\therefore$ 68% within one standard deviation

$$\mu_2 - \sigma_2 = 85 - 5 = 80$$

$$\mu_2 + \sigma_2 = 85 + 5 = 90$$

$\therefore$ Range of Q in one standard deviation is 80 to 90

Distribution of 'Q'

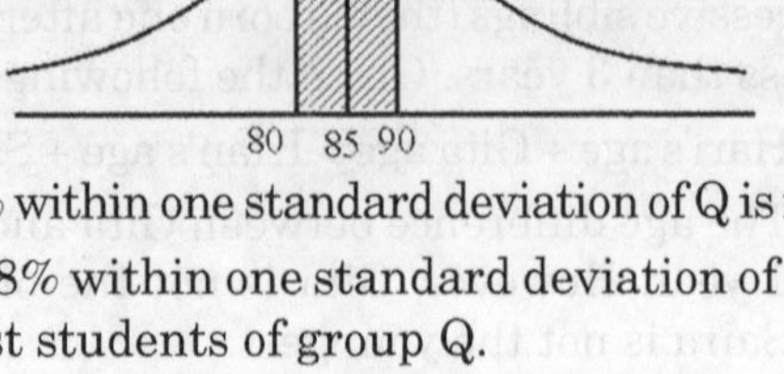

68% within one standard deviation of Q is narrower

$\therefore$ 68% within one standard deviation of Q means most students of group Q.

$\therefore$ Most students of group 'Q' scored marks in a narrower range than students in group 'P'

10. The following letter series is in the order of even letters series

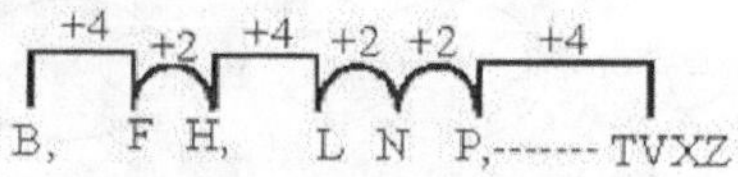

11. From given data, the following diagram can be possible.

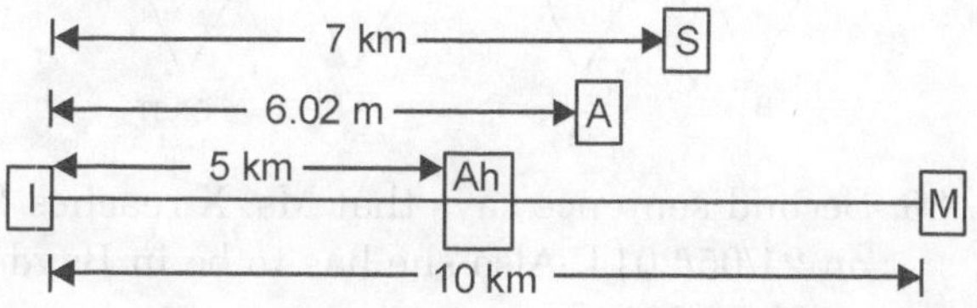

Here S → Susan

A → Arun

Ah → Ahmed

M → Michael

From the above diagram, Arun lives farthest away than Ahmed means more than 5 km but closer than Susan means less than 7 km, from the given alternatives.

So, only option 'c' is possible.

12. From the given question two statements will be followed.

For statment I

Arrange the given data according to their ages

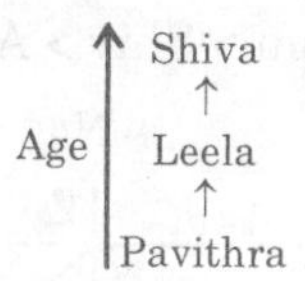

For statement II

Arrange the given data according to their winning

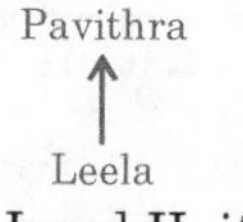

So, from statement I and II, it is clear that only option (*d*) is possible (i.e., statement I).

13.

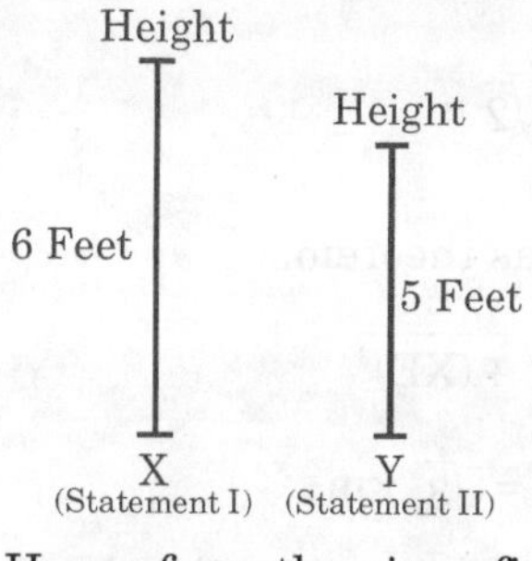

Hence from the given figure Mr. X is taller than Mr. Y by 1 foot.

14. Social science disciplines had a pre-colonial origin.

15.

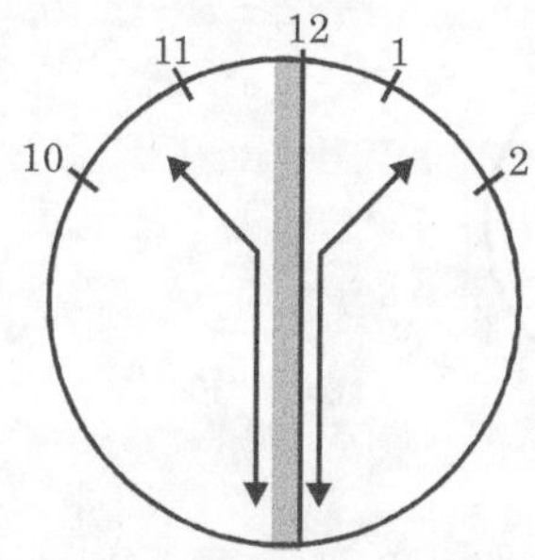

Mirror image of 1 : 20 is 10 : 30

10 : 30 was the time two and quarter hour back so time now will be 12 : 45

16.

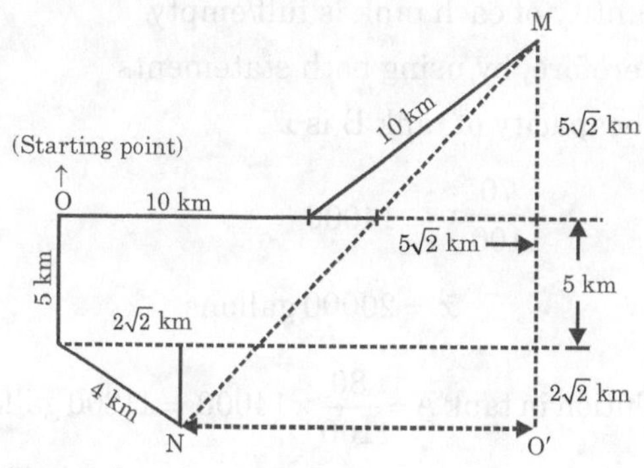

From the given figure

$$MN = \sqrt{(O'M)^2 + (O'N)^2}$$

$$O'M = 5\sqrt{2} + 5 + 2\sqrt{2} = 5 + 7\sqrt{2}$$

$$O'N = 10 + 5\sqrt{2} - 2\sqrt{2} = 10 + 3\sqrt{2}$$

$$MN = \sqrt{(5 + 7\sqrt{2})^2 + (10 + 3\sqrt{2})^2}$$

$$= \sqrt{25 + 98 + 70\sqrt{2} + 100 + 18 + 60\sqrt{2})}$$

$$\approx 20.61 \text{ km}$$

17.

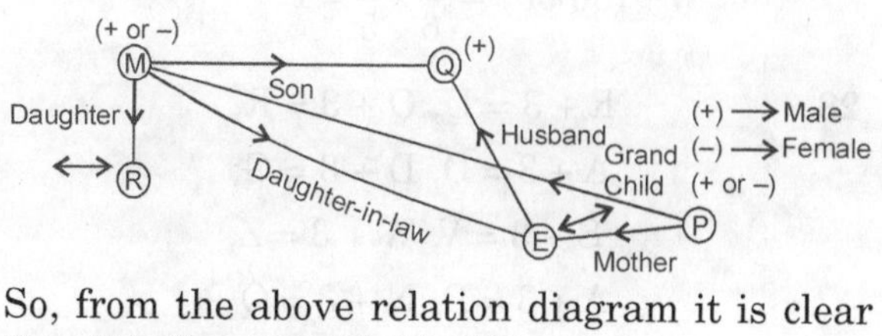

So, from the above relation diagram it is clear that P is the grandchild of M.

20. Manu age = sravan age + 2 months

Manu age = Trideep age – 3 months

Pavan age = Sravan's age + 1 month

From the above statement

Trideep age > Man > Pavan > Sravan

Hence, Trideep can occupy the extra space in the flat.

24.

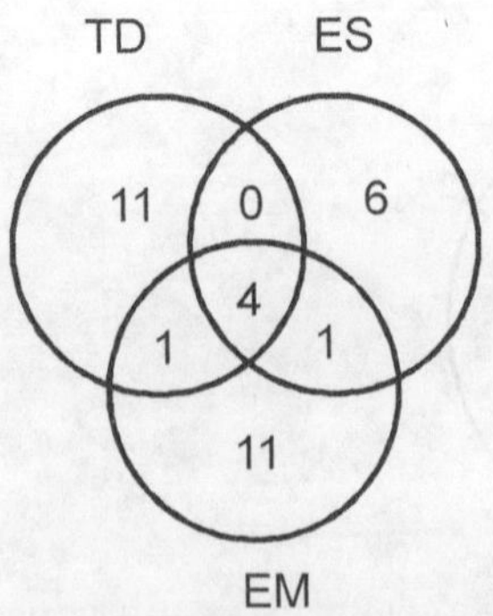

25. Statement-I can be used to solve the question if capacity of both tanks is already known

Statement-II can be used if it is known what quantity of each tank is full/empty.

Therefore, by using both statements

Let capacity of tank B is x

$$\frac{70}{100}x = 14000$$

$$\Rightarrow \qquad x = 20000 \text{ gallons}$$

$$\text{Solution in tank A} = \frac{80}{100} \times 14000 = 11200 \text{ gallons}$$

$$\text{Solution in tank B} = \frac{40}{100} \times 20000 = 8000 \text{ gallons}$$

$$\therefore \quad \text{Total solution} = 11200 + 8000 = 19200 \text{ gallons}$$

26. $66 \square 6 = \frac{66-6}{66+6} = \frac{60}{72} = \frac{5}{6}$

$$66 \diamond 6 = \frac{66+6}{66-6} = \frac{72}{60} = \frac{6}{5}$$

$$(66 \square 6) \rightarrow (66 \diamond 6) = \frac{5}{6} \times \frac{6}{5} = 1$$

28. R + 3 = U, O + 3 = R,

A + 3 = D, D + 3 = G;

S + 3 = V, W + 3 = Z,

A + 3 = D, N + 3 = Q

29. Community of people : Doctor

30. Since U does not want any portfolio, (*c*) and (*d*) are ruled out. R wants Home, or Finance or No portfolio, (*a*) is not valid.

Hence option (*b*) is correct.

34. S + 1: S + 2 :

Therefore concluding diagram can be

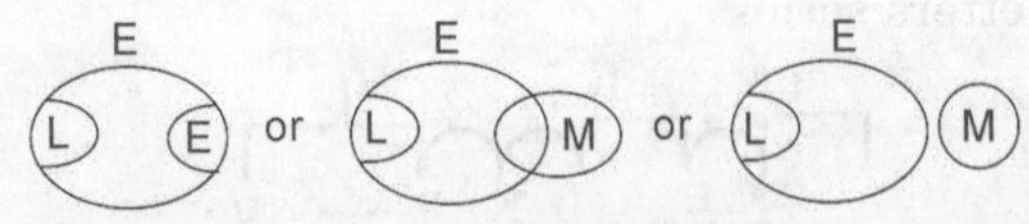

35.

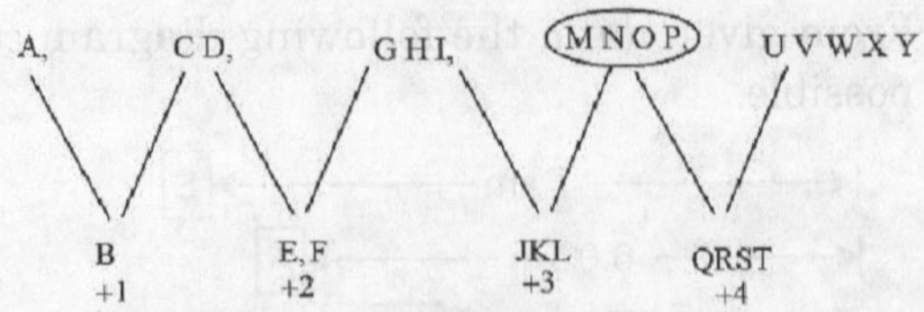

36. Second sentence says that Ms. X reaches Kochi on 21/05/2014. Also she has to be in Bagdogora on 22/05/2014.

$\therefore$ She stays in Kochi for only one day in may.

39.

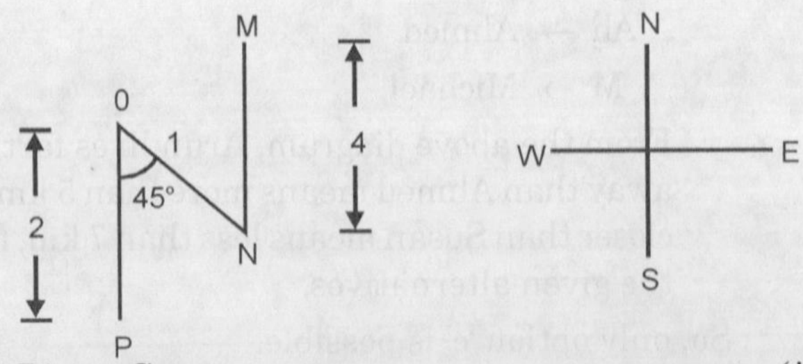

41. Riaz > Som ...(*i*)

Shiv > Ansu ...(*ii*)

Ansu is youngest in the group

$\therefore$ From equations (*i*) and (*ii*)

$\therefore$ Riaz > Som > Shiv > Ansu

42.

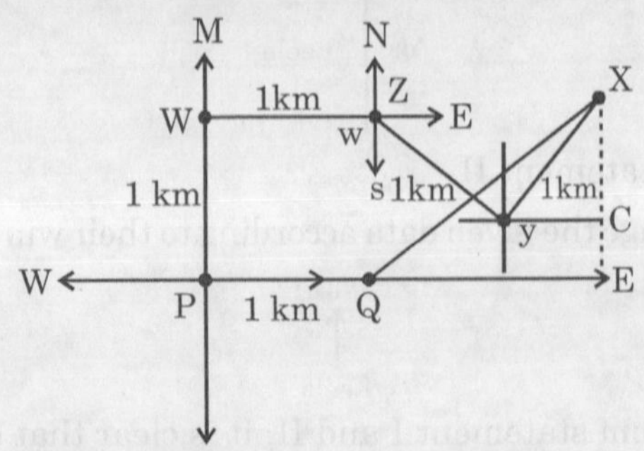

Here XE = 1 km

$$QE = \frac{2}{\sqrt{2}} = \sqrt{2}$$

$\therefore$ using Pythagoras theorem,

$$XQ = \sqrt{(QE)^2 + (XE)^2}$$

$$= \sqrt{2+1} = \sqrt{3} \text{ km}$$

$\therefore$ Statement I itself determines the eldest child which statement 2 is incorrect as Ansu is youngest in the group is not satisfied by statement 2

Hence option (*a*) is correct

44. ALRVX → only one vowel
EPVZB → only one vowel
ITZDF → only one vowel
OYEIK → three vowels

45. (*a*) Anuj: Even numbered floor (2, 4, 6)
(*b*) Bhola: Even numbered floor (2, 4, 6)
(*c*) Chandan lives on the floor above that of Faisal.
(*d*) Dilip: not on 2nd floor.
(*e*) Eswar: does not live immediately above or immediately below Bhola

From the options its clear, that only option (*b*) satisfies condition (*e*).

So, correct Ans is (*b*).

47. KCLFTSB: BST-Best, F-Of, LCK-Luck (Reverse order)

SHSWDG: GD-Good, WSHS-Wishes (Reverse order)

Similarly "ace the Exam'- C-Ace, T-The, XM-Exam

48. Diversity is shown in terms of difference language.

51. It is not mentioned that elephant is the largest animal.

52.
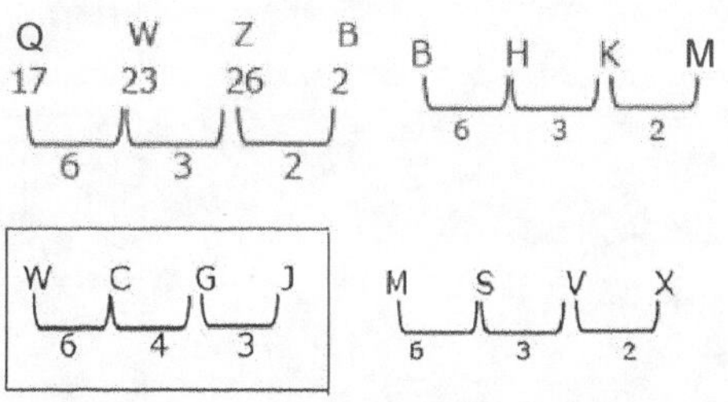

55.
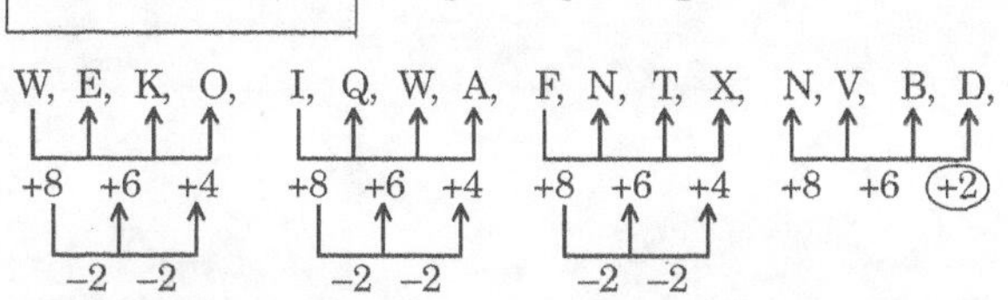

Hence the odd one from the following group is N, V, B, D.

56. By reading the instructions, it is clear that person who fills form T is a resident female with annual income of ₹ 9 Lakh.

59. Among given option, option (*d*) *i.e.* the number of people with malarial fever has increased this year do not contradic the conclusion.

70.
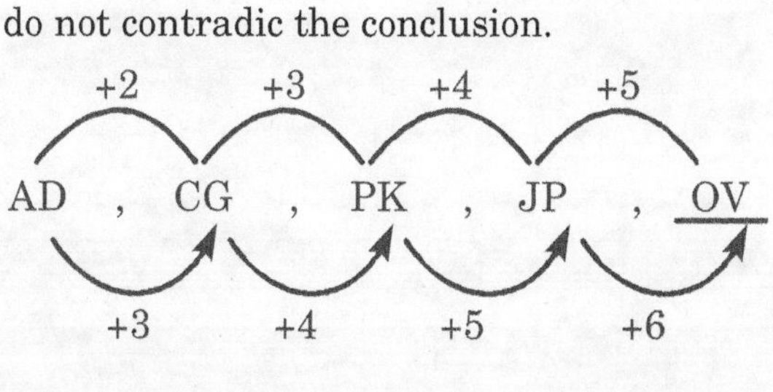

71. Let each truck carry 100 units.

$$2800 = 4n + e$$

where n = normal

$$3000 = 10n + e$$

where e = excess/pending

$$\therefore \quad n = \frac{100}{3}, \quad e = \frac{8000}{3}$$

$$5 \text{ days} \Rightarrow 500\,x = \frac{5100}{3} + \frac{8000}{3}$$

$$\Rightarrow \quad 500x = \frac{8500}{3} = 17$$

$$\Rightarrow \quad x > 5$$

Minimum possible = 6

72. The given passage clearly deals with how to deal with bereavement and grief and so after the tragedy occurs and not about precautions. Therefore, irrespective of the causes of death, a school student rarely gets into details of causes—which is beyond the scope of the context. Rest all are important in dealing with grief.

73. Given relationship is worker: workplace. A gladiator is

(*i*) a person, usually a professional combatant trained to entertain the public by engaging in mortal combat with another person or a wild.

(*ii*) A person engaged in a controversy or debate, especially in public.

74. From the passage it cannot be inferred that horses are given immunity as in option (*a*), since the aim is to develop medicine and in turn immunize humans. Option (*b*) is correct since it is given that horses develop immunity after some time. Refer "until their blood built up immunities". Even option(*c*) is invalid since medicine is not built till immunity is developed in the horses. Option (*d*) is incorrect since specific examples are cited to illustrate and this cannot capture the essence.

76.
$$H + G > I + S \quad ...(1)$$
and
$$G - S = 1 \quad ...(2)$$
G is not oldest, S is not youngest
$$\therefore \quad H + 1 > I$$
Irfan older than Hari
Gita older than Sarita
From given option SGHI

77. Use of chemical agents in warfare would be undesirable

78.

$$137 - \overline{001}\ \overline{011}\ \overline{111}$$
$$+276 - 010\ 111\ 110$$
$$\underline{100\ 011\ 101} = 435$$

Octal number system

$$\therefore \quad 731 - 111\ 011\ 001$$
$$672 - 110\ 111\ 110$$
$$\underline{100\ 010\ 011}$$

Hence 1623

NUMERICAL TYPE QUESTIONS

1. Middle number is the average of number on both sides

$\therefore$ Average of 3 and 3 is $\frac{3+3}{2} = \frac{6}{2} = 3$

2. $81 - 54 = 27; 27 \times \frac{2}{3} = 18$

$54 - 36 = 18; 18 \times \frac{2}{3} = 12$

$36 - 24 = 12; 12 \times \frac{2}{3} = 8$

$\therefore \quad 24 - 8 = 16$

3. 2, 3, 6, 15, **45**, 157.5, 630

Ratios: 1.5, 2, 2.5, 3, 3.5, 4

$\frac{\text{2nd number}}{\text{1st number}}$ is in increasing order as shown above

4. The given series is

12, 35, 81, 173, 357, ...

The given series follows the following pattern

$12 \times 2 + 11 = 35$

$35 \times 2 + 11 = 81$

$81 \times 2 + 11 = 173$

$173 \times 2 + 11 = 357$

$357 \times 2 + 11 = 725$

Numerical Ability

1

CHAPTER

Numbers and Algebra

MCQ TYPE QUESTIONS

2016

1. In a quadratic function, the value of the product of the roots (α, β) is 4. Find the value of

$$\frac{\alpha^n + \beta^n}{\alpha^{-n} + \beta^{-n}}$$

(*a*) n^4 (*b*) 4^n

(*c*) 2^{2n-1} (*d*) 4^{n-1}

2. Among 150 faculty members in an institute, 55 are connected with each other through Facebook® and 85 are connected through WhatsApp®. 30 faculty members do not have Facebook® or WhatsApp® accounts. The number of faculty members connected only through Facebook® accounts is ______________.

(*a*) 35 (*b*) 45

(*c*) 65 (*d*) 90

3. If $|9y-6| = 3$, then $y^2 - \frac{4y}{3}$ is ________.

(*a*) 0 (*b*) $+\frac{1}{3}$

(*c*) $-\frac{1}{3}$ (*d*) Undefined

4. If $q^{-a} = \frac{1}{r}$ and $r^{-b} = \frac{1}{s}$ and $s^{-c} = \frac{1}{q}$, the value of abc is .

(*a*) $(rqs)^{-1}$ (*b*) 0

(*c*) 1 (*d*) r^+q^+s

5. The sum of the digits of a two digit number is 12. If the new number formed by reversing the digits is greater than the original number by 54, find the original number.

(*a*) 39 (*b*) 57

(*c*) 66 (*d*) 93

6. Two finance companies, P and Q, declared fixed annual rates of interest on the amounts invested with them. The rates of interest offered by these companies may differ from year to year. Year-wise annual rates of interest offered by these companies are shown by the line graph provided below.

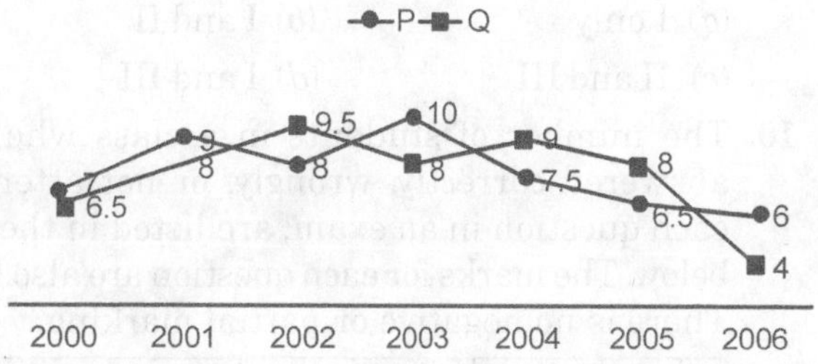

If the amounts invested in the companies, P and Q, in 2006 are in the ratio 8 : 9, then the amounts received after one year as interests from companies P and Q would be in the ratio:

(*a*) 2 : 3 (*b*) 3 : 4

(*c*) 6 : 7 (*d*) 4 : 3

2015

7. Read the following table giving sales data of five types of batteries for years 2006 to 2012

Year	Type I	Type I	Type I	Type I	Type I
2006	75	144	114	102	108
2007	90	126	102	84	126
2008	96	114	75	105	135
2009	105	90	150	90	75
2010	90	75	135	75	90
2011	105	60	165	45	120
2012	115	85	160	100	145

Out of the following, which type of battery achieved highest growth between the years 2006 and 2012?

(*a*) Type V (*b*) Type III

(*c*) Type II (*d*) Type I

8. If $a^2 + b^2 + c^2 = 1$, then $ab + bc + ac$ lies in the interval

(*a*) $\left[1, \frac{2}{3}\right]$

(*b*) $\left[\frac{-1}{2}, 1\right]$

(*c*) $\left[-1, \frac{1}{2}\right]$

(*d*) [2,–4]

9. If the list of letters, P, R, S, T, U is an arithmetic sequence, which of the following are also in arithmetic sequence?

I. 2P, 2R, 2S, 2T, 2U

II. P–3, R–3, S–3, T–3, U–3

III. P^2, R^2, S^2, T^2, U^2

(*a*) I only (*b*) I and II

(*c*) II and III (*d*) I and III

10. The number of students in a class who have answered correctly, wrongly, or not attempted each question in an exam, are listed in the table below. The marks for each question are also listed. There is no negative or partial marking.

Q. No.	Marks	Answered Correctly	Answered Wrongly	Not Attempted
1	2	21	17	6
2	3	15	27	2
3	1	11	29	4
4	2	23	18	3
5	5	31	12	1

What is the average of the marks obtained by the class in the examination?

(*a*) 2.290 (*b*) 2.970

(*c*) 6.795 (*d*) 8.795

2014

11. The monthly rainfall chart on 50 years of rainfall in Agra is shown in the following figure. Which of the following are true? (k percentile is the value such that k percent of the data fall below that value)

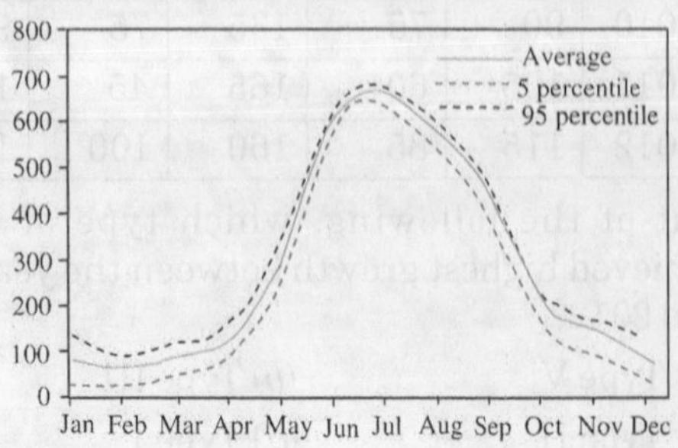

(*i*) On average, it rains more in July than in December

(*ii*) Every year, the amount of rainfall in August is more than that in January

(*iii*) July rainfall can be estimated with better confidence than February rainfall

(*iv*) In August, there is at least 500 mm of rainfall

(*a*) (*i*) and (*ii*)

(*b*) (*i*) and (*iii*)

(*c*) (*ii*) and (*iii*)

(*d*) (*iii*) and (*iv*)

12. The total exports and revenues from the exports of a country are given in the two charts shown below. The pie chart for exports shows the quantity of each item exported as a percentage of the total quantity of exports. The pie chart for the revenues shows the percentage of the total revenue generated through export of each item. The total quantity of exports of all the items is 500 thousand tonnes and the total revenues are 250 crore rupees. Which item among the following has generated the maximum revenue per kg?

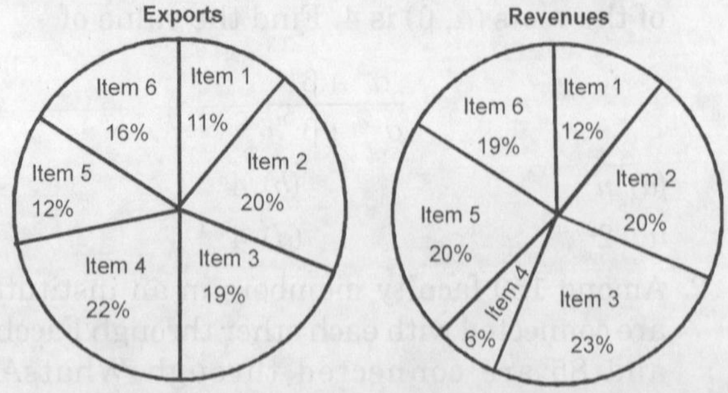

(*a*) Item 2 (*b*) Item 3

(*c*) Item 6 (*d*) Item 5

13. The statistics of runs scored in a series by four batsmen are provided in the following table. Who is the most consistent batsman of these four?

Batsman	Average	Standard deviation
K	31.2	5.21
L	46.0	6.35
M	54.4	6.22
N	17.9	5.90

(*a*) K (*b*) L

(*c*) M (*d*) N

14. The exports and imports (in crores of ₹) of a country from 2000 to 2007 are given in the following bar chart. If the trade deficit is defined as excess of imports over exports, in which year is the trade deficit 1/5th of the exports?

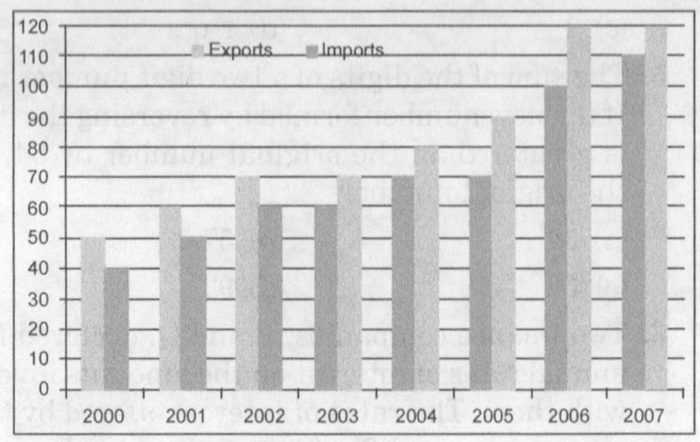

(*a*) 2005

(*b*) 2004

(*c*) 2007

(*d*) 2006

15. The table below has question-wise data on the performance of students in an examination. The marks for each question are also listed. There is no negative or partial marking in the examination.

Q.No.	Marks	Answered Correctly	Answered Wrongly	Not Attempted
1	2	21	17	6
2	3	15	27	2
3	2	23	18	3

What is the average of the marks obtained by the class in the examination?

(*a*) 1.34 (*b*) 1.74
(*c*) 3.02 (*d*) 3.91

16. The ratio of male to female students in a college for five years is plotted in the following line graph. If the number of female students in 2011 and 2012 is equal, what is the ratio of male students in 2012 to male students in 2011?

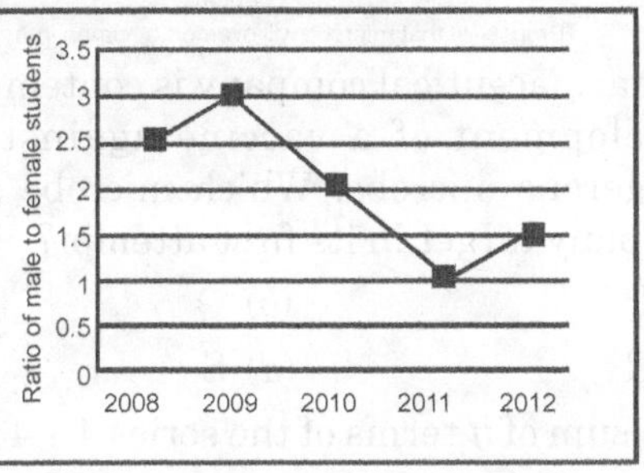

(*a*) 1 : 1 (*b*) 2 : 1
(*c*) 1.5 : 1 (*d*) 2.5 : 1

17. What is the average of all multiples of 10 from 2 to 198?

(*a*) 90 (*b*) 100
(*c*) 110 (*d*) 120

18. The value of $\sqrt{12+\sqrt{12+\sqrt{12+...}}}$ is

(*a*) 3.464 (*b*) 3.932
(*c*) 4.000 (*d*) 4.444

19. If x is real and $|x^2 - 2x + 3| = 11$, then possible values of $|-x^3 + x^2 - x|$ include

(*a*) 2, 4 (*b*) 2, 14
(*c*) 4, 52 (*d*) 14, 52

2013

20. A number is as much greater than 75 as it is smaller than 117. The number is

(*a*) 91
(*b*) 93
(*c*) 89
(*d*) 96

21. Following table provides figures (in rupees) on annual expenditure of a firm for two years - 2010 and 2011.

Category	2010	2011
Raw material	5200	6240
Power & fuel	7000	9450
Salary & wages	9000	12600
Plant & machinery	20000	25000
Advertising	15000	19500
Research & Development	22000	26400

In 2011, which of the following two categories have registered increase by same percentage?

(*a*) Raw material and Salary & wages
(*b*) Salary & wages and Advertising
(*c*) Power & fuel and Advertising
(*d*) Raw material and Research & Development

22. What will be the maximum sum of 44, 42, 40,... ?

(*a*) 502 (*b*) 504
(*c*) 506 (*d*) 500

23. The current erection cost of a structure is ₹ 13,200. If the labour wages per day increase by 1/5 of the current wages and the working hours decrease by 1/24 of the current period, then the new cost of erection in ₹ is

(*a*) 16,500 (*b*) 15,180
(*c*) 11,000 (*d*) 10,120

24. Find the sum of the expression

$$\frac{1}{\sqrt{1}+\sqrt{2}}+\frac{1}{\sqrt{2}+\sqrt{3}}+\frac{1}{\sqrt{3}+\sqrt{4}}+.....+\frac{1}{\sqrt{80}+\sqrt{81}}$$

(*a*) 7 (*b*) 8
(*c*) 9 (*d*) 10

25. In the summer of 2012, in New Delhi, the mean temperature of Monday to Wednesday was 41°C and of Tuesday to Thursday was 43°C. If the temperature on Thursday was 15% higher than that of Monday, then the temperature in °C on Thursday was

(*a*) 40 (*b*) 43
(*c*) 46 (*d*) 49

26. The set of values of p for which the roots of the equation $3x^2 + 2x + p(p - 1) = 0$ are of opposite sign is

(*a*) $(-\infty, 0)$ (*b*) $(0, 1)$
(*c*) $(1, \infty)$ (*d*) $(0, \infty)$

27. Find the sum to n terms of the series 10 + 84 + 734 + ...

(*a*) $\frac{9(9^n+1)}{10}+1$ (*b*) $\frac{9(9^n-1)}{8}+1$

(*c*) $\frac{9(9^n-1)}{8}+n$ (*d*) $\frac{9(9^n-1)}{8}+n^2$

28. If $3 \le X \le 5$ and $8 \le Y \le 11$ then which of the following options is TRUE?

(*a*) $\frac{3}{5} \le \frac{X}{Y} \le \frac{8}{5}$ (*b*) $\frac{3}{11} \le \frac{X}{Y} \le \frac{5}{8}$

(*c*) $\frac{3}{11} \le \frac{X}{Y} \le \frac{8}{5}$ (*d*) $\frac{3}{5} \le \frac{X}{Y} \le \frac{8}{11}$

29. Following table gives data on tourists from different countries visiting India in the year 2011.

Country	Number of Tourists
USA	2000
England	3500
Germany	1200
Italy	1100
Japan	2400
Australia	2300
France	1000

Which two countries contributed to the one third of the total number of tourists who visited India in 2011?

(*a*) USA and Japan

(*b*) USA and Australia

(*c*) England and France

(*d*) Japan and Australia

2012

30. If $(1.001)^{1259} = 3.52$ and $(1.001)^{2062} = 7.85$, then $(1.001)^{3321} =$

(*a*) 2.23 (*b*) 4.33

(*c*) 11.37 (*d*) 27.64

31. Raju has 14 currency notes in his pocket consisting of only ₹ 20 notes and ₹ 10 notes. The total money value of the notes is ₹ 230. The number of ₹ 10 notes that Raju has is

(*a*) 5 (*b*) 6

(*c*) 9 (*d*) 10

32. The data given in the following table summarizes the monthly budget of an average household.

Category	**Amount (Rs.)**
Food	4000
Clothing	1200
Rent	2000
Savings	1500
Other expenses	1800

The approximate percentage of the monthly budget **NOT** spent on savings is

(*a*) 10% (*b*) 14%

(*c*) 81% (*d*) 86%

2011

33. P, Q, R and S are four types of dangerous microbes recently found in a human habitat. The area of each circle with its diameter printed in brackets represents the growth of a single microbe surviving human immunity system within 24 hours of entering the body. The danger to human beings varies proportionately with the toxicity, potency and growth attributed to a microbe shown in the figure below :

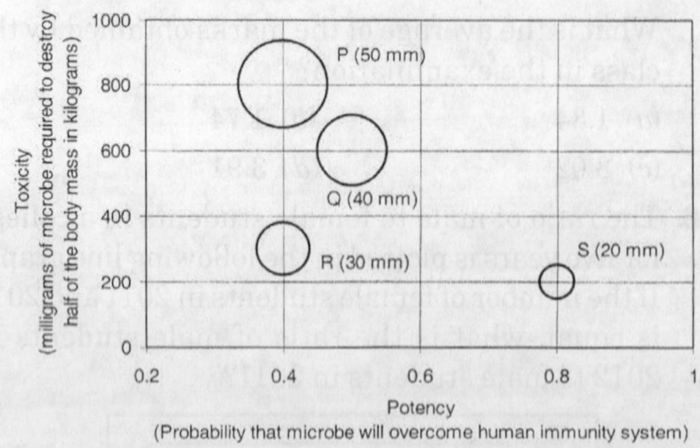

A pharmaceutical company is contemplating the development of a vaccine against the most dangerous microbe. Which microbe should the company target in its first attempt?

(*a*) P (*b*) Q

(*c*) R (*d*) S

34. The sum of n terms of the series $4 + 44 + 444 + \ldots$ is

(*a*) $(4/81)[10^{n+1} - 9n - 1]$

(*b*) $(4/81)[10^{n-1} - 9n - 1]$

(*c*) $(4/81)[10^{n+1} - 9n - 10]$

(*d*) $(4/81)[10^{n} - 9n - 10]$

35. Given that $f(y) = |y|/y$, and q is any non-zero real number, the value of $|f(q) - f(-q)|$ is

(*a*) 0 (*b*) –1

(*c*) 1 (*d*) 2

NUMERICAL TYPE QUESTIONS

2016

1. The following graph represents the installed capacity for cement production (in tonnes) and the actual production (in tonnes) of nine cement plants of a cement company. Capacity utilization of a plant is defined as ratio of actual production of cement to installed capacity. A plant with installed capacity of at least 200 tonnes is called a large plant and a plant with lesser capacity is called a small plant. The difference between total production of large plants and small plants, in tonnes is ______.

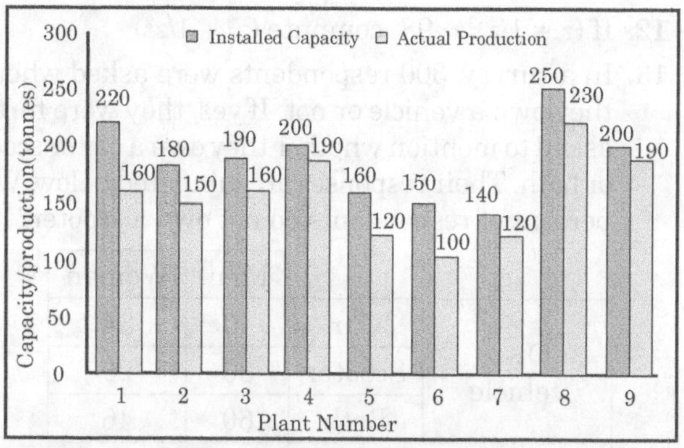

2. The numeral in the units position of $211^{870} + 146^{127} \times 3^{424}$ is ________.

2015

3. The exports and imports (in crores of Rs.) of a country from the year 2000 to 2007 are given in the following bar chart. In which year is the combined percentage increase in imports and exports the highest ?

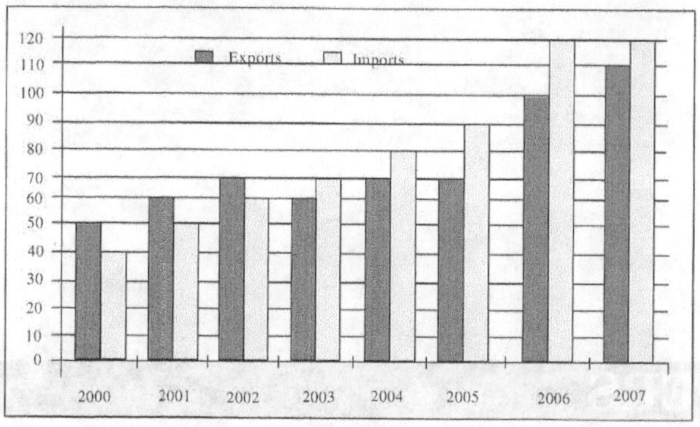

4. The pie chart below has the breakup of the number of students from different departments in an engineering college for the year 2012. The proportion of male to female students in each department is 5 : 4. There are 40 males in Electrical Engineering. What is the difference between numbers of female students in the Civil department and the female students in the Mechanical department?

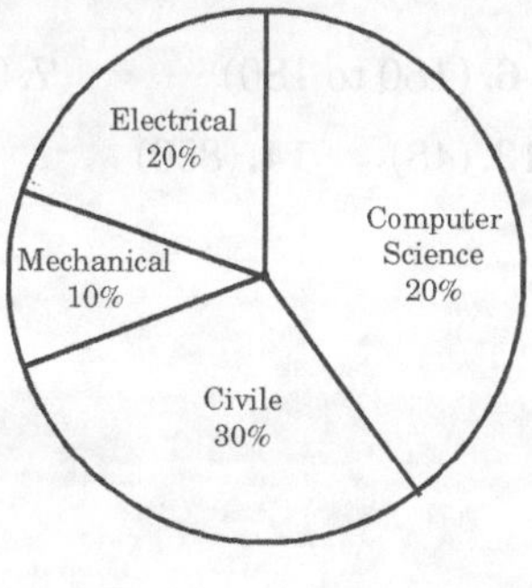

2014

5. A foundry has a fixed daily cost of ₹ 50,000 whenever it operates and variable cost of ₹ 800Q, where Q is the daily production in tonnes. What is the cost of production in ₹ per tonne for a daily production of 100 tonnes?

6. The smallest angle of a triangle is equal to two thirds of the smallest angle of a quadrilateral. The ratio between the angles of the quadrilateral is 3 : 4 : 5 : 6. The largest angle of the triangle is twice its smallest angle. What is the sum, in degrees, of the second largest angle of the triangle and the largest angle of the quadrilateral?

7. In a sequence of 12 consecutive odd numbers, the sum of the first 5 numbers is 425. What is the sum of the last 5 numbers in the sequence?

8. A firm producing air purifiers sold 200 units in 2012. The following pie chart presents the share of raw material, labour, energy, plant & machinery, and transportation costs in the total manufacturing cost of the firm in 2012. The expenditure on labour in 2012 is ₹ 4,50,000. In 2013, the raw material expenses increased by 30% and all other expenses increased by 20%. What is the percentage increase in total cost for the company in 2013?

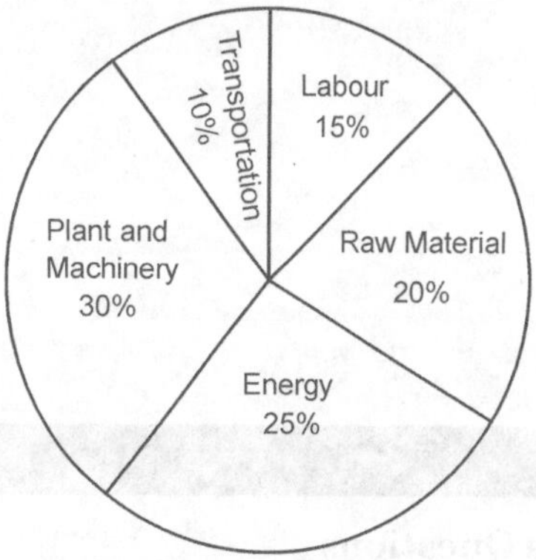

9. A firm producing air purifiers sold 200 units in 2012. The following pie chart presents the share of raw material, labour, energy, plant & machinery, and transportation costs in the total manufacturing cost of the firm in 2012. The expenditure on labour in 2012 is ₹ 4,50,000. In 2013, the raw material expenses increased by 30% and all other expenses increased by 20%. If the company registered a profit of ₹ 10 lakhs in 2012, at what price (in ₹) was each air purifier sold?

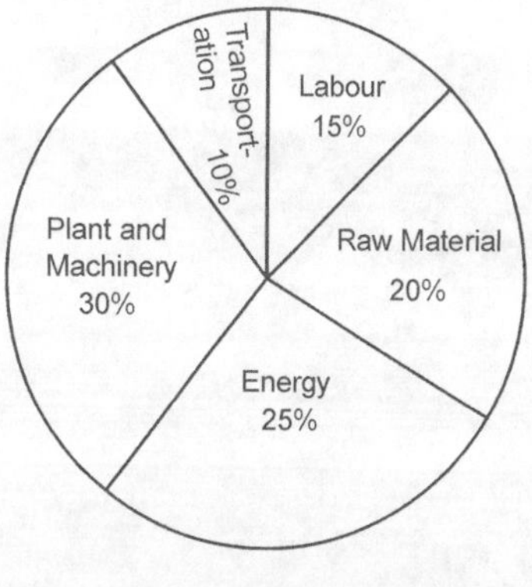

10. The sum of eight consecutive odd numbers is 656. The average of four consecutive even numbers is 87. What is the sum of the smallest odd number and second largest even number?

11. The ratio of male to female students in a college for five years is plotted in the following line graph. If the number of female students doubled in 2009, by what percent did the number of male students increase in 2009?

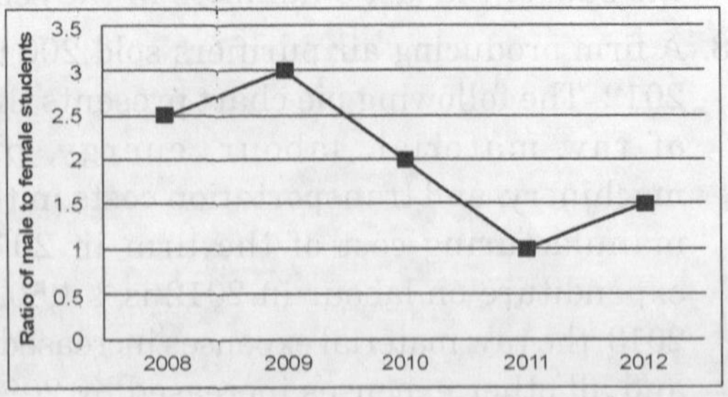

12. If $(z + 1/z)^2 = 98$, compute $(z^2 + 1/z^2)$

13. In a survey, 300 respondents were asked whether they own a vehicle or not. If yes, they were further asked to mention whether they own a car or scooter or both. Their responses are tabulated below. What percent of respondents do not own a scooter?

		Men	Women
Own vehicle	Car	40	34
	Scooter	30	20
	Both	60	46
Do not own vehicle		20	50

14. Round–trip tickets to a tourist destination are eligible for a discount of 10% on the total fare. In addition, groups of 4 or more get a discount of 5% on the total fare. If the one way single person fare is ₹ 100, a group of 5 tourists purchasing round–trip tickets will be charged ₹ ______.

ANSWERS

MCQ Type Questions

1. (*b*) 2. (*a*) 3. (*c*) 4. (*c*) 5. (*a*) 6. (*d*) 7. (*d*) 8. (*b*) 9. (*b*) 10. (*c*)
11. (*b*) 12. (*d*) 13. (*a*) 14. (*d*) 15. (*c*) 16. (*c*) 17. (*b*) 18. (*c*) 19. (*d*) 20. (*d*)
21. (*d*) 22. (*c*) 23. (*a*) 24. (*b*) 25. (*c*) 26. (*b*) 27. (*d*) 28. (*b*) 29. (*c*) 30. (*d*)
31. (*a*) 32. (*d*) 33. (*d*) 34. (*c*) 35. (*d*)

Numerical Type Questions

1. (120) 2. (7) 3. (2006) 4. (32) 5. (1300 to 1300) 6. (180 to 180) 7. (495) 8. (22)
9. (20,000) 10. (163) 11. (140 to 140) 12. (96) 13. (48) 14. (850)

EXPLANATIONS

MCQ TYPE QUESTIONS

1. $\dfrac{\alpha^n+\beta^n}{\alpha^{-n}+\beta^{-n}} = \dfrac{\alpha^n+\beta^n}{\dfrac{1}{\alpha^n}+\dfrac{1}{\beta^n}} = \dfrac{\alpha^n+\beta^n}{\dfrac{\beta^n+\alpha^n}{\alpha^n\times\beta^n}} = \alpha^n\times\beta^n$

$\alpha^n \times \beta^n = (\alpha\times\beta)^n = (4)^n$

So, the value of $\dfrac{\alpha^n+\beta^n}{\alpha^{-n}+\beta^{-n}}$ is 4^n.

2. From question

Total number of faculty members = 150

The number of faculty members having facebook account = (FB) = 55

The number of faculty members having whatsapp = (W) = 85

The number of faculty members do not have face book or Whats App accounts = 30

The number of faculty members having any account = 150 – 30 = 120

The number of faculty members having both the accounts = (FB + W) – 120 = (55 + 85) – 120 = 20

$\therefore$ The number of faculty members connected only through Facebook accounts = 55 – 20 = 35

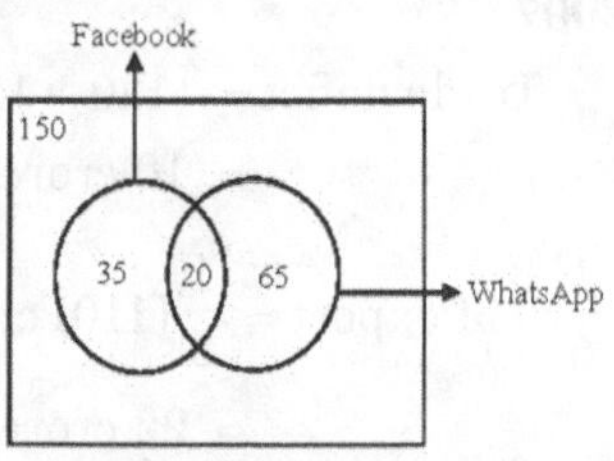

So the number of faculty members connected only through facebook accounts is 35.

3. Given, $|9y-6| = 3$

It has two values either

$(9y-6) = 3 \quad ...(i)$

Or $\quad -(9y-6) = 3$

$9y - 6 = -3 \quad ...(ii)$

From equation (*i*),

$9y = 9$

$\therefore \quad y = 1$

Or from equation (*ii*),

$9y = 3$

$\therefore \quad y = \dfrac{1}{3}$

Now $\quad y^2 = \dfrac{4y}{3}$

put $\quad y = 1,$

we get $\quad = 1-\dfrac{4}{3} = -\dfrac{1}{3}$

and also put $\quad y = \dfrac{1}{3}$

we get, $\quad \dfrac{1}{9}-\dfrac{4}{9} = \dfrac{-3}{9} = \dfrac{-1}{3}$

So, the required value is $\dfrac{-1}{3}$.

4. Given $\quad q^{-a} = \dfrac{1}{r}$

$\Rightarrow \quad \dfrac{1}{q^a} = \dfrac{1}{r}$

$\Rightarrow \quad q^a = r \quad ...(i)$

$r^{-b} = \dfrac{1}{s}$

$\Rightarrow \quad \dfrac{1}{r^b} = \dfrac{1}{s}$

$\Rightarrow \quad s = r^b \quad ...(ii)$

$s^{-c} = \dfrac{1}{q}$

$\Rightarrow \quad \dfrac{1}{s^c} = \dfrac{1}{q}$

$\Rightarrow \quad s^c = q \quad ...(iii)$

From equation (*i*),

$q^a = r$

$(s^c)^a = r \quad$ (from eq. (*iii*))

$S^{ac} = r$

$s^{ac} = (s)^{\frac{1}{b}} \quad$ (from eq. (*ii*))

$s^{abc} = s = s^1$

$\therefore \quad abc = 1$

So the value of *abc* is 1

5. Let the two digit number be xy

Now sum of the digit of a two digit number is 12

$\therefore \quad x + y = 12 \quad ...(i)$

Now according to question,

$(10y + x) - (10x + y) = 54$

$\therefore \quad 9y - 9x = 54$

$\therefore \quad y - x = 6 \quad ...(ii)$

Now solving equation (*i*) and (*ii*) we get,

$x = 3$ and $y = 9$

Hence the original number is 39.

$\therefore$ Option (*a*) is correct.

6. Amount invested by Company P in 2006 = ₹ 8x

and amount invested by Company Q in 2006 = ₹ 9x

Interest from Company P

$$= \frac{\frac{₹\,8x \times 6}{100}}{\frac{₹\,9x \times 4}{100}}$$

$$= \frac{4}{3}$$

∴ Option (d) is correct.

7. Type-I achieved a growth of 53% in the period which is higher than any other type of battery

10. Total number of students in class = 44

Average of marks,

$$= \frac{(2\times21)+(3\times15)+(1\times11)+(2\times23)+(5\times31)}{42}$$

$$= \frac{42+45+11+46+155}{44} = \frac{299}{44} = 6.795$$

∴ Option (c) is correct.

11. In the question the monthly average rainfall chart for 50 years has been given.

Let us check the options.

(i) On average, it rains more in July than in December ⇒ correct.

(ii) Every year, the amount of rainfall in August is more than that in January.

⇒ may not be correct because average rainfall is given in the question.

(iii) July rainfall can be estimated with better confidence than February rainfall.

⇒ From chart it is clear the gap between 5 percentile and 95 percentile from average is higher in February than that in July ⇒ correct.

(iv) In August at least 500 mm rainfall

⇒ May not be correct, because its 50 year average.

So correct option (b) (i) and (iii).

12. Item : 2

$$\frac{\frac{20}{100}\times250\times10^7}{\frac{20}{100}\times500\times10^3}$$

$0.5\times10^4 = 5\times10^3$ $\boxed{1}$ = Item 2

Item : 3

$$\frac{23\times250\times10^7}{19\times500\times10^3}$$

1.2 = Item 3

Item : 6

$$\frac{19}{16} = 1.18 = \text{Item 6}$$

Item : 5

$$\frac{20}{12} = \frac{5}{3} = 1.6 \Rightarrow \boxed{1.6 = \text{Item 5}}$$

14. For 2005,

trade defict = (90 – 70) crores = 20 crores

Now $\frac{1}{5}^{\text{th}}$ of export = $\frac{1}{5}$(70) crores

= 14 crores ≠ 20 crores.

Hence option (a) is wrong.

For 2004,

Trade defict =(80 – 70) crores = 10 crores

Now $\frac{1}{5}^{\text{th}}$ of export = $\frac{1}{5}$(70) crores

= 14 crores ≠ 20 crores.

Hence option (b) is wrong.

For 2007

Trade defict = (120 – 110) crores

= 10 crores

Now $\frac{1}{5}^{\text{th}}$ of export = $\frac{1}{5}$(110) crores

= 22 crores ≠ 10 crores.

Hence option (c) is wrong.

For 2006,

Trade defict = (120 – 100) crores

= 20 crores

Now $\frac{1}{5}^{\text{th}}$ of export = $\frac{1}{5}$(100) crores

= 20 crores = Trade deficts

Hence option (d) is correct.

15. Total question

44 × 2 = 88

44 × 3 = 132

$$\frac{144}{132} = \frac{88}{308}$$

Total marks obtained = (21×2) + (15×3) + (23×2)

= 133

Total Number of students = 44

$$\text{Average} = \frac{133}{44} = 3.02$$

16. Take number of female students in 2011 = 100

$\therefore$ Number of male in 2011 = 100

No. of female in 2012 = 100

No. of male in 2012 = 150

$$\text{Ratio} = \frac{150}{100}$$

17. All multiples of 10 from 2 to 198

$$= 10, 20, \ldots\ldots\ldots.190$$

Now $a_n = a + (n-1)d$

$$190 = 10 + (n-1) \times 10$$

$$180 = (n-1) \times 10$$

$$18 = n - 1$$

$$\therefore \quad n = 19$$

Hence total number of multiples are 19.

Now $S_n = \frac{n}{2}[2a + (n-1)d]$

$$= \frac{19}{2}[20 + (18)(10)]$$

$$= \frac{19}{2}[20 + 180] = 1900$$

Now average of all multiples of 10 from 2 to198

$$= \frac{10 + 20 + \ldots + 190}{19}$$

$$\therefore \quad = \frac{1900}{19} = 100$$

18. $\sqrt{12 + \sqrt{12 + \sqrt{12 + \ldots}}}$

For a particular type of question

i.e. $\sqrt{a + \sqrt{a + \sqrt{a + \ldots}}}$ we find a two consecutive number whose product is a. The greater number among the two consecutive number is the answer.

Here $a = 12$

i.e. $12 = 3 \times 4$

↓

greater number

Hence $\sqrt{12 + \sqrt{12 + \sqrt{12 + \ldots}}} = 4$

19. Here it is given that x is real

Now $|x^2 - 2x + 3| = 11$

$\Rightarrow$ $x^2 - 2x + 3 = 11$ or $x^2 - 2x + 3 = -11$

For $x^2 - 2x + 3 = 11 \Rightarrow x^2 - 2x - 8 = 0$

$$b^2 - 4ac = 4 - 4(1)(-8) = 36 > 0$$

Hence roots are real.

and for $x^2 - 2x + 14 = 0$

$$b^2 - 4ac = 4 - 4(1)(14) < 0$$

Hence roots are imaginary

$\therefore$ The value of x is the root of $x^2 - 2x - 8 = 0$

Now $x = \frac{+2 \pm \sqrt{4 - 4(1)(-8)}}{2}$

$$= \frac{2 \pm \sqrt{36}}{2}$$

$$= \frac{2 \pm 6}{2}$$

$$\therefore \quad x = 4, -2.$$

Now for $x = 4$,

$$|-x^3 + x^2 - x| = |-64 + 16 - 4| = 52.$$

and for $x = -2$

$$|-x^3 + x^2 - x| = |+8 + 4 + 2| = 14$$

Hence possible values of $|-x^3 + x^2 - x| = 14, 52$

22. $0 = 44 + (n-1)(-2)$

$$n - 1 = 22$$

$$n = 23$$

Sum $= \frac{n}{2}(a + l)$

$$= \frac{23}{2}(44 + 0)$$

$$= 23 \times 22 = 506$$

Hence (*c*) is the answer

23. Let '*d*' be the # of days required

Let '*h*' be the daily working hours.

$\therefore$ *dh* can be taken as a measure of the total work be done in order to erect the structure.

Now working hours per day $h' = \frac{23}{24}h$

$\therefore$ # of days $d' = \frac{dh}{h'} = \frac{dh}{\left(\frac{23}{24}h\right)} = \frac{24}{23}d$

Let c = daily wages earlier

$$\therefore \quad c = \frac{13,200}{d}$$

Now daily wages $c' = \frac{6}{5}c$

$\therefore$ Total new cost $= c' \times d'$

$$= \frac{6}{5} \times \frac{13200}{d} \times \frac{24}{23}d \approx 16,500$$

24. Given: The expression

$$S = \frac{1}{\sqrt{1}+\sqrt{2}} + \frac{1}{\sqrt{2}+\sqrt{3}} + ...+ \frac{1}{\sqrt{80}+\sqrt{81}}$$

To find: Sum of the expression.

Analysis: Each term of the expression is having irrational denominator. One can also see, that both the elements in the denominator of a term differ by only 1.

This gives an idea that on rationalising the denominator, each denominator will become 1, so the sum will be easier than earlier.

Rationalising the denominator,

$$S = \frac{\sqrt{2}-\sqrt{1}}{(\sqrt{2})^2-(\sqrt{1})^2} + \frac{\sqrt{3}-\sqrt{2}}{(\sqrt{3})^2-(\sqrt{2})^2} + ... + \frac{\sqrt{81}-\sqrt{80}}{(\sqrt{81})^2-(\sqrt{80})^2}$$

$$= \frac{\sqrt{2}-\sqrt{1}}{1} + \frac{\sqrt{3}-\sqrt{2}}{3-2} + ... + \frac{\sqrt{81}-\sqrt{80}}{81-80}$$

$$= \sqrt{2}-\sqrt{1}+\sqrt{3}-\sqrt{2}+....+81-\sqrt{80}$$

Now, one can see that the terms will cancel out.

Each term except $\sqrt{1}$ & $\sqrt{81}$

Occurs twice, once as positive and once a negative. So, after cancelling out, we are left with the following:

$$S = \sqrt{81}-\sqrt{1} = 9-1 = 8$$

So, the answer is (*b*).

25. $\frac{\text{Mon + Tues + Wed.}}{3} = 41$

Mon + Tues + Wed. = 123 ...(1)

$\frac{\text{Tues + Wed + Thurs.}}{3} = 43°$

Tue + Wed + Thu. = 129° ...(2)

(2) – (1)

Tues + Wed + Thu – (Mon + Tues. + Wed.)

= 129 – 123 = 6°

Thu. – Mon. = 6° ⇒ $\frac{115x}{100} - x = 6°$

Thus. = Mon × $\frac{115}{100} = \frac{15x}{100} = 6°$

Mon = x $x = 40°$

Thurs = $\frac{115x}{100}$

∴ Thurs = $\frac{115 \times 40}{100} = 46°$

30. Let $1.001 = x$

$x^{1259} = 3.52$, and $x^{2062} = 7.85$

∴ $x^{3321} = x^{1259} . x^{2062}$

$= 3.52 \times 7.85$

$= 27.64$

31. Let number of ₹ 20 notes be x and ₹ 10 notes be y

∴ $20x + 10y = 230$ ...(1)

and $x + y = 14$ ...(2)

Solving equations (1) and (2), we have

$x = 9$ and $y = 5$

Hence numbers of 10 rupee notes are 5.

32. Total budget = 10,500

Expenditure other than savings = 9000

Hence approximate percentage of monthly budget

$$= \frac{9000}{10500} = 86\%$$

33. By observation of the table, we can say S

	P	Q	R	S
Requirement	800	600	300	200
Potency	0.4	0.5	0.4	0.8

34. Let S = 4 (1 + 11 + 111 + ...)

$$= \frac{4}{9}(9 + 99 + 999 + ...)$$

$$= \frac{4}{9}\{(10-1)+(10^2-1)+(10^3-1)+........\}$$

$$= \frac{4}{9}\{(10+10^2+....+10^n)-n\}$$

$$= \frac{4}{9}\left\{10\frac{(10^n-1)}{9}-n\right\}$$

$$= \frac{4}{81}\{10^{n+1}-9n-10\}$$

35. Given : $f(y) = \frac{|y|}{y}$

$$\Rightarrow f(q) = \frac{|q|}{q}$$

$$f(-q) = \frac{|-q|}{-q}$$

$$= \frac{-|q|}{q}$$

$$|f(q)-f(q)| = \frac{|q|}{q} + \frac{|q|}{q}$$

$$= \frac{2|q|}{q} = 2$$

NUMERICAL TYPE QUESTIONS

1. From the given graph large plants which are having installed capacity of at least 200 tonnes are 1, 4, 8 and 9.

So, the total production of large plants

$$= 160 + 190 + 230 + 190$$
$$= 770$$

Now the remaining plants with installed capacity is less then 200 tonnes are 2, 3, 5, 6 and 7.

So, total production of small plants

$$= 150 + 160 + 120 + 100 + 120$$
$$= 650$$

Then difference $= 770 - 650$
$$= 120$$

2. Unit digit of $211^{870} + 146^{127} \times 3^{424}$ is $1 + 6 \times 1 = 7$

3. Increase in exports in 2006

$$= \frac{100-70}{70} = 42.8\%$$

Increase in imports in 2006

$$= \frac{120-90}{90} = 33.3\%$$

which is more than any other year.

4. According to question,

There are 40 males in Electrical Engg.

∴ Total number of females in Electrical Engg.

$$= 72 - 40 = 32$$

∴ Total number of students in Electrical Engg. = 72

⇒ 20% = 72

∴ 30% = Total number of students in Civil Engg.

∴ Total number of students in Civil Engg.

$$= \frac{30 \times 72}{20} = 36 \times 3 = 108$$

∴ Number of female students in Civil Engg.

$$= \frac{4 \times 108}{9} = 4 \times 12 = 48$$

Now, total number of students in Mechanical Engg.

$$= \frac{10 \times 72}{20} = 36$$

∴ Number of female students in Mechanical Engg.

$$= \frac{4 \times 36}{9} = 16$$

∴ Difference between female students in Civil and Mechanical departments = 48 – 16 = 32

5. Fixed cost = ₹ 50,000

Variable cost = ₹ 800Q

Q = daily production in tones

For 100 tonnes of production daily, total cost of production = 50,000 + 800 × 100 = 130,000

So, cost of production per tonne of daily production

$$= \frac{130{,}000}{100} = ₹\,1300.$$

6. Let the angles of quadrilateral are $3x$, $4x$, $5x$, $6x$

So, $3x + 4x + 5x + 6x = 360$

$$x = 20$$

Smallest angle of quadrilateral = 3 × 20 = 60°

Smallest angle of triangle $= \frac{2}{3} \times 60° = 40°$

Largest angle of triangle = 2 × 40° = 60°

Three angles of triangle are 40°, 60°, 80°

Largest angle of quadrilateral is 120°

Sum (2nd largest angle of triangle + largest angle of quadrilateral) = 60° + 120° =180°.

7. 8th observation is 7 × 2 = 14 more than 1st observation

9th observation is 14 more than 2nd observation

10th observation is 14 more than 3rd observation

11th observation is 14 more than 4th observation

12th observation is 14 more than 5th observation

Total 14 × 5 =70

Sum of the first five numbers = 425

Sum of last five numbers = 495

8.

	2012	2013
Transport (10%)	300,000	360,000
Labour (15%)	450,000	540,000
Raw material (20%)	750,000	780,000
Energy (25%)	750,000	900,000
Plant and Machinery (30%)	900,000	1,080,000
Total	**3,000,000**	**3,660,000**

Percentage increase in total cost = 22%

9. Total expenditure $= \frac{15}{100} x = 4{,}50{,}000$

$$x = 3 \times 10^6$$

Profit = 10 lakhs

So, total selling price = 40,00,000 ...(1)

Total purifies = 200 ...(2)

S.P. of each purifier = (1)/(2) = 20,000

10. Eight consecutive odd number = 656

$$a - 6, a - 1, a - 2, a, a + 2, a + 4, a + 6$$
$$a + 8 = 656$$
$$a = 81$$

Smallest $m = 75$ ...(1)

Average consecutive even numbers

$$\Rightarrow \frac{a-2+a+a+2+a+4}{4} = 87$$

$\Rightarrow a = 86$

Second largest number = 88 ...(2)

Adding equation (1) and (2),

$$75 + 88 = 163$$

11. Ratio of male to female student in 2008 = 5 : 2

In 2009, let the male student be $5x$ and female student be $2x$

Now if number of female students double in 2009

$\therefore$ Number of female student = $4x$

$\therefore$ Number of male student = $12x$

Gender	2008	2009
Male	$5x$	$12x$
Female	$2x$	$4x$

% of male student increase in 2009

$$= \left[\left(\frac{12x-5x}{5x}\right)\times 100\right]\%$$

$$= \left[\frac{7}{5}\times 100\right]\%$$

$$= 140\%$$

12. Expanding $\left(z+\frac{1}{z}\right)^2 = 98$

$$\therefore \quad z^2 + \frac{1}{z^2} + 2(z)\left(\frac{1}{z}\right) = 98$$

$$\therefore \quad z^2 + \frac{1}{z^2} = 96$$

13. Total respondents = 300

Now men respondents who do not have scooter

= 40 + 20 = 60

And women respondents who do not have scooter

= 34 + 50 = 84

$\therefore$ Total respondents who do not have scooter

= 60 + 84 = 144

$$\therefore \text{Required percent} = \left(\frac{144}{300}\times 100\right)\% = 48\%$$

14. One way single person fare = ₹ 100

$\therefore$ Two way fare for single person = ₹ 200

$\therefore$ For 5 persons two way fare = ₹ 1000

Now, total discount = (10 + 5)% = 15%

$$\text{Discount amount} = ₹\left[\frac{15}{100}\times 1000\right] = ₹\ 150$$

Amount to be paid = ₹(1000 – 150) = ₹ 850

2

CHAPTER

Percentage and Its Applications

MCQ TYPE QUESTIONS

2016

1. In a huge pile of apples and oranges, both ripe and unripe mixed together, 15% are unripe fruits. Of the unripe fruits, 45% are apples. Of the ripe ones, 66% are oranges. If the pile contains a total of 5692000 fruits, how many of them are apples?
   (a) 2029198 (b) 2467482
   (c) 2789080 (d) 3577422
2. A person moving through a tuberculosis prone zone has a 50% probability of becoming infected. However, only 30% of infected people develop the disease. What percentage of people moving through a tuberculosis prone zone remains infected but does not show symptoms of disease?
   (a) 15 (b) 33
   (c) 35 (d) 37
3. (x % of y) + (y % of x) is equivalent to ______.
   (a) 2 % of xy (b) 2 % of (xy/100)
   (c) xy % of 100 (d) 100 % of xy

2014

4. The population of a new city is 5 million and is growing at 20% annually. How many years would it take to double at this growth rate?
   (a) 3 – 4 years (b) 4 – 5 years
   (c) 5 – 6 years (d) 6 – 7 years
5. One percent of the people of country X are taller than 6 ft. Two percent of the people of country Y are taller than 6 ft. There are thrice as many people in country X as in country Y. Taking both countries together, what is the percentage of people taller than 6 ft?
   (a) 3.0 (b) 2.5
   (c) 1.5 (d) 1.25
6. Industrial consumption of power doubled from 2000-2001 to 2010-2011. Find the annual rate of increase in percent assuming it to be uniform over the years.
   (a) 5.6 (b) 7.2
   (c) 10.0 (d) 12.2
7. The Gross Domestic Product (GDP) in Rupees grew at 7% during 2012-2013. For international comparison, the GDP is compared in US Dollars (USD) after conversion based on the market exchange rate. During the period 2012-2013 the exchange rate for the USD increased from ₹ 50/ USD to ₹ 60/ USD. India's GDP in USD during the period 2012-2013
   (a) increased by 5% (b) decreased by 13%
   (c) decreased by 20% (d) decreased by 11%

2013

8. A firm is selling its product at ₹ 60 per unit. The total cost of production is ₹ 100 and firm is earning total profit of ₹ 500. Later, the total cost increased by 30%. By what percentage the price should be increased to maintained the same profit level.
   (a) 5 (b) 10
   (c) 15 (d) 30

2012

9. An automobile plant contracted to buy shock absorbers from two suppliers X and Y. X supplies 60% and Y supplies 40% of the shock absorbers. All shock absorbers are subjected to a quality test. The ones that pass the quality test are considered reliable. Of X's shock absorbers, 96% are reliable. Of Y's shock absorbers, 72% are reliable.
   The probability that a randomly chosen shock absorber, which is found to be reliable, is made by Y is
   (a) 0.288 (b) 0.334
   (c) 0.667 (d) 0.720

2011

10. There are two candidates P and Q in an election. During the campaign, 40% of the voters promised to vote for P, and rest for Q. However, on the day of election 15% of the voters went back on their promise to vote for P and instead voted for Q. 25% of the voters went back on their promise to vote for Q and instead voted for P. Suppose, P lost by 2 votes, then what was the total number of voters?
    (a) 100 (b) 110
    (c) 90 (d) 95

NUMERICAL TYPE QUESTION

2015

1. From a circular sheet of paper of radius 30 cm, a sector of 10% area is removed. If the remaining part is used to make a conical surface, then the ratio of the radius and height of the cone is ____.

ANSWERS

MCQ Type Questions

1. (*a*) **2.** (*c*) **3.** (*a*) **4.** (*a*) **5.** (*d*) **6.** (*b*) **7.** (*d*) **8.** (*a*) **9.** (*b*) **10.** (*a*)

Numerical Type Question

1. (2.06)

EXPLANATIONS

MCQ TYPE QUESTIONS

1. Given,

Total no. of fruits = 5692000

Unripe type of apples = 45% of 15% of 5692000

$$= \frac{45}{100} \times \frac{15}{100} \times 5692000$$

$$= 384210$$

Ripe type of apples $= (100-66)\% \times (100-15)\% \times 5692000$

$$= \frac{34}{100} \times \frac{85}{100} \times 5692000$$

$$= 1644988$$

$\therefore$ Total no. of apples = 384210 + 1644988 = 2029198

2. The percentage of people moving through a tuberculosis prone zone remains infected but does not show symptoms of disease.

$$= (100-30)\% \times 50\% = \frac{70}{100} \times \frac{50}{100} = \frac{35}{100} = 35\%$$

3. (x% of y) + (y% of x)

$$\frac{x}{100} \times y + \frac{y}{100} \times x = \frac{2xy}{100} = 2\% \text{ of } xy$$

$\therefore$ Option (*a*) is correct.

4. $A = P\left[1+\frac{r}{100}\right]^t$

$$10 \text{ million} = 5 \text{ million}\left[1+\frac{20}{100}\right]^t$$

$$2 = \left[1+\frac{1}{5}\right]^t$$

$$2 = \left(\frac{6}{5}\right)^t = \log 2 = t \log\left(\frac{6}{5}\right)$$

$$0.301 = t\ 0.0791$$

$$t = \frac{0.301}{0.0791} = 3.8 \text{ years}$$

$\therefore$ Hence around 3–4 years would it take to double the growth rate

5. Let number of people in country $y = 100$

So, number of people in country $x = 300$

Total number of people taller than 6ft in both the countries $= 300 \times \frac{1}{100} + 100 \times \frac{2}{100} = 5$

% of people taller than 6ft in both the countries

$$= \frac{5}{400} \times 100 = 1.25\%.$$

7. Per ₹ 100 final value ₹ 107

$\Rightarrow$ per $\frac{100}{50}$ Dollars final value $\frac{107}{60}$

for 100 dollars ____ ?

$$= \frac{100 \times 50}{100} \times \frac{107}{60} = 89.16$$

Descreased by 11%

9.

	X	*Y*
Supply	60%	40%
Reliable	96%	72%
Overall	0.576	0.288

$$\therefore \quad P(x) = \frac{0.288}{0.576+0.288} = 0.334$$

10.

P	Q
40%	60%
–6%	+6%
+15%	–15%
49%	51%

$\therefore$ 2% = 2 $\Rightarrow$ 100% = 100

NUMERICAL TYPE QUESTION

1. 90% of area of sheet = Cross sectional area of cone

$\Rightarrow 0.9 \times \pi \times 30 \times 30 = \pi \times r_1 \times 30$

($\because$ Slant height of cone (l) = 30)

$\Rightarrow$ 27 cm = r_1

$\therefore$ Height of the cone $= \sqrt{30^2 - 27^2} = 13.08$ cm

Hence ratio of radius and height is 2.06.

3 CHAPTER

Time and Work

MCQ TYPE QUESTIONS

2016

1. P, Q, R and S are working on a project. Q can finish the task in 25 days, working alone for 12 hours a day. R can finish the task in 50 days, working alone for 12 hours per day. Q worked 12 hours a day but took sick leave in the beginning for two days. R worked 18 hours a day on all days. What is the ratio of work done by Q and R after 7 days from the start of the project?

(*a*) 10:11 (*b*) 11:10

(*c*) 20:21 (*d*) 21:20

2. S, M, E and F are working in shifts in a team to finish a project. M works with twice the efficiency of others but for half as many days as E worked. S and M have 6 hour shifts in a day whereas E and F have 12 hours shifts. What is the ratio of contribution of M to contribution of E in the project?

(*a*) 1 : 1 (*b*) 1 : 2

(*c*) 1 : 4 (*d*) 2 : 1

3. It takes 10 s and 15 s, respectively, for two trains travelling at different constant speeds to completely pass a telegraph post. The length of the first train is 120 m and that of the second train is 150 m. The magnitude of the difference in the speeds of the two trains (in m/s) is ______.

(*a*) 2.0 (*b*) 10.0

(*c*) 12.0 (*d*) 22.0

4. The velocity V of a vehicle along a straight line is measured in m/s and plotted as shown with respect to time in seconds. At the end of the 7 seconds, how much will the odometer reading increase by (in m) ?

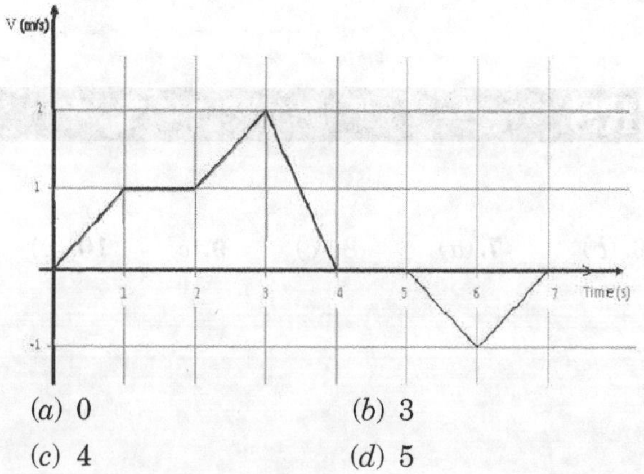

(*a*) 0 (*b*) 3

(*c*) 4 (*d*) 5

5. Ananth takes 6 hours and Bharath takes 4 hours to read a book. Both started reading copies of the book at the same time. After how many hours is the number of pages **to be** read by Ananth, twice that **to be** read by Bharath? Assume Ananth and Bharath read all the pages with constant pace.

(*a*) 1 (*b*) 2

(*c*) 3 (*d*) 4

2015

6. An electric bus has onboard instruments that report the total electricity consumed since the start of the trip as well as the total distance covered. During a single day of operation, the bus travels on stretches *M*, *N*, *O* and *P*, in that order. The cumulative distances travelled and the corresponding electricity consumption are shown in the table below.

Stretch	Comulative distance(km)	Electricity used (kWh)
M	20	12
N	45	25
O	75	45
P	100	57

The stretch where the electricity consumption per km is minimum is

(*a*) M

(*b*) N

(*c*) O

(*d*) P

2014

7. It takes 30 minutes to empty a half-full tank by draining it at a constant rate. It is decided to simultaneously pump water into the half-full tank while draining it. What is the rate at which water has to be pumped in so that it gets fully filled in 10 minutes?

(*a*) 4 times the draining rate

(*b*) 3 times the draining rate

(*c*) 2.5 times the draining rate

(*d*) 2 times the draining rate

2013

8. A tourist covers half of his journey by train at 60 km/h, half of the remainder by bus at 30 km/h and the rest by cycle at 10 km/h. The average speed of the tourist in km/h during his entire journey is

(*a*) 36 (*b*) 30
(*c*) 24 (*d*) 18

9. A car travels 8 km in the first quarter of an hour, 6 km in the second quarter and 16 km in the third quarter. The average speed of the car in km per hour over the entire journey is

(*a*) 30 (*b*) 36
(*c*) 40 (*d*) 24

2011

10. The fuel consumed by a motorcycle during a journey while traveling at various speeds is indicated in the graph below.

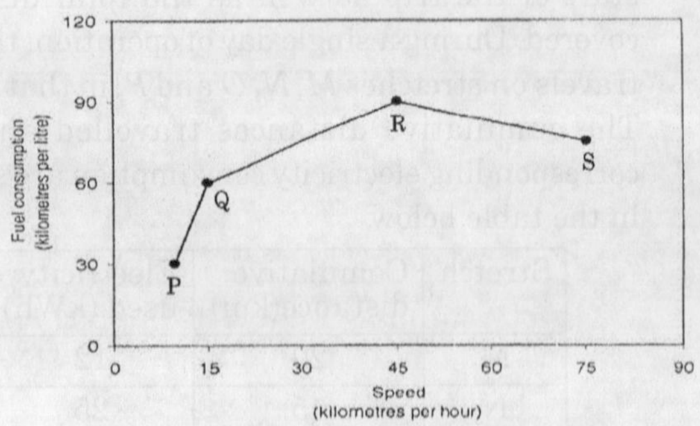

The distances covered during four laps of the journey are listed in the table below

Lap	Distance (kilometers)	Average speed (kilometers per hour)
P	15	15
Q	75	45
R	40	75
S	10	10

From the given data, we can conclude that the fuel consumed per kilometre was least during the lap

(*a*) P
(*b*) Q
(*c*) R
(*d*) S

2010

11. 5 skilled workers can build a wall in 20 days; 8 semi-skilled workers can build a wall in 25 days; 10 unskilled workers can build a wall in 30 days. If a team has 2 skilled, 6 semi-skilled and 5 unskilled workers, how long will it take to build the wall?

(*a*) 20 days
(*b*) 18 days
(*c*) 16 days
(*d*) 15 days

NUMERICAL TYPE QUESTIONS

2015

1. A tiger is 50 leaps of its own behind a deer. The tiger takes 5 leaps per minute to the deer's 4. If the tiger and the deer cover 8 metre and 5 metre per leap respectively, what distance in meters will the tiger have a run before it catches the deer?

2014

2. A man can row at 8 km per hour in still water. If it takes him thrice as long to row upstream, as to row downstream, then find the stream velocity in km per hour.

3. A train that is 280 metres long, travelling at a uniform speed, crosses a platform in 60 seconds and passes a man standing on the platform in 20 seconds. What is the length of the platform in metres?

ANSWERS

MCQ Type Questions

1. (*c*) **2.** (*b*) **3.** (*a*) **4.** (*d*) **5.** (*c*) **6.** (*b*) **7.** (*a*) **8.** (*c*) **9.** (*c*) **10.** (*a*)
11. (*d*)

Numerical Type Questions

1. (800) **2.** (4) **3.** (560 to 560)

EXPLANATIONS

MCQ TYPE QUESTIONS

1. Q can finish 1 day's 1 hour's work

$$= \frac{1}{25 \times 12} = \frac{1}{300}$$

R can finish 1 day's 1 hour's work

$$= \frac{1}{50 \times 12}$$

$$= \frac{1}{600}$$

Now Q working hours

$$= (7 - 2) \times 12$$

$$= 60 \text{ hr}$$

$\therefore$ R working hours = 7 × 18 = 126

After 7 days, the ratio of work done by Q & R is

$$Q : R$$

$$\frac{60}{300} : \frac{126}{600}$$

$$Q : R = 120 : 126$$

$$= 20 : 21$$

2. M is twice as efficient as E but worked for half as many days. So in this case they will do equal work if their shifts had same timings. But M's shift is for 6 hours, while E's shift for 12 hours. Hence, E will do twice the work as M.

Ratio of contribution of M : E in work is 1 : 2.

3. Here the speed of first train (S_1) = $\frac{120}{10}$ = 12 m/s

and the speed of second train (S_2) = $\frac{150}{15}$ = 10 m/s

$\therefore$ The magnitude of difference in the speeds of two trains = ($S_1 - S_2$) = (12 – 10) m/s = 2 m/s

4. From the given figure.

The odometer reading increases from starting point to end point.

Odometer reading = Area of the given diagram

Area of the velocity and time graph per second

1st sec ⇒ triangle

$$= \frac{1}{2} \times 1 \times 1 = \frac{1}{2}$$

2nd sec ⇒ square

$$= 1 \times 1 = 1$$

3rd sec ⇒ square + triangle

$$= 1 \times 1 + \frac{1}{2} \times 1 \times 1 = 1\frac{1}{2}$$

4th sec ⇒ triangle

$$= \frac{1}{2} \times 1 \times 2 = 1$$

5th sec ⇒ straight line

$$= 0$$

6th sec ⇒ triangle

$$= \frac{1}{2} \times 1 \times 1 = \frac{1}{2}$$

7th sec ⇒ triangle

$$= \frac{1}{2} \times 1 \times 1 = \frac{1}{2}$$

Total Odometer reading at 7 seconds

$$= \left(\frac{1}{2} + 1 + 1\frac{1}{2} + 1 + 0 + \frac{1}{2} + \frac{1}{2}\right) \text{m}$$

$$= 5 \text{ m}$$

So, the odometer reading is increased by 5 m.

6. For $M \Rightarrow \frac{12}{20} = 0.6$

$$N \Rightarrow \frac{25}{45} = 0.55$$

$$O \Rightarrow \frac{45}{75} = 0.6$$

$$P \Rightarrow \frac{57}{100} = 0.57$$

7. $V_{half} = 30(s)$ drawing rate = s

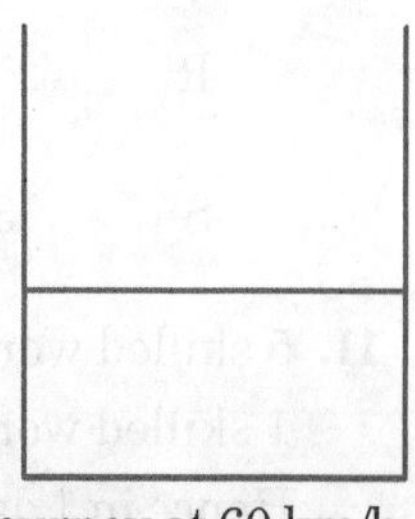

Total volume = 60 S tank

$(s^1)(10) - (s)10 = 30s$

$s^1(s) - s = 3s$

$s1 = 4s$

$s^1 = 4$ drawing rate

8. **Given:** A tourist covers half the journey at 60 km/h, one fourth at 30 km/h and the remaining one fourth at 10 km/h

To find: Avearge speed of the whole journey.

Analysis: One approach to this is to find individual time consumed in each of the three parts and use that value to get total time, and find average speed.

So,

Lets assume total length of journey = x km

$\therefore$ First phase at 60 km/h

$$\text{Time } t_1 = \frac{\text{Distance}}{\text{Speed}} = \frac{x}{2}\frac{1}{60} = \frac{x}{120} h$$

Second phase at 30 km/h

$$\text{Distance} = \frac{x}{4}\text{km}$$

$$\text{Time } t_2 = \frac{x}{4.30} = \frac{x}{120}$$

Third phase at 10 km/h

$$\text{Distance} = \frac{x}{4}\text{ km}$$

$$\text{Time } t_3 = \frac{x}{4.10} = \frac{x}{40}h$$

$$\therefore \quad \text{Total time} = \frac{x}{120}+\frac{x}{120}+\frac{x}{40} = \frac{x}{40}\left(\frac{1}{3}+\frac{1}{3}+1\right)$$

$$= \frac{x}{40}\frac{(1+1+3)}{3} = \frac{5x}{120}$$

$$\therefore \text{Average speed} = \frac{\text{Total Distance}}{\text{Total time}}$$

$$= \frac{x \cdot 120}{5x} = 24 \text{ km}/h$$

Hence, the answer is (c)

9. $\text{Average Speed} = \frac{\text{Total distance}}{\text{total time taken}}$

$$\text{Total Distance} = \frac{8+6+16}{15+15+15} = \frac{30\times 60}{45}$$

$= 40$km/h.

10.

	Fuel consumption	Actual
P	60 km/l	$\frac{15}{60} = \frac{1}{4}l$
Q	90 km/l	$\frac{75}{90} = \frac{5}{6}l$
R	75 km/l	$\frac{40}{75} = \frac{8}{15}l$
S	30 km/l	$\frac{10}{30} = \frac{1}{3}l$

11. 5 skilled workers build wall in 20 days

1 skilled worker build wall in 20 × 5 days

Hence in 1 day, part of work done by skilled work

$$= \frac{1}{100}$$

Similarly in 1 day part of work done by semi-skilled workers

$$= \frac{1}{25\times 8}$$

and in 1 day part of work done by un-skilled worker

$$= \frac{1}{30\times 10}$$

So part of work done in 1 day by 2 skilled, 6 semi-skilled and 5 unskilled

$$= \frac{2}{100}+\frac{6}{200}+\frac{5}{300} = \frac{1}{15}$$

So work done by given workers in days = 15

NUMERICAL TYPE QUESTIONS

1. Tiger – 1 leap ⇒ 8 meter

Speed = 5 leap/hr = 40m/min

Deer → 1 leap = 5 meter

speed = 4hr = 20m/min

Let at time 't' the tiger catches the deer.

∴ Distance travelled by deer + initial distance between them

50 × 8 ⇒ 400m = distance covered by tiger.

⇒ 40 × t = 400 + 20t

$$\Rightarrow t = \frac{400}{200} = 20 \text{ min}$$

⇒ total distance ⇒ 400 + 20 × t = 800 m

2. Speed of man = 8;

Left distance = d

$$\text{Time taken} = \frac{d}{8}$$

Upstream: Speed of stream = s

⇒ speed upstream = S' = (8 – s)

$$t' = \left(\frac{d}{8-s}\right)$$

Downstream:

$$\text{Given speed downstream} = t'' = \frac{d}{8+s}$$

$$\Rightarrow 3t' = t'' \Rightarrow \frac{3d}{8-s} = \frac{d}{8+s}$$

$$\Rightarrow \quad \frac{3d}{8-s} = \frac{d}{8+s}$$

$$\Rightarrow \quad s = 4 \text{ km/hr}$$

3. Let the length of platform = x.m

and train length = 280 m (given)

According to Question

$$\frac{x+280}{60} = \frac{280}{20}$$

$$\therefore \quad x = 560 \text{ m.}$$

■■

4 CHAPTER

Ratio, Proportion and Mixtures

MCQ TYPE QUESTIONS

2015

1. A cube of side 3 units is formed using a set of smaller cubes of side 1 unit. Find the proportion of the number of faces of the smaller cubes visible to those which are NOT visible.

(*a*) 1 : 4

(*b*) 1 : 3

(*c*) 1 : 2

(*d*) 2 : 3

2011

2. A container originally contains 10 litres of pure spirit. From this container 1 litre of spirit is replaced with 1 litre of water. Subsequently, 1 litre of the mixture is again replaced with 1 litre of water and this process is repeated one more time. How much spirit is now left in the container?

(*a*) 7.58 litres

(*b*) 7.84 litres

(*c*) 7 litres

(*d*) 7.29 litres

ANSWERS

MCQ Type Questions

1. (*c*) **2.** (*d*)

EXPLANATIONS

MCQ TYPE QUESTIONS

1.

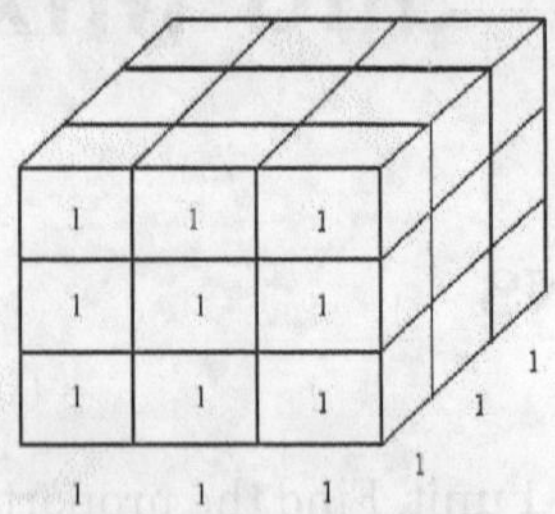

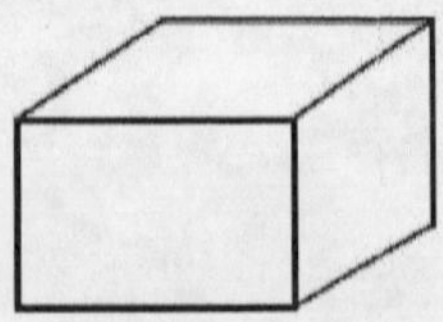

Number of faces per cube = 6

Total number of cubes = 9 × 3 = 27

∴ Total number of faces = 27 × 6 = 162

∴ Total number of non visible faces = 162 – 54 = 108

$$\therefore \frac{\text{Number of visible faces}}{\text{Number of non visible faces}} = \frac{54}{108} = \frac{1}{2}$$

2. $10\left(\frac{10-1}{10}\right)^3 = 10\left(\frac{9}{10}\right)^3 = \frac{729}{1000}$

$$\therefore \frac{729}{1000} \times 1 = 7.29 \text{ litres}$$

5 CHAPTER Permutations and Combinations & Probability

MCQ TYPE QUESTIONS

2016

1. Shaquille O' Neal is a 60% career free throw shooter, meaning that he successfully makes 60 free throws out of 100 attempts on average. What is the probability that he will successfully make exactly 6 free throws in 10 attempts?

(*a*) 0.2508 (*b*) 0.2816
(*c*) 0.2934 (*d*) 0.6000

2015

2. Four cards are randomly selected from a pack of 52 cards. If the first two cards are kings, what is the probability that the third card is a king?

(*a*) 4/52 (*b*) 2/50
(*c*) 1/52 × (1/52) (*d*) 1/52 × (1/52) × (1/50)

3. Five teams have to compete in a league, with every team playing every other team exactly once, before going to the next round. How many matches will have to be held complete the league round of matches?

(*a*) 20 (*b*) 10
(*c*) 8 (*d*) 5

4. Right triangle PQR is to be constructed in the xy – plane so that the right angle is at P and line PR is parallel to the-axis. The x and y coordinates of P, Q, and R are to be integers that satisfy the inequalities: $-4 \le x \le 5$ and $6 \le y \le 16$. How many different triangles could be constructed with these properties?

(*a*) 110 (*b*) 1,100
(*c*) 9,900 (*d*) 10,000

5. A coin is tossed thrice. Let X be the event that head occurs in each of the first two tosses. Let Y be the event that a tail occurs on the third toss. Let Z be the event that two tails occurs in three tosses. Based on the above information, which one of the following statements is TRUE?

(*a*) X and Y are not independent
(*b*) Y and Z are dependent
(*c*) Y and Z are independent
(*d*) X and Z independent

6. Ram and Ramesh appeared in an interview for two vacancies in the same department. The probability of Ram's selection is 1/6 and that of Ramesh is 1/8. What is the probability that only one of them will be selected?

(*a*) $\frac{47}{48}$ (*b*) $\frac{1}{4}$
(*c*) $\frac{13}{48}$ (*d*) $\frac{35}{48}$

7. Given set A = {2, 3, 4, 5} and Set B = {11, 12, 13, 14, 15}, two numbers are randomly selected, one from each set. What is probability that the sum of the two numbers equals 16 ?

(*a*) 0.20 (*b*) 0.25
(*c*) 0.30 (*d*) 0.33

8. The probabilities that a student passes in Mathematics, Physics and Chemistry are m, p and c respectively. Of these subjects, the student has 75% chance of passing in at least one, a 50% chance of passing in at least two and a 40% chance of passing in exactly two. Following relations are drawn in $m, p.c$:

I. $p + m + c = \frac{27}{20}$ II. $p + m + c = \frac{13}{20}$

III. $(p) \times (m) \times (c) = \frac{1}{10}$

(*a*) Only relation I is true.
(*b*) Only relation II is true.
(*c*) Relations II and III are true.
(*d*) Relations I and III are true.

2014

9. A five digit number is formed using the digits 1,3,5,7 and 9 without repeating any of them. What is the sum of all such possible five digit numbers?

(*a*) 6666660 (*b*) 6666600
(*c*) 6666666 (*d*) 6666606

10. You are given three coins: one has heads on both faces, the second has tails on both faces, and the third has a head on one face and a tail on the other. You choose a coin at random and toss it, and it comes up heads. The probability that the other face is tails is

(*a*) 1/4 (*b*) 1/3
(*c*) 1/2 (*d*) 2/3

2013

11. Out of all the 2-digit integers between 1 and 100, a 2-digit number has to be selected at random. What is the probability that the selected number is not divisible by 7?

(*a*) 13/90 (*b*) 12/90

(*c*) 78/90 (*d*) 77/90

12. What is the chance that a leap year, selected at random, will contain 53 Saturdays?

(*a*) 2/7 (*b*) 3/7

(*c*) 1/7 (*d*) 5/7

13. In a factory, two machines M1 and M2 manufacture 60% and 40% of the autocomponents respectively. Out of the total production, 2% of M1 and 3% of M2 are found to be defective. If a randomly drawn autocomponent from the combined lot is found defective, what is the probability that it was manufactured by M2?

(*a*) 0.35 (*b*) 0.45

(*c*) 0.5 (*d*) 0.4

2012

14. A and B are friends. They decide to meet between 1 PM and 2 PM on a given day. There is a condition that whoever arrives first will not wait for the other for more than 15 minutes. The probability that they will meet on that day is

(*a*) $\frac{1}{4}$ (*b*) $\frac{1}{16}$

(*c*) $\frac{7}{16}$ (*d*) $\frac{9}{16}$

2010

15. Given digits 2, 2, 3, 3, 3, 4, 4, 4, 4 how many distinct 4 digit numbers greater than 3000 can be formed?

(*a*) 50 (*b*) 51

(*c*) 52 (*d*) 54

NUMERICAL TYPE QUESTIONS

2015

1. How many four digit numbers can be formed with the 10 digits 0, 1, 2, ... 9 if no number can start with 0 and if repetitions are not allowed?

2014

2. In any given year, the probability of an earthquake greater than Magnitude 6 occurring in the Garhwal Himalayas is 0.04. The average time between successive occurrences of such earthquakes is _____ years.

3. 10% of the population in a town is HIV^+. A new diagnostic kit for HIV detection is available; this kit correctly identifies HIV^+ individuals 95% of the time, and HIV^- individuals 89% of the time. A particular patient is tested using this kit and is found to be positive. The probability that the individual is actually positive is _____

4. A batch of one hundred bulbs is inspected by testing four randomly chosen bulbs. The batch is rejected if even one of the bulbs is defective. A batch typically has five defective bulbs. The probability that the current batch is accepted is _____

ANSWERS

MCQ Type Questions

1. (*a*) **2.** (*b*) **3.** (*b*) **4.** (*c*) **5.** (*d*) **6.** (*b*) **7.** (*a*) **8.** (*d*) **9.** (*b*) **10.** (*b*)

11. (*d*) **12.** (*a*) **13.** (*c*) **14.** (*c*) **15.** (*b*)

Numerical Type Questions

1. 4536 **2.** (25 to 25) **3.** (0.47 to 0.48) **4.** (0.8145)

EXPLANATIONS

MCQ TYPE QUESTIONS

1. The question clearly explains that shaquille makes 60 free throws out of 100

Hence, Probability of free throw = $\frac{60}{100} = 0.6$

And probability of NOT free throw

$= 1 - 0.6 = 0.4$

So required probability of exactly 6 throws in 10 attempts will be given by

$^{10}C_6\, 0.6^6 \times 0.4^4 = 0.2508$

2. There are 4 kings in a pack of 52 cards.

If 2 cards are selected and both are kings, remaining cards will be 50 out of which 2 will be kings.

3. For a match to be played, we need 2 teams

No. of matches = No. of ways of selections 2 teams out of 5

$= {}^5C_2 = 10$

5. $x = \{HHT, HHH\}$

y depends on x

$z = \{TTH, TTT\}$

$\therefore$ 'd' is the correct choice.

6. P (Ram) = 1/6; p(Ramesh) = 1/8

p(only at) = p (Ram) × p (not ramesh) + p(Ramesh) × p(n_0 × R_{am})

$= \frac{1}{6} + \frac{7}{8} + \frac{1}{8} \times \frac{5}{6}$

$\Rightarrow \frac{12}{40} = \frac{1}{4}$

7. Here the given sets are

$A = \{2, 3, 4, 5\}$

and $B = \{11, 12, 13, 14, 15\}$

desired outcome = {(2, 14), (3, 13), (4, 12), (5, 11)}

$\therefore$ $n(E) = 4$

and total outcome = 4 × 5 = 20

$\therefore$ $n(s) = 20$

$\therefore$ Probability of an event

$= \frac{n(E)}{n(s)}$

$= \frac{4}{20} = \frac{1}{5} = 0.2$

9. The digit in unit place is selected in 4! Ways

The digit in tens place is selected in 4! Ways

The digit in hundreds place is selected in 4! Ways

The digit in thousands place is selected in 4! Ways

The digit in ten thousands place is selected in 4! Ways

Sum of all values for 1

$4! \times 1 \times (10^0 + 10^1 + 10^2 + 10^3 + 10^4)$

$= 4! \times 11111 \times 1$

Similarly for '3' $4! \times (11111) \times 3$

Similarly for '5' $4! \times (11111) \times 5$

Similarly for '7' $4! \times (11111) \times 7$

Similarly for '9' $4! \times (11111) \times 9$

$\therefore$ sum of all such numbers

$4! \times (11111) \times (1 + 3 + 5 + 7 + 9)$

$= 24 \times (11111) \times 25$

$= 6666600$

10. The probability that the other face is tail is $\frac{1}{3}$.

11. Given: All the 2-digit numbers

To find: Probability of selecting a number at random not divisible by 7

Analysis: Total number of 2-digit numbers

10, 11, ..., 99 → 90 numbers

Unfavourable numbers *i.e.* numbers divisible by 7 : 14, 21, ..., 98

Number of unfavourable numbers = 13

This can be hand counted or we can use the following properties:

Last term= First time + (no. – 1) Diff

98 = 14 + (no. – 1) 7

$\Rightarrow$ no. = 13

$\Rightarrow$ Probability = $\frac{\text{Total} - \text{unfavourable}}{\text{Total}}$

$= \frac{90 - 13}{90} = 77/90$

Hence, the solution is (*d*)

12. at a leap year

52 weeks, and 2 extra day they are (Mon, Tues) (Tues, Wed) (Wed Thu.) (Thus. Fri) (Fri. Sat) (Sat. Sun) (Sun Mond.)

$n(s) = 7$

$n(E) = 2 \quad P(E) = \frac{2}{7}$

14.

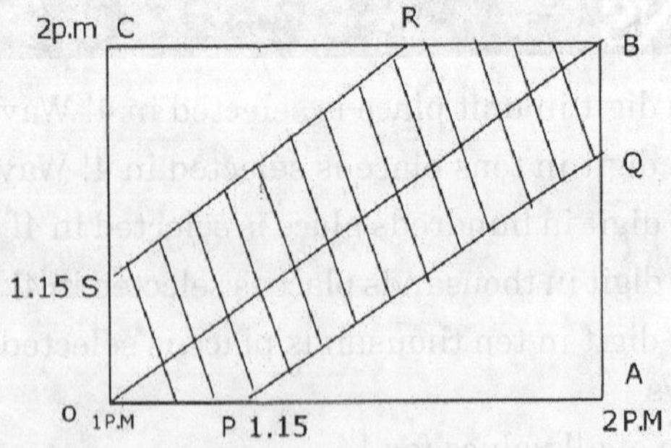

OB is the line when both A and B arrive at same time.

Total sample sapce = 60 × 60 = 3600

Favourable cases = Area of OABC – 2(Area of SRC)

$$= 3600 - 2 \times \left(\frac{1}{2} \times 45 \times 45\right)$$

$$= 1575$$

$$\therefore \text{ Required probability } = \frac{1575}{3600} = \frac{7}{16}$$

15. 2, 2, 3, 3, 4, 4, 4, 4 $x > 3000$

322 — 3, 4	344 — 2, 3, 4
323 — 2, 3, 4	422 — 3, 4
324 — 2, 3, 4	423 — 2, 3, 4
332 — 2, 3, 4	424 — 2, 3, 4
333 — 2, 4	432 — 2, 3, 4
334 — 2, 3, 4	433 — 2, 3, 4
342 — 2, 3, 4	434 — 2, 3, 4
343 — 2, 3, 4	442 — 2, 3, 4
	443 — 2, 3, 4
	444 — 2, 3, 4

NUMERICAL TYPE QUESTIONS

1. In thousands place, 9 digits except 0 can be placed

In hundreds place, 9 digits can be placed (including 0, excluding the one used in thousands place)

In tens place, 8 digits can be placed (excluding the ones used in thousands and hundreds place)

In ones place, 7 digits can be placed (excluding the one used in thousands, hundreds and tens place)

Total number of combinations

$$= 9 \times 9 \times 8 \times 7 = 4536$$

2. Given, in one year, probability of an earthquake > 6 magnitude = 0.04

So, average time between successive earthquakes = 1/0.04 = 25 years

3. Given, 10% of the population is HIV^+, so probability = 0.1 for HIV^+ and 0.9 for HIV^-

The diagnostic kit identifies → 0.95 HIV^+ and 0.89 HIV^- correctly

So, the required probability that the person is actually HIV^+ = 0.95/2 = 0.475

4. Probability for one bulb to be non-defective is

$$\frac{95}{100}$$

∴ Probabilities that none of the bulbs is

$$\text{defectives } \left(\frac{95}{100}\right)^4 = 0.8145$$

■■

6 CHAPTER

Miscellaneous

MCQ TYPE QUESTIONS

2016

1. In a 2 × 4 rectangle grid shown below, each cell is a rectangle. How many rectangles can be observed in the grid?

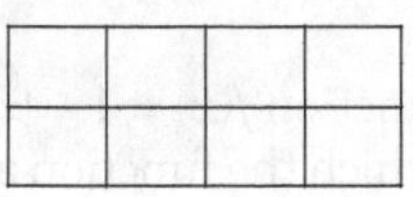

(a) 21 (b) 27
(c) 30 (d) 36

2.

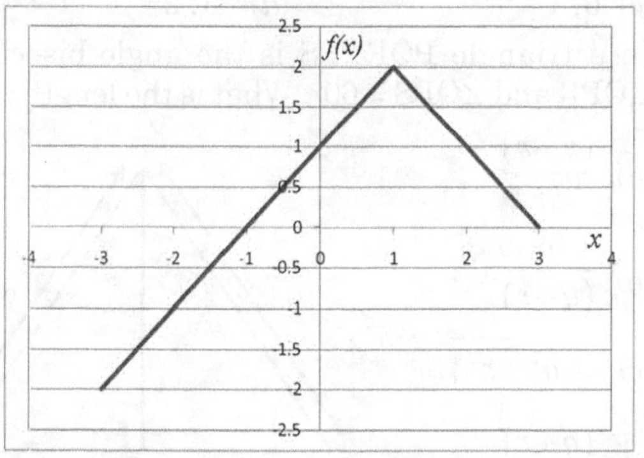

Choose the correct expression for $f(x)$ given in the graph.

(a) $f(x) = 1 - |x - 1|$ (b) $f(x) = 1 + |x - 1|$
(c) $f(x) = 2 - |x - 1|$ (d) $f(x) = 2 + |x - 1|$

3. The volume of a sphere of diameter 1 unit is ________ than the volume of a cube of side 1 unit.

(a) least (b) less
(c) lesser (d) low

4. A window is made up of a square portion and an equilateral triangle portion above it. The base of the triangular portion coincides with the upper side of the square. If the perimeter of the window is 6 m, the area of the window in m² is ________.

(a) 1.43 (b) 2.06
(c) 2.68 (d) 2.88

5. The binary operation □ is defined as $a \square b = ab + (a + b)$, where a and b are any two real numbers. The value of the identity element of this operation, defined as the number x such that $a \square x = a$, for any a, is ________.

(a) 0 (b) 1
(c) 2 (d) 10

6. Which of the following curves represents the function $y = \ln\left(\left|e^{\left[|\sin(|x|)|\right]}\right|\right)$ for $|x| < 2\pi$?

Here, x represents the abscissa and y represents the ordinate.

(a)

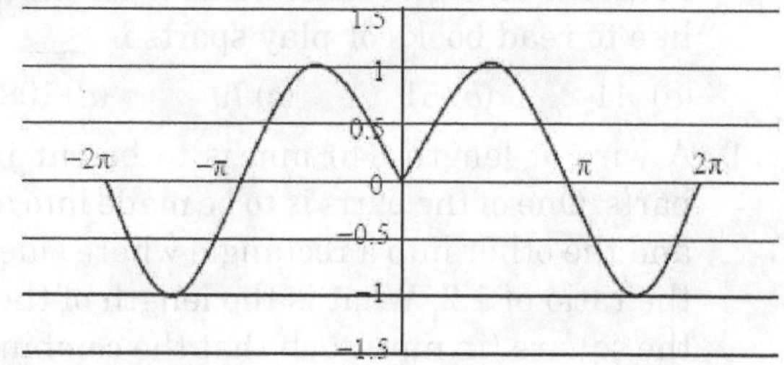

(b)

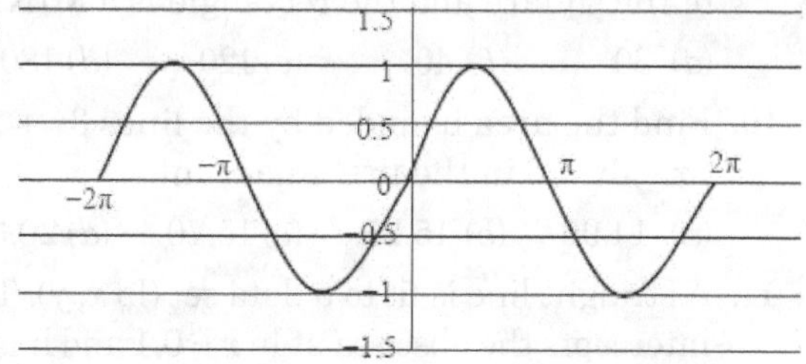

(c)

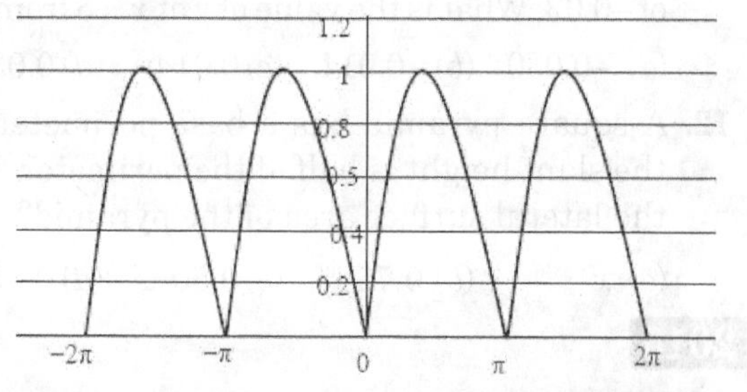

(d)

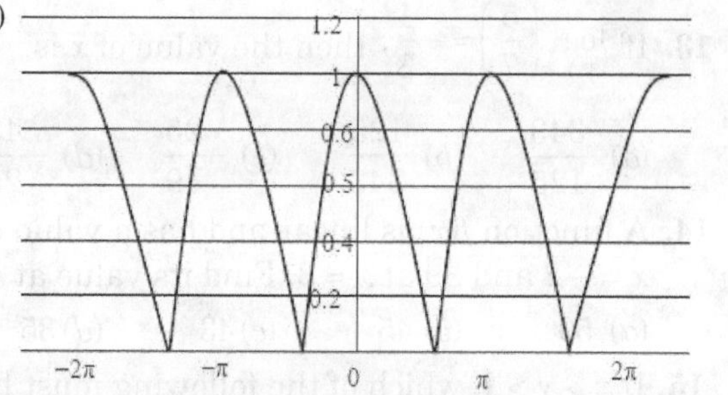

7. Given (9 inches)$^{1/2}$ = (0.25 yards)$^{1/2}$, which one of the following statements is TRUE?

(a) 3 inches = 0.5 yards
(b) 9 inches = 1.5 yards
(c) 9 inches = 0.25 yards
(d) 81 inches = 0.0625 yards

8. The Venn diagram shows the preference of the student population for leisure activities.

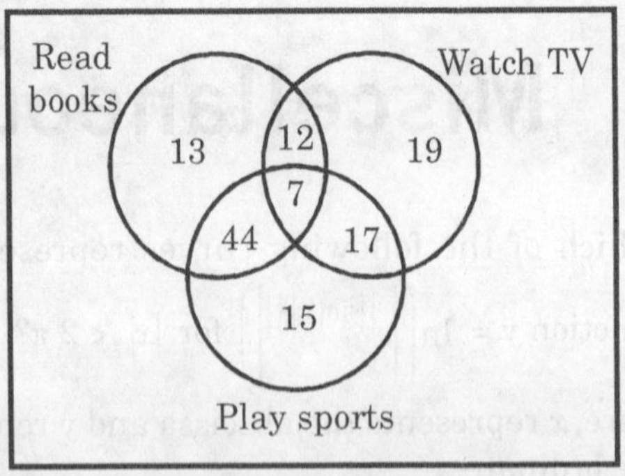

From the data given, the number of students who like to read books or play sports is ____.

(*a*) 44 (*b*) 51 (*c*) 79 (*d*) 108

9. A wire of length 340 mm is to be cut into two parts. One of the parts is to be made into a square and the other into a rectangle where sides are in the ratio of 1:2. What is the length of the side of the square (in mm) such that the combined area of the square and the rectangle is a MINIMUM?

(*a*) 30 (*b*) 40 (*c*) 120 (*d*) 180

10. Find the area bounded by the lines $3x + 2y = 14$, $2x - 3y = 5$ in the first quadrant

(*a*) 14.95 (*b*) 15.25 (*c*) 15.70 (*d*) 20.35

11. A straight line is fit to a data set ($\ln x, y$). This line intercepts the abscissa at $\ln x = 0.1$ and has a slope of -0.02. What is the value of y at $x = 5$ from the fit?

(*a*) -0.030 (*b*) -0.014 (*c*) 0.014 (*d*) 0.030

12. A square pyramid has a base perimeter x, and the slant height is half of the perimeter. What is the lateral surface area of the pyramid?

(*a*) x^2 (*b*) $0.75\,x^2$ (*c*) $0.50\,x^2$ (*d*) $0.25\,x^2$

2015

13. If $\log_x\left(\frac{5}{7}\right) = -\frac{1}{3}$, then the value of x is

(*a*) $\frac{343}{125}$ (*b*) $\frac{125}{343}$ (*c*) $-\frac{25}{49}$ (*d*) $-\frac{49}{25}$

14. A function $f(x)$ is linear and has a value of 29 at $x = -2$ and 39 at $x = 3$. Find its value at $x = 5$.

(*a*) 59 (*b*) 45 (*c*) 43 (*d*) 35

15. If $x > y > 1$, which of the following must be true?

(*i*) $\ln x > \ln y$ (*ii*) $e^x > e^y$

(*iii*) $y^x > x^y$ (*iv*) $\cos x > \cos y$

(*a*) *i* and *ii* (*b*) *i* and *iii*

(*c*) *iii* and *iv* (*d*) *ii* and *iv*

16. $\log \tan 1° + \log \tan 2°$ +-------+ $\log \tan 89°$ is ____.

(*a*) 1 (*b*) $1/\sqrt{2}$ (*c*) 0 (*d*) -1

17. Based on the given statements, select the most appropriate option to solve the given question.

What will be the total weight of 10 poles each of same weight?

Statements:

(I) One fourth of the weight of a pole is 5 kg

(II) The total weight of these poles is 160 kg more than the total weight of two poles.

(*a*) Statement II alone is not sufficient

(*b*) Statement II alone is not sufficient

(*c*) Either I or II alone is sufficient

(*d*) Both statements I and II together are not sufficient.

18. Consider a function $f(x) = 1 - |x|$ on $-1 \le x \le 1$. The value of x at which the function attains a maximum, and the maximum value of function are:

(*a*) 0, −1 (*b*) −1, 0

(*c*) 0, 1 (*d*) −1, 2

19. In a triangle PQR, PS is the angle bisector of ∠QPR and ∠QPS = 60°. What is the length of PS?

(*a*) $\frac{(q+r)}{qr}$

(*b*) $\frac{qr}{(q+r)}$

(*c*) $\sqrt{(q^2+r^2)}$

(*d*) $\frac{(q+r)^2}{qr}$

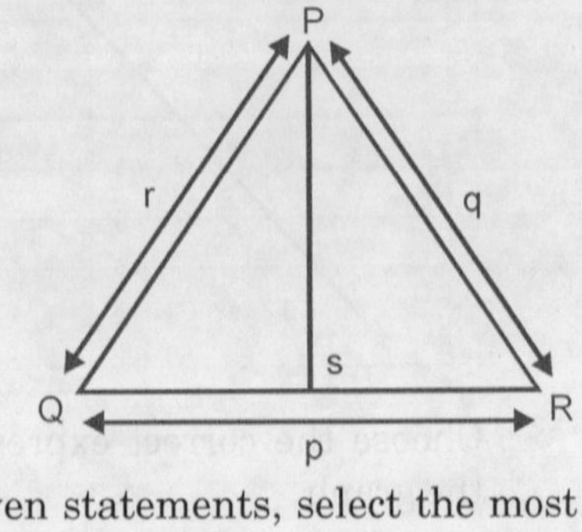

20. Based on the given statements, select the most appropriate option to solve the given question.

If two floors in a certain building are 9 feet apart, how many steps are there in a set of stairs that extends from the first floor to the second floor of the building?

Statements:

I. Each step is ¾ foot high.

II. Each step is 1 foot wide.

(*a*) Statement I alone is sufficient, but statement II alone is not sufficient.

(*b*) Statement II alone is sufficient, but statementI alone is not sufficient.

(*c*) Both statements together are sufficient, but neither statement alone is sufficient.

(*d*) Statement I and II together are not sufficient.

2014

21. If $y = 5x^2 + 3$, then the tangent at $x = 0$, $y = 3$

(*a*) passes through $x = 0$, $y = 0$

(*b*) has a slope of +1

(*c*) is parallel to the x-axis

(*d*) has a slope of −1

22. Let $f(x, y) = x^n y^m = P$. If x is doubled and y is halved, the new value of f is

(*a*) 2^{n-m} P (*b*) 2^{m-n} P
(*c*) $2(n-m)$ P (*d*) $2(m-n)$ P

23. A regular die has six sides with numbers 1 to 6 marked on its sides. If a very large number of throws show the following frequencies of occurrence: 1 → 0.167; 2 → 0.167; 3 → 0.152; 4 → 0.166; 5 → 0.168; 6 → 0.180. We call this die

(*a*) irregular (*b*) biased
(*c*) Gaussian (*d*) insufficient

24. Consider the equation : $(7526)_8 - (Y)_8 = (4364)_8$, where $(X)_N$ stands for X to the base N. Find Y.

(*a*) 1634 (*b*) 1737 (*c*) 3142 (*d*) 3162

25. At what time between 6 *a.m.* and 7 *a.m.* will the minute hand and hour hand of a clock make an angle closest to 60°?

(*a*) 6 : 22 *a.m.* (*b*) 6 : 27 *a.m.*
(*c*) 6 : 38 *a.m.* (*d*) 6 : 45 *a.m.*

26. The roots of $ax^2 + bx + c = 0$ are real and positive a, b and c are real. Then $ax^2 + b|x| + c = 0$ has

(*a*) No roots (*b*) 2 real roots
(*c*) 3 real roots (*d*) 4 real roots

2012

27. The cost function for a product in a firm is given by $5q^2$, where q is the amount of production. The firm can sell the product at a market price of ₹ 50 per unit. The number of units to be produced by the firm such that the profit is maximized is

(*a*) 5 (*b*) 10 (*c*) 15 (*d*) 25

28. Which of the following assertions are **CORRECT ?**

P : Adding 7 to each entry in a list adds 7 to the mean of the list

Q : Adding 7 to each entry in a list adds 7 to the standard deviation of the list

R : Doubling each entry in a list doubles the mean of the list

S : Doubling each entry in a list leaves the standard deviation of the list unchanged

(*a*) P, Q (*b*) Q, R (*c*) P, R (*d*) R, S

29. A political party orders an arch for the entrance to the ground in which the annual convention is being held. The profile of the arch follows the equation $y = 2x - 0.1x^2$ where y is the height of the arch in meters. The maximum possible height of the arch is

(*a*) 8 meters (*b*) 10 meters
(*c*) 12 meters (*d*) 14 meters

30. There are eight bags of rice looking alike, seven of which have equal weight and one is slightly heavier. The weighing balance is of unlimited capacity. Using this balance, the minimum number of weighings required to identify the heavier bag is

(*a*) 2 (*b*) 3 (*c*) 4 (*d*) 8

2011

31. If log (P) = (1/2) Log (Q) = (1/3) Log (R), then which of the following options is TRUE?

(*a*) $P^2 = Q^3 R^2$ (*b*) $Q^2 = PR$
(*c*) $Q^2 = R^3 P$ (*d*) $R = P^2 Q^2$

32. The variable cost (V) of manufacturing a product varies according to the equation V = 4q, where q is the quanity produced. The fixed cost (F) of production of same product reduces with q according to the equation F = 100/q. How many units should be produced to minimize the total cost (V + F)?

(*a*) 5 (*b*) 4 (*c*) 7 (*d*) 6

33. Three friends R, S and T shared toffee from a bowl. R took 1/3rd of the toffees, but returned four to the bowl. S took 1/4th of what was left but returned three toffees to the bowl. T took half of the remainder but returned two back into the bowl. If the bowl had 17 toffees left, how many toffees were originally there in the bowl?

(*a*) 38 (*b*) 31 (*c*) 48 (*d*) 41

NUMERICAL TYPE QUESTIONS

2015

1. In the given figure angle Q is a right angle, PS : QS = 3:1, RT : QT = 5:2 and PU : UR = 1:1. If area of triangle QTS is 20 cm², then the area of triangle PQR in cm² is______.

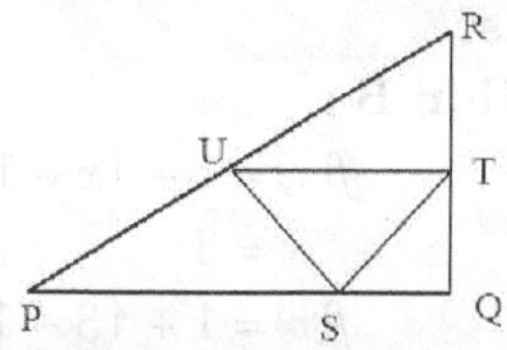

2. If p, q, r, s are distinct integers such that:

$f(p, q, r, s) = \max(p, q, r, s)$

$g(p, q, r, s) = \min(p, q, r, s)$

$h(p, q, r, s)$ = remainder of $(p \times q)/(r \times s)$ if $(p \times q) > (r \times s)$ or remainder of $(r \times s)/(p \times q)$ if $(r \times s) > (p \times q)$

Also a function $fgh(p, q, r, s) = f(p, q, r, s) \times g(p, q, r, s) \times h(p, q, r, s)$

Also the same operations are valid with two variable function of the form $f(p, q)$.

What is the value of $fg(h(2, 5, 7, 3), 4, 6, 8$

2014

3. When a point inside of a tetrahedron (a solid with four triangular surfaces) is connected by straight lines to its corners, how many (new) internal planes are created with these lines? ________.

ANSWERS

MCQ Type Questions

1. (*c*)	**2.** (*c*)	**3.** (*c*)	**4.** (*b*)	**5.** (*a*)	**6.** (*c*)	**7.** (*c*)	**8.** (*d*)	**9.** (*b*)	**10.** (*b*)
11. (*a*)	**12.** (*d*)	**13.** (*a*)	**14.** (*c*)	**15.** (*a*)	**16.** (*c*)	**17.** (*c*)	**18.** (*c*)	**19.** (*b*)	**20.** (*a*)
21. (*c*)	**22.** (*a*)	**23.** (*b*)	**24.** (*c*)	**25.** (*a*)	**26.** (*d*)	**27.** (*a*)	**28.** (*c*)	**29.** (*b*)	**30.** (*a*)
31. (*b*)	**32.** (*a*)	**33.** (*c*)							

Numerical Type Questions

1. 280 **2.** (8) **3.** (6)

EXPLANATIONS

MCQ TYPE QUESTIONS

1. In the given 2 × 4 rectangle grid, the following type of rectangles are present.

One figured rectangles = 8

Two figured rectangles = 10

Three figured rectangles = 4

Four figured rectangles = 5

Six figured rectangles = 2

Eight figured rectangles = 1

∴ Total number of rectangles

$$= 8 + 10 + 4 + 5 + 2 + 1 = 30$$

So the number of rectangles observed in the given grid is 30.

2. From the given graph function $f(x)$ must be equals to zero at $x = 3$.

From Option A :

$$f(x) = 1 - |x - 1|$$

at $x = 3$

$$f(x) = 1 - |3 - 1| = 1 - 2 = -1$$

So, it is false

From Option B :

$$f(x) = 1 + |x - 1|$$

at $x = 3$

$$f(x) = 1 + |3 - 1| = 1 + 2 = 3$$

So, it is false

From Option C :

$$f(x) = 2 - |x - 1|$$

at $x = 3$

$$f(x) = 2 \quad |3 - 1| = 2 - 2 = 0$$

So, it is true.

From Option D :

$$f(x) = 2 + |x - 1|$$

at $x = 3$

$$f(x) = 2 - |3 - 1| = 2 + 2 = 4$$

So, it is false.

Hence the correct expression for $f(x)$ given in the graph is $2 - |x - 1|$.

3. 'Lesser than' is apt because the sentence should be in comparative degree.

4. The possible window will be as follows :

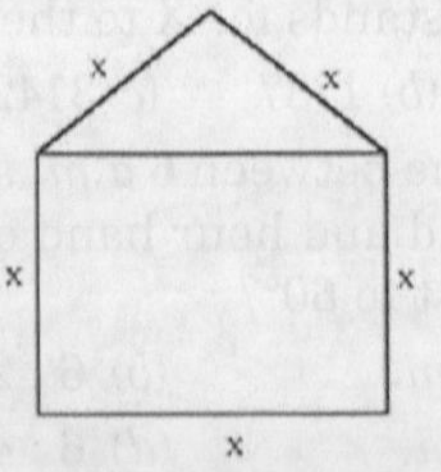

Perimeter of window = $5x$ = 6m

$$x = 1.2 \text{ m}$$

So area of window = Area of triangle + Area of square

$$= \frac{\sqrt{3}}{4}x^2 + x^2$$

$$= \frac{\sqrt{3}}{4}(1.2)^2 + (1.2)^2 = 2.06 \text{ m}^2$$

5. The binary operation is defined

$\Rightarrow a \quad b = ab + (a + b)$

$a \quad x = a$

∴ From the equation '*b*' is the variable

Option A: $x = 0$

$a \quad o = a \times 0 + (a + 0) = 0 + a = a$

Option B: $x = 1$

$a \quad 1 = a \times 1 + (a + 1) = a + a + 1 = 2a + 1$

Option C: $x = 2$

$a \quad 2 \Rightarrow a \times 2 + (a + 2) = 2a + a + 2 = 3a + 2$

Option D: $x = 10$

$a \quad 10 \Rightarrow a \times 10 + (a + 10) = 10a + a + 10 = 11a + 10$

∴ Option 'A' only True.

7. $(9 \text{ inches})^{1/2} = (0.25 \text{ yards})^{1/2}$,

Solving we get 9 inch = 0.25 yards

(since 1 inch = 0.028 yard)

8. Given Venn diagram is

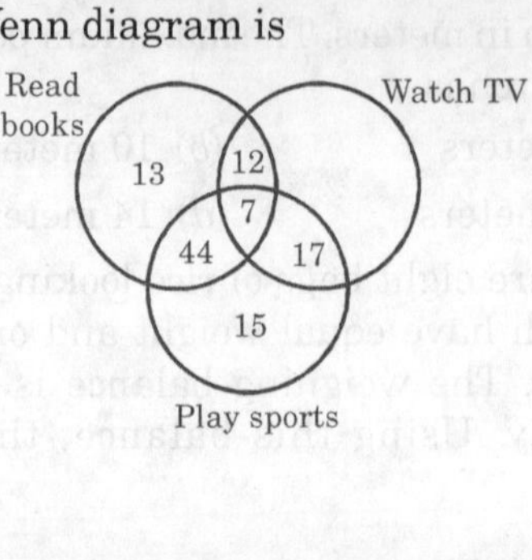

The number of students who like to read books or play sports

$$= 13 + 12 + 44 + 7 + 15 + 17 = 108$$

10. $A = \left[\frac{14}{3}, 0\right]$

$B = [0, 7]$,

$C = \left[\frac{5}{2}, 0\right]$

$D = \left[0, \frac{-5}{3}\right]$

$E = [4, 1]$,

$F = [0, 1]$

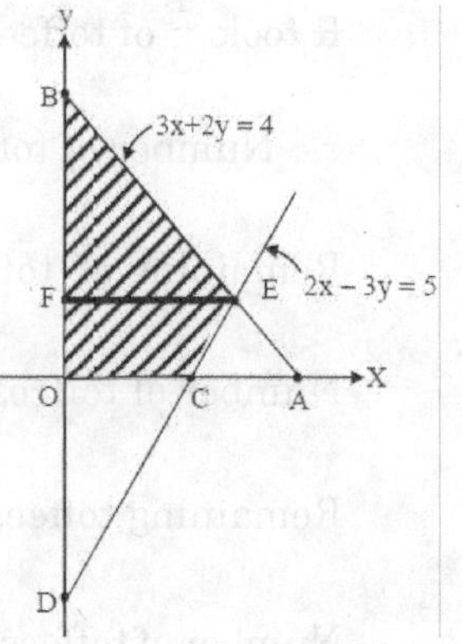

Required area = Area of Δ BFE + Area of quadrilateral FEOC

$$= \frac{1}{2}\times 4\times 6+\frac{1}{2}\times(4+2.5)\times 1$$

$$= 2\times 6+\frac{1}{2}\times 6.5$$

$$= 12 + 3.25 = 15.25 \text{ sq. units}$$

So, the area bounded by the given lines is 15.25 sq. units.

11. The equation of a straight line is

$$y = mx + c$$

$$m = \text{slope} = -0.02$$

set (log x, y)

Now, the line intercepts the abscissa at

$$\log x = 0.1$$

So, at abscissa $y = 0$

$\therefore$ $y = mx + c$

$0 = -0.02 \times 0.1 + c$

$\therefore$ $c = 0.002$

$y = mx + c$

$y = -0.02 \times \log x + c$

At $x = 5$

$\therefore$ $y = -0.02\times\log 5 + 0.002 = -0.030$

So, the value of y is – 0.030.

13. $\frac{5}{7} = x^{-1/3} \Rightarrow \frac{7}{5} = x^{1/3} \Rightarrow \left(\frac{7}{5}\right)^3 = x \Rightarrow x = \frac{343}{125}$

14. $f(x) = 2x + 33 = 2(5) + 33 = 43$

15. For whole numbers, greater the value greater will be its log.

Same logic for power of e.

16. $\log \tan 1° + \log \tan 89° = \log(\tan 1° \times \tan 89°)$

$= \log(\tan 1° \times \cot 1°)$

$= \log 1 = 0$

Using the same logic total sum is '0'.

18. The given function is

$$f(x) = 1 - |x| \text{ on } -1 \le x \le 1$$

In order to find maximum and minimum value of function in interval [–1, 1]

At $x = -1$, $f(-1) = 1 - |-1| = 1 - 1 = 0$

At $x = -0.5$, $f(-0.5) = 1 - |-0.5| = 1 - 0.5 = 0.5$

At $x = 0$, $f(0) = 1 - |0| = 1$

At $x = 0.5$, $f(0.5) = 1 - |0.5| = 0.5$

At $x = 1$, $f(1) = 1 - |1| = 1 - 1 = 0$

$\therefore$ Hence, maximum value occurs at $x = 0$ and its maximum value is 1.

$\therefore$ Option (c) is correct.

21. $y = 5x^2 + 3$, $\frac{dy}{dx} = 10x$

Slope of tangent $= \left(\frac{dy}{dx}\right)_{x=0,y=3} = 10 \times 0 = 0$

Slope = 0

$\Rightarrow$ tangent is parallel to x-axis.

22. $$P' = 2^n X^n \left(\frac{1}{2}\right)^m y^m$$

$$= 2^{n-m} X^n Y^m = 2^{n-m} P$$

23. For a very large number of throws, the frequency should be same for unbiased throw. As it not same, then the die is baised.

24. $(7526)_8 - (y)_8 = (4364)_8$

$\Rightarrow y_8 = (7526)_8 - (4364)_8$

```
     4  (8 + 2=10)
 7   5   2   6
 4   3   6   4
 -------------
 3   1   4   2
```

When we have base 8, we borrow 8 instead of 10 as done in normal subtraction.

25.

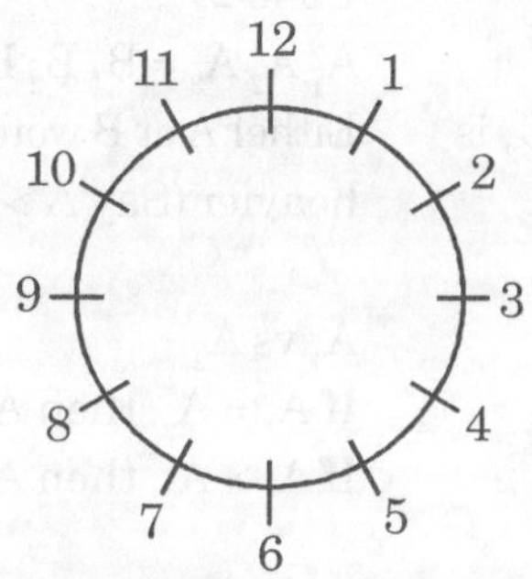

From option

At 6 a.m.

Both hand – 180° aparts,

At 6 : 20 a.m. – both hand apart

$\Rightarrow$ 180° + 10° – 120° = 70°

At 6 : 22 a.m. – both hand apart

$\Rightarrow$ 180° + 12° – 132° = 60° (closest)

26. $ax^2 + bx + c = 0$

Roots are real and positive

$\therefore \quad b^2 - 4ac > 0$

By the above condition we get two positive real roots

Now, $ax^2 + b\,|x| + c = 0$

This can be written as;

$ax^2 + bx + c$

$\Delta = b^2 - 4ac > 0$

Now for $ax^2 - bx + c$

$(-b)^2 - 4ac$

$\Rightarrow \quad b^2 - 4ac$

Again $b^2 - 4ac > 0$

So again we get real roots

Thus we have 4 real roots of $ax^2 + b\,|x| + c = 0$

27. $P = 50q - 5q^2$

$$\frac{dp}{dq} = 50 - 10q;\ \frac{d^2p}{dq^2} < 0$$

$\therefore$ p is maximum at $50 - 10q = 0$ or, $q = 5$

Else check with options

28. P and R always holds true

Else consider a sample set {1, 2, 3, 4} and check accordingly

29. $y = 2x - 0.1x^2;\ \dfrac{dy}{dx} = 2 - 0.2x;\ \dfrac{d^2y}{dx^2} < 0$

$\therefore$ y maximises at $2 - 0.2x = 0$

$\Rightarrow x = 10$

$\therefore$ $y = 20 - 10 = 10\text{m}$

30. Let us categorize bags in three groups as

$A_1\,A_2\,A_3 \qquad B_1\,B_2\,B_3 \qquad C_1\,C_2$

1st weighing A vs B

Case-1	**Case-2**
$A_1 A_2 A_3 = B_1 B_2 B_3$	$A_1 A_2 A_3 \neq B_1 B_2 B_3$
Then either C_1 or C_2 is heavier	Either A or B would be heavier (Say A > B)
2nd weighing	
C_1 vs C_2	A_1 vs A_2
If $C_1 > C_2$, then C_1	If $A_1 = A_2$, then A_3
If $C_1 < C_2$, then C_2	If $A_1 > A_2$, then A_1
If $A_1 < A_2$, then A_2	

31. $\log P = \dfrac{1}{2} \log Q$

$= \dfrac{1}{3} \log (R) = k$

$\therefore \quad P = b^k,\ Q = b^{2k},\ R = b^{3k}$

Hence, $Q^2 = b^{4k} = b^{3k}\,b^k = PR$

32. Checking with all options in formula: (4q+100/q) i.e. (V+F). Option (*a*) gives the minimum cost.

33. Let total number of toffees in bowl be x

R took $\dfrac{1}{3}$ of toffees and returned 4 to the bowl

$\therefore$ Number of toffees with R $= \dfrac{1}{3}x - 4$

Remaining of toffees in bowl $= \dfrac{2}{3}x + 4$

Number of toffees with S $= \dfrac{1}{4}\left(\dfrac{2}{3}x + 4\right) - 3$

Remaining toffees in bowl $= \dfrac{3}{4}\left(\dfrac{2}{3}x + 4\right) + 4$

Number of toffees with T $= \dfrac{1}{2}\left(\dfrac{3}{4}\left(\dfrac{2}{3}x + 4\right) + 4\right) + 2$

Remaining toffees in bowl

$$= \frac{1}{2}\left[\frac{3}{4}\left(\frac{2}{3}x + 4\right) + 4\right] + 2$$

Given : $\dfrac{1}{2}\left[\dfrac{3}{4}\left(\dfrac{2}{3}x + 4\right) + 4\right] + 2 = 17$

$\Rightarrow \quad \dfrac{3}{4}\left(\dfrac{2}{3}x + 4\right) = 27$

$\Rightarrow \quad x = 48$

NUMERICAL TYPE QUESTIONS

1. Let area of triangle PQR be 'A'

$$\frac{SQ}{PQ} = \frac{1}{1+3} = \frac{1}{4}$$

$$\frac{QT}{QR} = \frac{2}{2+5} = \frac{2}{7}$$

$\therefore$ Area of $\Delta QTS = \dfrac{1}{2} \times SQ \times QT$

$$= \frac{1}{2} \times \left(\frac{1}{4}PQ\right) \times \left(\frac{2}{7}QR\right)$$

$$= \frac{1}{4} \times \frac{2}{7} \times \left(\frac{1}{2} \times PQ \times QR\right)$$

$$= \frac{1}{14} \times \text{Area of } \Delta PQR$$

Given $20\text{ cm}^2 = \dfrac{1}{14} \times A$

$\therefore \quad A = 14\ \ 20 = 280\text{ cm}^2$

2. $fg(h(2, 5, 7, 3), 4, 6, 8) = fg(1, 4, 6, 8)$

$= f(1, 4, 6, 8) \times g(1, 4, 6, 8)$

$= 8 \times 1 = 8$

Engineering Mathematics

1

CHAPTER

Linear Algebra

MCQ TYPE QUESTIONS

2016

1. The solution to the system of equations

$$\begin{bmatrix} 2 & 5 \\ -4 & 3 \end{bmatrix}\begin{Bmatrix} x \\ y \end{Bmatrix} = \begin{Bmatrix} 2 \\ -30 \end{Bmatrix}$$

is

(*a*) 6, 2 (*b*) –6, 2

(*c*) –6, –2 (*d*) 6, –2

2. The condition for which the eigenvalues of the matrix

$$A = \begin{bmatrix} 2 & 1 \\ 1 & k \end{bmatrix}$$

are positive, is

(*a*) $k > \frac{1}{2}$ (*b*) $k > -2$

(*c*) $k > 0$ (*d*) $k < -\frac{1}{2}$

3. A real square matrix A is called skew-symmetric if

(*a*) $A^T = A$

(*b*) $A^T = A^{-1}$

(*c*) $A^T = -A$

(*d*) $A^T = A + A^{-1}$

2015

4. If any two columns of a determinant $P = \begin{vmatrix} 4 & 7 & 8 \\ 3 & 1 & 5 \\ 9 & 6 & 5 \end{vmatrix}$ are interchanged, which one of the following statements regarding the value of the determinant is CORRECT?

(*a*) Absolute value remains unchanged but sign will change

(*b*) Both absolute value and sign will change

(*c*) Absolute value will change but sign will not change

(*d*) Both absolute value and sign will remain unchanged

5. For a given matrix $\begin{bmatrix} 4+3i & -i \\ i & 4-3i \end{bmatrix}$, where $i = \sqrt{-1}$, the inverse of matrix P is

(*a*) $\frac{1}{24}\begin{bmatrix} 4-3i & i \\ -i & 4+3i \end{bmatrix}$

(*b*) $\frac{1}{25}\begin{bmatrix} i & 4-3i \\ 4+3i & -i \end{bmatrix}$

(*c*) $\frac{1}{24}\begin{bmatrix} 4+3i & -i \\ i & 4-3i \end{bmatrix}$

(*d*) $\frac{1}{25}\begin{bmatrix} 4+3i & -i \\ i & 4-3i \end{bmatrix}$

2014

6. Given that the determinant of the matrix $\begin{bmatrix} 1 & 3 & 0 \\ 2 & 6 & 4 \\ -1 & 0 & 2 \end{bmatrix}$ is – 12, the determinant of the matrix $\begin{bmatrix} 2 & 6 & 0 \\ 4 & 12 & 8 \\ -2 & 0 & 4 \end{bmatrix}$ is

(*a*) – 96

(*b*) – 24

(*c*) 24

(*d*) 96

7. The matrix from of the linear system $\frac{dx}{dt} = 3x - 5y$ and $\frac{dy}{dt} = 4x + 8y$ is

(*a*) $\frac{d}{dt} = \begin{Bmatrix} x \\ y \end{Bmatrix} = \begin{bmatrix} 3 & -5 \\ 4 & 8 \end{bmatrix}\begin{Bmatrix} x \\ y \end{Bmatrix}$

(*b*) $\frac{d}{dt}\begin{Bmatrix} x \\ y \end{Bmatrix} = \begin{bmatrix} 3 & 8 \\ 4 & -5 \end{bmatrix}\begin{Bmatrix} x \\ y \end{Bmatrix}$

(*c*) $\frac{d}{dt}\begin{Bmatrix} x \\ y \end{Bmatrix} = \begin{bmatrix} 4 & -5 \\ 3 & 8 \end{bmatrix}\begin{Bmatrix} x \\ y \end{Bmatrix}$

(*d*) $\frac{d}{dt}\begin{Bmatrix} x \\ y \end{Bmatrix} = \begin{bmatrix} 4 & 8 \\ 3 & -5 \end{bmatrix}\begin{Bmatrix} x \\ y \end{Bmatrix}$

8. One of the eigenvectors of the matrix $\begin{bmatrix} -5 & 2 \\ -9 & 6 \end{bmatrix}$ is

(*a*) $\begin{Bmatrix} -1 \\ 1 \end{Bmatrix}$ (*b*) $\begin{Bmatrix} -2 \\ 9 \end{Bmatrix}$

(*c*) $\begin{Bmatrix} 2 \\ -1 \end{Bmatrix}$ (*d*) $\begin{Bmatrix} 1 \\ 1 \end{Bmatrix}$

9. Consider a 3 × 3 real symmetric matrix S such that two of its eigenvalues are $a \neq 0$, $b \neq 0$

with respective eigenvectors $\begin{bmatrix} x_1 \\ x_2 \\ x_3 \end{bmatrix}, \begin{bmatrix} y_1 \\ y_2 \\ y_3 \end{bmatrix}$ · If $a \neq b$

then $x_1y_1 + x_2y_2 + x_3y_3$ equals

(*a*) a (*b*) b

(*c*) ab (*d*) 0

10. Which one of the following equations is a correct identity for arbitrary 3 × 3 real matrices P, Q and R?

(*a*) P(Q + R) = PQ + RP

(*b*) $(P - Q)^2 = P^2 - 2PQ + Q^2$

(*c*) det (P + Q) = det P + det Q

(*d*) $(P + Q)^2 = P^2 + PQ + QP + Q^2$

2013

11. The eigenvalues of a symmetric matrix are all

(*a*) complex with non-zero positive imaginary part

(*b*) complex with non-zero negative imaginary part

(*c*) real

(*d*) pure imaginary

2012

12. For the matrix $A = \begin{bmatrix} 5 & 3 \\ 1 & 3 \end{bmatrix}$, ONE of the normalized eigen vectors is given as

(*a*) $\begin{pmatrix} \frac{1}{2} \\ \frac{\sqrt{3}}{2} \end{pmatrix}$ (*b*) $\begin{pmatrix} \frac{1}{\sqrt{2}} \\ \frac{-1}{\sqrt{2}} \end{pmatrix}$

(*c*) $\begin{pmatrix} \frac{3}{\sqrt{10}} \\ \frac{-1}{\sqrt{10}} \end{pmatrix}$ (*d*) $\begin{pmatrix} \frac{1}{\sqrt{5}} \\ \frac{2}{\sqrt{5}} \end{pmatrix}$

13.
$$x + 2y + z = 4$$
$$2x + y + 2z = 5$$
$$x - y + z = 1$$

The system of algebraic equations given above has

(*a*) a unique solution of $x = 1, y = 1$ and $z = 1$.

(*b*) only, the two solutions of $(x = 1, y = 1, z = 1)$ and $(x = 2, y = 1, z = 0)$.

(*c*) infinite number of solutions.

(*d*) no feasible solution.

2011

14. Eigenvalues of a real symmetric matrix are always

(*a*) positive

(*b*) negative

(*c*) real

(*d*) complex

15. Consider the following system of equations :

$$2x_1 + x_2 + x_3 = 0,$$
$$x_2 - x_3 = 0,$$
$$x_1 + x_2 = 0.$$

This system has

(*a*) a unique solution

(*b*) no solution

(*c*) infinite number of solutions

(*d*) five solutions

(*d*) how to give support to a grieving friend

2010

16. One of the eigen vectors of the matrix

$$A = \begin{bmatrix} 2 & 2 \\ 1 & 3 \end{bmatrix} \text{ is}$$

(*a*) $\begin{Bmatrix} 2 \\ -1 \end{Bmatrix}$

(*b*) $\begin{Bmatrix} 2 \\ 1 \end{Bmatrix}$

(*c*) $\begin{Bmatrix} 4 \\ 1 \end{Bmatrix}$

(*d*) $\begin{Bmatrix} 1 \\ -1 \end{Bmatrix}$

2009

17. For a matrix $[M] = \begin{bmatrix} \frac{3}{5} & \frac{4}{5} \\ x & \frac{3}{5} \end{bmatrix}$, the transpose of the matrix is equal to inverse of the matrix, $[M]^T = [M]^{-1}$. The value of x is given by

(*a*) $-\frac{4}{5}$ (*b*) $-\frac{3}{5}$

(*c*) $\frac{3}{5}$ (*d*) $\frac{4}{5}$

2008

18. The matrix $\begin{bmatrix} 1 & 2 & 4 \\ 3 & 0 & 6 \\ 1 & 1 & p \end{bmatrix}$ has one eigenvalue equal to 3. The sum of the other two eigenvalues is

(*a*) p (*b*) p – 1

(*c*) p – 2 (*d*) p – 3

19. For what value of a, if any, will the following system of equations in x, y and z have a solution?

$$2x + 3y = 4$$
$$x + y + x = 4$$
$$x + 2y - z = a$$

(*a*) Any real number

(*b*) 0

(*c*) 1

(*d*) There is no such value

20. The eigenvectors of the matrix $\begin{bmatrix} 1 & 2 \\ 0 & 2 \end{bmatrix}$ are written in the form $\begin{bmatrix} 1 \\ a \end{bmatrix}$ and $\begin{bmatrix} 1 \\ b \end{bmatrix}$. What is a + b ?

(*a*) 0 (*b*) $\frac{1}{2}$

(*c*) 1 (*d*) 2

2007

21. If a square matrix A is real and symmetric, then the eigen values

(*a*) are always real

(*b*) are always real and positive

(*c*) are always real and non-negative

(*d*) occur in complex conjugate pairs

22. If $y = x + \sqrt{x + \sqrt{x + \sqrt{x + \ldots \infty}}}$, then $y(2) =$

(*a*) 4 or 1 (*b*) 4 only

(*c*) 1 only (*d*) undefined

23. The number of linearly independent eigenvectors of $\begin{bmatrix} 2 & 1 \\ 0 & 2 \end{bmatrix}$ is

(*a*) 0 (*b*) 1

(*c*) 2 (*d*) infinite

2006

24. Let x denote a real number. Find out the INCORRECT statement.

(*a*) $S = \{x : x > 3\}$ represents the set of all real numbers greater than 3

(*b*) $S = \{x : x^2 < 0\}$ represents the empty set

(*c*) $S = \{x : x \in A \text{ and } x \in B\}$ represents the union of set A and set B

(*d*) $S = \{x : a < x < b\}$ represents the set of all real numbers between a and b, where a and b are real numbers

25. Eigenvalues of a matrix $S \equiv \begin{bmatrix} 3 & 2 \\ 2 & 3 \end{bmatrix}$ are 5 and 1. What are the eigenvalues of the matrix $S^2 = SS$?

(*a*) 1 and 25 (*b*) 6 and 4

(*c*) 5 and 1 (*d*) 2 and 10

26. Match the items in columns I and II.

Column I	Column II
P. Singular matrix	1. Determinant is not defined
Q. Non-square matrix	2. Determinant is always one
R. Real symmetric matrix	3. Determinant is zero
S. Orthogonal matrix	4. Eigen values are always real
	5. Eigenvalues are not defined

Codes:

	P	Q	R	S
(*a*)	3	1	4	2
(*b*)	2	3	4	1
(*c*)	3	2	5	4
(*d*)	3	4	2	1

27. Multiplication of matrices E and F is G. Matrices E and G are

$$E \equiv \begin{bmatrix} \cos\theta & -\sin\theta & 0 \\ \sin\theta & \cos\theta & 0 \\ 0 & 0 & 1 \end{bmatrix} \text{ and } G \equiv \begin{bmatrix} 1 & 0 & 0 \\ 0 & 1 & 0 \\ 0 & 0 & 1 \end{bmatrix}.$$

What is the matrix F?

(a) $\begin{bmatrix} \cos\theta & -\sin\theta & 0 \\ \sin\theta & \cos\theta & 0 \\ 0 & 0 & 1 \end{bmatrix}$ (b) $\begin{bmatrix} \sin\theta & \cos\theta & 0 \\ -\cos\theta & \sin\theta & 0 \\ 0 & 0 & 1 \end{bmatrix}$

(c) $\begin{bmatrix} \cos\theta & \sin\theta & 0 \\ -\sin\theta & \cos\theta & 0 \\ 0 & 0 & 1 \end{bmatrix}$ (d) $\begin{bmatrix} \sin\theta & -\cos\theta & 0 \\ \cos\theta & \sin\theta & 0 \\ 0 & 0 & 1 \end{bmatrix}$

NUMERICAL TYPE QUESTIONS

2016

1. The number of linear independent eigenvectors of matrix $A = \begin{bmatrix} 2 & 1 & 0 \\ 0 & 2 & 0 \\ 0 & 0 & 3 \end{bmatrix}$ is ________.

2015

2. The lowest eigen value of the 2 × 2 matrix $\begin{bmatrix} 4 & 2 \\ 1 & 3 \end{bmatrix}$ is ________.

2014

3. The real root of the equation $5x - 2\cos x - 1 = 0$ (up to two decimal accuracy) is ________

ANSWERS

MCQ Type Questions

1. (*d*)	2. (*a*)	3. (*c*)	4. (*a*)	5. (*a*)	6. (*a*)	7. (*a*)	8. (*d*)	9. (*d*)	10. (*d*)
11. (*c*)	12. (*b*)	13. (*c*)	14. (*c*)	15. (*c*)	16. (*a*)	17. (*a*)	18. (*c*)	19. (*d*)	20. (*b*)
21. (*a*)	22. (*a*)	23. (*b*)	24. (*c*)	25. (*a*)	26. (*a*)	27. (*c*)			

Numerical Type Questions

1. (2) 2. (2) 3. (0.53 to 0.56)

EXPLANATIONS

MCQ TYPE QUESTIONS

1. Equations from this matrix are:

$2x + 5y = 2$ and $-4x + 3y = -30$

By solving these two equations, we get

$x = 6; y = -2$

2. $A = \begin{bmatrix} 2 & 1 \\ 1 & k \end{bmatrix}$

If λ_1, λ_2 are eigen values of $A_{2\times 2}$ then $|A| = \lambda_1 \lambda_2$

$$|A| = 2k - 1$$
$$\lambda_1, \lambda_2 > 0$$
$$2k - 1 > 0$$
$$\therefore \quad k > \frac{1}{2}$$

3. A real square matrix A is called skew symmetric if

$$A^T = -A$$

6. $\begin{vmatrix} 2 & 6 & 0 \\ 4 & 12 & 8 \\ -2 & 0 & 4 \end{vmatrix} = (2)^3 \begin{vmatrix} 1 & 3 & 0 \\ 2 & 6 & 4 \\ -1 & 0 & 2 \end{vmatrix}$

$$= 8 \times (-12) = -96$$

7. Given that $\frac{dx}{dt} = 3x - 5y$

$$\frac{dy}{dt} = 4x + 8y$$

Matrix term $\frac{d}{dt}\begin{Bmatrix} x \\ y \end{Bmatrix} = \begin{bmatrix} 3 & -5 \\ 4 & 8 \end{bmatrix}\begin{bmatrix} x \\ y \end{bmatrix}$

8. Eigen values of the matrix $\begin{bmatrix} -5 & 2 \\ -9 & 6 \end{bmatrix}$ are 4, – 3

$\therefore$ the eigen vector corresponding to eigen vector λ is $Ax = \lambda x$ (verify the options)

$\begin{pmatrix} 1 \\ 1 \end{pmatrix}$ is eigen vector corresponding to eigen value $\lambda = 3$

9. We know that the Eigen vectors corresponding to distinct Eigen values of real symmetric matrix are orthogonal.

$$\begin{bmatrix} x_1 \\ x_2 \\ x_3 \end{bmatrix}\begin{bmatrix} y_1 \\ y_2 \\ y_3 \end{bmatrix} = x_1 y_1 + x_2 y_2 + x_3 y_3 = 0$$

11. Suppose the eigen value of matrix A is n $(= \lambda + i B_B)$ say.

and the eigen vector is n whereas the conjugate pair of eign value and eign vector is λ and $\bar{n}$

So, $An = \lambda n$ ...(1)

And $A\bar{n} = \lambda\bar{x}$ ...(2)

Taking transpose of equation

$$x^{-T}A^T = \bar{x}^T\bar{\lambda} \quad ...(3)$$
$$\Rightarrow \quad x^{-T}A^T x = x^{-T}\bar{\lambda}n$$
$$\Rightarrow \quad x^{-T}An = x^{-T}\bar{\lambda}n$$
$$\Rightarrow \quad \lambda = \bar{\lambda}$$
$$\Rightarrow \quad \alpha + i\beta = \alpha - i\beta$$
$$\Rightarrow \quad 2i\beta = 0$$
$$\Rightarrow \quad \beta = 0$$

Hence eigen value of a symmetric matrix is real.

12. $|A - \lambda I| = 0$

$$\Rightarrow \quad \begin{vmatrix} 5-\lambda & 3 \\ 1 & 3-\lambda \end{vmatrix} = 0$$
$$\Rightarrow \quad (5-\lambda)(3-\lambda) = 3 = 0$$
$$\Rightarrow \quad \lambda^2 - 8\lambda + 15 - 3 = 0$$
$$\Rightarrow \quad \lambda^2 - 8\lambda + 12 = 0$$
$$\Rightarrow \quad \lambda = 2, \lambda = 6$$
$$(A - 2I)\lambda X = 0$$

At, $\lambda = 2$

$$\begin{pmatrix} 3 & 3 \\ 1 & 1 \end{pmatrix}\begin{pmatrix} x_1 \\ x_2 \end{pmatrix} = \begin{pmatrix} 0 \\ 0 \end{pmatrix}; \; x_1 + x_2 = 0, \; x_1 = -x_2$$

Hence the required vector is $\begin{bmatrix} \frac{1}{\sqrt{2}} \\ \frac{-1}{\sqrt{2}} \end{bmatrix}$

13. The given equations are

$$x + 2y + z = 4 \quad ...(i)$$
$$2x + y + 2z = 5 \quad ...(ii)$$
$$x + y + z = 1 \quad ...(iii)$$

Eliminating z from (*i*) & (*ii*), we have

$$y = 1 \quad ...(iv)$$

Eliminating x form (*ii*) & (*iii*), we have

$$y = 1 \quad ...(v)$$

Lines (*iv*) and (*v*) are coincident

Hence the given equation has infinite solutions.

14. Eigen values of a real symmetric matrix are always real.

15. Given equations are

$$2x_1 + x_2 + x_3 = 0 \quad ...(i)$$

$$x_2 - x_3 = 0 \quad ...(ii)$$

and $$x_1 + x_2 = 0 \quad ...(iii)$$

Eliminating x_3 from *(i)* and *(ii)*, we get

$$x_1 + x_2 = 0 \quad ...(iv)$$

Clearly *(iii)* and *(iv)* are coincident, i.e. they will meet at infinite points.

Hence given equations have infinite solutions.

16. Given : $A = \begin{bmatrix} 2 & 2 \\ 1 & 3 \end{bmatrix}$

Characteristic equation is

$$|A - \lambda I| = 0$$

$$\Rightarrow \begin{bmatrix} 2-\lambda & 2 \\ 1 & 3-\lambda \end{bmatrix} = 0$$

$$\Rightarrow (2-\lambda)(3-\lambda) - 2 = 0$$

$$\Rightarrow \lambda^2 - 5 + 4 = 0$$

$$\Rightarrow \lambda = 1, 4$$

Eigen vector for $\lambda = 1$ is

$$\begin{bmatrix} 2-\lambda_1 & 2 \\ 1 & 3-\lambda_1 \end{bmatrix}\begin{bmatrix} x_1 \\ x_2 \end{bmatrix} = 0$$

$$\Rightarrow \begin{bmatrix} 1 & 2 \\ 1 & 2 \end{bmatrix}\begin{bmatrix} x_1 \\ x_2 \end{bmatrix} = 0$$

$$\Rightarrow x_1 + 2x_2 = 0$$

$$\Rightarrow x_2 = -\frac{1}{2}x$$

Eigen vector set is

x_1	1	2	3	4
x_2	$-\frac{1}{2}$	-1	$-\frac{3}{2}$	-2

$\therefore$ Eigen vector is $\begin{Bmatrix} 2 \\ -1 \end{Bmatrix}$.

17. $$[M] = \begin{bmatrix} \frac{3}{5} & \frac{4}{5} \\ x & \frac{3}{5} \end{bmatrix}$$

$$[M]^T = \begin{bmatrix} \frac{3}{5} & x \\ \frac{4}{5} & \frac{3}{5} \end{bmatrix} \quad ....(i)$$

$$[M]^{-1} = \frac{1}{[M]} \text{ adj M}$$

$$|M| = \left(\frac{3}{5}\right)^2 - \frac{4}{5}x$$

$$\text{adj M} = [\text{cofactor } [M]]^T$$

$$= \begin{bmatrix} \frac{3}{5} & -x \\ -\frac{4}{5} & \frac{3}{5} \end{bmatrix}^T$$

$$= \begin{bmatrix} \frac{3}{5} & -\frac{4}{5} \\ -x & \frac{3}{5} \end{bmatrix} \quad .....(ii)$$

From equations *(i)* and *(ii)*, we get

$$\begin{bmatrix} \frac{3}{5} & x \\ \frac{4}{5} & \frac{3}{5} \end{bmatrix} = \frac{1}{\left(\frac{3}{5}\right)^2 - \frac{4}{5}x}\begin{bmatrix} \frac{3}{5} & -\frac{4}{5} \\ -x & \frac{3}{5} \end{bmatrix}$$

$$\therefore \quad x = \frac{1}{\left(\frac{3}{5}\right)^2 - \frac{4}{5}x} \times \left(-\frac{4}{5}\right)$$

$$\Rightarrow \left(\frac{3}{5}\right)^2 x - \frac{4}{5}x^2 = -\frac{4}{5}$$

$$\Rightarrow \frac{9}{25}x - \frac{4}{5}x^2 = -\frac{4}{5}$$

$$\Rightarrow \frac{4}{5}x^2 - \frac{9}{25}x - \frac{4}{5} = 0$$

$$20x^2 - 9x - 20 = 0$$

$$\Rightarrow 20x^2 - (25x - 16x) - 20 = 0$$

$$\Rightarrow 20x^2 - 25x + 16x - 20 = 0$$

$$\Rightarrow 5x(4x - 5) + 4(4x - 5) = 0$$

$$\Rightarrow (5x + 4)(4x - 5) = 0$$

$$\therefore \quad x = -\frac{4}{5} \text{ or } \frac{5}{4}$$

Alternately

$$A^T = A^{-1}$$

so, orthogonal matrix.

Now $$AA^{-1} = I = AA^T$$

$$\therefore \begin{bmatrix} \frac{3}{5} & \frac{4}{5} \\ x & \frac{3}{5} \end{bmatrix}\begin{bmatrix} \frac{3}{5} & x \\ \frac{4}{5} & \frac{3}{5} \end{bmatrix} = \begin{bmatrix} 1 & 0 \\ 0 & 1 \end{bmatrix}$$

$\Rightarrow \quad \frac{3}{5}x + \frac{3}{5} \times \frac{4}{5} = 0$

$\Rightarrow \quad \frac{3}{5}x = -\frac{12}{25}$

$\Rightarrow \quad x = -\frac{4}{5}$

18. Sum of diagonal vector = Sum of Eigen values

$\therefore \quad 1 + 0 + p = 3 + S$

$\Rightarrow \quad S = p - 2$

19.
$$2x + 3y = 4 \quad \text{...(i)}$$
$$x + y + z = 4 \quad \text{...(ii)}$$
$$x + 2y - z = a \quad \text{...(iii)}$$

$|A| = 0 \Rightarrow$ (i) is wrong.

We can see that addition of equations (*ii*) and (*iii*) produces

2x + 3y= 4 + a -(4)

Comparing it with equation (i)

i.e. $\quad 2x + 3y = 4$

We conclude that, the system of equations is soluble for a = 0

20. To find Eigen values of the matrix,

$$(1 - \lambda)(2 - \lambda) = 0$$

$\Rightarrow \quad \lambda = 1, 2$

For $\lambda = 1$,

$$\begin{bmatrix} 0 & 2 \\ 0 & 1 \end{bmatrix}_{2 \times 2} \begin{bmatrix} x \\ y \end{bmatrix}_{2 \times 1} = \begin{bmatrix} 0 \\ 0 \end{bmatrix}$$

$\Rightarrow \quad \begin{bmatrix} 2y \\ y \end{bmatrix} = \begin{bmatrix} 0 \\ 0 \end{bmatrix}$

$\Rightarrow \quad y = 0 = a$ and $x = 1$

$\Rightarrow$ *For* $\lambda = 2$,

$$\begin{bmatrix} -1 & 2 \\ 0 & 0 \end{bmatrix} \begin{bmatrix} x \\ y \end{bmatrix} = \begin{bmatrix} 0 \\ 0 \end{bmatrix}$$

$\Rightarrow \quad \begin{bmatrix} x + 2y \\ 0 \end{bmatrix} = \begin{bmatrix} 0 \\ 0 \end{bmatrix}$

$\Rightarrow \quad x = 2y$

For $x = 1$

$\Rightarrow \quad y = \frac{1}{2} = b$

$\therefore \quad a + b = 0 + \frac{1}{2}$

$= \frac{1}{2}$

21. $(A - \lambda I) = 0$

Solving for λ, we get the eigen values.

$$A = \begin{bmatrix} a & b \\ b & a \end{bmatrix}$$

Now $\quad \begin{bmatrix} a - \lambda & b \\ b & a - \lambda \end{bmatrix} = 0$

which gives $(a - \lambda)^2 - b^2 = 0$

$\Rightarrow \quad a - \lambda = \pm b$

$\Rightarrow \quad \lambda = a \pm b$

Hence real value of eigen.

22. Given $y = x + \sqrt{x + \sqrt{x + \sqrt{x + ... \infty}}}$

$\Rightarrow \quad (y - x) = \sqrt{xy}$

$\Rightarrow \quad (y - x)^2 = y$

$\Rightarrow \quad y^2 + x^2 - 2xy - y = 0$

For $x = 2$, $\quad y^2 - 5y + 1 = 0$

$\therefore \quad y = 4; y = 1$

23. $A = \begin{bmatrix} 2 & 1 \\ 0 & 2 \end{bmatrix}$

Now, $\quad |A - \lambda I| = 0$

$\Rightarrow \quad \begin{vmatrix} 2 - \lambda & 1 \\ 0 & 2 - \lambda \end{vmatrix} = 0$

$\Rightarrow \quad \lambda = 2$

Hence only 1 linearly independent eigen vector.

24. $\{X : x \in A \text{ and } x \in B\}$ represents the union of set A and set B.

The above symbol denotes intersection of set A and set B. Therefore the statement is incorrect

25. For S matrix, if eigen values are $\lambda_1, \lambda_2, \lambda_3$, ... then for S^2 matrix, the eigen values will be

$\lambda_1^2, \lambda_2^2, \lambda_3^3$, ...

For S matrix, if eigen values are 1 and 5, then for S^2 matrix, the eigen values are 1 and 25.

27. Given: $E \equiv \begin{bmatrix} \cos\theta & -\sin\theta & 0 \\ \sin\theta & \cos\theta & 0 \\ 0 & 0 & 1 \end{bmatrix}$

and $\quad G \equiv \begin{bmatrix} 1 & 0 & 0 \\ 0 & 1 & 0 \\ 0 & 0 & 1 \end{bmatrix}$

Now, $\quad E \times F = G$

$\therefore \quad F = [E]^{-1}$

NUMERICAL TYPE QUESTIONS

1.
$$A = \begin{pmatrix} 2 & 1 & 0 \\ 0 & 2 & 0 \\ 0 & 0 & 3 \end{pmatrix}$$

Eigen value $|A - \lambda I| = 0$

$$\begin{vmatrix} 2-\lambda & 1 & 0 \\ 0 & 2-\lambda & 0 \\ 0 & 0 & 3-\lambda \end{vmatrix} = 0$$

$(2 - \lambda)(2 - \lambda)(3 - \lambda) = 0$

$\lambda = 2, 2, 3$

For $\lambda = 2$, No. of Linearly independent eigenvector $= 3\text{-rank of }(A - 2I)$

$= 3 - 2 = 1$

For $\lambda = 3$, No. of Linearly independent eigen vectors $= 1$

so total linearly independent Eigen vector $= 2$

2. Let
$$A = \begin{bmatrix} 4 & 2 \\ 1 & 3 \end{bmatrix}$$

Characteristic equation of A is $|A - \lambda I| = 0$

$$\Rightarrow \begin{bmatrix} 4-\lambda & 2 \\ 1 & 3-\lambda \end{bmatrix} = 0$$

$\lambda^2 - 7\lambda + 10 = 0 \Rightarrow \lambda = 2, 5$

3. Let $f(x) = 5x - 2\cos x - 1$

$\Rightarrow f'(x) = 5 + 2\sin x$

$f(0) = -3; f(1) = 2.9$

By intermediate value theorem roots lie between 0 and 1.

Let $x_0 = 1 \text{ rad} = 57.32°$

By Newton Raphson method,

$$x_{n+1} = x_n - \frac{f(x_n)}{f'(x_n)}$$

$$\Rightarrow x_{n+1} = \frac{2x_n \sin x_n + 2\cos x_n + 1}{5 + 2\sin x_n}$$

$\Rightarrow x_1 = 0.5632$

$\Rightarrow x_2 = 0.5425$

$\Rightarrow x_3 = 0.5424$

2

CHAPTER

Calculus

MCQ TYPE QUESTIONS

2016

1. If $f(t)$ is a function defined for all $t \geq 0$, its Laplace transform F(s) is defined as

(a) $\int_0^\infty e^{st} f(t)dt$ (b) $\int_0^\infty e^{-st} f(t)dt$

(c) $\int_0^\infty e^{ist} f(t)dt$ (d) $\int_0^\infty e^{-ist} f(t)dt$

2. The value of the integral

$$\int_{-\infty}^{\infty} \frac{\sin x}{x^2 + 2x + 2} dx$$

evaluated using contour integration and the residue theorem is

(a) $-\pi \sin(1)/e$ (b) $-\pi \cos(1)/e$

(c) $\sin(1)/e$ (d) $\cos(1)/e$

3. The figure below represents a triangle PQR with initial coordinates of the vertices as P(1, 3), Q(4, 5) and R(5, 3.5). The triangle is rotated in the X-Y plane about the vertex P by angle θ in clockwise direction. If sinθ = 0.6 and cosθ = 0.8, the new coordinates of the vertex Q are

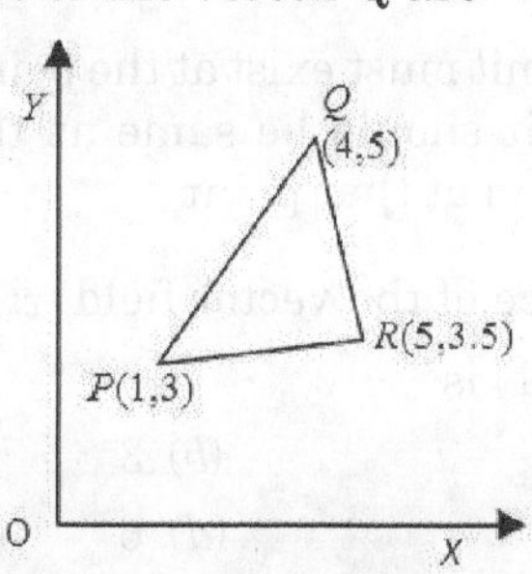

(a) (4.6, 2.8) (b) (3.2, 4.6)

(c) (7.9, 5.5) (d) (5.5, 7.9)

4. The values of x for which the function

$$f(x) = \frac{x^2 - 3x - 4}{x^2 + 3x - 4}$$

is **NOT** continuous are

(a) 4 and – 1 (b) 4 and 1

(c) – 4 and 1 (d) – 4 and – 1

5. Laplace transform of cos(ωt) is

(a) $\frac{s}{s^2 + \omega^2}$ (b) $\frac{\omega}{s^2 + \omega^2}$

(c) $\frac{s}{s^2 - \omega^2}$ (d) $\frac{\omega}{s^2 - \omega^2}$

6. The value of $\oint_\Gamma \frac{3z - 5}{(z-1)(z-2)} dZ$ along a closed path Γ is equal to (4 π i), where $z = x + iy$ and $i = \sqrt{-1}$. The correct path Γ is

(a) 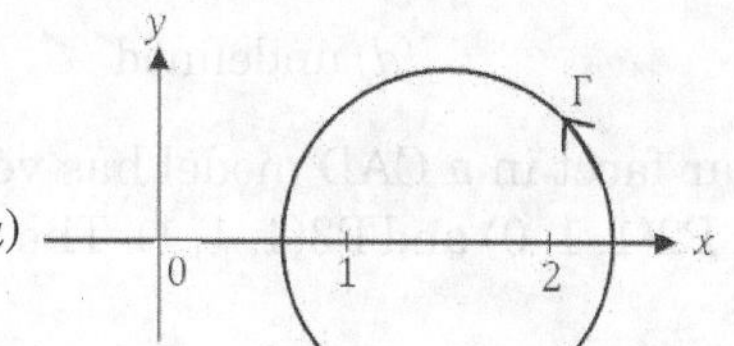

(b) 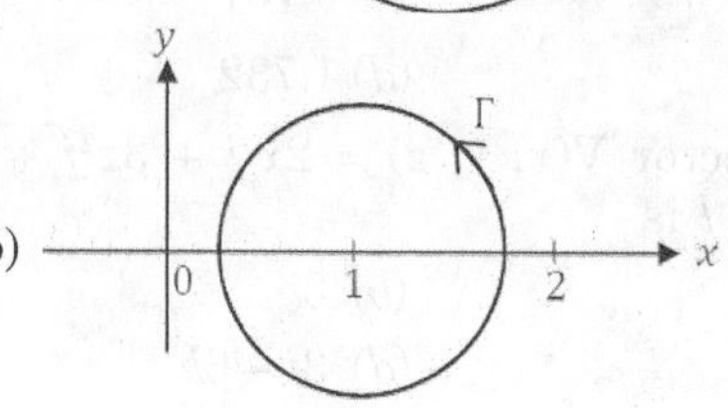

(c) 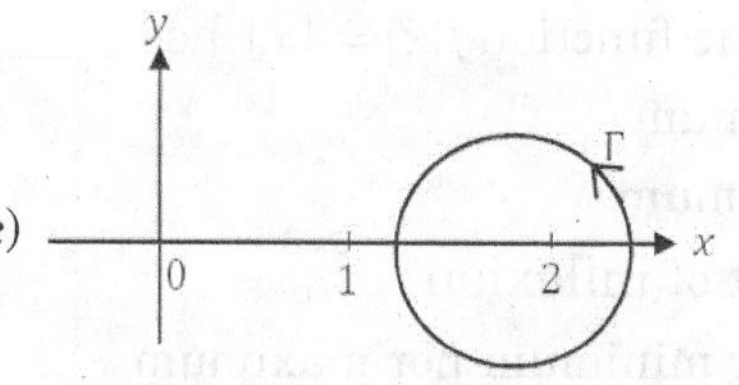

(d) 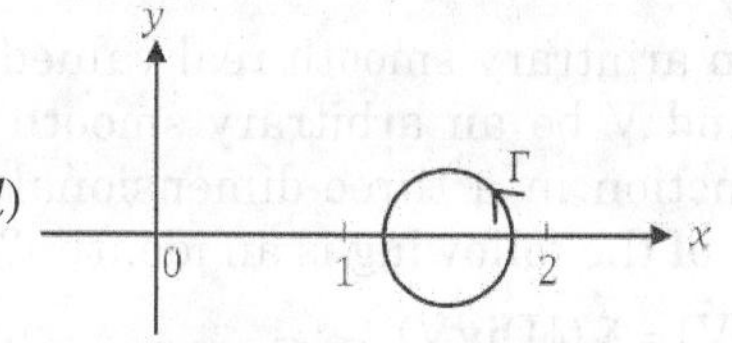

7. $\underset{x \to 0}{\text{Lt}} \frac{\log_e(1 + 4x)}{e^{3x} - 1}$ is equal to

(a) 0 (b) $\frac{1}{12}$

(c) $\frac{4}{3}$ (d) 1

8. Solutions of Laplace's equation having continuous second-order partial derivatives are called

(a) biharmonic functions

(b) harmonic functions

(c) conjugate harmonic functions

(d) error functions

9. $\lim_{x \to \infty} \sqrt{x^2 + x - 1} - x$ is

(a) 0 (b) ∞

(c) 1/2 (d) – ∞

10. A point P (1, 3, – 5) is translated by $2\hat{i}+3\hat{j}-4\hat{k}$ and then rotated counter clockwise by 90° about the *z*-axis. The new position of the point is

(*a*) (– 6, 3, – 9) (*b*) (– 6, – 3, – 9)

(*c*) (6, 3, – 9) (*d*) (6, 3, 9)

2015

11. The value of $\lim_{x\to\infty}\dfrac{1-\cos(x^2)}{2x^4}$ is

(*a*) 0 (*b*) $\dfrac{1}{2}$

(*c*) $\dfrac{1}{4}$ (*d*) undefined

12. A triangular facet in a CAD model has vertices: P1(0, 0, 0); P2(1, 1, 0) and P3(1, 1, 1). The area of the facet is

(*a*) 0.500 (*b*) 0.707

(*c*) 1.414 (*d*) 1.732

13. Curl of vector V(*x*, *y*, *z*) = $2x^2i + 3z^2j + y^3k$ at $x = y = z = l$ is

(*a*) – 3*i* (*b*) 3*i*

(*c*) 4*i* – 4*j* (*d*) 3*i* – 6*k*

14. At *x* = 0, the function *f*(*x*) = |*x*| has _______.

(*a*) A minimum

(*b*) A maximum

(*c*) A point of inflexion

(*d*) neither minimum nor maximum

15. Let ϕ be an arbitrary smooth real valued scalar function and V be an arbitrary smooth vector valued function in a three-dimensional space. Which one of the following is an identity?

(*a*) $\text{Curl}(\phi\overline{V}) = \nabla(\phi\text{Div}\overline{V})$

(*b*) $\text{Div}\,\overline{V} = 0$

(*c*) $\text{Div}\,\text{Curl}\,\overline{V} = 0$

(*d*) $\text{Div}(\phi\overline{V}) = \phi\text{Div}\,\overline{V}$

2014

16. $\underset{x\to 0}{\text{Lt}}\dfrac{x-\sin x}{1-\cos x}$ is

(*a*) 0 (*b*) 1

(*c*) 3 (*d*) not defined

17. Which one of the following describes the relationship among the three vectors, $\hat{i}+\hat{j}+\hat{k}$, $2\hat{i}+3\hat{j}+\hat{k}$ and $5\hat{i}+6\hat{j}+4\hat{k}$?

(*a*) The vectors are mutually perpendicular

(*b*) The vectors are linearly dependent

(*c*) The veators are linearly independent

(*d*) The vectors are unit vectors

18. The integral $\oint_C (ydx - xdy)$ is evaluated along the circle $x^2 + y^2 = \dfrac{1}{4}$ traversed in counter clockwise direction. The integral is equal to

(*a*) 0 (*b*) $\dfrac{-\pi}{4}$

(*c*) $\dfrac{-\pi}{2}$ (*d*) $\dfrac{\pi}{4}$

19. $\underset{x\to 0}{\text{Lt}}\left(\dfrac{e^{2x}-1}{\sin(4x)}\right)$ is equal to

(*a*) 0 (*b*) 0.5

(*c*) 1 (*d*) 2

20. Curl of vector $\bar{F} = x^2z^2\hat{i} - 2xy^2z\hat{j} + 2y^2z^3\hat{k}$ is

(*a*) $(4yz^3 + 2xy^2)\hat{i} + 2x^2z\hat{j} - 2y^2z\hat{k}$

(*b*) $(4yz^3 + 2xy^2)\hat{i} - 2x^2z\hat{j} - 2y^2z\hat{k}$

(*c*) $2xz^2\hat{i} - 4xyz\hat{j} + 6y^2z^2\hat{k}$

(*d*) $2xz^2\hat{i} + 4xyz\hat{j} + 6y^2z^2\hat{k}$

21. If a function is continuous at a point,

(*a*) The limit of the function may not exist at the point

(*b*) The function must be derivable at the point

(*c*) The limit of the function at the point tends to infinity

(*d*) The limit must exist at the point and the value of limit should be same as the value of the function at that point

22. Divergence of the vector field $x^2z\hat{i} + xy\hat{j} - yz^2\hat{k}$ at (1, –1, 1) is

(*a*) 0 (*b*) 3

(*c*) 5 (*d*) 6

23. The value of the integral

$$\int_0^2 \frac{(x-1)^2\sin(x-1)}{(x-1)^2+\cos(x-1)}dx \text{ is}$$

(*a*) 3 (*b*) 0

(*c*) –1 (*d*) –2

24. The value of the integral $\int_0^2\int_0^x e^{x+y}dydx$ is

(*a*) $\dfrac{1}{2}(e-1)$ (*b*) $\dfrac{1}{2}(e^2-1)^2$

(*c*) $\dfrac{1}{2}(e^2-e)$ (*d*) $\dfrac{1}{2}(e-\dfrac{1}{e})^2$

2013

25. The following surface integral is to be evaluated over a sphere for the given steady velocity vector field F = $xi + yj + zk$ defined with respect to a Cartesian coordinate system having, i, j and k as unit base vectors

$$\iint_s \frac{1}{4}(\text{F}.n)dA$$

where S is the sphere, $x^2 + y^2 + z^2 = 1$ and n is the outward unit normal vector to the sphere. The value of the surface integral is

(a) π (b) 2π

(c) $3\pi l 4$ (d) 4π

26. The value of the definite integral $\int_1^e \sqrt{x}\ \ln(x)\,dx$ is

(a) $\frac{4}{9}\sqrt{e^3}+\frac{2}{9}$ (b) $\frac{2}{9}\sqrt{e^3}-\frac{4}{9}$

(c) $\frac{2}{9}\sqrt{e^3}+\frac{4}{9}$ (d) $\frac{4}{9}\sqrt{e^3}-\frac{2}{9}$

2012

27. The area enclosed between the straight line $y = x$ and the parabola $y = x^2$ in the $x - y$ plane is

(a) 1/6 (b) 1/4

(c) 1/3 (d) 1/2

28. Consider the function $f(x) = |x|$ in the interval $-1 \le x \le 1$. At the point $x = 0$, $f(x)$ is

(a) continuous and differentiable.

(b) non–continuous and differentiable.

(c) continuous and non–differentiable.

(d) neither continuous nor differentiable.

29. $\lim_{x\to 0}\left(\frac{1-\cos x}{x^2}\right)$ is

(a) 1/4 (b) 1/2

(c) 1 (d) 2

30. At $x = 0$, the function $f(x) = x^3 + 1$ has

(a) a maximum value (b) a minimum value

(c) a singularity (d) a point of inflection

31. For the spherical surface $x^2 + y^2 + z^2 = 1$, the unit outward normal vector at the point $\left(\frac{1}{\sqrt{2}}, \frac{1}{\sqrt{2}}, 0\right)$ is given by

(a) $\frac{1}{\sqrt{2}}\hat{i}+\frac{1}{\sqrt{2}}\hat{j}$ (b) $\frac{1}{\sqrt{2}}\hat{i}-\frac{1}{\sqrt{2}}\hat{j}$

(c) $\hat{k}$ (d) $\frac{1}{\sqrt{3}}\hat{i}+\frac{1}{\sqrt{3}}\hat{j}+\frac{1}{\sqrt{3}}\hat{k}$

2011

32. A series expansion for the function sin θ is

(a) $1-\frac{\theta^2}{2!}+\frac{\theta^4}{4!}-....$ (b) $\theta-\frac{\theta^3}{3!}+\frac{\theta^5}{5!}-....$

(c) $1+\theta+\frac{\theta^2}{2!}+\frac{\theta^3}{3!}+....$ (d) $\theta+\frac{\theta^3}{3!}+\frac{\theta^5}{5!}+....$

33. If $f(x)$ is an even function and a is a positive real number, then $\int_{-a}^{a} f(x)\,dx$ equals

(a) 0 (b) a

(c) $2a$ (d) $2\int_0^a f(x)dx$

34. What is $\lim_{\theta\to 0}\frac{\sin\theta}{\theta}$ equal to?

(a) θ (b) sin θ

(c) 0 (d) 1

2010

35. The parabolic arc $y = \sqrt{x}$, $1 \le x \le 2$ is revolved around the x-axis. The volume of the solid of revolution is

(a) $\frac{\pi}{4}$ (b) $\frac{\pi}{2}$

(c) $\frac{3\pi}{4}$ (d) $\frac{3\pi}{2}$

36. The value of the integral $\int_{-\infty}^{\infty}\frac{dx}{1+x^2}$ is

(a) $-\pi$ (b) $-\frac{\pi}{2}$

(c) $\frac{\pi}{2}$ (d) π

37. The function $y = |2-3x|$ is

(a) continuous $\forall x \in$ R and differentiable $\forall x \in$ R

(b) is continuous $\forall x \in$ R and differentiable $\forall x \in$ R except at $x = 3/2$

(c) continuous $\forall x \in$ R and differentiable $\forall x \in$ R except at $x = 2/3$

(d) continuous $\forall x \in$ R except at $x = 3$ and differentiable $\forall\, x \in$ R

38. Velocity vector of a flow field is given as

$$\vec{V} = 2xy\hat{i} - x^2 z\,\hat{j}.$$

The vorticity vector at (1, 1, 1) is

(a) $4\hat{i} - \hat{j}$ (b) $4\hat{i} - \hat{k}$

(c) $\hat{i} - 4\hat{j}$ (d) $\hat{i} - 4\hat{k}$

2009

39. The divergence of the vector field $3xz\,\hat{i} + 2xy\,\hat{j} - yz^2\,\hat{k}$ at a point (1, 1, 1) is equal to

(a) 7 (b) 4
(c) 3 (d) 0

40. A path AB in the form of one quarter of a circle of unit radius is shown in the figure. Integration of $(x+y)^2$ on path AB traversed in a counter-clockwise sense is

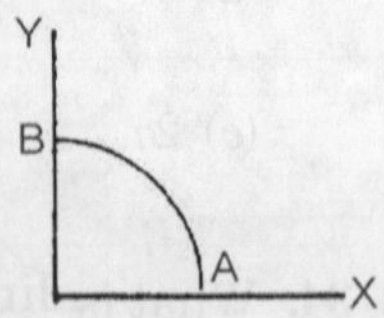

(a) $\frac{\pi}{2} - 1$
(b) $\frac{\pi}{2} + 1$
(c) $\frac{\pi}{2}$
(d) 1

41. The distance between the origin and the point nearest to it on the surface $z^2 = 1 + xy$ is

(a) 1 (b) $\frac{\sqrt{3}}{2}$
(c) $\sqrt{3}$ (d) 2

42. The area enclosed between the curves $y^2 = 4x$ and $x^2 = 4y$ is

(a) $\frac{16}{3}$ (b) 8
(c) $\frac{32}{3}$ (d) 16

2008

43. In the Taylor series expansion of e^x about $x = 2$, the coefficient of $(x-2)^4$ is

(a) $1/4!$ (b) $e^4/4!$
(c) $e^2/4!$ (d) $e^4/4!$

44. The value of $\lim_{x \to 8} \frac{x^{1/3} - 2}{(x-8)}$ is

(a) $\frac{1}{16}$ (b) $\frac{1}{12}$
(c) $\frac{1}{8}$ (d) $\frac{1}{4}$

45. The divergence of the vector field

$$(x-y)\,\hat{i} + (y-x)\,\hat{j} + (x+y+z)\,\hat{k} \text{ is}$$

(a) 0 (b) 1
(c) 2 (d) 3

46. Consider the shaded triangular region P shown in the figure. What is $\iint_P xy\,dx\,dy$?

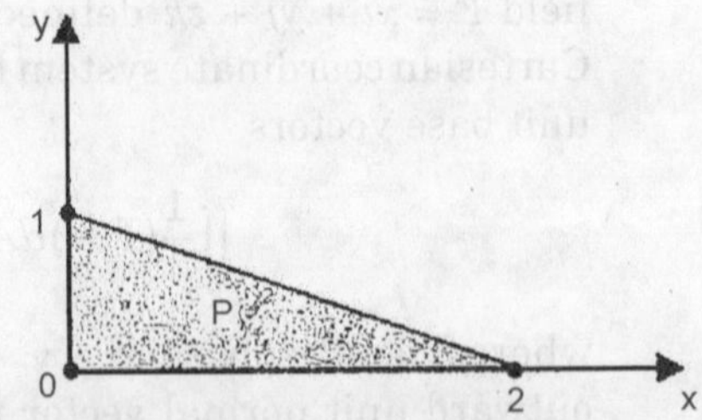

(a) $\frac{1}{6}$
(b) $\frac{2}{9}$
(c) $\frac{1}{16}$
(d) 1

47. The directional derivative of the scalar function $f(x, y, z) = x^2 + 2y^2 + z$ at the point P = (1, 1, 2) in the direction of the vector $\bar{a} = 3\hat{i} - 4\hat{j}$ is

(a) – 4 (b) – 2
(c) – 1 (d) 1

48. Which of the following integrals is unbounded?

(a) $\int_0^{\pi/4} \tan x\,dx$ (b) $\int_0^{\infty} \frac{1}{x^2+1}\,dx$
(c) $\int_0^{\infty} xe^{-x}\,dx$ (d) $\int_0^1 \frac{1}{1-x}\,dx$

49. The length of the curve $y = \frac{2}{3}x^{3/2}$ between $x = 0$ and $x = 1$ is

(a) 0.27 (b) 0.67
(c) 1 (d) 1.22

2007

50. The minimum value of function $y = x^2$ in the interval [1,5] is

(a) 0 (b) 1
(c) 25 (d) undefined

51. The area of a triangle formed by the tips of vectors $\bar{a}, \bar{b}$ and $\bar{c}$ is

(a) $\frac{1}{2}(\bar{a}-\bar{b}).(\bar{a}-\bar{c})$ (b) $\frac{1}{2}\left|(\bar{a}-\bar{b})\times(\bar{a}-\bar{c})\right|$
(c) $\frac{1}{2}\left|\bar{a}\times\bar{b}\times\bar{c}\right|$ (d) $\frac{1}{2}(\bar{a}\times\bar{b}).\bar{c}$

52. $$\lim_{x \to 0} \frac{e^x - \left(1 + x + \frac{x^2}{2}\right)}{x^3} =$$

(a) 0 (b) $\frac{1}{6}$
(c) $\frac{1}{3}$ (d) 1

2006

53. Equation of the line normal to function $f(x) = (x - 8)^{2/3} + 1$ at P (0, 5) is
(a) $y = 3x - 5$ (b) $y = 3x + 5$
(c) $3y = x + 15$ (d) $3y = x - 15$

54. If $f(x) = \dfrac{2x^2 - 7x + 3}{5x^2 - 12x - 9}$, then $\lim_{x \to 3} f(x)$ will be
(a) $-\dfrac{1}{3}$ (b) $\dfrac{5}{18}$
(c) 0 (d) $\dfrac{2}{5}$

NUMERICAL TYPE QUESTIONS

2016

1. Consider the function $f(x) = 2x^3 - 3x^2$ in the domain [–1, 2]. The global minimum of $f(x)$ is _______
2. If $y = f(x)$ satisfies the boundary value problem $y'' + 9y = 0$, $y(0) = 0$, $y(\pi/2) = \sqrt{2}$, then $y(\pi/4)$ is ________.
3. A scalar potential φ has the following gradient : $\nabla\varphi = yz\hat{i} + xz\hat{j} + xy\hat{k}$. = Consider the integral $\int_C \nabla\varphi . d\vec{r}$ on the curve $\vec{r} = x\hat{i} + y\hat{j} + z\hat{k}$.
The curve C is parameterized as follows :
$$\begin{cases} x = t \\ y = t^2 \text{ and } 1 \le t \le 3 \\ z = 3t^2 \end{cases}$$
The value of the integral is ________
4. The area (in percentage) under standard normal distribution curve of random variable Z within limits from – 3 to + 3 is ________
5. The value of the line integral $\int_C \bar{F} \cdot \bar{r}' \, ds$, where C is a circle of radius $\dfrac{4}{\sqrt{\pi}}$ units is ________.
Here, $\bar{F}(x, y) = y\hat{i} + 2x\hat{j}$ and $\bar{r}'$ is the **UNIT** tangent vector on the curve C at an arc length s from a reference point on the curve. $\hat{i}$ and $\hat{j}$ are the basis vectors in the x-y Cartesian reference. In evaluating the line integral, the curve has to be traversed in the counter-clockwise direction.

2015

6. Consider a spatial curve in three-dimensional space given in parametric form by
$x(t) = \cos t$, $y(t) = \sin t$, $z(t) = \dfrac{2}{\pi} t, 0 \le t \le \dfrac{\pi}{2}$
The length of the curve is ______
7. The surface integral $\iint_s \dfrac{1}{\pi}(9xi - 3yj)\, n dS$ over the sphere given by $x^2 + y^2 + z^2 = 9$ is ______.
8. The value of $\lim_{x \to 0}\left(\dfrac{-\sin x}{2\sin x + x\cos x}\right)$ is ______
9. The value of $\int_C \left[(3x - 8y^2)dx + (4y - 6xy)dy\right]$, (where C is boundary of the region bounded by $x = 0$, $y = 0$ and $x + y = 1$ is) ___________

ANSWERS

MCQ Type Questions

1. (b)	**2.** (a)	**3.** (a)	**4.** (c)	**5.** (a)	**6.** (b)	**7.** (c)	**8.** (b)	**9.** (c)	**10.** (a)
11. (c)	**12.** (b)	**13.** (a)	**14.** (a)	**15.** (c)	**16.** (a)	**17.** (b)	**18.** (c)	**19.** (b)	**20.** (a)
21. (d)	**22.** (c)	**23.** (b)	**24.** (b)	**25.** (a)	**26.** (c)	**27.** (a)	**28.** (c)	**29.** (b)	**30.** (d)
31. (a)	**32.** (b)	**33.** (d)	**34.** (d)	**35.** (d)	**36.** (d)	**37.** (d)	**38.** (d)	**39.** (c)	**40.** (b)
41. (a)	**42.** (a)	**43.** (c)	**44.** (b)	**45.** (d)	**46.** (a)	**47.** (b)	**48.** (d)	**49.** (d)	**50.** (b)
51. (b)	**52.** (b)	**53.** (b)	**54.** (b)						

Numerical Type Questions

1. – 5	**2.** –1	**3.** (726)	**4.** (99.73)	**5.** (16)	**6.** 1.86	**7.** (216)
8. (–0.333)	**9.** (1.60–1.70)					

EXPLANATIONS

MCQ TYPE QUESTIONS

1. According to the Laplace transformation of

$f(t)\ \forall \in t \geq 0$ is as $F(s)=L\{f(t)\}=\int_0^{\infty} e^{-st} f(t)dt$

2. $I = \int_{-\infty}^{\infty} \frac{\sin(x)}{x^2+2x+2}$

let $f(x) = \frac{I_m\left(e^{iz}\right)}{z^2+2z+2}$

Then poles of $f(z)$ are given by $z^2 + 2z + 2 = 0$

$\therefore\ z = -1 \pm i$

$R_1 = \text{Res}\left(f(z) : z = -1+i\right)$

$$= \underset{z\to -1\pm i}{\text{Lt}}\left[z-(-1+i)\right]\frac{e^{iz}}{\left[z-(-1+i)\right]\left[z-(-1-i)\right]}$$

$$= \frac{e^{i(-1+i)}}{-1+i+1+i} = \frac{e^{-i-1}}{2i}$$

$$\int_c f(z)dz = \int_c \frac{I_m\left(e^{iz}\right)}{z^2+2z+2}dz = I_m\left[2\pi i(R_1)\right]$$

$$= I_m\left[2\pi i\left(\frac{e^{-i-1}}{2i}\right)\right]$$

$$= I_m\left[\pi e^{-1}\left(\cos(1) - i\sin(1)\right)\right]$$

$$= -\frac{\pi\sin(1)}{e}$$

3.

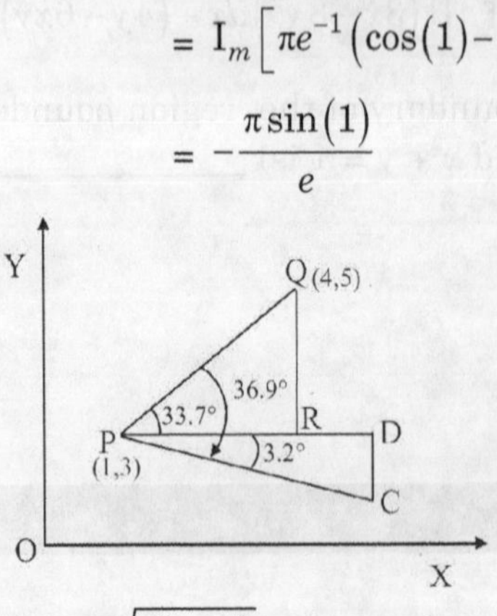

$PQ = \sqrt{2^2 + 3^2} = 3.6055 = PC$

PD = PC × cos 3.2 = 3.6

X co-ordinate of point C = 1 + 3.6 = 4.6

DC = 3.6 sin 3.2 = 0.2

Y co-ordinate of point C = 3.0 – 0.2 = 2.8

4. $f(x) = \frac{x^2-3x-4}{x^2+3x-4} = \frac{x^2-3x-4}{(x+4)(x-1)}$

The function is not defined at $x = -4, 1$ so function f(x) is discontinuous at $x = -4, 1$

5. $L\{\cos(\omega t)\} = \frac{s}{s^2+\omega^2}$ $\left(\because L\{\cos(at)\} = \frac{s}{s^2+a^2}\right)$

6. L The correct path is given in option (B).

$$\int_c \frac{\left(\frac{3z-5}{z-2}\right)}{z-1}dz = 2\pi i\left[\frac{3z-5}{z-2}\right]_{z=1} = 4\pi i$$

7. $\underset{x\to 0}{\text{lt}} \frac{\ln(1+x)}{e^{BC}-1} = \frac{0}{0}$

Apply in L-hospital rule

$$= \underset{x\to 0}{\text{lt}} \frac{\frac{1}{1+4x}\times 4}{e^{3x}\times 3} = \frac{4}{3}$$

8. The solution of Laplace's equation having continuous 2nd order partial derivatives is called a harmonic function.

9. $= \underset{x\to\infty}{\text{lt}}\left(\sqrt{2^2+x-1-x}\right)$

$$= \underset{x\to\infty}{\text{lt}}\left(\sqrt{x^2+x-1}-x\right)\times\frac{\left(\sqrt{x^2+x-1}+x\right)}{\left(\sqrt{x^2+x-1}+x\right)}$$

$$= \underset{x\to\infty}{\text{lt}}\frac{x^2+x-1-x^2}{\sqrt{x^2+x-1}+x}$$

$$= \underset{x\to 0}{\text{lt}}\frac{x\left(1-\frac{1}{x}\right)}{x\left[\sqrt{1+\frac{1}{x}-\frac{1}{x}}+1\right]} = \frac{1}{2}$$

10. "P" after translation = (1 + 2, 3 + 3, –5 –4)

= (3, 6, –9)

Rotation about z- axis means

$$\begin{bmatrix} x' \\ y' \\ z' \\ 1 \end{bmatrix} = \begin{bmatrix} \cos\theta & -\sin\theta & 0 & 0 \\ \sin\theta & \cos\theta & 0 & 0 \\ 0 & 0 & 1 & 0 \\ 0 & 0 & 0 & 1 \end{bmatrix}\begin{bmatrix} x \\ y \\ z \\ 1 \end{bmatrix}$$

$$= \begin{bmatrix} 0 & -1 & 0 & 0 \\ 1 & 0 & 0 & 0 \\ 0 & 0 & 1 & 0 \\ 0 & 0 & 0 & 1 \end{bmatrix}\begin{bmatrix} 3 \\ 6 \\ -9 \\ 1 \end{bmatrix}$$

$$= \begin{bmatrix} 0-6+0+0 \\ 3+0+0+0 \\ 0+0-9+0 \\ 0+0+0+1 \end{bmatrix}\begin{bmatrix} -6 \\ 3 \\ -9 \\ 1 \end{bmatrix}$$

Final point = [–6, 3, –9]

11. $\lim\limits_{x\to 0}\dfrac{1-\cos(x^2)}{2x^4} = \dfrac{0}{0}$

Using L Hospital Rule

$\lim\limits_{x\to 0}\dfrac{(\sin x^2)2x}{-8x^3} = \dfrac{0}{0}$

$\lim\limits_{x\to 0}\dfrac{(\cos x^2)2x2x+(\sin x^2)2}{24x^2}$

$\lim\limits_{x\to 0}\dfrac{(\cos x^2)4x^2+2\sin x^2}{24x^2}$

$= \lim\limits_{x\to 0}\dfrac{(-\sin x^2)4x^2+\cos x^2(8x)+2(\cos x^2)2x}{48x}$

$= \lim\limits_{x\to 0}\dfrac{(-\cos x^2)2x(4x^2)+(-\sin x^2)(8x)+(-\sin x^2)(2x)(8x)+12\cos x^2+(-\sin x^2)(2x)4x}{48}$

$= \dfrac{12}{48} = \dfrac{1}{4}$

12. Area $= \dfrac{1}{2}\times b\times h = \dfrac{1}{2}\times a\sqrt{2}\times a$

$a = 1$

$\Rightarrow \quad \dfrac{1}{\sqrt{2}} = 0.7071$

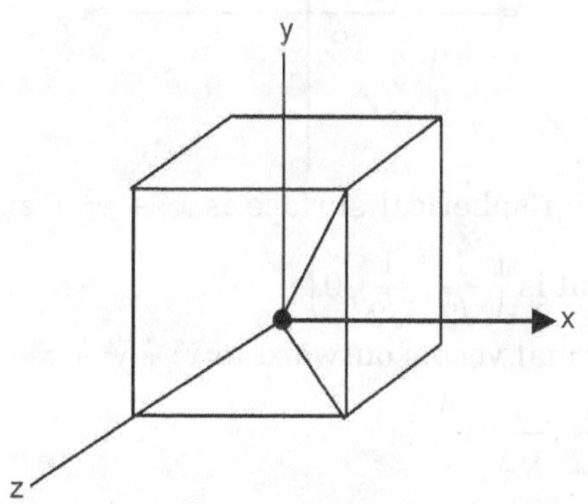

13. For condition regarding something which already happened, should have practiced is the correct choice.

14. For negative value of x, $f(x)$ will be positive

For positive values of x, $f(x)$ will be positive

$\therefore$ minimum value of $f(x)$ will occur at $x = 0$

16. $\lim\limits_{x\to 0}\dfrac{x-\sin x}{1-\cos x} = \left(\dfrac{0}{0}\right)$

Applying L. Hospital Rule,

$\lim\limits_{x\to 0}\dfrac{1-\cos x}{\sin x} = \left(\dfrac{0}{0}\right)$

Once again, L. Hospital rule

$\lim\limits_{x\to 0}\dfrac{\sin x}{\cos x} = \left(\dfrac{0}{1}\right) = 0$

17. Given vectors are

$i + j + k$, $2i + 3j + k$ and $5i + 6j + k$

$$\begin{vmatrix}1 & 1 & 1\\ 2 & 3 & 1\\ 5 & 6 & 1\end{vmatrix} = 0$$

$\therefore$ Vectors are linearly dependent.

18. Given integral $\oint_c (ydx - xdy)$

where C is $x^2 + y^2 = \dfrac{1}{4}$

Applying Green's theorem

$\oint_c \text{M}dx - \text{N}dy = \iint_R \left(\dfrac{\partial \text{N}}{\partial x} - \dfrac{\partial \text{M}}{\partial y}\right) dxdy$

where R is region included in c

$\oint_c ydx - xdy = \iint_R (-1-1)\,dxdy$

$= -2 \iint_R dxdy = -2 \times$ Region R

$= -2 \times$ area of circle with radius $\dfrac{1}{2}$

$= -2 \times \pi\left(\dfrac{1}{2}\right)^2 = \dfrac{-\pi}{2}$

19. $\lim\limits_{x\to 0}\dfrac{e^{2x}-1}{\sin 4x}$ is $\left(\dfrac{0}{0}\right)$

So, Applying L-Hospital Rule, $\lim\limits_{x\to 0}\dfrac{2e^{2x}}{4\cos 4x}$

$= \dfrac{2}{4} = 0.5$

20. Given $\vec{\text{F}} = x^2z^2i\;2xy^2z\text{J} + 2y^2z^3k$

$$\text{curl}\,\vec{\text{F}} = \nabla\times\vec{\text{F}} = \begin{vmatrix} i & j & k\\ \dfrac{\partial}{2x} & \dfrac{\partial}{2y} & \dfrac{\partial}{2z}\\ x^2z^2 & -2xy^2z & 2y^2z^3\end{vmatrix}$$

$= i\left(4yz^3 + 2xy^2\right) - j\left(0 - 2x^2z\right) + k\left(-2y^2z - 0\right)$

$= \left(4yz^2 + 2xy^2 + 2x^2zj\right) - 2y^2zk$

21. We know that $f(x)$ is continuous at $x = a$, if $\lim\limits_{x\to a} f(x)$ exists and equal to $f(a)$

22. Given $\vec{\text{F}} = x^2\,a\hat{i} + xy\hat{j} - yz^2\hat{k}$

$\text{div}\,\vec{\text{F}} = \nabla\cdot\vec{\text{F}}$

$= \dfrac{\partial}{\partial x}(x^2z) + \dfrac{\partial}{\partial y}(xy) + \dfrac{\partial}{\partial z}(-yz^2)$

$= 2xz + x - 2yz$

$\text{div}\,\vec{\text{F}}$ at $(1, -1, 1) = 2 + 1 + 2 = 5$

23. Let $I = \dfrac{\int_0^2 (x-1)^2 \sin(x-1)}{(x-1)^2 + \cos(x-1)} dx$

We know that, $\int_0^2 \dfrac{(2-x-1)^2 \sin(2-x-1)}{(2-x-1)^2 + \cos(2-x-1)} dx$

$$= \int_0^2 \frac{(1-x)^2 \sin(1-x)}{(1-x)^2 + \cos(1-x)} dx$$

$$= -\int_0^2 \frac{(x-1)^2 \sin(x-1)}{(x-1)^2 + \cos(x-1)} dx = -1$$

$\Rightarrow \quad I + I = 0$

$\Rightarrow \quad 2I = 0$

$\Rightarrow \quad I = 0$

24. $\int_0^2 \int_0^x e^{x+y} dy\, dx = \int_0^2 e^x \left(\int_0^x e^y dy\right) dx$

$$= \int_0^2 e^x \left(e^y\right)_0^x dx = \int_0^2 e^x \left(e^x - 1\right) dx$$

$$= \int_0^2 \left(e^{2x} - e^x\right) dx = \left(\frac{e^{2x}}{2} - e^x\right)_0^2$$

$$= \frac{e^4}{2} - e^2 - \frac{1}{2} + 1 = \frac{e^4}{2} - e^2 + \frac{1}{2}$$

$$= \frac{1}{2}(e^4 - 2e^2 + 1) = \frac{1}{2}(e^2 - 1)^2$$

25. Use Gauss-Divergence theorem,

$$I = \iint_s \frac{1}{4} \vec{F}.\hat{n} dA = \iiint_V \vec{\nabla}.\vec{F} dV$$

$$\vec{\nabla}.\vec{F} = 1 + 1 + 1 = 3$$

$$\therefore \quad I = \frac{1}{4} \times 3V = \frac{3}{4} \times \frac{4}{3}\pi(1)^3 = \pi$$

26. $\int_1^e \sqrt{x} \ln x dx = \left[\angle nx \int \sqrt{x} dx - \int \left(\frac{d}{dx} < nx\right)\left(\sqrt{x} dx\right) dx\right]_l^e$

(Integrate by parts)

27. Given curves are

$y = x$ and ...(1)

$y = x^2$ ...(2)

Solving equations (1) and (2), we have

$x = 0, x = 1$

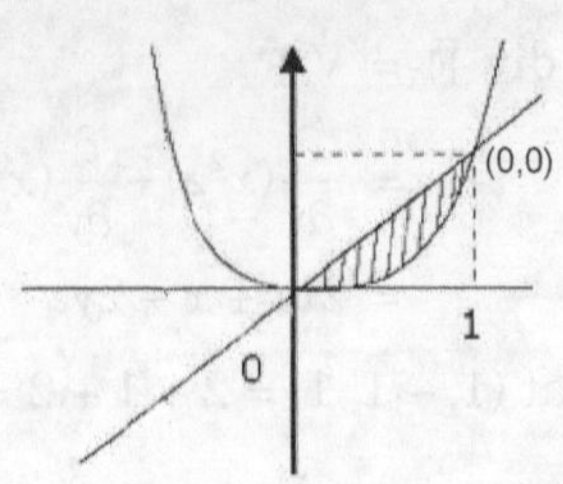

Area $= \int_0^1 (x - x^2) dx$

$$= \left(\frac{x^2}{2} - \frac{x^3}{3}\right)_0^1 = \frac{1}{2} - \frac{1}{3} = \frac{1}{6} \text{ sq units}$$

28. Given function is, $f(x) = |x|$

$|x|$ is continues at $x = 0$ but not differentiable.

29. $\underset{x \to 0}{\text{Lt}} \dfrac{1 - \cos x}{x^2}$

$$= \underset{x \to 0}{\text{Lt}} \frac{2\sin^2\left(\frac{x}{2}\right)}{x^2}$$

$$= 2 \underset{x \to 0}{\text{Lt}} \frac{\sin^2\left(\frac{x}{2}\right)}{\left(\frac{x^2}{2}\right) \times 4} = 2 \times \frac{1}{4} = \frac{1}{2}$$

30. The function $f(x) = x^3 + 1$ has a point of inflection at $x = 0$, since in the graph sign of the curvature (i.e., the concavity) is changed.

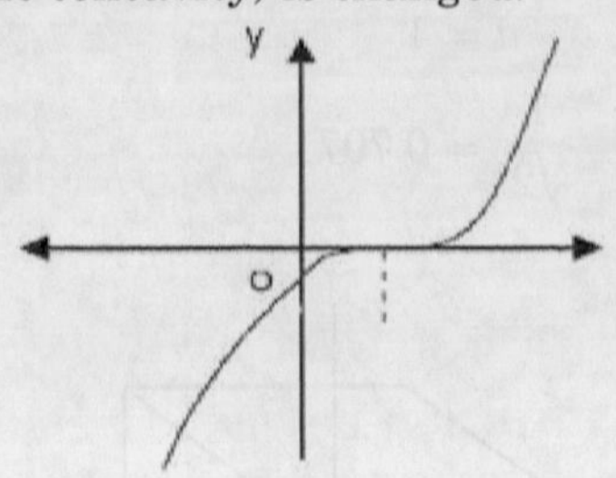

31. Given spherical surface is $x^2 + y^2 + z^2 = 1$ and point is $\left(\frac{1}{\sqrt{2}}, \frac{1}{\sqrt{2}}, 0\right)$

Normal vector outward to $x^2 + y^2 + z^2 - 1 = 0$ at $\left(\frac{1}{\sqrt{2}}, \frac{1}{\sqrt{2}}, 0\right)$ is

$$\frac{1}{\sqrt{2}}i + \frac{1}{\sqrt{2}}j + 0.k = \frac{1}{\sqrt{2}}i + \frac{1}{\sqrt{2}}j$$

Hence the outward unit normal vector

$$= \frac{\frac{1}{\sqrt{2}}i + \frac{1}{\sqrt{2}}j}{\left(\frac{1}{\sqrt{2}}\right)^2 + \left(\frac{1}{\sqrt{2}}\right)^2} = \frac{1}{\sqrt{2}}i + \frac{1}{\sqrt{2}}j \text{ ©}$$

32. $\sin x = x - \dfrac{x^3}{3!} + \dfrac{x^5}{5!} - \ldots$

33. $\int_{-a}^a f(x)dx = \begin{cases} 2\int_0^a f(x)dx; & f(x) \text{ is even} \\ 0; & f(x) \text{ is odd} \end{cases}$

34. Applying L' Hospitals rule, we have

$$\lim_{\theta \to 0} \frac{\cos\theta}{1} = \cos 0 = 1$$

35. Given: $y = \sqrt{x}$

$\Rightarrow$ $y^2 = x, 1 \le x \le 2$

$$\text{Required volume} = \pi\int_a^b y^2\,dx = \pi\int_1^2 x\,dx$$

$$= \pi\left(\frac{x^2}{2}\right)^2 = \pi\left(2-\frac{1}{2}\right) = \frac{3\pi}{2}$$

36. $$I = \int_{-\infty}^{\infty}\frac{dx}{1+x^2}$$

$$= \left[\tan^{-1}x\right]_{-\infty}^{\infty} = \tan^{-1}(\infty) - \tan^{-1}(-\infty)$$

$$= \frac{\pi}{2} - \left(-\frac{\pi}{2}\right) = \pi$$

37. Given : $y = |2-3x|$

At $x = \frac{2}{3}$, sharp point will be there.

Hence, function is continuous to all $\forall\ x \in R$ and also differentiable at all points, $\forall\ x \in R$ except at $x = \frac{2}{3}$.

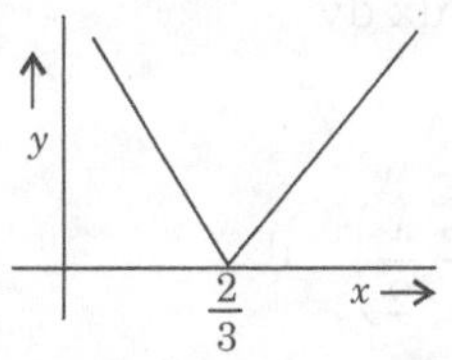

38. $\bar{v} = 2xyi - x^2zj$

Velocity of a vector

$$= \bar{\nabla}\times\bar{v}$$

$$= \begin{vmatrix} i & j & k \\ \frac{\partial}{\partial x} & \frac{\partial}{\partial y} & \frac{\partial}{\partial z} \\ 2xy & -x^2z & 0 \end{vmatrix}$$

$$= \hat{i}\left(0 - \frac{\partial}{\partial z}(-x^2z)\right)\hat{j}\left(0 - \frac{\partial}{\partial z}(2xy)\right)$$

$$+\hat{k}\left(\frac{\partial}{\partial x}(-x^2z) - \frac{\partial}{\partial y}(2xy)\right)$$

$$= \hat{i}(-x^2) + \hat{k}(-2\times z - 2x)$$

At (1, 1, 1),

$$\bar{\nabla}\times\bar{v} = \hat{i} + \hat{k}(-2-2) = i - 4k$$

39. $\vec{F} = 3xz\vec{i} + 2xy\vec{j} - yz^2\vec{k}$

$$\therefore \quad \text{div} = \vec{\nabla}.\vec{F} = \frac{\partial}{\partial x}(3xz) + \frac{\partial}{\partial y} - \frac{\partial}{\partial z}(yz^2)$$

$$= 3z + 2x - 2yz$$

At point (1, 1, 1),

divergence $= 3 + 2 - 2 = 3$

40. $x = \cos\theta$

$y = \sin\theta$

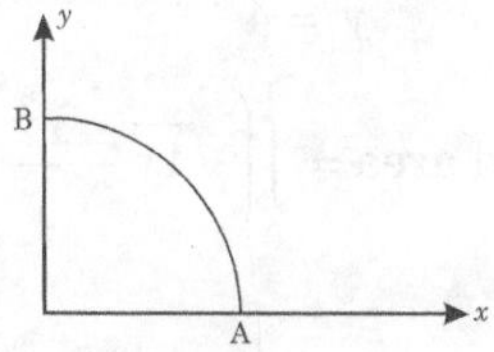

Path is $x^2 + y^2 = 1$

$R = e^{i\theta}$

$\therefore$ $(x + y)^2 = 1 + 2\sin\theta\cos\theta$

$$\int_0^{\pi/2}(1+\sin 2\theta)\,d\theta = \left[\theta - \frac{\cos 2\theta}{2}\right]_0^{\pi/2}$$

$$= \frac{\pi}{2} + \frac{1}{2} + \frac{1}{2} = \frac{\pi}{2} + 1$$

Alternately

Given : $x^2 + y^2 = 1$

Put $x = \cos\theta$, and $y = \sin\theta$

$$(x+y)^2 = \cos^2\theta + \sin^2\theta + 2\sin\theta\cos\theta$$

$$= 1 + \sin 2\theta$$

$$= \int_0^{\pi/2}(1+\sin 2\theta)\,d\theta$$

$$= \left[\theta - \frac{\cos 2\theta}{2}\right]_0^{\pi/2}$$

$$= \frac{\pi}{2} + \frac{1}{2} + \frac{1}{2} = \frac{\pi}{2} + 1$$

41. Not proper method

Short method :

Take a point on the curve z = 1, x = 0, y = 0

Length between origin and this point

$$l = \sqrt{(1-0)+(0-0)+(0-0)} = 1$$

This is minimum length because all options have length greater than l.

42. Given : $y^2 = 4x$

$x^2 = 4y$

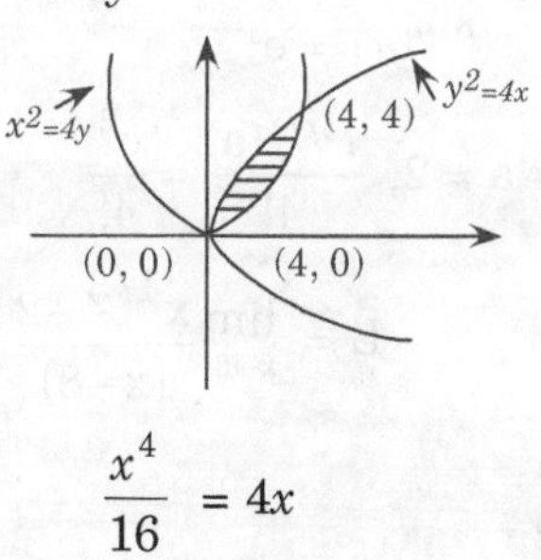

$$\therefore \quad \frac{x^4}{16} = 4x$$

$$\Rightarrow \quad x^4 = 64x$$

$$\Rightarrow \quad x(x^3 - 64) = 0$$

$\Rightarrow \quad x^3 = 64$

$\Rightarrow \quad x = 4$

$\therefore \quad y = 4$

$\therefore$ Required area $= \int_0^4 \left(\sqrt{4x} - \frac{x^2}{y}\right) dx$

$$= \left[2 \times x^{3/2} \times \frac{2}{3} - \frac{x^3}{12}\right]_0^4$$

$$= \frac{4}{3} \times (4)^{3/2} - \frac{64}{12}$$

$$= \frac{32}{3} - \frac{16}{3} = \frac{16}{3}$$

Alternately

For point where both parabolas cut each other

$y^2 = 4x$

$x^2 = 4y$

$\therefore \quad x^2 = 4\sqrt{4x}$

$\Rightarrow \quad x^2 = 8\sqrt{x}$

$\Rightarrow \quad x^4 = 16x$

$\Rightarrow \quad x^3 = 16$

$\therefore \quad x = (4, 0), (-4, 0)$

$\therefore$ Required area $= \int_0^4 \sqrt{4x} - \int_0^4 \frac{x^2}{4} dx$

$$= \left[2 \times \frac{2}{3} x^{3/2} - \frac{x^3}{12}\right]_0^4$$

$$= \frac{16}{3}$$

43. Taylor series expansion of f(x) about a is given by

$$f(x) = f(a) + \frac{(x-a)}{1!} f'(a) + \frac{(x-a)^2}{2!} f''(a) + \text{-----}$$

Coefficient of $(x - a)^4$ is $\frac{f''''(a)}{4!}$

Now $\quad f(x) = e^x$

$\Rightarrow \quad f''''(x) = e^x$

$\Rightarrow \quad f''''(a) = e^a$

Hence for a = 2, $\frac{f''''(a)}{4!} = \frac{e^2}{4!}$

44. $\quad L = \lim_{x\to 8} \frac{x^{1/3} - 2}{(x-8)}$

Let $x = a^3$

$\Rightarrow \quad a = 2$

$\therefore \quad \lim_{a\to 2} \frac{(a-2)}{a^3 - 8}$

By partial fractions,

$$(a^3 - 8) = (a - 2)(a^2 + 2a + 4)$$

$$\Rightarrow \quad L = \lim_{a\to 2} \frac{1}{a^2 + 2a + 4} = \frac{1}{12}$$

45. $\quad \nabla . A = \frac{\partial(x-y)}{\partial x} + \frac{\partial(y-x)}{\partial y} + \frac{\partial}{\partial z}(x+y+z)$

$$= 1 + 1 + 1 = 3$$

46. $I = \iint xy.\ dx\, dy$

The limit of y is from 0 to $\frac{1-\frac{x}{2}}{2}$ and limit of x from 0 to 2.

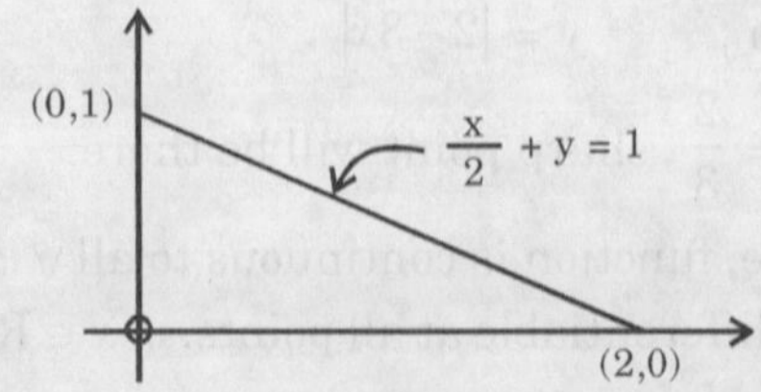

$$I = \int_0^2 \int_0^{1-\frac{x}{2}} xy \ dx\, dy$$

$$= \int_0^2 \frac{x}{2}\left(1 - \frac{x}{2}\right)^2 . dx$$

$$= \frac{1}{8}\int_0^2 x\ (x^2 + 4 - 4x).\ dx$$

$$= \frac{1}{8}\int_0^2 (x^3 + 4x - 4x^2).\ dx$$

$$= \frac{1}{8}\left(\frac{x^4}{4} + 2x^2 - \frac{4x^2}{3}\right)_0^2 = \frac{1}{6}$$

47. The directional derivative of a function ϕ along a vector $\vec{A}$ is gevin by

$$(\text{grad } \phi). \frac{\bar{A}}{|A|}$$

Hence directional derivative

$$= (\text{grad } (x^2 + 2y^2 + z)) . \frac{(3\hat{i} - 4\hat{j})}{\sqrt{3^2 + 4^2}}$$

$$= (2x\,\hat{i} + 4y\,\hat{j} + \hat{k}) \frac{(3\hat{i} - 4\hat{j})}{5}$$

$$= \frac{6x - 16y}{5}$$

Hence (1, 1, 2), directional derivative

$$= \frac{6 \times 1 - 16 \times 1}{5} = -2$$

48. To see whether integrals are bounded or unbounded, we need to see that the integrand doesn't reach infinity in the interval of integration. Let us write down range of the integrands in the 4 options,

$a \to (0, 1)$; $b \to (0, 1)$; $c \to (0, \frac{1}{e})$; $d \to (0, \infty)$

Thus, (*d*), i.e. $\int_0^1 \frac{1}{1-x}\,dx$ is unbounded.

49. Length, $L = \int \sqrt{(dy)^2 + (dx)^2}$

$$dx = \sqrt{x}\,.\,dx$$

$$\therefore \quad L = \int_0^1 \left(\sqrt{x+1}\right).\,dx$$

$$= \frac{2}{3}(x+1)^{3/2}\Big|_0^1 = 1.22$$

50. Since interval given is bounded, so minimum value of function is 1.

51. Let the vectors be $i, j, i + j$

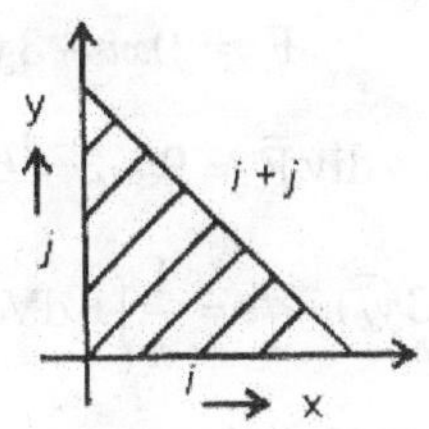

$(\vec{c})\ (\vec{b})\ (\vec{a})$

Area vector will be perpendicular to plane of x, y i.e. z.

k will be the required unit vector.

Now (*a*), (*d*) cannot give vector product

$$\therefore \quad \frac{1}{2}\left|(\vec{a}-\vec{b})\times(\vec{a}-\vec{c})\right| = \frac{1}{2}\left|(\vec{i})\times(\vec{j})\right|$$

$$= \frac{1}{2}k$$

52. We know, $e^x = 1 + \frac{x}{1!} + \frac{x^2}{2!} + \frac{x^3}{3!} + \frac{x^4}{4!} + \ldots.$

$$\text{Given : } \lim_{x\to 0}\frac{e^x - \left[1 + x + \frac{x^2}{2}\right]}{x^3}$$

$$= \lim_{x\to 0}\frac{\frac{x^3}{3!} + \frac{x^4}{4!} + \ldots}{x^3}$$

$$= \frac{1}{3!} = \frac{1}{6}$$

53. *Given:* $f(x) = (x - 8)^{2/3} + 1$

$$m_{\text{normal}} \times m_{\text{tangent}} = -1$$

where, $m_{\text{tangent}} = \frac{dy}{dx}$

and $m_{\text{normal}} = \frac{-1}{(dy/dx)}$

Here, $\frac{dy}{dx} = \frac{df(x)}{dx}$

$$= \frac{2}{3}(x-8)^{-1/3}$$

$$m_{\text{normal}} = \frac{-3}{2}(x-8)^{1/3}$$

$$\Rightarrow \quad [m_{\text{normal}}]_{(0,5)} = 3$$

Hence equation of normal is

$$(y - y_1) = m_{\text{normal}}(x - x_1)$$

$$\Rightarrow \quad (y - 5) = 3\,(x - 0)$$

$$\Rightarrow \quad y = 3x + 5$$

54.

$$I = \lim_{x\to 3}\frac{2x^2 - 7x + 3}{5x^2 - 12x - 9}$$

Applying L' Hospital rule, we get

$$I = \lim_{x\to 3}\frac{4x - 7}{10x - 12}$$

$$= \frac{4\times 3 - 7}{10\times 3 - 12}$$

$$= \frac{5}{18}$$

NUMERICAL TYPE QUESTIONS

1.

$$f(x) = 2x^3 - 3x^2$$

$$f'(x) = 0$$

$$\Rightarrow \quad 6x^2 - 6x = 0$$

$$6x(x - 1) = 0$$

$$x = 0 \text{ and } 1$$

$$f(-1) = -5,\ f(1) = -1,\ f(2) = -4$$

Global minimum = –5

2. $y'' + 9y = 0$

A.E. is $m^2 + 9 = 0$

$$m = \pm 3i$$

$$y = y_c + y_p$$

$$y = C_1\cos 3x + C_2\sin 3x \qquad (\because y_p = 0)$$

If $x = 0$, $y = 0$

$$0 = C_1(1) + C_2(0) \Rightarrow C_1 = 0$$

If $x = \pi/2$, $y = \sqrt{2}$

$$\sqrt{2} = C_1(0) + C_2 \sin\left(\tfrac{3\pi}{2}\right) \Rightarrow C_2 = -\sqrt{2}$$

$$\therefore \; y = -\sqrt{2}\sin 3x$$

If $x = \pi/4$

$$y\left(\tfrac{\pi}{4}\right) = -\sqrt{2}\sin\left(\tfrac{3\pi}{4}\right) = -\sqrt{2}\left(\frac{1}{\sqrt{2}}\right) = -1$$

3. $$\int_c \Delta\phi . d\bar{r} = \int_c (yz\hat{i} + xz\hat{j} + xy\hat{k}).(dx\hat{i} + dy\hat{j} + dz\hat{k})$$

$$= \int_c yzdx + xzdy + xydz$$

$x = t$, $y = t^2$, $z = 3t^2$

$$\begin{bmatrix} dx = dt \\ dy = 2t\,dt \\ dz = 6t\,dt \end{bmatrix}$$

$$= \int t^2 . 3t^2 \; dt + \int t.(3t^2)\; 2t\,dt + t(t^2)6t\;dt$$

$$= \int_1^3 [3t^4 + 6t^4 + 6t^4]dt$$

$$= 15\int_1^3 t^4\,dt = \frac{15}{5}[t^5]_1^3 = 3[3^5 - 1] = 726$$

4. In the standard normal curve the area between –3 and 3 is 0.9973

$\therefore$ Percentage of area is 99.73

5. By Greon's theorem

$$= \int_C \bar{F}\dot{r}ds = \int_C ydx + 2xdy$$

$$= \iint (2-1)dxdy$$

$$= \pi\left(\frac{4}{\sqrt{\pi}}\right)^2 = 16$$

6. The length of the curve

$$= \int_0^{\pi/2} \sqrt{\left(\frac{dx}{dt}\right)^2 + \left(\frac{dy}{dt}\right)^2 + \left(\frac{dz}{dt}\right)^2}\,dt$$

$$= \int_0^{\pi/2} \sqrt{(-\sin t)^2 + (\cos t)^2 + \left(\frac{2}{\pi}\right)^2}\,dt$$

$$= \int_0^{\pi/2} \sqrt{\sin^2 t + \cos^2 t + \frac{4}{\pi^2}}\,dt$$

$$= \int_0^{\pi/2} \sqrt{1 + \frac{4}{\pi^2}}\,dt$$

$$= \sqrt{1 + \frac{4}{\pi^2}} \cdot t\Big|_0^{\pi/2}$$

$$= \sqrt{1 + \frac{4}{\pi 2}} \times \frac{\pi}{2} = 1.8622$$

7. By Gauss divergence theorem,

$$\int_S \bar{F}.\bar{n}\;ds = \int_V \text{div}\,\bar{F}dV$$

Here $\bar{F} = 9x\bar{i} - 3y\hat{j}$

$$\text{div}\,\bar{F} = 9 - 3 = 6$$

$$\therefore \int_s \frac{1}{\pi}(9x\bar{i} - 3y\bar{j}).\bar{n}ds = \frac{1}{\pi}\int_V 6\,dV = \frac{1}{\pi}6V$$

$$= \frac{1}{\pi}6\left(\frac{4}{3}\pi r^3\right)$$

$$= 8(3)^3 = 216$$

3

CHAPTER

Differential Equations

MCQ TYPE QUESTIONS

2015

1. Fine the solution of $\frac{d^2y}{dx^2} = y$ which passes through the origin and the point $\left(\text{in } 2, \frac{3}{4}\right)$,

(a) $y = \frac{1}{2}e^x - e^{-x}$ (b) $y = \frac{1}{2}(e^x + e^{-x})$

(c) $y = \frac{1}{2}(e^x - e^{-x})$ (d) $y = \frac{1}{2}e^x + e^{-x}$

2. The Laplace transform of e^{i5t} where $i = \sqrt{-1}$ is

(a) $\frac{s-5i}{s^2-25}$ (b) $\frac{s+5i}{s^2+25}$

(c) $\frac{s+5i}{s^2-25}$ (d) $\frac{s-5i}{s^2+25}$

3. Consider the following differential equation:

$\frac{dy}{dt} = -5y$; initial condition : $y = 2$ at $t = 0$.

The value of y at $t = 3$ is

(a) $-5e^{-10}$ (b) $2e^{-10}$

(c) $2e^{-15}$ (d) $-15e^{2}$

4. Laplace transform of the function $f(t)$ is given by $F(s) = L\{f(t)\} = \int_0^\infty f(t)e^{-st}dt$. Laplace transform of the function shown below is given by

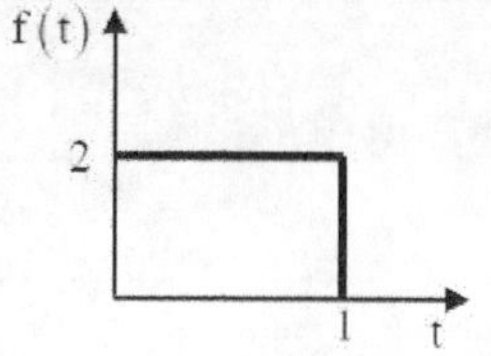

(a) $\frac{1-e^{-2s}}{s}$

(b) $\frac{1-e^{-s}}{2s}$

(c) $\frac{2-2e^{-s}}{s}$

(d) $\frac{1-2e^{-s}}{s}$

2014

5. The general solution of the differential equation $\frac{dy}{dx} = \cos(x+y)$, with c as a constant, is

(a) $y + \sin(x+y) = x + c$ (b) $\tan\left(\frac{x+y}{2}\right) = y + c$

(c) $\cos\left(\frac{x+y}{2}\right) = x + c$ (d) $\tan\left(\frac{x+y}{2}\right) = x + c$

6. Consider two solutions $x(t) = x_1(t)$ and $x(t) = x_2(t)$ of the differential equation

$\frac{d^2x(t)}{dt^2} + x(t) = 0,\ t > 0,$

Such that

$$x_1(0) = 1, \left.\frac{dx_1(t)}{dt}\right|_{t=0} = 0,\ x_2(0) = 0, \left.\frac{dx_2(t)}{dt}\right|_{t=0} = 1.$$

The Wronskian $W(t) = \begin{vmatrix} x_1(t) & x_2(t) \\ \frac{dx_1(t)}{dt} & \frac{dx_2(t)}{dt} \end{vmatrix}$ at $t = \pi/2$ is

(a) 1 (b) –1

(c) 0 (d) $\pi/2$

7. The solution of the initial value problem $\frac{dy}{dx} = -2xy$; $y(0) = 2$ is

(a) $1 + e^{-x^2}$ (b) $2e^{-x^2}$

(c) $1 + e^{x^2}$ (d) $2e^{x^2}$

8. Laplace transform of cos (ω t) is $\frac{s}{s^2+\omega^2}$. The laplace transform of $e^{-2t}\cos(4t)$ is

(a) $\frac{s-2}{(s-2)^2+16}$ (b) $\frac{s+2}{(s-2)^2+16}$

(c) $\frac{s-2}{(s+2)^2+16}$ (d) $\frac{s+2}{(s+2)^2+16}$

9. Consider an ordinary differential equation. $\frac{dx}{dt} = 4t + 4$. If $x = x_0$ at $t = 0$, the increment in x calculated using Runge-Kutta fourth order multi-step method with a step size of $\Delta t = 0.2$ is

(a) 0.22 (b) 0.44

(c) 0.66 (d) 0.88

2013

10. The partial differential equation

$$\frac{\partial u}{\partial t} + u\frac{\partial u}{\partial \lambda} = \frac{\partial^2 u}{\partial x^2} \text{ is a}$$

(a) linear equation of order 2

(b) non-linear equation of order 1

(c) linear equation of order 1

(d) non-linear equation of order 2

11. The function $f(t)$ satisfies the differential equation $\frac{d^2 f}{dt^2} + f = 0$ and the auxiliary conditions, $f(0) = 0$, $\frac{df}{dt}(0) = 4$. The Laplace transform of $f(t)$ is given by

(a) $\frac{2}{s+1}$ (b) $\frac{4}{s+1}$

(c) $\frac{4}{s^2+1}$ (d) $\frac{2}{s^4+1}$

12. The solution to the differential equation $\frac{d^2u}{dx^2} - k\frac{du}{dx} = 0$ where k is a constant subjected to the boundary conditions $u(0) = 0$ and $u(L) = U$ is

(a) $u = U\frac{x}{L}$ (b) $u = U\left(\frac{1-e^{kx}}{1-e^{kL}}\right)$

(c) $u = U\left(\frac{1-e^{-kx}}{1-e^{-kL}}\right)$ (d) $u = U\left(\frac{1+e^{kx}}{1+e^{kL}}\right)$

2012

13. The inverse Laplace transform of the function $F(s) = \frac{1}{s(s+1)}$ is given by

(a) $f(t) = \sin t$ (b) $f(t) = e^{-t}\sin t$

(c) $f(t) = e^{-t}$ (d) $f(t) = 1 - e^{-t}$

14. Consider the differential equation $x^2\frac{d^2y}{dx^2} + x\frac{dy}{dx} - 4y = 0$ with the boundary conditions of $y(0) = 0$ and $y(1) = 1$. The complete solution of the differential equation is

(a) x^2 (b) $\sin\left(\frac{\pi x}{2}\right)$

(c) $e^x \sin\left(\frac{\pi x}{2}\right)$ (d) $e^{-x}\sin\left(\frac{\pi x}{2}\right)$

2011

15. Consider the differential equation $\frac{dy}{dx} = (1 + y^2)x$. The general solution with constant c is

(a) $y = \tan\frac{x^2}{2} + \tan c$

(b) $y = \tan^2\left(\frac{x}{2} + c\right)$

(c) $y = \tan^2\left(\frac{x}{2}\right) + c$

(d) $y = \tan\left(\frac{x^2}{2} + c\right)$

2010

16. The Blasius equation, $\frac{d^3 f}{d\eta^3} + \frac{f d^2 f}{2\, d\eta^2} = 0$, is a

(a) second order nonlinear ordinary differential equation

(b) third order nonlinear ordinary differential equation

(c) third order linear ordinary differential equation

(d) mixed order nonlinear ordinary differential equation

17. The Laplace transform of a function $f(t)$ is $\frac{1}{s^2(s+1)}$. The function $f(t)$ is

(a) $t - 1 + e^{-t}$ (b) $t + 1 + e^{-t}$

(c) $-1 + e^{-t}$ (d) $2t + e^t$

2009

18. The inverse Laplace transform of $\frac{1}{(s^2+s)}$ is

(a) $1 + e^t$

(b) $1 - e^t$

(c) $1 - e^{-t}$

(d) $1 + e^{-t}$

19. The solution of $x\frac{dy}{dx} + y = x^4$ with the condition $y(1) = \frac{6}{5}$ is

(a) $y = \frac{x^4}{5} + \frac{1}{x}$ (b) $y = \frac{4x^4}{5} + \frac{4}{5x}$

(c) $y = \frac{x^4}{5} + 1$ (d) $y = \frac{x^5}{5} + 1$

2008

20. Let f = y^x. What is $\frac{\partial^2 f}{\partial x \partial y}$ at x = 2, y = 1 ?

(a) 0 (b) In 2

(c) 1 (d) $\frac{1}{\text{In } 2}$

21. Given that y″ + 2y′ + y = 0, y(0) = 0, y(1) = 0. What is y(0.5) ?

(a) 0 (b) 0.37

(c) 0.62 (d) 1.13

2007

22. The solution of $\frac{dy}{dx} = y^2$ with initial value $y(0) = 1$ is bounded in the interval

(a) $-\infty \le x \le \infty$ (b) $-\infty \le x \le 1$

(c) $x < 1, x > 1$ (d) $-2 \le x \le 2$

23. If F (s) is the Laplace transform of function $f(t)$, then Laplace transform of $\int_0^t f(\tau)\, d\tau$ is

(a) $\frac{1}{s}$ F (s) (b) $\frac{1}{s}$ F (s) $- f(0)$

(c) sF (s) $- f(0)$ (d) $\int$ F (s) ds

2006

24. The solution of the differential equation $\frac{dy}{dx} + 2xy = e^{-x^2}$ with $y(0) = 1$ is

(a) $(1 + x)\, e^{+x^2}$ (b) $(1 + x)\, e^{-x^2}$

(c) $(1 - x)\, e^{+x^2}$ (d) $(1 - x)\, e^{x^2}$

25. For $\frac{d^2y}{dx^2} + 4\frac{dy}{dx} + 3y = 3e^{2x}$, the particular integral is

(a) $\frac{1}{15}e^{2x}$ (b) $\frac{1}{5}e^{2x}$

(c) $3e^{2x}$ (d) $C_1e^{-x} + C_2e^{-3x}$

NUMERICAL TYPE QUESTION

2014

1. If $y = f(x)$ is the solution of $\frac{d^2y}{dx^2} = 0$ with the boundary conditions $y = 5$ at $x = 0$, and $\frac{dy}{dx} = 2$ at $x = 10$, $f(15)$ = _____

ANSWERS

MCQ Type questions

1. (c)	**2.** (b)	**3.** (c)	**4.** (c)	**5.** (d)	**6.** (a)	**7.** (b)	**8.** (d)	**9.** (d)	**10.** (d)
11. (c)	**12.** (b)	**13.** (d)	**14.** (a)	**15.** (d)	**16.** (b)	**17.** (a)	**18.** (c)	**19.** (a)	**20.** (c)
21. (a)	**22.** (c)	**23.** (b)	**24.** (b)	**25.** (b)					

Numerical Type Question

1. (34 to 36)

EXPLANATIONS

MCQ TYPE QUESTIONS

1. $\frac{d^2y}{dx^2} = y \Rightarrow (D^2 - 1)\,y = 0$

$D^2 - 1 = 0$

$\Rightarrow \quad D = \pm 1$

$y = c_1 e^x + c_2 e^{-x}$

Passes through (0, 0) and $\left(142, \frac{3}{4}\right)$ (0, 0)

$\Rightarrow \quad 0 = C_1 + C_2 \quad$ ...(1)

$\left(142, \frac{3}{4}\right)$

$\frac{3}{4} = C_1 e^{142} + C_2 e^{-142} = C_1 2 + \frac{C_2}{2}$

$\Rightarrow 2C_1 + \frac{1}{2}C_2 = \frac{3}{4} \quad$ ...(2)

Solving (1) and (2)

$\Rightarrow \quad C_1 = \frac{1}{2}$

$C_2 = -\frac{1}{2}$

$\therefore \quad y = \frac{1}{2}e^x - \frac{1}{2}e^{-x}$

$= \frac{1}{2}(e^x - e^{-x})$

2. $\mathcal{L}\{e^{i5t}\} = \mathcal{L}\,(\cos 5t + i \sin 5t)$

$= \frac{s+5i}{s^2+25}$

3. Here, $\frac{dy}{dt} = -5y$;

$\therefore \quad \frac{dy}{y} = -5dt$

Taking integration on both sides

$\int \frac{dy}{y} = \int -5dt$

$\therefore \quad \log y = -5t + C \quad$ ...(*i*)

Now at $t = 0$, $y = 2$

$\therefore \quad C = \log 2$

Thus equation (*i*) changes to

$\log y = -5t + \log 2$

$\therefore \quad \log\left(\frac{y}{2}\right) = -5t$

$\therefore \quad y = 2e^{-5t}$

Now at $t = 3$

$\therefore \quad y = 2e^{-15}$

4. $f(t) = 2;\ 0 < t < 1 > 0$; otherwise

$\therefore \quad L[f(t)] = \int_0^1 2e^{-st}dt = 2\frac{e^{-st}}{-s}\Big|_0^1$

$= \frac{2-2e^{-s}}{s}$

5. $\frac{dy}{dx} = \cos(x+y)$

Put $\quad x + y = v \Rightarrow 1 + \frac{dy}{dx}$

$= \frac{dv}{dx}$

$\Rightarrow \quad \frac{dy}{dx} = \frac{dv}{dx} - 1$

$\Rightarrow \quad \frac{dv}{dx} - 1 = \cos v$

$\Rightarrow \quad \frac{1}{1+\cos v}dv = dx$ (variable separable)

$\int \frac{1}{2\cos^2\frac{v}{2}}dv = \int dx + c \quad \left[\begin{array}{l} 1+\cos 2\theta = 2\cos^2\theta \\ 1+\cos\theta = 2\cos^2\frac{\theta}{2} \end{array}\right]$

$\frac{1}{2}\int \sec^2\frac{v}{2}\,dv = x + c$

$\tan\frac{v}{2} = x + c$

$\Rightarrow \quad \tan\left(\frac{x+y}{2}\right) = x + c$

6. Given Differential equation is

$\frac{d^2x(t)}{dt^2} + x(t) = 0$

Auxiliary equation is $m^2 + 1 = 0$

$m = 0 \pm i$

Complementary y solution is

$$x_c = c_1 \cos t + c^2 \sin t$$

Particular solution $x_p = 0$

$\therefore$ General solution $x = c_1 \cos t + c_2 \sin t$

Let $x_1(t) = \cos t$ $x_2(t) = \sin t$

Clearly $x_1(0) = 1 \frac{dx_1}{dt} = 0$

and $x_2(0) = 0, \frac{dx_2}{dt} = 1$

$$W = \begin{vmatrix} x_1(t) & x_2(t) \\ \frac{dx_1}{dt} & \frac{dx_2}{dt} \end{vmatrix}_{x=\pi/2}$$

$$= \begin{vmatrix} \cos t & \sin t \\ -\sin t & \cos t \end{vmatrix}_{t=\pi/2} = \cos^2 t + \sin^2 t = 1$$

7. Given $\frac{dy}{dx} = -2y, y(0) = 2$ is

$\Rightarrow$ $\frac{dy}{dx} + 2xy = 0$

comparing with $\frac{dy}{dx} + py = Q$

$$P = 2x;\ Q = 0$$

I.F. $= e^{\int pdx} = e^{\int 2xdx} = e^{x^2}$

Solution is $y(\text{IF}) = \int Q(\text{IP})dx + C$

$$ye^{x^2} = 0 + C$$

$\therefore$ $y = Ce^{-x^2}$

Given $y(0) = 2$

$\Rightarrow$ $2 = C$

$\therefore$ $y = 2e^{-x^2}$

8. We know that if $L\{f(t)\} = F(s)$

Then $L\{e^{at} f(t)\} = F(s-a)$

$\therefore$ $L\{e^{-2t} \cos 4t\} = \frac{s+2}{(s+2)^2 + 4^2}$

$$= \frac{s+2}{(s+2)^2 + 16}$$

9. Given $\frac{dx}{dt} = 4t + 4$

$x = x_0$ at $t = 0$

$n = 0.2$

Calculate $x(0.2)$ value

$K_1 = f(t_0, x_0) = f(0, x_0) = 4$

$$K_2 = f\left(t_0 + \frac{h}{2}, x_0 + \frac{1}{2}K_2 h\right)$$

$= f(0 + 0.1, x_0 + 0.4)$

$= f(001, x_0 + 0.4)$

$= 4(0.1) + 4$

$= 4.4$

$$K_3 = f\left(x_0 + \frac{h}{2}, x_0 + \frac{K_2 h}{2}\right)$$

$= f(t_0 + 0.1, x_0 + (2.2)(0.2))$

$= f(0.1, x_0 + 0.44)$

$= 4(0.1) + 4$

$= 4.4$

$K_4 = f(t_0 + h, x_0 + K_3 h)$

$= f(0 + 0.2, x_0 + 0.88)$

$= f(0.2, x_0 + 0.88)$

$= 4(0.2) + 4$

$= 4.8$

$$x(0.2) = x_1 = x_0 + \frac{h}{6}(K_1 + 2K_2 + 2K_3 + K_4)$$

$$= x_0 + \frac{0.2}{6}(4 + 2(4.4) + 2(4.4) + (4.8))$$

$$= x_0 + \frac{0.2}{6}(4 + 8.8 + 8.8 + 4.8)$$

$= x_0 + 0.88$

Increment as $x = x_1 - x_0 = x_0 + 0.88 - x_0 = 0.88$

10. Order is index of a derivative present in a PDE (i.e.)maximum there.

Hence, $\frac{\partial^2 u}{\partial n^2} \Rightarrow$ highest index 2 $\Rightarrow$ order = 2

if y, y_2 solution of a PDE, then PDE is linearly independent.

$$\frac{\partial y_1}{\partial t} + \frac{y_1 \partial y_1}{\partial tx} = \frac{\partial y_1^2}{\partial x^2} \quad ...(1)$$

$$\frac{\partial y_2}{\partial t} + \frac{y_2 \partial y_2}{\partial x} = \frac{\partial^2 y_2}{\partial x^2} \quad ...(2)$$

If $y = ay_1 + a_1 y_2$ is a sum of the PDE

$$\frac{\partial}{\partial t}(ay_1 + a_1 y_1) + (ay_1 + a_2 y_2)\frac{\partial}{d_2}(ay_1 + ay_2) = \frac{\partial^2}{d2^2}(ay_1 + ay_2)$$

Hence PDE is non linear of order 2

11. Check by options

$$L^{-1}\left\{\frac{4}{s^2+1}\right\} = 4 \text{ sint satisfies } \frac{d^2f}{dt^2}+f=0$$

12. Since $\frac{d^2u}{dx^2}-\frac{kdu}{dx}=0$ and check by options.

13. $$L^{-1}\left(\frac{1}{S(S+1)}\right) = \frac{1}{S}-\frac{1}{S+1}$$

$$L^{-1}\left(\frac{1}{S}-\frac{1}{S+1}\right) = 1-e^{-t}$$

14. $$x^2\frac{d^2y}{dx^2}+x\frac{dy}{dx}-4y = 0$$

$\Rightarrow x^2\,y'' + xy' - xy = 0$

From Euler-Cauchy's Equation

$a_0\,x^2y''+a_1xy'+a_2y = 0$

Given:

$a_0 = 1, a_1=1, a_2=-4$

$\therefore \quad a_0m(m-1)+a_1m+a_2 = 0$

$m(m-1)+m-4 = 0$

$m^2 = 4$

$\Rightarrow \quad m = \pm 2$

$y(x) = Ax^2 + Bx^{-2}$

$= Ax^2 + \frac{B}{x^2}$

Applying Boundary condition

$y(0) = A0 + \frac{B}{0} = 0$

$\therefore \quad B = 0$

and $y(1) = A(1) = 1$

$\therefore \quad A = 1$

$y(x) = x^2$

15. Given differential equation is

$$\frac{dy}{dx} = (1+y^2)\,x$$

$$\Rightarrow \quad \frac{dy}{1+y^2} = xdx$$

Integrating, we have

$$\Rightarrow \quad \tan^{-1} y = \frac{x^2}{2}+c$$

$$\Rightarrow \quad y = \tan\left(\frac{x^2}{2}+c\right)$$

16. According to Blasius equation,

$$f''' + \frac{f}{2}.f'' = 0$$

is a third order linear ordinary differential equation.

17. Given : $L[f(t)] = \left[\frac{1}{s^2(s+1)}\right]$

$$\therefore \quad f(t) = L^{-1}\left[\frac{1}{s^2(s+1)}\right]$$

Let $$\frac{1}{s^2(s+1)} = \frac{As+B}{s^2}+\frac{C}{(s+1)}$$

$$\therefore \quad 1 = (As+B)(s+1)+C(s^2)$$

Put $s = 0$

$\Rightarrow \quad B = 1$

Put $s = -1$,

$\Rightarrow \quad C = 1$

Equating coefficients of s^2 on both sides, we have

$0 = A + C$

Put $C = 1$

$\Rightarrow \quad A = -1$

$$\therefore \quad \frac{1}{s^2(s+1)} = \frac{-s}{s^2}+\frac{1}{s^2}+\frac{1}{(s+1)}$$

$$= -\frac{1}{s}+\frac{1}{s^2}+\frac{1}{(s+1)}$$

$$\therefore \quad \mathcal{L}^{-1}\left[\frac{1}{s^2(s+1)}\right] = \mathcal{L}^{-1}\left[-\frac{1}{s}\right]+\mathcal{L}^{-1}\left[\frac{1}{s^2}\right]+\mathcal{L}^{-1}\left[\frac{1}{s+1}\right]$$

$$= -1+t+e^{-t}$$

18. $$\frac{1}{(s^2+s)} = \frac{1}{s(s+1)}$$

$$= \frac{1}{s}-\frac{1}{(s+1)}$$

$\therefore$ Laplace transform $= (1-e^{-t})$

19. Given : $x\frac{dy}{dx}+y = x^4$

$$\Rightarrow \quad \frac{dy}{dx}+\frac{y}{x} = x^3$$

Its a linear differential equation of the form

$$\frac{dy}{dx}+Py = Q$$

where P, Q are only functions of x

$$P = \frac{1}{x}, Q = x^3$$

Solution is given by

$$ye^{\int Pdx} = \int\left(Qe^{\int Pdx}\right) dx + C$$

$\Rightarrow$ $$y.e^{\int \frac{1}{x}dx} = \int\left(Q.e^{\int \frac{1}{x}dx}\right)dx + C$$

$\Rightarrow$ $$ye^{\ln x} = \int x^3.e^{\ln x} + C$$

$\Rightarrow$ $$yx = \int x^3.xdx + C$$

$\Rightarrow$ $$yx = \int x^4dx + C$$

$\Rightarrow$ $$yx = \frac{x^5}{5} + C$$

$\Rightarrow$ $$y = \frac{x^4}{5} + \frac{C}{x}$$

Given condition : $y(1) = \frac{6}{5}$

$\therefore$ $$\frac{6}{5} = \frac{(1)^4}{5} + \frac{C}{1}$$

$\Rightarrow$ $$C = \frac{6}{5} - \frac{1}{5} = 1$$

$\therefore$ $$y = \frac{x^4}{5} + \frac{1}{x}$$

20. Given: $f = y^x$

$\Rightarrow$ $\ln f = x \ln y$

$\Rightarrow$ $$\frac{1}{f}\frac{\partial f}{\partial y} = \frac{x}{y}$$

$\Rightarrow$ $$\frac{\partial f}{\partial y} = \frac{x}{y}.\ y^x$$

Now, $$\ln\frac{\partial f}{\partial y} = \ln\frac{x}{y} + x \ln y$$

$\Rightarrow$ $$\frac{1}{\frac{\partial f}{\partial y}}\cdot\frac{\partial^2 f}{\partial x\,\partial y} = \left[\frac{y}{x}.\frac{1}{y} + \ln y\right]$$

$\Rightarrow$ $$\frac{\partial^2 f}{\partial x\,\partial y} = \left[\frac{1}{x} + \ln y\right]\frac{x}{y}.\ y^x$$

$$= \left[\frac{1}{2} + 0\right] \quad 2\ 1 = 1$$

21. *Given:* $y'' + 2y' + y = 0$

A.E is, $D^2 + 2D + 1 = 0$

$\Rightarrow$ $(D + 1)^2 = 0$

$\Rightarrow$ $m = -1$

C.F. is $(C_1 + C_2 x)e^{-x}$

P.I. = 0

Now, $y(0) = 0$

$\Rightarrow$ $C_1 = 0$

and $y(1) = 0$

$\Rightarrow$ $C_2e^{-1} = 0$

$\Rightarrow$ $C_2 = 0$

$\therefore$ C.F. = 0

$\Rightarrow$ $y(0.5) = 0$

22. Given : $$\frac{dy}{dx} = y^2$$

$\Rightarrow$ $$\frac{dy}{dy^2} = dx$$

Integrating $$-\frac{1}{y} = x + c$$

At $y = 0$, $c = -1$

$\therefore$ $$x = 1 - \frac{1}{y}$$

$\Rightarrow$ $$y = \frac{-1}{x-1}$$

Only (*c*) $x \neq 0$

23. From definition, we know

$$\int_0^t f(t)dt = \frac{1}{s}\ F(s)$$

24. First order equation is

$$\frac{dy}{dx} + Py = Q,$$

where $P = 2x$ and $Q = e^{-x^2}$

Since P and Q are functions of x

$\therefore$ Integrating factor $= e^{+\int Pdx} = e^{+x^2}$

$$e^{x^2}\frac{dy}{dx} + (2xe^{x^2})y = e^{-x^2}.e^{+x^2}$$

$$\frac{d}{dx}(ye^{x^2}) = 1$$

$$ye^{x^2} = x + C$$

Since $y(0) = 1$

$\Rightarrow$ $C = 1$

Hence its solution is

$$y = (1 + x)\, e^{-x^2}$$

25. Given differential equation may be written as

$$D^2y + 4Dy + 3y = 3e^{2x}$$

$\therefore$ $$\text{P.I.} = \frac{3e^{2x}}{(D^2 + 4D + 3)}$$

Substituting D = 2, we get

$$\text{P.I.} = 3e^{2x}\left(\frac{1}{15}\right) = \frac{e^{2x}}{5}$$

NUMERICAL TYPE QUESTION

1. Given $\frac{d^2y}{dx^2} = 0$, $y = 5$ at $x = 0$, $\frac{dy}{dx} = 2$ at $x = 10$

$\Rightarrow$ Auxilliary equation is $m^2 = 0$

$\Rightarrow$ $m = 0, 0$

$y_c = (c_1 + c_2x)\, e^x = c_1 + c_2x$

$y_p = 0$

General solution $y = y_c + y_p$

$\Rightarrow$ $y = c_1 + c_2x$

$y = 5$ at $x = 0$ $\quad\Rightarrow\quad c_1 = 5$

$\frac{dy}{dx} = 2$ at $x = 10$ $\quad\Rightarrow\quad 2 = c_2$

$y(x) = 5 + 2x$ $\quad\Rightarrow\quad y\,(15) = 5 + 30 = 35$

4
CHAPTER

Probability and Statistics

MCQ TYPE QUESTIONS

2016

1. Consider a Poisson distribution for the tossing of a biased coin. The mean for this distribution is μ. The standard deviation for this distribution is given by

(a) $\sqrt{\mu}$ (b) μ^2
(c) μ (d) 1/μ

2. Three cards were drawn from a pack of 52 cards. The probability that they are a king, a queen, and a jack is

(a) $\frac{16}{5525}$ (b) $\frac{64}{2197}$
(c) $\frac{3}{13}$ (d) $\frac{8}{16575}$

2015

3. Given two complex numbers $Z_1 = 5+(\sqrt{3})i$ and $z_2 = \frac{2}{\sqrt{3}} + 2i$, the argument of $\frac{z_1}{z_2}$ in degrees is

(a) 0
(b) 30
(c) 60
(d) 90

2014

4. The argument of the complex number $\frac{1+i}{1-i}$, where $i = \sqrt{-1}$, is

(a) $-\pi$ (b) $-\frac{\pi}{2}$
(c) $\frac{\pi}{2}$ (d) π

5. An analytic function of a complex variable $z = x + i\,y$ is expressed as $f(z) = u(x, y) + i\,v(x, y)$, where $i = \sqrt{-1}$. If $u(x, y) = 2\,x\,y$, then $v(x, y)$ must be

(a) $x^2 + y^2$ + constant
(b) $x^2 - y^2$ + constant
(c) $-x^2 + y^2$ + constant
(d) $-x^2 - y^2$ + constant

6. An analytic function of a complex variable $z = x + iy$ is expressed as $f(z) = u(x, y) + iv(x, y)$, where $i = \sqrt{-1}$. If $u(x, y) = x^2 - y^2$, then expression for $v(x, y)$ in terms of x, y and a general constant c would be

(a) $xy + c$ (b) $\frac{x^2 + y^2}{2} + c$
(c) $2xy + c$ (d) $\frac{(x - y)^2}{2} + c$

7. If z is a complex variable, the value of $\int_5^{3i} \frac{dz}{z}$ is

(a) $-0.511 - 1.57i$ (b) $-0.511 + 1.57i$
(c) $0.511 - 1.57i$ (d) $0.511 + 1.57i$

2011

8. The product of two complex numbers $1 + i$ and $2 - 5i$ is

(a) $7 - 3i$ (b) $3 - 4i$
(c) $-3 - 4i$ (d) $7 + 3i$

2010

9. The modulus of the complex number $\left(\frac{3+4i}{1-2i}\right)$ is

(a) 5 (b) $\sqrt{5}$
(c) $\frac{1}{\sqrt{5}}$ (d) $\frac{1}{5}$

2009

10. An analytic function of a complex variable $z = x + iy$ is expresed as

$$f(z) = u(x, y) + iv(x, y), \text{ where } i = \sqrt{-1}.$$

If $u = xy$, the expression of v should be

(a) $\frac{(x+y)^2}{2} + k$

(b) $\frac{x^2 - y^2}{2} + k$

(c) $\frac{y^2 - x^2}{2} + k$

(d) $\frac{(x-y)^2}{2} + k$

2007

11. If $\varphi(x, y)$ and $\psi(x, y)$ are functions with continuous second derivatives, then

$\varphi(x, y) + i\psi(x, y)$

can be expressed as an analytic function of

$x + iy$ ($i = \sqrt{-1}$), when

(a) $\frac{\partial \varphi}{\partial x} = -\frac{\partial \psi}{\partial x}; \frac{\partial \varphi}{\partial y} = \frac{\partial \psi}{\partial y}$

(b) $\frac{\partial \varphi}{\partial y} = -\frac{\partial \psi}{\partial x}; \frac{\partial \varphi}{\partial x} = \frac{\partial \psi}{\partial y}$

(c) $\frac{\partial^2 \varphi}{\partial x^2} + \frac{\partial^2 \varphi}{\partial y^2} = \frac{\partial^2 \psi}{\partial x^2} + \frac{\partial^2 \psi}{\partial y^2} = 1$

(d) $\frac{\partial \varphi}{\partial x} + \frac{\partial \varphi}{\partial y} = \frac{\partial \psi}{\partial x} + \frac{\partial \psi}{\partial y} = 0$

2006

12. Assuming $i = \sqrt{-1}$ and t is a real number, $\int_0^{p/3} e^{it}\,dt$ is

(a) $\frac{\sqrt{3}}{2} + i\frac{1}{2}$ (b) $\frac{\sqrt{3}}{2} - i\frac{1}{2}$

(c) $\frac{1}{2} + i\frac{\sqrt{3}}{2}$ (d) $\frac{1}{2} + i\left(1 - \frac{\sqrt{3}}{2}\right)$

NUMERICAL TYPE QUESTION

2016

1. The probability that a screw manufactured by a company is defective is 0.1. The company sells screws in packets containing 5 screws and gives a guarantee of replacement if one or more screws in the packet are found to be defective. The probability that a packet would have to be replaced is ________.

ANSWERS

MCQ Type Questions

1. (a) **2.** (a) **3.** (a) **4.** (c) **5.** (c) **6.** (c) **7.** (b) **8.** (a) **9.** (b) **10.** (c)
11. (b) **12.** (a)

Numerical Type Question

1. (0.41)

EXPLANATIONS

MCQ TYPE QUESTIONS

1. In the case of Poisson distribution mean = variance

Given, mean = μ

∴ variance = μ

⇒ standard deviation = $\sqrt{\mu}$

2. $= \frac{{}^4C_1 \times {}^4C_1 \times {}^4C_1}{{}^{52}C_3}$

$= \frac{16}{5525}$

3. $z_1 = 5+(5\sqrt{3})i$

$\arg z_1 = \tan^{-1}\left(\frac{5\sqrt{3}}{5}\right) = \tan^{-1}(\sqrt{3}) = 60°$

$z_2 = \frac{2}{\sqrt{3}} + 2i$

$\arg z_2 = \tan^{-1}\left(\frac{2}{\frac{2}{\sqrt{3}}}\right) = \tan^{-1}(\sqrt{3}) = 60°$

$\arg\left(\frac{z_1}{z_2}\right) = \arg(z_1) - \arg(z_2)$

$= 60° - 60° = 0$

4. Given: $z = \frac{1+i}{1-i} \Rightarrow z = \frac{(1+i)(1+i)}{(1-i)(1+i)}$

$= \frac{(1+i)^2}{(1^2-i^2)} = \frac{1+2i+i^2}{(1\ +1)}$

$= \frac{1+2i-1}{2} = i$

Arg (z) = Arg (i)

$= \tan^{-1}\left(\frac{y}{x}\right) = \tan^{-1}\infty = \frac{\pi}{2}$

5. Given $f(z) = u + iv$ is analytic and $u = 2xy$

We know that if $f(z)$ is analytic then CR equations will be satisfied.

i.e., $\frac{\partial u}{\partial x} = \frac{\partial v}{\partial y}$ and $\frac{\partial u}{\partial y} = -\frac{\partial v}{\partial x}$

Verify the options which satisfy the above C-R equations.

$v = -x^2 + y^2$ + constant, satisfies the C-R equation.

6. Given $f(z) = \mu x(x, y) + iv\,(x, y)$ is analytic and $x = x^2 - y^2$

We know that if $f(z) = \mu + iv$ is analytic then C-R equations will be satisfied.

i.e., $\frac{\partial\mu}{\partial x} = \frac{\partial v}{\partial xy}$ and $\frac{\partial\mu}{\partial y} = \frac{-\partial v}{\partial x}$

∴ $v = 2xy + c$ is correct answer

7. $\int_5^{3i}\frac{dz}{z} = \ln(z)\Big|_5^{3i} = \ln 3i - \ln 5$

$= \ln 3 + i\frac{\pi}{2} - (\ell n 5 + i0)$

$\left[\therefore \ln z = \ln r = i\theta \text{ where } r = \sqrt{x^2+y^2}, \theta = \arg z\right]$

$= \ln 3 - \ln 5 + i\frac{22}{14}$

$= -0.511 + 157i$

8. $(1 + i)(2 - 5i) = 2 - 5i + 2i + 5 = 7 - 3i$

9. Given : $z = \frac{3+4i}{1-2i}$

Rationalizing the denominator, we get

$z = \frac{3+4i}{1-2i} \times \frac{1+2i}{1+2i}$

$= \frac{3+6i+4i-8}{5}$

$= \frac{10i-5}{5} = 2i - 1$

∴ Modulus of $|z| = |2i-1| = \sqrt{2^2+(-1)^2} = \sqrt{5}$

10. Given u = x,y

For analytic function

$\frac{\partial u}{\partial x} = \frac{\partial v}{\partial y}$

and $\frac{\partial u}{\partial y} = -\frac{\partial v}{\partial x}$

By Milne Thomson method

Let w = u + zv

∴ $\frac{dw}{dz} = \frac{\partial y}{\partial x} + i\frac{\partial v}{\partial x}$

$= \frac{\partial u}{\partial x} - i\frac{\partial u}{\partial y}$

$$\Rightarrow \qquad \frac{dw}{dz} = y - ix$$

Replacing x by z and y by 0, we get

$$\frac{dw}{dz} = 0 - iz$$

where, $z = x + iy$

$$\therefore \qquad dw = -iz\, dz$$

Integrating, $\quad w = -i\dfrac{z^2}{2} + C$

where C is a constant,

$$\therefore \qquad V = \text{Im}\left[-i\frac{z^2}{2} + C\right]$$

$$= \text{Im}\left[-i\frac{(x^2 - y^2 + 2ixy)}{2}\right]$$

$$\Rightarrow \qquad v = \frac{y^2 - x^2}{2}$$

11. Let analytical function be

$$f(x + iy) = (x + iy)^2$$

$$= (x^2 - y^2) + 2ixy$$

Now, $\quad \varphi(x, y) = x^2 - y^2$

$\qquad\quad \psi(x, y) = 2xy$

Clearly $\quad \dfrac{\partial \varphi}{\partial y} = -\dfrac{\partial \psi}{\partial x}; \; \dfrac{\partial \varphi}{\partial x} = \dfrac{\partial \psi}{\partial y}$

12. $$\int_0^{\pi/3} e^{it}\, dt = \left[\frac{e^{it}}{i}\right]_0^{\pi/2}$$

$$= \frac{1}{i}\left[e^{i\pi/3} - 1\right]$$

$$= -i\left[\cos\frac{\pi}{3} + i\sin\frac{\pi}{3} - 1\right]$$

$$= -i\left[\frac{1}{2} + i\frac{\sqrt{3}}{2} - 1\right] = \frac{\sqrt{3}}{2} + i\frac{1}{2}$$

NUMERICAL TYPE QUESTION

1. we require $P(x \geq 1) = 1 - P(x = 0)$

$$= 1 - 5C_0\,(0.1)^0\,(0.9)^5$$

$$= 0.4095 \approx 0.41$$

■■

5 CHAPTER

Numerical Methods

MCQ TYPE QUESTIONS

2016

1. Numerical integration using trapezoidal rule gives the best result for a single variable function, which is

(a) linear (b) parabolic
(c) logarithmic (d) hyperbolic

2. The root of the function $f(x) = x^3 + x - 1$ obtained after first iteration on application of Newton-Raphson scheme using an initial guess of $x_0 = 1$ is

(a) 0.682
(b) 0.686
(c) 0.750
(d) 1.000

2015

3. Among the four normal distributions with probability density functions as shown below, which one has the lowest variance?

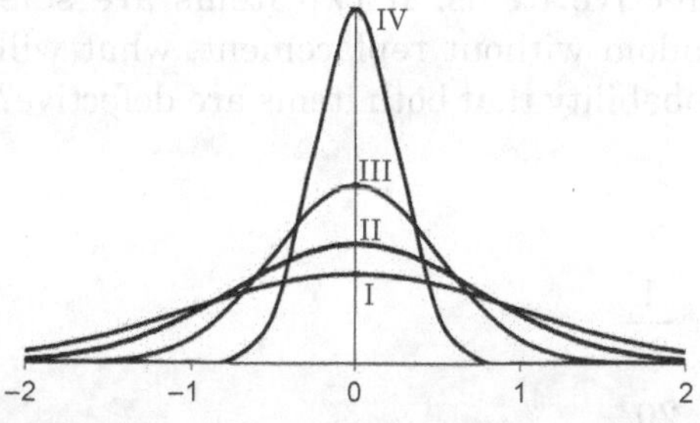

(a) I (b) II
(c) III (d) IV

4. If P(X) = 1/4, P(Y) = 1/3, and P(X∩Y) = 1/2, the value of P(Y/X) is

(a) $\frac{1}{4}$
(b) $\frac{4}{25}$
(c) $\frac{1}{3}$
(d) $\frac{29}{50}$

2014

5. In the following table, x is a discrete random variable and $p(x)$ is the probability density. The standard deviation of x is

x	1	2	3
p(x)	0.3	0.6	0.1

(a) 0.18 (b) 0.36
(c) 0.54 (d) 0.6

6. A box contains 25 parts of which 10 are defective. Two parts are being drawn simultaneously in a random manner from the box. The probability of both the parts being good is

(a) $\frac{7}{20}$ (b) $\frac{42}{125}$
(c) $\frac{25}{29}$ (d) $\frac{5}{9}$

7. A machine produces 0, 1 or 2 defective pieces in a day with associated probability of 1/6, 2/3 and 1/6, respectively. The mean value and the variance of the number of defective pieces produced by the machine in a day, respectively, are

(a) 1 and 1/3
(b) 1/3 and 1
(c) 1 and 4/3
(d) 1/3 and 4/3

8. The number of accidents occurring in a plant in a month follows Poisson distribution with mean as 5.2. The probability of occurrence of less than 2 accidents in the plant during a randomly selected month is

(a) 0.029 (b) 0.034
(c) 0.039 (d) 0.044

2013

9. Let x be a normal random variable with mean 1 and variance 4. The probability P{X < 0} is

(a) 0.5
(b) greater than zero and less than 0.5
(c) greater than 0.5 and less than 1.0
(d) 1.0

10. The probability that a student knows the correct answer to a multiple choice question is $\frac{2}{3}$. If the student does not know the answer, then the student guesses the answer. The probability of the guessed answer being correct is $\frac{1}{4}$. Given that the student has answered the question correctly, the conditional probability that the student knows the correct answer is

(a) $\frac{2}{3}$ (b) $\frac{3}{4}$

(c) $\frac{5}{6}$ (d) $\frac{8}{9}$

2011

11. An unbiased coin is tossed five times. The outcome of each toss is either a head or a tail. The probability of getting at least one head is

(a) $\frac{1}{32}$ (b) $\frac{13}{32}$

(c) $\frac{16}{32}$ (d) $\frac{31}{32}$

2010

12. A box contains 2 washers, 3 nuts and 4 bolts. Items are drawn from the box at random one at a time without replacement. The probability of drawing 2 washers first followed by 3 nuts and subsequently the 4 bolts is

(a) $\frac{2}{315}$ (b) $\frac{1}{630}$

(c) $\frac{1}{1260}$ (d) $\frac{1}{2520}$

2009

13. If three coins are tossed simultaneously, the probability of getting atleast one head is

(a) $\frac{1}{8}$

(b) $\frac{3}{8}$

(c) $\frac{1}{2}$

(d) $\frac{7}{8}$

14. The standard deviation of a uniformly distributed random variable between 0 and 1 is

(a) $\frac{1}{\sqrt{12}}$ (b) $\frac{1}{\sqrt{3}}$

(c) $\frac{5}{\sqrt{12}}$ (d) $\frac{7}{\sqrt{12}}$

2008

15. A coin is tossed 4 times. What is the probability of getting heads exactly 3 times?

(a) $\frac{1}{4}$ (b) $\frac{3}{8}$

(c) $\frac{1}{2}$ (d) $\frac{3}{4}$

2007

16. Let X and Y be two independent random variables. Which one of the relations between expectation (E), variance (Var) and covariance (*Cov*) given below is **FALSE** ?

(a) $E(XY) = E(X)E(Y)$

(b) $Cov(X, Y) = 0$

(c) $Var(X + Y) = Var(X) + Var(Y)$

(d) $E(X^2 Y^2) = (E(X))^2 (E(Y))^2$

2006

17. A box contains 20 defective items and 80 non-defective items. If two items are selected at random without replacement, what will be the probability that both items are defective?

(a) $\frac{1}{5}$

(b) $\frac{1}{25}$

(c) $\frac{20}{99}$

(d) $\frac{19}{495}$

18. The number of customers arriving at a railway reservation counter is Poisson distributed with an arrival rate of eight customers per hour. The reservation clerk at this counter takes six minutes per customer on an average with an exponentially distributed service time. The average number of the customers in the queue will be

(a) 3 (b) 3.2

(c) 4 (d) 4.2

NUMERICAL TYPE QUESTIONS

2016

1. Solve the equation $x = 10\cos(x)$ using the Newton-Raphson method. The initial guess is $x = \pi/4$. The value of the predicted root after the first iteration, up to second decimal, is ________
2. Gauss-Seidel method is used to solve the following equations (as per the given order):

$$x_1 + 2x_2 + 3x_3 = 5$$
$$2x_1 + 3x_2 + x_3 = 1$$
$$3x_1 + 2x_2 + x_3 = 3$$

Assuming initial guess as $x_1 = x_2 = x_3 = 0$, the value of x_3 after the first iteration is ________.

3. The error in numerically computing the integral $\int_0^{\pi}(\sin x + \cos x)dx$ using the trapezoidal rule with three intervals of equal length between 0 and π is ________.

2014

4. Consider an unbiased cubic dice with opposite faces coloured identically and each face coloured red, blue or green such that each colour appears only two times on the dice. If the dice is thrown thrice, the probability of obtaining red colour on top face of the dice at least twice is ________
5. A group consists of equal number of men and women. Of this group 20% of the men and 50% of the women are unemployed. If a person is selected at random from this group, the probability of the selected person being employed is ________

ANSWERS

MCQ Type Questions

1. (*a*)	**2.** (*c*)	**3.** (*d*)	**4.** (*c*)	**5.** (*d*)	**6.** (*a*)	**7.** (*a*)	**8.** (*b*)	**9.** (*b*)	**10.** (*d*)
11. (*d*)	**12.** (*c*)	**13.** (*d*)	**14.** (*a*)	**15.** (*a*)	**16.** (*d*)	**17.** (*d*)	**18.** (*b*)		

Numerical Type Questions

1. 1.5641 **2.** –6 **3.** (0.1862) **4.** (0.25 to 0.27) **5.** (0.64 to 0.66)

EXPLANATIONS

MCQ TYPE QUESTIONS

1. $f(x)$ is a linear function

2. $$x_{n+1} = x_n - \frac{f(x_n)}{f'(x_n)}$$

for n = 0, $$x_1 = x_0 - \frac{f(x_0)}{f'(x_0)}$$

$$f(x) = x^3 + x + 1$$

$$f'(x) = 3x^2 + 1$$

given $x_0 = 1$

$$\therefore \quad f(x_0) = f(1) = 1$$

$$f'(x_0) = f'(1) = 4$$

$$x_1 = 1 - \frac{1}{4} = \frac{3}{4} = 0.75$$

3. We have Probability distribution function of Normal Distribution

$$f(x) = \frac{1}{\sigma\sqrt{x}} e^{-\frac{(x-\mu)^2}{2\sigma^2}} \quad \text{...(1)}$$

Variance = σ^2 is lowest

$\Rightarrow$ σ also lowest

$\Rightarrow$ If σ decreases $\Rightarrow$ $f(x)$ increases ($\because$ from (1))

$\Rightarrow$ Curve will have highest peak

4. $$P(Y/X) = \frac{P(X \cap Y)}{P(X)} = \frac{\frac{1}{12}}{\frac{1}{4}} = \frac{1}{3}$$

5. Given

x	1	2	3
$p(x)$	0.3	0.6	0.1

mean $(\mu) = \exp(x) = 1 \times 0.3 + 2 \times 0.6 + 3 \times 0.1$

$= 0.3 + 1.2 + 0.3 = 1.8$

$E(x^2) = \sum x^2 P(x)$

$= 1 \times 0.3 + 4 \times 0.6 + 9 \times 0.1$

$= 0.3 + 2.4 + 0.9 = 3.6$

Variance $v(x) = E(x^2) - \mu^2 = 3.6 - (1.8)^2$

S. D $(\sigma) = +\sqrt{v(x)}$

$= +\sqrt{3.6 - (1.8)^2} = \sqrt{0.36} = 0.6$

6. Two parts can be selected in the following way $15C_2 \times 10C_0$

Good	Bad
15	10
2	0
1	1
0	2

Total no. of parts = 25.

Two parts can be selected in $25C_2$ ways

$$\text{Required probability} = \frac{15C_2 \times 10C_0}{25C_2}$$

$$= \frac{105}{300} = \frac{7}{20}$$

7. Let 'x' be no. of defective pieces.

x	0	1	2
$P(x)$	$1/6$	$2/3$	$1/6$

$$\text{Mean } (\mu) = E(x) = \Sigma\, x\, P(x)$$

$$= \left(0 \times \frac{1}{6}\right) + \left(1 \times \frac{1}{3}\right) + \left(2 \times \frac{1}{6}\right)$$

$$= 0 + \frac{2}{3} + \frac{1}{3} = 1$$

$$E(x^2) = \Sigma\, x^2\, P(x)$$

$$= \left(0 \times \frac{1}{6}\right) + \left(1 \times \frac{2}{3}\right) + \left(4 \times \frac{1}{6}\right)$$

$$= 0 + \frac{2}{3} + \frac{2}{3} = \frac{4}{3}$$

Variance, $$V(x) = E(x^2) - \{E(x)\}^2$$

$$= \frac{4}{3} - 1 = 1/3$$

8. Given $\lambda = 5.2$

Let x be random variable which follows Poisson's distribution

$$P(x < 2) = P(x = 0) + P(x = 1)$$

$$= \frac{e^{-\lambda}\lambda^0}{0!} + \frac{e^{-\lambda}}{1!}\lambda^1 = e^{-5.2}(6.2)$$

$$= 0.0055 \times 6.2$$

$$= 0.034$$

9. $$P(x < 0) = P\left\{\frac{(X - \bar{X})}{\sigma} < \frac{0 - \bar{X}}{\sigma}\right\}$$

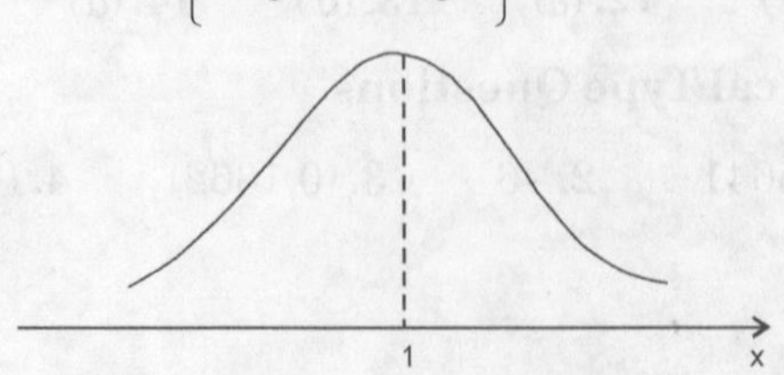

$\because$ $\bar{x} = 1, \sqrt{V} = \sigma$ or $\sigma = \sqrt{4} = 2$

$\therefore$ $P\left\{\frac{x-1}{2} < -\frac{1}{2}\right\}$ or $P\left\{z < -\frac{1}{2}\right\}$

$$P\left(z < -\frac{1}{2}\right) = P\left(z > 1/2\right)$$

$\because$ $P(x > a)$ represents area under curve for $x > a$

10. Probability = $\dfrac{2/3\times 1}{(2/3\times 1)+\left(\dfrac{1}{3}\times\dfrac{1}{4}\right)} = 8/9$

11. P (at least one head) = 1 – P (no heads)

$$= 1 - \frac{1}{2^5} = \frac{31}{32}$$

12. *Given :* Washers – 2, Nuts – 3, Bolts – 4,

Total = 9 objects

Hence probability of drawing two washers from 9 objects

$$= \frac{^2C_2}{^9C_2} = \frac{1}{36}$$

Probability of drawing, 3 nuts out of remaining 7 objects

$$= \frac{^3C_3}{^7C_3}\ \frac{1}{^7C_3} = \frac{1}{35}$$

Probability of drawing 4 bolts out of remaining 4 bolts

$$= \frac{^4C_4}{^4C_4} = 1$$

$\therefore$ Required probability = $\dfrac{1}{36}\times\dfrac{1}{35}\times 1 = \dfrac{1}{1260}$

13. Let probability of getting atleast one head = P(H)

$\therefore$ P(at least one head) = 1 –P (no head)

$\Rightarrow$ P(H) = 1– P(all tails)

But in all cases, $2^3 = 8$

$\therefore$ $P(H) = 1 - \dfrac{1}{8} = \dfrac{7}{8}$

Alternately

Probability of getting at least one head

$$= 1 - {}^3C_0\left(\frac{1}{2}\right)^0\left(\frac{1}{2}\right)^3$$

$$= 1 - \frac{1}{8} = \frac{7}{8}$$

Alternately

From Binomial theorem

p^r q^{h-r}

Probability of getting at least one head

$$= P\,{}^nC_r p^r q^{n-r}$$

$$= 3C_1\left(\frac{1}{2}\right)^3 + 3C_2\left(\frac{1}{2}\right)^3 + 3C_3\left(\frac{1}{2}\right)^3$$

$$= (3+3+1)\left(\frac{1}{2}\right)^3 = \frac{7}{8}$$

14. A uniform distribution and density function

$$f(x) = \int_{-0}^{x} f(x)dx \begin{cases} 0, & x\le a \\ \dfrac{x-a}{b-a}, & a<x<b \\ 0, & x\ge b \end{cases}$$

Density function

$$f(x) = \begin{cases} \dfrac{1}{b-a} & a,x<b \\ 0 & a>x,x>b \end{cases}$$

$\therefore$ Mean $E(x) = \sum_{x=a}^{b} x(F(x) = \dfrac{a+b}{2}$

$\therefore$ Variance $= F(x)^2 - [f(x)]^2$

$$= \sum_{x=a}^{b} x^2F(x) - \left\{\sum_{x=a}^{b} xF(x)\right\}^2$$

Put the value of F(x), we get

Variance

$$= \sum_{x=a}^{b} x^2\frac{1}{b-a}dx - \left\{\sum_{x=a}^{b} x.\frac{1}{b-a}dx\right\}^2$$

$$= \left[\frac{x^3}{3(b-a)}\right]_a^b - \left[\frac{xL}{2(b-a)}\right]^2$$

$$= \frac{b^3-a^3}{3(b-a)} - \frac{(b^2-a^2)^L}{4(b-a)^2}$$

$$= \frac{(b-a)(b^2+ab+a^2)}{3(b-a)} - \frac{(b-a)^2(b+a)^2}{4(b-a)^2}$$

$$= \frac{4b^2+4ab+4a^2-3a^2-3b^2-6ab}{12}$$

$$= \frac{b^2+a^2-2ab}{12} = \frac{(b-a)^2}{12}$$

Standared deviance

$= \sqrt{\text{variance}}$

$$= \sqrt{\frac{(b-a)^2}{12}} = \frac{b-a}{\sqrt{12}}$$

Given : $b = 1, a = 0$

$\therefore$ Standard variance = $\dfrac{1-0}{\sqrt{12}} = \dfrac{1}{\sqrt{12}}$

15. Number of favourable cases are given by

H H H T
H H T H
H T H H
T H H H

Total number of cases = ${}^2C_1\ {}^2C_1\ {}^2C_1\ {}^2C_1$

= 16

$\therefore$ Probability = $\dfrac{4}{16} = \dfrac{1}{4}$

16. Since X and Y are independent hence (*a*), (*b*), (*c*) are true Only (*d*) is odd one.

17. $$P\text{ (Both defective)} = \frac{N\text{(Both defective)}}{\text{Sample space}} = \frac{{}^{20}C_2}{{}^{100}C_2} = \frac{19}{495}$$

18. Average number of customer in queue

$$= \frac{\lambda^2}{\mu(\mu - A)} = \frac{8 \times 8}{10 \times (10-8)} = 3.2$$

NUMERICAL TYPE QUESTIONS

1. Let $f(x) = x - 10\cos(x)$

and $x_0 = \frac{\pi}{4}$

$f'(x) = 1 + 10\sin(x)$

$$x_1 = x_0 - \frac{f(x_0)}{f'(x_0)} = \frac{\pi}{4} - \frac{\left(\frac{\pi}{4} - \frac{10}{\sqrt{2}}\right)}{\left(1 + \frac{10}{\sqrt{2}}\right)}$$

$$\Rightarrow \frac{\pi}{4} + \frac{(6.2857)}{(8.0711)}$$

$\therefore\ x_1 = 1.5641$

2. Let x + 2y + 3z = 5

2x + 3y + z = 1

3x + 2y + z = 3 and $x_0 = 0, y_0 = 0, z_0 = 0$

Then first iteration will be

$x_1 = x_1 = 5 - 2y_0 - 3z_0 = 5 - 0 - 0 = 5$

$$x_0 = x_0 = \frac{1}{3}(1 - 2x_1 - z_0)$$

$$= \frac{1}{3}(1 - 10 - 0) = -3$$

$x_3 = z_1 = 3 - 3x_1 - 2y_1$

$= 3 - 15 + 6 = -6$

$\therefore\ x_3 = -6$

3. $$f(x) = \int_o^\pi (\sin x + \cos x)dx$$

x	0	$\frac{\pi}{3}$	$\frac{2\pi}{3}$	π
$f(x)$	1	$\frac{\sqrt{3}}{2}$	$\frac{\sqrt{3}}{2}$	–1

$$h = \frac{b-a}{n} = \frac{\pi - 0}{3} = \frac{\pi}{3}$$

Trapezoidal Rule

$$\int_o^\pi f(x)dx = \frac{h}{2}[(y_0 + y_3) + 2(y_1 + y_2)]$$

$$= \frac{\pi}{6}\left[(1-1) + 2\left(\frac{\sqrt{3}}{2} + \frac{\sqrt{3}}{2}\right)\right]$$

$= 1.8137$

By Exact solution

$$f(x) = \int_o^\pi (\sin x + \cos x)dx$$

$$= [-\cos x + \sin x]_0^\pi$$

$= 1 - (-1) = 2$

Error = Exact value – approximate value

$= 2 - 1.8137$

$= 0.1863$

4. $$p = \frac{2}{6} = \frac{1}{3}$$

$$q = 1 - \frac{1}{3} = \frac{2}{3}$$

Using Binomial distribution

$$p(x \geq 2) = 3_{C_2}\left(\frac{1}{3}\right)^2\left(\frac{2}{3}\right)^1 + 3_{C_3}\left(\frac{1}{3}\right)^3\left(\frac{2}{3}\right)^0$$

$$= \frac{6}{27} + \frac{1}{27}$$

$$= \frac{7}{27}$$

5. Let M → men

W → women

E → employed

U → unemployed

Given P(M) = 0.5

P(W) = 0.5

$P(U/M) = 0.20$

$P(U/W) = 0.50$

By Total probability,

$P(U) = P(M)\ P(U/M) + P(W)\ P(U/M)$

$= 0.5 \,.\, 0.20 + 0.5 \,.\, 0.50 = 0.35$

Required probability = P(E) = 1 – P(U)

$= 1 - 0.35 = 0.65$

6
CHAPTER

Complex Variables

MCQ TYPE QUESTIONS

2016

1. $f(z) = u(x, y) + i\, v(x, y)$ is an analytic function of complex variable $z = x + i\, y$ where $i = \sqrt{-1}$. If $u(x, y) = 2\, xy$, then $v(x, y)$ may be expressed as

(a) $-x^2 + y^2 +$ constant

(b) $x^2 - y^2 +$ constant

(c) $x^2 + y^2 +$ constant

(d) $-(x^2 + y^2) +$ constant

2013

2. Match the CORRECT pairs

Numerical Integration Scheme	Order of Fitting Polynomial
P Simpson's 3/8 Rule	1 First
Q Trapezoidal Rule	2 Second
R Simpson's 1/3 Rule	3 Third

(a) P-2, Q-1, R-3 (b) P-3, Q-2, R-1

(c) P-1, Q-2, R-3 (d) P-3, Q-1, R-2

2012

3. A box contains 4 red balls and 6 black balls. Three balls are selected randomly from the box one after another, without replacement. The probability that the selected set contains one red ball and two black balls is

(a) 1/20 (b) 1/12

(c) 3/10 (d) 1/2

2011

4. The integral $\int_1^3 \frac{1}{x} dx$, when evaluated by using Simpson's $\frac{1}{3}$ rule on two equal subintervals each of length l, equals

(a) 1.000

(b) 1.098

(c) 1.111

(d) 1.120

2010

5. Torque exerted on a flywheel over a cycle is listed in the table. Flywheel energy (in J per unit cycle) using Simpson's rule is

Angle (degree)	0	60	120	180	240	300	360
Torque (Nm)	0	1066	–323	0	323	–355	0

(a) 542

(b) 993

(c) 1444

(d) 1986

2007

6. A calculator has accuracy up to 8 digits after decimal place. The value of $\int_0^{2\pi} \sin x\, dx$ when evaluated using this calculator by trapezoidal method with 8 equal intervals, to 5 significant digits is

(a) 0.00000

(b) 1.0000

(c) 0.00500

(d) 0.00025

2006

7. Match the items in Lists I and II.

List I	List II
P. Gauss–Seidel method	1. Interpolation
Q. Forward Newton – Gauss method	2. Non–linear differential equations
R. Runge–Kutta method	3. Numerical integration
S. Trapezoidal Rule	4. Linear algebraic equations

Codes:

	P	Q	R	S
(a)	1	4	3	2
(b)	1	4	2	3
(c)	1	3	2	4
(d)	4	1	2	3

NUMERICAL TYPE QUESTIONS

2016

1. A function f of the complex variable $z = x + i\,y$, is given as $f(x, y) = u(x, y) + i\,v(x, y)$, where $u(x, y) = 2\,kxy$ and $v(x, y) = x^2 - y^2$. The value of k, for which the function is analytic, is ____

2015

2. Simpson's $\frac{1}{3}$ rule is used to integrate the function $f(x) = \frac{3}{5}x^2 + \frac{9}{5}$ between $x = 0$ and $x = 1$ using the least number of equal sub-intervals. The value of the integral is ________

3. Using a unit step size, the value of integral $\int_1^2 x \ln x\, dx$ by trapezoidal rule is ________

4. Newton-Raphson method is used to find the roots of the equation, $x^3 + 2x^2 + 3x - 1 = 0$. If the initial guess is $x_0 = 1$, then the value of x after 2nd iteration is ________.

2014

5. Using the trapezoidal rule, and dividing the interval of integration into three equal subintervals, the definite integral $\int_{-1}^{+1} |x|\, dx$ is ______

6. The best approximation of the minimum value attained by $e^{-x} \sin(100x)$ for $x \geq 0$ is ______

7. The value of $\int_{2.5}^{4} \ln(x)\, dx$ calculated using the Trapezoidal rule with five subintervals is ______

8. The definite integral $\int_1^3 \frac{1}{x} dx$ is evaluated using Trapezoidal rule with a step size of 1. The correct answer is ______

ANSWERS

MCQ Type Questions

1. (*a*) 2. (*d*) 3. (*d*) 4. (*c*) 5. (*) 6. (*a*) 7. (*d*)

Numerical Type Questions

1. (–1) 2. (2) 3. 0.69 4. 0.30 5. (1.10 to 1.12) 6. (1.00 to –0.94) 7. (1.74 to 1.76)
8. (1.1 to 1.2)

EXPLANATIONS

MCQ TYPE QUESTIONS

1. For a function of complex variable to be an analytic function Cauchy-Riemann equation has to satisfies which states as

$$\frac{\partial u}{\partial x}=\frac{\partial v}{\partial y} \text{ and } \frac{\partial u}{\partial y}=-\frac{\partial v}{\partial x}$$

Only option (A) $-x^2+y^2+$constant fulfills this.

2. Trapezoidal rule fits linear polynomial (Ist order) simplex 1/3 rule fits a parabolic curve (2nd order polynomial slip on 3/5 rule fit a able curve (3rd order polynomial

$\therefore$ Correct Answer $\Rightarrow d$

3. Given, $\frac{\text{Red}}{4}\ \frac{\text{Black}}{6}$

The selection will be RBB or BBR of BRB

Probability of selecting RBB = $\frac{4}{10}\times\frac{6}{9}\times\frac{5}{8}$

Probability of selecting BBR = $\frac{6}{10}\times\frac{5}{9}\times\frac{4}{8}$

Probability of selecting BRB = $\frac{6}{10}\times\frac{4}{9}\times\frac{5}{8}$

P (Red = 1) = sum of above three probabilities = 0.5

4. Given: $\int_1^3 \frac{1}{x}dx$

Here, $a = 1$, $b = 3$, $n = 2$

$$h = \frac{b-a}{n} = 1$$

$x_0 = 1$	$x_1 = 2$	$x_2 = 3$
$y_0 = 1$	$y_2 = \frac{1}{2} = 0.5$	$y_3 = \frac{1}{3} = 0.33$

By Simpson's rule

$$\int_1^3 \frac{1}{x}dx = \frac{1}{3}h[(y_1+y_3)+4(y_2)]$$

$$= \frac{1}{3}(1)[(1+0.33)+4(0.51)] = 1.11$$

5. Given : $h = 60° - 0° = 60°$

$$= 60\times\frac{\pi}{180} = 1.047$$

$y_0 = 0,\ y_1 = 1066,$

$y_2 = -323,\ y_4 = 323,$

$y_5 = -355,\ y_6 = 0$

By Simpson's rule, flywheel energy

$$= \frac{h}{3}[(y_0+y_6)+4(y_1+y_3+y_5)+2(y_2+y_4)]$$

$$= \frac{1.047}{3}[(0+0)+4(1066+0-355)+2(-323+323)]$$

= 993 Nm rad = 993 Joules/cycle

6. (b), (c), (d) cannot be possible.

NUMERICAL TYPE QUESTIONS

1. Given, $u = 2kxy$

$v = x^2 - y^2$

$u_x = v_y$

Condition for which given function to be analytical is

$$\frac{\partial u}{\partial x} = \frac{\partial v}{\partial y}$$

$$2ky = -2y$$

$$k = -1$$

2.

x	0	$\frac{1}{2}$	1
$y = f(x) = \frac{3}{5}x^2 + \frac{9}{5}$	$\frac{9}{5}$	$\frac{39}{20}$	$\frac{12}{5}$

$$\int_0^1 y\,dx = \frac{\left(\frac{1}{2}\right)}{2}\left[\left(\frac{9}{5}+\frac{12}{5}\right)+4\left(\frac{39}{20}\right)\right]$$

3.

x	1	2
$y = 1hx$	0	$21h2$

By Trapezoidal Rule,

$$\int_1^2 x\ln x\,dx = \frac{1}{2}[0+2\ln 2] = 0.69$$

4. By Newton-Raphson Method,

1st iteration, $x_1 = x_0 - \frac{f(x_0)}{f'(x_0)}$

$$= 1 - \frac{f(1)}{f'(1)} = 1 - \frac{5}{10} = \frac{1}{2}$$

Where $f(x) = x^3 + 2x^2 + 3x - 1 \Rightarrow f(1) = 5$

$f'(x) = 3x^2 + 4x + 3 \Rightarrow f'(1) = 10$

2nd iteration, $x_2 = x_1 \frac{f(x_1)}{f'(x_1)} = 0.5 - \frac{f(0.5)}{f'(0.5)} = 0.3043$

5. $\int_{-1}^{+1}|x|\ dx$

Let $y = |x|$

n = no. of subintervals = 3

$$h = \frac{x_n - n_o}{n} = \frac{1-(-1)}{3} = \frac{2}{3}$$

Values of x are,

$$-1, -1 + \frac{2}{3}, -1 + 2\left(\frac{2}{3}\right), -1 + 3\left(\frac{2}{3}\right)$$

x	-1	$-\frac{1}{3}$	$\frac{1}{3}$	1
$y = \lvert x \rvert$	1	$\frac{1}{3}$	$\frac{1}{3}$	1

Trapezodial rule

$$= \int_{x_o}^{x^n} f(x)dx = \frac{h}{2}\ [(y_o + y_n) + 2\ (y_1 + ...+ y_{n-1})]$$

$$= \int_{-1}^{1} |x| dx = \frac{1}{3}\left[(1+1)+2\left(\frac{1}{3}+\frac{1}{3}\right)\right]$$

$$= \frac{1}{3}\left[2+\frac{4}{3}\right] = \frac{1}{3} \times \frac{10}{3} = \frac{10}{9} = 1.1111$$

7. Given $\int_{2.5}^{4} \ln x\ dx \quad n = 5$

Let $y = \ln x \qquad n = \frac{4-2.5}{5} = \frac{1.5}{5} = 0.30$

x	2.5	2.8	3.1	3.4	3.7	4.0
$y = \ln x$	0.9163	1.0296	1.1314	1.2238	1.3083	1.3863

Trapezoidal Rule is $\int_{x_0}^{x_n} f(x)dx$

$$= \frac{h}{2}[(y_0 + y_n) + 2y_1 + ...y_{n-1}]$$

$$\int_{2.5}^{4} \ln x\ dx$$

$$= \frac{0.3}{2}[(0.9163+1.3863)+(1.0296+1.314+1.2238+1.3083)]$$

$= 0.15\ (2.3026 + 9.3862)$

$= 1.7533$

8. Given, $\int_{1}^{3} \frac{1}{x} dx$

h = step size 1

n = no. of sub intervals

$$= \frac{3-1}{1} = 2$$

Let $y =$

x	1	2	3
y	1	$\frac{1}{2}$	$\frac{1}{3}$

By trapezoidal rule

$$\int_{x_0}^{x_n} f(x)dx = \frac{h}{2}[(y_0 + y_n) + 2(y_1 + ... + y_{n-1})]$$

$$\int_{1}^{3} \frac{1}{x} dx = \frac{1}{2}\left[\left(1+\frac{1}{3}\right)+2\left(\frac{1}{2}\right)\right]$$

$$= \frac{1}{2}\left[\frac{4}{3}+1\right] = \frac{1}{2} \times \frac{7}{3} = 1.1667$$

Technical Section

1 CHAPTER

Engineering Mechanics

MCQ TYPE QUESTIONS

2016

1. A point mass M is released from rest and slides down a spherical bowl (of radius R) from a height H as shown in the figure below. The surface of the bowl is smooth (no friction). The velocity of the mass at the bottom of the bowl is

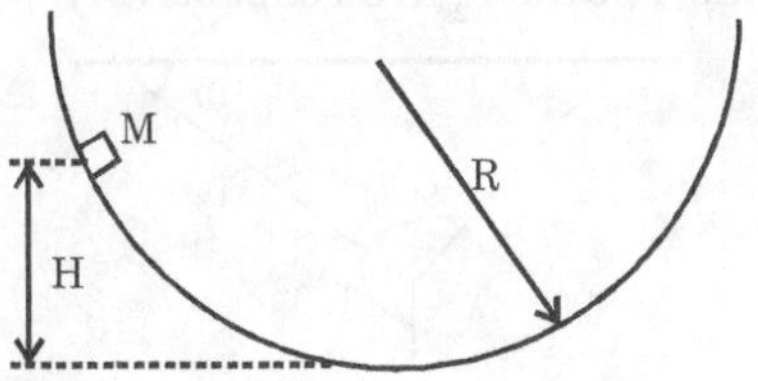

(a) $\sqrt{gH}$ (b) $\sqrt{2gR}$

(c) $\sqrt{2gH}$ (d) 0

2. A two-member truss PQR is supporting a load W. The axial forces in members PQ and QR are respectively

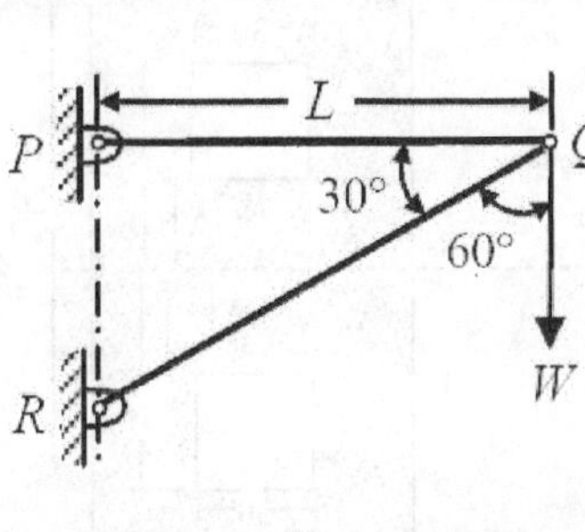

(a) 2W tensile and $\sqrt{3}$W compressive

(b) $\sqrt{3}$W tensile and 2W compressive

(c) $\sqrt{3}$W compressive and 2W tensile

(d) 2W compressive and $\sqrt{3}$W tensile

3. A point mass having mass M is moving with a velocity V at an angle θ to the wall as shown in the figure. The mass undergoes a perfectly elastic collision with the smooth wall and rebounds. The total change (final minus initial) in the momentum of the mass is

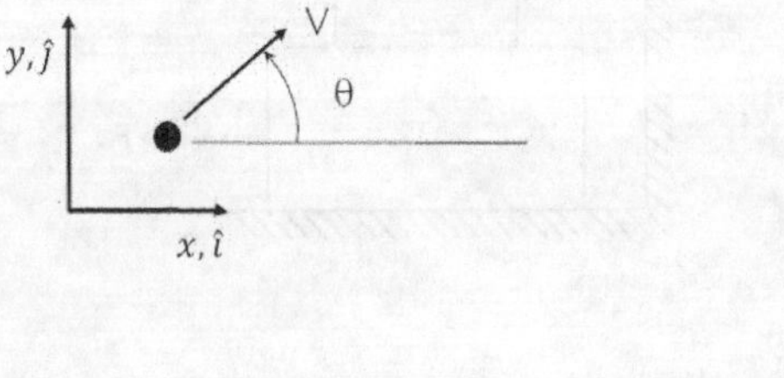

(a) $-2MV\cos\theta\,\hat{j}$ (b) $2MV\sin\theta\,\hat{j}$

(c) $2MV\cos\theta\,\hat{j}$ (d) $-2MV\sin\theta\,\hat{j}$

4. A system of particles in motion has mass center G as shown in the figure. The particle i has mass m_i and its position with respect to a fixed point O is given by the position vector r_i. The position of the particle with respect to G is given by the vector ρ_i. The time rate of change of the angular momentum of the system of particles about G is (The quantity $\ddot{\rho}_i$ indicates second derivative of ρ_i with respect to time and likewise for γ_i).

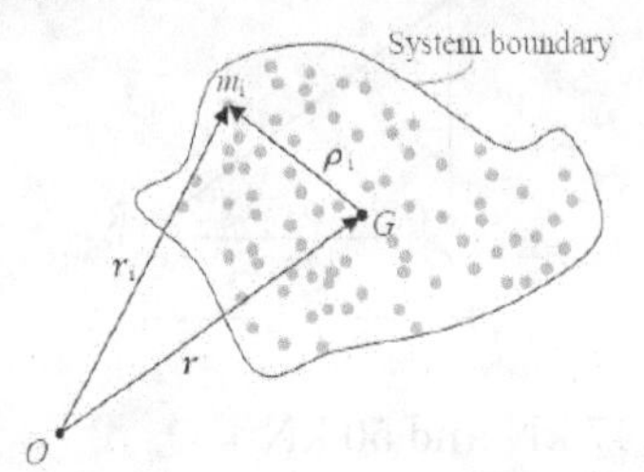

(a) $\sum_i r_i \times m_i\ddot{\rho}_i$ (b) $\sum_i \rho_i \times m_i\ddot{r}_i$

(c) $\sum_i r_i \times m_i\ddot{r}_i$ (d) $\sum_i \rho_i \times m_i\ddot{\rho}_i$

5. For the situation shown in the figure below the expression for H in terms of r, R and D is

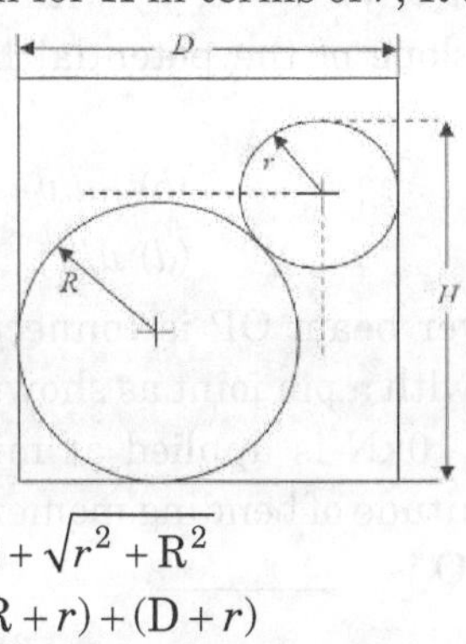

(a) $H = D + \sqrt{r^2 + R^2}$

(b) $H = (R + r) + (D + r)$

(c) $H = (R + r) + \sqrt{D^2 - R^2}$

(d) $H = (R + r) + \sqrt{2D(R + r) - D^2}$

6. A force F is acting on a bent bar which is clamped at one end as shown in the figure.

The CORRECT free body diagram is

(a) 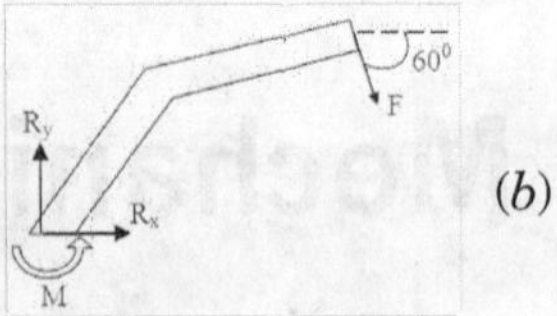

(b) 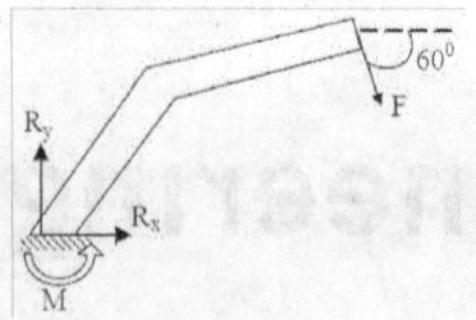

(c) 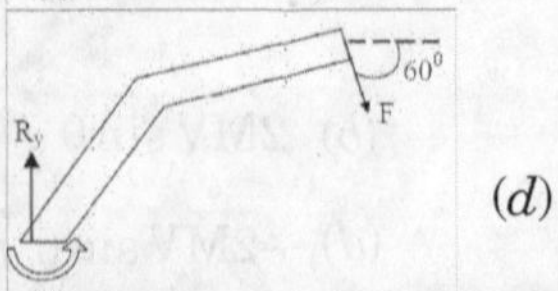

(d) 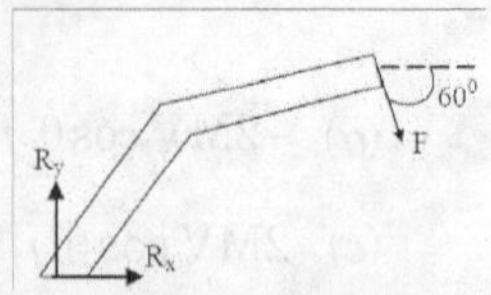

2015

7. For the truss shown in figure, the magnitude of the force in member PR and the support reaction at R are respectively

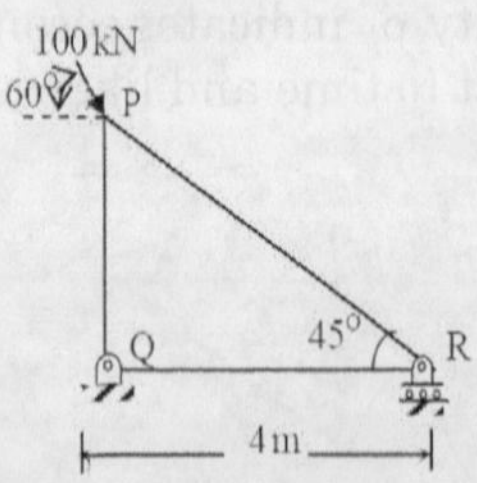

(a) 122.47 kN and 50 kN

(b) 70.71 kN and 100 kN

(c) 70.71 kN and 50 kN

(d) 81.65 kN and 100 kN

8. If the velocity for a potential flow is given by V(x, y) = u(x, y)i + v(x, y)j with usual notation then the slope of the potential line at (x, y) is ______.

(a) v/u (b) $-u/v$

(c) v^2/u^2 (d) u/v

9. A cantilever beam OP is connected to another beam PQ with a pin joint as shown in the figure. A load of 10kN is applied at midpoint of PQ. The magnitude of bending moment (in kN-m) at fixed end O is ______.

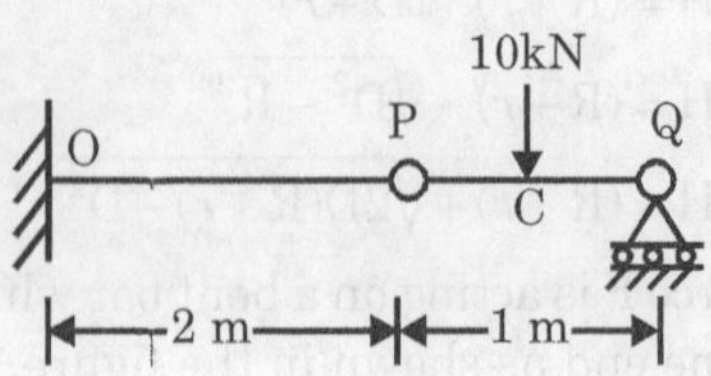

(a) 2.5 (b) 5

(c) 10 (d) 25

10. For the truss shown in the figure, the magnitude of the force in (kN) in the member SR is

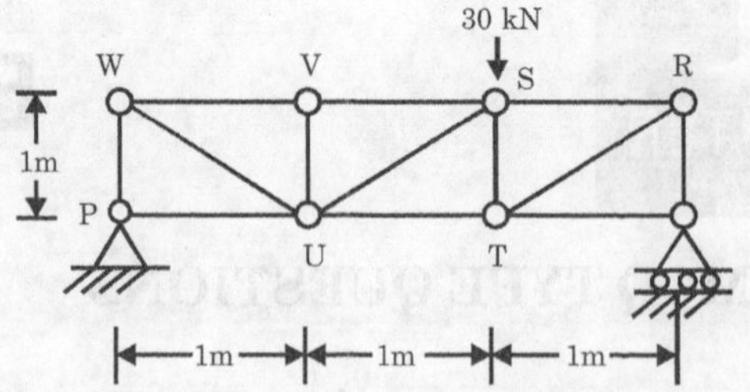

(a) 10 (b) 14.14

(c) 20 (d) 28.28

11. A weight of 500 N is supported by two metallic ropes as shown in the figure. The values of tensions T_1 and T_2 are respectively

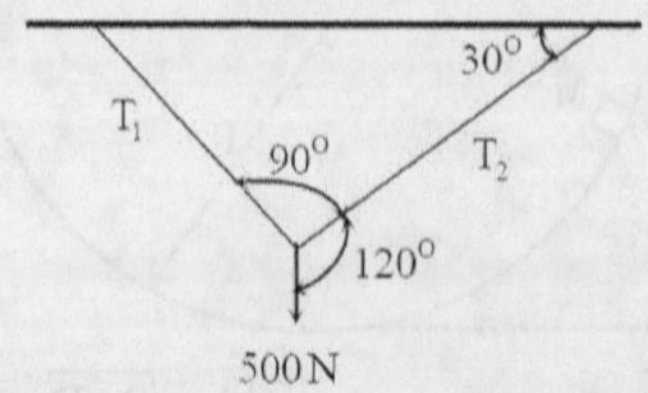

(a) 433 N and 250 N (b) 250 N and 433 N

(c) 353.5 N and 250 N (d) 250 N and 353.5 N

12. The value of moment o inertia of the section shown in the figure about the X-axis is

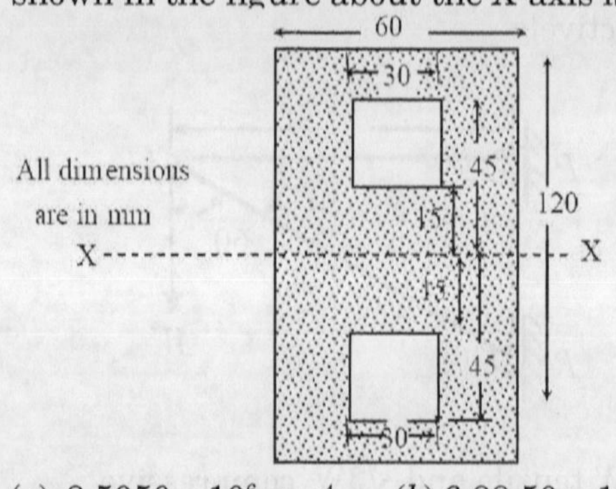

(a) 8.5050×10^6 mm^4 (b) $6.88.50 \times 10^5$ mm^4

(c) 7.7625×10^6 mm^4 (d) 8.5725×10^6 mm^4

2014

13. A block R of mass 100 kg is placed on a block S of mass 150 kg shown in the figure. Block R is tied to the wall by a massless and inextensible string PQ. If the coefficient of static friction for all surfaces is 0.4, the minimum force F (in kN) needed to move the block S is

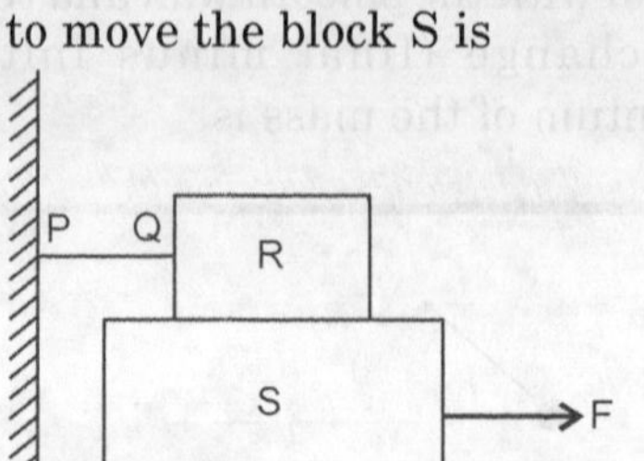

(a) 0.69 (b) 0.88
(c) 0.98 (d) 1.37

14. Consider a single degree-of-freedom system with viscous damping excited by a harmonic force. At resonance, the phase angle (in degree) of the displacement with respect to the exciting force is

(a) 0 (b) 45
(c) 90 (d) 135

15. A mass m_1 of 100 kg travelling with a uniform velocity of 5 m/s along a line collides with a stationary mass m_2 of 1000 kg. After the collision, both the masses travel together with the same velocity. The coefficient of restitution is

(a) 0.6 (b) 0.1
(c) 0.01 (d) 0

16. An annular disc has a mass m, inner radius R and outer radius 2R. The disc rolls on a flat surface without slipping. If the velocity of the centre of mass is v, the kinetic energy of the disc is

(a) $\frac{9}{16}mv^2$ (b) $\frac{11}{16}mv^2$
(c) $\frac{13}{16}mv^2$ (d) $\frac{15}{16}mv^2$

17. A force P is applied at a distance x from the end of the beam as shown in the figure. What would be the value of x so that the displacement at 'A' is equal to zero?

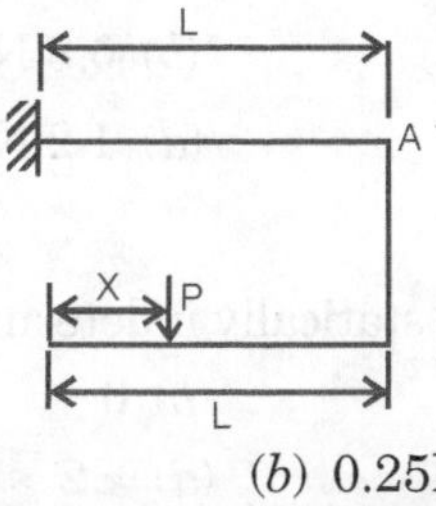

(a) 0.5L (b) 0.25L
(c) 0.33L (d) 0.66L

18. Consider a rotating disk cam and a translating roller follower with zero offset. Which one of the following pitch curves, parameterized by t, lying in the interval 0 to 2π, is associated with the maximum translation of the follower during one full rotation of the cam rotating about the center at $(x, y) = (0, 0)$?

(a) $x(t) = \cos t$, $y(t) = \sin t$
(b) $x(t) = \cos t$, $y(t) = 2\sin t$
(c) $x(t) = \frac{1}{2} + \cos t$, $y(t) = 2\sin t$
(d) $x(t) = \frac{1}{2} + \cos t$, $y(t) = \sin t$

19. In a statically determinate plane truss, the number of joints (j) and the number of members (m) are related by

(a) $j = 2m - 3$ (b) $m = 2j + 1$
(c) $m = 2j - 3$ (d) $m = 2j - 1$

20. For the truss shown in the figure, the forces F_1 and F_2 are 9 kN and 3 kN, respectively. The force (in kN) in the member QS is

All dimensions are in m

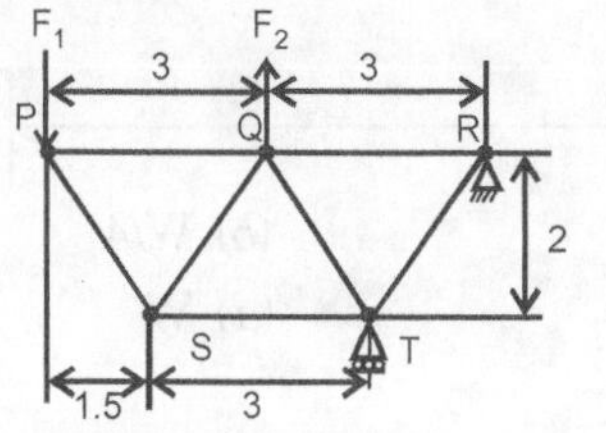

(a) 11.25 tension
(b) 11.25 compression
(c) 13.5 tension
(d) 13.5 compression

21. A frame is subjected to a load P as shown in the figure. The frame has a constant flexural rigidity EI. The effect of axial load is neglected. The deflection at point A due to the applied load P is

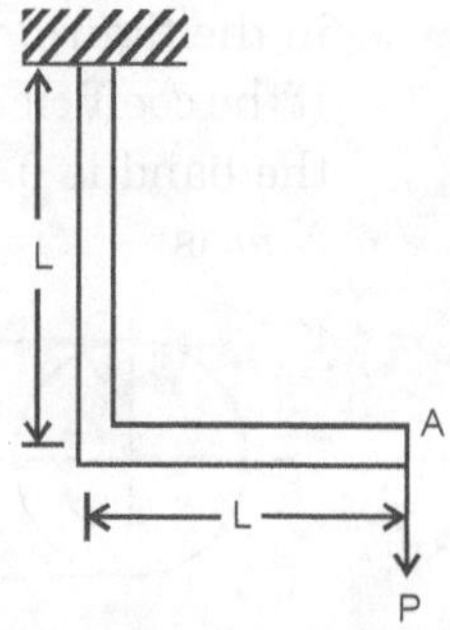

(a) $\frac{1}{3}\frac{PL^3}{EI}$
(b) $\frac{2}{3}\frac{PL^3}{EI}$
(c) $\frac{PL^3}{EI}$
(d) $\frac{4}{3}\frac{PL^3}{EI}$

22. A wardrobe (mass 100 kg, height 4 m, width 2 m, depth 1 m), symmetric about the Y-Y axis, stands on a rough level floor as shown in the figure. A force P is applied at mid-height on the wardrobe so as to tip it about point Q without slipping. What are the minimum values of the force (in Newton) and the static coefficient of friction μ between the floor and the wardrobe, respectively?

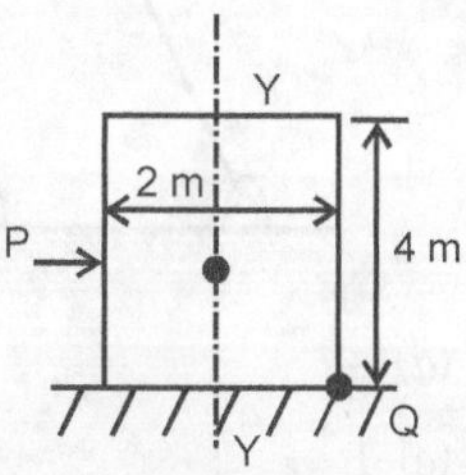

(a) 490.5 and 0.5
(b) 981 and 0.5
(c) 1000.5 and 0.15
(d) 1000.5 and 0.25

2013

23. A pin Jointed uniform rigid rod of weight W and length L is supported horizontally by an external force F as shown in the figure below. The force F is suddenly removed. At the instant of force removal, the magnitude of vertical reaction developed at the support is

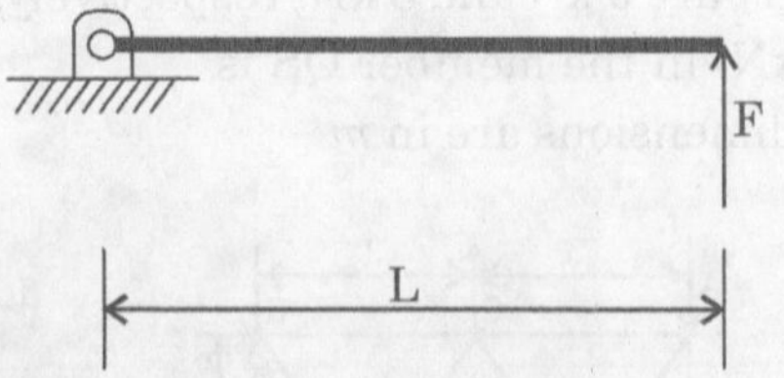

(*a*) zero (*b*) W/4
(*c*) W/2 (*d*) W

2012

24. A solid disc of radius r rolls without slipping on a horizontal floor with angular velocity ω and angular acceleration α. The magnitude of the acceleration of the point of contact on the disc is

(*a*) zero (*b*) $r\alpha$
(*c*) $\sqrt{(r\alpha)^2 + (r\omega^2)^2}$ (*d*) $r\omega^2$

25. A force of 400 N is applied to the brake drum of 0.5 m diameter in a band–brake system as shown in the figure, where the wrapping angle is 180°. If the coefficient of friction between the drum and the band is 0.25, the braking torque applied, in *N.m* is

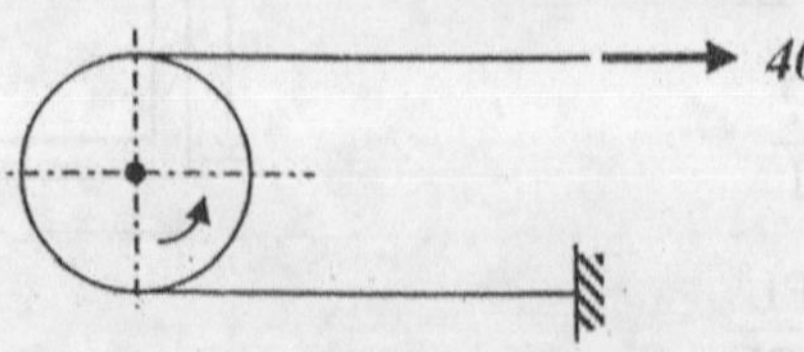

(*a*) 100.6 (*b*) 54.4
(*c*) 22.1 (*d*) 15.7

2011

26. A double-parallelogram mechanism is shown in the figure. Note that PQ is a single link. The mobility of the mechanism is

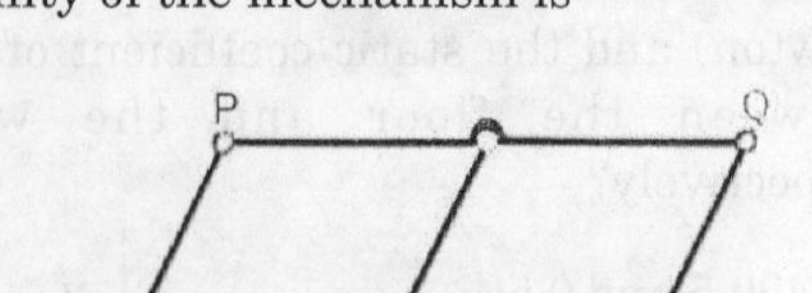

(*a*) –1 (*b*) 0
(*c*) 1 (*d*) 2

27. The coefficient of restitution of a perfectly plastic impact is

(*a*) 0 (*b*) 1
(*c*) 2 (*d*) ∞

28. A stone with mass of 0.1 kg is catapulted as shown in the figure. The total force F_x (in N) exerted by the rubber band as a function of distance x (in m) is given by $F_x = 300\,x^2$. If the stone is displaced by 0.1 m from the un-stretched position ($x = 0$) of the rubber band, the energy stored in the rubber band is

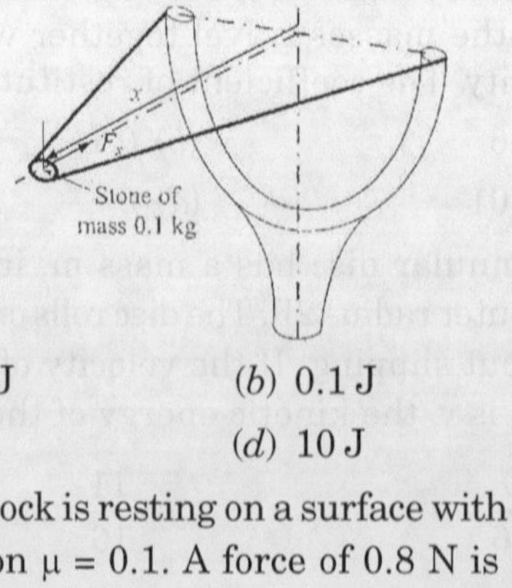

(*a*) 0.01 J (*b*) 0.1 J
(*c*) 1 J (*d*) 10 J

29. A 1 kg block is resting on a surface with coefficient of friction μ = 0.1. A force of 0.8 N is applied to the block as shown in the figure. The friction force is

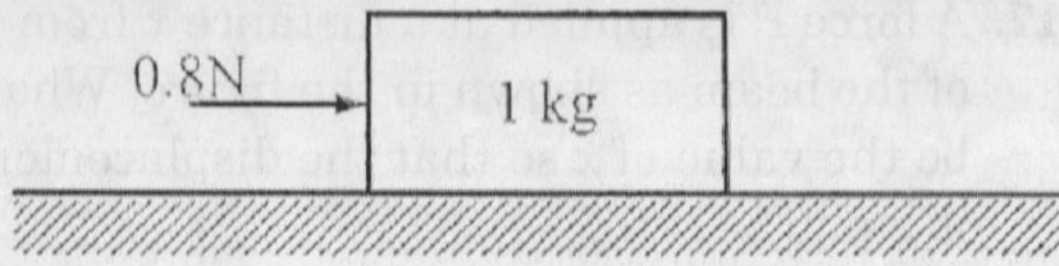

(*a*) 0 (*b*) 0.8 N
(*c*) 0.98 N (*d*) 1.2 N

2010

30. Mobility of a statically indeterminate structure is

(*a*) ≤ –1 (*b*) 0
(*c*) 1 (*d*) ≥ 2

31. There are two points P and Q on a planar rigid body. The relative velocity between the two points

(*a*) should always be along PQ
(*b*) can be oriented along any direction
(*c*) should always be perpendicular to PQ
(*d*) should be along QP when the body undergoes pure translation

32. A band brake having band-width of 80 mm, drum diameter of 250 mm, coefficient of friction of 0.25 and angle of wrap of 270 degrees is required to exert a friction torque of 1000 Nm. The maximum tension (in kN) developed in the band is

(*a*) 1.88 (*b*) 3.56
(*c*) 6.12 (*d*) 11.56

2009

33. A block weighing 981 N is resting on a horizontal surface. The coefficient of friction between the block and the horizontal surface is $\mu = 0.2$.A vertical cable attached to the block provides partial support as shown. A man can pull horizontally with a force of 100 N. What wil be the tension, T (in N) in the cable if the man is just able to move the block to the right ?

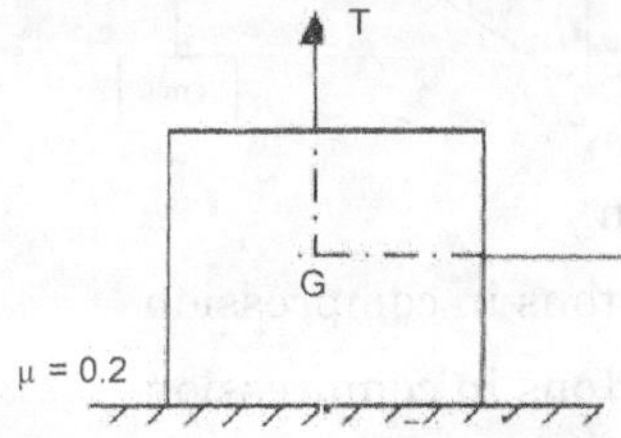

(*a*) 176.2 (*b*) 196.0

(*c*) 481.0 (*d*) 981.0

2008

34. A straight rod of length L(t), hinged at one end and freely extensible at the other end, rotates through an angle θ(t) about the hinge. At time t, L(t) = 1 m, L(t) = 1 m/s, θ(t) = $\frac{\pi}{4}$ rad and θ(t) = 1 rad/s. The magnitude of the velocity at the other end of the rod is

(*a*) 1 m/s (*b*) $\sqrt{2}$ m/s

(*c*) $\sqrt{3}$ m/s (*d*) 2 m/s

2007

35. During inelastic collision of two particles, which one of the following is conserved ?

(*a*) Total linear momentum only

(*b*) Total kinetic energy only

(*c*) Both linear momentum and kinetic energy

(*d*) Neither linear momentum nor kinetic energy

36. A block of mass M is released from point P on a rough inclined plane with inclination angle θ, shown in the figure below. The coefficient of friction is μ. If $\mu < \tan\theta$, then the time taken by the block to reach another point Q on the inclined plane, where PQ = *s*, is

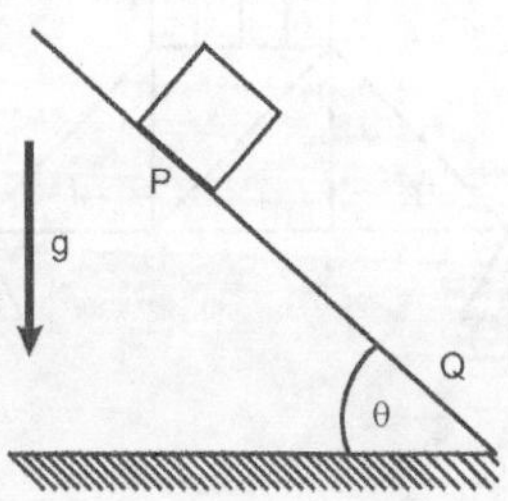

(*a*) $\sqrt{\dfrac{2s}{g\cos\theta(\tan\theta-\mu)}}$

(*b*) $\sqrt{\dfrac{2s}{g\cos\theta(\tan\theta+\mu)}}$

(*c*) $\sqrt{\dfrac{2s}{g\sin\theta(\tan\theta-\mu)}}$

(*d*) $\sqrt{\dfrac{2s}{g\sin\theta(\tan\theta+\mu)}}$

2006

37. If a system is in equilibrium and the position of the system depends upon many independent variables, the principle of virtual work states that partial derivatives of its total potential energy with respect to each of the independent variable must be

(*a*) –1.0 (*b*) 0

(*c*) 1.0 (*d*) ∞

38. If point A is in equilibrium under the action of the applied forces, the values of tension T_{AB} and T_{AC} are respectively

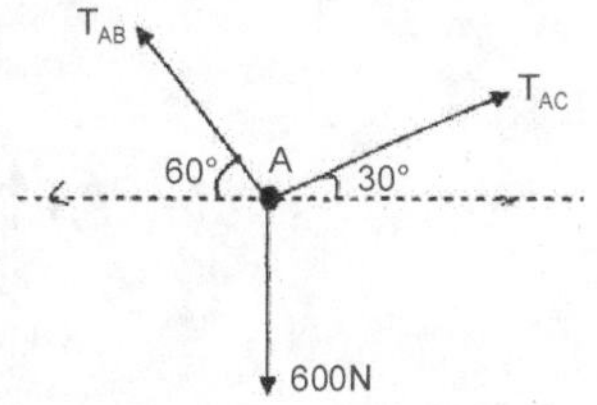

(*a*) 520 N and 300 N (*b*) 300 N and 520 N

(*c*) 450 N and 150 N (*d*) 150 N and 450 N

2005

39. A simple pendulum of length 5 m, with a bob of mass 1 kg, is in simple harmonic motion. As it passes through its mean position, the bob has a speed of 5 m/s. The net force on the bob at the mean position is

(*a*) zero (*b*) 2.5 N

(*c*) 5 N (*d*) 25 N

40. Two books of mass 1 kg each are kept on a table, one over the other. The coefficient of friction on every pair of contacting surfaces is 0.3. the lower book is pulled with a horizontal force F. The minimum value of F for which slip occurs between the two books is

(*a*) zero (*b*) 1.06 N

(*c*) 5.74 N (*d*) 8.83 N

41. A shell is fired from a connon. At the instant the shell is just about to leave the barrel, its velocity relative to the barrel is 3m/s, while the barrel is swinging upwards with a constant angular velocity of 2 rad/s. The magnitude of the absolute velocity of the shell is

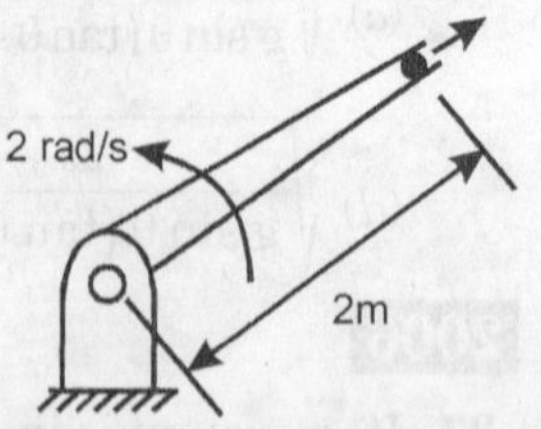

(*a*) 3 m/s

(*b*) 4 m/s

(*c*) 5 m/s

(*d*) 7 m/s

42. An elevator (lift) consists of the elevator cage and a counter weight, of mass m each. The cage and the counterweight are connected by a chain that passes over a pulley. The pulley is coupled to a motor. It is desired that the elevator should have a maximum stopping time of t seconds from a peak speed v. If inertias of the pulley and the chain are neglected, the minimum power that the motor must have is

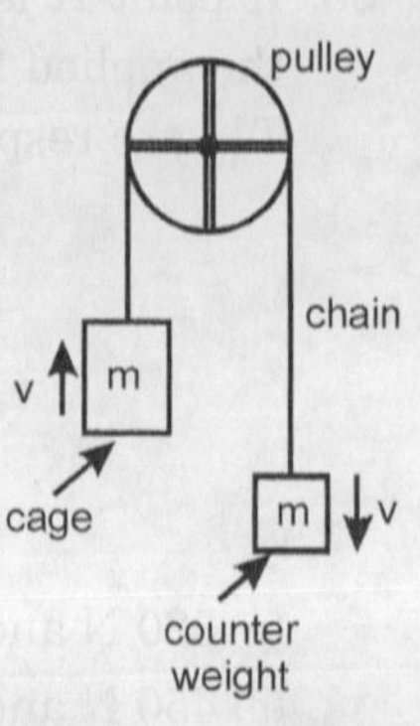

(*a*) $\frac{1}{2}mv^2$

(*b*) $\frac{mv^2}{2t}$

(*c*) $\frac{mv^2}{t}$

(*d*) $\frac{2mv^2}{t}$

43. A 1 kg mass of clay, moving with a velocity of 10 m/s, strikes a stationary wheel and sticks to it. The solid wheel has a mass of 20 kg and a radius of 1m. Assuming that wheel and the ground are both rigid and that the wheel is set into pure rolling motion, the angular velocity of the wheel immediately after the impact is approximately

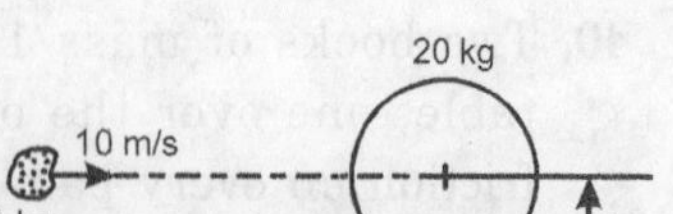

(*a*) zero

(*b*) $\frac{1}{3}$ rad / s

(*c*) $\sqrt{\frac{10}{3}}$ rad / s

(*d*) $\frac{10}{3}$ rad/s

2004

44. The figure shows a pin-jointed plane truss loaded at the point M by hanging a mass of 100 kg. The member LN of the truss is subjected to a load of

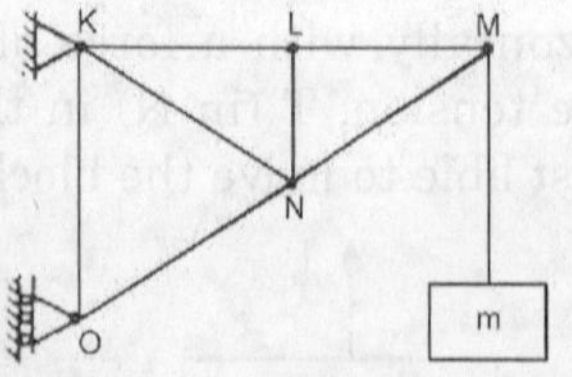

(*a*) 0 Newton

(*b*) 490 Newtons in compression

(*c*) 981 Newtons in compression

(*d*) 981 Newtons in tension

45. The figure shows a pair of pin-jointed gripper-tongs holding an object weighing 2000 N. The co-efficient of friction μ at the gripping surface is 0.1. XX is the line of action of the input force and YY is the line of application of gripping force. If the pin-joint is assumed to be frictionless, then magnitude of force F required to hold the weight is

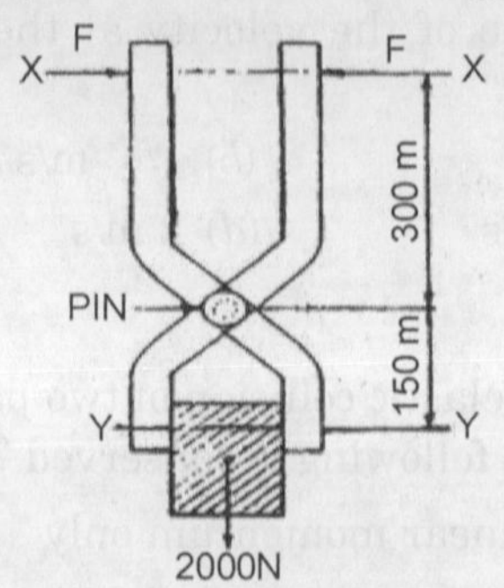

(*a*) 1000 N (*b*) 2000 N

(*c*) 2500 N (*d*) 5000 N

2003

46. A truss consists of horizontal members (AC, CD, DB and EF) and vertical members (CE and DE) having length l each. The members AE, DE and BF are inclined at 45° to the horizontal. For the uniformly distributed load "P" per unit length on the member EF of the truss shown in figure given below, the force in the member CD is

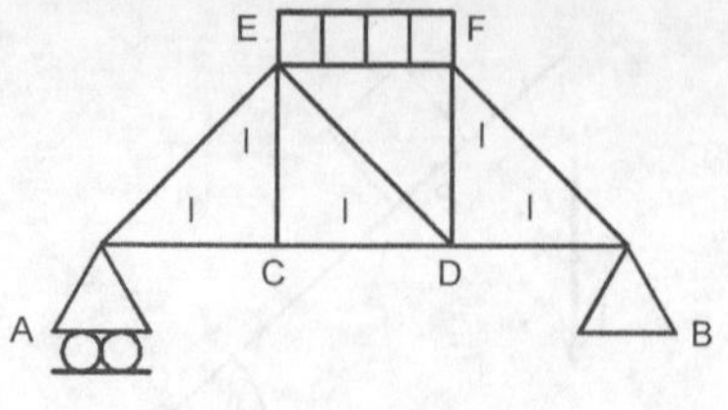

(a) $\frac{pl}{2}$ (b) pl

(c) 0 (d) $\frac{2pl}{3}$

47. A bullet of mass "m" travels at a very high velocity v (as shown in the figure) and gets embedded inside the block of mass "M" initially at rest on a rough horizontal floor. The block with the bullet is seen to move a distance "s" along the floor. Assuming   to be the coefficient of kinetic friction between the block and the floor and "g" the acceleration due to gravity what is the velocity v of the bullet?

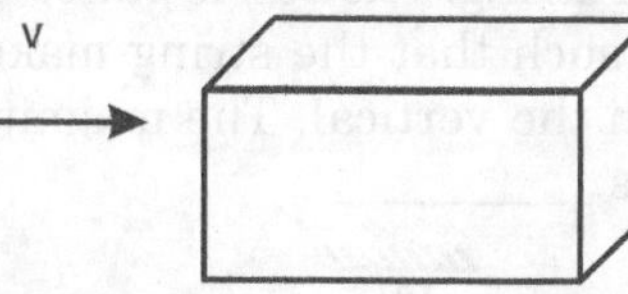

(a) $\frac{M+m}{m}\sqrt{2\mu\, gs}$

(b) $\frac{M-m}{m}\sqrt{2\mu\, gs}$

(c) $\frac{\mu(M+m)}{m}\sqrt{2\, gs}$

(d) $\frac{M}{m}\sqrt{2\mu\, gs}$

Common Data Q. (48 –49)

A reel of mass "m" and radius of gyration "k" is rolling down smoothly from rest with one end of the thread wound on it held in the ceiling as depicted in the figure. Consider the thickness of the thread and its mass negligible in comparison with the radius "r" of the hub and the reel mass "m". Symbol "g" represents the acceleration due to gravity.

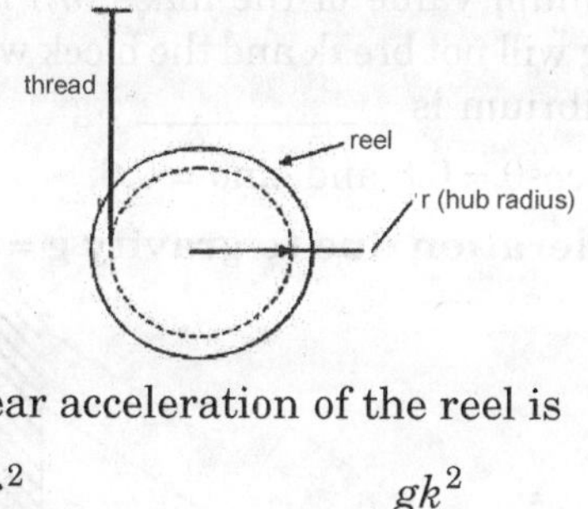

48. The linear acceleration of the reel is

(a) $\frac{gr^2}{(r^2+k^2)}$ (b) $\frac{gk^2}{(r^2+k^2)}$

(c) $\frac{grk}{(r^2+k^2)}$ (d) $\frac{mgr^2}{(r^2+k^2)}$

49. The tension in the thread is

(a) $\frac{mgr^2}{(r^2+k^2)}$ (b) $\frac{mgrk}{(r^2+k^2)}$

(c) $\frac{mgk^2}{(r^2+k^2)}$ (d) $\frac{mg}{(r^2+k^2)}$

2002

50. A toy cylinder of weight w, length l, and radius r rolls without slipping on the inner cylindrical surface of radius R, which is fixed.

(a) Determine differential equation of motion of the toy for large oscillations about its lowest point using energy method.

(b) Write down equation of the motion and circular frequency of oscillation of the toy for small oscillations.

The centre of gravity of the joy is assumed to be at the centre of the toy cylinder.

2001

51. Mass M slides in a frictionless slot in the horizontal dircection and the bob of mass m is hinged to mass M at C, through a rigid massless rod. This system is released from rest with $\theta = 30°$. At the instant when $\theta = 0$, the velocities of m and M can be determined using the fact that, for the system (*i.e.*, m and M together)

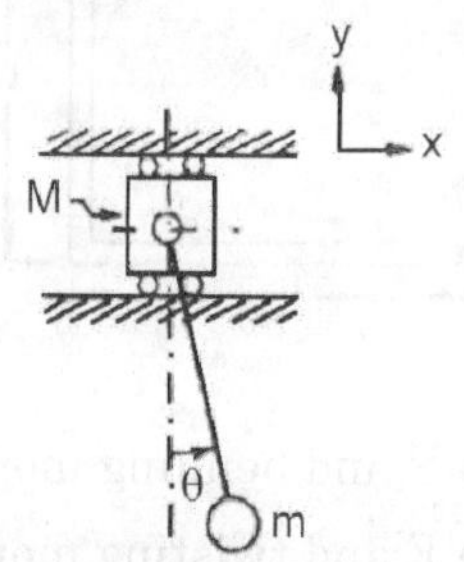

(a) the linear momentum in x and y directions are conserved but the energy is not conserved

(b) the linear momentum in x and y directions are conserved and the energy is also conserved

(c) the linear momentum in x direction is conserved and the energy is also conserved

(d) the linear momentum in y direction is conserved and the energy is also conserved

52. The 2 kg mass C moving horizontally to the right, with a velocity of 5 m/s, strickes the 8 kg mass B at the lower end of the rigid massless rod AB. The rod is suspended from a frictionless hinge at A and is initially at rest. If coefficient of restitution between mass C and B is one, then angular velocity of the rod AB immediately after impact will be

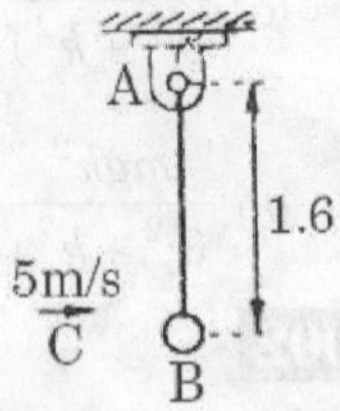

(a) 0.425 rad/s

(b) 0.625 rad/s

(c) 0.825 rad/s

(d) 0.925 rad/s

2000

53. A steel wheel of 600 mm diameter rolls on a horizontal steel rail. It carries a load of 500 N. The coefficient of rolling resistance is 0.3 mm. The force in N, necessary to roll the wheel along the rail is

(a) 0.5 (b) 5

(c) 15 (d) 150

1999

54. A concentrated force, F is applied (perpendicular to the plane of the figure) on the tip of the bent bar shown in figure. The equivalent load at a section close to the fixed end is

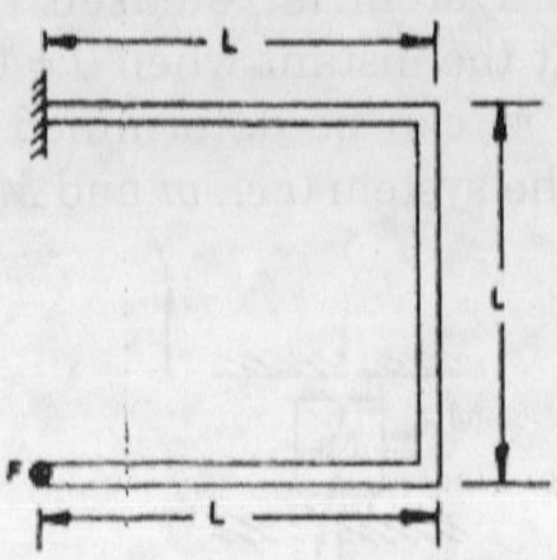

(a) Force F

(b) Force F and bending moment FL

(c) Force F and twisting moment FL

(d) Force F, bending moment F L, and twisting moment FL

1998

55. A car moving with uniform acceleration covers 450 m in a 5 second interval, and covers 700 m in the next 5 second interval. The acceleration of the car is

(a) 7 m/s^2 (b) 50 m/s^2

(c) 25 m/s^2 (d) 10 m/s^2

56. A cannon ball fired from the ground attained a maximum height of 1 km. At this maximum height, the radius of curvature of its path was also equal to 1 km. With this data, determine the initial velocity and angle of projection from the the ground. Neglect air resistance.

1997

57. The efficiency of a power screw is maximum when the lead angle is

(a) $\frac{\pi}{2} - \frac{r}{2}$ (b) $\frac{\pi}{2} - r$

(c) $\pi - \frac{r}{2}$ (d) $\pi - r$

Where r is the friction angle.

NUMERICAL TYPE QUESTIONS

2016

1. A rigid ball of weight 100 N is suspended with the help of a string. The ball is pulled by a horizontal force F such that the string makes an angle of 30° with the vertical. The magnitude of force F (in N) is ______

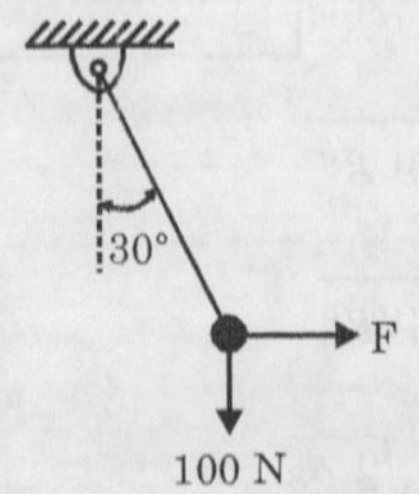

2. A car is moving on a curved horizontal road of radius 100 m with a speed of 20 m/s. The rotating masses of the engine have an angular speed of 100 rad/s in clockwise direction when viewed from the front of the car. The combined moment of inertia of the rotating masses is 10 kg-m^2. The magnitude of the gyroscopic moment (in N-m) is ______.

3. A block of mass m rests on an inclined plane and is attached by a string to the wall as shown in the figure. The coefficient of static friction between the plane and the block is 0.25. The string can withstand a maximum force of 20 N. The maximum value of the mass (m) for which the string will not break and the block will be in static equilibrium is ______ kg.

Take cosθ = 0.8 and sinθ = 0.6.

Acceleration due to gravity g = 10 m/s^2

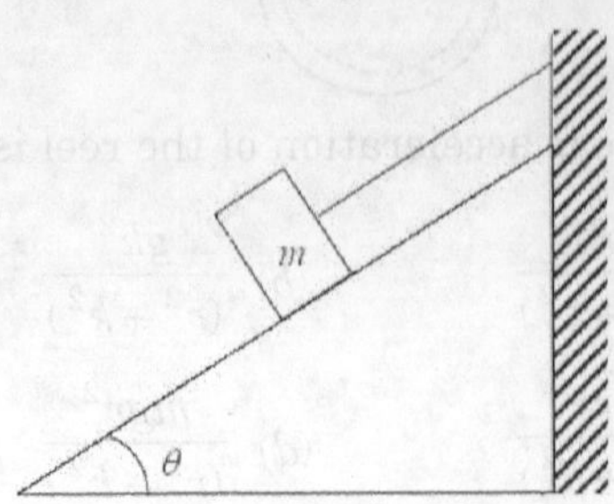

4. A horizontal bar with a constant cross-section is subjected to loading as shown in the figure. The Young's moduli for the sections AB and BC are 3E and E, respectively.

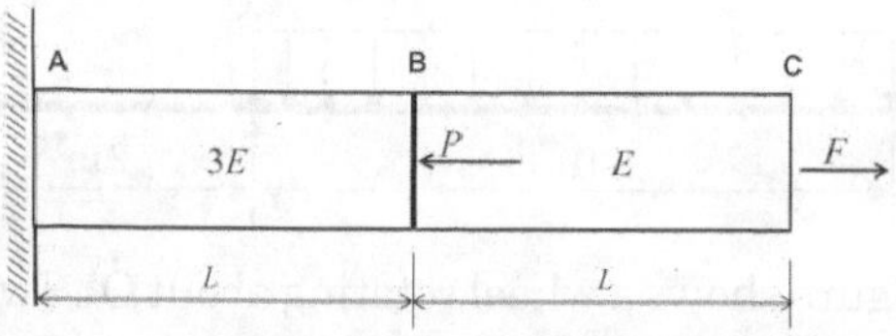

For the deflection at C to be zero, the ratio P/Fis ______.

5. The forces F_1 and F_2 in a brake band and the direction of rotation of the drum are as shown in the figure. The coefficient of friction is 0.25. The angle of wrap is $3\pi/2$ radians. It is given that R = 1 m and F_2 = 1 N. The torque (in N-m) exerted on the drum is ______

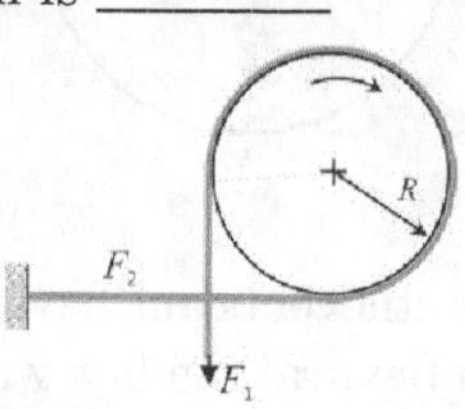

6. An inextensible massless string goes over a frictionless pulley. Two weights of 100 N and 200 N are attached to the two ends of the string. The weights are released from rest, and start moving due to gravity. The tension in the string (in N) is ______.

7. A circular disc of radius 100 mm and mass 1 kg, initially at rest at position A, rolls without slipping down a curved path as shown in figure. The speed v of the disc when it reaches position B is ______ m/s.

Acceleration due to gravity g = 10 m/s².

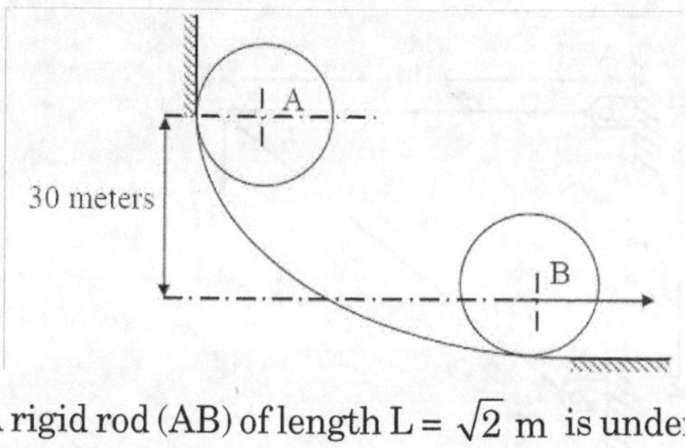

8. A rigid rod (AB) of length L = $\sqrt{2}$ m is undergoing translational as well as rotational motion in the x-y plane (see the figure). The point A has the velocity $V_1 = \hat{i} + 2\hat{j}$ m/s. The end B is constrained to move only along the x direction.

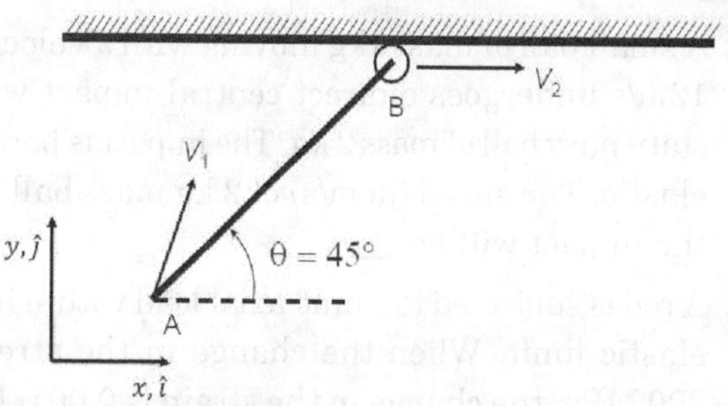

The magnitude of the velocity V_2 (in m/s) at the end B is ______

2015

9. A wheel of radius r rolls without slipping on a horizontal surface shown below. If the velocity of point P is 10 m/s in the horizontal direction, the magnitude of velocity of point Q (in m/s) is ______.

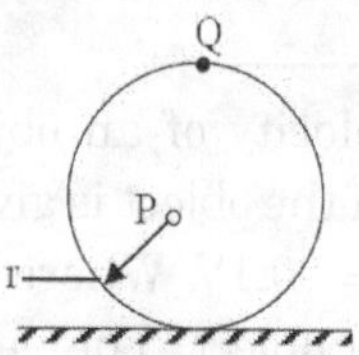

10. Two identical trusses support a load of 100 N as shown in the figure. The length of each truss is 1.0 m, cross-sectional area is 200 mm²; Young's modulus E = 200 GPa. The force in the truss AB (in N) is ______

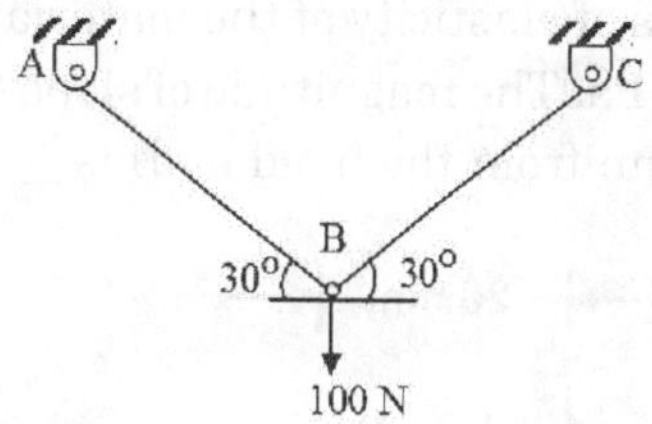

11. A swimmer can swim 10 km in 2 hours when swimming along the flow of a river. While swimming against the flow, she takes 5 hours for the same distance. Her speed in still water (in km/h) is ______.

12. A ball of mass 0.1 kg, initially at rest, is dropped from height of 1 m. Ball hits the ground and bounces off the ground. Upon impact with the ground, the velocity reduces by 20%. The height (in m) to which the ball will rise is ______

13. A cantilever beam with flexural rigidity of 200 Nm² is loaded as shown in the figure. The deflection (in mm) at the tip of the beam is ______.

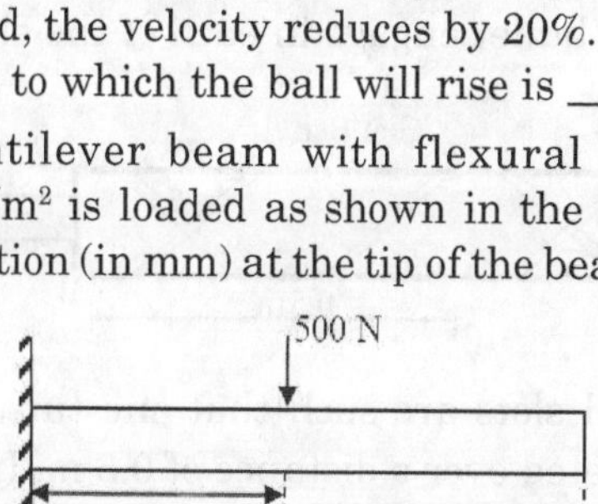

14. A small ball of mass 1kg moving with a velocity of 12m/s undergoes a direct central impact with a stationary ball of mass 2 kg. The impact is perfectly elastic. The speed (in m/s) of 2 kg mass ball after the impact will be ______.

15. A rod is subjected to a unit-axial load within linear elastic limit. When the change in the stress is 200 MPa, the change in the strain is 0.001. If the Poisson's ratio of the rod is 0.3, the modulus of rigidity (in GPA) is ________

16. The flow stress (in MPa) of a material is given by $\sigma = 500\ \varepsilon^{0.1}$. Where ε is true strain. The Young's modulus of elasticity of the material is 200 GPa. A block of thickness 100 mm made of this material is compressed to 95 mm thickness and then the load is removed. The final dimension of the block (in mm) is ________.

17. The initial velocity of an object is 40 m/s. The acceleration of the object is given by the following expression $a = -0.1V$. Where V is instantaneous velocity of the object. The velocity of the object after 3 seconds will be ________.

18. A cantilever beam with square cross section of 6mm side is subjected to a load of 2 kN normal to the top surface as shown in the figure. The young's modulus of elasticity of the material of the beam is 210 GPa. The magnitude of slope (in radian) at Q (20 mm from the fixed end) is ________.

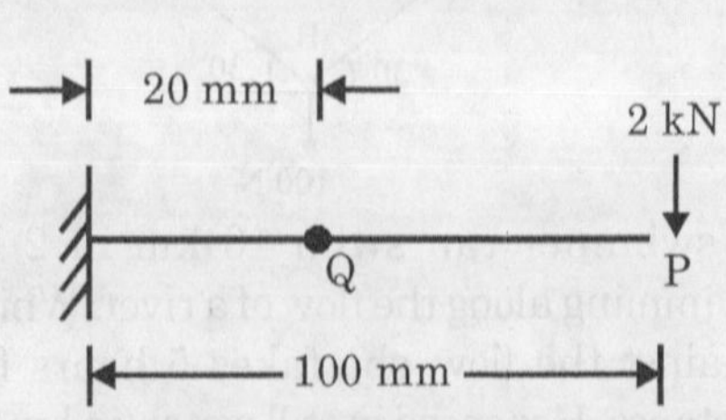

19. A bullet spins as the shot is fired from a gun. For this purpose, two helical slots as shown in the figure are cut in the barrel. Projections A and B on the bullet engage in each of the slots

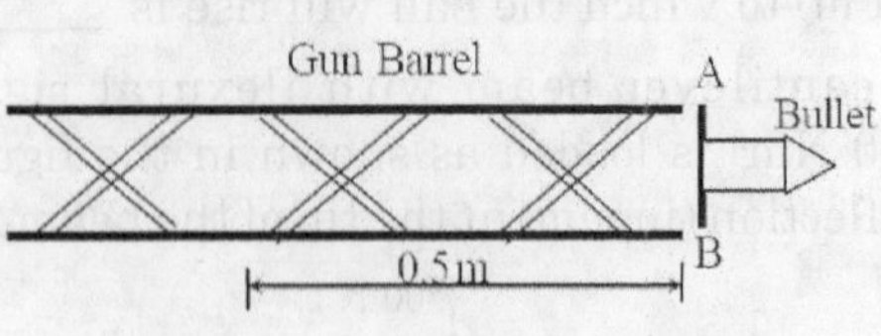

Helical slots are such that one turn of helix is completed over a distance of 0.5 m. If velocity of bullet when it exits the barrel is 20 m/s, its spinning speed in rad/s is ______.

20. For the overhanging beam shown in figure, the magnitude of m aximum bending moment (in kN-m) is ________.

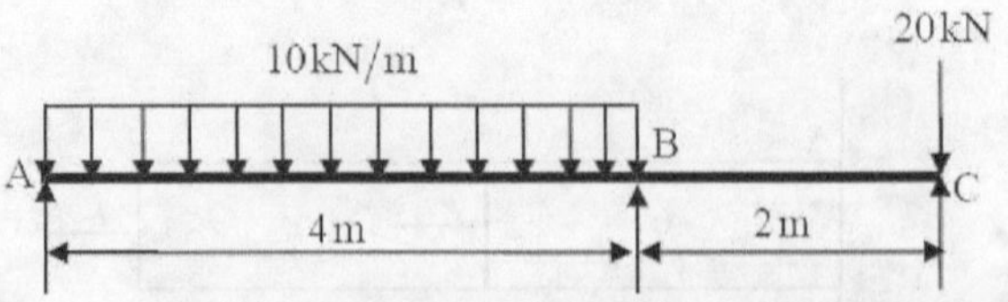

21. Figure shows a wheel rotating about O_2. Two points A and B located along the radius of wheel have speeds of 80 m/s and 140 m/s respectively. The distance between the points A and B is 300 mm. The diameter of the wheel (in mm) is ________

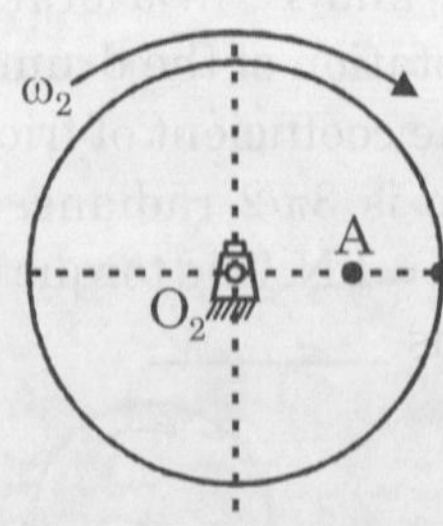

2014

22. Consider a cantilever beam, having negligible mass and uniform flexural rigidity, with length 0.01 m. The frequency of vibration of the beam, with a 0.5 kg mass attached at the free tip, is 100 Hz. The flexural rigidity (in N.m^2) of the beam is ____

23. A block weighing 200 N is in contact with a level plane whose coefficients of static and kinetic friction are 0.4 and 0.2, respectively. The block is acted upon by a horizontal force (in newton) $p=10t$, where t denotes the time in seconds. The velocity (in m/s) of the block attained after 10 seconds is ____

24. A two member truss ABC is shown in the figure. The force (in kN) transmitted in member AB is ________

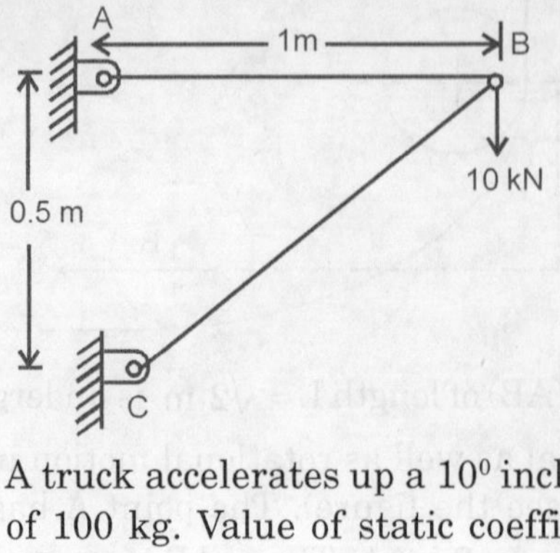

25. A truck accelerates up a 10° incline with a crate of 100 kg. Value of static coefficient of friction between the crate and the truck surface is 0.3. The maximum value of acceleration (in m/s^2) of the truck such that the crate does not slide down is ________

26. Maximum fluctuation of kinetic energy in an engine has been calculated to be 2600 J. Assuming that the engine runs at an average speed of 200 rpm, the polar mass moment of inertia (in kg.m^2) of a flywheel to keep the speed fluctuation within ± 0.5% of the average speed is ______

27. A body of mass (M) 10 kg is initially stationary on a 45° inclined plane as shown in figure. The coefficient of dynamic friction between the body and the plane is 0.5. The body slides down the plane and attains a velocity of 20 m/s. The distance travelled (in meter) by the body along the plane is ______

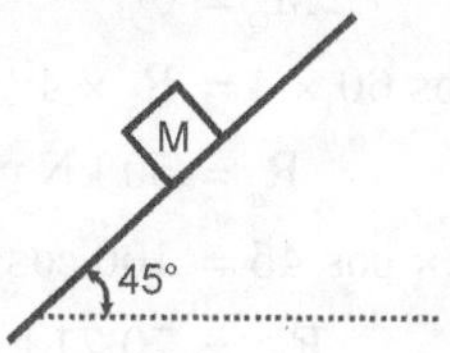

28. A uniform slender rod (8 m length and 3 kg mass) rotates in a vertical plane about a horizontal axis 1 m from its end as shown in the figure. The magnitude of the angular acceleration (in rad/s^2) of the rod at the position shown is ______

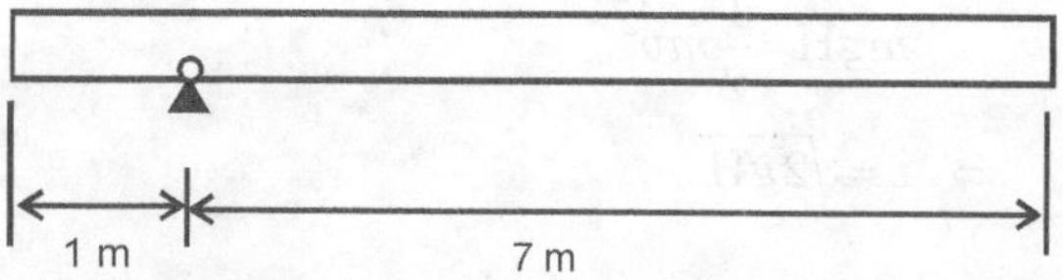

29. A ladder AB of length 5 m and weight (W) 600 N is resting against a wall. Assuming frictionless contact at the floor (B) and the wall (A), the magnitude of the force P (in Newton) required to maintain equilibrium of the ladder is ______

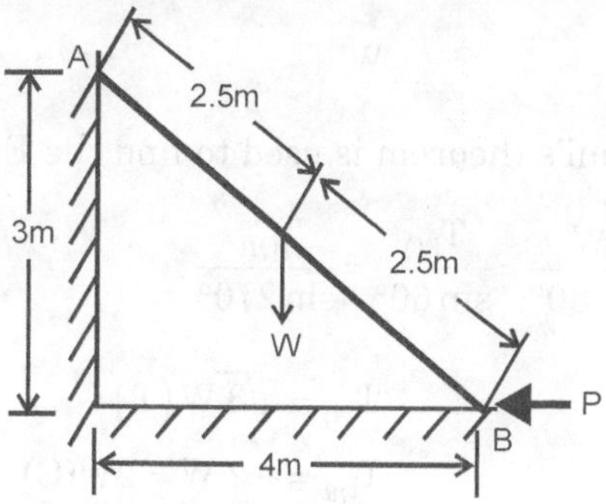

ANSWERS

MCQ Type Questions

1. (c)	**2.** (b)	**3.** (d)	**4.** (b)	**5.** (d)	**6.** (a)	**7.** (c)	**8.** (b)	**9.** (c)	**10.** (c)
11. (a)	**12.** (b)	**13.** (d)	**14.** (c)	**15.** (d)	**16.** (c)	**17.** (c)	**18.** (c)	**19.** (c)	**20.** (a)
21. (d)	**22.** (a)	**23.** (b)	**24.** (a)	**25.** (b)	**26.** (c)	**27.** (a)	**28.** (b)	**29.** (b)	**30.** (a)
31. (b)	**32.** (b)	**33.** (c)	**34.** (b)	**35.** (a)	**36.** (a)	**37.** (b)	**38.** (a)	**39.** (a)	**40.** (d)
41. (c)	**42.** (c)	**43.** (b)	**44.** (a)	**45.** (b)	**46.** (a)	**47.** (a)	**48.** (a)	**49.** (c)	**50.** (*)
51. (c)	**52.** (b)	**53.** (d)	**54.** (c)	**55.** (d)	**56.** (*)	**57.** (a)			

"*" Refers to Subjective Question.

Numerical Type Questions

1. 57.735 N	**2.** 200	**3.** 5	**4.** 4	**5.** 2.248
6. (133.33)	**7.** (20)	**8.** (3)	**9.** (20)	**10.** 100
11. (3.5)	**12.** 0.64	**13.** 0.26	**14.** (8)	**15.** (77)
16. (95.18)	**17.** (29.632)	**18.** (0.1587)	**19.** 251.3	**20.** (40)
21. 1400	**22.** (0.064 to 0.067)	**23.** (4.8 to 5.0)	**24.** (18 to 22)	**25.** (1.0 to 1.3)
26. (590 to 595)	**27.** (56 to 59)	**28.** (1.9 to 2.1)	**29.** (400 N)	

EXPLANATIONS

MCQ TYPE QUESTIONS

1. In the system energy remains conserve

$$mgH = \frac{1}{2}mv^2$$

$$\Rightarrow v = \sqrt{2gH}$$

2.

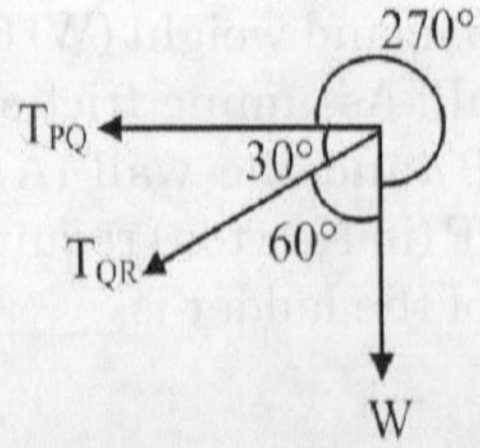

Lami's theorem is used to find the F

$$\frac{W}{\sin 30°} = \frac{T_{PQ}}{\sin 60°} = \frac{T_{QR}}{\sin 270°}$$

$$\Rightarrow \quad T_{PQ} = \sqrt{3}\,W(T)$$

$$\Rightarrow \quad T_{QR} = -2\,W = 2W(C)$$

3. Vertical component of velocity – V sin θ

Horizontal component of velocity – V cos θ

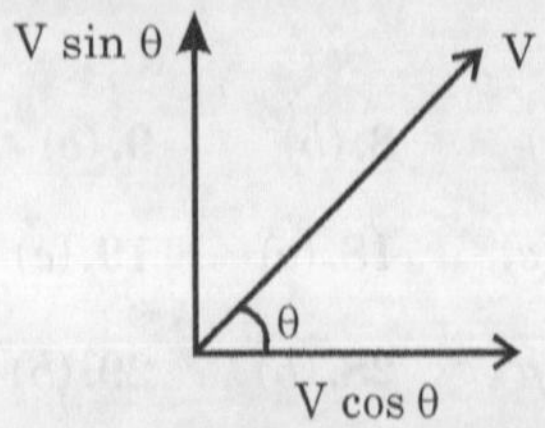

Only vertical component is responsible for change in velocity so change in momentum

= mass × change in velocity

= m(– V sin θ – V sin θ)

= – 2 mV sin θ [in y direct]

So change in momentum = – 2 mV sin θ

5.

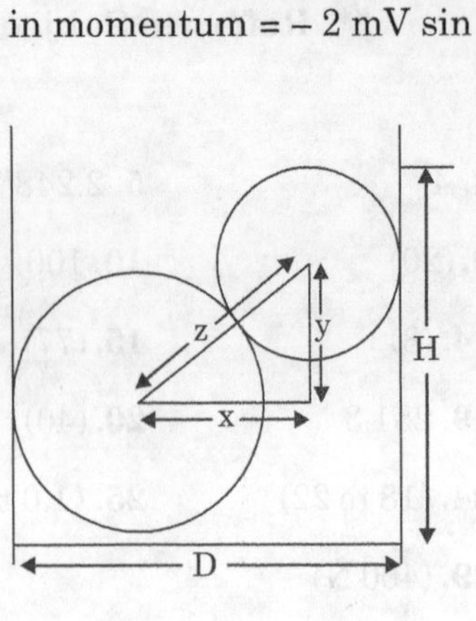

$$z = R + r$$

$$x = D - (R + r)$$

$$y = \sqrt{z^2 - x^2}$$

$$= \sqrt{(R+r)^2 - [D - (R+r)]^2}$$

$$H = r + R + y$$

$$= (R+r) + \sqrt{(R+r)^2 - [D-(R+r)]^2}$$

$$H = (R+r) + \sqrt{2D(R+r) - D^2}$$

7.

$$\Sigma M_Q = 0$$

$$100 \times \cos 60 \times 4 = R_a \times 4$$

$$\Rightarrow \quad R_a = 50 \text{ kN}$$

$$F_{PR} \times \cos 45 = 100 \cos 60$$

$$\Rightarrow \quad F_{PR} = 70.71 \text{ kN}$$

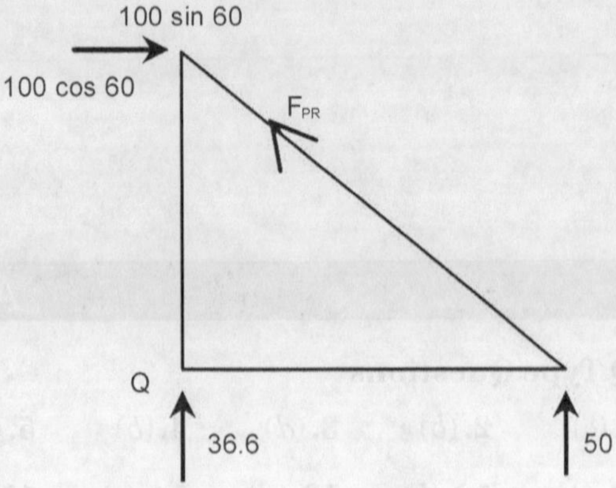

8. Here $V(x, y) = u(x, y)\,\hat{i} + v(x, y)\,\hat{j}$

As we know that $u = \frac{-\partial\phi}{\partial x}$

$$\therefore \quad \frac{-\partial\phi}{\partial x} = -u(x, y) \qquad ...(i)$$

Similarly $v = \frac{-\partial\phi}{\partial y}$

$$\therefore \quad \frac{\partial\phi}{\partial y} = -v(x, y) \qquad ...(ii)$$

From equations (*i*) and (*ii*) we get,

$$\frac{\partial y}{\partial x} = \frac{\partial\phi}{\partial x} \times \frac{\partial y}{\partial\phi} = \frac{+u(x,y)}{-v(x,y)} = \frac{-u}{v}$$

9.

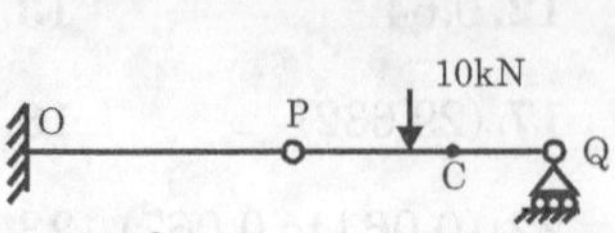

P-being internal hinge

$$\Sigma M_P = 0 \quad ...(1)$$

Condition 'PQ', reaction at Q i.e. $R_Q = 10$ kN

Now from (1), $\Sigma M_P = 0$

$M_0 - R_Q \times 1 + R_C \times 0.5 = 0$

$\Rightarrow \quad M_0 = 10$kN

10.

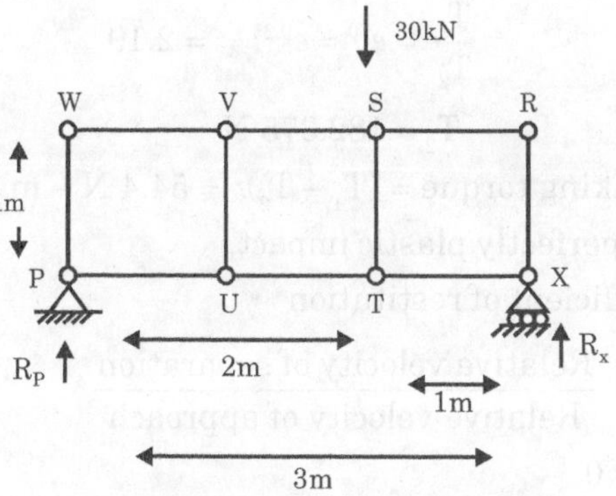

$R_p + R_x = 30000$

$\Sigma M_P = 0$

$R_x \times 3 = 2 \times 3000$

$\Rightarrow \quad R_x = 20$ kN

and $\quad R_p = 10$ kN

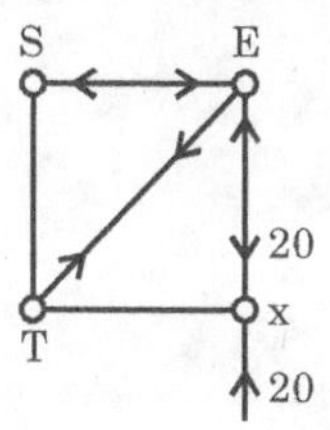

For balance at 'x' $\rightarrow F_{Rx} = 20$ kN

at 'RT' $\rightarrow F_{RT} \cos 45 = 20$

$$\Rightarrow \quad F_{RT} = \frac{20}{\cos 45} \quad ...(i)$$

Also $\quad F_{SR} = F_{RT} \cos 45 \quad ... (ii)$

From equations (*i*) and (*ii*),

$F_{SR} = 20$ kN

11. Using sine rule

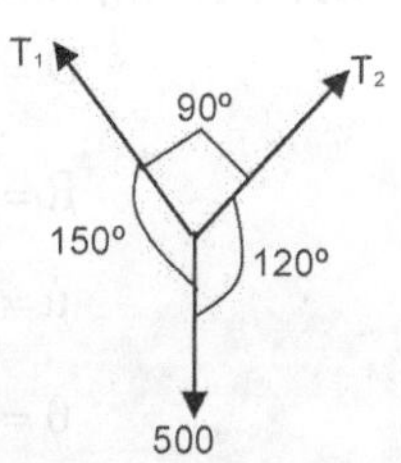

$$\frac{T_1}{\sin 120°} = \frac{T_2}{\sin 150°} = \frac{500}{\sin 90°}$$

$T_1 = 500 \times \sin 120°$

$T_2 = 500 \sin 150°$,

$T_1 = 433$ N, $T_2 = 250$N

12. Moment of Inertia,

$$I_{xx} = \frac{1}{12}\left[(120)^3 \times 60\right] - 2\left[\frac{1}{12} \times (30)^4 + 30 \times 30 \times 30\right]$$

$= 6.885 \times 10^6$ mm^4

13.

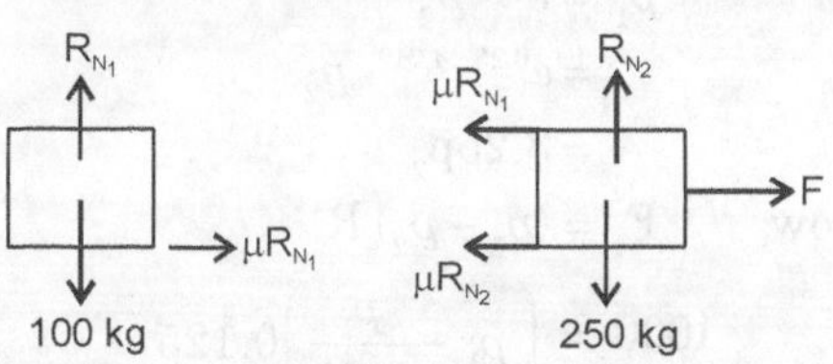

F = 0.4 × 100 × 9.81 + 0.4 × 250 × 9.81

= 1.37 kN.

15. Coefficient of restitution = $\dfrac{\text{Velocity of separation}}{\text{Velocity of approach}}$

$$= \frac{V_2 - V_1}{u_1 - u_2} = \frac{V - V}{5 - 0} = 0$$

$\because V_2 = V_1 = V$ final velocity is same.

16. $\therefore \; I = \frac{1}{2}m(2R)^2 - \frac{1}{2}mR^2 = \frac{3}{2}mR^2$

$$(K.E)_{\text{Rotation}} = \frac{1}{2}I \cdot \omega^2 = \frac{1}{2} \cdot \frac{3}{2} \cdot mR^2 \cdot \omega^2 = \frac{3}{4} \cdot \frac{1}{4} \cdot m(2R\omega)^2$$

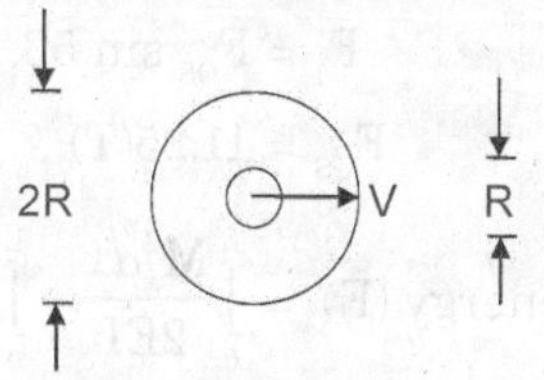

$$\therefore \quad (K{\cdot}E)_R = \frac{3}{16} \cdot mv^2$$

$$(K{\cdot}E)_{\text{Translation}} = \frac{1}{2}mv^2$$

$$(K{\cdot}E)_T = (K{\cdot}E)_R + (K{\cdot}E)_T$$

$$= \frac{3}{16}mv^2 + \frac{1}{2}mv^2 = \frac{11}{16}mv^2.$$

17.

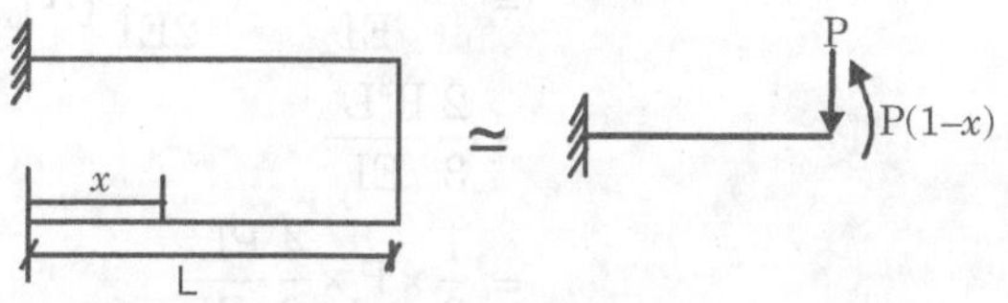

$\delta_1 = \dfrac{P1^3}{3EI}$ due to point load (downward)

$\delta_2 = \dfrac{ML^2}{2EI}$ due to moment (upward)

$$\Rightarrow \; \delta = \delta_1 - \delta_2 = \frac{P1^3}{3EI} - \frac{P(1-x)1^2}{2EI} = 0$$

$$\Rightarrow \; \frac{P1^3}{3EI} - \frac{P1^3}{2EI} + \frac{P \times 1^2 x}{2EI} = 0$$

$$\Rightarrow \; \left[\frac{-1}{3}1 + x\right]\frac{P1^2}{2EI} = 0$$

$$\Rightarrow \; x = \frac{1}{3}1 = 0.33L.$$

18. From all the four options the maximum amplitudes is in point 'C' as t = 0.

$(x)_{t=0} \quad (y)_{t=0}$

(A) $x = 1 \quad y = 0$

(B) $x = 1 \quad y = 0$

(C) $x = \dfrac{3}{2} \quad y = 0$

(D) $x = \dfrac{3}{2} \quad y = 0$

Option 'C' has maximum net amplitude.

20. By method of sections

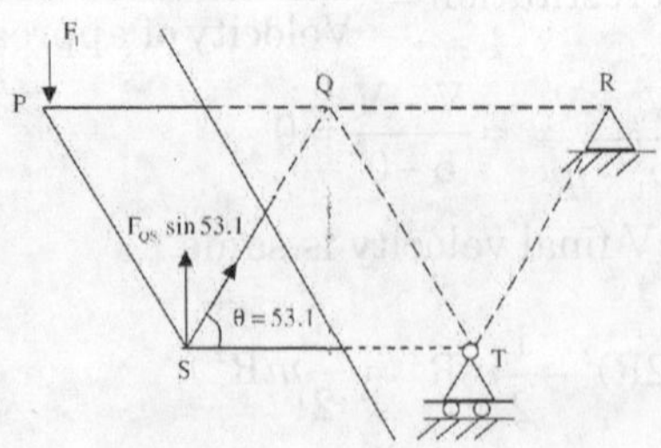

$$\sum V = 0$$

$$F_1 = F_{QS} \sin 53.1$$

$$\Rightarrow \quad F_{QS} = 11.25(T)$$

21. $$\text{Strain energy (E)} = \int_0^L \frac{M_x^2 dx}{2EI} + \int_0^L \frac{M_y^2 dx}{2EI}$$

$$M_X = PX$$

$$M_Y = PL$$

$$\frac{1}{2} P\delta = \int_0^L \frac{(PX)^2 dx}{2EI} + \int_0^L \frac{(PL)^2 dx}{2EI}$$

$$= \frac{P^2\left[\frac{X^3}{6}\right]_0^L}{EI} + \frac{P^2L^2}{2EI}[X]_0^L$$

$$= \frac{2}{3}\frac{P^2L^2}{EI}$$

$$= \frac{1}{2} \times P \times \frac{4}{3}\frac{PL^2}{EI}$$

$$\delta = \frac{4}{3}\frac{PL^2}{EI}$$

22. Taking moments about Q, $\Sigma M_Q = 0$

$$\Rightarrow \quad W \times 1m = P \times 2m$$

$$\Rightarrow \quad P = \frac{100 \times 9.81}{2} = 490.5 \text{ N}$$

$$\Sigma H = 0$$

$$\Rightarrow \quad F_F = P = 490.5 \text{ N}$$

$$\Sigma V = 0$$

$$\Rightarrow \quad R_N = mg = 981 \text{ N}$$

$$\text{Friction Force} = \mu R_N$$

$$\mu = \frac{490}{981} = 0.5$$

23.

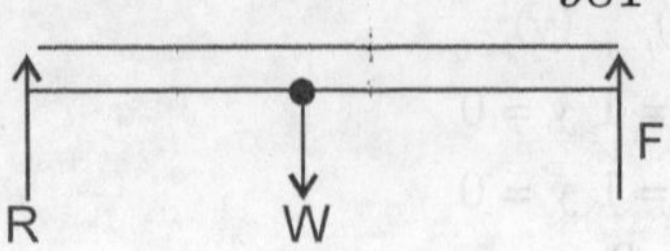

$$R + F = W$$

$$\text{For } F = 0 \Rightarrow R = W$$

25.

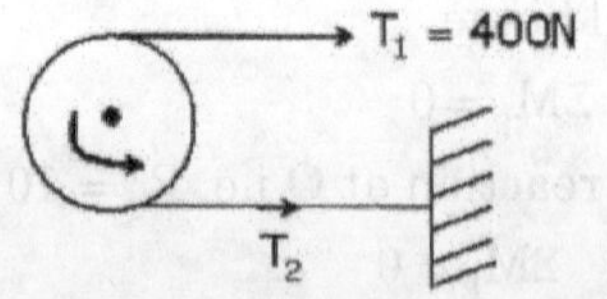

As the drum is rotating in anti clock wise direction, T_1 will be right side & T_2 will be clock side.

$$\& \quad \frac{T_1}{T_2} = e^{\mu\theta} = e^{0.25 \times \pi} = 2.19$$

$$T_2 = 182.375 \text{ N}$$

Breaking torque = $(T_1 - T_2)r$ = 54.4 N – m

27. For perfectly plastic impact,

Coefficient of restitution

$$= \frac{\text{Relative velocity of separation}}{\text{Relative velocity of approach}}$$

$$= 0$$

since both bodies clinge together after impact.

28. Energy stored in the bar

$$= \text{Work done by the stone}$$

$$= \int_0^{0.1} F_x \, dx$$

$$= \int_0^{0.1} 300 \times x^2 dx$$

$$= 300 \times \frac{x^3}{3} = 100 \times 0.1^3 = 0.1\text{J}$$

29. Limiting friction force between block and the surface is 0.98 N. But applied force is 0.8 N which is less than the limiting friction force. Therefore friction force for the given case is 0.8 N.

31. Given : b = 80 mm = 0.08 m

D = 250 mm = 0.25 m

$$R = \frac{D}{2}$$

$$\mu = 0.25$$

$$\theta = 270° = 270 \times \frac{\pi}{180} = 4.712 \text{ radians}$$

F_t = 1000 N-m.

$$\text{Now} \quad \frac{p_1}{p_2} = e^{\mu\theta}$$

$$\Rightarrow \quad p_1 = e^{\mu\theta} \cdot p_2$$

$$= e^{0.25 \times 4.712} \cdot p_2$$

$$= 3.25 p_2$$

$$\text{Now} \quad F_t = (p_1 - p_2) R$$

$$\therefore \quad 1000 = \left(p_1 - \frac{p_1}{3.25}\right) 0.125$$

Solving, we get maximum tension in the belt,

$$p_1 = 11.56 \text{ kN}$$

33.

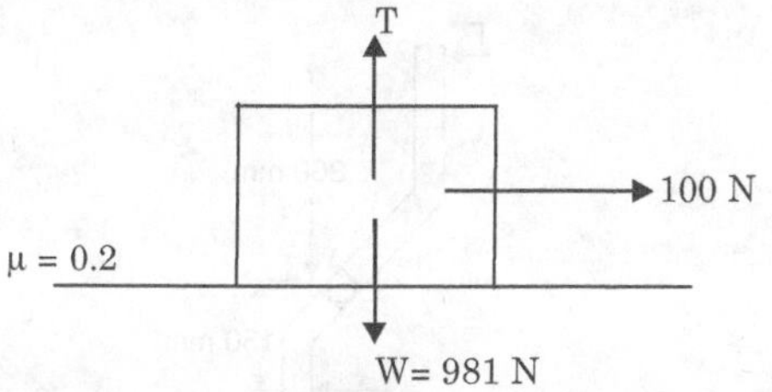

Net normal force on block

$$N = W - T$$
$$= 981 - T \quad ...(i)$$

Frictional force, $f = \mu N$

Under equilibrium, i.e. when man is just able to move the block

$$\mu N = 100$$

From equation (*i*),

$$\mu(981 - T) = 100$$
$$\Rightarrow \quad 0.2\,(981 - T) = 100$$
$$T = -\frac{100}{0.2} + 981$$
$$= 981 - 500 = 481N$$

34.

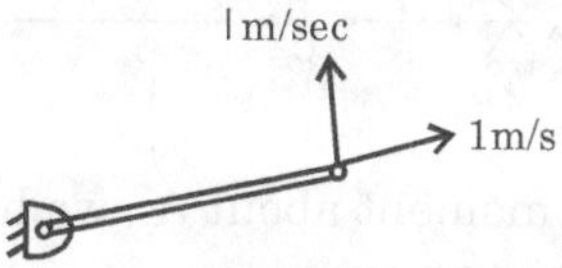

$$v = \sqrt{1^2 + 1^2} = \sqrt{2} \text{ m/s}$$

35. Linear momentum and Kinetic energy is conserved in elastic collision and in inelastic collision.

Kinetic energy (heat, sound etc.) is conserved.

36. $Mg \sin\theta - \mu\, Mg \cos\theta = Ma$

$$\Rightarrow \quad a = g \sin\theta - \mu\, g \cos\theta$$

Now, $s = ut + \frac{1}{2}at^2$

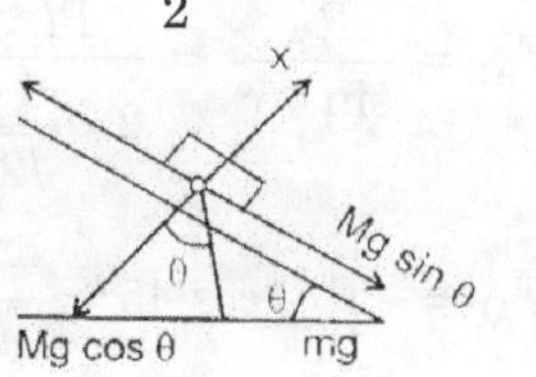

But $u = 0$

$$\therefore \quad t = \sqrt{\frac{2s}{a}} = \sqrt{\frac{2s}{g\cos\theta(\tan\theta - \mu)}}$$

37. Total potential energy = *f*(independent variable)

Hence for a system in equilibrium,

Total potential energy = constant

Thus, partial derivatives of its total potential energy with respect to each of independent variable must be zero.

38.

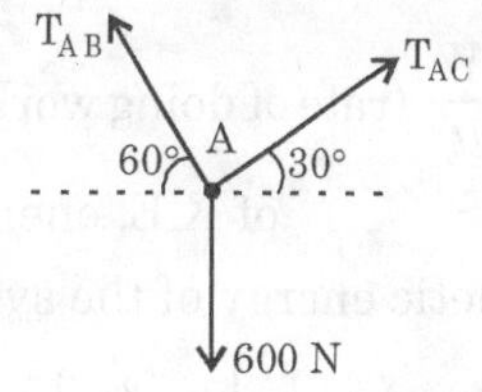

$$T_{AB} \cos 60° = T_{AC} \cos 30° \quad ...(i)$$
$$T_{AB} \sin 60° + T_{AC} \sin 30° = 600 \quad ...(ii)$$

Solving equations (*i*) and (*ii*), we get

$$T_{AB} = 520N$$

and $T_{AC} = 300N$

39. Force at mean position is zero.

40. μ for each surface is same = 0.3

Its free body diagram ($m_1 = m_2 = 1$ kg)

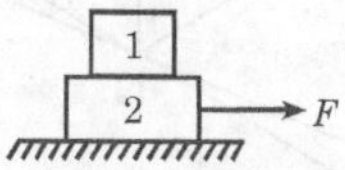

$$N = mg = g \text{ N}$$
$$N' = 2mg = 2g \text{ N}$$

Equation of motion of 1

$$\mu N = 1 \times a_1$$
$$\Rightarrow \quad a_1 = \mu g \quad ....(i)$$

towards right

Equation of motion of 2

$$F - (-\mu N + \mu N') = 1 \times a_2$$
$$\Rightarrow \quad F - (3\mu g) = a_2 \quad ....(ii)$$

towards right

Now for relative motion between two blocks

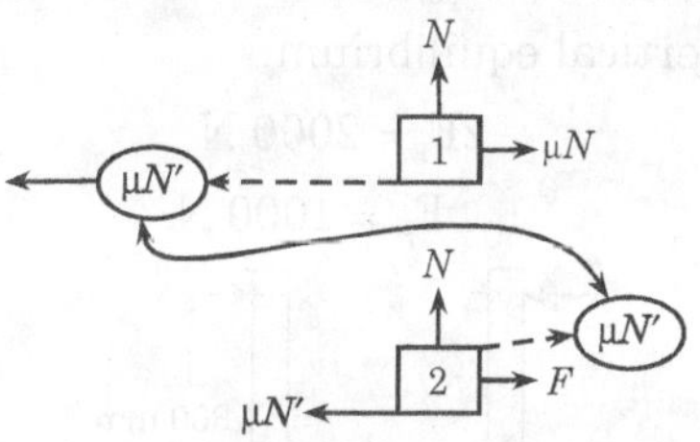

$$a_2 > 0$$
$$\therefore \quad F - 3\mu g > 0$$
$$\Rightarrow \quad F > 3\mu g$$
$$\Rightarrow \quad F > 3 \times 0.3 \times 9.8$$
$$\Rightarrow \quad F > 8.82 \text{ N}$$
$$\therefore \quad F = 8.83 \text{ N}$$

41.

$$\omega r = 2 \times 2$$
$$= 4 \text{m/sec}$$
$$v_{net} = \sqrt{3^2 + 4^2}$$
$$= 5 \text{ m/s}$$

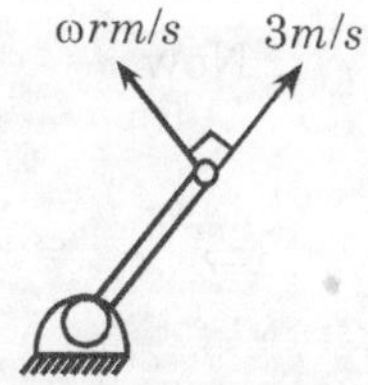

42. Power = $\frac{dW}{dt}$ (rate of doing work or rate of change of K.E. energy of the system)

Initial Kinetic energy of the system

$$= \frac{1}{2}mv^2 + \frac{1}{2}mv^2$$

$$= mv^2$$

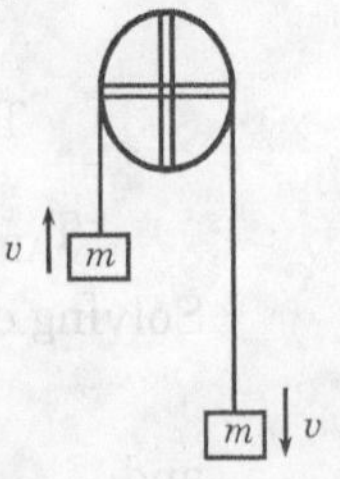

Final kinetic energy = 0

Time duration it occurs = t

$$\Rightarrow \quad \text{Power} = \frac{mv^2}{t}$$

44.

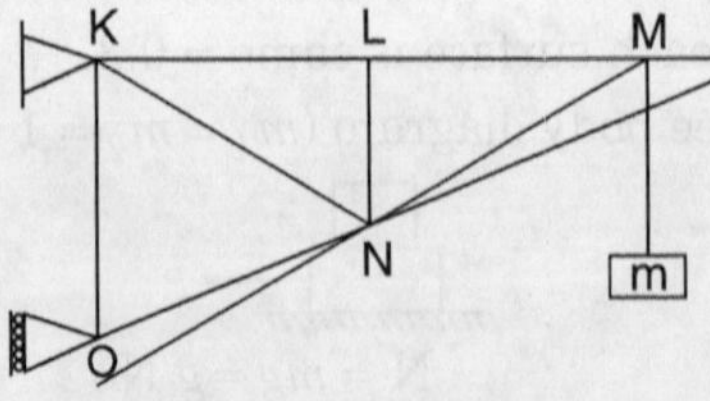

Consider all forces at point L

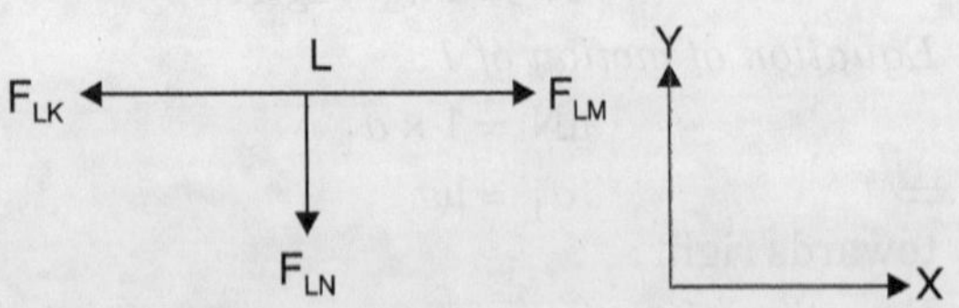

Summation of all forces along x-direction,

$$F_{LK} = F_{LM}$$

Summation of all forces along y-direction,

$$F_{LN} = 0.$$

45. For vertical equilibrium

$$2F_r = 2000 \text{ N}$$

$$\Rightarrow \quad F_r = 1000 \text{ N}$$

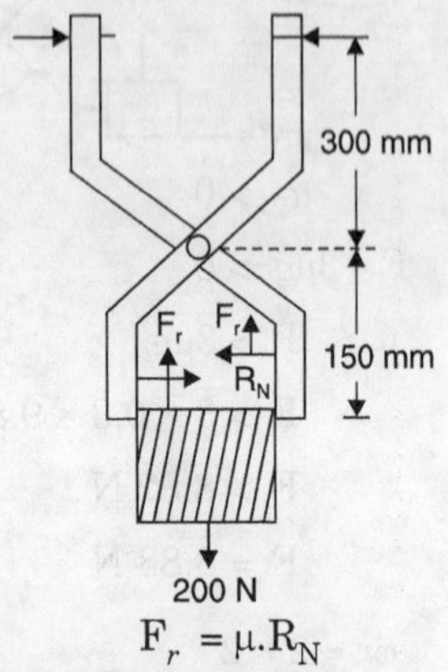

Now $\quad F_r = \mu.R_N$

$\therefore \quad 1000 = 0.1\ R_N$

$$\Rightarrow \quad R_N = \frac{1000}{0.1} = 10000 \text{ N.}$$

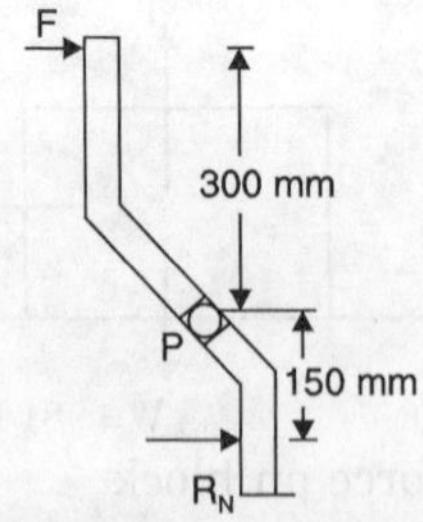

Taking moment about P, we have

$$0.3 \times F = R_N \times 0.150$$

$$\Rightarrow \quad 0.3\ F = 10000 \times 0.150$$

$$\Rightarrow \quad F = \frac{10000 \times 0.15}{0.3} = \frac{10{,}000 \times 15}{30}$$

$$= 5000 \text{ N.}$$

46. Total load on EF member = Pl

where, l = length of EF

For horizontal equilibrium, $R_A + R_B = Pl$

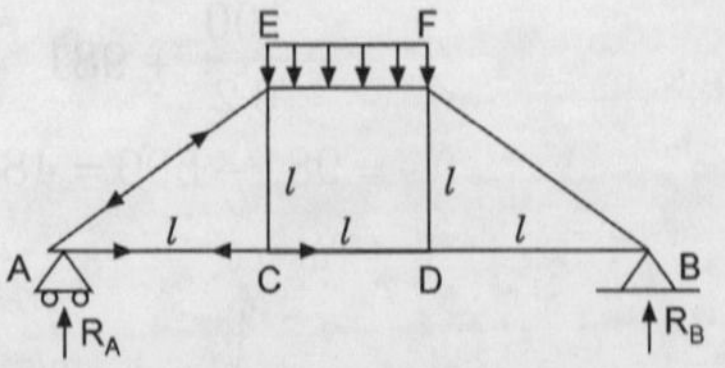

Taking moment about A., we have

$$Pl \times \frac{3l}{2} = R_B \times 3l$$

$$\Rightarrow \quad R_B = \frac{Pl}{2}$$

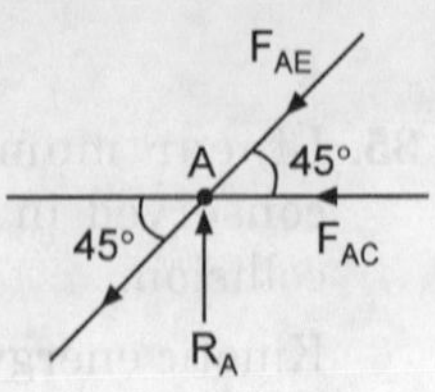

Considernig a point A.

For horizontal equilibrium

$$F_{AC} + F_{AE} \cos 45° = 0$$

For vertical equilibrium

$$R_A - F_{AE} \sin 45° = 0$$

$$\Rightarrow \quad F_{AE} = \frac{R_A}{\sin 45°} = \frac{Pl}{2 \times \frac{1}{\sqrt{2}}} = \frac{Pl}{\sqrt{2}}$$

$$\therefore \quad F_{AC} = -F_{AE} \cos 45° = \frac{-Pl}{\sqrt{2}} \cdot \frac{1}{\sqrt{2}} = \frac{-Pl}{2}$$

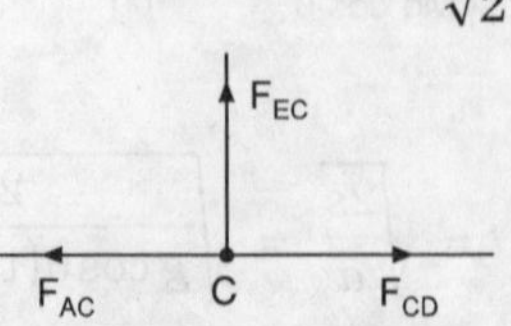

–ve sign shows that force on member AC is opposite of assumed.

Now consider a point C

For horizontal equilibrium, $F_{AC} = F_{CD} = \frac{Pl}{2}$

47. Let, m = mass of the bullet

v = velocity of bullet

M = mass of the block

u = final velocity of system (block + bullet)

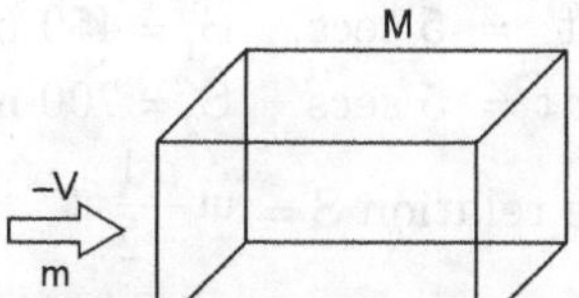

Form conservation of linear momentum

$$mv = (M + m)\,u$$

$$\Rightarrow \quad u = \frac{mv}{m + M} \qquad ...(i)$$

For horizontal equilibrium of body

$$(M + m)\,g = N$$

Frictional force, $F_r = N = u(M + m)\,g$

$$\text{Frictional retardation} = \frac{\mu\,(M+m)g}{(M+m)} = -\,g$$

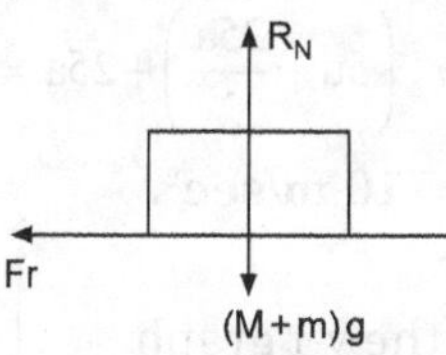

(–ve sign shows that acceleration is opposite to motion)

Let v_f = final velocity of system = 0

From $\quad v_f^2 = u^2 + 2as$

$$\Rightarrow \quad u^2 + 2as = 0$$

$$\Rightarrow \quad u^2 - 2\ gs = 0$$

From equation (i)

$$\left(\frac{mv}{M+m}\right)^2 = 2\ \ gs$$

$$\frac{mv}{M+m} = \sqrt{2\ \mu gs}$$

$$\Rightarrow \quad v = \frac{M+m}{m}\sqrt{2\mu gs}$$

48. *Given:* m = mass of the reel

k = radius of gyration

r = radius of the hub

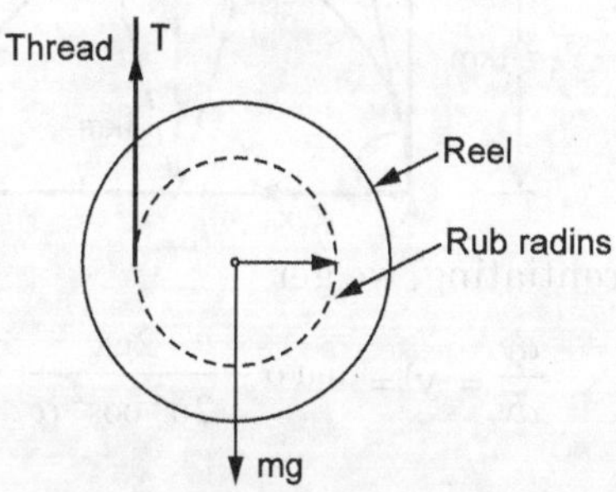

For vertical translational motion

$$mg - T = ma_t \qquad ...(i)$$

where T = tension in thread

a_t = tangential acceleration of the point of contact of thread and hub.

For rotation motion

$$T \times r = I\alpha$$

$$\Rightarrow \quad T = \frac{I\alpha}{r} = I\,\frac{a_t}{r^2} = mk^2 \cdot \frac{a_t}{r^2} \qquad ...(ii)$$

From equations (ii) and (iii), we have

$$mg - \frac{mk^2 a_t}{r^2} = ma_t$$

$$\Rightarrow \quad mg = ma_t\left(1+\frac{k^2}{r^2}\right) = ma_t\left(\frac{r^2+k^2}{r^2}\right)$$

$$\Rightarrow \quad a_f = \frac{g\,r^2}{(k^2+r^2)}$$

49. Tension in thread, $T = \dfrac{mk^2}{r^2} \times \dfrac{gr^2}{(k^2+r^2)} = \dfrac{mgk^2}{(k^2+r^2)}$

50. Translational velocity of the centre of toy,

$$v = (R - r)\,\dot{\theta}$$

Angular velocity of toy,

$$\omega = (\dot{\beta} - \dot{\theta})$$

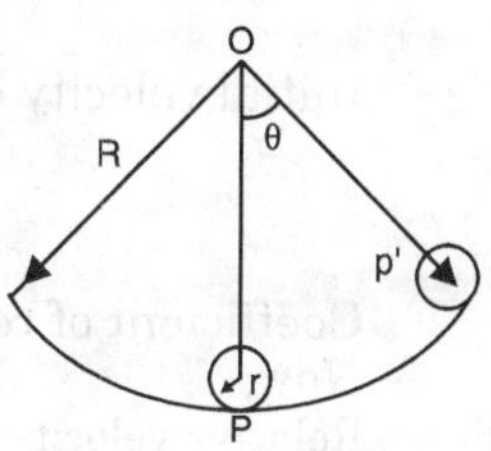

Now, $\quad R\theta = r\beta$

$$\Rightarrow \quad \beta = \frac{R}{r}\theta$$

∴ Angular velocity of toy,

$$\omega = (\dot{\beta} - \dot{\theta}) = \left(\frac{R}{r} - 1\right)\dot{\theta}$$

Kinetic energy of toy,

$$T = \frac{1}{2}\,\frac{w}{g}\cdot v^2 + \frac{1}{2}\,\frac{w}{g}\cdot \omega^2$$

$$= \frac{1}{2}\,\frac{w}{g}\,(R-r)^2\dot{\theta}^2 + \frac{1}{2}\,\frac{w}{g}\,\frac{r^2}{2}\left[\left(\frac{R}{r}-1\right)^2\right]^2 \dot{\theta}^2$$

Potential energy of toy,

$$v = \frac{w}{g}\cdot g\cdot (R-r)(1-\cos\theta)$$

$$= w\,(R-r)(1-\cos\theta)$$

(a) **Equation of Motion for large oscillation**

Applying energy principle,

$$\frac{d}{dt}\,(T + U) = 0$$

$$\Rightarrow \frac{1}{2}\,\frac{w}{g}\,(R-r)^2\,2\cdot\ \dot{\theta}\,\ddot{\theta} + \frac{1}{2}\,\frac{w}{g}\cdot\frac{r^2}{2}\,\frac{(R-r)^2}{r^2}$$

$$\cdot\ 2\,\dot{\theta}\,\ddot{\theta} + w\,(R-r)\sin\theta\cdot\dot{\theta} = 0$$

$$\Rightarrow \frac{w}{g}(R-r)\ddot{\theta}+\frac{1}{2}\frac{w}{g}(R-r)\ddot{\theta}+w\sin\theta=0$$

$$\Rightarrow \ddot{\theta}=-\frac{2gw}{3w(R-r)}\sin\theta=-\frac{2g}{3(R-r)}\sin\theta$$

(*b*) **Equation of motion for small oscillation**

$$\frac{w}{g}(R-r)^2\ddot{\theta}+\frac{1}{2}\frac{w}{g}(R-r)^2\ddot{\theta}+w(R-r)\theta=0$$

$$\Rightarrow \frac{3}{2}\frac{w}{g}(R-r)^2\ddot{\theta}+w(R-r)\theta=0$$

$$\Rightarrow \frac{3}{2}\frac{w}{g}(R-r)\ddot{\theta}+w\theta=0$$

$$\Rightarrow \ddot{\theta}+\frac{2}{3}\frac{g}{(R-r)}\theta=0$$

$\therefore$ Natural frequency, $\omega_n=\sqrt{\dfrac{2g}{3(R-r)}}$

52.

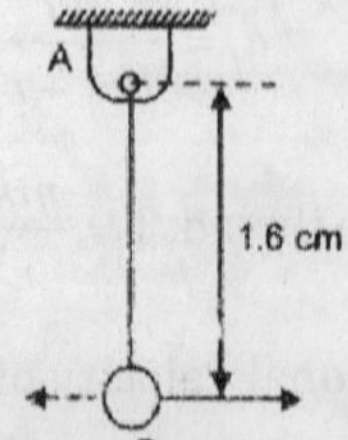

Initial velocity of C, $V_C = 5$ m/s
$m_C = 2$kg
$V_B = 0$,
$m_B = 8$ kg

Coefficient of restitution = 1

Now,
Relative velocity of approach
= Relative velocity of separation

Approach velocity $= V_C$

Let after impect,
velocity of C $= V_C'$
and velocity of B $= V_B'$

$\therefore \quad V_C = V_B' - V_C'$

Applying conservation of linear momentum,

$$m_C V_C = m_B V_B' - m_C V_C'$$

$$\therefore \quad 10 = 4V_B' - V_C'$$

$$\Rightarrow \quad V_B' - V_C' = 2$$

$$\Rightarrow \quad 3V_B' = 3$$

$$V_B' = 1 \text{ m/s}$$

$$\therefore \quad \omega=\frac{V_B'}{1.6}=\frac{1}{1.6}=0.625 \text{ rad/s}$$

53.

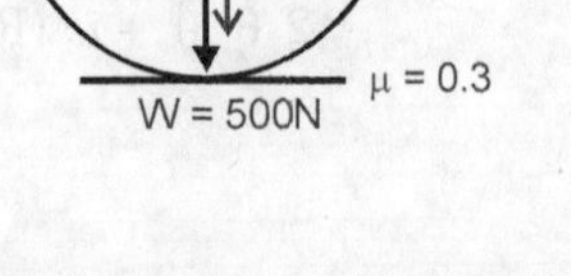

Here, $r=\dfrac{600}{2}=300$ mm.

W = 500 N.

$\therefore \quad F = \mu \times W = 0.3 \times 500 = 150 \times$ N.

55. For $t_1 = 5$ secs, $S_1 = 450$ m
and for $t_2 = 5$ secs $S_2 = 700$ m

From the relation $S = ut+\frac{1}{2}at^2$

$$450 = 5u+\frac{1}{2}\times a\times 25 \text{ [since initial velocity } = u]$$

or $$450 = 5u+\frac{25a}{2} \qquad \text{......}(i)$$

Also after first 5 seconds, velocity of car,

$$v = u + at = u + 5a.$$

After first 5 seconds, u + 5a will be the initial velocity for next 5 seconds, therefore

$$700 = (4+5a)5+\frac{1}{2}a\times 25 = 20+25a+\frac{25a}{2}$$

From equation (i), we have

$$700 = \left(5u+\frac{25a}{2}\right)+25a = 450+25a,$$

or a = **10 m/sec²**.

Alternately

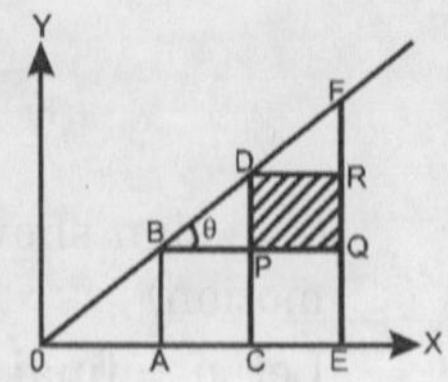

Considering the v-t graph

Area ACDB = 450

Area CEFD = 700

Difference is the shaded region

DPQR = 250

$$\therefore \quad PD = \frac{250}{5} = 50$$

$$a = \tan\theta = \frac{PD}{BP} = \frac{50}{5} = \mathbf{10}$$

56. Maximum height obtained, $y_1 = 1000$ m.

Radius of curvature at maximum height,

$$\rho = 1000 \text{ m}$$

Equation of path is,

$$y = x\tan\alpha - \frac{gx^2}{2u^2\cos^2\alpha}$$

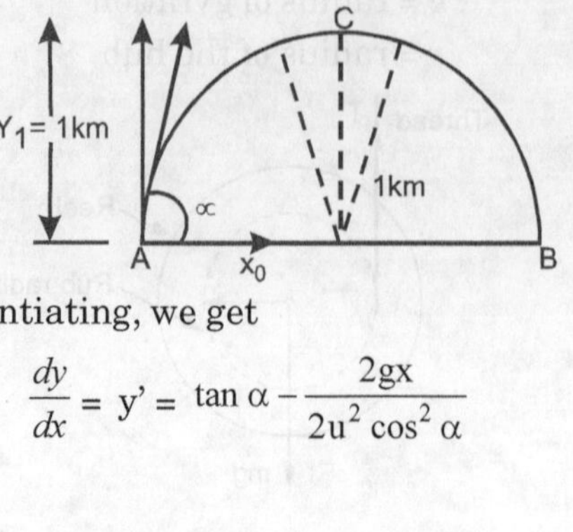

Differentiating, we get

$$\frac{dy}{dx} = y' = \tan\alpha - \frac{2gx}{2u^2\cos^2\alpha}$$

Again differentiating, we get

$$\frac{d^2y}{dx^2} = y'' = 0 - \frac{2g}{2u^2\cos^2\alpha} \quad ...(1)$$

$$\text{Radius of curvature, } \rho = \left| \frac{\left[1+\left(\frac{dy}{dx}\right)^2\right]^{3/2}}{\left(\frac{d^2y}{dx^2}\right)} \right| \quad ...(2)$$

$$\text{At C (maximum height), } \frac{dy}{dx} = 0 \quad ...(3)$$

Substituting equations (1) and (3) in equation (2), we get

$$\rho = \left| \frac{2u^2\cos^2\alpha}{-2g} \right|$$

$$\text{or} \quad 1000 = \frac{u^2\cos^2\alpha}{2g} \quad ...(4)$$

Equation of maximum height is

$$y = \frac{y_0^2}{2g} = \frac{(u\sin\alpha)^2}{2g}$$

where $u\sin\alpha$ is the vertical component of initial velocity.

$$\therefore \quad y = 1000 = \frac{u^2\sin^2\alpha}{2g} \quad ...(5)$$

Dividing equation (5) by equation (4), we get

$$\tan^2\alpha = 2$$

or $\alpha = \tan^{-1} =$ **54.74°**

$$\text{From equation (5), } \frac{u^2\sin^2\alpha}{2g} = 1000$$

$$\text{or Initial velocity, } u = \sqrt{\frac{1000 \times 2 \times 9.8}{\sin^2 54.74°}}$$

= **171.45 m/sec**

and angle of projection = **54.74°**

NUMERICAL TYPE QUESTIONS

1. 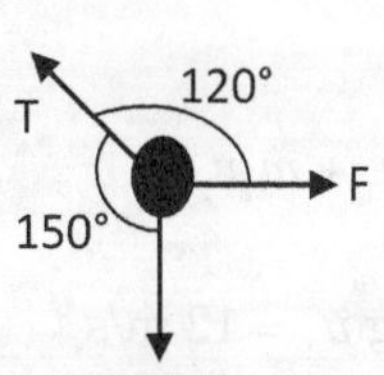

Lami's theorem is used to find the F

$$\frac{F}{\sin 150°} = \frac{T}{\sin 120°}$$

$$\Rightarrow \quad F = 57.753$$

2. R = 100m, v = 200 m/sec,

$$\omega_p = \frac{V}{R} = 0.2 \text{ rad/sec}$$

$$\omega_s = 100 \text{ rad/sec}$$

$$I = 10 \text{ kg-m}^2$$

$$\text{Gyroscopic moment} = I\omega_s\omega_p = 10 \times 0.2\ 100 = 200 \text{ N-m}$$

3. 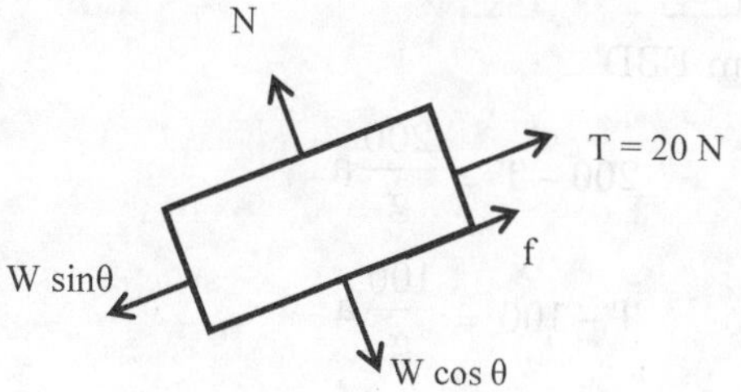

$$\sum F_y = 0$$

$$\Rightarrow \quad N = W\cos\theta = 0.8\text{ W}$$

$$f = \mu N = 0.2\text{ W}$$

$$\sum F_x = 0$$

$$\Rightarrow \quad 0.6\text{ W} = 20 + 0.2\text{ W}$$

$$\Rightarrow \quad W = 50\text{ N}$$

$$\Rightarrow \quad m = 5\text{ kg}$$

4. 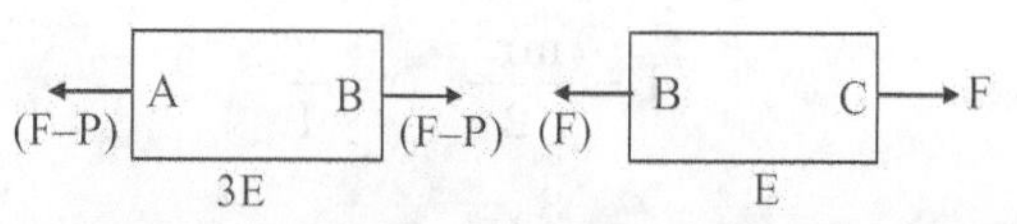

$$\delta_{AB} + \delta_{BC} = 0$$

$$\frac{(F-P)L}{A(3E)} + \frac{(F)L}{A(E)} = 0$$

$$\frac{(F-P)}{3} + F = 0$$

$$F - P + 3F = 0$$

$$4F = P$$

$$\frac{P}{F} = 4$$

5. $\frac{F_1}{F_2} = e^{\mu\theta} = e^{0.25\times\frac{3\pi}{2}} = 3.248$

$\therefore$ $F_1 = 3.248N$

Torque, $T = (F_1 - F_2) \times r$

$= (3.248 - 1) \times 1$

$= 2.248$ N-m

6.

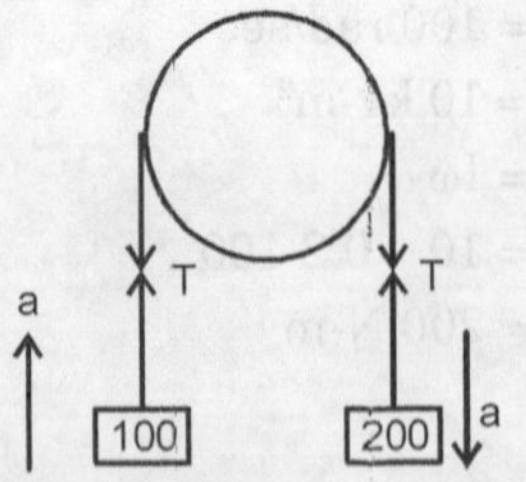

From FBD

$$200 - T = \frac{200}{g}a$$

$$T - 100 = \frac{100}{g}a$$

On adding

$$100 = \frac{300}{g}a \Rightarrow a = \frac{g}{3}$$

$$T = 133.3N$$

7. Conservation of energy

Total (KE + PE) = const.

(KE + PE)A = (KE + PE)B

$$0 + mg30 = \frac{1}{2}mv^2 + \frac{1}{2}I\omega^2 + 0$$

$$30 \times 10 = \frac{1}{2}\times 1 \times v^2 + \frac{1}{2}\times\frac{1}{2}\times 1 \times v^2$$

$$I = \frac{mr^2}{2},\ \omega = \frac{V}{I}$$

$$30 \times 10 = \frac{3}{4}V^2$$

$$V = 20 \text{ m/s}$$

8. $V_1 = i + 2j$

$|V_1| = \sqrt{2^2 + 1^2} = \sqrt{5}$

By drawing velocity diagram

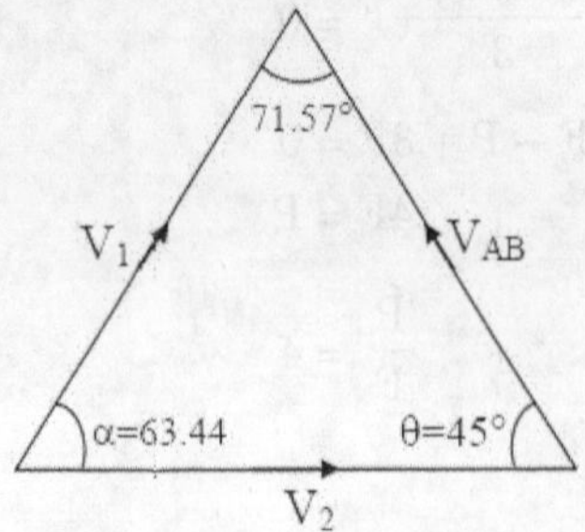

$$\alpha = \tan^{-1}\left(\frac{2}{1}\right) = 63.44$$

By applying sine rule

$$\frac{\sqrt{5}}{\sin 45} = \frac{V_2}{\sin 71.57}$$

$\therefore$ $V_2 = 3$ m/sec

9. At p^t 'A', the velocity

$\Rightarrow 10 + 10 = 20$ m/s

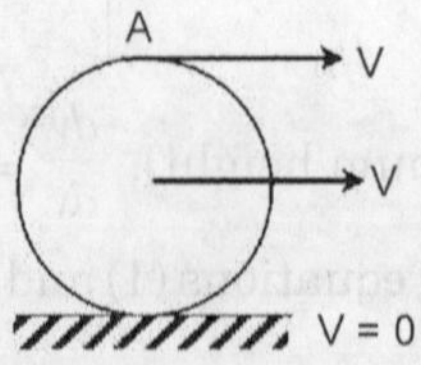

10. $\Rightarrow$ $2f \sin 30 = 100$

$\Rightarrow$ $f = 100$ N

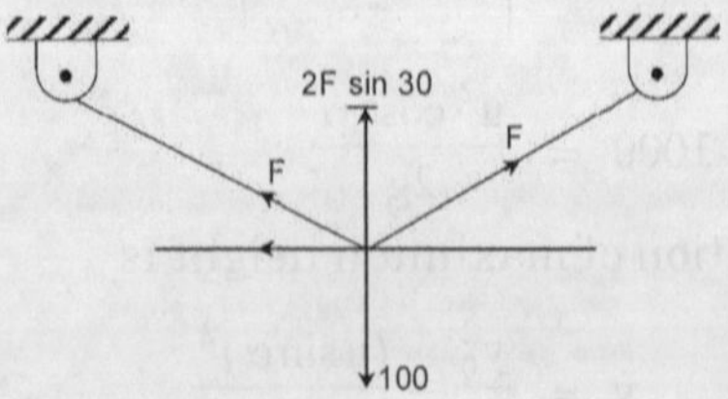

11. Let Swimmer = x

River = y

$$2 = \frac{10}{x + y}$$

$$5 = \frac{10}{x - y}$$

On solving we get $x = 3.5$ km/h

12. $v = \sqrt{2gh} = \sqrt{2 \times 9.01 \times 1} \Rightarrow 4.4294$ m/s

$v' = 0 \times v = 3.5432$ m/s

$h = \frac{v^2}{2g} = 0.64$m

13. Deflection = Deflection at load + Slope × Distance

$$\Rightarrow \frac{W1^3}{3EI} + \frac{W1^2}{2EI} \times .05$$

W = 500

1 = .05

EI = 200

δ = 0.2604 mm

14. For elastic collision

$$m_1u_1 + m_1u_2 = m_1v_1 + m_2v_2 \quad ...(i)$$

moment conservation

$m_1 = 1$ kg $u_1 = 12$ m/s

$m_2 = 2$ kg $u_2 = 0$ m/s

$$\frac{1}{2}m_1u_1^2 + \frac{1}{2}m_2u_2^2 = \frac{1}{2}m_1v_1^2 + \frac{1}{2}m_2v_2^2 \quad ...(ii)$$

energy conservation

From equation (i)

$$12 = v_1 + 2v_2 \quad ...(iii)$$

From equation (ii)

$$\frac{1}{2}\times 1\times 144+\frac{1}{2}\times 2\times 0 = \frac{1}{2}\times 1\times v_1^2+\frac{1}{2}\times 2\times v_2^2$$

$$\Rightarrow \quad 144 = v_1^2 + 2v_2^2 \quad ...(iv)$$

From equations (iii) and (iv)

$$144 = 144 + 4v_2^2 - 48v_2 + 2v_2^2$$

$$\Rightarrow \quad 6v_2^2 - 48v_2 = 0$$

$$6v_2(v_2 - 8) = 0$$

$$\Rightarrow \quad v_2 = 8 \text{ m/s}$$

15. Modulus of rigidity (G)

$$G = \frac{E}{2(1+\mu)}$$

It has given change in stress = 200MPa

Change in strain = 0.001

Here, $200 = E \times 0.003$

$$\Rightarrow \quad E = \frac{200\text{MPa}}{0.002} = 200\times 10^3 \text{ MPa}$$

$$= 200\text{GPa}$$

$$G = \frac{200}{2(1+0.3)} = \frac{200}{2\times 1.3} = \frac{100}{1.3} = \frac{1000}{13} = 77\text{GPa}$$

16. True strain $= ln\left(\frac{100}{95}\right) = 0.05129$

$$\sigma = 500 \times (0.05129)^{0.1} = 371.51$$

Upto elastic limits using hooks law,

$$\varepsilon = \left(\frac{\sigma}{E}\right)$$

$$\varepsilon = \Delta l/l = \left(\frac{\sigma}{E}\right)$$

$$\Delta l = \left(\frac{l.\sigma}{E}\right)$$

$$\Delta l = 100\times\left(\frac{371.51}{200000}\right)$$

$$\Delta l = 0.1857 \text{ mm}$$

– considering this for elastic recovery

∴ This will be added to 95 mm

∴ Final dimension = 95.18575 mm

17.

$$a = -0.1v$$

$$\frac{dv}{dt} = -0.1v$$

$$\ln v = -0.1t + \ln k$$

$$v = ke^{-0.1t}$$

At $t = 0$; $v = 40$

∴ $k = 40$

$$v = 40e^{-0.1t}$$

At t = 3 seconds

$$v = 40e^{-0.1\times 3} = 29.6327 \text{ m/s}$$

18. Slope, $\theta_Q = \frac{-Px}{2EI}(2L - x)$

where x is measured from the fixed end

$$\therefore \quad \theta_Q = \frac{-2000(0.02)}{2\times 210\times 10^9\times\frac{(0.006)^4}{12}}(2\times 0.1 - 0.02)$$

$$= -0.1587$$

The magnitude of slope is 0.1587 radian

19. Time taken force revolution $= \frac{0.5}{20} = 0.025$ sec.

The spinning speed is $\frac{2\pi}{0.02s}$ rad/sec = 251.3 rad/sec

20. BMD:

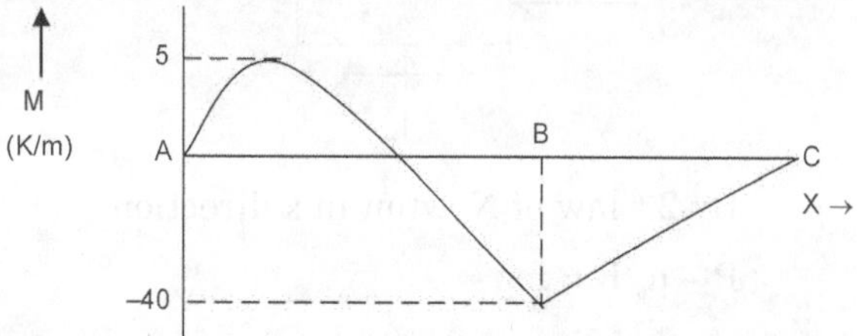

$$R_A = 10\text{kN}$$

$$R_B = 50\text{kN}$$

Maximum bending momentum occurs at reaction B and has a magnitude of 40 kN-m.

21.

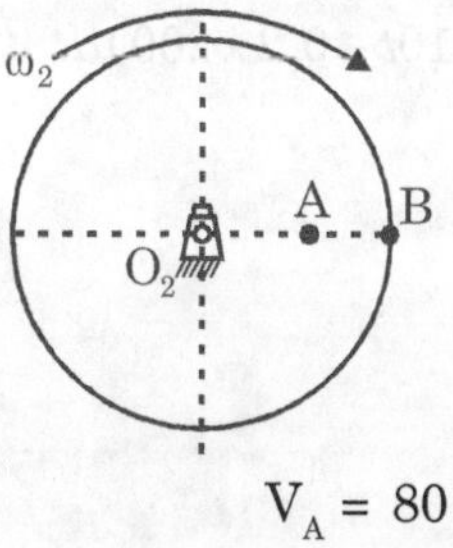

$$V_A = 80 \text{ m/s}, V_B = 140 \text{ m/s}$$

$$r_B - r_A = 300 \quad ...(i)$$

$$\omega \times r_A = 80$$

$$\omega \times r_B = 140$$

$$\therefore \quad \frac{r_B}{r_A} = 1.75 \quad ...(ii)$$

Solving (i) and (ii), $r_B = 700$ mm.

∴ Diameter of wheel is 1400 mm.

22. $S = \frac{FL^3}{3EI}$

$$k = \frac{F}{S} = \frac{3EI}{L^3}$$

$$k = \frac{3EI}{(0.01)^3}$$

$$k = 3,000,000\ EI$$

$$\omega_n = \sqrt{\frac{k}{m}} = \sqrt{\frac{3000,000EI}{0.5}}$$

$$= 2449.48\sqrt{EI}$$

$$f_n = \frac{\omega_n}{2\pi} \Rightarrow 100 = \frac{2449.48\sqrt{EI}}{2\pi}$$

$$EI = 0.065\ N.m^2.$$

23.

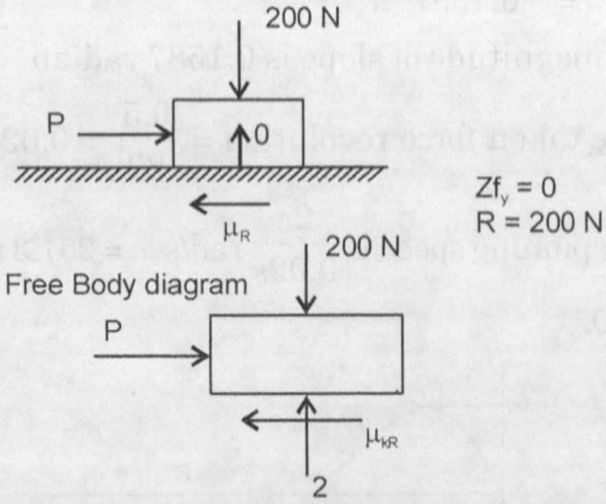

By 2nd law of Newton in x direction

$$P - \mu_k R = ma$$

$$\Rightarrow \quad P - \mu_k R = m\frac{dv}{dt}$$

$$\Rightarrow v_2 - v_1 = \int_1^2 dv = \frac{1}{m}\int_{t_1}^{2}(P - \mu_k R)dt$$

$$= \frac{1}{\left(\frac{200}{9.81}\right)}\int_0^{10}(10t - 0.2\times 200)dt$$

$$v_2 - 0 = 4.90\ m/s.$$

$$\therefore\ v_2 = 4.90 m/s$$

24.

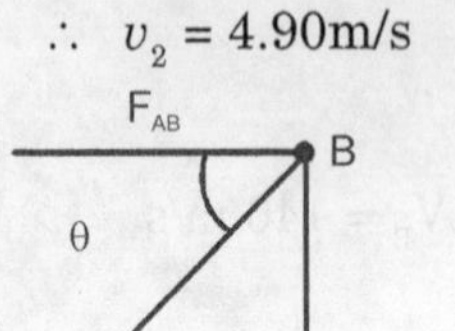

FBD of joint B

$$\theta = Tan^{-1}\left(\frac{0.5}{1}\right) = 26.5$$

$$\Sigma V = 0$$

$$\frac{F_{BC}\sin\theta}{F_{BC}\cos\theta} = \frac{10\ kN}{F_{AB}}$$

$$F_{AB} = \frac{10}{Tan26.2} = 20\ kN.$$

25.

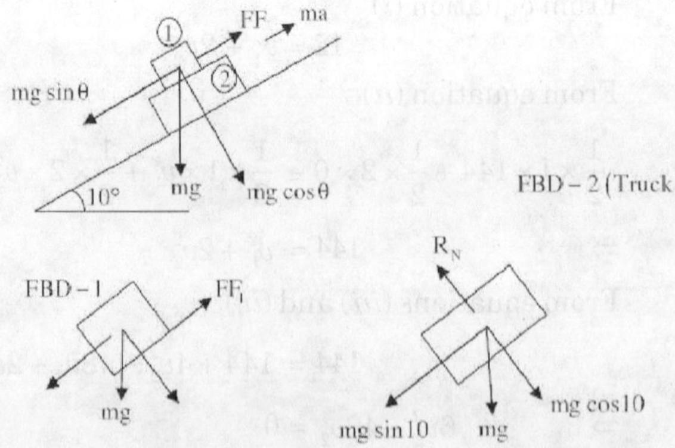

$$\Sigma F_{along} = ma$$

$$F_F - mg\sin\theta = ma$$

$$\Rightarrow \mu\, mg\cos 10 - mg\sin 10 = ma$$

$$\Rightarrow a = [0.3\cos 10 - \sin 10]\ 9.81 = 1.19.$$

26.

$$\Delta E = I\omega^2 C_s \frac{\pi}{4}$$

$$\Rightarrow 2600 = I\times\left(\frac{2\pi\times 200}{60}\right)^2\times\left(\frac{0.5}{100} - \left(\frac{-0.5}{100}\right)\right)$$

$$\Rightarrow I = 595.$$

27. FBD of block

$$\text{Net force} = ma$$

$$mg\sin 45 - \mu mg\cos 45 = ma$$

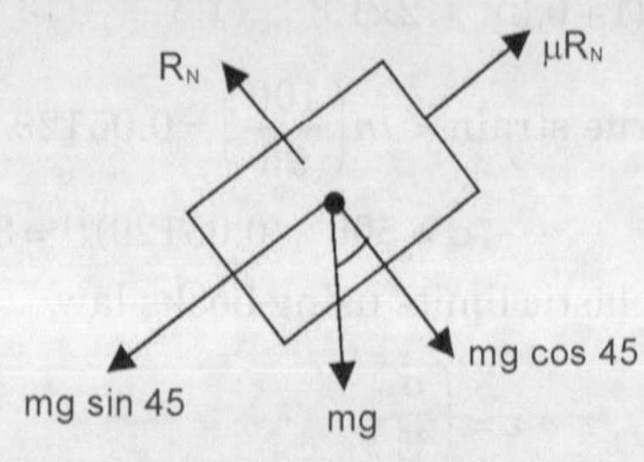

$$\Rightarrow \quad a = 3.46\ m/s^2$$

$$\Rightarrow \quad V^2 - u^2 = 2as$$

$$\Rightarrow \quad S = \frac{V^2}{2a} = \frac{20^2}{2\times 3.46} = 57.8\ m.$$

29. Taking moments about B

$$\Sigma M_B = 0$$

$$\Rightarrow \quad W\times 2 = R_A\times 3$$

$$\Rightarrow \quad R_A = \frac{600\times 2}{3} = 400$$

$$\sum H = 0$$

$$\Rightarrow \quad R_A = P = 400\ N$$

2

CHAPTER

Strength of Materials

MCQ TYPE QUESTIONS

2016

1. The cross sections of two hollow bars made of the same material are concentric circles as shown in the figure. It is given that $r_3 > r_1$ and $r_4 > r_2$, and that the areas of the cross-sections are the same. J_1 and J_2 are the torsional rigidities of the bars on the left and right, respectively. The ratio J_2/J_1 is

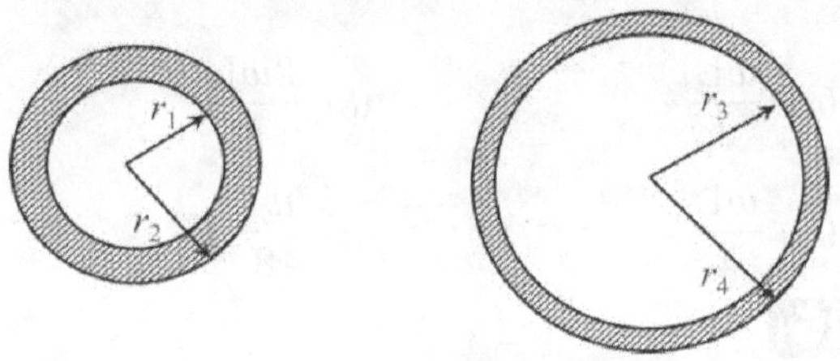

(a) > 1 (b) < 0.5

(c) =1 (d) between 0.5 and 1

2. A cantilever beam having square cross-section of side a is subjected to an end load. If a is increased by 19%, the tip deflection decreases approximately by

(a) 19% (b) 29%

(c) 41% (d) 50%

3. The principal stresses at a point inside a solid object are σ_1 = 100 MPa, σ_2 = 100 MPa and σ_3 = 0 MPa. The yield strength of the material is 200 MPa. The factor of safety calculated using Tresca (maximum shear stress) theory is n_T and the factor of safety calculated using von Mises (maximum distortional energy) theory is n_V. Which one of the following relations is TRUE?

(a) $n_T = (\sqrt{3}/2)n_V$ (b) $n_T = (\sqrt{3})n_V$

(c) $n_T = n_V$ (d) $n_V = (\sqrt{3})n_T$

4. A shaft with a circular cross-section is subjected to pure twisting moment. The ratio of the maximum shear stress to the largest principal stress is

(a) 2.0 (b) 1.0

(c) 0.5 (d) 0

5. A thin cylindrical pressure vessel with closed-ends is subjected to internal pressure. The ratio of circumferential (hoop) stress to the longitudinal stress is

(a) 0.25 (b) 0.50

(c) 1.0 (d) 2.0

6. The value of true strain produced in compressing a cylinder to half its original length is

(a) 0.69 (b) – 0.69

(c) 0.5 (d) – 0.5

7. A rigid horizontal rod of length 2L is fixed to a circular cylinder of radius R as shown in the figure. Vertical forces of magnitude P are applied at the two ends as shown in the figure. The shear modulus for the cylinder is G and the Young's modulus is E.

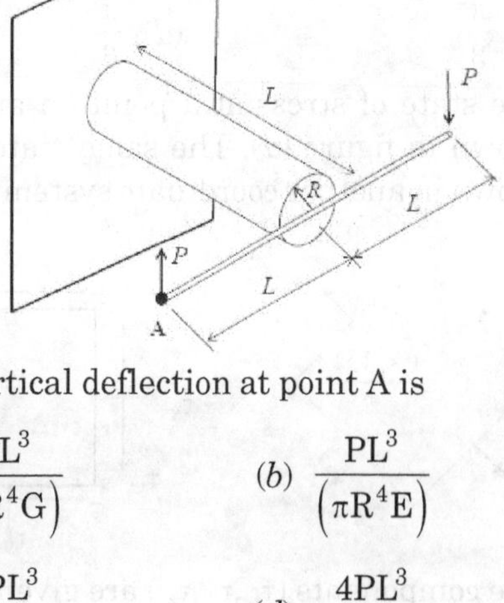

The vertical deflection at point A is

(a) $\frac{PL^3}{(\pi R^4 G)}$ (b) $\frac{PL^3}{(\pi R^4 E)}$

(c) $\frac{2PL^3}{(\pi R^4 E)}$ (d) $\frac{4PL^3}{(\pi R^4 G)}$

8. A simply supported beam of length 2L is subjected to a moment M at the mid-point x = 0 as shown in the figure. The deflection in the domain $0 \le x \le L$ is given by

$$W = \frac{-Mx}{12EIL}(L - x)(x + c),$$

where E is the Young's modulus, I is the area moment of inertia and c is a constant (to be determined).

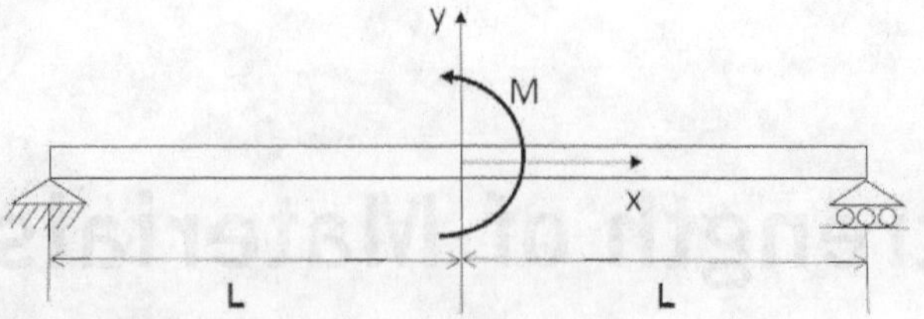

The slope at the center $x = 0$ is

(a) $\frac{ML}{(2EI)}$ (b) $\frac{ML}{(3EI)}$

(c) $\frac{ML}{(6EI)}$ (d) $\frac{ML}{(12EI)}$

9. The cross-sections of two solid bars made of the same material are shown in the figure. The square cross-section has flexural (bending) rigidity I_1, while the circular cross-section has flexural rigidity I_2. Both sections have the same cross-sectional area. The ratio I_1/I_2 is

(a) $\frac{1}{\pi}$ (b) $\frac{2}{\pi}$

(c) $\frac{\pi}{3}$ (d) $\frac{\pi}{6}$

10. The state of stress at a point on an element is shown in figure (a). The same state of stress is shown in another coordinate system in figure (b).

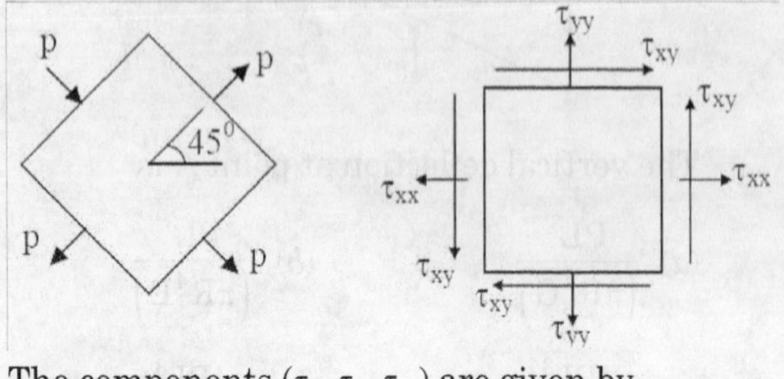

The components $(\tau_{xx}, \tau_{yy}, \tau_{xy})$ are given by

(a) $\left(p/\sqrt{2}, -p/\sqrt{2}, 0\right)$

(b) $(0, 0, p)$

(c) $\left(p, -p, p/\sqrt{2}\right)$

(d) $\left(0, 0, p/\sqrt{2}\right)$

11. Engineering strain of a mild steel sample is recorded as 0.100%. The true strain is

(a) 0.010% (b) 0.055%

(c) 0.099% (d) 0.101%

12. Two circular shafts made of same material, one solid (S) and one hollow (H), have the same length and polar moment of inertia. Both are subjected to same torque. Here, θ_S is the twist and τ_S is the maximum shear stress in the solid shaft, whereas θ_H is the twist and τ_H is the maximum shear stress in the hollow shaft. Which one of the following is TRUE?

(a) $\theta_S = \theta_H$ and $\tau_S = \tau_H$ (b) $\theta_S > \theta_H$ and $\tau_S > \tau_H$

(c) $\theta_S < \theta_H$ and $\tau_S < \tau_H$ (d) $\theta_S = \theta_H$ and $\tau_S < \tau_H$

13. A beam of length L is carrying a uniformly distributed load w per unit length. The flexural rigidity of the beam is EI. The reaction at the simple support at the right end is

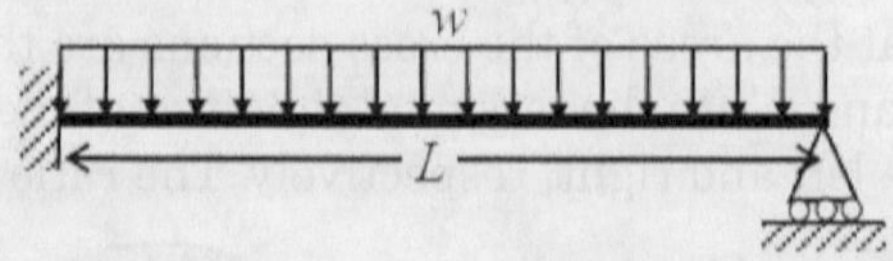

(a) $\frac{wL}{2}$ (b) $\frac{3wL}{8}$

(c) $\frac{wL}{4}$ (d) $\frac{wL}{8}$ z

2015

14. Which one of the following types of stress-strain relationship best describes the behavior of brittle materials, such as ceramics and thermosetting plastics, (σ = stress and ε = strain)?

(a)

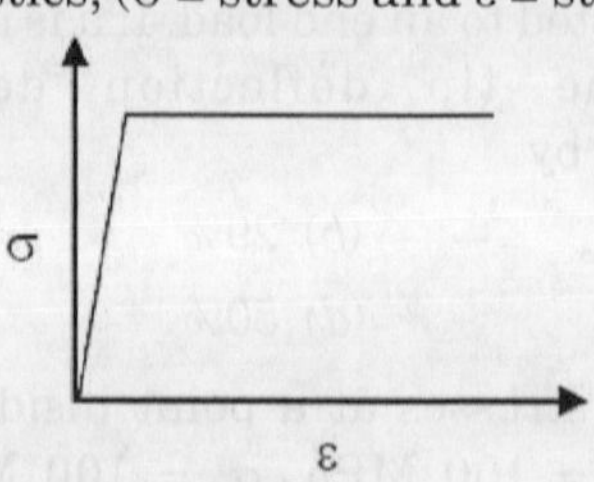

(b)

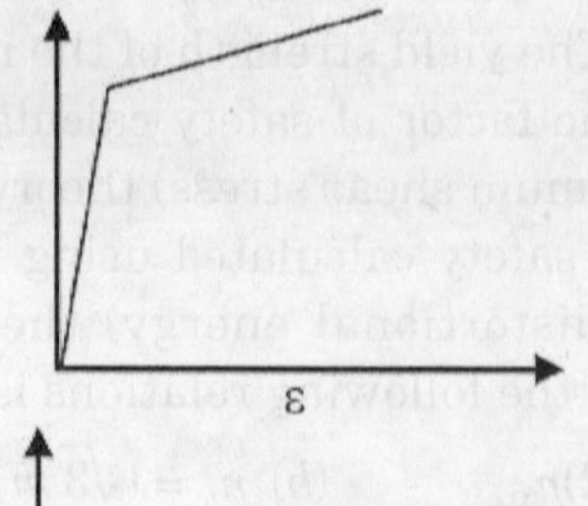

(c)

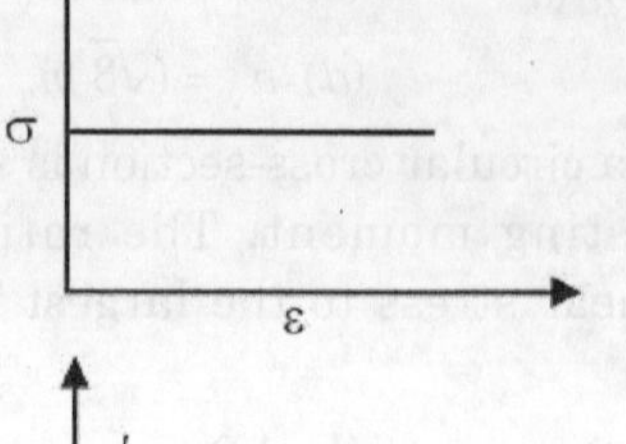

(d)

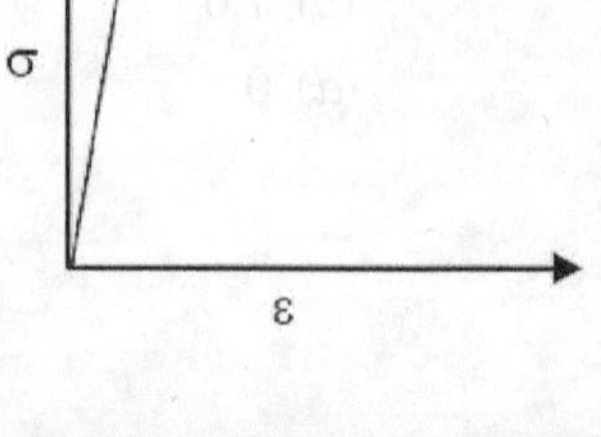

15. Which one of the following is the most conservative fatigue failure criterion?
(a) Soderberg (b) Modified Goodman
(c) ASME Elliptic (d) Gerber

16. For flow through a pipe of radius R, the velocity and temperature distribution are as follows:

$u(r,x) = C_1$, and $T(r,x) = C_2\left(1-\left(\frac{r}{R}\right)^3\right)$, where C_1 and C_2 are constants

The bulk mean temperature is given by $T_m = \frac{2}{u_m R^2}\int_0^R u(r,x)T(r,x)rdr$,

with U_m bing the mean velocity of flow. The value of T_m is
(a) $\frac{0.5C_2}{U_m}$ (b) $0.5C_2$
(c) $0.6C_2$ (d) $\frac{0.6C_2}{U_m}$

17. A gas is stored in a cylindrical tank of inner radius 7 m and wall thickness 50 mm. The gage pressure of the gas is 2MPa. The maximum shear stress (in MPa) in the wall is
(a) 35 (b) 70
(c) 140 (d) 280

18. In a plane stress condition, the components of stress at a point are $\sigma_x = 20$ MPa, $\sigma_y = 80$MPa and $\tau_{xy} = 40$MPa. The maximum shear stress (in MPa) at the point is ________.
(a) 20 (b) 25
(c) 50 (d) 100

19. The strain hardening exponent n of stainless steel SS 304 with distinct yield and UTS values undergoing plastic deformation is
(a) $n < 0$ (b) $n = 0$
(c) $0 < n < 1$ (d) $n = 1$

20. For the given fluctuating fatigue load, the values of stress amplitude and stress ratio are respectively

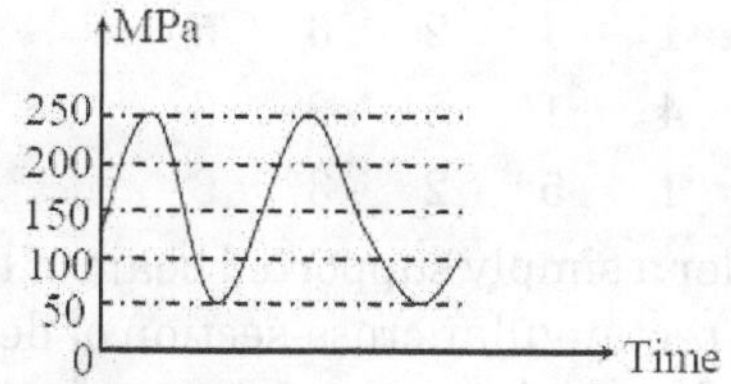

(a) 100 MPa and 5
(b) 250 MPa and 5
(c) 100 MPa and 0.20
(d) 250 MPa and 0.20

2014

21. A circular rod of length 'L' and area of cross-section 'A' has a modulus of elasticity 'E' and coefficient of thermal expansion 'α'. One end of the rod is fixed and other end is free. If the temperature of the rod is increased by ΔT, then
(a) stress developed in the rod is E α ΔT and strain developed in the rod is α ΔT
(b) both stress and strain developed in the rod are zero
(c) stress developed in the rod is zero and strain developed in the rod is α ΔT
(d) stress developed in the rod is E α ΔT and strain developed in the rod is zero

22. A steel cube, with all faces free to deform, has Young's modulus, E, Poisson's ratio, v, and coefficient of thermal expansion, α. The pressure (hydrostatic stress) developed within the cube, when it is subjected to a uniform increase in temperature, ΔT, is given by
(a) 0 (b) $\frac{\alpha(\Delta T)E}{1-2v}$
(c) $-\frac{\alpha(\Delta T)E}{1-2v}$ (d) $\frac{\alpha(\Delta T)E}{3(1-2v)}$

23. A thin plate of uniform thickness is subject to pressure as shown in the figure below

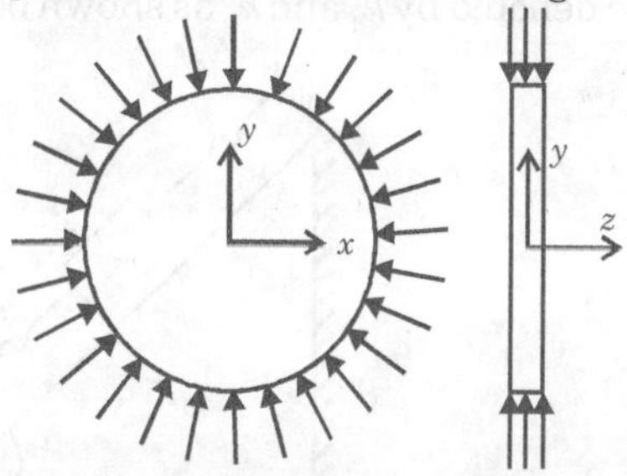

Under the assumption of plane stress, which one of the following is correct?
(a) Normal stress is zero in the z-direction
(b) Normal stress is tensile in the z-direction
(c) Normal stress is compressive in the z-direction
(d) Normal stress varies in the z-direction

24. The relationship between true strain (ε_T) and engineering strain (ε_E) in a uniaxial tension test is given as
(a) $\varepsilon_E = \ln(1 + \varepsilon_T)$
(b) $\varepsilon_E = \ln(1 - \varepsilon_T)$
(c) $\varepsilon_T = \ln(1 + \varepsilon_E)$
(d) $\varepsilon_T = \ln(1 - \varepsilon_E)$

25. The flexural rigidity (EI) of a cantilever beam is assumed to be constant over the length of the beam shown in figure. If a load P and bending moment PL/2 are applied at the free end of the beam then the value of the slope at the free end is

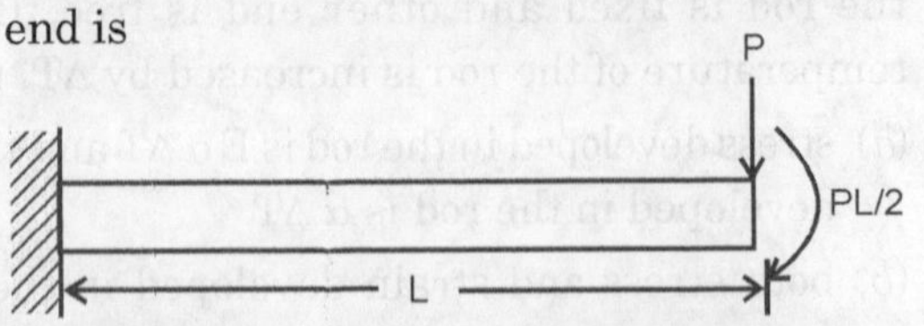

(*a*) $\frac{1}{2}\frac{PL^2}{EI}$ (*b*) $\frac{PL^2}{EI}$

(*c*) $\frac{3}{2}\frac{PL^2}{EI}$ (*d*) $\frac{5}{2}\frac{PL^2}{EI}$

26. A cantilever beam of length, L, with uniform cross-section and flexural rigidity, EI, is loaded uniformly by a vertical load, w per unit length. The maximum vertical deflection of the beam is given by

(*a*) $\frac{wL^4}{8EI}$ (*b*) $\frac{wL^4}{16EI}$

(*c*) $\frac{wL^4}{4EI}$ (*d*) $\frac{wL^4}{24EI}$

27. What is the natural frequency of the spring mass system shown below? The contact between the block and the inclined plane is frictionless. The mass of the block is denoted by m and the spring constants are denoted by k_1 and k_2 as shown below:

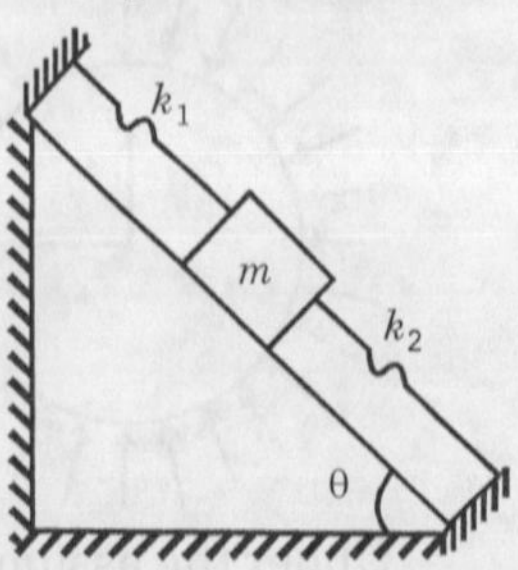

(*a*) $\sqrt{\frac{k_1 + k_2}{2m}}$

(*b*) $\sqrt{\frac{k_1 + k_2}{4m}}$

(*c*) $\sqrt{\frac{k_1 - k_2}{m}}$

(*d*) $\sqrt{\frac{k_1 + k_2}{m}}$

28. Consider the two states of stress as shown in configurations I and II in the figure below. From the standpoint of distortion energy (von-Mises) criterion, which one of the following statements is true?

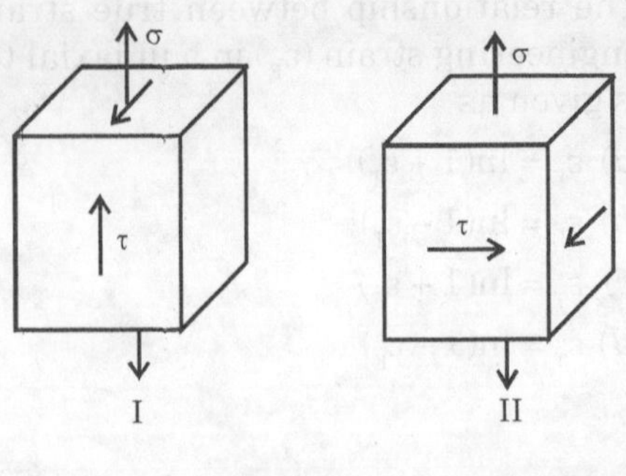

(*a*) I yields after II

(*b*) II yields after I

(*c*) Both yield simultaneously

(*d*) Nothing can be said about their relative yielding

29. Which one of following is **NOT** correct?

(*a*) Intermediate principal stress is ignored when applying the maximum principal stress theory

(*b*) The maximum shear stress theory gives the most accurate results amongst all the failure theories

(*c*) As per the maximum strain energy theory, failure occurs when the strain energy per unit volume exceeds a critical value

(*d*) As per the maximum distortion energy theory, failure occurs when the distortion energy per unit volume exceeds a critical value

30. The stress-strain curve for mild steel is shown in the figure given below. Choose the correct option referring to both figure and table.

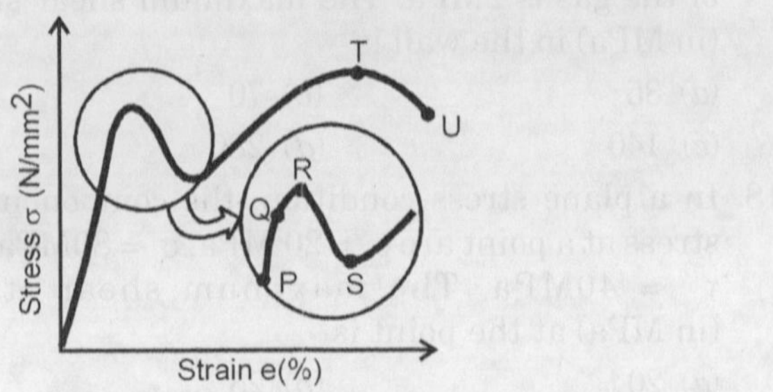

Point on the graph	Description of the point
P	1. Upper Yield Point
Q	2. Ultimate Tensile Strength
R	3. Proportionality Limit
S	4. Elastic Limit
T	5. Lower Yield Point
U	6. Failure

Codes:

	P	Q	R	S	T	U
(*a*)	1	2	3	4	5	6
(*b*)	3	1	4	2	6	5
(*c*)	3	4	1	5	2	6
(*d*)	4	1	5	2	3	6

31. Consider a simply supported beam of length, $50h$, with a rectangular cross-section of depth, h, and width, $2h$. The beam carries a vertical point load, P, at its mid-point. The ratio of the maximum shear stress to the maximum bending stress in the beam is

(*a*) 0.02 (*b*) 0.10

(*c*) 0.05 (*d*) 0.01

2013

32. A rod of length L having uniform cross-sectional area A is subjected to a tensile force P as shown in the figure below If the Young's modulus of the material varies linearly from E_1 to E_2 along the length of the rod the normal stress developed at the section-SS is

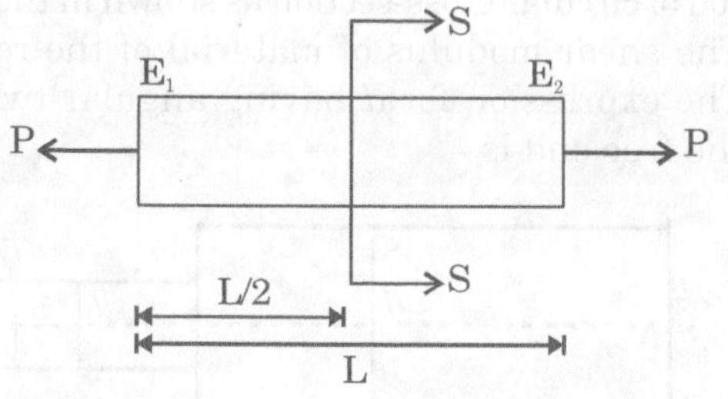

(*a*) $\frac{P}{A}$ (*b*) $\frac{P(E_1 - E_2)}{A(E_1 + E_2)}$

(*c*) $\frac{PE_2}{AE_1}$ (*d*) $\frac{PE_1}{AE_2}$

33. Two threaded bolts A and B of same material and length are subjected to identical tensile load. If the elastic strain energy stored in bolt A is 4 times that of bolt B and the mean diameter of bolt A is 12 *mm*, the mean diameter of bolt B in *mm* is

(*a*) 16 (*b*) 24

(*c*) 36 (*d*) 48

34. For a ductile material, toughness is a measure of

(*a*) resistance to scratching

(*b*) ability to absorb energy up to fracture

(*c*) ability to absorb energy till elastic limit

(*d*) resistance to indentation

35. If two nodes are observed at a frequency of 1800 rpm during whirling of a simply supported long slender rotating shaft, the first critical speed of the shaft in rpm is

(*a*) 200 (*b*) 450

(*c*) 600 (*d*) 900

36. A simply supported beam of length L is subjected to a varying distributed load sin $(3\pi x/L)$ Nm^{-1}, where the distance *x* is measured from the left support. The magnitude of the vertical reaction force in N at the left support is

(*a*) zero

(*b*) L/3π

(*c*) L/π

(*d*) 2L/π

2012

37. A thin walled spherical shell is subjected to an internal pressure. If the radius of the shell is increased by 1% and the thickness is reduced by 1%, with the internal pressure remaining the same, the percentage change in the circumferential (hoop) stress is

(*a*) 0 (*b*) 1

(*c*) 1.08 (*d*) 2.02

38. A cantilever beam of length *L* is subjected to a moment *M* at the free end. The moment of inertia of the beam cross section about the neutral axis is *I* and the Young's modulus is *E*. The magnitude of the maximum deflection is

(*a*) $\frac{ML^2}{2EI}$ (*b*) $\frac{ML^2}{EI}$

(*c*) $\frac{2ML^2}{EI}$ (*d*) $\frac{4ML^2}{EI}$

39. For a long slender column of uniform cross section, the ratio of critical buckling load for the case with both ends clamped to the case with both ends hinged is

(*a*) 1 (*b*) 2

(*c*) 4 (*d*) 8

40. The homogeneous state of stress for a metal part undergoing plastic deformation is

$$T = \begin{pmatrix} 10 & 5 & 0 \\ 5 & 20 & 0 \\ 0 & 0 & -10 \end{pmatrix}$$

where the stress component values are in *MPa*. Using von Mises yield criterion, the value of estimated shear yield stress, in *MPa* is

(*a*) 9.50 (*b*) 16.07

(*c*) 28.52 (*d*) 49.41

41. The state of stress at a point under plane stress condition is

$\sigma_{xx} = 40MPa$, $\sigma_{yy} = 100MPa$ and $\tau_{xy} = 40MPa$.

The radius of the Mohr's circle representing the given state of stress in MPa is

(*a*) 40 (*b*) 50

(*c*) 60 (*d*) 100

42. A solid steel cube constrained on all six faces is heated so that the temperature rises uniformly by ΔT. If the thermal coefficient of the material is α, Young's modulus is *E* and the Poisson's ratio is υ, the thermal stress developed in the cube due to heating is

(*a*) $-\frac{\alpha(\Delta T)E}{(1-2\upsilon)}$ (*b*) $-\frac{2\alpha(\Delta T)E}{(1-2\upsilon)}$

(*c*) $-\frac{3\alpha(\Delta T)E}{(1-2\upsilon)}$ (*d*) $-\frac{\alpha(\Delta T)E}{3(1-2\upsilon)}$

43. A solid circular shaft needs to be designed to transmit a torque of 50 *N.m.* If the allowable shear stress of the material is 140 MPa, assuming a factor of safety of 2, the minimum allowable design diameter in *mm* is

(*a*) 8 (*b*) 16
(*c*) 24 (*d*) 32

2011

44. A column has a rectangular cross-section of 10 mm × 20 mm and a length of 1 m. The slenderness ratio of the column is close to

(*a*) 200 (*b*) 346
(*c*) 477 (*d*) 1000

45. A simply supported beam PQ is loaded by a moment of 1 kN-m at the mid-span of the beam as shown in the figure. The reaction forces R_P and R_Q at supports P and Q respectively are

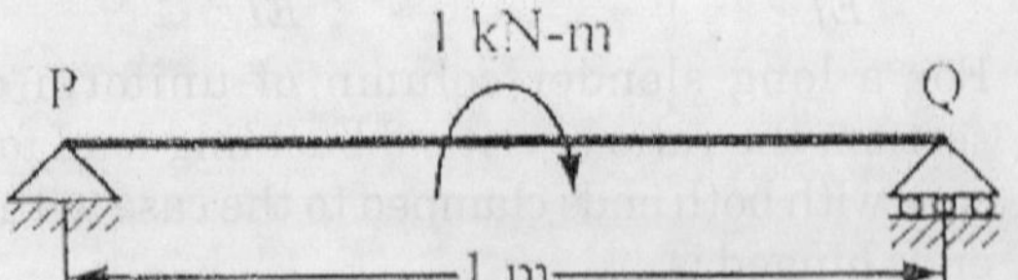

(*a*) 1 kN downward, 1 kN upward
(*b*) 0.5 kN upward, 0.5 kN downward
(*c*) 0.5 kN downward, 0.5 kN upward
(*d*) 1 kN upward, 1 kN upward

46. A thin cylinder of inner radius 500 mm and thickness 10 mm is subjected to an internal pressure of 5 MPa. The average circumferential (hoop) stress in MPa is

(*a*) 100 (*b*) 250
(*c*) 500 (*d*) 1000

47. Match the following criteria of material failure, under biaxial stresses σ_1 and σ_2 and yield stress σ_y, with their corresponding graphic representations :

P. Maximum-normal-stress criterion
Q. Maximum-distortion-energy criterion
R. Maximum-shear-stress criterion

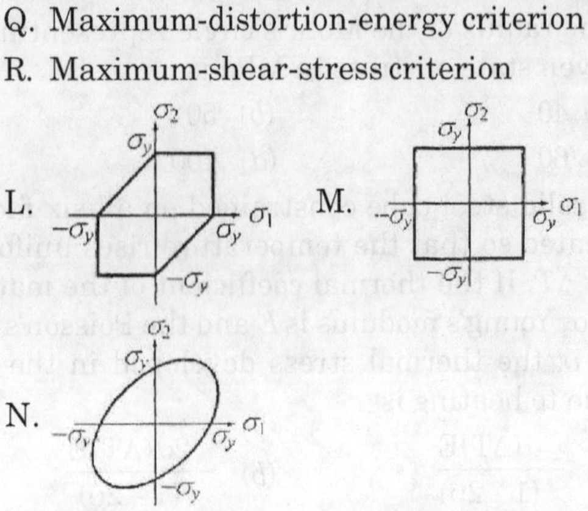

(*a*) P-M, Q-L, R-N (*b*) P-N, Q-M, R-L
(*c*) P-M, Q-N, R-L (*d*) P-N, Q-L, R-M

48. The shear strength of a sheet metal is 300 MPa. The blanking force required to produce a blank of 100 mm diameter from a 1.5 mm thick sheet is close to

(*a*) 45 kN (*b*) 70 kN
(*c*) 141 kN (*d*) 3500 kN

49. A torque T is applied at the free end of a stepped rod of circular cross-section as shown in the figure. The shear modulus of material of the rod is G. The expression for *d* having angular twist θ at the free end is

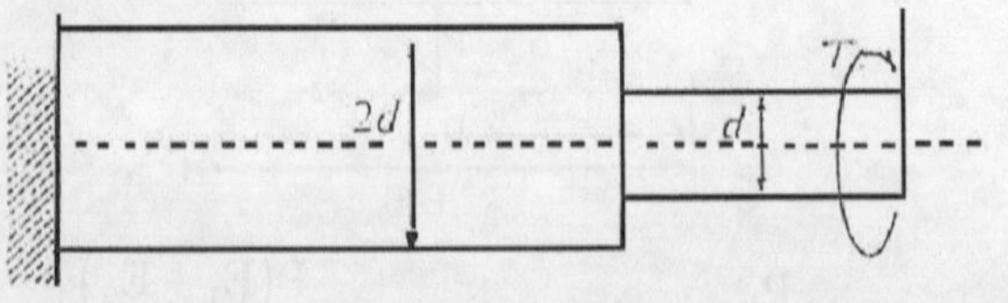

(*a*) $\left(\frac{32TL}{\pi\theta G}\right)^{\frac{1}{4}}$ (*b*) $\left(\frac{18TL}{\pi\theta G}\right)^{\frac{1}{4}}$

(*c*) $\left(\frac{16TL}{\pi\theta G}\right)^{\frac{1}{4}}$ (*d*) $\left(\frac{2TL}{\pi\theta G}\right)^{\frac{1}{4}}$

50. A cubic casting of 50 mm side undergoes volumetric solidification shrinkage and volumetric solid contraction of 4% and 6% respectively. No riser is used. Assume uniform cooling in all directions. The side of the cube after solidification and contraction is

(*a*) 48.32 mm (*b*) 49.90 mm
(*c*) 49.94 mm (*d*) 49.96 mm

2010

51. The state of plane-stress at a point is given by $\sigma_x = -200$MPa, $\sigma_y = 100$ MPa and $\tau_{xy} = 100$MPa. The maximum shear stress (in MPa) is

(*a*) 111.8 (*b*) 150.1
(*c*) 180.3 (*d*) 223.6

52. A bracket (shown in figure) is rigidly mounted on wall using four rivets. Each rivet is 6 mm in diameter and has an effective length of 12 mm.

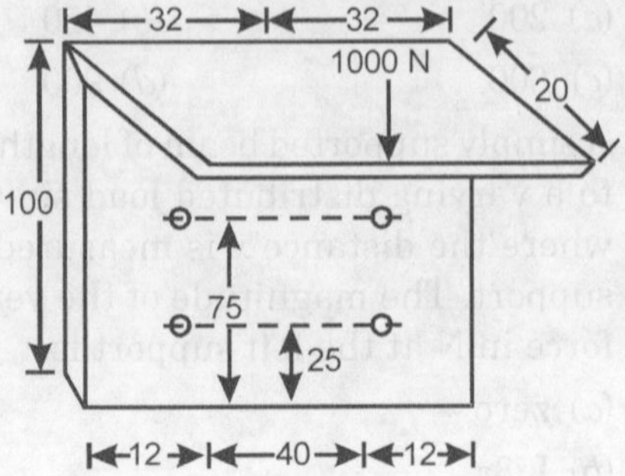

Direct shear stress (in MPa) in the most heavily loaded rivet is

(*a*) 4.4 (*b*) 8.8
(*c*) 17.6 (*d*) 35.2

2009

53. If principal stresses in a plane stress problem are $\sigma_1 = 100$ MPa, $\sigma_2 = 40$ MPa, then magnitude of the maximum shear stress (in MPa) will be

(*a*) 176.2 (*b*) 196.0
(*c*) 481.0 (*d*) 981.0

54. A solid shaft of diameter, d and length, L is fixed at both the ends. A torque, T_0 is applied at a distance $\frac{L}{4}$ from the left end as shown in the figure given below.

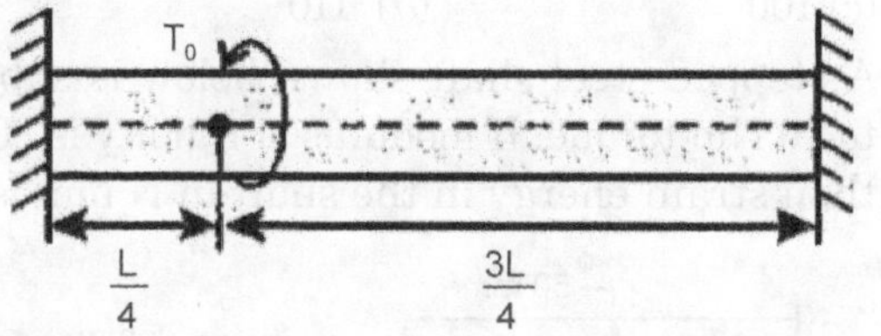

The maximum shear stress in the shaft is

(*a*) $\frac{16T_0}{\pi d^3}$ (*b*) $\frac{12T_0}{\pi d^3}$

(*c*) $\frac{8T_0}{\pi d^3}$ (*d*) $\frac{4T_0}{\pi d^3}$

55. A frame of two arms of equal length L is shown in the figure. The flexural rigidity of each arm of the frame is EI. The vertical deflection at the point of application of load P is

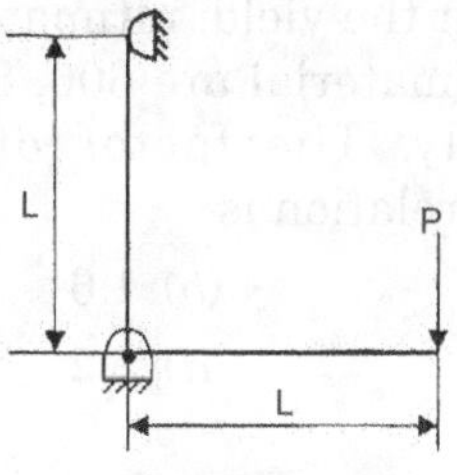

(*a*) $\frac{PL^3}{3EI}$ (*b*) $\frac{2PL^3}{3EI}$

(*c*) $\frac{2PL^3}{EI}$ (*d*) $\frac{4PL^3}{3EI}$

56. A uniform rigid rod of mass M and length L is hinged at one end as shown in the figure. A force P is applied at a distance of 2L/3 from the hinge so that the rod swings to the right. The reaction at the hinge is

(*a*) –P

(*b*) 0

(*c*) $\frac{P}{3}$

(*d*) $\frac{2P}{3}$

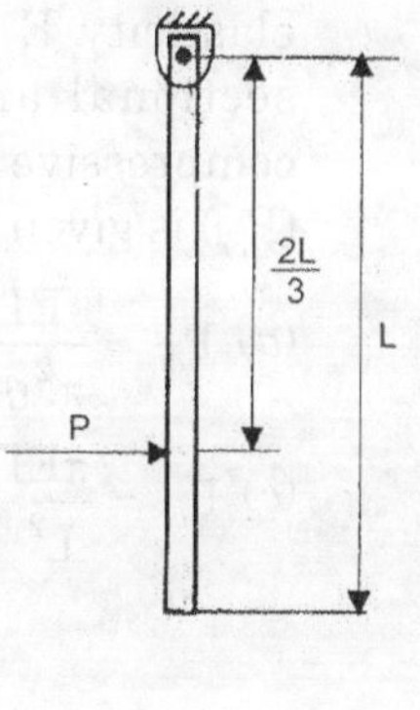

2008

57. The transverse shear stress acting in a beam of rectangular cross-section, subjected to a transverse shear load, is

(*a*) variable with maximum at the bottom of the beam

(*b*) variable with maximum at the top of the beam

(*c*) uniform

(*d*) variable with maximum of the neutral axis

58. A rod of length L and diameter D is subjected to a tensile load P. Which of the following is sufficient to calculate resulting change in diameter?

(*a*) Young's modulus

(*b*) Shear modulus

(*c*) Poisson's ratio

(*d*) Both Young's modulus and shear modulus

59. A cantilever type gate hinged at Q is shown in the figure. P and R are the centers of gravity of the cantilever part and the counterweight respectively. The mass of the cantilever part is 75 kg. The mass of the counterweight, for static balance, is

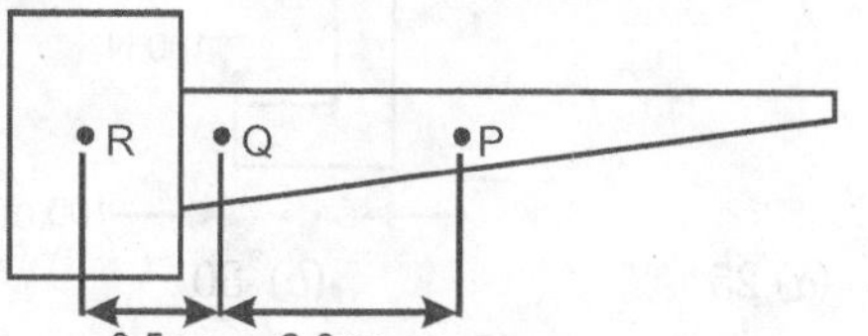

(*a*) 75 kg (*b*) 150 kg
(*c*) 225 kg (*d*) 300 kg

60. For the component loaded with a force F as shown in the figure, the axial stress at the corner point P is

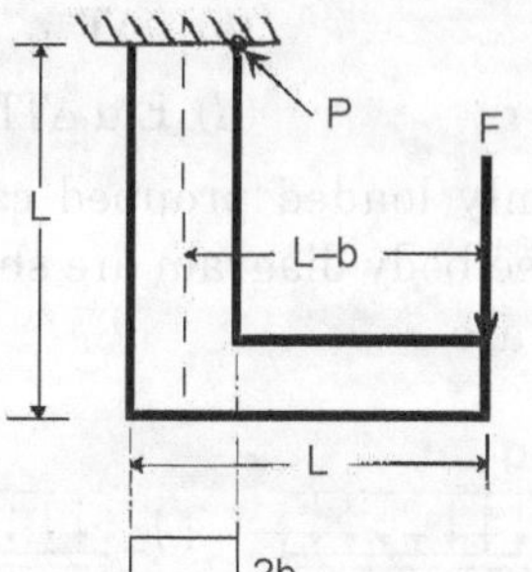

(*a*) $\frac{F(3L-b)}{4b^3}$ (*b*) $\frac{F(3L+b)}{4b^3}$

(*c*) $\frac{F(3L+4b)}{4b^3}$ (*d*) $\frac{F(3L-2b)}{4b^3}$

61. A solid circular shaft of diameter 100 mm is subjected to an axial stress of 50 MPa. It is further subjected to a torque of 10 kNm. The maximum principal stress experienced on the shaft is closest to

(*a*) 41 MPa (*b*) 82 MPa

(c) 164 MPa (*d*) 204 MPa

62. Consider a truss PQR loaded at P with a force F as shown in the figure.

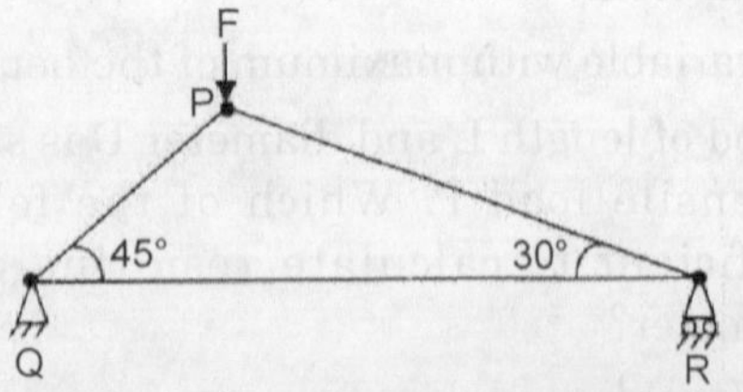

The tension in the member QR is

(*a*) 0.5 F (*b*) 0.63 F

(c) 0.73 F (*d*) 0.87 F

2007

63. In a simply-supported beam loaded as shown below, the maximum bending moment in Nm is

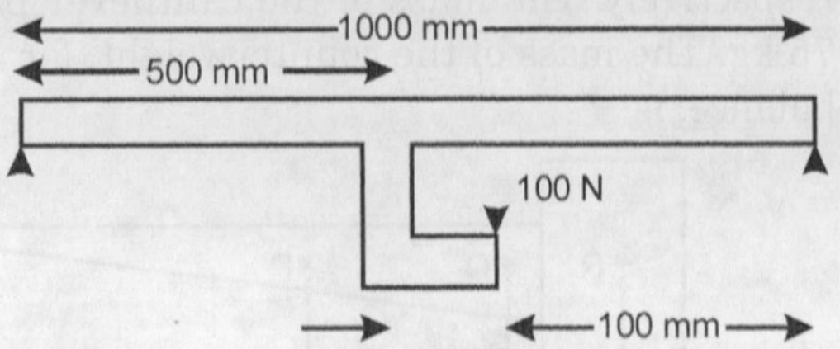

(*a*) 25 (*b*) 30

(c) 35 (*d*) 60

64. A steel rod of length L and diameter D, fixed at both ends, is uniformly heated to a temperature rise of ΔT. Young's modulus is E and co-efficient of linear expansion is α. The thermal stress in the rod is

(*a*) 0 (*b*) $\alpha \Delta T$

(c) $E \alpha \Delta T$ (*d*) $E \alpha \Delta TL$

65. A uniformly loaded propped cantilever beam and its free body diagram are shown below. The reactions are

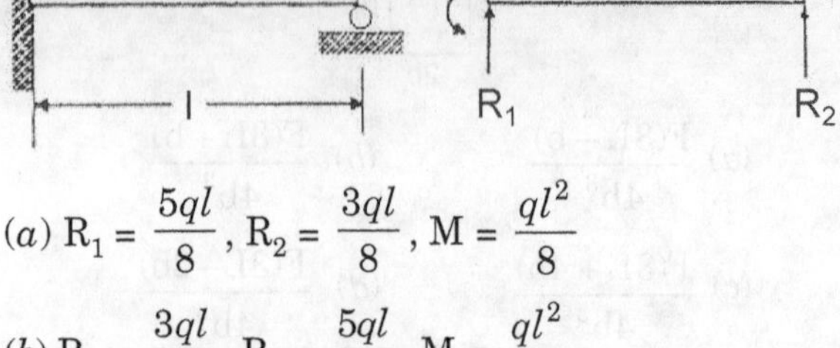

(*a*) $R_1 = \frac{5ql}{8}, R_2 = \frac{3ql}{8}, M = \frac{ql^2}{8}$

(*b*) $R_1 = \frac{3ql}{8}, R_2 = \frac{5ql}{8}, M = \frac{ql^2}{8}$

(*c*) $R_1 = \frac{5ql}{8}, R_2 = \frac{3ql}{8}, M = 0$

(*d*) $R_1 = \frac{3ql}{8}, R_2 = \frac{5ql}{8}, M = 0$

66. A 200 × 100 × 50 mm steel block is subjected to a hydrostatic pressure of 15 MPa.

The Young's modulus and Poisson's ratio of the material are 200 GPa and 0.3 respectively. The change in the volume of the block in mm^3 is

(*a*) 85 (*b*) 90

(c) 100 (*d*) 110

67. A stepped steel shaft shown below is subjected to 10 Nm torque. If modulus of rigidity is 80 GPa, then strain energy in the shaft in N mm is

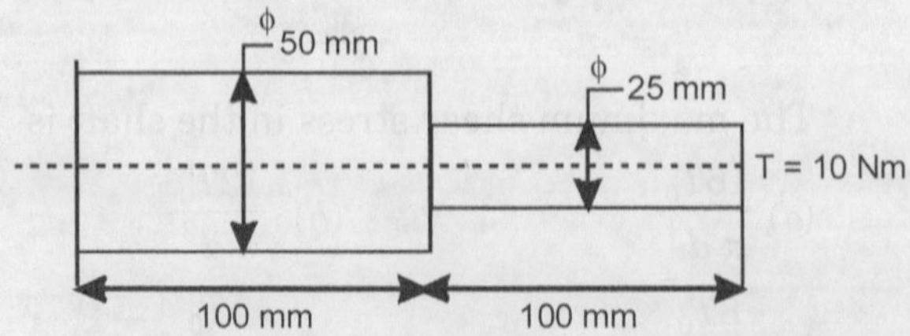

(*a*) 4.12 (*b*) 3.46

(c) 1.73 (*d*) 0.86

68. A thin spherical pressure vessel of 200 mm diameter and 1 mm thickness is subjected to an internal pressure varying from 4 to 8 MPa. Assume that the yield, ultimate, and endurance strength of material are 600, 800 and 400 MPa respectively. The factor of safety as per Goodman's relation is

(*a*) 2.0 (*b*) 1.6

(c) 1.4 (*d*) 1.2

2006

69. For a circular shaft of diameter d subjected to torque T, the maximum value of the shear stress is

(*a*) $\frac{64T}{\pi d^3}$ (*b*) $\frac{32T}{\pi d^3}$

(c) $\frac{16T}{\pi d^3}$ (*d*) $\frac{8T}{\pi d^3}$

70. A pin-ended column of length L, modulus of elasticity E and second moment of the cross-sectional area I is loaded centrically by a compressive load P. The critical buckling load (P_{cr}) is given by

(*a*) $P_{cr} = \frac{EI}{\pi^2 L^2}$ (*b*) $P_{cr} = \frac{\pi^2 EI}{3L^2}$

(c) $P_{cr} = \frac{\pi EI}{L^2}$ (*d*) $P_{cr} = \frac{\pi^2 EI}{L^2}$

71. According to Von-Mises' distortion energy theory, the distortion energy under three dimensional stress state is represented by

(*a*) $\frac{1}{2E}\left[\sigma_1^2+\sigma_2^2+\sigma_3^2-2\nu(\sigma_1\sigma_2+\sigma_3\sigma_2+\sigma_1\sigma_3\right]$

(*b*) $\frac{1-2\nu}{6E}\left[\sigma_1^2+\sigma_2^2+\sigma_3^2+2(\sigma_1\sigma_2+\sigma_3\sigma_2+\sigma_1\sigma_3)\right]$

(c) $\frac{1+\nu}{3E}\left[\sigma_1^2+\sigma_2^2+\sigma_3^2-(\sigma_1\sigma_2+\sigma_3\sigma_2+\sigma_1\sigma_3)\right]$

(*d*) $\frac{1}{3E}\left[\sigma_1^2+\sigma_2^2+\sigma_3^2-\nu(\sigma_1\sigma_2+\sigma_3\sigma_2+\sigma_1\sigma_3)\right]$

72. A steel bar of 40 mm × 40 mm square cross-section is subjected to an axial compressive load of 200 kN. If length of the bar is 2 m and E = 200 GPa, then elongation of the bar will be

(*a*) 1.25 mm (*b*) 2.70 mm

(*c*) 4.05 mm (*d*) 5.40 mm

73. A bar having a cross-sectional area of 700mm^2 is subjected to axial loads at the positions indicated. The value of stress in the segment QR is

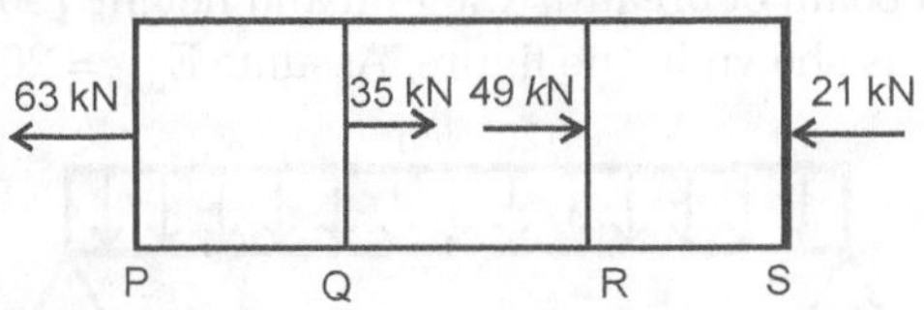

(*a*) 40 MPa (*b*) 50 MPa

(*c*) 70 MPa (*d*) 120 MPa

2005

74. A uniform, slender cylindrical rod is made of a homogeneous and isotropic material. The rod rests on a frictionless surface. The rod is heated uniformly. If radial and longitudinal thermal stresses are represented by σ_r and σ_z, respectively, then

(*a*) $\sigma_r = 0, \sigma_z = 0$ (*b*) $\sigma_r \neq 0, \sigma_z = 0$

(*c*) $\sigma_r = 0, \sigma_z \neq 0$ (*d*) $\sigma_r \neq 0, \sigma_z \neq 0$

75. Two identical cantilever beams are supported as shown, with their free ends in contact through a rigid roller. After the load P is applied, the free ends will have

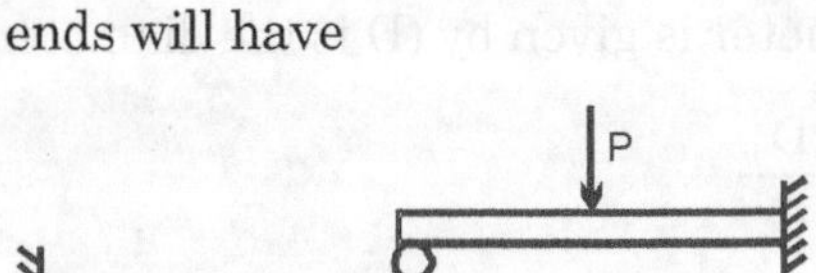

(*a*) equal deflections but not equal slopes

(*b*) equal slopes but not equal deflections

(*c*) equal slopes as well as equal deflections

(*d*) neither equal slopes nor equal deflections

76. The two shafts AB and BC, of equal length and diameters *d* and 2*d*, are made of the same material. They are joined at B through a shaft coupling, while ends A and C are built-in (cantilevered). A twisting moment T is applied to the coupling. If T_A and T_C represent the twisting moments at the ends A and C, respectively, then

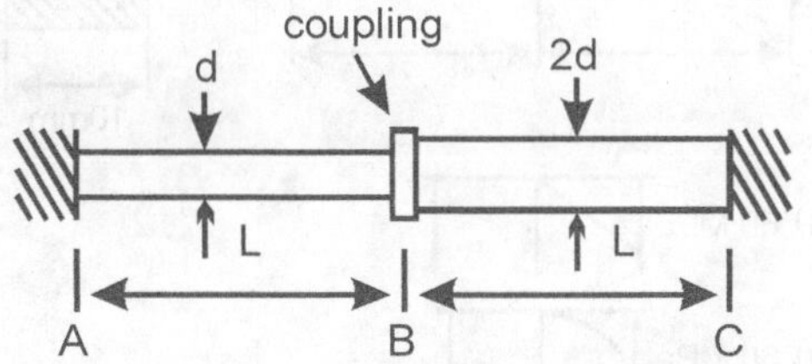

(*a*) $T_C = T_A$

(*b*) $T_C = 8T_A$

(*c*) $T_C = 16\ T_A$

(*d*) $T_A = 16\ T_C$

77. A beam is made up of two identical bars AB and BC, by hinging them together at B. The end A is built-in (cantilevered) and the end C is simply-supported. With the load P acting as shown, the bending moment at A is

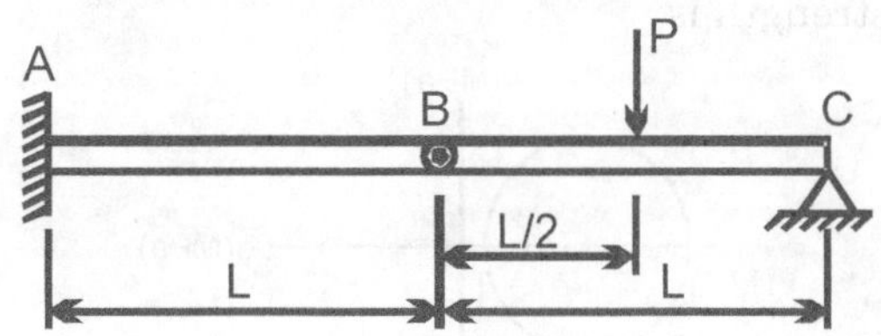

(*a*) zero (*b*) $\frac{PL}{2}$

(*c*) $\frac{3PL}{2}$ (*d*) indeterminate

78. A cantilever beam carries the anti-symmetric load shown, where w_0 is peak intensity of the distributed load. Qualitatively, the correct bending moment diagram for this beam is

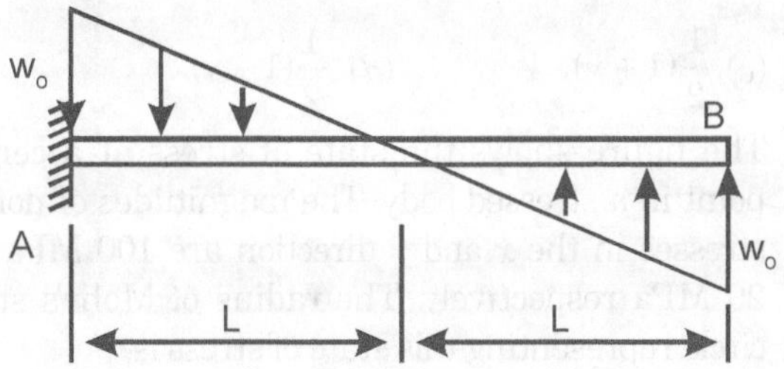

(*a*) 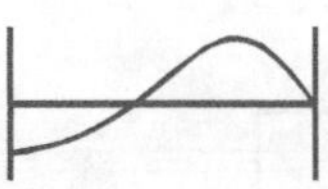 (*b*) 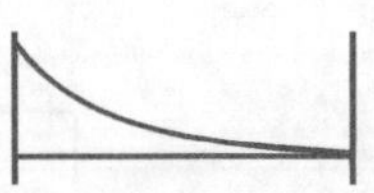

(*c*) 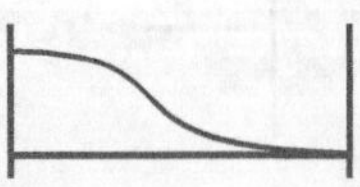 (*d*) 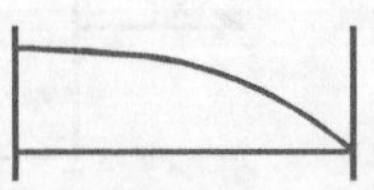

79. A cantilever beam has the square cross section 10 mm × 10 mm. It carries a transverse load of 10 N. Considering only bottom fibres of the beam, the correct representation of the longitudinal variation of the bending stress is

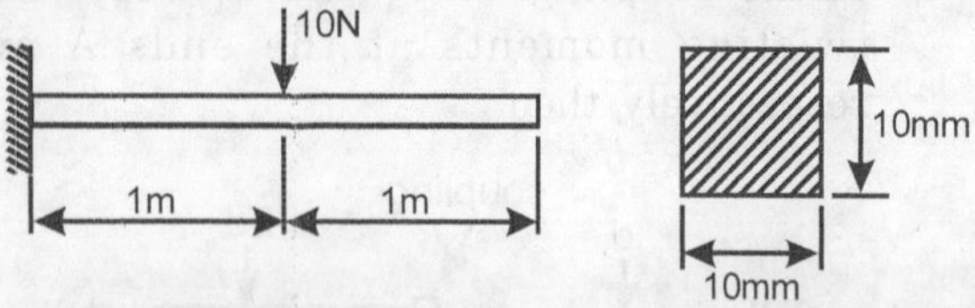

(a) 60 MPa 

(b) 60 MPa 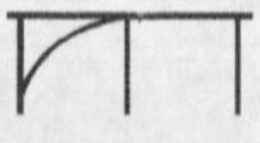

(c) 400 MPa 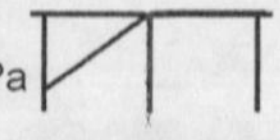

(d) 400 MPa 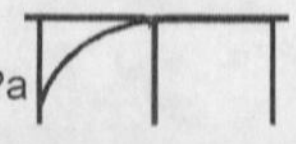

80. The Mohr's circle of plane stress for a point in a body is shown. The design is to be done on the basis of the maximum shear stress theory for yielding. Then, yielding will just begin if the designer chooses a ductile material whose yield strength is

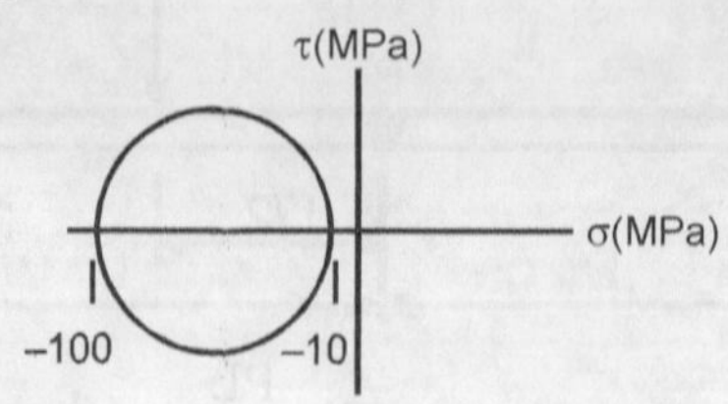

(a) 45 MPa (b) 50 MPa
(c) 90 MPa (d) 100 MPa

2004

81. In terms of Poisson's ratio (ν), the ratio of Young's modulus (E) to Shear modulus (G) of elastic materials is

(a) $2(1+\nu)$ (b) $2(1-\nu)$

(c) $\frac{1}{2}(1+\nu)$ (d) $\frac{1}{2}(1-\nu)$

82. The figure shows the state of stress at a certain point in a stressed body. The magnitudes of normal stresses in the x and y direction are 100 MPa and 20 MPa respectively. The radius of Mohr's stress circle representing this state of stress is

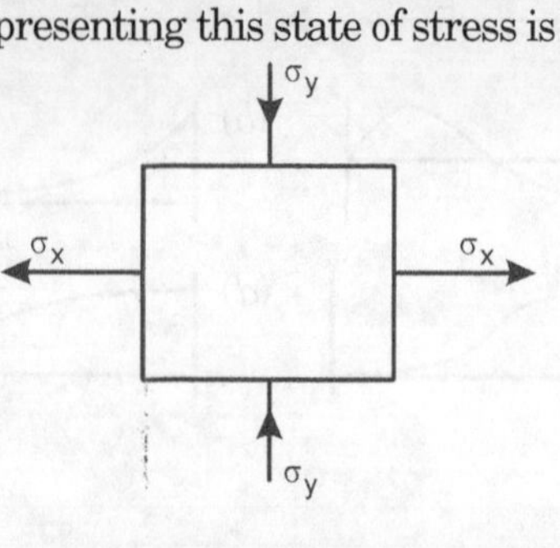

(a) 120 (b) 80
(c) 60 (d) 40

83. A torque of 10 Nm is transmitted through a stepped shaft as shown in figure. The torsional stiffnesses of individual sections of lengths MN, NO and OP are 20 Nm/rad, 30 Nm/rad and 60 Nm/rad respectively. The angular deflection between the ends M and P of the shaft is

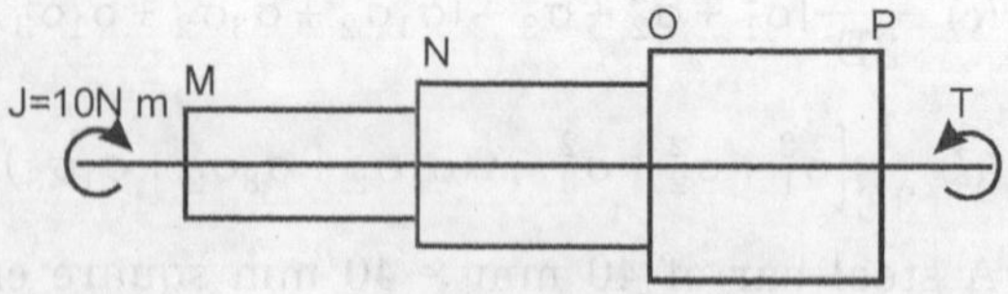

(a) 0.5 rad (b) 1.0 rad
(c) 5.0 rad (d) 10.0 rad

84. A solid circular shaft of 60 mm diameter transmits a torque of 1600 N.m. The value of maximum shear stress developed is

(a) 37.72 MPa (b) 47.72 MPa
(c) 57.72 MPa (d) 67.72 MPa

Common Data Q. 85-86

A steel beam of breadth 120 mm and height 750 mm is loaded as shown in the figure. Assume E_{steel} = 200 GPa.

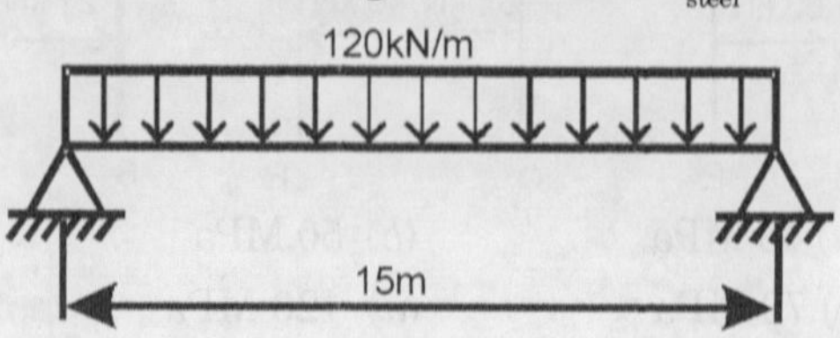

85. The beam is subjected to a maximum bending moment of

(a) 3375 kNm
(b) 4750 kNm
(c) 6750 kNm
(d) 8750 kNm

86. The value of maximum deflection of the beam is

(a) 93.75 mm (b) 83.75 mm
(c) 73.75 mm (d) 63.75 mm

2003

87. The second moment of a circular area about the diameter is given by (D is the diameter).

(a) $\frac{\pi D^4}{4}$

(b) $\frac{\pi D^4}{16}$

(c) $\frac{\pi D^4}{32}$

(d) $\frac{\pi D^4}{64}$

88. A concentrated load of P acts on a simply supported beam of span L at a distance $\frac{L}{3}$ from the left support. The bending moment at the point of application of the load is given by

(*a*) $\frac{PL}{3}$ (*b*) $\frac{2PL}{3}$

(*c*) $\frac{PL}{9}$ (*d*) $\frac{2PL}{9}$

89. Two beams, one having square cross section and another circular cross-section, are subjected to the same amount of bending moment. If cross sectional area as well as material of both the beams are the same, then

(*a*) maximum bending stress developed in both the beams is the same

(*b*) circular beam experiences more bending stress than the square one

(*c*) square beam experiences more bending stress than the circular one

(*d*) both the beams will experience same deformation

90. Two identical circular rods of same diameter and same length are subjected to same magnitude of axial tensile force. One of the rods is made out of mild steel having the modulus of elasticity of 206 GPa. The other rod is made out of cast iron having the modulus of elasticity of 100 GPa. Assume both the materials to be homogeneous and isotropic and the axial force causes the same amount of uniform stress in both the rods. The stresses developed are withn the proportional limit of the respective materials. Which of the following observations is correct ?

(*a*) Both rods elongate by the same amount

(*b*) Mild steel rod elongates more than the cast iron rod

(*c*) Cast iron rod elongates more than the mild steel rod

(*d*) As the stresses are equal strains are also equal in both the rods

91. Maximum shear stress developed on the surface of a solid circular shaft under pure torsion is 240 MPa. If shaft diameter is doubled, then maximum shear stress developed corresponding to the same torque will be

(*a*) 120 MPa (*b*) 60 MPa

(*c*) 30 MPa (*d*) 15 MPa

92. A simply supported laterally loaded beam was found to deflect more than a specified value. Which of the following measures will reduce the deflection ?

(*a*) Increase area moment of inertia

(*b*) Increase span of the beam

(*c*) Select a different material having lesser modulus of elasticity

(*d*) Magnitude of the load to be increased

93. A shaft subjected to torsion experiences a pure shear stress τ on the surface. The maximum principal stress on the surface which is at 45° to the axis will have a value

(*a*) $\tau \cos 45°$ (*b*) $2\tau \cos 45°$

(*c*) $\tau \cos^2 45°$ (*d*) $2\,\tau \sin 45° \cos 45°$

Common Data Q. 94-95

The state of stress at a point "P" in a two dimensional loading is such that the Mohr's circle is a point located at 175 MPa on the positive normal stress axis.

94. Determine maximum and minimum principal stresses respectively from the Mohr's circle

(*a*) + 175 MPa, – 175MPa

(*b*) +175 MPa, + 175 MPa

(*c*) 0, –175 MPa

(*d*) 0, 0

95. Determine directions of maximum and minimum principal stresses at the point "P" from he Mohr's circle.

(*a*) 0, 90° (*b*) 90°, 0

(*c*) 45°, 135° (*d*) all directions

2002

96. The total area under the stress-strain curve of a mild steel specimen tested up to failure under tension is a measure of

(*a*) ductility (*b*) ultimate strength

(*c*) stiffness (*d*) toughness

97. The number of components in a stress tensor defining stress at a point in three dimensions is

(*a*) 3 (*b*) 4

(*c*) 6 (*d*) 9

98. The relationship between Young's modulus (E), Bulk modulus (K) and Poisson's ratio (μ) is given by

(*a*) $E = 3K(1-2\mu)$ (*b*) $K = 3E(1-2\mu)$

(*c*) $E = 3K(1-\mu)$ (*d*) $K = 3E(1-\mu)$

99. The ratio of Euler's buckling loads of columns with the same parameters having (*i*) both ends fixed, and (*ii*) both ends hinged is

(*a*) 2 (*b*) 4

(*c*) 6 (*d*) 8

100. A chalk stick is twisted to failed to failure by applying opposite torques T at the two ends. Take a square element ABCD with two sides parallel to the longitudinal axis of the stick.

(*i*) The principal stresses σ_1 and σ_2 will be

(*a*) $\frac{16T}{\pi d^3}, -\frac{16T}{\pi d^3}$

(*b*) $-\frac{16T}{\pi d^3}, \frac{16T}{\pi d^3}$

(*c*) $\frac{16T}{\pi d^3}, \frac{16T}{\pi d^3}$

(*d*) $-\frac{16T}{\pi d^3}, -\frac{16T}{\pi d^3}$

(*ii*) The angles of principal plane will be

(*a*) 45°, 135° (*b*) 55°, 125°

(*c*) 65°, 115° (*d*) 75°, 105°

101. A cantilever AB of length 'L' has fixed end A and free end B. It is loaded by applying a concentrated load W at the mid point C of the cantilever.

(*a*) Determine the deflection and slope at points C and B.

(*b*) Show deflections and slopes on the cantilever.

2001

102. The shapes of the bending moment diagram for a uniform cantilever beam carrying a uniformly distributed load over its length is

(*a*) a suraight line (*b*) a hyperbola

(*c*) an elipse (*d*) a parabola

103. The area moment of inertia of a square of size 1 unit about its diagonal is

(*a*) $\frac{1}{3}$ (*b*) $\frac{1}{4}$

(*c*) $\frac{1}{12}$ (*d*) $\frac{1}{6}$

104. Bars AB and BC, each of negleble mass, support load P as shown in the figure. In this arrangement,

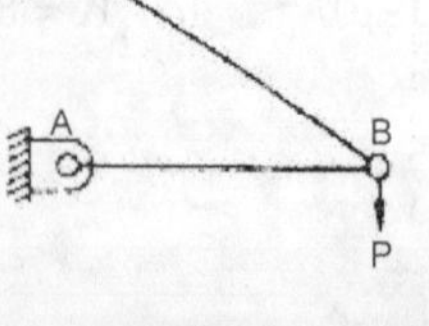

(*a*) bar AB is subjected to bending but bar BC is not subjected to bending

(*b*) bar AB is not subjected to bending but bar BC is subjected to bending

(*c*) neither bar AB nor bar BC is subjected to bending

(*d*) both bars AB and BC are subjected to bending

105. Two helical tensile springs of the same material and also having identical mean coil diameter and weight, have wire diameters *d* and *d*/2. The ratio of their stiffnesses is

(*a*) 1 (*b*) 4

(*c*) 64 (*d*) 128

106. The maximum principle stress for the stress state shown in the figure is

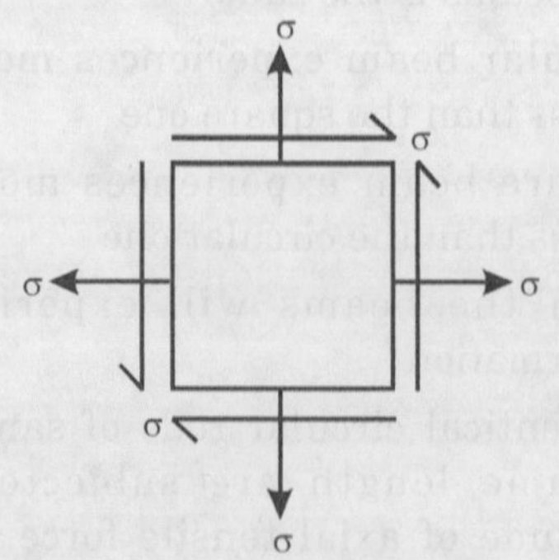

(*a*) σ (*b*) 2σ

(*c*) 3σ (*d*) 1.5σ

107. For the loading on truss shown in the figure, the force in member CD is

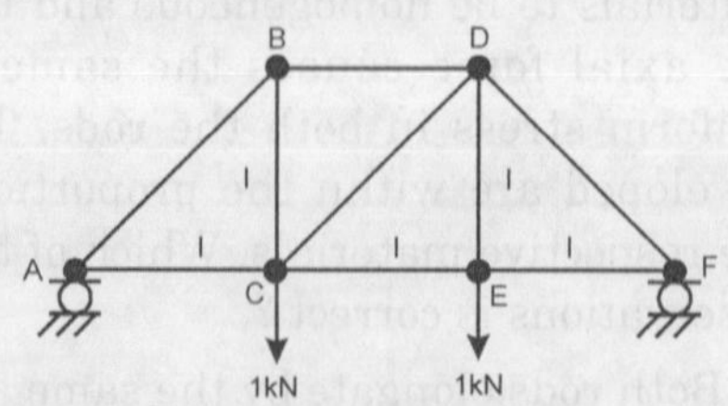

(*a*) 0 kN (*b*) 1 kN

(*c*) $\sqrt{2}$ kN (*d*) $1\sqrt{2}$ kN

Common Data Q. 108-109

Using area moment, the vertical deflection of the uniform beam AB at the flexural rigidity of the beam is EI.

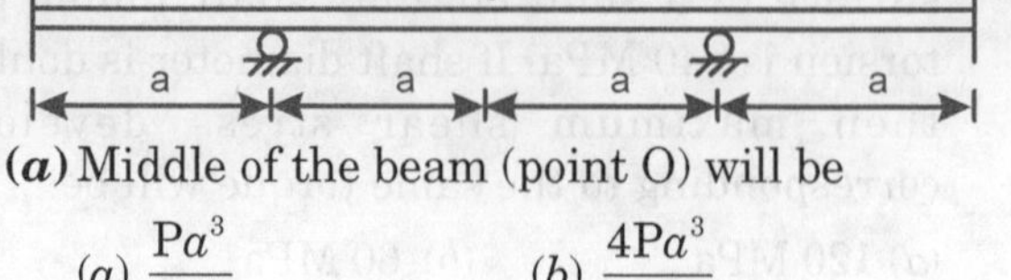

108. (*a*) Middle of the beam (point O) will be

(*a*) $\frac{Pa^3}{2EI}$ (*b*) $\frac{4Pa^3}{2EI}$

(*c*) $\frac{4Pa^3}{3EI}$ (*d*) $\frac{Pa^3}{3EI}$

109. (*b*) Left end of the beam (point A) will be

(*a*) $\frac{Pa^3}{2EI}$ (*b*) $\frac{4Pa^3}{2EI}$

(*c*) $\frac{4Pa^3}{3EI}$ (*d*) $\frac{Pa^3}{3EI}$

110. A belt drive shown in Fig. (*a*) has an angle of wrap of 160° on the smaller pulley. Adding an idler as shown in Fig. (*b*), increases the wrap angle to 200°. The slack side tension is the same in both cases and the centrifugal force is negligible. By what percentage is the torque capacity of the belt drive increased by adding the idler? (Use coefficient of friction μ = 0.3.)

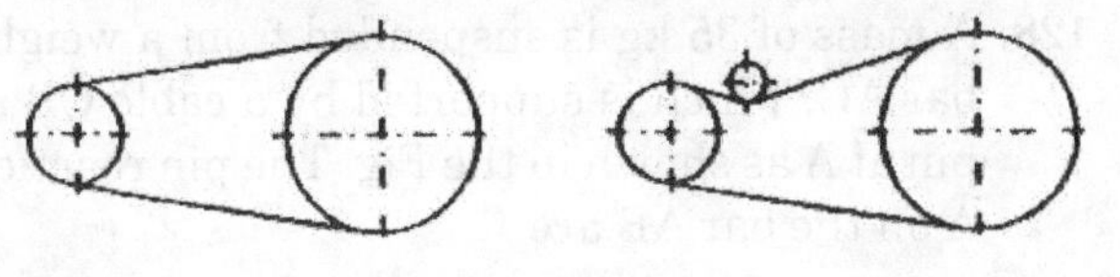

(a) (b)

(*a*) 32.25% (*b*) 36.25%

(*c*) 42.25% (*d*) 46.25%

2000

111. In a bi-axial stress problem, the stresses in *x* and *y* directions are σ_x = 200 MPa and σ_y= 100 MPa. The maximum principal stress in MPa, is

(*a*) 50 (*b*) 100

(*c*) 150 (*d*) 200

112. A steel shaft 'A' of diameter '*d*' and length'*l*' is subjected to a torque 'T'. Another shaft 'B' made of aluminium of the same diameter '*d*' and length 0.5 *l* is also subjected to the same torque 'T'. The shear modulus of steel is 2.5 times the shear modulus of aluminium. The shear stress in the steel shaft is 100 MPa. The shear stress in the aluminium shaft, in MPa, is

(*a*) 40 (*b*) 50

(*c*) 100 (*d*) 250

113. A simply supported beam carries a load 'P' through a bracket, as shown in the figure. The maximum bending moment in the beam is

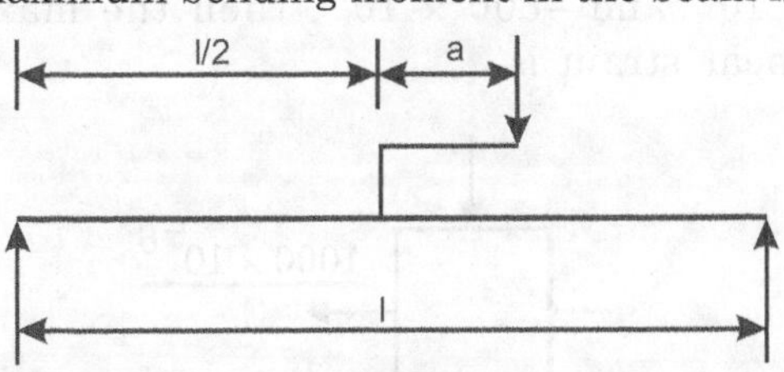

(*a*) $\frac{Pl}{2}$ (*b*) $\frac{Pl}{2} + \frac{aP}{2}$

(*c*) $\frac{Pl}{2} + aP$ (*d*) $\frac{Pl}{2} - aP$

1999

114. Which theory of failure will you use for aluminium components under steady loading

(*a*) Principal stress theory

(*b*) Principal strain theory

(*c*) Strain energy theory

(*d*) Maximum shear stress theory

115. A thin walled cylindrical vessel of wall thickness t and diameter *d* is filled with gas to a gauge pressure of *p*. The maximum shear stress on the vessel wall will then be

(*a*) $\frac{pd}{t}$ (*b*) $\frac{pd}{2t}$

(*c*) $\frac{pd}{4t}$ (*d*) $\frac{pd}{8t}$

116. A lean elastic beam of given flexural rigidity, EI, is loaded by a single force F as shown in the figure. How many boundary conditions are necessary to determine the deflected centre line of the beam?

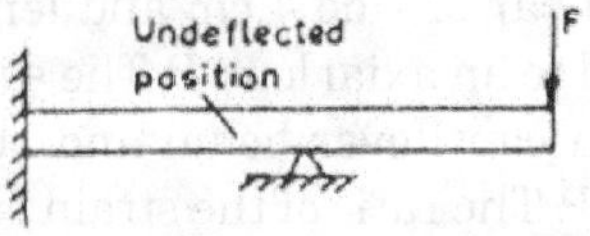

(*a*) 5 (*b*) 4

(*c*) 3 (*d*) 2

117. Consider the signboard mounting shown in figure given. The wind load acting perpendicular to the plane of the figure is F = 100 N. We wish to limit the deflection, due to bending, at point A of the hollow cylindrical pole of outer diameter 150 mm to 5 mm. Find the wall thickness for the pole. Assume E = 2.0×10^{11} N/m^2.

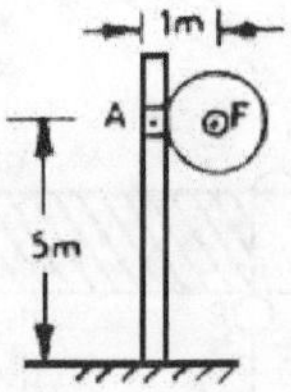

118. The peak bending stress at critical section of a component varies between 100 MN/m^2 and 300 MN/m^2. The ultimate tensile strength of the material is 700 MN/m^2, yield point in tension is 500 MN/m^2, endurance limit for reversed bending is 350 MN/m^2. Find the factor of safety.

1998

119. The buckling load for a column pinned at both ends is 10 kN. If the ends are fixed, the buckling load changes to

(*a*) 40kN (*b*) 2.5 kN

(*c*) 5kN (*d*) 20 kN

120. The normal stesses at a point are σ_x = 10 MPa and, σ_y = 2 MPa; the shear stress at this point is 4MPa. The maximum principal stress at this point is

(*a*) 16 MPa (*b*) 14 MPa
(*c*) 11 MPa (*d*) 10 MPa

121 The ratio of average shear stress to the maximum shear stress in a beam with a square cross-section is

(*a*) 1 (*b*) $\frac{2}{3}$
(*c*) $\frac{3}{2}$ (*d*) 2

122. The maximum principal strain in a thin cylindrical tank, having a radius of 25 cm and wall thickness of 5 mm when subjected to an internal pressure of 1 MPa, is (taking Young's midulus as 200 GPa and Poisson's ratio as 0.2)

(*a*) 2.25×10^{-4} (*b*) 2.25
(*c*) 2.25×10^{-6} (*d*) 22.5

123. A square bar of side 4 cm and length 100 cm is subjected to an axial load P. The same bar is then used as a cantilever beam and subjected to an end load P. The ratio of the strain energies, stored in the bar in the second case to that stored in the first case, is

(*a*) 16 (*b*) 400
(*c*) 1000 (*d*) 2500

124. For the composite beam shown in figure, flexural rigidities EI of AB and DC are equal to 10^5 N-cm², add EI of BD is 2×10^5 N-cm². Using moment - area theorem, determine the loacation and magnitude of maximum deflection between B and C.

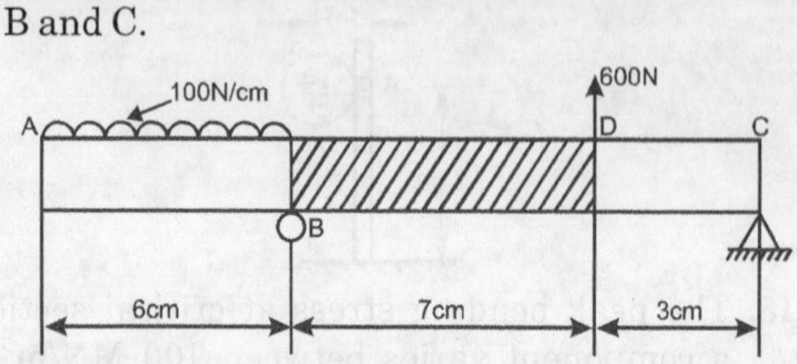

125. For the piping system shown in figure, $(EI)_{AB} = 10^{11}$ N-cm² and $(EI)_{BC} = 8 \times 10^{11}$ N-cm². The axial rigidity of DC, $(EA)_{DC} = 10^7$ N. Determine the load on DC.

126. A flat road has a curve segment with a radius of 100 m. While negotiating this curve, a vehicle slipped on its tyres as well as tried to rollover at a particular speed. Calculate this speed assuming a friction coefficient of 0.5. Also calculate the height of CG of the vehicle above ground if the tread (distance between the tyres at the front or rear) is 1.2 m.

1997

127. The alloying element mainly used to improve the endurance strength of steel materials is

(*a*) Nickel (*b*) Vanadium
(*c*) Molybdenum (*d*) Tungston

128. A mass of 35 kg is suspended from a weightless bar AC, which is supported by a cable CB and a pin at A as shown in the Fig. The pin reactions at A on the bar AB are

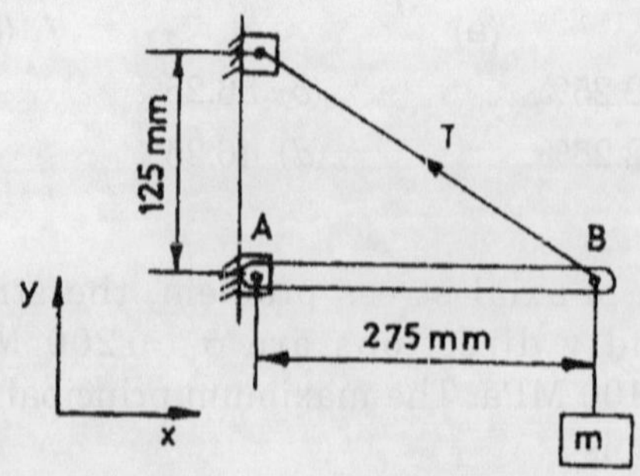

(*a*) R_x = 343.4 N, R_y = 755.4 N (*b*) R_x = 343.4 N, R_y = 0
(*c*) R_x = 755.4 N, R_y = 343.4 N (*d*) R_x = 755.5 N, R_y = 0

129. A small element at the critical section of a component is in a bi-axial state of stress with the two principal stresses being 360 MPa and 140 MPa. The maximum working stress according to Distortion Energy Theory is

(*a*) 220 MPa (*b*) 110 MPa
(*c*) 314 MPa (*d*) 330 MPa

1996

130. If the two principal strains at a point are 1000×10^{-6} and -600×10^{-6}, then the maximum shear strain is

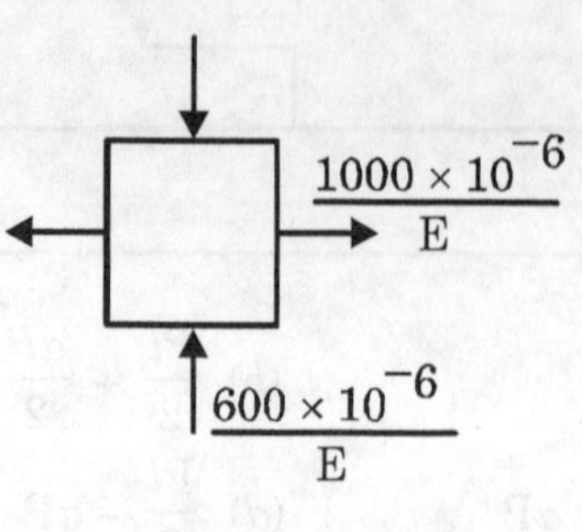

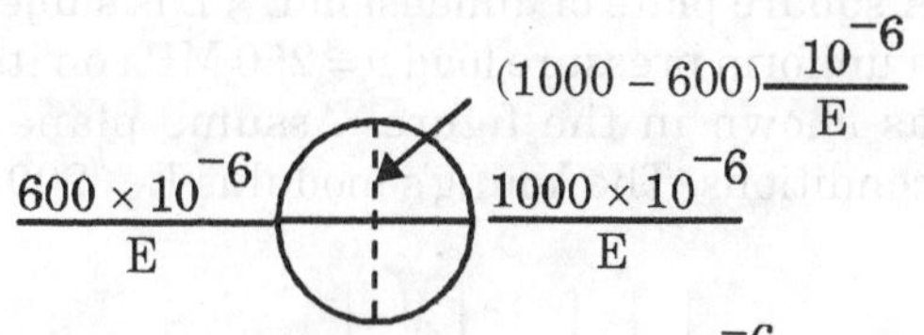

Maximum shear strain = 200×10^{-6}

(a) 800×10^{-6} (b) 500×10^{-6}

(c) 1600×10^{-6} (d) 200×10^{-6}

131. If the ratio of the diameter of rivet hole to the pitch of rivets is 0.25, then the tearing efficiency of the joint is

(a) 0.50 (b) 0.75

(c) 0.25 (d) 0.87

132. A solid shaft can resist a bending moment of 3.0 kNm and a twisting moment of 4.0 kNm together, then the maximum torque that can be applied is

(a) 7.0 kNm (b) 3.5 kNm

(c) 4.5 kNm (d) 5.0 kNm

133. A thick cylinder is subjected to an internal pressure of 60 MPa. If the hoop stress on the outer surface is 150 MPa, then the hoop stress on the internal surface is

(a) 105 MPa (b) 180 MPa

(c) 210 MPa (d) 135 MPa

134. AB and CD two uniform and identical bars of mass 10 kg each, as shown in the figure. The hinges at A and B are frictionless. The assembly is released from rest and motion occurs in the vertical plane. At the instant that the hinge B passes the point B, the angle between the two bars will be

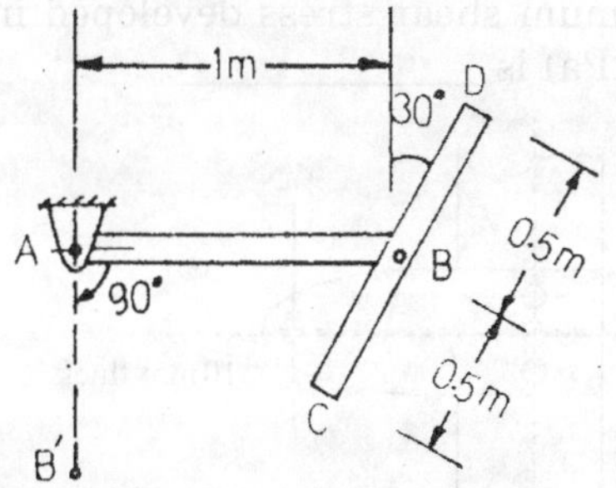

(a) 60 degrees (b) 37.4 degrees

(c) 30 degrees (d) 45 degrees

135. A thin cylinder of 100 mm internal diameter and 5mm thickness is subjected to an internal pressure of 10MPa and a torque of 2000 Nm. Calculate the magnitudes of the princial stresses.

136. Two bars AB and BC are connected by a friction less hinge at B. The assembly is supported and loaded as shown in Figure. Draw the shear force and bending moment diagrams for the combined beam AC, clearly labelling the important values. Also indicate your sign convention.

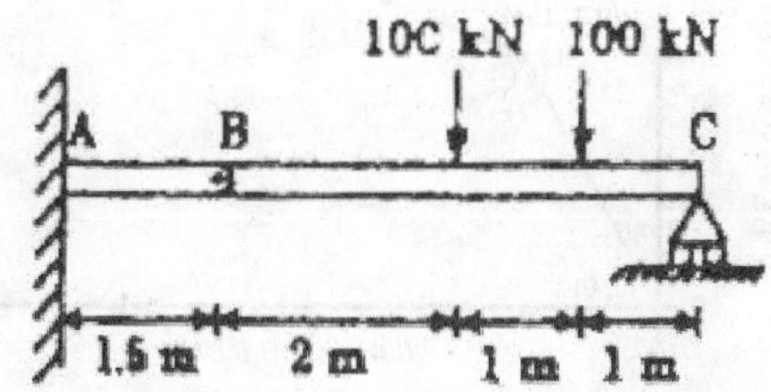

137. A simply supported beam is subjected to a single force P at a distance 'b' from one of the supports. Obtain the expression for the deflection under the load using Castigliano's theorem. How do you calculate deflection at the mid-point the beam ?

NUMERICAL TYPE QUESTIONS

2016

1. The figure shows cross-section of a beam subjected to bending. The area moment of inertia (in mm^4) of this cross-section about its baseis _________.

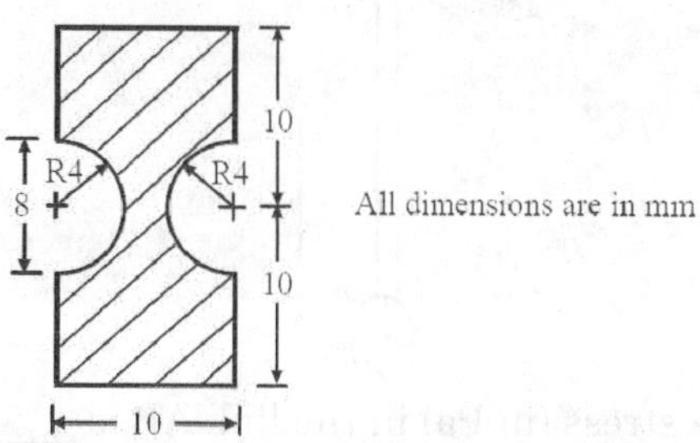

2. A simply-supported beam of length 3L is subjected to the loading shown in the figure.

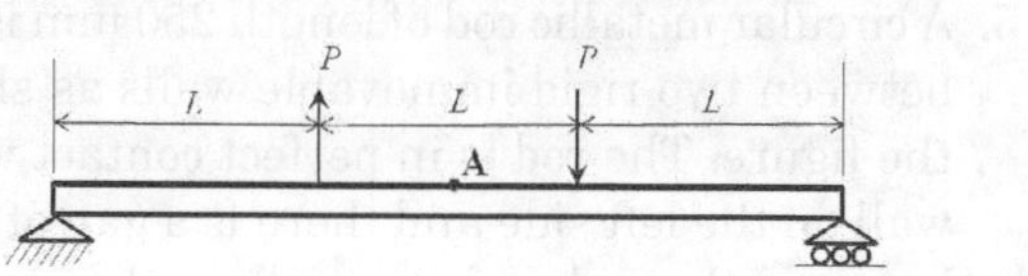

It is given that P = 1 N, L = 1 m and Young's modulus E = 200 GPa. The cross-section is a square with dimension 10 mm × 10 mm. The bending stress (in Pa) at the point A located at the top surface of the beam at a distance of 1.5L from the left end is _________.

(Indicate compressive stress by a negative sign and tensile stress by a positive sign.)

3. A hypothetical engineering stress-strain curve shown in the figure has three straight lines PQ, QR, RS with coordinates P(0,0), Q(0.2,100), R(0.6,140) and S(0.8,130). 'Q' is the yield point, 'R' is the UTS point and 'S' the fracture point.

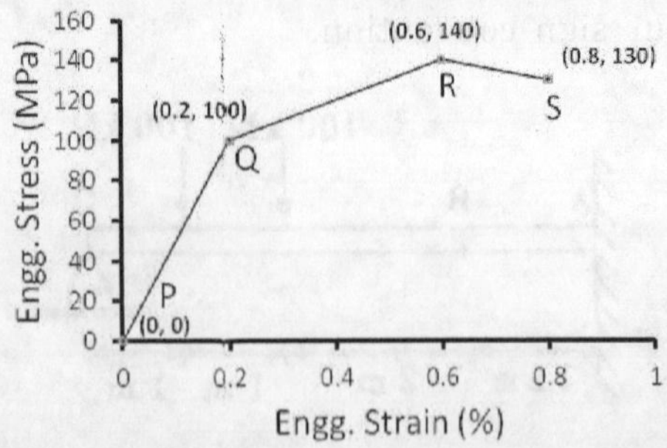

The toughness of the material (in MJ/m^3) is ______.

4. In the figure, the load P = 1 N, length L = 1 m, Young's modulus E = 70 GPa, and the cross-section of the links is a square with dimension 10 mm × 10 mm. All joints are pin joints.

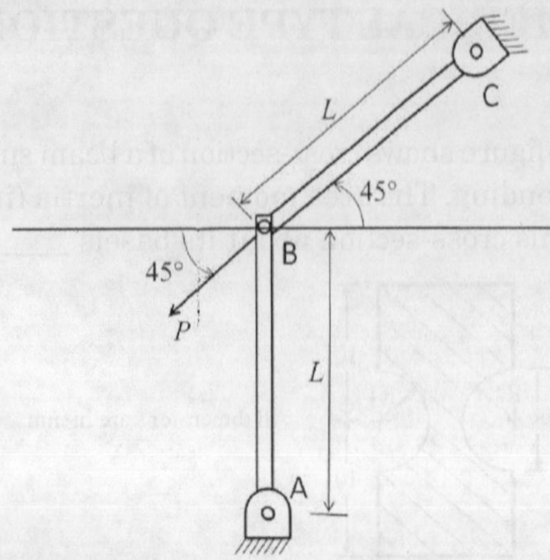

The stress (in Pa) in the link AB is ______ (Indicate compressive stress by a negative sign and tensile stress by a positive sign.)

5. A circular metallic rod of length 250 mm is placed between two rigid immovable walls as shown in the figure. The rod is in perfect contact with the wall on the left side and there is a gap of 0.2 mm between the rod and the wall on the right side. If the temperature of the rod is increased by 200° C, the axial stress developed in the rod is ______ MPa.

Young's modulus of the material of the rod is 200 GPa and the coefficient of thermal expansion is 10^{-5} per °C.

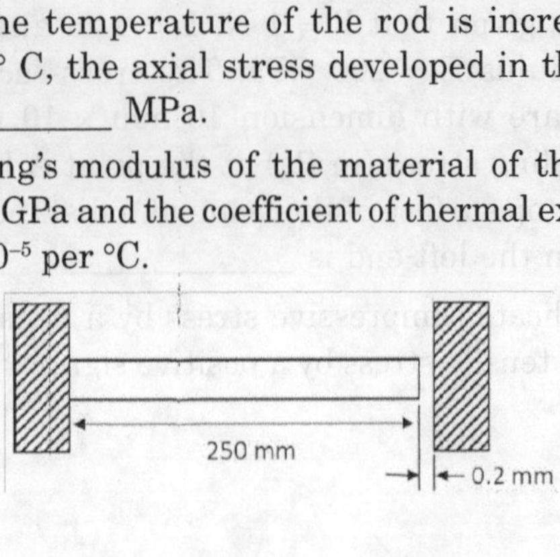

6. A square plate of dimension L × L is subjected to a uniform pressure load p = 250 MPa on its edges as shown in the figure. Assume plane stress conditions. The Young's modulus E = 200 GPa.

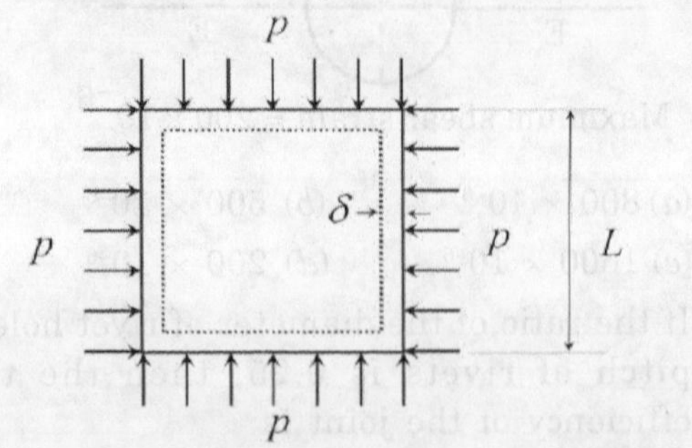

The deformed shape is a square of dimension L − 2 δ. If L = 2 m and δ = 0.001 m, the Poisson's ratio of the plate material is ______.

2015

7. Consider a steel (Young's modulus E = 200 GPa) column hinged on both sides. Its height is 1.0m and cross-section is 10 mm × 20 mm. The lowest Euler critical buckling load (in N) is ______.

8. Consider a stepped shaft subjected to a twisting moment applied at B as shown in the figure. Assume shear modulus, G = 77 GPa. The angle of twist at C (in degrees) is ______

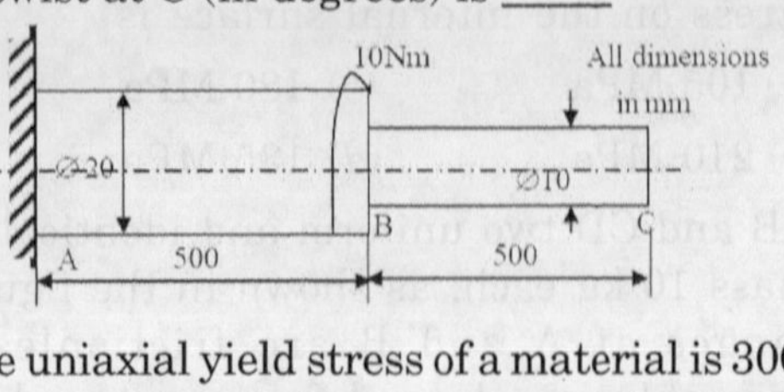

9. The uniaxial yield stress of a material is 300 MPa. According to von Mises criterion, the shear yield stress (in MPa) of the material is ______.

10. A cantilever bracket is bolted to a column using three M12 × 1.75 bolts, P, Q and R. The value of maximum shear stress developed in the bolt P (in MPa) is ______.

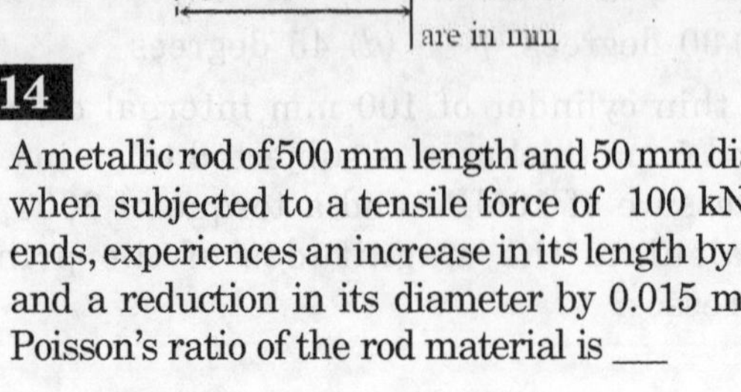

2014

11. A metallic rod of 500 mm length and 50 mm diameter, when subjected to a tensile force of 100 kN at the ends, experiences an increase in its length by 0.5 mm and a reduction in its diameter by 0.015 mm. The Poisson's ratio of the rod material is ___

12. The state of stress at a point is given by $\sigma_x = -6$ MPa, $\sigma_y = 4$ MPa and $\tau_{xy} = -8$ MPa. The maximum tensile stress (in MPa) at the point is ______

13. A 200 mm long, stress free rod at room temperature is held between two immovable rigid walls. The temperature of the rod is uniformly raised by 250°C. If the Young's modulus and coefficient of thermal expansion are 200 GPa and 1×10^{-5} /°C, respectively, the magnitude of the longitudinal stress (in MPa) developed in the rod is ______.

14. A metal rod of initial length L_0 is subjected to a drawing process. The length of the rod at any instant is given by the expression, $L(t) = L_0(1 + t^2)$, where t is the time in minutes. The true strain rate (in min^{-1}) at the end of one minute is ______

15. During pure orthogonal turning operation of a hollow cylindrical pipe, it is found that the thickness of the chip produced is 0.5 mm. The feed given to the zero degree rake angle tool is 0.2 mm/rev. The shear strain produced during the operation is ______

16. In a structure subjected to fatigue loading, the minimum and maximum stresses developed in a cycle are 200 MPa and 400 MPa respectively. The value of stress amplitude (in MPa) is ______

17. A disc clutch with a single friction surface has coefficient of friction equal to 0.3. The maximum pressure which can be imposed on the friction material is 1.5 MPa. The outer diameter of the clutch plate is 200 mm and its internal diameter is 100 mm. Assuming uniform wear theory for the clutch plate, the maximum torque (in N.m) that can be transmitted is ______

18. A rotating steel shaft is supported at the ends. It is subjected to a point load at the centre. The maximum bending stress developed is 100 MPa. If the yield, ultimate and corrected endurance strength of the shaft material are 300 MPa, 500 MPa and 200 MPa, respectively, then the factor of safety for the shaft is ______

19. Two solid circular shafts of radii R_1 and R_2 are subjected to same torque. The maximum shear stresses developed in the two shafts are τ_1 and τ_2. If $R_1/R_2 = 2$, then τ_2/τ_1 is ______

20. If the Poisson's ratio of an elastic material is 0.4, the ratio of modulus of rigidity to Young's modulus is ______

21. The number of independent elastic constants required to define the stress-strain relationship for an isotropic elastic solid is ______

22. A shaft is subjected to pure torsional moment. The maximum shear stress developed in the shaft is 100 MPa. The yield and ultimate strengths of the shaft material in tension are 300 MPa and 450 MPa, respectively. The factor of safety using maximum distortion energy (von-Mises) theory is ______

ANSWERS

MCQ Type Questions

1. (*a*)	**2.** (*d*)	**3.** (*c*)	**4.** (*b*)	**5.** (*d*)	**6.** (*b*)	**7.** (*d*)	**8.** (*c*)	**9.** (*c*)	**10.** (*b*)
11. (*c*)	**12.** (*d*)	**13.** (*b*)	**14.** (*d*)	**15.** (*a*)	**16.** (*c*)	**17.** (*c*)	**18.** (*c*)	**19.** (*c*)	**20.** (*c*)
21. (*c*)	**22.** (*a*)	**23.** (*a*)	**24.** (*c*)	**25.** (*b*)	**26.** (*a*)	**27.** (*d*)	**28.** (*c*)	**29.** (*b*)	**30.** (*c*)
31. (*d*)	**32.** (*a*)	**33.** (*b*)	**34.** (*b*)	**35.** (*c*)	**36.** (*b*)	**37.** (*d*)	**38.** (*a*)	**39.** (*c*)	**40.** (*b*)
41. (*b*)	**42.** (*a*)	**43.** (*b*)	**44.** (*b*)	**45.** (*a*)	**46.** (*b*)	**47.** (*c*)	**48.** (*c*)	**49.** (*b*)	**50.** (*a*)
51. (*c*)	**52.** (*b*)	**53.** (*c*)	**54.** (*a*)	**55.** (*d*)	**56.** (*c*)	**57.** (*d*)	**58.** (*d*)	**59.** (*d*)	**60.** (*d*)
61. (*b*)	**62.** (*b*)	**63.** (*b*)	**64.** (*c*)	**65.** (*a*)	**66.** (*b*)	**67.** (*c*)	**68.** (*c*)	**69.** (*c*)	**70.** (*d*)
71. (*c*)	**72.** (*a*)	**73.** (*a*)	**74.** (*a*)	**75.** (*c*)	**76.** (*c*)	**77.** (*b*)	**78.** (*c*)	**79.** (*a*)	**80.** (*c*)
81. (*a*)	**82.** (*c*)	**83.** (*b*)	**84.** (*a*)	**85.** (*a*)	**86.** (*a*)	**87.** (*d*)	**88.** (*d*)	**89.** (*b*)	**90.** (*c*)
91. (*c*)	**92.** (*a*)	**93.** (*d*)	**94.** (*b*)	**95.** (*d*)	**96.** (*d*)	**97.** (*c*)	**98.** (*a*)	**99.** (*b*)	**100.** (*)
101. (*)	**102.** (*d*)	**103.** (*c*)	**104.** (*c*)	**105.** (*c*)	**106.** (*b*)	**107.** (*a*)	**108.** (*a*)	**109.** (*c*)	**110.** (*c*)

111. (*d*)	**112.** (*c*)	**113.** (*c*)	**114.** (*d*)	**115.** (*d*)	**116.** (*d*)	**117.** (*)	**118.** (*)	**119.** (*a*)	**120.** (*c*)
121. (*b*)	**122.** (*a*)	**123.** (*d*)	**124.** (*)	**125.** (*)	**126.** (*)	**127.** (*b*)	**128.** (*d*)	**129.** (*c*)	**130.** (*d*)
131. (*b*)	**132.** (*d*)	**133.** (*c*)	**134.** (*c*)	**135.** (*)	**136.** (*)	**137.** (*)			

Numerical Type Questions

1. 21439.07	**2.** Zero	**3.** 0.85	**4.** (0)	**5.** (239.8)	**6.** (0.2)
7. 3285-3295	**8.** 0.236	**9.** (173.28)	**10.** 341	**11.** (0.29 to 0.31)	
12. (8.4 to 8.5)	**13.** (499 to 501)	**14.** (0.9 to 1.1)	**15.** (2.8 to 3.0)	**16.** (99 to 101)	
17. (529 to 532)	**18.** (1.9 to 2.1)	**19.** (7.9 to 8.1)	**20.** (0.35 to 0.36)	**21.** (1.9 to 2.1)	
22. (1.7 to 1.8)					

"*" Refers to Subjective Question.

EXPLANATIONS

MCQ TYPE QUESTIONS

1. Cross-sectional area is same so

$$A_1 = A_2$$

$$\Rightarrow \pi\left[r_2^2 - r_1^2\right] = \pi\left[r_4^2 - r_3^2\right]$$

$$\therefore r_2^2 - r_1^2 = r_4^2 - r_3^2$$

$$\frac{J_2}{J_1} = \frac{\frac{\pi}{2}\left[r_4^4 - r_3^4\right]}{\frac{\pi}{2}\left[r_2^4 - r_1^4\right]} = \frac{\left(r_4^2 + r_3^2\right)\left(r_4^2 - r_3^2\right)}{\left(r_2^2 + r_1^2\right)\left(r_2^2 - r_1^2\right)}$$

$$\frac{J_2}{J_1} = \frac{\left(r_4^2 + r_3^2\right)}{\left(r_2^2 + r_1^2\right)}$$

$$\frac{J_2}{J_1} > 1 \qquad \left[\because r_4 > r_2 \;\&\; r_3 > r_1\right]$$

2. 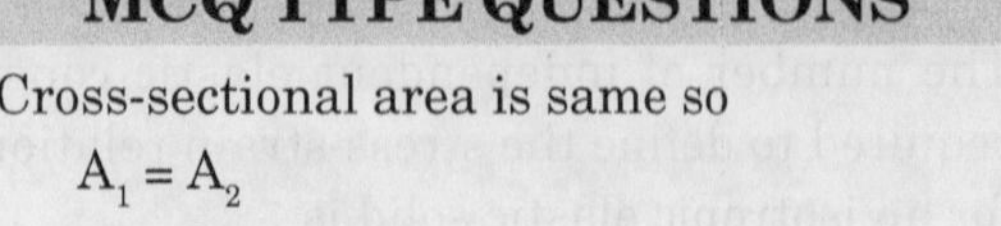

$$\Delta = \frac{wl^3}{3EI}$$

$$\frac{\Delta_2}{\Delta_1} = \frac{\left(\frac{a^4}{12}\right)}{\left(\frac{(1.19a)^4}{12}\right)} = \frac{1}{1.19^4} = 0.5$$

$\Delta_2 = 0.5\Delta_1$, reduced by 50 %.

3. According to Distortion Energy theory

$$\sqrt{\sigma_1^2 + \sigma_2^2 - \sigma_1\sigma_2} = \frac{S_{yt}}{n_V}$$

According to maximum shear stress theory

$$\sigma_1 - \sigma_2 = \frac{S_{yt}}{n_T}$$

$$\Rightarrow \quad n_T = \frac{200}{100} = 2$$

But $\sigma_1 = \sigma_2$, let it is σ_1

$$\sqrt{\sigma_1^2 + \sigma_2^2 - \sigma_1\sigma_2} = \frac{S_{yt}}{n_V}$$

$$\Rightarrow \quad n_V = \frac{200}{100} = 2$$

$$\therefore \quad n_T = n_V$$

4. 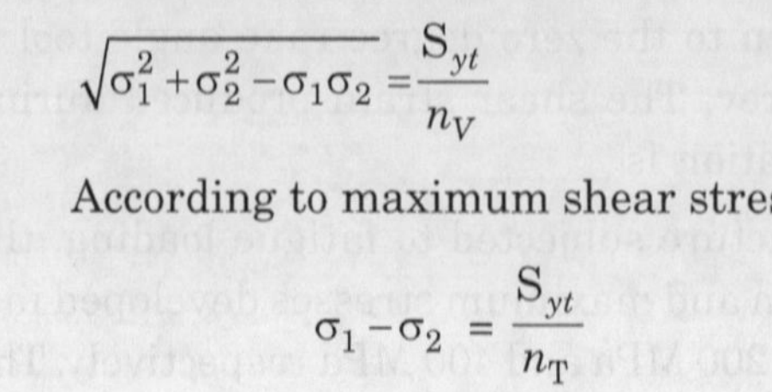

Mohr's circle

$$\tau_{max} = \tau_{xy}$$

$$\sigma_1 = \tau_{xy}$$

$$\therefore \quad \frac{\tau_{max}}{\sigma_1} = 1$$

5. Hoopstress $(\sigma_h) = \dfrac{Pd}{2t}$

Longitudional stress $(\sigma_\ell) = \dfrac{Pd}{4t}$

6. Engg. strain $(\varepsilon) = \dfrac{\text{Change in Length}}{\text{Original Length}}$

$$= \frac{\left(\frac{L}{2} - L\right)}{L} = -0.5$$

True strain $= \ln(1 + \varepsilon) = \ln(1 - 0.5)$

$= -0.69$

7. $T = P(2L) = 2PL$

$$\theta = \frac{TL}{GJ} = \frac{(2PL)(L)}{(G)\frac{\pi}{32}(2R)^4} = \frac{32 \times 2PL^2}{\pi GR^4(16)}$$

$$Q = \frac{4PL^2}{\pi GR^4}$$

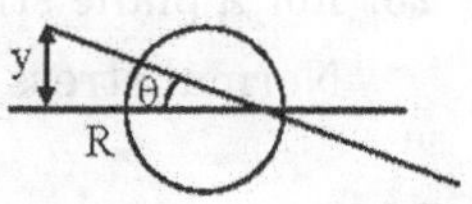

$$y = R\theta = \frac{4PL^3}{\pi GR^4}$$

8.

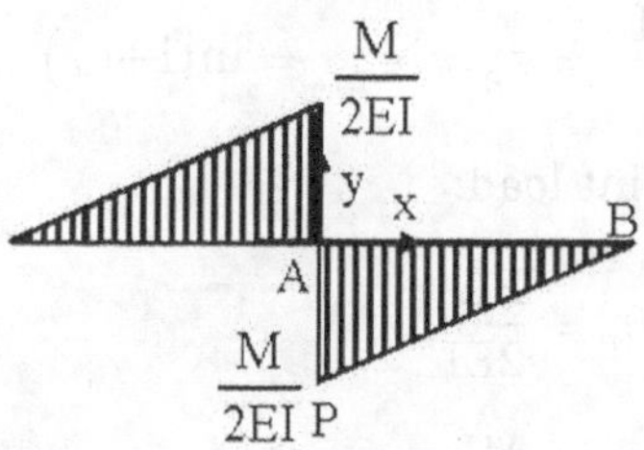

$$\Delta_{BA} = -\frac{1}{EI}\left(\begin{array}{l}\text{First moment of area}\\ \text{APB about B}\end{array}\right)$$

$$= -\frac{1}{EI}\left(\frac{M}{2} \times \frac{1}{2} \times L \times \frac{2L}{3}\right) = \frac{ML^2}{6EI}$$

Slope at 'A' $= \dfrac{\Delta_{BA}}{L} = \dfrac{ML}{6EI}$

$\because$ (at 'A' x = 0, W = 0 so slope $= \dfrac{\Delta_{BA}}{L}$)

9. Flexurd rigidity = EI

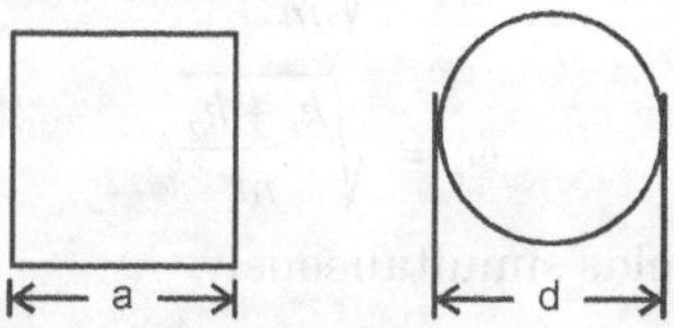

Both have same cross-section

$$a^2 = \frac{\pi}{4}d^2$$

$$\frac{I_1}{I_2} = \frac{E_1 \frac{a^2}{1^2}}{E_2 \frac{\pi d^4}{64}}$$

$E_1 = E_2$ both section of same material

$$\frac{I_1}{I_2} = \frac{\frac{\pi^2 d^4}{16} \times 12}{\frac{\pi}{64}d^4} = \frac{\pi}{3}$$

10.

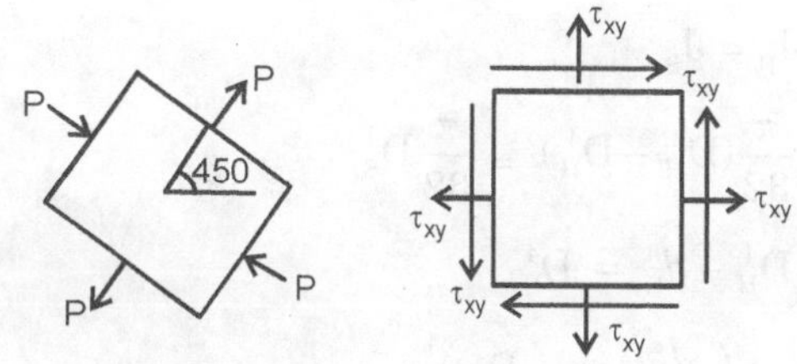

$$\sigma_\theta = \left(\frac{\sigma_x + \sigma_y}{2}\right) + \left(\frac{\sigma_x - \sigma_y}{2}\right)\cos 2\theta + T_{xg} \sin 2\theta$$

Here $\theta = -45$

$\sigma_\theta = Txy$

$\sigma_x = p$

$\sigma_y = -p$

$$\sigma_\theta = T_{xy} = \left(\frac{P - P}{2}\right) + \left(\frac{P + P}{2}\right)\cos 90°$$

$= 0$

When $\theta = +45$

$$\sigma_\theta = \left(\frac{P - P}{2}\right) + \left(\frac{P + P}{2}\right)\cos 90°$$

When $\theta = 45°$, $T_\theta = T_{xy}$

$$T_\theta = \left(\frac{\sigma_x - \sigma_y}{2}\right)\sin 2\theta - T_{xy} \cos 2\theta$$

$$T_\theta = T_{xy} = \left(\frac{P + P}{2}\right)\sin 90° = P$$

$\therefore$ $T_{xx}, T_{yy}, T_{xy} = 0, 0, p$

11. Engg. strain $\varepsilon = \dfrac{0.1}{100}$

True strain $\varepsilon_T = \ln(1 + \varepsilon)$

$$= \ln\left(1 + \frac{0.1}{100}\right)$$

$= 0.0009995$

ε_T in %age = 0.0995%

12. $\theta_S = \dfrac{TL}{CJ}$ (solid)

$\theta_H = \dfrac{TL}{CJ}$ (hollow)

$\theta_S = \theta_H = \theta$

Since material length , polar moment of inertia and applied torque all are same.

$$\tau_S = \frac{T}{J} \times r_S \text{ (or) } \tau_S \propto r_S$$

$$\tau_H = \frac{T}{J} \times r_H \text{ (or) } \tau_H \propto r_H$$

$$\therefore J_H = J_S$$

$$\frac{\pi}{32}(D_H^4 - D_H^4) = \frac{\pi}{32} D_S^4$$

$$D_H^4 - d_S^4 = D_S^4$$

$$1 - \left(\frac{d_S}{D_H}\right) = \left(\frac{D_S}{D_H}\right)^4$$

Since $\frac{d_S}{d_H} < 1$

$\therefore D_S < D_H$

$\therefore r_H > r_S$

$\therefore \tau_H > \tau_S$

13. $\frac{w\ell^4}{8EI} = \frac{R\ell^3}{3EI} \quad R = \frac{3w\ell}{8}$

16.

$$T_m = \frac{2}{u_m R^2} \int_0^R C_1 C_2 \left(1 - \left(\frac{r}{R}\right)^3\right) r\, dr$$

$$= \frac{2C_1C_2}{u_m R^2} \int_0^R \left[r - \frac{r^4}{R^3}\right] dr$$

$$= \frac{2C_1C_2}{u_m R^2}\left(\frac{3}{10}R^2\right) = \frac{0.6C_1C_2}{u_m}$$

Since $u(r, x)$ is constant, $u_m = C_1$

$$\therefore \quad t_m = 0.6C_2$$

17. $\sigma_c = \frac{pd}{2t} = \frac{2 \times 14}{2 \times 0.05} = 280\text{MPa}$

$\sigma_1 = \frac{pd}{4t} = \frac{2 \times 14}{4 \times 0.05} = 140\text{MPa}$

Maximum shear stress,

$$\tau_{max} = \frac{\sigma_1 - \sigma_2}{2} = \frac{\sigma_c - \sigma_1}{2} = 70 \text{ MPa}$$

Note : This stress τ_{max} is the maximum in-plane shear stress.

Absolute τ_{max} will be $\frac{\sigma_c}{2} = 140$ MPa

18. $\tau_{max} = \sqrt{\left(\frac{\sigma_y - \sigma_x}{2}\right)^2 + \tau_{xy}^2}$

$$= \sqrt{\left(\frac{80 - 20}{2}\right)^2 + 40^2} = 50 \text{ MPa}$$

19. n lies between 0 and 1.0 means material is a perfectly plastic solid, while 1 represents a 100% elastic solid.

20. Stress amplitude $= \frac{\sigma_{max} - \sigma_{min}}{2}$

$$= \frac{250 - 50}{2} = 100 \text{ Mpa}$$

Stress ratio $= \frac{\sigma_{min}}{\sigma_{max}} = \frac{50}{250} = 0.2$

21. Since one end of the rod is fixed and other is free to expand. Hence the Temperature stresses

$$= 0 \;\&\; \epsilon = \frac{\delta L}{L} \frac{\alpha \Delta T L}{L} = \alpha \Delta T$$

22. Since all the faces are free to expand the stresses due to temperature rise is equal to 0.

23. For a plane stress criteria.

Normal stress in Z direction = 0.

24.

$$\epsilon_{True} = \int_{L_o}^{L_f} \frac{dL}{L} = \ln\left[1 + \frac{\Delta L}{L_o}\right]$$

But $\frac{\Delta L}{L_o} = \epsilon_E \Rightarrow \epsilon_{True} = \ln(1 + \epsilon_E)$

25. Slope due to point load

$$\theta_1 = \frac{PL^2}{2EI}$$

$$\theta_2 = \frac{ML}{EI}$$

$$\theta = \frac{PL^2}{2EI} + \frac{PL}{2}\frac{L}{EI} = \frac{PL^2}{EI}$$

26. Maximum deflection for cantilever subjected to UDL is $\frac{Wl^4}{8EI}$.

27. It is parallel

$$k_{eq} = k_1 + k_2$$

$$\omega_n = \sqrt{\frac{k_{eq}}{m}}$$

$$\omega_n = \sqrt{\frac{k_1 + k_2}{m}}$$

28. Both yields simultaneously.

31. Given, $b = 2h$, $d = h$, $l = 50h$, force $= p$

$$\text{shear stress} = \frac{3}{2} \times \frac{\text{shear force}}{d \times b}$$

$$= \frac{3}{2} \times \frac{p}{h \times 2h} = \frac{3p}{4h^2}$$

$$\text{Bending stress} = \frac{M}{1} \times y = \frac{p \times 1/2}{\frac{bd^3}{12}} \times \frac{d}{2}$$

$$= \frac{p \times 50h/2 \times \frac{h}{2}}{\frac{(2h)h^3}{12}}$$

$$= \frac{\frac{25}{2} \times 6ph^2}{h^2} = \frac{75p}{h^2}$$

$$\frac{\text{Shear stress}}{\text{Bending stress}} = \frac{\frac{3P}{4h^2}}{\frac{75p}{h^2}} = \frac{3}{4 \times 75} = 0.01.$$

32. At section – 55

The left side of rod

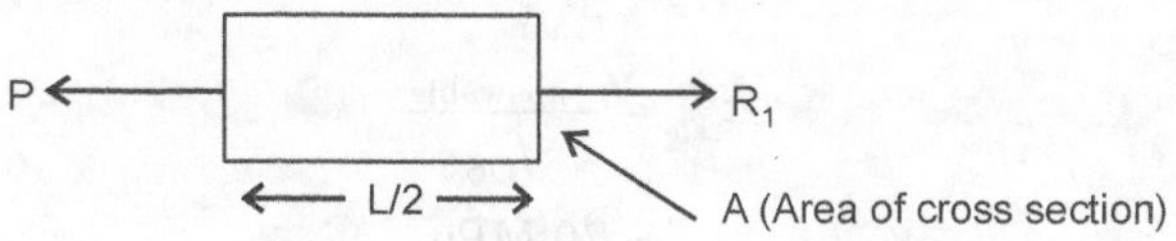

$$\text{Normal stress } 6_n = \frac{R}{A} = \frac{P}{A} \left(\because \begin{matrix} \varepsilon Fn = 0 \\ R - P = 0 \end{matrix} \right)$$

The right side of rod

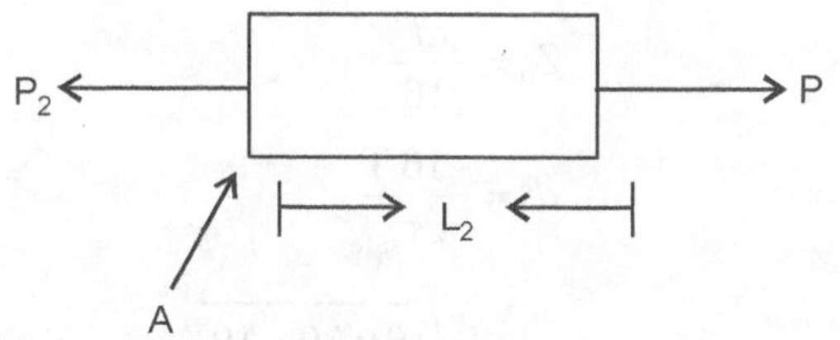

$$\Sigma fx = 0$$

$$\Rightarrow \quad R_2 - P = 0$$

$$\Rightarrow \quad R_2 = P$$

$$6n = \frac{P_2}{A}$$

$$= \frac{P}{A}$$

$$\therefore \text{ Hence normal} = \frac{P}{A}$$

33. $d_B = 2d_A \left[\because U \alpha \frac{1}{A} \right]$

34. For a ductive material toughness is a measure of ability to absorb energy up to fracture

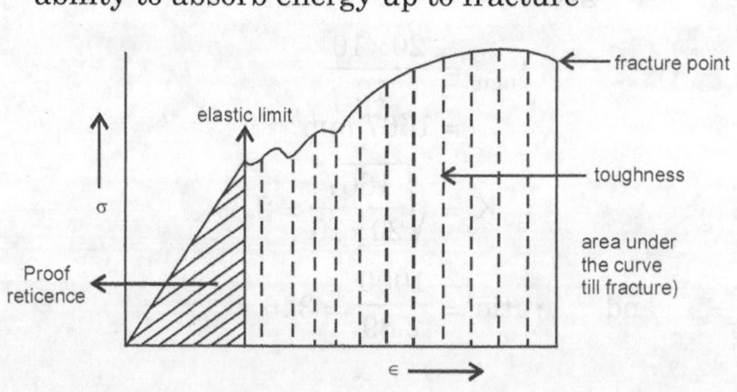

Verification:

(*a*) Resistance to scrateling is resistance to abrasion.

(*b*) Toughness is the ability to absorb energy upto fracture.

(*c*) Proof resilence is the ability to absorb energy till elastic limit.

(*d*) Hardness the resistance to indentation.

35.

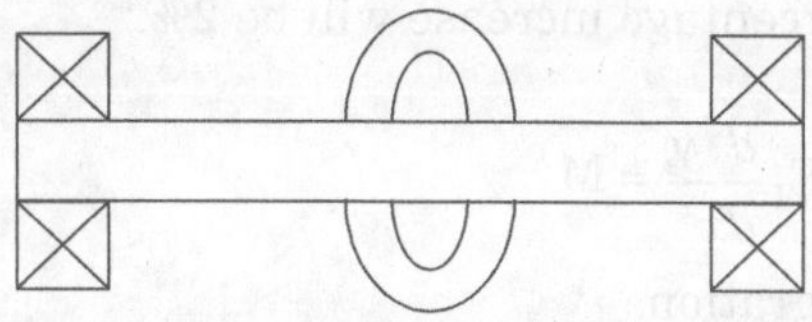

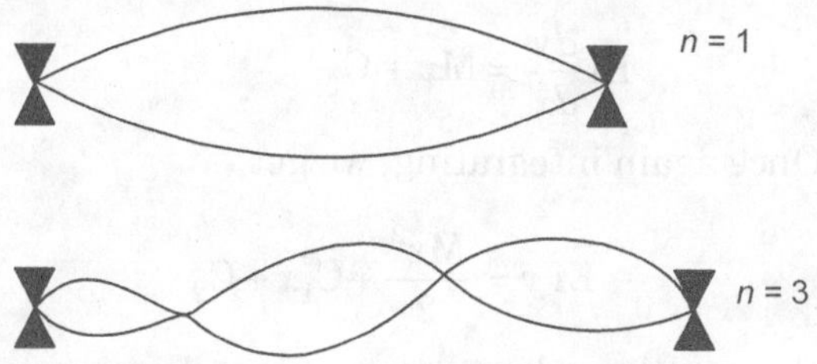

The whiring frequency of shaft is given by

$$f = \frac{\pi}{2} n^2 \sqrt{\frac{gEI}{wL^4}}$$

$$f_1 = \frac{\pi}{2} \sqrt{\frac{gEI}{wL^4}}$$

$n \Rightarrow$ node number

frequency of 1st mode

2 nodes are present in 3rd mode.

$$f_3 = 3^2 f_1 = 1800 \text{ (rpm)}$$

$$\Rightarrow \quad f_1 = \frac{1800}{9} = 200 (rpm)$$

36.

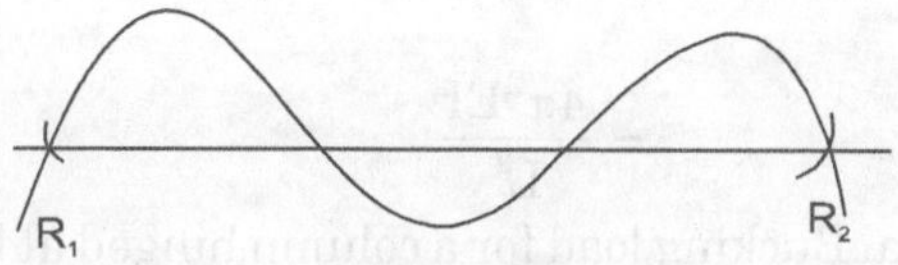

$$\text{Total load} = \int_0^2 \sin\left(\frac{3nx}{L}\right) dx$$

$$P = -\frac{L}{3\pi} \left[\cos\left(\frac{3\pi x}{L}\right) \right]_0^L = \frac{2L}{3\pi}$$

By symmetry, $R_1 = R_2 = P/2$

37. Hoop stress for a thin spherical shell

$$\sigma_h = \frac{Pr}{2t}$$

Applying logarithm on both sides, we get

$$\log(\sigma_h) = \log\left(\frac{P}{2}\right) + \log(r) - \log(t)$$

Differentiating above equation,

$$\frac{d\sigma_h}{\sigma_h} = 0 + \frac{dr}{r} - \frac{dt}{t}$$

Given : $\frac{dr}{r} = 0.01$ and $\frac{dt}{t} = -0.01$

Substituting, we get

$$\frac{d\sigma_h}{\sigma_h} = 0.02,$$

Hence percentage increase will be 2%.

38. $$E_I \frac{d^2y}{dx^2} = M$$

upon integration,

$$EI\frac{dy}{dx} = Mx + C_1$$

Once again integrating, we get

$$EI\,y = \frac{Mx^2}{2} + C_1x + C_2$$

For cantilever beam at $x = 0$,

$$\frac{dy}{dx} = 0$$

& $x = 0$

$y = 0$

From this we get $C_1 = C_2 = 0$, Hence

$$y = \frac{Mx^2}{2EI};$$

maximum deflection

$$y_{max} = \frac{ML^2}{2EI}$$

39. Critical Buckling load for column fixed at both ends

$$= \frac{4\pi^2 EI}{L^2}$$

Critical Bucking load for a column hinged at both ends $\frac{\pi^2 EI}{L^2}$

Hence, $\frac{Pcr_1}{Pcr_2} = 4$

40. $$\sigma_{eq} = \sqrt{\frac{1}{2}\{(\sigma_{11}-\sigma_{22})^2+(\sigma_{11}-\sigma_{33})^2+(\sigma_{22}-\sigma_{33})^2+6\left[\sigma_{12}^2+\sigma_{23}^2+\sigma_{13}^2\right]\}}$$

We know, $\sigma_{11} = 10, \sigma_{22} = 20,$

$\sigma_{33} = -10;$

$\sigma_{12} = 5 = \sigma_{21}$

$\sigma_{23} = \sigma_{13} = 0 = \sigma_{32} = \sigma_{31}$

$\therefore$ $\sigma_{eq} = 27.839$ MPa

Shear stress at yield

$$\tau_y = \frac{\sigma_{eq}}{\sqrt{3}}$$

$$= 16.07 \text{ MPa}$$

41. $\sigma_{xx} = 40$ MPa; $\sigma_{yy} = 100$ MPa,

$\tau_{xy} = 40$ MPa

$$\text{Radius of Mohr's circle} = \sqrt{\left[\frac{\sigma_{xx}-\sigma_{yy}}{2}\right]^2 + (\tau xy)^2}$$

$$= \sqrt{30^2 + 40^2}$$

$$= 50 \text{ MPa}$$

43. $T = 50N - n;$

$\tau_{allowable} = 140$ MPa

$f_{OS} = 2;$

$$\tau_{safe} = \frac{\tau_{allowable}}{f_{OS}}$$

$$= 70 \text{ MPa}$$

We know, $\frac{T}{Z_p} = \tau_{safe}$

$$Z_p = \frac{\pi d^3}{16}$$

$$\Rightarrow \quad d^3 = \frac{16\,T}{\pi\,\tau_{safe}}$$

$$d = \sqrt{\frac{16 \times 50 \times 10^3}{\pi \times 70}}$$

$$d = 15.38 \text{ mm} \approx 16 \text{ mm}$$

44. $$\text{Slenderness ratio} = \frac{\text{length of column}}{\text{least radius of gyration}}$$

$$= \frac{L}{K}$$

But $K = \sqrt{\frac{I_{min}}{A}}$

where I_{min} = minimum area moment of inertia

i.e. I_{xx} or I_{yy}, whichever is less.

For given section

$$I_{min} = \frac{20 \times 10^3}{12}$$

$$= 1667 \text{ mm}^3$$

$$\therefore \quad K = \sqrt{\frac{1667}{20 \times 10}} = 2.89$$

and $$\text{ratio} = \frac{1000}{2.89} = 346$$

45. Take moments about 'Q'

$$R_Q \times 1 - 1 = 0$$

$\Rightarrow$ $R_Q = 1 \text{ kN} \uparrow$

But $R_P + R_Q = 0$

$\therefore$ $R_P = -R_Q = -1 \text{ kN}$

$\Rightarrow$ $R_P = 1 \text{ kN} \downarrow$

46. Given : $p = 5$ MPa

$d = 1000$ mm

$t = 10$ mm

Hoop stress, $\sigma_{Hoop} = \dfrac{pd}{2t}$

$= 250$ MPa

48. Blanking force $= \tau \times A_s$

$= 300 \times \pi dt$

$= 300 \times \pi \times 100 \times 1.5$

$= 141$ kN

49.

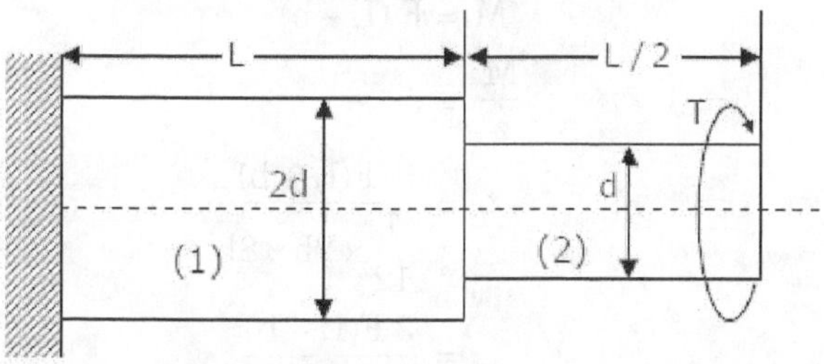

Angular twist at the free end

$$\theta = \Sigma\left(\frac{TL}{GJ}\right)$$

$$= \left(\frac{TL}{GJ}\right)_1 + \left(\frac{TL}{GJ}\right)_2$$

$$= \frac{T}{G}\left[\left(\frac{L}{J}\right)_1 + \left(\frac{L}{J}\right)_2\right]$$

$$= \frac{T}{G}\left[\frac{L}{\frac{\pi}{32}(2d)^4} + \frac{L/2}{\frac{\pi}{32}d^4}\right]$$

$$= \frac{TL}{\pi G d^4}\left[\frac{32}{16} + \frac{32}{2}\right]$$

$$= \frac{18TL}{\pi G d^4}$$

$\Rightarrow$ $d = \left[\dfrac{18TL}{\pi G\theta}\right]^{1/4}$

50. Volumetric solidification shrinkage and volumetric solid contraction cause decrease in dimensions.

Volume of cube $= 50^3 = 125000 \text{ mm}^3$

After considering both the allowances,

$V = 125000 \times 0.96 \times 0.94$

$= 112800 \text{ mm}^3$

$\therefore$ Side of cube $= \sqrt[3]{112800} = 48.32$ mm

51. Given : $\sigma_x = -200$ MPa

$\sigma_y = 100$ MPa

and $\tau_{xy} = 100$ MPa

$$\therefore \quad \tau_{max} = \sqrt{\left(\frac{\sigma_x - \sigma_y}{2}\right)^2 + \tau_{xy}^2}$$

$$= \sqrt{\left(\frac{-200 - 100}{2}\right)^2 + 100^2}$$

$= 180.2$ MPa

52. Given : Number of rivets = 4

Diameter of each rivet, $d = 6 \times 10^{-3}$ m

Direct load on the bracket = 1000 N

Since there is no eccentricity, therefore

load on each rivet $= \dfrac{1000}{4} = 250$ N

$\therefore$ Shear stress $= \dfrac{P}{A}$

$$= \frac{250}{\frac{\pi}{4}(6)^2} = 8.8 \text{ Pa}$$

53. Maximum shear stress,

$$\tau_{max} = \frac{\sigma_1 - \sigma_2}{2}$$

$$= \frac{100 - 40}{2}$$

$= 30$ MPa

54. $\tau_{max} = \dfrac{16}{\pi d^3}\left[\sqrt{M^2 + T^2}\right]$

But M = 0

$\therefore$ $\tau_{max} = \dfrac{16T_0}{\pi d^3}$

Alternately

We know, $\dfrac{T}{J} = \dfrac{\tau}{r}$

where, $J = \frac{\pi d^4}{32}$

$\therefore \quad \tau = \frac{Tr}{J}$

$= \frac{T_o \times R \times 32}{\pi d^4} = \frac{16T_o}{\pi d^3}$

55. $EI \frac{d^2 y}{dn^2} = -M_x$

$\Rightarrow \quad M_x = -P(L - x)$

$\therefore \quad EI \frac{d^2 y}{dx^2} = P(L - x)$

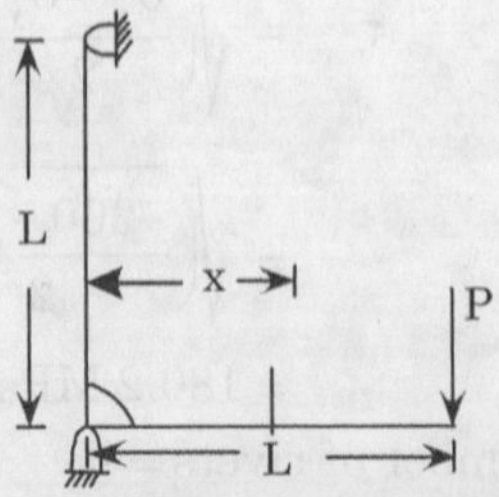

Integrating twice and puting x = L, we get

$$y = \frac{4PL^3}{3EI}$$

Alternately

For BC, $\quad M_x = P \times x$

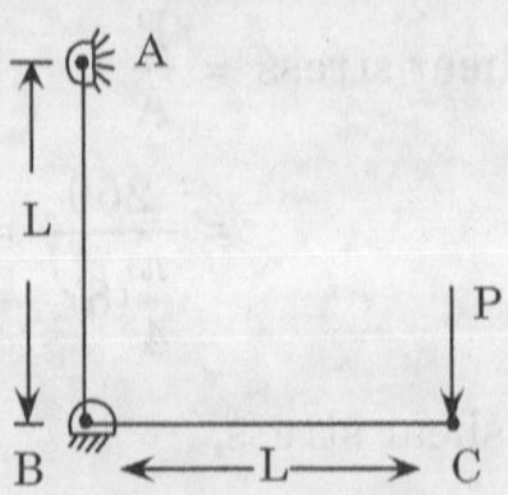

Strain energy, $U_{BC} = \int_0^L \frac{M_x^2\, dx}{2EI} = \frac{P^2 L^3}{6EI}$

For AB, $\quad M_y = P \times L$

Strain energy, $U_{AB} = \int_0^L \frac{M_y^2\, dy}{2EI} = \frac{P^2 L^3}{2EI}$

Total strain energy, $U = \frac{P^2 L^3}{2EI} + \frac{P^2 L^3}{6EI}$

$= \frac{P^2 L^3}{6EI}$

$\therefore \quad \frac{1}{2} \times P \times \text{deflection} = \frac{4P^2 L^3}{6EI}$

$\Rightarrow \quad \text{Deflection} = \frac{4PL^3}{3EI}$

56. Moment about A is zero

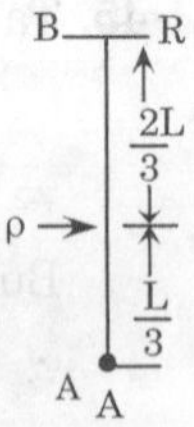

$\therefore \quad R \times L = P \times \frac{L}{3}$

$\Rightarrow \quad R = \frac{P}{3}$

57.

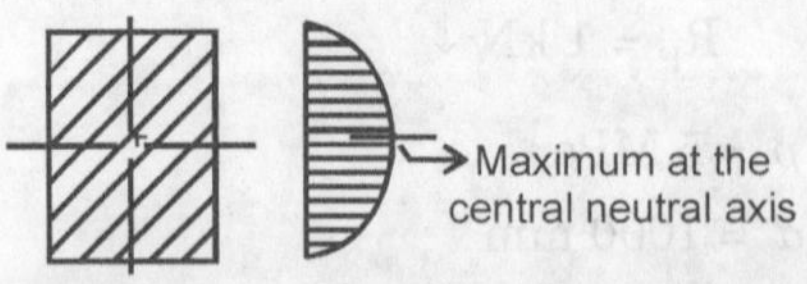

58. Both Poisson's ratio and Young's modulus are required to calculate lateral strain in rod.

Now m, E and N are related by

$$N = \frac{mE}{2(m + 1)}$$

59. For balancing moments about the hinge Q,

$75 \times 2 = m_R \times 0.5$

$\Rightarrow \quad m_R = 300$ kg

60. Binding moment about P is given by

$M = F(L - b)$

$\Rightarrow \quad \frac{M}{z} = f$

$\Rightarrow \quad f_P = \frac{F(L - b)}{\frac{1}{12} \times 2b \times 8b^2}$

$= \frac{3F(L - b)}{4b^3}$

Axial stress due to direct load F,

$$\sigma = \frac{F}{4b^2}$$

$\therefore$ Axial stress $= f_P + \sigma$

$\therefore \quad f_P + \sigma = \frac{3F(L - b)}{4b^3} + \frac{F}{4b^2}$

$= \frac{1}{4b^3}[3FL - 2Fb]$

$= \frac{F(3L - 2b)}{4b^3}$

61. Maximum stress, $\sigma_{max} = \frac{p}{2} + \sqrt{\frac{p^2}{4} + q^2}$

$\therefore \quad q_{max} = \frac{T}{\frac{\pi}{16} D^3}$

$= \frac{10 \times 10^3 \times 10^3}{\frac{\pi}{16} \times (100)^3}$

$= \frac{160}{3.14}$ N/mm²

$= 51$ MPa

$$\therefore \quad \sigma_{max} = \frac{50}{2} + \sqrt{(25)^2 + (51)^2}$$

$$= 25 + 57$$

$$= 82 \text{ MPa}$$

62. $T_{PQ} \cos 40° + T_{PR} \cos 60° + F = 0$ ...(*i*)

and $T_{PQ} \sin 40° = T_{PR} \sin 60°$ ...(*ii*)

From these equations, T_{PQ} and T_{PR} can be found in terms of F.

Now at Q, $T\ T_{QR} = T_{PQ} \cos 45°$

Solving these equations, we get

$$T_{QR} = 0.63 \text{ F}$$

63.

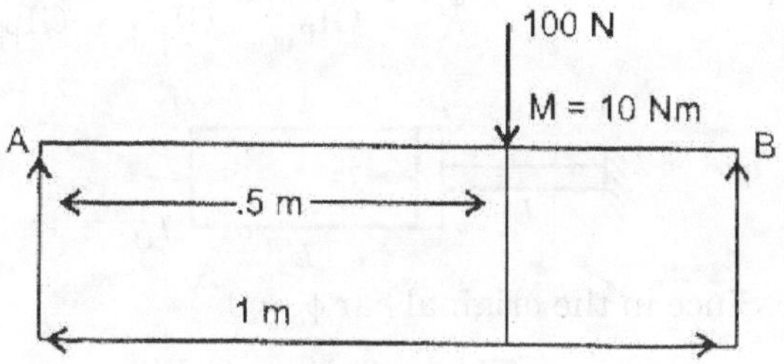

Taking moment about A,

$\Rightarrow \quad 100 \times 0.5 + 10 = R_B$

$R_B = 60$ N

$\therefore \quad R_A = 100 - 60 = 40$ N

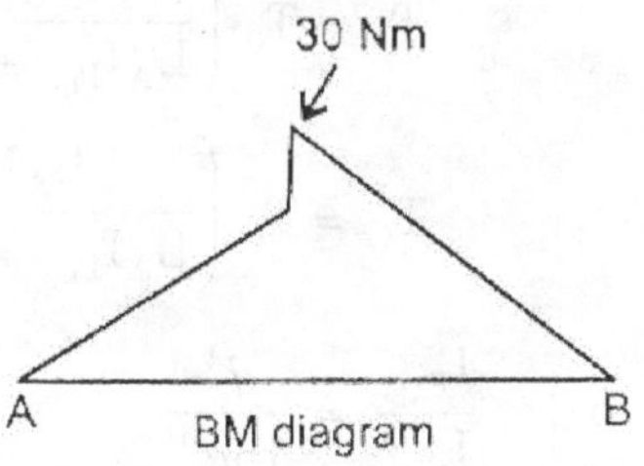

64. Since rod is free to expand, therefore elongation, $\Delta L = L\,\alpha\,\Delta t$

$$\therefore \quad \text{strain} = \frac{\Delta L}{L} = \alpha\,\Delta t$$

Thermal stress = $E\,\alpha\,\Delta t$

Since rod is fixed at both ends, so thermal strain will be zero but there will be thermal stresses.

65.

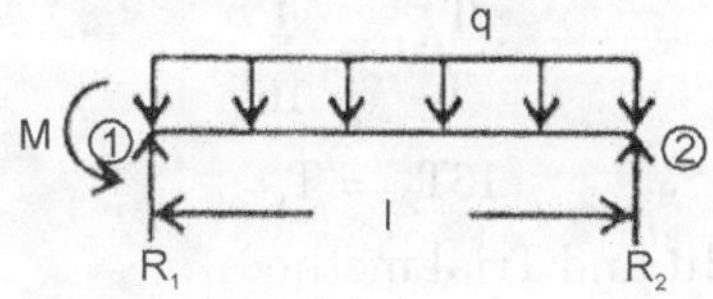

$R_1 + R_2 = ql$ ...(*i*)

Moment about ①,

$$R_2 l + M - \frac{ql^2}{2} = 0 \quad ...(ii)$$

Moment about ②,

$$R_1 l - \frac{ql^2}{2} - M = 0 \quad ...(iii)$$

From equations (*i*), (*ii*) and (*iii*), we get

we get $R_1 = \dfrac{5ql}{8}$

$$R_2 = \frac{3ql}{8}$$

and $M = \dfrac{ql^2}{8}$

66. Here, $P_x = P_y = P_z = -15 \times 10^6$ Pa (compression)

Now, $e_x = \dfrac{P_x}{E} - \dfrac{P_y}{E.m} - \dfrac{P_z}{Em}$

$$= 3 \times 10^{-5}$$

where, m = Poisson's ratio

$$e_x = e_y = e_z = 3 \times 10^{-5}$$

$$e_x + e_y + e_z = 9 \times 10^{-5}$$

$\therefore$ Change in volume = Total strain × Original

$= 90\text{m}^3$

67.

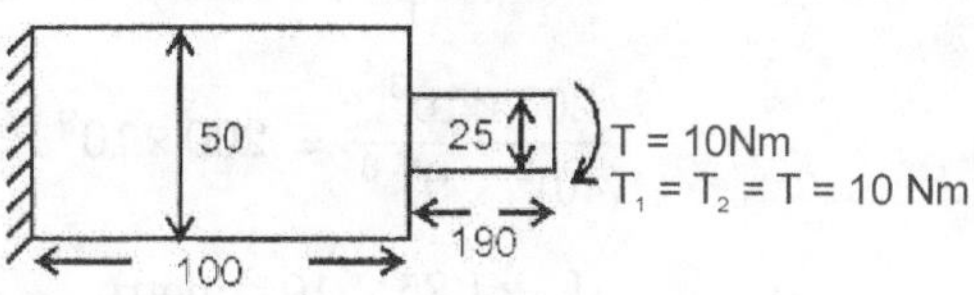

Here, $J_1 = \dfrac{\pi}{32} d^4$

$$= \left(\frac{\pi}{32}\right) \times (25)^4 \times 10^{-12}$$

$$G_1 = 80 \times 10^9$$

$$l = 0.1/\text{m}$$

$$\theta_1 = 0.33 \times 10^{-3}$$

Now $T_1 = \dfrac{J_1 G_1 \theta_1}{l_1}$

$\Rightarrow \quad \theta_1 = \dfrac{T_1 l_1}{J_1 G_1}$

$$\therefore \quad \frac{\theta_1}{\theta_2} = \left(\frac{J_2}{J_1}\right)\left(\frac{l_1}{l_2}\right)\left(\frac{G_2}{G_1}\right)$$

Now $\theta_1 = 4\theta_2$

$\therefore \quad \theta = \theta_1 + \theta_2$

$$= \frac{5}{4} \times 0.33 \times 10^{-3}$$

$$\text{Strain energy} = \frac{1}{2}T\theta$$

$$= \frac{1}{2} \times 10 \times \frac{5}{4} \times 0.33 \times 10^{-3}$$

$$= 1.73$$

69. Let T = torque;

d = diameter of shaft

and τ = maximum value of shear stress,

$$\therefore \quad T = \frac{\pi}{16}d^3 \times \tau$$

$$\Rightarrow \quad \tau = \frac{16T}{\pi d^3}$$

70. Critical buckling load (P_{cr}) for a column having length L, modulus of elasticity E and second moment of cross sectional area I is loaded centrically.

Condition :

Both ends are having pin joint. *i.e.*, hinged

so, $n = 1$

$$\therefore \quad P_{cr} = n\frac{\pi^2 EI}{L^2} = \frac{\pi^2 EI}{L^2}$$

72. We know, $\sigma = E\dfrac{\Delta L}{L}$

$$\therefore \quad \frac{200 \times 10^3}{(40)^2 \times 10^{-6}} = 200 \times 10^9 \times \frac{\Delta L}{2}$$

$$\Rightarrow \quad \Delta L = 1.25 \times 10^{-3} \text{ metre}$$

$$= 1.25 \text{ mm}$$

73.

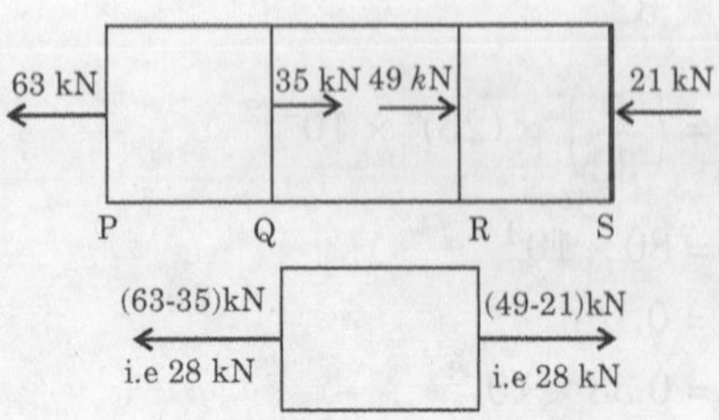

$$\text{Stress} = \frac{\text{Force}}{\text{Area}}$$

$$= \frac{28 \times 10^3}{700 \times 10^{-6}} \text{N/m}^2$$

$$= 4 \times 10^{-5} \text{ N/m}^2$$

$$= 40 \text{ MPa}$$

74. Rod is not restrained but completely free

76. Twisting moment T is applied to the coupling.

Now, $\dfrac{T}{I_J} = \dfrac{\tau}{r} = \dfrac{G\phi}{l}$

Indeterminate structure due to fixed end.

Let T_A and T_C are twisting moments felt at A and C.

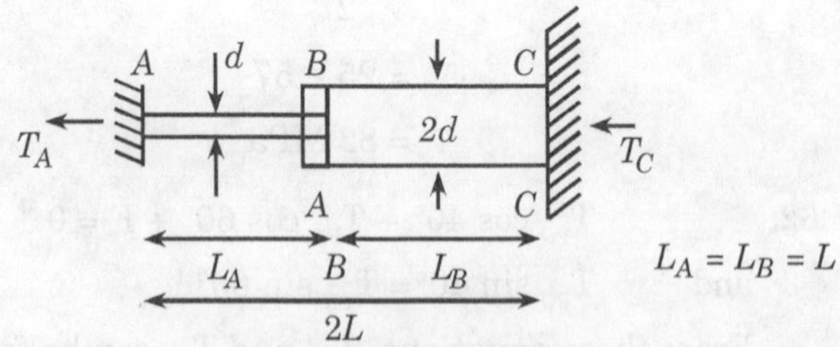

Hence from static equilibrium

$$T_A + T_C = T \qquad ....(i)$$

Now selecting redundant torque T_C and analyzing corresponding released structure

$$\phi_c = \frac{TL_A}{GI_{P_{AB}}} - \frac{T_C L_A}{GI_{P_{AB}}} - \frac{T_C L_B}{GI_{P_{BC}}}$$

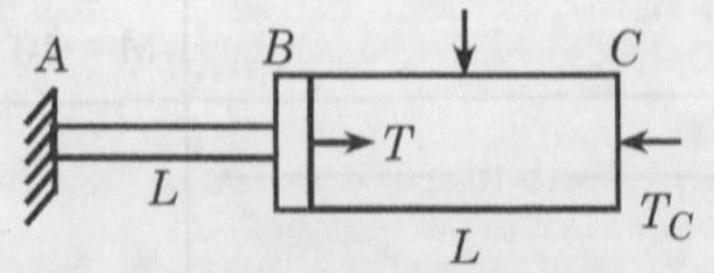

Since in the original bar $\phi_b = 0$

$$\therefore \quad \frac{TL_A}{GI_{P_{AB}}} - \frac{T_C L_A}{GI_{P_{AB}}} - \frac{T_C L_C}{GI_{P_{BC}}} = 0 \qquad ....(ii)$$

Solving equations (*i*) and (*ii*), we get

$$T_A = T_0\left(\frac{L_B I_{P_{AB}}}{L_A I_{P_{CB}} + L_B I_{P_{AB}}}\right),$$

$$T_C = T_0\left(\frac{L_A I_{P_{BC}}}{L_A I_{P_{BC}} + L_B I_{P_{AB}}}\right)$$

Now, $\dfrac{I_{P_{AB}}}{I_{P_{BC}}} = \dfrac{d^4}{16d^4}$

$$\Rightarrow \quad \boxed{I_{P_{BC}} = 16 I_{P_{AB}}} \text{ and } \boxed{L_A = L_B = L}$$

$$\therefore \quad T_A = T_0\left[\frac{LI_{P_{AB}}}{17LI_{P_{AB}}}\right]$$

$$\text{and} \quad T_C = T_0\left(\frac{16LI_{P_{AB}}}{17LI_{P_{AB}}}\right)$$

$$\Rightarrow \quad \frac{T_A}{T_C} = \frac{1}{16}$$

$$\Rightarrow \quad 16T_A = T_C$$

78. Use Hit and Trial method.

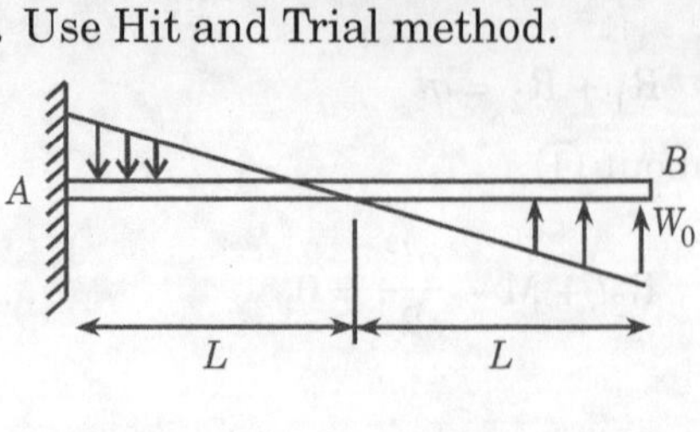

$$V = \frac{dm}{dn}$$

Consider first option (*a*) of moment diagram.

At right end B, slope must be zero, as there is no shear force at B. So, option (*a*) is wrong.

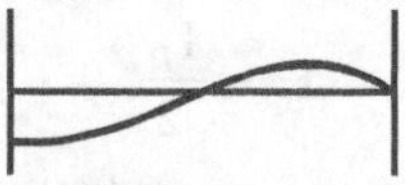

1

Now consider option (*b*).

Due to symmetrical load intensity at left side too the shear force equal to zero so slope at left side must be zero. So option (*b*) is wrong.

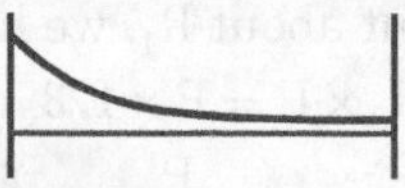

2

Now, consider option (*c*).

Here at both the ends, slope is zero means shear force at both the end is zero and also when we move from right to left, rate of increase of shear force decrease due to triangular shape of load intensity and at middle, slope should be maximum and there after decreases. So in option (*c*), all these criterion fulfills. Hence option (*c*) is the correct.

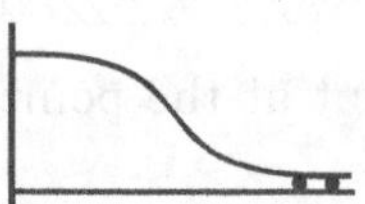

3

Now, consider option (*d*). Its middle part slope and right most part slope is strictly not agreeing with the load shown.

79. Here, $I = \dfrac{bh^3}{12} = \dfrac{0.01 \times 0.01^3}{12} m^4$

$y = 0.005$ m

$M_x = 10x$ N-m

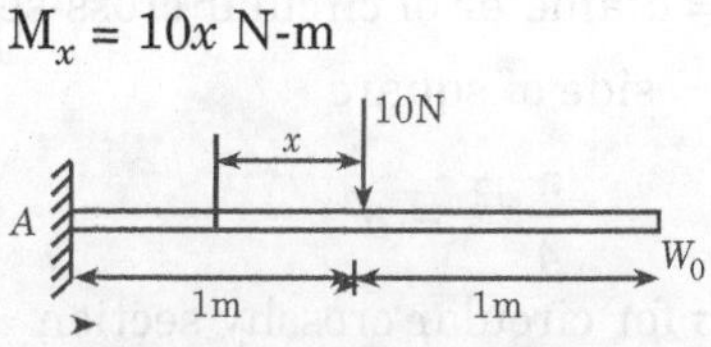

At left extreme of cantilever

$$\sigma_{bending} = 60 \times 1 \text{ MPa}$$

After 1 *m* $= 60$ MPa

$$\sigma_{bending} = 60 \times 0 \text{ MPa} = 0 \text{ MPa}$$

and it will continue over whole span.

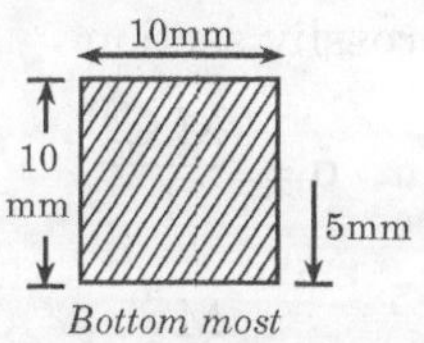

Bottom most fiber

$$\sigma_{(bending\ stress)} = \frac{Mxy}{I}$$

$$= \frac{(10x)(0.005)}{(0.01)^4/12} \text{N/m}^2$$

$$= 60 \times \text{MPa}$$

(straight line equation)

80. Applying maximum shear stress theory

$$\tau_{max} = \frac{\sigma_{Y.P}}{2} = \frac{\sigma_x - \sigma_y}{2}$$

$$\therefore \quad \sigma_{Y.P.} = \sigma_x - \sigma_y$$

$$= -100 - (-10)$$

$$= -90 \text{ MPa}$$

81. We know $\quad E = 2\,G[1 + \nu)$

$$\Rightarrow \quad \frac{E}{G} = 2\,[1 + \nu]$$

82.

$$r = \frac{\sigma_x + \sigma_y}{2}$$

$$= \frac{100 + 20}{2}$$

$$= 60 \text{ MPa}$$

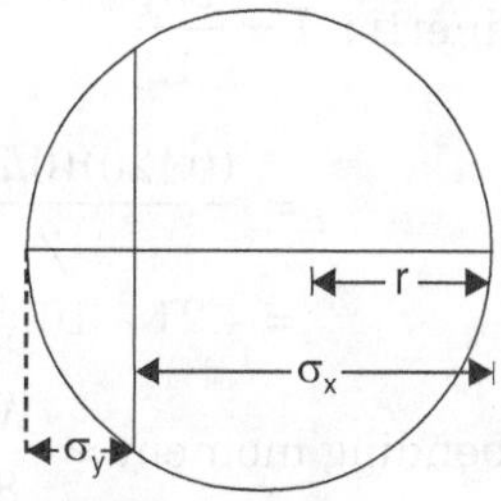

$\therefore$ Radius of Mohr's circle = 60 MP*a*

83. Torque will be same on the part MN, NO, and OP as they are connected in series.

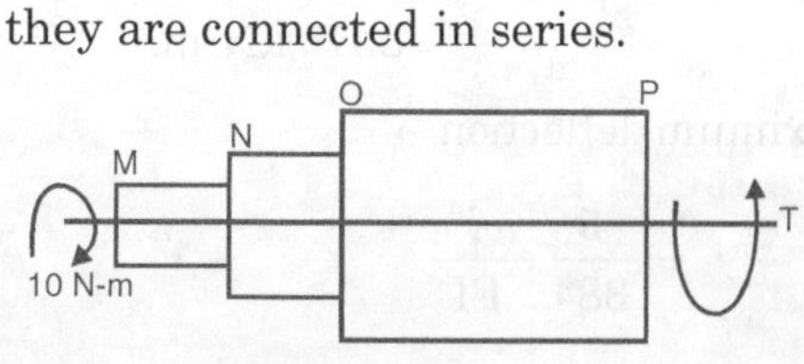

$$(\delta)_{MN} = \frac{10 \text{ N} - \text{m}}{20 \text{ N} - \text{m/rad.}}$$

$$= 0.5 \text{ radian}$$

$$(\delta)_{NO} = \frac{10}{30} = \frac{1}{3} \text{ radian}$$

$$(\delta)_{OP} = \frac{10}{60} = \frac{1}{6} \text{ radian}$$

$$(\delta)_{MP} = \frac{1}{2} + \frac{1}{3} + \frac{1}{6}$$

$$= \frac{3 + 2 + 1}{6} = 1 \text{ radian}$$

84. $f_s = \dfrac{16T}{\pi d^3}$

$$= \frac{16 \times 1600}{\pi \times (0.06)^3}$$

$= 37.72$ MPa.

85. *Given* : $b = 120$ mm

$h = 750$ mm

$E_{steel} = 200 \times 10^9$ Pa

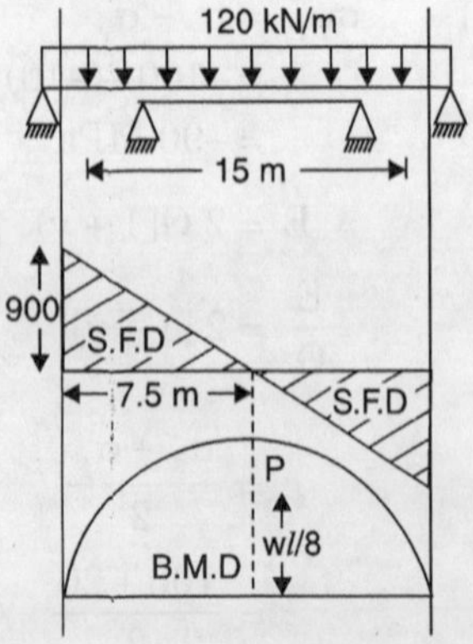

Moment of inertia, $I = \dfrac{bh^3}{12}$

$$= \frac{(0.120)(0.750)^3}{12}$$

$= 4.21 \times 10^{-3}\ m^4$

Maximum bending moment $= \dfrac{Wl}{8}$

$$= \frac{(120 \times 15) \times 15}{8}$$

$= 3375$ kN-m.

86. Maximum deflection

$$= \frac{5}{384}\ \frac{wl^4}{EI}$$

$$= \frac{5}{384} \times \frac{(120)\,(15)^4 \times 1000}{200 \times 10^9 \times 4.21 \times 10^{-3}}$$

$= 0.0939$ m

$= 93.94$ mm.

87. Polar Moment of Inertia perpendicular to the plane of paper,

$$I_P = \frac{\pi D^4}{32}$$

Applying "*Perpendicular axes*" theorem

$$I_x + I_y = I_P$$

$\Rightarrow \quad 2I_x = I_P$

$\Rightarrow \quad I_x = \dfrac{I_P}{2} \qquad \ldots[\because I_x = I_y]$

$$= \frac{\pi D^4}{64}$$

88. For horizontal equilibrium

$$R_1 + R_2 = P$$

Taking moment about R_1, we have

$$R_2 \times L = P \times L/3$$

$$R_2 = \frac{P}{3}$$

$\therefore \quad R_1 = \dfrac{2P}{3}$

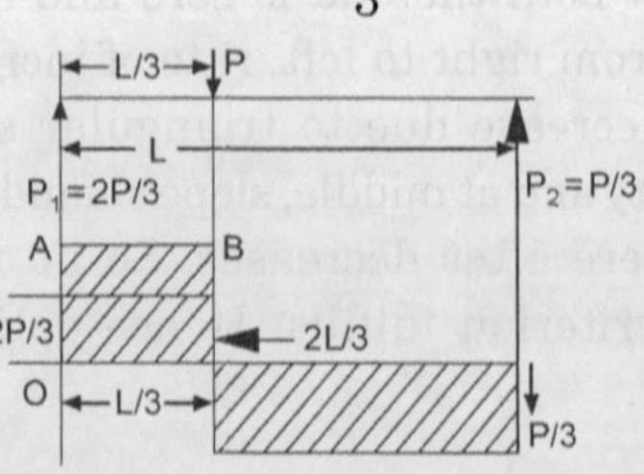

Bending moment at the point of application of load

= Area under OABC

$$= \frac{2P}{3} \times \frac{L}{3}$$

$$= \frac{2PL}{9}$$

89. Cross-sectional area of square and circular cross-sections are equal

Let d = diameter of circular cross-section

a_2 = side of square

$\therefore \quad \dfrac{\pi}{4} d^2 = a^2$

Now, σ for circular crosshy section

$$= \frac{M.C.}{I}$$

$$= \frac{M.\dfrac{d}{2}}{\dfrac{\pi d^3}{64}} = \frac{32\,M}{\pi d^3}$$

For square crosshy section

$$\sigma = \frac{M.C.}{I}$$

$$= \frac{M\dfrac{a}{2}}{\dfrac{a^4}{12}} = \frac{6M}{a^3}$$

$$\frac{\sigma \text{ for circular cross-section}}{\sigma \text{ for square cross-section}} = \frac{\frac{32M}{\pi d^3}}{\frac{6M}{d^3}} = \frac{32\,M}{\pi\,d^3} \times \frac{a^3}{6M}$$

$$= \frac{32}{\pi}\cdot\left(\frac{a^3}{d^3}\right)$$

Since, $a = \frac{\pi}{4}d^2 = 7$

$\therefore$ σ for circular cross-section > σ for square cross-section

90. Elongation in cost iron,

$$\Delta_{CI} = \left(\frac{Pl}{AE}\right)_{C.I.}$$

Elongation in mild steel,

$$\Delta_{m.s.} = \left(\frac{Pl}{AE}\right)_{M.S.}$$

$$\Rightarrow \quad \frac{\Delta_{C.I}}{\Delta_{M.S.}} = \frac{\frac{Pl}{AE_{C.I.}}}{\frac{Pl}{AE_{M.S.}}}$$

$$= \frac{E_{M.S.}}{E_{C.I.}}$$

$$= \frac{206}{100} = 2.06$$

$$\therefore \quad \Delta_{CI} > \Delta_{M.S.}$$

91. *Given:* $\tau_1 = 240$ MPa

When staff diameter is doubled, maximum share stress = τ_2

Let d be the original diameter.

Since torque in both the cases is constant

$$\therefore \quad \frac{\tau_1}{\tau_2} = \frac{(T_{r_1}/I_{P_1})}{(T_{r_2}/I_{P_2})}$$

$$= \left(\frac{r_1}{I_{P_1}} \times \frac{I_{P_2}}{r_2}\right)$$

$$= \frac{d}{2\times\left(\frac{\pi d^2}{32}\right)} \times \frac{\pi\,(2d)^4}{32\times\frac{2d}{2}}$$

$$= \frac{\pi d\times 16d^4}{\pi d^4\times d\times 2} = 8$$

$$\Rightarrow \quad \tau_2 = \frac{\tau_1}{8}$$

$$= \frac{240}{8} = 30 \text{ MPa}$$

92. From differential equation of elastic curve

$$EI\frac{d^2y}{dx^2} = M$$

$$\Rightarrow \quad y = \frac{1}{EI}\iint M dx\,dx + C_1x + C_2$$

where y = deflection of beam

As area moment increases, deflection decreases.

93. Force on plane BC is clockwise and is taken as positive.

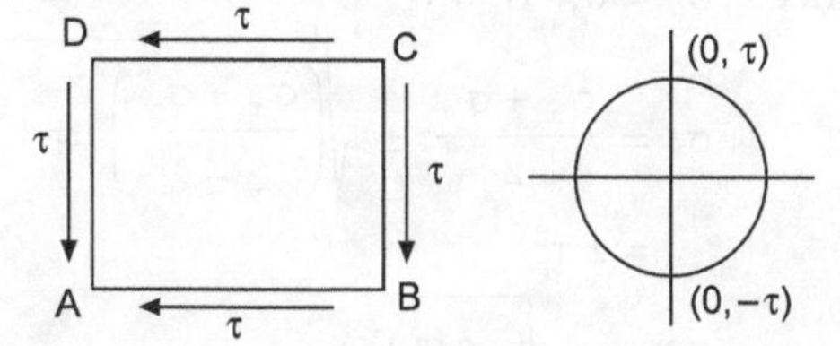

Thus maximum principal stress on the surface which is at 45° to the axis

$$= 2\tau \sin 45° \cos 45°$$

$$= 2\tau\times\frac{1}{\sqrt{2}}\times\frac{1}{\sqrt{2}}$$

$$= \tau$$

since angle in Mohr's circle is taken as double of the required a angle.

94. Mohr's circle equation

$$\left(\sigma - \frac{\sigma_x+\sigma_y}{2}\right)^2 + (\tau)^2 = \left(\sqrt{\left(\frac{\sigma_x-\sigma_y}{2}\right)^2 + (\tau_{xy})^2}\right)^2$$

where, σ = principal stress

τ = shear stress

According to question,

$$\frac{\sigma_x+\sigma_y}{2} = 175$$

$$\Rightarrow \quad \sigma_x = \sigma_y = 175 \text{ MPa}$$

95. Since Mohr's circle is a point, hence direction of maximum and minimum principal stresses of point P is in all direction.

99. $$P_{cr} = \frac{\pi^2 EI}{l_e^2}$$

For both ends fixed, $l_e = \frac{l}{2}$

and for both ends hinged, $l_e = l$

$$\therefore \quad \frac{(P_{cr})_i}{(P_{cr})_{ii}} = \frac{\dfrac{\pi^2 EI}{\left(\frac{l}{2}\right)^2}}{\dfrac{\pi^2 EI}{l}}$$

$$= \frac{4\pi^2 EI \times l^2}{l^2 \times \pi^2 EI} = 4$$

100. In the given problem, only torque 3 is applied. So only shear stress is produced. Magnitude of shear stress is given by

$$\tau = \frac{16\,T}{\pi d^3}$$

Since $\sigma_y = 0, \ \sigma_x = 0$

$$\therefore \quad \sigma_1 = \frac{\sigma_x + \sigma_y}{2} \pm \sqrt{\left(\frac{\sigma_x - \sigma_y}{2}\right)^2 + \tau^2}$$

$$= \pm \tau^{2\backslash}$$

$$\sigma_1, \sigma_2 = \pm \sqrt{\left(\frac{16\,T}{\pi d^3}\right)^2}$$

$$\sigma_1 = + \frac{16\,T}{\pi d^3}, \text{ and } \sigma_2 = -\frac{16\,T}{\pi d^3}$$

$$\tan 2\theta = \frac{2\tau_{xy}}{\sigma_x - \sigma_y}$$

$$\therefore \quad \theta_1 = 45^\circ \text{ and } \theta_2 = 135^\circ$$

101. (*a*)

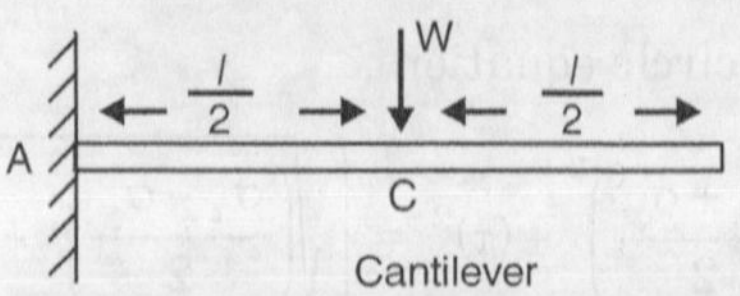

Cantilever

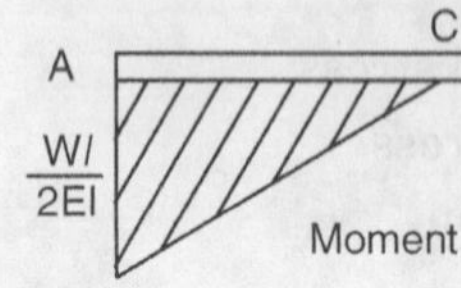

Moment area diagram

Slopes.

$$\theta_B = \text{area of } \frac{M}{EI} \text{ diagram between A and B}$$

$$= \frac{Wl}{2EI} \times \frac{l}{2} = \frac{Wl^2}{4EI}$$

$$\theta_C = \text{area of } \frac{M}{EI} \text{ diagram between A and C}$$

$$= \frac{Wl^2}{4EI}$$

Deflection.

$$y_B = \text{Moment of area of } \frac{M}{EI} \text{ diagram about B}$$

$$= \frac{Wl^2}{4EI}\left[\frac{l}{2} + \frac{2}{3} \times \frac{l}{2}\right]$$

$$= \frac{Wl^2}{4EI} \times \frac{5}{6} \times l = \frac{5\,Wl^3}{24\,EI}$$

$$y_C = \text{Moment of area of } \frac{M}{EI} \text{ diagram about C}$$

$$= \frac{Wl^2}{4EI} \times \frac{2}{3} \times \frac{l}{2} = \frac{Wl^3}{12EI}$$

(*b*)

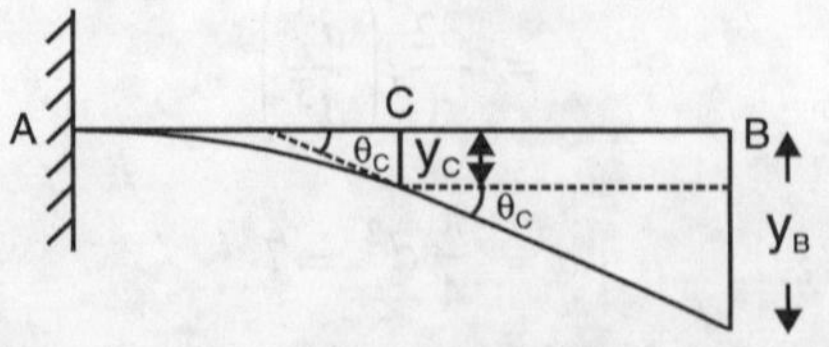

102. The equation of parabola is

$$M = \frac{qx^2}{2}$$

105.

$$\delta = \frac{64\,P\,R^3 n}{G_1 d^4}$$

$$\Rightarrow \quad \frac{P}{\delta} = \frac{G_1 d^4}{64\,R^3 n}$$

$$\Rightarrow \quad K = k\frac{d^4}{n}$$

Since wire weight is the same, therefore

$$\frac{\pi}{4} d_3^2\, n_1 = \frac{\pi}{4} d_2^2\, n_2$$

$$\Rightarrow \quad \frac{n_1}{n_2} = \frac{1}{4}$$

$$\therefore \quad \frac{k_1}{k_2} = \frac{d^4}{(d/2)^4} \times \frac{n_2}{n_1}$$

$$= 16 \times 4 = 64$$

106.

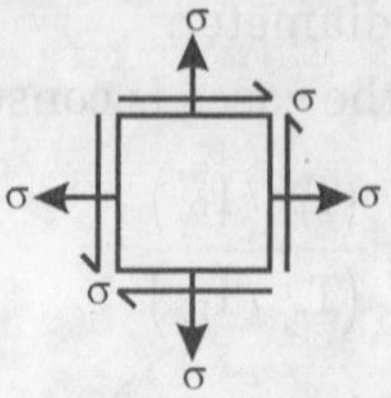

Maximum principal stress,

$$\sigma_1 = \frac{\sigma_x + \sigma_y}{2} + \sqrt{\left(\frac{\sigma_x - \sigma_y}{2}\right)^2 + \tau^2}$$

$$= \frac{\sigma + \sigma}{2} + \sqrt{\left(\frac{\sigma - \sigma}{2}\right)^2 + \sigma^2}$$

$$= 2\sigma$$

107.

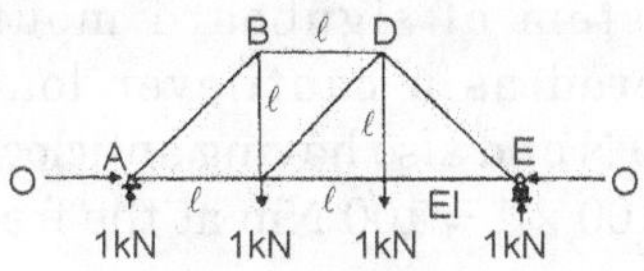

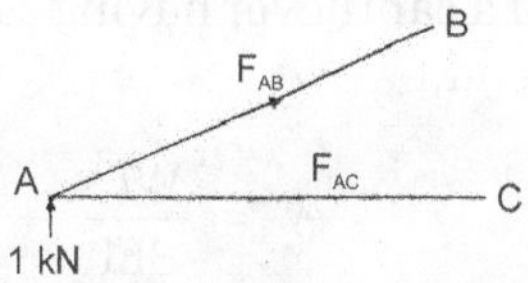

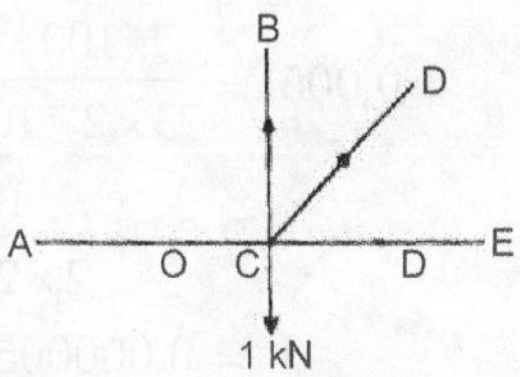

$$F_{AB}\cos 45° = 1\text{kN}$$

$$\Rightarrow \quad F_{AB} = \sqrt{2}\ \text{kN}.$$

$$F_{AC} = 0$$

Similarly, $F_{EF} = 0$

$$F_{CE} = 0$$

$$F_{BC} + F_{FD}\cos 45° = 1\text{kN}$$

$$\Rightarrow F_{CD}\sin 45° + 0 + 0 = 0$$

$$\Rightarrow \quad F_{CD} = 0$$

108–109.

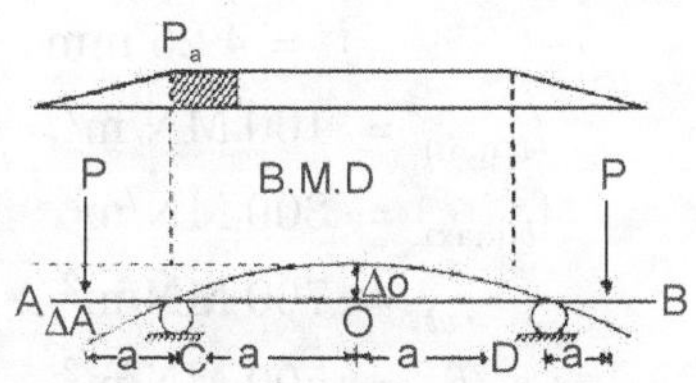

108. Deflection of point O = Deviation of point C

$$\Delta t_{c/o} = \frac{(\text{area of B.M}) \times \bar{x}}{\text{EI}}$$

$$= \frac{(\text{P}.a \times a) \times \frac{a}{2}}{\text{EI}} = \frac{\text{P}a^3}{2\text{EI}}$$

109. Deflection of point A

$$= \text{Deviation of C} - \Delta_0$$

$$\Delta_A = t_{c/o} - \Delta_0$$

$$= \frac{\left(\frac{1}{2} \times a \times \text{P}a\right) \times \left(\frac{2}{3a}\right)}{\text{EI}} + \frac{\left(\text{P}a^2 \times \left(\frac{3}{2a}\right)\right)}{\text{EI}} - \frac{\text{P}.a^3}{\text{EI}}$$

$$= \frac{\text{P}a^3}{3\text{EI}} + \frac{3\text{P}a^3}{2\text{EI}} - \frac{\text{P}a^3}{2\text{EI}} = \frac{4\text{P}a^3}{3\text{EI}}$$

110. Torque carrying capacity = $(T_1 - T_2) \times r$

where, r = radius of pulley (is same)

T_2 = tension in the slack side

$$Q_1 = 160 \times \frac{\pi}{180} = 2.7925$$

$$Q_2 = 200 \times \frac{\pi}{180} = 3.490$$

Using the relation, $\frac{T_1}{T_2} = e^{\mu\theta}$

For first case $\frac{T_1}{T_2} = e^{\mu\theta_1} = e^{0.3 \times 2.790} = 2.309$

For second case

$$\frac{T_1'}{T_2} = e^{0.3 \times 3.490} = 2.8490\ (T_2 - T_2'\ \text{given})$$

Now, $\frac{T_1 - T_2}{T_2} = 1.309$

$$\Rightarrow \quad T_1 - T_2 = 1.309\ T_2$$

Now, $\frac{T_1' - T_2}{T_2} = 1.8490$

$$\Rightarrow \quad T_1' - T_2 = 1.8490\ T_2$$

Hence % increases in torque carrying capacity.

$$= \frac{(T_1' - T_2) - (T_1' - T_2)}{T_1 - T_2}$$

$$= \frac{1.8490\,T_2 - 1300\,T_2}{1.309\,T_2} = 41.25\ \%$$

111.

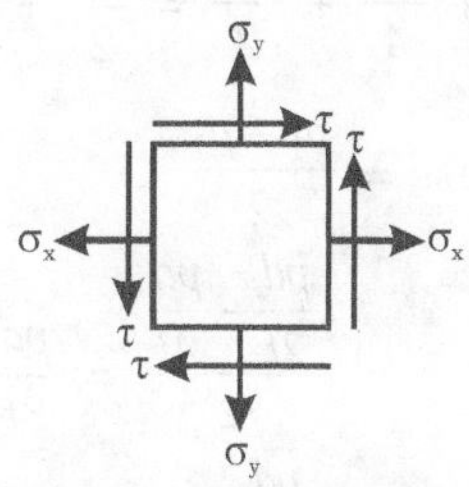

Given: $\sigma_x = 200$ MPa

$\sigma_y = 100$ MPa

$$\sigma_{1,2} = \frac{\sigma_x + \sigma_y}{2} \pm \frac{1}{2}\sqrt{(\sigma_x - \sigma_y)^2 + 4\tau^2}$$

$$= \left(\frac{200 + 100}{2}\right) \pm \frac{1}{2}\sqrt{(200 - 100)^2}$$

$$= 150 \pm 50 = 200\ \text{MPa},\ 100\ \text{MPa}$$

$$\therefore \quad \sigma_{max} = 200\ \text{MPa}$$

112. $\frac{\text{T}}{j} = \frac{\tau}{d/2}$

113. $R_1 + R_2 = P$

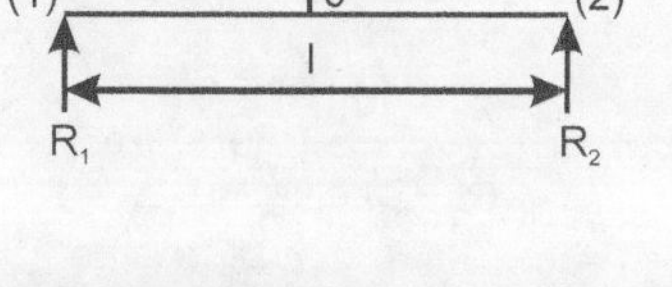

Taking moment about (1),

$$\Sigma M_1 = 0$$

$$\therefore \quad R_2 \times \ell = P \times \left(\frac{\ell}{2} + a\right)$$

$$\Rightarrow \quad R_2 = \frac{P\left(\frac{\ell}{2}+a\right)}{\ell}$$

$$\therefore \quad R_1 = P - \frac{P\left(\frac{\ell}{2}+a\right)}{\ell}$$

$$= P\ell - \frac{P\ell}{2} - P.a. = \frac{\frac{P\ell}{2} - Pa}{\ell}$$

Maximum bending moment will be at the center, *i.e.* at O.

Taking moment about O, we have

$$\Sigma M_0 = \left(R_1 \times \frac{\ell}{2}\right) + (P \times a) + \left\{\frac{\left(\frac{Pl}{2} + P.a.\right)}{\ell} \times \frac{\ell}{2}\right\}$$

$$= \left(\frac{\frac{P\ell}{2} - Pa}{\ell}\right) \times \frac{\ell}{2} + (P.a) + \frac{\left(\frac{P\ell}{2} + Pa\right)}{\ell} \times \frac{\ell}{2}$$

$$= \frac{P\ell}{4} - \frac{Pa}{2} + Pa + \frac{P\ell}{4} + \frac{Pa}{2} = \frac{P\ell}{2} + Pa$$

119. Maximum shear stress $= \dfrac{p_1 - p_2}{2}$

$$= \frac{\frac{pd}{2t} - \frac{pd}{4t}}{2} = \frac{pd}{8t}$$

$$\therefore \quad p_1 = \frac{pd}{2t} \text{ and } p_2 = \frac{pd}{4t}$$

120. $$M = EI\frac{d^2y}{dx^2}$$

Since the above equation is second order differential equation, therefore two boundary conditions are needed to solve y.

121. Given $F = 100$ N,

$d_o = 150$ mm

$\delta = 5$ mm,

$E = 2 \times 10^{11} N/m^2$

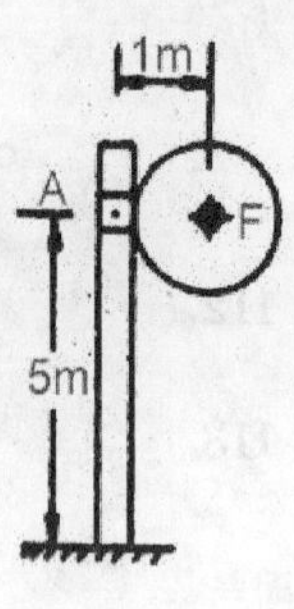

Thickness of pole = ?

The system of signboard mounting can be considered as a cantilever loaded at A i.e, W = 100 N and also having anticlockwise moment of M = 100 × 1 = 100 Nm at the free end.

Deflection of a cantilever having concentrated load at the free end,

$$y = \frac{Wl^3}{3EI} + \frac{MI^2}{2EI}$$

or, $$0.005 = \frac{100 \times 5^3}{3 \times 2 \times 10^{11} \times I} + \frac{100 \times 5^2}{2 \times 2 \times 10^{11} \times I}$$

or, $$I = 0.0000053$$

or, $$\frac{\pi}{64}\left(d_0^4 - d_i^4\right) = 0.0000053$$

or, $$(0.15)^4 - d_i^4 = \frac{0.0000053 \times 64}{\pi}$$

or, $$d_i = 0.1415 \text{ m}$$

∴ Wall thickness of pole

$$= \frac{d_i - d_0}{2} = \frac{150 - 141.5}{2}$$

$$= 4.25 \text{ mm}$$

122. Given $f_{b(min)} = 100 \text{ MN/m}^2$,

$f_{b(max)} = 300 \text{ MN/m}^2$

$f_{ut} = 700 \text{ MN/m}^2$,

$f_{yt} = 500 \text{ MN/m}^2$,

$f_e = 350 \text{ MN/m}^2$

Now, $$f_m = \frac{f_{b(min)} + f_{b(max)}}{2}$$

$$= \frac{100 + 300}{2} = 200 \text{ MN/m}^2$$

$$f_v = \frac{f_{b(max)} - f_{b(min)}}{2}$$

$$= \frac{300 - 100}{2} = 100 \text{ MN/m}^2.$$

For design based on ultimate tensile strength, using Goodman's criterion, we have

$$\frac{1}{F.S.} = \frac{f_m}{f_{ut}} + \frac{f_u}{f_e}$$

$$= \frac{200}{700} + \frac{100}{350} = 0.572$$

or, $$F.S = 1.75.$$

For design based on yield strength, using Soderberg equation,

or $$F.S. = \frac{f_m}{f_{yt}} + \frac{f_u}{f_e}$$

$$= \frac{200}{500} + \frac{100}{350} = 0.686$$

or, F.S. = 1.46.

Taking larger of the two values, we have

F.S. = 1.75.

119. $P_{cr} = \dfrac{\pi^2 EI}{\ell^2}$

I case ℓ_I = L (for both ends pinned)

II case $\ell_{II} = \dfrac{L}{2}$ (for both ends fixed)

$$\frac{P_{crI}}{P_{crII}} = \frac{\ell_{II}^2}{\ell_I^2} = \frac{\left(\frac{L}{2}\right)^2}{L^2} = \frac{1}{4}$$

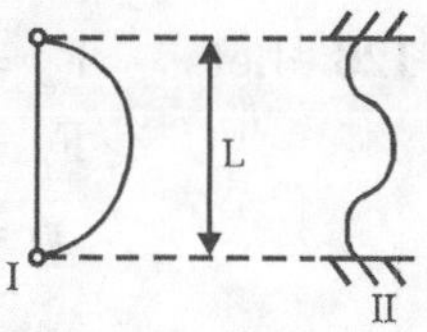

Given, P_{crI} = **10 kN**

$\therefore \quad P_{cr\,II} = 4\,P_{cr\,II}$ = **40 kN**

120. Maximum principal stress

$$= \left(\frac{\sigma_x + \sigma_y}{2}\right) + \sqrt{\left(\frac{\sigma_x - \sigma_y}{2}\right)^2 + \sigma^2}$$

$$= \left(\frac{10+2}{2}\right) + \sqrt{\left(\frac{10-2}{2}\right)^2 + 4^2}$$

= **11.66 MPa**

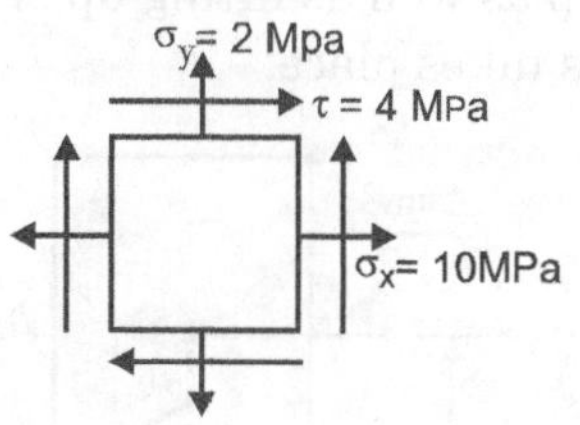

121. shear stress, $q_{max.} = \dfrac{Fd^2}{8I} = \dfrac{F.d^2}{8.\dfrac{d^4}{12}} = \dfrac{3F}{2d^2}$

where F = shear load

d = side of the square section

$$q_{av} = \frac{F}{d^2}$$

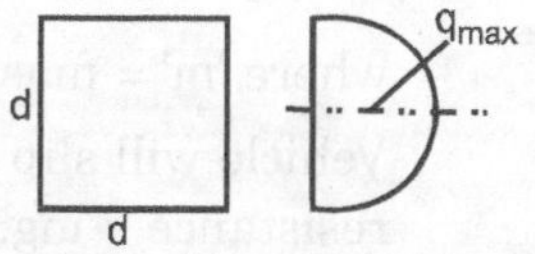

$$\therefore \quad \frac{q_{av}}{q_{max.}} = \frac{\frac{F}{d^2}}{\frac{3F}{2d^2}} = \frac{2}{3}$$

122. Principal strain $= \dfrac{pd}{2tE} - \dfrac{pd}{4tEm}$

$$= \frac{pd}{4tE}\left[2 - \frac{1}{m}\right]$$

$$= \frac{1\times10^6 \times 50\times10^{-2}}{4\times5\times10^{-3}\times200\times10^9}(2-0.2)$$

$= \mathbf{2.25 \times 10^{-4}}$

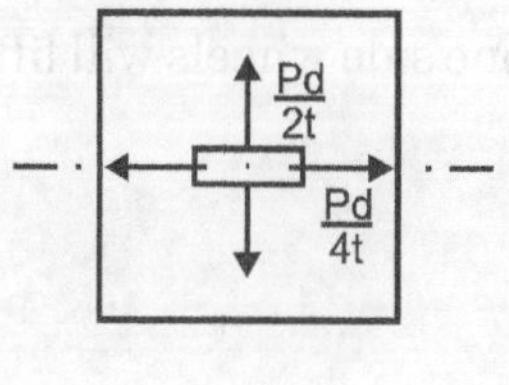

123. $U_1 = \dfrac{1}{2}(P)\left(\dfrac{PL}{AE}\right) = \dfrac{P^2\times100}{2\times4\times4\times E} = \dfrac{25P^2}{8E}$

$$U_2 = \frac{1}{2}(P)\times\left(\frac{PL^3}{3EI}\right)$$

$$= \frac{1}{2}\cdot\frac{P^2\times100^3}{E\dfrac{1}{12}\times4\times4^3} = \frac{15625}{2}\frac{P^2}{E}$$

$$\therefore \quad \frac{U_2}{U_1} = \frac{15625}{2}\times\frac{8}{25} = \mathbf{2500}$$

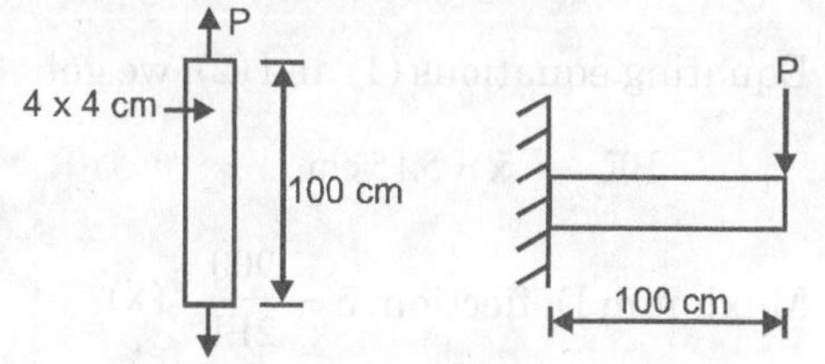

124. $EI_{AB} = EI_{CD} = 10^5$ N-cm²

Taking moments about 'D', we have ($R_B \times 10$)

$$= 100\times6\times\left(\frac{6}{2}+10\right) - 600\times3$$

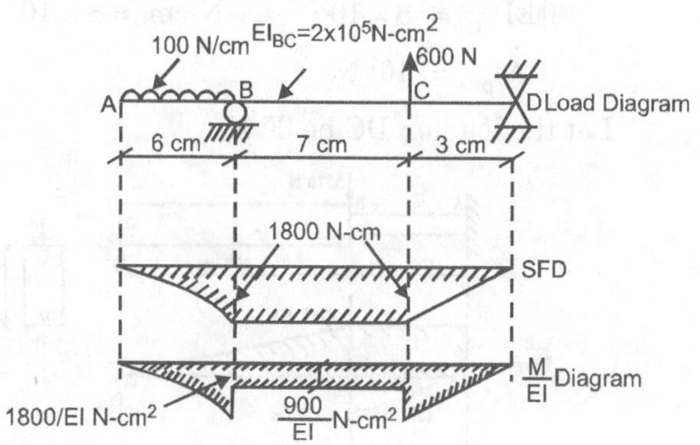

$\therefore \quad R_B$ = 600 N; R_A = 600 N

Let the maximum deflection occur at point F at a distance $\bar{x}$ from B.

Slope of the deflected beam is zero at F.

δ = EF

= Moment of $\dfrac{M}{EI}$ diagram between B and E about B

(i.e.) $EF = \dfrac{900}{EI}\times\bar{x}\times\dfrac{\bar{x}}{2} = \dfrac{900}{2EI}(\bar{x})^2$ ...(1)

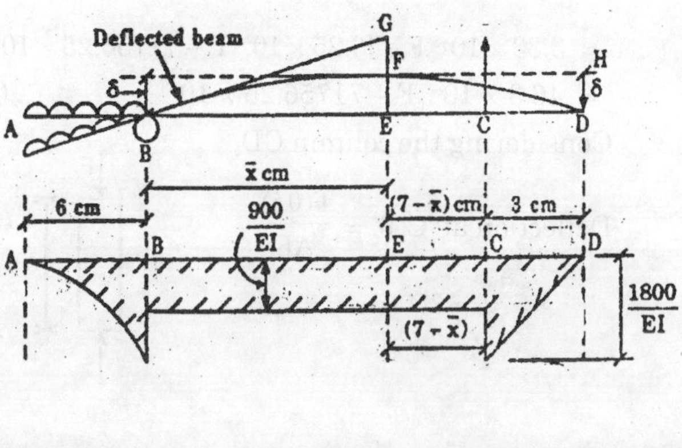

Same deflection can be obtained by finding the moment of $\frac{M}{EI}$ diagram between E and D about 'D'.

$\delta = EF$

$$= \frac{1}{2}(3) \times \left(\frac{1800}{EI}\right) \times \left(\frac{2}{3} \times 3\right) + (7-\bar{x})\left(\frac{900}{EI}\right)\left[3 + \left(\frac{7-\bar{x}}{2}\right)\right]$$

$$= \frac{900}{EI} \times 6 + \left[\frac{900}{EI}\right]\left[(7-\bar{x})\left(\frac{13-\bar{x}}{2}\right)\right] \quad ...(2)$$

Equating equations (1) and (2), we get

$$BE = \bar{x} = 5.15 \text{ cm}$$

$$\text{Maximum Deflection, } \delta = \frac{900}{2EI} \times (\bar{x})^2$$

$$= \frac{900}{2 \times 10^5} \times (5.15)^2$$

$$= \mathbf{0.119 \text{ cm}}$$

125. Explanation : $(EI)_{AB} = 10^{11}$ N-cm^2 = 10^7 Nm^2

$(EI)_{BC} = 8 \times 10^{11}$ N-cm^2 = 8×10^7 Nm^2

$(EA)_{DC} = 10^7$ N.

Let the load on DC be 'F' N.

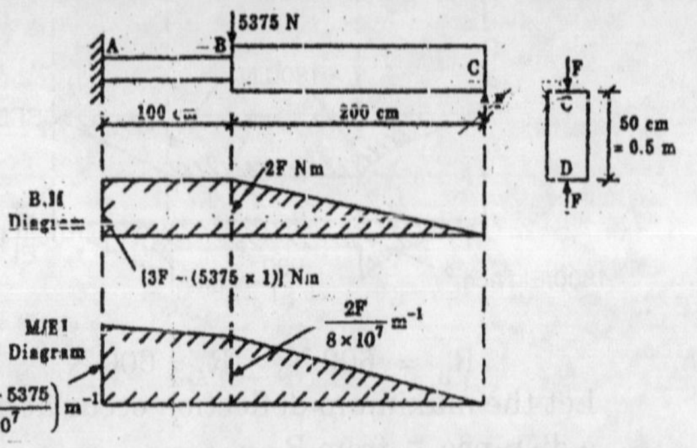

$$\text{Deflection at C, } \delta = \text{Area of } \frac{M}{EI} \text{ diagram about 'C'}$$

$$= \left(\frac{1}{2} \times 2 \times \frac{2F}{8 \times 10^7}\right)\left(\frac{2}{3} \times 2\right) + \left\{1 \times \frac{2F}{8 \times 10^7} \times \left(\frac{1}{2} + 2\right)\right.$$

$$\left. + \frac{1}{2} \times 1 \times \left(\frac{2}{3} \times 1 + 2\right)\left(\frac{3F - 5375}{10^7} - \frac{2F}{8 \times 10^7}\right)\right\}$$

$= 3.33 \times 10^{-8}$ F + {6.25×10^{-8} F – 71756.25×10^{-8}}

$= 46.3 \times 10^{-8}$ F – 71756.25×10^{-8} ...(1)

Considering the column CD,

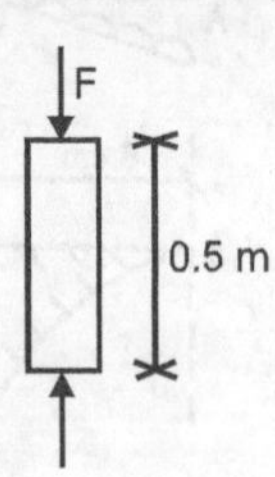

$$\text{Deflection at 'C' } \delta_C = \frac{F(0.5)}{AE}$$

$$\delta_C = \frac{0.5F}{10^7}$$

$$= 5 \times 10^{-8}\text{F} \quad ...(2)$$

Equating equations (1) and (2), we get

5×10^{-8} F= 46.3×10^{-8} F – 71756.25×10^{-8}

Solving we get,

Load on CD F = **1737.44 N**

126. Let F_e = centrifugal force acting outwards

F_f = Frictional force acting inwards

r = radius of curve = 100 m.

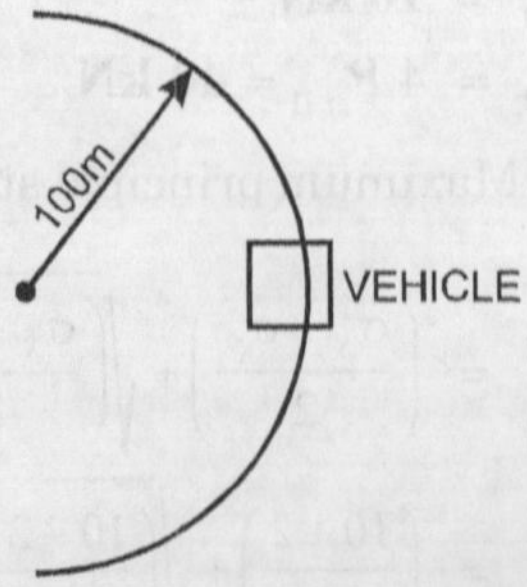

Let 'v' be the velocity of the vehicle at which both the slip as well as lifting up of a pair of one side wheels takes place.

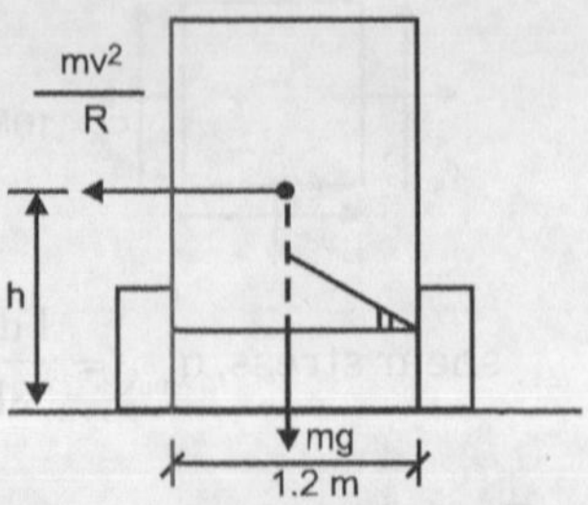

$$\text{Centrifugal force, } F_e = \frac{mv^2}{g}$$

where 'm' = mass of the vehicle.

Vehicle will slip if this is equal to the frictional resistance μmg.

$$\therefore \quad \frac{mv^2}{R} = \mu mg$$

$$v = \sqrt{\mu g R}$$

$$= \sqrt{0.5 \times 9.8 \times 100}$$

$$= \mathbf{22.1 \text{ m/s}}$$

Let 'h' be the height of CG above ground

Turning couple = $F_c \times h$

A pair of one side wheels will lift from the ground if

$$F_c \times h \geq mg \times \frac{1.2}{2}$$

$$h \geq mg \times \frac{0.6}{mv^2 / R}$$

$$\geq \frac{100 \times 9.8 \times 0.6}{(22.1)^2} = \mathbf{1.2\ m}$$

128. $\cos\theta = \frac{125}{302}$

$\Rightarrow \quad \theta = 65.5°$

Resolving the forces vertically,

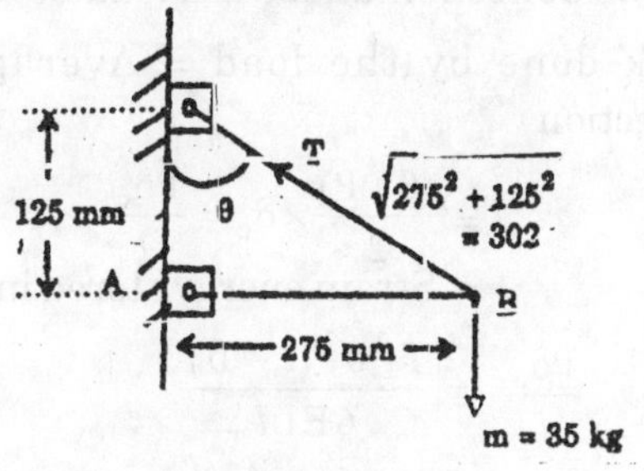

$$35 \times 9.81 = T\cos 65.5°$$

or $\quad T = 830$ N

$R_y = 0$

Resolving the forces horizontally,

$T \sin 65.5° = R_x$

or $\quad 830 \times 0.91 = R_x$

or $\quad R_x = \mathbf{755.4\ N}$

129. According to distorsion energy theorem

$$f_1^2 + f_2^2 - f_1 f_2 = (\text{Working stress})^2$$

$$\text{Working stress} = \sqrt{(360)^2 + (140)^2 - (360 \times 140)}$$

$$= \mathbf{314.32\ MPa}$$

130.

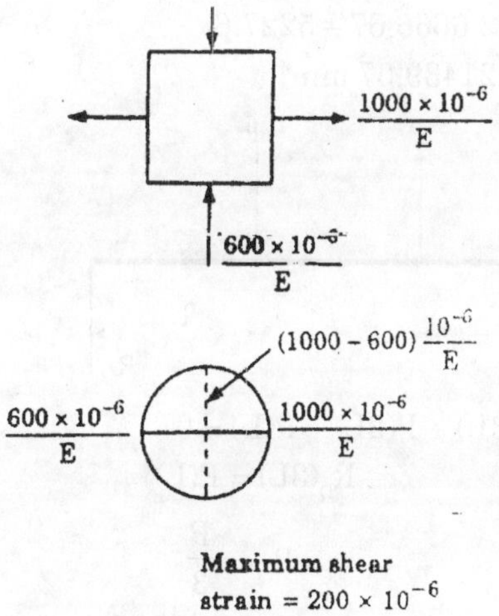

131. $\frac{d}{p} = 0.25$

$$\eta_{tearing} = \left(\frac{P-d}{P}\right) \times 100 = 75\%$$

132. Maximum tensile stress due to bending,

$$f_t = \frac{M}{Z} = \frac{3 \times 10^3}{\left(\frac{\frac{\pi}{64} \times d^4}{\frac{d}{2}}\right)} = \frac{30.558}{d^3} \text{ kPa.}$$

Maximum shear stress due to torsion, $f_s = \frac{TR}{J}$

$$= \frac{4 \times 10^3 \times d/2}{\frac{\pi}{32} \times d^4} = \frac{20.7}{d^3} \text{ kPa.}$$

When both the stress acts simultaneously, maximum induced shear stress,

$$f_{s\,max.} = \frac{1}{2}\left[\sqrt{(f_t)^2 + 4(f_s)^2}\right]$$

Maximum torque that can be applied,

$$T = \frac{\pi d^3}{16} f_{s\,max.}$$

$$= \frac{\pi d^3}{16} \times \frac{1}{2}\sqrt{\left(\frac{30.558}{d^3}\right)^2 + 4\left(\frac{20.37}{d^3}\right)^2}$$

$$= \mathbf{5\ kNm}$$

135. *Given*, Internal diameter = 0.1 m

Thickness = 0.005 m

Longitudinal stress, $p_2 = \frac{pd}{4t}$

$$= \frac{10 \times 10^6 \times 0.1}{4 \times 0.005} = 50 \times 10^7 \text{ N / m}^2$$

$$= 50 \text{ MN / m}^2$$

Internal pressure, $p = 10 \times 10^6$ Pa $= 10^7$ N/m²

Hoop stress, $\quad p = \frac{pd}{2t} = 100 \text{ MN / m}^2$

Force at the rim of cylinder

$$= \frac{\text{Torque}}{\text{Radius}} = \frac{2000}{0.05} = 40000 \text{ N}$$

$$\text{Shear stress, } q = \frac{\text{Force}}{\pi dt}$$

$$= \frac{40000}{\pi \times 0.1 \times 0.005} = 2.5464 \times 10^6 \text{ N / m}^2$$

$$\text{Principal stresses} = \frac{p_1 - p_2}{2} \pm \sqrt{\left(\frac{p_1 - p_2}{2}\right)^2 + q^2}$$

$$= \frac{10 + 50}{2} \pm \sqrt{\left(\frac{100 - 50}{2}\right)^2 + (2.5464)^2}$$

$$= 75 \pm 25.129$$

$$= 100.129 \text{ MN/m}^2 \text{ and } 49.871 \text{ MN/m}^2$$

136. Taking moment about C

$$R_B \times 4 = 100 \times 2 + 200 \times 1$$

or $R_B = 100$ kN

$\therefore$ $R_C = 200$ kN

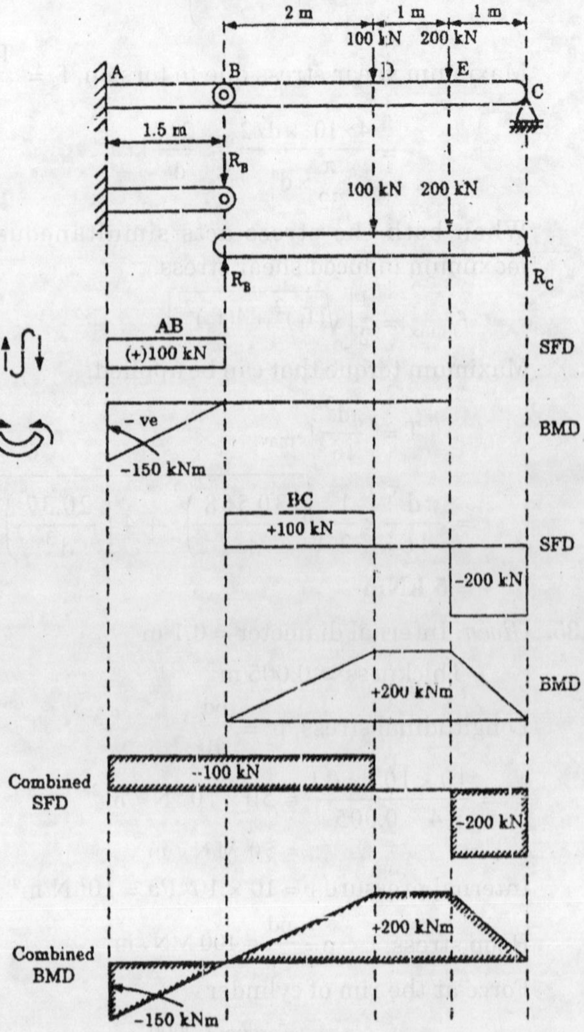

137. Let the load P acts at a distance b from the support B and l be the total length of the beam.

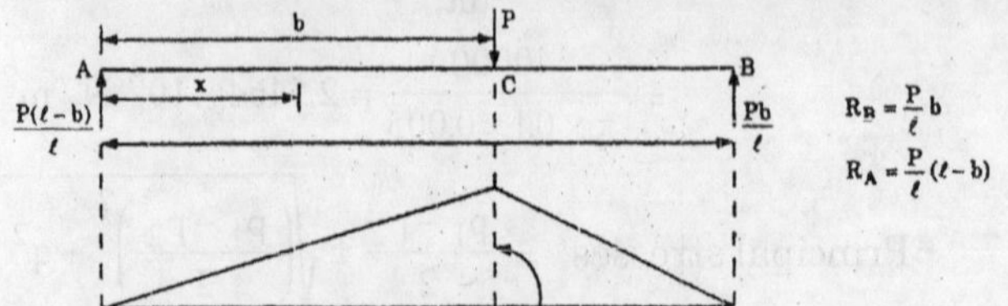

Strain energy stored in the beam due to bending.

U = Strain energy stored by the length AC + Strain energy stored by the length BC

$$= \int_0^b \left[\frac{P}{l}(l-b)\,x\right]^2 \frac{dx}{2EI} + \int_0^{(l-b)} \left[\left(\frac{Pn}{l}\right)x\right]^2 \frac{dx}{2EI}$$

$$= \frac{P^2(l-b)^2}{2EI\,l^2}\left[\frac{x^3}{3}\right]_0^b + \frac{P^2b^2}{2EI\,l^2}\left[\frac{x^3}{3}\right]_0^{l-b}$$

$$= \frac{P^2(l-b)^2}{2EI\,l^2}\left[(l-b)^2b^3 + b^2(l-b)^3\right]$$

$$= \frac{P^2}{6EI\,l^2}(l-b)^2b^2(b+l-b)$$

$$= \frac{P^2b^2(l-b)^2}{6EI\,l}$$

Let the deflection under the load be 'δ_c'.

Work done by the load = Average load × Deflection

$$= \frac{(0+P)}{2} \times \delta_c = \frac{P\delta_c}{2}$$

= Strain energy stored in the beam.

$$\therefore \quad \frac{P\delta_c}{l} = \frac{P^2b^2(l-b)}{6EI\,l}$$

$$\delta_c = \frac{Pb^2(l-b)^2}{3EI\,l}$$

Deflection at the mid point, $b = \frac{l}{2}$

$$\text{dmid point} = \frac{P\left(\frac{l}{2}\right)^2\left(l-\frac{l}{2}\right)^2}{3EI\,l} = \frac{Pl^3}{48EI}$$

NUMERICAL TYPE QUESTIONS

1. $I = \frac{10\times 20^3}{12} + 10\times 20(10)^2 - \left[\frac{\pi}{64}\times(8^4) + \frac{\pi}{4}\times(8^2)(10^2)\right]$

$= 2.6666.67 - 5227.6$

$= 21439.07$ mm^4

2.

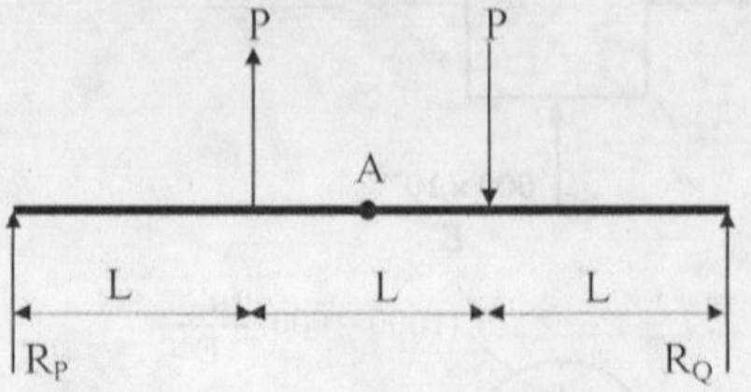

$$-R_Q(3L) + P(2L) - P(L) = 0$$

$$R_Q(3L) = P(L)$$

$$R_Q = \frac{P}{3}$$

$$M_A = R_Q(1.5L) - P(0.5L)$$

$$\Rightarrow \frac{P}{3}\left(\frac{3}{2}L\right) - \frac{P}{2}L$$

$$M_A = 0$$

$$\Rightarrow \qquad \sigma_A = 0 \qquad [\because \sigma \propto M]$$

3. Toughness = Area under diagram

$$= \frac{1}{2}\times\frac{0.2}{100}\times 100 + \frac{1}{2}\left(\frac{0.2}{100}\right)(100+140)$$

$$+\frac{1}{2}\left(\frac{0.2}{100}\right)(140+130)$$

T = 0.1 + 0.48 + 0.27 = 0.85 MJ/m³

4.

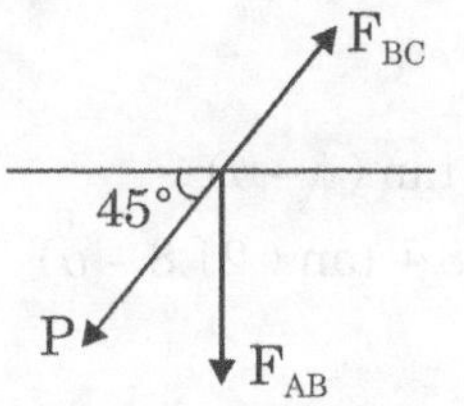

Applying lome's Theorem

$$\frac{F_{AB}}{\sin 180} = \frac{P}{\sin 135} = \frac{F_{BC}}{\sin 45}$$

$$F_{AB} = 0$$

5. $\ell \alpha t = 250 \times (10^{-5})(200)$

$$= \frac{1}{2} = 0.5 \text{ mm} > \text{gap}$$

Deformation prevented $= 0.5 - 0.2 = 0.3 = \frac{P\ell}{AE}$

$$\Rightarrow \qquad 0.3 = \frac{\sigma(250+0.2)}{200\times 10^3}$$

$$\sigma = 239.8 \text{ MPa}$$

6. $\frac{\Delta V}{V} = \frac{\sigma}{E}(1-\nu)$

$$\varepsilon_v = \varepsilon_x + \varepsilon_y + \varepsilon_z$$

$$\varepsilon_v = \left(\frac{\sigma}{E} - \frac{\mu\sigma}{E}\right) + \left(\frac{\sigma}{E} - \frac{\mu\sigma}{E}\right) - \frac{\mu(\sigma+\sigma)}{E}$$

Due to plane stress condition $\varepsilon_z = 0$

$$\varepsilon_v = \frac{\sigma}{E}\times 2 \times (1-\mu)$$

$$\frac{(2-2\times 0.001)^2 - 2^2}{2^2} = -\frac{P}{E}\times 2(1-\mu)$$

$$-2\times 10^{-3} = \frac{250}{200\times 10^3}\times 2(1-\mu)$$

$$\frac{250}{200} = 1-\mu$$

$$0.8 = 1-\mu$$

$$\therefore \qquad \mu = 0.2$$

7. Euler's critical load $= \frac{\pi^2 EI}{l^2}$

$$\Rightarrow \qquad \frac{\pi^2 \times 200\times 10^9 \times .02\times .01^3}{12} = 3289.8681 \text{ N}$$

8. Angle of twist at (C) = Angle of twist at (B)

$$\Rightarrow \quad \theta = \frac{TL}{GJ}$$

$$\Rightarrow \quad \frac{10\times 0.5\times 32}{77\times 10^9 \times \pi \times .024}$$

$$\Rightarrow \quad 0.236050$$

9. If there is uniaxial loading yield stress is σ_y

As per Von Mises failure theory

$$\sigma_y^2 = \sigma_1^2 + \sigma_1\sigma_2 + \sigma_2^2$$

Under pure shear stress loading

$$\sigma_1 = -\sigma_2 = \tau$$

then $\qquad \sigma_y^2 = \tau^2 + \tau^2 + \tau^2 = 3\tau^2$

$$\Rightarrow \qquad \tau = \frac{\sigma_y}{\sqrt{3}}$$

Hence $\qquad \tau = \frac{\sigma_y}{\sqrt{3}} = 0.5577\sigma_y = 173.28$

where τ is shear yield stress

10. P = 9 kN, e = 250 mm

Primary shear force

$$P_p' = P_q' = P_r' = \frac{p}{3} = \frac{q}{3} = 3 \text{ kN}$$

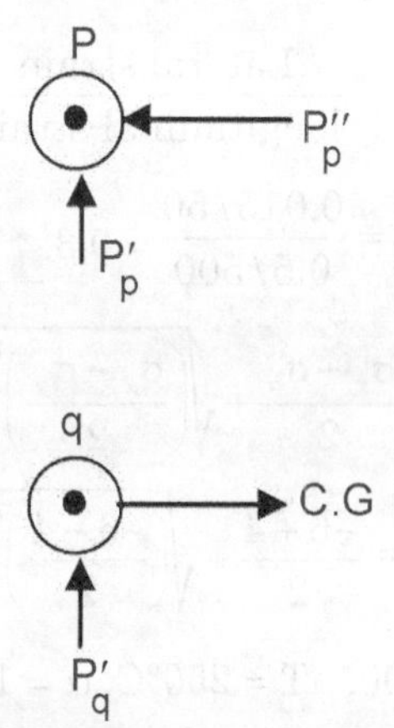

Secondary shear force:

By symmetry C.G lies at the centre of bolt Q.

$$\therefore \qquad r_p = 30 \text{ mm}$$

$$r_r = 30 \text{ mm}$$

$$r_q = 0$$

$$C = \frac{Pe}{r_q^2 + r_q^2 + r_r^2}$$

$$= \frac{9000 \times 250}{30^2 + 0^2 + 30^2}$$

$$= 1250$$

$$P''_p = P''_r = C$$

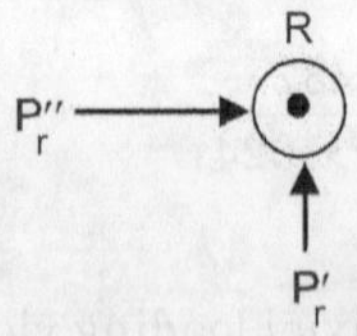

$$P''_p = Cr_p = 37.5 \text{ kN}$$

$$P'_r = Cr_r = 37.5 \text{ kN}$$

$$P''_q = 0$$

Resultant shear force

Due to symmetry stress in P Q R will be equal in magnitude

$$P_p = P_r = \sqrt{(3)^2 + (37.5)^2}$$

$$= 38.5648 \text{ kN}$$

$$\tau = \frac{P}{A} = \frac{38.5648}{\frac{\pi}{4}(0.012)^2}$$

$$= 340.987 \text{ MP}a \simeq 341 \text{ MP}a$$

11. l = 500 mm, d = 50 mm, p = 100 KN

δl = 0.5 mm, δd = 0.015

Poission's Ratio

$$\left(\frac{l}{m}\right) = \frac{\text{Lateral strain}}{\text{longitudinal strain}}$$

$$\frac{\delta d/d}{\delta l/l} = \frac{0.015/50}{0.5/500} = 0.3.$$

12. $$\sigma_1 = \frac{\sigma_x + \sigma_y}{2} + \sqrt{\left(\frac{\sigma_x - \sigma_y}{2}\right)^2 + (\tau_{xy})^2}$$

$$= \frac{-6+4}{2} + \sqrt{\left(\frac{-6-4}{2}\right)^2 + (-8)^2} = 8.43.$$

13. l = 200, ΔT = 250°C, $\alpha = 1 \times 10^{-5}$ /°C

E = 200 GPa = 200 × 10^3 MPa

$\sigma = \alpha\,\Delta$TE = 1 × 10^{-5} × 250 × 200 × 10^3 = 500 MPa.

14. $$\varepsilon = ln\ \frac{l_i}{l_0}$$

$$\frac{d\varepsilon}{dt} = \frac{l_0}{l_i} \times \frac{1}{l_0}\frac{dl_i}{dt} = \frac{2t}{(1+t^n)}$$

$$\frac{d\varepsilon}{dt} = \frac{2\times 1}{1+1} = 1.$$

15. Chip thickness ratio $r = \frac{0.2}{0.5} = \frac{t_1}{t_2}$

$$\tan\phi = \frac{r\cos\alpha}{1 - r\sin\alpha} = 0.4$$

$$\phi = 21.8°$$

Shear strain = cot ϕ + tan ($\phi - \alpha$)

= cot 21.8 + tan (21.8 − α)

= 2.9

16. Stress Amplitude $= \frac{(\sigma_{\max} - \sigma_{\min})}{2}$

$$= \frac{400 - 200}{2} = 100.$$

17. As per uniform wear theory

$$\text{Torque} = \frac{\pi\mu P_a d}{8}\left[D^2 - d^2\right]$$

$$= \frac{\pi \times 0.3 \times 1.5 \times 10^6}{8}\left[0.2^2 - 0.1^2\right]$$

$$= 530.14 \text{ N} - \text{m}$$

18. σ = min of $\frac{(S_{yt}, S_{ut}, S_e)}{F_0 S}$

$$\text{FOS} = \frac{200}{100} = 2.$$

19. $$\tau = \frac{16\,T}{\pi d^3}$$

$$\Rightarrow \frac{\tau_2}{\tau_1}\left(\frac{d}{d_2}\right)^3 = \left(\frac{R_1}{R_2}\right)^3 = 2^3 = 8.$$

20. From E = 2G(1 + μ)

$$\frac{G}{E} = \frac{1}{2(1+\mu)}$$

$$= \frac{1}{2(1+0.4)} = 0.357$$

21. $$\tau_{\max} = \frac{\sigma_u}{2\text{FOS}}$$

$$\Rightarrow \quad \text{FOS} = \frac{300}{2\times 100} = 1.5$$

■■

3

CHAPTER

Theory of Machines

MCQ TYPE QUESTIONS

2016

1. In the gear train shown, gear 3 is carried on arm 5. Gear 3 meshes with gear 2 and gear 4. The number of teeth on gear 2, 3, and 4 are 60, 20, and 100, respectively. If gear 2 is fixed and gear 4 rotates with an angular velocity of 100 rpm in the counterclockwise direction, the angular speed of arm 5 (in rpm) is

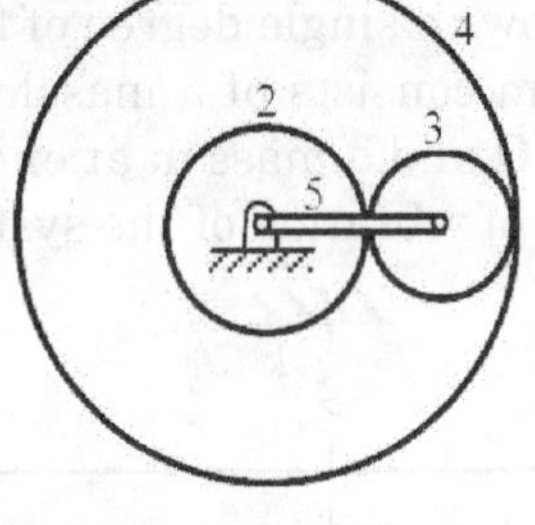

(*a*) 166.7 counterclockwise

(*b*) 166.7 clockwise

(*c*) 62.5 counterclockwise

(*d*) 62.5 clockwise

2. For the brake shown in the figure, which one of the following is TRUE?

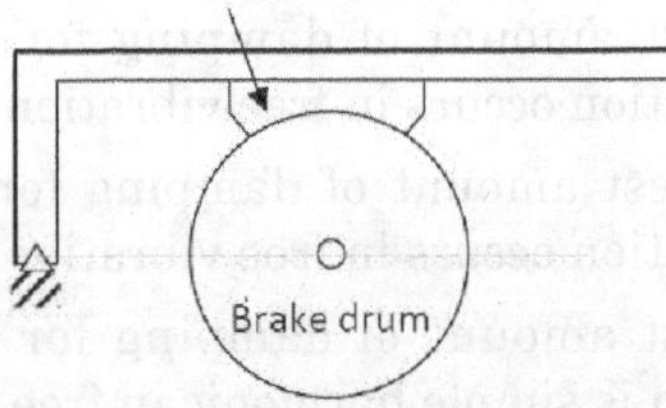

(*a*) Self energizing for clockwise rotation of the drum

(*b*) Self energizing for anti-clockwise rotation of the drum

(*c*) Self energizing for rotation in either direction of the drum

(*d*) Not of the self energizing type

3. A rigid link PQ is undergoing plane motion as shown in the figure (V_P and V_Q are non-zero). V_{QP} is the relative velocity of point Q with respect to point P.

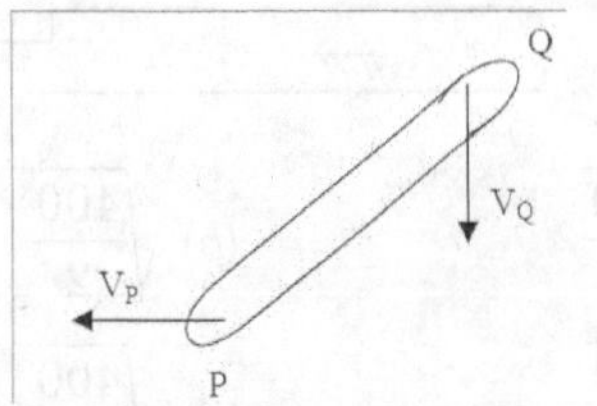

Which one of the following is TRUE?

(*a*) V_{QP} has components along and perpendicular to PQ

(*b*) V_{QP} has only one component directed from P to Q

(*c*) V_{QP} has only one component directed from Q to P

(*d*) V_{QP} has only one component perpendicular to PQ

4. The number of degrees of freedom in a planner mechanism having *n* links and *j* simple hinge joints is

(*a*) $3(n-3)-2j$ (*b*) $3(n-1)-2j$

(*c*) $3n-2j$ (*d*) $2j-3n+4$

2015

5. Consider a slider crank mechanism with nonzero masses and inertia. A constant torque τ is applied on the crank as shown in the figure. Which of the following plots best resembles variation of crank angle, θ versus time

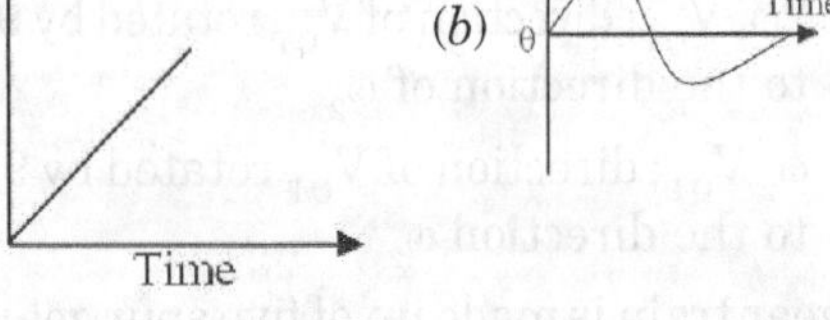

(*a*)

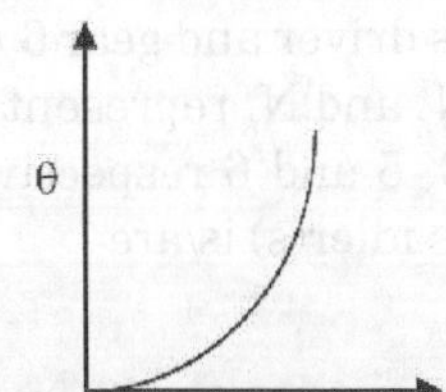

(*b*)

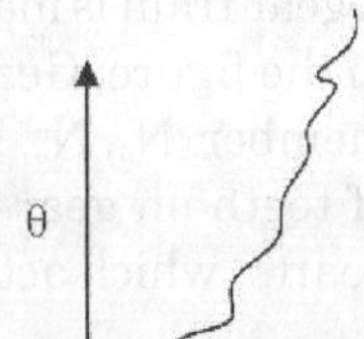

(*c*)

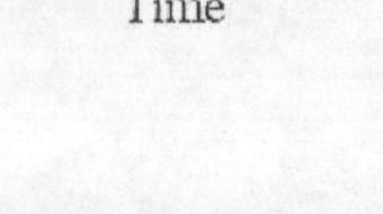

(*d*)

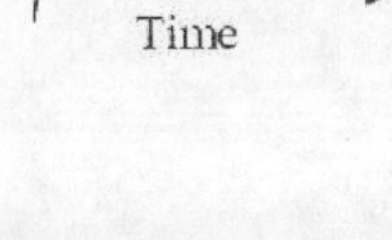

6. Considering massless rigid rod and small oscillations, the natural frequency (in rad/s) of vibration of the system shown in the figure is

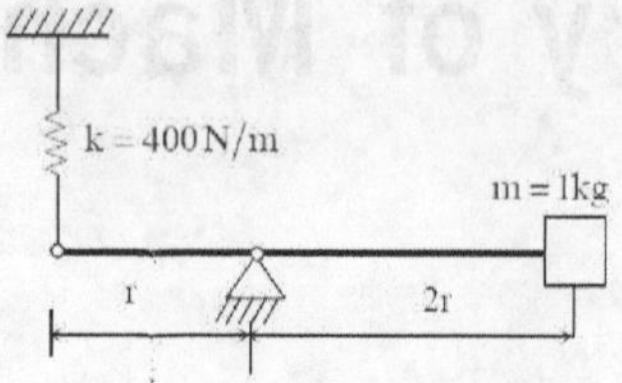

(*a*) $\sqrt{\dfrac{400}{1}}$ (*b*) $\sqrt{\dfrac{400}{2}}$

(c) $\sqrt{\dfrac{400}{3}}$ (*d*) $\sqrt{\dfrac{400}{4}}$

7. A single-degree. Freedom spring-mass system is subjected to a sinusoidal force of 10 N amplitude and frequency ω along the axis of the spring. The stiffness of the spring is 150 N/m, damping factor is 0.2 and the undamped natural frequency is 10.ω At steady state, the amplitude of vibration (in m) is approximately

(*a*) 0.05 (*b*) 0.07

(*c*) 0.70 (*d*) 0.90

8. In the figure, link 2 rotates with constant angular velocity ω_2. A slider link 3 moves outwards with a constant relative velocity $V_{Q/P,}$ where Q is a point on slider 3 and P is a point on link 2. The magnitude and direction of Coriolis component of acceleration is given by

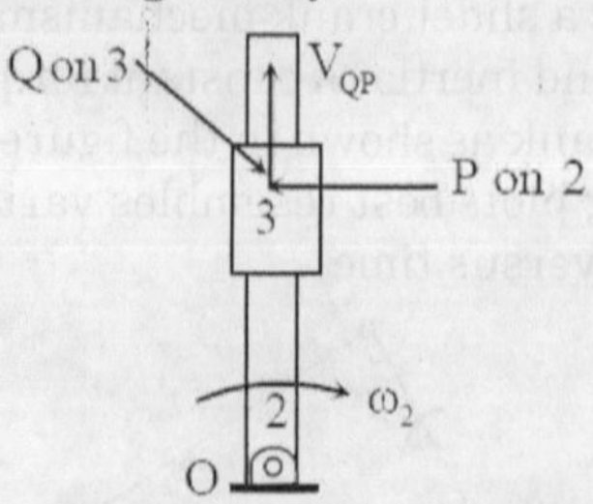

(*a*) $2\omega_2 V_{Q/P}$; direction of $V_{Q/P}$ rotated by 90° in the direction ω_2

(*b*) $\omega_2 V_{Q/P;}$ direction of $V_{Q/P}$ rotated by 90° in the direction ω_2

(*c*) $2\omega_2 V_{Q/P}$; direction of $V_{Q/P}$ rotated by 90° opposite to the direction of ω_2

(*d*) $\omega_2 V_{Q/P}$; direction of $V_{Q/P}$ rotated by 90° opposite to the direction ω_2

9. A gear train is made up of five spur gears as shown in the figure. Gear 2 is driver and gear 6 is driven member. N_2, N_3, N_4, N_5 and N_6 represent number of teeth on gears 2, 3, 5 and 6 respectively. The gear(s) which act(s) as idler(s) is/are

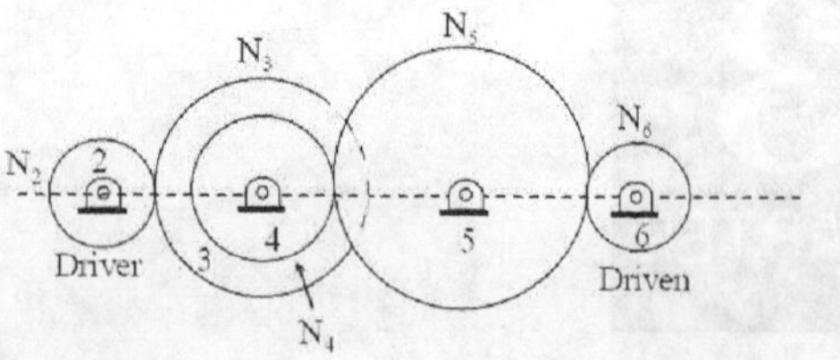

(*a*) only 3 (*b*) only 4

(*c*) only 5 (*d*) Both 3 and 5

10. The number of degrees of freedom of the linage shown in the figure is

(*a*) -3 (*b*) -0

(*c*) 1 (*d*) 2

11. Figure shows a single degree of freedom system. The system consists of a massless rigid bar OP hinged at O and a mass m at end P. The natural frequency of vibration of the system *i*

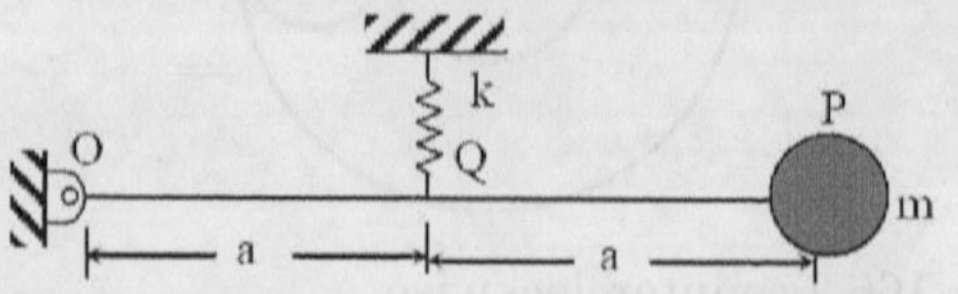

(*a*) $f_n = \dfrac{1}{2\pi}\sqrt{\dfrac{k}{4m}}$ (*b*) $f_n = \dfrac{1}{2\pi}\sqrt{\dfrac{k}{2m}}$

(*c*) $f_n = \dfrac{1}{2\pi}\sqrt{\dfrac{k}{m}}$ (*d*) $f_n = \dfrac{1}{2\pi}\sqrt{\dfrac{2k}{m}}$

2014

12. Critical dampimg is the

(*a*) largest amount of damping for which no oscillation occurs in free vibration

(*b*) smallest amount of damping for which no oscillation occurs in free vibration

(*c*) largest amount of damping for which the motion is simple harmonic in free vibration

(*d*) smallest amount of damping for which the motion is simple harmonic in free vibration

13. A circular object of radius *r* rolls without slipping on a horizontal level floor with the center having velocity V. The velocity at the point of contact between the object and the floor is

(*a*) zero

(*b*) V in the direction of motion

(*c*) V opposite to the direction of motion

(*d*) V vertically upward from the floor

14. A rigid link PQ is 2 m long and oriented at 20° to the horizontal as shown in the figure. The magnitude and direction of velocity V_Q, and the direction of velocity V_P are given. The magnitude of V_P (in m/s) at this instant is

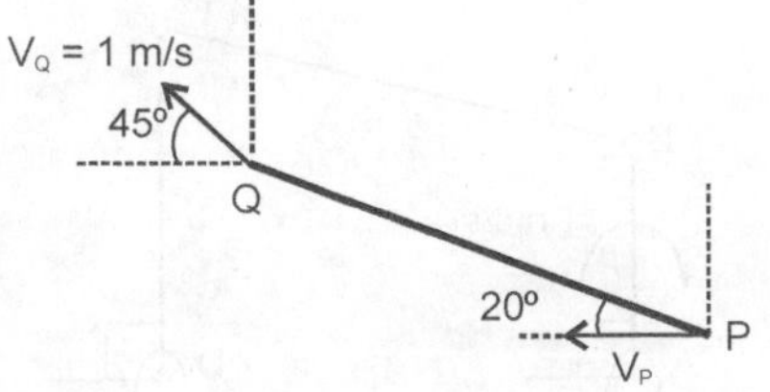

(*a*) 2.14 (*b*) 1.89
(*c*) 1.21 (*d*) 0.96

15. A rigid uniform rod AB of length L and mass m is hinged at C such that AC = L/3, CB = 2L/3. Ends A and B are supported by springs of spring constant *k*. The natural frequency of the system is given by

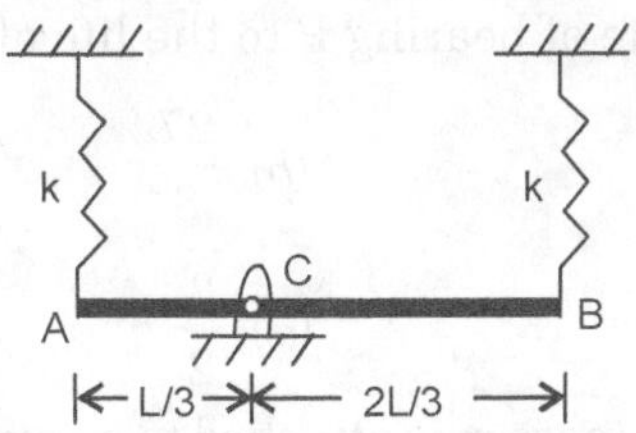

(*a*) $\sqrt{\frac{k}{2m}}$ (*b*) $\sqrt{\frac{k}{m}}$
(*c*) $\sqrt{\frac{2k}{m}}$ (*d*) $\sqrt{\frac{5k}{m}}$

16. Which one of the following is a CFC refrigerant?
(*a*) R744 (*b*) R290
(*c*) R502 (*d*) R718

17. A 4-bar mechanism with all revolute pairs has link lengths l_f = 20 mm, l_{in} = 40 mm, l_{co} = 50 mm and l_{out} = 60 mm. The suffixes '*f*', '*in*', '*co*' and '*out*' denote the fixed link, the input link, the coupler and output link respectively. Which one of the following statements is true about the input and output links?
(*a*) Both links can execute full circular motion
(*b*) Both links cannot execute full circular motion
(*c*) Only the output link cannot execute full circular motion
(*d*) Only the input link cannot execute full circular motion

18. In vibration isolation, which one of the following statements is **NOT** correct regarding Transmissibility (T)?
(*a*) T is nearly unity at small excitation frequencies
(*b*) T can be always reduced by using higher damping at any excitation frequency
(*c*) T is unity at the frequency ratio of $\sqrt{2}$
(*d*) T is infinity at resonance for undamped systems

2013

19. A link OB is rotating with a constant angular velocity of 2 *rad/s* in counter clockwise direction and a block is sliding radially outward on it with an uniform velocity of 0.75 *m/s* with respect to the rod, as shown in the figure below. If OA = 1 *m* the magnitude of the absolute acceleration of the block at location A in *m/s*² is

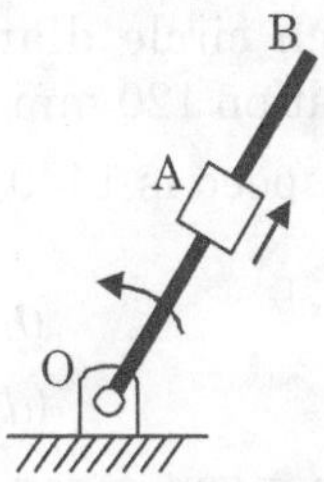

(*a*) 3 (*b*) 4
(*c*) 5 (*d*) 6

20. A planar closed kinematic chain is formed with rigid links PQ = 2.0 m, QR = 3.0 m, RS = 2.5 m and SP = 2.7 m with all revolute joints. The link to be fixed to obtain a double rocker (rocker-rocker) mechanism is
(*a*) PQ (*b*) QR
(*c*) RS (*d*) SP

21. A flywheel connected to a punching machine has to supply energy of 400 Nm while running at a mean angular speed of 20 rad/s. If the total fluctuation of speed is not to exceed ±2%, the mass moment of inertia of the flywheel in kg/m² is
(*a*) 25 (*b*) 50
(*c*) 100 (*d*) 125

22. A compound gear train with gears P, Q, R and S has number of teeth 20, 40, 15 and 20, respectively. Gears Q and R are mounted on the same shaft as shown in the figure below. The diameter of the gear Q is twice that of the gear R. If the module of the gear R is 2 mm, the center distance in mm between gears P and S is

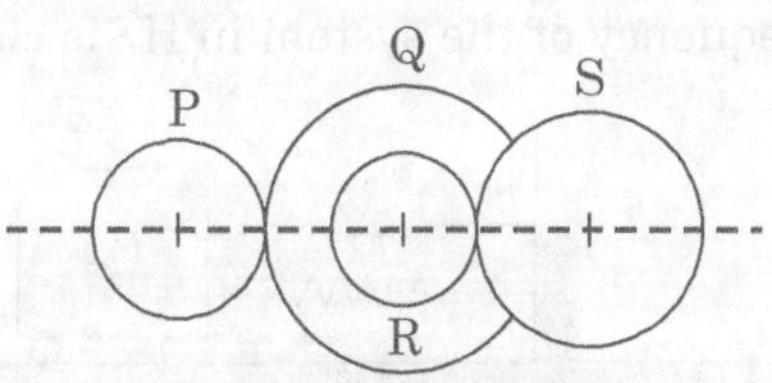

(*a*) 40 (*b*) 80
(*c*) 120 (*d*) 160

23. A single degree of freedom system having mass 1 kg and stiffness 10 kN/m initially at rest is subjected to an impulse force of magnitude 5 kN for 10^{-4} seconds. The amplitude in mm of the resulting free vibration is

(*a*) 0.5 (*b*) 1.0

(*c*) 5.0 (*d*) 10.0

2012

24. The following are the data for two crossed helical gears used for speed reduction:

Gear I : Pitch circle diameter in the plane of rotation 80 mm and helix angle 30°

Gear II: Pitch circle diameter in the plane of rotation 120 mm and helix angle 22.5°

If the input speed is 1440 *rpm*, the output speed in *rpm* is

(*a*) 1200 (*b*) 900

(*c*) 875 (*d*) 720

25. In the mechanism given below, if the angular velocity of the eccentric circular disc is 1 *rad/s*, the angular velocity (*rad*/s) of the follower link for the instant shown in the figure is

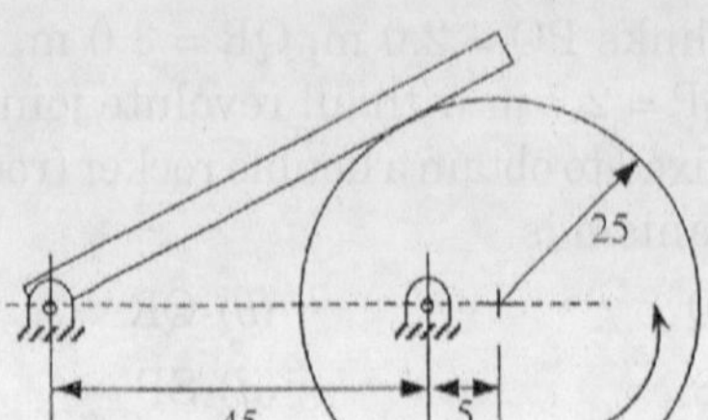

Note : All dimensions are in *mm*.

(*a*) 0.05 (*b*) 0.1

(*c*) 5.0 (*d*) 10.0

26. A circular solid disc of uniform thickness 20 *mm*, radius 200 *mm* and mass 20 *kg*, is used as a flywheel. If it rotates at 600 *rpm*, the kinetic energy of the flywheel, in *Joules* is

(*a*) 395 (*b*) 790

(*c*) 1580 (*d*) 3160

2011

27. A mass of 1 kg is attached to two identical springs each with stiffness k = 20 kN/m as shown in the figure. Under frictionless conditions, the natural frequency of the system in Hz is close to

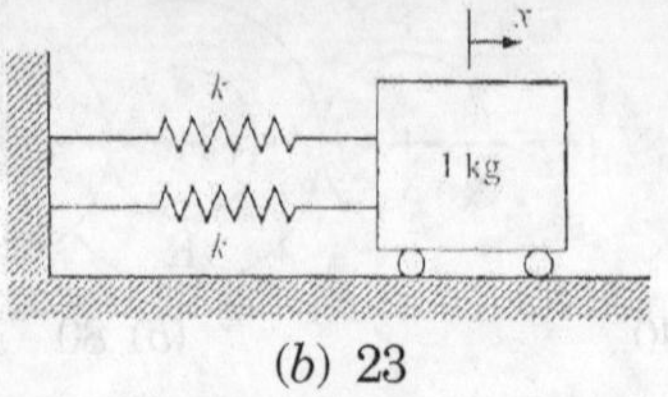

(*a*) 32 (*b*) 23

(*c*) 16 (*d*) 11

28. For the four-bar linkage shown in the figure, the angular velocity of link AB is 1 rad/s. The length of link CD is 1.5 times the length of link AB. In the configuration shown,the angular velocity of link CD in rad/s is

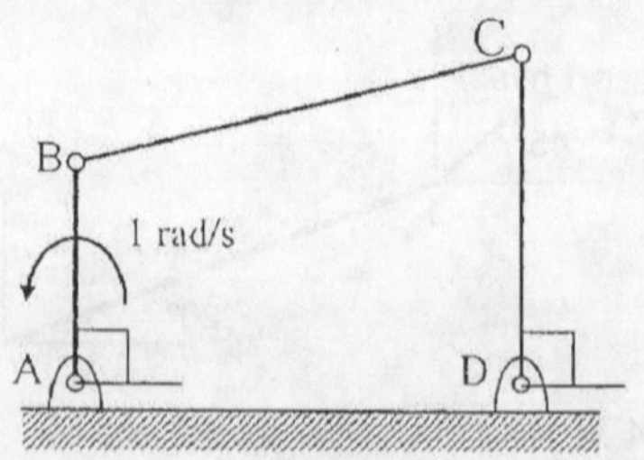

(*a*) 3 (*b*) $\frac{3}{2}$

(*c*) 1 (*d*) $\frac{2}{3}$

29. Two identical ball bearings P and Q are operating at loads 30 kN and 45 kN respectively. The ratio of the life of bearing P to the life of bearing Q is

(*a*) $\frac{81}{16}$ (*b*) $\frac{27}{8}$

(*c*) $\frac{9}{4}$ (*d*) $\frac{3}{2}$

30. A disc of mass m is attached to a spring of stiffness k as shown in the figure. The disc rolls without slipping on a horizontal surface. The natural frequency of vibration of the system is

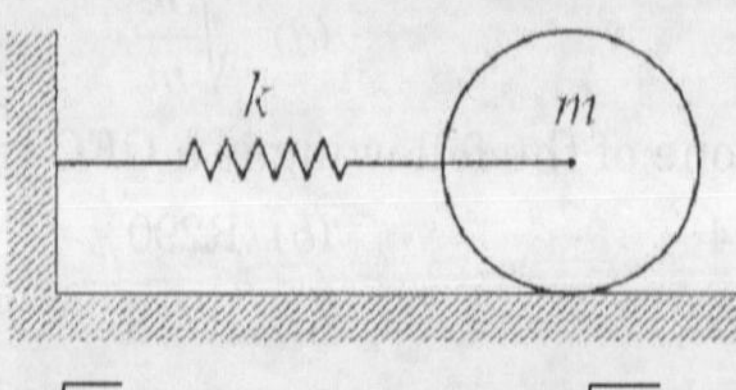

(*a*) $\frac{1}{2\pi}\sqrt{\frac{k}{m}}$ (*b*) $\frac{1}{2\pi}\sqrt{\frac{2k}{m}}$

(*c*) $\frac{1}{2\pi}\sqrt{\frac{2k}{3m}}$ (*d*) $\frac{1}{2\pi}\sqrt{\frac{3k}{2m}}$

2010

31. Which of the following statements is INCORRECT ?

(*a*) Grashof's rule states that for a planar crank-rocker four bar mechanism, the sum of the shortest and longest link lengths cannot be less than the sum of the remaining two link lengths

(*b*) Inversions of a mechanism are created by fixing different links one at a time

(*c*) Geneva mechanism is an intermittent motion device

(*d*) Gruebler's criterion assumes mobility of a planar mechanism to be one

2009

32. A simple quick return mechanism is shown in the figure. The forward to return ratio of the quick return mechanism is 2 : 1. If radius of the crank O_1P is 125 mm, then distance 'd' (in mm) between crank centre to lever pivot centre point should be

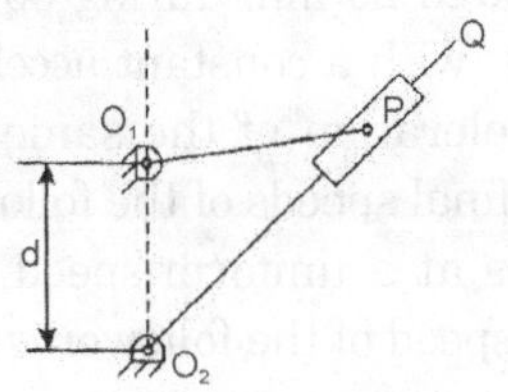

(*a*) 144.3 (*b*) 216.5
(*c*) 240.0 (*d*) 250.0

33. An epicyclic gear train is shown schematically in the given figure.

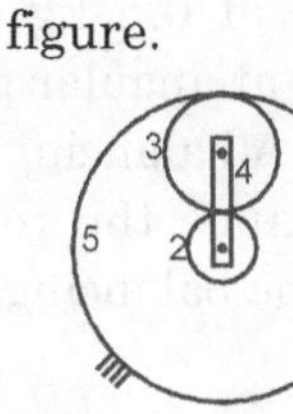

The sun gear 2 on the input shaft is a 20 teeth external gear. The planet gear 3 is a 40 teeth external gear. The ring gear 5 is a 100 teeth internal gear. The ring gear 5 is fixed and the gear 2 is rotating at 60 rpm ccw (ccw = counter-clockwise and cw = clockwise). The arm 4 attached to the output shaft will rotate at

(*a*) 10 rpm ccw (*b*) 10 rmp cw
(*c*) 12 rpm cw (*d*) 12 rpm ccw

34. Match the approaches given below to perform stated kinematics/ dynamics analysis of machine.

List I — *Analysis*

P. Continous relative rotation
Q. Velocity and acceleration
R. Mobility
S. Dynamic-static analysis

List II — *Approach*

1. D' Alembert's principle
2. Grubler's criterion
3. Grashoff's law
4. Kennedy's theorem

Codes :

	P	Q	R	S
(*a*)	1	2	3	4
(*b*)	3	4	2	1
(*c*)	2	3	4	1
(*d*)	4	2	1	3

2008

35. A planar mechanism has 8 links and 10 rotary joints. The number of degrees of freedom of the mechanism, using Gruebler's criterion, is

(*a*) 0 (*b*) 1
(*c*) 2 (*d*) 3

36. A circular disk of radius R rolls without slipping at a velocity v. The magnitude of the velocity at point P (see figure) is

(*a*) $\sqrt{3}$ v
(*b*) $\sqrt{3}$ v/2
(*c*) v/2
(*d*) 2v/ $\sqrt{3}$

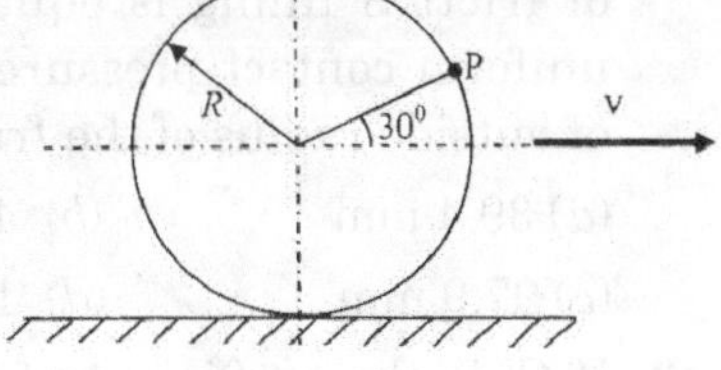

2007

37. The input link O_2P of a four bar linkage is rotated at 2 rad/s counter clockwise direction as shown below. The angular velocity of the coupler PQ in rad/s, at an instant when $\angle O_4O_2P$ = 180°, is

(*a*) 4
(*b*) $2\sqrt{2}$
(*c*) 1
(*d*) $\frac{1}{\sqrt{2}}$

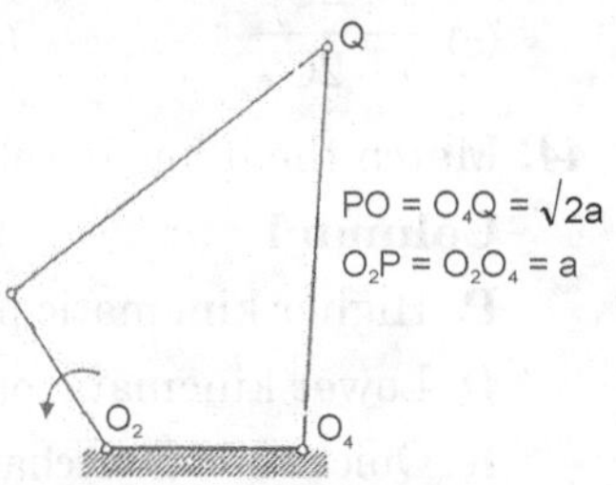

38. The speed of an engine varies from 210 rad/s to 190 rad/s. During a cycle, the change in kinetic energy is found to be 400 Nm. The inertia of the flywheel in kgm^2 is

(*a*) 0.10 (*b*) 0.20
(*c*) 0.30 (*d*) 0.40

2006

39. For a four-bar linkage in toggle position, the value of mechanical advantage is

(*a*) 0.0 (*b*) 0.5
(*c*) 1.0 (*d*) ∞

40. The number of inversions for a slider crank mechanism is

(*a*) 6 (*b*) 5
(*c*) 4 (*d*) 3

41. Match the items in columns I and II.

Column I	Column II
P. Addendum	1. Cam
Q. Instantaneous center of velocity	2. Beam
R. Section modulus	3. Linkage
S. Prime circle	4. Gear

Codes:

	P	Q	R	S
(a)	4	2	3	1
(b)	4	3	2	1
(c)	3	2	1	4
(d)	3	4	1	2

42. A disk clutch is required to transmit 5 kW at 2000 rpm. The disk has a friction lining with coefficient of friction equal to 0.25. Bore radius of friction lining is equal to 25 mm. Assume uniform contact pressure of 1 MPa. The value of outside radius of the friction lining is

(a) 39.4 mm (b) 49.5 mm

(c) 97.9 mm (d) 142.9 mm

43. If C_f is the coefficient of speed fluctuation of a flywheel, then ratio of $\omega_{max}/\omega_{min}$ will be

(a) $\dfrac{1-2C_f}{1+2C_f}$ (b) $\dfrac{2-C_f}{1+2C_f}$

(c) $\dfrac{1+2C_f}{1-2C_f}$ (d) $\dfrac{2+C_f}{2-C_f}$

44. Match the items in columns I and II

Column I	Column II
P. Higher kinematic pair	**1.** Grubler's equation
Q. Lower kinematic pair	**2.** Line contact
R. Quick return mechanism	**3.** Euler's equation
S. Mobility of a linkage	**4.** Planer
	5. Shaper
	6. Surface contact

Codes:

	P	Q	R	S
(a)	2	6	4	3
(b)	6	2	4	1
(c)	6	2	5	3
(d)	2	6	5	1

45. A machine of 250 kg mass is supported on springs of total stiffness 100 kN/m. Machine has an unbalanced rotating force of 350 N at speed of 3600 rpm. Assuming a damping factor of 0.15, the value of transmissibility ratio is

(a) 0.0531 (b) 0.9922

(c) 0.0162 (d) 0.0028

46. In a four-bar linkage, S denotes the shortest link length, L is the longest link length, P and Q are lengths of other two links. Atleast one of the three moving links will rotate by 360° if

(a) S + L ≤ P + Q (b) S + L > P + Q

(c) S + P ≤ L + Q (d) S + P > L + Q

2005

47. The number degrees of freedom of a planar linkage with 8 links and 9 simple revolute joints is

(a) 1 (b) 2

(c) 3 (d) 4

48. In a cam-follower mechanism, the follower needs to rise through 20 mm during 60° of cam rotation, the first 30° with a constant acceleration and then with a deceleration of the same magnitude. The initial and final speeds of the follower are zero. The cam rotates at a uniform speed of 300 rpm. The maximum speed of the follower is

(a) 0.60 m/s (b) 1.20 m/s

(c) 1.68 m/s (d) 2.40 m/s

49. A rotating disc of 1 m diameter has two eccentric masses of 0.5 kg each at radii of 50 mm and 60 mm at angular positions of 0° and 150°, respectively. A balancing mass of 0.1 kg is to be used to balance the rotor. What is the radial position of the balancing mass?

(a) 50 mm

(b) 120 mm

(c) 150 mm

(d) 280 mm

50. A weighing machine consists of a 2 kg pan resting on spring. In this condition, with the pan resting on the spring, the length of the spring is 200 mm. When a mass of 20 kg is placed on the pan, the length of the spring becomes 100 mm. For the spring, the un-deformed length l0 and the spring constant k (stiffness) are

(a) l_0 = 220 mm, k = 1862 N/m

(b) l_0 = 210 mm, k = 1960 N/m

(c) l_0 = 200 mm, k = 1960 N/m

(d) l_0 = 200 mm, k = 2156 N/m

2004

51. For a mechanism shown below, the mechanical advantage for the given configuration is

(a) 0 (b) 0.5

(c) 1.0 (d) ∞

52. In the figure shown, the relative velocity of link 1 with respect to link 2 is 12 m/sec. Link 2 rotates at a constant speed of 120 rpm. The magnitude of Coriolis component of acceleration of link 1 is

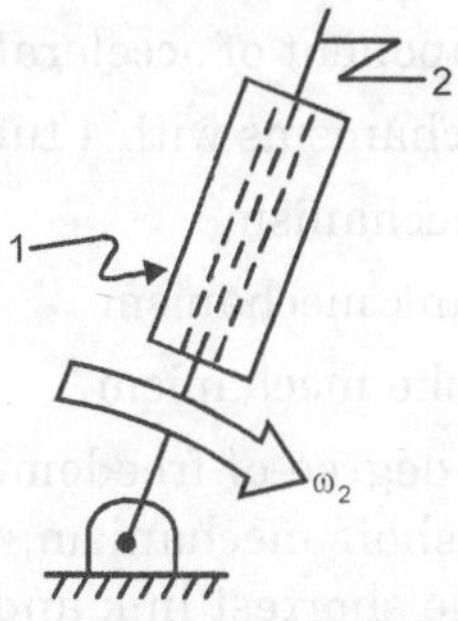

(*a*) 302 m/s² (*b*) 604 m/s²

(*c*) 906 m/s² (*d*) 1208 m/s²

53. **The figure below shows a planar mechanism with single degree of freedom. The instant center 24 for the given configuration is located at a position**

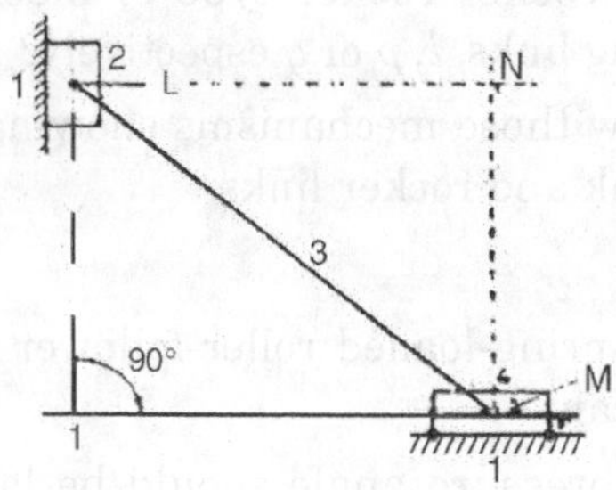

(*a*) L (*b*) M

(*c*) N (*d*) ∞

54. **A uniform stiff rod of length 300 mm and having a weight of 300 N is pivoted at one end and connected to a spring at the other end. For keeping the rod vertical in a stable position, the minimum value of spring constant K needed is**

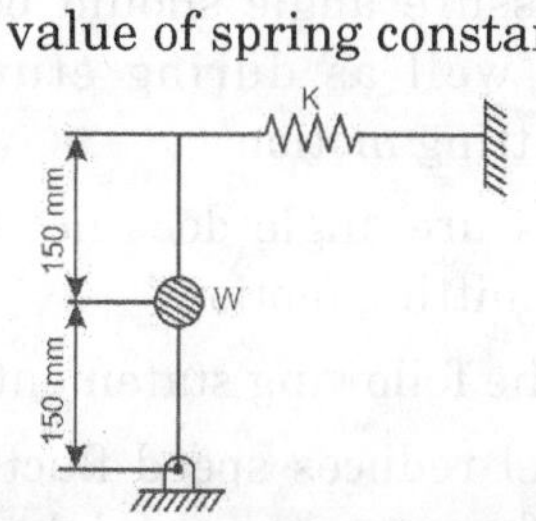

(*a*) 300 N/m (*b*) 400 N/m

(*c*) 500 N/m (*d*) 1000 N/m

55. **An ejector mechanism consists of a helical compression spring having a spring constant of K = 981 × 10³ N/m. It is pre-compressed by 100 mm from its free state. If it is used to eject a mass of 100 kg held on it, the mass will move up through a distance of**

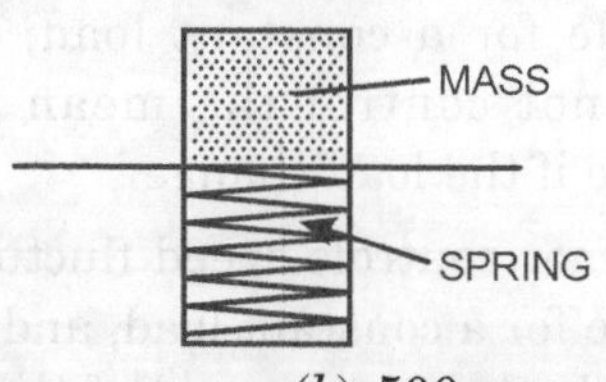

(*a*) 100 mm (*b*) 500 mm

(*c*) 981 mm (*d*) 1000 mm

56. **A rigid body shown in the Fig. (*a*) has a mass of 10 kg. It rotates with a uniform angular velocity 'ω'. A balancing mass of 20 kg is attached as shown in Fig. (*b*). The percentage increase in mass moment of inertia as a result of this addition is**

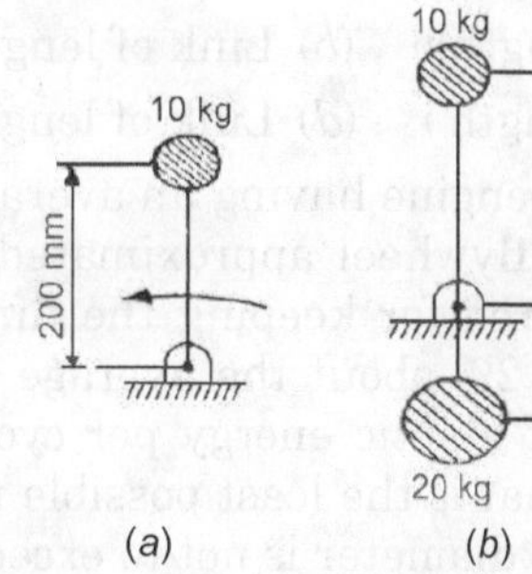

(*a*) (*b*)

(*a*) 25% (*b*) 50%

(*c*) 100% (*d*) 200%

57. **Match the following**

List I	List II
Type of Mechanism	*Motion achieved*
P. Scott – Russel mechanism	**1.** Intermittent motion
Q. Geneva mechanism	**2.** Quick return motion
R. Off-set slider-crank mechanism	**3.** Simple harmonic motion
S. Scotch Yoke mechanism	**4.** Straight line motion

Codes:

	P	Q	R	S
(*a*)	2	3	1	4
(*b*)	3	2	4	1
(*c*)	4	1	2	3
(*d*)	4	3	1	2

58. **Match the following with respect to spatial mechanisms.**

List I	List II
Type of Joint	*Degrees of constraint*
P. Revolute	**1.** Three
Q. Cylindrical	**2.** Five
R. Spherical	**3.** Four
	4. Two
	5. Zero

Codes:

	P	Q	R
(*a*)	1	3	3
(*b*)	5	4	3
(*c*)	2	3	1
(*d*)	4	5	3

2003

59. The lengths of the links of a 4-bar linkage with revolute pairs only are p, q, r, and s units. Given that $p < q < r < s$. Which of these links should be the fixed one, for obtaining a "*double crank*" mechanism ?

(*a*) Link of length p (*b*) Link of length q

(*c*) Link of length r (*d*) Link of length s

60. For a certain engine having an average speed of 1200 rpm, a flywheel approximated as a solid disc, is required for keeping the fluctuation of speed within 2% about the average speed. The fluctuation of kinetic energy per cycle is found to be 2kJ. What is the least possible mass of the flywheel if its diameter is not to exceed 1m?

(*a*) 40 kg (*b*) 51 kg

(*c*) 62 kg (*d*) 73 kg

Common Data Q. 61-62

The circular disc shown in its plan view in the figure rotates in a plane parallel to the horizontal plane about the point O at a uniform angular velocity ω. Two other points A and B are located on the line OZ at distances r_A and r_B from O respectively.

61. The velocity of point B with respect to point A is a vector of magnitude

(*a*) 0

(*b*) $\omega\,(r_B - r_A)$ and direction opposite to the direction of motion of point B

(*c*) $\omega\,(r_B - r_A)$ and direction same as the direction of motion of point B

(*d*) $\omega\,(r_B - r_A)$ and direction being from O to Z.

62. The acceleration of point B with respect to point A is a vector of magnitude

(*a*) 0

(*b*) $\omega\,(r_B - r_A)$ and direction same as the direction of motion of point B

(*c*) $\omega^2\,(r_B - r_A)$ and direction opposite to be direction of motion of point B

(*d*) $\omega^2\,(r_B - r_A)$ and direction being from Z to O.

2002

63. The minimum number of teeth on the pinion to operate without interference in standard full height involute teeth gear mechanism with 20 pressure angle is

(*a*) 14 (*b*) 12

(*c*) 18 (*d*) 32

64. The minimum number of links in a single degree-of-freedom planar mechanism with both higher and lower kinematic pairs is

(*a*) 2 (*b*) 3

(*c*) 4 (*d*) 5

65. Coriolis component of acceleration is present in

(*a*) 4-bar mechanisms with 4 turning pairs

(*b*) shaper mechanism

(*c*) slider-crank mechanism

(*d*) Scotch Yoke mechanism

66. In a single degree-of-freedom, 4-link, revolute jointed Grashoff mechanism, '*l*' is the longest link, 's' is the shortest link and p, q are the two remaining links. These links may be in any order.

(*a*) Write down the condition for feasibilty of Grashoff mechanism.

(*b*) When will you get crank-crank, crank rocker and rocker-rocker type of mechanisms by fixing links, l, p or q espectively?

(*c*) Draw these mechanisms showing motions of crank and rocker links.

2001

67. For a spring-loaded roller-follower driven with a disc cam

(*a*) the pressure angle should be larger during rise than that during return for ease of trasmitting motion

(*b*) the pressure angle should be smaller during rise than that during return for ease of transmitting motion

(*c*) the pressure angle should be large during rise as well as during eturn for ease of transmitting motion

(*d*) the pressure angle does not affect the ease of transmitting motion

68. Which of the following statement is correct?

(*a*) Flywheel reduces speed fluctuations during a cycle for a constant load, but flywheel does not control the mean speed of the engine if the load changes

(*b*) Flywheel does not educe speed fluctuations during a cycle for a constant load, but flywheel does control the mean speed of the engine if the load changes

(*c*) Governor controla speed fluctuations during a cycle for a constant load, but governor does not control the mean speed of the engine if the load change

(*d*) Governor controls speed fluctuations during a cycle for a constant load, and governor also controls the mean speed of the engine if the load changes

2000

69. A punch press, fitted with a flywheel having a radius of gyration of 0.5 m, runs at 260 rev min.The press is capable of punching 600 holes per hour. Each punching operation takes 1.5 seconds and requires a work of 10,000 Nm. The rating of the electric motor used is 2 kW. What is the mass of the flywheel, if speed of the flywheel should not drop below 240 rev/min ?

(*a*) 240 kg (*b*) 280 kg

(*c*) 340 kg (*d*) 420 kg

1999

70. For the planar mechanism shown in figure select the most appropriate choice for the motion of link 2 when link 4 is moved upwards.

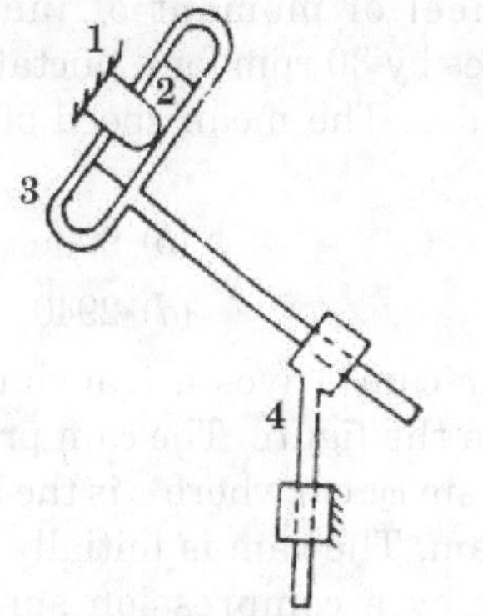

(*a*) Link 2 rotates clockwise

(*b*) Link 2 rotates counter - clockwise

(*c*) Link 2 does not move

(*d*) Link 2 motion cannot be determined

71. As shown in Figure. a person A is standing at the centre of a rotating platform facing person B who is riding a bicycle, heading East. The relevant speeds and distances are shown in given figure. person a bicycle, heading East. At the instant under consideration, what is the apparent velocity of B as seen by A?

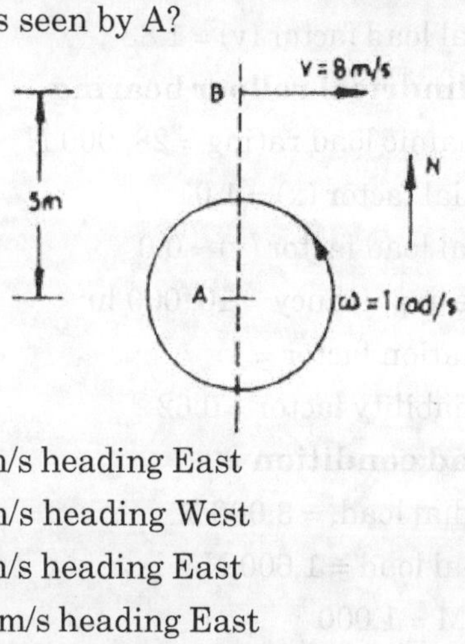

(*a*) 3 m/s heading East

(*b*) 3 m/s heading West

(*c*) 8 m/s heading East

(*d*) 13 m/s heading East

72. For the audio cassette mechanism shown in figure given below where is the instantaneous centre of rotation (point) of the two spools?

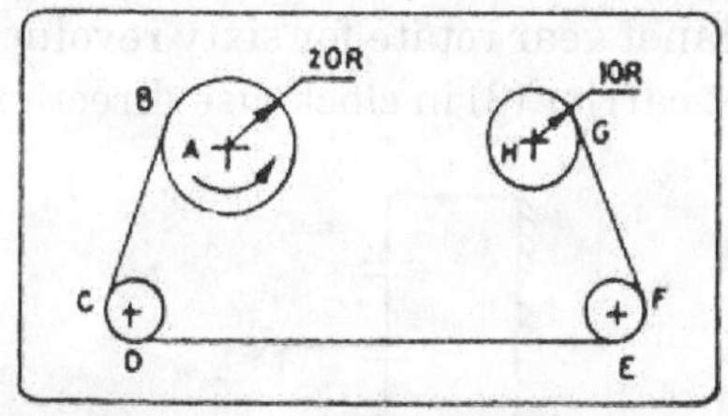

(*a*) Point P lies to the left of both the spools but at infinity along the line joining A and H

(*b*) Point P lies in between the two spools on line joining A and H, such that PH = 2 AP

(*c*) Point P lies to the right of both the spools on the line joining A and H, such that AH = HP

(*d*) Point P lies at the intersection of the line joining B and C and the line joining G and F

73. With regard to belt drives with given pulley diameters, centre distance and coefficient of friction between the pulley and the belt materials, which of the statement below are FALSE?

(*a*) A crossed flat belt configuration can transmit more power than an open flat belt configuration

(*b*) A "V" belt has greater power transmission capacity than an open flat belt

(*c*) Power transmission is greater when belt tension is higher due to centrifugal effects than the same belt drive when centrifugal afects are absent.

(*d*) Power transmission is the greatest just before the point of slipping is reached

74. Given figure shows a 2 degree of freedom manipulator consisting of a rotary base and a sliding arm which slides radially with respect to the base. The instantaneous position, angular acceleartion and relative velocity of arm with respect to the base are shown in figure. The radial relative acceleration of the arm with respect to the base is zero. Obtain the magnitude and direction of the absolute acceleration of the point P on the sliding arm.(5)

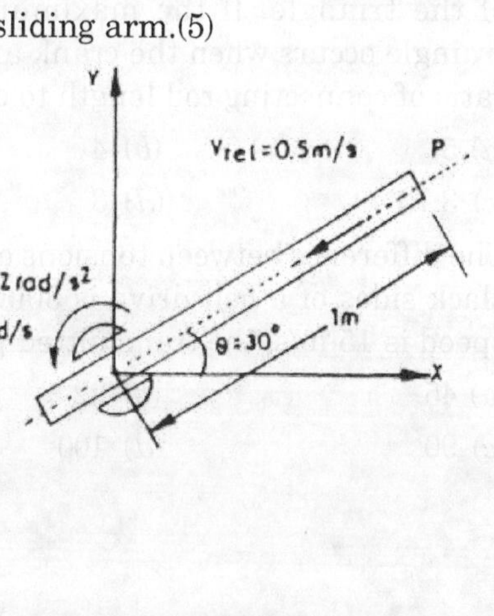

75. A planetary gear train is shown in Figure. Internal gear (1) has 104 teeth and is held fixed and planet gear (2) has 96 teeth. How much does the planet gear rotate for sixty revolutions of the planet carrier (3) in clockwise direction?

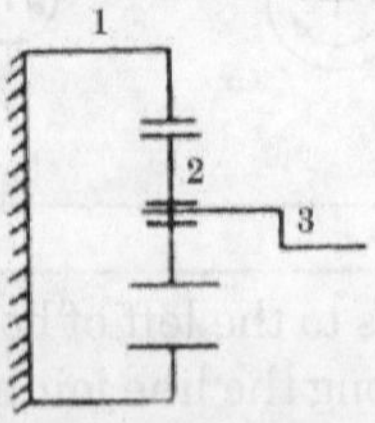

76. A flat belt opon drive transmits 1.5 kW of power. The coefficient of friction between the belt and the drive pulley is 0.25 and the lap angle is 159°. The drive pulley is rotating in the clockwise direction with the linear belt velocity of 3.75 m/s. Determine the x and y components of the reaction force on the drive pulley shaft.

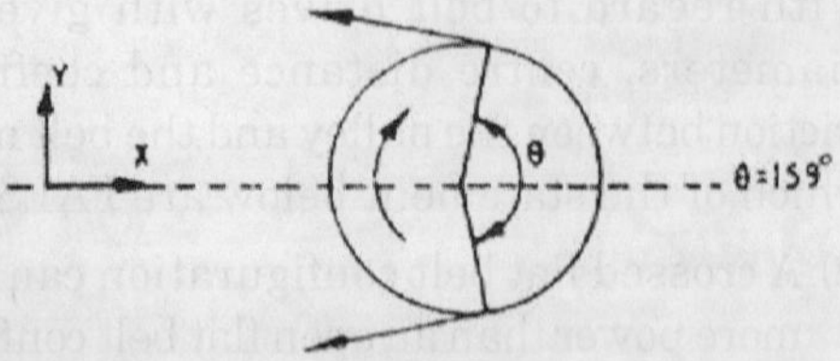

1998

77. The basic load rating of a ball bearing is

(*a*) the maximum static radial load that can be applied without causing any plastic deformation of bearing components.

(*b*) the radial load at which 90% of the group of apparently indentical bearings run for one million revolutions before the first evidence of failure.

(*c*) the maximum radial load that can be applied during operation without any plastic deformation of bearing components.

(*d*) a combination of radial and axial loads that can be applied without any plastic deformation.

78. Consider the triangle formed by the connecting rod and the crank of an IC engine as the two sides of the triangle. If the maximum area of this traingle occurs when the crank angle is 75°, the ratio of connecting rod length to crank radius is

(*a*) 5 (*b*) 4

(*c*) 3.73 (*d*) 3

79. The difference between tensions on the tight and slack sides of a belt drive is 3000 N. If the belt speed is 15 m/s, the transmitted power in kW is

(*a*) 45 (*b*) 22.5

(*c*) 90 (*d*) 100

80. The profile of a cam in a particular zone is given by $x = \sqrt{3}\cos\theta$ and $y = \sin\theta$. The normal to the cam profile at $\theta = \pi/4$ is at an angle (with respect to x axis)

(*a*) $\frac{\pi}{4}$ (*b*) $\frac{\pi}{2}$

(*c*) $\frac{\pi}{3}$ (*d*) 0

81. A strip with a cross-section 150 mm × 4.5 mm is being rolled with 20% reduction of area using 450 mm diameter rolls. The angle subtended by the deformation zone at the roll centre is (in radian)

(*a*) 0.01 (*b*) 0.02

(*c*) 0.03 (*d*) 0.06

82. A flywheel of moment of inertia 9.8 kg m² fluctuates by 30 rpm for a fluctation in energy of 1936 joules. The mean speed of the flywheel is (in rpm)

(*a*) 600 (*b*) 900

(*c*) 968 (*d*) 2940

83. A planar cam drives a translating follower as shown in the figure. The cam profile is given by $r = 5 + 2 \sin \omega t$ cm where ω is the angular velocity of the cam. The cam is initially pressed by the follower by a compression spring of stiffness 2 kN/cm when r = 5 cm. Calculate the maximum operating speed of this system for the follower to maintain contact with the cam profile. The mass of the follower can be taken as 1 kg. The initial spring compression is 5 mm.

84. Select one of the following two bearings and justify why did you reject the other :

(*a*) Deep groove ball bearing

Dynamic load rating = 47, 000 N

radial factor (x) = 0.56

axial load factor (y) = 1.83

(*b*) Cylindrical rolloer bearing

dynamic load rating = 28, 000 N

radial factor (x) = 1.0

axial load factor (y) = 0.0

Life expectancy = 10, 000 hr

Rotation factor = 1

Reliability factor = 0.62

Load condition

Radial load, = 3,000 N

Axial load = 1,600 N

RPM = 1,000

1997

85. To restore stable operating condition in a hydrodynamic journal bearing, when it encounters higher magnitude loads,

(*a*) oil viscosity is to be decreased

(*b*) oil viscosity is to be increased

(*c*) oil viscosity index is to be increased

(*d*) oil viscosity index is to be decreased

86. Total slip will occur in a belt drive when

(*a*) angle of rest is zero

(*b*) angle of creep is zero

(*c*) angle of rest is greater than angle of creep

(*d*) angle of creep is greater than angle of rest

87. To make a worm drive reversible, it is necessary to increase

(*a*) centre distance

(*b*) worm diameter factor

(*c*) number of starts

(*d*) reduction ratio

88. The cross head velocity in the slider crank mechanism, for the position shown in Figure is,

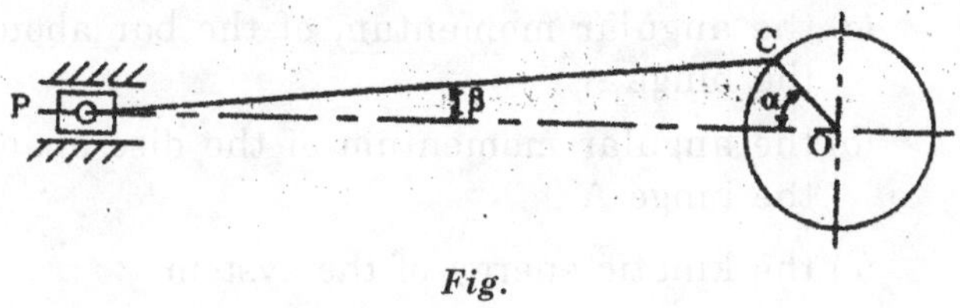

Fig.

(*a*) $V_C \cos(90-\overline{\alpha+\beta})\cos\beta$

(*b*) $V_C \cos(90-\overline{\alpha+\beta})\sec\beta$

(*c*) $V_C \cos(90-\overline{\alpha-\beta})\cos\beta$

(*d*) $V_C \cos(90-\alpha-\beta)\sec\beta$

where V_C is the linear velocity of the crank pin.

89. The dynamic load capacity of 6306 bearing is 22 kN. The maximum radial load it can sustain to operate at 600 rev/min, for 2000 hours is

(*a*) 4.16 kN (*b*) 3.60 kN

(*c*) 6.25 kN (*d*) 5.29 kN

90. The percentage improvement in power capacity of a flat belt drive, when the wrap angle at the driving pulley is increased from 150° to 210° by an idler arrangement for a friction coefficient of 0.3, is

(*a*) 25.21 (*b*) 33.92

(*c*) 40.17 (*d*) 67.85

91.

List I	List II
(*a*) Automobile wheel mounting on axle	1. Magneto bearing
(*b*) High speed grinding	2. Angular contact bearing spindle
(*c*) I.C. Engine connecting rod	3. Taper roller bearing
(*d*) Leaf spring eye mounting	4. Hydrodynamic journal bearing
	5. Sintered metal bearing
	6. Taflon/Nylon bush.

92. A multiple disc clutchis to transmit 4 kW at 750 rev/min. Available steel and bronze discs of 40 mm inner radius and 70 mm outer radius are to be assembled alternately in appropriate numbers. The clutch is to operate in oil with an expected coefficient of friction of 0.1 and the maximum allowable pressure is not to exceed 350 kPa. Assume uniform wear condition to prevail and specify the number of steel (Driving) and bronze (driven) discs required. Also determine what axial force is to be applied to develop the full torque.

93. A single stage spur gear reducer is found to have a pinion with 15 teeth and tip (addendum) diameter of 85 mm. The mating gear has 33 teeth and 175 mm tip diameter. If the gears are supposed to have standard 20° pressure angle full depth involute teeth, what should be the module and centre distance between them. Supported by due calculations state whether there is interference or not.

94. A metal strip is to be rolled from an initial wrought thickness of 3.5 mm to a final rolled thickness of 2.5 mm in a single pass rolling mill having rolls of 250 mm diameter. The strip is 450 mm wide. The average coefficient of friction in the roll gap is 0.08. Taking plane-strain flow stress of 140 MPa, for the metal, and assuming negligible spreading, estimate the roll separating force.

1996

95. In thick film hydrodynamic journal bearings, the coefficient of friction

(*a*) increases with increases in load

(*b*) is independent of load

(*c*) decreases with increase in load

(*d*) may increase or decrease with increase in load

96. A wheel of mass m and radius r is in accelerated rolling motion without slip under a steady axle torque T. If the coefficient of kinetic friction is μ, the friction force from the ground on the wheel is

(*a*) μ mg (*b*) T/r

(*c*) zero (*d*) none these

97. A rod of length 1 m is sliding in a corner as shown in figure. At an instant when the rod makes an angle of 60 degrees with the horizontal plane, the velocity of point A on the rod is 1 m/s. The angular velocity of the rod at this instant

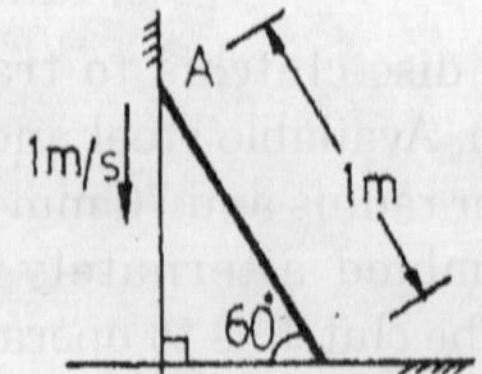

(*a*) 2 rad/s (*b*) 1.5 rad/s

(*c*) 0.5 rad/s (*d*) 0.75 rad/s

98. Five holes of diameter 10 mm each are to be punched in a sheet 3 mm thick at a pitch of 25 mm. What should be the minimum capacity of the press (in tonnes) if the yield point of material is 50 MPa and

(*a*) one hole is punched per stroke,

(*b*) five holes are punched in a single stroke.

Show with a neat sketch the arrangement of punches so as to avoided eccentric loading (i.e. provide balanced loading). It may be noted that no shear is provided on the punches.

99. Figure below show a planetary gear train. Gears 2, 4 and 5 have 24, 40 and 144 teeth respectively. Gear 5 is fixed. Gear 2 is rotating clockwise at 700 rpm. What will be the rpm of the arm and gear 4 ?

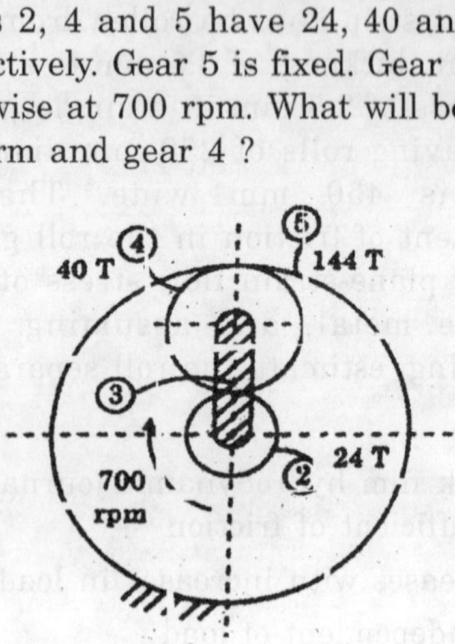

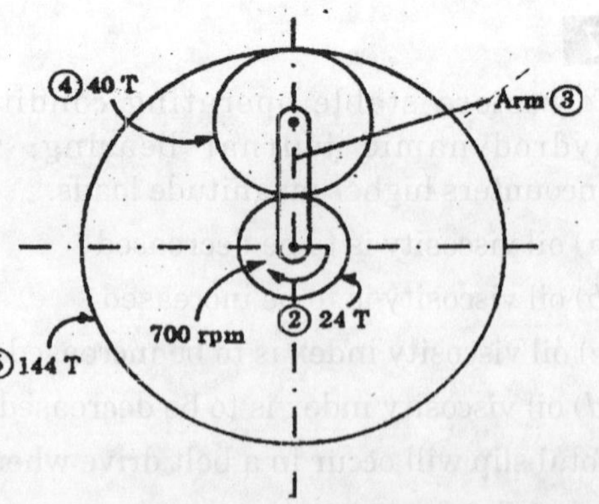

100. A bar AB is hinged to a fixed support at A and the centre of a uniform disc at B. Both hinges are frictionless. At a certain instant, the system is observed to be moving in the horizontal plane, when for the bar has an angular velocity ω while disc has an anglar velocity of ω relative to the bar. The bar and the disc are each of mass m. Determine

(*a*) the angular momentum of the bar about the hinge A

(*b*) the angular momentum of the disc about the hinge A

(*c*) the kinetic energy of the system

NUMERICAL TYPE QUESTIONS

2016

1. A slider crank mechanism with crank radius 200 mm and connecting rod length 800 mm is shown. The crank is rotating at 600 rpm in the counterclockwise direction. In the configuration shown, the crank makes an angle of 90° with the sliding direction of the slider, and a force of 5 kN is acting on the slider. Neglecting the inertia forces, the turning moment on the crank (in kN-m) is ________.

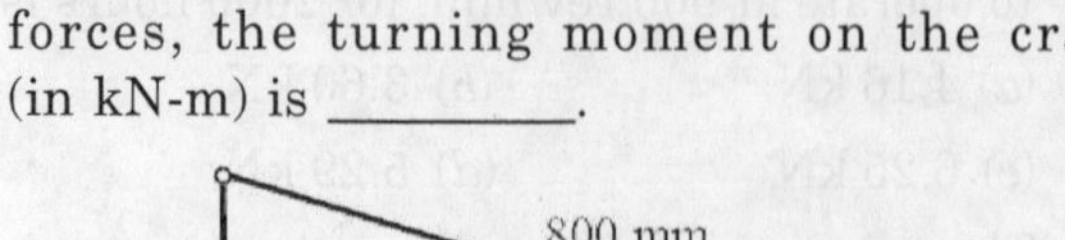

2. A mass of 2000 kg is currently being lowered at a velocity of 2 m/s from the drum as shown in the figure. The mass moment of inertia of the drum is 150 kg-m^2. On applying the brake, the mass is brought to rest in a distance of 0.5 m. The energy absorbed by the brake (in kJ) is ________

3. The rod AB, of length 1 m, shown in the figure is connected to two sliders at each end through pins. The sliders can slide along QP and QR. If the velocity V_A of the slider at A is 2 m/s, the velocity of the midpoint of the rod at this instant is ________ m/s.

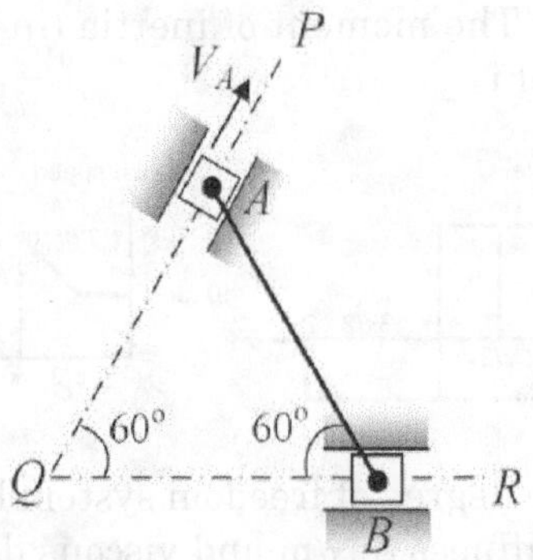

4. Two masses m are attached to opposite sides of a rigid rotating shaft in the vertical plane. Another pair of equal masses m_1 is attached to the opposite sides of the shaft in the vertical plane as shown in figure. Consider m = 1 kg, e = 50 mm, e_1 = 20 mm, b = 0.3 m, a = 2 m and a_1 = 2.5 m. For the system to be dynamically balanced, m_1 should be ________ kg.

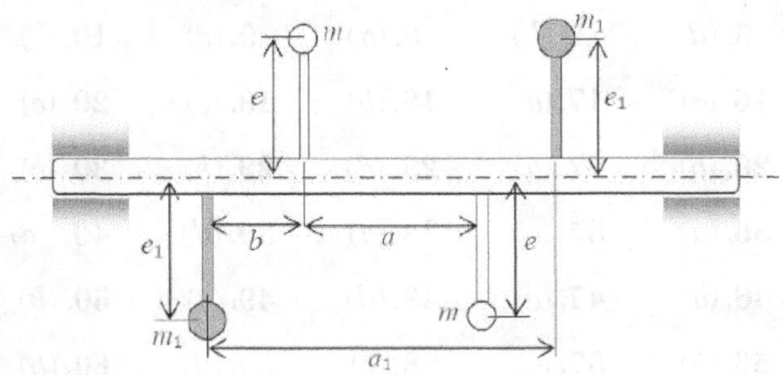

2015

5. Consider an ant crawling along the curve $(x - 2)^2 + y^2 = 4$, where x and y are in meters. The ant starts at the point (4, 0) and moves counter-clockwise with a speed of 1.57 meters per second. The time taken by the ant to reach the point (2, 2) is (in seconds) ________.

6. In a spring mass system, the mass in 'm' and the spring constant is K. The critical damping coefficient of the system is 0.1 kg/s. In another spring mass system, the mass is 2m and the spring constant is 8 K. The critical damping coefficient of the system (in kg/s) of this system is ______.

7. In a certain slider-crank mechanism, length of crank and connecting rod are equal. If the crank rotates with the uniform angular speed of 14 rad/s and the crank length is 300 mm, the maximum acceleration of the slider (in m/s²) is ______.

8. The torque (in N-m) exerted on the crank shaft of a two stroke engine can be described as T = 10000 + 1000 sin 1θ – 1200 cos 2θ, where θ is the crank angle as measured from inner dead center position. Assuming the resisting torque to be constant, the power (in kW) developed by the engine at 100 rpm is ____________.

2014

9. A slider crank mechanism has slider mass of 10 kg, stroke of 0.2 m and rotates with a uniform angular velocity of 10 rad/s. The primary inertia forces of the slider are partially balanced by a revolving mass of 6 kg at the crank, placed at a distance equal to crank radius. Neglect the mass of connecting rod and crank. When the crank angle (with respect to slider axis) is 30°, the unbalanced force (in newton) normal to the slider axis is ______

10. An offset slider-crank mechanism is shown in the figure at an instant. Conventionally, the Quick Return Ratio (QRR) is considered to be greater than one. The value of QRR is______

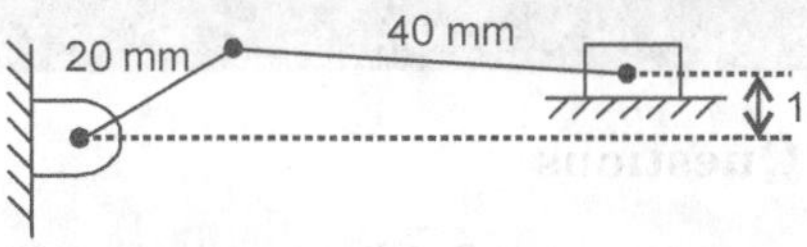

11. A rigid link PQ of length 2 m rotates about the pinned end Q with a constant angular acceleration of 12 rad/s². When the angular velocity of the link is 4 rad/s, the magnitude of the resultant acceleration (in m/s²) of the end P is ______

12. The damping ratio of a single degree of freedom spring-mass-damper system with mass of 1 kg, stiffness 100 N/m and viscous damping coefficient of 25 N.s/m is ______

13. A four-wheel vehicle of mass 1000 kg moves uniformly in a straight line with the wheels revolving at 10 rad/s. The wheels are identical, each with a radius of 0.2 m. Then a constant braking torque is applied to all the wheels and the vehicle experiences a uniform deceleration. For the vehicle to stop in 10 s, the braking torque (in N.m.) on each wheel is ________

14. A slider-crank mechanism with crank radius 60 mm and connecting rod length 240 mm is shown in figure. The crank is rotating with a uniform angular speed of 10 rad/s, counter clockwise. For the given configuration, the speed (in m/s) of the slider is ______

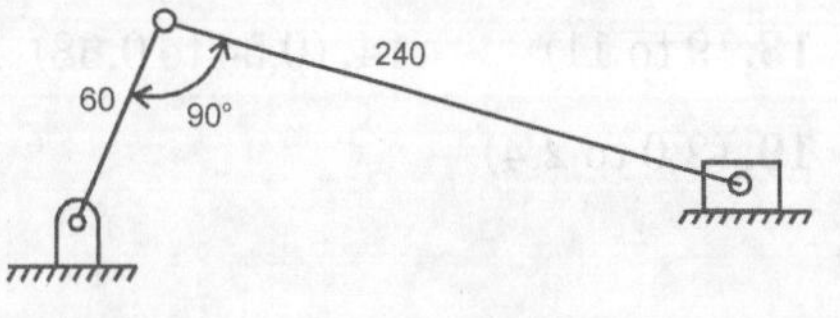

15. A mass-spring-dashpot system with mass $m = 10$ kg, spring constant $k = 6250$ N/m is excited by a harmonic excitation of 10 cos(25t) N. At the steady state, the vibration amplitude of the mass is 40 mm. The damping coefficient (c, in N.s/m) of the dashpot is ______

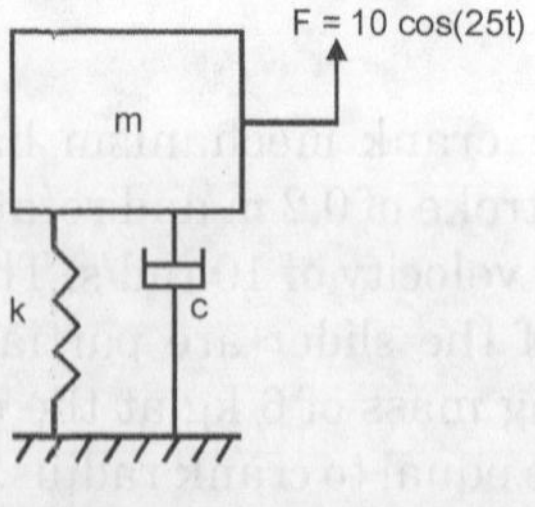

16. An amount of 100 kW of heat is transferred through a wall in steady state. One side of the wall is maintained at 127°C and the other side at 27°C. The entropy generated (in W/K) due to the heat transfer through the wall is ______

17. A point mass is executing simple harmonic motion with an amplitude of 10 mm and frequency of 4 Hz. The maximum acceleration (m/s^2) of the mass is ______

18. Torque and angular speed data over one cycle for a shaft carrying a flywheel are shown in the figures. The moment of inertia (in kg.m^2) of the flywheel is ____

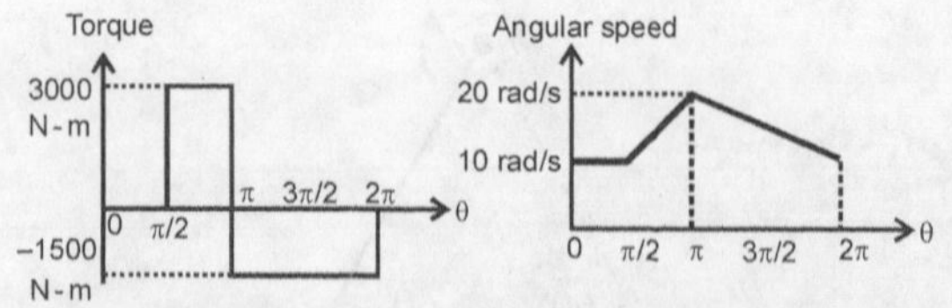

19. A single degree of freedom system has a mass of 2 kg, stiffness 8 N/m and viscous damping ratio 0.02. The dynamic magnification factor at an excitation frequency of 1.5 rad/s is ______

ANSWERS

MCQ Type Questions

1. (c)	**2.** (a)	**3.** (d)	**4.** (b)	**5.** (d)	**6.** (d)	**7.** (b)	**8.** (a)	**9.** (c)	**10.** (c)
11. (a)	**12.** (b)	**13.** (a)	**14.** (d)	**15.** (d)	**16.** (c)	**17.** (a)	**18.** (b)	**19.** (c)	**20.** (c)
21. (a)	**22.** (b)	**23.** (c)	**24.** (b)	**25.** (b)	**26.** (b)	**27.** (a)	**28.** (d)	**29.** (b)	**30.** (c)
31. (b)	**32.** (d)	**33.** (a)	**34.** (b)	**35.** (b)	**36.** (a)	**37.** (c)	**38.** (a)	**39.** (d)	**40.** (c)
41. (b)	**42.** (a)	**43.** (d)	**44.** (d)	**45.** (d)	**46.** (a)	**47.** (c)	**48.** (d)	**49.** (c)	**50.** (b)
51. (a)	**52.** (a)	**53.** (c)	**54.** (c)	**55.** (a)	**56.** (b)	**57.** (c)	**58.** (c)	**59.** (d)	**60.** (b)
61. (c)	**62.** (d)	**63.** (c)	**64.** (c)	**65.** (b)	**66.** (*)	**67.** (c)	**68.** (a)	**69.** (*)	**70.** (b)
71. (d)	**72.** (d)	**73.** (c)	**74.** (*)	**75.** (*)	**76.** (*)	**77.** (a)	**78.** (c)	**79.** (a)	**80.** (c)
81. (d)	**82.** (a)	**83.** (*)	**84.** (*)	**85.** (b)	**86.** (a)	**87.** (c)	**88.** (b)	**89.** (d)	**90.** (d)
91. (a) –3, (b) –1, (c) –4, (d) –6	**92.** (*)	**93.** (*)	**94.** (*)	**95.** (c)	**96.** (a)	**97.** (a)	**98.** (*)	**99.** (*)	**100.** (*)

Numerical Type Questions

1. 1	**2.** (14.11)	**3.** (1)	**4.** (2)	**5.** (2)	**6.** (0.38 to 0.42)
7. (117.6)	**8.** (104 to 105)	**9.** (29 to 31)	**10.** (1.2 to 1.3)	**11.** (39 to 41)	**12.** (1.24 to 1.26)
13. (9 to 11)	**14.** (0.54 to 0.68)	**15.** (9 to 11)	**16.** (80 to 85)	**17.** (6.3 to 6.4)	**18.** (20 to 32)
19. (2.0 to 2.4)					

"*" Refers to Subjective Question.

EXPLANATIONS

MCQ TYPE QUESTIONS

1. Given $T_2 = 60$

$N_2 = 0$

$T_3 = 20$

$T_4 = 100$

$N_2 = 100$ rpm (ccw)

Relative velocity equation

$$\frac{N_4 - N_a}{N_2 - N_a} = \frac{T_2}{T_4} = \frac{100 - N_a}{0 - N_a} = \frac{-60}{100}$$

$$1.6N_a = 100$$

$$N_a = 62.5 \text{ rpm (ccw)}$$

2.

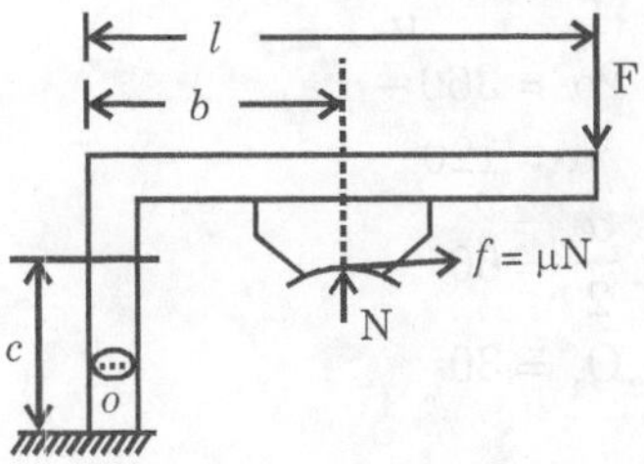

Taking moment about hinge in clockwise

$$N_b - Fl - \mu Nc = 0$$

$$F = \frac{N(b - \mu c)}{l}$$

Self energizing if $b > \mu c$ so for clockwise rotation of drum the brake is self energizing.

4. $DOF = 3(n - 1) - 2j$

6. Form $\Rightarrow dx = 2r\theta$

$$\frac{d^2x}{dt^2} = 2r.\ddot{\theta}$$

Taking moments

$$2r\,\theta.m.\,2r.\ddot{\theta} + 400 \times (r\theta) \times r\theta = 0$$

$$\Rightarrow 4m.\ddot{\theta} + 400 \times \theta = 0$$

$$\Rightarrow \omega = \sqrt{\frac{400}{4}}$$

7. Amplitude of vibration,

$$A = \frac{f_0/s}{\sqrt{1 - \left(\frac{\omega}{\omega_n}\right)^2 + \left(\frac{2\xi\omega}{\omega_n}\right)^2}}$$

$$= \frac{10/150}{\sqrt{\left[1 - \left(\frac{1}{10}\right)\right]^2 + \left(2 \times 0.2 \times \frac{1}{10}\right)^2}} = 0.06605 \simeq 0.07$$

8. Direction is obtained by rotating velocity vector through 90° in the direction of rotation of the link.

9. $$\frac{\omega_2}{\omega_6} = \frac{\omega_2}{\omega_3}\frac{\omega_3}{\omega_5}\frac{\omega_5}{\omega_6} = \frac{N_3}{N_2}\frac{N_5}{N_4}\frac{N_6}{N_5} = \frac{N_3N_6}{N_2N_4}$$

$\therefore$ only Gear(5) is Idle

10. Number of links, N = 6

Total number as binary joints, $j = 7$

$$F = 3(N - 1) - 2j = 15 - 14 = 1.$$

11. Force in the spring F = 2mg [from equilibrium]

Deflection as mass at P,

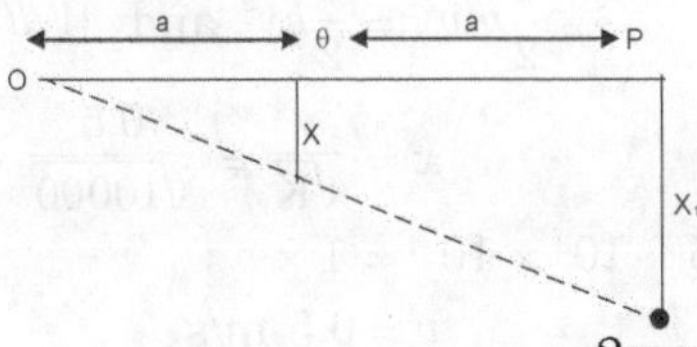

$$x_1 = \frac{x}{a} \times 2a = 2x = 2 \times \frac{2mg}{k} = \frac{4mg}{k}$$

$$w_n = \sqrt{\frac{g}{s}} = \sqrt{\frac{g}{x_1}} = \sqrt{\frac{k}{4m}};$$

$$f_n = \frac{1}{2\pi}\omega_n = \frac{1}{2\pi}\sqrt{\frac{k}{4m}}$$

13. Velocity at point of contact = R ω

(R = Radius of point from Instantaneous centre)

$\because$ The instantaneous centre is at intersection of object and floor, hence radius R = 0

$\therefore$ Velocity at point is zero.

14. By Instantaneous center method; say ω = angular velocity of rod

$v_q = r_q.\omega$ ___ (i) $\quad v_p = r_p.\omega$ ___ (ii)

Dividing equation (i) with equation (ii) we get

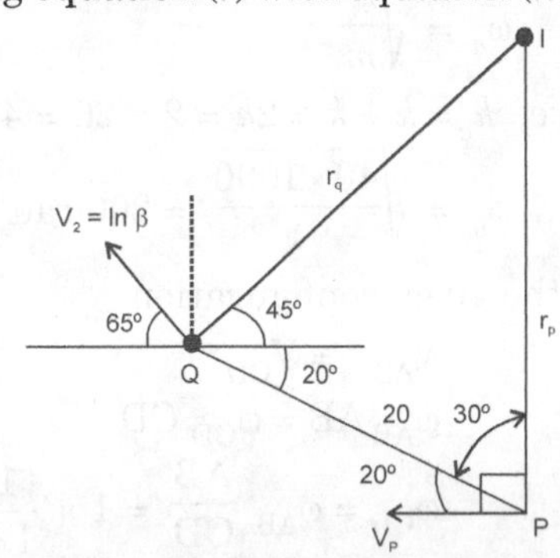

$$\frac{v_q}{v_p} = \frac{r_q}{r_p}; v_p = \frac{r_p}{r_q}.v_q$$

Now by sine law, $\frac{r_p}{\sin(45+20)} = \frac{r_q}{\sin(70°)}$

$$\frac{r_p}{r_q} = \frac{\sin(65°)}{\sin(70°)}$$

So, $$v_p = \frac{\sin(65)}{\sin(70)}v_q = 0.96.$$

16. Among all refrigerants R502 is the only CFC refrigerant.

17. S + l < P + Q

20 + 60 < 40 + 50

$\Rightarrow$ 80 < 90

If smaller link is fixed both input and Output link execute full circular motion.

19. $2vw = 2 \times 0.75 \times 2 = 3 \text{ m/s}^2$

20. The link opposite to shortest link is fixed.

21. $$\Delta E = I\omega_m^2 c_f$$

22. $$m = \frac{D}{T} \text{ and } m_p = m_Q$$

$m_R = m_s$ for gear meshing condition

23. $$\frac{1}{2}mv^2 = \frac{1}{2}kx^2 \text{ and } \int F\,dt = m(v_2 - v_1)$$

$$\Rightarrow \quad x = \frac{v}{\sqrt{K}} = \frac{0.5}{\sqrt{10000}}$$

$$\Rightarrow \quad 5 \times 10^3 \times 10^{-4} = 1 \times v$$

$$\Rightarrow \quad v = 0.5 \text{ m/s}$$

24. For helical gears

$$\text{velocity ratio} = \frac{d_1 \cos\Phi}{d_2 \cos\Phi} = \frac{80\cos 30^\circ}{120\cos 22.5} = \frac{N_2}{N_1}$$

$$\therefore \quad N_2 = 1440 \times 0.625 = 900 \text{ rpm}$$

26. K.E. of fly wheel $= \frac{1}{2}I\omega^2$

$$\text{where, } I = \frac{mR^2}{2} = \frac{20 \times 0.2^2}{2} = 0.4 \text{ kg} - \text{m}^2$$

$$\omega = \frac{2\pi N}{60} = \frac{2 \times \pi \times 600}{60} = 62.83 \text{ rad/s}$$

$$\therefore \quad \text{K.E.} = \frac{1}{2} \times 0.4 \times 62.83^2 = 790 \text{ J}$$

27. Natural frequency of the system

$$\omega_n = \sqrt{\frac{k_e}{m}}$$

where, $k_e = k + k = 2k = 2 \times 20 = 40$ kN/m

$$\therefore \quad \omega_n = \sqrt{\frac{40 \times 1000}{1}} = 200 \text{ rad/s} = 32 \text{ Hz}$$

28. For the given configuration

$$V_{AB} = V_{CD}$$

$$\therefore \quad \omega_{AB}\,AB = \omega_{CD}\,CD$$

$$\Rightarrow \quad \omega_{CD} = \omega_{AB}\frac{AB}{CD} = 1 \times \frac{1}{1.5} = \frac{2}{3} \text{ rad/s}$$

29. For ball bearing

$$P.(L)^{1/3} = C$$

Given: $P_P = 30$ kN

$P_2 = 45$ kN

$$\frac{L_P}{L_2} = \left(\frac{P_2}{P_P}\right)^3 = \left(\frac{45}{30}\right)^3 = \left(\frac{3}{2}\right)^3 = \frac{27}{8}$$

30.

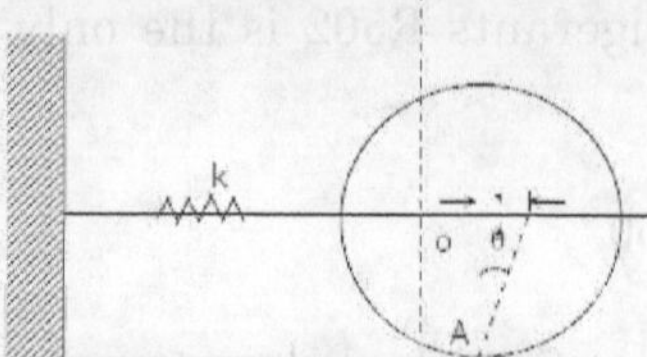

Taking moments about instantaneous centre 'A',

$$I_a\ddot{\theta} + (kx)\,r = 0$$

$$\Rightarrow \quad (I_O + mr^2)\ddot{\theta} + kx\,(\theta r)r = 0$$

$$\Rightarrow \quad \left(\frac{1}{2}mr^2 + mr^2\right)\ddot{\theta} + k\,(\theta r^2) = 0$$

$$\Rightarrow \quad \ddot{\theta} + \frac{kr^2}{\frac{3}{2}mr^2}\theta = 0$$

$$\Rightarrow \quad \ddot{\theta} + \frac{2k}{3m}\theta = 0$$

$$\therefore \quad \omega_n = \frac{1}{2\pi}\sqrt{\frac{2k}{3m}}$$

32. $$\frac{2}{1} = \frac{360 - \alpha}{\alpha}$$

$$\Rightarrow \quad 2\alpha = 360 - \alpha$$

$$\Rightarrow \quad \alpha = 120^\circ$$

$$\therefore \quad \frac{\alpha}{2} = 60^\circ$$

i.e. $\angle PO_2O_1 = 30^\circ$

In ΔPO_2O_1

$$\tan 30^\circ = \frac{O_1P}{d}$$

$$\Rightarrow \quad d = \frac{125}{\tan 30^\circ} = 219.29$$

and $$\sin 30^\circ = \frac{125}{x}$$

$$\Rightarrow \quad x = \frac{125}{1/2} = 250$$

33. $T_2 = 20$; $T_3 = 40$; $T_5 = 100$

Condition	Arm	Gear 2	Gear 3	Gear 5
1. Arm fixed – gear 2 rotates with + 1 rotation	0	+1	$-\frac{T_2}{T_3}$	$-\frac{T_2}{T_3} \times \frac{T_3}{T_5}$
2. Arm fixed – gear 2 rotates with + x rotations	0	$+x$	$-x\frac{T_2}{T_3}$	$-x\frac{T_2}{T_5}$
3. Adding + y rotations	y	$y + x$	$y - x\frac{T_2}{T_3}$	$y - x\frac{T_2}{T_5}$

Given: $$y - x\frac{T_2}{T_5} = 0$$

$$\Rightarrow \quad x = 5y \qquad \text{...(i)}$$

and $$y + x = 60 \qquad \text{...(ii)}$$

Solving equations (*i*) and (*ii*), we get

$$y + 5y = 60$$

$$\Rightarrow \quad y = 10 \text{ rpm ccw}$$

35. By Gruebler's criterion,

$$F = 3(L-1) - 2j = 3(8-1) - 2 \times 10 = 1$$

36. Instantaneous centre of rotation of the disk is 0, i.e. its point of contact with ground every other point on the disk is rotating with respect to this center with an angular velocity.

$$\omega = \frac{V}{R}$$

$\therefore$ Velocity of P = ω

$$OP = \frac{V}{R} \times \sqrt{3}R = \sqrt{3}v$$

38. We know, $\Delta E = I\omega^2 C_S$

Here, $\omega = 200$

$\Delta E = 400$

$$C_S = \frac{\omega_{max} - \omega_{min}}{\omega_{max} + \omega_{min}} = 3\frac{20}{200} = 0.1$$

$\Delta E = 400$ N – m

$$\therefore \quad I = \frac{\Delta E}{\omega^2 C_S} = \frac{400}{(200)^2 \times .1} = 0.10$$

39. Mechanical advantage $= \frac{\text{Load (output force)}}{\text{Effort (input force)}}$

For a four bar linkage in toggle position,

effort = 0

$\therefore$ Mechanical advantage = ∞

40. There are four number of inversions for a slider crank mechanism.

42. Torque, $T = \frac{P \times 60}{2\pi \times N} = 23.87$ N m

Axial thrust, $W = P \times \pi\,(r_1^2 - r_2^2)$

But $$T = \frac{2}{3}\mu \times P \times \pi (r_1^2 - r_2^2)\frac{(r_1^3 - r_2^3)}{(r_1^2 - r_2^2)} = \mu w r$$

$\therefore \quad r_2 = 39.4$ mm

43. $$C_f = \frac{\omega_{max} - \omega_{min}}{\frac{\omega_{max} + \omega_{min}}{2}}$$

$$\Rightarrow \quad \frac{\omega_{max}}{\omega_{min}} = \frac{2 + C_f}{2 - C_f}$$

45. $$\omega_n = \sqrt{\frac{S}{M}} = 20$$

$$\omega = \frac{2\pi \times N}{60} = 377$$

$\therefore$ Transmissibility ratio $= \frac{\omega_n^2}{\omega^2 - \omega_n^2} = 0.0028$

46. According to Grashoff's Criteria,

$$S + L \le P + Q$$

47. Number of degree of freedom,

$$n = 3(l - 1) - 2J - h$$
$$= (3 \times 7) - (2 \times 9) - 0 = 3$$

48. $$\omega = \frac{2\pi N}{60}$$

Time taken to move 30° $= \frac{\frac{\pi}{180} \times 30}{\omega}$

$$= \frac{\pi}{6} \times \frac{60}{2\pi N} = \frac{5}{N}\text{sec}$$

During this period, follower move 20 mm = 0.02 m and u = 0

From $$s = ut + \frac{1}{2}at^2$$

$$0.02 = 0 + \frac{1}{2}a \times \left(\frac{5}{N}\right)^2$$

$\Rightarrow \quad a = 144$ m/sec^2

$\therefore \quad v = u + at$

$$= 0 + \left(144 \times \frac{5}{300}\right)$$

$= 2.4$ m/sec

49. Along x-axis,

$$0.5\,(-60 \times 10^{-3} \cos 30° + 50 \times 10^{-3})\,\omega^2 = 0.1\omega^2 \times x \times 10^{-3}$$

$\therefore \quad x = -9.8076$ mm

Along y-axis,

$$0.5\,(60 \times 10^{-3} \sin 30°)\,\omega^2 = 0.1\omega^2\, y$$

$\Rightarrow \quad y = 150$ mm

$\therefore \quad r = \sqrt{x^2 + y^2} = 150.32$ mm

50. Let initial length is δ without load, and stiffness is k.

$\therefore \quad 2g = k(\delta - 0.2)$ ...(i)

and $\quad 22g = k(\delta - 0.1)$ ...(ii)

Solving equations (i) and (ii), we get

$\delta = 210$ mm

$k = 1960$ N/m

52. Velocity of link 1 with respect to link 2,

$$V_{1/2} = 12 \text{ m/s}$$

$$\omega = \frac{2\pi N}{60} = \frac{2\pi \times 120}{60} = 4\pi$$

$\therefore$ Corioli's component of acceleration

$$= 2V_{1/2}\omega = 2 \times 12 \times 4\pi$$
$$= 301.59 \approx 302 \text{ m/s}^2.$$

54. Let weight be displaced by x from equilibrium position.

$$\therefore \quad \tan\theta = \frac{x}{300} = \frac{y}{150}$$

$$\Rightarrow \quad y = \frac{x}{2}$$

Taking moment about hinged point

$$kx.(0.3) = W.y = W \times \frac{x}{2}$$

$$\therefore \quad k = \frac{300}{2 \times 0.3} = 500 \text{ N/m}.$$

58. For revolute joint, degree of freedom = 1
For cylinderical joint, degree of freedom = 2
For spherical joint, degree of freedom = 3
Degree of constraints = 6 – Degree of freedom

59. To obtain a DOUBLE CRANK MECHANISM, shortest link is always fixed. While obtaining a DOUBLE LEVER MECHANISM, the link opposite to the SHORTEST LINK is fixed.

60. Average speed, N = 1200 rpm
Coefficient of fluctuation of speed,

$$c_s = \frac{\omega_1 - \omega_2}{\omega} = 2\% = 0.02$$

Fluctuation of kinetic energy, $\Delta E = 2 \times 10^3$ J

Now $\Delta E = \frac{1}{2} I\omega_1^2 - \frac{1}{2} I\omega_2^2 = \frac{1}{2} I(\omega_1^2 - \omega_2^2)$

Since $\frac{\omega_1 + \omega_2}{2} = \omega$

$$\therefore \quad \Delta E = I\left(\frac{\omega_1 + \omega_2}{\omega}\right)(\omega_1 - \omega_2)$$

$$= I\omega \frac{(\omega_1 - \omega_2)}{\omega} . \omega = I\omega^2 c_s$$

$$\Rightarrow \quad 2 \times 10^2 = \frac{1}{2} MR^2 \times \omega^2 \times {c_s}^2$$

where R = radius of disc

$$\therefore \quad 2 \times 10^2 = \frac{1}{2} M \times \left(\frac{1}{2}\right)^2 \times \left(\frac{2176200}{60}\right) \times 0.02$$

$$\therefore \quad M = \frac{2 \times 10^3 \times 60 \times 60 \times 8}{0.02 \times (2 \times \pi \times 1200)^2} = 50.65 \approx 51 \text{ kg}$$

61. Linear velocity of B = $\omega_B r_B$
Linear velocity of A = $\omega_A r_A$

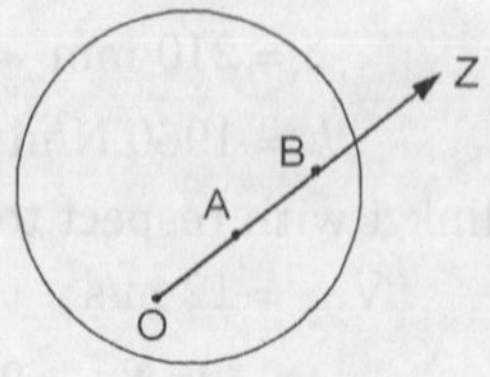

Obviously, $\omega_B r_B > \omega_A r_A$
because $\omega_B = \omega_B = \omega$ and $r_B > r_A$
$\therefore \quad \omega r_B > \omega r_A$
So, relative velocity = $\omega (r_B - r_A)$
in the direction of motion of point B and, it will experience a centripetal acceleration of $\omega^2 (r_B - r_A)$ towards "O".

62. Acceleration of point B with respect to point A

$$a_{B/A} = \overline{\omega} \times \overline{v}_{B/A} = \omega . \omega (r_B - r_A)$$

$$= \omega^2 (r_B - r_A)$$

and direction from z to O.

66. (***a***) *Condition for feasibility of Grashoff Mechanism*
Sum of the shortest and longest link lengths should not be greater than the sum of the remaining two links lengths, if there is continuous relative motion between the two links.

i.e. $\quad l + s < p + q$

(***b***) When shortest link, i.e. s is fixed, we will get crank-crank meachanism. To obtain crank-rocker mechanism we have to fix the link p. When shortest link s is made coupler and the link opposite to it is fixed, then other two links p_1 and q would oscillate. The mechanism is called *rocker-rocker mechanism*.

(***c***)

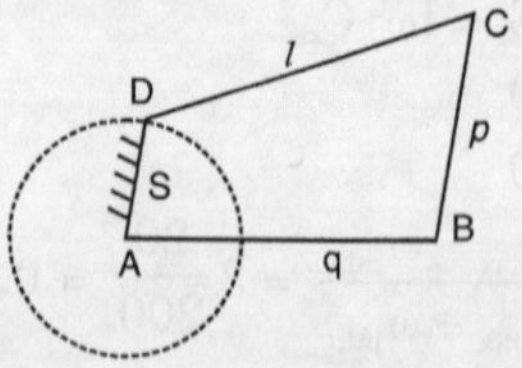

I

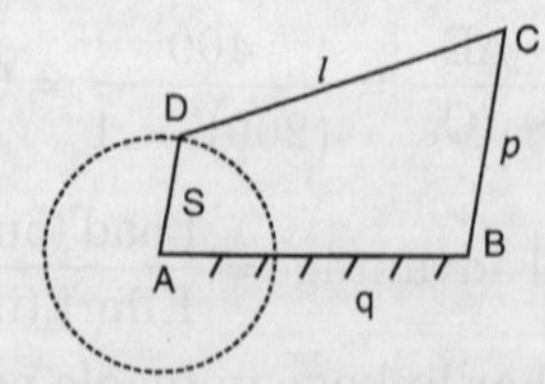

II

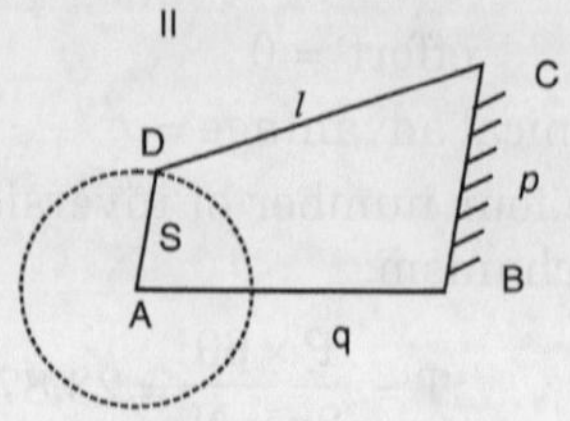

III

69. *Given:* $k = 0.53$

N = 260 rev/min = 600 holes/hr

Energy required for punching,

ε_1 = 10000 N.m

Motor power supplied, P = 2kW

N_{min} = 240 rev/min

Energy supplied by motor,

ε_2 = 2kW = 2000 W = 2000 N-m/s

Energy absorbed during one punching operation which takes 1.5 seconds = 10,000 N-m

$$\text{Energy required} = \frac{1000}{1.5} = 6666.67 \text{ N-m/s}$$

Maximum fluctuation of energy per second
= 6666.67 – 2000 = 4666.67 N-m/s

Now, $\Delta\varepsilon = \frac{1}{2} mk^2 [{\omega_1}^2 - {\omega_2}^2]$

$$\therefore \quad 4666.67 = \frac{1}{2} m (0.5)^2$$

$$\left[\left(\frac{2\pi \times 260}{60}\right)^2 - \left(\frac{259 \times 240}{60}\right)^2\right] = 0.125\, m\, [109.56]$$

$$\therefore \quad m = 340 \text{ kg}$$

71. Relative velocity = 8 – (–1 × 5)

= 13 m/s towards East.

74. Given,

$\omega = 1$ rad/s,

$\omega^2 = -0.732$ rad/s²,

$V_{rel} = 0.5$ m/s.

$V_{po} = r \times \omega = 1 \times 1 = 1$ m/s

$V_{rel} = 0.5 \text{ m/s} = V_p$

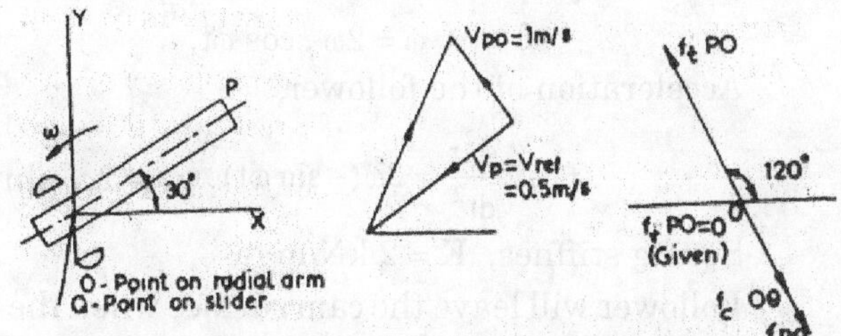

f_{COQ} = Component = $2\,\omega v$

$= 2 \times 1 \times 1 = 2$ m/s²

Tangential acceleration of arm with respect to base

$= \omega^2 r$

$= f_t(PO) = -0.732 \times 1$

$= -0.732$ m/sec²

Radial acceleration of arm with respect to base = 0

∴ Magnitude of absolute acceleration of the point P on the sliding arm,

$f_{Po} = 2 - (-0.732)$

$= 2.732$ m/sec²

75. Given, $T_1 = 104$ T

$T_2 = 96$ T

N = 60 revolutions of planet carrier.

T_1 is fixed.

S.No.	Conditions of motion	Arm (3)	Wheel (2)	Wheel (1)
1.	Arm fixed-wheel (2) rotates through + T revolution (i.e., 1 revolution anti-clockwise	0	+1	$+\frac{T_2}{T_1}$
2.	Arm fixed wheel (2) rotates through + x revolution	0	+ x	$+\frac{T_2}{T_1}x$
3.	Add + y revolution to all elements.	+ y	+ y	+ y
4.	Total motion	y	$x + y$	$y + \frac{T_2}{T_1}x$

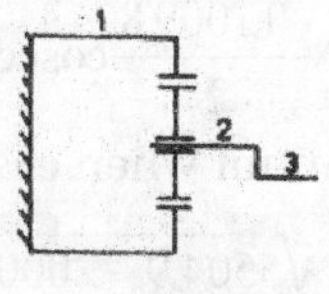

Now, $y + \frac{T_2}{T_1}x = 0$, where, $y = -60$

$\therefore \quad -60 + \frac{T_2}{T_1}x = 0,$

or, $\frac{T_2}{T_1}x = 60$

or, $x = -\frac{60T_2}{T_1}$

$= +60 \times \frac{104}{96}$

= 65 revolutions anti-clockwise.

∴ Ratation of planet gear (2) for 60 revolutions of planet carrier (3) clockwise,

$x + y = 65 - 60$

= 5 revolutions anti clockwise.

76. Given

$P = 1.5$ kW

$\mu = 0.25$

$\theta = 159°$

$v = 375$ m/s

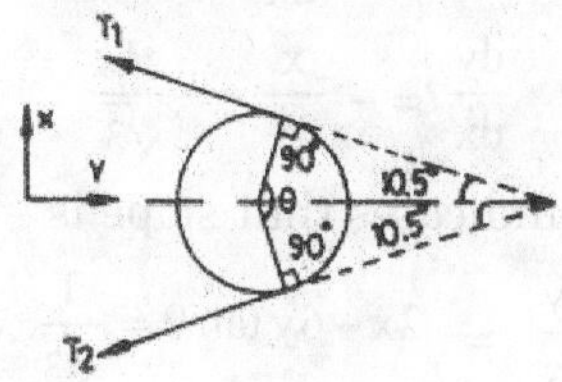

From $P = (T_1 - T_2)\,v$

$1.5 \times 1000 = (T_1 - T_2) \times 3.75$

or $T_1 - T_2 = 400$ .....(*i*)

Now, $\frac{T_1}{T_2} = e^{\mu\theta} = e^{0.25} \times \frac{159 \times \pi}{180} = 2$

or, $T_1 = T_2$ .....(*ii*)

From equations (*i*) and (*ii*), we get

$T_2 = 400$ N

and $T_1 = 800$N

X-Component $-(T_1 \cos 10.5° + T_2 \cos 10.5°)$

$= -(800 \cos 10.5° + 400 \sin 10.5°)$

$= -1179.9$ N

Y - Component : $T_1 \sin 10.5° - T_2 \sin 10.5°$

$= 800 \sin 10.5° - 400 \sin 10.5°$

$= 72.89$ N

78. $\Delta = \frac{1}{2}(PQ)(PR) \sin A$

Area will be maximum when A = 90°

i.e. PQR is a right angled triangle.

∴ Ratio of connecting rot length to crank radius,

$\frac{\ell}{r} = \tan 75° = 3.732;$

$\ell = 3.732\, r$

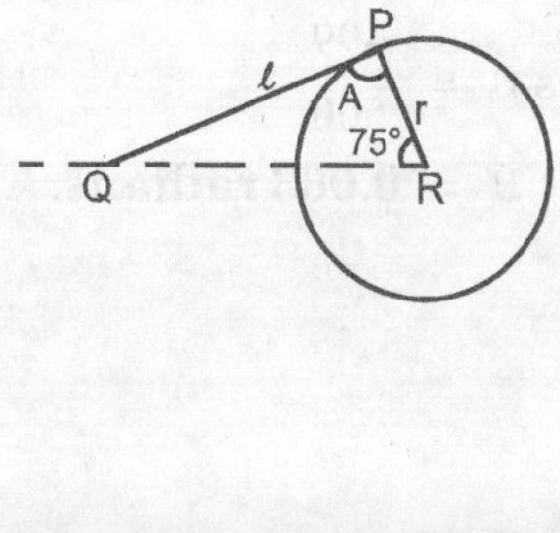

79. Given, $T_1 - T_2 = 3000$ N

where T_1, T_2 = tensions on tight and slack side respectively

v = belt speed = 15 m/sec

Power = $(T_1 - T_2)\, v$

= 3000 × 45000 watt = **45 kW.**

80. $\sin^2\theta + \cos^2\theta = 1$

Equation of curve is $y^2 + \frac{x^2}{3} = 1$

or $x^2 + 3y^2 = 3$

To find Slope at required point $\left(\frac{\sqrt{3}}{2}, \frac{1}{\sqrt{2}}\right)$,

differentiating we get

$$2x + 6y\frac{dy}{dx} = 0$$

or $\frac{dy}{dx} = -\frac{x}{3y} = -\frac{1}{\sqrt{3}}$

–ve sign indicates that slope is – ve.

$\therefore\ \frac{dy}{dx} = 2x + 6y \tan\theta = \frac{1}{\sqrt{3}}$

or $\theta = 30°$

∴ Angle made by normal = 60° = **π/3 radians.**

81. Given, h_0 = initial thickness = 4.5 mm

Sectional area = 150 × 4.5 mm²

Since area is reduced by 20%, therefore

t = 0.8 × 4.5 = 3.60

final area = 0.8 × 1.50 × 4.5 sq. mm.

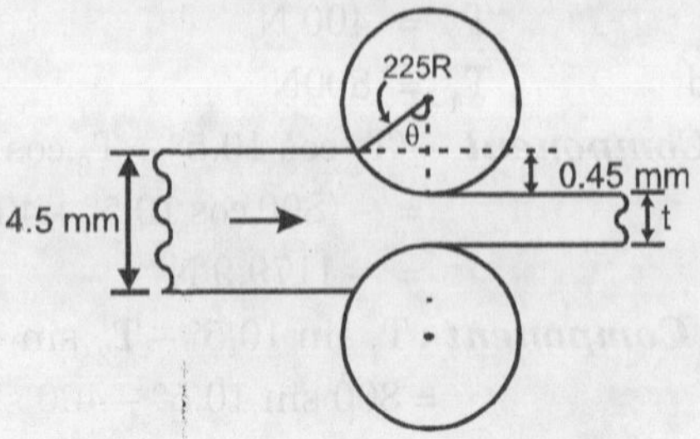

Keeping width constant, final thickness

$$h_1 = \frac{0.8 \times 150 \times 4.5}{150} = 0.8 \times 4.5 \text{ mm}$$

$\therefore\ R - R\cos\theta = \frac{4.5 - (0.8 \times 4.5)}{2}$

or $R(1 - \cos\theta) = \frac{4.5 \times 0.2}{2}$

or $1 - \cos\theta = \frac{4.5 \times 0.2}{2 \times 225}$

or $\cos\theta = \frac{499}{500}$

$\therefore\ \theta$ = **0.063 radians.**

82. $\Delta E = mR^2\omega^2\Delta S = mR^2\omega(\omega_1 - \omega_2)$

$\therefore\ 1936 = 9.8 \times \omega\left(\frac{2\pi}{60} \times 30\right)$

or Mean speed of flywheel, ω = **600 rpm**

83. Displacement of the follower, r = 5 + 2 sin ωt

∴ Velocity of the follower, $V = \frac{dr}{dt} = 2(\cos\omega t)$.

ω = 2ω . cos ωt.

Acceleration of the follower,

$$f = \frac{d^2r}{dt^2} = 2\omega(-\sin\omega t).\omega = -2\omega^2 \sin\omega t$$

Spring stiffnes, K = 2 kN/m

Follower will leave the cam surface when the net force acting through is is zero.

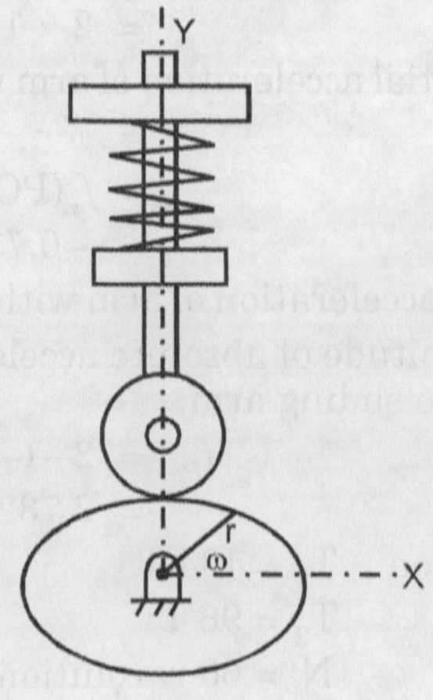

Initial spring force = $\left(\frac{5}{10}\right)$ 2000 N

Force due to compression of spring = (5 + 2 sin ωt) 2000 N

Weight due to the follower = 1 × 9.8 = 9.8 N

Force due to acceleration = – (1) (– 2 sin ωt) (ω²)

Now for any value of rotation, θ of the cam, algebraic sum of the spring force, weight of the follower and the accelerating force is equal to the vertical reaction between cam and follower. When this reaction is zero, then this follower will just begin to leave the cam.

$\therefore\ S + K \times x + W + 2\,m\,\omega^2 \sin\omega t = 0$

1000 + 10,000 + 2000 (2 sin ωt) + 1 × 9.8 + 2 × 2 × ω² sin ωt = 0

or 11,000 + 4000 sin ωt + 9.8 + 4ω² sin ωt = 0

or, $\omega^2 = \frac{-11009.8}{2}\text{cosec}\,\omega t - 2000$

ω will be maximum when cosec ωt = – 1

or $\omega = \sqrt{5504.9 - 2000} = 59.20$ rad / sec.

∴ Maximum allowable cam shaft speed,

$$N = \frac{\omega \times 60}{2\pi} = \frac{59.20 \times 60}{2\pi} = \mathbf{565.32\ rpm.}$$

84. Load conditions

Radial load, F_r = 3000 N

Axial load F_a = 1600 N

RPM = 1,000

First Case : Deep groove ball bearing

Radial load = X = 0.56

Axial load factor = Y = 1.83

Minimum dynamic load required

$$P = X \times F_r + Y \times F_a$$
$$= 0.56 \times 3000 + 1.83 \times 1600$$
$$= 4608 \text{ N}$$

To select the bearing which can bear maximum load, determine a factor

$$\text{Factor} = \frac{\text{Given dynamic load rating}}{\text{Working dynamic load}}$$
$$= \frac{47000}{4608} = 10.2$$

For second case : Cylindrical roller bearing

$$\text{Dynamic load} = 1 \times 300 + 0 \times 1600$$
$$= 3000 \text{ N.}$$
$$\text{Factor} = \frac{28000}{3000} = 9.33$$

Hence select the first bearing, which has bigger factor.

89. Number of revolutions in life = 2000 × 60 × 600

= 72 × 10^5 revolutions

$\therefore$ L = 72

$$\text{Maximum radial load} = \frac{22}{\sqrt[3]{L}} = \frac{22}{\sqrt[3]{72}}$$
$$= \mathbf{5.29\ kN}$$

92. *Given*: Power transmited = 4kW,

Speed = 750 rpm

Inner radius, r_1 = 40 mm,

Outer radius, r_2 = 70 mm.

Coefficient of friction of oil, = 0.1

Maximum allowable pressure,

$$p = 350 \text{ kPa}$$
$$= 350 \times 10^3 \text{ N/m}^2$$
$$= 0.35 \text{ N/mm}^2$$

Now $$\frac{2\pi NT}{60} = 4000 \text{ W}$$

$$\therefore \quad T = \frac{4000}{25\pi} = 51 \text{ N/m}$$

Axial force, $$W = p_{ave.}\pi(\gamma_1^2 - \gamma_2^2)$$

$$= 350 \times 10^3 \pi\left[\left(\frac{7}{100}\right)^2 - \left(\frac{4}{100}\right)^2\right] = 3620 \text{ N}$$

Also $$T = n\ W\gamma$$

or $$51 = n \times 0.1 \times 3620 \times \frac{5.5}{100}$$

$\therefore$ Number of pairs of contact surfaces, n = 2.51

Take *n* = 4 since 4 is the nearest even number i.e. **3 steel and 2 bronze discs** may be used.

(n = total number of discs – 1)

$$T' = \frac{T}{n} = \frac{52}{4} = 13 \text{ N-m}$$

$$\therefore \quad 13 = \mu W'\gamma$$
$$= 0.1W'\frac{5.5}{100} = \frac{0.55\,W'}{100}$$

$\therefore$ Actual axial force,

$$W' = \frac{1300}{0.55} = \mathbf{2360\ N}$$

93. *Given* : Number of teeth on gear,

T = 33

Addendum circle radius of gear

$$= \frac{D_A}{2} = \frac{175}{2} = 87.5 \text{ mm}$$

Number of teeth on pinion,

t = 15

Addendum circle radius of pinion

$$= \frac{d_A}{2} = \frac{85}{2} = 42.5 \text{ mm}$$

Pressure angle, $\phi = 20°$

Now, rubbing velocity,

$$v_R = \omega_1 \quad r = \omega_2 r$$

where r, R = pitch circle radii of pinion and gear respectively.

Now, $$\frac{\omega_1}{\omega_2} = \frac{T}{t} = \frac{33}{15} = 2.2$$

$\therefore$ $$\omega_1 = 2.2\,\omega_2$$

Now $$V_R = 2.2\,\omega_2 \times r = \omega_2 \times R$$

or $$R = 2.2\ r.$$

$$\text{Gear ratio} = \frac{T}{t} = 2.2 = \frac{R}{r} = \frac{\left(\frac{D_A}{2} - m\right)}{\left(\frac{D_A}{2} - m\right)}$$

or $$2.2 = \frac{87.5 - m}{42.5 - m}$$

or $(42.5 - m) \times 2.2 = 87.5 - m$

or $93.5 - 2.2\ m = 87.5 - m,$

$\therefore$ $m = 5.$

Now, centre resistance = R + r

where $$R = R_A - m = \left(\frac{D_A}{2} - m\right)$$
$$= \left(\frac{175}{2} - 5\right) = 82.5 \text{ mm}$$

$$r = r_A - m = \left(\frac{d_A}{2} - m\right)$$

$$= \left(\frac{85}{2} - 5\right) = 37.5 \text{ mm}$$

$\therefore$ Centre resistance = 82.5 + 37.5 = 120 mm

Minimum number of teeth required on pinion in order to aviod interference,

$$t = \frac{2\,A_w}{G\sqrt{1 + \frac{1}{G}\left(\frac{1}{G} + 2\right)\sin^2 20° - 1}}$$

$$= \frac{2 \times 5}{2.2\sqrt{1 + \frac{1}{2.2}\left(\frac{1}{1.22} + 2\right)\sin^2 20° - 1}} = 7.46$$

Since actual number of teeth on pinion = 15 > 7.46

So there will be no interference.

94.

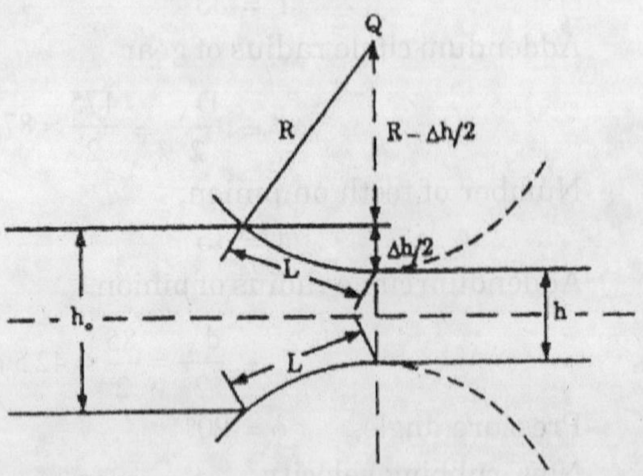

$$\text{Roll gap pressure, } P_a = \left(\frac{h}{\mu L}\right)\left[e^{\frac{\mu L}{h}} - 1\right]\sigma_p$$

where,
$$h = \frac{3.5 + 2.5}{2} = 3 \text{ mm}$$

$$\Delta h = 3.5 - 2.5 = 1 \text{ mm}$$

Length of contact,

$$L = \sqrt{R^2 - \left(R - \frac{\Delta h}{2}\right)^2}$$

$$= \sqrt{R.\Delta h} = \sqrt{125 \times 1} = 11.18\text{mm}$$

$$\therefore \quad P_a = \left[\frac{3}{0.08 \times 11.18}\right]\left[e^{\frac{0.08 \times 11.18}{3}} - 1\right] \times 140$$

$$= 163.11 \text{ MPa}$$

Roll separating force,

$$F_R = 163.11 \times 11.18 \times 450$$

$$= 820606 \text{ N or } \mathbf{820.6\ kN}$$

98. Explanation : Given, Diameter of each hole, d = 10 mm = 0.01 m

Number of holes to be punched = 5

Thickness of sheet, t = 3 mm = 0.003 m

Pitch of holes = 25 mm

Yield point of material = 50 MPa = 50×10^6 N/m².

$$\text{Area resistimg shear/hole} = \pi \times \left(\frac{10}{1000}\right) \times \left(\frac{3}{1000}\right) \text{ m}^2$$

Force required/hole

$$= \frac{50 \times 10^6 \times \pi \times 10 \times 13}{10^3 \times 10^3} = 20420 \text{ N.}$$

(a) If one hole is to be punched/stroke, minimum capacity of the press

$$= \frac{20420}{9.8 \times 1000} = \mathbf{2.08\ T}$$

(b) If five holes are to be punched/stroke and if all the holes are punched together minimum capacity of the press

$$= 2.06 \times 5 = \mathbf{10.4\ T}$$

If the punching time is staggered without any overlap, capacity of the press required is only 2.08 Tonnes.

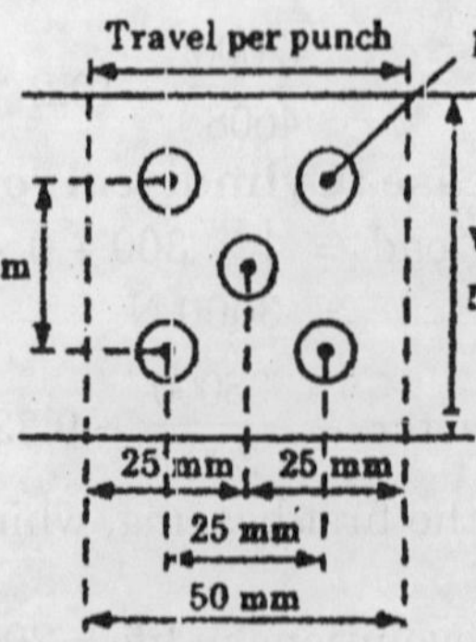

Fig. Arrangement of punches to avoid eccentric loading

99.

Action	**3**	**2**	**4**	**5**
Fix arm '3' gear 2 in clock wise direction	0	+ 1	$-\frac{24}{40}$	$-\frac{24}{40} \times \frac{40}{144}$
Multiply by x	0	+ x	$-\frac{24}{40}$x	$-\frac{40}{144}$x
Add y	+ y	x + y	$y - \frac{24}{40}$x	$y - \frac{40}{144}$x

Wheel (5) is fixed,

$$y - \frac{40}{144}x = 0$$

$$\therefore \quad y = \frac{x}{6}$$

rpm of gear (2) = 700 rpm,

i.e. $\quad x + y = 700$

or $\quad x + \frac{x}{6} = 700$

or $\quad x = 600$

$\therefore \quad y = 100$

rpm of arm (3) = + y = 100 rpm in clockwise direction

$$\text{rpm of gear (4)} = y - \frac{24}{40}x$$

$$= 100 - \frac{24}{40} \times 600 = \mathbf{-260\ rpm}$$

Negative sign indicates that rpm of gear 4 is 260 rpm anticlockwise.

100. ***(a)*** Angular momentum of bar about hinge A,

$$\Rightarrow \quad L_{bar_A} = L_{bar_A} \times \omega$$

$$= \frac{1}{3} m_b (2a)^2 \omega = \frac{4}{3} ma^2 \omega$$

(b) Angular momentum of the disc about the hinge A

$$L_{disc_A} = L_{disc_B} + L_{AB}$$

$$= \frac{1}{2} m_d a^2 \omega + \left\{ \left(\frac{1}{2} m_d a^2 \right) + m_d (2a)^2 \right\}$$

$$= ma^2\omega + m(2a)^2\omega = \mathbf{5\ ma^2\omega}$$

(c) Kinetic energy of the system

$$= \frac{1}{2} I_{bar_A} \times \omega^2 + \frac{1}{2} I_{disc_B} \times \omega^2 + \frac{1}{2} I_{AB} \times \omega^2$$

$$= \frac{1}{2} \frac{m(2a)^2}{3} \omega^2 + \frac{1}{2} \frac{ma^2}{2} \omega^2 + \frac{1}{2} \left[\frac{ma^2}{2} + m(2a)^2 \right] \omega^2$$

$$= \frac{2ma^2\omega^2}{3} + \frac{ma^2\omega^2}{4} + \frac{9}{4} ma^2\omega^2$$

$$= \frac{38}{12} ma^2 \omega^2$$

NUMERICAL TYPE QUESTIONS

1. Given $F_p = 5$ kN

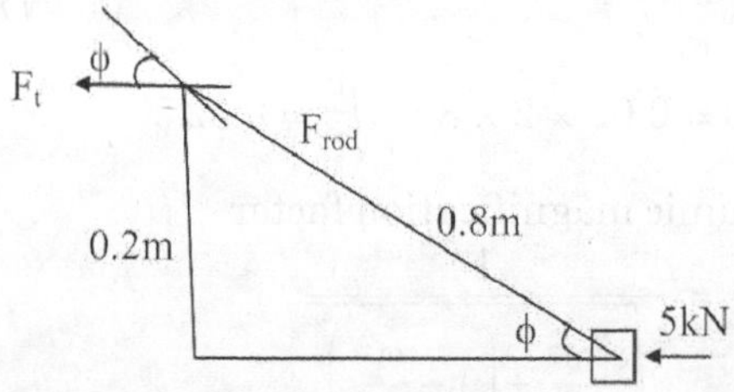

$$OF_{rod} = \frac{F_p}{\cos\phi}$$

$$F_t = F_{rod} \cos\phi$$

$$\therefore \quad F_t = 5 \text{ kN}$$

Turning moment = $F_t \cdot r = 5 \times 0.2 = 1$ kN-m

2.

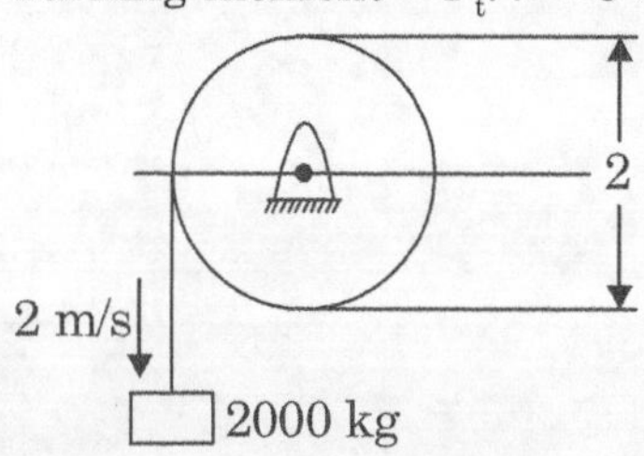

$$v = r\omega$$

$$\omega_i = \frac{V_u}{r} = \frac{2}{1} = 2 \text{ rad/sec}$$

$$\text{Loss in KE of drum} = \frac{1}{2} J(\omega_i^2 - \omega_f^2) = \frac{1}{2} \times 150(2^2 - 0^2)$$

$$= 300 \text{ Joule}$$

3. Locate the I-centre for the link AB as shown in figureM is the mid point of AB

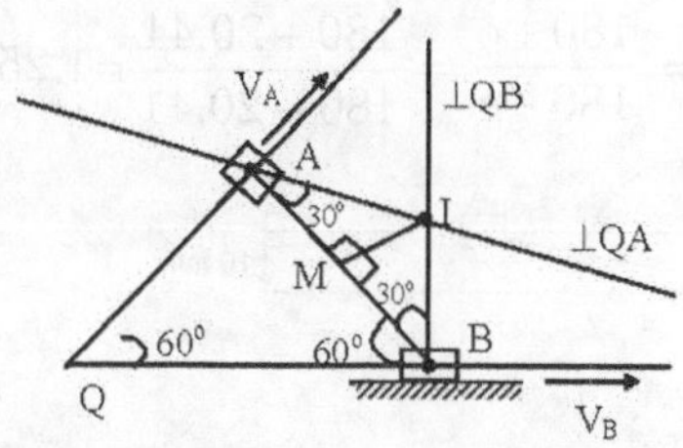

Given $V_A = 2$ m/sec

$$V_A = IA.\omega \Rightarrow \omega = \frac{V_A}{IA}$$

$$V_M = IM.\omega = IM \frac{V_A}{IA} = \frac{IM}{IA}.V_A$$

$$= \sin 30°.\ V_A = \frac{1}{2}.2 = 1 \text{ m/sec}$$

4. By symmetric two system is in dynamic balance when

$$mea = m_1 e_1 a_1$$

$$m_1 = m \frac{e}{e_1} \cdot \frac{a}{a_1} = 1 \times \frac{50}{20} \frac{2}{2.5} = 2 \text{ kg}$$

5.

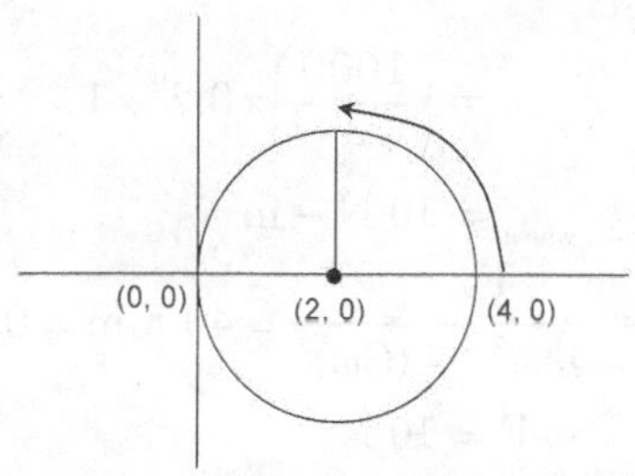

$$\frac{1}{4} \times \text{circumference}$$

$$\Rightarrow \frac{1}{4} \times \quad \pi \times 4$$

$$\text{time} = \frac{\pi}{1.5 \rightarrow} = 2\text{sec}$$

7. $a_{max} = 2r\omega^2$ (where θ = 0) i.e at Inner dead centre

$$= 2 \times 0.3 \times 14^2$$

$$a_{max} = 117.6 \text{ m/s}^2$$

9. $r = \frac{0.2}{2} = 0.1$

Now m= 6kg

$$F = mr\,(10^2) \sin\theta$$

$$= 6 \times 0.1 \times 100 \times \sin 30° = 30 \text{ N}$$

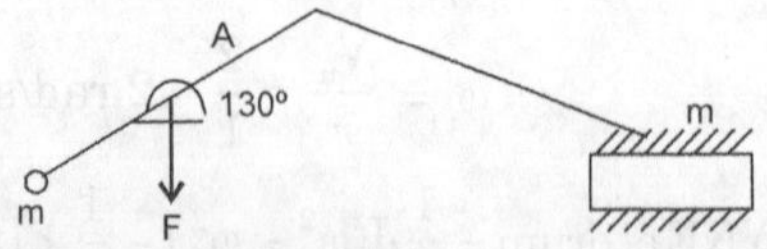

10. AB = stroke length

AO = 40 – 20 = 20

BO = 40 + 20 = 60

$\angle AOB = \angle BOC - \angle AOC = 80.41 - 60 = 20.41$

$$QRR = \frac{180+\phi}{180-\phi} = \frac{180+20.41}{180-20.41} = 1.255.$$

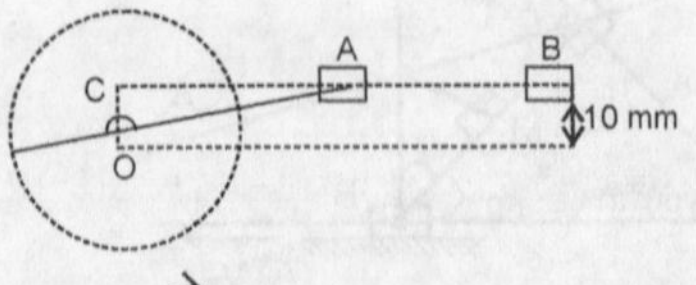

11.

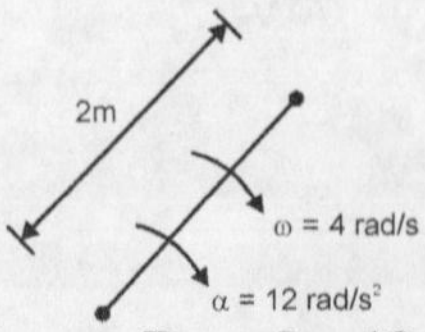

$a_t = R\alpha = 2 \times 12 = 24$

$a_R = \omega^2 R = 16 \times 2 = 32$

$$a = \sqrt{a_t^2 + a_R^2} = \sqrt{24^2 + 32^2} = 40 \text{ m/s}^2.$$

13. m = 1000 kg, ω = 10 Rad/sec, R = 0.2 m, t = 10 sec when ω = 0

$\therefore \quad \omega_F = \omega_i - \alpha t$

$\therefore \quad 0 = 10 - \alpha \ 10$

$\therefore \quad \alpha = 1 \text{ rad/sec}^2$

Now T = I·α

$$\therefore \quad (T)_{\text{each wheel}} = \left(\frac{1000}{4}\right) \times 0.2^2 \times 1$$

$\therefore \quad (T)_{\text{each wheel}} = 10 \text{ N} - \text{m}$

15.
$$X = \frac{F}{\sqrt{(k - m\omega^2)^2 + (C\omega)^2}} = 40 \text{ mm} = 0.04$$

F = 10 N

ω = 25

$$\Rightarrow \quad 0.04 = \frac{10}{\sqrt{(6250 - 10 \times 25^2)^2 + (C \times 25)^2}}$$

C = 10Ns/m

16.
$$\Delta S_1 = \frac{Q}{T_1}$$

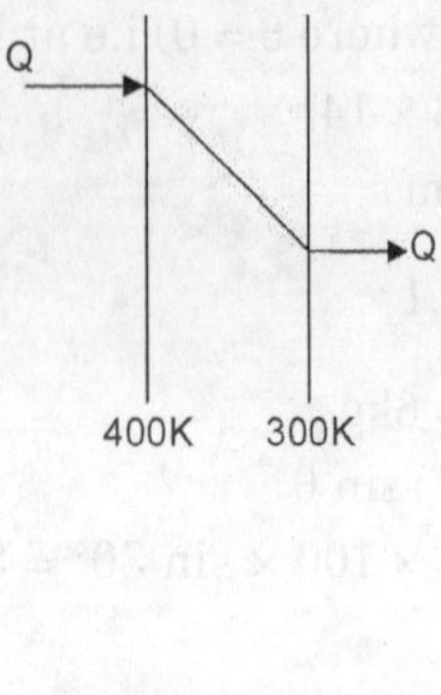

$$\Delta S_2 = \frac{Q}{T_2}$$

$$(\Delta S)_{\text{generated}} = \Delta S_1 + \Delta S_2$$

$$= \frac{Q}{400} - \frac{Q}{300} = \frac{100 \times 10^3}{100}\left(\frac{1}{4} - \frac{1}{3}\right)$$

$$= 10^3 \times -0.0833 = -83.33 \text{ W/K}$$

Entropy generated = 83.33 W/K.

17. Given A = 10 mm = 0.01 m

f = 4 Hz

$x = A \cos \omega t$

$$v = \frac{dx}{dt} = -A\omega \sin \omega t$$

$$\frac{d^2x}{dt^2} = a = -A\omega^2 \cos \omega t$$

For maximum, t = 0

$\therefore \quad a = A\omega^2 = 0.01(2\pi f)^2$

$\because \quad \omega = 2\pi f = 0.01(2 \times \pi \times 4)^2 = 6.32 \text{ m/s}^2$

18.
$$\omega = \frac{\omega_{max} + \omega_{min}}{2} = \frac{20+0}{2} = 10$$

$$C_s = \frac{\omega_{max} - \omega_{min}}{\omega} = \frac{20-0}{10} = 2$$

ΔE = Area of T – θ diagram

$$= \frac{\pi}{2} \times 3000 + 1500 \times \pi = 3000\pi$$

$\Delta E = I\omega^2 C_s$

$\Rightarrow 3000\pi = I(10)^2 \times 2 \qquad \Rightarrow I = 47.12$

19.
$$\text{Damping ratio} = 0.02 = \frac{C}{2m\omega_n} = \frac{C}{2 \times 8 \times \sqrt{\frac{S}{m}}}$$

$$\Rightarrow C = 0.02 \times 2 \times 8 \times \sqrt{\frac{8}{8}} = 0.32$$

Dynamic magnification factor

$$= \frac{1}{\sqrt{\frac{c^2\omega^2}{s^2} + \left(1 - \frac{\omega^2}{\omega_n^2}\right)^2}}$$

$$= \frac{1}{\sqrt{\frac{(0.32)^2 \times (1.5)^2}{(8)^2} + \left[1 - \frac{(1.5)^2}{(1)^2}\right]}} = 0.799$$

4

CHAPTER

Vibrations

MCQ TYPE QUESTIONS

2016

1. A solid disc with radius a is connected to a spring at a point d above the center of the disc. The other end of the spring is fixed to the vertical wall. The disc is free to roll without slipping on the ground. The mass of the disc is M and the spring constant is K. The polar moment of inertia for the disc about its centre is $J = Ma^2/2$.

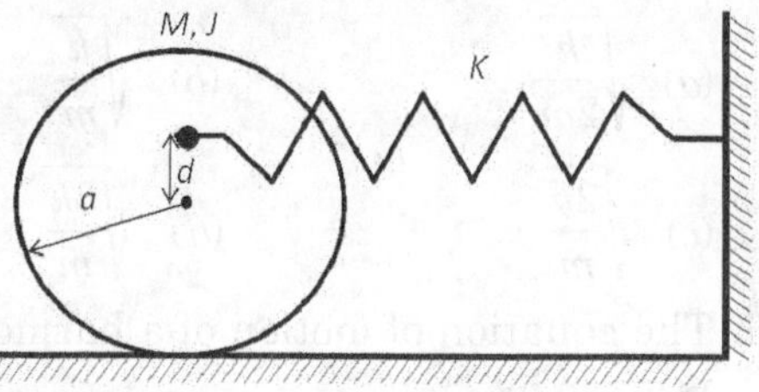

The natural frequency of this system in rad/s is given by

(a) $\sqrt{\dfrac{2K(a+d)^2}{3Ma^2}}$ (b) $\sqrt{\dfrac{2K}{3M}}$

(c) $\sqrt{\dfrac{2K(a+d)^2}{Ma^2}}$ (d) $\sqrt{\dfrac{K(a+d)^2}{Ma^2}}$

2. A single degree of freedom mass-spring-viscous damper system with mass m, spring constant k and viscous damping coefficient q is critically damped. The correct relation among m, k, and q is

(a) $q = \sqrt{2km}$ (b) $q = 2\sqrt{km}$

(c) $q = \sqrt{\dfrac{2k}{m}}$ (d) $q = 2\sqrt{\dfrac{k}{m}}$

2013

3. A metric thread of pitch 2 *mm* and thread angle 60° is inspected for its pitch diameter using 3-wire method. The diameter of the best size wire in *mm* is

(a) 0.866 (b) 1.000

(c) 1.154 (d) 2.000

4. A long thin walled cylindrical shell, closed at both the ends, is subjected to an internal pressure. The ratio of the hoop stress (circumferential stress) to longitudinal stress developed in the shell is

(a) 0.5 (b) 1.0

(c) 2.0 (d) 4.0

5. A bar is subjected to fluctuating tensile load from 20 *k*N to 100 *k*N. The material has yield strength of 240 MPa and endurance limit in reversed bending is 160 MPa. According to the Soderberg principle, the area of cross-section in mm² of the bar for a factor of safety of 2 is

(a) 400 (b) 600

(c) 750 (d) 1000

2012

6. A concentrated mass m is attached at the centre of a rod of length $2L$ as shown in the figure. The rod is kept in a horizontal equilibrium position by a spring of stiffness k. For very small amplitude of vibration, neglecting the weights of the rod and spring, the undamped natural frequency of the system is

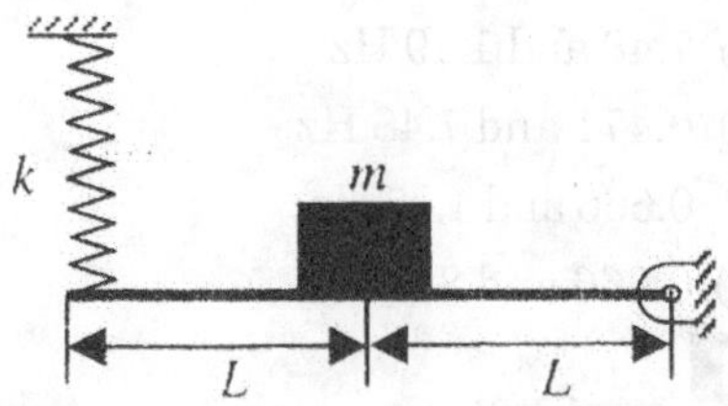

(a) $\sqrt{\dfrac{k}{m}}$ (b) $\sqrt{\dfrac{2k}{m}}$

(c) $\sqrt{\dfrac{k}{2m}}$ (d) $\sqrt{\dfrac{4k}{m}}$

2010

7. The natural frequency of a spring-mass system on earth is ω_n. The natural frequency of this system on the moon ($g_{moon} = g_{earth}/6$) is

(a) ω_n (b) $0.408\,\omega_n$

(c) $0.204\,\omega_n$ (d) $0.167\,\omega_n$

8. A mass m attached to a spring is subjected to a harmonic force as shown in figure. The amplitude of the forced motion is observed to be 50 mm. The value of m (in kg) is

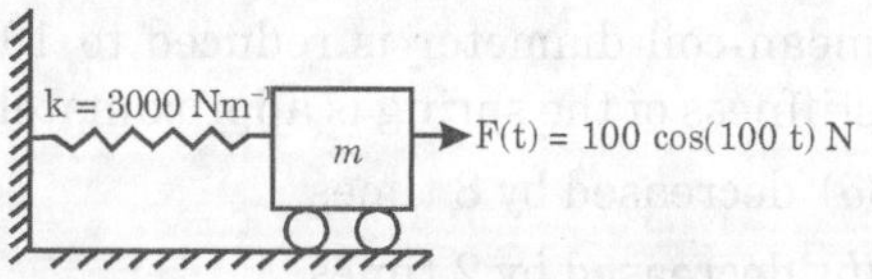

(a) 0.1 (b) 1.0

(c) 0.3 (d) 0.5

2009

9. The rotor shaft of a large electric motor supported between short bearings at both the ends shows a deflection of 1.8 mm in the middle of the rotor. Assuming the rotor to be perfectly balanced and supported at knife edges at both the ends, the likely critical speed (in rpm) of the shaft is

(*a*) 350 (*b*) 705

(*c*) 2810 (*d*) 4430

10. An automotive engine weighing 240 kg is supported on four springs with linear characteristics. Each of the front two springs have a stiffness of 16 MN/m while the stiffness of each rear spring is 32 MN/m. The engine speed (in rpm), at which resonance is likely to to occur, is

(*a*) 6040 (*b*) 3020

(*c*) 1424 (*d*) 955

11. A vehicle suspension system consists of a spring and a damper. The stiffness of the springs is 3.6 kN/m and damping constant of the damper is 400 Ns/m. If mass is 50 kg, then damping factor (d) and damped natural frequency (f_n) respectively are

(*a*) 7.48 and 1.19 Hz

(*b*) 0.471 and 7.48 Hz

(*c*) 0.666 and 1.35 Hz

(*d*) 0.666 and 8.50 Hz

2008

12. The natural frequency of the spring mass system shown in the figure is closest to

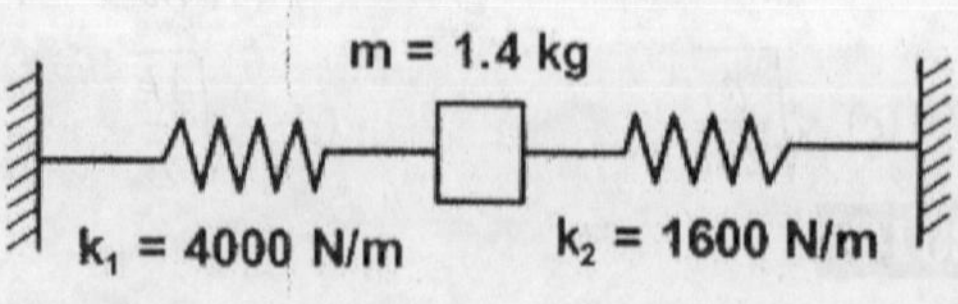

(*a*) 8 Hz

(*b*) 10 Hz

(*c*) 12 Hz

(*d*) 14 Hz

13. A compression spring is made of music wire of 2 mm diameter having a shear strength and shear modulus of 800 MPa and 80 GPa respectively. The mean coil diameter is 20 mm, free length is 40 mm and the number of active coils is 10. If mean coil diameter is reduced to 10 mm, then stiffness of the spring is approximately

(*a*) decreased by 8 times

(*b*) decreased by 2 times

(*c*) increased by 2 times

(*d*) increased by 8 times

2007

14. For an under damped harmonic oscillator, resonance

(*a*) occurs when excitation frequency is greater than undamped natural frequency

(*b*) occurs when excitation frequency is less than undamped natural frequency

(*c*) occurs when excitation frequency is equal to undamped natural frequency

(*d*) never occurs

15. Natural frequency of the system shown below is

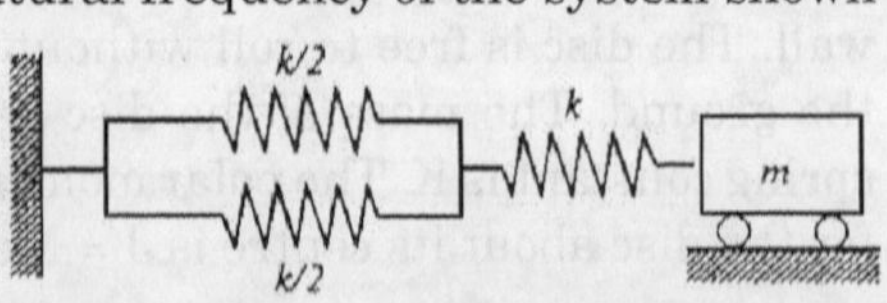

(*a*) $\sqrt{\frac{k}{2m}}$ (*b*) $\sqrt{\frac{k}{m}}$

(*c*) $\sqrt{\frac{2k}{m}}$ (*d*) $\sqrt{\frac{3k}{m}}$

16. The equation of motion of a harmonic oscillator is given by

$$\frac{d^2x}{dt^2} + 2\zeta\omega_n\frac{dx}{dt} + \omega_n^2 x = 0$$

and the initial conditions at $t = 0$ are $x(0) = X$, $\frac{dx}{dt}(0) = 0$. The amplitude of $x(t)$ after n complete cycles is

(*a*) $Xe^{-2n\pi\left(\frac{\zeta}{\sqrt{1-\zeta^2}}\right)}$ (*b*) $Xe^{2n\pi\left(\frac{\zeta}{\sqrt{1-\zeta^2}}\right)}$

(*c*) $Xe^{-2n\pi\left(\frac{\sqrt{1-\zeta^2}}{\zeta}\right)}$ (*d*) X

2006

17. The differential equation governing the vibrating system is

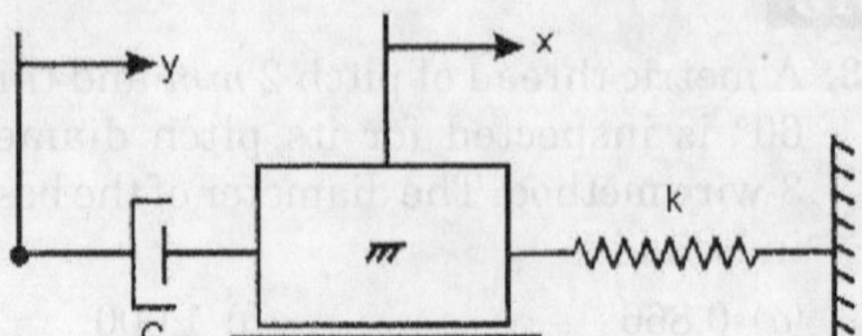

(*a*) $m\ddot{x} + c\dot{x} + k(x - y) = 0$

(*b*) $m(\ddot{x} - \ddot{y}) + c(\dot{x} - \dot{y}) + kx = 0$

(*c*) $m\ddot{x} + c(\dot{x} - \dot{y}) + kx = 0$

(*d*) $m(\ddot{x} - \ddot{y}) + c(\dot{x} - \dot{y}) + k(x - y) = 0$

2005

18. There are four samples P, Q, R and S, with natural frequencies 64, 96, 128 and 256 Hz, respectively. They are mounted on test setups for conducting vibration experiments. If a loud pure note of frequency 144 Hz is produced by some instrument, which of the samples will show the most perceptible induced vibration?

(*a*) P (*b*) Q
(*c*) R (*d*) S

19. In a spring-mass system, the mass is 0.1 kg and the stiffness of the spring is 1 kN/m. By introducing a damper, the frequency of oscillation is found to be 90% of the original value. What is the damping coefficient of the damper?

(*a*) 1.2 N.s/m (*b*) 3.4 N.s/m
(*c*) 8.7 N.s/m (*d*) 12.0 N.s/m

2004

20. A vibrating machine is isolated from the floor using springs. If ratio of excitation frequency of vibration of machine to the natural frequency of the isolation system is equal to 0.5, then transmissibility of ratio of isolation is

(*a*) $\frac{1}{2}$ (*b*) $\frac{3}{4}$
(*c*) $\frac{4}{3}$ (*d*) 2

21. A mass M, of 20 kg is attached to the free end of a steel cantilever beam of length 1000 mm having a cross-section of 25 × 25 mm. Assume mass of the cantilever to be neglible and E_{steel} = 200 GPa. If lateral vibration of this system is critically damped using a viscous damper, then damping constant of the damper is

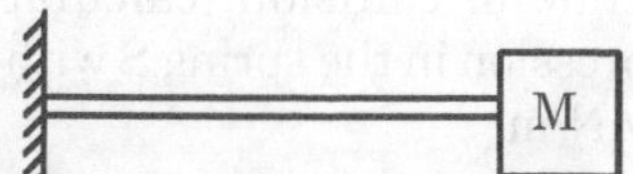

(*a*) 1250 Ns/m (*b*) 625 Ns/m
(*c*) 312.50 Ns/m (*d*) 156.25 Ns/m

2003

22. Consider the arrangement shown in the figure below, where J is combined polar mass moment of inertia of the disc and the shafts. K_1, K_2, K_3 are torsional stiffness of the respective shafts. The natural frequency of torsional oscillation of the disc is given by

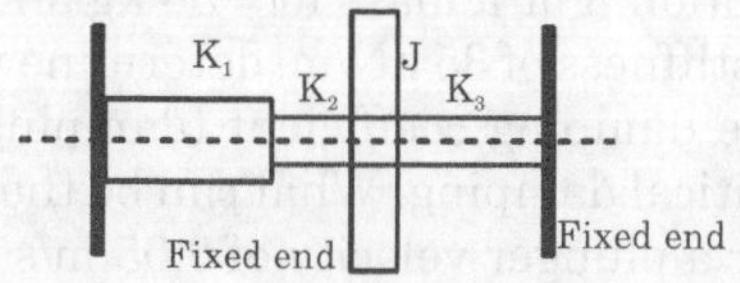

(*a*) $\sqrt{\frac{K_1+K_2+K_3}{J}}$

(*b*) $\sqrt{\frac{K_1K_2+K_2K_3+K_3K_1}{J(K_1+K_2)}}$

(*c*) $\sqrt{\frac{K_1K_2K_3}{J(K_1K_2+K_2K_3+K_3K_1)}}$

(*d*) $\sqrt{\frac{K_1K_2+K_2K_3+K_3K_1}{J(K_2+K_3)}}$

Common Data Q. 23 – 24

A uniform rigid slenderbar of mass 10 kg, hinged at the left end is suspended with the help of spring and damper arrangement as shown in the figure where k = 2 kN/m, C = 500 Ns/m and stiffness of the torsional spring K_t is 1 kN/m/rad. Ignore the hinge dimensions.

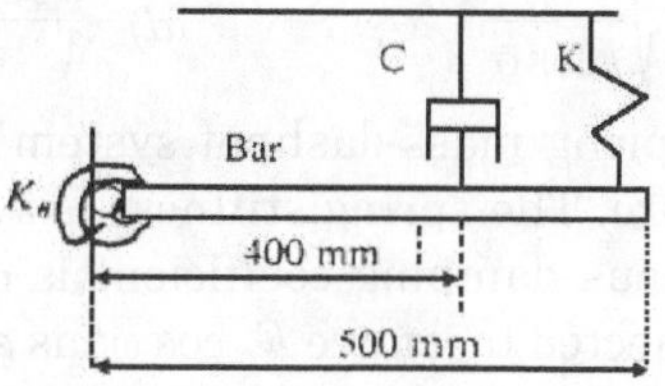

23. The un-damped natural frequency of oscillations of the bar about the hinge point is

(*a*) 42.43 rad/s (*b*) 30 rad/s
(*c*) 17.32 rad/s (*d*) 14.14 rad/s

24. The damping coefficient in the vibration equation is given by

(*a*) 500 Nms/rad (*b*) 500 N/(m/s)
(*c*) 80 Nms/rad (*d*) 80 N/(m/s)

2002

25. If length of the cantilever beam is halved, then natural frequency of the mass M at the end of this cantilever beam of negligible mass is increased by a factor of

(*a*) 2 (*b*) 4
(*c*) $\sqrt{8}$ (*d*) 8

2001

26. In the figure shown, the spring deflects by δ to position A (equilibrium position) when a mass *m* is kept on it. During free vibration, the mass is at position B at some instant. The change in potential energy of the spring-mass system from position A to position B is

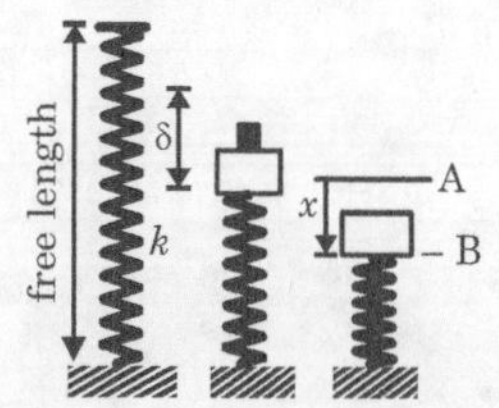

(a) $\frac{1}{2}kx^2$ (b) $\frac{1}{2}kx^2 - mgx$

(c) $\frac{1}{2}k(x+\delta)^2$ (d) $\frac{1}{2}kx^2 + mgx$

27. The assembly shown in the figure is composed of two massless rods of length l with two particles, each of mass m. The natural frequency of this assembly for small oscillations is

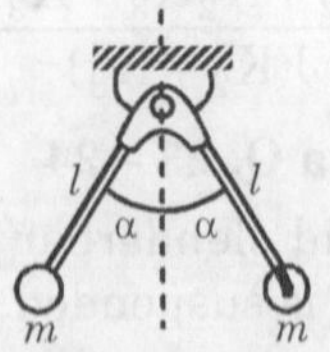

(a) $\sqrt{\frac{g}{l}}$ (b) $\sqrt{\frac{2g}{l\cos\alpha}}$

(c) $\sqrt{\frac{g}{l\cos\alpha}}$ (d) $\sqrt{\frac{g\cos\alpha}{l}}$

28. A spring mass-dashpot system is shown in the figure. The spring stiffness is k, mass is m, and viscous damping coefficient is c. The system is subjected to a force $F_0 \cos \omega t$ as shown. Write the equations of motion which are needed to determine x. (No need to determine x.)

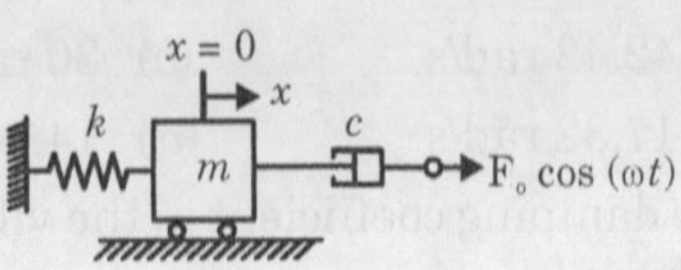

2000

29. The natural frequency of an undamped vibrating system is 100 rad/s. A damper with a damping factor of 0.8 is introduced into the system. The frequency of vibration of the damped system, in rad/s, is

(a) 60 (b) 75

(c) 80 (d) 100

30. As shown in the figure, a mass of 100 kg is held between two springs. The natural frequency of vibration of the system, in cycles/s, is

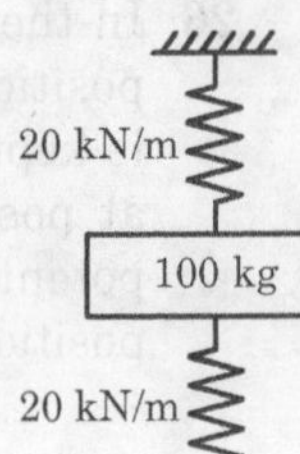

(a) $\frac{1}{2\pi}$

(b) $\frac{5}{\pi}$

(c) $\frac{10}{\pi}$

(d) $\frac{20}{\pi}$

1999

31. $\frac{\partial^2 u}{\partial t^2} = C^2 \frac{\partial^2 u}{\partial x^2}$ represents the equation for

(a) Vibration of a stretched string

(b) Motion of a projectile in a gravitational field

(c) Heat flow in thin rod

(d) Oscillation of a simple pendulum

32. Consider the system of two wagons shown in the figure. The natural frequencies of this system are

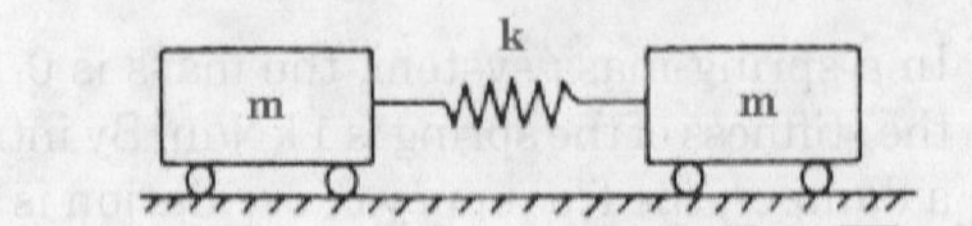

(a) 0, $\frac{\sqrt{2k}}{m}$ (b) $\frac{\sqrt{k}}{m}, \frac{\sqrt{2k}}{m}$

(c) $\frac{\sqrt{k}}{m}, \frac{\sqrt{k}}{2m}$ (d) 0, $\frac{\sqrt{k}}{2m}$

33. A proper base isolation is to be designed for mounting a sensitive instrument as shown in the figure. At the point of mounting, the base vibration due to other disturbances is indicated in the figure. If the permitted absolute displacement amplitude on the rigid mounting pad is 0.001 mm, find the stiffness of the spring. Assume that the total mass of the mounting pad and the instrument is 50 kg.

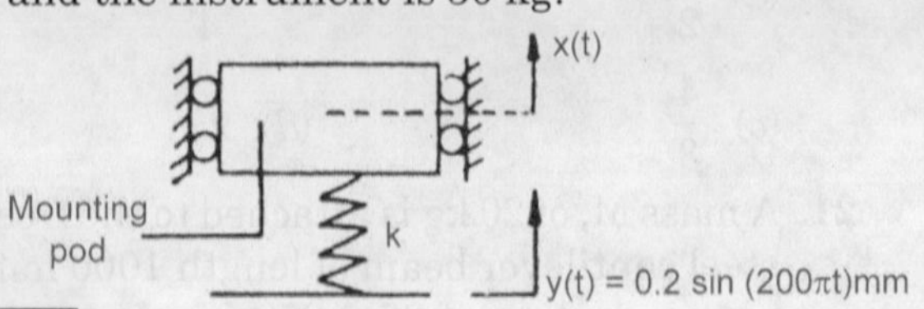

1998

34. A railway wagon (A) of mass 10,000 kg collides with another identical wagon (B) as shown in the figure. If A is moving at 5 m/s and B is at rest at the time of collision, calculate the maximum compression in the spring S with a spring constant of 2MN/m.

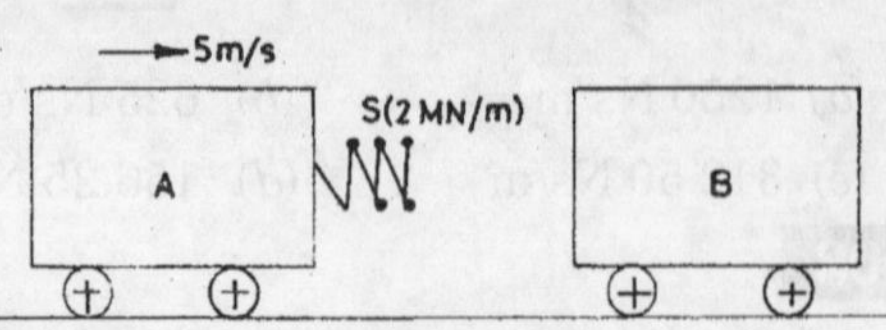

1997

35. The suspension system of a two-wheeler can be equated to a single spring-mass system with a viscous damper connected in series. Sketch the free body diagram and give the equations of motion. For a mass m = 50 kg and a spring with a stiffness of 35 kN/m, determine what should be the damping coefficient (damping constant) for critical damping. What can be the damping force for a plunger velocity of 0.05 m/s ?

1996

36. A ball A of mass m falls under gravity from a height h and strikes another ball B of mass m which is supported at rest on a spring of stiffness k. Assume perfeetly elastic impact. Immediately after the impact

(*a*) the velocity of ball A is $\frac{1}{2}\sqrt{2gh}$

(*b*) the velocity of ball A is zero

(*c*) the velocity of both balls is $\frac{1}{2}\sqrt{2gh}$

(*d*) none of the above

37. A mass of 1 kg is suspended by means of 3 springs as shown in Fig 2.8. The spring constants K_1, K_2 and K_3 are respectively 1 kN/m, 3kN/m and 2 kN/m. The natural frequency of the system is approximately

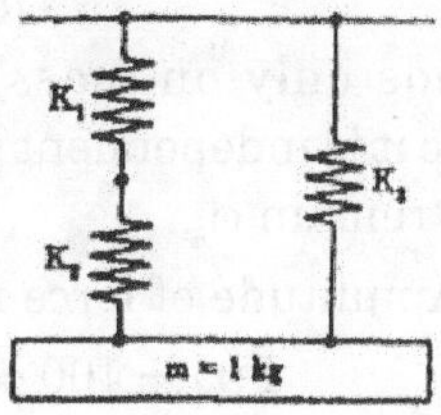

(*a*) 46.90 Hz (*b*) 52.44 Hz

(*c*) 60.55 Hz (*d*) 77.46 Hz

38. A cylinder of mass 1 kg and radius 1 m is connected by two identical springs at a height of 0.5m above the center as shown in Fig. The cylinder rolls without sliping. If the spring constant is 30 kN/m for each spring, find the natural frequency of the system for small oscillations.

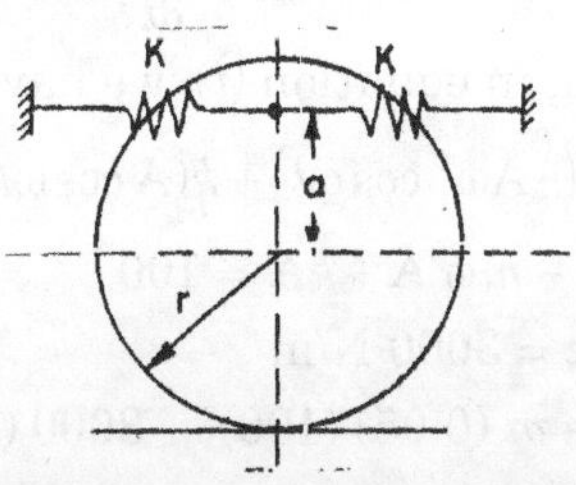

NUMERICAL TYPE QUESTIONS

2016

1. A single degree of freedom spring mass system with viscous damping has a spring constant of 10 kN/m. The system is excited by a sinusoidal force of amplitude 100 N. If the damping factor (ratio) is 0.25, the amplitude of steady state oscillation at resonance is ________mm.

2. The system shown in the figure consists of block A of mass 5 kg connected to a spring through a massless rope passing over pulley B of radius *r* and mass 20 kg. The spring constant *k* is 1500 N/m. If there is no slipping of the rope over the pulley, the natural frequency of the system is ___ rad/s.

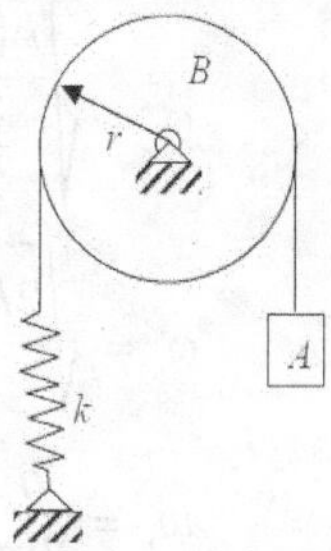

3. The static deflection of a spring under gravity, when a mass of 1 kg is suspended from it, is 1 mm. Assume the acceleration due to gravity g =10 m/s². The natural frequency of this spring-mass system (in rad/s) is ____________.

4. A single degree of freedom spring-mass system is subjected to a harmonic force of constant amplitude. For an excitation frequency of $\sqrt{\frac{3k}{m}}$, the ratio of the amplitude of steady state response to the static deflection of the spring is _______

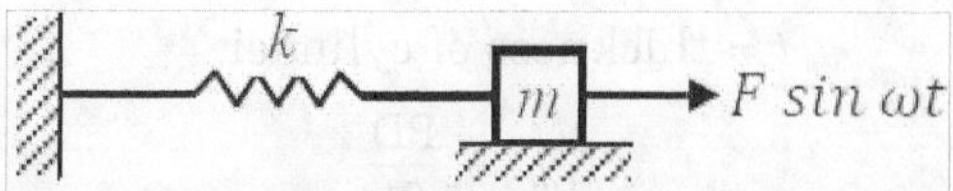

ANSWERS

MCQ Type Questions

1. (*a*)	**2.** (*b*)	**3.** (*c*)	**4.** (*c*)	**5.** (*d*)	**6.** (*d*)	**7.** (*a*)	**8.** (*d*)	**9.** (*b*)	**10.** (*a*)
11. (*a*)	**12.** (*b*)	**13.** (*d*)	**14.** (*c*)	**15.** (*a*)	**16.** (*a*)	**17.** (*c*)	**18.** (*a*)	**19.** (*c*)	**20.** (*c*)
21. (*a*)	**22.** (*b*)	**23.** (*a*)	**24.** (*c*)	**25.** (*c*)	**26.** (*b*)	**27.** (*d*)	**28.** (*)	**29.** (*a*)	**30.** (*c*)
31. (*a*)	**32.** (*c*)	**33.** (*)	**34.** (*)	**35.** (*)	**36.** (*b*)	**37.** (*b*)	**38.** (*)		

"*" Refers to Subjective Question.

Numerical Type questions

1. 20 **2.** (10) **3.** (100) **4.** (0.5)

EXPLANATIONS

MCQ TYPE QUESTIONS

1.

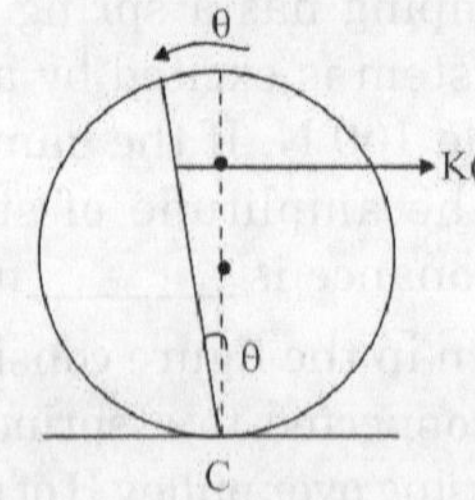

$$I_c = \frac{3}{2}ma^2$$

$$\omega_c = \sqrt{\frac{k(a+d)^2}{\frac{3}{2}ma^2}}$$

$$\omega_n = \sqrt{\frac{2k(a+d)^2}{3ma^2}}$$

2. $$w_n = \frac{q}{2m}$$

$$q = 2mw_n = 2m\sqrt{\frac{k}{m}}$$

$$q = 2\sqrt{km}$$

3. $d = \dfrac{p}{\tan\theta}$

4. $\underline{6h} = \dfrac{\text{PD}}{2t}$ where P = internal pressure

$6l = \dfrac{\text{PD}}{4t}$ D = inner dia of cylinder.

t = thickness of cylinder.

$$\therefore \quad \frac{6h}{6l} = \frac{\frac{\text{PD}}{2t}}{\frac{\text{PD}}{4t}} = 2$$

5. Soderberg criteria is:

$$\frac{\sigma_a}{\sigma_e} + \frac{\sigma_m}{s_{y+}} = \frac{1}{n} \text{ and } \sigma_a = \frac{100-20}{2\text{A}}$$

$$\Rightarrow \quad \frac{40}{160} + \frac{60}{240} = \frac{1}{2} \times \text{A} \times 10^3 \quad \sigma_m = \frac{100+20}{2\text{A}}$$

$$\text{A} = 1000 \text{ mm}^2$$

6.

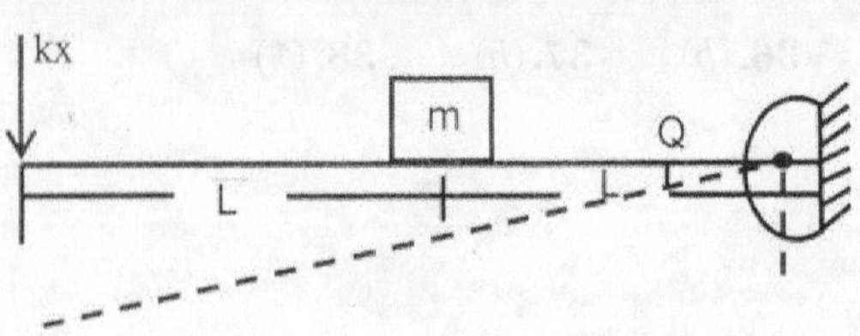

Equation of motion.

$$m\left(L\ddot{\theta}\right).L\ddot{\theta} + kx \times 2\text{ L} = 0$$

$$m\left(L\ddot{\theta}\right).L\bar{\theta} + \text{k}(2\text{L}.\theta) \times 2\text{L} = 0$$

$$mL^2\ddot{\theta} + 4\text{k}L^2\theta = 0$$

$$\ddot{\theta} + \sqrt{\frac{4k}{m}}\,\theta = 0$$

$$\omega_n = \sqrt{\frac{4k}{m}}$$

7. Natural frequency, $\omega_n = \sqrt{\dfrac{\text{K}}{m}}$

It depends only on mass but not on weight. Therefore it is independent of gravity of a system, so it will remain ω_n.

8. Given : Amplitude of force motion, A = 50 mm

$$\therefore \quad \text{F}(t) = 100\cos(100t) \quad ...(i)$$

Also, $$m\ddot{x} + kx = 100\cos(100t)$$

$$\omega = 100$$

General solution is,

$$x = \text{A}\cos\omega t$$

$$\Rightarrow \quad \dot{x} = \frac{\partial x}{\partial t} = -\text{A}\omega\sin\omega t$$

and $$\ddot{x} = \frac{\partial^2 x}{\partial t^2} = -\text{A}\omega^2\cos\omega t$$

Putting in equation (i), we have

$$m\left(-\text{A}\omega^2\cos\omega t\right) + k(\text{A}\cos\omega t) = 100\cos(\omega t)$$

$$\Rightarrow \quad -m\omega^2\text{A} + k\text{A} = 100$$

Here k = 3000 N/m

$$\therefore \quad -m(0.05)(100)^2 + 3000(0.05) = 100$$

$$\Rightarrow \quad m = \frac{50}{100} = 0.5$$

9. $$\text{N}_c = \frac{0.4985}{\sqrt{\delta}}\text{ rps} = \frac{0.4985 \times 60}{\sqrt{1.8 \times 10^{-3}}}\text{ rpm} = 705\text{ rpm}$$

Alternately

When speed of the shaft (ω) is equal to natural speed of vibration, it is called *critical speed* (ω_n)

i.e. $$\omega = \omega_n = \sqrt{\frac{g}{\text{A}}} = \sqrt{\frac{9.81}{1.8 \times 10^{-3}}}$$

$$= 73.82 = \frac{60 \times 73.82}{2\pi} = 705$$

10. At resonance, $\omega!! = \omega_n = \sqrt{\dfrac{k_{eq}}{m}}$

$$k_{eq} = 2k_1 + 2k_2 = 2(k_1 + k_2)$$
$$= 2(32 + 16) = 96 \times 10^6 \text{ N/m}$$
$$\omega = \sqrt{\frac{k}{m}} \text{ rad/s} = \frac{60}{2\pi}\sqrt{\frac{k}{m}} \; rpm$$
$$= \frac{30}{\pi}\sqrt{\frac{96 \times 10^6}{240}} = 6019.1 \; rpm$$

11. Damping factor $= \dfrac{C}{2\sqrt{sm}}$

$$= \frac{400}{2\sqrt{50 \times 3.6 \times 10^3}} = 0.741$$
$$\omega_d = \sqrt{1 - \Sigma^2 \omega_n} \qquad \ldots\left[\omega_n = \sqrt{\frac{s}{m}} = 8.48\sigma\right]$$
$$= \sqrt{1 - (471)^2} \times 8.481 = 7.48$$

Now $\omega_d = 2\text{pNd}$

$$\therefore \quad N_d = \frac{\omega_l}{2\pi} = \frac{7.48}{2 \times 3.14} = 1.19$$

12.
$$k_{eq} = k_1 + k_2$$
$$= 5600 \text{ N/m}$$
$$\therefore \quad f = \frac{1}{2\pi}\sqrt{\frac{k}{m}}$$
$$= 10 \text{ Hz}$$

13.
$$k_1 = \frac{G\, d_1^{\,4}}{8D_1^{\,3}\, n_1}$$

and
$$k_2 = \frac{G\, d_2^{\,4}}{8D_2^{\,3}\, n_2}$$

Here, $d_1 = d_2 = 2$ mm (diameter of spring wire)

$G = 80$ GPa

$D_1 = 20$,

$D_2 = 10$

$n_1 = n_2 = 10$

$$\therefore \quad \frac{k_1}{k_2} = \left(\frac{D_2}{D_1}\right)^3 = \frac{1}{8}$$
$$\Rightarrow \quad k_2 = 8\, k_1$$

14. For underdamped harmonic oscillator,

$$\omega_d = \omega_n$$

15. Equivalent k, $k_{\text{equ}} = \dfrac{1}{k} + \dfrac{1}{\dfrac{k}{2} + \dfrac{k}{2}} = \dfrac{k}{2}$

$$\therefore \quad \omega_n = \sqrt{\frac{k}{2m}}$$

17.

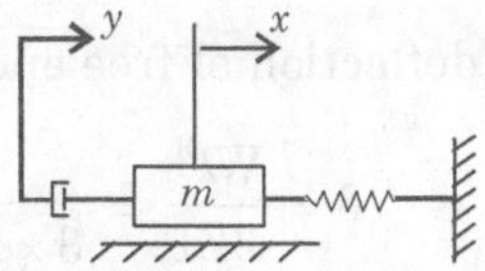

$$\therefore \quad m\frac{d^2x}{dt^2} + c\left(\frac{dx}{dt} - \frac{dy}{dt}\right) + kx = 0$$

This is the differential equation governing the above vibrating system.

19.
$$f_d = 0.9 f_n$$
$$\Rightarrow \quad \omega_d = 0.9\, \omega_n$$
$$\Rightarrow \quad \sqrt{\omega_n^2\; a^2} = 0.9\, \omega_n$$

Squaring $\omega_n^2 - a^2 = 0.81\, \omega_n^2$

$$\Rightarrow \quad 0.19\omega_n^2 = a^2$$
$$\Rightarrow \quad 0.19 \times \frac{s}{m} = \frac{c^2}{(2m)^2}$$
$$\Rightarrow \quad c^2 = 0.76\; sm$$
$$\Rightarrow \quad c = \sqrt{0.76 \times 1 \times 10^3 \times 0.1}$$
$$= 8.717 \text{ Ns/m}$$

20. Transmissibility ratio,

$$\xi = \frac{\sqrt{1 + \left(\dfrac{2c\omega}{\omega_n c_c}\right)^2}}{\sqrt{\left(\dfrac{2c\omega}{\omega_n c_c}\right)^2 + \left(1 - \dfrac{\omega^2}{\omega_n^2}\right)^2}}$$

If $c = 0$, then

$$\xi = \frac{1}{\left(1 - \dfrac{\omega^2}{\omega_n^2}\right)}$$
$$= \frac{1}{1 - (0.5)^2} = \frac{4}{3}.$$

21. Given : length of cantilever beam,

$$l = 1000 \text{ mm} = 1 \text{ m}$$

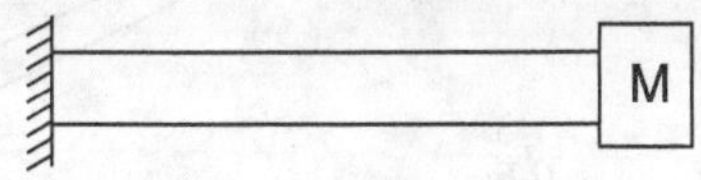

Moment of inertia of the shaft,

$$I = \frac{1}{12} bd^3$$
$$= \frac{25 \times (25)^3}{12} = 3.25 \times 10^{-8} \text{ m}^4$$
$$E_{\text{steel}} = 200 \times 10^9 \text{ Pa.}$$

Mass, $M = 20$ kg

$\therefore$ Weight, $W = 20 \times 9.81 = 196.2$

Now, deflection of free end,

$$\delta = \frac{Wl^3}{3EI} = \frac{196.2 \times (1)^3}{3 \times 200 \times 10^9 \times 3.25 \times 10^{-8}}$$
$$= 0.01 \text{ m}$$

Now, $\omega_n = \sqrt{\frac{g}{\delta}} = \sqrt{\frac{9.81}{0.01}} = 31.22 \text{ rad/s}$

As system is critically damped,

$$\left(\frac{C_c}{2m}\right)^2 = \omega_n^2$$

Critical damping constant,

$$C_c = 2m\omega_n = 2 \times 20 \times 31.22$$
$$= 1248.99 \text{ N-s/m}^2.$$

22.

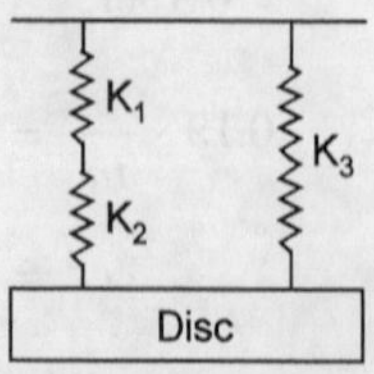

$$K_{eq} = \frac{1}{\frac{1}{K_1} + \frac{1}{K_2}} + \frac{1}{K_3} = \frac{K_1K_2}{K_1K_2} + K_3$$
$$= \frac{K_1K_2 + K_1K_3 + K_2K_3}{K_1 + K_2}$$

$\therefore$ Natural frequency $= \sqrt{\frac{K_1K_2 + K_1K_3 + K_2K_3}{J(K_1 + K_2)}}$

23. $x_1 = l_1 Q$

$x_2 = Q.l_2$

For small deflection, after equilibrium taking momentum about O

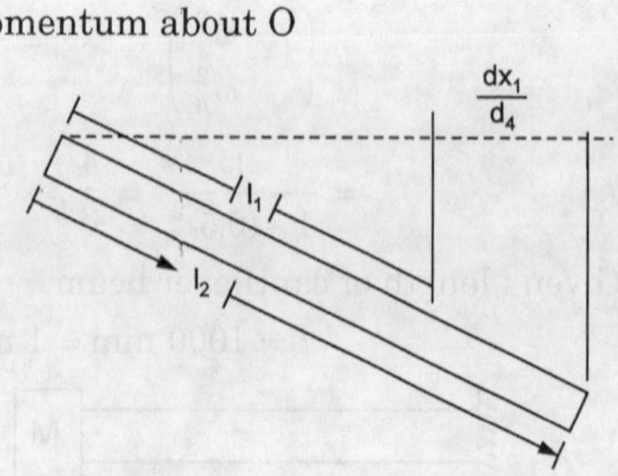

$$C\frac{dx_1}{dt}.l_1 + kx_2l_2 + I\alpha + k_0.Q = 0 \quad ...(i)$$

$$\Rightarrow \quad \omega_n = \sqrt{\frac{kl_2^2 + kw}{\frac{ml^2}{3}}}$$
$$= \sqrt{\frac{2 \times 10^3 \times (0.5)^2 \times 1 \times 10^3}{\frac{10 \times (0.05)^2}{3}}}$$
$$= \sqrt{\frac{1.5 \times 10^3 \times 3}{10 \times (.5)^2}} = 42.43 \text{ rad/sec.}$$

24. For static equilibrium

$$k_t . \theta_s + k_t . \theta_s L^2 = mgl^2$$

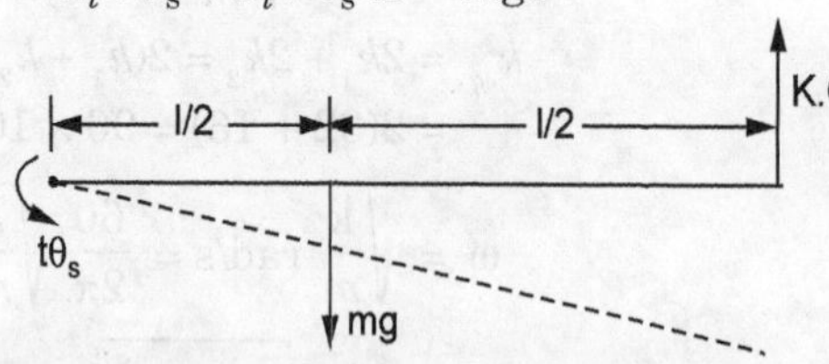

For additional deflection θ, equation will be

$$I.\ddot{\theta} = \frac{mgl}{2} - k_t(\theta_s + \theta) - C.l_1^2\theta - kl^2(\theta_s + \theta)$$

$$\Rightarrow \quad I.\ddot{\theta} + C.l_1^2\theta + (k_t + kl^2)\theta = 0$$

where, $I = ml^2/3$

$\therefore$ Un-damped frequency,

$$\omega = \sqrt{\frac{(k_t + kl^2) \times 3}{ml^2}}$$
$$= \sqrt{\frac{(1000 + 2000 \times 0.5)^2 \times 3}{10 \times 0.5^2}}$$
$$= 42.426 = 42.43 \text{ rad/sec.}$$

Equivalent damping coefficient

$$= Cl_1^2 \text{ Nm/rad} = 500 \times (0.04)^2$$
$$= 80 \text{ Nms/rad}$$

26. Potential energy at A $= mg(l - \delta)$

Total energy at B $= mg[l - (\delta + x)] + \frac{1}{2}kx^2$

$\therefore$ Change in energy $= mgl - mg(\delta + x) + \frac{1}{2}kx^2 - mgl + mg\delta.$

$$= \frac{1}{2}kx^2 - mgx.\delta$$

27. Net restoring torque when displaced by a small angle θ,

$$\tau = mg\cos(\alpha - \theta)l - mg(\alpha + \theta)l$$
$$= 2\,mgl\cos\alpha\sin\theta$$

For very small θ, $\sin\theta \approx \theta$

$$\therefore \quad \tau = 2mgl\cos\alpha\,\theta \text{ (restorative)}$$

Now, $I\frac{d^2\theta}{dt^2} + 2\,mgl\cos\alpha\,\theta = 0$

But $I = 2ml^2$

$$\therefore \quad 2ml^2\frac{d^2\theta}{dt^2} + 2mgl\cos\alpha\,\theta = 0$$

$$\Rightarrow \quad \frac{d^2\theta}{dt^2} + \frac{g\cos\alpha}{l}\theta = 0$$

$$\therefore \quad \omega_n = \sqrt{\frac{g\cos\alpha}{l}}$$

28. Equating forces acting on the body, we get

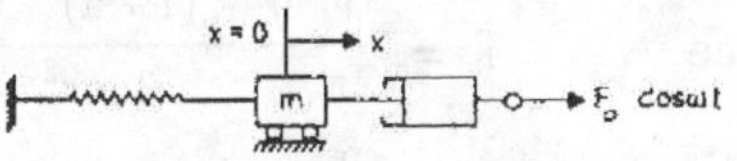

$$m\frac{d^2x}{dt^2} = F_0 \cos \omega t - c\frac{dx}{dt} - Kx.$$

$$\Rightarrow \quad m\frac{d^2x}{dt^2} + c\frac{dx}{dt} + Kx = F_0 \cos \omega t$$

29. $\xi < 1$, hence it is underdamped vibration case.

∴ Frequency of the system,

$$\omega_d = \sqrt{1-\xi^2} \times \omega_n$$

$$= \sqrt{1-0.64} \times 100 = 60$$

30. $S = S_i + S_2$

$= 20 + 20 = 40$ kN/m $= 40{,}000$ N/m

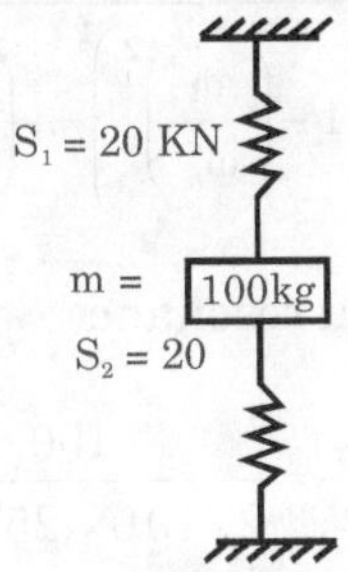

∴ Natural frequency of vibration of the system,

$$f_n = \frac{1}{2\lambda}\sqrt{\frac{S}{m}}$$

$$= \frac{1}{2\lambda}\sqrt{\frac{40 \times 1000}{100}} = \frac{20}{2\lambda} = \frac{10}{\lambda}$$

33. Given, Absolute displacement of mounting pad

$= 0.001$ mm

$m = 50$ kg

Displacement of base = 0.2 mm

Displacement of manufacturing pad = 0.001 mm

Net displacement $= 2 - 0.001 = 0.199$ mm

Force transmitted $= kx = 0.199$ K.

Natural frequency,

$$\omega_n = 200\,\pi = \frac{d^2x}{dt^2}$$

Now, $\quad m\frac{d^2x}{dt^2} = S(y-x)$

or, $\quad m\,\omega_n^2 x = S \times 0.199$

or, $\quad \frac{50 \times (200\pi)^2 \times 0.001}{1000} = S \times 0.199$

or $\quad$ S (or k) $= 99.12$ N/mm

34. Given, Mass of wagon, A = 10,000 kg

Initial velocity of moving wagon, A = 5 m/s

Stiffness of spring $= 2 \times 10^6$ N/m

Let 'v' be the velocity of both wagons when the spring is compressed to maximum.

According to momentum conservation

$$m_1u_1 + 0 = m_1v + m_2v$$

Since $\quad m = m_2,$

$$v = \frac{u_1}{2} = \frac{5}{2} = 2.5 \text{ m/s}.$$

$$\text{K.E. before impact} = \frac{1}{2}m_1u_1^2 + 0 = \frac{1}{2} \times 10000 \times 5^2$$

(When both are moving with velocity) v = 2.5 m/s

K.E. after impact

$$= \frac{1}{2} \times 10000 \times 2.5^2 + \frac{1}{2} \times 10000 \times 2.5^2$$

$$= 10000 \times 2.5^2$$

If δ is the deflection of spring and 'K' be the stiffness,

$$\text{energy stored in the spring} = \frac{1}{2}K\delta^2$$

Change in K.E. = Energy stored by spring

$$\text{or} \quad 10000 \times \frac{(5)^2}{2} - 10000 \times (2.5)^2 = \frac{1}{2} \times 2 \times 10^6 \times \delta^2$$

Solving, we get

$\delta = 0.25$

m = **25 cm**

35.

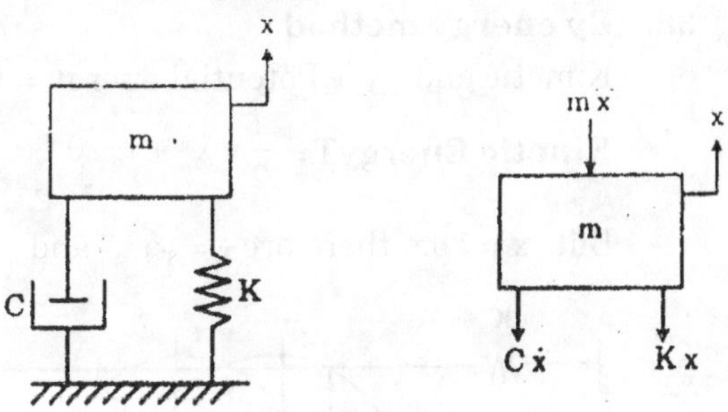

Free body diagram

Considering the system in **Figure** for free vibrations of the system, we get

$$m\ddot{x} + C\dot{x} + kx = 0 \quad \text{...(A)}$$

$$x = e^{st}$$

satisfies the above equation, giving

$$ms^2 + Cs + k = 0$$

The two roots are,

$$s_{1,2} = \frac{C}{2m} \pm \sqrt{\left(\frac{C}{2m}\right)^2 - \frac{k}{m}}$$

$$x = A\,.\,e^{s_1t} + B\,.\,e^{s_2t}$$

where A and B are constants of integration that depend on the initial conditions with which the motion of mass, m is started. This depends on the two roots s_1 and s_2 which may be real or complex depending on the quantity under the square root is + ve or –ve. A special case exists when the term under the square root is zero, then the system is said to possess critical damping.

$$\left(\frac{C_c}{2m}\right)^2 = \frac{k}{m} = \omega_n^2$$

where ω_n = natural circular frequency,

C_c = critical damping coefficient.

$C_c = 2m\,\omega_n = 2\sqrt{k.m}$

It is to be noted that critical damping is a system property that depends on the mass and stiffness and it has nothing to do with the acutal damping, C present in the system.

$$\omega_n = \sqrt{\frac{k}{m}} = \sqrt{\frac{35000}{50}} = 26.4 \text{ rad/s}$$

$\therefore$ $C_c = 2m\,\omega_n = 2 \times 50\ 26.4$

$= \mathbf{2640\ N\text{-}s/m}$

36. In a perfectly elastic collision between equal masses of two bodies, velocities exchange on impact.

Velocity just before impect = Velocity immediately after impact.

Before impact	**After impact**
$V_A = \sqrt{2gh}$	$V_A = 0$
$V_B = 0$	$VB = \sqrt{2gh}$

38. By energy method

Kimetic energy + Potential energy = 0

Kinetic EnergyT= $\frac{1}{2}m\dot{x}^2 + I_o\,.\,\theta$

But $x = r\,\theta$, therefore $\dot{x} = r\dot{\theta}$ and $I_o = \frac{1}{2}mr^2$

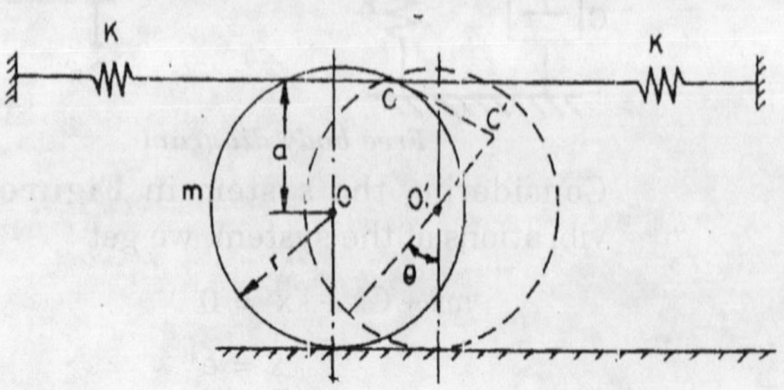

Thus $T = \frac{1}{2}mr^2\dot{\theta}^2 + mr^2\dot{\theta}^2 = mr^2\dot{\theta}^2$

Potential Energy

End of the spring is moved from C to C', which for small angles

$x + a \text{ sim } \theta \approx x + a\theta = r\theta + a\theta = (r + a)\,\theta$

Force exerted by a spring = k(r + n) θ,

Work done on a spring = Average force × Deflection

$\therefore$ $U = (2) \times \frac{1}{2}k(r+a)^2\,\theta^2.$

and $\frac{d}{dt}(T+U) = \frac{3}{2}mr^2\dot{\theta}\,\ddot{\theta} + 2k\,(r+a)^2\,\theta.\dot{\theta} = 0$

or $\ddot{\theta} + \frac{4k\,(r+a)^2}{3\text{ mr}}\theta = 0$

Hence $f_n = \frac{1}{2\pi}\frac{\sqrt{4k\,(1+a)^2\ 3\,mr^2}}{3\,mr^2}$ cps.

or Given m = 1 kg, r = 1 m

a = 0.5 m, k = 30000 N/m

$\therefore$ Natural frequency of the system,

$$f_n = \frac{1}{2\pi}\frac{\sqrt{4 \times 30{,}000 \times (1+0.5)^2}}{3 \times 1 \times (1)^2}$$

$= \mathbf{47.746\ cps}$

Numerical Type questions

1. k = 10 kN/m

F_0 = 100 N

$\xi = 0.25$

$$X = \frac{(F_0/k)}{\sqrt{\left(1-\left(\frac{\omega}{\omega_n}\right)^2\right)^2 + \left(2\zeta\frac{\omega}{\omega_n}\right)^2}}$$

$\frac{\omega}{\omega_n} = 1$ at resonance

$$X = \frac{F_0}{2k\zeta} = \frac{100}{2 \times 100 \times .25 \times 10^3} = 20\text{ mm}$$

2. $KE = \frac{1}{2}m\dot{x}^2 + \frac{1}{2}I\dot{\theta}^2$

$m = 5$ kg, $\theta = \frac{x}{r}$

$I = \frac{20 \times r^2}{2} = 10r^2$

$KE = \frac{1}{2}5\dot{x}^2 + \frac{1}{2}10r^2.\frac{\dot{x}^2}{r^2} = \frac{1}{2}(15)\dot{x}^2$

$\therefore$ $m_{eq} = 15$

$PE = \frac{1}{2}kx^2$

$k_{eq} = k = 1500$ N/m

Natural frequency

$$\omega_n = \sqrt{\frac{k_{eq}}{m_{eq}}} = \sqrt{\frac{1500}{15}} = 10 \text{ rad/sec}$$

3. $\omega_n = \sqrt{\frac{g}{\delta}}$

$\delta = 10^{-3}$m

$$\omega_n = \sqrt{\frac{10}{10-3}} = 100 \text{ rad/sec}$$

4. $MF = \pm\frac{1}{1-\left(\frac{\omega}{\omega_n}\right)^2} = \pm\frac{1}{1-3} = 0.5$

5 CHAPTER

Design of Machine Elements

MCQ TYPE QUESTIONS

2016

1. The spring constant of a helical compression spring DOES NOT depend on

(*a*) coil diameter

(*b*) material strength

(*c*) number of active turns

(*d*) wire diameter

2. A machine element XY, fixed at end X, is subjected to an axial load P, transverse load F, and a twisting moment T at its free end Y. The most critical point from the strength point of view is

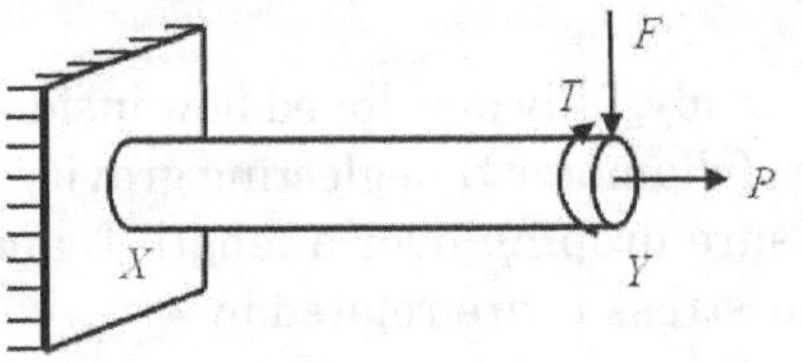

(*a*) a point on the circumference at location Y

(*b*) a point at the center at location Y

(*c*) a point on the circumference at location X

(*d*) a point at the center at location X

3. In a structural member under fatigue loading, the minimum and maximum stresses developed at the critical point are 50 MPa and 150 MPa, respectively. The endurance, yield, and the ultimate strengths of the material are 200 MPa, 300 MPa and 400 MPa, respectively. The factor of safety using modified Goodman criterion is

(*a*) $\frac{3}{2}$ (*b*) $\frac{8}{5}$

(*c*) $\frac{12}{7}$ (*d*) 2

4. Which of the bearings given below SHOULD NOT be subjected to a thrust load?

(*a*) Deep groove ball bearing

(*b*) Angular contact ball bearing

(*c*) Cylindrical (straight) roller bearing

(*d*) Single row tapered roller bearing

2015

5. A pinion with radius r_1, and inertia I_1 is driving a gear with radius r_2 and inertia I_2. Torque τ_1 is applied on pinion. The following are free body diagrams of pinion and gear showing important forces (F_1 and F_2) of interaction. Which of the following relations hold true?

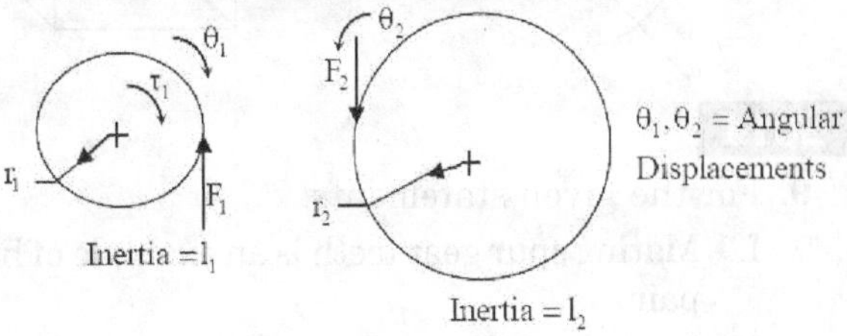

(*a*) $F_1 \neq F_2; \tau_1 = I_1\ddot{\theta}_1 : F_2 = I_2 \frac{r_1}{r_2^2}\ddot{\theta}_1$

(*b*) $F_1 = F_2; \tau_1 = \left[I_1 + I_2\left(\frac{r_1}{r_2}\right)^2\right]\ddot{\theta}_1 : F_2 = I_2 \frac{r_1}{r_2^2}\ddot{\theta}_1$

(*c*) $F_1 = F_2; \tau_1 = I_1\ddot{\theta}_1 : F_2 = I_2 \frac{1}{r_2}\ddot{\theta}_2$

(*d*) $F_1 \neq F_2; \tau_1 = \left[I_1 + I_2\left(\frac{r_1}{r_2}\right)^2\right]\ddot{\theta}_1 ; F_2 = I_2 \frac{1}{r_2}\ddot{\theta}_2$

6. The number of degrees of freedom of the planetary gear train shown in the figure is

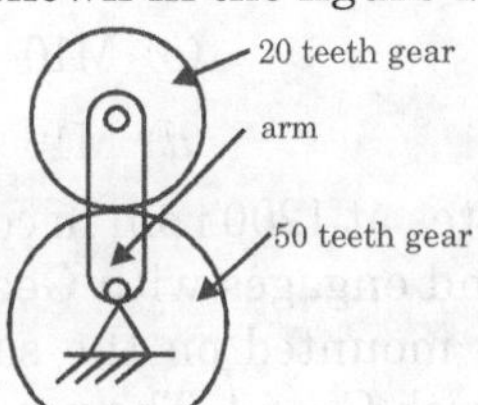

(*a*) 0 (*b*) 1

(*c*) 2 (*d*) 3

7. A hollow shaft ($d_o = 2d_i$, where d_o & d_i are the outer and inner diameter respectively) needs to transmit 20kW power at 3000 RPM. If the maximum permissible shear stress is 30 MPa, d_o is

(*a*) 11.29 mm (*b*) 22.58 mm

(*c*) 33.87 mm (*d*) 45.16 mm

8. For the same material and the mass, which of the following configurations of flywheel will have maximum mass moment of inertia about the axis of rotation OO' passing through the center of gravity

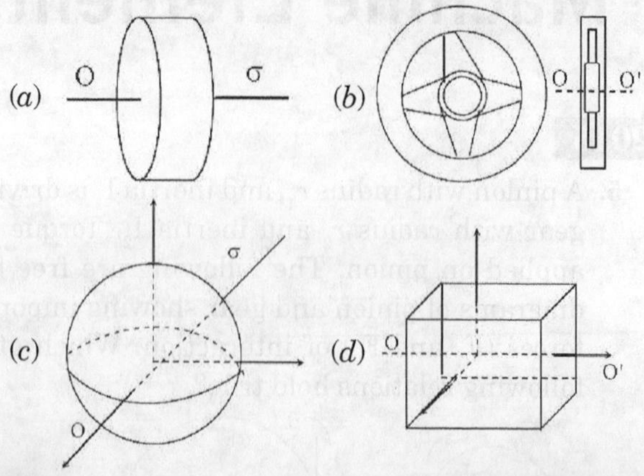

(a) 300 rpm counterclockwise

(b) 300 rpm clockwise

(c) 4800 rpm counterclockwise

(d) 4800 rpm clockwise

12. Which one of the following is used to convert a rotational motion into a translational motion?

(a) Bevel gears

(b) Double helical gears

(c) Worm gears

(d) Rack and pinion gears

13. It is desired to avoid interference in a pair of spur gears having a 20° pressure angle. With increase in pinion to gear speed ratio, the minimum number of teeth on the pinion

(a) increases

(b) decreases

(c) first increases and then decreases

(d) remains unchanged

2014

9. For the given statements:

I. Mating spur gear teeth is an example of higher pair

II. A revolute joint is an example of lower pair

Indicate the correct answer.

(a) Both I and II are false

(b) I is true and II is false

(c) I is false and II is true

(d) Both I and II are true

10. For the three bolt system shown in the figure, the bolt material has shear yield strength of 200 MPa. For a factor of safety of 2, the minimum metric specification required for the bolt is

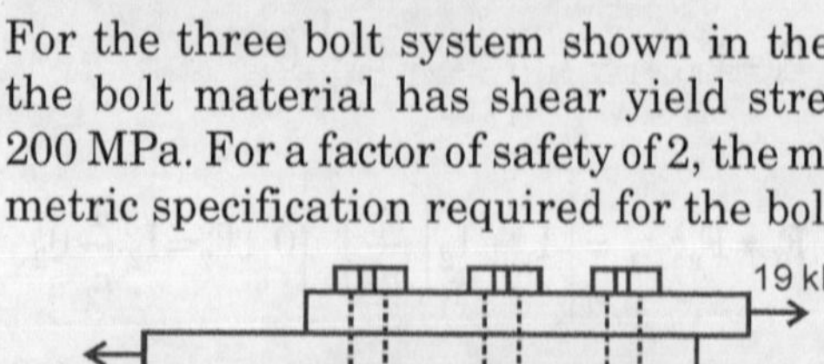

(a) M8 (b) M10

(c) M12 (d) M16

11. Gear 2 rotates at 1200 rpm in counter clockwise direction and engages with Gear 3. Gear 3 and Gear 4 are mounted on the same shaft. Gear 5 engages with Gear 4. The numbers of teeth on Gears 2, 3, 4 and 5 are 20, 40, 15 and 30, respectively. The angular speed of Gear 5 is

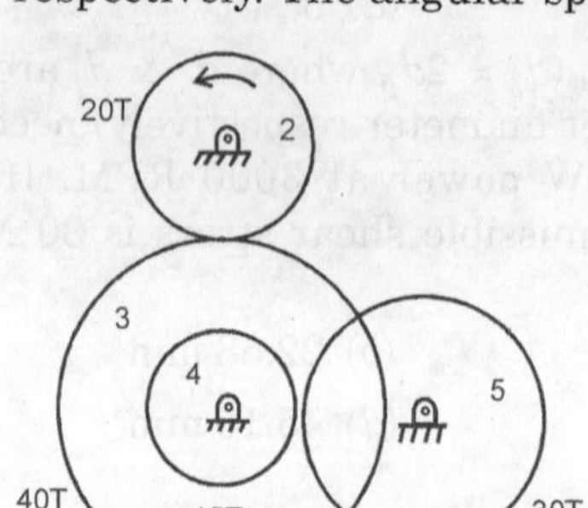

2013

14. For steady, fully developed flow inside a straight pipe of diameter D neglecting gravity effects the pressure drop Δp over a length L and the wall shear stress τ_w are related by

(a) $\tau_w = \dfrac{\Delta p D}{4L}$ (b) $\tau_w = \dfrac{\Delta p D^2}{4L^2}$

(c) $\tau_w = \dfrac{\Delta p D}{2L}$ (d) $\tau_w = \dfrac{4\Delta p L}{D}$

15. In order to have maximum power from a Pelton turbine, the bucket speed must be

(a) equal to the jet speed

(b) equal to half of the jet speed

(c) equal to twice the jet speed

(d) independent of the jet speed

16. Water is coming out from a tap and falls vertically downwards. At the tap opening, the stream diameter is 20 mm with uniform velocity of 2 m/s. Acceleration due to gravity is 9.81 m/s². Assuming steady, inviscid flow, constant atmospheric pressure everywhere and neglecting curvature and surface tension effects, the diameter in mm of the stream 0.5 m. below the tap is approximately.

(a) 10 (b) 15

(c) 20 (d) 25

17. A hinged gate of length 5 m, inclined at 30° with the horizontal and with water mass on its left, is shown in the figure below. Density of water is 1000 kg/m³. The minimum mass of the gate in kg per unit width (perpendicular to the plane of paper) required to keep it closed is

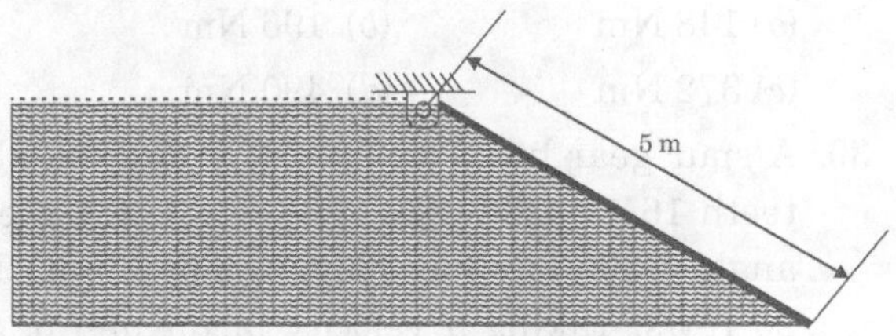

(*a*) 5000 (*b*) 6600
(*c*) 7546 (*d*) 9623

2011

18. A streamline and an equipotential line in a flow field
(*a*) are parallel to each other
(*b*) are perpendicular to each other
(*c*) intersect at an acute angle
(*d*) are identical

19. Figure shows the schematic for the measurement of velocity of air (density = 1.2 kg/m³) through a constant-area duct using a pitot tube and a water-tube manometer. The differential head of water (density = 1000 kg/m³) in the two columns of the manometer is 10 mm. Take acceleration due to gravity as 9.8 m/s². The velocity of air in m/s is

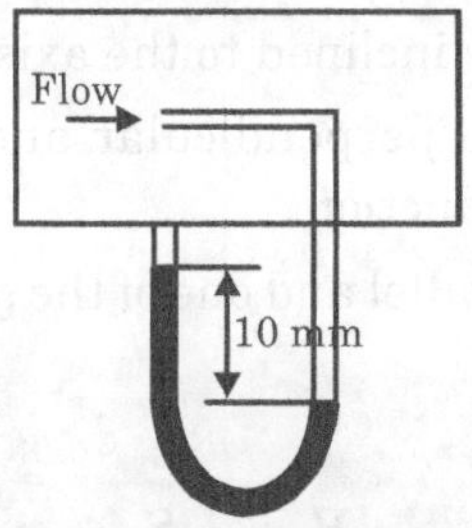

(*a*) 6.4 (*b*) 9.0
(*c*) 12.8 (*d*) 25.6

20. The ratios of the laminar hydrodynamic boundary layer thickness to thermal boundary layer thickness of flows of two fluids P and Q on a flat plate are $\frac{1}{2}$ and 2 respectively. The Reynolds number based on the plate length for both the flows is 10^4. The Prandtl and Nusselt numbers for P are $\frac{1}{8}$ and 35 respectively. The Prandtl and Nusselt number, for Q are respectively
(*a*) 8 and 140 (*b*) 8 and 70
(*c*) 4 and 40 (*d*) 4 and 35

2010

21. Tooth interference in an external involute spurgear pair can be reduced by
(*a*) decreasing center distance between gear pair
(*b*) decreasing module
(*c*) decreasing pressure angle
(*d*) increasing number of gear teeth

22. For the epicyclic gear arrangement shown in the figure, ω_2 = 100 rad/s clockwise (CW) and ω_{arm} = 80 rad/s counter clockwise (CCW). The angular velocity ω_5 (in rad/s) is

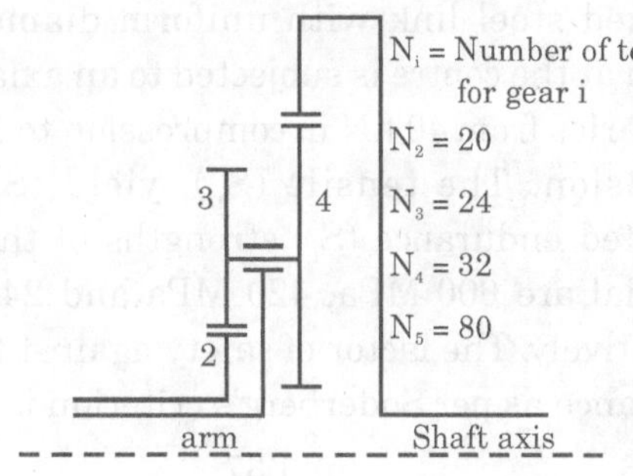

(*a*) 0
(*b*) 70 CW
(*c*) 140 CCW
(*d*) 140 CW

23. A lightly loaded full journal bearing has journal diameter of 50 mm, bush bore of 50.05 mm and bush length of 20 mm. If rotational speed of journal is 1200 rpm and average viscosity of liquid lubricant is 0.03 Pa s, the power loss (in W) will be
(*a*) 37 (*b*) 74
(*c*) 118 (*d*) 237

24. For the configuration shown, the angular velocity of link AB is 10 rad/s counterclockwise. The magnitude of the relative sliding velocity (in ms^{-1}) of slider B with respect to rigid link CD is

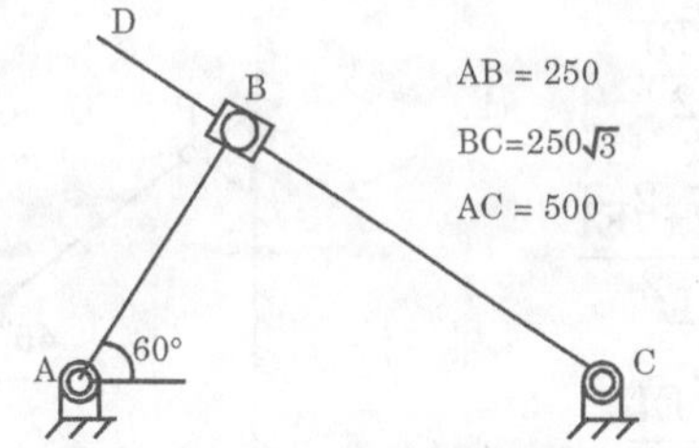

(*a*) 0 (*b*) 0.86
(*c*) 1.25 (*d*) 2.50

2009

25. A solid circular shaft of diameter d is subjected to a combined bending moment M and torque, T. The material property to be used for designing the shaft using the relation $\frac{16}{\pi d^3}\sqrt{M^2 + T^2}$ is

(*a*) ultimate tensile strength (S_u)

(*b*) tensile yield strength (S_y)

(*c*) torisonal yield strength (S_{sy})

(*d*) endurance strength (S_e)

26. A forged steel link with uniform diameter of 30 mm at the centre is subjected to an axial force that varies from 40 kN in compression to 160 kN in tension. The tensile (S_u), yield (S_y) and corrected endurance (S_e) strengths of the steel material are 600 MPa, 420 MPa and 240 MPa respectively. The factor of safety against fatigue endurance as per Soderberg's criterion is

(*a*) 1.26 (*b*) 1.37

(*c*) 1.45 (*d*) 2.00

2008

27. An axial residual compressive stress due to a manufacturing process is present on the outer surface of a rotating shaft subjected to bending. Under a given bending load, the fatigue life of the shaft in the presence of the residual compressive stress is

(*a*) decreased

(*b*) increased or decreased, depending on the external bending load

(*c*) neither decreased nor increased

(*d*) increased

28. The rod PQ of length L and with flexural rigidity EI is hinged at both ends. For what minimum force F, is it expected to buckle?

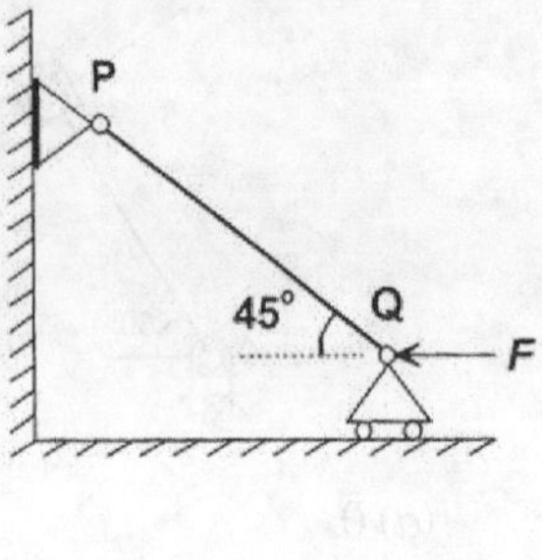

(*a*) $\frac{\pi^2 EI}{L^2}$

(*b*) $\frac{\sqrt{2}\pi^2 EI}{L^2}$

(*c*) $\frac{\pi^2 EI}{\sqrt{2}L^2}$

(*d*) $\frac{\pi^2 EI}{2L^2}$

29. A clutch has outer and inner diameters 100 mm and 40 mm respectively. Assuming a uniform pressure of 2 MPa and coefficient of friction of liner material 0.4, the torque carrying capacity of the clutch is

(*a*) 148 Nm (*b*) 196 Nm

(*c*) 372 Nm (*d*) 490 Nm

30. A spur gear has a module of 3 mm, number of teeth 16, a face width of 36 mm and a pressure angle of 20°. It is transmitting a power of 3 kW at 20 rev/s. Taking a velocity factor of 1.5, and a form factor of 0.3, the stress in the gear tooth is about

(*a*) 32 MPa (*b*) 46 MPa

(*c*) 58 MPa (*d*) 70 MPa

31. Match the type of gears with their most appropriate description.

List-I (*Type of gear*)

P. Helical

Q. Spiral Bevel

R. Hypoid

S. Rack and pinion

List-II (*Description*)

1. Axes non parallel and non intersectin

2. Axes parallel and teeth are inclined to the axis

3. Axes parallel and teeth are parallel to the axis

4. Axes are perpendicular and intersecting, and teeth are inclined to the axis

5. Axes are perpendicular and used for large speed reduction

6. Axes parallel and one of the gears has infinite radius

Codes:

	P	Q	R	S
(*a*)	2	4	1	6
(*b*)	1	4	5	6
(*c*)	2	6	4	2
(*d*)	6	3	1	5

2007

32. A ball bearing operating at a load F has 8000 hours of life. The life of the bearing, in hours, when the load is doubled to 2F is

(*a*) 8000

(*b*) 6000

(*c*) 4000

(*d*) 1000

33. A natural feed journal bearing of diameter 50 mm and length 50 mm operating at 20 revolution/second carries a load of 2.0 kN. The lubricant used has a viscosity of 20 mPa s. The radial clearance is 50 μm. The Sommerfeld number for the bearing is

(*a*) 0.062 (*b*) 0.125
(*c*) 0.250 (*d*) 0.785

34. A block-brake shown below has a face width of 300 mm and a mean co-efficient of friction of 0.25. For an activating force of 400 N, the braking torque in Nm is

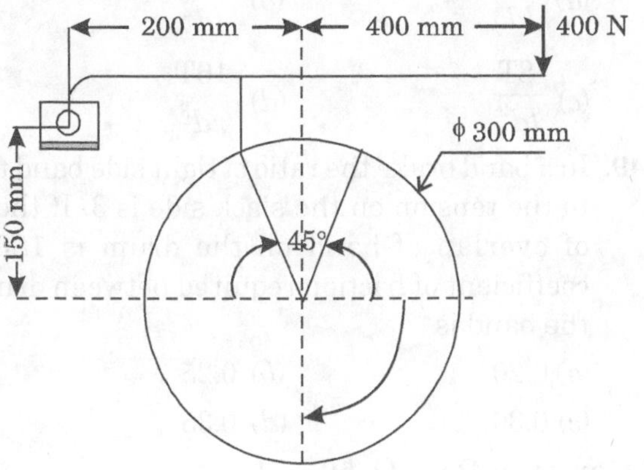

(*a*) 30 (*b*) 40
(*c*) 45 (*d*) 60

35. The piston rod of diameter 20 mm and length 700 mm in a hydraulic cylinder is subjected to a compressive force of 10 kN due to the internal pressure. The end conditions for the rod may be assumed as guided at the piston end and hinged at the other end. The Young's modulus is 200 GPa. The factor of safety for the piston rod is

(*a*) 0.68 (*b*) 2.75
(*c*) 5.62 (*d*) 11.0

2006

36. Twenty degree full depth involute profiled 19-tooth pinion and 37-tooth gear are in mesh. If the module is 5 mm, then center distance between the gear pair will be

(*a*) 140 mm (*b*) 150 mm
(*c*) 280 mm (*d*) 300 mm

37. A 60 mm long and 6 mm thick fillet weld carries a steady load of 15 kN along the weld. The shear strength of the weld material is equal to 200 MPa. The factor of safety is

(*a*) 2.4 (*b*) 3.4
(*c*) 4.8 (*d*) 6.8

2005

38. Which one of the following is a criterion in the design of hydrodynamic journal bearings?

(*a*) Sommerfeld number
(*b*) rating life
(*c*) Specific dynamic capacity
(*d*) Rotation factor

2004

39. Two mating spur gears have 40 and 120 teeth respectively. The pinion rotates at 1200 rpm and transmits a torque of 20 N.m. The torque transmitted by the gear is

(*a*) 6.6 Nm (*b*) 20 Nm
(*c*) 40 Nm (*d*) 60 Nm

40. In terms of theoretical stress concentration factor (K_t) and fatigue stress concentration factor (K_f), then notch sensitivity '*q*' is expressed as

(*a*) $\frac{(K_f - 1)}{(K_t - 1)}$ (*b*) $\frac{(K_f - 1)}{(K_t + 1)}$
(*c*) $\frac{(K_t - 1)}{(K_f - 1)}$ (*d*) $\frac{(K_f + 1)}{(K_t + 1)}$

41. In a bolted joint two members are connected with an axial tightening force of 2200 N. If the bolt used has metric threads of 4 mm pitch, then torque required for achieving the tightening force is

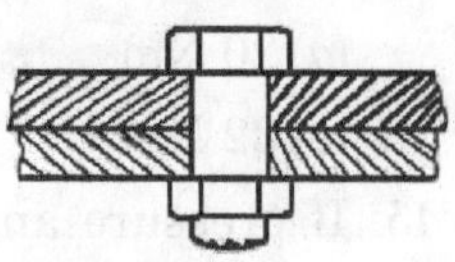

(*a*) 0.7 Nm
(*b*) 1.0 Nm
(*c*) 1.4 Nm
(*d*) 2.8 Nm

42. The figure given below shows a steel rod of 25 mm² cross sectional area. It is loaded at four points, K, L, M and N. Assume E_{steel} = 200 GPa. The total change in length of the rod due to loading is

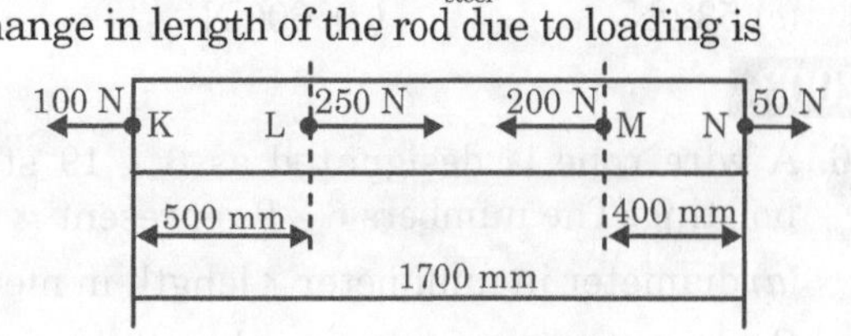

(*a*) 1 μm (*b*) – 10 μm
(*c*) 16 μm (*d*) – 20 μm

43. Match the following

List-I *Type of gears*	List-II *Arrangement of shafts*
P. Bevel gears	1. Non-parallel off-set shafts
Q. Worm gears	2. Non-parallel intersecting shafts
R. Herringbone gears	3. Non-parallel, non-intersecting shafts
S. Hypoid gears	4. Parallel shafts

Codes:

	P	Q	R	S
(a)	4	2	1	3
(b)	2	3	4	1
(c)	3	2	1	4
(d)	1	3	4	2

Common Data Q. 44-45

A compacting machine shown in the figure below is used to create a desired thrust force by using a rack and pinion arrangement. The input gear is mounted on the motor shaft. The gears have involute teeth of 2 mm module.

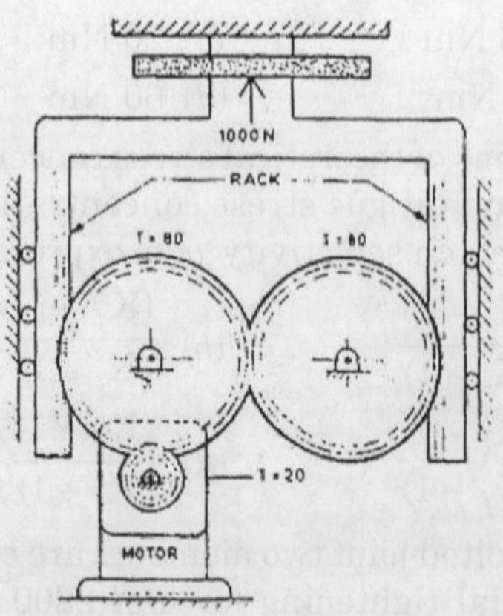

44. If drive efficiency is 80%, then torque required on the input shaft to create 1000 N output thrust is

(a) 20 Nm (b) 25 Nm
(c) 32 Nm (d) 50 Nm

45. If pressure angle of the rack is 20°, then force acting along the line of action between rack and the gear teeth is

(a) 250 N (b) 342 N
(c) 532 N (d) 600 N

2003

46. A wire rope is designated as 6 × 19 standard hoisting. The numbers 6 × 9 represent

(a) diameter in millimeter × length in meter
(b) diameter in centimeter × length in meter
(c) number of strands × number of wires in each strand
(d) number of wires in each strand × number of strands

47. A flexible rotor-shaft system comprises of a 10 kg rotor disc placed in the middle of a mass-less shaft of diameter 30mm and length 500 mm between bearings (shaft is being taken mass-less as the equivalent mass of the shaft is included in the rotor mass) mounted at the ends. The bearings are assumed to simulate simply supported boundary conditions. The shaft is made of steel for which the value of E is 2.1×10^{11} Pa. What is the critical speed of rotation of the shaft ?

(a) 60 Hz (b) 90 Hz
(c) 135 Hz (d) 180 Hz

48. Square key of side "d/4" each and length l is used to transmit torque "T" from the shaft of diameter "d" to the hub of a pulley. Assuming length of the key to be equal to thickness of the pulley, the average shear stress developed in the key is given by

(a) $\frac{4T}{ld}$ (b) $\frac{16T}{ld^2}$
(c) $\frac{8T}{ld^2}$ (d) $\frac{16T}{\pi d^3}$

49. In a band brake the ratio of tight side band tension to the tension on the slack side is 3. If the angle of overlap of band on the drum is 180°, the coefficient of friction required between drum and the band is

(a) 0.20 (b) 0.25
(c) 0.30 (d) 0.35

Common Data Q. 50 – 51

The overall gear ratio in a 2 stage speed reduction gear box (with all spur gears) is 12. The input and output shafts of the gear box are collinear. The countershaft which is parallel to the input and output shafts has a gear (Z_2 teeth) and pinion (Z_3 = 15 teeth) to mesh with pinion (Z_1 = 16 teeth) on the input shaft and gear (Z_4 teeth) on the output shaft respectively. It was decided to use a gear ratio of 4 with 3 module in the first stage and 4 module in the second stage.

50. Z_2 and Z_4 are

(a) 64 and 45 (b) 45 and 64
(c) 48 and 60 (d) 60 and 48

51. The centre distance in the second stage is

(a) 90 mm (b) 120 mm
(c) 160 mm (d) 240 mm

2002

52. A lead-screw with half nuts in a lathe, free to rotate in both directions has

(a) V-threads (b) Whitworth threads
(c) Buttress threads (d) Acme threads

53. The coupling used to connect two shafts with large angular misalignment is

(a) a Flange coupling
(b) an Oldham's coupling
(c) a Flexible bush coupling
(d) a Hooke's joint

54. A static load is mounted at the centre of a shaft rotating at uniform angular velocity. This shaft will be designed for

(*a*) maximum compressive stress (static)

(*b*) maximum tensile stress (static)

(*c*) maximum bending moment (static)

(*d*) fatigue loading

55. Large speed reductions (greated than 20) in one stage of a gear train are possible through

(*a*) Spur gearing (*b*) Worm gearing

(*c*) Bevel gearing (*d*) Helical gearing

56. If the wire diameter of a closed coil helical spring subjected to compressive load is increased from 1 cm to 2 cm, other parameters remaining same, then deflection will decrease by a factor of

(*a*) 16 (*b*) 8

(*c*) 4 (*d*) 2

57. A hollow mild steel shaft is subjected to torsional and bending moments T and M respectively. The outside and inside diameters of the shaft are D and *d* respectively.

(*a*) Derive the expressions for the maximum shear stress, and the maximum normal stress induced due to combined T and M.

(*b*) Name any one theory of failure, which may be used to design this shaft using these stresses.

2001

58. The spur gear in the figure is driven clockwise at 100 rpm. The ring gear is held stationary. For the number of teeth shown on the gears, the arm rotates at

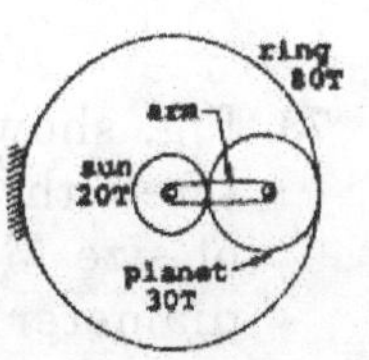

(*a*) 0 rpm

(*b*) 20 rpm

(*c*) 33.33 rpm

(*d*) 66.67 rpm

2000

59. A 1.5 kW motor is running at 1440 rev/min. It is to be connected to a stirrer running at 36rev/min. The gearing arrangement suitable for this application is

(*a*) differential gear (*b*) helical gear

(*c*) spur gear (*d*) worm gear

60. The life of a ball bearing at a load of 10 kN is 8000 hours. Its life in hours, if the load is increased to 20 kN, keeping all other conditions the same, is

(*a*) 4000 (*b*) 2000

(*c*) 1000 (*d*) 500

61. A steel bolt having a nominal diameter of 20 mm and a pitch of 2.4 mm is used to connect two plates of 10 mm thickness each. An aluminium tube of inner diameter 22 mm and outer diameter of 40 mm is separating the plates as shown in the figure. The nut is pulled snug (just tight) and then given a one-third additional turn. Find the resulting stresses in the bolt and the tube neglecting the deformation of the plates. Young's modulus of steel and aluminium are 207×10^3 MPa and 67.5×10^3 MPa respectively.

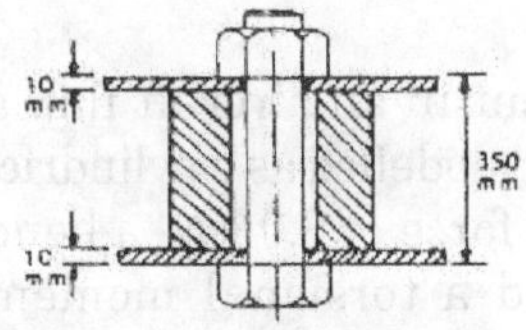

62. A single block brake with a 300 mm diameter brake drum shown in the figure is used to absorb a torque of 75 Nm. The coefficient of friction between drum and the lining is 0.35. The pressure on the block is uniform. The force P will be

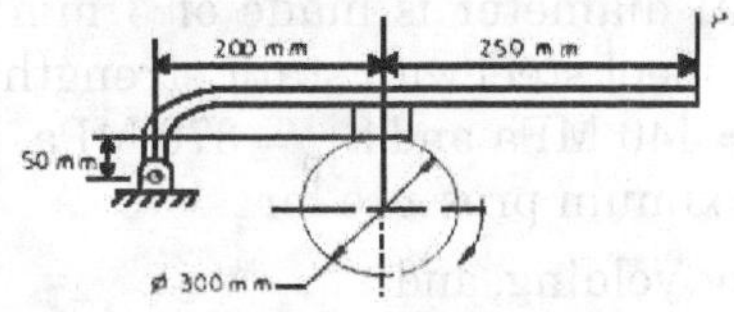

(*a*) 480 N (*b*) 580 N

(*c*) 680 N (*d*) 880 N

63. A full journal bearing with a journal of 75 mm diameter and a bearing of 75 mm length is subjected to a radial load of 2500 N at 400 rev/min. The lubricant is SAE 30 at 75° C having a viscosity of 16.5×10^{-3} kg/ms. Radial clearance is 0.03 mm. Eccentricity of the bearing is 0.27. The Sommerfeld number will be

(*a*) 12.4×10^6 (*b*) 18.4×10^6

(*c*) 23.4×10^6 (*d*) 28.4×10^6

1999

64. Bolts in the flanged end of pressure vessel are usually pre - tensioned Indicate which of the following statments in NOT TRUE ?

(*a*) Pre - tensioning helps to seal the pressure vessel

(*b*) Pre - tensioning increases the fatigue life of the bolts

(*c*) Pre - tensioning reduces the maximum tensile stress in the bolts

(*d*) Pre - tensioning helps to reduce the effect of pressure pulsations in the pressure vessel

1998

65. A component used in the Mars pathfinder can be idealized as a circular bar clamped at its ends. The bar should withstand a torque of 1000 Nm. The component is assembled on earth when the temperature is 30°C. Temperature on Mars at the site of landing is –70°C. The material of the bar has an allowable shear stress of 300 MPa and its Young's modulus is 200 GPa. Design the diameter of the bar taking a factor of safety of 1.5 and assuming a coefficient of thermal expansion for the material of the bar as 12×10^{-6}/°C.

1997

66. A component in an aircraft flap actuator can be adequately modelled as a cylindrical bar subjected to an axial force of 8.0 kN, a bending moment of 55 Nm and a torsional moment of 30 Nm. A 20 mm diameter solid bar of 7075 – T 6 aluminium having S_{ut} = 591 MPa, S_{yp} = 542 MPa and S_{syp} = 271 MPa is recommended for its use. Determine the factor of safety available as per stress strain theory.

67. A spherical pressure vessel of 600 mm internal diameter is made of 3 mm thick cold drawn sheet steel with static strength properties of S_{ut} = 440 MPa and S_{yp} = 370 MPa. Determine the maximum pressure for

(a) static yelding, and

(b) eventual fatigue failure, when the pressure fluctuates between 0 and p_{max}, when yielding is also not permitted (Do not apply any factor of safety).

1996

68. The bolts in a rigid flanged coupling connecting two shafts transmitting power are subjected to

(*a*) shear force and bending moment

(*b*) axial force

(*c*) torsion

(*d*) torsion and bending moment.

69. In spur gears, the circle on which the involute is generated is called the

(*a*) pitch circle (*b*) clearance circle

(*c*) base circle (*d*) addendum circle

70. Design a helical spring for a maximum compression load of 1200 N and a deflection of 30 mm. The permissible shear stress in the spring material is 400 MPa and the modulus of rigidity is 80GPa.

Take the spring index C as 5 and the Whal's factor, $K = \dfrac{4C-1}{4C-4} + \dfrac{9{,}615}{C}$

71. A power screw moves a nut against a force of 10 kN at a speed of 6 mm/s. The screw has a single square thread of 6mm pitch on a major diameter of 45 mm. Calculate the power of the driving motor. The coefficient of friction at the screw thread is 0.15.7.Five holes of diameter 10 mm each are to be punched in a sheet 3 mm thick at a pitch of 25 mm. What should be the minimum capacity of the press (in tonnes) if the yield point of material is50 MPa and

(*a*) one hole is punched per stroke,

(*b*) five holes are punched in a single stroke.

Show with a neat sketch the arrangement of punches so as to avoided eccentric loading (i.e. provide balanced loading). It may be noted that no shear is provided on the punches.

72. Fifty flat pieces 1 mm thick and initial dimensions as shown in Fig. (*a*) are to be milled in a single cut to the final dimension shown in Fig. (*b*) using end milling. If the cutter of diameter 25 mm has 10 teeth and rotates at 100 rpm, find

(*a*) the maximum uncut chip thickness if the horizontal feed of the table is 10 mm/min, assuming single tooth in contact.

(*b*) the material removal rate.

(*c*) the maximum torque if the yield strength of the material is 150 MPa.

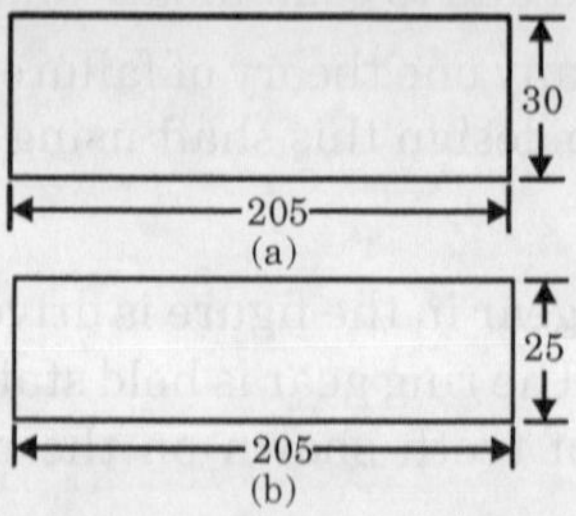

Fig. 21

73. Fig. shows the 3-wire method of inspecting screw threads. Assuming ISO metric threads of size M16 with pitch = 2mm and effective diameter = 14.701 mm, determine the over write measurement S when the diameter of the roller used for the measurement is 1.155 mm. The diameter of the roller corresponds to the best wire size for which the roller touches the thread flanks at points located along the effective diameter.

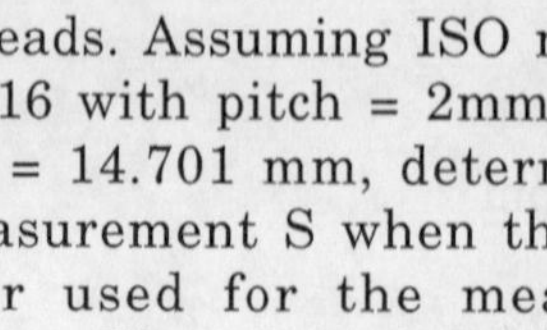

74. Alloy steel forged gear blank shown in Fig. (*a*) is to be machined to produce precision spur gear with a module of 5 mm. The finished gear is shown in Fig. (*b*). Write down the operation sequence to machine this component assuming a medium batch size of 300 gears. The operations should be listed in proper sequence as 1, 2, 3, ...

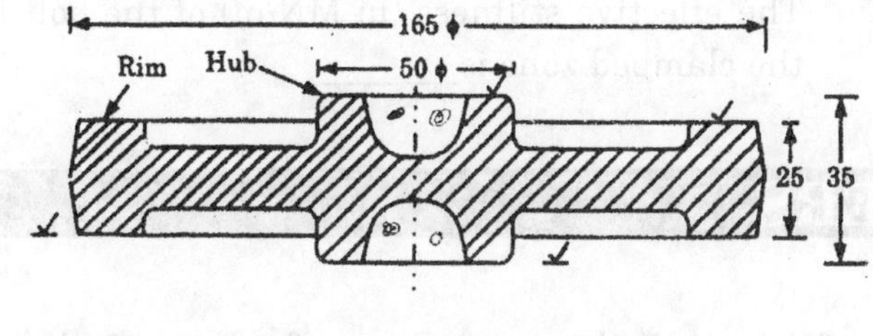

(*a*)

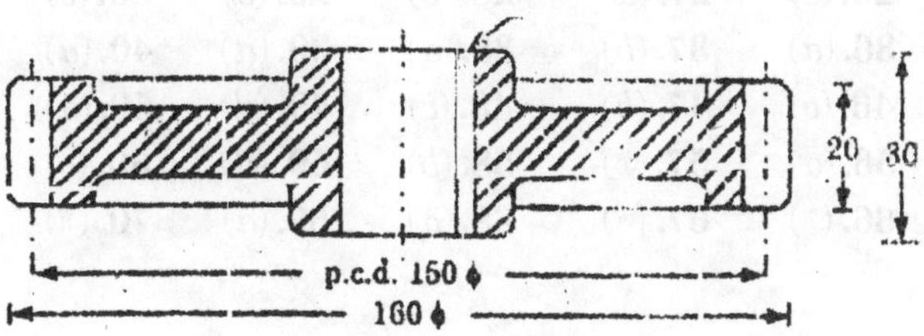

(*b*) All diamensions in mm

NUMERICAL TYPE QUESTIONS

2016

1. A bolted joint has four bolts arranged as shown in figure. The cross sectional area of each bolt is 25 mm^2. A torque T = 200 N-m is acting on the joint. Neglecting friction due to clamping force, maximum shear stress in a bolt is _____ MPa.

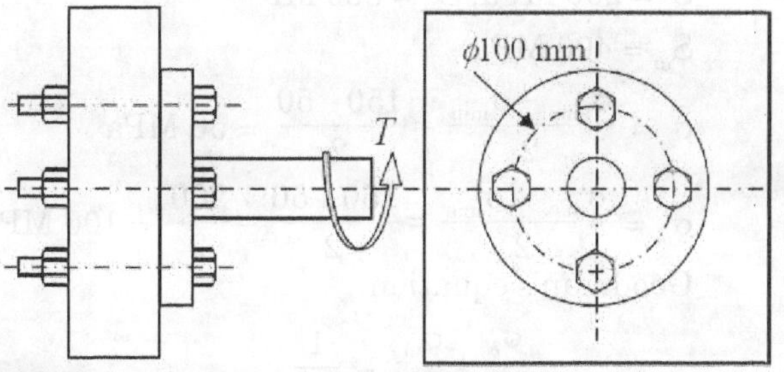

2015

2. A rope-brake dynamometer attached for the crank shaft I.C. engine measures a brake power of 10kW when the speed of rotation of the shaft is 400 rad/s. The shaft torque (in N-m) sensed by the dynamometer is _______.
3. A hallow shaft of 1 m length is designed to transmit a power of 30 kW at 700 rpm. The maximum permissible angle of twist in the shaft is 1°. The inner diameter of the shaft is 0.7 times the outer diameter. The modulus of rigidity is 80 GPa. The outside diameter (in mm) of the shaft is _________.
4. For ball bearings, the fatigue life L measured in number of revolutions and the radial load F are related by $FL^{1/3}$ = K, where K is a constant. It withstands a radial load of 2 kN for a life of 540 million revolutions. The load (in kN) for a life of one million revolutions is _______

2014

5. A pair of spur gears with module 5 mm and a center distance of 450 mm is used for a speed reduction of 5:1. The number of teeth on pinion is _____
6. A hydrodynamic journal bearing is subject to 2000 N load at a rotational speed of 2000 rpm. Both bearing bore diameter and length are 40 mm. If radial clearance is 20 μm and bearing is lubricated with an oil having viscosity 0.03 Pa.s, the Sommerfeld number of the bearing is _____
7. Consider a flywheel whose mass M is distributed almost equally between a heavy, ring-like rim of radius R and a concentric disk-like feature of radius R/2. Other parts of the flywheel, such as spokes, etc, have negligible mass. The best approximation for α, if the moment of inertia of the flywheel about its axis of rotation is expressed as αMR^2, is _______
8. A spur pinion of pitch diameter 50 mm rotates at 200 rad/s and transmits 3 kW power. The pressure angle of the tooth of the pinion is 20°. Assuming that only one pair of the teeth is in contact, the total force (in Newton) exerted by a tooth of the pinion on the tooth on a mating gear is _______
9. The diameter of a recessed ring was measured by using two spherical balls of diameter d_2 = 60 mm and d_1 = 40 mm as shown in the figure.

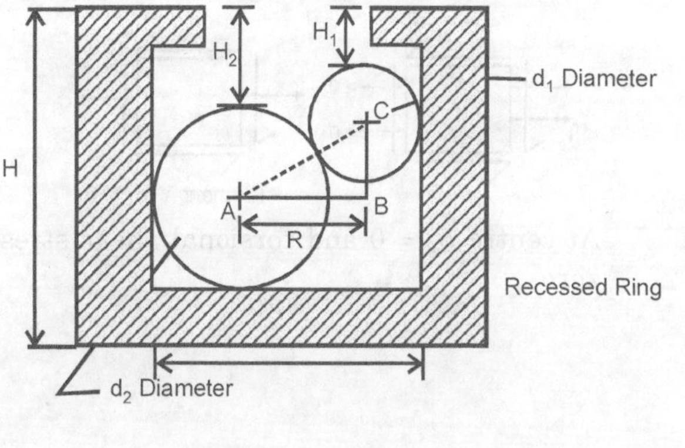

The distance H_2 = 35.55 mm and H_1 = 20.55 mm. The diameter (D, in mm) of the ring gauge is ______

10. Ball bearings are rated by a manufacturer for a life of 10^6 revolutions. The catalogue rating of a particular bearing is 16 kN. If the design load is 2 kN, the life of the bearing will be $p \times 10^6$ revolutions, where p is equal to

11. A bolt of major diameter 12 mm is required to clamp two steel plates. Cross sectional area of the threaded portion of the bolt is 84.3 mm². The length of the threaded portion in grip is 30 mm, while the length of the unthreaded portion in grip is 8 mm. Young's modulus of material is 200 GPa. The effective stiffness (in MN/m) of the bolt in the clamped zone is______

ANSWERS

MCQ Type Questions

1. (*b*)	**2.** (*c*)	**3.** (*d*)	**4.** (*c*)	**5.** (*b*)	**6.** (*c*)	**7.** (*b*)	**8.** (*b*)	**9.** (*d*)	**10.** (*b*)
11. (*a*)	**12.** (*d*)	**13.** (*a*)	**14.** (*a*)	**15.** (*b*)	**16.** (*b*)	**17.** (*d*)	**18.** (*b*)	**19.** (*c*)	**20.** (*a*)
21. (*b*)	**22.** (*d*)	**23.** (*a*)	**24.** (*d*)	**25.** (*c*)	**26.** (*c*)	**27.** (*d*)	**28.** (*b*)	**29.** (*b*)	**30.** (*c*)
31. (*a*)	**32.** (*d*)	**33.** (*c*)	**34.** (*c*)	**35.** (*a*)	**36.** (*a*)	**37.** (*b*)	**38.** (*a*)	**39.** (*d*)	**40.** (*a*)
41. (*c*)	**42.** (*b*)	**43.** (*b*)	**44.** (*b*)	**45.** (*c*)	**46.** (*c*)	**47.** (*b*)	**48.** (*c*)	**49.** (*d*)	**50.** (*a*)
51. (*b*)	**52.** (*d*)	**53.** (*d*)	**54.** (*d*)	**55.** (*b*)	**56.** (*a*)	**57.** (*)	**58.** (*b*)	**59.** (*d*)	**60.** (*c*)
61. (*)	**61.** (*b*)	**63.** (*c*)	**64.** (*c*)	**65.** (*)	**66.** (*)	**67.** (*)	**68.** (*a*)	**69.** (*a*)	**70.** (*)
71. (*)	**72.** (*)	**73.** (*)	**74.** (*)						

Numerical Type Questions

1. (160)	**2.** 25	**3.** 44.52	**4.** 16.286	**5.** 29 to 31	**6.** 0.75 to 0.85
7. 0.55 to 0.57	**8.** 638 to 639	**9.** 92 to 94	**10.** 500 to 540	**11.** 460 to 470	

"*" Refers to Subjective Question.

EXPLANATIONS

MCQ TYPE QUESTIONS

1. Spring constant for a helical spring is given by following formula

$$k = \frac{Gd^4}{64R^3 n}$$

This is independent of material strength.

2.

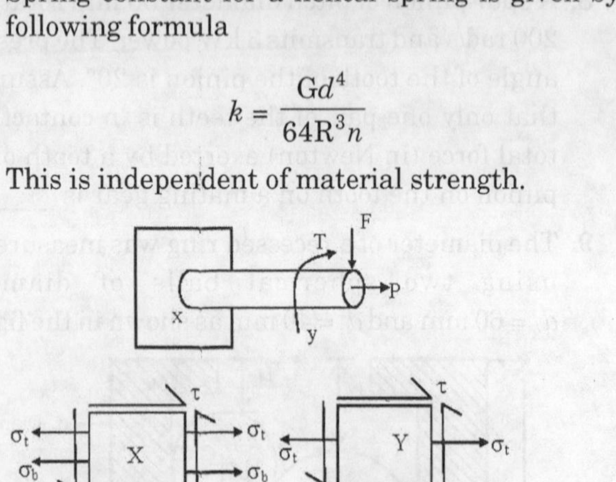

Element Y(σ_b = 0)

At centre σ_b = 0 and torsional shear stress are zero.

3. σ_{max} = 150 MPa

σ_{min} = 50 MPa

σ_m = 100 MPa

S_e = 200 MPa, S_{yt} = 300 MP

S_{ut} = 400 MPa

$$\sigma_a = \frac{\sigma_{max} - \sigma_{min}}{2} = \frac{150 - 50}{2} = 50 \text{ MPa}$$

$$\sigma_m = \frac{\sigma_{max} + \sigma_{min}}{2} = \frac{150 + 50}{2} = \frac{200}{2} = 100 \text{ MPa}$$

Goodman's equation

$$\frac{\sigma_a}{S_e} + \frac{\sigma_m}{S_{ut}} = \frac{1}{FS}$$

$$\Rightarrow \quad \frac{50}{200} + \frac{100}{400} = \frac{1}{FS}$$

$$\Rightarrow \quad FS = 2$$

5. $F_1 = F_2$ [Vertical equilibrium]

For Gear 1   For Gear 2

$\tau_1 - F_1 r_1 = I_{-1}\ddot{\theta}_1$ $F_2 r_2 = I_2 \ddot{\theta}_2$

$$\tau_1 - I_2\ddot{\theta}_2 \frac{r_1}{r_2} = I_1\ddot{\theta}_1 \text{ ... using } F_2 = \frac{I_2\ddot{\theta}_2}{r_2} \text{ and}$$

$$F_1 = \tau_1 = I_1\ddot{\theta}_1 + I = \ddot{\theta}_2 \frac{r_1}{r_2}$$

$$= \left[I_1 + I_2 \frac{\ddot{\theta}_2}{\ddot{\theta}_1}\frac{r_1}{r_2}\right]\ddot{\theta}_1$$

$$\tau_1 = \left[I_1 + I_2\left(\frac{r_1}{r_2}\right)^2\right]\ddot{\theta}_1 \ldots\ldots \left[\frac{\ddot{\theta}_1}{\ddot{\theta}_2} = \frac{r_2}{r_1}\right]$$

and clearly $F_2 = \dfrac{I_2 r_1 \ddot{\theta}_1}{r^2}$

6. A planetary gear train has 2 DOF and hence requires two input to get desired output.

7. $$P = T\omega$$

$$20 \times 10^3 = T \times \frac{2\pi \times 3000}{60}$$

$\therefore$ $T = 63.662$ N– m

Now $$\frac{T}{I_p} = \frac{\tau}{r},$$

So, $$\frac{63.662}{\frac{\pi}{32}(15d_1^4)} = \frac{30 \times 10^6}{d_1} \quad (r_0 = d),$$

$\therefore$ $d_1 = 11.295$ mm

$\therefore$ $d_0 = 2d_1 = 22.59$ mm

8. Rim med wheel has maximum mass located away from the axis of rotation. Thus will have maximum moment of inertia.

9. Since higher pair has a line or point contact and lower pair has a surface of Area contact. Hence both are true.

i.e., spur gear has line contact (Higher pair) and Revolute joint has surface contact (lower pair).

10. $\sigma_y = 200$ MPa

$P = 19 \times 10^3$ N

$$\sigma = \frac{P}{\frac{\pi}{4}d^2} \Rightarrow d = 10.9 \text{ mm}$$

M10

11. $$\frac{N_5}{N_2} = \frac{T_2 \times T_4}{T_3 \times T_5} = \frac{20 \times 15}{40 \times 30}$$

$\Rightarrow$ $N_5 = 1200 \times 0.25 = 300$ rpm ccw.

14. Assumption:

(*i*) Flow is steady *y*(*i.e.*) $\frac{\partial}{\partial t}() = 0$

(*ii*) Fully developed the $\frac{\partial}{\partial t}(\ ) - 0$; properties are not changing in the direction of the flow.

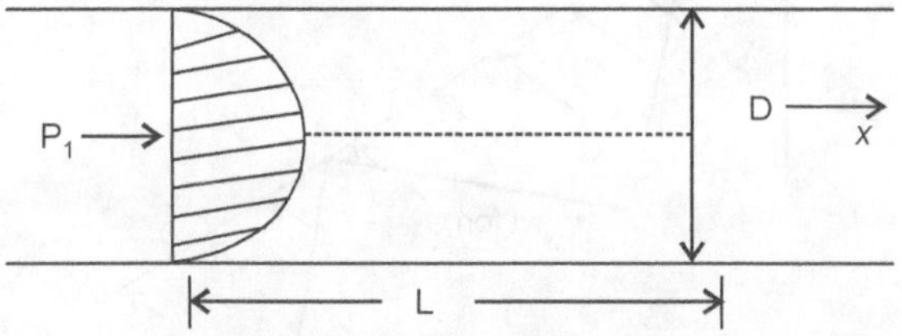

→ Pressure is constant along the vertical axis.

→ Pressure along horizontal axis does change.

$\Delta P.\ P_2.\ P_1 < 0$

Apply N2M (2nd) over the length l

$\Rightarrow P_1\pi r^2 - (P_1 - |\Delta P|)\pi r^2 - 2\pi r\, l_{ic}$

$$\frac{\Delta P}{L} = \frac{2\tau}{6L_1}$$

Neither P nor l depend as on r.

So, $\frac{\tau}{\varepsilon}$ is independ at then ($t = r$ where (is on stat)

At center $r = 0$, $\tau = c \times 0 = 0$

$$\tau = \frac{2\tau wr}{D}$$

$$\frac{\Delta P}{L} = \frac{2}{r} \times \frac{\tau w \times r}{D} \Rightarrow TW = \frac{\Delta PD}{4L}$$

15. We know $\Rightarrow$ [Power = Force × Velocity]

$$f = -m[v_f - v_i]$$

$$f = -fQ[(-v_i + 2u) - v_i]$$

$$f = -fQ(-2v_i + 2u)$$

$$\boxed{\therefore f = 2fQ[-v_i - u]} \quad \ldots(1)$$

So, $P = 2PQ(v_i - u)$.

To find speed at maximum power:

$$\frac{dp}{du} = 0, \frac{dp}{du} = 2fQ[(v_i - u) + u(-1)]$$

$$= 2fQ[v_i - 2u]$$

$$\frac{dp}{du} = 0,$$

$$\Rightarrow \boxed{v_i = 2n} \Rightarrow \boxed{u = \frac{v_i}{2}}$$

16. $$Q = A_1 V_1 = A_2 V_2$$

$$\Rightarrow \frac{\pi}{4}(20)^2 \times 2 = \frac{\pi}{4} \times d^2 \times \sqrt{2 \times 9.8 \times 0.5}$$

$$d \approx 15\text{mm}$$

17. For equilibrium:

Torque due to pressure of water = Torque due to *wf* of gate.

M → max of gate.

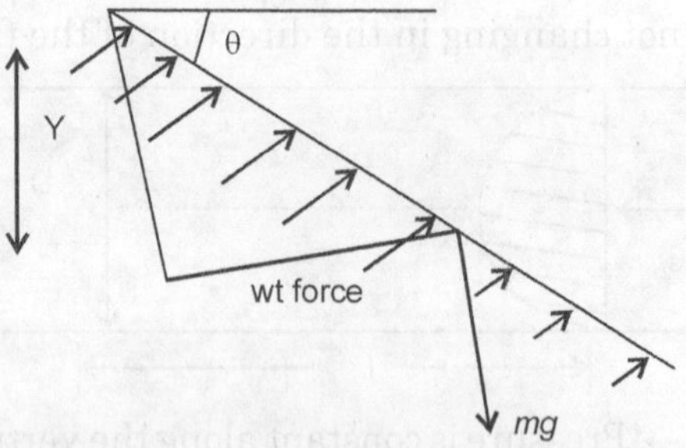

$$\int_0^L eg\sin\theta s^2 d = mg\frac{1}{2}\cos\theta$$

$$m = \frac{2eL^2\tan\theta}{3}$$

$$= \frac{2}{3}\times 10^3\times 5^2\times\tan 30^\circ = 9653xy.$$

18. $\left(\frac{dy}{dx}\right)_\phi \times \left(\frac{dy}{dx}\right)_\psi = -1$

Slope of equipotential Line

× slope of stream function = – 1

They are orthogonal to each other.

19. From Bernoulli's equation

$$\frac{V_1^2 - V_2^2}{2g} = \frac{p_2 - p_1}{\rho_a g}$$

$$\Rightarrow \quad 1\,V_1 = \sqrt{\frac{2(p_2 - p_1)}{\rho_a}}$$

But $p_2 - p_1 = (\rho g h)_{water}$

$= 9810 \times 10 \times 10^{-3}$

$= 98.1\ N/m^2$

$$\therefore \quad V_1 = \sqrt{\frac{2\times 98.1}{1.2}} = 1.2\ m/s$$

20. $\frac{\delta_t}{\delta} = \frac{1}{1.026}\times P_r^{-1/3}$

For fluid Q : $\frac{1}{2} = \frac{\delta_t}{\delta} = \frac{1}{1.026}\times P_r^{-1/3}$

$\Rightarrow \quad P_r = 8$

For fluid P : Laminar flow over flat plate

$N_u = 0.664\ Re_L^{1/2}\ P_r^{1/3}$

$= 35$

Similarly for fluid Q:

$Nu = 0.664\ Re_L^{1/2}\ P_r^{1/3}$

$= 0.664\ (10^4)^{1/2}\ 8^{1/3}$

$\simeq 140$

22.

Condition	Arm/*a*	2	3/4	5
Arm fixed 2-rotates + 1 revolution	0	1	$-\frac{20}{26} = \frac{5}{6}$	$-\frac{20}{26}\times\frac{32}{80} = \frac{1}{3}$
Arm fixed 2-rotates + X revolution	0	X	$-\frac{5}{6}X$	$-\frac{X}{3}$
Add Y (for arm rotation)	Y	X + Y	$-\frac{5}{6}X + Y$	$-\frac{X}{3} + Y$

Given : $\omega_{arm} = -80$ rad/sec = Y (CCW)

$\omega_2 = 100$ rad/sec = X + Y (CW)

∴ X = 100 – Y = 100 + 8

= 108 rad/sec (CW)

Also, $\omega_s = -\frac{1}{3}X + Y$

$= -\frac{1}{3}(180) + (-80)$

= – 140 rad/sec (CCW)

∴ $\omega_s = 140$ rad/sec (CW)

Note : CCW→ Counter clockwise

CW → Clockwise

23. *Given* : $d_j = 50$ mm = 0.05 m

$d_B = 50.05 = 0.05005$ m

$l_B = 20$ mm = 0.02 m

N = 1200 rpm

μ = 0.03 Pa–sec

Frictional factor, *f* = Shear stress × Area

$= \frac{\mu U}{h}\times\pi d_j l_B$

Here, $h = r_B - r_j$

= 25.025 – 25 = 0.025 = 0.000025m

$$U = \frac{\pi d_j N}{60} = \frac{\pi\times 0.05\times 1200}{60}$$

= 3.141

$$\therefore \quad f = \frac{0.03\times 3.141}{0.000025}\times\pi(0.05)\,(0.02)$$

= 11.841

Torque, $T = f\times r_j$

= 11.841 × 0.025 = 0.2960

$$\text{Power loss} = T\times\omega = \frac{2\pi NT}{60}$$

$$= \frac{2\pi\times 1200\times 0.2960}{60}$$

= 37.196 W ≈ 37.2 W

24. *Given* : ω_{AB} = 10 rad/sec (CCW)

AB = 250 mm, BC = $250\sqrt{3}$, AC = 500

Direction of velocity of AB is perpendicular to link AB and direction of velocity of AB is parallel to line CD.

Also, direction of relative velocity of slides B with respect to C is in line with link BC.

$\therefore \quad V_C = 0$

But $V_B = r_{AB} \times W_{AB} = 250 \times 10 \times 10^{-3} = 2.5$ m/sec

$\therefore \quad V_{BC} = V_B - V_C = 2.5$ m/sec

25. For a circular shaft of diameter *d*,

$$\frac{T'}{J} = \frac{\tau}{\left(\frac{d}{2}\right)}$$

where, T′ = net torsional moment

τ = shear stress (torsional shear strength)

J = polar moment of inertia

$$\therefore \quad t = \frac{T'\left(\frac{d}{2}\right)}{J}$$

But for a solid circular shaft

$$J = \frac{\pi}{32} d^4$$

$$\therefore \quad \tau = \frac{16T'}{\pi d^3}$$

But $\quad T' = \sqrt{M^2 + T^2}$

$$\therefore \quad \tau = \frac{16}{\pi d^3}\sqrt{M^2 + T^2}$$

26. Fatigue endurance, σ_{en} = 240 MPa

$$A = \frac{\pi}{4} \times (30)^2 = 706.5 \times 10^{-6}\, m^2$$

Soderberg's equation is

$$\frac{1}{h} = \frac{\sigma_m}{\sigma_y} + \frac{\sigma_a}{\sigma_{en.f}}$$

Now $\quad F_m = \frac{160 + 40}{2} = 100$ kN

$$\therefore \quad \sigma_m = \frac{F_m}{A} = \frac{100kN}{706.5 \times 10^{-6}} = 141.54 \text{ MPa}$$

and $\quad F_a = \frac{160 - 40}{2} = 60$ kN

$$\therefore \quad \sigma_a = \frac{60 \times 10^3}{706.5} = 84.93 \text{ MPa}$$

$$\therefore \quad \frac{1}{h} = \frac{141.54}{420} + \frac{84.93}{240}$$

$\Rightarrow \quad h = 1.45$

27. From Gerber's parabola, the endurance strength remains constant at maximum value irrespective of the compressive load, whereas for tensile load it decreases sharply with increases in σm and stress in the shaft, its strength in fatigue increases.

28. Since both ends hinged, therefore

$$L_e = L$$

$$\therefore \text{ Buckling load}, W = \frac{\pi^2 EI}{L^2}$$

Also $\quad W = F \cos 45°$

$$\therefore \quad F = \frac{\pi^2 EI}{L^2} / \cos 45°$$

$$= \frac{\sqrt{2}\,\pi^2 EI}{L^2}$$

29. $T = \int r.\, \mu.\, p.\, 2\pi r.\, dr = \int r.\, 0.4.\, 2\, .\, 2\pi r.\, dr$

$$= 1.6\,\pi \int_{20}^{50} r^2\, .\, dr$$

$$= \frac{1.6\,\pi}{3} \frac{\left[50^3 - 20^3\right]}{10^3} = 196 \text{ Nm}$$

30. Lewis equation is given as

$$F = \sigma_d\, C_v\, b\pi\, ym$$

F can be calculated as

$$F = \frac{P}{\omega r} \qquad \text{where, } \omega = 20 \times 2\pi \text{ rad/sec}$$

$$D = mZ = 3 \times 16 = 48\text{mm}$$

$$C_v = 1.5,\ b = 36$$

$$y = 0.3,\ m = 3$$

Now $\quad F = \frac{3 \text{ kW}}{40\pi \times 24}$

$$\therefore \quad \frac{3 \times 10^3 \times 10^3}{960\,\pi} = \sigma_d \times 1.5 \times 36 \times \pi \times 0.3 \times 3$$

$\Rightarrow \quad \sigma_d \approx 65$

Hence nearest stress in gear tooth taking in account some F.O.S., σ = 58

31.

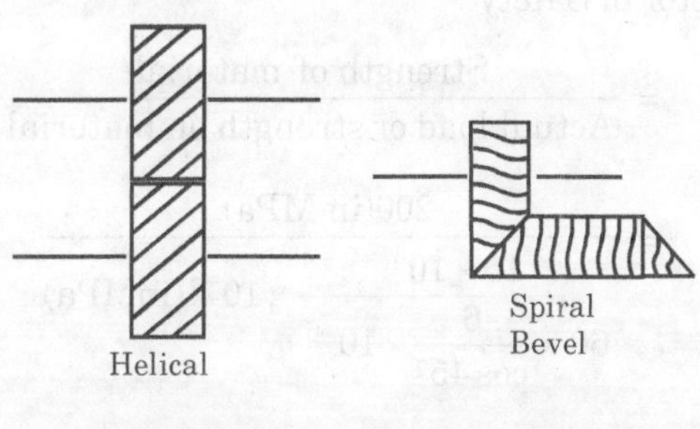

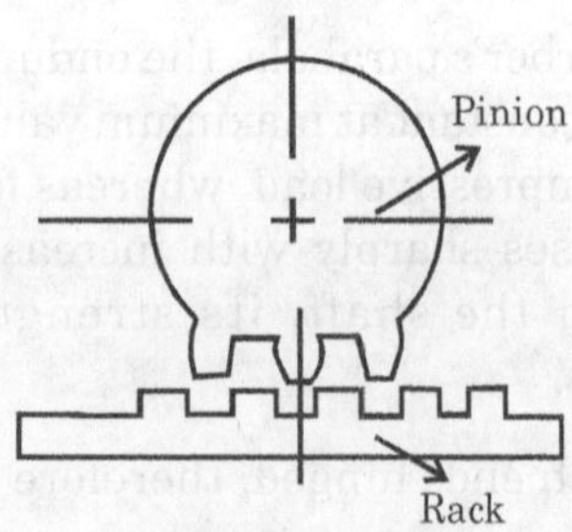

32. $$\text{Life} \propto \frac{1}{(\text{Load})^3}$$

$\therefore$ $$\text{New life} = \frac{8000}{(2)^3} = 1000 \text{ MS}$$

33. Sommerfeld Number $= \dfrac{ZN}{P}\left(\dfrac{d}{c}\right)^2$

Given: $\lambda = 50\ \mu m$

$P = 2kN$

$d = 50$ mm

$N = 20$ rps

$Z = 20$ mPas

34. Taking moment about A,

$$R_N \times 200 = 400 \times 600$$

$$\Rightarrow \quad R_N = 1200 \text{ N}$$

$$\therefore \quad \text{Braking torque} = \mu \,.\, R_N \times \frac{D}{2} = 45 \text{ Nm}$$

35. $$\text{Factor of safety} = \frac{E}{\sigma}$$

$$= \frac{\dfrac{200 \times 10^5}{10 \times 1000}}{\dfrac{\pi}{4} \times \left(\dfrac{20}{1000}\right)^2} = 0.68$$

36. $$\text{Centre distance} = \frac{D_1 + D_2}{2} = \frac{mT_1 + mT_2}{2}$$

$$= \frac{5(19+37)}{2} = 140 \text{ mm}$$

37. Factor of Safety

$$= \frac{\text{Strength of material}}{\text{Actual load or strength on material}}$$

$$= \frac{200(\text{in MPa})}{\dfrac{\dfrac{15 \times 10^3}{6}}{60 \times \dfrac{6}{\cos 45°} \times 10^{-6}} \times 10^{-6} (\text{inMPa})}$$

$$= \frac{200(\text{in MPa})}{58.91(\text{in MPa})} = 3.4$$

38. Sommerfield Number, also called Bearing Characteristic Number,

$$S = \frac{Z_n}{P} \cdot \left(\frac{D}{C_d}\right)^2$$

39. We know $\dfrac{N_p}{N_G} = \dfrac{T_G}{T_p}$

where, N_p = speed of pinion,

N_G = speed of gear wheel

T_G = number of teeth of gear,

T_p = number of teeth of pinion

$$\therefore \quad \frac{1200}{N_G} = \frac{120}{40}$$

$$\Rightarrow \quad N_G = 400 \text{ r.p.m.}$$

Since power transmitted by both gear will be equal

i.e. $T_p\, \omega_p = T_G\, \omega_G$

where, T_p = torque transmitted by pinion,

T_G = torque transmitted by gear wheel

$$\therefore \quad \frac{20 \times 2\pi \times 1200}{60} = \frac{T_G \times 2\pi \times 400}{60}$$

$\therefore$ Torque transmitted by gear,

$$T_G = 60 \text{ N-m.}$$

41. Tightening force = 2200 N

Pitch = 4 mm = 0.004 m

$$\text{Torque} = \text{Tightening force} \times \frac{0.004}{2\pi}$$

$$= 1.4 \text{ N-m.}$$

42. Total change in length

$$= \frac{\Sigma Pl}{AE}$$

$$= \frac{P_1 l_1 + P_2 + P_3 l_3}{AE}$$

$$= \frac{100 \times 0.5}{AE} + \frac{(-150) \times (0.8)}{AE} + \frac{0.5 \times 0.4}{AE}$$

$$= \frac{1}{25 \times 10^{-6} \times 200 \times 10^9} [50 - 120 + 20]$$

$$= -1 \times 10^{-5} \text{m}$$

$$= -10\ \mu\text{m.}$$

44. Given : Module, m = 2

$$\frac{D}{T} = 2$$

$$\Rightarrow \quad D = 80 \times 2$$

$$= 160 \text{ mm}$$

$$2F = 1000$$

$$\Rightarrow \quad F = 500 \text{ N}$$

Let T_1 be the torque applied by motor and T_2 be the torque applied by gear.

$\therefore$ Power transmission = 80%

Now, $$T_1\omega_1 = \frac{2T_2 \times \omega_2}{0.8}$$

$$\Rightarrow \quad T_1 = \frac{2 \times F \times (D/2)}{0.8} \times \frac{\omega_2}{\omega_1}$$

$$= 2 \times 500 \times \frac{0.16}{2} \times \frac{1}{0.8} \times \frac{1}{4}$$

$$= 25 \text{ N-m.}$$

45. $\because \quad P \cos\phi = F$

$\therefore$ Force acting along the line of action,

$$P = \frac{F}{\cos\phi}$$

$$= \frac{500}{\cos 20°} = 532 \text{ N.}$$

47. Here, mass of rotar, $m = 10$ kg

diameter of shaft, $d = 30 \times 10^5$ m

length of shaft, $l = 500 \times 10^{-3}$ m

E for steel $= 2.1 \times 10^{11}$ N/m²

Deflection of shaft, $$\Delta = \frac{mgl^3}{48EI}$$

Here, $$I = \frac{\pi}{64} d^4$$

$$= \frac{\pi}{64} \times (30 \times 10^{-3})^4$$

$$= 3.976 \times 10^{-8} \text{ m}^4$$

$$= \frac{10 \times 9.81 \times (500 \times 10^{-3})^3}{48 \times 2.1 \times 10^{11} \times 3.976 \times 10^{-8}}$$

$$= 3.06 \times 10^{-5} \text{ m}$$

$$\omega_n = \sqrt{\frac{g}{\Delta}}$$

$$= \sqrt{\frac{9.81}{3.06 \times 10^{-5}}}$$

$$= 566.24 \text{ rad/s}$$

$$f_n = \frac{\omega_n}{2\pi} = \frac{566.24}{2 \times 3.142}$$

$$= 90 \text{ Hz.}$$

48. Force of the shaft circumference is given by

$$P = \frac{T}{d/2}$$

$$\Rightarrow \quad P = \frac{2T}{d}$$

Shearing area = width × length = $\frac{d}{4} \times l = \frac{ld}{4}$

Average shear stress = $\frac{P}{\text{shearing area}}$

$$= \frac{\frac{2T}{d}}{\frac{ld}{4}} = \frac{8T}{ld^2}$$

49. Given: $$\frac{\text{Tension in tight side}}{\text{Tension is slack side}} = \frac{T_1}{T_2} = 3$$

$$\frac{T_1}{T_2} = e^{\mu\theta}$$

where, θ = angle of overlap

= 180°

= π radians

$$\therefore \quad 3 = e^{\mu\theta}$$

$$\Rightarrow \quad \ln(3) = \mu\pi.$$

$$\Rightarrow \quad \mu = \frac{1}{\pi} \ln(3) = 0.35$$

50. Given: $$\frac{N_1}{N_4} = 12$$

$$\frac{N_1}{N_2} = 4 = \frac{D_2}{D_1}$$

$$m_1 = 3$$

$$m_2 = 4$$

Now, $$\frac{D_1}{Z_1} = \frac{D_2}{Z_2}$$

$$\Rightarrow \quad \frac{Z_1}{Z_2} = \frac{D_1}{D_2} = \frac{N_2}{N_1} = \frac{1}{4}$$

$$\Rightarrow \quad Z_2 = Z_1 \times 4 = 64$$

and $$\frac{N_1}{N_4} = \frac{D_2 \times D_4}{D_1 \times D_3}$$

$$\Rightarrow \quad 12 = \frac{D_4}{D_3}$$

$$\Rightarrow \quad \frac{D_4}{D_3} = 3$$

Also, $$\frac{Z_3}{Z_4} = \frac{D_3}{D_4}$$

$$\Rightarrow \quad Z_4 = Z_3 . \frac{D_4}{D_3}$$

$$= Z_3 \times 3$$

$$= 15 \times 3$$

$$= 45$$

51. Now, $x = r_4 + r_3 = \dfrac{D_4 + D_3}{2}$

But $\dfrac{D_4}{Z_4} = \dfrac{D_3}{Z_3} = 4$

$\Rightarrow \quad D_4 = 180$

$D_3 = 60$

$\therefore \quad x = \dfrac{180 + 60}{2} = 120 \text{ mm}$

56. $\delta = \dfrac{64\,PR^3 n}{4d^4}$

$\therefore \quad \delta \propto \dfrac{1}{d^4}$

$\Rightarrow \quad \dfrac{\delta_1}{\delta} = \dfrac{d^4}{d_1^4} = \dfrac{1^4}{2^4}$

$\Rightarrow \quad \delta_1 = \dfrac{\delta}{16}$

57. Outside diameter = D

Inside diameter = d

(*a*) Maximum shear stress due to frequency,

$$\frac{\tau_{max}}{1} = \frac{T}{J}\rho$$

$$= \frac{T}{\frac{\pi}{32}\left(D^4 - d^4\right)} \times \frac{D}{2}$$

$$= \frac{16\,TD}{\pi\left(D^4 - d^4\right)}$$

and maximum bending stress due to bending moments only,

$$\sigma_b = \frac{My}{I}$$

$$= \frac{M}{\frac{\pi}{64}\left(D^4 - d^4\right)} \times \frac{D}{2}$$

$$= \frac{32\,MD}{\pi\left(D^4 - d^4\right)}$$

$\therefore$ Maximum shear stress due to torque T and bending moment M, i.e.

$$\tau_{max} = \sqrt{\left(\frac{\sigma_b}{2}\right)^2 + \tau^2_{max}}$$

$$= \sqrt{\left\{\frac{16\,MD}{\pi\left(D^4 - d^4\right)}\right\}^2 + \left\{\frac{16\,TD}{\pi\left(D^4 - d^4\right)}\right\}^2}$$

$$= \frac{16\,D\sqrt{M^2 + T^2}}{\pi\left(D^4 - d^4\right)}$$

Maximum normal stress,

$$\sigma_1 = \frac{\sigma_b}{2} + \frac{32MD}{\pi\left(D^2 - d^4\right)}$$

$$= \frac{32MD}{\pi\left(D^2 - d^4\right)} \times \frac{1}{2}$$

$$+ \sqrt{\left\{\frac{32\,MD}{2\pi\left(D^4 - d^4\right)}\right\}^2 + \left\{\frac{16\,TD}{\pi\left(D^4 - d^4\right)}\right\}^2}$$

$$= \frac{16\,MD}{\pi\left(D^4 - d^4\right)}$$

$$+ \sqrt{\left\{\frac{16\,MD}{\pi\left(D^4 - d^4\right)}\right\}^2 + \left\{\frac{16\,TD}{\pi\left(D^4 - d^4\right)}\right\}^2}$$

$$= \frac{16\,D}{\pi\left(D^4 - d^4\right)}\left\{M + \sqrt{M^2 + T^2}\right\}$$

(*b*) Using maximum shear stress theory, we may design the shaft.

58.

Arm	**Sun**	**Planet**	**Ring**
+1	+1	+1	+1
0	$\frac{80}{30} \times \frac{30}{20}$	$-\frac{80}{30}$	–1
1	5	$-\frac{5}{3}$	0

For 5 revolutions of Sun, Arm rotates by 1.

Hence for 100 revolutions of Sun, Arm rotates by

$$\frac{100}{5} = 20$$

60. $\text{Life} \propto \left(\dfrac{1}{P}\right)^3$

$\therefore \quad L_2 = L_1\left(\dfrac{P_1}{P_2}\right)^3$

$= 8000\left(\dfrac{10}{20}\right)^3$

$= 1000 \text{ hrs.}$

61. In $\dfrac{1}{3}$ rd additional turn, *i.e.*

α movement of unit $= \dfrac{2.4}{3} = 0.8$ mm

Nut in tension let have force P_1 and tube is in compression let have P_2 force

In equilibrium,

pull on the rod = push on the tube

$$\therefore \quad \sigma_s\left(\frac{\pi}{4}\right)(20)^2 = \sigma_c\left(\frac{\pi}{4}\right)\left[(44)^2 - (22)^2\right]$$

$$\Rightarrow \quad \frac{\sigma_s}{\sigma_c} = 3.63 \quad ...(i)$$

Total movement of nut = Expansion in rod + Compression in tube

$$\therefore \quad 0.8 \text{ mm} = \left.\frac{\sigma_s \ell}{E}\right|_{rod} + \left.\frac{\sigma_s \ell}{E}\right|_{tube}$$

$$0.8 \text{ mm} = \frac{\sigma_s \times 350\text{mm}}{207\times 10^{-2}\text{N/mm}^2} + \frac{\sigma_c \times 330 \text{ mm}}{67.5 \times 10^{-2}\text{N/mm}^2}$$

$$\Rightarrow 1.69 \times 10^{-2}\,\sigma_s + 4.89 \times 10^2\,\sigma_c = 0.8 \quad ...(ii)$$

From equations (i) and (ii), we get

$$\sigma_c = \frac{0.8}{11.02 \times 10^2} = 7.2 \times 10^{-4} \text{ N/mm}^2$$

and $\sigma_s = 26.3 \times 10^{-4} \text{ N/mm}^2$

62.

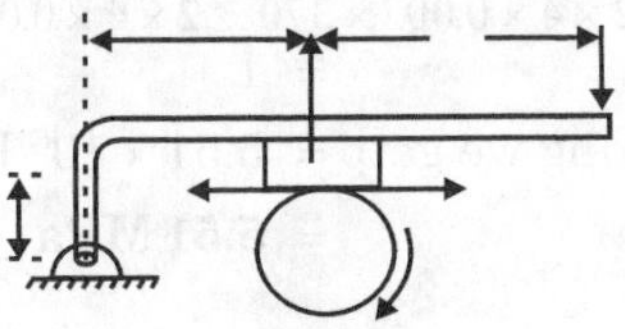

Here, D = 300 mm = 0.3 m

Torque = 75 N.m

Friction reaction on drum

$$= \mu R = \frac{\text{Torque}}{D/2} = \frac{75}{0.3/2} = 500 \text{ N}$$

But $\mu = 0.35$

$$\therefore \quad R = \frac{500}{0.35}$$

Taking moment about hence, we get

$$P \times (200 + 250) + \mu R \times 50 = R \times 200$$

$$\Rightarrow P \times 450 + 500 \times 50 = \frac{500}{0.35} \times 200$$

$$\Rightarrow P = \frac{260\,714}{450} = 579.4 \text{ N}$$

63. *Given:*

d = 75 mm

l = 75mm

w = 2500 N

N = 400 rev/min

Z = 16.5×10^{-3} kg/ms

at 75°C of SAE 30

$c_1 = 0.03$ mm

Eccentricity, e = 0.27 mm

Diametral clearance, $c = 2c_1 = 2 \times 0.03 = 0.06$ mm

$$\text{Sommerfield Number} = \frac{ZN}{P}\left(\frac{d}{c}\right)^2$$

$$\text{Bearing pressure, } p = \frac{W}{ld} = \frac{2500}{75 \times 75} = 0.44 \text{ N/mm}^2$$

$\therefore$ Sommerfield Number

$$= \frac{ZN}{P}\left(\frac{d}{c}\right)^2 = \frac{16.5 \times 10^{-3} \times 400}{0.44}\left(\frac{75}{0.06}\right)^2 = 23.4 \times 10^6$$

65. Given, $T_{max.}$ = 1000 Nm

T_e = 30°C

T_m = –70°C.

$f_{s\ allow}$ = 300 MPa

E = 200 GPa = 200×10^9 Pa

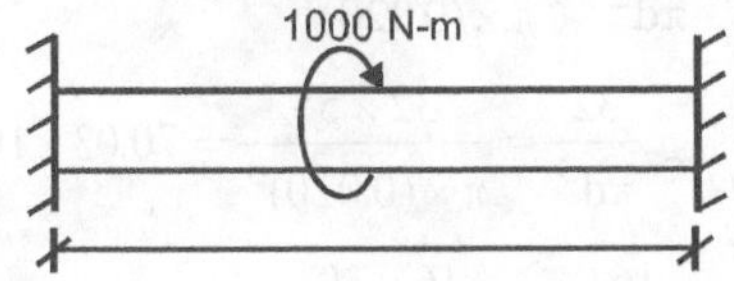

Factor of safety = 1.5

$\alpha = 12 \times 10^{-6}$/°C.

Change in length $\Delta l = \alpha\, l\,(\Delta T)$

where, l = original length of bar (metres).

Change in length at Mars.

$$= 12 \times 10^{-6} \times l \times [30 - (-70)] = 12 \times 10^{-4} \times l \text{ metres.}$$

$$\text{Linear strain} = \frac{\text{Change in length}}{\text{Original length}} = \frac{12\times 10^{-4} \times l}{l} = 12 \times 10^{-4}$$

Axial stress, f_a = E × Linear strain

$$= 200 \times 10^9 \times 12 \times 10^{-4} \text{ N/m}^2.$$

$$= \mathbf{2.4 \times 10^9\ N/m^2}.$$

From maximum shear stress equation,

$$f_{s\ max.} = \left[\left(\frac{K_s \times 16T}{\pi d^3}\right)^2 + \frac{1}{4}\left(\frac{4F_a}{\pi d^2}\right)^2\right]^{1/2}$$

$$\text{where, } f_{s\,max} = \frac{\text{Allowable shear stress}}{\text{F.O.S.}}$$

$$= \frac{300}{1.5} = 200 \text{ MPa}$$

K_s = 1 for steady loads

$$f_a = \frac{F_a}{(\pi d^2/4)} = 2.4 \times 10^8 \text{ N/m}^2.$$

$$\therefore 200 \times 10^6 = \left[\left(\frac{16 \times 1000}{\pi d^3}\right)^2 + \frac{1}{4}(2.4 \times 10^8)^2\right]^{1/2}$$

$$\text{or } 4 \times 10^6 = \left(\frac{16 \times 1000}{\pi d^3}\right)^2 + 1.44 \times 10^{16}$$

$$\text{or } \left(\frac{1000}{\pi d^3}\right)^2 = 2.56 \times 10^{16}$$

$$\text{or } \frac{1000}{\pi d^3} = 1.6 \times 10^8$$

$$\therefore d = 0.03169 \text{ m} = \mathbf{3.169\ cm.}$$

66. *Given*, Axial force, F_a = 8kN = 8×10^3 N

Bending moment, M_b = 55 Nm

Torsional moment, M_t = 30 Nm

Diameter of bar = 20 mm = 0.02 m,

Ultimate strength = 591 MPa,

Yield point strength = 542 MPa

$$f_{direct} = \frac{4F}{\pi d^2} = \frac{4 \times 8000}{\pi \times (0.020)^2} = 25.465 \times 10^6 \text{ Pa}$$

$$f_{bending} = \frac{32}{\pi d^3} = \frac{32 \times 55}{\pi \times (0.020)^3} = 70.03 \times 10^6 \text{ Pa}$$

$$f_{shear} = \frac{16T}{\pi d^3} = \frac{16 \times 30}{\pi \times (0.020)^3} = 191 \times 10^6 \text{ Pa}$$

$$\text{Maximum stress} = \sqrt{(f_d + f_b)^2 + 4(f_s)^2}$$

$$= 10^6\sqrt{(25.465 + 70.03)^2 + 4(19.1)^2}$$

$$= 102.85 \text{ MPa}$$

$$f_{max} \le \frac{\text{Utimate strength, } f_{yp}}{\text{factor of safety}}$$

$$\therefore \text{Factor of safety} = \frac{f_{yp}}{f_{max}}$$

$$= \frac{542}{102.85} = \mathbf{5.27}$$

67. *Given*, Internal diameter of vessel = 600 mm

Thickness of shell = 3 mm

Ultimate strength of steel, S_{ut} = 400 MPa

Yield point strength, S_{yp} = 370 MPa.

$$S_h = \frac{pD}{4t}$$

where p = internal pressure

D = diameter

t = thickness

$$s_t = s_{yp} = \frac{p_{max} D}{4t}$$

(Neglecting factor of safety)

$$\therefore p_{max.} = \frac{s_{yp} \times 4t}{D}$$

$$= \frac{370 \times 10^6 \times 4 \times 0.003}{0.6}$$

$$= 7.4 \times 10^5 \text{ Pa} = 7.4 \text{ MPa}$$

When pressure fluctuates between 0 and $p_{max.}$

$$s_m = s_a$$

As no yielding is also allowed, Soderberg's equation is to be used.

$$\textit{i.e. } \frac{s_m}{s_{yp}} + \frac{s_8}{s_e} = \frac{1}{F.S}$$

$$\text{or } \frac{p \times 0.6}{2 \times 4 \times 0.003 \times 370} + \frac{p \times 0.6}{2 \times 4 \times 0.003 \times (0.5 \times 440)} = 1$$

Solving we get p = 5.51×10^6 Pa

= **5.51 MPa**

72.

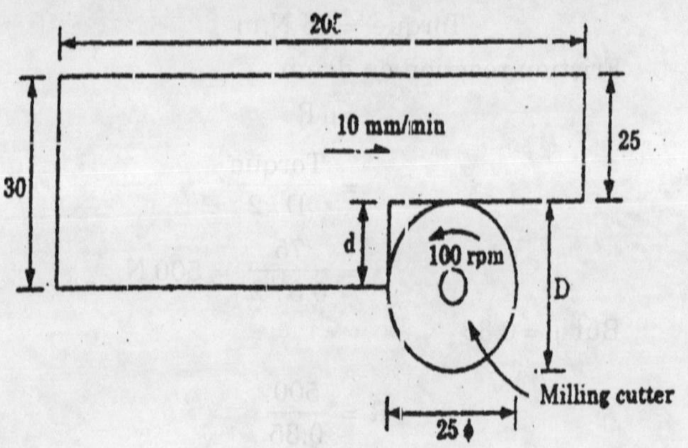

Plan view

Number of teeth = 10

(a) Maximum uncut chip thickness = $\frac{2f}{NZ}\sqrt{\frac{d}{D}}$

where f = feed of table in mm/min = 10 mm/min

N = Cutter rpm = 100

Z = Number of teeth in the cutter = 10

d = depth of cut in mm = 30 – 25 = 5 mm

D = diameter of the cutter in mm = 25 mm.

Maximum uncut chip thickness = $\frac{2\times10}{100\times10}\sqrt{\frac{5}{25}}$

$= \mathbf{8.94\times10^{-3}\ mm.}$

(b) Material removal rate= f × d × w,

where w is the width of the job

$= \frac{10}{60}\times5\times50$

$= \mathbf{41.7\ mm^3/sec}$

(c) According to the power law,

Power required for milling

$= 1.4 \times (t_{\text{uncut maximum}})^{-0.4}\ J/mm^3$

$= 1.4 \times (8.3 \times 10^{-3})^{-0.4}$

$= 9.5\ J/mm^3$

Material removal rate = 41.7 mm³/sec

Power required = 9.5 × 41.7 W

= 396.9 W

Also Power required = $\frac{2\pi NT}{60}$

$\therefore \quad T = \frac{396.9\times60}{2\times\pi\times100}$

$= \mathbf{37.9\ Nm.}$

73. *Given* Pitch, p = 2mm

Effective diameter, E = 14.701 mm

Thread type = M 16

Diameter of wire, d = 1.155 mm

Angle of thread, θ = 60° for metric threads

Now $S = D_{effective} + 2\,[AO + OB]$

$= 14.701+2\left[\frac{1.155}{2}\cos60°+1.155\right]$

= 17.589 mm

Hence over write measurement

= 17.589 mm

74.

	Machine Required	Operation
(1)	Lathe Hold the job by the hub. Center it	(A) Turn OD to 150 (B) Machine the flange face to a depth of 7.5 mm with respect to hub face (C) Machine the hub face to 5 mm with respect to the flange face (D) Hole to be machined to $25^{-0.10}_{\phi-0.05}$
(2)	Lathe Hold the job by the machined hub. Center it	(A) Machine the flange face to a thickness of 20 mm (B) Machine the hub face to 5 mm with respect to flange face
(3)	Grinding Hold by the flange and grind	(A) Grind the ID to $25^{+0.21}_{\phi-0.000}$
(4)	Shaping Hold by the flange	(A) Cutting the key way 8 W × 3.5 deep
(5)	Milling Hold by the ID of hub	(A) Gear cutting 150 PCD and 5 module with proper indexing

NUMERICAL TYPE QUESTIONS

1. Let Resisting force in each bolt = FN

Net Resisting torque = TR = $4F\times50\times10^{-3}$

Applied torque (T) = Resisting torque

$T = 4F\times50\times10^{-3}$

$F = \frac{200}{4\times50}\times1000$

F = 1000 N

$F = \tau\times A_R$

$\tau = \frac{F}{AR}$

$= \frac{1000}{25} = 40$ mPa

2. P = Tω

$\therefore\ T = \frac{P}{\omega} = \frac{10000}{400} = 25$ N–m

3. P = Tω $\quad 30\times1000 = T\times\frac{2\pi\times700}{60}$

∴ T = 409.256 N–m

$\frac{T}{I_p} = \frac{G\theta}{l}$

$\frac{409.256}{\frac{\pi}{32}(1-0.7^4)d_0^4} = \frac{80\times10^9}{1}\times\frac{\pi}{180}\left(\begin{matrix} l=1m \\ \theta=\frac{\pi}{180}\text{radians}\end{matrix}\right)$

Solving, we get , $d_0 = 44.5213$ mm

4. $FL^{\frac{1}{3}} = k$

$F_1L_1^{\frac{1}{3}} = F_2L_2^{\frac{1}{3}}$

$2 \times 540^{\frac{1}{3}} = F_2(1)^{\frac{1}{3}}$

$\therefore \quad F_2 = 16.286$ kN

5. Given speed Ratio = 5 : 1

$\frac{5}{1} = \frac{T_2}{T_1} = \frac{d_2}{d_1}$

$\Rightarrow d_2 = 5d_1$

centre distance = $\frac{d_1 + d_2}{2} = 450$

$\Rightarrow \; d_1 + d_2 = 900$

$\Rightarrow \; 5d_1 + d_1 = 900$

$\Rightarrow \; d_1 = 150$

$m = \frac{d_1}{T_1} \Rightarrow T_1 = \frac{150}{5} = 30.$

6. $S = \left(\frac{r}{c}\right)^2 \frac{\mu D_s}{P}$

$d = 40$ mm $\quad \Rightarrow \; r = 20$ mm

$c = 20\ \mu$m $\quad \Rightarrow \; c = 20 \times 10^{-3}$ mm

$\mu = 0.03$ PaS $\quad \Rightarrow \; \mu = 0.03 \times 10^{-6}$ MPa.S

$D_s = 2000$ rpm $\quad \Rightarrow \; D_s = \frac{2000}{60}$ rps

$P = \frac{p}{1 \times d} = \frac{2000}{40 \times 40} = 1.25$ MPa

$S = \left(\frac{20}{20 \times 10^{-3}}\right)^2 \times \frac{0.03 \times 10^{-6} \times \frac{2000}{60}}{1.25} = 0.8$

7. Moment of Inertia of fly wheel

$I = m_r R^2$

R = mean Radius of Rim

$= M\left(\frac{R + R/2}{2}\right)^2$

$= 0.56\ MR^2$

$\Rightarrow \quad \alpha = 0.56.$

8. Diameter = 50 mm = 0.05 m

$\omega = 200$ rad/s

$\Rightarrow \quad N = \frac{\omega \times 60}{2\pi}$

$\alpha = 20°$

P = 3 kW

$T = \frac{p \times 10^6 \times 60}{2\pi N}$

$= \frac{3 \times 10^6 \times 60}{2\pi \times \frac{\omega \times 60}{2\pi}}$

$T = 15000$ N – mm

Tangential component of force

$F_t = \frac{2T}{d} = \frac{2 \times 1500}{50} = 600$ N

Total force = $\frac{F_t}{\cos 20} = 638.5$ N

9. $D_{ring} = d_1 \sec\frac{\theta}{2} + 2(h_1 + r_1)\tan\frac{\theta}{2} \quad \therefore \theta = 60º$

$= 4 - \sec 30 + 2\,(20.55 + 20) \tan 30$

$D_{ring} = 93$ mm

10. $L = \left(\frac{c}{w}\right)^k \times 10^6 \quad k = 3$ (Ball bearing)

$= \left(\frac{16}{2}\right)^3 \times 10^6 = 8^3 \times 10^6 = P \times 10^6$

P = 512

11. $d_1 = 12$ mm;

$l_1 = 8$ mm

$A_2 = 84.33\ mm^2$

$A_1 = \frac{\pi}{4}(d_1)^2$

$l_2 = 30$ mm

$K_1 = \frac{A_1E_1}{l_1};$

$K_2 = \frac{A_2E_2}{l_2}$

$= \frac{84.33 \times 200}{30} = 562.2$

$= \frac{\frac{\pi}{4}(12)^2 \times 200}{8} = 2827.4$

$\frac{1}{K} = \frac{1}{K_1} + \frac{1}{K_2}$

$= \frac{\pi}{2} \times 3000 + 1500 \times \pi$

$= \frac{1}{2827.4} + \frac{1}{562.2}$

$\Rightarrow \quad K = 468.9$ MN/m

6

CHAPTER

Fluid Mechanics

MCQ TYPE QUESTIONS

2016

1. The instantaneous stream-wise velocity of a turbulent flow is given as follows:

$$u(x, y, z, t) = \bar{u}(x, y, z) + u'(x, y, z, t)$$

The time-average of the fluctuating velocity $u'(x, y, z, t)$ is

(*a*) $\frac{u'}{2}$ (*b*) $-\frac{\bar{u}}{2}$

(*c*) zero (*d*) $\frac{\bar{u}}{2}$

2. For a floating body, buoyant force acts at the

(*a*) centroid of the floating body

(*b*) center of gravity of the body

(*c*) centroid of the fluid vertically below the body

3. The volumetric flow rate (per unit depth) between two streamlines having stream functions ψ_1 and ψ_2 is

(*a*) $|\psi_1 + \psi_2|$ (*b*) $\psi_1\psi_2$

(*c*) ψ_1/ψ_2 (*d*) $|\psi_1 - \psi_2|$

4. Assuming constant temperature condition and air to be an ideal gas, the variation in atmospheric pressure with height calculated from fluid statics is

(*a*) linear (*b*) exponential

(*c*) quadratic (*d*) cubic

(*d*) centroid of the displaced fluid

5. Consider a frictionless, massless and leak-proof blocking a rectangular hole of dimensions 2 R × L at the bottom of an open tank as shown in the figure. The head of the plug has the shape of a semi-cylinder of radius R. The tank is filled with a liquid of density ρ up to the tip of the plug. The gravitational acceleration is *g*. Neglect the effect of the atmospheric pressure.

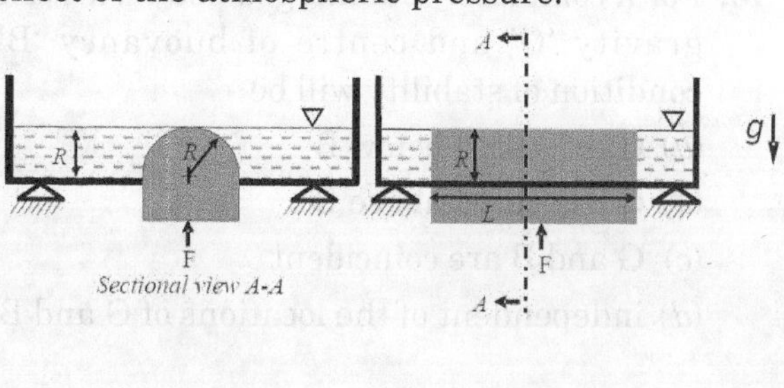

Sectional view A-A

The force F required to hold the plug in its position is

(*a*) $2\rho R^2 gL\left(1-\frac{\pi}{4}\right)$ (*b*) $2\rho R^2 gL\left(1+\frac{\pi}{4}\right)$

(*c*) $\pi R^2 \rho gL$ (*d*) $\frac{\pi}{2}\rho R^2 gL$

6. For a certain two-dimensional incompressible flow, velocity field is given by $2xy\,\hat{i} - y^2\hat{j}$. The streamlines for this flow are given by the family of curves

(*a*) x^2y^2 = constant

(*b*) xy^2 = constant

(*c*) $2xy - y^2$ = constant

(*d*) xy = constant

7. For a two-dimensional flow, the velocity field is $\vec{u} = \frac{x}{x^2+y^2}\hat{i} + \frac{y}{x^2+y^2}\hat{j}$, where $\hat{i}$ and $\hat{j}$ are the basis vectors in the *x*-*y* Cartesian coordinate system. Identify the **CORRECT** statements from below.

(1) The flow is incompressible.

(2) The flow is unsteady.

(3) *y*-component of acceleration, $a_y = \frac{-y}{\left(x^2+y^2\right)^2}$

(4) *x*-component of acceleration, $a_x = \frac{-(x+y)}{\left(x^2+y^2\right)^2}$

(*a*) (2) and (3) (*b*) (1) and (3)

(*c*) (1) and (2) (*d*) (3) and (4)

2015

8. Consider fully developed flow in a circular pipe with negligible entrance length effects. Assuming the mass flow rate, density and friction factor to be constant, if the length of the pipe is doubled and the diameter is halved, the head loss due to friction will increase by a factor of

(*a*) 4 (*b*) 16

(*c*) 32 (*d*) 64

9. The Blasius equation related to boundary layer theory is a

(*a*) third-order linear partial differential equation

(*b*) third-order nonlinear partial differential equation

(*c*) second-order nonlinear ordinary differential equation

(*d*) third-order nonlinear ordinary differential equation

10. For flow of viscous fluid over a flat plate, if the fluid temperature is the same as the plate temperature, the thermal boundary layer is

(*a*) thinner than the velocity boundary layer

(*b*) thicker than the velocity boundary layer

(*c*) of the same thickness as the velocity boundary layer

(*d*) not formed at all

11. Match the following pairs:

Equation	Physical Interpretation
(P) $\nabla \times \vec{V} = 0$	(I) Incompressible continuity equation
(Q) $\nabla . \vec{V} = 0$	(II) Steady flow
(R) $\frac{D\vec{V}}{Dt} = 0$	(III) Irrotational flow
(S) $\frac{\partial \vec{V}}{\partial t} = 0$	(IV) Zero acceleration of fluid particle

(*a*) P – IV, Q – I, R – II, S – III

(*b*) P – IV, Q – III, R – I, S – II

(*c*) P – III, Q – I, R – IV, S – II

(*d*) P – III, Q – I, R – II, S – IV

12. Within a boundary layer for a steady incompressible flow, the Bernoulli equation

(*a*) holds because the flow is steady

(*b*) holds because the flow is incompressible

(*c*) holds because the flow is transitional

(*d*) does not hold because the flow is frictional

13. In the laminar flow of air (Pr = 0.7) over a heated plate if δ and δ_T denote, respectively, the hydrodynamic and thermal boundary layer thicknesses, then

(*a*) $\delta = \delta_T$ (*b*) $\delta > \delta_T$

(*c*) $\delta < \delta_T$ (*d*) $\delta = 0$ but $\delta_T \neq 0$

14. The head loss for a laminar incompressible flow through a horizontal circular pipe is h_1. Pipe length and fluid remaining the same, if the average flow velocity doubles and the pipe diameter reduces to half its previous value, the head loss is h_2. The ratio h_2/h_1 is

(*a*) 1 (*b*) 4

(*c*) 8 (*d*) 16

15. Three parallel pipes connected at the two ends have flow-rates Q_1, Q_2 and Q_3 respectively, and the corresponding frictional head losses are hL_1, h_{L2} and h_{L3} respectively. The correct expressions for total flow rate (Q) and frictional head loss across the two ends (h_L) are

(*a*) $Q = Q_1 + Q_2 + Q_3$; $h_L = h_{L1} + h_{L2} + h_{L3}$

(*b*) $Q = Q_1 + Q_2 + Q_3$; $h_L = h_{L1} = h_{L2} = h_{L3}$

(*c*) $Q = Q_1 = Q_2 = Q_3$; $h_L = h_{L1} + h_{L2} + h_{L3}$

(*d*) $Q = Q_1 = Q_2 = Q_3$; $h_L = h_{L1} = h_{L2} = h_{L3}$

16. Couette flow is characterized by

(*a*) steady, incompressible, laminar flow through a straight circular pipe

(*b*) fully developed turbulent flow through a straight circular pipe

(*c*) steady, incompressible, laminar flow between two fixed parallel plates

(*d*) steady, incompressible, laminar flow between one fixed plate and the other moving with a constant velocity

17. A rigid container of volume 0.5 m³ contains 1.0 kg of water at 120°C (v_f = 0.00106 m³/kg, v_g = 0.8908 m³/kg). The state of water is

(*a*) Compressed liquid

(*b*) Saturated liquid

(*c*) A mixture of saturated liquid and saturated vapor

(*d*) Superheated vapor

2014

18. Consider a two-dimensional laminar flow over a long cylinder as shown in the figure below.

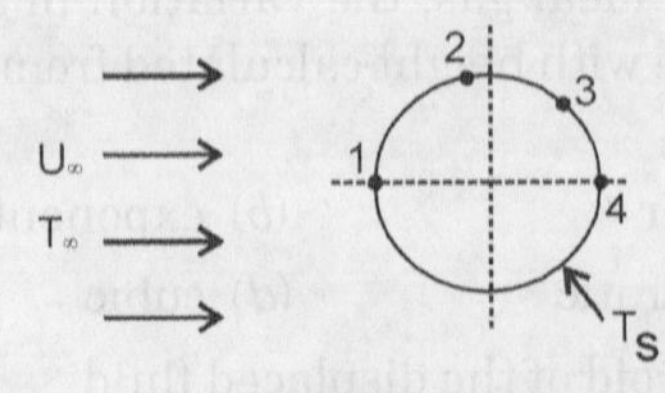

The free stream velocity is U_∞ and the free stream temperature T_∞ is lower than the cylinder surface temperature T_s. The local heart transfer coefficient is minimum at point

(*a*) 1 (*b*) 2

(*c*) 3 (*d*) 4

19. For a completely submerged body with centre of gravity 'G' and centre of buoyancy 'B', the condition of stability will be

(*a*) G is located below B

(*b*) G is located above B

(*c*) G and B are coincident

(*d*) independent of the locations of G and B

20. For laminar forced convection over a flat plate, if the free stream velocity increases by a factor of 2, the average heat transfer coefficient

(a) remains same

(b) decreases by a factor of $\sqrt{2}$

(c) rises by a factor of $\sqrt{2}$

(d) rises by a factor of 4

21. For an incompressible flow field, $\vec{v}$, which one of the following conditions must be satisfied?

(a) $\nabla.\vec{V} = 0$ (b) $\nabla \times \vec{V} = 0$

(c) $(\vec{V}.\nabla) \times \vec{V} = 0$ (d) $\frac{\partial V}{\partial t} + (\vec{V}.\nabla)\vec{V} = 0$

22. Consider the turbulent flow of a fluid through a circular pipe of diameter, D. Identify the correct pair of statements.

I. The fluid is well-mixed

II. The fluid is unmixed

III. $Re_D < 2300$

IV. $Re_D > 2300$

(a) I, III (b) II, IV

(c) II, III (d) I, IV

23. For a gas turbine power plant, identify the correct pair of statements.

P. Smaller in size compared to steam power plant for same power output

Q. Starts quickly compared to steam power plant

R. Works on the principle of Rankine cycle

S. Good compatibility with solid fuel

(a) P, Q (b) R, S

(c) Q, R (d) P, S

24. A siphon is used to drain water from a large tank as shown in the figure below. Assume that the level of water is maintained constant. Ignore frictional effect due to viscosity and losses at entry and exit. At the exit of the siphon, the velocity of water is

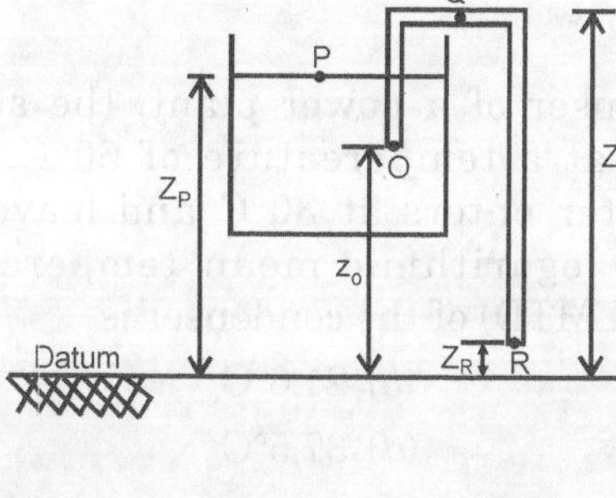

(a) $\sqrt{2g(Z_Q - Z_R)}$ (b) $\sqrt{2g(Z_P - Z_R)}$

(c) $\sqrt{2g(Z_O - Z_R)}$ (d) $\sqrt{2gZ_Q}$

25. A flow field which has only convective acceleration is

(a) a steady uniform flow

(b) an unsteady uniform flow

(c) a steady non-uniform flow

(d) an unsteady non-uniform flow

26. Kaplan water turbine is commonly used when the flow through its runner is

(a) axial and the head available is more than 100 m

(b) axial and the head available is less than 10 m

(c) radial and the head available is more than 100 m

(d) mixed and the head available is about 50 m

27. Consider the following statements regarding streamline(s):

(i) It is a continuous line such that the tangent at any point on it shows the velocity vector at that point

(ii) There is no flow across streamlines

(iii) $\frac{dx}{u} = \frac{dy}{v} = \frac{dz}{w}$ is the differential equation of a streamline, where u, v and w are velocities in directions x, y and z, respectively

(iv) In an unsteady flow, the path of a particle is a streamline

Which one of the following combinations of the statements is true?

(a) (i), (ii), (iv) (b) (ii), (iii), (iv)

(c) (i), (iii), (iv) (d) (i), (ii), (iii)

28. Consider a velocity field $\vec{V} = K(y\hat{i} + x\hat{k})$, where K is a constant. The vorticity, Ω_z, is

(a) – K (b) K

(c) – K/2 (d) K/2

29. Two infinite parallel plates are placed at a certain distance apart. An infinite radiation shield is inserted between the plates without touching any of them to reduce heat exchange between the plates. Assume that the emissivities of plates and radiation shield are equal. The ratio of the net heat exchange between the plates with and without the shield is

(a) 1/2 (b) 1/3

(c) 1/4 (d) 1/8

2013

30. Consider one-dimensional steady state heat conduction, without heat generation, in a plane wall with boundary conditions as shown in the figure below. The conductivity of the wall is given by $k = k_0 + bT$, where k_0 and b are positive constants, and T is temperature

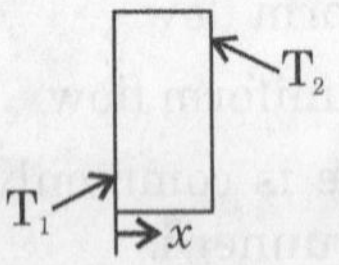

where $T_2 > T_1$

As x increases, the temperature gradient (dT/dx) will

(*a*) remain constant (*b*) be zero

(*c*) increase (*d*) decrease

31. Consider one-dimensional steady state heat conduction along x-axis ($0 \le x \le L$), through a plane wall with the boundary surfaces ($x = 0$ and $x = L$) maintained at temperatures of 0°C and 100°C. Heat is generated uniformly throughout the wall. Choose the CORRECT statement

(*a*) The direction of heat transfer will be from the surface at 100°C to the surface at 0°C

(*b*) The maximum temperature inside the wall must be greater than 100°C

(*c*) The temperature distribution is linear within the wall

(*d*) The temperature distribution is symmetric about the mid-plane of the wall

32. A steel ball of diameter 60 mm is initially in thermal equilibrium at 1030°C in a furnace. It is suddenly removed from the furnace and cooled in ambient air at 30°C, with convective heat transfer coefficient = 20 W/m²K. The thermo-physical properties of steel are density ρ = 7800 kg/m³, conductivity k = 40 W/mK and specific heat c = 600 J/kg/K. The time required in seconds to cool the steel ball in air from 1030°C to 430°C is

(*a*) 519 (*b*) 931

(*c*) 1195 (*d*) 2144

33. Two large diffuse gray parallel plates, separated by a small distance, have surface temperatures of 400 K and 300 K. If the emissivities of the surfaces are 0.8 and the Stefan-Boltzmann constant is 5.67×10^{-8} W/m²K⁴, the net radiation heat exchange rate in kW/m² between the two plates is

(*a*) 0.66 (*b*) 0.79

(*c*) 0.99 (*d*) 3.96

2012

34. Oil flows through a 200 *mm* diameter horizontal cast iron pipe (friction factor, $f = 0.0225$) of length 500 *m*. The volumetric flow rate is 0.2 *m*³/*s*. The head loss (in *m*) due to friction is (assume $g = 9.81\ m/s^2$)

(*a*) 116.18 (*b*) 0.116

(*c*) 18.22 (*d*) 232.36

35. An incompressible fluid flows over a flat plate with zero pressure gradient. The boundary layer thickness is 1 *mm* at a location where the Reynolds number is 1000. If the velocity of the fluid alone is increased by a factor of 4, then the boundary layer thickness at the same location, in *mm* will be

(*a*) 4 (*b*) 2

(*c*) 0.5 (*d*) 0.25

36. A large tank with a nozzle attached contains three immiscible, inviscid fluids as shown. Assuming that the changes in h_1, h_2 and h_3 are negligible, the instantaneous discharge velocity is

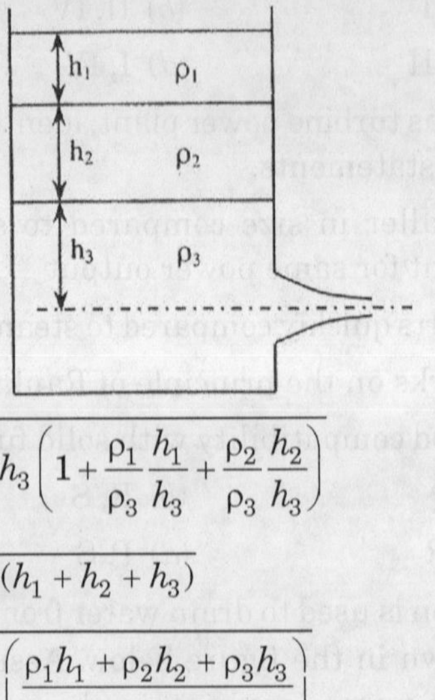

(*a*) $\sqrt{2gh_3\left(1+\frac{\rho_1}{\rho_3}\frac{h_1}{h_3}+\frac{\rho_2}{\rho_3}\frac{h_2}{h_3}\right)}$

(*b*) $\sqrt{2g(h_1+h_2+h_3)}$

(*c*) $\sqrt{2g\left(\frac{\rho_1 h_1+\rho_2 h_2+\rho_3 h_3}{\rho_1+\rho_2+\rho_3}\right)}$

(*d*) $\sqrt{2g\left(\frac{\rho_1 h_2 h_3+\rho_2 h_3 h_1+\rho_3 h_1 h_2}{\rho_1 h_1+\rho_2 h_2+\rho_3 h_3}\right)}$

2011

37. In a condenser of a power plant, the steam condenses at a tempereature of 60°C. The cooling water enters at 30°C and leaves at 45°C. The logarithmic mean temperature difference (LMTD) of the condenser is

(*a*) 16.2°C (*b*) 21.6°C

(*c*) 30°C (*d*) 37.5°C

38. A pipe of 25 mm outer diameter carries steam. The heat transfer coefficient between the cylinder and surroundings is 5 W/m^2K. It is proposed to reduce the heat loss from the pipe by adding insulation having a thermal conductivity of 0.05 W/mK. Which one of the following statements is TRUE ?

(*a*) The outer radius of the pipe is equal to the critical radius.

(*b*) The outer radius of the pipe is less than the critical radius.

(*c*) Adding the insulation will raduce the heat loss.

(*d*) Adding the insulation will increase the heat loss.

39. A spherical steel ball of 12 mm diameter is initially at 1000 K. It is slowly cooled in a surronding of 300 K. The heat transfer coefficient between the steel ball and the surrounding is 5 W/m^2K. The thermal conductivity of steel is 20 W/mK. The temperature difference between the centre and the surface of the steel ball is

(*a*) large because conduction resistance is far higher than the convective resistance.

(*b*) large because conduction resistance is far less than the convective resistance.

(*c*) small because conduction resistance is far higher than the convective resistance.

(*d*) small because conduction resistance is far less than the convective resistance.

2010

40. For the stability of a floating body, under the influence of gravity alone, which of the following is TRUE?

(*a*) Metacentre should be below centre of gravity

(*b*) Metacentre should be above centre of gravity

(*c*) Metacentre and centre of gravity must lie on the same horizontal line

(*d*) Metacentre and centre of gravity must lie on the same vertical line

41. The maximum velocity of a one-dimensional incompressible fully developed viscous flow, between two fixed parallel plates, is 6 ms^{-1}. The mean velocity (in ms^{-1}) of the flow is

(*a*) 2 (*b*) 3

(*c*) 4 (*d*) 5

42. A smooth pipe of diameter 200 mm carries water. The pressure in the pipe at section S_1 (elevation: 10 m) is 50 kPa. At section S_2 (elevation: 12 m) the pressure is 20 kPa and velocity is 2 ms^{-1}. Density of water is 1000 kgm^{-3} and acceleration due to gravity is 9.8 ms^{-2}. Which of the following is TRUE ?

(*a*) Flow is from S_1 to S_2 and head loss is 0.53 m

(*b*) Flow is from S_2 to S_1 and head loss is 0.53 m

(*c*) Flow is from S_1 to S_2 and head loss is 1.06 m

(*d*) Flow is from S_2 to S_1 and head loss is 1.06 m

43. Match the following

P: Compressible flow	U: Reynolds number
Q: Free surface flow	V: Nusselt number
R: Boundary layer flow	W: Weber number
S: Pipe flow	X: Froude number
T: Heat convection	Y: Match number
	Z: Skin friction coefficient

(*a*) P - U; Q - X; R - V; S - Z; T - W

(*b*) P - W; Q - X; R - Z; S - U; T - V

(*c*) P - Y; Q - W; R - Z; S - U; T - X

(*d*) P - Y; Q - W; R - Z; S - U; T - V

2009

44. A coolant fluid at 30 °C flows over a heated flat plate maintained at a constant temperature of 100°C. The boundary layer temperature distribution at a given location on the plate may be approximated as

$$T = 30 + 70 \exp(-y)$$

where y (in m) is the distance normal to the plate and T is in °C.

If thermal conductivity of the fluid is 1.0 W/mK, then local convective heat transfer coefficient (in W/m^2K) at that location will be

(*a*) 0.2 (*b*) 1

(*c*) 5 (*d*) 10

45. Consider steady, incompressible and irrotational flow through a reducer in a horizontal pipe where the diameter is reduced from 20 cm to 10 cm. The pressure in the 20 cm pipe just upstream of the reducer is 150kPa. The fluid has a vapour pressure of 50 kPa and a specific weight of 5 kN/m^3. Neglecting frictional effects, the maximum discharge (in m^3/s) that can pass through the reducer without causing cavitation is

(*a*) 0.05 (*b*) 0.16

(*c*) 0.27 (*d*) 0.38

46. You are asked to evaluate assorted fluid flows for their suitability in a given laboratory application. The following three flow choices, expressed in terms of the two-dimensional velocity fields in the xy-plane, are made available.

P. u = 2y, v = – 3x

Q. u = 3xy, v = 0

R. u = 2x, v = 2y

Which flows (s) should be recommended when the application requires the flow to be incompressible and irrotational ?

(*a*) P and R (*b*) Q

(*c*) Q and R (*d*) R

47. Water at 25 °C is flowing through a 1.0 km long G.I. pipe of 200 mm diameter at the rate of 0.07 m^3/s. If value of Darcy friction factor for this pipe is 0.02 and density of water is 1000 kg/m^3, the pumping power (in kW) required to maintain the flow is

(*a*) 1.8 (*b*) 17.4

(*c*) 20.5 (*d*) 41.0

48. The velocity profile of a fully developed laminar flow in a straight circular pipe, as shown in the figure, is given by the expression

$$u(r) = -\frac{R^2}{4\mu}\left(\frac{dp}{dx}\right)\left(1-\frac{r^2}{R^2}\right)$$

where $\frac{dp}{dx}$ is a constant.

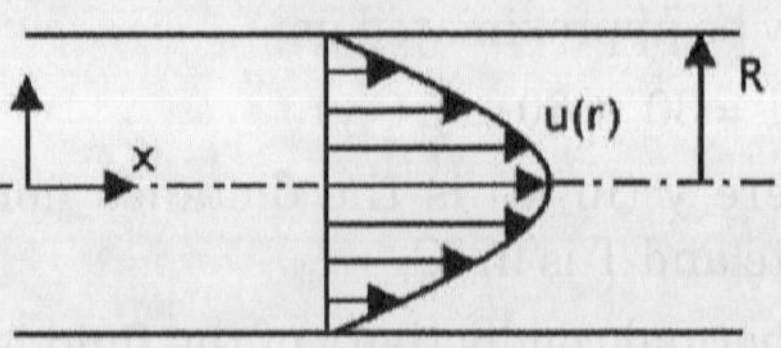

The average velocity of fluid in the pipe is

(*a*) $-\frac{R^2}{8\mu}\left(\frac{dp}{dx}\right)$ (*b*) $-\frac{R^2}{4\mu}\left(\frac{dp}{dx}\right)$

(*c*) $-\frac{R^2}{2\mu}\left(\frac{dp}{dx}\right)$ (*d*) $-\frac{R^2}{\mu}\left(\frac{dp}{dx}\right)$

2008

49. For the continuity equation given by $\vec{\nabla}\bullet\vec{V} = 0$ to be valid, where $\vec{V}$ is the velocity vector, which one of the following is a necessary condition?

(*a*) Steady flow (*b*) Irrotational flow

(*c*) Inviscid flow (*d*) Incompressible flow

50. A two dimensional fluid element rotates like a rigid body. At a point within the element, the pressure is 1 unit. Radius of the Mohr's circle, characterizing the state of stress at that point, is

(*a*) 0.5 unit (*b*) 0 unit

(*c*) 1 unit (*d*) 2 unit

51. Water, having a density of 1000 kg/m^3, issues from a nozzle with a velocity of 10 m/s and the jet strikes a bucket mounted on a Pelton wheel. The wheel rotates at 10 rad/s. The mean diameter of the wheel is 1 m. The jet is split into two equal streams by the bucket, such that each stream is deflected by 120°, as shown in the figure. Friction in the bucket may be neglected. Magnitude of the torque exerted by the water on the wheel, per unit mass flow rate of the incoming jet, is

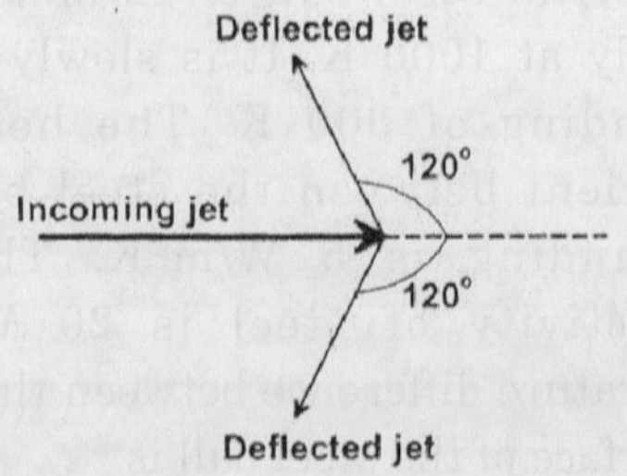

(*a*) 0 (N.m)/(kg/s) (*b*) 1.25 (N.m)/(kg/s)

(*c*) 2.5 (N.m)/(kg/s) (*d*) 3.75 (N.m)/(kg/s)

2007

52. Water has a critical specific volume of 0.003155 m^3/kg. A closed and rigid steel tank of volume 0.025 m^3 contains a mixture of water and steam at 0.1 MPa. The mass of the mixture is 10 kg. The tank is now slowly heated. The liquid level inside the tank

(*a*) will rise

(*b*) will fall

(*c*) will remain constant

(*d*) may rise or fall depending on the amount of heat transferred

53. Consider an incompressible laminar boundary layer flow over a flat plate of length L, aligned with the direction of an oncoming uniform free stream. If F is the ratio of the drag force on the front half of the plate to the drag force on the rear half, then

(*a*) F < 1/2 (*b*) F =1 2

(*c*) F = 1 (*d*) F > 1

54. In a steady flow through a nozzle, the flow velocity on the nozzle axis is given by

$$v = u_0\,(1 + \frac{3x}{L})\,i,$$

where x is distance along the axis of the nozzle from its inlet plane and L is length of the nozzle.

The time required for a fluid particle on the axis to travel from the inlet to the exit plane of the nozzle is

(a) $\frac{L}{u_0}$ (b) $\frac{L}{3u_0}\ln 4$

(c) $\frac{L}{4u_0}$ (d) $\frac{L}{2.5\,u_0}$

55. Consider steady laminar incompressible axi-symmetric fully developed viscous flow through a straight circular pipe of constant cross-sectional area at a Reynolds number of 5. The ratio of inertia force to viscous force on a fluid particle is

(a) 5 (b) $\frac{1}{5}$

(c) 0 (d) ∞

56. A model of a hydraulic turbine is tested at a head of 1/4th of that under which the full scale turbine works. The diameter of the model is half of that of the full scale turbine. If N is RPM of the full scale turbine, then RPM of the model will be

(a) N/4 (b) N/2

(c) N (d) 2N

57. Which combination of the following statements about steady incompressible forced vortex flow is correct ?

P : Shear stress is zero at all points in the flow.

Q : Vorticity is zero at all points in the flow.

R : Velocity is directly proportional to the radius from the centre of the vortex.

S : Total mechanical energy per unit mass is constant in the entire flow field.

(a) P and Q (b) R and S

(c) P and R (d) P and S

2006

58. For a Newtonian fluid

(a) shear stress is proportional to shear strain

(b) rate of shear stress is proportional to shear strain

(c) shear stress is proportional to rate of shear strain

(d) rate of shear stress is proportional to rate of shear strain

59. In a two-dimensional velocity field with velocities u and v along the x and y directions respectively, the convective acceleration along the x-direction is given by

(a) $u\frac{\partial u}{\partial x}+v\frac{\partial u}{\partial y}$

(b) $u\frac{\partial u}{\partial x}+v\frac{\partial v}{\partial y}$

(c) $u\frac{\partial v}{\partial x}+v\frac{\partial u}{\partial y}$

(d) $v\frac{\partial u}{\partial x}+u\frac{\partial u}{\partial y}$

60. A two-dimensional flow field has velocities along the x and y directions given by

$u = x^2t$ and $v = -2xyt$

respectively, where t is time.

The equation of streamlines is

(a) $x^2 y$ = constant

(b) $x y^2$ = constant

(c) $x y$ = constant

(d) not possible to determine

61. The velocity profile in fully developed laminar flow in a pipe of diameter D is given by

$u = u_0(1 - 4r^2/D^2)$,

where r is radial distance from the center.

If viscosity of the fluid is μ, the pressure drop across a length L of the pipe is

(a) $\frac{\mu u_0 L}{D^2}$ (b) $\frac{4\mu u_0 L}{D^2}$

(c) $\frac{8\mu u_0 L}{D^2}$ (d) $\frac{16\mu u_0 L}{D^2}$

62. A siphon draws water from a reservoir and discharges it out at atmospheric pressure. Assuming ideal fluid and the reservoir is large, the velocity at point P in the siphon tube is

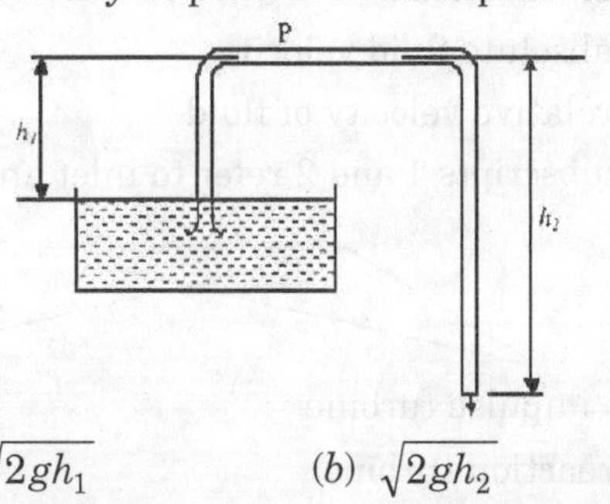

(a) $\sqrt{2gh_1}$ (b) $\sqrt{2gh_2}$

(c) $\sqrt{2g(h_2-h_1)}$ (d) $\sqrt{2g(h_2+h_1)}$

2005

63. The velocity components in the x and y directions of a two dimensional potential flow are u and v, respectively. Then $\frac{\partial u}{\partial x}$ is equal to

(a) $\frac{\partial v}{\partial x}$ (b) $-\frac{\partial v}{\partial x}$

(c) $\frac{\partial v}{\partial y}$ (d) $-\frac{\partial v}{\partial y}$

64. A venturimeter of 20 mm throat diameter is used to measure the velocity of water in a horizontal pipe of 40 mm diameter. If the pressure difference between the pipe and throat sections is found to be 30 kPa, then, neglecting frictional losses, the flow velocity is

(a) 0.2 m/s (b) 1.0 m/s

(c) 1.4 m/s (d) 2.0 m/s

65. A U-tube manometer with a small quantity of mercury is used to measure the static pressure difference between two locations A and B in a conical section through which an incompressible fluid flows. At a particular flow rate, the mercury column appears as shown in the figure. The density of mercury is 13600 kg/m³ and $g = 9.81$ m/s². Which of the following is correct ?

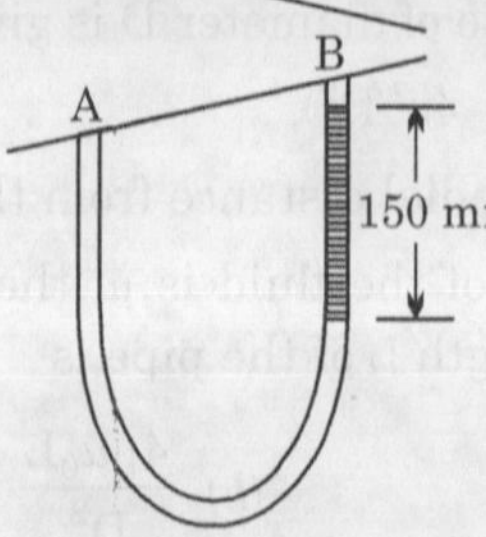

(a) Flow direction is A to B and $p_A - p_B = 20$ kPa

(b) Flow direction is B to A and $p_A - p_B = 1.4$ kPa

(c) Flow direction is A to B and $p_B - p_A = 20$ kPa

(d) Flow direction is B to A and $p_B - p_A = 1.4$ kPa

66. In the velocity diagram shown below

u = blade velocity

C = absolute fluid velocity

w = relative velocity of fluid

and subscripts 1 and 2 refer to inlet and outlet

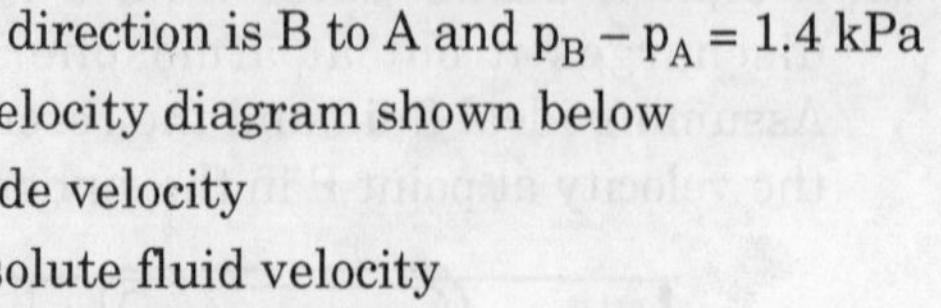

(a) an impulse turbine

(b) a reaction turbine

(c) a centrifugal compressor

(d) an axial flow compressor

67. A leaf is caught in a whirlpool. At a given instant, the leaf is at a distance of 120 m from the centre of the whirlpool. The whirlpool can be described by the following velocity distribution:

$$V_r = -\left(\frac{60 \times 10^3}{2\pi r}\right) \text{ m/s}$$

and $$V_\theta = \frac{300 \times 10^3}{2\pi r} \text{ m/s}$$

where r (in metres) is the distance from the centre of the whirlpool.

What will be the distance of the leaf from the centre when it has moved through half a revolution?

(a) 48 m (b) 64 m

(c) 120 m (d) 142 m

2004

68. An incompressible fluid (kinematic viscosity, 7.4×10^{-7} m²/s, specific gravity, 0.88) is held between two parallel plates. If the top plate is moved with a velocity of 0.5 m/s while the bottom one is held stationary, then fluid attains a linear velocity profile in the gap of 0.5 mm between these plates ; the shear stress in Pascals on the surface of top plate is

(a) 0.651×10^{-3} (b) 0.651

(c) 6.51 (d) 0.651×10^3

69. A fluid flow is represented by the velocity field $\vec{V} = ax\,\vec{i} + ay\,\vec{j}$, where a is a constant. The equation of stream line passing through a point (1, 2) is

(a) $x - 2y = 0$ (b) $2x + y = 0$

(c) $2x - y = 0$ (d) $x + 2y = 0$

70. For a fluid flow through a divergent pipe of length L having inlet and outlet radii of R_1 and R_2 respectively and a constant flow rate of Q, assuming the velocity to be axial and uniform at any cross-section, the acceleration at the exit is

(a) $\frac{2Q(R_1 - R_2)}{\pi L R_2^3}$ (b) $\frac{2Q^2(R_1 - R_2)}{\pi L R_2^3}$

(c) $\frac{2Q^2(R_1 - R_2)}{\pi^2 L R_2^5}$ (d) $\frac{2Q^2(R_2 - R_1)}{\pi^2 L R_2^5}$

71. A closed cylinder having a radius R and height H is filled with oil of density ρ. If the cylinder is rotated about its axis at an angular velocity of ω, then thrust at the bottom of the cylinder is

(*a*) $\pi R^2 \rho g H$ (*b*) $\pi R^2 \dfrac{\rho\omega^2 R^2}{4}$

(*c*) $\pi R^2 (\rho\omega^2 R^2 + \rho g H)$ (*d*) $\pi R^2 \left(\dfrac{\rho\omega^2 R^2}{4} + \rho g H\right)$

72. For air flow over a flat plate, velocity (U) and boundary layer thickness (δ) can be expressed respectively, as

$$\frac{U}{U_\alpha} = \frac{3}{2}\frac{y}{\delta} - \frac{1}{2}\left(\frac{y}{\delta}\right)^3 ; \delta = \frac{4.64x}{\sqrt{Re_x}}$$

If the free stream velocity is 2 m/s, and air has kinematic viscosity of 1.5×10^{-5} m²/s and density of 1.23 kg/m³, then wall shear stress at x = 1 m, is

(*a*) 2.36×10^2 N/m² (*b*) 43.6×10^{-3} N/m²

(*c*) 4.36×10^{-3} N/m² (*d*) 2.18×10^{-3} N/m²

73. A centrifugal pump is required to pump water to an open water tank situated 4 km away from the location of the pump through a pipe of diameter 0.2 m having Darcy's friction factor of 0.01. The average speed of water in the pipe is 2 m/s. If it is to maintain a constant head of 5 m in the tank, neglecting other minor losses, then absolute discharge pressure at the pump exit is

(*a*) 0.449 bar (*b*) 5.503 bar

(*c*) 44.911 bar (*d*) 55.203 bar

2003

74. A cylindrical body of cross-sectional area A, height H and density ρ_s, is immersed to a depth h in a liquid of density ρ, and tied to the bottom with a string. The tension in the string is

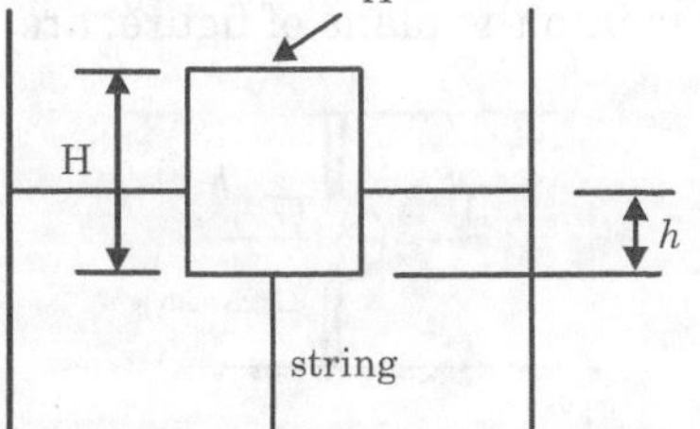

(*a*) $\rho g h A$ (*b*) $(\rho_s - \rho) g h A$

(*c*) $(\rho - \rho_s) g h A$ (*d*) $(\rho h - \rho_s H) g A$

75. A water container is kept on a weighing balance. Water from a tap is falling vertically into the container with a volume flow rate of Q; the velocity of the water when it hits the water surface is U. At a particular instant of time, the total mass of the container and water is m. The force registered by the weighing balance at this instant of time is

(*a*) $mg + \rho QU$

(*b*) $mg + 2\rho QU$

(*c*) $mg + \rho QU^2/2$

(*d*) $\rho QU^2/2$

76. Air flows through a venturi and into atmosphere. Air density is ρ; atmospheric pressure is P_a; throat diameter is D_t; exit diameter is D and exit velocity is U. The throat is connected to a cylinder containing a frictionless piston attached to a spring. The spring constant is k. The bottom surface of the piston is exposed to atmosphere. Due to the flow, the piston moves by distance x. Assuming incompressible frictionless flow, x is

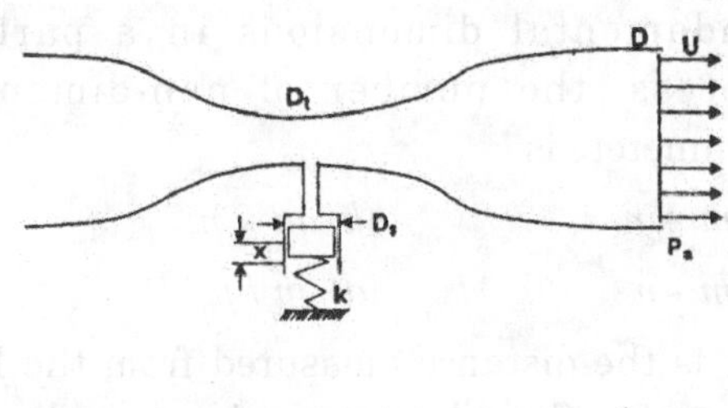

(*a*) $\left(\dfrac{\rho U^2}{2k}\right)\pi D_s^2$ (*b*) $\dfrac{\rho U^2}{8k}\left(\dfrac{D^2}{D_t^2} - 1\right)\pi D_s^2$

(*c*) $\dfrac{\rho U^2}{2k}\left(\dfrac{D^2}{D_t^2} - 1\right)\pi D_s^2$ (*d*) $\dfrac{\rho U^2}{8k}\left(\dfrac{D^4}{D_t^4} - 1\right)\pi D_s^2$

77. Consider a laminar boundary layer over a heated flat plate. The free stream velocity is U_∞. At some distance x from the leading edge, the velocity boundary layer thickness is δ_v and the thermal boundary layer thickness is δ_T. If the Prandtl number is greater than 1, then

(*a*) $\delta_v > \delta_T$ (*b*) $\delta_T > \delta_v$

(*c*) $\delta_v \approx \delta_T \sim (U_\infty x)^{-1/2}$ (*d*) $\delta_v \approx \delta_T \sim x^{-1/2}$

Common Data Q. 78-79

A syringe with a frictionless plunger contains water and has at its end a 100 mm long needle of 1 mm diameter. The internal diameter of the syringe is 10 mm. Water density is 1000 kg/m³.

The plunger is pushed in at 10 mm/s and the water comes out as a jet

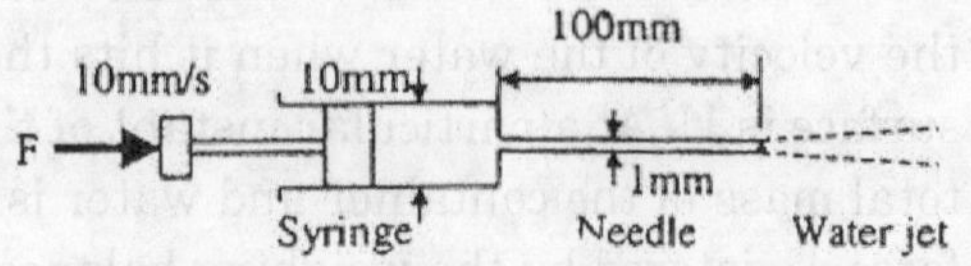

78. Assuming ideal flow, the force F in newtons required on the plunger to push out the water is

(*a*) 0 (*b*) 0.04

(*c*) 0.13 (*d*) 1.15

79. Neglect losses in the cylinder and assume fully developed laminar viscous flow throughout the needle; the Darcy friction factor is 64/Re, where Re is the Reynolds number. Given that the viscosity of water is 1.0×10^{-3} kg/s m, the force F in newtons required on the plunger is

(*a*) 0.13 (*b*) 0.16

(*c*) 0.3 (*d*) 4.4

2002

80. If there are *m* physical quantities and *n* fundamental dimensions in a particular process, the number of non-dimensional parameters is

(*a*) $m + n$ (*b*) $m \times n$

(*c*) $m - n$ (*d*) m / n

81. If x is the distance measured from the leading edge of a flat plate, then laminar boundary layer thickness varies as

(*a*) $1/x$ (*b*) $x^{4/5}$

(*c*) x^2 (*d*) $x^{1/2}$

82. Flow separation in flow past a solid object is caused by

(*a*) a reduction of pressure to vapour pressure

(*b*) a negative pressure gradient

(*c*) a positive pressure gradient

(*d*) the boundary layer thickness reducing to zero

83. A correctly designed convergent-divergent nozzle working at a designed load is

(*a*) always isentropic

(*b*) always choked

(*c*) never choked

(*d*) never isentropic

84. The value of Biot number is very small (less than 0.01) when

(*a*) convective resistance of the fluid is negligible

(*b*) conductive resistance of the fluid is negligible

(*c*) conductive resistance of the solid is negligible

(*d*) none of these

85. The properties of mercury at 300 K are:

density = 13529 kg/m^3

specific heat at constant pressure = 0.1393 kJ/kg-K

dynamic viscosity = 0.1523.10^{-2} N.s/m^2

thermal conductivity = 8.540 W/m-K.

The Prandtl number of the mercury at 300 K is

(*a*) 0.0248 (*b*) 2.48

(*c*) 24.8 (*d*) 248

86. What is the value of the view factor for two inclined flat plates having common edge of equal width, and with an angle of 20 degrees?

(*a*) 0.83 (*b*) 1.17

(*c*) 0.66 (*d*) 1.34

2001

87. The SI unit of kinematic viscosity (υ) is

(*a*) m^2/sec (*b*) *kg*/m–sec

(*c*) *m*/sec^2 (*d*) m^3/sec^2

88. A static fluid can have

(*a*) non-zero normal and shear stress

(*b*) negative normal stress and zero shear stress

(*c*) positive normal stress and zero shear stress

(*d*) zero normal stress and non-zero shear stress

89. The horizontal and vertical hydrostatic forces F_x and F_y on the semi-circular gate, having a width *w* into the plane of figure, are

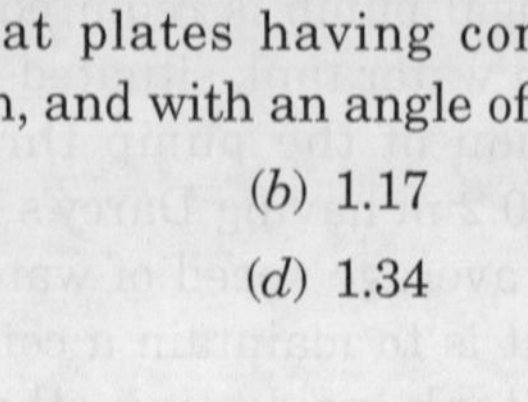

(*a*) $F_x = pghrw$ and $F_y = 0$

(*b*) $F_x = 2pghrw$ and $F_y = 0$

(*c*) $F_x = 2pghrw$ and $F_y = pgwr^2/2$

(*d*) $F_x = 2pghrw$ and $F_y\ \pi pgwr^2/2$

90. The 2-D flow with, velocity

$\vec{v} = (x + 2y + 2)\ \vec{i} + (4 - y)\ \vec{j}$ is

(*a*) comprsssible and irrotational
(*b*) compressible and not irrotational
(*c*) incompressible and irrotational
(*d*) incompressible and not irrotational

91. Water (Prandtl number ~ 6) flows over a flat plate which is heated over the entire length. Which one of the following relationship between the hydrodynamic boundary layer thickness (δ) and the thermal boundary layer thickness (δ_t) is true?

(*a*) $\delta_t > \delta$ (*b*) $\delta_t < \delta$
(*c*) $\delta_t = \delta$ (*d*) Can not be predicted

92. Water (ρ = 1000 kg/m³) flows horizontally through a nozzle into the atmosphere under the conditions given below. (Assume steady state flow).

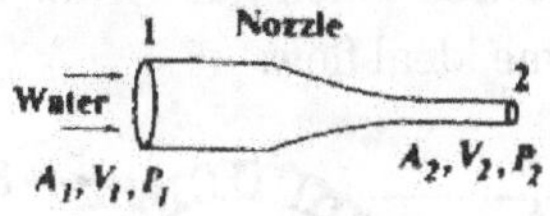

At inlet :

$A_1 = 10^{-3}\ m^2$

$V_1 = 2$ m/sec

$P_1 = 3 \times 10^5$ Pa (gauge)

At outlet :

$A_2 = 10^{-4}\ M^2$

$P_2 = P_{atm}$

Determine the external horizontal force needed to keep the nozzle in place.

(*a*) 168 N (*b*) 192 N
(*c*) 224 N (*d*) 254 N

93. Water flows through a 0.6 m diameter, 1000 m long pipe from a 30 overhead tank to a village. Find the discharge (in liters/sec) at the village (at ground level), assuming a Fanning friction factor *f* = 0.04 and ignoring minor losses due to bends etc.

(*a*) 232 (*b*) 324
(*c*) 362 (*d*) 418

2000

94. Navier Stoke's equation represents the conservation of

(*a*) energy (*b*) mass
(*c*) pressure (*d*) momentum

95. In given figure, if the pressure of gas in bulb A is 50 cm Hg vacuum and p_{atm} = 76 cm Hg, then height of column H is equal to

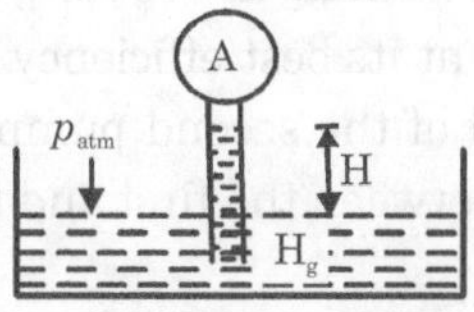

(*a*) 26 cm (*b*) 50 cm
(*c*) 76 cm (*d*) 126 cm

1999

96. If '*p*' is the guage pressure within a spherical droplet, then guage pressure within a bubble of the same fluid and of same size will

(*a*) $\frac{p}{4}$ (*b*) $\frac{p}{2}$
(*c*) *p* (*d*) 2*p*

97. If velocity of water inside a smooth tube is doubled, the n turbulent flow heat transfer coefficient between the water and the tube will

(*a*) remain unchanged
(*b*) increase to double its value
(*c*) increase but will not reach double its value
(*d*) increase to more than double its value

98. Kinematic viscosity of air at 20°C is given to be 1.6×10^{-5} m²/s. Its kinematic viscosity at 70°C will be vary approximately

(*a*) 2.2×10^{-5} m²/s (*b*) 1.6×10^{-5} m²/s
(*c*) 1.2×10^{-5} m²/s (*d*) 3.2×10^{-5} m²/s

99. Water flows through a vertical contraction from a pipe of diameter *d* to another of diameter *d*/2 (see figure). The flow velocity at the inlet to the contraction is 2m/s and pressure 200 kN/m² . If the height of the contraction measures 2m , then pressure at the exit of the contraction will be very nearly

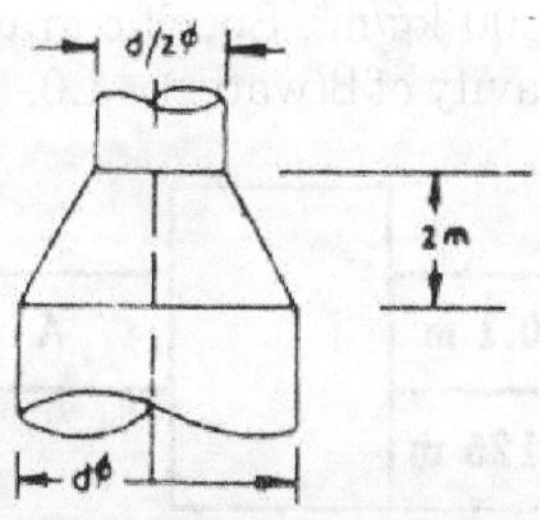

(*a*) 168 kN/m² (*b*) 192 kN/m²
(*c*) 150 kN/m² (*d*) 174 kN/m²

100. A centrifugal pump running at 750 rpm discharges water at 0.1 m^3/s against a head of 10 m at its best efficiency. A second pump of the same homologous series, when working at 500 rpm, is to deliver water at 0.05 m^3/s at its best efficiency. What will be the design head of the second pump and what is the scale ratio between the first and the second?

1998

101. The dicharge velocity at the pipe exit in figure is

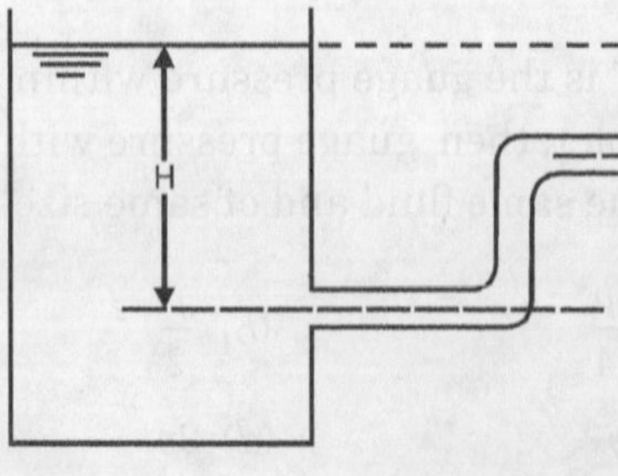

(*a*) $\sqrt{2gH}$ (*b*) $\sqrt{2gh}$

(*c*) $\sqrt{g(H+h)}$ (*d*) 0

102. List I

(*a*) High head, low flow

(*b*) Low head, high flow

(*c*) Heat transfer

(*d*) Low drag

List II

(1) Streamlined body rate

(2) Boundary layer rate

(3) Orifice meter

(4) Centrifulgal pump

(5) Axial flow pump

(6) Nusselt number

103. A cylinder of mass 10 kg and area of cross-section 0.1 m^2 is tied down with string in a vessel containing two luquids as shown in Figure. Calculate gauge pressure on the cylinder bottom and the tension in the string. Density of water = 1000 kg/m^3. Specific gravity of A = 0.8. Specific gravity of B(water) = 1.0.

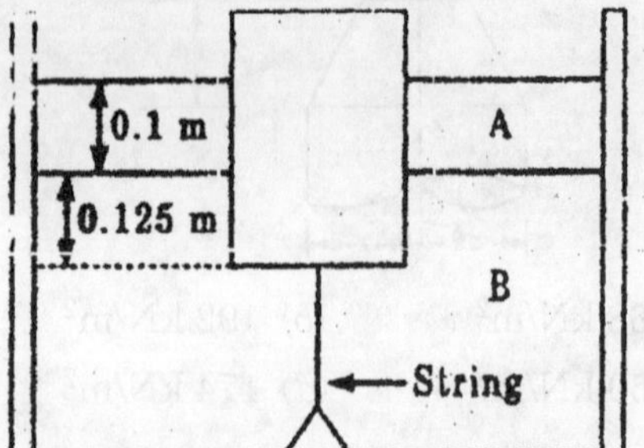

104. A farmer wishes to connect two pipes of different lengths and diameters to a common header supplied with 8×10^3 m^3/s of water from a pump. One pipe is 100 m long and 5 cm in diameter. The other pipe is 800 m long. Determine the diameter of the second pipe such that both pipes have the same flow rate. Assume the pipes to be laid on level ground and friction factor for both pipes as 0.02. Also determine the head loss in metres of water in the pipes.

Note : Pressure drop = $f\frac{1}{2}\rho v^2 \frac{L}{D}$, where f is friction factor.

105. A vertical jet of water is able to keep a hemispherical shell as shown in Figure. The water jet is issuing from a nozzle of area A_N with a velocity V_N at the nozzle exit. Using Bernoulli's equation and momentum theorem, develop expression for jet velocity at the shell and the height H at which the shell is balanced. Shell mass = m. Water density is ρ, and asume ideal flow.

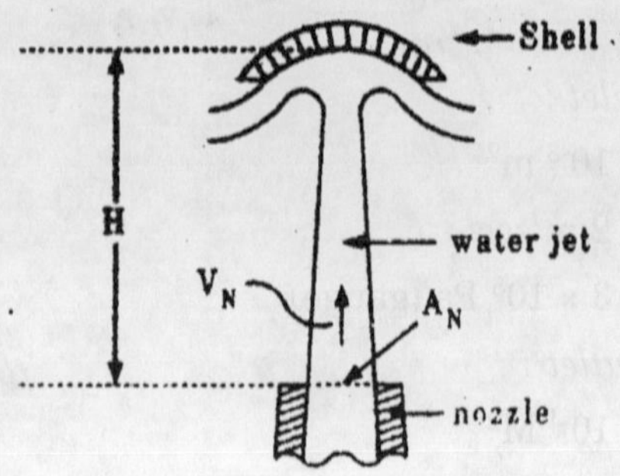

1997

106. The specific speed of a centrifugal comperssor is grenerally

(*a*) higher than that of an axial comperssor

(*b*) less than that of a reciprocating compressor

(*c*) independant of the type of compressor,but depends only on the size of the compressor

(*d*) more than the specific speed of the reciprocating compressor but less that of the axial compressor

107. The Reynolds number for flow of a certain fluid in a circular tube is specified as 2500. What will be the Reynolds number when the tube diameter is increased by 20% and the fluid velocity is decreased by 40% keeping fluid the same ?

(*a*) 1200 (*b*) 1800

(*c*) 3600 (*d*) 200

108. List I | **List II**

List I	List II
(*a*) Steam nozzle	1. Mach Number
(*b*) Compressible flow	2. Reaction Turbine
(*c*) Surface tension	3. Biot Number
(*d*) Heat conduction	4. Nusselt Number
	5. Supersaturation
	6. Weber Number

109. A water turbine delivering 10 MW power is to be tested with the help of a geometrically similar 1:8 model, which runs at the same speed as the proto-type.

(*a*) Find the power developed by the model assuming the efficiencies of the model and the proto-type are equal.

(*b*) Find the ratio of the heads and the ratio of mass flow rates between the proto-typ and the model.

110. A full journal bearing, 200 mm in diameter and 200 mm long, is to support a radial load of 45 kN at an operating speed of 960 rev/min. If it is to operate at a Sommerfeld number of 0.08 when the radial clearance is 0.2 mm, determine the viscosity of the oil to be used. Estimate also the power loss in the bearing if the resulting coefficient of friction value is 0.00315.

(Note : Sommerfeld number is bearing characteristic number with speed in rev/s)

1996

111. In flow through a pipe, the transition from laminar to turbulent flow does not depend on

(*a*) velocity of the fluid

(*b*) density of the fluid

(*c*) diameter of the pipe

(*d*) length of the pipe

112. For a given heat flow and for the same thickness, the temperature drop across the material will be maximum for

(*a*) copper (*b*) steel

(*c*) glass-wool (*d*) refractory brick

113. The dimenation of surface tension is

(*a*) ML^{-1} (*b*) L^2T^{-1}

(*c*) $ML^{-1}T^1$ (*d*) MT^{-2}

114. A mercury manometer is used to measure the static pressure at a point in a water pipe as shown in figure. The level difference of mercury in the two limbs is 10 mm. The gauge pressure at that point is

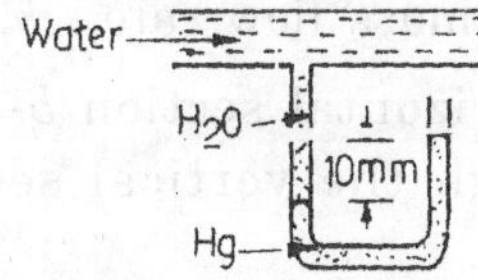

(*a*) 1236 Pa (*b*) 1333 Pa

(*c*) zero (*d*) 98 Pa

115. For laminar flow through a long pipe, the pressure drop per unit length increases

(*a*) in linear proportion to the cross-sectional area

(*b*) in proportion to the dimeter of the pipe

(*c*) in inverse proportion to the cross-sectional area

(*d*) in inverse proportion to the square of cross-sectional area

116.

List I	List II
(*a*) Fourier number	1. Surface tension
(*b*) Weber number	2. Forced convection
(*c*) Grashoff number	3. Natural convection
(*d*) Schmidt number	4. Radiation
	5. Transient heat conduction
	6. Mass diffusion

117. In a hydroelectric station, water is available at the rate of 175 m³/s under a head of 18m. The turbines run at a speed of 150 rpm with overall efficiency of 82%. Find the number of turbines required if they have the maximum specific speed of 460.

NUMERICAL TYPE QUESTIONS

2016

1. Oil (kinematic viscosity, $\upsilon_{oil} = 1.0 \times 10^{-5}$ m²/s) flows through a pipe of 0.5 m diameter with a velocity of 10 m/s. Water (kinematic viscosity, $\upsilon_w = 0.89 \times 10^{-6}$ m²/s) is flowing through a model pipe of diameter 20 mm. For satisfying the dynamic similarity, the velocity of water (in m/s) is ______.

2. A steady laminar boundary layer is formed over a flat plate as shown in the figure. The free stream velocity of the fluid is U_o. The velocity profile at the inlet *a*-*b* is uniform, while that at a downstream location c-d is given by $u = U_o\left[2\left(\frac{y}{\delta}\right) - \left(\frac{y}{\delta}\right)^2\right]$.

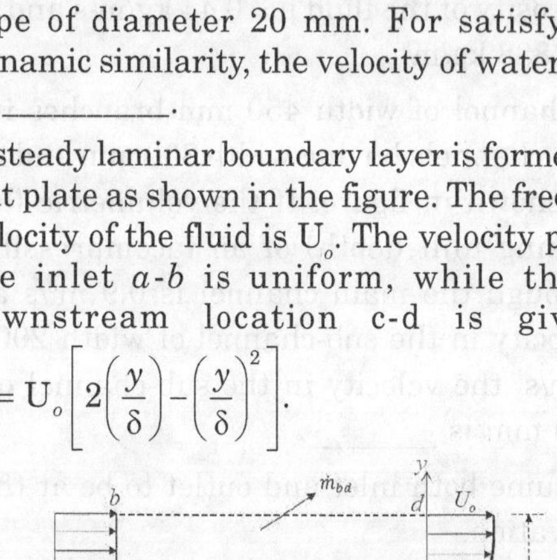

The ratio of the mass flow rate, $\dot{m}_{bd}$, leaving through the horizontal section *b-d* to that entering through the vertical section *a-b* is ________.

3. A fluid (Prandtl number, Pr = 1) at 500 K flows over a flat plate of 1.5 m length, maintained at 300 K. The velocity of the fluid is 10 m/s. Assuming kinematic viscosity, $\nu = 30 \times 10^{-6}$ m²/s, the thermal boundary layer thickness (in mm) at 0.5 m from the leading edge is ________.

4. The large vessel shown in the figure contains oil and water. A body is submerged at the interface of oil and water such that 45 percent of its volume is in oil while the rest is in water. The density of the body is ________ kg/m³.

The specific gravity of oil is 0.7 and density of water is 1000 kg/m³.

Acceleration due to gravity $g = 10$ m/s².

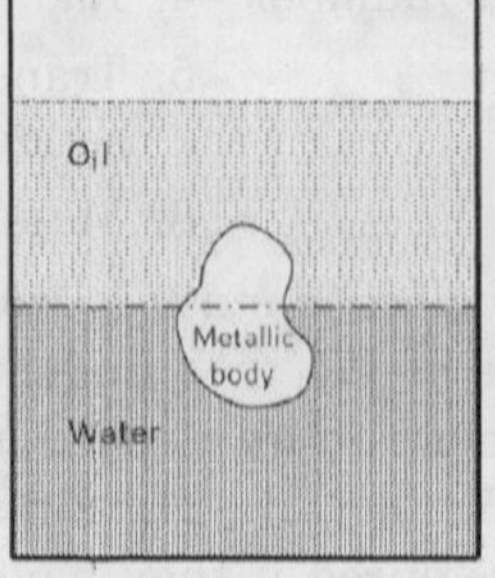

5. Consider fluid flow between two infinite horizontal plates which are parallel (the gap between them being 50 mm). The top plate is sliding parallel to the stationary bottom plate at a speed of 3 m/s. The flow between the plates is solely due to the motion of the top plate. The force per unit area (magnitude) required to maintain the bottom plate stationary is ________ N/m².

Viscosity of the fluid $\mu = 0.44$ kg/m-s and density $\rho = 888$ kg/m³.

6. A channel of width 450 mm branches into two sub-channels having width 300 mm and 200 mm as shown in figure. If the volumetric flow rate (taking unit depth) of an incompressible flow through the main channel is 0.9 m³/s and the velocity in the sub-channel of width 200 mm is 3 m/s, the velocity in the sub-channel of width 300 mm is ________.

Assume both inlet and outlet to be at the same elevation.

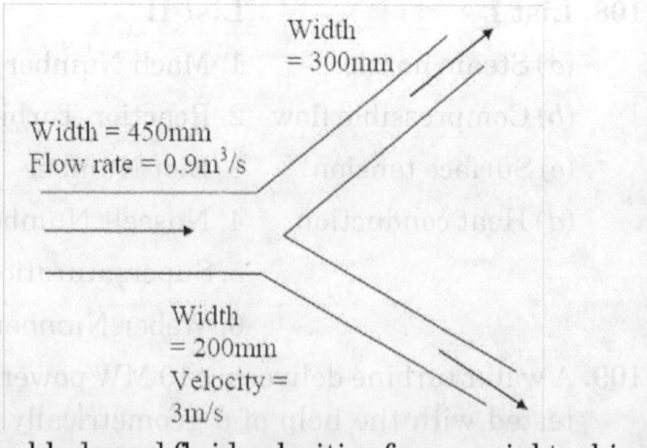

7. The blade and fluid velocities for an axial turbine are as shown in the figure.

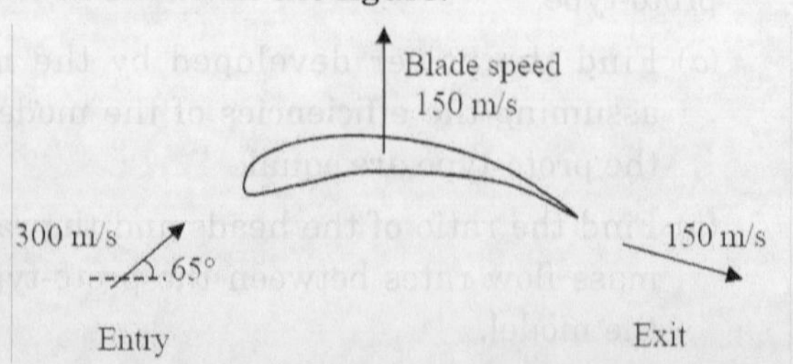

8. Consider a fully developed steady laminar flow of an incompressible fluid with viscosity μ through, a circular pipe of radius R. Given that the velocity at a radial location of R/2 from the centerline of the pipe is U_1, the shear stress at the wall is $K\mu U_1/R_1$ where K is ________.

9. The water jet exiting from a stationary tank through a circular opening of diameter 300 mm impinges on a rigid wall as shown in the figure. Neglect all minor losses and assume the water level in the tank to remain constant. The net horizontal force experienced by the wall is ________ kN.

Density of water is 1000 kg/m³.

Acceleration due to gravity g = 10 m/s².

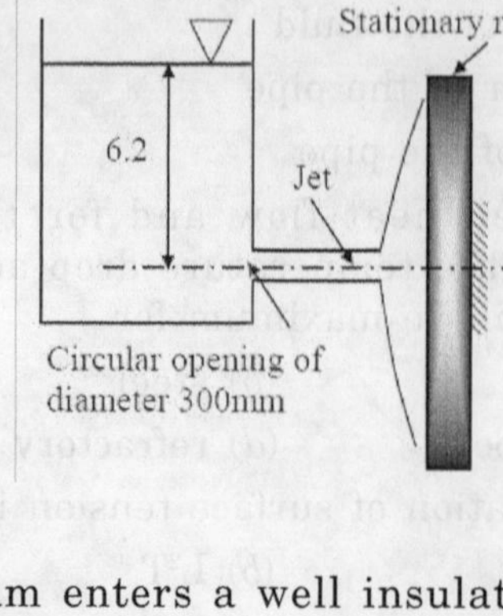

2015

10. Steam enters a well insulated turbine and expands isentropically throughout. At an intermediate pressure, 20 percent of the mass is extracted for process heating and the remaining steam expands isentropically to 9 kPa.

Inlet to turbine: P = 14 MPa, T = 560°C, h = 3486 kJ/kg, s = 6.6 kJ/(kg.K)

Intermediate stage: h= 27776 kJ/kg

Exit of turbine: P= 9 kPa, h_f = 174 kJ/kg, h_g = 2574 kJ/kg,

s_f = 0.6 kJ/(kg.K), s_g = 8.1 kJ/(kg.K)

If the flow rate of steam entering the turbine is 100 kg/s, then the work output (in MW) is _______.

11. Water (ρ = 1000 kg/m³) flows through a venturimeter with inlet diameter 80 mm and throat diameter 40 mm. The inlet and throat gauge pressures are measured to be 400 kPa and 130 kPa respectively. Assuming the venturimeter to be horizontal and neglecting friction, the inlet velocity (in m/s) is _______.

12. Air (ρ = 1.2 kg/m³ and kinematic viscosity, $v = 2 \times 10^{-5}\ m^2/s$) with a velocity of 2 m/s flows over the top surface of a flat plate of length 2.5 m. If the average value of friction coefficient is $C_f = \dfrac{1.328}{\sqrt{Re_x}}$, the total drag force (in N) per unit width of the plate is _______.

13. The velocity field of an incompressible flow is given by

$V = (a_1x + a_2y + a_3z)i + (b_1x + b_2y + b_3z)j + (c_1x + c_2y + c_3z)k$, where $a_1 = 2$ and $c_3 = -4$. The value of b_2 is _______.

14. For a fully developed laminar flow of water (dynamic viscosity 0.001 Pa-s) through a pipe of radius 5 cm, the axial pressure gradient is –10Pa/m. The magnitude of axial velocity (in m/s) at the radial location of 0.2 cm is _______.

2014

15. An ideal water jet with volume flow rate of 0.05 m³/s strikes a flat plate placed normal to its path and exerts a force of 1000 N. Considering the density of water as 1000 kg/m³, the diameter (in mm) of the water jet is ____

16. Water flows through a pipe having an inner radius of 10 mm at the rate of 36 kg/hr at 25°C. The viscosity of water at 25°C is 0.001 kg/m.s. The Reynolds number of the flow is _______

17. For a fully developed flow of water in a pipe having diameter 10 cm, velocity 0.1 m/s and kinematic viscosity 10^{-5} m²/s, the value of Darcy friction factor is ____

18. In a simple concentric shaft-bearing arrangement, the lubricant flows in the 2 mm gap between the shaft and the bearing. The flow may be assumed to be a plane Couette flow with zero pressure gradient. The diameter of the shaft is 100 mm and its tangential speed is 10 m/s. The dynamic viscosity of the lubricant is 0.1 kg/m.s. The frictional resisting force (in newton) per 100 mm length of the bearing is _______.

19. An aluminium alloy (density 2600 kg/m³) casting is to be produced. A cylindrical hole of 100 mm diameter and 100 mm length is made in the casting using sand core (density 1600 kg/m³). The net buoyancy force (in newton) acting on the core is _______

20. The difference in pressure (in N/m²) across an air bubble of diameter 0.001 m immersed in water (surface tension = 0.072 N/m) is _______

21. Water flows through a 10 mm diameter and 250 m long smooth pipe at an average velocity of 0.1 m/s. The density and the viscosity of water are 997 kg/m³ and 855×10^{-6} N.s/m², respectively. Assuming fully-developed flow, the pressure drop (in Pa) in the pipe is _______

22. Consider laminar flow of water over a flat plate of length 1 m. If the boundary layer thickness at a distance of 0.25 m from the leading edge of the plate is 8 mm, the boundary layer thickness (in mm), at a distance of 0.75 m, is _______

23. A fluid of dynamic viscosity 2×10^{-5} kg/m.s and density 1 kg/m³ flows with an average velocity of 1 m/s through a long duct of rectangular (25 mm × 15 mm) cross-section. Assuming laminar flow, the pressure drop (in Pa) in the fully developed region per meter length of the duct is _______

24. At the inlet of an axial impulse turbine rotor, the blade linear speed is 25 m/s, the magnitude of absolute velocity is 100 m/s and the angle between them is 25°. The relative velocity and the axial component of velocity remain the same between the inlet and outlet of the blades. The blade inlet and outlet velocity triangles are shown in the figure. Assuming no losses, the specific work (in J/kg) is _______

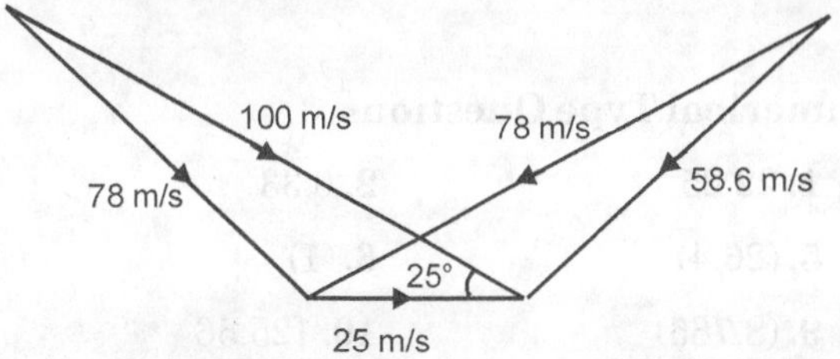

25. Water flows through a tube of diameter 25 mm at an average velocity of 1.0m/s. The properties of water are ρ =1000kg/m³ , μ = 7.25×10^{-4} N.s /m², k = 0.625 W/mK, Pr = 4.85.

Using Nu = $0.023Re^{0.8}\ Pr^{0.4}$, the convective heat transfer coefficient (in W/m².K) is _______

Using $Nu = 0.023 Re^{0.8} Pr^{0.4}$, the convective heat transfer coefficient (in $W/m^2.K$) is ______

26. Two identical metal blocks L and M (specific heat = 0.4 kJ/kg.K), each having a mass of 5 kg, are initially at 313 K. A reversible refrigerator extracts heat from block L and rejects heat to the block M until the temperature of block L reaches 293 K. The final temperature (in K) of block M is ______

27. Steam with specific enthalpy (h) 3214 kJ/kg enters an adiabatic turbine operating at steady state with a flow rate 10 kg/s. As it expands, at a point where h is 2920 kJ/kg, 1.5 kg/s is extracted for heating purposes. The remaining 8.5 kg/s further expands to the turbine exit, where h = 2374 kJ/kg. Neglecting changes in kinetic and potential energies, the net power output (in kW) of the turbine is ______

28. In a compression ignition engine, the inlet air pressure is 1 bar and the pressure at the end of isentropic compression is 32.42 bar. The expansion ratio is 8. Assuming ratio of specific heats (γ) as 1.4, the air standard efficiency (in percent) is ______

ANSWERS

MCQ Type Questions

1. (c)	**2.** (d)	**3.** (d)	**4.** (a)	**5.** (a)	**6.** (b)	**7.** (b)	**8.** (d)	**9.** (d)	**10.** (d)
11. (c)	**12.** (d)	**13.** (c)	**14.** (c)	**15.** (b)	**16.** (d)	**17.** (c)	**18.** (b)	**19.** (a)	**20.** (c)
21. (a)	**22.** (d)	**23.** (a)	**24.** (b)	**25.** (c)	**26.** (b)	**27.** (d)	**28.** (a)	**29.** (a)	**30.** (c)
31. (b)	**32.** (d)	**33.** (a)	**34.** (a)	**35.** (c)	**36.** (a)	**37.** (b)	**38.** (c)	**39.** (d)	**40.** (b)
41. (c)	**42.** (c)	**43.** (d)	**44.** (b)	**45.** (b)	**46.** (d)	**47.** (b)	**48.** (a)	**49.** (d)	**50.** (b)
51. (d)	**52.** (b)	**53.** (d)	**54.** (b)	**55.** (a)	**56.** (c)	**57.** (b)	**58.** (c)	**59.** (a)	**60.** (a)
61. (d)	**62.** (c)	**63.** (d)	**64.** (d)	**65.** (a)	**66.** (b)	**67.** (b)	**68.** (b)	**69.** (c)	**70.** (c)
71. (d)	**72.** (c)	**73.** (b)	**74.** (d)	**75.** (a)	**76.** (d)	**77.** (a)	**78.** (b)	**79.** (c)	**80.** (c)
81. (d)	**82.** (c)	**83.** (b)	**84.** (b)	**85.** (a)	**86.** (c)	**87.** (a)	**88.** (c)	**89.** (d)	**90.** (d)
91. (b)	**92.** (d)	**93.** (d)	**94.** (d)	**95.** (a)	**96.** (d)	**97.** (c)	**98.** (a)	**99.** (c)	**100.** (*)
101. (b)	**102.** (a) – 4 (b) – 5 (c) – 6 (d) – 1	**103.** (*)	**104.** (*)	**105.** (*)	**106.** (d)	**107.** (b)	**108.** (a) – 5 (b) – 1 (c) – 6 (d) – 3		
109. (*)	**110.** (*)	**111.** (c)	**112.** (c)	**113.** (d)	**114.** (b)	**115.** (c)	**116.** (a) – 5 (b) – 1 (c) – 3 (d) – 6	**117.** (*)	

Numerical Type Questions

1. 22.25	**2.** 0.33	**3.** 6.12	**4.** (865)
5. (26.4)	**6.** (1)	**7.** (52.807)	**8.** (2.667)
9. (8.765)	**10.** 125.56	**11.** (6)	**12.** 0.0159
13. (2)	**14.** (6.24)	**15.** (56 to 57)	**16.** (635 to 638)
17. (0.06 to 0.07)	**18.** (15 to 16)	**19.** (7 to 8)	**20.** (287 to 289)
21. (6800 to 6900)	**22.** (13.5 to 14.2)	**23.** (1.7 to 2.0)	**24.** (3250 to 3300)
25. (4600 to 4625)	**26.** (333 to 335)	**27.** (7580 to 7582)	**28.** (59 to 61)

"*" Refers to Subjective Question.

EXPLANATIONS

MCQ TYPE QUESTIONS

1. Time average of fluctuating velocity is zero.

2. For floating body Buoyancy force acts through the center of buoyancy which is C.G for displaced volume.

5.

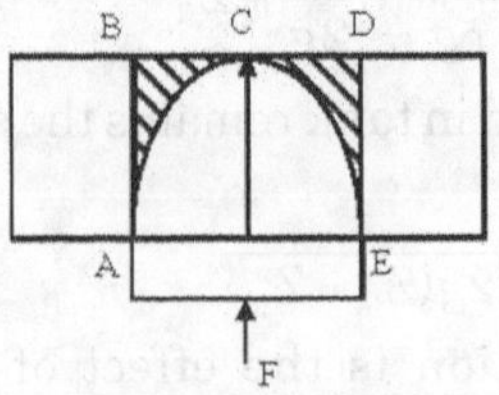

Force required to balance hydrostatic force

$$F_V = [(\text{Volume})_{\text{Rectangle}} - (\text{Volume})_{\text{Semicircle}}] \times \rho g$$

$$F_V = \left[2R \times R \times L - \frac{\pi R^2}{2}.L\right] \times \rho g$$

$$= 2\rho R^2 gL\left[1 - \frac{\pi}{4}\right]$$

6.

$$v = 2xy\,i - y^2 j$$

$$u = \frac{\partial x}{\partial y}, \; v = -\frac{\partial x}{\partial x}$$

$$2xy = \frac{\partial x}{\partial y} = 2$$

$$2xy\partial y = \partial \psi$$

on integrating

$$\psi = xy^2 + f(x)$$

Now

$$\frac{\partial x}{\partial x} = y^2 + f'(x)$$

$$-v = y^2 + f'(x)$$

$$-(-y^2) = y^1 + f'(x)$$

$$f'(x) = 0$$

$\Rightarrow$ $f(x) = \text{constant}$

so $\psi = xy^2 + \text{constant}$

7.

$$a_x = u\frac{\partial u}{\partial x} + v\frac{\partial u}{\partial y}$$

$$= \frac{x}{x^2 + y^2}\left(\frac{x^2 + y^2 - x \times 2x}{(x^2 + y^2)^2}\right) - \frac{y}{(x^2 + y^2)} \times x \times \frac{1}{(x^2 + y^2)} \times 2y$$

$$= \frac{x(x^2 + y^2 - 2x^2) - 2xy^2}{(x^2 + y^2)(x^2 + y^2)} = \frac{-x^3 - xy^2}{(x^2 + y^2)}$$

$$\therefore \; a_x = \frac{x}{(x^2 + y^2)^2}$$

$$a_y = u\frac{\partial v}{\partial x} + v\frac{\partial v}{\partial y}$$

$$= \frac{x}{(x^2 + y^2)} \times \frac{-y}{(x^2 + y^2)^2} \times 2x \times \frac{y}{(x^2 + y^2)} \times \left(\frac{(x^2 + y^2) - y \times 2y}{(x^2 + y^2)^2}\right)$$

$$= \frac{-2x^2 y + yx^2 - y^3}{(x^2 + y^2)^3}$$

$$= \frac{-y}{(x^2 + y^2)^2}$$

The velocity components are not functions of time, so flow is steady according to continuity equation,

$$\frac{\partial u}{\partial x} + \frac{\partial v}{\partial y} = \frac{-(x^2 - y^2)}{(x^2 + y^2)^2} + \frac{(x^2 - y^2)}{(x^2 + y^2)^2} = 0$$

Since it satisfies the above continuity equation for 2D and incompressible flow.

$\therefore$ The flow is incompressible.

8.

$$\text{head loss} = \frac{fLv^2}{2gd} = \frac{fl\left(\frac{Q}{A}\right)^2}{2gd}$$

$$\Rightarrow \quad \text{head loss} \propto \frac{L}{d^5}$$

$$h_1^* \alpha \frac{L}{d^5}$$

$$h_2^* \alpha \frac{2.L}{\left(d/2\right)^5} = \frac{L}{d^5} \times 64$$

$$\frac{h_1}{h_2} = 1/64 \Rightarrow \frac{h_2}{h_1} = 64$$

12. Bernoulli equation does not hold because the flow is frictional.

13. When

$P_r < 1 \;\Rightarrow\; \delta_T > \delta$

$P_r > 1 \;\Rightarrow\; \delta_T < \delta$

$P_r = 1 \Rightarrow \delta_T = \delta$

14. Head loss, $h \propto \dfrac{u_{\text{avg}}}{D^2}$

$$\frac{h_2}{h_1} = \left(\frac{D_1}{D_2}\right)^2 \frac{u_{\text{avg},2}}{u_{\text{avg},1}} = 2^2 \times 2$$

$$\frac{h_2}{h_1} = 8$$

15. Total flow rate $Q = Q_1 + Q_2 + Q_3$

Head loss $h = h_{L_1} = h_{L_2} = h_{L_3}$

16. Coutte flow is steady incompressible, laminar flow between one fixed plate and other moving with constant velocity.

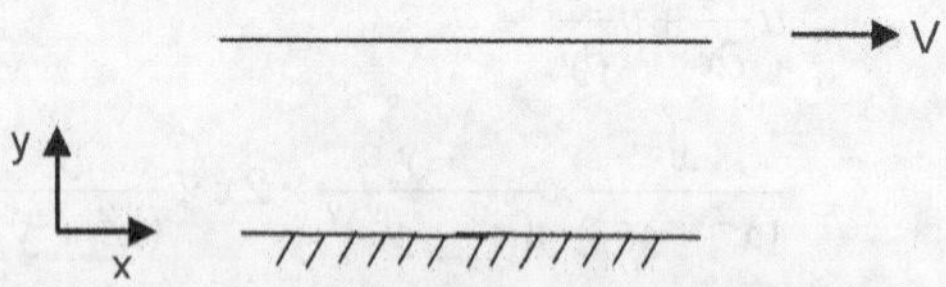

17.
$$V = \frac{0.5}{1} \text{ m}^3/\text{kg} = 0.5 \text{ m}^3/\text{kg}$$

Since $V_f < V < V_g$ the state of water is mixture of saturated water and saturated vapour.

18. For laminar flow, the heat transfer coefficient is minimum where the boundary layer thickness is maximum and vice versa. For turbulent-region boundary layer thickness is maximum at 3 but for laminar boundary layer thickness is maximum at 2 so minimum heat transfer coefficient.

19. A body in a liquid is said to be stable, when given small displacement, it returns to its original position.

Stability of completely submerged Bodies

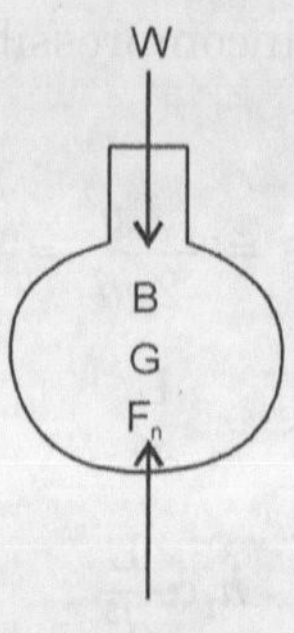

The center of gravity 'G' is below the center of Buoyancy 'B'.

20. For laminar flow,

$$Nu = 0.664\,(Re)^{0.5}\,(Pr)^{0.33}$$

$$\frac{hL}{k} = 0.664\left(\frac{\rho VD}{\mu}\right)^{0.5}(Pr)^{0.33}$$

$$h \propto V^{0.5};\; h \propto \sqrt{V}$$

So when free stream velocity increases by a factor of 2, then the average heat transfer coefficient rises by a factor of $\sqrt{2}$.

21. Incompressible flow condition

$$\nabla \cdot \vec{V} = 0$$

22. $Re_D > 2300$ means it is a turbulent flow. In turbulent flow, the fluid is well mixed. The fluid is unmixed, for a very-low Reynolds number laminar flow.

23. Steam power plants are bulky due to presence of boiler and condenser. Gas turbines are compact, as compressors and turbines are coupled on a common shaft. In steam power plants, boiler takes lot of time to get started, as compared to Gas Turbines.

24. Applying Bernoulli's equation between the points 'P' and 'R'.

$$\frac{P_{atm}}{\rho g} + \frac{V_P^2}{2g} + Z_p = \frac{P_{atm}}{\rho g} + \frac{V_R^2}{2g} + Z_R$$

Since the level of liquid in tank remains the same, $V_P = 0$

$$\therefore \quad V_R = \sqrt{2g(Z_P - Z_R)}.$$

25. Convective acceleration is the effect of time independent acceleration of fluid with respect to space that means flow is steady non-uniform flow.

26. Kaplan turbine is an axial flow turbine and works at low heads (<50).

28. $\Omega_z = \dfrac{\partial v}{\partial x} - \dfrac{\partial u}{\partial y}$

$(v = 0)$

$$\therefore \quad \Omega_z = \frac{\partial(ky)}{\partial y} = -k.$$

29.
$$\frac{q_{\text{with sheilds}}}{q_{\text{without sheilds}}} = \frac{1}{n+1}$$

here, $n = 1$

$$\therefore \quad \frac{q}{q_{\text{without}}} = \frac{1}{2}.$$

30. $k = k_o + b_T$, $k = f(T)$

From energy equation:

$$\frac{d}{dx}\left(\frac{kdT}{dx}\right) = 0$$

$$\int \frac{d}{an}\left(\frac{kdT}{dx}\right) = \int 0 \;\Rightarrow\; \frac{kdT}{dx} \cdot C \qquad ...(1)$$

$$\int K(T)\,dT = \int C.dx$$

$$\int (k_0 + b_T)\,dT = Cx + A$$

A is integration constant

$$\left(K_o T + \frac{bT^2}{2}\right) = Cx + A$$

Using B.C's

$$x = 0,\ T = 0$$

$$x = 1,\ T = 100$$

A = 0

$$k_0 \times (100) + \frac{b(10)^4}{2} = C$$

$$(100\,k_0 + 5000b) = C$$

From equation (1)

$$\boxed{\frac{dT}{dx} = \left[\frac{100k_0 + 5000b}{k_0 + b_T}\right]} \quad ...(2)$$

31. $$\frac{d^2t}{dx^2} + \frac{q}{x} = 0$$

$$\frac{dT}{dx} = \frac{-q}{k}x + c_1$$

$$1 = \frac{-q}{k}\frac{x^2}{2} + c_1x + c_2 \quad \text{(parabolic)}$$

at $x = 0, T_1 = 0$

$x = l, T_2 = 100$

$$0 = 0 + c_2 \Rightarrow \boxed{c_2 = 0}$$

$$100 = \frac{-qt^2}{2k} + c_1 l$$

$$\Rightarrow \boxed{\frac{100}{L} - \frac{qL}{2K} = c_1}$$

$$\therefore \quad T = \frac{-q}{k}\frac{x^2}{2} + \left(\frac{100}{l} - \frac{ql}{2k}\right)x$$

For maximum temp:

$$\frac{dT}{dx} - 0 \Rightarrow \frac{-q \times Lx}{2x} + \frac{100}{L}\frac{qL}{2x} = 0$$

32. $$\frac{T - T_\infty}{T_i - T_\infty} = e^{-\frac{hA}{\rho Vc}t} \text{ or } 0.4 = e$$

$$\frac{hA}{\rho Vc} = \frac{h(b)}{\rho(D)c} \because \frac{V}{A} = \frac{4/3\pi r^3}{4\pi r^2} = \frac{r}{3} = \frac{D}{6}$$

33. $$q_{T-2} = \frac{0(T_1^4 - T_2^4)}{1/\varepsilon_1 + 1/\varepsilon_2 - 1}$$

$$= \frac{5.67 \times 10^{-8}(400^4 - 300^4)}{\left(2/.8 - 1\right)} = 0.66\frac{kw}{m^2}$$

34. $$H = \frac{fLV^2}{2gD}$$

$$= \frac{0.0225 \times 500 \times \left(\frac{0.2}{\Pi \times 0.2^2/4}\right)^2}{2 \times 9.81 \times 0.4}$$

$= 116.18$ m

35. Boundary layer thickness $(\delta) \alpha \frac{1}{\sqrt{Re}}$

$$Re \,\alpha\, V \Rightarrow \delta \,\alpha\, \frac{1}{\sqrt{V}}$$

If velocity of fluid is increased by 4 times then boundary layer thickness reduces by half

36. Applying Bernoulli's equation, at exit we get,

$$\frac{P_1}{\rho_3} + gz_1 + \frac{V_1^2}{2} = \frac{P_2}{\rho_3} + gz_2 + \frac{V_2^2}{2}$$

We know that $Z_1 = Z_2$, $V_1 = 0$ & $P_2 = P_{atm}$

Hence it reduce to

$$\frac{P_1}{\rho_3} = \frac{V_2^2}{2}$$

$$V_2 = \sqrt{\frac{2P_1}{\rho_3}}$$

But $P_1 = \rho_1 gh_1 + \rho_2 gh_2 + \rho_3 gh_3$

Upon substituting we get,

$$V_2 = \sqrt{2gh_3\left[1 + \frac{\rho_1 h_1}{\rho_3 h_3} + \frac{\rho_2 h_2}{\rho_3 h_3}\right]}$$

37. Flow configuration in condenser as shown below.

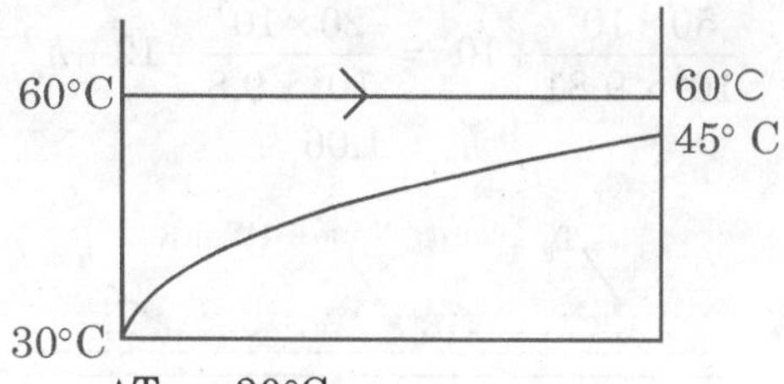

$\Delta T_1 = 30°C$

$\Delta T_2 = 15°C$

$$LMTD = \frac{\Delta T_1 - \Delta T_2}{\ln\left(\frac{\Delta T_1}{\Delta T_2}\right)} = \frac{30 - 15}{\ln\left(\frac{30}{15}\right)}$$

$= 21.6°$ C

38. Critical radius of insulation

$$= \left(\frac{k}{h}\right) = \left(\frac{0.05}{5}\right) \text{m} = 10 \text{ mm}$$

$$(r_{outer}) > r_{critical}$$

Thus, adding insulation shall decrease H.T. Rate.

39. $$Bi = \frac{hL}{K} = \frac{5 \times 0.002}{20} = 0.0005$$

For the given condition, Biot number tends to zero, i.e. conduction resistance is far less than convection resistance. Therefore temperature between centre and surface is very small.

40. For *stable* floating body metacentre should be above centre of gravity.

For *unstable* floating body, metacentre should be below centre of gravity.

For *neutral* equilibrium, both metacentre and centre of gravity should coincide.

41. For fully developed incompressible viscous flow,

$$V_{mean} = \frac{2}{3} V_{max}$$

$$= \frac{2}{3} \times 6 = 4 \text{ m/s}$$

42. At section S_1 — **At section S_2**

$P_1 = 50$ kPa — $P_2 = 20$ kPa

$z_1 = 10$ m — $z_2 = 12$ m

D = 200 mm, — $l = 1000 \text{ kg/m}^3$

V = 2 m/s — $g = 9.8 \text{ m/s}^2$

Applying Bernoulli's equation at two sections, we have

$$\frac{P_1}{\rho g} + z_1 = \frac{P_2}{\rho g} + z_2 + h_2$$

where h_2 = Loss of head from s_1 to s_2

$$\therefore \quad \frac{50 \times 10^3}{10^3 \times 9.81} + 10 = \frac{20 \times 10^3}{10^3 \times 9.8} + 12 + h_2$$

$$\Rightarrow \quad h_2 = 1.06$$

44.

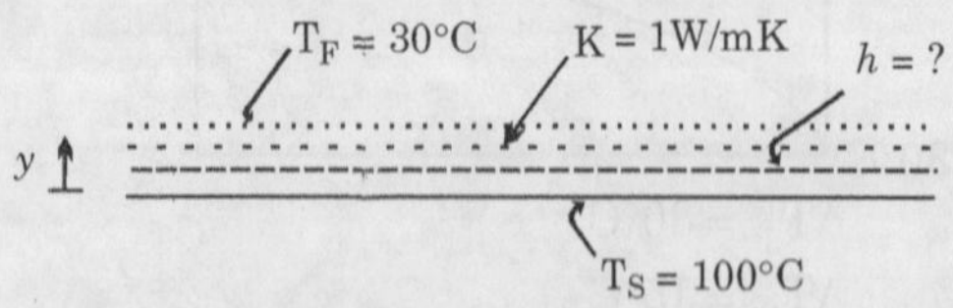

Under steady state conditions

Heat transfer by conduction

= Heat transfer by convection

$$\therefore \quad -kA\frac{dT}{dy} = h\,A\Delta T$$

$$\Rightarrow \quad -k\frac{dT}{dy} = h\,\Delta T$$

$$\Rightarrow \quad -k\frac{dT}{dy}(30 + 70e^{-y}) = h\,(100 - 30^{\circ})$$

$$\Rightarrow \quad -k\frac{\partial}{\partial y}(-70e^{-y}) = h \times 70$$

But $y = 0$

$$\therefore \quad -k(-70e^{-0}) = 70h$$

$$\Rightarrow \quad 70k = 70h$$

But $k = 1$

$$\therefore \quad h = 1 \text{W/m}^2\text{k}$$

Alternately

$$hA\Delta t = kA\frac{dT}{dy}$$

$$\therefore \quad h(100 - 30) = k \times (100 - 30)$$

$$\Rightarrow \quad h = k = 1 \text{ W/m}^2\text{K}$$

45. Considering potential head difference = 0,

i.e $z_1 = z_2$

Apply Bernoulli's theorem

$$\frac{p}{\rho g} + \frac{v^2}{2g} = C$$

$$\therefore \quad \frac{p_1}{w_1} + \frac{v_1^2}{2g} = \frac{p_2}{w_2} + \frac{v_2^2}{2g}$$

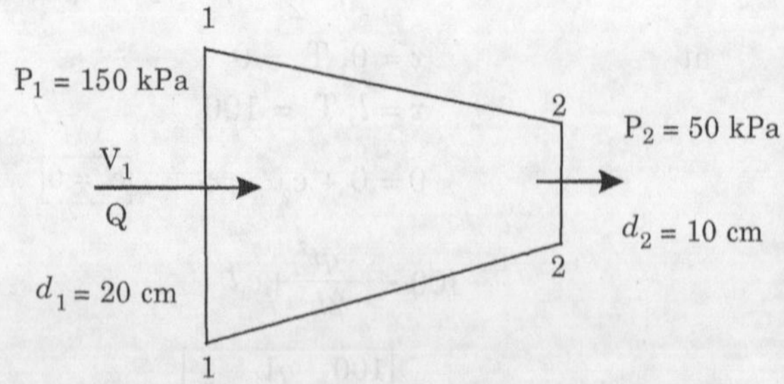

But $w_1 = w_2 = w = 5$ (incompressible flow)

$$\therefore \quad \frac{150}{5} + \frac{v_1^2}{2g} = \frac{50}{5} + \frac{v_2^2}{2g}$$

$$\Rightarrow \quad \frac{v_2^2 - v_1^2}{2g} = \frac{150 - 5}{5}$$

$$\Rightarrow \quad \frac{v_2^2 - v_1^2}{2g} = 20\,m \quad \text{...}(i)$$

Also, discharge, $Q = \left(\frac{\pi}{4}d_1^2\right)v_1$

$$= \left(\frac{\pi}{4}d_2^2\right)v_2$$

$$\Rightarrow \quad \frac{v_1}{v_2} = \left(\frac{d_2}{d_1}\right)^2 = \left(\frac{10}{20}\right)^2$$

$$\Rightarrow \quad v_1 = \frac{v_2}{4} \quad \text{...}(ii)$$

From equations (*i*) and (*ii*), we get

$$\frac{v_2^2 - \frac{v_2^2}{16}}{2g} = 20$$

$$\Rightarrow \quad \frac{15v_2^2}{32g} = 20$$

$$\Rightarrow \quad v_2 = \sqrt{\frac{20 \times 32g}{15}}$$

$$= 20.45 \text{ m/s}$$

$$\therefore \quad \text{Discharge, } Q = \left(\frac{\pi}{4}d_2^2\right)v_2$$

$$= \frac{\pi}{4}(0.1)^2 \times 20.45$$

$$= 0.16 \text{ m}^3\text{/sec}$$

46. For irrotational flow,

$$\frac{\partial^2 \psi}{\partial x^2} + \frac{\partial^2 \psi}{\partial y^2} = 0$$

For R : $\quad \frac{\partial \psi}{\partial x} = \text{v}$

$\Rightarrow \quad \frac{\partial \psi}{\partial x} = 2y$

$\Rightarrow \quad \frac{\partial^2 \psi}{\partial x^2} = \frac{\partial}{\partial x}(2\text{y}) = 0$

Now $\quad \frac{\partial \psi}{\partial y} = -\text{u}$

$\Rightarrow \quad \frac{\partial^2 \psi}{\partial y^2} = \frac{\partial}{\partial y}(\text{u})$

$\quad = -\frac{\partial}{\partial y}(-2\text{x}) = 0$

$\therefore \quad \frac{\partial^2 \psi}{\partial x^2} + \frac{\partial^2 \psi}{\partial y^2} = 0$

Hence it is irrotational flow.

For P : $\quad \frac{\partial^2 \psi}{\partial x^2} = -3$

and $\quad \frac{\partial^2 \psi}{\partial y^2} = -2$

$\therefore \quad \frac{\partial^2 \psi}{\partial y^2} + \frac{\partial^2 \psi}{\partial y^2} = -5 \neq 0$

Hence no irrotational.
Similarly Q is also not irrotational.

47. $\quad v = \dfrac{0.07}{\frac{\pi}{4} \times (0.2)^2} = 2.23\text{m/s}$

Head loss, $\text{h}_\text{f} = \dfrac{4f\,\text{L}v^2}{d \times 2g}$

$= \dfrac{4 \times 0.02 \times 1000 \times (2.23)^2}{0.2 \times 2 \times 9.81}$

$= 25.35$

$\therefore \quad$ Power $= \rho g h_f$

$= 9.81 \times 0.07 \times 25.35$

$= 17.4 \text{ kW}$

Alternately

Head loss due to friction in pipe,

$$\text{h}_\text{f} = \frac{\lambda l Q^2}{12.1 d^5}$$

where, l = length of pipe
Q = discharge
d = diameter of pipe
λ = Darcy friction coefficient

$\therefore \quad h_f = \dfrac{0.02 \times 1000 \times (0.07)^2}{12.1 \times (0.2)^5}$

$= 25.31\text{m}$

$\therefore \quad$ Power $= \rho g \text{ Q } h_f$

$= 1000 \times 9.80 \times 0.01 \times 25.37$

$= 17362.66 \text{ W}$

$= 17.36 \text{ kW}$

$\approx 17.4 \text{ kW}$

48.

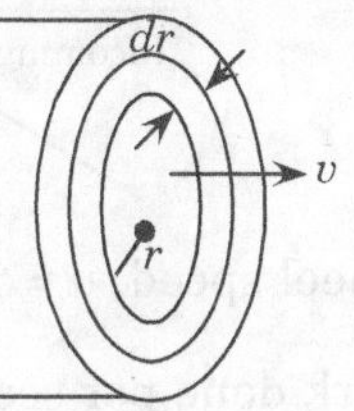

$$\text{u} = -\frac{R^2}{4\mu}\left(\frac{dp}{dx}\right)\left(1 - \frac{r^2}{\text{R}}\right)$$

$dQ = \text{u}(2\pi r dr)$

$\therefore \quad \text{Q} = 2\pi \int_o^R 4r dr$

$= 2\pi \int_o^R \frac{1}{4\mu} \text{R}^2 \left(-\frac{dp}{dx}\right)\left(1 - \frac{r^2}{\text{R}^2}\right) \text{rdr}$

Solving, we get

$$\text{Q} = \frac{\pi R^4}{8\mu}\left(-\frac{dp}{dx}\right)$$

Now $\quad$ Q = Area × average velocity

$\therefore \quad u_{\text{av}} = \dfrac{Q}{\text{Area}} = \dfrac{Q}{\pi \text{R}^2}$

or $\quad u_{\text{av}} = \dfrac{\text{R}^2}{8\mu}\left(\dfrac{dp}{dx}\right)$

49. Basic equation of continuity for fluid flow is given by

$$\frac{\partial \rho \text{u}}{\partial \text{u}} + \frac{\partial \rho \text{v}}{\partial \text{y}} + \frac{\partial \rho \text{w}}{\partial \text{z}} = \frac{\partial \rho}{\partial \text{t}}$$

Now if ρ remains constant, then only we can write

$$\vec{\nabla} \bullet \vec{\text{V}} = 0$$

i.e. $\quad \dfrac{\partial \text{u}}{\partial \text{x}} + \dfrac{\partial \text{v}}{\partial \text{y}} + \dfrac{\partial \text{w}}{\partial \text{z}} = 0$

hence incompressible flow

50. Since pressure in fluid is of hydrodynamic type

$\therefore \quad p_1 = p_2 = p_3 = 1$

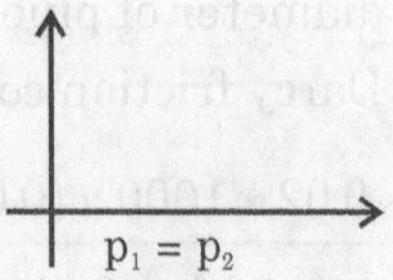

Hence radius of Mohr's circle is zero.

51. *Given* :

$$\phi = 60°$$

$$k = 1 \text{ (No friction)}$$

$$v = 10 \text{ m/sec}$$

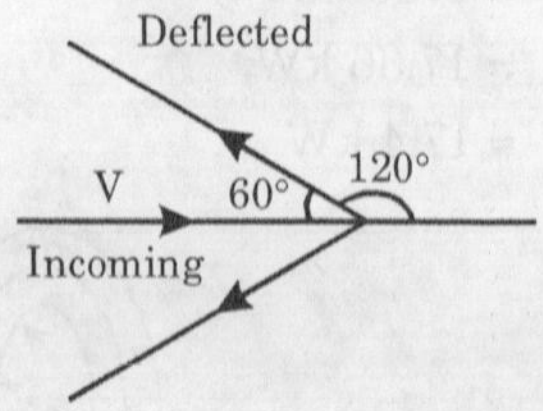

Wheel speed, $u = 10 \times \frac{1}{2} = 5$ m/sec

Work done per sec by wheel

$$= \frac{W}{g} (v - u)(1 + k \cos \phi).\ u$$

$$\text{Power} = \frac{m.g}{g} (v - u)(1 + K \cos \phi).\ u$$

Now $\quad P = T.\ \omega$

$$\therefore \quad \frac{T}{m} = \frac{(v - u)(1 + k \cos f)\ u}{w}$$

$$= \frac{(10 - 5)(1 + \cos 60°).5}{10}$$

$$= 3.75 \text{ N.m/(kg/sec)}$$

53. A, B, C can be used out easily.

54.

$$v = u_0\left(1 + \frac{3x}{L}\right)$$

$$\therefore \quad \frac{dx}{dt} = u_0\left(1 + \frac{3x}{L}\right)$$

$$\Rightarrow \quad \frac{dx}{u_0\left(1 + \frac{3x}{L}\right)} = dt$$

Integrating

$$\int_0^L \frac{dx}{u_0\left(1 + \frac{3x}{L}\right)} = \int_0^T dt$$

$$\Rightarrow \quad \frac{L}{3u_0}\left[\ln\left(1 + \frac{3x}{L}\right)\right]_0^L = T$$

$$\Rightarrow T = \frac{L}{3u_0} \ln(4)$$

55. Reynold's number $= \dfrac{\text{Inertia force}}{\text{Viscous force}} = 5$

56. From similitude,

$$\left(\frac{gH}{N^2D^2}\right)_{model} = \left(\frac{gH}{N^2D^2}\right)_{actual}$$

$$\therefore \quad \left(\frac{H \times 4}{4 \times N^2 \times D^2}\right)_{model} = \left(\frac{g.H}{N^2D^2}\right)_{actual}$$

$$\Rightarrow \quad N_{model} = N_{actual}$$

57. Clearly zero shear stress and vortex.

58. For a Newtonian fluid, shear stress is proportional to rate of shear strain.

59. Two dimensional velocity field with velocities u, v and along x and y direction.

$\therefore$ Acceleration along x direction,

$$a_x = a_{convective} + a_{temporalor\ local}$$

$$= \underbrace{u\frac{\partial u}{\partial x} + v\frac{\partial u}{\partial y} + \omega\frac{\partial u}{\partial z}}_{\text{Convective acceleration}} + \underbrace{\frac{\partial u}{\partial t}}_{\text{temporal acceleration}}$$

Since, $\dfrac{\partial u}{\partial z} = 0$ for 2-dimensional field

$\therefore$ Convective acceleration $= u\dfrac{\partial u}{\partial x} + v\dfrac{\partial u}{\partial y}$

60. *Given* : $\quad u = x^2t$

and $\quad v = -2xyt$

We know, $\quad \dfrac{\partial \psi}{\partial y} = v = -2xyt \quad$ ...(*i*)

$\quad \dfrac{\partial \psi}{\partial y} = -u = -x^2t \quad$ ...(*ii*)

Integrating equation (*i*), we get

$\quad \psi = -x^2yt + f'(y) \quad$ ...(*iii*)

Differentiating (*iii*) with respect to y, we get

$\quad \dfrac{\partial \psi}{\partial y} = -x^2t + f'(y) \quad$ ...(*iv*)

Equating the value of $\dfrac{\partial \psi}{\partial y}$ from equations (*ii*) and (*iv*), we get

$$-x^2t = -x^2t + f'(y)$$

Since $f'(y) = 0$

$\therefore \quad f(y) = C$ where, 'C' is constant of integration

$\therefore \quad \psi = -x^2yt + C$

C is a numerical constant, so it can be taken as zero.

$\therefore \quad \psi = -x^2yt$

For equation of stream lines,

$$\psi = \text{constant}$$

$$\therefore \quad -x^2yt = \text{constant}$$

For a particular instance,

$$x^2y = \text{constant}$$

61. By Hagen - Poiseuille law, for steady laminar flow in circular pipes

$$\tau = -\mu\frac{\partial u}{\partial r}$$

$$\tau = \frac{-\partial P}{\partial x}.\frac{r}{2}$$

$$\mu\frac{\partial u}{\partial r} = \frac{\partial P}{\partial x}.\frac{r}{2}$$

$$\mu u_0\left(\frac{-8r}{D^2}\right) = \frac{P}{L}.\frac{r}{2} \quad ...\left[\because u = u_0\left(1-\frac{4r^2}{D^2}\right)\right]$$

$$\therefore \quad P = \frac{-16\mu L u_0}{D^2} \quad [(-) \text{ sign is due to drop}]$$

62. By energy conservation, velocity at point

$$Q = \sqrt{2g(h_2 - h_1)}$$

As there is a continuous and uniform flow, so velocity of liquid at point Q and P is same.

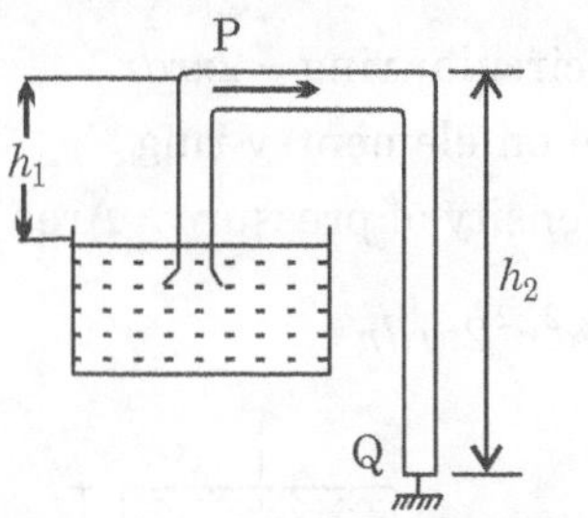

$$V_P = \sqrt{2g(h_2 - h_1)}$$

63.

$$\frac{\partial^2\phi}{\partial x^2} + \frac{\partial^2\phi}{\partial y^2} = 0 \quad ...(i)$$

(Laplace Equation should be satisfied)

$$\therefore \quad u = -\frac{\partial\phi}{\partial x}$$

and

$$v = -\frac{\partial\phi}{\partial y}$$

$$\Rightarrow \quad \frac{\partial u}{\partial x} = -\frac{\partial^2\phi}{\partial x^2}$$

and

$$\frac{\partial v}{\partial y} = -\frac{\partial^2\phi}{\partial y^2}$$

Putting values in equation (i), we get

$$\left(-\frac{\partial u}{\partial x}\right) + \left(-\frac{\partial v}{\partial y}\right) = 0$$

$$\Rightarrow \quad \frac{\partial u}{\partial x} = -\frac{\partial v}{\partial y}$$

64. We know, $A_1V_1 = A_2V_2$

$$\Rightarrow \quad V_2 = \frac{D_1^2}{D_2^2}.V_1$$

$$= \frac{16}{4}V_1$$

$$\Rightarrow \quad V_2 = 4V_1$$

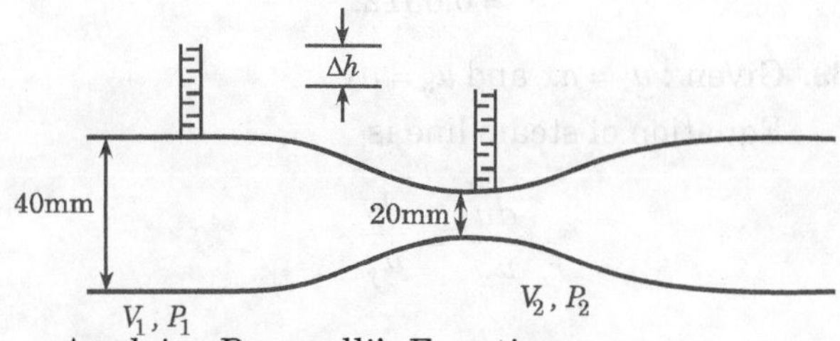

Applying Bernoulli's Equation

$$\frac{P_1}{\rho g} + \frac{V_1^2}{2g} + z_1 = \frac{P_2}{\rho g} + \frac{V_2^2}{2g} + z_2$$

$$\Rightarrow \quad \frac{P_1 - P_2}{eg} = \frac{V_2^2 - V_1^2}{2g}$$

$$\Rightarrow \quad \frac{15V_1^2}{2} = \frac{30 \times 10^3}{1000}$$

$$\Rightarrow \quad V_1^2 = 4$$

$$\Rightarrow \quad V_1 = 2.0 \text{ m/s}$$

So velocity of flow is 2.0 m/sec

65.

$$P_A - P_B = \rho g \times \Delta h$$

$$= 13600 \times 9.81 \times 0.15$$

$$= 20 \text{ kPa}$$

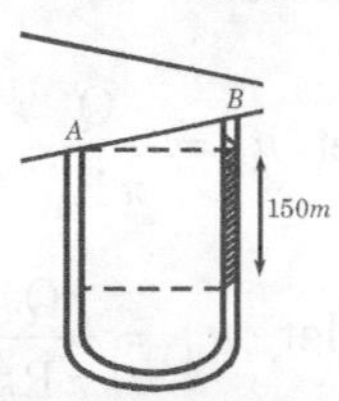

As pressure is decreasing from A to B, hence flow direction is A to B.

66. As $u_1 = u_2$ and $w_2 >> w_1$

68. *Given* : $\nu = 7.4 \times 10^{-7}$ m²/s

$$\rho = 0.88 \times 1000 = 880$$

$$v = 0.5 \text{ m/s}$$

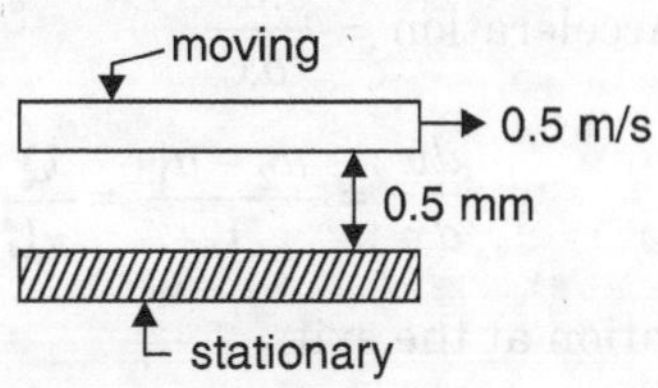

Shear stress, $\tau = \mu \times \dfrac{du}{dy}$

$$= (v \times \rho) \times \frac{du}{dy}$$

$$= 7.4 \times 10^{-7} \times 880 \times \frac{0.5}{(0.0005)}$$

$$= 0.6512.$$

69. Given : $u_x = ax$ and $u_y = ay$

Equation of steam line is

$$\frac{du}{u_x} = \frac{dy}{u_y}$$

$$\Rightarrow \quad \frac{du}{ax} = \frac{dy}{ay}$$

Integrating both sides, we have

$$\log(ax) = \log(ay) + \log c$$

$$\Rightarrow \quad ax = c \,.\, ay$$

$$\Rightarrow \quad x = cy$$

Since steam line is passed through point (1, 2)

$$\therefore \quad 1 = 2c$$

$$\Rightarrow \quad c = \frac{1}{2}$$

$$\therefore \quad x = \frac{y}{2}$$

Hence equation of steam line is

$$2x - y = 0$$

70. Velocity at inlet, $u_1 = \dfrac{Q}{\pi R_1^2}$

Velocity at outlet, $u_2 = \dfrac{Q}{\pi R_2^2}$

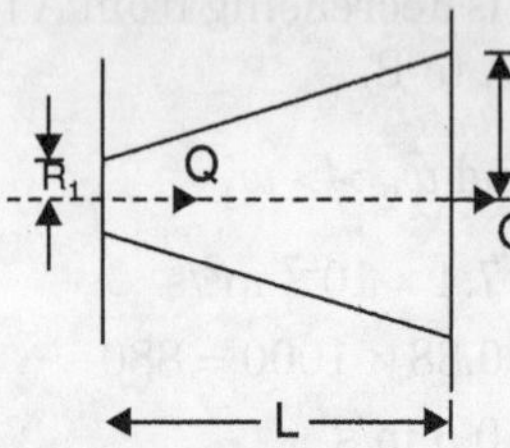

Acceleration $= \dfrac{v.dv}{dx}$

$$\therefore \quad \frac{dv}{dx} = \frac{u_2 - u_1}{L} = \frac{Q}{\pi L}\left(\frac{1}{R_2^2} - \frac{1}{R_1^2}\right)$$

Acceleration at the exit

$$= u_2 \,.\, \frac{dv}{dx}$$

$$= \frac{Q}{\pi R_2^2}\left[\frac{Q}{\pi L}\left(\frac{R_1^2 - R_2^2}{R_1^2 R_2^2}\right)\right]$$

$$= \frac{Q^2}{\pi^2 R_2^2 L}\left[\frac{(R_1 - R_2)(R_1 + R_2)}{R_1^2 R_2^2}\right]$$

Consider limiting case, *i.e.* $R_1 \to R_2$, we have

Acceleration at the exit

$$= \frac{Q^2}{\pi^2 R_L^2 L}\left[\frac{(R_1 - R_2)(2R_2)}{R_2^4}\right]$$

$$= \frac{2Q^2(R_1 - R_2)}{\pi^2 R_2^5 L}$$

71. $$\frac{\partial p}{\partial r} = \frac{\rho v^2}{r}$$

$$= \frac{\rho.\,\omega^2 r^2}{r}$$

$$= \rho\omega^2 r. \qquad [\because \quad v = \omega \times r]$$

$$\therefore \quad \int_0^p \partial p = \int_0^r \rho\omega^2 r \, dr$$

$$\Rightarrow \quad p = \frac{\rho}{2}\omega^2 r^2$$

Area of circular ring $= 2\pi r dr$

$\therefore$ Force on elementry ring

= Intensity of pressure × Area of ring

$$= \frac{\rho}{2}\omega^2 r^2 2\pi r dr$$

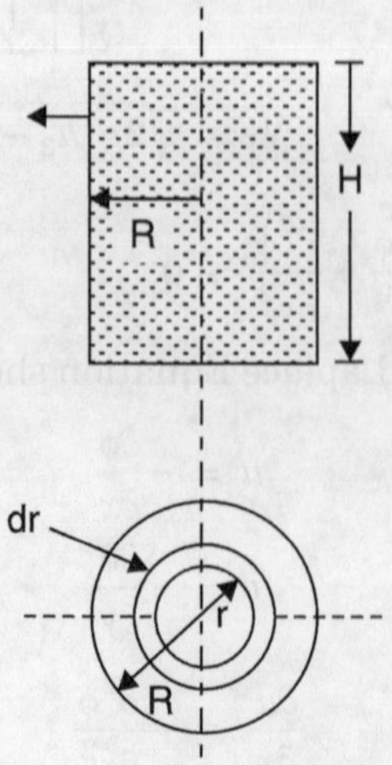

$\therefore$ Total force on the top of the cylinder

$$= \int_0^R \frac{\rho}{2}\omega^2 r^2 2\pi r \, dr$$

$$= \frac{\rho}{2}\omega^2 2\pi \int_0^R r^3 \, dr$$

$$= \frac{\rho}{2} . \omega^2 \, 2\pi \frac{R^4}{4}$$

$$= \frac{\rho}{4} \omega^2 \times \pi R^4$$

Thrust at the bottom of the cylinder

= Weight of water in cylinder

+ Total force on the top of cylinder

$$= \rho g \times \pi R^2 \times H + \frac{\rho}{4} \omega^2 \times \pi R^4$$

$$= \pi R^2 \left[\frac{\rho \omega^2 R^2}{4} + \rho g H \right]$$

72. Given : $\rho = 1.23 \text{ kg/m}^3$

$$\nu = \frac{\mu}{\rho} = 1.5 \times 10^{-5} \text{ m}^2/\text{s}$$

$u = 2$ m/s

$x = L = 1$ m

$$\therefore \quad R_{ex} = \frac{\rho u L}{\mu}$$

$$= \frac{2 \times 1}{\left(\frac{\mu}{\rho}\right)}$$

$$= \frac{2 \times 1}{1.5 \times 10.5}$$

$$= 1.34 \times 10^5$$

Now, shear stress,

$$\tau_0 = \mu \left(\frac{du}{dy}\right)_{y=0}$$

where, $$\frac{u}{U} = \frac{3}{2}\frac{y}{\delta} - \frac{y^3}{2\delta^3}$$

$$\frac{du}{dy} = U\left[\frac{3}{2\delta} - \frac{3}{2}\frac{y^2}{\delta^3}\right]$$

Hence $$\left(\frac{du}{dy}\right)_{y=0} = \frac{3U}{2\delta}$$

Given : $$\delta = \frac{4.64x}{\sqrt{Re_x}}$$

$$= \frac{4.64 \times x}{\sqrt{\frac{\rho U x}{\mu}}}$$

Puting $x = 1$, $$(\delta)_{x=1} = \frac{4.64}{\sqrt{\frac{2 \times 1}{1.5 \times 10^{-5}}}}$$

$$= 0.0127$$

$$\therefore \quad \tau_0 = \mu \cdot \frac{du}{dy}$$

$$= \frac{3}{2} \frac{\mu U}{\delta}$$

$$= \frac{3}{2} \times \frac{(1.5 \times 10^{-5} \times 1.23) \times 2}{0.0127}$$

$$= 4.355 \times 10^{-3} \times \text{N/m}^2$$

73. *Given* : $d = 0.2$ m

$L = 4000$ m

$f = 0.01$

$v = 2$ m/s

Head loss due to friction,

$$h_f = \frac{fLv^2}{2gd}$$

$$= \frac{0.01 \times 4000 \times (2)^2}{2 \times 9.81 \times 0.2} = 40.77 \text{ m}$$

Pressure corresponding to this head

$= \rho g \, (h_f + h + h_{atm})$

$= 1000 \times 9.81 \, (40.77 + 5 + 10.3)$

$= 5.50 \times 10^5 \text{ N/m}^2 = 5.50$ bar.

74. $T_1 + Mg$ = Buoyancy force

$\tau_1 + \rho_s \, H.A. \, g = \rho. \, h \, A.g$

$\therefore \quad \tau_1 = g.A \, (\rho h - \rho_s.H)$

75. Volume flow rate = Q

Mass of water strike = ρQ

Velocity of water when it hit water surface = U

Force on weighing balance due to water strike

= Initial momentum – final momentum

= $\rho QU - 0$

= ρQU

(since final velocity is perpendicular to initial velocity)

Now total force on weighing balance

= $mg + \rho QU$

76. Applying Bernoulli's equation at points (1) and (2), we have

$$\frac{P_1}{\rho g} + \frac{v_1^2}{2g} + z_1 = \frac{P_2}{\rho g} + \frac{v_2^2}{2g} + z_2$$

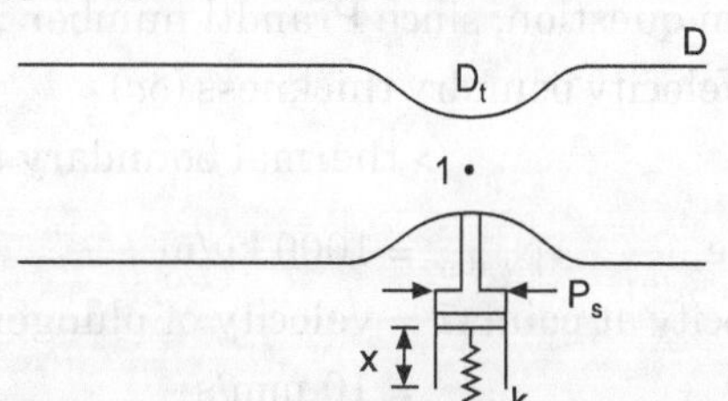

Since venturi is horizontal, therefore

$$z_1 = z_2$$

Now, $$\left(\frac{P_1}{\rho g} - \frac{P_2}{\rho g}\right) = \frac{v_2^2}{2g} - \frac{v_1^2}{2g}$$

$$\Rightarrow \quad (P_1 - P_2) = \frac{\rho g}{2g}(v_2^2 - v_1^2)$$

$$= \frac{\rho}{2}(v_2^2 - v_1^2)$$

Since $P_2 = P_a$ = atmospheric pressure

$$\therefore \quad (P_1 - P_a) = \frac{\rho}{2}(v_2^2 - v_1^2) \quad ...(i)$$

Applying continuity equation at points (1) and (2), we have

$$A_1 v_1 = A_2 v_2$$

$$\Rightarrow \quad v_1 = \left(\frac{A_2}{A_1}\right) v_2$$

Since $v_2 = U$

$$\therefore \quad v_1 = \left(\frac{\frac{\pi}{4}D^2}{\frac{\pi}{4}D_t^2}\right) U$$

$$\Rightarrow \quad v_1 = \left(\frac{D}{D_t}\right)^2 U$$

From equation (*i*),

$$P_1 - P_a = \frac{\rho}{2}\left[v^2 - \left(\frac{D}{D_t}\right)^2 U^2\right]$$

$$= \frac{\rho}{2} U^2 \left[1 - \frac{D^4}{D_t^4}\right]$$

At point P

Spring force = pressure force due air

$$\therefore \quad -kx = \frac{\pi}{4} D_s^2 \times \frac{\rho U^2}{2}\left[1 - \frac{D^4}{D_t^4}\right]$$

$$\Rightarrow \quad x = -\frac{\pi}{8}\frac{D_s^2 \rho U^2}{k}\left[1 - \frac{D^4}{D_t^4}\right]$$

77. Prandtl number $= \dfrac{\text{Molecular diffusivity of mom}}{\text{Molecular diffusivity of heat}}$

From question, since Prandtl number >1

∴ Velocity boundry thickness (δv)

> thermal boundary thickness

78. Here, $\rho_{water} = 1000 \text{ kg/m}^3$

Velocity at points 1 = velocity of plunger

= 10 mm/s

= 0.01 m/s

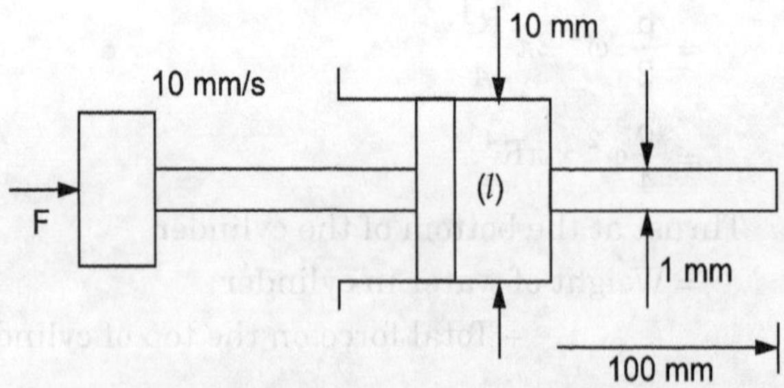

Applying Bernoulli's equation at points 1 and 2, we have

$$\frac{P_1}{\rho g} + \frac{v_1^2}{2g} + z_1 = \frac{P_2}{\rho g} + \frac{v_2^2}{2g} + z_2$$

Since $z_1 = z_2$ and $P_2 = 0$

$$\therefore \quad \frac{P_1}{\rho g} = \frac{v_2^2}{2g} - \frac{v_1^2}{2g}$$

$$\Rightarrow \quad P_1 = \frac{\rho}{2}(v_2^2 - v_1^2) \quad ...(i)$$

Applying continuity equation at points (1) and (2), we have

$$A_1 v_1 = A_2 v_2$$

$$\Rightarrow \quad v_2 = \left(\frac{A_1}{A_1}\right) v_1$$

$$= \frac{\frac{\pi}{4} \times (0.01)^2}{\frac{\pi}{4} \times (0.001)^2} v_1$$

$$= 100\, v_1$$

$$= 100 \times 0.01$$

$$= 1 \text{ m/s}$$

Now from equation (*i*),

$$P_1 = \frac{\rho}{2}[v_2^2 - v_1^2]$$

$$= \frac{1000}{2}[(1)^2 - (0.01)^2]$$

$$= 499.95 \text{ N/m}^2$$

Force required on plunger

$$= P_1 \times v_1$$

$$= 499.95 \times \frac{11}{4} \times (0.01)^2$$

$$= 0.04 \text{ N.}$$

79. Given : v = velocity of water

$= 10 \times 10^{-3}$ kg/sm

Now, $$\text{Re} = \frac{\rho v_2 d}{\upsilon}$$

$$= \frac{1000 \times 1 \times 0.001}{1 \times 10^{-3}} \quad \text{... (since } v_2 = 1)$$

$$= 1000$$

Now, Darcy's friction factor,

$$4f = \frac{64}{Re}$$

$$= \frac{64}{1000} = 0.064$$

Head loss in needle,

$$h_l = \frac{flv^2}{2gD}$$

$$= \frac{0.064 \times 0.1 \times (1)^2}{2 \times 9.8 \times 0.001}$$

$$= 0.3265$$

Applying Bernoulli's equation at points 1 and 2, we have

$$\frac{P_1}{\rho g} + \frac{v_1^2}{2g} + z_1 = \frac{P_2}{\rho g} + \frac{v_2^2}{2g} + z_2 + h_l$$

Since $z_1 = z_2$ and $P_2 = 0$

$$\therefore \quad \frac{P_1}{\rho g} = \frac{v_2^2 - v_1^2}{2g} + h_l$$

$$P_1 = \frac{\rho}{2}(v_2^2 - v_1^2) + \rho g\, h_l$$

$$= \frac{1000}{2}[(1)^2 - (0.01)^2] + 1000 \times 9.8 \times 0.3265$$

$$= 499.95 + 3199.7$$

$$= 3699.65$$

Now force required on plunger

$$= P_1 \times A_1$$

$$= 3699.65 \times \frac{\pi}{4} \times (0.01)^2 = 0.3 \text{ N}$$

85. Prandtl number

$$= \frac{\mu C_p}{K}$$

$$= 0.1523 \times 10^{-2} \frac{Ns}{m^2} \times \frac{0.1393 \times 1000}{8.54 / mK} \frac{kJ}{kgK}$$

$$= 0.0248$$

89. $F_x = 2pghrw$

$$F_y = w_2 - w_1 = \frac{\pi r^2}{2}.pg \times w = \frac{\pi r^2 pgw}{2}$$

90. Here, $\bar{v} = (x + 2y + 2)\,\hat{i} + (4 - y)\,\hat{j}$

$$u = x + 2y + 2,$$

$$v = 4 - y$$

$$\therefore \quad \frac{\partial v}{\partial x} = 1, \frac{\partial v}{\partial y} = -1$$

$$\therefore \quad \frac{\partial v}{\partial x} + 1, \frac{\partial v}{\partial y} = 0,$$

hence incompressible.

Again, $\omega = \frac{1}{2}\left(\frac{\partial v}{\partial x} - \frac{\partial v}{\partial y}\right) = \frac{1}{2}(0 - 2) = -1$

Hence not irrotational.

91. $\frac{\delta}{\delta t} = (P_r)^n$

Here, $P_r = 6 > 1$

$$\therefore \quad \frac{\delta}{\delta t} > 1$$

$$\Rightarrow \quad \delta > \delta t$$

92. *Given :* $P = 1000 \text{ kg/m}^3$

$A_1 = 10^{-5} \text{ m}^2$

$A_2 = 10^{-4} \text{m}^2$

$V_1 = 2$ m/sec.

$P_2 = P_{atn}$

$P_1 = 3 \times 10^{-5}$ Pa(gange)

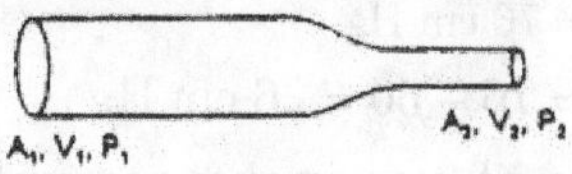

Now, $A_1 V_1 = A_2 V_2$

$$\Rightarrow \quad V_2 = \frac{10^{-3} \times 2}{10^{-4}} = 20 \text{ m/sec.}$$

Writing momentum equation,

$$p_1 A_1 - R_x - p_2 A_2 = pQ(V_2 - V_1)$$

$$\Rightarrow \quad 3 \times 10^5 \times 10^{-3} - R_x - 1.01325 \times 10^5 \times 10^{-4}$$

$$= 1000 \times 10^{-3} \times 2\,(20 - 2)$$

$$\Rightarrow \quad 300 - R_x - 10.1325 = 36$$

$\therefore$ External horizontal force,

$$F_{ext} = 253.87 \text{ N}$$

93. Apply Bernoulies equation at (1) and (2), we get

$$\frac{p_1}{lg} + \frac{v_1^2}{2g} + z_1 = \frac{p_2}{lg} + \frac{v_2^2}{2g} + z_2 + h_f$$

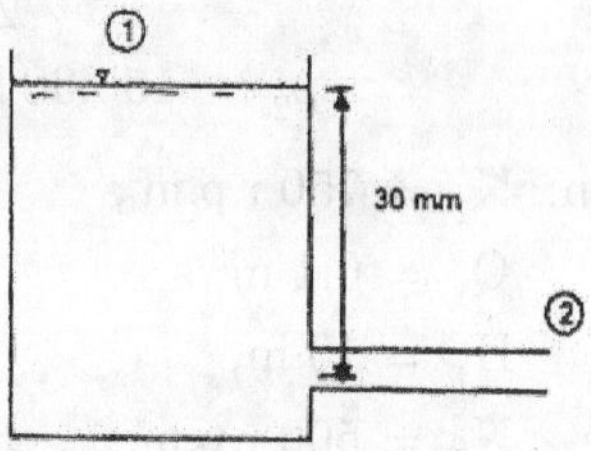

Neglecting the minor losses, we get

$$0 + 0 + 30 = 0 + \frac{v_2^2}{2g} + 0 + \frac{4flv_2^2}{d \times 2g}$$

$$\Rightarrow \quad 30 = \frac{v_2^2}{2g} + \frac{4 \times .04 \times 1000 \times v_2^2}{0.6 \times 2g}$$

$$\Rightarrow \quad 30 \times 2g = v_2^2\,[267.67]$$

$$\Rightarrow \quad v_2^2 = \frac{30 \times 2 \times 9.81}{267.67}$$

$$\Rightarrow \quad v_2 = 1/48 \text{ m/sec}$$

Now discharge in pipe,

$$Q = \text{Area} \times \text{velocity}$$

$$= \frac{\pi}{4} d^2 \times v_2$$

$$= \frac{\pi}{4} \times (0.6)^2 \times 1.48$$

$$= 0.418 \text{ m}^3\text{/sec or 418 lit/sec.}$$

95.

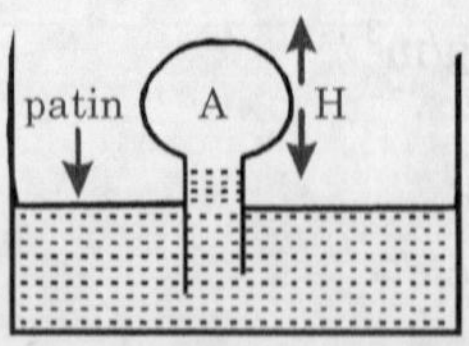

$$p_A = 50 \text{ cm Hg vacuum}$$

$$P_{atm} = 76 \text{ cm Hg}$$

$$H = 76 - 50 = 26 \text{ cm Hg}$$

98. Viscosity of gas increases with increasing temperature.

99. From continuity equation,

$$A_1 v_1 = A_2 v_2$$

or $$\frac{\pi}{4} \times d^2 \times 2 = \frac{\pi}{4}\left(\frac{d}{2}\right)^2 \times v_2$$

or $$v_2 = 8 \text{ m/s.}$$

Applying Bernoulli's theorem,

$$\frac{p_1}{w} + \frac{v_1^2}{2g} + z_1 = \frac{p_2}{w} + \frac{v_2^2}{2g} + z_2$$

or $$\frac{200 \times 1000}{9810} + \frac{(2)^2}{2 \times 9.81} + 0 = \frac{p_2}{w} + \frac{(8)^2}{2 \times 9.81} + 2$$

or $$p_2 = 150.38 \text{ kN/m}^2$$

100. Given, $N_1 = 750$ r.p.m.,

$Q_1 = 0.1$ m³/s,

$H_1 = 10$ m,

$N_2 = 500$ r.p.m.,

$Q_2 = 0.05$ m³/s.

We know

$$\left(\frac{N\sqrt{Q}}{H_1^{3/4}}\right) = \left(\frac{N\sqrt{Q}}{H_2^{3/4}}\right)_2$$

or, $$\frac{750 \times \sqrt{0.1}}{(10)^{3/4}} = \frac{500 \times \sqrt{0.05}}{H_2^{3/4}}$$

or, $$H_2^{3/4} = \frac{500 \times \sqrt{0.05}}{42.18}$$

$$\therefore \quad H_2 = 3.67 \text{ m}$$

Scale-ratio : $$L_r^{2.5} = \frac{Q_1}{Q_2} = \frac{0.1}{0.05}$$

$$\therefore \quad L_r = 1.32$$

103. *Given,* Mass of cylinder = 10 kg

= Sectional area, X = 0.1 m²

Density of water, P_w = 1000 kg/m³

Density of liquid, A = 0.8 × 1000 = 800 kg/m³

Density of liquid, B = 1 × 1000 = 1000 kg/m³

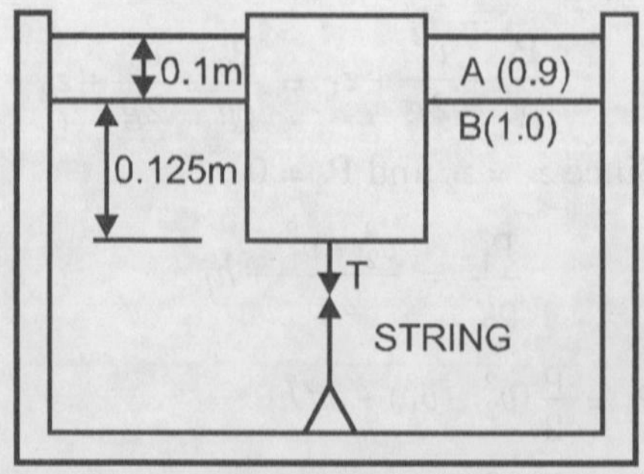

Volume of liquid A displaced = 0.1 × 0.1 = 0.01 m³

Mass of liquid A displaced = 0.01 × 800 = 8 kg

Volume of luquid B displaced = 0.125 × 0.1

= 0.0125 m³

Mass of liquid B displaced = 0.0125 × 1000

= 12.5 kg

Total mass of liquid displaced = 8 + 12.5

= 20.5 kg

Upward thrust = 20.5 × 9.81 = 201.1 N

Weight of cylinder = mg

= 10 × 9.81 = 98.1 N

Net upward thrust = 201.1 – 98.1 = 103 N

∴ Tension in string = 103 N

Pressure at the bottom of the cyclinder

= 0.1 × 800 × 9.81 + 0.125 × 1000 × 9.81

= **2011 Pa (gauge)**

104. Given, $Q = 8 \times 10^{-3}$ m³/sec

D_1 = 5 cm = 0.05 m

l_2 = 800 m

friction factor, f = 0.02

For the flow to be same, h_f should be same in both pipes.

$$h_{f_1} = \frac{fL_1V_1^2}{2gD_1} \quad ......(1)$$

$$h_{f_2} = \frac{fL_2V_2^2}{2gD_2} \quad ......(2)$$

$$Q_1 = \frac{\pi}{4}D_1^2V_1 \quad ......(3)$$

$$Q_2 = \frac{\pi}{4}D_2^2V_2 \quad ......(4)$$

Substituting equations (3) and (4) in (1) and (2) and equating friction losses, we get

$$\frac{0.02\times 100}{2gD_1}\times\left(\frac{Q_1^2\times 16}{\pi^2 D_1^4}\right) = \frac{0.02\times 800}{2gD_2}\times\left(\frac{Q_2^2\times 16}{\pi^2 D_2^4}\right)$$

But $D = 5$ cm, and $Q_1 = Q_2$.

Therefore, we get D_2 = **7.57 cm**

$$V = \frac{4\times 10^{-3}}{\frac{\pi}{4}\times\left(\frac{5}{100}\right)^2} = 2.04 \text{ m/s}$$

Total head loss of water in the pipes $= 2\times\frac{f\,1\,V_1^2}{2g\,D}$

$$= \frac{2\times 0.02\times 100\times 2.04^2}{2\times 9.81\times 5} = \mathbf{0.17\ m}$$

105. Let, V_N = Velocity of jet at nozzle

V_2 = Velocity of jet at shell

A_N = Area of nozzle

Applying Bernoulli's theorem between sections (1) and (2), we have

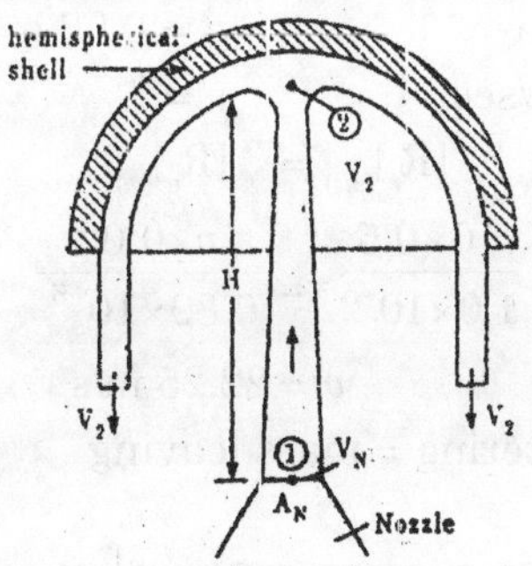

$$\frac{P_{atm}}{\gamma} + \frac{V_N^2}{2g} + 0 = \frac{P_{atm}}{\gamma} + \frac{V_2^2}{2g} + H$$

(For flow rate to be unity)

$\therefore$ Jet velocity at the shell, $V_2 = \sqrt{V_N^2 - 2gH}$

Rate of change of momentum $= \dot{m}V_2 - (-\dot{m}V_2)$
$= mg$

where $\dot{m}$ = mass flow rate of water, and
m = mass of the shell.

$$2\dot{m}V_2 = mg$$

or $\dot{m} = A_N \cdot V_N$

Now $2(A_N V_N)\left(\sqrt{V_N^2 - 2gH}\right) = mg$

or $$H = \frac{1}{2g}\left[V_N^2 - \left(\frac{mg}{2A_N V_N}\right)^2\right]$$

107. $$Re = \frac{\rho VL}{\mu}$$

$= 2500 \times 1.2 \times 0.6$

$=$ **1800**

109. *Given* : Output of turbine = 10 MW

$$\frac{L_m}{L_p} = \frac{1}{8},$$

where m, p are subscripts used for model and prototype respectively.

$$\frac{P_p}{P_m} = \frac{N_p^3 D_p^3}{N_m^3 D_m^5}$$

$$= \left(\frac{D_p}{D_m}\right)^5 \text{ as } N_p = N_m$$

i.e., $\frac{P_p}{P_m} = (8)^5$

Power developed by the model,

$$P_m = \frac{P_p}{8^5}$$

$$= \frac{10\times 10^6}{8^5} = \mathbf{305.2\ W}$$

(***b***) Specific speed of a turbine,

$$N_s = \frac{N\sqrt{P}}{H^{5/4}}$$

i.e., $$\frac{N_p\sqrt{P_p}}{H_p^{5/4}} = \frac{N_m\sqrt{P_m}}{H_m^{5/4}}$$

..... where $N_p = N_m$

$$\therefore \frac{H_p}{H_m} = \left(\frac{P_p}{P_m}\right)^{\frac{2}{5}} = 8^2 = \mathbf{64}$$

Ratio of mass flow rates

$$= \frac{m_p}{m_m} = \frac{\rho Q_p}{\rho Q_m}$$

$$= \frac{N_p D_p^3}{N_m D_m^3}$$

..... where $N_p = N_m$

$$= \left(\frac{D_p}{D_m}\right)^3 = 8^3 = \mathbf{512}$$

110. Given :Diameter of bearing = 200 mm = 0.2 m

Length of bearing = 200 mm = 0.2 m

Radial load to be supported = 45 kN

Operating speed = 960 rpm

$$= \frac{960}{60} = 16\text{rps.}$$

Sommerfeld number = 0.8

Radial clearance = 0.2 mm

Coefficient of friction, f = 0.00315

$$\text{Bearing pressure, p} = \frac{\text{Load}}{\text{Projected area}}$$

$$= \frac{45 \times 10^3}{0.2 \times 0.2}$$

$$= 11.25 \times 10^5 \text{ N/m}^2$$

$$= 1.125 \text{ MN/m}^2$$

$$\text{Sommerfeld number, S} = \frac{Zn'}{p}\left(\frac{D}{C_d}\right)^2$$

where D = Diameter of bearing

D_d = Diameter clearance

= 2 × 0.2 = 0.4 mm

n´ = rev./sec of shaft

Z = Dynamic viscosity in Ns/m²

$$\therefore \quad 0.08 = \frac{Z \times 16}{11.25 \times 10^5}\left(\frac{200}{0.4}\right)^2$$

or Dynamic viscosity,

Z = 0.0225 Ns/m²

or = 0.00225 Pa sec.

Now,Rubbing velocity,

$$V = \frac{\pi Dn}{60} = \frac{\pi \times 0.2 \times 960}{60}$$

= 10 m/sec.

Power loss in bearing= Heat generated

or, $H_g = f\,F\,V$

= 0.00314 × 45 × 10³ × 10

= **1425.02 Watt.**

112. $$q = \frac{1}{2} kA \frac{\partial T}{\partial X}$$

'k' should be least for ∂T to be maximum for a given q, A and ∂X.

114. Pressure = ρgh

$$= 13{,}600 \times 9.81 \times \frac{10}{1000}$$

= **1333 Pa**

117. *Given*,Discharge, Q = 175 m³/sec

Head, H = 18 m

Speed of turbine, N = 150 rpm

Overall efficiency, η = 82%

Maximum specific speed, Ns = 460

$$\text{Specific speed of turbine, Ns} = \frac{N\sqrt{P}}{H^{5/4}}$$

$$\therefore \quad 460 = \frac{150 \times \sqrt{p}}{(18)^{5/4}}$$

or,Power available at turbine shaft,

P = 12927.5 kW

Power developed by a turbine = $P_t \times \eta$

= 12927.5 × 0.82

= 10600 kW

$$\text{Power available} = \frac{\rho g Q H}{1000}$$

$$= \frac{10^3 \times 9.8 \times 175 \times 18}{1000}$$

= 30870 kW

Number of turbine units required,

$$= \frac{30870}{10600} \quad \textbf{2.91, i.e. 3.}$$

Thus three turbine units are to be installed.

NUMERICAL TYPE QUESTIONS

1. The Reynold's number remain same so,

oil	**water**
$v = 1.0 \times 10^{-5}$ m²/s	$v = 0.89 \times 10^{-6}$ m²/s
$d = 0.5$ m	$d = 0.02$ m
$v = 10$ m/sec	$v = ?$

$$[R_e]_{(oil)} = [R_e]_{(w)}$$

$$\frac{10 \times 0.5}{1.0 \times 10^{-5}} = \frac{v \times 0.02}{0.89 \times 10^{-6}}$$

$$\Rightarrow \quad v = 22.25 \text{ m/s}$$

2. mass entering = mass leaving

$$\dot{m}_{ab} = \dot{m}_{bd} + \int_0^\delta \rho dy$$

$$\int_0^\delta u dy = u_0 \int_0^\delta 2\left(\frac{y}{\delta}\right) - \left(\frac{y}{\delta}\right)^2$$

$$= u_0\left[\frac{y^2}{\delta} - \frac{y^3}{3\delta^2}\right]_0^\delta$$

$$= u_0\left[\delta - \frac{\delta}{3}\right]$$

$$\int_0^{\delta} u dy = \frac{2}{3} u_0 \rho\delta$$

$$\dot{m}_{ab} = \dot{m}_{bd} + \frac{2}{3} u_0 \rho\delta$$

$$\rho u_0 \delta = \frac{\dot{m}_{bd}}{\dot{m}_{ab}} + \frac{2u_0 \rho\delta}{3u_0 \rho\delta}$$

$$\frac{\dot{m}_{bd}}{\dot{m}_{ab}} = 1 - \frac{2}{3} = \frac{1}{3}$$

3. $V = 10$ m/s, $\quad x = 0.5$ m,

$$V = 30 \times 10^{-6} \text{ m}^2/\text{s}$$

$$Re_x = \frac{V \times x}{\nu} = \frac{10 \times 0.5}{30 \times 10^{-6}}$$

$$= 166666.66$$

$$= 1.667 \times 10^5 < 5 \times 10^5$$

$\therefore$ Flow is laminar

$$\frac{\delta_h}{\delta_t} = (\Pr)^{\frac{1}{3}} = 1$$

$$\therefore \quad \delta_h = \delta_t$$

$$\delta_h = \frac{5x}{\sqrt{Re_x}} = \frac{5 \times 0.5}{\sqrt{1.667 \times 10^5}}$$

$$= 6.123 \times 10^{-3} \text{ m}$$

$$= 6.12 \text{ mm}$$

4. $\rho_{oil} V_{oil} g + \rho_W V_W g = \rho_{body} V_{body} g$

$$700 \times 0.45 V_{body} + 1000 \times 0.55 V_{body} = \rho_{body} V_{body}$$

$$\rho_{body} = 865 \text{ kg/m}^3$$

5. $v = 3$ m/s

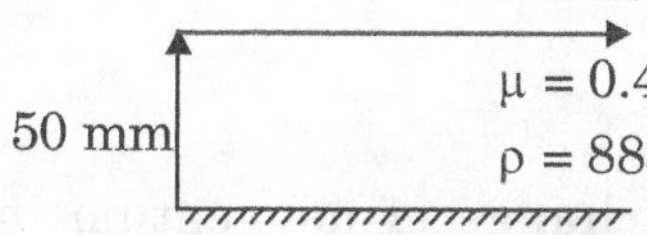

$$\mu = 0.44 \text{ kg}$$

$$\rho = 888 \text{ kg/m}^3$$

$$\frac{du}{dy} = 7\frac{8.0}{0.05} = 60$$

$$\text{Shear strees} = T = \mu \frac{du}{dy}$$

$$= 0.44 = 60 = 26.4 \text{ N/m}^2$$

6.

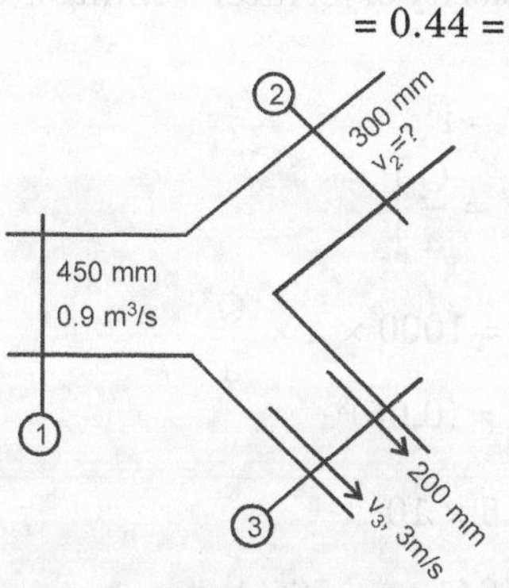

Apply conservation of mass

$$A_1V_1 = A_2V_2 + A_3V_3$$

$$0.9 = (.3)^2 \frac{1}{2} + (0.2)^2 \times 3$$

$$V_2 = 1 \text{ m/s}$$

7. Given: $\quad V_1 = 300$ m/sec

$$u = 150 \text{ m/sec}$$

$$V_{f1} = V_{f2}$$

$$(\alpha = 25°)$$

$$V_2 = 150 \text{ m/sec}$$

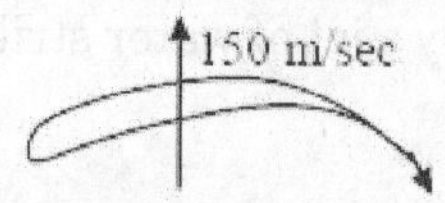

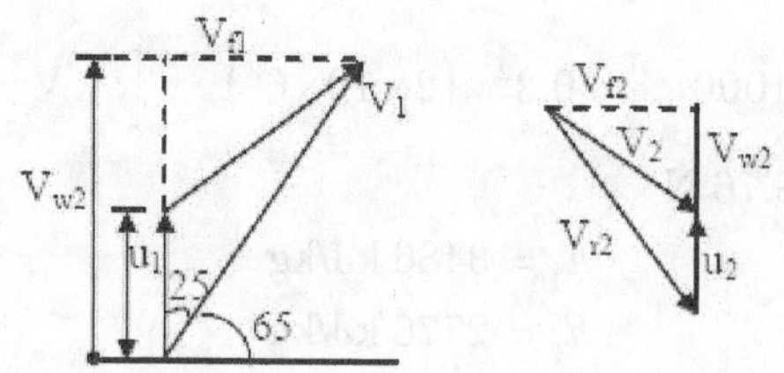

$$\text{Specific work} = [V_{w1} + V_{w2}] \,.\, u$$

$$V_{w1} = V_1 \cos 25$$

$$V_{w1} = 300 \cos 25$$

$$= 271.89 \text{ m/s}$$

$$V_{f1} = V_{f2} = V_1 \sin 25$$

$$V_{f2} = 300 \sin 25$$

$$= 126.78 \text{ m/s}$$

$$V_{w2} = \sqrt{V_2^2 - V_{f2}^2}$$

$$= \sqrt{150^2 - 126.78^2}$$

$$= \sqrt{6426.83}$$

$$V_{w2} = 80.16 \text{ m/s}$$

$$\text{Specific work} = [V_{w1} + V_{w2}] \,.\, u$$

$$= [271.89 + 80.76] \times 150$$

$$= 52807.5 \text{ J/kg}$$

$$= 52.81 \text{ kJ/kg}$$

8. $\tau_w = -\frac{\partial P}{\partial x}\left(\frac{r}{2}\right)$

$$u = -\frac{R^2}{4\mu}\left(\frac{\partial p}{dx}\right)\left(1 - \frac{r^2}{R^2}\right)$$

$$u_1 = -\frac{R^2}{4\mu}\left(\frac{\partial p}{dx}\right)\left(1 - \frac{R^2}{4 \times R^2}\right)$$

$$= -\frac{R^2}{4\mu}\left(\frac{\partial p}{dx}\right) \times \frac{3}{4}$$

$$\left(-\frac{\partial p}{dx}\right) = \frac{16\mu u_1}{3R^2}$$

$$\tau_w = -\frac{\partial p}{dx} \times \frac{R}{2} = \frac{k \times \mu \times u_1}{R}$$

$$\frac{16 \times \mu \times u_1}{3R^2} \times \frac{R}{2} = \frac{k \times \mu \times u_1}{R}$$

$$k = \frac{16\mu \times u_1}{3R^2} \times \frac{R}{2} \times \frac{R}{\mu u_1}$$

$$= \frac{8}{3} = 2.66$$

9. Force exerted by a jet of water striking fixed wall $= \rho a v^2$

$$= 1000 \times \frac{\pi}{4} \times 0.3^2 \times V^2$$

$$= 1000 \times \frac{\pi}{4} \times 0.3^2 \times (2 \times 10 \times 6^2) \qquad [\because V = \sqrt{2gh}]$$

$= 8.76$ N

10.
$$h_1 = 3486 \text{ kJ/kg}$$
$$h_2 = 2776 \text{ kJ/kg}$$
$$s_1 = s_3 = 6.6$$
$$6.6 = .6 + x(0.1 - .6)$$
$$x = 0.8$$

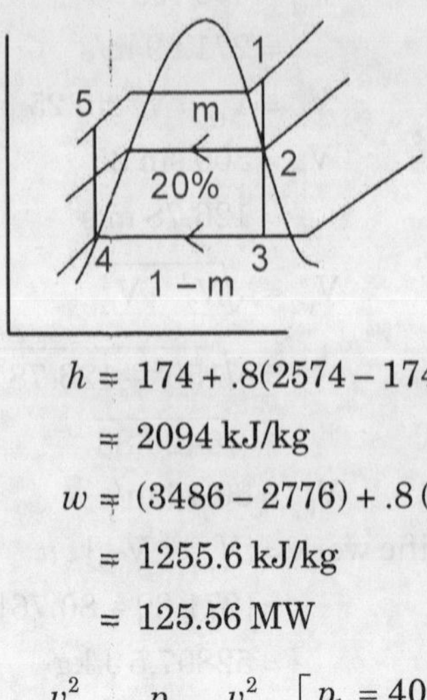

$$h = 174 + .8(2574 - 174)$$
$$= 2094 \text{ kJ/kg}$$
$$w = (3486 - 2776) + .8(2776 - 2094)$$
$$= 1255.6 \text{ kJ/kg}$$
$$= 125.56 \text{ MW}$$

11.
$$\frac{p_1}{\rho g} + \frac{v_1^2}{2g} = \frac{p_2}{\rho g} + \frac{v_2^2}{2g} \qquad \begin{bmatrix} p_1 = 400000 \\ p_2 = 130000 \end{bmatrix}$$

$$v_1 \times 80^2 = v_2 \times 40^2$$
$$v_2 = 4v_1$$

Substituting v_2 and solving for v_1 we get

$$v_1 = 6 \text{m/s}$$

12.
$$C_f = \frac{1.320}{\sqrt{Re_x}}$$

$$R_{ex} = \frac{\rho v d}{\mu} = \frac{vd.}{v}$$

$$v = 2 \text{m/s}$$
$$1 = 2.5 \text{m}$$
$$\gamma = 2 \times 10^{-5} \text{ m}^2/\text{s}$$
$$F = \frac{1}{2} C_f \rho A v^2$$
$$A = 2.5 \times 1$$

On substituting we get

$$F = 0.0159 \text{N}$$

13.
$$\frac{\partial u}{\partial x} + \frac{\partial v}{\partial y} + \frac{\partial w}{\partial z} = 0$$
$$a_1 + b_2 + c_3 = 0$$
$$2 - 4 + b_2 = 0$$
$$b_2 = 2$$

14.
$$u = \frac{-1}{4\mu} \frac{\partial P}{\partial x} (R^2 - r^2)$$

$$= \frac{-1}{4 \times 0.001} \times (-10)((0.05)^2 - (0.002)^2)$$

$$u = 6.24 \text{ m/s}$$

15.

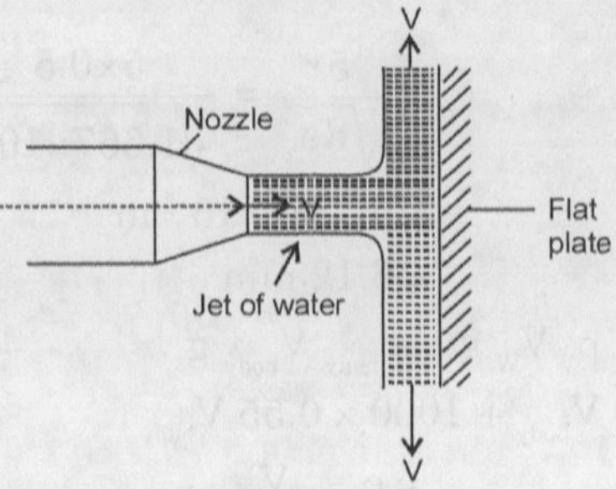

Given : $Q = 0.05$ m³/s

$F_x = 1000$ N

$\rho = 1000$ kg/m³

$D = ?$

F_x = Rate of change of momentum in the direction of force

$$= \frac{\text{Mass}}{\text{Time}} \times [\text{Initial velocity} - \text{Final velocity}]$$

$\frac{\text{Mass}}{\text{Sec}} = \rho a v$, velocity of jet after striking is equal to zero

$\therefore \quad \rho a V (V - 0) = F_x$

$$\left[\because Q = aV \Rightarrow V = \frac{Q}{a}\right]$$

$$1000 = 1000 \times a \times \frac{Q^2}{a^2}$$

$$Q^2 = a; \; a = (0.05)^2$$

$$\frac{\pi}{4} d^2 = 2.5 \times 10^{-3}$$

$$d = 0.05641 \text{ m} = 56.41 \text{ mm.}$$

16. Given Q = 36 kg /hr

$1\ m^3/hr = 1000\ kg/hr$

So converting kg /hr to m^3/s

$Q = 10^{-5}\ m^3/s$

$$R_e = \frac{\rho VD}{\mu} = \frac{\rho D}{\mu} \times \frac{Q}{A}$$

$$= \frac{\rho D}{\mu} \times \frac{Q}{\frac{\pi}{4}D^2} = \frac{4\rho Q}{\mu D\pi} = \frac{4 \times 1000 \times 10^{-5}}{0.001 \times 20 \times 10^{-3} \times \pi}$$

$R_e = 636.62.$

17. Given, D = 10 cm = 0.1 m

V = 0.1 m/s

$\upsilon = 10^{-5}\ m^2/s$

$$R_e = \frac{VD}{\upsilon} = \frac{0.1 \times 0.1}{10^{-5}}$$

$R_e = 1000$

$\therefore$ flow is laminar

Darcy friction factor $= \frac{64}{R_e}$ (for laminar flow)

$$= \frac{64}{1000} = 0.064$$

18.

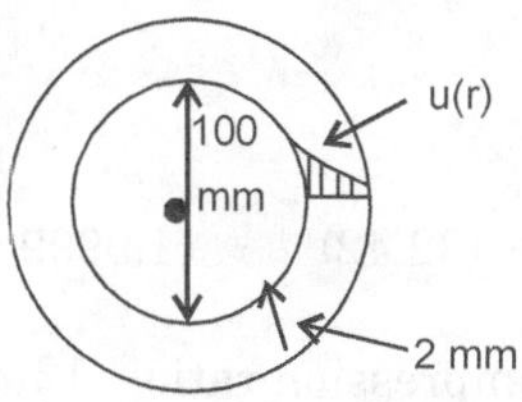

$$\tau_w = \tau_{cylinder} = \mu \left.\frac{du}{dr}\right|$$

$$F = A \times \mu\ \frac{du}{dr} = \frac{\pi D\mu[u_t - u_w]}{(\Delta t)}$$

u_t = tangential velocity

u_w = velocity at bearing

$$F = \pi \times 0.1 \times 0.1 \times \frac{0.1[10-0]}{2 \times 10^{-3}}$$

F = 15.707 N.

19. d = 1600 kg/m³

ρ = 2600 kg/m³

Net buouancy force = weight of liquid displaced – weight of solid body

$= \upsilon \times \rho \times g - \upsilon \times d \times g = \upsilon g\ (\rho - d)$

$= \frac{\pi}{4} d^2\ L \times g\ (\rho - d)$

$= \frac{\pi}{4} \times 0.1^2 \times 0.1 \times 9.81\ (2600 - 1600) = 7.7\ N.$

20.

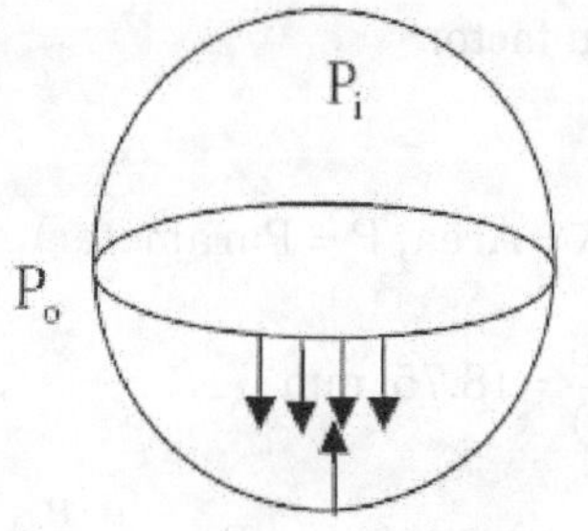

Surface tension in a bubble $= \frac{\Delta p.r}{4}$

we know, $P_i - P_o = \frac{4T}{r}$

$$\Delta p = \frac{4 \times 0.072}{0.001} = 288\ N/m^2.$$

21.

$$Re = \frac{\rho VD}{\mu}$$

$$= \frac{997 \times 0.1 \times 10 \times 10^{-3}}{855 \times 10^{-6}}$$

= 1166.081 (laminar flow)

$$f = \frac{64}{Re} = \frac{64}{Re}$$

$$= \frac{64}{1166.081} = 0.054$$

$$\Delta p = \frac{fl\rho V^2}{2D}$$

$$= \frac{0.054 \times 250 \times 997 \times (0.1)^2}{2 \times 10 \times 10^{-3}}$$

= 6840 pa.

22. $\therefore \quad \frac{\delta}{x} = \frac{k}{\sqrt{x}}$, here '$k$' is constant

$$\therefore \quad \frac{\delta_1}{\sqrt{x_1}} = \frac{\delta_2}{\sqrt{x_2}}$$

$$\therefore \quad \frac{8}{\sqrt{0.25}} = \frac{\delta_2}{\sqrt{0.75}} \qquad \therefore\ \delta_2 = 13.86\ mm.$$

$$\because \quad \frac{\delta}{x} = \frac{0.60}{\sqrt{Re_x}}$$

$$\text{or} \quad \frac{\delta}{x} = \frac{0.60}{\sqrt{xu_\infty/\upsilon}} = \frac{0.60\sqrt{\upsilon}}{\sqrt{x} \cdot \sqrt{u_\infty}}$$

$\because\ \sqrt{\upsilon}$ and $\sqrt{u_\infty}$ is constant in both the cases

23. Given,

$\mu = 2\ \ 10^{-5}$ kg/m.s, $\rho = 1\ kg/m^3$, $u_{av} = 1 m/sec$

Duct area = 25 mm 15 mm

$$\therefore \quad \frac{\Delta P}{L} = \frac{4 \times \lambda \times \rho \times U_{av}^2}{2 \times D_n} \qquad ....(1)$$

Here λ = Friction factor

D_n = Dia

$\therefore \; D_n = \frac{4 \times A}{P}$ (A = Area, P = Parameter)

$\lambda = \frac{4 \times 25 \times 15}{2(25+15)} = 18.75$ mm ...(2)

$\lambda = \frac{16}{Re} \qquad \therefore Re = \frac{\rho \cdot u_{av} \cdot D_n}{\pi}$

$\therefore \; Re = \frac{1 \times 1 \times 18.75 \times 10^{-3}}{2 \times 10^{-5}} = 937.5$

$\therefore \; \lambda = \frac{16}{937.5} = 1.707 \times 10^{-2}$ ...(3)

Here from equation (3),

$\frac{\Delta P}{L} = \frac{4 \times 1.707 \times 10^{-2} \times (1)^2 \times 2}{2 \times 18.75 \times 10^{-3}} = 1.8208$ pa/m.

24.

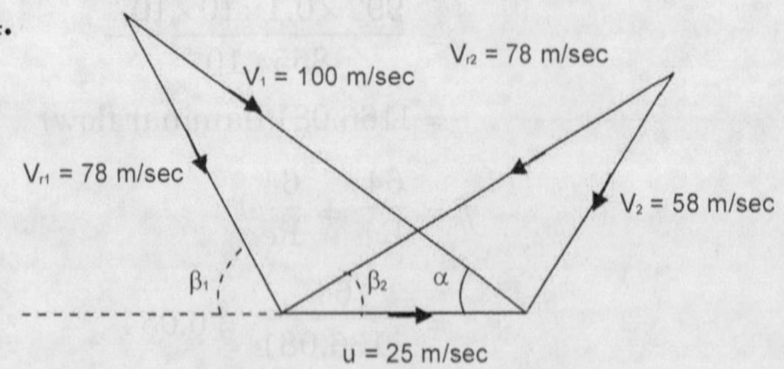

Given $\alpha = 25^\circ$

$V_{r_1} \sin\beta_1 = V_1 \sin\alpha$

$\therefore \; 78 \sin\beta_1 = 100 \sin 25^\circ$

$\therefore \; \beta_1 = 32.81^\circ$

$\because \; V_{r_1} = V_{r_2}$

$\therefore \; \beta_1 = \beta_2 = 32.81^\circ$

Now $\therefore \Delta V_w = V_{r_1}\cos\beta_1 + V_{r_2}\cos\beta_2$

$= 2V_{r_1}\cos\beta_1$

$= 2 \times 78 \times \cos 32.81^\circ$

$\therefore \; \Delta V_w = 131.1136$ m/sec

$\therefore$ Sp. power $= \Delta V_w \cdot u = 131.1136 \times 25$

$= 3277.84$ J/kg.

25. $Re = \frac{\rho V D}{\mu} = \frac{1000 \times 1 \times 25 \times 10^{-3}}{7.25 \times 10^{-4}}$

$= 34482.758$

Pr = 4.85

Nu = 0.023

$Re^{0.8} Pr^{0.4} = 184.5466 = \frac{hD}{k}$

$\therefore \; h = \frac{184.5466 \times 0.625}{25 \times 10^{-3}}$

$= 4613.6659$ W/m²k.

26. Given $m_1 = m_2 = 5$Kg

$c_1 = c_2 = 0.4$KJ / Kg.K

$T_{L,i} = T_{M,i} = 313$K

$T_{M,f} = 293$K, $T_{L,f} = ?$

$m_1 c_1 \Delta T_1 = m_2 c_2 \Delta T_2$

$\Delta T_1 = \Delta T_2$

$T_{L,i} - T_{L,f} = T_{M,i} - T_{M,f}$

$313 - T_{L,f} = 313 - 293$

$T_{L,f} = 333$k

27. $\dot{w} = \dot{m}_{tot}(h_1 - h_2) + (\dot{m}_{tot} - \dot{m}_{out})(h_2 - h_3)$

$\dot{w} = 10(3214 - 2920) + 8.5(2920 - 2374)$

$\dot{w} = 7581$ w.

28. Compression ratio,

$r_k = \frac{v_1}{v_2} = \left(\frac{p_2}{p_1}\right)^{\frac{1}{\gamma}} = (32.42)^{\frac{1}{1.4}} = 11.9999 \simeq 12$

cutoff ratio, $r_k = \frac{\text{compression ratio}}{\text{expansion ratio}} = \frac{12}{8} = 1.5$

$\eta = 1 - \frac{1}{r_k^{\gamma-1}}\left(\frac{r_e^{\gamma} - 1}{\gamma(r_c - 1)}\right)$

$= 1 - \frac{1}{12^{0.4}}\left(\frac{(15)^{1.4} - 1}{1.4(1.5-1)}\right)$

$\eta = 59.599\%$.

■■

7 CHAPTER Heat Transfer

MCQ TYPE QUESTIONS

2016

1. A plastic sleeve of outer radius $r_0 = 1$ mm covers a wire (radius $r = 0.5$ mm) carrying electric current. Thermal conductivity of the plastic is 0.15 W/m-K. The heat transfer coefficient on the outer surface of the sleeve exposed to air is 25 W/m²-K. Due to the addition of the plastic cover, the heat transfer from the wire to the ambient will

(a) increase (b) remain the same

(c) decrease (d) be zero

2. Which of the following statements are **TRUE** with respect to heat and work?

(i) They are boundary phenomena

(ii) They are exact differentials

(iii) They are path functions

(a) both (i) and (ii)

(b) both (i) and (iii)

(c) both (ii) and (iii)

(d) only (iii)

3. A hollow cylinder has length L, inner radius r_1, outer radius r_2, and thermal conductivity k. The thermal resistance of the cylinder for radial conduction is

(a) $\frac{\ln(r_2/r_1)}{2\pi k L}$ (b) $\frac{\ln(r_1/r_2)}{2\pi k L}$

(c) $\frac{2\pi k L}{\ln(r_2/r_1)}$ (d) $\frac{2\pi k L}{\ln(r_1/r_2)}$

4. Consider the radiation heat exchange inside an annulus between two very long concentric cylinders.

The radius of the outer cylinder is R_o and that of the inner cylinder is R_i. The radiation view factor of the outer cylinder onto itself is

(a) $1-\sqrt{\frac{R_i}{R_o}}$ (b) $\sqrt{1-\frac{R_i}{R_o}}$

(c) $1-\left(\frac{R_i}{R_o}\right)^{1/3}$ (d) $1-\frac{R_i}{R_o}$

5. Grash of number signifies the ratio of

(a) inertia force to viscous force

(b) buoyancy force to viscous force

(c) buoyancy force to inertia force

(d) inertial force to surface tension force

6. For a heat exchanger, ΔT_{max} is the maximum temperature difference and ΔT_{min} is the minimum temperature difference between the two fluids. LMTD is the log mean temperature difference. C_{min} and C_{max} are the minimum and the maximum heat capacity rates. The maximum possible heat transfer (Q_{max}) between the two fluids is

(a) C_{min} LMTD (b) $C_{min}\ \Delta T_{max}$

(c) $C_{max}\ \Delta T_{max}$ (d) $C_{max}\ \Delta T_{min}$

2015

7. A 10 mm diameter electrical conductor is covered by an insulation of 2 mm thickness. The conductivity of the insulation is 0.08 W/m-K and the convection coefficient at the insulation surface is 10 W/m²-K. Addition of further insulation of the same material will

(a) increase heat loss continuously

(b) decrease heat loss continuously

(c) increase heat loss to a maximum and then decrease heat loss

(d) decrease heat loss to a minimum and then increase heat loss

8. The total emissive power of a surface is 500 W/m² at a temperature T_1 and 1200 W/m² at a temperature T_2, where the temperatures are in Kelvin. Assume the emissivity of the surface to be constant, the ratio of the temperatures $\frac{T_1}{T_2}$ is

(a) 0.308 (b) 0.416

(c) 0.803 (d) 0.874

9. A balanced counter flow heat exchanger has a surface area of 20m² and overall heat transfer coefficient of 20 W/m² – K Air (C_P = 1000 J/kg – K) entering at 0.4 kg/s and 280 K is to be preheated by the air leaving the system at 0.4 kg/s and 300 K. The outlet temperature (in K) of the preheated air is

(a) 290 (b) 300

(c) 320 (d) 350

10. A cylindrical uranium fuel rod of radius 5 mm in a nuclear reactor is generating heat at the rate of 4×10^7 W/m³. The rod is cooled by a liquid (convective heat transfer coefficient 1000 W/ m² – K) at 25°C. At steady state the surface temperature (in K) of the rod is

(a) 308 (b) 398
(c) 418 (d) 448

11. The ratio of momentum diffusivity (v) to thermal diffusivity (α), is called

(a) Prandtl number (b) Nusselt number
(c) Biot number (d) Lewis number

12. A brick wall $\left(k = 0.9 \dfrac{W}{m.k}\right)$ of thickness 0.18 m separates the warm air in a room from the cold ambient air. On a particular winter day, the outside air temperature is – 5°C and the room needs to be maintained at 27°C. The heat transfer coefficient associated with outside air is $20 \dfrac{W}{m^2 K}$. Neglecting the convective resistance of the air inside the room, the heat loss, in $\left(\dfrac{W}{m^2}\right)$ is

(a) 88 (b) 110
(c) 128 (d) 160

2014

13. Biot number signifies the ratio of

(a) convective resistance in the fluid to conductive resistance in the solid
(b) conductive resistance in the soild to convective resistance in the fluid
(c) inertia force to viscous force in the fluid
(d) buoyancy force to viscous force in the fluid

14. The non-dimensional fluid temperature profile near the surface of a convectively cooled flat plate is given by $\dfrac{T_W - T}{T_W - T_\infty} = a + b\dfrac{y}{L} + c\left(\dfrac{y}{L}\right)^2$, where y is measured perpendicular to the plate, L is the plate length, and a, b and c are arbitrary constants. T_W and T_∞ are wall and ambient temperatures, respectively. If the thermal conductivity of the fluid is k and the wall heat flux is q'', the Nusselt number $Nu = \dfrac{q''}{T_W - T_\infty}\dfrac{L}{k}$ is equal to

(a) a (b) b
(c) $2c$ (d) $(b + 2c)$

15. In a heat exchanger, it is observed that $\Delta T_1 = \Delta T_2$, where ΔT_1 is the temperature difference between the two single phase fluid streams at one end and ΔT_2 is the temperature difference at the other end. This heat exchanger is

(a) a condenser
(b) an evaporator
(c) a counter flow heat exchanger
(d) a parallel flow heat exchanger

16. Consider a long cylindrical tube of inner and outer radii, r_i and r_o, respectively, length, L and thermal conductivity, k. Its inner and outer surfaces are maintained at T_i and T_o, respectively ($T_i > T_o$). Assuming one-dimensional steady state heat conduction in the radial direction, the thermal resistance in the wall of the tube is

(a) $\dfrac{1}{2\pi k L}\ln\left(\dfrac{r_i}{r_o}\right)$ (b) $\dfrac{1}{2\pi r_i k}$

(c) $\dfrac{1}{2\pi k L}\ln\left(\dfrac{r_o}{r_i}\right)$ (d) $\dfrac{1}{4\pi k L}\ln\left(\dfrac{r_o}{r_i}\right)$

17. A solid sphere of radius r_1 = 20 mm is placed concentrically inside a hollow sphere of radius r_2 = 30 mm as shown in the figure.

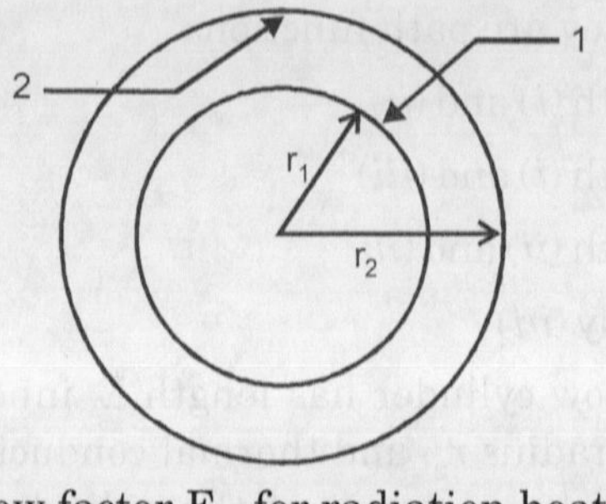

The view factor F_{21} for radiation heat transfer is

(a) $\dfrac{2}{3}$ (b) $\dfrac{4}{9}$

(c) $\dfrac{8}{27}$ (d) $\dfrac{9}{4}$

18. Match Group A with Group B :

Group A
P . Biot number
Q . Grashof number
R . Prandtl number
S . Reynolds number

Group B
1 . Ratio of buoyancy to viscous force
2. Ratio of inertia force to viscous force
3. Ratio of momentum to thermal diffusivities
4. Ratio of internal thermal resistance to boundary layer thermal resistance

Codes :

	P	Q	R	S
(a)	4	1	3	2
(b)	4	3	1	2
(c)	3	2	1	4
(d)	2	1	3	4

2013

19. The pressure, dry bulb temperature and relative humidity of air in a room are 1 bar, 30°C and 70%, respectively. If the saturated steam pressure at 30°C is 4.25 kPa, the specific humidity of the room air in kg water vapour/kg dry air is

(a) 0.0083 (b) 0.0101
(c) 0.0191 (d) 0.0232

20. A cylinder contains 5 m^3 of an ideal gas at a pressure of 1 bar. This gas is compressed in a reversible isothermal process till its pressure increases to 5 bar The work in *kJ* required for this process is

(a) 804.7 (b) 953.2
(c) 981.7 (d) 1012.2

21. Specific enthalpy and velocity of steam at inlet and exit of a steam turbine, running under steady state, are as given below

	Specific enthalpy (kJ/kg)	Velocity (m/s)
Inlet steam condition	3250	180
Exit steam condition	2360	5

The rate of heat loss from the turbine per kg of steam flow rate is 5 kW Neglecting changes potential energy of steam, the power developed in kW by the steam turbine per kg of steam flow rate, is

(a) 901.2 (b) 911.2
(c) 17072.5 d) 17082.5

22. The pressure, temperature and velocity of air flowing in a pipe are 5 bar, 500 K and 50 m/s, respectively. The specific heats of air at constant pressure and at constant volume are 1.005 kJ/kgK and 0.718 *kJ/kg*K, respectively. Neglect potential energy. If the pressure and temperature of the surroundings are 1 bar and 300 K respectively the available energy in *kJ/kg* of the air stream is

(a) 170 (b) 187
(c) 191 (d) 213

2012

23. For an opaque surface, the absorptivity (α), transmissivity (τ) and reflectivity (ρ) are related by the equation :

(a) $\alpha + \rho = \tau$ (b) $\rho + \alpha + \tau = 0$
(c) $\alpha + \rho = 1$ (d) $\alpha + \rho = 0$

24. Which one of the following configurations has the highest fin effectiveness?

(a) Thin, closely spaced fins
(b) Thin, widely spaced fins
(c) Thick, widely spaced fins
(d) Thick, closely spaced fins

25. Consider two infinitely long *thin* concentric tubes of circular cross section as shown in the figure. If D_1 and D_2 are the diameters of the inner and outer tubes respectively, then the view factor F_{22} is given by

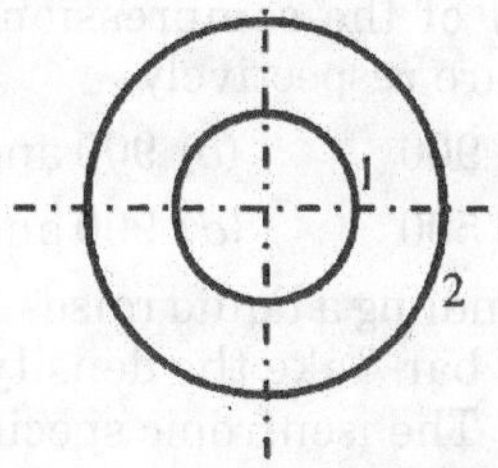

(a) $\left(\frac{D_2}{D_1}\right)-1$ (b) zero
(c) $\left(\frac{D_1}{D_2}\right)$ (d) $1-\left(\frac{D_1}{D_2}\right)$

26. Water (C_p =4.18 *kJ/kg.K*) at 80°*C* enters a counterflow heat exchanger with a mass flow rate of 0.5 kg/s. Air (C_p = 1 kJ/kg.K) enters at 30°*C* with a mass flow rate of 2.09 *kg/s*. If the effectiveness of the heat exchanger is 0.8, the LMTD (in °C) is

(a) 40 (b) 20
(c) 10 (d) 5

2011

27. Heat and work are

(a) intensive properties
(b) extensive properties
(c) point functions
(d) path functions

28. The contents of a well-insulated tank are heated by a resistor of 23 Ω in which 10 A current is flowing. Consider the tank along with its contents as a thermodynamic system. The work done by the system and the heat transfer to the system are positive. The rates of heat (Q), work (W) and change in internal energy (ΔU) during the process in kW are

(a) Q = 0, W = – 2.3, ΔU = + 2.3
(b) Q = + 2.3, W = 0, ΔU = + 2.3
(c) Q = –2.3, W = 0, ΔU = – 2.3
(d) Q = 0, W = + 2.3, ΔU = – 2.3

29. The values of enthalpy of steam at the inlet and outlet of a steam turbine in a Rankine cycle are 2800 kJ/kg and 1800 kJ/kg respectively. Neglecting pump work, the specific steam consumption in kg/kW-hour is

(*a*) 3.60 (*b*) 0.36

(*c*) 0.06 (*d*) 0.01

30. An ideal Brayton cycle, operating between the pressure limits of 1 bar and 6 bar, has minimum and maximum temperature of 300 K and 1500 K. The ratio of specific heats of the working fluid is 1.4. The approximate final temperatures in Kelvin at the end of the compression and expansion processes are respectively

(*a*) 500 and 900 (*b*) 900 and 500

(*c*) 500 and 500 (*d*) 900 and 900

31. A pump handling a liquid raises its pressure from 1 bar to 30 bar. Take the density of the liquid as 990 kg/m^3. The isentropic specific work done by the pump in kJ/kg is

(*a*) 0.10 (*b*) 0.30

(*c*) 2.50 (*d*) 2.93

32. The crank radius of a single-cylinder I.C. engine is 60 mm and the diameter of the cylinder is 80 mm. The swept volume of the cylinder in cm^3 is

(*a*) 48 (*b*) 96

(*c*) 302 (*d*) 603

2010

33. A phenomenon is modeled using n dimensional variables with k primary dimensions. The number of non-dimensional variables is

(*a*) k (*b*) n

(*c*) $n - k$ (*d*) $n + k$

34. A fin has 5 mm diameter and 100 mm length. The thermal conductivity of fin material is 400 Wm^{-1} K^{-1}. One end of the fin is maintained at 130°C and its remaining surface is exposed to ambient air at 30°C. If the convective heat transfer coefficient is 40 Wm^{-2} K^{-1}, the heat loss (in W) from the fin is

(*a*) 0.08 (*b*) 5.0

(*c*) 7.0 (*d*) 7.8

2009

35. In a parallel flow heat exchanger operating under steady state, the heat capacity rates (product of specific heat at constant pressure and mass flow rate) of the hot and cold fluid are equal. The hot fluid, flowing at 1 kg/s with C_p = 4 kJ/kgK, enters the heat exchanger at 102 °C while the cold fluid has an inlet temperature of 15 °C. The overall heat transfer coefficient for the heat exchanger is estimated to be 1 kW/m^{2}K and the corresponding heat transfer surface area is 5 m^2. Neglect heat transfer between the heat exchanger and the ambient. The heat exchanger is characterized by the following relation:

$$2\varepsilon = -\exp(-2\text{NTU})$$

The exit temperature (in °C) for the cold fluid is

(*a*) 45 (*b*) 55

(*c*) 65 (*d*) 75

36. Consider steady-state heat conduction across the thinkness in a plane composite wall (as shown in the figure) exposed to convection conditions on both sides.

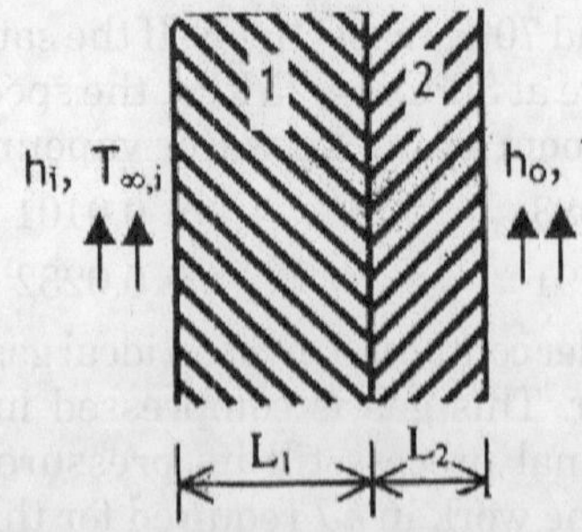

Given: h_1 = 20 W/m^{2}K; h_o = 50 W/m^{2}K; $T_{\infty,i}$ = 20 °C; $T_{\infty,o}$ = 20 °C; k_1 = 20 W/mK; k_2 = 50 W/mK; L_1 = 0.30 m and L_2 = 0.15 m. Assuming negligible contact resistance between the wall surfaces, the interface temperature, T (in °C), of the two walls will be

(*a*) – 0.50 (*b*) 2.75

(*c*) 3.75 (*d*) 4.50

2008

37. A cyclic device operates between three thermal reservoirs, as shown in the figure. Heat is transferred to/form the cyclic device. It is assumed that heat transfer between each thermal reservoir and the cyclic device takes place across negligible temperature difference. Interactions between the cyclic device and the respective thermal reservoirs that are shown in the figure are all in the form of heat transfer.

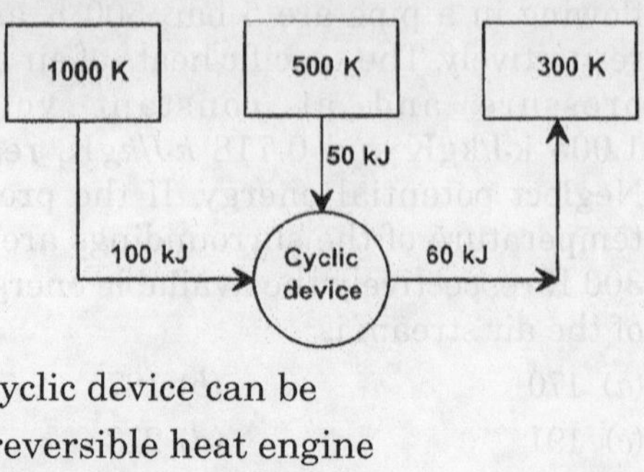

The cyclic device can be

(*a*) a reversible heat engine

(*b*) a reversible heat pump or a reversible refrigerator

(*c*) an irreversible heat engine

(*d*) an irreversible heat pump or an irreversible refrigerator

38. For the three-dimensional object shown in the figure below, five faces are insulated. The sixth face (PQRS), which is not insulated, interacts thermally with the ambient, with a convective heat transfer coefficient of 10 W/m^2.K. The ambient temperature is 30°C. Heat is uniformly generated inside the object at the rate of 100 W/m^3. Assuming the face PQRS to be at uniform temperature, its steady state temperature is

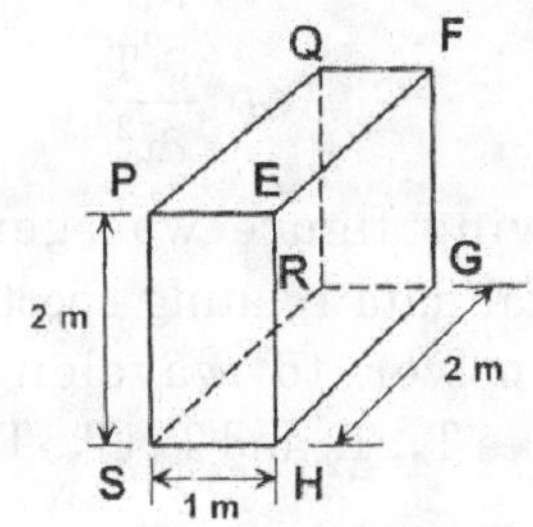

(*a*) 10°C (*b*) 20°C

(*c*) 30°C (*d*) 40°C

39. A hollow enclosure is formed between two infinitely long concentric cylinders of radii 1 m and 2 m, respectively. Radiative heat exchange takes place between the inner surface of the larger cylinder (surface-2) and the outer surface of the smaller cylinder (surface-1). The radiating surfaces are diffuse and the medium in the enclosure is non-participating. The fraction of the thermal radiation leaving the larger surface and striking itself is

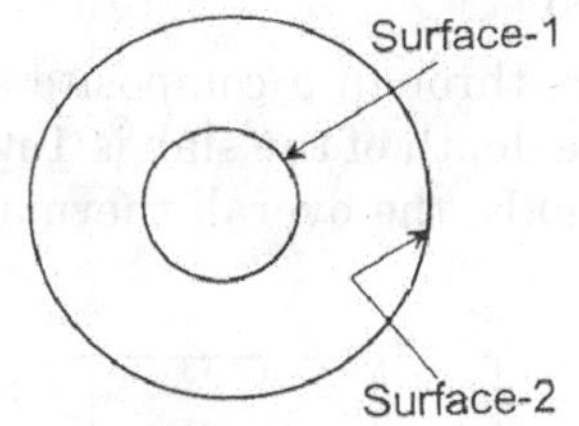

(*a*) 0.25 (*b*) 0.5

(*c*) 0.75 (*d*) 1

40. Steady two-dimensional heat conduction takes place in the body shown in the figure below. The normal temperature gradients over surfaces P and Q can be considered to be uniform. The temperature gradient $\frac{\partial T}{\partial x}$ at surface Q is equal to 10 K/m. Surfaces P and Q are maintained at constant temperatures as shown in the figure, while the remaining part of the boundary is insulated. The body has a constant thermal conductivity of 0.1 W/m.K. The values of $\frac{\partial T}{\partial x}$ and $\frac{\partial T}{\partial y}$ at surface P are

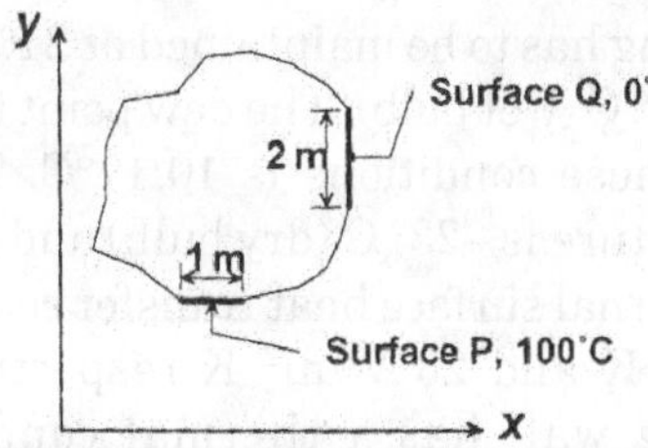

(*a*) $\frac{\partial T}{\partial x} = 20$ K/m, $\frac{\partial T}{\partial y} = 0$ K/m

(*b*) $\frac{\partial T}{\partial x} = 0$ K/m, $\frac{\partial T}{\partial y} = 10$ K/m

(*c*) $\frac{\partial T}{\partial x} = 10$ K/m, $\frac{\partial T}{\partial y} = 10$ K/m

(*d*) $\frac{\partial T}{\partial x} = 0$ K/m, $\frac{\partial T}{\partial y} = 20$ K/m

2007

41. The temperature distribution within the thermal boundary layer over a heated isothermal flat plate is given by $\frac{T - T_w}{T_\infty - T_w} = \frac{3}{2}\left(\frac{y}{\delta_t}\right) - \frac{1}{2}\left(\frac{y}{\delta_t}\right)^3$, where T_w and T_∞ are the temperatures of plate and free stream respectively, and y is the normal distance measured from the plate. The local Nusselt number based on the thermal boundary layer thickness δ_t is given by

(*a*) 1.33 (*b*) 1.50

(*c*) 2.0 (*d*) 4.64

42. In a counter flow heat exchanger, hot fluid enters at 60°C and cold fluid leaves at 30°C. Mass flow rate of the hot fluid is 1 kg/s and that of the cold fluid is 2 kg/s. Specific heat of the hot fluid is 10 kJ/kgK and that of the cold fluid is 5 kJ/kgK. The Log Mean Temperature Difference (LMTD) for the heat exchanger in °C is

(*a*) 15 (*b*) 30

(*c*) 35 (*d*) 45

43. The average heat transfer co-efficient on a thin hot vertical plate suspended in still air can be determined from observations of the change in plate temperature with time as it cools. Assume the plate temperature to be uniform at any instant of time and radiation heat exchange with the surroundings negligible. The ambient temperature is 25°C, the plate has a total surface area of 0.1 m^2 and a mass of 4 kg. The specific heat of the plate material is 2.5 kJ/kgK. The convective heat transfer co-efficient in W/m^{2}K, at the instant when the plate temperature is 225°C and the change in plate temperature with time dT/dt = –0.02 K/s, is

(*a*) 200 (*b*) 20

(*c*) 15 (*d*) 10

44. A building has to be maintained at 21° C (dry bulb) and 14.5° C (wet bulb). The dew point temperature under these conditions is 10.17°C. The outside temperature is –23° C (dry bulb) and the internal and external surface heat transfer coefficients are 8 W/m^2 K and 23 W/m^2 K respectively. If the building wall has a thermal conductivity of 1.2 W/mK, the minimum thickness (in m) of the wall required to prevent condensation is

(*a*) 0.471 (*b*) 0.407
(c) 0.321 (*d*) 0.125

45. A heat transformer is a device that transfers a part of the heat, supplied to it at an intermediate temperature, to a high temperature reservoir while rejecting the remaining part to a low temperature heat sink. In such a heat transformer, 100 kJ of heat is supplied at 350 K. The maximum amount of heat in kJ that can be transferred to 400 K, when the rest is rejected to a heat sink at 300 K is

(*a*) 12.50 (*b*) 14.29
(c) 33.33 (*d*) 57.14

2006

46. In a composite slab, the temperature at the interface (Tinter) between two materials is equal to the average of the temperatures at the two ends. Assuming steady one-dimensional heat conduction, which of the following statements is true about the respective thermal conductivities?

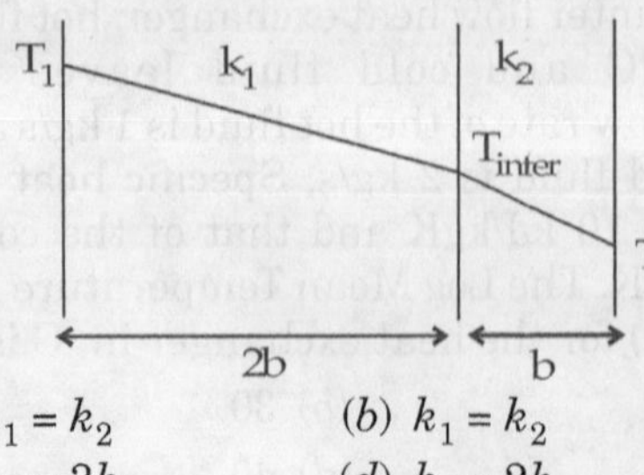

(*a*) $2k_1 = k_2$ (*b*) $k_1 = k_2$
(c) $2k_1 = 3k_2$ (*d*) $k_1 = 2k_2$

47. A 100 W electric bulb was switched on in a 2.5 m × 3 m × 3 m size thermally insulated room having a temperature of 20°C. The room temperature at the end of 24 hours will be

(*a*) 321°C (*b*) 341°C
(c) 450°C (*d*) 470°C

48. With an increase in the thickness of insulation around a circular pipe, heat loss to surroundings due to

(*a*) convection increases, while that due to conduction decreases

(*b*) convection decreases, while that due to conduction increases

(*c*) convection and conduction decreases

(*d*) convection and conduction increases

2005

49. In a case of one dimensional heat conduction in a medium with constant properties, T is the temperature at position x, at time t. Then $\frac{\partial T}{\partial t}$ is proportional to

(*a*) $\frac{T}{x}$ (*b*) $\frac{\partial T}{\partial x}$
(*c*) $\frac{\partial^2 T}{\partial x \partial t}$ (*d*) $\frac{\partial^2 T}{\partial x^2}$

50. The following figure was generated from experimental data relating spectral black body emissive power to wavelength at three temperatures T_1, T_2 and T_3 ($T_1 > T_2 > T_3$).

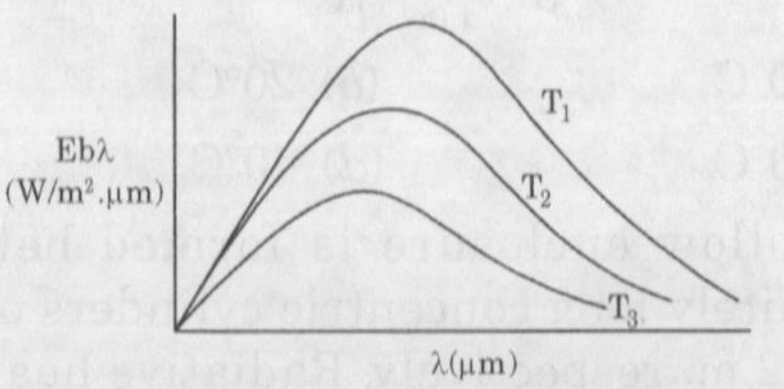

The conclusion is that the measurements are

(*a*) correct because the maxima in $E_{b\lambda}$ show the correct trend

(*b*) correct because Planck's law is satisfied

(*c*) wrong because the Stefan Boltzmann law is not satisfied

(*d*) wrong because Wien's displacement law is not satisfied

51. Heat flows through a composite slab, as shown below. The depth of the slab is 1 m. The *k* values are in W/m.K. the overall thermal resistance in K/W is

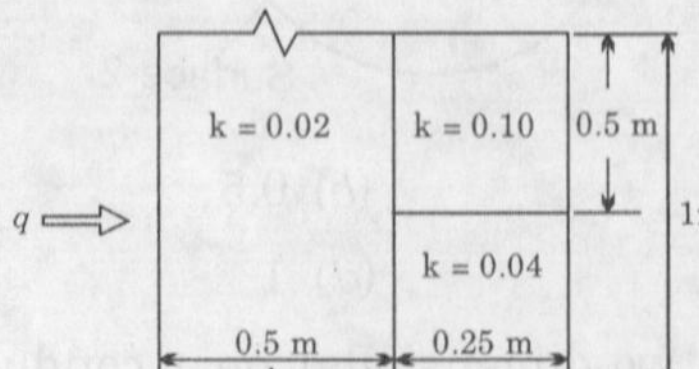

(*a*) 17.2 (*b*) 21.9
(c) 28.6 (*d*) 39.2

52. A small copper ball of 5 mm diameter at 500 K is dropped into an oil bath whose temperature is 300 K. The thermal conductivity of copper is 400 W/m.K, its density 9000 kg/m^3 and its specific heat 385 J/kg.K. If the heat transfer coefficient is 250 W/m^2.K and lumped analysis is assumed to be valid, the rate of fall of the temperature of the ball at the beginning of cooling will be, in K/s,

(*a*) 8.7 (*b*) 13.9
(c) 17.3 (*d*) 27.7

53. A solid cylinder (surface 2) is located at the centre of a hollow sphere (surface 1). The diameter of the sphere is 1 m, while the cylinder has a diameter and length of 0.5 m each. The radiation configuration factor F_{11} is

(*a*) 0.375 (*b*) 0.625
(*c*) 0.75 (*d*) 1

54. Hot oil is cooled from 80 to 50° C in an oil cooler which uses air as the coolant. The air temperature rises from 30 to 40° C. The designer uses a LMTD value of 26° C. The type of heat exchanger is

(*a*) parallel flow (*b*) double pipe
(*c*) counter flow (*d*) cross flow

Linked Answer Q. (55-56)

An un-insulated air conditioning duct of rectangular cross section 1m × 0.5 m, carrying air at 20°C with a velocity of 10 m/s, is exposed to an ambient of 30°C. Neglect the effect of duct construction material. For air n the range of 20-30°C, data are as follows: thermal conductivity = 0.025 W/m.K: velocity = 18 μPa.s; Prandtl number = 0.73; density = 1.2 kg/m³. The laminar flow Nussch number is 3.4 for constant wall temperature conditions and, for turbulent flow, Nu = 0.023 Re0.8Pr0.8

55. The Reynolds number for the flow is

(*a*) 444 (*b*) 890
(*c*) 4.44×10^5 (*d*) 5.33×10^5

56. The heat transfer per metre length of the duct, in watts, is

(*a*) 3.8 (*b*) 5.3
(*c*) 89 (*d*) 769

2004

57. A solar collector receiving solar radiation at the rate of 0.6 kW/m² transforms it to the internal energy of a fluid at an overall efficiency of 50%. The fluid heated to 350 K is used to run a heat engine which rejects heat at 313 K. If the heat engine is to deliver 2.5 kW power, then minimum area of the solar collector required would be

(*a*) 83.33 m² (*b*) 16.66 m²
(*c*) 39.68 m² (*d*) 79.36 m²

58. A steel billet of 2000 kg mass is to be cooled from 1250 K to 450 K. The heat released during this process is to be used as a source of energy. The ambient temperature is 303 K and specific heat of steel is 0.5 kJ/kg K. The available energy of this billet is

(*a*) 490.44 MJ (*b*) 30.95 MJ
(*c*) 10.35 MJ (*d*) 0.10 MJ

59. A stainless steel tub (k_s = 19 W/mK) of 2 cm ID and 5 cm OD is insulated with 3 cm thick asbestos (k_a = 0.2 W/mK). If the temperature difference between the innermost and outermost surfaces is 600°C, the heat transfer rate per unit length is

(*a*) 0.94 W/m (*b*) 9.44 W/m
(*c*) 944.88 W/m (*d*) 9447.21 W/m

60. A spherical thermocouple junction of diameter 0.706 mm is to be used for the measurement of temperature of a gas stream. The convective heat transfer co-efficient on the bead surface is 400 W/m²K. Thermophysical properties of thermocouple material are k = 20 W/mK, C = 400 J/kg K and ρ = 8500 kg/m³. If the thermocouple initially at 30°C is placed in a hot stream of 300°C, then time taken by the bead to reach 298°C, is

(*a*) 2.35 s (*b*) 4.9 s
(*c*) 14.7 s (*d*) 29.4 s

61. In a condenser, water enters at 30°C and flows at the rate 1500 kg/hr. The condensing steam is at a temperature of 120°C and cooling water leaves the condenser at 80°C. Specific heat of water is 4.187 kJ/kg K. If the overall heat transfer coefficient is 2000 W/m² K, then heat transfer area is

(*a*) 0.707 m² (*b*) 7.07 m²
(*c*) 70.7 m² (*d*) 141.4 m²

2003

62. A 2 kW, 40 litre water heater is switched on for 20 minutes. The heat capacity C_p for water is 4.2 kJ/kg K. Assuming all the electrical energy has gone into heating the water, increase of the water temperature in degree centrgrade is

(*a*) 2.7 (*b*) 4.0
(*c*) 14.3 (*d*) 25.25

63. An industrial heat pump operates between the temperatures of 27°C and –13°C. The rates of heat addition and heat rejection are 750 W and 1000 W, respectively. The COP for the heat pump is

(*a*) 7.5 (*b*) 6.5
(*c*) 4.0 (*d*) 3.0

64. A plate having 10 cm² area each side is hanging in the middle of a room of 100 m² total surface area. The plate temperature and emissivity are respectively 800 K and 0.6. The temperature and emissivity values for the surfaces of the room are 300 K and 0.3 respectively. Boltzmann's constant $\sigma = 5.67 \times 10^{-8}$ W/m² K⁴. The total heat loss from the two surfaces of the plate is

(*a*) 13.66 W (*b*) 27.32 W
(*c*) 27.87 W (*d*) 13.66 MW

65. In a counter flow heat exchanger, for the hot fluid the heat capacity = 2 kJ/kg K, mass flow rate = 5 kg/s, inlet temperature = 150°C, outlet temperature = 100 °C. For the cold fluid, heat capacity = 4 kJ/kg K, mass flow rate = 10 kg/s, inlet temperature = 20 °C. Neglecting heat transfer to the surroundings, the outlet temperature of the cold fluid in °C is

(*a*) 7.5 (*b*) 32.5

(*c*) 45.5 (*d*) 70.0

Common Data Q. 66-67

Heat is being transferred by convection from water at 48°C to a glass plate whose surface that is exposed to the water is at 40 °C. The thermal conductivity of water is 0.6 W/mK and the thermal conductivity of glass is 1.2 W/mK. The spatial gradient of temperature in the water at the water-glass inerface is $dT/dy = 1 \times 10^4$ K/m.

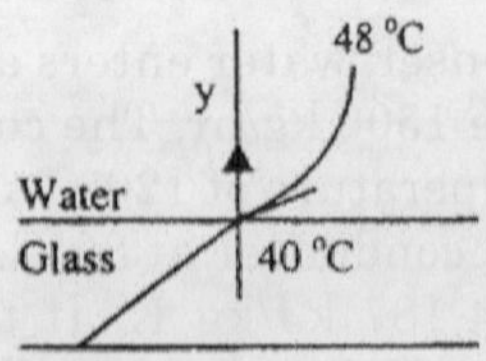

66. The value of the temperature gradient in the glass at the water-glass interface in K/m is

(*a*) -2×10^4 (*b*) 0.0

(*c*) 0.5×10^4 (*d*) 2×10^4

67. The heat transfer coefficient *h* in W/m² K is

(*a*) 0.0 (*b*) 4.8

(*c*) 6 (*d*) 750

2002

68. For the same inlet and outlet temperatures of hot and cold fluids, the Log Mean Temperature Difference (LMTD) is

(*a*) greater for parallel flow heat exchanger than for counter flow heat exchanger.

(*b*) greater for counter flow heat exchanger than for parallel flow heat exchanger.

(*c*) same for both parallel and counter flow heat exchangers.

(*d*) dependent on the properties of the fluids.

69. A copper tube of 20 mm outer diameter, 1 mm thickness and 20 m long (Thermal conductivity = 400 W/m²–K). is carrying saturated steam at 150° C (Convective heat transfer coefficient = 150 W/m² – K). The tube is exposed to an ambient temperature of 27° C. The convective heat transfer coefficient of air is 5 W/m²–K. Glass wool is used for insulation (Thermal conductivity = 0.075 W/m–K). If the thickness of the insulation used is 5 mm higher than the critical thickness of insulation, calculate the rate of heat lost by the steam and the rate of steam condensation in kg/hr. (The enthalpy of condensation of steam = 2230 kJ/kg.)

2001

70. For the circular tube of equal length and diameter shown below, the view factor F_{13} is 0.17. The view factor F_{12} in this case will be

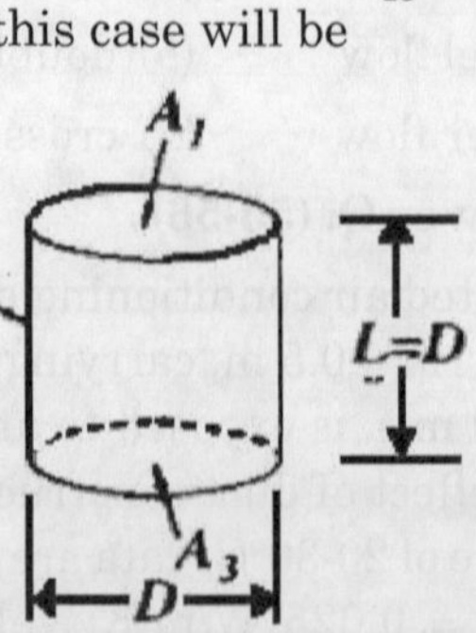

(*a*) 0.17 (*b*) 0.21

(*c*) 0.79 (*d*) 0.83

71. In descending order of magnitude, the thermal conductivity of (*a*) pure iron, (*b*) liquid water, (*c*) saturated water vapour, and (*d*) pure aluminum can be arranged as

(*a*) *a b c d* (*b*) *b c a d*

(*c*) *d a b c* (*d*) *d c b a*

72. A composite wall, having unit length normal to the plane of paper, is insulated at the top and bottom as shown in figure. It is comprised of four different materials A, B, C and D.

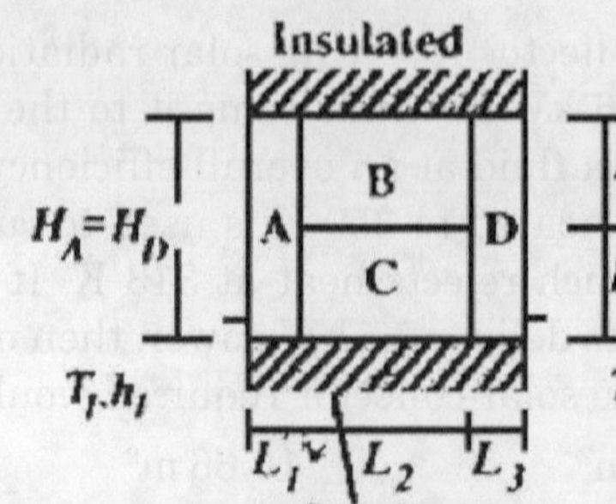

The dimensions are :

$H_A = H_D = 3$m, $H_B = H_C = 1.5$ m,

$L_1 = L_3 = 0.05$ m, $L_2 = 0.1$ m.

The thermal conductivity of the materials are:

$K_A = K_D = 50$ W/m-K,

$K_B = 10$ W/m-K,

$K_C = 1$ W/m-k.

The fluid temperatures and heat transfer coefficients (see figure) are: $T_1 = 200° C$,

$h_1 = 50\ W/m^2\text{-}K$,

$T_2 = 25° C$ $h_2 = 10\ W/m^2\text{-}K$.

Assuming one-dimensional conduction,

(*a*) sketch the thermal circuit of the system, and

(*b*) determine the rate of heat transfer through the wall.

2000

73. A steam turbine receives steam steadily at 10 bar with an enthalpy of 3000 kJ/kg and discharges at 1 bar with an enthaply of 2700 kJ/kg. The work output is 250 kJ/kg. The changes in kinetic and potential energies are negligible. The heat transfer from the turbine casing to the surroundings is equal to

(*a*) 0 kJ (*b*) 50 kJ

(*c*) 150 kJ (*d*) 250 kJ

74. A steel steam pipe 10 cm inner diameter and 11 cm outer diameter is covered with an insulation having a thermal conductivity of 1 W/mK. If the convective heat transfer coefficient between the surface of insulation and the surrounding air is 8 W/m^2K, then critical radius of insulation is

(*a*) 10 cm (*b*) 11 cm

(*c*) 12.5 cm (*d*) 15 cm

75. Air enters a counterflow heat exchanger at 70° C and leaves at 40° C. Water enters at 30° C and leaves at 50° C. The LMTD in degree C is

(*a*) 5.65 (*b*) 14.43

(*c*) 19.52 (*d*) 20.17

76. A thin metal plate is exposed to solar radiation. The air and the surroundings are at 30° C. The heat transfer coefficient by free convection from the upper surface of the plate is 17.4 W/m^2K. The plate has an absorptivity of 0.9 at solar wavelength and an emissivity of 0.1 at the long wavelength. Neglecting any heat loss from the lower surface, determine the incident solar radiation intensity in kW/m^2, if the measured equilibrium temperature of the plate is 50° C. Stefan Boltzmann constant is $5.67 \times 10^{-8}\ W/m^2K^4$.

1999

77. In a pulverised-fuel-fired large power boiler, then heat transfer from the burning fuel to the walls of the furnace is

(*a*) by conduction only

(*b*) By convection only

(*c*) by conduction and convection

(*d*) predominantly by radiation

78. It is proposed to coat a 1 mm diameter wire with enamel paint (k = 0.1 W/mK) to increase heat transfer with air. If the air side heat transfer coefficient is 100 W/m^2 K, then optimum thickness of enamel paint should be

(*a*) 0.25 mm (*b*) 0.5 mm

(*c*) 1 mm (*d*) 2 mm

79. Two fluids, A and B exchange heat in a counter-current heat exchanger. Fluid A enters at 420° C and has a mass flow rate of 1 kg/s. Fluid B enters at 20° C and also has a mass flow rate of 1 kg/s. Effectiveness of heat exchanger is 75%. Determine the heat transfer rate and exit temperature of Fluid B.

(specific heat Fluid A = 1 kJ/kgK and that of Fluid B = 4 kJ/kgK).

1998

80. Heat transfer coefficients for free convection in gases, forced convection in gases and vapours, and for boiling water lie, respectively in the ranges of

(*a*) 5 – 15, 20 – 200 and 3000 – 50,000 W/m^2K.

(*b*) 20 – 50, 200 – 500, and 50,000 – 10^5 W/m^2K.

(*c*) 50 – 100, 500 – 1000, and 10^5 – 10^6 W/m^2K.

(*d*) 20 – 100, 200 – 1000, and a constant 10^6 W/m^2K.

81. A steel ball of mas 1 kg and specific heat 0.4 kJ/kg is at a temperature of 60°C. It is dropped into 1 kg water at 20°C. The final steady state temperature of water is

(*a*) 23.5°C (*b*) 30°C

(*c*) 35°C (*d*) 40°C

82. The temperature variation under steady heat conduction across a composite slab of two materials with thermal conductivities K_1 and K_2 is shown in figure. Then, which one of the following statements holds ?

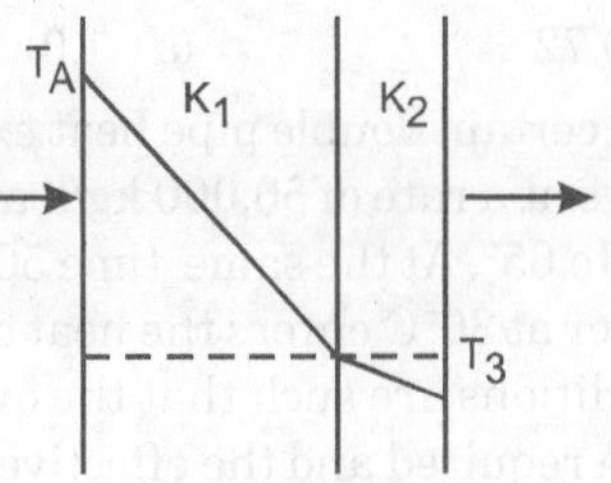

(*a*) $K_1 > K_2$ (*b*) $K_1 = K_2$

(*c*) $K_1 = 0$ (*d*) $K_1 < K_2$

83. A gas filled tube has 2 mm inside diameter and 25 cm length. The gas is heated by an electrical wire of diameter 50 microns (0.05 mm) located along the axis of the tube. Current and voltage drop across the heating element are 0.5 amps and 4 volts, respectively. If the measured wire and inside tube wall temperatures are 175°C and 150°C respectively, find the thermal conductivity of the gas filling the tube.

84. A hot fluid at 200°C enters a heat exchanger at a mass flow rate of 10^4 kg/hr. Its specific heat is 2000J/kg K. It is to be cooled by another fluid entering at 25°C with a mass flow rate 2500 kg/hr and specific heat 400 J/kg K. The overall heat transfer coefficient based on outside area of 20m² is 250 W/m²K. Find the exit temperature of the hot fluid when the fluids are in parallel flow.

85. Hot water flows with a velocity of 0.1 m/s in a 100 m long, 0.1 m diameter pipe. Heat loss from the pipe outer wall is uniform and equal to 420 W/m². If the inlet water temperature is 80°C, calculate the water temperature at the exit. Neglect effect of pipe wall thickness. C_p (water) = 4.2 kJ/kg °C and density of water = 1000 kg/m³.

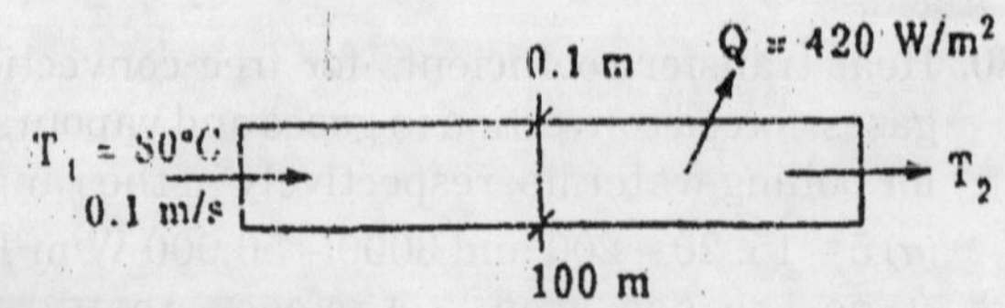

1997

86. In radiative heat transfer, a gray surface is one

(*a*) which appears gray to the eye

(*b*) whose emissivity is independant of wavelength

(*c*) which has reflectivity equal to zero

(*d*) which appears equally bright from all directions.

87. In a certain heat exchanger, both the fluids have identical mass flow rate-specific heat product. The hot fluid enters at 76°C and leaves at 47°C and the cold fluid entering at 26°C leaves at 55°C. The effectiveness of the heat exchanger is

(*a*) 0.16 (*b*) 0.58

(*c*) 0.72 (*d*) 1.0

88. In a certain double pipe heat exchanger, hot water flows at a rate of 50,000 kg/h and gets cooled from 95° to 65°. At the same time 50,000 kg/h of cooling water at 30°C enters the heat exchanger. The flow conditions are such that the overall heat transfer area required and the effectiveness, assuming the two streams are in parallel flow. Assume for both the streams C_p = 4.2 kJ/kg K.

1996

89. For a current wire of 20 mm diameter exposed to air (h = 20 W/m² K), maximum heat dissipation occurs when thickness of insulation (k = 0.5 W/mK) is

(*a*) 20 mm (*b*) 25 mm

(*c*) 20 mm (*d*) 10 mm

90. Show that in a double-pipe counterflow heat exchanger if $m_h c_h = m_c C_c$, the temperature profiles of the two fluids along its length are parallel straight lines.

NUMERICAL TYPE QUESTIONS

2016

1. A steel ball of 10 mm diameter at 1000 K is required to be cooled to 350 K by immersing it in a water environment at 300 K. The convective heat transfer coefficient is 1000 W/m²-K. Thermal conductivity of steel is 40 W/m-K. The time constant for the cooling process τ is 16 s. The time required (in *s*) to reach the final temperature is ________

2. An infinitely long furnace of 0.5 m × 0.4 m cross-section is shown in the figure below. Consider all surfaces of the furnace to be black. The top and bottom walls are maintained at temperature $T_1 = T_3$ = 927 °C while the side walls are at temperature $T_2 = T_4$ = 527 °C. The view factor, F_{1-2} is 0.26. The net radiation heat loss or gain on side 1 is ______ W/m.

Stefan-Boltzmann constant = 5.67×10^{-8} W/m²-K⁴

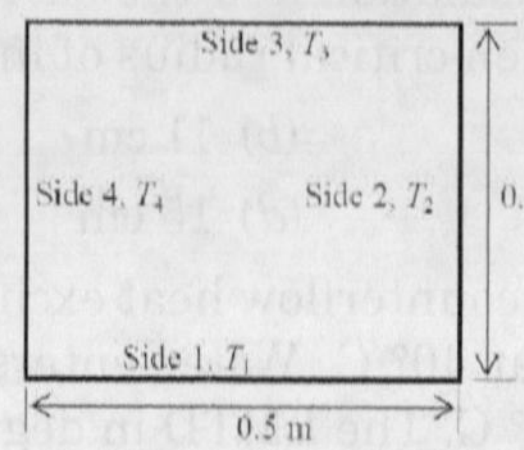

3. Consider a parallel-flow heat exchanger with area A_p and a counter-flow heat exchanger with area A_c. In both the heat exchangers, the hot stream flowing at 1 kg/s cools from 80°C to 50°C. For the cold stream in both the heat exchangers, the flow rate and the inlet temperature are 2 kg/s and 10 °C, respectively. The hot and cold streams in both the heat exchangers are of the same fluid. Also, both the heat exchangers have the same overall heat transfer coefficient. The ratio A_c/A_p is ______.

4. Two cylindrical shafts A and B at the same initial temperature are simultaneously placed in a furnace. The surfaces of the shafts remain at the furnace gas temperature at all times after they are introduced into the furnace. The temperature variation in the axial direction of the shafts can be assumed to be negligible. The data related to shafts A and B is given in the following Table.

Quantity	Shaft A	Shaft B
Diameter (m)	0.4	0.1
Thermal conductivity (W/m-K)	40	20
Volumetric heat capacity (J/m³-K)	2×10^6	2×10^7

The temperature at the centerline of the shaft A reaches 400°C after two hours. The time required (in hours) for the centerline of the shaft B to attain the temperature of 400°C is ______

5. Steady one-dimensional heat conduction takes place across the faces 1 and 3 of a composite slab consisting of slabs A and B in perfect as shown in the figure, where k_A, k_B denote the respective thermal conductivities. Using the data as given in the figure, the interface temperature T_2 (in °C) is ______

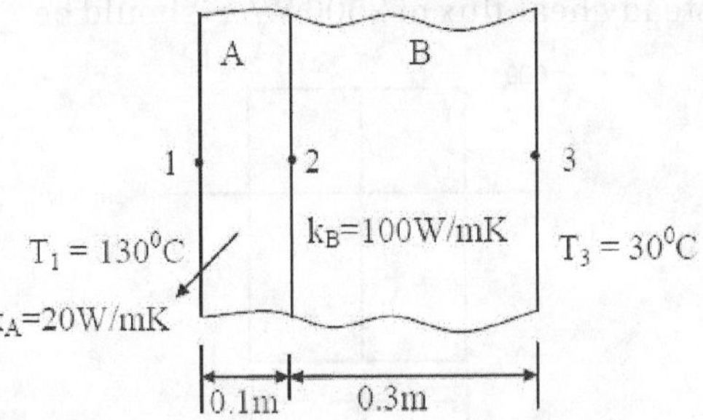

6. Two large parallel plates having a gap of 10 mm in between them are maintained at temperatures T_1 = 1000 K and T_2 = 400 K. Given emissivity values, $\varepsilon_1 = 0.5$, $\varepsilon_2 = 0.25$ and Stefan-Boltzmann constant $\sigma = 5.67 \times 10^{-8}$ W/m²-K⁴, the heat transfer between the plates (in kW/m²) is ______.

7. A cylindrical steel rod, 0.01 m in diameter and 0.2 m in length is first heated to 750 °C and then immersed in a water bath at 100 °C. The heat transfer coefficient is 250 W/m²-K. The density, specific heat and thermal conductivity of steel are ρ = 7801 kg/m³, c = 473 J/kg-K, and k = 43 W/m-K, respectively. The time required for the rod to reach 300 °C is ______ second

2015

8. A well insulated rigid container of volume 1 m³ contains 1.0 kg of an ideal gas [C_p = 1000 J/(kg.K) and C_v = 800 J/(kg.K)] at a pressure of 10^5 Pa. A stirrer is rotated at constant rpm in the container for 1000 rotations and the applied torque is 100 N-m. The final temperature of the gas (in K) is ______.

9. If a foam insulation is added to a 4cm outer diameter pipe as shown in the figure, the critical radius of insulation (in cm) is ______.

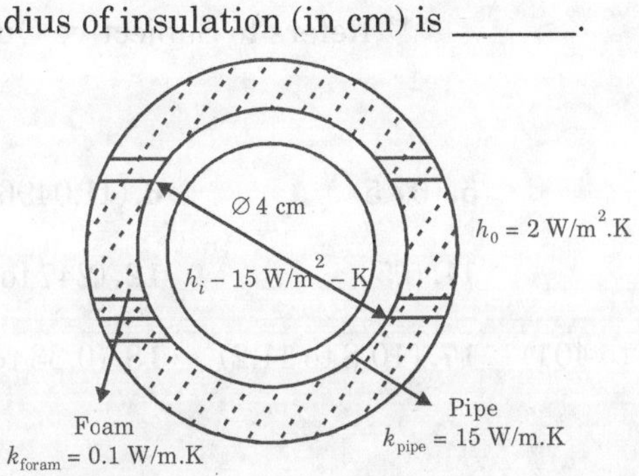

10. A Prandtl tube (Pitot-static tube with C = 1) is used to measure the velocity of water. The differential manometer reading is 10 mm of liquid column with a relative density of 10. Assuming g = 9.8 m/s², the velocity of water (in m/s) is ______.

11. Refrigerant vapor enters into the compressor of a standard vapor compression cycle at – 10°C (h = 402 kJ/kg) and leaves the compression at 50°C(h = 432 kJ/kg). It leaves the condenser at 30°C (h = 237 kJ/kg). The COP of the cycle is ______.

12. One side of a wall is maintained at 400 K and the other at 300 K. The rate of heat transfer through the wall is 1000 W and the surrounding temperature is 25°C. Assuming no generation of heat within the wall, the irreversibility (in W) due to heat transfer through the wall is ______.

2014

13. Consider one dimensional steady state heat conduction across a wall (as shown in figure below) of thickness 30 mm and thermal conductivity 15 W/m.K. At x = 0, a constant heat flux, q'' = 1×10⁵ W/m² is applied. On the other side of the wall, heat is removed from the wall by convection with a fluid at 25°C and heat transfer coefficient of 250 W/m².K. The temperature (in °C), at x = 0 is ______

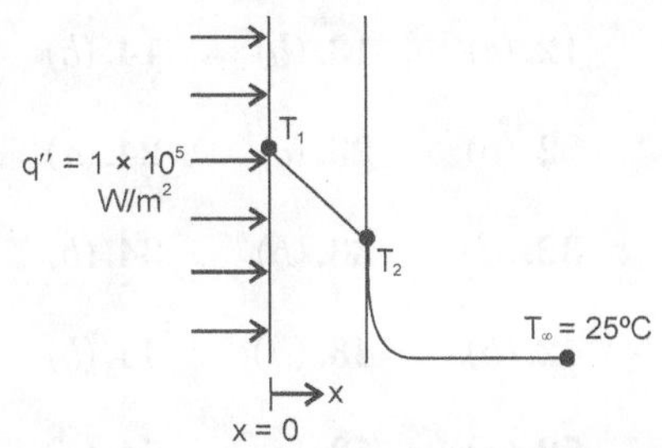

14. A hemispherical furnace of 1 m radius has the inner surface (emissivity, ε = 1) of its roof maintained at 800 K, while its floor (ε = 0.5) is kept at 600 K. Stefan-Boltzmann constant is 5.668 × 10⁻⁸ W/m².K⁴. The net radiative heat transfer (in kW) from the roof to the floor is ______

15. A material P of thickness 1 mm is sandwiched between two steel slabs, as shown in the figure below. A heat flux 10 kW/m² is supplied to one of the steel slabs as shown. The boundary temperatures of the slabs are indicated in the figure. Assume thermal conductivity of this steel is 10 W/m.K. Considering one-dimensional steady state heat conduction for the configuration, the thermal conductivity (k, in W/m.K) of material P is ______

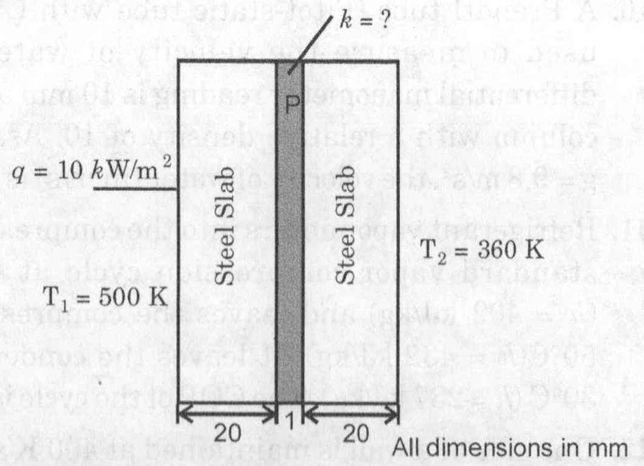

16. Heat transfer through a composite wall is shown in figure. Both the sections of the wall have equal thickness (l). The conductivity of one section is k and that of the other is $2k$. The left face of the wall is at 600 K and the right face is at 300 K. The interface temperature T_i (in K) of the composite wall is ______

17. A double-pipe counter-flow heat exchanger transfers heat between two water streams. Tube side water at 19 liter/s is heated from 10°C to 38°C. Shell side water at 25 liter/s is entering at 46°C. Assume constant properties of water; density is 1000 kg/m^3 and specific heat is 4186 J/kg K. The LMTD (in °C) is ______

18. A plane wall has a thermal conductivity of 1.15 W/m.K. If the inner surface is at 1100°C and the outer surface is at 350°C, then the design thickness (in meter) of the wall to maintain a steady heat flux of 2500 W/m^2 should be ______

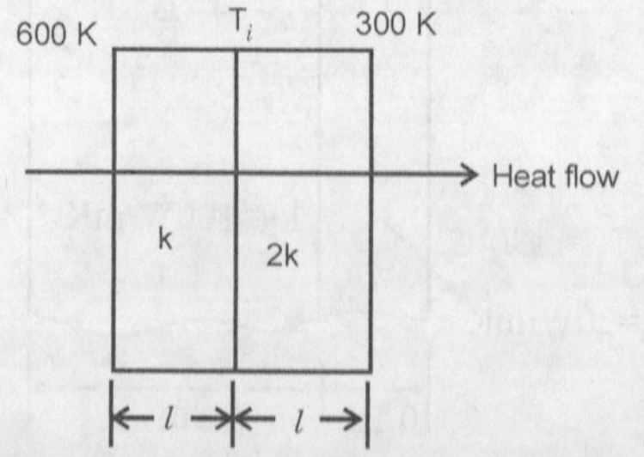

ANSWERS

MCQ Type Questions

1. (*a*)	**2.** (*b*)	**3.** (*a*)	**4.** (*d*)	**5.** (*b*)	**6.** (*b*)	**7.** (*c*)	**8.** (*c*)	**9.** (*a*)	**10.** (*b*)
11. (*a*)	**12.** (*c*)	**13.** (*b*)	**14.** (*b*)	**15.** (*c*)	**16.** (*c*)	**17.** (*b*)	**18.** (*a*)	**19.** (*c*)	**20.** (*a*)
21. (*a*)	**22.** (*b*)	**23.** (*c*)	**24.** (*a*)	**25.** (*d*)	**26.** (*c*)	**27.** (*d*)	**28.** (*a*)	**29.** (*a*)	**30.** (*a*)
31. (*d*)	**32.** (*d*)	**33.** (*b*)	**34.** (*b*)	**35.** (*b*)	**36.** (*c*)	**37.** (*a*)	**38.** (*d*)	**39.** (*b*)	**40.** (*d*)
41. (*b*)	**42.** (*b*)	**43.** (*d*)	**44.** (*b*)	**45.** (*d*)	**46.** (*d*)	**47.** (*c*)	**48.** (*a*)	**49.** (*d*)	**50.** (*c*)
51. (*c*)	**52.** (*a*)	**53.** (*c*)	**54.** (*d*)	**55.** (*c*)	**56.** (*c*)	**57.** (*a*)	**58.** (*a*)	**59.** (*c*)	**60.** (*b*)
61. (*a*)	**62.** (*c*)	**63.** (*a*)	**64.** (*b*)	**65.** (*b*)	**66.** (*c*)	**67.** (*d*)	**68.** (*b*)	**69.** (*)	**70.** (*d*)
71. (*c*)	**72.** (*)	**73.** (*b*)	**74.** (*c*)	**75.** (*b*)	**76.** (*)	**77.** (*c*)	**78.** (*b*)	**79.** (*)	**80.** (*a*)
81. (*a*)	**82.** (*d*)	**83.** (*)	**84.** (*)	**85.** (*)	**86.** (*b*)	**87.** (*b*)	**88.** (*)	**89.** (*b*)	**90.** (*)

"*" Refers to Subjective Question.

Numerical Type Questions

1. 42.22 sec	**2.** 24530.688W/m	**3.** (0.9278)	**4.** (5)	**5.** (67.5)	**6.** (11.0496)
7. (43.49)	**8.** 1283.4-1287.4	**9.** (5)	**10.** 1.32	**11.** 5.5	**12.** (247 to 249)
13. (620 to 630)	**14.** (24.0 to 25.2)	**15.** (0.09 to 0.11)	**16.** (399 to 401)	**17.** (10.8 to 11.2)	**18.** (0.33 to 0.35)

EXPLANATIONS

MCQ TYPE QUESTIONS

1.

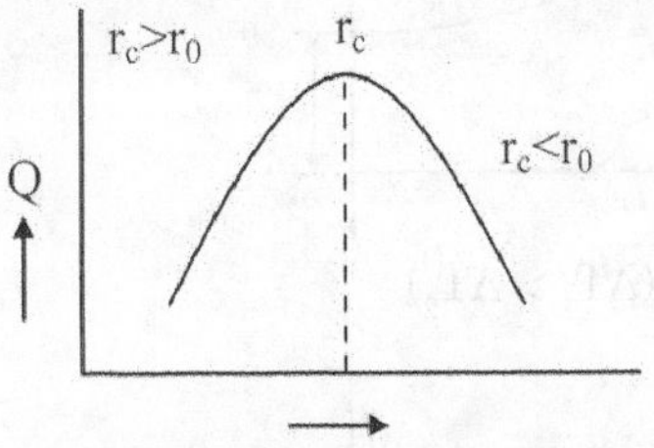

r_0 = 1 mm, k = 0.15 W/m-k

h = 25 W/m²-K

$$r_c = \frac{k}{h_0} \text{ for cylindrical shape}$$

$$= \frac{0.15}{25} \times 1000 = 0.15 \times 40 = 6 \text{ mm}$$

$\therefore$ $r_c > r_0$

$\Rightarrow$ The heat transfer from the wire to the ambient will increase

3.

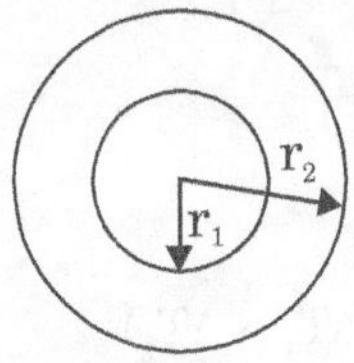

For a hollow cylinder

$$Q = \frac{2\pi k L(T_1 - T_2)}{\ln\left(\frac{r_2}{r_1}\right)}$$

$$= \frac{T_1 - T_2}{\frac{\ln\left(\frac{r_2}{r_1}\right)}{2\pi k L}} = \frac{T_1 - T_2}{R_{th}}$$

$$\therefore \quad R_{th} = \frac{\ln\left(\frac{r_2}{r_1}\right)}{2\pi k L}$$

4.

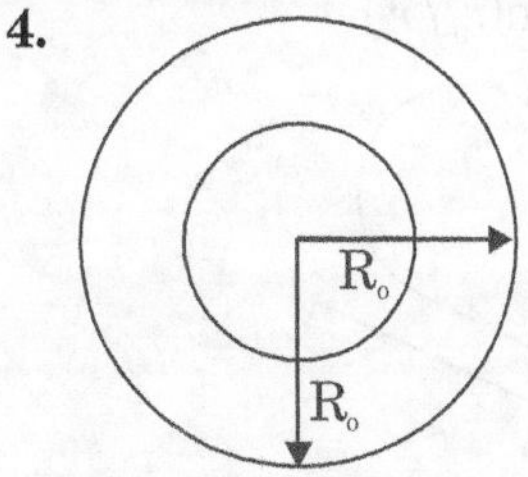

View factor

$$F_{1-1} = 0$$

$$F_{1-2} = 1$$

Reciprocacy theorem

$$A_2F_{2-1} = A_1F_{1-2}$$

$$\Rightarrow \quad F_{2-1} = \frac{A_1}{A_2}F_{1-2} = \frac{2\pi R_i L}{2\pi R_o L} = \frac{R_i}{R_o}$$

$$F_{2-1} + F_{2-2} = 1$$

$$\therefore \quad F_{2-2} = 1 - F_{2-1} = 1 - \frac{R_i}{R_o}$$

5. Grashof number = inertia force × $\frac{\text{Buoyancy force}}{(\text{viscous force})^2}$

6. The temperature difference is not for a given fluid but across the fluids and maximum heat transfer occurs for C_{min} and the temperature difference is equal to $\left(T_{h_i} - T_{c_i}\right)$.

7. r_c = 8mm

$\therefore$ the heat lost increases to maximum and then decreases.

8. Emissive power $\propto T^4$

$$\Rightarrow \quad \frac{500}{1200} = \left(\frac{T_1}{T_2}\right)^4$$

$$\therefore \quad \frac{T_1}{T_2} = 0.8034$$

9. Counter flow heat exchanged

Surface Area A = 20 m² mass flowrate = 0.4 kg/s

$$u = \frac{20W}{m^2K} \quad \text{Temperature } T_{c}i = 280K$$

$$C_p \text{ of air} = 1000\frac{1}{kgK}$$

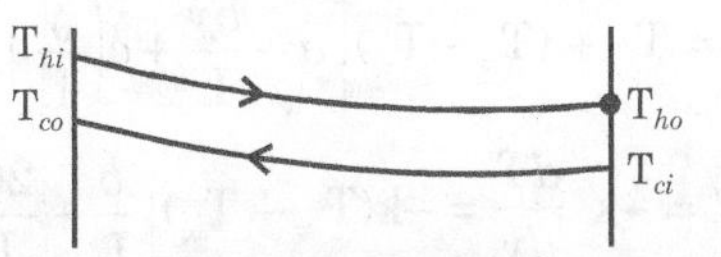

Since m is same for both flow = 0.4 kg/s

Assume C_p is same = 100 J/kg.K

Hence $\quad \Delta T_1 = T_i - T_{c0} = \Delta T_2 = T_{h0} - T_{ci}$

$$\Delta T_1 = 300 - T_{co} = T_{ho} - 280$$

$$\Delta T_m = \Delta T_1 = \Delta T_2$$

$$uA\Delta T_m = mC_p(T_{co} - T_{ci})$$

$$20 \times 20 \times (300 - T_{co}) = 0.4 \times 1000(T_{co} - 280)$$

$$2T_{co} = 300 + 280$$

$$T_{co} = \frac{580}{2} = 290 \text{ K}$$

10. The surface temperature,

$$(T_w) = T_a + \frac{q_g}{2h}R = 298 + \frac{4\times10^7}{2\times1000}\times5\times10^{-3}$$

$$T_w = 398 \text{ K}$$

11. $Pr = \dfrac{\mu C_p}{k} = \dfrac{\rho V C_p}{k} = \dfrac{V}{\dfrac{k}{\rho c_p}} = \dfrac{V}{\alpha}$

12.

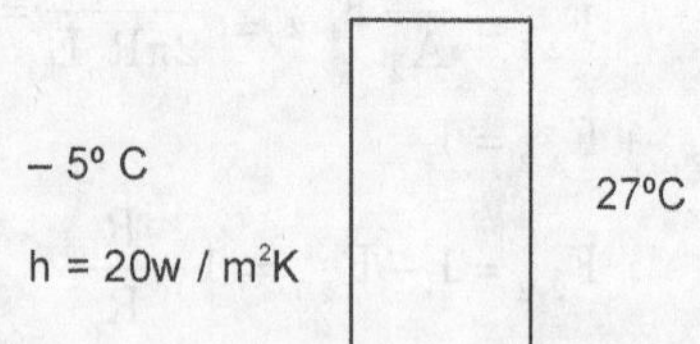

Total thermal resistance $\dfrac{1}{R_{th}} = \dfrac{1}{\dfrac{1}{h} + \dfrac{1}{k}}$

$$\frac{1}{R_{th}} = \frac{1}{\dfrac{1}{20} + \dfrac{0.18}{0.9}} = 4 \text{ W / m}^2\text{K}$$

$$Q = \frac{\Delta T}{R_{th}} = [27 - (-5)] \times 4 = 128 \text{ W / m}^2$$

13. Biot - number :

Biot number provides a way to compare the conduction resistance within a solid body to the convection resistance external to the body (offered by the surrounding fluid) for heat transfer:

$$Bi = \frac{hs}{k}; \quad s = \frac{\text{Volume of the body}}{\text{Surface area}}$$

Where 's' is a characteristic dimension of the solid
'h' is convective heat transfer coefficient
'k' is thermal conductivity of the body.

14. $\dfrac{T_w - T}{T_w - T_\infty} = a + \dfrac{by}{L} + c\left(\dfrac{y}{L}\right)^2$

$$T = T_w + (T_\infty - T_w)\left[a + \frac{by}{L} + c\left(\frac{y}{L}\right)^2\right]$$

$$q'' = -k\frac{dT}{dy} = -k(T_\infty - T_w)\left[\frac{b}{L} + \frac{2Cy}{L^2}\right]$$

at y = 0

$$\Rightarrow \frac{q''L}{(T_w - T_\infty)k} = b = N_u$$

$\therefore \quad N_u = b.$

15. <u>Temperature distribution</u>

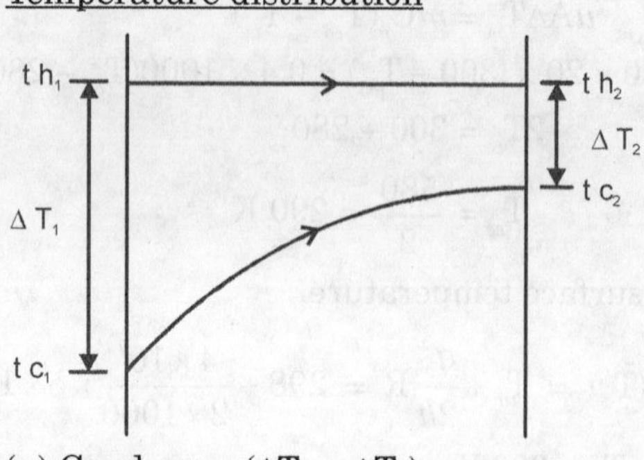

(*a*) Condenser ($\Delta T_1 > \Delta T_2$)

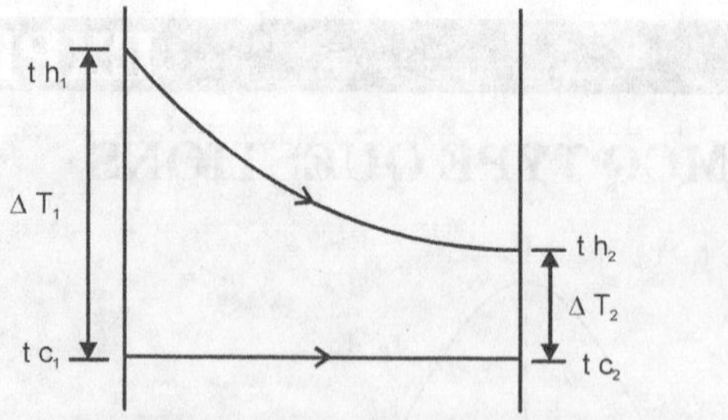

(*b*) Evaporator ($\Delta T_1 > \Delta T_2$)

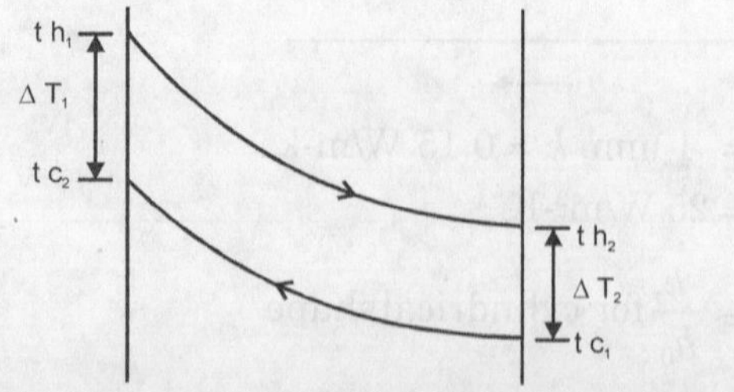

(*c*) Counter flow Heat Exchanger ($\Delta T_1 = \Delta T_2$)

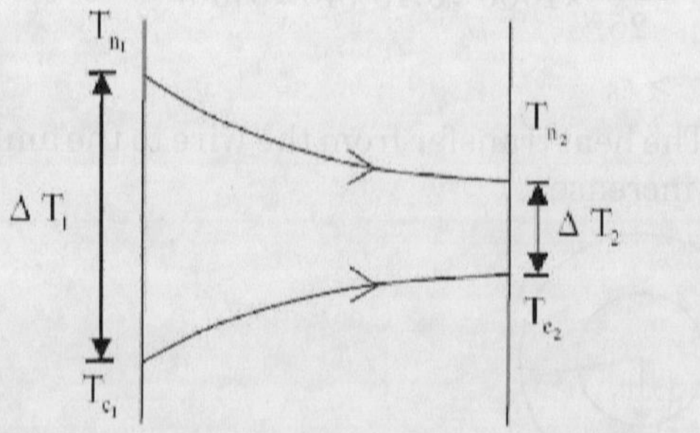

(*d*) Parallel flow Exchanger ($\Delta T_1 > \Delta T_2$)

From the above temperature distribution, we can say option '*c*' is the correct answer.

16. $A_r = 2\pi rL$

From Fourier 's Law

$$q_r = kA_r\frac{dT}{dr}$$

$$q_r = 2\pi kr L\frac{dT}{dr}$$

Boundary conditions:

$T = T_i$ at $r = r_i$

$T = T_0$ at $r = r_0$

$$q = \frac{2\pi kL(T_i - T_0)}{\ln(r_0 / r_i)}$$

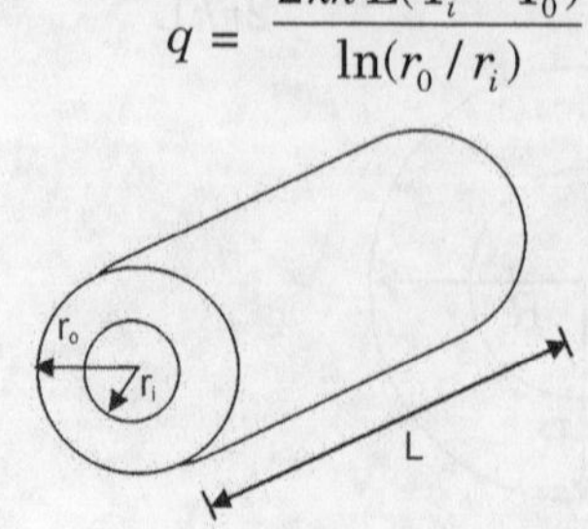

$$= \frac{T_i - T_0}{\dfrac{\ln(r_0 / r_i)}{2\pi kL}} = \frac{T_i - T_0}{R_{th}}$$

$$R_{th} = \frac{\ln(r_0 / r_i)}{2\pi kL}$$

17. $F_{11} + F_{12} = 1$, here $F_{11} = 0$

$\therefore \quad F_{12} = 1$

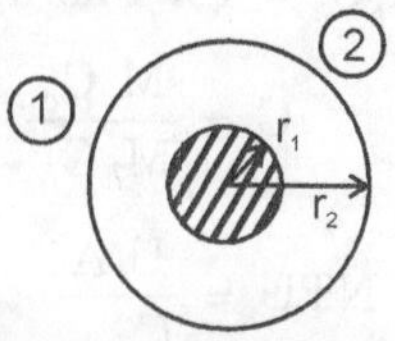

Now from reciprocating law

$$A_1 F_{12} = A_2 F_{21}$$

$$\therefore \quad 2\pi(20)^2 \times 1 = 2\pi(30)^2 \times F_{21}$$

$$\Rightarrow \therefore F_{21} = \left(\frac{10}{30}\right)^2 = \frac{4}{9}.$$

19. $RH = 0.7 = \dfrac{pv}{4.25}$ and $w = \dfrac{0.622pv}{100 - pv}$

20. $P_1 V_1 \ln p_2/p_1 = W_{isothermal}$

21. $$\dot{Q} - \dot{W} = \dot{m}\left[(h_2 - h_1) + \left(\frac{V_2^2 - V_1^2}{2}\right)\right]$$

and $\dot{\theta} = -5$ kW $\dot{m} = 1 kg/s$

22. For flow stream,

$$A.E. = (h_2 - h_1) + K.E. - T_0(S_2 - S_1)$$
$$= c_p(T_2 - T = 187 \text{ kJ/kg}$$

25. $$F_{22} = 1 - F_{21}$$
$$= 1 - \frac{A_1}{A_2} = 1 - \frac{D_1}{D_2}$$

26. $$\varepsilon = \frac{C_h}{C_{min}} \times \frac{(t_{h_1} - t_{h_2})}{(t_{h_1} - t_{c_1})}$$

$$0.8 = \frac{4.18 \times 0.5}{2.09 \times 1} \times \frac{(80 - t_{h_2})}{(80 - 30)}$$

$$t_{h_2} = 40°C$$

$$m_h c_{p_h}(t_{h_1} - t_{h_2}) = m_c c_{p_c}(t_{c_2} - t_{c_1})$$

$$0.5 \times 4.18 \times 40 = 2.09 \times 1 \times (t_{c_2} - 30)$$

$$t_{c_2} = 70°C$$

$$\theta_1 = \theta_2 = 10°C$$

$\therefore \quad$ LMTD = 10°C

27. Heat and work are path functions.

Since δQ and δW are dependent on path followed between two given end states of a thermodynamic process undergone by system.

28. $$W_{electric} = i^2R$$

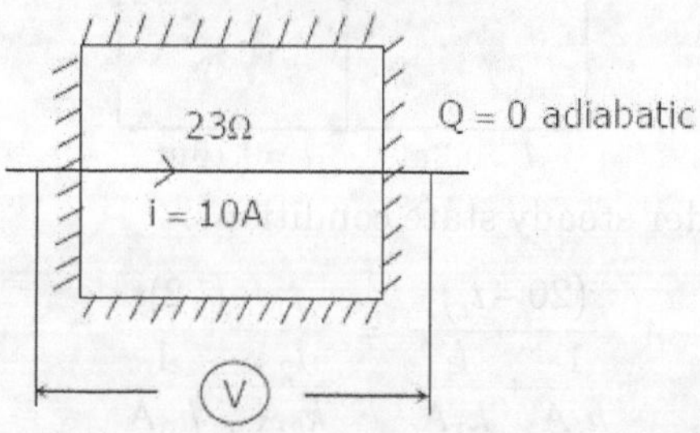

$$W_{electric} = i^2R$$
$$= (10^2 \times 20) \text{ watts}$$
$$= -2.3 \text{ kW (on system)}$$

First law : $Q - w = \Delta U$

$$\therefore \quad 0 - (-W_{electric}) = \Delta u$$

29. Work done by the turbine

$$W = 2800 - 1800$$
$$= 1000 \text{ kJ/kg}$$
$$= 1000 \text{ kW-s/kg}$$

$$\text{Specific fuel consumption} = \frac{1}{1000} \times 3600$$
$$= 3.6 \text{ kg/kW} - \text{hr}$$

30. Ideal Brayton cycle,

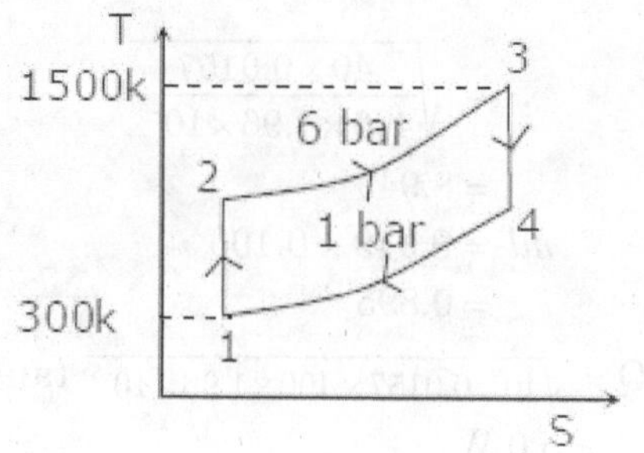

At the end of compression,

$$\text{Temperature, } T_2 = T_1\left(\frac{P_3}{P_4}\right)^{\frac{\gamma-1}{\gamma}}$$

$$= 300 \times 6^{\frac{0.4}{1.4}} = 500 \text{ K}$$

At the end of expansion;

$$\text{Temperature, } T_4 = \frac{T_3}{\left(\frac{P_3}{P_4}\right)^{\frac{\gamma-1}{\gamma}}} = \frac{1500}{6^{\frac{0.4}{1.4}}} = 900 \text{ K}$$

31. Work done by pump $= \upsilon(p_2 - p_1) = \dfrac{(30-1) \times 100}{990}$

$$= 2.93 \text{ kJ/kg}$$

32. Stroke of the cylinder, $l = 2r = 2 \times 60 = 120$ mm

$$\text{Swept volume} = \frac{\pi}{4} d^2 \times l$$
$$= \frac{\pi}{4} \times 80^2 \times 120$$
$$= 603 \text{ cm}^3$$

34. Fin diameter, d = 5mm

Fin length, L = 100 mm

k = 400 W/mK

T_0 = 130°C

T_∞ = 30°C

h = 40 W/m²K

∴ Heat loss from fin

$$Q = \sqrt{kAhP}\,(T_o - T_\infty)\ \tanh ml$$

where $P = \pi d$

$= 3.14 \times 5 \times 10^{-3}$

$= 0.015$

$$A = \frac{\pi}{4} d^2$$

$$= \frac{\pi}{4} \times (5 \times 10^{-3})^2$$

$$= 1.96 \times 10^{-5}\ m^2$$

$$\therefore \quad m = \sqrt{\frac{hp}{kA}}$$

$$= \sqrt{\frac{40 \times 0.0157}{400 \times 1.96 \times 10^{-5}}}$$

$= 8.949$

$\therefore \quad ml = 8.949 \times 0.100$

$= 0.895$

$$\therefore \quad Q = \sqrt{40 \times 0.0157 \times 400 \times 1.96 \times 10^{-5}}\ \tanh(0.895)$$

$= 5.0$ W

35. Given : $t_{h_1} = 102°$ C

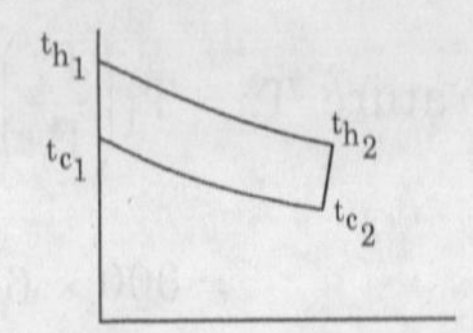

$t_{c_2} = 15°$C

$$NTU = \frac{UA}{C_{min}}$$

$$= \frac{1 \times 10^3 \times 5}{1 \times 4 \times 10^3}$$

$= 1.25$

$$\in = \frac{1 - \exp(-2NTU)}{2}$$

$$= \frac{1 - \exp(-2.5)}{2}$$

$= 0.458$

$$\therefore \quad 0.458 = \frac{t_{c_1} - t_{c_2}}{t_{h1} - t_{c_2}} = \frac{t_{c_1} - 15}{102 - 15}$$

or $t_{c_1} = 0.458 \times (102 - 15) + 15$

$= 54.846°$ C

$\approx 55°$C

Alternately

$C = M_c C_c = M_h C_h = 1 \times 4$ kJ/kg K

$$\therefore \quad C = \frac{M_c C_c}{M_h C_h} = 1$$

$$NTU = \frac{U.A}{C_{min}}$$

$$= \frac{1 \times 5}{4} = \frac{5}{4}$$

From the relation

$$\in = \frac{1 - e^{-(2 \times 5/4)}}{2} = 0.46$$

Now, $$\in = \frac{t_{c_2} - 15}{102 - 15}$$

$\therefore \quad 0.46 = t_{c_2}$

$\Rightarrow \quad t_{c_2} = 54.9° \approx 55°$C

36. $$R_{eq} = \frac{1}{h_2 A} + \frac{l_1}{k_1 A} + \frac{l_2}{k_2 A} + \frac{1}{h_2 A}$$

$$= \frac{1}{A}\left(\frac{1}{20} + \frac{0.30}{20} + \frac{0.15}{50} + \frac{1}{50}\right)$$

$$= \frac{0.088}{A}$$

Now $$Q_2 = \frac{(t_1 + t_2)}{R_{eq}}$$

$$= \frac{t_1 - t}{\frac{1}{h_i A} + \frac{L_1}{k_1 A}}$$

$$\therefore \quad \frac{22}{\frac{0.88}{A}} = \frac{t_1 - t}{\frac{0.065}{A}}$$

$$\Rightarrow \quad \frac{2.95 \times 0.065}{0.088} = 293 - t$$

$\Rightarrow \quad t = 3.75$ °C

Alternately

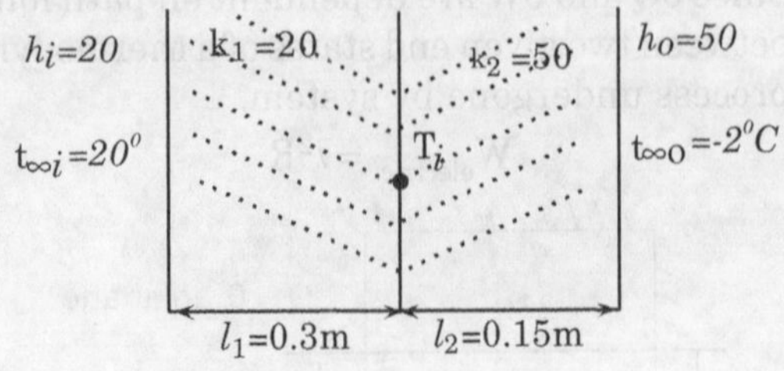

Under steady state conditions.

$$\frac{(20 - t_i)}{\frac{1}{h_i A} + \frac{l_1}{k_1 A}} = \frac{t_1 - (-2)}{\frac{l_2}{k_2 A} + \frac{1}{h_0 A}}$$

$$\therefore \quad \frac{20 - t_i}{\frac{1}{20} + \frac{0.3}{20}} = \frac{t_i + 2}{\frac{0.15}{50} + \frac{1}{50}}$$

$$\Rightarrow \quad \frac{20 - t_i}{\frac{1.3}{20}} = \frac{T_i + 2}{\frac{1.15}{50}}$$

$$\Rightarrow \quad 20 - t_i = \frac{1.3}{1.15} \times \frac{50}{20} \{t_i + 2\}$$

$$\Rightarrow \quad 20 - t_i = 2.826\,(t_i + 2)$$

$$\Rightarrow \quad 3.826\,t_i = 20 - 2 \times 2.826$$

$$\Rightarrow \quad t_i = 3.75°C$$

37. Since heat is taken from the high temperature sources and rejected to low temperature sink, hence device is a heat engine not heat pump. Since, the temperature differences are negligible, engine is reversible.

38. Since heat would be flowing from the block to the surroundings through face PQRS, hence jump would decrease from centre to PQRS to the surroundings

$$\Rightarrow \quad T_s \geq 30°$$

$$\therefore \quad T_s = 40°$$

39. $F_{11} + F_{12} = 1$

$F_{21} + F_{22} = 1$

$F_{11} = 0$ due to concave surface

$F_{12} = 1$

Now $A_1 F_{12} = A_2 F_{21}$

$$\Rightarrow \quad F_{21} = F_{12} \frac{A_1}{A_2}$$

$$\Rightarrow \quad F_{12} = 1$$

$$\therefore \quad F_{21} = \frac{A_1}{A_2} = \frac{2\pi r_1 l}{2\pi r_2 l}$$

$$= \frac{r_1}{r_2} = \frac{1}{2} = 0.5$$

$$\therefore \quad F_{22} = 1 - F_{21} = 0.5$$

40. Direction of heat flow is always normal to surface of constant temperature

So for surface P_1, $\frac{dT}{dx} = 0$

From energy conservation,

Heat rate at P = Heat rate at Q

$$0.1 \times 1 \left(\frac{dT}{dy}\right)_P = 0.1 \times 2 \times \left(\frac{dT}{dx}\right)_Q$$

$$\Rightarrow \quad \frac{dT}{dy} = 20 \text{ K/m}$$

41. Only (*b*) satisfies the conditions.

42. Heat capacity of hot fluid = 1 × 10 = 10 kJ/k – s

Heat capacity of cold fluid = 2 × 5 = 10 kJ/k – s

Since heat capacity is same, so LMTD is difference of temperature at either end

i.e., LMTD = 60° – 30° = 30°C

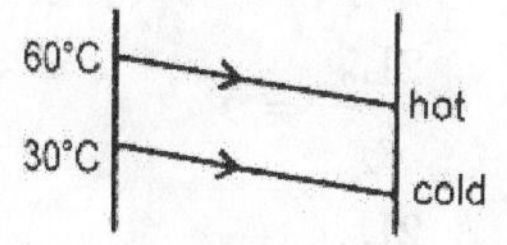

43. From heat balance equation,

$$-hA\,(T - T_0) = mc \frac{dt}{dz}$$

where, T_0 = Ambient temperature

$$\therefore \quad h = \frac{4 \times 2.5 \times 10^3 \times 0.2}{0.1 \times 200}$$

$$= 10 \text{ w/m}^2\text{k}$$

44. Calculated easily by formula.

45. $$\frac{Q_H}{Q_L} = \frac{400}{300} - \frac{4}{3}$$

$$Q_H + Q_L = Q = 100$$

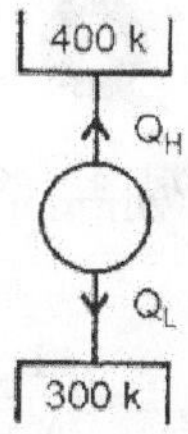

$$\therefore \quad \frac{4}{3} Q_L + Q_L = 100$$

$$\Rightarrow \quad Q_L = 42.85 \text{ kJ}$$

$$\therefore \quad Q_H = 100 - 42.85 = 57.14 \text{ kJ}$$

46. $$Q = \frac{-k_1 A}{2b}(T_{inter} - T_1)$$

$$= \frac{-k_2 A(T_2 - T_{inter})}{b}$$

$$T_{inter} = \frac{T_1 + T_2}{2}$$

$$\therefore \quad k_1 = 2k_2$$

47. Heat generated by bulb = 100 × 24 × 60 × 60 Joule.

$\therefore$ Heat dissipated $= \rho \times V \times C_v \times (T - 20)$

$\Rightarrow 100 \times 24 \times 60 \times 60 = 1.24 \times 9 \times 2.5 \times 3 \times C_v (T - 20)$

$$\Rightarrow \quad T = 450°$$

49. One dimensional heat conduction equation is

$$\frac{\partial^2 T}{\partial x^2} = \frac{1}{\alpha} \frac{\partial T}{\partial t}$$

$$\therefore \quad \frac{\partial^2 T}{\partial x^2} \propto \frac{\partial T}{\partial t}$$

50. Wien's displacement law is not satisfied, *i.e.*,

$$\lambda_{max} T = C$$

which tells lower λ_{max} for higher temperature.

51. $$\frac{L_1}{k_1} = \frac{0.5}{0.02} = 25$$

$$\frac{2L_2}{k_2} = 5$$

$$\frac{2L_3}{k_3} = 12.5$$

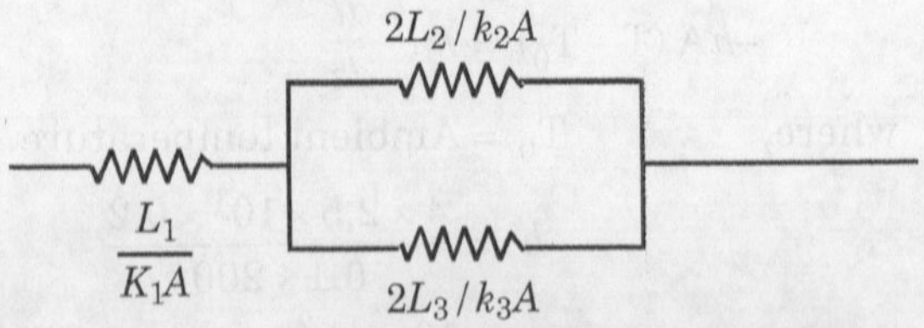

$\therefore$ $$R_{total} = 25 + 3.6 = 28.6 \text{ K/W}$$

53. $$A_1F_{12} + A_1F_{11} = 1$$

$$A_2F_{21} + A_2F_{22} = 1$$

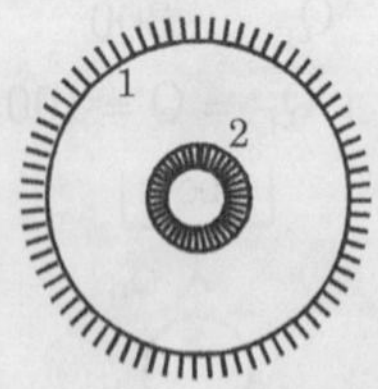

Since, $F_{22} = 0$

therefore $$F_{21} = \frac{1}{A_2}$$

$\Rightarrow$ $$F_{12} = \frac{1}{A_2}$$

Now, $$A_1F_{11} = 1 - A_1F_{12} = \left(1 - \frac{A_1}{A_2}\right)$$

$\therefore$ $$F_{11} = \frac{1}{A_1} - \frac{1}{A_2}$$

54. $$\theta = \frac{\theta_1 - \theta_2}{\ln\frac{\theta_1}{\theta_2}}$$

80° ⟶ 50°

30° ⟶ 40°

parallel flow

$\theta_1 = 50°$, $\theta_2 = 10°$

80° ⟶ 50

40° ⟵ 30°

Counter flow

$\theta_1 = 40°$, $\theta_2 = 20°$

As given value is between these two, so cross flow is the best option.

55. $$L_C = \frac{\Delta A}{D}$$

$$= 4 \times \frac{1 \times 0.5}{2 \times 1.5}$$

$$= 0.666$$

Reynolds number,

$$Re = \frac{\rho L_C v}{u}$$

$$= \frac{1.2 \times 0.66 \times 10}{18 \times 10^{-6}} = 4.4 \times 10^5$$

57. Given: Receiving solar radiation at the rate of 0.6 kW/m^2

Internal energy of fluid after absorbing solar radiation

$$= 0.6 \times \frac{1}{2} \text{ kW/m}^2$$

$$= 0.3 \text{ kW/m}^2$$

$$\eta_{engine} = 1 - \frac{315}{350} = 0.1$$

$$0.1 = \frac{W}{Q_1}$$

$\therefore$ $$Q_1 = \frac{2.5}{0.1} = 25 \text{ kW}$$

Let A be minimum area of collector.

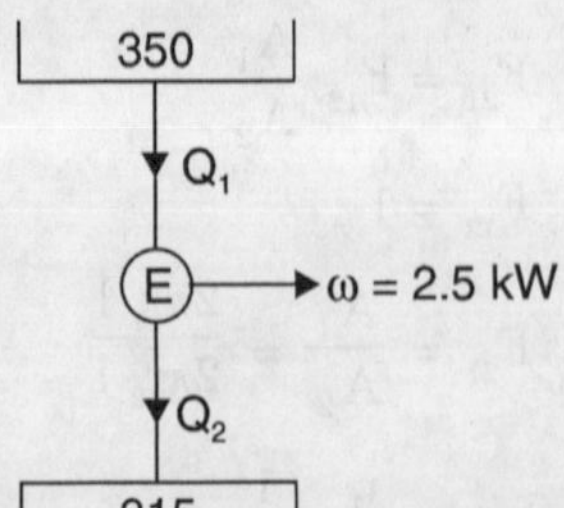

$\therefore$ $$Q_1 = 0.3 \times A$$

or $$25 \text{ kW} = 0.3 \text{ kW/m}^2$$

or $$A = \frac{25}{0.3}$$

$$= 83.33 \text{ m}^2.$$

58. Given : $T_1 = 1250$ K

$T_2 = 450$ K

Specific heat, $c = 0.5$ kJ/kgK

Ambient temperature, $T_0 = 303$ K

Mass of billet, $m = 2000$ kg.

Available energy $= m(h_1 - h_2) - T_0 \Delta s$

$= 490.439$ MJ

$$= m(h_1 - h_2) - T_0\, mc.\ln\frac{T_1}{T_2}$$

$$= mc(T_1 - T_2) - T_0\, mc.\ln\frac{T_1}{T_2}$$

$$= 2000 \times 0.5 \times (1250 - 450)$$

$$- 303 \times 0.5 \times 2000 \times \ln\left(\frac{1250}{450}\right)$$

Alternative method

Given : $T_0 = 303$ K

$T_2 = 450$ K

$m = 2000$ kg

$T_1 = 1250$ K

$c = 0.5$ kJ/kg.K

$$\text{Available energy} = mc\int_{T_1}^{T_2}\left(1 - \frac{T_0}{T}\right)dT$$

$$= mc\,[T - T_0 \ln T]_{T_1}^{T_2}$$

$$= mc\left[(T_2 - T_1) - T_0 \ln\frac{T_2}{T_1}\right]$$

$$= 2000 \times 0.5 \times 10^3 \left[450 - 1250 - 303\ln\frac{450}{1250}\right]$$

$$\omega_{max} = -490.44 \text{ MJ.}$$

59. Given : $d_1 = 2$cm

$r_1 = 1$ cm

$d_2 = 5$ cm

$r_2 = 2.5$ cm

$d_3 = 5 + 2 \times 3 = 11$

$r_3 = 5.5$ cm

Equivalent thermal resistance

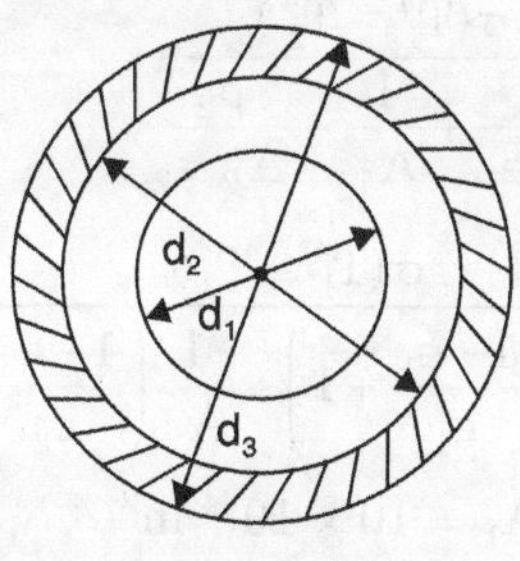

$$R_t = \frac{1}{2\pi k_{st}}\ln\frac{r_2}{r_1} + \frac{1}{2\pi k_a}\ln\frac{r_3}{r_2}$$

$$= \frac{1}{2\pi \times 19}\ln\frac{2.5}{1} + \frac{1}{2\pi \times 0.2} \times \ln\frac{5.5}{2.5}$$

$$= 7.67 \times 10^{-3} + 0.627$$

$$= 0.635$$

$\therefore$ Heat transfer rate per unit length

$$= \frac{t_1 - t_2}{R_t}$$

$$= \frac{600}{0.635} = 944.88 \text{ W/m.}$$

60. $$L_C = \frac{V}{A} = \frac{\frac{4}{3}\pi r^3}{4\pi r^2}$$

$$= \frac{r}{3} = \frac{0.706}{6}$$

$$= 0.1176 \text{ mm}$$

$$\alpha = \frac{k}{\rho c} = \frac{20}{400 \times 8500}$$

$$= 5.88 \times 10^{-6}$$

$$\text{Fourier number} = \frac{\alpha\tau}{(L_C)^2} = \frac{5.88 \times 10^{-6}\tau}{(0.1176 \times 10^{-3})^2}$$

$$= 425.17\,\tau$$

$$\text{Biot number} = \frac{hL_C}{k}$$

$$= \frac{400 \times 0.1176 \times 10^{-3}}{20}$$

$$= 0.00235$$

Now $$\frac{t - t_\infty}{t_i - t_\infty} = e^{-B_iFo}$$

$\therefore$ $$\frac{298 - 30}{300 - 30} = e^{-0.00235 \times 425.17\tau}$$

$\Rightarrow$ $\tau = 4.9$ sec.

Alternately

Given $h = 400$ W/m²K

$k = 20$ W/mK

$c = 400$ J/kg.K

$\rho = 8500$ kg/m³

$T(t) = 298$°C

$T_i = 30$°C

$T_\infty = 300$°C

Now, $$L = \frac{V}{A} = \frac{\frac{1}{6}\pi D^3}{\pi D^2} = \frac{1}{6}D$$

$$= \frac{1}{6} \times 0.706 \times 10^{-3} \text{ m}$$

$$= 1.76 \times 10^{-4} \text{ m.}$$

$$\text{Biot Number} = \frac{hL}{k}$$

$$= \frac{400 \times 1.176 \times 10^{-4}}{20}$$

$$= 0.0023.$$

Since $B_i < 0.1$, lumped system analysis can be used

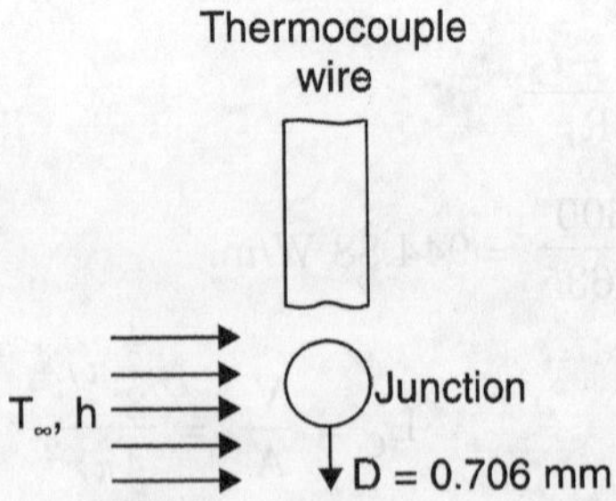

$$\frac{T(t)}{T_i - T_\infty} = e^{-\frac{hAT}{\rho cV}}$$

$$\frac{298-300}{30-300} = e^{-\frac{hAT}{\rho cV}}$$

$$\Rightarrow \quad \frac{-2}{270} = e^{-\frac{hAT}{\rho cV}}$$

$$\Rightarrow \quad \ln\left(\frac{270}{2}\right) = \frac{hAT}{\rho cV}$$

$$\Rightarrow T = \frac{\rho cV}{cA} \ln\left(\frac{270}{2}\right)$$

$$= \frac{\rho c}{h}\left(\frac{V}{A}\right) \ln 135$$

$$= \frac{8500 \times 400 \times 1.176 \times 10^{-4} \times 4.90}{400}$$

$\approx$ 4.9 sec.

61. Here, $\theta_1 = 120 - 30 = 90°C$

$\theta_2 = 120 - 80 = 40°C$

Log mean temperature difference,

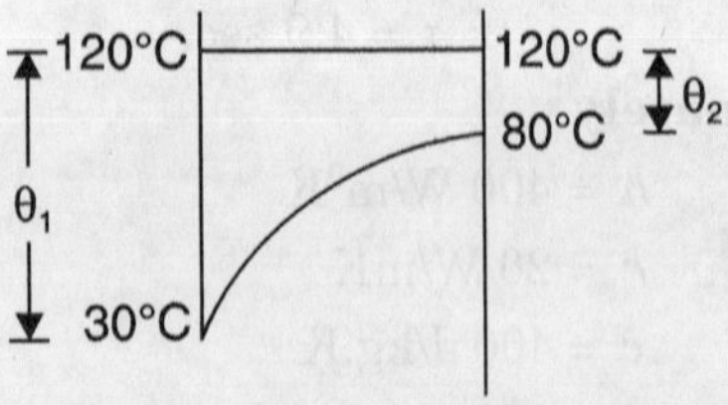

$$\theta_m = \frac{\theta_1 - \theta_2}{\ln\frac{\theta_1}{\theta_2}}$$

$$= \frac{(90-40)}{\ln\frac{90}{40}} = 61.65$$

$$Q = mc_{pw}\,\Delta T$$

$$= 1500 \times \frac{1}{3600} \times 4.187 \times (50) \times 10^3$$

$$= 87.22 \times 10^3 \text{ Joule/s}$$

Again $\quad Q = UA\theta_m$

$\therefore$ $87.22 \times 10^3 = 2000 \times A \times 61.65$

$\Rightarrow \quad A = 0.707 \text{ m}^2$

62. Heat supplied, $\quad Q = mc_p\,\Delta T$

$$= 2 \times 1000 \times 20 \times 60$$

$$= 2400 \times 1000 \text{ W}$$

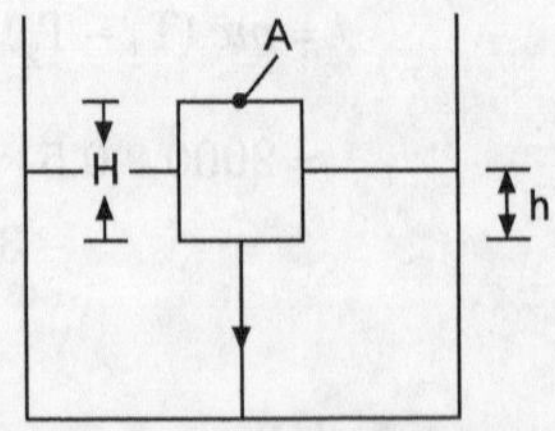

$\therefore$ Increase of water temperature,

$$\Delta T = \frac{Q}{mc_p} = \frac{2400 \times 1000}{40 \times 4.2 \times 1000}$$

$$= 14.3$$

63. Heat addition, $Q_2 = 750$ W

Heat rejected, $Q_1 = 1000$ W

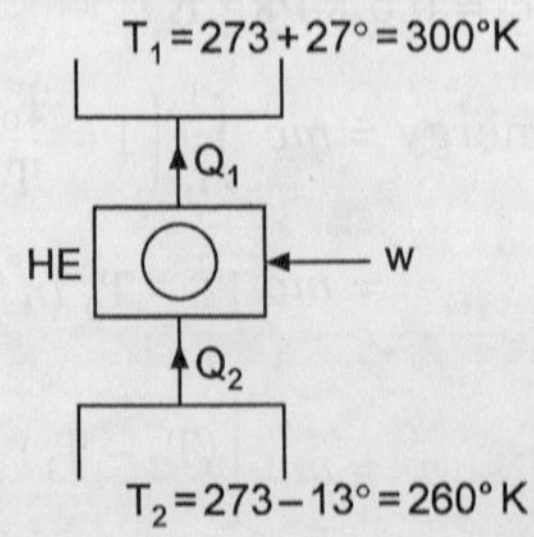

$$[\text{C.O.P}]_{\text{H.P.}} = \frac{Q_1}{W} = \frac{Q_1}{Q_1 - Q_2}$$

$$= \frac{1000}{1000-750} = 4$$

$$[\text{C.O.P.}]_{\text{H.P}} = \frac{T_1}{T_1 - T_2}$$

$$= \frac{300}{300-260} = 7.5$$

For better design, we take minimum C.O.P of heat pump.

64. Heal loss from one surface of the plate,

$$\theta_{net} = \frac{\sigma(T_1^4 - T_2^4)}{\frac{\rho_1}{A_1 \epsilon_1} + \frac{1}{A_1} + \frac{\rho_2}{A_2 \epsilon_2}}$$

$$= \frac{\sigma(T_1^4 - T_2^4)}{\frac{1}{A_1}\left(\frac{1-\epsilon_1}{\epsilon_1} + 1\right) + \frac{1}{A_2}\left(\frac{1-\epsilon_2}{\epsilon_2}\right)}$$

Here, $\quad A_1 = 10 \times 10^{-4} \text{ m}^2$

$T_1 = 800$

$\epsilon_1 = 0.6$

$T_2 = 300$

$A_2 = 100 \text{ m}^2$

$\epsilon_2 = 0.3$

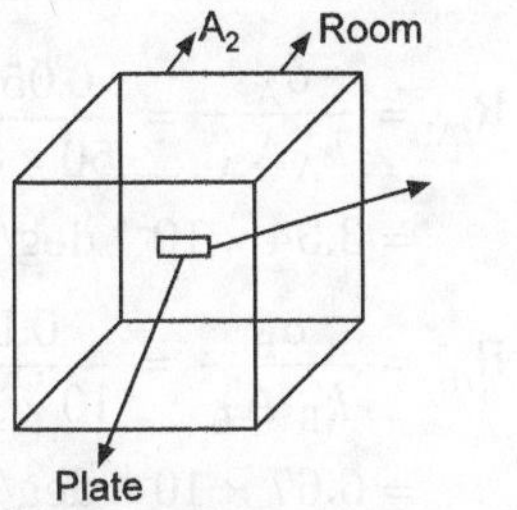

$$\therefore Q = \frac{5.67 \times 10^{-8}(800^4 - 300^4)}{\frac{10^4}{10}\left(\frac{1}{0.6}\right) + \frac{1}{100}\left(\frac{1}{0.3} - 1\right)} = 13.66 \text{ W}$$

$\therefore$ Heat loss from both the surfaces = 2 Q_{net}

= 2 × 13.66 = 27.32 W.

65. For hot fluid

Heat capacity, c_h = 2 kJ/kg K

Mass flow rate, m_h = 5 kg/s

Inlet temperature = $T_{P_{11}}$ = 150°C

Outlet temperature = $T_{P_{12}}$ = 100°C

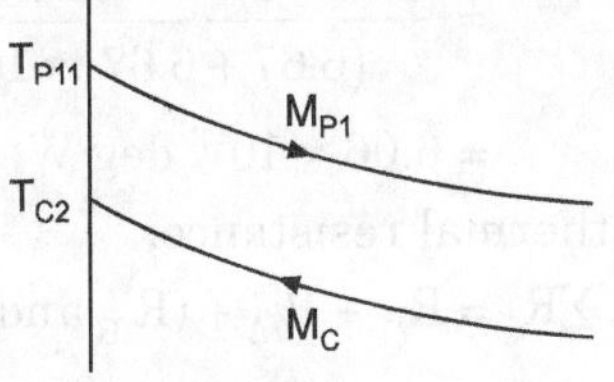

Similarly for cold fluid

Heat capacity, c_C = 4 KJ/kg.K

Mass flow rate, $\dot{m}_C$ = 10 kg/s

Inlet temperature, T_{C_1} = 20°C

Outlet temperature, T_{C_2} = ?

Since heat gain = Heat loss

$\Rightarrow 2 \times 5 \times (150 - 100) = 4 \times 10 \times (T_{C2} \times 20)$

$$\Rightarrow \quad T_{C2} = 20 + \frac{2 \times 5 \times 50}{4 \times 10}$$

= 32.5°C

66. Temperature of water = 48°C

Temperature of glass plate = 40°C

Thermal conductivity of water = $k_w = 0.6\ \omega/mK$

Thermal conductivity of glass = $k_G = 1.2\ \omega/mK$

The spatial gradient of temeprature in water at the water-glass interface = $\left(\frac{dT}{dy}\right)_w = 1 \times 10^5$ K/m

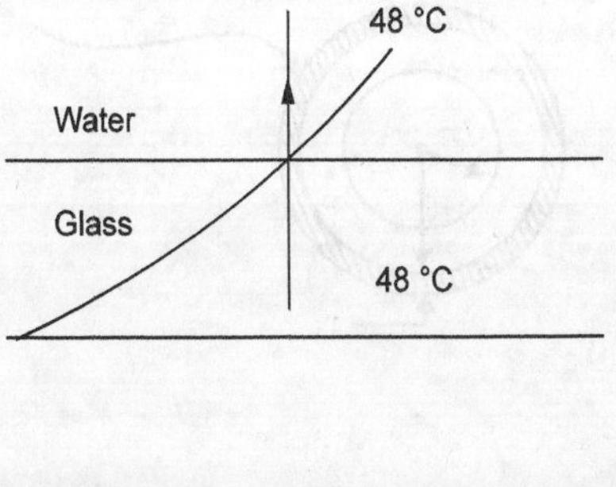

At water glass interface

$$q = k_w \left(\frac{dT}{dy}\right)_w = k_G \left(\frac{dT}{dy}\right)_G$$

$$\Rightarrow \quad \left(\frac{dT}{dy}\right)_G = \frac{k_w}{k_G} \times \left(\frac{dT}{dy}\right)_w$$

$$= \frac{0.6}{1.1} \times 1 \times 10^4 = 0.5 \times 10^4 \text{ K/m}$$

67. For convection

Heat flux, $q = h(T_f - T_i) = k_w \left(\frac{dT}{dy}\right)_w$

$= 0.6 \times 1 \times 10^4 = 6000$ W/m

$$\therefore \quad h = \frac{q}{(T_f - T_i)} = \frac{6000}{8}$$

= 750 W/m^2K

69. Given : $D_1 = 20 - 2 \times 1 = 18$ mm

r_1 = 9 mm

r_2 = 10 mm

D_2 = 20 mm

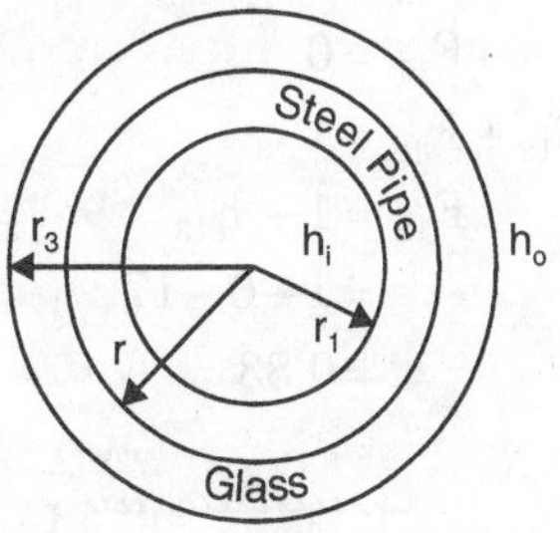

h_i = 150 W/m² – K

h_o = 5 W/m² – K

K_{st} = 400 W/m–K

$K_{G/am}$ = 0.075 W/m–K

Critical thickness of insulation

$$r_c = \frac{K}{h_o} = \frac{0.075}{5} = 0.015 \text{ m} = 15 \text{ cm}$$

Hence $r_3 = r_c + 5 = 20$ mm

Total thermal resistance per unit length of the pipe

$$R_t = \frac{1}{2\pi r_1 h_i} + \frac{1}{2\pi K_{st}} \ln \frac{r_2}{r_1} + \frac{1}{2\pi K_{G/am}} \ln \frac{r_3}{r_2} + \frac{1}{2\pi r_3 h_o} \text{ m}$$

$$= \frac{1}{2\pi \times 9 \times 10^{-3} \times 150} + \frac{1}{2\pi \times 400} \ln \frac{10}{9} + \frac{1}{2\pi \times 0.75} \ln \frac{20}{10} + \frac{1}{2\pi \times 20 \times 10^{-3} \times 5}$$

= 3.18 deg/watt

Hence heat loss per unit length of the pipe

$$= \frac{t_i - t_o}{R_t}$$

$$= \frac{150 - 27}{3.18}$$

$$= 38.679 \text{ W/m}$$

Hence total heat loss in one hour per metre length of the pipe

$$= 38.679 \times 3600 \text{ J/sec-m}$$

$$= 139.245 \text{ kJ/hour-m}$$

Hence rate of steam condensation

$$= \frac{139.245}{2230} \text{ kg/hr}$$

$$= 0.06244 \text{ kg/hr-m}$$

70. Given : $\frac{l}{r_1} = 2$

$$\frac{r_3}{l} = 0.5$$

$$F_{13} = 0.17$$

$$F_{11} = 0$$

$$F_{11} + F_{12} + F_{13} = 1$$

$$\therefore \quad F_{12} = 1 - F_{13}$$

$$= 1 - 0 - 17$$

$$= 0.83$$

71.

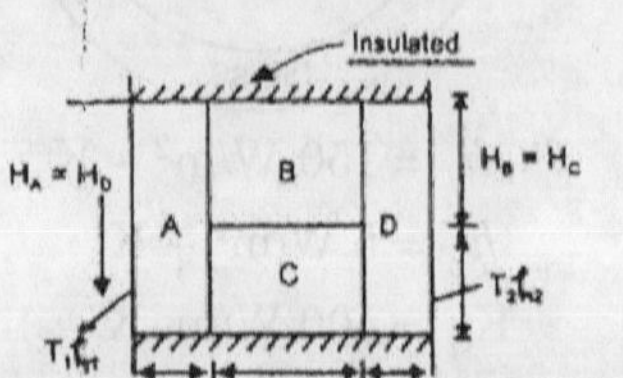

(a)

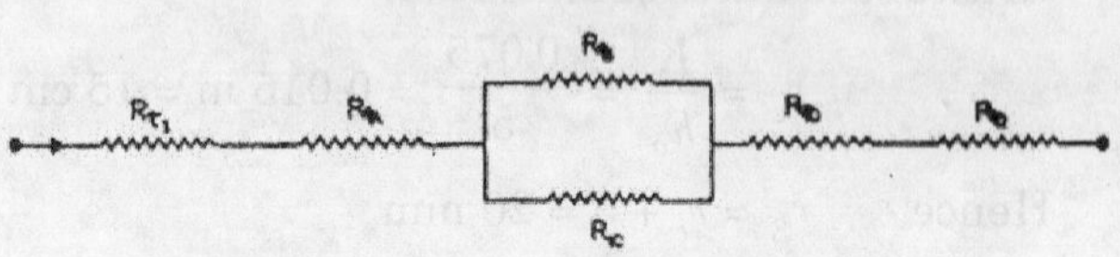

This is thermal circuit of the system.

(b) Now we have to determine the rate of heat transfer through the wall.

$$A_1 = A_2 = 3 \times 1 = 3\text{m}^2 = A_A = A_D$$

and $A_B = A_C = 1.5 \times 1 = 1.5 \text{ m}^2$

(i) $R_{t1} = \frac{1}{h_1 A_1}$

$$= \frac{1}{50 \times 3}$$

$$= \frac{1}{150} = 6.67 \times 10^{-3} \text{ deg/W}$$

(ii) $R_{tA} = \frac{\delta_A}{k_A A_A} = \frac{0.05}{50 \times 3}$

$$= 3.34 \times 10^{-4} \text{ deg/W} = R_{tD}$$

(iii) $R_{tB} = \frac{\delta_B}{k_B A_B} = \frac{0.1}{10 \times 1.5}$

$$= 6.67 \times 10^{-3} \text{ deg/W}$$

(iv) $R_{tC} = \frac{\delta_c}{k_c A_c} = \frac{0.1}{10 \times 1.5}$

$$= 66.67 \times 10^{-3} \text{ deg/W}$$

(v) $R_{t2} = \frac{1}{h_2 A_2} = \frac{1}{10 \times 3}$

$$= 33.34 \times 10^{-3} \text{ deg/W}$$

Since R_{tB} and R_{tc} are in parallel, therefore equivalent resistance

$$= \frac{R_{t_B} \times R_{t_c}}{R_{t_B} + R_{t_c}}$$

$$= \frac{6.67 \times 10^{-3} \times 6.67 \times 10^{-3}}{(6.67 + 6.67) \times 10^{-3}}$$

$$= 6.06 \times 10^{-3} \text{ deg/W}$$

$\therefore$ Total thermal resistance,

$$\Sigma R_t = R_{t1} + R_{tA} + (R_{tB} \text{ and } R_{tC})_{\text{Equivalent}} + R_{tD} + R_{t2}$$

$$= [6.67 \times 10^{-3} + 0.334 \times 10^{-3} + 6.06 \times 10^{-3} + 0.334 \times 10^{-3}] \text{ deg/W}$$

$$= 46.73 \times 10^{-3} \text{ deg/W}$$

Given : $T_1 = 200°C$,

$$T_2 = 25°C$$

$\therefore$ Heat transfer through the wall,

$$Q = \frac{\Delta t}{\Sigma R_t} = \frac{200 - 25}{46.73 \times 10^{-3}}$$

$$= 3744.91 \text{W}$$

73. $\Delta h = \Delta Q - \Delta W$

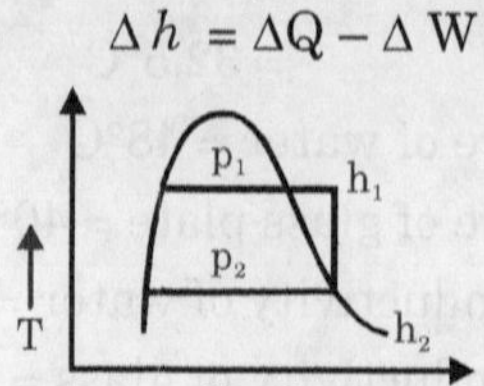

$$\therefore \quad \Delta Q = (-300 + 250) \text{ kT/kg}$$

$$= -50 \text{ kT/kg}$$

74. $h = 8 \text{ W/m}^2\text{k}$,

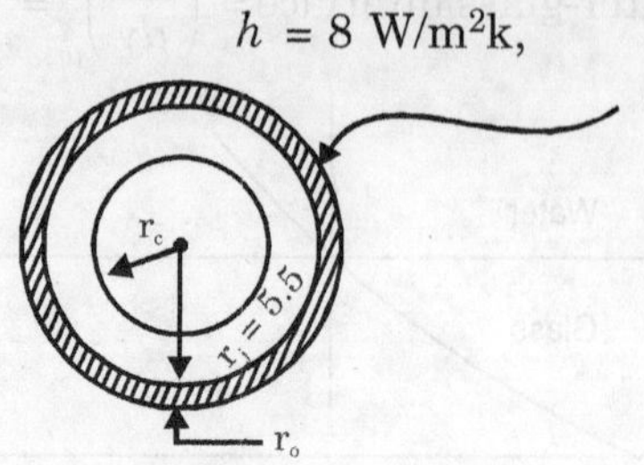

$$k = 1 \text{ W/mk}$$

$$r_i = \frac{10}{2} = 5 \text{ cm and}$$

$$r_o = \frac{11}{2} = 5.5 \text{ cm}$$

$\therefore$ Critical radius, $r_c = \frac{1}{8} = 12.5$ cm

75. $\therefore$ $\Delta T_1 = 70 - 50$

$= 20°C,$

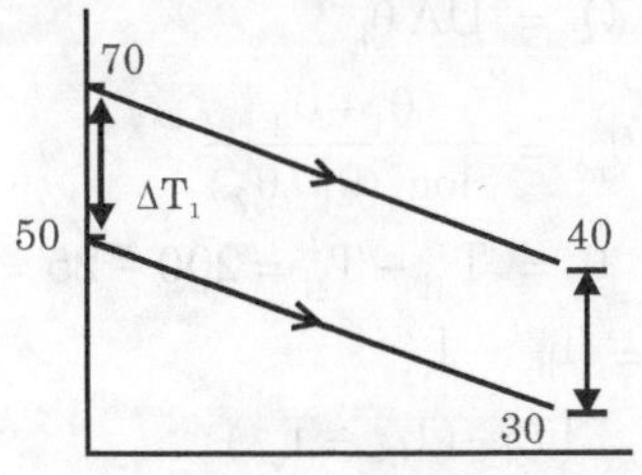

$$\Delta T_2 = 40 - 30 = 10°C$$

$$\text{LMTD} = \frac{\Delta T_1 \ \Delta T_2}{\log\left(\frac{\Delta T_1}{\Delta T_2}\right)} = \frac{20 - 10}{\log\left(\frac{20}{10}\right)}$$

$$= \frac{10}{\log 2} = 14.43$$

76. Given : $t_\infty = 30°C = 30 + 273$

$= 303$ K

$t_s = 50°C = 50 + 273$

$= 323°$ K

$h = 17.4 \text{ W/m}^2 \text{ k}$

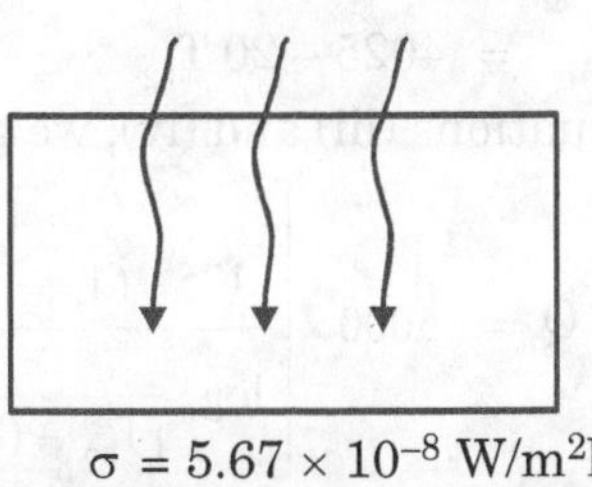

$\sigma = 5.67 \times 10^{-8} \text{ W/m}^2\text{k}^4$

$\alpha = 0.9$

$\varepsilon = 0.1$

In equilibrium condition,

Heat input to plate = Heat output from the plate

$\Rightarrow$ $\alpha I = h(t_s - t_\infty) + \sigma\varepsilon(T_s^4 - T_\infty^4)$

$\Rightarrow$ $0.9 I = 17.4 \times 20 + 5.67 \times 10^{-8} \times 0.1 \times [(323)^4 - (303)^4]$

$= 348 + 13.92 = 361.92$

$\therefore$ Incident radiation,

$$I = \frac{361.92}{0.9} = 402.13 \ \frac{W}{m^2}$$

$$= 0.402 \ \frac{kW}{m^2}$$

78. Critical radius of insulation,

$$r_c = \frac{k}{h} = \frac{0.1}{100} = 0.001 = 1 \text{ mm.}$$

Radius of min = 0.5 mm.

$\therefore$ Thickness of coating = 1 – 0.5 = 0.5 mm.

79. Given, $t_{1A} = 420°$ C,

$m_A = 1$ kg/s,

$t_{1B} = 20°$

$m_B = 1$ kg/s

$\varepsilon = 75\%,$

$C_{pA} = 1$ kJ/kgK,

$C_{pB} = 4$ kJ/kgk.

We know, $\varepsilon = \frac{t_{A1} - t_{A2}}{t_{A1} - t_{B1}} = 0.75$

[since ${}^mC_{ph} > {}^mC_{pc}$]

where, $t_{A_2} = t_{A_1} - 0.75\,(420 - 20)$

$= 420 - 0.75\,(420 - 20) = 120°$ C.

Heat transfer rate,

$$Q = mC_{pc}(t_{B_2} - t_{B_1})$$

$$= 1 \times 4\,(t_{B_2} - 20°)$$

$$= mC_{pn}\,(t_{A_1} - t_{A_2})$$

or $4(t_{B_2} - 20°) = 1 \times 1\,(420 - 120°) = 300$

$\therefore$ $t_{B_2} = \frac{300}{4} + 20° = 95°$ C

Heat transfer rate = 300 kJ/s = 300 kW

81. Heat lost by ball = Heat gained by water

or $m_b \times C_b \times (T_b - T_s) = m_w \times C_w \times (T_s - T_w)$

where subscript w stands for water and b stands for ball.

and T_s = steady state temperature

$\therefore$ $1 \times 0.4 \times (60 - T_s) = 1 \times 4.18 \times (T_s - 20)$

[since $C_w = 4.18$ kJ/kg]

or $24 \times 0.4\,T_s = 4.18\,T_s - 83.6$

$\therefore$ $T_s = 23.49°C \approx$ **23.5°C**

82. $Q = KA\frac{dt}{dx}$

where Slope $= \frac{dt}{dx} = \frac{Q}{KA}$

i.e., slope is more for less value of K, therefore

$$\left(\frac{dt}{dx}\right)_{\text{I material}} > \left(\frac{dt}{dx}\right)_{\text{II material}}$$

But $Q = Q_2$

$\therefore$ $K < K_2$

83. Given, Current = 0.5 amp.

Voltage = 4 volts.

Inside tube radius, $R_t = 1$ mm = 0.001 m.

Wire radius, $R_w = 0.025$ mm.

Inside tube temperature,

$$T_1 = 150°C = 150 + 273 = 423°K.$$

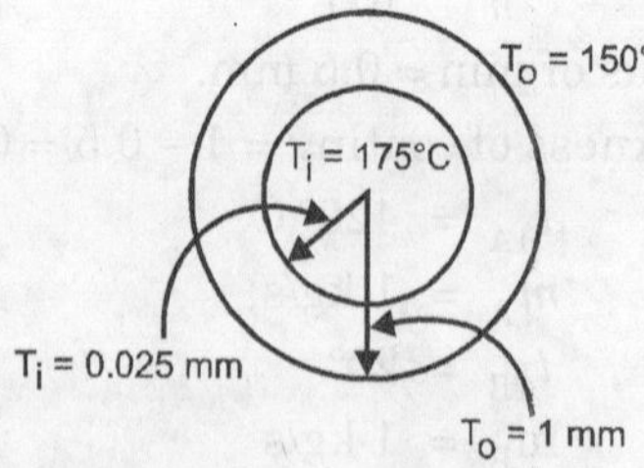

Wire temperature,

$$T_2 = 175 + 273 = 448°K$$

Length of the tube,

$$l = 25 \text{ cm}$$

Heat given by electrical wire = VI

$$= 4 \times 0.5 = 2W$$

Heat conducted by the tube $= \dfrac{2\pi kL(T_i - T_o)}{ln\left(\dfrac{r_o}{r_i}\right)}$

$$\therefore \quad VI = \frac{2\pi kL(T_i - T_o)}{ln\left(\dfrac{r_o}{r_i}\right)}$$

$$\therefore \text{ Thermal conductivity, } k = \frac{VI \,.ln\left(\dfrac{r_o}{r_i}\right)}{2\pi L(T_i - T_o)}$$

$$= \frac{2 \times ln\left(\dfrac{1}{0.025}\right)}{2\pi \times 25 \times 10^{-2}(175 - 150)}$$

$$= \mathbf{0.188 \ W/mK}$$

84. Given, Area, A = 20 m²

Overall heat transfer coefficient, U = 250 W/m²K

$$C_h = 2000 \text{ J/kg°K}$$

$$T_{c1} = 25°C$$

$$m_c = 2500 \text{ kg/hr} = 0.694 \text{ kg/sec}$$

$$C_c = 400 \text{ J/kg° K}$$

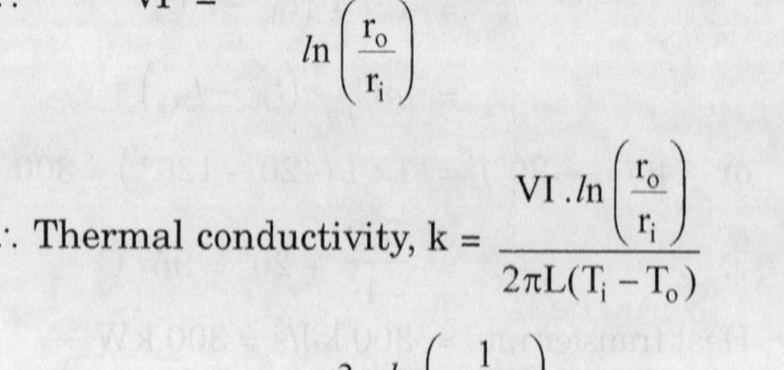

Heat transferred by hot fluid,

$$Q = m_h C_h (T_{h1} - T_{h2})$$

$$= 2.777 \times 2000 \times (200 - T_{h2})$$

$$= 5554 \times (200 - T_{h2}) \quad \text{..... (i)}$$

Heat transferred by cold fluid,

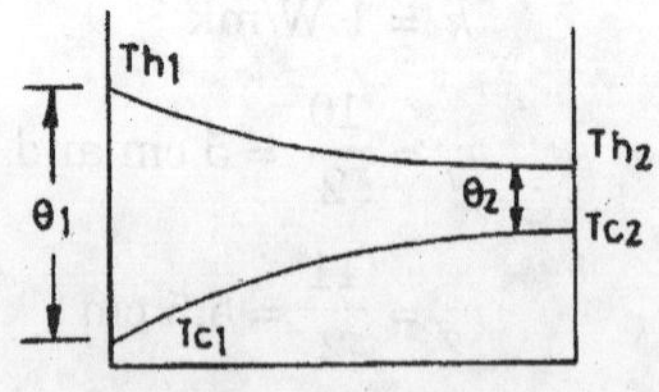

$$Q = m_c C_c (T_{c2} - T_{c1})$$

$$= 0.694 \times 400 \times (T_{c2} - 25)$$

$$= 277.6 \times (Tc_2 - 25) \quad \text{..... (ii)}$$

Also heat transferred is given by,

$$Q = UA\theta_m$$

$$\text{where } \theta_m = \frac{\theta_1 - \theta_2}{\log_e(\theta_1/\theta_2)}$$

$$\theta = T_{h1} - T_{c1} = 200 - 25 = 175°C$$

$$\theta_2 = T_{h_2} - T_{c_2}$$

$$\theta_m = \frac{175 - (T_{h_2} - T_{c_2})}{\log_e\left(\dfrac{175}{T_{h_2} - T_{c_2}}\right)}$$

$$\therefore \quad Q = UA\theta_m = 250 \times 20 \times \left[\frac{175 - (T_{h2} - T_{c2})}{\log_e\left(\dfrac{175}{T_{h2} - T_{c2}}\right)}\right] \quad \text{...(iii)}$$

From equations (i) and (ii), we have

$$5554 \times (200 - T_{h2}) = 277.6 \times (T_{c2} - 25)$$

$$\text{or} \quad 20 \times (200 - T_{h2}) = T_{c2} - 25$$

$$\text{or} \quad T_{c2} = 25 + 20 \times (200 - T_{h2})$$

$$= 4025 - 20\,T_{h2} \quad \text{......(iv)}$$

From equations (iii) and (iv), we get

$$Q = 5000 \times \left[\frac{175 - (T_{h_2} - (4025 - T_{h_2}))}{\log_e \dfrac{175}{[T_{h_2} - (4025 + 20T_{h_2})]}}\right]$$

$$= 5000\left[\frac{175 - (T_{h_2} - 4025 + 20T_{h_2})}{\log_e \dfrac{175}{[T_{h_2} - 4025 + 20T_{h_2}]}}\right]$$

$$= 5000\left[\frac{175 - (21T_{h_2} - 4025)}{\log_e \dfrac{175}{[(21T_{h_2} - 4025)]}}\right] \quad \text{......(v)}$$

Equating equations (i) and (v), we get

$$5554 \times (200 - T_{h2})$$

$$= 5000\left[\frac{175 - (21T_{h_2} - 4025)}{\log_e \dfrac{175}{[(21T_{h_2} - 4025)]}}\right]$$

By trial and error method, we get

$$T_{h2} = \mathbf{191.666°C}$$

∴ Temperature of hot fluid when it leaves the exchanger= **191.666°C**

85. Velocity of water, V = 0.1 m/sec

Length of pipe, l = 100

Diameter of pipe, d = 0.1 m

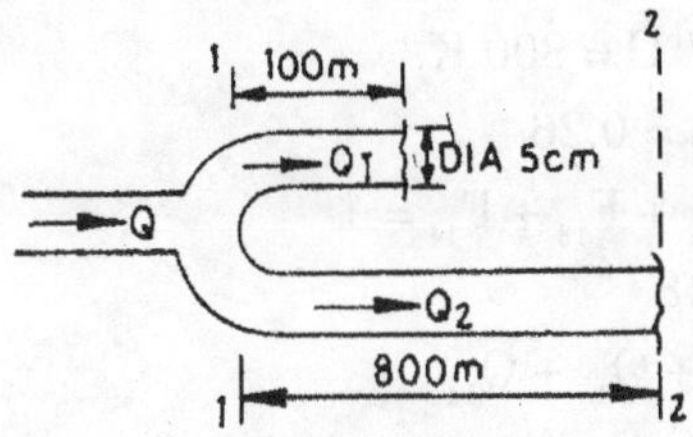

Flow rate, Q = A × V

$$= \frac{\pi d^2}{4} \times V$$

$$= \frac{\pi \times (0.1)^2}{4} \times 0.1$$

$$= 0.7854 \times 10^{-3} m^3/sec.$$

Mass flow rate, m = ρ × Q

where ρ = 1000 kg/m³.

$$\therefore \quad m = 1000 \times 0.7854 \times 10^{-3}$$

$$= 0.7854 \text{ kg/sec.}$$

Heat loss from the wall of pipe = 420 W/m²

Surface area of pipe = πd × l

$$= \pi \times 0.1 \times 100 = 31.415$$

∴ Heat loss, Q = 420 × 31.415 = 13194.3 Watt.

AlsoHeat loss, $Q = m \times C_p \times (T_i - T_o)$

where T_o = Outlet temperature, and

T_i = Inlet temperature = 80°C

$$\therefore \quad Q = 0.7854 \times 4.2 \times 10^3 \times (80 - T_o) = 1319.3.$$

or T_o = **76°C.**

87.

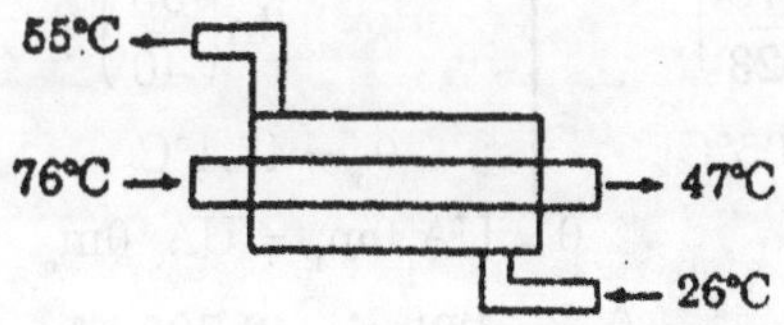

Effectiveness of the heat exchanger

$$= \frac{\text{Heat transferred}}{\text{Maximum possible heat transfer}}$$

$$= \frac{m\,s\,(55-26)}{m\,s\,(76-26)} = \mathbf{0.58}$$

88. *Given :* Flow rate of hot water,

$$m_h = 50000 \text{ kg/hour}$$

Flow rate of cold water,

$$m_c = 50{,}000 \text{ kg/hour}$$

$$t_{h1} = 95° C$$

$$t_{h2} = 65°C$$

$$t_{C1} = 30°C$$

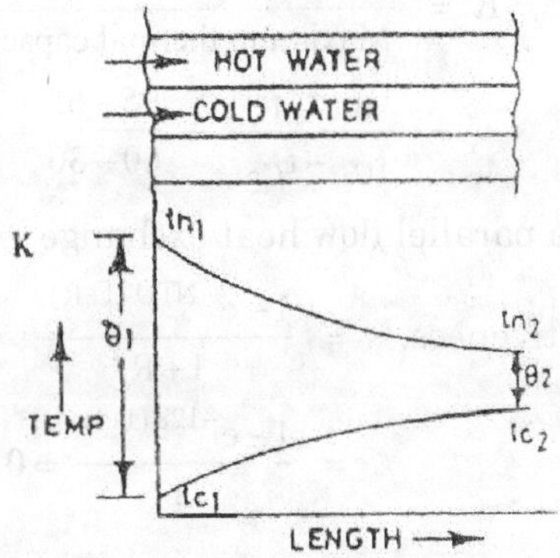

Overall heat transfer coefficient,

$$U = 2270 \text{ W/m}^2\text{K}$$

For both streams, C_p = 4.2 kJ/kg°K

Double Pipe Parallel Flow Heat Exchanger

$$95° \rightarrow 65°C$$

$$30°C \rightarrow t°C$$

$$95°C \rightarrow 65°C$$

∴ Hot fluid gives up heat,

$$q_h = m_h C_h (t_{h1} - t_{h2})$$

$$= 50{,}000 \times 4.2 \times (95 - 65) \text{ kJ/hour}$$

$$= 63 \times 10^5 \text{ kJ/hour} = 1750 \text{ kWatt.}$$

Now, Cold fluid pick up heat,

$$= m_c \cdot c_c (t_{c1} - t_{c2})$$

= Heat given by hot fluid.

or 50,000 × 4.2 × (95 − 65) = 50000 × 4.2 × (t − 30)

$$\therefore \quad t = 60°C$$

Now, Log Mean Temperature Difference

$$\text{LMTD} = \frac{\theta_1 - \theta_2}{ln\frac{\theta_1}{\theta_2}} = \frac{(95-30)-(65-60)}{ln\left(\frac{95-30}{65-60}\right)} = \frac{60}{ln 13}$$

= 23.4 degrees

Heat exchange,Q = U.A. LMTD

$$= (m\,C_p)(T_{h,in} - T_{h,out})$$

∴Heat transfer area required,

$$A = \frac{mC_p\,(T_{h,in} - T_{h,out})}{U.\,LMTD}$$

$$= \left(\frac{50{,}000}{60 \times 60}\right)\frac{4.2 \times 10^3 (95-65)}{2270 \times 23.4}$$

$$= \mathbf{32.9\ m^2}$$

∴ Effectiveness of heat exchanger,

$$\varepsilon = \frac{\text{Actual heat transfer}}{\text{Maximum possible heat transfer}}$$

$$= \frac{T_{h,in} - T_{h,out}}{T_{h,in} - T_{h,in}} = \frac{95-65}{95-30} = \mathbf{0.462}$$

ALTERNATELY

$$NTU = \frac{95-65}{23.4} = 1.28$$

$$R = \frac{\text{Minimum thermal capacity}}{\text{Maximum thermal capacity}}$$

$$= \frac{t_{H_1}-t_{H_2}}{t_{C_1}-t_{C_2}} = \frac{95-65}{60-30} = 1$$

For a parallel flow heat exchange

$$\text{effectiveness, } \varepsilon = \frac{1-e^{-NTU(1+R)}}{1+R}$$

$$= \frac{1-e^{-128(1+1)}}{2} = \mathbf{0.462}$$

90.

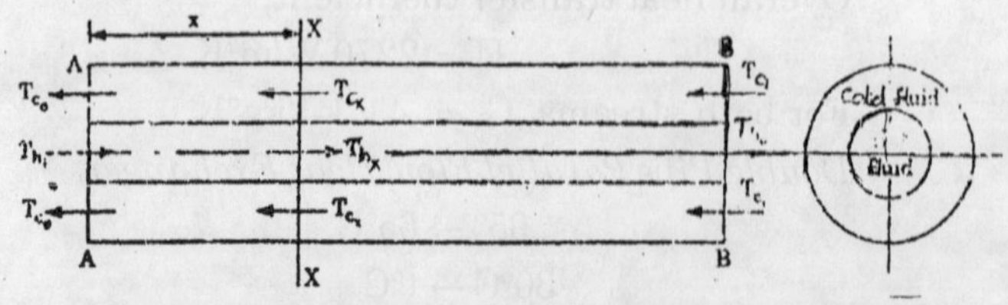

Consider a section XX at a distance x from the end AA.

Let T_{h_X} and T_{c_X} be the temperatures of hot and cold fluid in XX plane.

Heat given out by hot fluid in the zone AA, XX = $m_h \cdot c_c\ (T_{h_i} - T_{h_X})$.

This is equal to the heat gained by the cold fluid during its travel from XX to AA.

Heat received by cold fluid = $m_c \cdot c_c\ (T_{c_O} - T_{c_X})$

$$\therefore m_h \cdot c_h\ (T_{h_i} - T_{h_X}) = m_c c_c\ (T_{c_O} - T_{c_X})$$

$$\Rightarrow \quad (T_{h_i} - T_{h_X}) = (T_{c_o} - T_{c_x}) \quad \text{as } m_h c_h = m_c c_c$$

$$\text{or} \quad (T_{h_i} - T_{c_O}) = (T_{h_X} - T_{c_X})$$

Temperature difference between hot and cold fluid is constant at any cross-section. It means the temperature profiles of the two fluids along its length are parallel straight lines.

NUMERICAL TYPE QUESTIONS

1. $$\text{Biot Number} = \frac{hL_C}{K}$$

$$\text{For sphere } L_C = \frac{\text{Volume}}{\text{surface area}} = \frac{d}{6}$$

$$\therefore \quad Bi = \frac{hd}{6K} = \frac{1000\times0.01}{6\times40}$$

$$= 0.0416 < 0.1$$

Hence lumped heat analysis is used

$$\frac{T-T_\infty}{T_i-T_\infty} = e^{\frac{-hA_St}{\rho VC_p}} = e^{\frac{-t}{t^*}}$$

Thermal time constant,

$$t^* = \frac{\rho VC_p}{hA_S} = 16\,\text{sec}$$

$$\therefore \quad \frac{350-300}{1000-300} = e^{\frac{-t}{16}}$$

$$\Rightarrow \quad t = 42.2249\ \text{sec}$$

2. $T_1 = 927°C = 1200$ K,

$T_2 = 527°C = 800$ K,

$F_{12} = F_{14} = 0.26$

$F_{11} + F_{12} + F_{13} + F_{14} = 1$

$F_{13} = 0.48$

$Q = Q_{12} + Q_{13} + Q_{14}$

$Q_{13} = 0$ since the temperatures are same

$$Q = Q_{12} + Q_{14} = 2\times\sigma_b\times A\times F_{12}\times\left(T_1^4 - T_2^4\right)$$

$$Q = 2\times5.67\times10^{-8}\times(0.5\times1)\times0.26\times(1200^4-800^4)$$

$$= 24530.688$$

3.

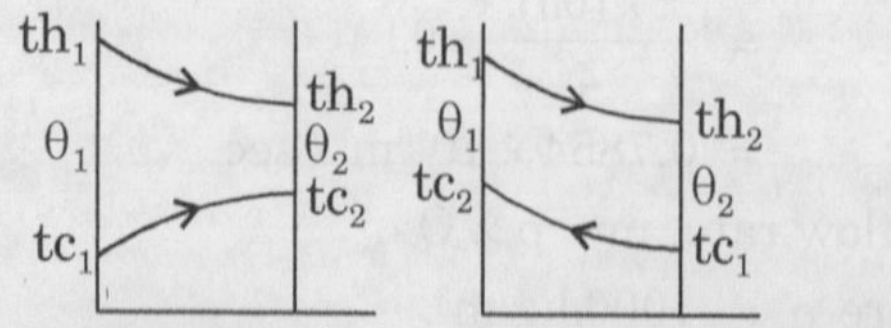

Parallel flow	Counter flow
$\theta_1 = th_1 - tc_1$	$\theta_1 = th_1 - tc_2$
$= 80 - 10$	$= 80 - 25$
$= 70°C$	$= 55°C$
$\theta_2 = th_2 - tc_2$	$\theta_2 = th_2 - tc_1$
$= 50° - 25°$	$= 50 - 10$
$= 25°$	$= 40°$
$\theta_m = \frac{\theta_1-\theta_2}{\ln\left(\frac{\theta_1}{\theta_2}\right)}$	$\theta_m = \frac{\theta_1-\theta_2}{\ln\left(\frac{\theta_1}{\theta_2}\right)}$
$\theta_m = \frac{70-25}{\ln\left(\frac{70}{23}\right)}$	$\theta_m = \frac{50-40}{\ln\left(\frac{55}{40}\right)}$
$\theta_m = 43.705°C$	$\theta_m = 47.1°C$

$$\theta = UA_P\theta m_P = UA_c\,\theta m_c$$

$$\frac{A_C}{A_P} = \frac{\theta m_P}{\theta m_c} = \frac{43.705}{47.1} = 0.927$$

4. $$\frac{T-T_\infty}{T_i-T_\infty} = e^{\frac{-hA_st}{\rho VC_p}}$$

$(\rho C_P)_A = 2\times10^6\ J/m^3K$,

$(\rho C_P)_B = 2\times10^7\ J/m^3K$,

$d_A = 0.4$ m, $d_B = 0.1$ m

$t_A = 2hrs = 2\times3600$

$= 7200$ sec

$h_A = h_B$

$$\left(\frac{hA_s}{\rho VC_p}\times t\right)_A = \left(\frac{hA_s}{\rho VC_p}\times t\right)_B$$

$$\Rightarrow \left(\frac{\pi\times 0.4\times L\times 7200}{2\times 10^6\times\frac{\pi}{4}\times(0.4)^2\times L}\right)_A$$

$$= \left(\frac{\pi\times 0.1\times L\times t}{2\times 10^7\times\frac{\pi}{4}\times(0.1)^2\times L}\right)_B$$

$$\Rightarrow t_B = 18000 \text{ secs} = 5 \text{ hrs}$$

5.

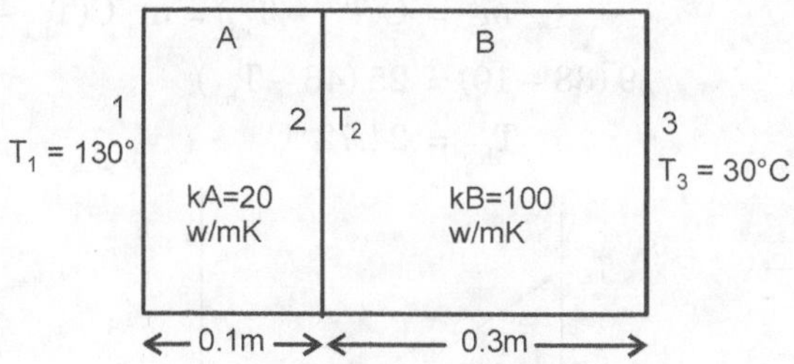

$$\frac{Q}{A} = \frac{kA.(T_1 - T_2)}{LA}$$

$$= \frac{kB(T_2 - T_3)}{LB}$$

$$\frac{20(130 - T_2)}{0.1} = \frac{100(T_2 - 30)}{0.3}$$

$$\therefore \quad T_2 = 67.5°C$$

6.

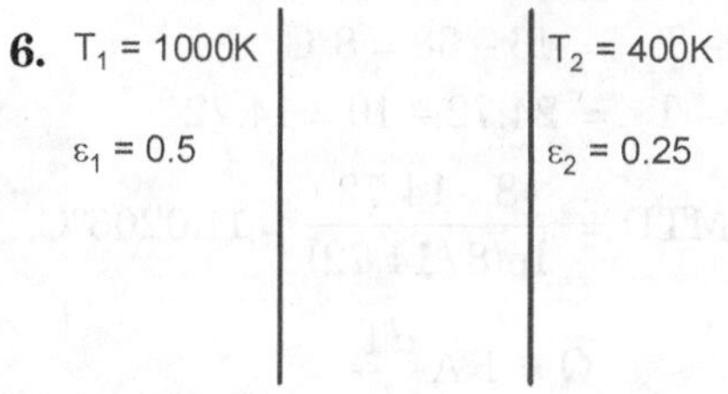

$$Q+2 = \frac{\sigma\left(T_1^4 - T_2^4\right)}{\frac{1}{\varepsilon_1}+\frac{1}{\varepsilon_2}-1}$$

$$= \frac{5.76\times 10^{-8}\left(1000^4 - 400^4\right)}{\frac{1}{0.5}+\frac{1}{0.25}-1}$$

$$= 11049.696 \text{ W/m}^2$$

$$= 11.049 \text{ kW/m}^2$$

7. d = 0.01 m

L = 0.2 m, R = 7801 Kg/m^3

t_i = 750°C, t_a = 100°C

C = 473 J/Kg K, Kj k = 43 W/mK

h = 250 W/m^{2}K

$$L_c = \frac{V}{A} = \frac{\frac{\pi}{4}d^2l}{\pi dl} = \frac{d}{4}$$

$$\left(\frac{t - t_a}{t_i - t_a}\right) = \exp\left[\frac{-hA\tau}{\rho VC}\right] = \exp\left[\frac{-h\tau}{\rho CL_c}\right]$$

$$\frac{300 - 100}{750 - 100} = \exp\left[\frac{-250\times 4\times\tau}{7801\times 473\times 0.01}\right]$$

$$\Rightarrow \quad \tau = 43.49 \text{ s}$$

9. Critical radius $r_c = \frac{k}{h_0} = \frac{0.1}{2} = 05 \text{ m} = 5 \text{ cm}$

10. Velocity as water = $C_v\sqrt{2gh}$

C_v = 1 (Given)

$$h = x\left[\frac{s_g}{s_0}-1\right]$$

$$= 0.01\,(10 - 1) = 0.09 \text{ m}$$

$$\therefore \text{ velocity of flow } = \sqrt{2\times 9.8\times 0.09} = 1.328 \text{ m/s}$$

11. Work done = 432 – 402 = 30 kJ/kg

Refrigerating effect = 402 – 237 = 165 kJ/kg.

$$\text{COP} = \frac{165}{30} = 5.5$$

13. $$Q = \frac{T - T_\infty}{\frac{L}{kA}+\frac{1}{hA}}$$

$$q = \frac{Q}{A} = \frac{T - T_\infty}{\frac{L}{k}+\frac{1}{h}}$$

$$1\times 10^5 = \frac{T - 25}{\frac{30\times 10^{-3}}{15}+\frac{1}{250}}$$

$$600 = T - 25°C$$

$$T = 625°C.$$

14. Let the base be (1) and hemispherical furnace be (2)

$$\therefore \quad F_{11} + F_{12} = 1 \quad ...(1)$$

$$F_{21} + F_{22} = 1 \quad ...(2)$$

$$\because \quad F_{11} = 0$$

$$\therefore \quad F_{12} = 2$$

$$A_1F_{12} = A_2F_{21}$$

$$\therefore \quad F_{21} = \frac{A_1}{A_2}F_{12} = \frac{\pi R^2}{2\pi R^2}F_{12} = 0.5F_{12}$$

$$\therefore \quad F_{21} = 0.5\times 2.0.5$$

$$\therefore \quad F_{22} = 0.5$$

So, $F_{11} = 0$, $F_{12} = 1$, $F_{21} = 0.5 < F_{22} = 0.5$

Now Radiative heat transfer,

$\Rightarrow \theta = A_1 F_{12} \varepsilon_2 \times 6(800^4 - 600^4)$ watt

$\therefore \theta = \pi \times 1^2 \times 1 \times 0.5 \times 5.668 \times 10^{-8}(800^4 - 600^4)$ watt

or $\theta = 24.9$ kW,

15. $t_1 = 20$ mm; $k_1 = 10$ W/mk

$t_2 = 1$ mm; $k_2 = ?$

$t_3 = 20$ mm; $k_3 = 1010$ W/mk

$T_1 = 500\ k$; $T_2 = 360\ k$; $q_{in} = 1010$ kW/m²

Diagram

$$q_{in} = \frac{T_1 - T_2}{\frac{t_1}{k_1 A_1} + \frac{t_2}{k_2 A_2} + \frac{t_3}{k_3 A_3}}$$

Taking unit Area

$$10 \times 10^3 \text{ W/m}^2 = \frac{(500-360)}{\left(\frac{20\times10^{-3}}{10} + \frac{1\times10^{-3}}{k_2} + \frac{20\times10^{-3}}{10}\right)}$$

$$10\left(4 + \frac{1}{k_2}\right) = 140$$

$$\frac{1}{k_2} = 10$$

$$\Rightarrow \quad k_2 = 0.10 \text{ W/m.k.}$$

16. $R = R_1 + R_2$

$$= \frac{L}{KA} + \frac{L}{KA} = \frac{L}{KA} + \frac{L}{2KA} = \frac{3L}{2KA}$$

$$Q = \frac{T_L - T_R}{T_R}$$

$$= \frac{300}{\frac{3L}{2KA}} = 200 \times \frac{KA}{L} \rightarrow \quad ...(1)$$

At interface,

$$Q = \frac{T_L - T_i}{R_i}$$

$$= \frac{600 - T_i}{\frac{L}{KA}} = (600 - T_i)\frac{KA}{L} \rightarrow \quad ...(2)$$

Equating 1 and 2,

$$200\frac{KA}{L} = (600 - T_i)\frac{KA}{L}$$

$$200 = 600 - T_i$$

$$T_i = 400 \text{ K}$$

17. Given: $\dot{m}_h = 25$ L/S; $T_{h,i} = 46°C$, $T_{h,o} = ?$

$\dot{m}_c = 19$ L/S; $T_{c,i} = 10°C$, $T_{c,o} = 38°C$

$\rho = 1000$ kg/m³, C = 4186 J/kg.k

Energy balance,

$$\dot{m}_c = C(T_{c,o} - T_{c,i}) = m_n\ C(T_{h,i} - T_{h,o})$$

$$19\,(38 - 10) = 25\,(46 - T_{h,o})$$

$$T_{h,o} = 24.72\ °C$$

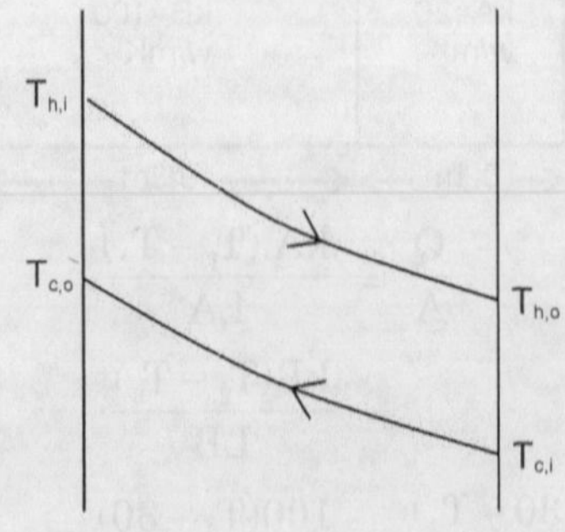

$$\text{LMTD} = \frac{\theta_1 - \theta_2}{\ln \theta_1 / \theta_2}$$

$$\theta_1 = T_{h,i} - T_{c,o} = 46 - 38 = 8°C$$

$$\theta_2 = T_{h,o} - T_{c,i} = 24.72 - 10 = 14.72$$

$$\text{LMTD} = \frac{8 - 14.72}{\ln(8/14.72)} = 11.0206°C.$$

18.

$$Q = KA\frac{dT}{dx}$$

$$q = \frac{Q}{A} = 2500 \text{ W/m}^2$$

$$2500 = K\frac{dT}{dx}$$

$$2500 = 1.15 \times \frac{(1100 - 350)}{x}$$

$$x = 0.345 \text{ m.}$$

8
CHAPTER

Thermodynamics

MCQ TYPE QUESTIONS

2016

1. The INCORRECT statement about regeneration in vapor power cycle is that

(*a*) it increases the irreversibility by adding the liquid with higher energy content to the steam generator

(*b*) heat is exchanged between the expanding fluid in the turbine and the compressed fluid before heat addition

(*c*) the principle is similar to the principle of Stirling gas cycle

(*d*) it is practically implemented by providing feed water heaters

2. The internal energy of an ideal gas is a function of

(*a*) temperature and pressure

(*b*) volume and pressure

(*c*) entropy and pressure

(*d*) temperature only

3. The INCORRECT statement about the characteristics of critical point of a pure substance is that

(*a*) there is no constant temperature vaporization process

(*b*) it has point of inflection with zero slope

(*c*) the ice directly converts from solid phase to vapor phase

(*d*) saturated liquid and saturated vapor states are identical

2015

4. For an ideal gas with constant values of specific heats, for calculation of the specific enthalpy,

(*a*) it is sufficient to know only the temperature

(*b*) both temperature and pressure are required to be known

(*c*) both temperature and volume are required to be known

(*d*) both temperature and mass are required to be known

5. The Vander Waals equation of state is $\left(p+\frac{a}{v^2}\right)(v-b)=\text{RT}$, where p is pressure, v is specific volume, T is temperature and R is characteristic gas constant. The SI unit of a is

(*a*) $\frac{\text{J}}{kg.\text{K}}$ (*b*) $\frac{m^3}{kg}$

(*c*) $\frac{m^5}{kgs^2}$ (*d*) $\frac{\text{Pa}}{kg}$

6. Which of the following statements regarding a Rankine cycle with reheating are TRUE?

(*i*) increase in average temperature of heat addition

(*ii*) reduction in thermal efficiency

(*iii*) drier steam at the turbine exit

(*a*) only (*i*) and (*ii*) are correct

(*b*) only (*ii*) and (*iii*) are correct

(*c*) only (*i*) and (*iii*) are correct

(*d*) (*i*), (*ii*) and (*iii*) are correct

7. For the same values of peak pressure, peak temperature and heat rejection, the correct order of efficiency for Otto, Dual and Diesel cycles is

(*a*) $\eta_{\text{Otto}} < \eta_{\text{Dual}} < \eta_{\text{Diesel}}$

(*b*) $\eta_{\text{Diesel}} < \eta_{\text{Dual}} > \eta_{\text{Otto}}$

(*c*) $\eta_{\text{Dual}} < \eta_{\text{Diesel}} < \eta_{\text{Otto}}$

(*d*) $\eta_{\text{Diesel}} > \eta_{\text{Otto}} < \eta_{\text{Dual}}$

8. The thermodynamic cycle shown in figure (T/s diagram) indicates

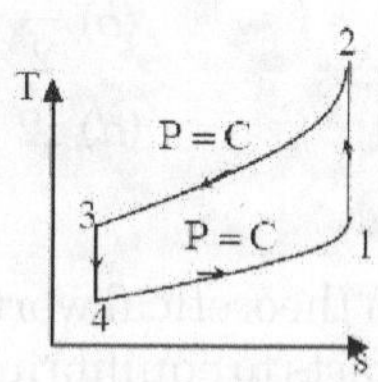

(*a*) Reversed Cannot cycle

(*b*) Reversed Brayton cycle

(*c*) Vapor compression cycle

(*d*) Vapor absorption cycle

9. An air-standard Diesel cycle consists of the following processes:

1-2: Air is compressed isentropically.

2-3: Heat is added at constant pressure.

3-4: Air expands isentropically to the original volume.

4-1: Heat is rejected at constant volume.

If γ and T denotes the specific heat ratio and temperature, respectively the efficiency of the cycle is

(*a*) $1-\dfrac{T_4-T_1}{T_3-T_2}$ (*b*) $1-\dfrac{T_4-T_1}{\gamma(T_3-T_2)}$

(*c*) $1-\dfrac{\gamma(T_4-T_1)}{T_3-T_2}$ (*d*) $1-\dfrac{T_4-T_1}{(\gamma-1)(T_3-T_2)}$

10. Saturated vapor is condensed to saturated liquid in condenser. The heat capacity ratio is $C_r=\dfrac{c_{min}}{c_{max}}$. The effectiveness ($\varepsilon$) of the condenser is

(*a*) $\dfrac{1-\exp[-NTU(1+C_r)]}{1+C_r}$

(*b*) $\dfrac{1-\exp[-NTU(1-C_r)]}{1-C_r\exp[-NTU(1-C_r)]}$

(*c*) $\dfrac{NTU}{1+NTU}$

(*d*) $1-\exp(-NTU)$

11. Which of the following statements are TRUE, when the cavitation parameter $\sigma = 0$?

i. the local pressure is reduced to vapor pressure
ii. cavitation starts
iii. boiling of liquid starts
iv. cavitation stops

(*a*) *i*, *ii* and *iv* (*b*) only *ii* and *iii*
(*c*) only *i* and *iii* (*d*) *i*, *ii* and *iii*

12. A solid sphere 1 of radius '*r*' is placed inside a hollow, closed hemispherical surface 2 of radius '4*r*'. The shape factor $F_{2\text{-}1}$ is

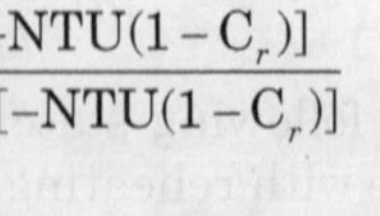

(*a*) $\dfrac{1}{12}$ (*b*) $\dfrac{1}{2}$
(*c*) 2 (*d*) 12

2014

13. The maximum theoretical work obtainable, when a system interacts to equilibrium with a reference environment, is called

(*a*) Entropy (*b*) Enthalpy
(*c*) Exergy (*d*) Rothalpy

14. In a power plant, water (density = 1000 kg/m^3) is pomped from 80 kPa to 3 MPa. The pump has an isentropic efficiency of 0.85. Assuming that the temperature of the water remains the same, the specific work (in kJ/kg) supplied to the pump is

(*a*) 0.34 (*b*) 2.48
(*c*) 2.92 (*d*) 3.43

15. A reversible heat engine receives 2 kJ of heat from a reservoir at 1000 K and a certain amount of heat from a reservoir at 800 K. It rejects 1 kJ of heat to a reservoir at 400 K. The net work output (in kJ) of the cycle is

(*a*) 0.8 (*b*) 1.0
(*c*) 1.4 (*d*) 2.0

16. The thermal efficiency of an air-standard Brayton cycle in terms of pressure ratio r_p and $\gamma(=c_p/c_v)$ is given by

(*a*) $1-\dfrac{1}{r_p^{\gamma-1}}$ (*b*) $1-\dfrac{1}{r_p^{\gamma}}$

(*c*) $1-\dfrac{1}{r_p^{1/\gamma}}$ (*d*) $1-\dfrac{1}{r_p^{(\gamma-1)/\gamma}}$

17. Which one of the following pairs of equations describes an irreversible heat engine?

(*a*) $\oint\delta Q>0$ and $\oint\dfrac{\delta Q}{T}<0$

(*b*) $\oint\delta Q<0$ and $\oint\dfrac{\delta Q}{T}<0$

(*c*) $\oint\delta Q>0$ and $\oint\dfrac{\delta Q}{T}>0$

(*d*) $\oint\delta Q<0$ and $\oint\dfrac{\delta Q}{T}>0$

18. The process utilizing mainly thermal energy for removing material is

(*a*) Ultrasonic Machining
(*b*) Electrochemical Machining
(*c*) Abrasive Jet Machining
(*d*) Laser Beam Machining

19. A certain amount of an ideal gas is initially at a pressure P_1 and temperature T_1. First, it undergoes a constant pressure process 1-2 such that $T_2 = 3T_1/4$. Then, it undergoes a constant volume process 2-3 such that $T_3 = T_1/2$. The ratio of the final volume to the initial volume of the ideal gas is

(*a*) 0.25 (*b*) 0.75
(*c*) 1.0 (*d*) 1.5

20. As the temperature increases, the thermal conductivity of a gas

(*a*) increases
(*b*) decreases
(*c*) remains constant
(*d*) increases up to a certain temperature and then decreases

21. Moist air at 35°C and 100% relative humidity is entering a psychometric device and leaving at 25°C and 100% relative humidity. The name of the device is

(*a*) Humidifier (*b*) Dehumidifier

(*c*) Sensible heater (*d*) Sensible cooler

2012

22. An ideal gas of mass *m* and temperature T_1 undergoes a reversible isothermal process from an initial pressure P_1 to final pressure P_2. The heat loss during the process is Q. The entropy change ΔS of the gas is

(*a*) $mR\ln\left(\frac{P_2}{P_1}\right)$ (*b*) $mR\ln\left(\frac{P_1}{P_2}\right)$

(*c*) $mR\ln\left(\frac{P_2}{P_1}\right)-\frac{Q}{T_1}$ (*d*) zero

2011

23. If a mass of moist air in an airtight vessel is heated to a higher temperature, then

(*a*) specific humidity of the air increases

(*b*) specific humidity of the air decreases

(*c*) relative humidity of the air increases

(*d*) relative humidity of the air decreases

2010

24. One kilogram of water at room temperature is brought into contact with a high temperature thermal reservoir. The entropy change of the universe is

(*a*) equal to entropy change of the reservoir

(*b*) equal to entropy change of water

(*c*) equal to zero

(*d*) always positive

25. A mono-atomic ideal gas (γ = 1.67, molecular weight = 40) is compressed adiabatically from 0.1 MPa, 300 K to 0.2 MPa. The universal gas constant is 8.314 kJ $kmol^{-1}$ K^{-1}. The work of compression of the gas (in kJ kg^{-1}) is

(*a*) 29.7

(*b*) 19.9

(*c*) 13.3

(*d*) 0

26. Consider the following two processes:

(*i*) A heat source at 1200 K loses 2500 kJ of heat to a sink at 800 K

(*ii*) A heat source at 800 K loses 2000 kJ of heat to a sink at 500 K

Which of the following statements is true?

(*a*) Process I is more irreversible than Process II

(*b*) Process II is more irreversible than Process I

(*c*) Irreversibility associated in both the processes are equal

(*d*) Both the processes are reversible

27. A moist air sample has dry bulb temperature of 30°C and specific humidity of 11.5 g water vapour per kg dry air. Assume molecular weight of air as 28.93. If the saturation vapour pressure of water at 30°C is 4.24 kPa and the total pressure is 90 kPa, then relative (in %) of air sample is

(*a*) 50.5 (*b*) 38.5

(*c*) 56.5 (*d*) 68.5

2009

28. If a closed system is undergoing an irreversible process, the entropy of the system

(*a*) must increase

(*b*) always remains constant

(*c*) must decrease

(*d*) can increase, decrease or remain constant

29. A frictionless piston-cylinder device contains a gas initially at 0.8 MPa and 0.015 m^3. It expands quasi-statically at constant temperature to a final volume of 0.030 m^3. The work output (in kJ) during this process will be

(*a*) 8.32 (*b*) 12.00

(*c*) 554.67 (*d*) 8320.00

2008

30. 2 moles of oxygen are mixed adiabatically with another 2 moles of oxygen in mixing chamber, so that the final total pressure and temperature of the mixture become same as those of the individual constituents at their initial states. The universal gas constant is given as R. The chang in entropy due to mixing, per mole of oxygen, is given by

(*a*) –R ln2 (*b*) 0

(*c*) R ln2 (*d*) R ln4

31. A balloon containing an ideal gas is initially kept in an evacuated and insulated room. The balloon ruptures and the gas fills up the entire room. Which one of the following statements is TRUE at the end of above process?

(*a*) The internal energy of the gas decreases from its initial value, but the enthalpy remains constant

(*b*) The internal energy of the gas increases from its initial value, but the enthalpy remains constant

(*c*) Both internal energy and enthalpy of the gas remain constant

(*d*) Both internal energy and enthalpy of the gas increase

32. A thermal power plant operates on a regenerative cycle with a single open feedwater heater, as shown in the figure.

For the state points shown, the specific enthalpies are:

$$h_1 = 2800 \text{ kJ/kg and } h_2 = 200 \text{ kJ/kg.}$$

The bleed to the feedwater heater is 20% of the boiler steam generation rate. The specific enthalpy at state 3 is

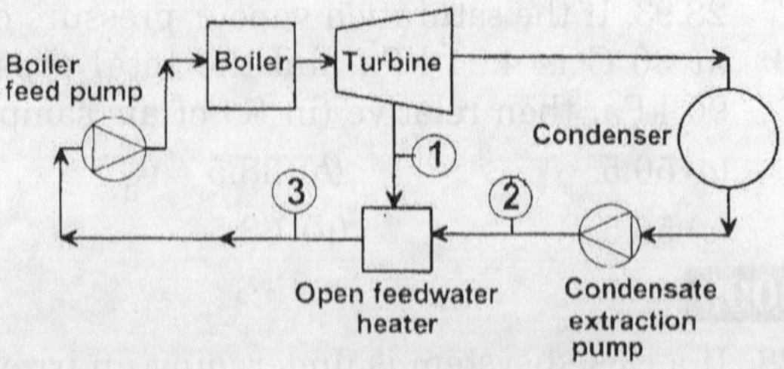

(*a*) 720 kJ/kg (*b*) 2280 kJ/kg

(*c*) 1500 kJ/kg (*d*) 3000 kJ/kg

33. In a steady state steady flow process taking place in a device with a single inlet and a single outlet, the work done per unit mass flow rate is given by

$$w = -\int_{inlet}^{outlet} vdp,$$

where v is specific volume and p is pressure. The expression for w given above

(*a*) is valid only if the process is both reversible and adiabatic

(*b*) is valid only if the process is both reversible and isothermal

(*c*) is valid for any reversible process

(*d*) is incorrect; it must be $w = \int_{inlet}^{outlet} pdv$

2007

34. Which of the following relationships is valid only for reversible processes undergone by a closed system of simple compressible substance (neglect changes in kinetic and potential energy ?

(*a*) $\delta Q = dU + \delta W$

(*b*) $T\,dS = dU + p\,dV$

(*c*) $T\,dS = dU + \delta W$

(*d*) $\delta Q = dU + p\,dV$

35. Which combination of the following statements is correct ?

P. A gas cools upon expansion only when its Joule-Thomson co-efficient is positive in the temperature range of expansion.

Q. For a system undergoing a process, its entropy remains constant only when the process is reversible.

R. The work done by a closed system in an adiabatic process is a point function.

S. A liquid expands upon freezing when the slope of its fusion curve on Pressure.

Temperature diagram is negative.

(*a*) R and S (*b*) P and Q

(*c*) Q, R and S (*d*) P, Q and R

2006

36. Dew point temperature is the temperature at which condensation begins when the air is cooled at constant

(*a*) volume (*b*) entropy

(*c*) pressure (*d*) enthalpy

37. The statements concern psychrometric chart.

1. Constant relative humidity lines are uphill straight lines to the right
2. Constant wet bulb temperature lines are downhill straight lines to the right
3. Constant specific volume lines are downhill straight lines to the right
4. Constant enthalpy lines are concident with constant wet bulb temperature lines

Which of the statements are correct?

(*a*) 2 and 3

(*b*) 1 and 2

(*c*) 1 and 3

(*d*) 2 and 4

38. Match items from groups I, II, III, IV and V.

Group I	Group II	Group III	Group IV	Group V
	When added to the system, is	Differential	Function	Phenomenon
E Heat	G Positive	I Exact	K Path	M Transient
F Work	H Negative	J Inexact	L Point	N Boundary

(*a*) F – G – J – K – M
E – G – I – K – N

(*b*) E – G – I – K – M
F – H – I – K – N

(*c*) F – H – J – L – N
E – H – I – L – M

(*d*) E – G – J – K – N
F – H – J – K – M

2005

39. The following four figures have been drawn to represent a fictitious thermodynamic cycle, on the *p-v* and T-*s* planes.

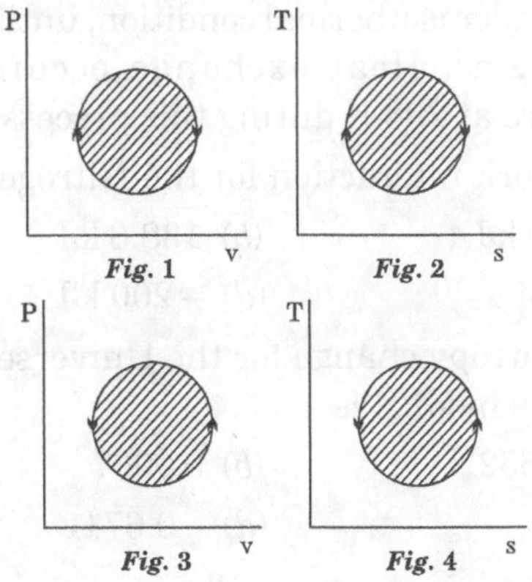

Fig. 1 *Fig. 2* *Fig. 3* *Fig. 4*

According to the first law of thermodynamics, equal areas are enclosed by

(*a*) figures 1 and 2 (*b*) figures 1 and 3

(*c*) figures 1 and 4 (*d*) figures 2 and 3

40. A *p-v* diagram has been obtained from a test on a reciprocating compressor. Which of the following represents that diagram?

(*a*)

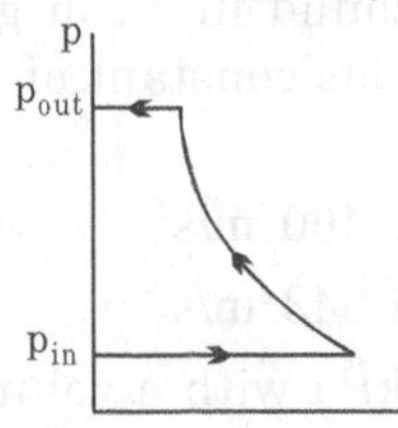

(*b*)

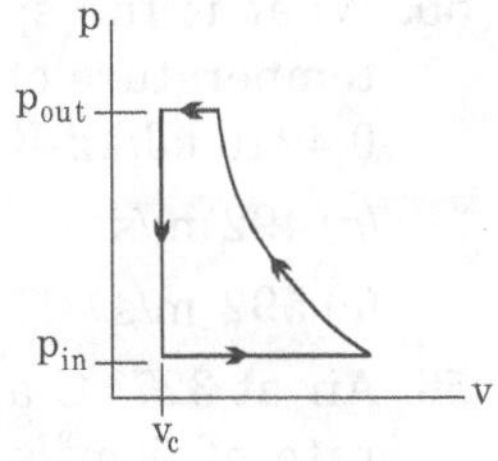

(*c*)

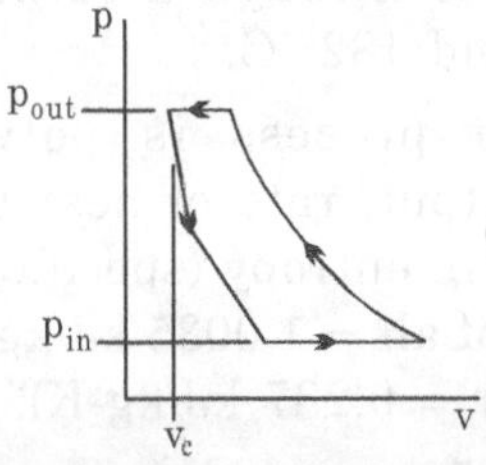

(*d*)

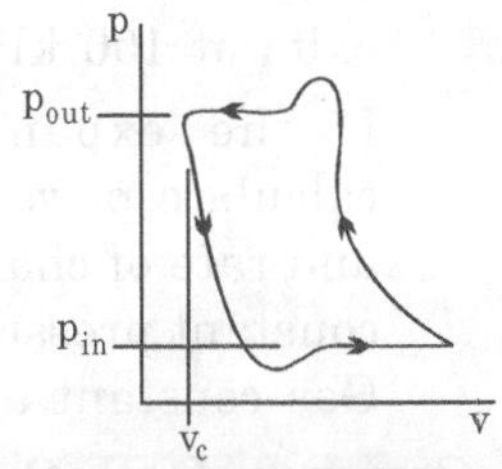

41. For a typical sample of ambient air (at 35° C, 75% relative humidity and standard atmospheric pressure), the amount of moisture in kg per kg of dry air will be approximately

(*a*) 0.002 (*b*) 0.027

(*c*) 0.25 (*d*) 0.75

42. Water at 42° C is sprayed into a stream of air at atmospheric pressure, dry bulb temperature of 40° C and a wet bulb temperature of 20° C. The air leaving the spray humidifier is not saturated. Which of the following statements is true?

(*a*) Air gets cooled and humidified

(*b*) Air gets heated and humidified

(*c*) Air gets heated and dehumidified

(*d*) Air gets cooled and dehumidified

43. A reversible thermodynamic cycle containing only three processes and producing work is to be constructed. The constraints are

(*i*) there must be one isothermal process,

(*ii*) there must be one isentropic process,

(*iii*) maximum and minimum cycle pressures and the clearance volume are fixed, and

(*iv*) polytropic processes are not allowed.

Then the number of possible cycles are

(*a*) 1 (*b*) 2

(*c*) 3 (*d*) 4

44. Nitrogen at an initial state of 10 bar, 1 m^3 and 300 K is expanded isothermally to a final volume of 2 m^3. The *p-v*-T relation is

$$\left(p + \frac{a}{v^2}\right)v = RT, \text{ where } a > 0.$$

The final pressure

(*a*) will be slightly less than 5 bar

(*b*) will be slightly more than 5 bar

(*c*) will be exactly 5 bar

(*d*) cannot be ascertained in the absence of the value of a.

Linked Answer Q. 45-46

The following table of properties was printed out for saturated liquid and saturated vapour of ammonia. The titles for only the first two columns are available. All that we know is that the other columns (columns 3 to 8) contain data on specific properties, namely, internal energy (kJ/kg), enthalpy (kJ/kg) and entropy (kJ/kg. K).

t(°C)	*p* (kPa)						
–20	190.2	88.76	0.3657	89.05	5.6155	1299.5	1418.0
0	429.6	179.69	0.7114	180.36	5.3309	1318.0	1442.2
20	857.5	272.89	1.0408	274.30	5.0860	1332.2	1460.2
40	1554.9	368.74	1.3574	371.43	4.8662	1341.0	1470.2

45. The specific enthalpy data are in columns

(*a*) 3 and 7 (*b*) 3 and 8

(*c*) 5 and 7 (*d*) 5 and 8

46. When saturated liquid at 40° C is throttled to –20° C, the quality at exit will be

(*a*) 0.189

(*b*) 0.212

(*c*) 0.231

(*d*) 0.788

2004

47. A gas contained in a cylinder is compressed, the work required for compression being 5000 kJ. During the process, heat interaction of 2000 kJ causes the surroundings to the heated. The change in internal energy of the gas during the process is

(*a*) – 7000 kJ (*b*) – 3000 kJ

(*c*) + 3000 kJ (*d*) + 7000 kJ

48. The pressure gauges G_1 and G_2 installed on the system show pressures of P_{G1} = 5.00 bar and P_{G2} = 1.00 bar. Value of unknown pressure P is

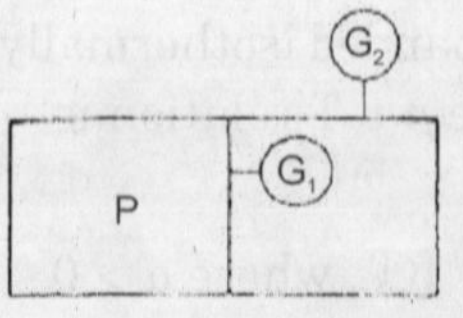

(*a*) 1.01 bar (*b*) 2.01 bar

(*c*) 5.00 bar (*d*) 7.01 bar

49. Dew point temperature of air at one atmospheric pressure (1.013 bar) is 18°C. The air dry bulb temperature is 30°C. The saturation pressure of water at 18°C and 30°C are 0.02062 bar and 0.04241 bar respectively. The specific heat of air and water vapour respectively are 1.005 and 1.88 kJ/kg K and the latent heat of vaporization of water at 0°C is 2500 kJ/kg. The specific humidity (kJ/kg of dry air) and enthalpy (kJ/kg of dry air) of this moist air respectively, are

(*a*) 0.01051, 52.64 (*b*) 0.01291, 63.15

(*c*) 0.01481, 78.60 (*d*) 0.01532, 81.40

2003

50. For air with a relative humidity of 80%

(*a*) dry bulb temperature is less than the wet bulb temperature

(*b*) dew point temperature is less than wet bulb temperature

(*c*) dew point and wet bulb temperatures are equal

(*d*) dry bulb and dew point temperatures are equal

51. Considering the relationship T*d*S = *d*U + *pd*V between the entropy (S), internal energy (U), pressure (*p*), temperature (T) and volume (V), which of the following statements is correct ?

(*a*) It is applicable only for a reversible process

(*b*) For an irreversible process, T*d*S > *d*U + *pd*V

(*c*) It is valid only for an ideal gas

(*d*) It is equivalent to 1 law, for a reversible process

Common Data Q. 52-53

Nitrogen gas (molecular weight 28) is enclosed in a cylinder by a piston, at the initial condition of 2 bar, 298 K and 1 m^3. In a particular process, the gas slowly expands under isothermal condition, until the volume becomes 2m^3. Heat exchange occurs with the atmosphere at 298 K during this process.

52. The work interaction for the Nitrogen gas is

(*a*) 200 kJ (*b*) 138.6 kJ

(*c*) 2 kJ (*d*) – 200 kJ

53. The entropy change for the Universe during the process in kJ/K is

(*a*) 0.4652 (*b*) 0.0067

(*c*) 0 (*d*) – 0.6711

2002

54. Carnot cycle is having an efficiency of 0.75. If temperature of the high temperature reservoir is 727° C, what is the temperature of low temperature reservoir?

(*a*) 23° C (*b*) –23° C

(*c*) 0° C (*d*) 250° C

55. What is the speed of sound in Neon gas at a temperature of 500K (Gas constant of Neon is 0.4210 kJ/kg-K)?

(*a*) 492 m/s (*b*) 460 m/s

(*c*) 592 m/s (*d*) 543 m/s

56. Air at 327° C and 400 kPa with a volume flow rate of 5 m^3/s flows through a turbine and exits at 100 kPa and 182° C.

If the expansion process is polytropic, calculate power output, rate of heat transfer and rate of change in entropy (specific heat at constant pressure of air = 1.0035 kJ/kg-K, and Gas constant of air = 0.287 kJ/kg-K).

Common Data Q. 57-58

An ice making plant refrigerant R-12 is having an evaporator saturation temperature of –25° C and the condenser saturation temperature of 35° C. The vapour is leaving the compressor at 65° C. Following table shows the properties of the refrigerant.

Temp°C	Pressure, kPa	Saturation Enthalpy kJ/kg	
		Liquid	Vapour
–25	123.7	13.3	176.5
35	850.0	69.6	201.5
Enthalpy of superheated refrigerant at 850 kPa and 65°C = 225.5 kJ/kg			

57. **(*i*)** Coefficient of performance will be

(*a*) 1.28 (*b*) 1.58

(*c*) 1.92 (*d*) 2.18

58. **(*ii*)** Power consumption will be

(*a*) 1.23 kW (*b*) 1.63 kW

(*c*) 1.93 kW (*d*) 2.23 kW

2001

59. A cyclic heat engine does 50 kJ of work per cycle. If efficiency of the heat engine is 75%, then heat rejected per cycle is

(*a*) $16\frac{2}{3}$ kJ (*b*) $33\frac{1}{3}$ kJ

(*c*) $37\frac{1}{2}$ kJ (*d*) $66\frac{2}{3}$ kJ

60. A small steam whistle (perfectly insulated and doing no shaft work) causes a drop of 0.8 kJ/kg in the enthalpy of steam from entry to exit. If kinetic energy of the steam at entry is negligible, then velocity of the steam at exit is

(*a*) 4 m/s (*b*) 40 m/s

(*c*) 80 m/s (*d*) 120 m/s

61. For air at a given temperature, as the relative humidity is increased isothermally, the wet bulb temperature

(*a*) and specific enthaplpy increase

(*b*) and specific enthalpy decrease

(*c*) increases and specific enthalpy decreases

(*d*) decreases and specific enthalpy increases

2000

62. Availability of a system at any given state is

(*a*) a property of the system

(*b*) the maximum work obtainable as the system goes to dead state

(*c*) total energy of the system

(*d*) the maximum useful work obtainable as the system goes to dead state

63. When an ideal gas with constant specific heats is throttled adiabatically, with negligible changes in kinetic and potential energies

(*a*) $\Delta h = 0, \Delta T = 0$

(*b*) $\Delta h > 0, \Delta T = 0$

(*c*) $\Delta h > 0, \Delta S > 0$

(*d*) $\Delta h = 0, \Delta S > 0$

where *h*, T and S represent respectively, enthalpy, temperature and entropy, temperature and entropy.

64. When atmospheric air is heated at constant pressure, its

(*a*) humidity ratio does not change

(*b*) relative humidity increases

(*c*) dew point temperature does not change

(*d*) wet bulb temperature increases

65. A certain mass of a pure substance undergoes an irreversible process form state 1 to state 2, the path of the process being a straight line on the T–s diagram. If T_1 = 330 K, T_2 = 440 K; U_1 = 170 kJ, U_2 = 190 kJ; H_1 = 220 kJ, H_2 = 247 kJ and S_1 = 0.23 kJ/K and S_2 = 0.3 kJ/K where T, U, H and S represent temperature, calculate internal energy, enthalpy and entropy respectively.

1999

66. A Stirling cycle and a Carnot cycle operate between 50°C and 350°C. Their efficiencies are η_s and η_c respectively. In this case, which of the following statements is true?

(*a*) $\eta_s > \eta_c$

(*b*) $\eta_s = \eta_c$

(*c*) $\eta_s < \eta_c$

(*d*) The sign of ($\eta_s - \eta_c$) depends on the working fluids used

67. A Carnot cycle refrigerator operates between 250°K and 300°K. Its coefficient of performance is

(*a*) 6.0 (*b*) 5.0

(*c*) 1.2 (*d*) 0.8

68. A gas turbine power plant has a specific output of 350 kJ/ kg and an efficiency of 34%. A regenerator is installed and the efficiency increases to 51% . The specific output will be closest to

(*a*) 350 kJ/kg (*b*) 468 kJ/kg

(*c*) 525 kJ/kg (*d*) 700 kJ/kg

69. An aeroplane is cruising at a speed of 800 kmph at an altitude, where the air temperature is 0°C. The flight Mach number at this speed is nearly

(*a*) 1.5 (*b*) 0.254

(*c*) 0.67 (*d*) 2.04

70. An isolated thermodynamic system executes a process. Choose the correct statement (*s*) from the following

(*a*) No heat is transferred

(*b*) No work is done

(*c*) No mass flows across the boundary of the system

(*d*) No chemical reaction takes place within the system

71. Select statements from List II matching the processes in List I. Enter your answer as A,B if the correct choice for (1) is (A) and that for (2) is (B)

List I	List II
(1) Intercooling	(A) No heat transfer during compression
(2) Isothermal compression	(B) Reduces low pressure compressor work
	(C) Heat rejection during compression
	(D) Reduces high pressure compressor work

72. An isentropic air turbine is used to supply 0.1 kg/s of air at 0.1 MN/m^2 and at 285° K to a cabin. The pressure at inlet to the turbine is 0.4 MN/m^2. Determine the temperature at turbine inlet and the power developed by the turbine. Assume C_p = 1.0 kJ/kg K.

73. An adiabatic steam turbine receives dry saturated steam at 1.0 MN/m^2 and discharges it at 0.1 MN/m^2. The steam flow rate is 3 kg/s and the moisture at exit in negligible. If the ambient temperature is 300°K, determine the rate of entropy production and the lost power.

Steam properties :

p	T_{sat}	h_f	h_g	s_f	s_g
MN/m^2	°C	kJ/kg	kJ/kg	kJ/kg K	kJ/kg K
10	179.9	762.8	2778.1	2.139	6.586
0.1	99.6	417.5	2675.5	1.303	7.359

1998

74. If V_N and α are the nozzle exit velocity and nozzle angle in an impulse turbine, the optimum blade velocity is given by

(*a*) $V_N \cos 2\alpha$ (*b*) $V_N \sin 2\alpha$

(*c*) $\frac{V_N \cos\alpha}{2}$ (*d*) $\frac{V_N 2}{2}$

75. A Curtis stage, Rateau stage and a 50% reaction stage in a steam turbine are examples of

(*a*) different types of impulse stages

(*b*) different types of reaction stages

(*c*) a simple impulse stage, a velocity compounded impulse stage and reaction stage

(*d*) a velocity compounded impulse stage, a simple impulse stage and a reaction stage

76. The isentropic heat drop in the nozzle of an impulse steam turbine with a nozzle efficiency 0.9, blade velocity rati 0.5, and mean blade velocity 150 m/s in kJ/kg is

(*a*) 50 (*b*) 40

(*c*) 60 (*d*) 75

77. Air (C_p = 1 kJ/kg, γ = 1.4) enters a compressor at a temperature of 27°C. The compressor pressure ratio is 4. Assuming an efficiency of 80%, the compressor work required in kJ/kg is

(*a*) 160 (*b*) 172

(*c*) 182 (*d*) 225

78. Pertaining to a steam boiler, pick the correct statement among the following

(*a*) Primary boiler heat surfaces include evaporator section, economizer and air preheater

(*b*) Primary boiler heat transfer surfaces include evaporator section, super heater section and economizer.

(*c*) Secondary boiler heat transfer surfaces include super heater, economizer and air preheater.

(*d*) Primary boiler heat transfer surfaces include evaporator section, super heater section and reheat section

Directions : Match 4 correct pairs between List I and List II for following questions. No credit will be given for partial matching. Write your answer using only the letters *a* to *d* and numbers 1 to 6.

79. List I

(*a*) Heat to work

(*b*) Heat to lift weight

(*c*) Heat to strain energy

(*d*) Heat to electromagnetic energy

List II

(1) Nozzle

(2) Endothermic chemical reaction

(3) Heat engine

(4) Hot air balloon/evaporation

(5) Thermal radiation

(6) Bimetallic strips

80. A Rankinge cycle operates between pressures of 80 bar and 0.1 bar. The maximum cycle temperature is 600°C. If the steam turbine and condensate pump efficiencies are 0.9 and 0.8, respectively, calculate the specific work and thermal efficiency. Relevant steam table extract is given below :

P bar T°C Specific volume m³/kg
Specific enthalpy kJ/kg
Specific entropy kJ/kg K

V_f V_g

h_f h_{fg} h_g S_f

S_{fg} S_g

0.1 45.84 0.0010103 14.68
191.9 2392.3 2584.2 0.6488
7.5006 8.1494

80 295.1 0.001385 0.0235
1317 1440.5 2757.5 3.2073
2.5351 5.7424

80 bar – 600°C **v**

0.486

Superheat table **h**

3642

s 7.0206

81. The minimum pressure and temperature in an Otto cycle are 100 kPa and 27°C. The amount of heat added to the air per cycle is 1500 kJ/kg. Determine the pressures and temperatures at all points of the air standard Otto cycle. Also calculate the specific work and the thermal efficiency of the cycle for a compression ratio of 8 : 1 [Take C_v(air) as 0.72 kJ/kg K and $\left(\frac{C_p}{C_v}\right)_{air} = 1.4$].

82. Find the required air-fuel ratio in a gas turbine whose turbine and compressor efficiencies are 85% and 80% respectively. Maximum cycle temperature is 875 C The working fluid can be taken as air (C_p = 1.00 kJ/kg K, γ = 1.4) which enters the compressor at 1 atm and 27°C. The pressure ratio is 4. The fuel used has calorific value of 42000 kJ/kg. There is a loss of 10% of calorific value in the combustion chamber.

1997

83. Which among the following is the boiler mounting?

(*a*) Blow off cock (*b*) Feed pump
(*c*) Economiser (*d*) Superheater

84. For an ideal gas the expression

$\left[T\left(\frac{\partial s}{\partial T}\right)p - T\left(\frac{\partial s}{\partial T}\right)v\right]$ is always equal to

(*a*) zero (*b*) $\frac{C_p}{C_v}$
(*c*) R (*d*) RT

85. A system undergoes a state change from 1 to 2. According the second law of thermodynamics for the process to be feasible, the entropy change, $S_2 - S_1$ of the system

(*a*) is positive or zero
(*b*) is negative or zero
(*c*) is zero
(*d*) can be positive, negative or zero

86. Kaplan turbine is

(*a*) a high head mixed flow turbine
(*b*) a low axial flow turbine
(*c*) an outward flow reaction turbine
(*d*) an impulse inward flow turbine

87. Refer to figure, the absolute pressure of gas A in the bulb is

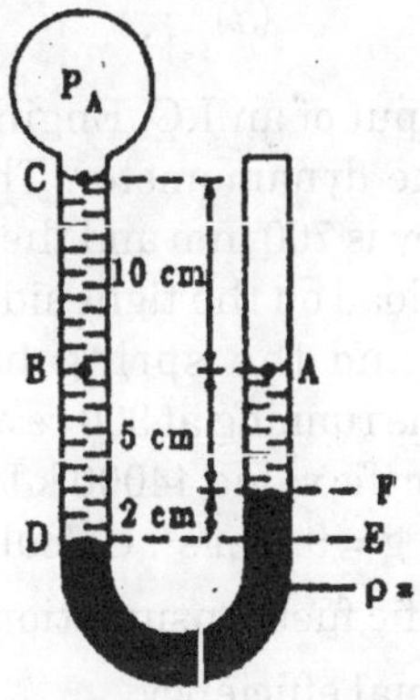

(*a*) 771.2 mm Hg (*b*) 752.65 mm Hg
(*c*) 767.35 mm Hg (*d*) 748.8 mm Hg

88. The following data pertain to a single stage impulse steam turbine :

Nozzle angle = 20°, Blade velocity = 200 m/s
Relative steam velocity at entry = 350 m/s
Blade inlet = 30° Blade exit angle = 25°.

If blade friction is neglected, the work done per kg steam is

(*a*) 124 kJ (*b*) 164 kJ
(*c*) 169 kJ (*d*) 174 kJ

89. Consider an actual regenerative Rankine cycle with one open feed water heater. For each kg steam entering the turbine, if m kg steam with a specific enthalpy of h_1 is bled from the turbine, and the specific enthalpy of liquid water entering the heater is h_2, then h_3 specific enthaply of saturated liquid leaving the heater is equal to

(*a*) $mh_1 - (h_2 - h_1)$
(*b*) $h_1 - m(h_2 - h_1)$
(*c*) $h_2 - m(h_2 - h_1)$
(*d*) $mh_2 - (h_2 - h_1)$

90. Consider a two stage reciprocating air compressor with a perfect intercooler operating at the best intermediate pressure. Air enters the low pressure cylinder at 1 bar, 27°C and leaves the high pressure cylinder at 9 bar. Assume the index of compression and expansion in each stage is 1.4, and that for air R = 286.7 J/kg K, the work done per kg air in the high pressure cylinder is

(*a*) 111 kJ (*b*) 222 kJ

(c) 37 kJ (*d*) 74 kJ

91. Clearance volume of a reciprocating compressor is 100 ml, and the volume of the cylinder at bottom dead centre is 1.0 litre. The clearance ratio of the compressor is

(*a*) $\frac{1}{11}$ (*b*) $\frac{1}{10}$

(c) $\frac{1}{9}$ (*d*) $\frac{1}{11}$

92. The power output of an I.C. Engine is measured by a rope brake dynamometer. The diameter of the brake pulley is 700 mm and the rope diameter is 25 mm. The load on the tight side of the rope is 50 kg. mass and the spring balance reads 50 N. The engine running at 900 rev/min consumes the fuel, of calorific value 44000 kJ/kg, at a rate of 4kg/h. Assume g = 9.8 m/s^2. Calculate

(a) brake specific fuel consumption.

(b) brake thermal efficiency.

93. At a place where the surroundings are at 1 bar 27°C, a closed rigid thermally insulated tank contains 2 kg air at 2 bar, 27°C. This air is then churned for a while, by a paddle wheel connected to an external motor. If it is given that the irreversibilityof the process is 100 kJ, find the final temperature and the increase in availability of air. Assume for air, c_v = 0.718 kJ/kg K.

94. In an air-standard regenerative gas turbine cycle, the pressure ratio is 5. Air enters the compressor at 1 bar, 300°K and leaves at 490°K. The maximum temperature in the cycle is 1000°K. Calculate the cycle efficiency, given that the efficiency of the regenerator and the adiabatic efficiency of the turbine are each 80%. Assume for air, the ratio of specific heats is 1.4. Also, show the cycle on a T– s diagram.

95. Air enters a frictionless adiabatic converging nozzle at 10 bar, 500°K with negligible velocity. The nozzle discharges to a region at 2 bar. If the exit area of the nozzle is 2.5 cm^2, find the flow rate of air through the nozzle. Assume for air, C_p = 1005 J/kgK and C_v = 718J/kg K.

1996

96. The specific haeat of an ideal gas depend on its

(*a*) temperature

(*b*) pressure

(c) volume

(*d*) molecular weight and structure

97. For reversible adiabatic compression in a steady flow process, the work transfer per unit mass is

(*a*) $\int$ p dv (*b*) $\int$ v dp

(c) $\int$ T ds (*d*) $\int$ s dT

98. A cycle consisting of two reversible isothermal processes and two reversible isobaric processes is known as

(*a*) Atkinson cycle (*b*) Stirling cycle

(c) Brayton cycle (*d*) Ericsson cycle

99. When wet steam flows through a throttle valve and remains wet at exit

(*a*) its temperature and quility increases

(*b*) its temperature decreases but quality increases

(c) its temperature increases but quality decreases

(*d*) its temperature and quality decreases

100. A solar energy based heat engine which receives 80 kJ of heat at 100 degree C and rejects 70 kJ of heat to the ambient at 30°C is be designed. The thermal efficiency of the heat engine is

(*a*) 70% (*b*) 1.88%

(c) 12.5% (*d*) indeterminate

101. For two cycles coupled in series, the topping cycle has an efficiency of 30% and the bottoming cycle has an efficiency of 20%. The overall combined cycle efficiency is

(*a*) 50% (*b*) 44%

(c) 38% (*d*) 55%

102. A steam plant has the boiler efficiency of 92%, turbine efficiency (mechanical) of 94%, generator efficiency of 95% and cycle efficiency of 44%. If 6% of the generated power is used to run the auxilliaries, the overall plant efficiency is

(*a*) 34% (*b*) 39%

(c) 45% (*d*) 30%

103. An iron cube at a temperature of 400 degree C is dropped into an insulated bath containing 10 kg water at 25 degree C. The water finally reaches a temperature of 50 degree C at steady state. Given that the specific heat of water is equal to 4186 J/kg K. Find the entropy changes for the iron cube and the water. Is the process irreversible ? If so, why ?

104. The velocity of steam exiting the nozzles of the impulse stage of a turbine is 400m/s. The blades operate close to the maximum blading efficiency. The nozzle angle is 20°. Considering equiangular blades and neglecting blade friction, calculate for a steam flow of 0.6 kg/s, the diagram power and the diagram efficiency.

105. Air enters the compressor of a gas turbine plant operating on Brayton cycle at 1 bar, 27°C. The pressure ratio in the cycle is 6. If $W_T = 2.5\ W_C$, where W_T and W_C are the turbine and compressor work respectively, calculate the maximum temperature and the cycle efficiency.

NUMERICAL TYPE QUESTIONS

2016

1. Consider two hydraulic turbines having identical specific speed and effective head at the inlet. If the speed ratio (N_1/N_2) of the two turbines is 2, then the respective power ratio (P_1/P_2) is ________.

2. For water at 25 °C, dp_s/dp_s = 0.189 kPa/K (p_s is the saturation pressure in kPa and T_s is the saturation temperature in K) and the specific volume of dry saturated vapour is 43.38 m³/kg. Assume that the specific volume of liquid is negligible in comparison with that of vapour. Using the Clausius-Clapeyron equation, an estimate of the enthalpy of evaporation of water at 25 °C (in kJ/kg) is ________.

3. An ideal gas undergoes a reversible process in which the pressure varies linearly with volume. The conditions at the start (subscript 1) and at the end (subscript 2) of the process with usual notation are: p_1 = 100 kPa, V_1 = 0.2 m³ and p_2 = 200 kPa, V_2 = 0.1 m³ and the gas constant, R = 0.275 kJ/kg-K. The magnitude of the work required for the process (in kJ) is ________.

4. In a steam power plant operating on an ideal Rankine cycle, superheated steam enters the turbine at 3 MPa and 350°C. The condenser pressure is 75 kPa. The thermal efficiency of the cycle is ________ percent.

Given data:

For saturated liquid, at P = 75 kPa, h_f = 384.39 kJ/kg, v_f = 0.001037 m³/kg, s_f = 1.213 kJ/kg-K

At 75 kPa, h_{fg} = 2278.6 kJ/kg, s_{fg} = 6.2434 kJ/kg-K

At P = 3 MPa and T = 350 °C (superheated steam), h = 3115.3kJ/kg, s = 6.7428 kJ/kg-K

5. The heat removal rate from a refrigerated space and the power input to the compressor are 7.2 kW and 1.8 kW, respectively. The coefficient of performance (COP) of the refrigerator is ________

6. Consider a simple gas turbine (Brayton) cycle and a gas turbine cycle with perfect regeneration. In both the cycles, the pressure ratio is 6 and the ratio of the specific heats of the working medium is 1.4. The ratio of minimum to maximum temperatures is 0.3 (with temperatures expressed in K) in the regenerative cycle. The ratio of the thermal efficiency of the simple cycle to that of the regenerative cycle is ________.

7. A piston-cylinder device initially contains 0.4 m³ of air (to be treated as an ideal gas) at 100 kPa and 80°C. The air is now isothermally compressed to 0.1 m³. The work done during this process is ____ kJ.

(Take the sign convention such that work done on the system is negative)

8. A refrigerator uses R-134 a as its refrigerant on an ideal vapour-compression refrigeration cycle between 0.14 MPa and 0.8 MPa. If the mass flow rate of the refrigerant is 0.05 kg/s, the rate of heat rejection to the environment is ________ kW.

Given data :

At P = 0.14 Mpa,
h = 236.04 kJ/kg,
s = 0.9322 kJ/kg-K

At P = 0.8 MPa,
h = 272.05 kJ/kg (superheated vapour)

At P = 0.8 MPa,
h = 93.42 kJ (saturated liquid)

9. The partial pressure of water vapour in a moist air sample of relative humidity 70% is 1.6 kPa, the total pressure being 101.325 kPa. Moist air may be treated as an ideal gas mixture of water vapour and dry air. The relation between saturation temperature (T_s in K) and saturation pressure (p_s in kPa) for water is given by $\ln(p_s/p_o)$ = 14.317 – 5304 T_s, where p_o = 101.325 kPa. The dry bulb temperature of the moist air sample (in °C) is ________.

10. Steam at an initial enthalpy of 100 kJ/kg and inlet velocity of 100 m/s, enters an insulated horizontal nozzle. It leaves the nozzle at 200 m/s. The exit enthalpy (in kJ/kg) is ______.

11. In a mixture of dry air and water vapor at a total pressure of 750 mm of Hg, the partial pressure of water vapor is 20 mm of Hg. The humidity ratio of the air in grams of water vapor per kg of dry air (g_w/kg_{da}) is ______.

12. In a 3-stage air compressor, the inlet pressure is p_1, discharge pressure is p_4 and the intermediate pressures are p_2 and p_3 ($p_2 < p_3$). The total pressure ratio of the compressor is 10 and the pressure ratios of the stages are equal. If p_1 = 100 kPa, the value of the pressure p_3 (in kPa) is ______.

13. In the vapour compression cycle shown in the figure, the evaporating and condensing temperatures are 260 K and 310 K, respectively. The compressor takes in liquid-vapour mixture (state 1) and isentropically compresses it to a dry saturated vapour condition (state 2). The specific heat of the liquid refrigerant is 4.8 kJ/kg-K and may be treated as constant. The enthalpy of evaporation for the refrigerant at 310 K is 1054 kJ/kg.

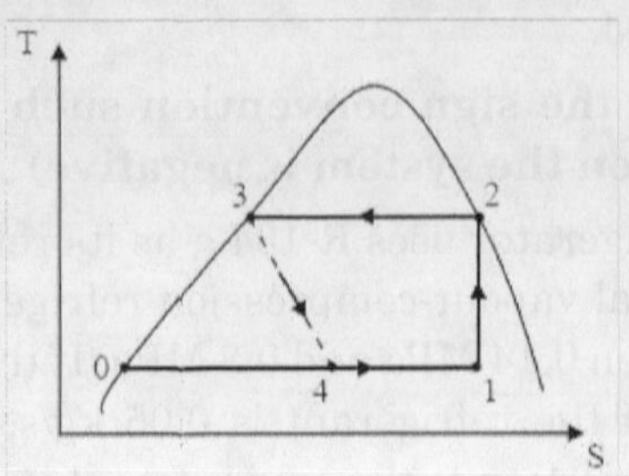

The difference between the enthalpies at state points 1 and 0 (in kJ/kg) is ______

2015

14. A Carnot engine (CE-1) works between two temperature reservoirs A and B, where T_A = 900 K and T_B = 500 K. A second Carnot engine (CE-2) works between temperature reservoirs B and C, where T_C = 300 K. In each cycle of CE-1 and CE-2, all the heat rejected by CE-1 to reservoir B is used by CE-2. For one cycle of operation, if the net Q absorbed by CE-1 from reservoir A is 150 MJ, the net heat rejected to reservoir C by CE-2 (in MJ) is ______.

15. A stream of moist air (mass flow rate = 10.1 kg/s) with humidity ratio of $0.01 \frac{\text{kg}}{\text{kg dry air}}$ mixes with a second stream of superheated water vapour flowing at 0.1 kg/s. Assuming proper and uniform mixing with no condensation, the humidity ratio of the final stream $\left(\text{in} \frac{\text{kg}}{\text{kg dry air}}\right)$ is ______.

16. Air enters a diesel engine with a density of 1.0 kg/m3. The compression ratio is 21. At steady state, the air intake is 30×10^{-3} kg/s and the net work output is 15 kW. The mean effective pressure (kPa) is ______.

17. Temperature of nitrogen in a vessel of volume 2 m³ is 288 K. A U-tube manometer connected to the vessel shows a reading of 70 cm of mercury (level higher in the end open to atmosphere). The universal gas constant is 8314 J/kmol-K, atmospheric pressure is 1.01325 bar, acceleration due to gravity is 9.81 m/s² and density of mercury is 13600 kg/m³. The mass of nitrogen (in kg) in the vessel is ______.

18. The COP of a carnot heat pump operating between 6°C and 37°C is ______.

19. In a Rankine cycle, the enthalpies at turbine entry and outlet are 3159 kJ/kg. and 2187 kJ/kg respectively. If the specific pump work is 2 kJ/kg, the specific steam consumption (in kg/kW-h) of the cycle based on the net output is ______.

20. One kg of air (R = 287 J/kg.K)undergoes an irreversible process between equilibrium state1(20°C,0.9m³)and equilibrium state 2 (20°C,0.6m³). The change in entropy $S_2 - S_1$ (in J/kg.K) is ______.

21. Work is done on adiabatic system due to which its velocity changes from 10m/s to 20m/s, elevation increases by 20 m and temperature increases by 1K. The mass of the system is 10 kg, C_v = 100 J / (kg.K) and gravitational acceleration is 10m/s². If there is no change in any other component of the energy of the system, the magnitude of total work done (in kJ) on the system is ______.

22. A cylindrical tank with closed ends is filled with compressed air at a pressure of 500 kPa. The inner radius of the tank is 2m, and it has wall thickness of 10 mm. The magnitude of maximum in-plane shear stress (in MPa) is ______.

23. Air in a room is at 35° and 60% relative humidity (RH). The pressure in the room is 0.1 MPa. The saturation pressure of water at 35°C is 5.63 kPa. The humidity ratio of the air (in gram/kg of dry air) is ______.

24. Steam enters a turbine at 30 bar, 300°C (u = 2750 kJ/kg, h = 2993 kJ/kg) and exits the turbine as saturated liquid at 15 kPa (u = 225 kJ/kg, h = 226 kJ/kg). Heat loss to the surrounding is 50 kJ/kg of steam flowing through the turbine. Neglecting changes in kinetic energy and potential energy, the work output of the turbine (in kJ/kg of steam) is ______

25. A mixture of ideal gases has the following composition by mass:

N_2	O_2	CO_2
60%	30%	10%

If the Universal gas constant is 8314 J/mol-K, the characteristic gas constant of the mixture (inJ/kg.K) is ______.

2014

26. 1.5 kg of water is in saturated liquid state at 2 bar (v_f = 0.001061 m³/kg, μ_f = 504.0 *k*J/kg, h_f = 505 kJ/kg). Heat is added in a constant pressure process till the temperature of water reaches 400°C (v = 1.5493 m³/kg, u = 2967.0 kJ/kg, h = 3277.0 kJ/kg). The heat added (in kJ) in the process is ______.

27. An ideal reheat Rankine cycle operates between the pressure limits of 10 kPa and 8 MPa, with reheat being done at 4 MPa. The temperature of steam at the inlets of both turbines is 500°C and the enthalpy of steam is 3185 kJ/kg at the exit of the high pressure turbine and 2247 kJ/kg at the exit of low pressure turbine. The enthalpy of water at the exit from the pump is 191 kJ/kg. Use the following table for relevant data.

Superheated steam temperature (°C)	Pressure (MPa)	v (m³/kg)	h (kJ/kg)	s (kJ/kg.K)
500	4	0.08644	3446	7.0922
500	8	0.04177	3399	6.7266

Disregarding the pump work, the cycle efficiency (in percentage) is ______

28. In an air-standard Otto cycle, air is supplied at 0.1 MPa and 308 K. The ratio of the specific heats (γ) and the specific gas constant (R) of air are 1.4 and 288.8 J/kg.K, respectively. If the compression ratio is 8 and the maximum temperature in the cycle is 2660 K, the heat (in kJ/kg) supplied to the engine is ______

29. A pure substance at 8 MPa and 400 °C is having a specific internal energy of 2864 kJ/kg and a specific volume of 0.03432 m³ / kg . Its specific enthalpy (in kJ/kg) is ______

30. A spherical balloon with a diameter of 10 m, shown in the figure below is used for advertisements. The balloon is filled with helium (R_{He} = 2.08 kJ/kg.K) at ambient conditions of 15°C and 100 kPa. Assuming no disturbances due to wind, the maximum allowable weight (in newton) of balloon material and rope required to avoid the fall of the balloon (R_{air} = 0.289 kJ/kg.K) is ______

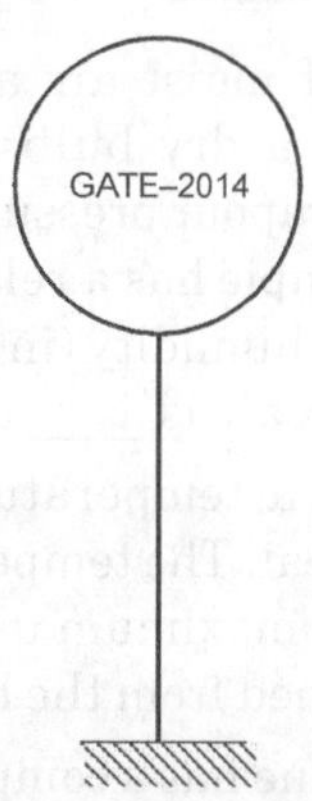

31. In an ideal Brayton cycle, atmospheric air (ratio of specific heats, c_p/c_v = 1.4, specific heat at constant pressure = 1.005 kJ/kg.K) at 1 bar and 300 K is compressed to 8 bar. The maximum temperature in the cycle is limited to 1280 K. If the heat is supplied at the rate of 80 MW, the mass flow rate (in kg/s) of air required in the cycle is ______

32. Steam at a velocity of 10 m/s enters the impulse turbine stage with symmetrical blading having blade angle 30°. The enthalpy drop in the stage is 100 kJ. The nozzle angle is 20°. The maximum blade efficiency (in percent) is ______

33. In a concentric counter flow heat exchanger, water flows through the inner tube at 25°C and leaves at 42°C. The engine oil enters at 100°C and flows in the annular flow passage. The exit temperature of the engine oil is 50°C. Mass flow rate of water and the engine oil are 1.5 kg/s and 1 kg/s, respectively. The specific heat of water and oil are 4178 J/kg.K and 2130 J/kg.K, respectively. The effectiveness of this heat exchanger is ______

34. A heat pump with refrigerant R22 is used for space heating between temperature limits of –20°C and 25°C. The heat required is 200 MJ/h. Assume specific heat of vapour at the time of discharge as 0.98 kJ/kg.K. Other relevant properties are given below. The enthalpy (in kJ/kg) of the refrigerant at isentropic compressor discharge is ______

Saturation Specific enthalpy temperature	**Pressure Specific entropy**
T_{sat}(°C)	P(MN/m²)
h_f(kJ/kg)	h_g(kJ/kg)
S_f(kJ/kg.K)	S_g(kJ/kg.K)
–20	0.2448
177.21 397.53	0.9139 1.7841
25	1.048 230.07
413.02 1.1047	1.7183

35. A sample of moist air at a total pressure of 85 kPa has a dry bulb temperature of 30°C (saturation vapour pressure of water = 4.24 kPa). If the air sample has a relative humidity of 65%, the absolute humidity (in gram) of water vapour per kg of dry air is ______

36. A source at a temperature of 500 K provides 1000 kJ of heat. The temperature of environment is 27°C. The maximum useful work (in kJ) that can be obtained from the heat source is ______

37. A diesel engine has a compression ratio of 17 and cut-off take place at 10% of the stroke. Assuming ratio of specific heats (γ) as 1.4, the air-standard efficiency (in percent) is ______

38. A reversed Carnot cycle refrigerator maintains a temperature of –5°C. The ambient air temperature is 35°C. The heat gained by the refrigerator at a continuous rate is 2.5 kJ/s. The power (in watt) required to pump this heat out continuously is ______

39. A thin gas cylinder with an internal radius of 100 mm is subject to an internal pressure of 10 MPa. The maximum permissible working stress is restricted to 100 MPa. The minimum cylinder wall thickness (in mm) for safe design must be ______

40. A closed system contains 10 kg of saturated liquid ammonia at 10°C. Heat addition required to convert the entire liquid into saturated vapour at a constant pressure is 16.2 MJ. If the entropy of the saturated liquid is 0.88 kJ/kg.K, the entropy (in kJ/kg.K) of saturated vapour is ______

ANSWERS

MCQ Type Questions

1. (c)	**2.** (d)	**3.** (c)	**4.** (a)	**5.** (c)	**6.** (c)	**7.** (b)	**8.** (b)	**9.** (b)	**10.** (d)
11. (d)	**12.** (a)	**13.** (c)	**14.** (d)	**15.** (c)	**16.** (d)	**17.** (a)	**18.** (d)	**19.** (b)	**20.** (a)
21. (b)	**22.** (b)	**23.** (d)	**24.** (d)	**25.** (a)	**26.** (b)	**27.** (b)	**28.** (a)	**29.** (a)	**30.** (b)
31. (c)	**32.** (a)	**33.** (c)	**34.** (c)	**35.** (a)	**36.** (c)	**37.** (a)	**38.** (d)	**39.** (a)	**40.** (d)
41. (b)	**42.** (b)	**43.** (a)	**44.** (b)	**45.** (d)	**46.** (d)	**47.** (c)	**48.** (b)	**49.** (b)	**50.** (b)
51. (d)	**52.** (b)	**53.** (a)	**54.** (b)	**55.** (d)	**56.** (*)	**57.** (d)	**58.** (d)	**59.** (a)	**60.** (b)
61. (a, c)	**62.** (d)	**63.** (d)	**64.** (a, c)	**65.** (*)	**66.** (c)	**67.** (b)	**68.** (a)	**69.** (c)	**70.** (a,b,c)
71. (c, b)	**72.** (*)	**73.** (*)	**74.** (d)	**75.** (d)	**76.** (a)	**77.** (c)	**78.** (c)	**79.** (a) – 3, (b) – 4, (c) – 6, (d) – 5	**80.** (*)
81. (*)	**82.** (*)	**83.** (a)	**84.** (c)	**85.** (d)	**16.** (b)	**17.** (a)	**88.** (a)	**89.** (c)	**90.** (a)
91. (c)	**92.** (*)	**93.** (*)	**94.** (*)	**95.** (*)	**96.** (a)	**97.** (b)	**98.** (d)	**99.** (b)	**100.** (c)
101. (b)	**102.** (a)	**103.** (*)	**104.** (*)	**105.** (*)					

"*" Refers to Subjective Question.

Numerical Type Questions

1. 0.25 **2.** 2443.25 kJ/kg **3.** 15 **4.** 26 **5.** (4) **6.** (0.803)

7. (–55.65) **8.** (8.9315) **9.** (19.89) **10.** (85) **11.** (17.0410) **12.** (464.151) **13.** (1103.44)

14. 50 **15.** 0.02 **16.** 525 **17.** 4.55 **18.** 10 **19.** 3.780 **20.** –116.36

21. 4.5 **22.** 25 **23.** 21.74 **24.** 2717 **25.** 274.99 **26.** 4155 to 4160 **27.** 40 to 42

28. 1400 to 1420 **29.** 3135 to 3140 **30.** 5300 to 5330 **31.** 105 to 112

32. 85.1 to 89.9 **33.** 0.65 to 0.67 **34.** 430 to 440 **35.** 19 to 22

36. 399 to 401 **37.** 58 to 62 **38.** 370 to 375 **39.** 9.8 to 10.6

40. 6.4 to 6.7

EXPLANATIONS

MCQ TYPE QUESTIONS

5. $p+\frac{a}{v^2}$ both term should gave same unit since they are getting added

$$\Rightarrow \quad \frac{N}{m^2} = a\left(\frac{kg}{m^3}\right)^2$$

$$\Rightarrow \quad a(\text{unit}) = \frac{m^6}{kg^2}.kg.\frac{m}{s^2m^2}$$

$$= \frac{m^5 kg}{kg^2 s^2} = \frac{m^5}{kgs^2}$$

6. With reheat average temperature of heat addition increases. Hence, efficiency of cycle also increases. Further, the quality of steam is higher at turbine exit.

7. For same values of peak pressure and temperature. Diesel cycle is most efficient and Otto cycle is least. Efficiency of dual cycle lies in between.

$\eta_{Diesel} > \quad \eta_{Dual} > \eta_{Otto}$

9. Heat applied, $Q_s = c_p(T_3 - T_2)$

Heat rejected, $Q_r = c_r(T_4 - T_1)$

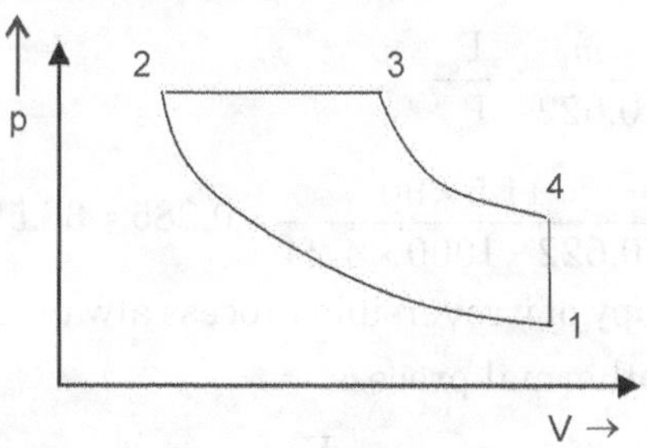

$$\eta = 1 - \frac{Q_r}{Q_s} = 1 - \frac{1}{\gamma}\frac{(T_4 - T_1)}{(T_3 - T_2)}$$

10. E of condenser is given by 1-exp(-NTU)

Because $\quad C_r = \frac{C_{min}}{C_{max}} = 0.\ (\text{as } C_{max} \to \infty)$

12. $\quad f_{11} + f_{12} = 1$

$\therefore \quad f_{12} = 1$

$f_{21}A_2 = f_{12}A_1$

$\therefore \quad f_{21} = \frac{f_{12}A_1}{A_2}$

$$= \frac{1 \times 4\pi r^2}{\frac{1}{2}4\pi(4r)^2 + \pi(4r)^2} = \frac{1}{12}$$

13. Energy (or) Available Energy :

The maximum portion of energy which could be converted into useful work by ideal processes which reduce the system to dead state (a state in equilibrium with the earth and its atmosphere).

14. Specific volume $= \frac{\text{volume of fluid}}{\text{Mass of fluid}}$

$$= \frac{1}{\rho} = \frac{1}{1000} = 10^{-3}\ m^3/kg$$

$$\eta = \frac{\text{Isentropic compressor work}}{\text{Actual compressor work}}$$

$$\text{Actual compressor work} = \frac{V(\Delta P)}{\eta}$$

$$= \frac{10^{-3} \times (3000 - 80)KPa}{0.85}$$

$$= \frac{2.92}{0.85} = 3.43\ kJ/kg.$$

15.

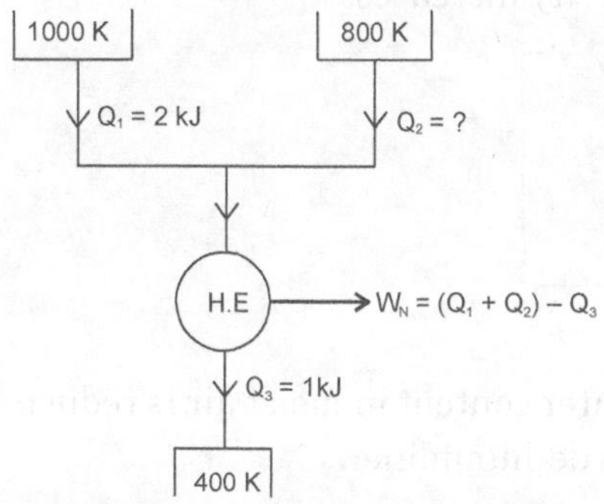

We know that for reversible heat engine, change in entropy is always zero

That is $\Delta S = 0$

$$\frac{Q_3}{T_3} - \left(\frac{Q_1}{T_1} + \frac{Q_2}{T_2}\right) = 0$$

$$\frac{1}{400} - \frac{2}{1000} - \frac{Q_2}{800} = 0$$

$Q_2 = 0.4$ kJ

$W_N = (Q_1 + Q_2) - Q_3$

$= (2 + 0.4) - 1 = 1.4$ kJ.

16. Thermal efficiency of air standard efficiency

$$= 1 - \frac{1}{(r_p)^{\frac{\gamma-1}{\gamma}}}$$

19. For (1 – 2) process : Constant pressure process $\{P_2 = P_1\}$

$$\frac{T_1}{V_1} = \frac{T_2}{V_2} \Rightarrow V_2 = \frac{T_2}{T_1} \times V_1$$

For (2 – 3) process : Constant volume process $\{V_3 = V_2\}$

Given $\quad T_2 = \frac{3T_1}{4}$

$$V_2 = \frac{3T_1}{4T_1} \times V_1 = \frac{3}{4}V_1$$

$$\Rightarrow \quad \frac{V_3}{V_1} = \frac{V_2}{V_1} = \frac{3}{4} = 0.75$$

20. $K = \dfrac{nV\lambda C_V}{3N_A}$

Where K is thermal conductivity

V is mean particle speed

λ is mean free path

C_V is molar head capacity

N_A is Avogadro's number

n is particles per unit volume

Gases transfer heat by direct collisions between molecules. As the temperature increases, the thermal conductivity increases due to increase in speed, movement and collisions in the molecules. From the above expression, by increasing mean particle speed, the thermal conductivity increases.

21.

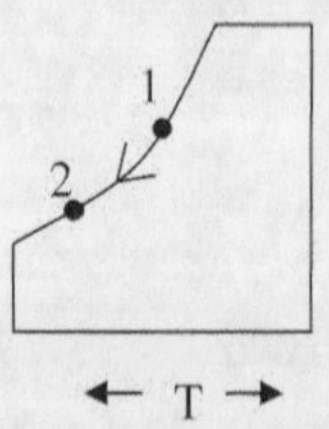

Since water content in moist air is reducing. The device is de-humidifier.

25. Given : $\gamma = 1.67$

Molecular weight, M = 40

$P_1 = 0.10$ MPa

$R_u = 8.314$ kJ/k mol.K

$T_1 = 300$ K

$P_2 = 0.2$ MPa

Characteristic gas constant,

$$r = \frac{R_u}{M} = \frac{8.314}{40} = 0.208\ kJ/kgK$$

From $$\frac{T_2}{T_1} = \left(\frac{P_2}{P_1}\right)^{\frac{\gamma-1}{\gamma}} = 2^{\frac{0.64}{1.67}}$$

$$T_2 = 396.18 \text{ K}$$

Work done in compression

$$= \frac{r(T_2 - T_1)}{-1} = \frac{0.208(396.18 - 300)}{0.67} = 29.8 \text{ kJ/kg}$$

26. From Clausius inequality,

$$\oint \frac{\partial Q}{T} < 0 \text{ for irreversible process}$$

i.e. $$\frac{Q_1}{T_1} + \frac{Q_2}{T_2} + \frac{Q_3}{T_3} + ... < 0$$

For process I

$$\oint \frac{\partial Q}{T} = \frac{2500}{1200} - \frac{2500}{800} = -1.042 < 0$$

For process II

$$\oint \frac{\partial Q}{T} = \frac{2000}{800} - \frac{2000}{500} = -1.5 < 0$$

Hence process II is more irreversible than process I.

27. Given : DBT = 30°C

$w = 11.5$ g/kg

$\therefore$ Relative humidity

$$= \frac{\text{Specific humidity}}{0.622} \times \frac{\text{Total pressure}}{\text{Saturation vapour pressure}}$$

$$= \frac{w}{0.622} \times \frac{P_a}{P_{vs}}$$

$$= \frac{11.5 \times 90}{0.622 \times 1000 \times 4.24} = 0.385 \approx 38.5\%$$

28. Entropy of irreversible process always increases.

29. In isothermal process

$$W = P_1V_1\, ln\frac{V_1}{V_2}$$

$$= 0.8 \times 10^3 \times 0.015\, ln\left(\frac{0.015}{0.030}\right) \text{kJ}$$

$$= -8.317 \approx -8.32 \text{ kJ}$$

negative sign shows work done by the system.

$\therefore$ Work output = 8.32 kJ

Alternately

$$P_1V_1 = P_2V_2$$

$\therefore$ $$0.8 \times 0.015 = P_2 \times 0.030$$

$\Rightarrow$ $$P_2 = \frac{0.8 \times 0.015}{0.030} = 0.4$$

$\therefore$ $$W = P_1V_1 \log\frac{P_1}{P_2}$$

$$= 0.8 \times 0.015\, ln2 = 8.31 \text{ kJ}$$

30. Since, $\Delta Q = 0$, therefore

$$S = \int \frac{dQ}{T} = 0$$

31. Since kinetic energy of the molecules won't change in vacuum, hence temperature and internal energy remains constant.

Now $\frac{pV}{T}$ = constant

$\Rightarrow$ pV = constant

$\Rightarrow$ H = U + pV = constant

32. $h_3 = 0.2h_1 + h_2 \times 0.8$

$= 560 + 160$

$= 720$ kJ/kg

33. For steady flow process,

$$\Delta W = -\Delta H + \Delta Q$$
$$= -\Delta H + T.\ \Delta S$$

But for reversible process,

$$T.\ \Delta S = \Delta H - V.\ \Delta p$$

$\Rightarrow$ $\Delta W - \Delta H + \Delta H = -V.\ \Delta p$

$\Rightarrow$ $\Delta W = -V.\ \Delta p$

For steady flow reversible process,

$$W = -\int_{\text{inlet}}^{\text{outlet}} v\ dp$$

34. For reversible processes, by first law

$$\delta Q = \delta U + \delta W$$

Also by definition of entropy,

$$dS = \frac{\delta Q}{T}$$

$\therefore$ $TdS = dU + \delta W$

36.

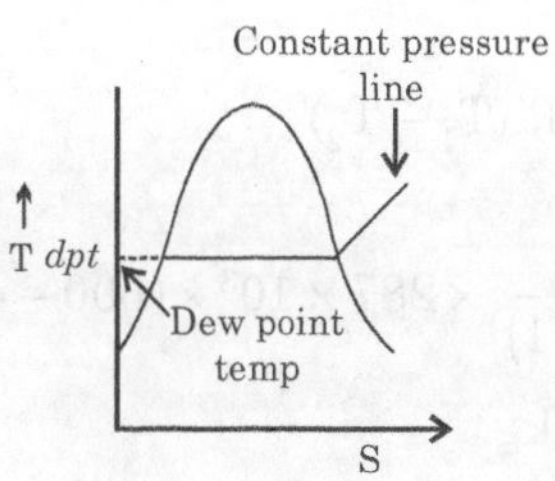

Air is cooled at constant pressure to make unsaturated air to saturated one.

38.

When added to the system		Differential	Function	Phenomenon
Heat	Positive	Inexact	Path	Boundary
Work	Negative	Inexact	Path	Transient

40. *At the time of intake,* intake valve operates only when pressure inside the cylinder is lower than the atmospheric pressure, and appreciable large pressure difference at the time of suction starting, causing fluttering at initial.

At the time of compression, process followed is polytorpic and continues till pressure exceeds the delivery line pressure for opening the delivery valve.

41. Here, $\phi = \frac{p_v}{p_s}$

$\Rightarrow$ $p_v = \phi \cdot p_s$

$(p_s)_{35°C} = 0.05628$ bar

$\omega = 0.622\ p_v$

42. *Given:* $t_{DBT} = 40°$

$t_{WBT} = 20°$

Water sprayed at temperature = 42°

Since, $t_{\text{water spray}} > t_{DBT}$

so heating and humidification.

44. $\left(p + \frac{a}{v^2}\right)V = RT$

$\therefore$ $\left(10 + \frac{a}{1}\right) \times 1 = \left(p + \frac{a}{4}\right) \times 2$

$\Rightarrow$ $p = 5 + \frac{a}{2} - \frac{a}{4} = 5 + \frac{a}{4}$

47. $dQ = du + dw$

$\therefore$ $-2000 = du + (-5000\text{ kJ})$

$\Rightarrow$ $du = 3000$ kJ.

48.

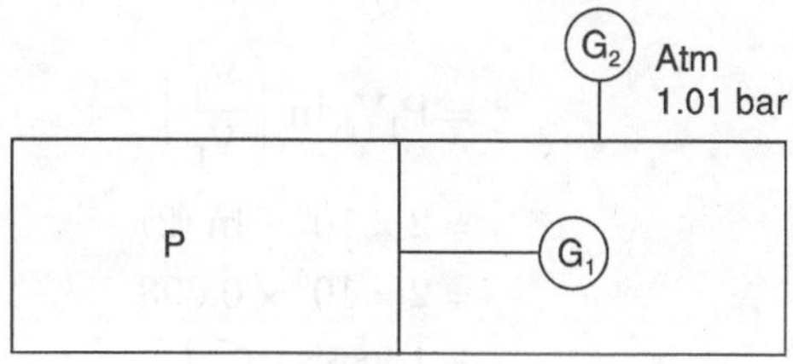

$P = P_{G_2}$ + Atmospheric pressure

$= 1.00 + 1.0$

$= 2.01$ bar.

49. Given : $p = 1.013$ bar

$(p_v)_{tdp} = 0.02062$

Specific humidity

$$= 0.622\ \frac{p_v}{p - p_v}$$

$$= \frac{0.622 \times 0.02062}{1.013 - 0.02062}$$

$= 0.01291$ kg/kg of dry air

$h = 1.022\ t_d + w\ (h_{fgdp} + 2.3t_{dp})$

$= 63.15$ kJ/kg of dry air.

Alternative method

p_v = Saturation pressure = 0.02062 bar

p_a = Partial pressure of air = 0.04241 bar

t_d = Dry bulb temperature = 30°

t_w = Wet bulb temperature = 18°

Specific humidity,

$$w = \frac{p_w}{p - p_v} \times 0.622$$

$$= 0.01292 \text{ kg/kg of dry air.}$$

Total enthalpy,

$$h = Cp_a\, t_d + wh_s + w_d p_s (t_d - t_\omega)$$

$$= 1.005 \times 30 + 0.0129 \times 2500$$

$$+ 0.01292 \times 1.88(30 - 18)$$

$$= 63.15 \text{ kJ/kg of dry air.}$$

50.

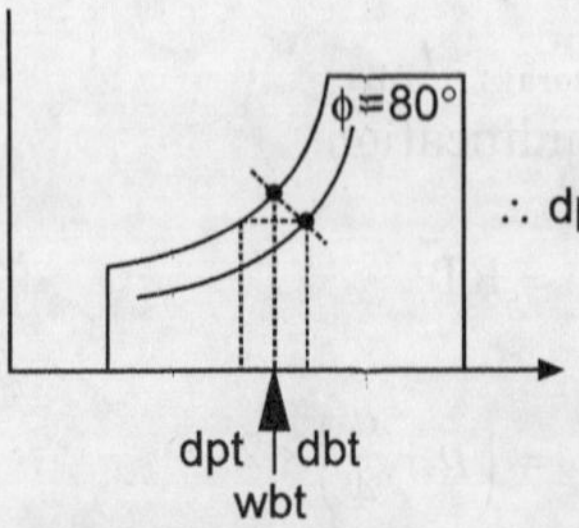

$\therefore$ dpt < wbt

52. Work interaction,
$$W = \int_{V_1}^{V_2} P dV$$

$$= \int_{V_1}^{V_2} \frac{c}{V} dV = c \ln\left(\frac{V_2}{V_1}\right)$$

$$= P_1V_1 \ln\left(\frac{V_2}{V_1}\right)$$

$$= 2 \times 10^5 \times \ln(2)$$

$$= 2 \times 10^5 \times 0.693$$

$$= 1.386 \times 10^5 \text{ J}$$

$$= 138.6 \text{ kJ}$$

53.
$$dQ = dU + dW$$

For isothermal process, $dU = 0$

$$\therefore \quad dQ = dW$$

$$ds = \frac{dQ}{T} = \frac{dW}{T}$$

$$= \frac{138.6 \times 10^3}{298}$$

$$= 0.4652 \text{ kJ/K.}$$

54.
$$\eta = 1 - \frac{T_2}{T_1}$$

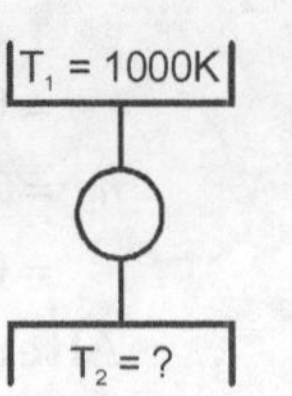

$$\therefore \quad 0.75 = 1 - \frac{T_2}{1000}$$

$$\Rightarrow \quad T_2 = 250 \text{ K}$$

$$= 250 - 273$$

$$= -23°\text{C}$$

55.
$$C = \sqrt{\rho RT}$$

$$= \sqrt{1.4 \times .4210 \times 1000 \times 500}$$

$$= 542.86$$

$$\simeq 543 \text{ m/s}$$

56. Given: $p_1 = 400$ kPa

$p_2 = 100$ kPa

$T_1 = 327°C = 600$ K

$T_2 = 182°C$

$= 455$ K

Now,
$$\left(\frac{T_2}{T_1}\right) = \left(\frac{P_2}{P_1}\right)^{\frac{n-1}{n}}$$

$$\therefore \quad \left(\frac{455}{600}\right) = \left(\frac{100}{400}\right)^{\frac{n-1}{n}}$$

Taking log, we get

$$\frac{n-1}{n} \log \frac{1}{4} = \log\left(\frac{455}{600}\right)$$

$$\Rightarrow \quad n = 1.25$$

Power output

$$= \frac{p_1 v_1}{n-1}\left[1 - \left(\frac{p_2}{p_1}\right)^{\frac{n-1}{n}}\right]$$

$$= \frac{400 \times 10^3 \times 5}{1.25 - 1}\left[1 - \left(\frac{1}{4}\right)^{\frac{0.25}{1.25}}\right]$$

$$= 1937.13 \text{ kW}$$

Rate of heat transfer,

$$Q_R = \frac{(\gamma - n)}{(\gamma - 1)(n - 1)} R (T_1 - T_2)$$

$$= \frac{1.4 - 1.25}{(1.4 - 1)(1.25 - 1)} \times 287 \times 10^3 \times (600 - 455)$$

$$= 104.037 \text{ kJ/sec/kg}$$

$$= 104.037 \times 11.6134 \text{ kJ/sec}$$

$$= 1208.33 \text{ kJ/sec}$$

Rate of change of enthalpy,

$$\Delta s = \frac{\eta - \gamma}{(\gamma - 1)(n - 1)} R \ln \frac{T_2}{T_1}$$

$$= \left(\frac{-1.4 + 1.25}{0.4 \times 0.25}\right) \times 287 \ln\left(\frac{455}{600}\right)$$

$$= 0.119 \text{ kJ}$$

57-58. *Given :* $h_2 = 225.5$ kJ/kg

From chart, $h_1 = 176.5$ kJ/kg

$h_3 = h_4$

$= 69.6$ kJ/kg

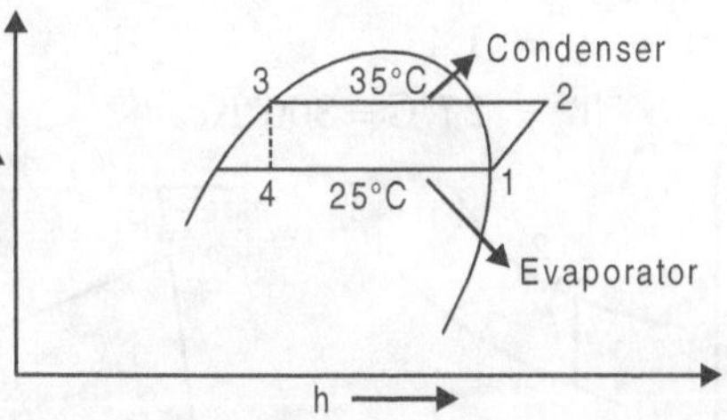

57. Coefficient of performance,

$$COP = \frac{h_1 - h_4}{h_2 - h_1}$$

$$= \frac{\text{Refrigerating effect}}{\text{compressor work}}$$

$$= \frac{176.5 - 69.6}{225.5 - 176.5}$$

$$= 2.18$$

58. Since capacity of the plant = 5 kW

$\therefore \quad m\,(h_1 - h_4) = 5$ kW

where m in kg/s

$\therefore \quad m\,(176.5 - 96.6) = 5$

$\Rightarrow \quad m = 0.467$ kg/s

Power consumption $= m\,(h_2 - h_1)$

$= 0.0467\,(225.5 - 176.5)$

$= 2.288$ kW

59.

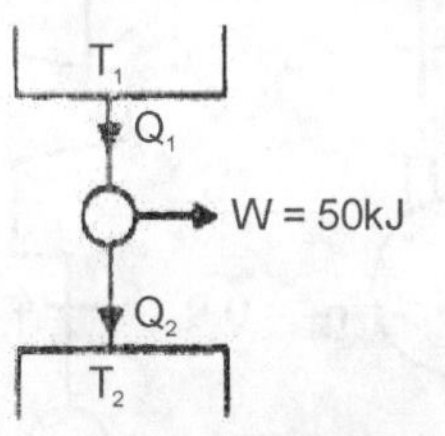

$\eta = 0.75$

$$\therefore \quad \eta_{HE} = \frac{W}{Q_1} = 0.75$$

$$\text{Net heat input, } Q_1 = \frac{50}{0.75} = \frac{200}{3}$$

Also, $W = Q_1 - Q_2$

$$\therefore \quad 50 = \frac{200}{3} - Q_2$$

$$\therefore \text{ Net heat required, } Q_2 = \frac{50}{3}$$

$$= 16.66 \text{ kJ}$$

60. $$\frac{v_1^2}{2} + h_1 = \frac{v_2^2}{2} + h_2$$

Here $v_1 = 0$

$$\therefore \quad h_1 = \frac{v_2^2}{2} + h_2$$

$$\Rightarrow \quad v_2 = \sqrt{2(h_1 + h_2)}$$

$$= \sqrt{2 \times 0.8 \times 1000}$$

$= 40$ m/sec.

62.

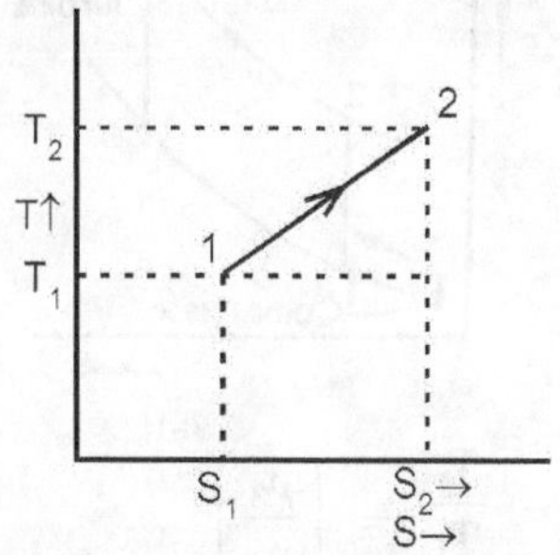

Given:

$T_1 = 330$ K, $T_2 = 440$ K

$U_1 = 170$ *k*J, $U_2 = 190$ *k*J

$H_1 = 220$ *k*J, $H_2 = 247$ *k*J

$S_1 = 0.23$ *k*J/K, $S_2 = 0.3$ kJ/K

Now, $\Delta U = \Delta Q - \Delta W$

where, ΔQ = area under curve an T-S diagram

$$= T_1\,(S_2 - S_1) + \frac{1}{2}(S_2 - S_1)(T_2 - T_1)$$

$$= 330\,(0.3 - 0.23) + \frac{1}{2}(0.3 - 0.23) \times (440 - 330)$$

$= 26.95$ kJ

$$\therefore \quad \Delta W_{1-2} = \Delta Q - \Delta U$$

$$= 26.95 - (190 - 170)$$

$$= 6.95 \text{ kJ}$$

66. Efficiency η of Stirling cycle is equal to Carnot cycle only when regenerative arrangement is employed.

67. Coefficient of performance (C.O.P)

$$= \frac{T}{T_1 - T_2} = \frac{250}{300 - 250}$$

$$= 5$$

69. $$M = \frac{v}{C} = \frac{800 \text{ kmph}}{C}$$

$$= \frac{800 \times 1000}{3600 \times 331.2} = 0.67$$

$$C = \sqrt{\gamma RT}$$

$$= \sqrt{1.4 \times 0.287 \times 273 \times 1000}$$

$$= 331.2 \text{ m/s.}$$

70. For an isolated system, Q = 0, W = 0 and mass does not flow through the system.

72. Given,
$$m = 0.1 \text{ kg/s}$$
$$P_4 = 0.1 \text{ MN/m}^2$$
$$T_4 = 285 \text{ K},$$
$$P_3 = 0.4 \text{ MN/m}^2$$
$$C_p = 1.0 \text{ kJ/kgK}$$

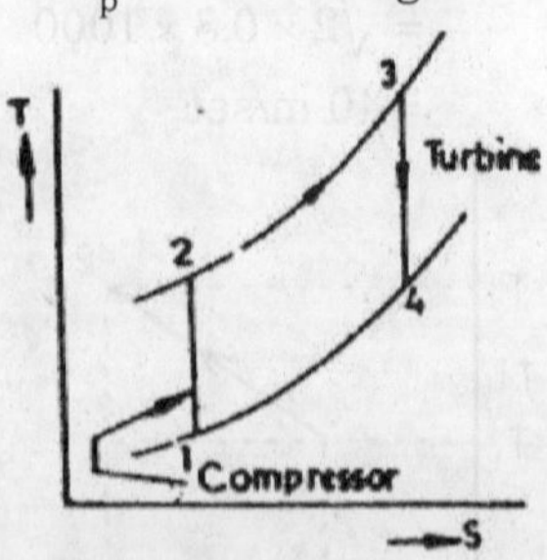

Now, $$\frac{T_3}{T_4} = \left(\frac{p_3}{p_4}\right)^{\frac{\gamma-1}{\gamma}}$$

or $$T_3 = 285\left(\frac{0.4}{0.1}\right)^{0.286}$$
$$= 423.68 \text{ K} \qquad [\text{since } \gamma = 1.4]$$

Power developed $= mC_p(T_3 - T_4)$
$$= 0.1 \times 1 \times (423.68 - 285)$$
$$= 13.87 \text{ kW}$$

73. Given,
$$P_1 = 1.0 \text{ MN/m}^2,$$
$$P_2 = 0.1 \text{ MN/m}^2,$$
$$m = 3 \text{ kg/s},$$
$$T_0 = 300 \text{ K}$$
$$S_{g1} = 6.586 \text{ kJ/kgK},$$
$$S_{g2} = 7.359 \text{ kJ/kgK}.$$

∴ Rate of Entropy production
$$= m(s_{g2} - S_{g1})$$
$$= 3(7.359 - 6.586)$$
$$= 2.319 \text{ kJ/K/s}$$

76. Blade velocity ratio $= \frac{u}{v_1} = 0.5$

or $u = 0.5\, v_1$

where u = mean blade velocity, and

v_1 = Absolute velocity of steam at inlet
$$= \frac{150}{0.5} = 300 \text{ m/s}$$

$$\text{K.E. at inlet} = \frac{1}{2} \times m \times v_1^2 = \frac{m}{2}(300)^2 \text{ J}$$

Nozzle efficiency = 0.9

∴ Isentropic heat drop in nozzle
$$= \frac{(300)^2}{2 \times 0.9} = \mathbf{50 kJ/kg.}$$

77. For air,
$$C_p = 1 \text{ kJ/kg}$$
$$\gamma = 1.4$$
$$T = 27°C = 300°K$$

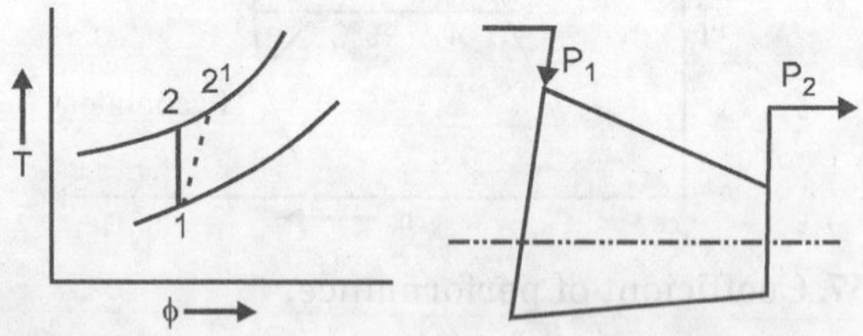

$$\text{Pressure ratio} = \frac{4P_2}{P_1}$$

$$\therefore \frac{T_2}{T_1} = \left(\frac{P_2}{P_1}\right)^{\frac{\gamma-1}{\gamma}} = (4)^{\frac{0.4}{1.4}} = 1.48$$

or $$T_2 = 1.48 \times 300 = 445.8°K$$

$$T'_2 - T_1 = \frac{T_2 - T_1}{0.8} = \frac{445.8 - 300}{0.8} = 182.24.$$

Compressor work $= C_p(T'_2 - T_1)$
$$= 1 \times 182.24 \text{ kJ/kg}$$
$$\approx \mathbf{182 \text{ kJ/kg}}.$$

78. Economizer, preheater, Super heater are all accessories of the boiler and hence forms secondary boiler heat transfer surface.

80.

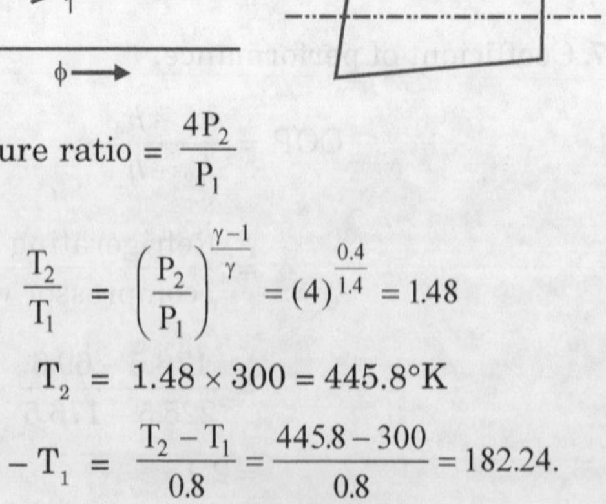

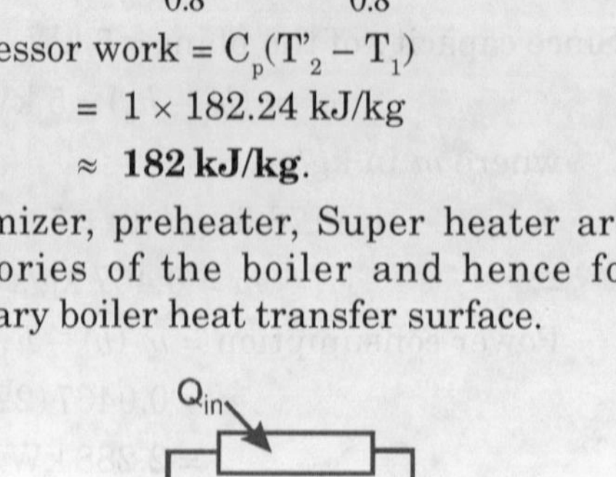

State 1 $P = 0.1 \text{ bar} = 0.81 \text{ kN/m}^2$,

$$h = 191.9 \text{ kJ/kg}$$
$$v = 0.0010103 \text{ m}^3\text{/kg}$$

State 2 $P_2 = 80 \text{ bar} = 7840 \text{ kN/m}^2$

$$W_{pump\ in} = \frac{v_1(P_2 - P_1)}{\eta_p}$$
$$= \frac{0.0010103\,(7840 - 9.81)}{0.9}$$
$$= 8.7898 \text{ kJ/kg}$$
$$h_{2a} = h_1 + W_{pump\ in}$$
$$= 191.9 + 8.79 = 200.69 \text{ kJ/kg.}$$

State 3 $P_3 = 80 \text{ bar} = 7840 \text{ kN/m}^2$

$$T_3 = 600°C,$$
$$h_3 = 3642 \text{ kJ/kg},$$
$$S_3 = 7.0206 \text{ kJ/kg}$$

State 4 $P_4 = 0.1 \text{ bar} = 9.81 \text{ kN/m}^2$

$$S_3 = S_{4s} = 7.0206$$
$$x_4 = \frac{S_{4s} - S_f}{S_{fg}}$$
$$= \frac{7.0206 - 0.6488}{7.5006} = 0.85$$
$$h_{4s} = h_f + 0.85\ h_{fg}$$
$$= 191.9 + 0.85 \times 2392.3 = 2225.3 \text{ kJ/kg}$$
$$\eta_T = \frac{h_3 - h_{4a}}{h_3 - h_{4s}}$$

or
$$h_{4a} = h_3 - \eta_T(h_3 - h_{4s})$$
$$= 3642 - 0.8(3642 - 2225.3)$$
$$= 2508.64 \text{ kJ/kg.}$$

Specific work output = $W_{Turbine} - W_{Pump}$

$$= (h_3 - h_{4a}) - 200.69$$
$$= [3642 - 2508.64] - 200.69$$
$$= \mathbf{932.67\ kJ/kg}$$

$$\text{Thermal efficiency} = \frac{\text{Work output}}{\text{Heat input}} \times 100$$
$$= \frac{932.67}{(h_3 - h_{2a})} \times 100$$
$$= \frac{932.67}{(3642 - 200.69)} \times 100$$
$$= \mathbf{27.1\%}$$

81. $P = 100 \text{ kPa}$

$$T = 27 + 273 = 300 \text{ K}$$

Heat added per cycle,

$$Q_{in} = 1500 \text{ kJ/kg}$$
$$\frac{V_1}{V_2} = 8$$
$$C_v = 0.72 \text{ kJ/kg K}$$
$$\frac{C_p}{C_v} = 1.4$$

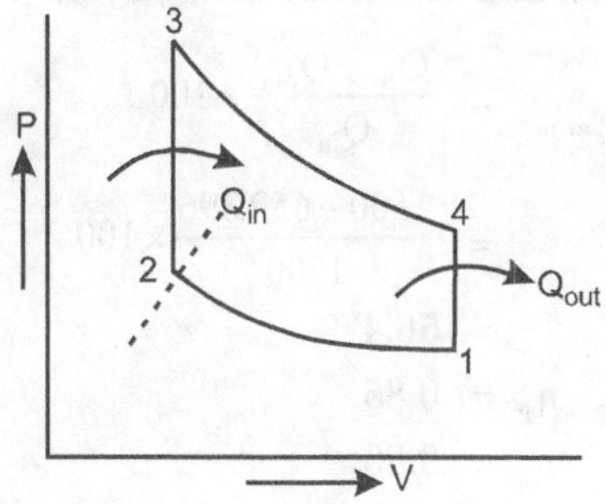

Process 1 – 2

$$\frac{P_2}{P_1} = \left(\frac{V_1}{V_2}\right)^{\gamma}$$
$$\therefore \quad P_2 = 100(8)^{1.4} = \mathbf{1837.92\ kPa}$$
$$T_2 = \left(\frac{P_2}{P_1}\right)^{\frac{\gamma-1}{\gamma}} \times T_1$$
$$= \left(\frac{1837.92}{100}\right)^{\frac{0.4}{1.4}} \times 300 = \mathbf{689.22\ K}$$

Process 2 – 3 (Constant volume process)

$$\frac{P_2}{P_1} = \frac{P_3}{T_3}$$

Heat added per cycle = 1500 kJ/kg

$$C_v(T_3 - T_2) = 1500$$
$$\therefore \quad T_3 = \left(\frac{1500}{0.72}\right) + 689.22$$
$$= \mathbf{2772.6\ K}$$
$$P_3 = \left(\frac{P_2}{T_2}\right) T_3 = \mathbf{7393.5\ kPa}$$

Process 3 – 4 (Adiabatic expansion process)

$$\frac{P_4}{P_3} = \left(\frac{V_3}{V_4}\right)^{\gamma} = \left(\frac{V_2}{V_1}\right)^{\gamma}$$
$$= \left(\frac{1}{8}\right)^{1.4} = 0.0544$$
$$P_4 = 7393.5 \times 0.0544 = \mathbf{402.3\ kPa}$$
$$\frac{P_4}{P_3} = \left(\frac{T_4}{T_3}\right)^{\frac{\gamma}{\gamma-1}}$$
$$\therefore \quad T_4 = \left(\frac{402.3}{7393.5}\right)^{\frac{0.4}{1.4}} \times 2772.6 = \mathbf{1206.8\ K}$$

(i) Heat rejected/kg of mass = $C_v(T_4 - T_1)$

$= 0.72(1206.8 - 300)$

$= \mathbf{652.896\ kJ/kg}$

(ii) Specific work = Work done/kg of air taken in

$= (1500 - 652.896)$

$= \mathbf{847.104\ kJ/kg}$

(iii) Thermal efficiency of the cycle

$$\eta_{thermal} = \frac{Q_{in} - Q_{out}}{Q_{in}} \times 100$$

$$= \frac{1500 - 652.896}{1500} \times 100$$

$= \mathbf{56.47\%}$

82. Given, $\eta_T = 0.85$

$\eta_c = 0.80$

$T_3 = 875°C = 875 + 273 = 1148° K$

For air

$C_p = 1.00$ kJ/kg K

$\gamma = 1.4$

$P_1 = 1$ atm

$T_1 = 27°C = 273 + 27 = 300° K$

Pressure ratio, $\frac{P_2}{P_1} = 4$

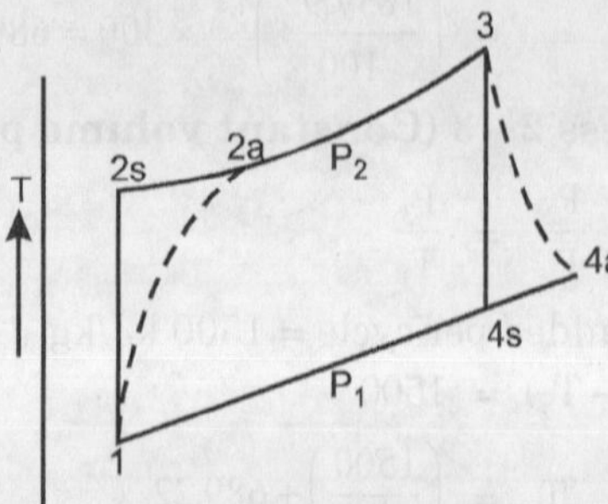

Calorific value of fuel used = 42000 kJ/kg

Loss of Calorific Value (C.V.) = 10%

Work done during compression = $mC_v(T_{2a} - T_1)$

$$\frac{P_2}{P_1} = \left(\frac{T_{2s}}{T_1}\right)^{\frac{\gamma}{\gamma-1}}$$

$$T_{2s} = (4)^{\frac{0.4}{1.4}} \times 300$$

$= 445.80$ K

$$\eta_c = 0.8 = \left(\frac{T_{2s} - T_1}{T_{2a} - T_1}\right)$$

or $$T_{2a} = \left(\frac{T_{2s} - T_1}{0.8}\right) + T_1$$

$= 482.25°C$

Heat added in the combustion chamber

$= mC_p(T_3 - T_{2a})$ kJ

$= m \times 1[(875 + 273) - 482.25]$

$= 665.75$ m kJ

$= 665.75$ kJ/kg of air

Mass of fuel burnt/kg of air

$$= \frac{665.75}{42{,}000 \times 0.9}$$

$= 0.0176$ kg

$$\text{A/F ratio} = \frac{1}{0.0176} = \mathbf{56.80}$$

84. We know, $\left(\frac{\partial s}{\partial T}\right)_v = \frac{C_v}{T}$

and $$T\left(\frac{\partial s}{\partial T}\right) - T\left(\frac{\partial s}{\partial T}\right)_v = C_p - C_v = R$$

87. $P_{atm} = 760$ mm Hg

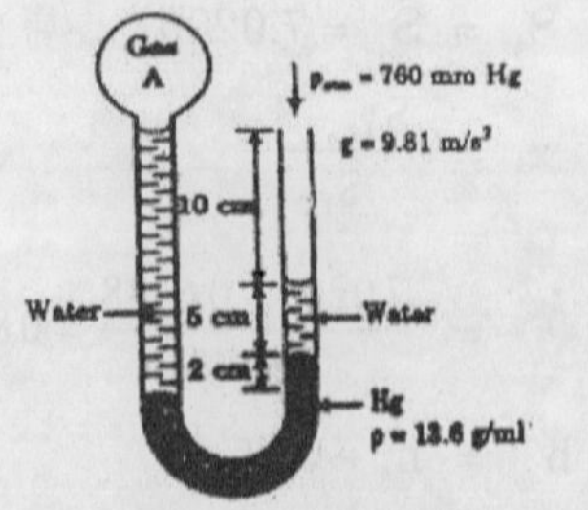

Let P_A be the absolute pressure at 'A'

Pressure at D = Pressure at E

or $P_A + P_{CD} = P_{atm} + P_{AF} + P_{FE}$

or $$P_A + \frac{170}{13.6} = 760 + \frac{50}{13.6} + 20$$

$P_A = \mathbf{771.2\ mm\ Hg}$

88.

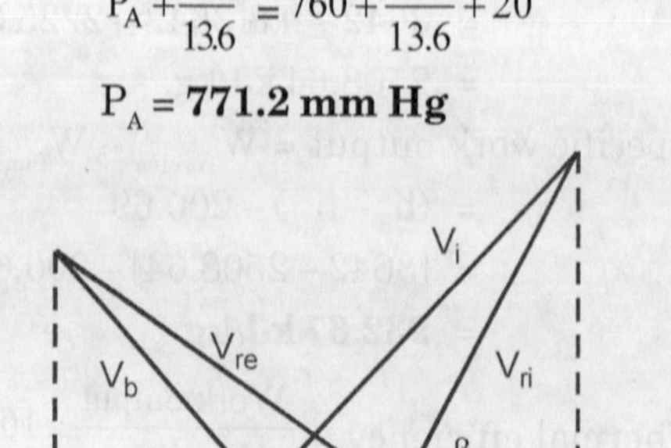

Given : $\alpha = 20°$

$\beta = 30°$

$x = 350 \cos 30° = 303$ m/s

$\therefore$ $V_{wi} = 303 + V_b$

$= 303 + 200 = 503$ m/s

$$\cos 25° = \frac{200 + V_{wo}}{350} = 0.9063$$

$V_{wo} = 118$ m/s

Work done/kg of steam

$$= \frac{(V_{wi} + V_{wo})V_b}{1000} \text{ kJ}$$

$$= \frac{(503+118)\,200}{1000} = \mathbf{124\ kJ}$$

90. $p_2 = \sqrt{p_1 \cdot p_3} = \sqrt{1 \times 9} = 3$ bar

Work done in either cylinder

$$= \frac{n}{n-1} RT_1 (3^{\frac{n-1}{n}} - 1)$$

$$= \frac{14}{0.4} \times 286.7 \times 300 (3^{\frac{0.4}{1.4}} - 1)$$

$= 3.5 \times 286.7 \times 300\ (1.37 - 1) =$ **111 kJ**

91. Clearance ratio $= \frac{\text{Clearance volume}}{\text{Swept volume}}$

$$= \frac{100}{(1000 - 100)} = \mathbf{\frac{1}{9}}$$

92. Brake power $= \frac{(W-S)\,\pi\,(D+d)\,N}{6000}$

where, W = weight on tight side of rope

= 50 × 9.81 = 490.5 N

S = spring balance reading = 50 N

D = diameter of pulley = 0.7 m

d = diameter of rope = 0.025 m

N = shaft speed in rpm = 900

$\therefore$ Brake power $= \frac{(490.5 - 50)\,\pi\,(0.7 + 0.025)\,900}{60{,}000}$

= 15.03 kW

Brake specific fuel consumption

$$= \frac{\text{kg of fuel / hr}}{\text{BHP in kW}}$$

$$= \frac{4}{15.049}$$

$= 0.265$ kg./BHP / hr. or kg / kWh

Brake thermal efficiency

$$= \frac{\text{Heat equivalent of brake power}}{\text{Heat energy supplied by fuel}}$$

$$= \frac{\text{BHP}}{m_f \times \text{C.V.}}$$

where, BHP = Brake power in kW = 15.049

m_f = mass of fuel supplied to engine per minute

$$= \frac{4}{60} = \frac{1}{15} \text{ kg / mt.}$$

C.V. = calorific value

= 44000 kJ/kg

$\therefore$ Brake thermal efficiency $= \frac{15.049 \times 60}{(1/15) \times 44000}$

$= 0.3078 =$ **30.78%**

93. Irreversibility, $I = T_0\ (\Delta s_{system} + \Delta s_{surroundings})$

where, T_0 = temperature of surroundings = 300 K

From data, $\Delta s_{surroundings} = 0$

$\therefore \quad I = T_0 \Delta s_{system}$

or $\quad \Delta s_{system} = \frac{I}{T_0}$

$$= \frac{100}{300} = 0.333 \text{ kJ / K}$$

Also, $\quad \Delta s_{system} = m\left[C_v \ln\frac{T_2}{T_1} + R \ln\frac{V_2}{V_1}\right]$

Here $\quad V_1 = V_2$

$\therefore \quad \Delta s_{system} = mC_v \ln\frac{T_2}{T_1}$

$$0.333 = 2 \times 0.718 \ln\frac{T_2}{300}$$

$\therefore \quad T_2 = 3783$ K

= **105.3°C**

Increase in availability $= \phi_2 - \phi_1$

$= (E_2 - E_1) - T_0 (s_2 - s_1)$

$$= m\left[C_v(T_2 - T_1) - T_0 C_v \ln\frac{T_2}{T_1}\right]$$

$$= 2 \times 0.718 \left[(378.3 - 300) - 300 \ln\frac{378.3}{300}\right]$$

= **12.54 kJ**

94. Figure shows the process on T–s chart with regeneration.

Assume compressor efficiency to be 100%

Turbine efficiency, $\eta_T = 0.8$

Regenerator efficiency = 0.8

Pressure ratio, $\frac{p_2}{p_1} = 5$

Given : $p_1 = 1$

$T_1 = 300°$ K

$T_2 = 490°$ K

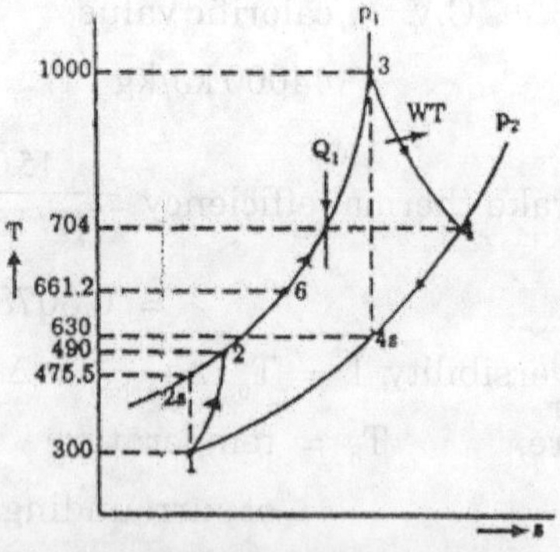

Maximum temperature of cycle,

$$T_3 = 1000° \text{ K}$$

Ratio of specific heats,

$$\gamma = 1.4$$

Now $\quad \dfrac{T_{2s}}{T_1} = \left(\dfrac{p_2}{p_1}\right)^{(\gamma-1)/\gamma} = (5)^{0.4/1.4}$

or $\quad T_{2s} = 300 \times (5)^{0.4/1.4} = 475.14 \text{ K}$

Also, $\quad \dfrac{T_3}{T_{4s}} = \left(\dfrac{p_3}{p_4}\right)^{(\gamma-1)/\gamma}$

$$= (5)^{0.4/1.4} = 1.5838$$

$\therefore \quad T_{4s} = \dfrac{1000}{1.5838} = 630 \text{ K}$

With regenerator,

$$T_3 - T_4 = \eta_T(T_3 - T_{4s})$$
$$= 0.8\,(1000 - 630) = 296$$
$$T_4 = T_3 - 296$$
$$= 1000 - 296 = 704 \text{ K}$$
$$W_T = h_3 - h_4$$
$$= C_p\,(T_3 - T_4)$$
$$= 1.005\,(296) = 297 \text{ kJ/kg}$$
$$W_C = h_1 - h_2$$
$$= C_p(T_2 - T_1)$$
$$= 1.005\,(490 - 300) = 191 \text{ kJ/kg}$$
$$W_{net} = W_T - W_C$$
$$= 297 - 191 = 106 \text{ kJ/kg}$$

Regenerator efficiency

$$= 0.8 = \frac{T_6 - T_2}{T_4 - T_2}$$

or $\quad T_6 - 490 = 0.8\,(704 - 490)$

or $\quad T_6 = 661.2$

Work consumed by compressor

$$= C_p(T_{2s} - T_1)$$
$$= 1.005\,(475.14 - 300)$$
$$= 176.01 \text{ kJ.kg}$$

Work done by turbine $= C_p(T_3 - T'_4)$

$$= 1.005\,(1000 - 705.11)$$
$$= 296.36 \text{ kJ/kg}$$

Net work = Turbine work – Comperssor work

$$= 296.36 - 1760.01$$
$$= 120.35 \text{ kJ/kg}$$

Heat supplied, $\quad Q_1 = h_3 - h_6$

$$= C_p(T_3 - T_6)$$
$$= 1.005(1000 - 661.2)$$
$$= 340 \text{ kJ/kg}$$

$\therefore$ Cycle efficiency, $\eta = \dfrac{\text{Net work}}{\text{Heat supplied}}$

$$= \frac{W_{net}}{Q_1}$$
$$= \frac{100.35}{340} = \mathbf{0.35 \text{ or } 35\%}$$

95. $\quad p_i = 10$ bar

Let $\quad$ i = inlet cross section

T = 500 K $\quad$ e = exit cross section

As velocity at entry is negligible,

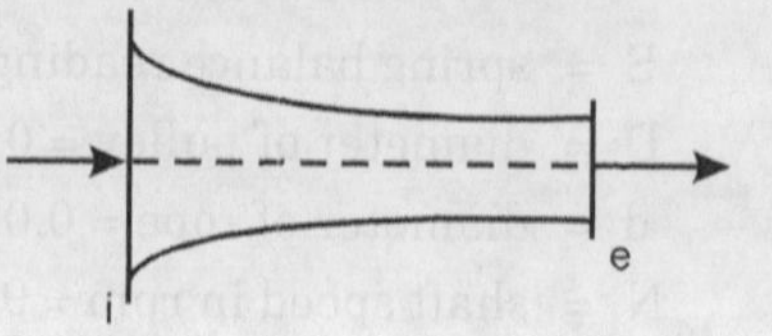

$p_i = p_e = 10$ bar = stagnation pressure at inlet

We know,

Critical pressure, $\quad p^* = p_o\left(\dfrac{2}{\gamma+1}\right)^{\gamma/\gamma-1}$

$$= 10\left(\frac{2}{2.4}\right)^{\frac{1.4}{0.4}} = 5.28 \text{ bar}$$

Back pressure = 2 bar < p*

i.e. the converging nozzle is choked

$$p_e = p^* = 5.28 \text{ bar}$$

$$T_e = T^* = T_o\left(\frac{2}{\gamma+1}\right)$$

$$= 500\left(\frac{2}{24}\right) = 416.67 \text{ K}$$

$$v_e = \sqrt{2C_p(T_o - T_e)}$$
$$= \sqrt{2 \times 1005(500 - 416.67)}$$
$$= \mathbf{409 \text{ m/s}}$$

Alternatively

$$v_e = V^* = \sqrt{\gamma RT}$$

$$= \sqrt{14 \times 287 \times 416.67} = \mathbf{409\ m/s}$$

$$\text{Flow rate} = \frac{RT_e}{p_e}$$

$$= \frac{287 \times 416.67}{5.28 \times 10^5} = 0.2265\ m^3/kg$$

$$m = \frac{A_e v_e}{\text{flow rate}}$$

$$= \frac{2.5 \times 10^{-4} \times 409}{0.2265} = \mathbf{0.451\ kg/s}$$

100. $\eta_{th} = \frac{(80-70)}{80} = \frac{1}{8} = 12.5\ \%$

101. $\eta_{\text{Topping cycle}} = \frac{W_T}{Q_1} = \frac{Q_1 - Q_2}{Q_1} = 30\%$, *i.e.* 0.3

or $\frac{Q_2}{Q_1} = 0.7$ ...*(i)*

$\eta_{\text{Bottoming cycle}} = \frac{W_B}{Q_2} = \frac{Q_2 - Q_3}{Q_2} = 20\%$, *i.e.* 0.2

or $\frac{Q_3}{Q_2} = 0.8$ ...*(ii)*

From equation *(i)* and *(ii)*

$$\text{Combined cycle efficiency} = \frac{(W_T + W_B)}{Q_1}$$

$$= \frac{Q_1 - Q_3}{Q_1}$$

$$= \frac{\frac{Q_2}{0.7} - 0.8\ Q_2}{\frac{Q_2}{0.7}} = 44\%$$

103. *Given*, Temperature of iron = 400° C = 673° K,

Temperature of iron cube = 25° C = 298° K

Mass of water = 10 kg

Temperature of water and cube after equilibrium = 50°C = 323°K

Specific heat of water, C_{pw} = 4186 J/kg°K

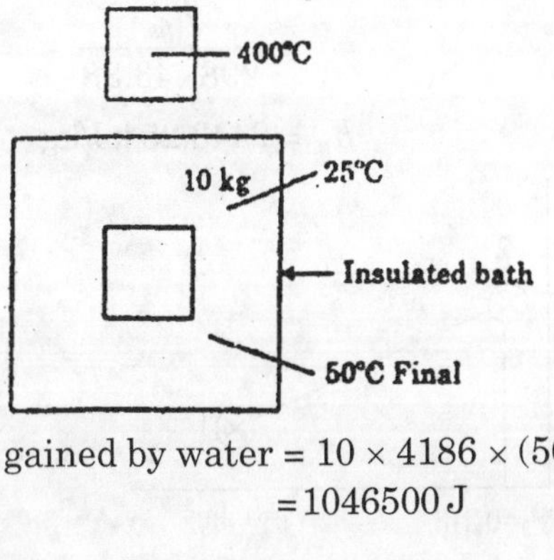

Heat gained by water = 10 × 4186 × (50 – 25) J = 1046500 J

Heat lost by iron cube = 1046500 = m × s × (400 – 50)

Now, Heat lost by iron cube = Heat gained by water

or $(m_i s_i)_{\text{iron cube}} \times 350 = 1046500\ J$

or $(m_i s_i)_{\text{iron cube}} = 2990\ J/K$

where m = mass of iron cube in kg, and

s = specific heat of iron in J/kg °K

$$\text{Entropy of iron at 673°K} = m_i s_i \log_e \left(\frac{673}{273}\right)$$

$$= 2990 \times \log_e \left(\frac{673}{273}\right)$$

= 2697.7 kJ/°K [Taking 0°C as datum]

$$\text{Entropy of water at 298°K} = 10 \times 4186 \times \log_e \left(\frac{298}{273}\right)$$

= 3667.8 kJ/°K

$$\text{Entropy of iron at 323°K} = 2990 \times \log_e \left(\frac{323}{273}\right)$$

= 502.8 kJ/°K

Entropy of water at 323°K

$$= 10 \times 4186 \times \log_e \left(\frac{323}{273}\right)$$

= 7070.03 kJ/°K

Change in entropy of iron = 502.8 – 2697.7 = –2194.9 kJ/°K

Change in entropy of water = 7040.03 – 3667.8 = + 3372.23 kJ/°K

Net change in entropy = 3372.23 – 2194.9 = 1177.33 kJ/°K.

Since Δ S > 0, hence the process is irreversible

104. *Given*, Velocity of steam, C_1 = 400 m/sec

α = 20° (nozzle angle)

$\theta = \phi$ (equiangular blades)

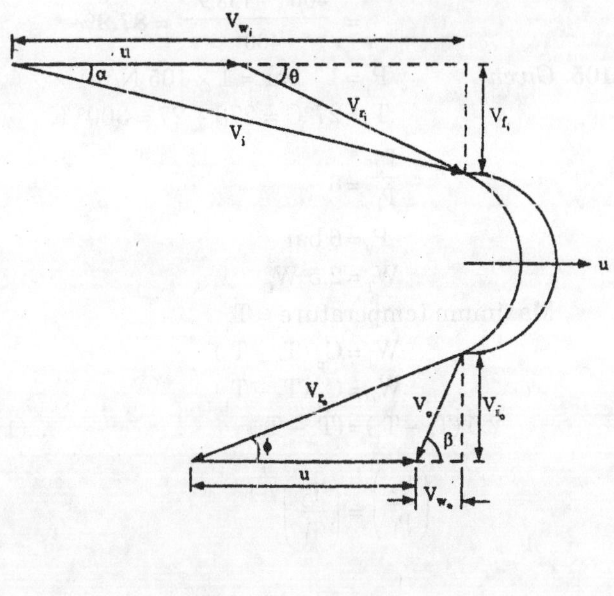

Steam inlet velocity, $V_i = 400$ m/s

For maximum blading efficiecy,

$$\text{Blade speed} = \frac{V_i}{2} = 200 \text{ m/s}$$

From inlet velocity triangle

$$V_{w_i} = 400 \cos 20° = 375.9 \text{ m/s}$$

$$V_{f_i} = 400 \sin 20° = 136.8 \text{ m/s}$$

$$\theta = \tan^{-1}\left(\frac{136.8}{375.9 - 200}\right) = 37.87°$$

$$V_{r_i} = \frac{V_{f_i}}{\sin 37.87°}$$

$$= \frac{136.8}{\sin 37.87°} = 22.8 \text{ m/s}$$

$$V_{r_i} = V_{r_o} \text{ and } \theta = \phi$$

From outlet velocity triangle

$$V_{w_o} = (222.8 \cos 37.87° - 200)$$

$$= -24.1 \text{ m/s}$$

$$V_{f_o} = V_{f_i} = 136.8 \text{ m/s}$$

$$V_o = \sqrt{V_{f_o}^2 + V_{w_o}^2} = 138.9 \text{ m/s}$$

$$\text{Blade power/Diagram power} = \frac{m(V_{w_i} - V_{w_o})\,u}{1000}$$

$$= \frac{0.6[375.9 - (-24.1)]}{1000} \times 200$$

$$= \mathbf{48\ kW}$$

$$\text{Diagram efficiency} = \frac{\text{Change in kinetic energy per kg}}{\text{Energy supplied per kg}}$$

$$= \frac{V_i^2 - V_o^2}{V_i^2}$$

$$= \frac{400^2 - 138.9^2}{400^2} = \mathbf{87.9\%}$$

105. *Given,* $P_1 = 1$ bar $= 1 \times 105$ N/m³

$T_1 = 27°C = 273 + 27 = 300°$ K

$$\frac{P_2}{P_1} = 6$$

$\therefore$ $P_2 = 6$ bar

$W_T = 2.5\ W_C$

Maximum temperature = T_3

$W_T = C_p\,(T_3 - T_2)$

$W_C = C_p\,(T_2 - T_1)$

$$2.5\,(T_2 - T_1) = (T_3 - T_2) \quad ...(1)$$

$$\left(\frac{P_2}{P_1}\right) = \left(\frac{T_2}{T_1}\right)^{\frac{\gamma}{\gamma-1}}$$

or $$6 = \left(\frac{T_2}{300}\right)^{\frac{1.4}{0.4}}$$

or $$T_2 = 500.6 \text{ K} \quad ...(2)$$

Substituting equation (2) in (1), we get

$$2.5\,(500.6 - 300) = T_3 - 500.6$$

$\therefore$ Maximum temperature, $T_3 = 1452.1$ K

$$\text{Cycle efficiency} = \frac{W_{out}}{Q_{in}}$$

$$= \frac{C_p(T_3 - T_2) - C_p(T_2 - T_1)}{C_p(T_3 - T_2)}$$

$$= \frac{1.5(500.6 - 300)}{(1452.1 - 500.6)}$$

$$= \mathbf{31.6\%}$$

NUMERICAL TYPE QUESTIONS

1. $$N_s = \frac{N\sqrt{P}}{H^{\frac{5}{4}}}$$

Given $N_{s1} = N_{s2}$, $H_1 = H_2$

$$\frac{N_1}{N_2} = 2,\ \frac{P_1}{P_2} = ?$$

$$N_1\sqrt{P_1} = N_2\sqrt{P_2}$$

$$\Rightarrow \frac{P_1}{P_2} = \left[\frac{N_1}{N_2}\right]^2 = \left(\frac{1}{2}\right)^2 = 0.25$$

2. $$\frac{dP_s}{dT_s} = 0.189 \text{ kPa/K}$$

$$T_{sat} = 273 + 25 = 298 \text{ K}$$

$$v_g = 43.38 \text{ m}^3\text{/kg}$$

$$v_f = 0$$

$$v_{fg} = v_g - v_f$$

$$= 43.38 - 0$$

$$= 43.38 \text{ m}^3\text{/kg}$$

$$\frac{dP_s}{dT_s} = \frac{h_{fg}}{T_{sat} \times v_{fg}}$$

$$= \frac{h_{fg}}{298 \times 43.38}$$

$$h_{fg} = 2443.25 \text{ kJ/kg}$$

3.

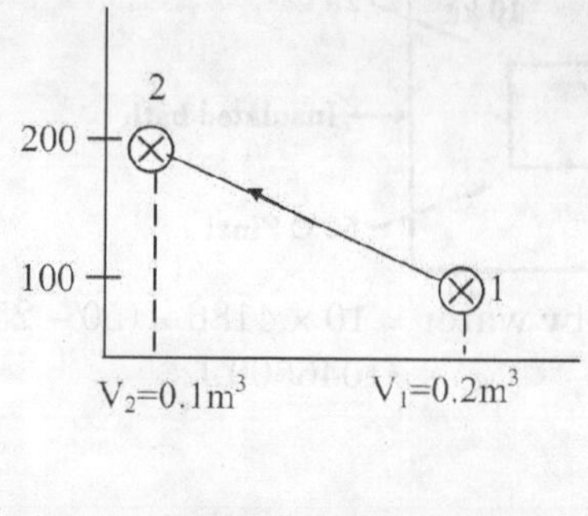

$$_1W_2 = \frac{1}{2}(P_1 + P_2)(V_1 - V_2)$$

$$= \frac{1}{2}(100+200)(0.2-0.1)$$

$$= 15 \text{ kJ}$$

4.

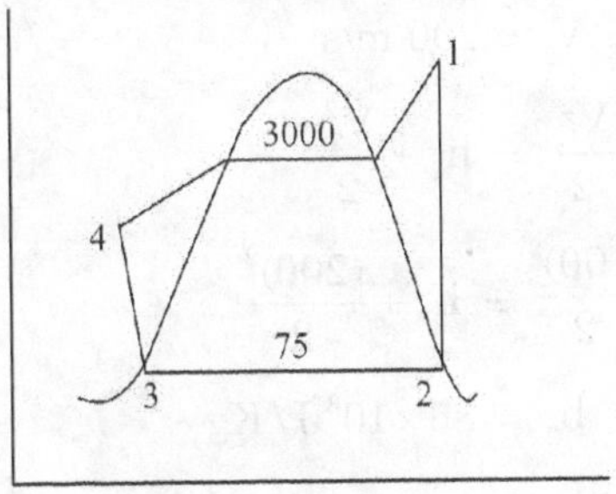

$h_1 = 3115.3$ kJ/kg

$s_1 = 6.7428$ kJ/kg

$s_1 = s_2 = s_f + xs_{fg}$

$$x = \frac{6.7428 - 1.213}{6.2434} = \frac{5.5298}{6.2434} = 0.8857$$

$$h_2 = h_f + xh_{fg}$$
$$= 348.39 + 0.8857 \times 2278.6$$
$$= 348.39 + 2018.16$$
$$= 2402.55 \text{ kJ/kg}$$

Pump work $= W_p = v_f(p_4 - p_3) = 3.033$ kJ/kg

$$h_4 = h_3 + v_f(p_4 - p_3)$$
$$= 384.34 + 0.001037(3000 - 75)$$
$$= 384.34 + 3.033$$
$$= 387.37 \text{ kJ/kg}$$

$$W_{net} = W_T - W_P$$
$$= (h_1 - h_2) - W_P$$
$$= (3115.3 - 2402.55) - 3.033$$
$$= 709.72 \text{ kJ/kg}$$

Q_s = HEAT SUPPLIED $= h_1 - h_4$

$$= 3115.3 - 387.37 = 2727.93 \text{ kJ/kg}$$

$$\eta_{th} = \frac{W_{net}}{Q_s}$$

$$= \frac{709.72}{2727.93} \times 100 = 0.26 \text{ or } 26\%$$

5.

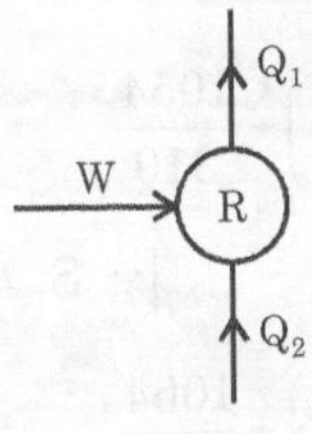

$$\text{COP} = \frac{Q_2}{W} = \frac{7.2}{1.8}$$

$$\text{COP} = 4$$

6. Brayton cycle eH

$$\eta_{Brayton} = 1 - \frac{1}{(r_p)^{\frac{\gamma-1}{\gamma}}}$$

given $r_p = 6, \gamma = 1.4$

$$\therefore \quad \eta_{Brayton} = 1 - \frac{1}{(6)^{\frac{1.4-1}{1.4}}} = 0.4006$$

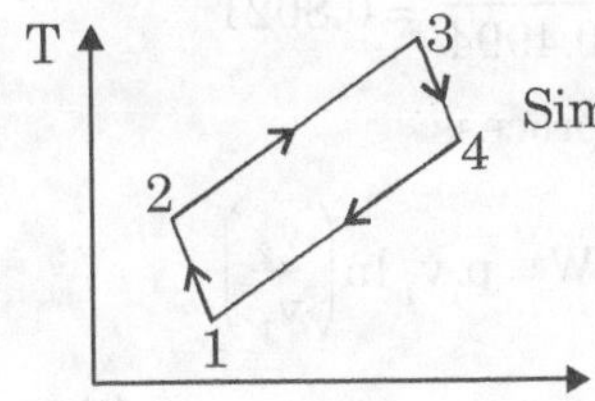

Simple Brayton cycle

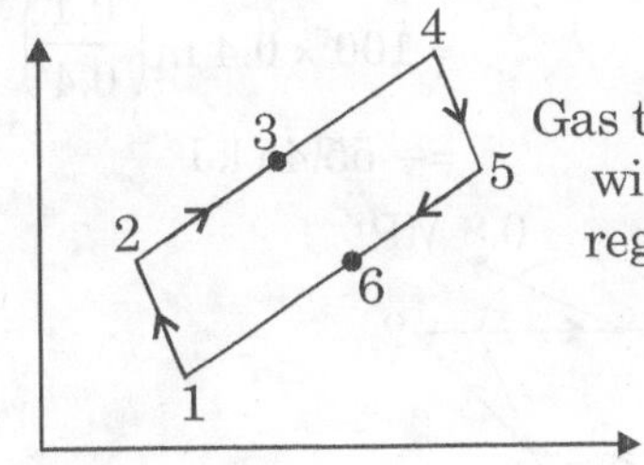

Gas turbine cycle with perfect regeneration

$r_p = 6, \gamma = 1.4$

$$\text{effectiveness } \varepsilon = \frac{C_p(T_5 - T_2)}{C_p(T_3 - T_2)} = 1$$

$$T_5 = T_3$$

$$\text{Given } \frac{T_{min}}{T_{max}} = 0.3 = \frac{T_1}{T_4}$$

$$W_{net} = W_T - W_C$$
$$= C_p(T_4 - T_5) - C_p(T_2 - T_1)$$
$$= C_p(T_4 - T_3) - C_p(T_2 - T_1) \quad [\because T_4 = T_3]$$

$$\eta_{Regenerative} = \frac{W_{net}}{\text{Heat-supplied}}$$

$$= \frac{C_p[(T_4 - T_3) - (T_2 - T_1)]}{C_p(T_4 - T_3)}$$

$$= 1 - \left(\frac{T_2 - T_1}{T_4 - T_3}\right)$$

$$= \frac{1 - T_1\left(\frac{T_2}{T_1} - 1\right)}{1 - T_4\left(1 - \frac{T_3}{T_4}\right)}$$

$$\eta_{Regenerative} = 1 - \frac{T_1}{T_4}\left(\frac{\frac{T_2}{T_1} - 1}{1 - \frac{T_5}{T_4}}\right)$$

$$= 1 - \frac{T_1}{T_4}\left(\frac{(r_p)^{\frac{\gamma-1}{\gamma}} - 1}{1 - \frac{1}{(r_p)^{\frac{\gamma-1}{\gamma}}}}\right) = 1 - \frac{T_1}{T_4}(r_p)^{\frac{\gamma-1}{\gamma}}$$

$$= 1 - 0.3(6)^{\frac{1.4-1}{1.4}} = 0.4994$$

$$\frac{\eta_{brayton}}{\eta_{Regenerative}} = \frac{0.4006}{0.4994} = 0.8021$$

7. For isothermal process

$$W = p_1 v_1 \ln\left(\frac{v_2}{v_1}\right)$$

$$= 100 \times 0.4 \ln\left(\frac{0.1}{0.4}\right)$$

$$= -55.45 \text{ kJ}$$

8.

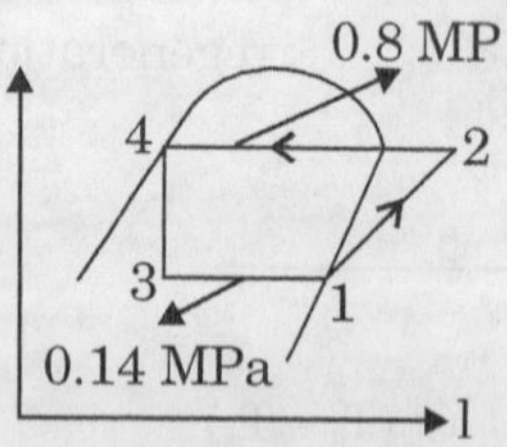

$h_1 = 236.04$ kJ/kg

$h_2 = 272.05$ kJ/kg

$h_3 = 93.42$ kJ/kg

$s_1 = 0.9322$ kJ/kg

m = 0.05 kg/s

Heat Rejection = $m(h_2 - h_3)$

$= 0.05(272.05 - 93.42)$

$= 8.931$ kW

9. $\phi = 0.7$, $\quad P_0 = 101.325$

$$\phi = \frac{P_v}{P_{sat}} \Rightarrow P_{sat} = \frac{P_v}{\phi}$$

$$P_{sat} = \frac{1.6}{0.7} = 2.28571 \text{ kPa}$$

Corresponding to saturation pressure whatever is the temperature is the DBT

$$\ln\left(\frac{P_S}{P_0}\right) = 14.317 - \frac{5304}{T_{sat}}$$

$$\ell n\left(\frac{2.28571}{101.325}\right) = 14.317 - \frac{5304}{T_{sat}}$$

$$-3.7916 = 14.317 - \frac{5304}{T_{sat}}$$

$$T_{sat} = \frac{5304}{18.1086} = 292.898 \text{ K} = 19.89°\text{C}$$

10.

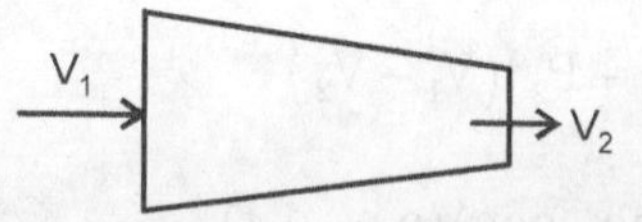

$h_1 = 100$ KJ/Kg

$V_1 = 100$ m/s

$V_2 = 200$ m/s

$$h_1 + \frac{V_1^2}{2} = h_2 + \frac{V_2^2}{2}$$

$$100 \times 10^3 + \frac{(100)^2}{2} = h_2 + \frac{(200)^2}{2}$$

$$h_2 = 85 \times 10^3 \text{ J/Kg}$$

11.

$$w = 0.622 \frac{Pv}{Pt - Pv}$$

$$= 0.622 \times \left(\frac{20}{750 - 20}\right)$$

$= 0.017$ Kgw/Kgda

$= 17$ gw/Kgda

12. $(r_P)_{OPT} = \left(\frac{P_4}{P_1}\right)^{\frac{1}{3}} = (10)^{\frac{1}{3}} = 2.1544$

$$\frac{P_2}{P_1} = \frac{P_3}{P_2} = \frac{P_4}{P_3} = 2.1544$$

$$P_2 = 2.1544 \times 100 = 215.44$$

$$\frac{P_3}{P_2} = 2.1544$$

$$P_3 = 2.1544 \times 215.44 = 464.15 \text{ kPa}$$

13.

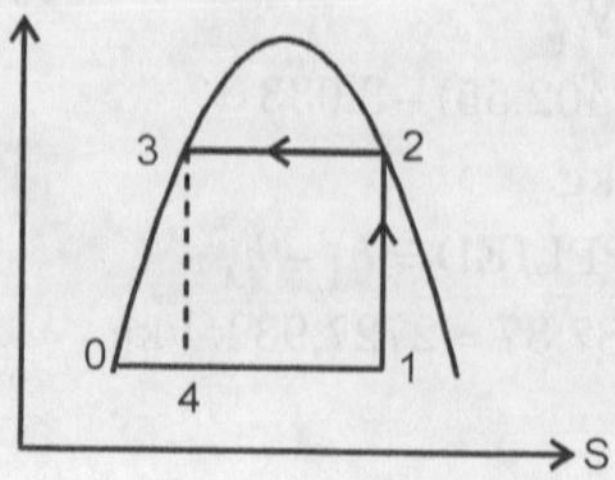

$$dQ = CpdT$$

$$S_2 - S_0 = \int_0^3 \frac{CpdT}{T} + \frac{hfg}{310}$$

$$S_1 - S_0 = 4.8\ln\left(\frac{310}{260}\right) + \frac{1054}{310}$$

$[\because S_1 = S_2]$

$$\frac{h_1 - h_0}{260} = 4.8\ln\left(\frac{310}{260}\right) + \frac{1054}{310}$$

$$h_1 - h_0 = 1103.51 \text{ KJ/Kg}$$

14.

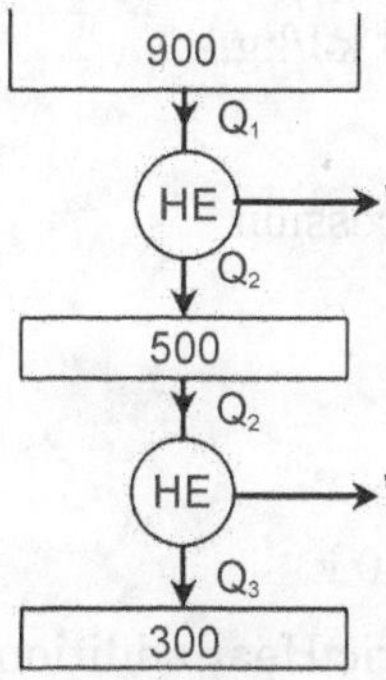

$$\eta_1 = \frac{1-T_1}{T_1} = \frac{1-500}{900} = 4.44$$

$$Q_2 = (1-\eta_1)\times Q_1$$
$$= 53.33 \text{ MJ}$$

$$\eta_2 = \frac{1-T_3}{T_2} = 1-\frac{300}{500} = 0.4$$

$$Q_3 = (1-\eta_2)\times Q_2 = 50 \text{ MJ}$$

15. $$\omega_{new} = \frac{m_1\omega_1 + m_2\omega_2}{m_1+m_2}$$

$$\Rightarrow \frac{0.1\times 10.1 + .1\times 1}{10.1+.1} = 0.02 \text{ kg/kg dry air}$$

16. $$\text{mep} = \frac{\text{work output}}{\text{swept volume}}$$

$$= \frac{12}{30\times 10^{-3}\left(1-\frac{1}{21}\right)}$$

$$= 525 \text{ kPa}$$

17. $$p = \rho gh$$
$$= 0.7 \times 1360 \times 9.81$$
$$= 9339.12\text{pa}$$

Actual pressure = atmospheric pressure

= 110664.12 pa

pv = mrT ⇒ m pv/RT

$$= \frac{110664.12\times 2}{288\times 8314} \Rightarrow 4.5539\text{kg}$$

18. $$(\text{COP})_{c.p.} = \frac{T_1}{T_1-T_2} = \frac{310}{31} = 10$$

19. Specific steam consumption = $\frac{3600}{W_T - W_P}$

Now, $W_T = h_2 - h_1$

$= 3159 - 2187$ kJ/kg

∴ $W_T = 972$ kJ/kg

and $W_P = 20$ kJ/kg

Thus specific steam consumption

$$= \frac{3600}{972-20}\text{kg/kW-h} = 3.7815 \text{ kg/kW.h}$$

20. $$s_2 - s_1 = R\ln\frac{V_2}{V_1} = 287\ln\frac{0.6}{0.9}$$

$$= -116.368 \text{ J/kgK}$$

21. Using SFEE

$$m\left[\left(\frac{v_1^2}{2}-\frac{v_2^{\ 2}}{2}\right)+(z_1-z_2)g+(h_1-h_2)\right] = W$$

$$10\left[\frac{10^2}{2}-\frac{20^2}{2}+(-20)\times 10+(100\times -1)\right] = W$$

W = – 4.5 kJ

∴ Work done on the system is 4.5 kJ.

22. Maximum in-plane shear stress

$$\tau_{max} = \frac{pd}{8t} = \frac{500\times 4}{8\times 10}\text{MPa} = 25 \text{ MPa}$$

23. $$\phi = \frac{P_w}{P_s} = 0.6 = \frac{P_w}{5.63}$$

∴ $P_w = 3.378$ kPa

Humidity Ratio,

$$w = 0.622\frac{P_w}{P_a - P_w}$$

$$= 0.622\times\frac{3.378}{100-3.378}$$

$$= 0.021745 \text{ kg/kg of dry air}$$

or 21.745 g/kg of dry air

24. Work output = (2993 – 226.50) kJ/kg = 2717 kJ/kg

25. Gas constant of mixture,

$$R_m = \frac{\text{Universal gas constant}}{\text{Average molar mass}}$$

Average molar mass

$$= \frac{100}{\frac{60}{28}+\frac{30}{32}+\frac{10}{44}} = 30.233 \text{ kg/ kmol}$$

$$R_m = \frac{8314}{30.233} = 274.996 \text{ J / kg – K}$$

26. Given, m = 1.5 kg

$h_1 = h_f = 505$ kJ/kg $\qquad h_2 = 3277.0$

From I[st] Law,

$$dQ = du + pdv = dh - vdp$$

$$dQ = dh \text{ (as } vdp = 0)$$

$$Q_{add} = dQ = m\,(h_2 - h_1) = (3277.0 - 505)\times 1.5$$

$$Q_{added} = 4158 \text{ kJ.}$$

27. $W_{HP} = h_2 - h_3$

$W_{LP} = h_4 - h_5$

Net work

$W_T = W_{HP} + W_{LP}$

$= (h_2 - h_3) + (h_4 - h_5)$

Given :

$h_1 = 191$ kJ/kg

$h_2 = 3399$ kJ/kg

$h_3 = 3185$ kJ/kg

$h_4 = 3446$ kJ/kg

$h_5 = 2247$ kJ/kg

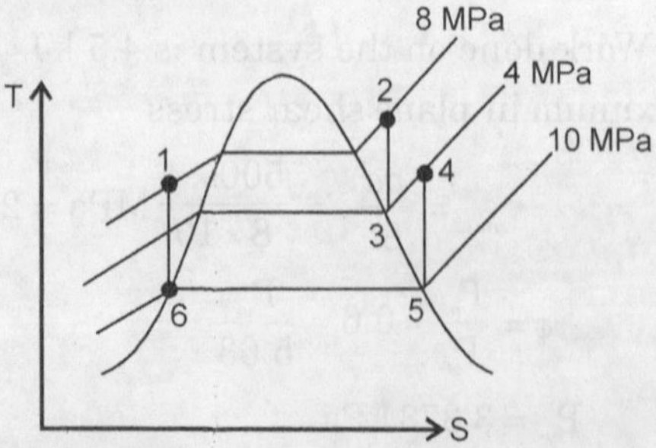

$$\eta_{cycle} = \frac{w_{net}}{Q_{added}}$$

$$\eta_{cycle} = \frac{w_1 - w_{pump}}{Q_{added}}$$

$$\eta_{cycle} = \frac{(h_2 - h_3) + (h_4 - h_5)}{(h_2 - h_1) + (h_4 - h_3)}$$

$$\eta_{cycle} = 0.407 = 40.7\%.$$

28. Otto cycle

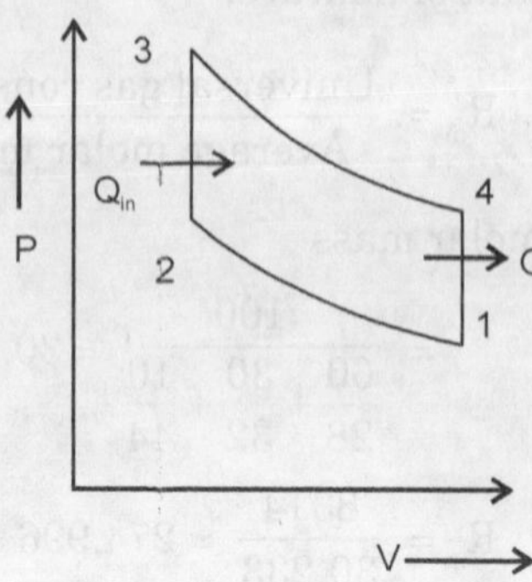

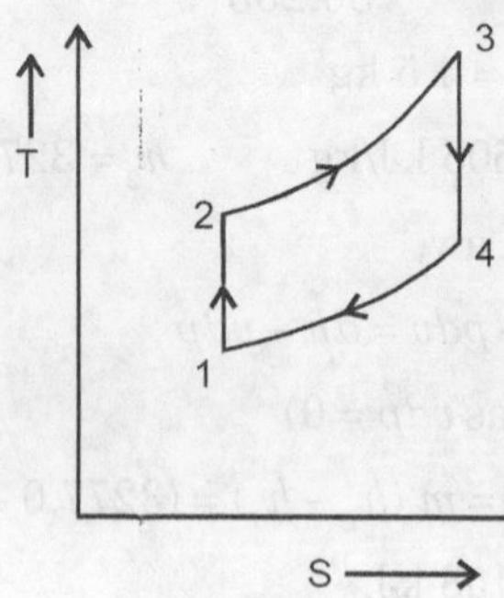

Given : $P_1 = 0.1$ MPa

$T_1 = 308$ K

$r = 1.4$

R = 0.2888 kJ/kg.k

$T_3 = 2660$ K

(1-2) Isentropic compression

PV^r = constant

$$\frac{T_2}{T_1} = (r)^{\gamma - 1}$$

$T_2 = 308 \times 8^{0.4} = 698.40$ k

(2-3) Isochoric Heat addition process

$$\frac{C_p}{C_v} = 1.4$$

R = 0.2888 kJ/kg

$C_p - C_v = 0.2888$

$$C_v\left(\frac{C_p}{C_v} - 1\right) = 0.2888$$

$$C_v = \frac{0.2888}{1.4 - 1} = 0.722 \text{ kJ/kg}$$

$Q_{in} = C_v (T_3 - T_2) = 0.722 (2660 - 698.40)$

$= 1416.27$ kJ/kg

29. $h = u + pv$

$h = 2864 + (8000 \text{ kPa}) \times (0.03432)$

$= 2864 + 274.56$

$= 3138.56$

$= 3139$

30.

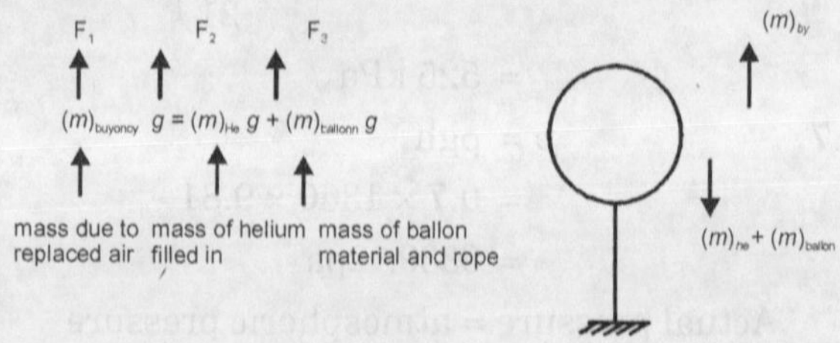

$F_1 = (m)_{buyoncy} \cdot g$

$$= \frac{p \cdot vg}{R_a . T}$$

$$= \frac{100 \times 10^3 \times \frac{4}{3} \times \pi \times 5^3 \times 9.81}{0.289 \times 10^3 \times 288}$$

$\therefore$ $F_1 = 6171.3$ N

$$F_2 = (m)_{Helium} \cdot g = \frac{p \cdot v \cdot g}{R_{He} . T}$$

$$= \frac{100 \times 10^3 \times \frac{4}{3} \times \pi \times 5^3 \times 9.81}{2.08 \times 10^3 \times 288}$$

$\therefore$ $F_2 = 857.45$ N

$\therefore$ $(F)_3 = F_1 - F_2$

$= 6171.3 - 857.45$

$= 5313.85$ N.

31.

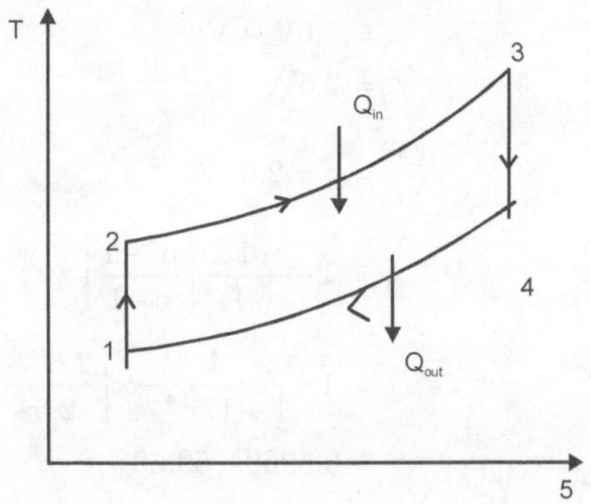

Given : $\gamma = 1.4,$

$c_p = 1.005$ kJ/kg.k

$P_1 = 1$ bar,

$P_2 = 8$ bar

$T_1 = 300$ K,

$T_3 = 1280$ K

$$r = \frac{P_2}{P_1} = 8$$

$Q_{in} = 80000$ kW.

(1-2) isentropic process

$$\frac{T_2}{T_1} = (r_p)^{r-\frac{1}{r}}$$

$$T_2 = 300 \times (8)^{\frac{1.4-1}{1.4}}$$

$$= 543.43 \text{ k}$$

$$Q_{in} = mc_p\,(T_3 - T_2)$$

$$80,000 = m \times 1.005\,(1280 - 300)$$

$$m = 108.07 \text{ kg/s.}$$

32. $(\eta)_{max}$ maximum blading efficiency = $\cos^2 \alpha$

$$= \cos^2 20° = 0.8830$$

$\therefore \quad (\eta_b)_{max} = 88.30\%$

33. $T_{c,i} = 25°C;$

$T_{c,o} = 42°C;$

$m_c = 1.5$ kg/s;

$C_c = 4.178$ kJ/kg.k

$T_{h,i} = 100°C;$

$T_{h,o} = 50°C;$

$m_h = 1$ kg/s;

$C_h = 2.130$ kJ/kg.k

$$m_c C_c = 1.5 \times 4.178$$

$$= 6.267 \text{ kW/°C} = C_{max}$$

$$m_h C_h = 1 \times 2.130$$

$$= 2.130 \text{ kW/°C} = C_{min}$$

$$\text{effectiveness} = E = \frac{C_h\left(T_{h,i} - T_{h,o}\right)}{C_{min}\left(T_{h,i} - T_{c,i}\right)}$$

$$= \frac{100 - 50}{100 - 25}$$

$$= 0.666$$

34.

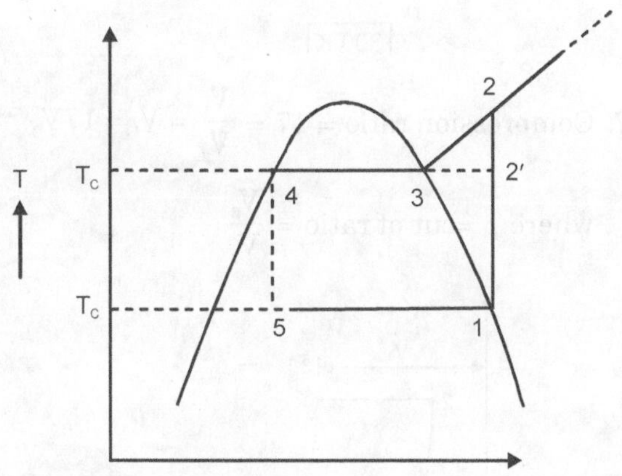

Given : $T_1 = 253$ K

$T_2' = 298$ K

$$S_1 = S_2' = S_2^2$$

$$\therefore \quad S_1 = S_2 = S_2' + C_P \ln \frac{T_2}{T_2'}$$

$$1.7841 = 1.7183 + 0.98 \ln \frac{T_2}{298}$$

$$\therefore \quad T_2 = 318.69 \text{ K}$$

$\therefore$ Enthalpy of discharge of compressor

$$\Rightarrow \quad h_2 = h_2' + C_P\,(T_2' - T_1')$$

$$\therefore \quad h_2 = 413.02 + 0.98\,(318.69 - 298)$$

$$\therefore \quad h_2 = 433.3 \frac{kJ}{kg}.$$

35. P_T =85 KPa, DBT = 30ºC

$P_{w.s} = 4.24$ KPa, RH = 65%

$$\therefore \quad P_w = P_{w.s} \times \frac{RH}{100} = 4.24 \times 0.65$$

$$= 2.756 \text{ KPa}$$

now $\quad \omega = 622 \dfrac{P_w}{P_T}$ gm (or water) / mg.d.a

$$\therefore \quad \omega = \frac{622 \times 2.756}{85} = 20.17 \text{ gm}$$

(of water)/mg.d.a

36. $$\eta = \frac{W_{net}}{Q_{in}} = 1 - \frac{T_{sink}}{T_{source}}$$

$$1 - \frac{300}{500} = \frac{W_{net}}{1000}$$

$$W_{net} = 1000 \times 0.4 = 400 \text{ kJ.}$$

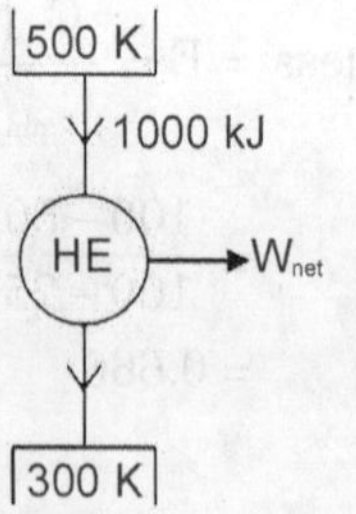

37. Compression ratio = 17 = $\frac{V_1}{V_2} = V_1 = 17V_2$

where ρ =cut of ratio = $\frac{V_3}{V_2}$

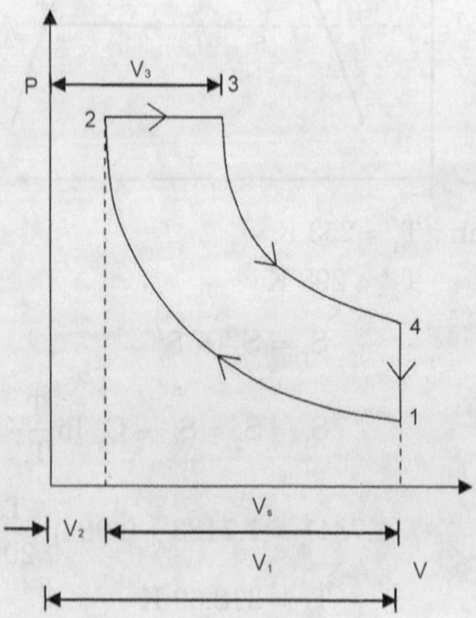

$$V_3 - V_2 = \frac{10}{100} V_s$$

$$V_3 - V_2 = 0.1(V_1 - V_2)$$

$$= 0.1(17V_2 - V_2)$$

$$= 0.1(16V_2)$$

$$V_3 = 1.6V_2 + V_2$$

$$V_3 = 2.6V_2$$

$$\frac{V_3}{V_2} = \rho = 2.6$$

$$\eta = 1 - \frac{1}{r(r)^{r-1}}\left[\frac{\rho^r - 1}{\rho - 1}\right]$$

$$= 1 - \frac{1}{1.4(17)^{1.4-1}} \times \left[\frac{2.6^{1.4} - 1}{2.6 - 1}\right]$$

$$= 0.5960 = 59.60\%.$$

38. $\frac{Q_1}{268} = \frac{Q_2}{308}$

$$Q_2 = \frac{2.5 \times 308}{268}$$

$$Q_2 = 2.873 \text{ kw}$$

$$w = Q_2 - Q_1 = 0.373 \text{ kw} = 373.13 \text{ watt.}$$

39. $\sigma_1 = \frac{pd}{2t}$

$$\Rightarrow \quad 100 = \frac{100 \times 2 \times 100}{2 \times t}$$

$$\Rightarrow \quad t = 10 \text{ mm}$$

40. Tds = du + pdv (closed system)

$$283.15 \times 10(S_2 - S_1) = 16.2 \times 10^3$$

$$\therefore \quad S_2 = 0.88 + \frac{16.2 \times 10^3}{283.15 \times 10}$$

$$= 6.6013 \text{ kJ/kg}$$

■■

9

CHAPTER

Engineering Materials

MCQ TYPE QUESTIONS

2016

1. The "Jominy test" is used to find
   (*a*) Young's modulus (*b*) hardenability
   (*c*) yield strength (*d*) thermal conductivity
2. In the phase diagram shown in the figure, four samples of the same composition are heated to temperatures marked by a, b, c and d.

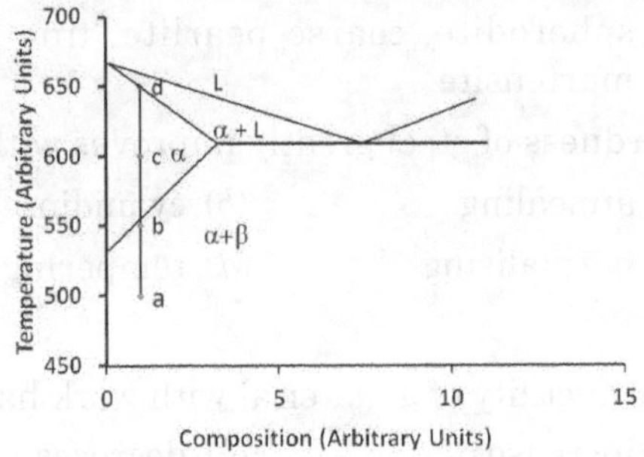

At which temperature will a sample get solutionized the fastest?
   (*a*) a (*b*) b
   (*c*) c (*d*) d
3. In a binary system of A and B, a liquid of 20% A (80%B) is coexisting with a solid of 70%. A (30%B). For an overall composition having 40% A, the fraction of solid is
   (*a*) 0.40 (*b*) 0.50
   (*c*) 0.60 (*d*) 0.75

2011

4. The crystal structure of austenite is
   (*a*) body centered cubic
   (*b*) face centered cubic
   (*c*) hexagonal closed packed
   (*d*) body centered tetragonal
5. During *normalizing* process of steel, the specimen is heated
   (*a*) between the upper and lower critical temperature and cooled in still air.
   (*b*) above the upper critical temperature and cooled in furnace.
   (*c*) above the upper critical temperature and cooled in still air.
   (*d*) between the upper and lower critical temperature and cooled in furnace.

2010

6. The material property which depends only on the basic crystal structure is
   (*a*) fatigue strength
   (*b*) work hardening
   (*c*) fracture strength
   (*d*) elastic constant

2009

7. The effective number of lattice points in the unit cell of simple cubic, body centered cubic, and face centered cubic space lattices, respectively, are
   (*a*) 1, 2, 2 (*b*) 1, 2, 4
   (*c*) 2, 3, 4 (*d*) 2, 4, 4

2007

8. If a particular Fe-C alloy contains less than 0.83% carbon, it is called
   (*a*) high speed steel
   (*b*) hypoeutectoid steel
   (*c*) hypereutectoid steel
   (*d*) cast iron
9. Which of the following engineering materials is the most suitable candidate for hot chamber die casting ?
   (*a*) Low carbon steel (*b*) Titanium
   (*c*) Copper (*d*) Tin

2006

10. The main purpose of spheroidising treatment is to improve
    (*a*) hardenability of low carbon steels
    (*b*) machinability of low carbon steels
    (*c*) hardenability of high carbon steels
    (*d*) machinability of high carbon steels
11. The ultimate tensile strength of a material is 400 MPa and the elongation up to maximum load is 35%. If the material obeys power law of hardening, then true stress-true strain relation (stress in MPa) in the plastic deformation range is
    (*a*) $\sigma = 540\ \varepsilon^{0.30}$
    (*b*) $\sigma = 775\ \varepsilon^{0.30}$
    (*c*) $\sigma = 540\ \varepsilon^{0.35}$
    (*d*) $\sigma = 775\ \varepsilon^{0.35}$

12. Match the items in columns I and II.

Column I	Column II
P. Charpy test	1. Fluidity
Q. Knoop test	2. Microhardness
R. Spiral test	3. Formability
S. Cupping test	4. Toughness
	5. Permeability

Codes :

	P	Q	R	S
(*a*)	4	5	3	2
(*b*)	3	5	1	4
(*c*)	2	4	3	5
(*d*)	4	2	1	3

2005

13. When the temperature of a solid metal increases,
(*a*) strength of the metal decreaes but ductility increases
(*b*) both strength and ductility of the metal decrease
(*c*) both strength and ductility of the metal increase
(*d*) strength of the metal increases but ductility decreases

2004

14. The S-N curve for steel becomes asymptotic nearly at
(*a*) 10^3 cycles (*b*) 10^4 cycles
(*c*) 10^6 cycles (*d*) 10^9 cycles

15. The percentage of carbon in gray cast iron is in the range of
(*a*) 0.25 to 0.75 percent (*b*) 1.25 to 1.75 percent
(*c*) 3 to 4 percent (*d*) 8 to 10 percent

16. From the lists given below, choose the most appropriate set of heat treatment process and the corresponding process characteristics

Process	Characteristics
P. Tempering	1. Austenite is converted into bainite
Q. Austempering	2. Austenite is converted into martensite
R. Martempering	3. Cementite is converted into globular structure
	4. Both hardness and brittleness are reduced
	5. Carbon is absorbed into the metal

Codes :

	P	Q	R
(*a*)	3	1	5
(*b*)	4	3	2
(*c*)	4	1	2
(*d*)	1	5	4

2003

17. During heat treatment of steel, the hardness of various structures in increasing order is
(*a*) martensite, fine pearlite, coarse pearlite, spherodite
(*b*) fine pearlite, coarse pearlite, spherodite, martensite
(*c*) martensite, coarse pearlite, fine pearlite, spherodite
(*d*) spherodite, coarse pearlite, fine pearlite, martensite

18. Hardness of steel greatly improves with
(*a*) annealing (*b*) cyaniding
(*c*) normalising (*d*) tempering

2002

19. The ductility of a material with work hardening
(*a*) increases (*b*) decreses
(*c*) remains unaffected (*d*) unpredictable

2000

20. Abrasive material used in grinding wheel selected for grinding ferrous alloys is
(*a*) silicon carbide
(*b*) diamond
(*c*) aluminium oxide
(*d*) boron carbide

21. Cast steel crankshaft surface is hardened by
(*a*) nitriding (*b*) normalising
(*c*) carburising (*d*) induction heating

22. The stress-strain behaviour of a material is shown in Figure. Its resilience and toughness, in Nm/m^3, are respectively

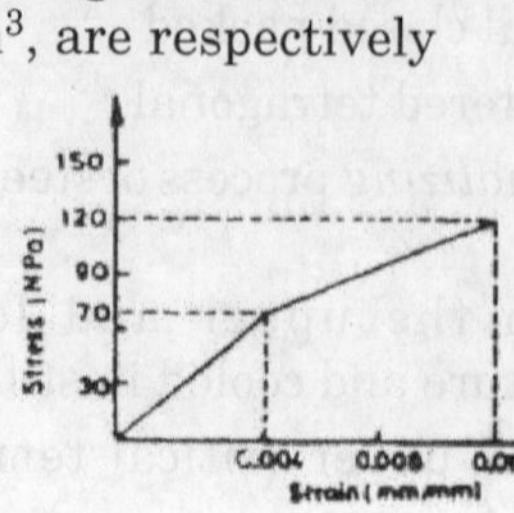

(*a*) $28 \times 10^4, 76 \times 10^4$
(*b*) $28 \times 10^4, 48 \times 10^4$
(*c*) $14 \times 10^4, 90 \times 10^4$
(*d*) $76 \times 10^4, 104 \times 10^4$

1999

23. Under repeated loading a material has the stress-strain curve shown in the figure, which of the following statements is true?

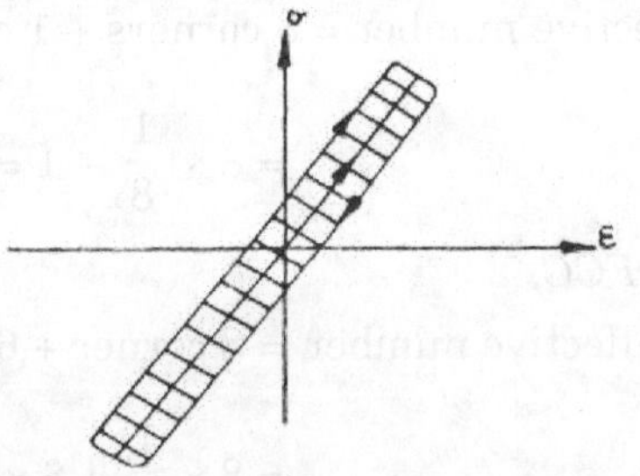

(*a*) The smaller the shaded area, the better the material damping

(*b*) The larger the shaded area, the better the material damping

(*c*) Material damping is an independent material property and does not depend on this curve

(*d*) None of these

1998

24. Decreasing grain size in a polycrystalline material

(*a*) increases yield strength and corrosion resistance.

(*b*) decreases yield strength and corrosion resistance

(*c*) decreases yield strength but increases corrosion resistance

(*d*) increases yield strength but decreases corrosion resistance.

1997

25. On completion of heat treatment, the resulting structure will have retained Austenite if

(*a*) rate of cooling is greater than the critical cooling rate

(*b*) rate of cooling is less than the critical cooling rate

(*c*) martensite formation starting temperature is above the room temperature

(*d*) martensite formation finish temperature is below the room temperature

1996

26. The iron-carbon diagram and the TTT curves are determined under

(*a*) equilibrium and non-equilibrium conditions respectively

(*b*) non-equilibrium and equilibrium conditions respectively

(*c*) equilibrium conditions for both

(*d*) non-equilibrium conditions for both.

ANSWERS

MCQ Type Questions

1. (*b*)	**2.** (*c*)	**3.** (*a*)	**4.** (*b*)	**5.** (*c*)	**6.** (*a*)	**7.** (*b*)	**8.** (*c*)	**9.** (*b*)	**10.** (*d*)
11. (*c*)	**12.** (*d*)	**13.** (*a*)	**14.** (*c*)	**15.** (*c*)	**16.** (*c*)	**17.** (*d*)	**18.** (*d*)	**19.** (*b*)	**20.** (*c*)
21. (*c*)	**22.** (*c*)	**23.** (*a*)	**24.** (*d*)	**25.** (*d*)	**26.** (*a*)				

EXPLANATIONS

MCQ TYPE QUESTIONS

1. The depth up to which the required hardness is obtained is called as hardenability and it is determined by using Jomney end quench test.
2. For aluminium alloy solution hardening process will be used to increase strength and hardness. In this process component will be heated to 550°C above temperature so that solute particles can penetrate into the lattice easily.
3. Weight of the liquid of A = $W_{LA} = l_a \times W_l = 0.2 \times W_l$ (Where W_l is weight of total liquid)

Weight of solid of A = $W_{SA} = S_a (W_T - W_l)$

$= 0.7 \times (W_T - W_l)$

$$W_A = 0.2\, W_l + 0.7\, (W_T - W_l)$$

$$W_A = 0.7\, W_T - 0.5 W_l$$

$$\frac{W_A}{W_T} = 0.7 - 0.5\, \frac{W_\ell}{W_T}$$

$$0.4 = 0.7 - 0.5\, (\%\, W_l)$$

$$(0.5)\, (\%\, W_l) = 0.3$$

$$(\%\, W_l) = \frac{3}{5}$$

$$(\%\, W_s) = 1 - \frac{3}{5}$$

$$= 0.4$$

7. *For a simple cubic unit cell*

8 corners have 8 atoms and each atom at one corner contributes to 8 unit cells.

$\therefore$ Effective number $= 8 \times \frac{1}{8} = 1$

For body centred cubic,

effective number = 8 corners + 1 centre

$$= 8 \times \frac{1}{8} + 1 = 2$$

For FCC,

effective number = 8 corner + 6 faces

$$= 8 \times \frac{1}{8} + 6 \times \frac{1}{2}$$

$$= 1 + 3 = 4$$

Hence answer is (1, 2, 4)

10. Machinability of high carbon steels.

11. $$\sigma = \left(\sigma_u + \frac{\Delta L}{L}\sigma_u\right)\varepsilon^{\Delta L/L}$$

17.

ROCKWELL	HARDNES STRUCTURE
RC 15	→ Coarse Pearlite
RC 25	→ Fine peralite
RC 65	→ Martensite

22. Resilience $= \frac{1}{2}(70 \times 0.004) \times 10^6$

$= 14 \times 10^4\ \text{Nm/m}^3$

Toughness

$$= \frac{1}{2}[(70 \times 0.004 + 50 \times 0.008) + 70 \times 0.008] \times 10^6$$

$= \mathbf{90 \times 10^4}$

10 CHAPTER Metal Casting

MCQ TYPE QUESTIONS

2016

1. Equal amounts of a liquid metal at the same temperature are poured into three moulds made of steel, copper and aluminium. The shape of the cavity is a cylinder with 15 mm diameter. The size of the moulds are such that the outside temperature of the moulds do not increase appreciably beyond the atmospheric temperature during solidification. The sequence of solidification in the mould from the fastest to slowest is

   (Thermal conductivities of steel, copper and aluminum are 60°5, 401 and 237 W/m-K, respectively.

   Specific heats of steel, copper and aluminium are 434, 385 and 903 J/kg-K, respectively.

   Densities of steel, copper and aluminium are 7854, 8933 and 2700 kg/m^3, respectively.)

   (*a*) Copper - Steel - Aluminium

   (*b*) Aluminium - Steel - Copper

   (*c*) Copper - Aluminium - Steel

   (*d*) Steel - Copper - Aluminium

2010

2. In a gating system, the ratio 1 : 2 : 4 represents

   (*a*) sprue base area : runner area : ingate area

   (*b*) pouring basin area : ingate area : runner area

   (*c*) sprue base area : ingate area : casting area

   (*d*) runner area : ingate area : casting area

3. Two pipes of inner diameter 100 mm and outer diameter 110 mm each are joined by flash-butt welding using 30 V power supply. At the interface, 1 mm of material melts from each pipe which has a resistance of 42.4 Ω. If the unit melt energy is 64.4 MJ m^{-3}, then time required for welding (in s) is

   (*a*) 1

   (*b*) 5

   (*c*) 10

   (*d*) 20

2009

4. Two streams of liquid metal which are not hot enough to fuse properly result into a casting defect known as

   (*a*) cold shut

   (*b*) swell

   (*c*) sand wash

   (*d*) scab

5. Match the items in Column I and Column II.

   Column I

   P. Metallic Chills

   Q. Metallic Chaplets

   R. Riser

   S. Exothermic Padding

   Column II

   1. Support for the core
   2. Reservoir of the molten metal
   3. Control cooling of critical sections
   4. Progressive solidification

   Codes :

	P	Q	R	S
(*a*)	1	3	2	4
(*b*)	1	4	2	3
(*c*)	3	4	2	1
(*d*)	4	1	2	3

2008

6. While cooling, a cubical casting of side 40 mm undergoes 3%, 4% and 5% volume shrinkage during the liquid state, phase transition and solid state, respectively. The volume of metal compensated from the riser is

   (*a*) 2%

   (*b*) 7%

   (*c*) 8%

   (*d*) 9%

2007

7. Volume of a cube of side 'l' and volume of a sphere of radius 'r' are equal. Both the cube and the sphere are solid and of same material. They are being cast. The ratio of the solidification time of the cube to the same of the sphere is

(*a*) $\left(\frac{4\pi}{6}\right)^3\left(\frac{r}{l}\right)^6$

(*b*) $\left(\frac{4\pi}{6}\right)\left(\frac{r}{l}\right)^2$

(*c*) $\left(\frac{4\pi}{6}\right)^2\left(\frac{r}{l}\right)^3$

(*d*) $\left(\frac{4\pi}{6}\right)^2\left(\frac{r}{l}\right)^4$

8. A 200 mm long down sprue has an area of cross-section of 650 mm^2 where the pouring basin meets the down sprue (i.e. at the beginning of the down sprue). A constant head of molten metal is maintained by the pouring basin. The molten metal flow rate is 6.5×10^5 mm^3/s. Considering the end of down sprue to be open to atmosphere and an acceleration due to gravity of 10^4 mm/s^2, the area of the down sprue in mm^2 at its end (avoiding aspiration effect) should be

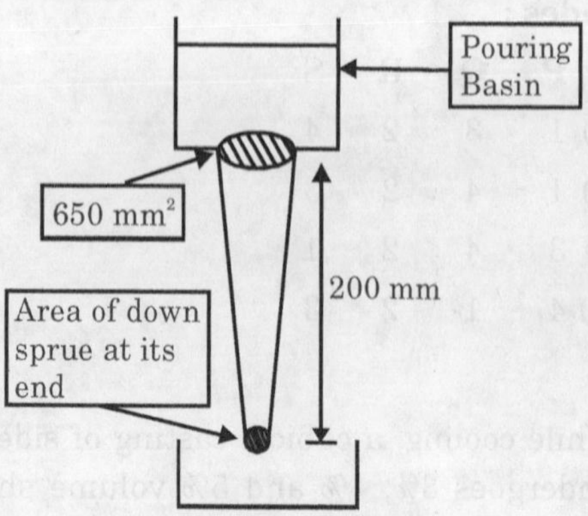

(*a*) 650.0 (*b*) 350.0

(*c*) 290.7 (*d*) 190.0

2006

9. An expendable pattern is used in

(*a*) slush casting

(*b*) squeeze casting

(c) centrifugal casting

(*d*) investment casting

10. In a sand casting operation, the total liquid head is maintained constant such that it is equal to the mould height. The time taken to fill the mould with a top gate is t_A. If the same mould is filled with a bottom gate, then the time taken is t_B. Ignore the time required to fill the runner and frictional effects. Assume atmospheric pressure at the top molten metal surfaces. The relation between t_A and t_B is

(*a*) $t_B = \sqrt{2}\ t_A$

(*b*) $t_B = 2\ t_A$

(*c*) $t_B = \frac{t_A}{\sqrt{2}}$

(*d*) $t_B = 2\sqrt{2}\ t_A$

2005

11. Match the itemsvg of List I (Equipment) with the items of List II (Process) and select the correct answer using the given codes.

List I (*Equipment*)	List II (*Process*)
P. Hot Chamber Machine	**1.** Cleaning
Q. Muller	**2.** Core making
R. Dielectric Baker	**3.** Die casting
S. Sand Blaster	**4.** Annealing
	5. Sand mixing

Codes :

	P	Q	R	S
(*a*)	2	1	4	5
(*b*)	4	2	3	5
(*c*)	4	5	1	2
(*d*)	3	5	2	1

12. A mould has a downsprue whose length is 20 cm and the cross sectional area at the base of the downsprue is 1 cm^2. The downsprue feeds a horizontal runner leading into the mould cavity of volume 1000 cm^3. The time required to fill the mould cavity will be

(*a*) 4.05 *s* (*b*) 5.05 *s*

(*c*) 6.05 *s* (*d*) 7.25 *s*

2004

13. Misrun is a casting defect which occurs due to

(*a*) very high pouring temperature of the metal

(*b*) insufficient fluidity of the molten metal

(*c*) absorption of gases by the liquid metal

(*d*) improper alignment of the mould flasks

14. Gray cast iron blocks 200 × 100 × 10 mm are to be cast in sand moulds. Shrinkage allowance for pattern making is 1%. The ratio of the volume of pattern to that of the casting will be

(*a*) 0.97 (*b*) 0.99

(*c*) 1.01 (*d*) 1.03

15. Match the following

Product	**Process**
P. Molded luggage	**1.** Injection molding
Q. Packaging containers for liquid	**2.** Hot rolling
R. Long structural shapes	**3.** Impact extrusion
S. Collapsible tubes	**4.** Transfer molding
	5. Blow molding
	6. Coining

Codes :

	P	Q	R	S
(*a*)	1	4	6	3
(*b*)	4	5	2	3
(*c*)	1	5	3	2
(*d*)	5	1	2	2

2003

16. Hardness of green sand mould increases with

(*a*) increase in moisture content beyond 6 percent

(*b*) increase in permeability

(*c*) decrease in permeability

(*d*) increase in both moisture content and permeability

17. With a solidification factor of 0.97×97^6 s/m^2, the solidification time (in seconds) for a spherical casting of 200 mm diameter is

(*a*) 539 (*b*) 1078

(*c*) 4311 (*d*) 3233

2002

18. The primary purpose of a sprue in a casting mould is to

(*a*) feed the casting at a rate consistent with the rate of solification

(*b*) act as a reservoir for molten metal

(*c*) feed molen metal from the pouring basin to the gate

(*d*) help feed the casting until all solidification takes place

19. In centrifugal casting, the impurities are

(*a*) uniformly distributed

(*b*) forced towards the outer surface

(*c*) trapped near the mean radius of the casting

(*d*) collected at the centre of the casting

2001

20. The height of the down-squre is 175 mm and its cross-sectional area at the base is 200 mm^2. The cross-sectional area of the horizontal runner is also 200 mm^2. Assuming no losses, indicate the correct choice for the time (in seconds) required to fill a mold cavity of volume 10^6 mm^3. (Use g = m/s^2).

(*a*) 2.67

(*b*) 8.45

(*c*) 26.72

(*d*) 84.50

2000

21. Two castings, a cube and a slab of the same material solidify under identical mould conditions. The volumes of the castings are equal but the slab dimensions are in the ratio of 1 : 2 : 4. Find ratio of the solidification time of the cube to that of the slab.

(*a*) 4 : 3

(*b*) 8 : 3

(*c*) 9 : 3

(*d*) 14 : 3

1999

22. Identify the stress - state in the FLANGE portion of a PARTIALLY DRAWN CYLINDRICAL CUP when deep - drawing without a blank holder

(*a*) Tensile in all three directions

(*b*) No stress in the flange at all, because there is no blank-holder

(*c*) Tensile stress in one direction and compressive in the one other direction

(*d*) Compressive in two directions and tensile in the third direction

23. Which of the following materials requires the largest shrinkage allowance, while making a pattern for casting?

(*a*) Aluminium

(*b*) Brass

(*c*) Cast Iron

(*d*) Plain Carbon Steel

1998

24. Chills are used in moulds to
    (*a*) achieve directional solidification
    (*b*) reduce the possibility of blowholes
    (*c*) reduce freezing time
    (*d*) smoothen metal flow for reducing splatter.

NUMERICAL TYPE QUESTIONS

2016

1. Heat is removed from a molten metal of mass 2 kg at a constant rate of 10 kW till it is completely solidified. The cooling curve is shown in the figure.

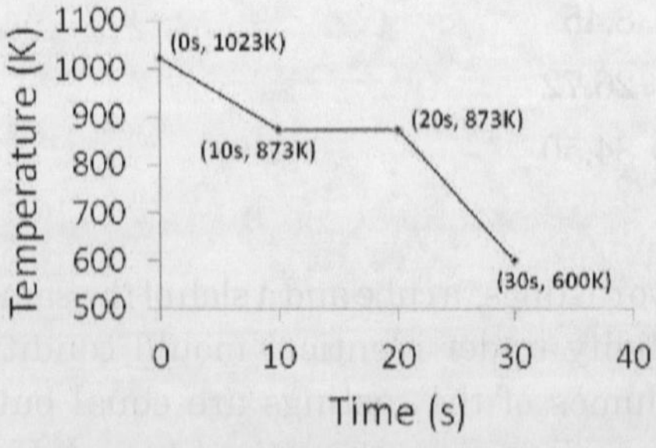

Assuming uniform temperature throughout the volume of the metal during solidification, the latent heat of fusion of the metal (in kJ/kg) is _____

2. Gray cast iron blocks of size 100 mm × 50 mm × 10 mm with a central spherical cavity of diameter 4 mm are sand cast. The shrinkage allowance for the pattern is 3%. The ratio of the volume of the pattern to volume of the casting is _______.

2001

3. Two solid workpieces,
   (*i*) a sphere with radius R and
   (*ii*) a cylinder with diameter equal to its height, have to be sand cast.

   Both workpieces have the same volume. Show that the cylindrical workpiece will solidify faster than the spherical workpiece.

1999

4. In a particular mould design, the down sprue has an area of cross section of 6.45 cm^2 where the pouring basin leads into the sprue. The sprue is 20 cm long. The required metal flow rate at the top section of the sprue is 820 cm^3/s. Determine the pouring height necessary above the sprue top. Also determine the area of cross-section of the sprue at its bottom to avoid aspiration of liquid metal.

1998

5. **List I**
   (*a*) Sand casting
   (*b*) Plaster mould casting
   (*c*) Shell mould casting
   (*d*) Investment casting

   List II
   (1) Symmetrical and circular shapes only
   (2) Parts have hardened skins and soft interior
   (3) Minimum post-casting processing
   (4) Parts have a tendency to warp
   (5) Parts have soft skin and hard interior
   (6) Suitable only for non-ferrous metals

6. **List I**
   (*a*) Aluminium brake shoe
   (*b*) Plastic water bottle
   (*c*) Stainless steel cups
   (*d*) Soft drink can (aluminium)

   List II
   (1) Deep drawing
   (2) Blow moulding
   (3) Sand casting
   (4) Centrifugal casting
   (5) Impact extrusion
   (6) Upset forging

7. Two castings of the same metal have the same surface area. One casting is in the form of a sphere and the other is a cube. What is the ratio of the solidification time for the sphere to that of the cube?

1996

8. Light impurities in the molten metal are prevented from reaching the mould cavity by providing a
   (*a*) strainer (*b*) button well
   (*c*) skim bob (*d*) all of the above

9.
List I	**List II**
(*a*) Rivets for aircraft body	1. Forging
(*b*) Carburettor body	2. Cold heading
(*c*) Crankshafts	3. Aluminium-based alloy
(*d*) Nails	4. Pressure die casting
	5. Investment casting

10. The pouring basin is a sand mould is 200 mm in diameter and 100 mm in height. The mould cavity is a cube of 125 mm side and must be filled in 25 seconds. The maximum velocity must ensure laminar flow into the mould cavity and in the gates of circular corss-section. Design the sprue to avoid aspiration effects.

Assume

(i) a constant temperature,

(ii) no loss in the velocity head, and

(iii) diameter of gate = 8 times the diameter at the sprue base.

Properties of molten metal : kinematic viscosity = 0.9 mm²/s,

density = 7000 kg/m³, and

c_p = 33.6J/mole –K.

11. (*a*) A steel cup of height 30 mm and internal diameter 40 mm, with a flange of width 10 mm is to be deep drawn from a sheet 1 mm thick. Determine the diameter of the blank and the drawing force. What is the draw ratio ? Can the cup be drawn in a single operation ?

The properties of steel are :

Yield point = 150 MPa, Ultimate tensile strength = 350 MPa, Limiting draw ratio = 1.9.

Neglect entry radius and blank holder effects.

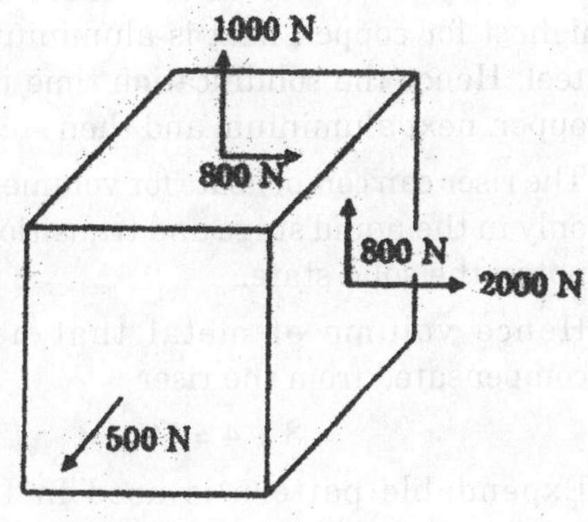

(*b*) A cube of 5 mm side is loaded as shown in Fig.

(i) Determine the principal stresses σ_1, σ_2 and σ_3

(ii) Will the cube yield if the yield strength of the material is 70 MPa ? Use Von Mises theory.

ANSWERS

MCQ Type Questions

1. (*c*)	**2.** (*c*)	**3.** (*d*)	**4.** (*a*)	**5.** (*d*)	**6.** (*b*)	**7.** (*d*)	**8.** (*c*)	**9.** (*d*)	**10.** (*c*)
11. (*d*)	**12.** (*b*)	**13.** (*b*)	**14.** (*d*)	**15.** (*b*)	**16.** (*c*)	**17.** (*b*)	**18.** (*c*)	**19.** (*d*)	**20.** (*a*)
21. (*d*)	**22.** (*b*)	**23.** (*a*)	**24.** (*a*)						

Numerical Type Questions

1. 50 **2.** (0.913) **3.** (*) **4.** (*)

5. $\begin{pmatrix} (a)-4 \\ (b)-6 \\ (c)-5 \\ (d)-3 \end{pmatrix}$ **6.** $\begin{pmatrix} (a)-3 \\ (b)-2 \\ (c)-5 \\ (d)-1 \end{pmatrix}$ **7.** (*) **8.** (*d*)

9. $\begin{pmatrix} (a)-3 \\ (b)-5 \\ (c)-4 \\ (d)-2 \end{pmatrix}$ **10.** (*) **11.** (*)

"*" Refers to Subjective Question.

EXPLANATIONS

MCQ TYPE QUESTIONS

1. Solidification time is inversely proportional to diffusivity and based on the values diffusivity is highest for copper, next is aluminium and then steel. Hence the solidification time is lowest for copper, next aluminium and then steel.

6. The riser can compensate for volume shrinkange only in the liquid stage and transition stage and not in the solid state.

Hence volume of metal that needs to be compensated from the riser

$$= 3 + 4 = 7\%$$

9. Expendable pattern is used in Investment casting.

10. Let 'h' be height of mould and same will be liquid head.

Top gating, $t_A = \frac{cV}{\sqrt{2gh}}$

where, c = constant

V = volume of mould

Bottom gating, $t_B = \frac{c \times V}{\sqrt{2g \times 2h}}$

Shear strain = cotϕ + tan (ϕ – α) = 1.65

$$\therefore \quad t_B = \frac{t_A}{\sqrt{2}}$$

12.
$$v_{\text{at base}} = \sqrt{2gh}$$
$$= \sqrt{2 \times 9.8 \times .2} \text{ m/sec}$$
$$= 1.9798 \text{ m/sec} = 197.98 \text{ cm/sec}$$

Area at base = 1 cm²

Volume to be filled = 1000 cm³

$$\therefore \text{Time required} = \frac{1000}{197.98 \times 1} = 5.050 \text{ sec}$$

14. Given : L = 200 mm

B = 100 mm

H = 10 mm

Volume of the casting,

$$V_c = 200 \times 100 \times 10 \text{ mm}^3 = 2 \times 10^5 \text{ mm}^3.$$

Shrinkage allowance = 1%

Hence, Shrinkage allowance given along length

$$= 200 \times 0.01 = 2 \text{ mm}$$

Shrinkage allowance given along breadth

$$= 100 \times 0.01 = 1 \text{ mm}$$

Shrinkage allowance given along height

$$= 10 \times 0.01 = 0.1 \text{ mm}$$

Volume of the pattern,

$$V_p = (200 + 2)(100 + 1)(10 + 0.1)$$
$$= 2.06 \times 10^5 \text{ mm}^3$$

$$\therefore \quad \frac{V_p}{V_c} = \frac{2.06 \times 10^5}{2 \times 10^5} = 1.03.$$

17. Solidification factor, $q = 0.97 \times 10^6$ S/m²

Diameter of spherical casting = 200 mm = 0.2 m

According to Caine's relation

Solidification time

$$= q.\left(\frac{V}{A}\right)^2 = 0.97 \times 10^6 \left[\frac{4/3\,\pi\,(0.1)^3}{4\pi(0.1)^2}\right]^2$$

$$= 0.97 \times 10^6 \left(\frac{.1}{3}\right)^2 = 1077.78 \approx 1078 \text{ sec.}$$

20. Time required to fill the mould

$$t = \frac{V}{V_g . A_g}$$

$$= \frac{10^6 mm^3}{200mm^3\sqrt{2gh_t}}$$

$$= \frac{10^6}{200\sqrt{2 \times 10 \times 0.175}} = 2.67$$

21. Solidification time, $t_s \propto v\left(\frac{V}{A}\right)^2$

where, v = constant

V = volume

A = surface area of heat dissipation

Let slab dimensions are a, $2a$, $4a$.

Then cube volume = $a \;\; 2a \;\; 4a = 8a^3$

∴ Side of cube = $2a$

$$\frac{t_{\text{cube}}}{t_{\text{slab}}} = \frac{\left(\frac{V}{A}\right)_{\text{cube}}}{\left(\frac{V}{A}\right)_{\text{slab}}} = \frac{A_{\text{slab}}}{A_{\text{cube}}}$$

$$= \frac{2(a \times 2a + 2a \times 4a + a \times 4a)}{6a^2}$$

$$= \frac{28a^2}{6a^2} = \frac{14}{3}$$

$$\therefore \quad t_{\text{cube}} : t_{\text{slab}} = \mathbf{14 : 3}$$

NUMERICAL TYPE QUESTIONS

1. $m = 2$ kg, Q = 10 kW

Time taken for removing latent heat

$$= 20 - 10 = 10 \text{ sec}$$

$$\text{Time} = \frac{\text{Latent heat}}{Q}$$

$$\text{Latent heat} = \text{time} \times Q = 10 \times 10 = 100 \text{ kJ}$$

$$\text{Latent heat/kg} = \frac{100}{2} = 50 \text{ kJ/kg}$$

2. Ratio of volume of pattern to casting

$$\frac{[(0.97 \times 100)(0.97 \times 50)(0.97 \times 10)]}{(100 \times 50 \times 10)} = 0.913$$

Volume of cavity is assumed to be small and negligible.

3. Volume of sphere, $V_s = \frac{4}{3}\pi R^3$

Surface area of sphere, $A_g = 4\pi R^2$

Volume of cylinder, $V_C = \frac{\pi}{4} d^2 h \qquad (\because h = d)$

$$= \frac{\pi}{4} d^3$$

Area of cylinder,

$$A_c = \pi d h + 2 \times \frac{\pi}{4} d^2$$

$$= \pi d^2 + \frac{\pi}{2} d^2 = \frac{3\pi}{2} d^2$$

Now, $\frac{V_s}{A_s} = \frac{4}{3} \frac{\pi R^3}{4\pi R^2} = \frac{R}{3}$

$$\frac{V_c}{A_c} = \frac{\pi}{4} \frac{d^3}{3\pi/2} \times d^2 = \frac{d}{6}$$

But $\frac{4}{3}\pi R^3 = \frac{\pi}{4} d^3$

$$\therefore \quad d = \left(\frac{16}{3}\right)^{1/3} R.$$

Thus, $\frac{V_c}{A_c} = \frac{1}{6}\left(\frac{16}{3}\right)^{1/3} R.$

Solidfication time of sphere,

$$T_s = K \frac{V_s}{A_s} = K. \frac{R}{3} = 0.333 \text{ KR}$$

Similarly, solidfication time of cylinder,

$$t_c = K \frac{1}{6}\left(\frac{16}{3}\right)^{1/3}. R = 0.2962 \text{ KR}$$

Since solidfication time of cylinder is less than that of sphere, hence it solidfies faster.

4.

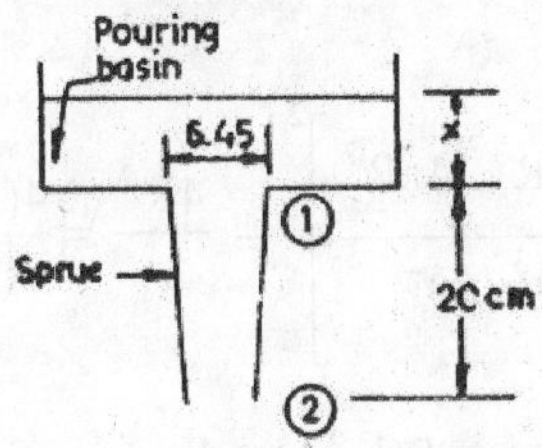

Given, $A_1 = 6.45$ cm^2

$h = 20$ cm

$Q = 820$ cm^3/s

From continuity equation,

$$Q = A \times V$$

or, $820 = A \times \sqrt{2gh}$

$$= 6.45 \times \sqrt{2gx}$$

or, $x = 8.246 \approx 8.5$ cm

To avoid aspiration effects,

$$\frac{A_2}{A_1} = \sqrt{\frac{x}{20 + x}} = \sqrt{\frac{8.5}{28.5}}$$

or, $A_2 = 6.45\sqrt{\frac{8.5}{28.5}} = 3.52$ cm^2

7. According to Chvorinov's rule,

Solidificatin time, $\tau_s = K\,(V/A)^2$

where V = Volume of casting

A = Surface area of casting

K = constant.

For sphere and cube of given case, $A_s = A_c$

Volume of sphere, $V_s = \frac{4}{3}\pi R^3$

where, R = Radius of sphere and volume of cube

$V_c = a^3$

where a = side of cube

$\therefore \quad 6a^2 = A$

or $\quad a = \sqrt{\frac{A}{6}}$

and $\quad A = 4\pi R^2$

or $\quad R = \sqrt{\frac{A}{4\pi}}$

Ratio of time taken

$$= \frac{\text{Solidification time of sphere}}{\text{Solidification time of cube}} = \left[\frac{K(V_{sphere}/A_{sphere})}{K(V_{cube}/A_{cube})}\right]^2$$

$$= \left[\frac{\frac{4}{3}\pi R^3/4\pi R^2}{a^3/6a^2}\right]^2 = \frac{4R^2}{a^2} = \frac{4(A/4\pi)}{(A/6)} = \frac{\mathbf{6}}{\boldsymbol{\pi}}$$

11. Diameter of the blank

$$D_o = \sqrt{d^2 + 4dh}$$

$$= \sqrt{40^2 + 4 \times 40 \times 30} = 80 \text{ mm}$$

$$\text{Drawing force} = \sigma_t \pi dt \left[\frac{D_o}{d} - 1\right]$$

$$= 150 \times 10^6 \times \pi \times \frac{40}{1000} \times \frac{1}{1000}[1.9 - 1] = \mathbf{16964.6\ N}$$

Draw ratio: It is the ratio of initial blank diameter to the diameter of the cup drawn from the blank (usually approximated by the punch diameter).

$$\text{Draw ratio} = \frac{D_o}{d} = \frac{80}{40} = 2$$

As the limiting draw ratio is only 1.9 for the material used (given in the problem), the cup cannot be drawn in single operation.

Minimum number of operations required are two.

(b) *(i)* Stress due to 1000 N load

$$= \frac{1000}{5 \times 5 \times 10^{-6}} \text{ Pa} = 40 \text{ MPa}$$

Stress due to 2000 N load = 80 MPa

Stress due to 500 N load = 20 MPa

and stress due to 800 N shear = 32 MPa

Principal stress due to 40 MPa tensile, 80 MPa tensile and 32 MPa shear stress can be calculated using Mohr's circle.

OP = 40 MPa, OQ = 80 MPa

PS = 32 MPa, RQ = 32 MPa

Radius of Mohr's circle = RV

$$= \sqrt{(VQ)^2 + (QR)^2}$$

$$= \sqrt{(20)^2 + (32)^2} = 37.74 \text{ MPa}$$

Maximum tensile stress,

σ_1 = OU = OV + VU = 60 + 37.74 = 97.74 MPa

Minimum tensile stress,

σ_2 = OT = OV – VT = 60 – 37.74 = 23.26 MPa

Tensile stress acting in the third mutually perpendicular plane is the third principal stress.

$$\sigma_3 = \mathbf{20\ MPa}$$

(ii) *According to Von Mises theory*

Yield stress,

$$\sigma_o = \frac{1}{\sqrt{2}}[(\sigma_1 - \sigma_2)^2 + (\sigma_2 - \sigma_3)^2 + (\sigma_3 - \sigma_1)^2]^{1/2}$$

$$= \frac{1}{\sqrt{2}}[(97.74 - 22.26)^2 + (22.26 - 20)^2 + (20 - 97.74)^2]^{1/2}$$

= **76.63 MPa**

Yield strength of the material is given as 70 MPa. Hence the cube will yield according to Von Mises theory, under the action of these forces.

11 CHAPTER

Forming Processes

MCQ TYPE QUESTIONS

2012

1. Match the following metal forming processes with their associated stresses in the workpiece.

Metal forming process	**Type of stress**
1. Coining	P. Tensile
2. Wire Drawing	Q. Shear
3. Blanking	R. Tensile and compressive
4. Deep Drawing	S. Compressive

(*a*) 1–S, 2–P, 3–Q, 4–R (*b*) 1–S, 2–P, 3–R, 4–Q
(*c*) 1–P, 2–Q, 3–S, 4–R (*d*) 1–P, 2–R, 3–Q, 4–S

2. A solid cylinder of diameter 100 *mm* and height 50 *mm* is forged between two frictionless flat dies to a height of 25 *mm*. The percentage change in diameter is

(*a*) 0 (*b*) 2.07
(*c*) 20.7 (*d*) 41.4

3. In a single pass rolling process using 410 *mm* diameter steel rollers, a strip of width 140 *mm* and thickness 8 *mm* undergoes 10 % reduction of thickness. The *angle of bite in radians* is

(*a*) 0.006 (*b*) 0.031
(*c*) 0.062 (*d*) 0.600

2007

4. In open-die forging, a disc of diameter 200 mm and height 60 mm is compressed without any barreling effect. The final diameter of the disc is 400 mm. The true strain is

(*a*) 1.986 (*b*) 1.686
(*c*) 1.386 (*d*) 0.602

5. The thickness of a metallic sheet is reduced from an initial value of 16 mm to a final value of 10 mm in one single pass rolling with a pair of cylindrical rollers each of diameter of 400 mm. The bite angle in degree will be

(*a*) 5.936
(*b*) 7.936
(*c*) 8.936
(*d*) 9.936

6. Match the correct combination for following metal working processes.

Processes	**Associated state of stress**
P. Blanking	1. Tension
Q. Stretch Forming	2. Compression
R. Coining	3. Shear
S. Deep Drawing	4. Tension and Compression
	5. Tension and Shear

Codes :

	P	**Q**	**R**	**S**
(*a*)	2	1	3	4
(*b*)	3	4	1	5
(*c*)	5	4	3	1
(*d*)	3	1	2	4

7. The force requirement in a blanking operation of low carbon steel sheet is 5.0 kN. The thickness of the sheet is '*t*' and diameter of the blanked part is '*d*'. For the same work material, if the diameter of the blanked part is increased to 1.5 *d* and thickness is reduced to 0.4 *t*, the new blanking force in kN is

(*a*) 3.0 (*b*) 4.5
(*c*) 5.0 (*d*) 8.0

2006

8. A 4 mm thick sheet is rolled with 300 mm diameter rolls to reduce thickness without any charge in its width. The friction coefficient at the work-roll interface is 0.1. The minimum possible thickness of the sheet that can be produced in a single pass is

(*a*) 1.0 mm (*b*) 1.5 mm
(*c*) 2.5 mm (*d*) 3.7 mm

2004

9. 10 mm diameter holes are to be punched in a steel sheet of 3 mm thickness. Shear strength of the material is 400 N/mm^2 and penetration is 40%. Shear provided on the punch is 2 mm. The blanking force during the operation will be

(*a*) 22.6 kN (*b*) 37.7 kN
(*c*) 61.6 kN (*d*) 94.3 kN

10. Through holes of 10 mm diameter are to be drilled in a steel plate of 20 mm thickness. Drill spindle speed is 300 rpm, feed 0.2 mm/rev and drill point angle is 120°. Assuming drill over travel of 2 mm, the time for producing a hole will be
(*a*) 4 seconds (*b*) 25 seconds
(*c*) 100 seconds (*d*) 110 seconds

11. In a rolling process, sheet of 25 mm thickness is rolled to 20 mm thickness. Roll is of diameter 600 mm and it rotates at 100 rpm. The roll strip contact length will be
(*a*) 5 mm (*b*) 39 mm
(*c*) 78 mm (*d*) 120 mm

2003

12. Cold working of steel is defined as working
(*a*) at its recrystallisation temperature
(*b*) above its recrystallisation temperature
(*c*) below its recrystallisation temperature
(*d*) at two thirds of the melting temperature of the metal

13. A shell of 100 mm diameter and 100 mm height with the corner radius of 0.4 mm is to be produced by cup drawing. The required blank diameter is
(*a*) 118 mm (*b*) 161 mm
(*c*) 224 mm (*d*) 312 mm

14. A brass billet is to be extruded from its initial diameter of 100 mm to a final diameter of 50 mm. The working temperature of 700 °C and the extrusion constant is 250 MPa. The force required for extrusion is
(*a*) 5.44 MN (*b*) 2.72 MN
(*c*) 1.36 MN (*d*) 0.36 MN

15. A metal disc of 20 mm diameter is to be punched from a sheet of 2 mm thickness. The punch and the die clearance is 3%. The required punch diameter is
(*a*) 19.88 mm
(*b*) 19.94 mm
(*c*) 20.06 mm
(*d*) 20.12 mm

2002

16. Hot rolling of mild steel is carried out
(*a*) at recrystallisation temperature
(*b*) between 100° C to 150° C
(*c*) below recrystallisation temperature
(*d*) above recrystallisation temperature

2001

17. The cutting force in punching and blanking operations mainly depends on
(*a*) modulus of elasticity of metal
(*b*) shear strength of metal
(*c*) bluk modulus of metal
(*d*) yield strength of metal

18. For rigid perfectly-plastic work material, negligible interface friction and no redundant work, the theoretically maximum possible reduction in the wire drawing operation is
(*a*) 0.36 (*b*) 0.63
(*c*) 1.00 (*d*) 2.72

19. A number of cold rolling passes are required in a two-high rolling mill to reduce the thickness of a plate from 50 mm to 25 mm. The roll diameter is 700 mm and the coefficient of friction at the roll-work interfacCCe is 0,1. It is required that the draft in each pass must be the same. Assuming no front and back tensions.
(*i*) Minimum number of passes required are
(*a*) 4 (*b*) 5
(*c*) 6 (*d*) 7
(*ii*) The draft in each pass will be
(*a*) 1.5 mm (*b*) 2.5 mm
(*c*) 3.5 mm (*d*) 4.5 mm

2000

20. A 1.5 mm thick sheet is subject to unequal biaxial stretching and the true strains in the directions of stretching are 0.05 and 0.09. The final thickness of the sheet in mm is
(*a*) 1.414 (*b*) 1.304
(*c*) 1.362 (*d*) 289

1996

21. A wire of 0.1 mm diameter is drawn from a rod of 15mm diameter. Dies giving reductions of 20%, 40% and 80% are available. For minimum error in the final size, the number of stages and reduction at each stage respectively would be
(*a*) 3 stages and 80% reduction for all three stages
(*b*) 4 stages and 80% reduction for first three stages followed by a finishing stage of 20% reduction
(*c*) 5 stages and reduction of 80%, 80%, 40%, 40%, 20% in a sequence
(*d*) none of the above

NUMERICAL TYPE QUESTIONS

2016

1. A 300 mm thick slab is being cold rolled using roll of 600 mm diameter. If the coefficient of friction is 0.08, the maximum possible reduction (in mm) is ____

1999

2. The shape ABCDE shown in Figure. is to be produced by blanking operation from a larger sheet of 4 mm thickness. The material is an alloy steel having ultimate shear strength, USS of 600 MPa.

(*a*) Calculate the force required in the blanking operation.

(*b*) Determine the coordinates of the centre of pressure during the operation with origin at A.

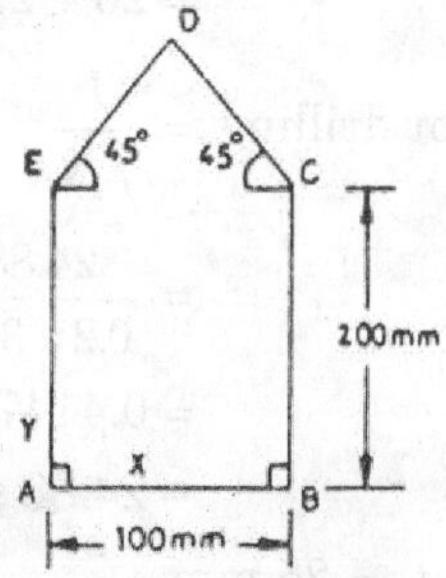

ANSWERS

MCQ Type Questions

1. (*a*)	**2.** (*d*)	**3.** (*c*)	**4.** (*c*)	**5.** (*d*)	**6.** (*d*)	**7.** (*a*)	**8.** (*c*)	**9.** (*a*)	**10.** (*b*)
11. (*b*)	**12.** (*c*)	**13.** (*c*)	**14.** (*b*)	**15.** (*a*)	**16.** (*d*)	**17.** (*b*)	**18.** (*b*)	**19.** (*i-(d)* *ii-(c)*)	**20.** (*a*)
21. (*d*)									

Numerical Type Questions

1. 1.92 **2.** (*)

"*" Refers to Subjective Question.

EXPLANATIONS

MCQ TYPE QUESTIONS

2. From incompressibility, we get

$$\pi d_1^2 h_1 = \pi d_2^2 h_2$$

$$d_2 = d_1 \times \sqrt{\frac{h_1}{h_2}}$$

$$d_2 = 100 \times \sqrt{\frac{50}{25}} = 141.42$$

Percentage change in diameter

$$= \frac{d_2 - d_1}{d_1} \times 100 = 41.42\%$$

3. $$\Delta H = D(1 - \cos\alpha)$$

$$\cos\alpha = 1 - \frac{\Delta H}{D} = 1 - \frac{0.1 \times 8}{410}$$

$$\boxed{\alpha = 3.57°}$$

$$\alpha = 3.57 \times \frac{\pi}{180} \text{ radians}$$

$$\boxed{\alpha = 0.062 \text{ radians}}$$

8. $$\mu = \sqrt{\frac{t_i - t_f}{R}}$$

$\Rightarrow$ $$t_f = 2.5 \text{ mm}$$

9. Given : Shear strength, $\tau = 400$ N/mm^2

Sheet thickness, $t = 3$ mm

Diameter of hole, $d = 10$ mm,

Shear on punch, $t_1 = 2$ mm

Penetration of punch on fraction,

$$2p = 0.4$$

Blanking force when shear is applied on the punch

$$= \pi d t \left(\frac{t_p}{t_1}\right) \tau$$

$$= \pi \times 10 \times 3 \times 400 \times \left(\frac{3 \times 0.4}{2}\right)$$

$$= \mathbf{22.61 \text{ kN.}}$$

10. Given : Diameter of hole, $d = 10$ mm

Thickness of steel plate, $t = 20$ mm

$$\alpha = \frac{120}{2} = 60°$$

Break through distance,

$$A = \frac{d}{2\tan\alpha} = \frac{10}{2\tan 60°}$$

$$= 2.8867 \text{ mm}$$

Total length of tool travel,

$$l = t + A + 2$$

$$= 20 + 2.8867 + 2 = 24.88$$

$$\text{Time for drilling} = \frac{l}{fN}$$

$$= \frac{24.88}{0.2 \times 300}$$

$$= 0.4147 \text{ min}$$

$$= 24.88 \text{ sec} \approx \mathbf{25 \text{ sec.}}$$

11. Given : t_i = 25 mm

t_f = 20 mm

R = 300 mm

N = 100 r.p.m.

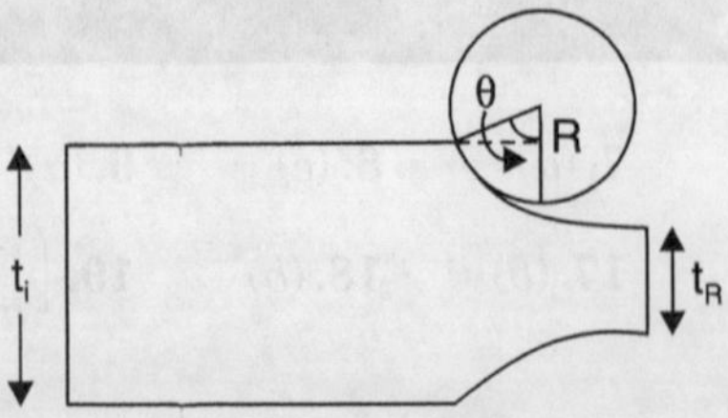

Angle subtended by the deformed zone at the roll center,

$$Q_i = \sqrt{\frac{t_i - t_R}{R}} = \sqrt{\frac{25-20}{300}} = 0.129 \text{ rad.}$$

∴ Roll strip contact length

$$= \theta.R = 0.129 \times 300 = 38.72 \text{ mm}$$

13. Let d = diameter of shell = 100 mm

h = height of shell = 100 mm

r = corner radius = 0.4 mm

Since $d \geq 20r$

$$\therefore \text{ Black diameter} = \sqrt{d^2 + 4dh}$$

$$= \sqrt{(100)^2 + 4 \times 100 \times 100}$$

$$= 223.6 \simeq 224 \text{ mm}$$

14. Force required for extrusion $= k.A_0 \ln\left(\frac{A_0}{A_f}\right)$

where, k = extrusion constant = 250×10^6 N/m^2

A_0 = initial area $= \frac{\pi}{4} \times (0.1)^2$

A_f = final area $= \frac{\pi}{4} \times (0.05)^2$

$$\text{Now, } P = 250 \times 10^6 \times \frac{\pi}{4} \times (0.1)^2 \ln\left[\frac{\frac{\pi}{4} \times (0.1)^2}{\frac{\pi}{4} \times (0.05)^2}\right]$$

$$= 2.72 \times 10^6 \text{ N/m} = 2.72 \text{ MN}$$

15. Punch diameter = 20 – 2 × 3% of thickness

$$= 20 - 2 \times \frac{3}{100} \times 2$$

$$= 20 - 0.1^2 = 19.88 \text{ mm}$$

19. Reduction in thickness in one pass,

$$\Delta h = \mu^2 R = (0.1)^2 \times 350 \text{ mm} = 3.5 \text{ mm}$$

and we have to reduce 25 mm.

(*i*) Minimum number of passes required

$$= \frac{50 - 25}{3.5} = 7.14 \approx 7 \text{ passes.}$$

(*b*) Draft in each pass,

$$\Delta h = \mu^2 R = (0.1)^2 \times 350 \text{ mm} = 3.5 \text{ mm}$$

NUMERICAL TYPE QUESTIONS

1. Maximum possible reduction

$$= \Delta H/\text{pass} = \mu^2 R$$

$$= 0.08^2 \times 300 = 1.92 \text{ mm}$$

2. Given, t = 4 mm

f_{su} = 600 MPa = 600 N/mm^2

(*a*) Required force,

$$F = P.t.f_{su}$$

where P = Perimeter ABCDE

$$= 100 + 200 + \frac{2 \times 100}{\sin 45 \times 2} + 200$$

$$= 641.42 \text{ mm.}$$

$$\therefore \quad F = 641.42 \times 4 \times 600 = 1539.41 \text{ kN}$$

(*b*) Co-ordinates of centre of pressure

X-coordinate = 50 mm.

$$\bar{y} = \frac{(100 \times 100 \times 200) + \left(200 + \frac{50}{3}\right)\left(\frac{1}{2} \times 100 \times 50\right)}{100 \times 200 + \frac{1}{2} \times 100 \times 50}$$

$$= 112.96 \text{ mm}$$

∴ Co-ordinates of centre of pressure with origin A are (50, 112.96)

■■

12 CHAPTER

Metal Joining Processes

MCQ TYPE QUESTIONS

2016

1. Under optimal conditions of the process the temperatures experienced by a copper work piece in fusion welding, brazing and soldering are such that

(*a*) $T_{welding} > T_{soldering} > T_{brazing}$

(*b*) $T_{soldering} > T_{welding} > T_{brazing}$

(*c*) $T_{brazing} > T_{welding} > T_{soldering}$

(*d*) $T_{welding} > T_{brazing} > T_{soldering}$

2. The welding process which uses a blanket of fusible granular flux is

(*a*) tungsten inert gas welding

(*b*) submerged arc welding

(*c*) electroslag welding

(*d*) thermit welding

2012

3. In abrasive jet machining, as the distance between the nozzle tip and the work surface increases, the material removal rate

(*a*) increases continuously.

(*b*) decreases continuously.

(*c*) decreases, becomes stable and then increases.

(*d*) increases, becomes stable and then decreases.

4. A fillet welded joint is subjected to transverse loading *F* as shown in the figure. Both legs of the fillets are of 10 *mm* size and the weld length is 30 *mm*. If the allowable shear stress of the weld is 94 *MPa*, considering the minimum throat area of the weld, the maximum allowable transverse load in *kN* is

(*a*) 14.44

(*b*) 17.92

(*c*) 19.93

(*d*) 22.16

5. In a DC arc welding operation, the voltage–arc length characteristic was obtained as $V_{arc} = 20 + 5l$ where the arc length l was varied between 5 *mm* and 7 *mm*. Here V_{arc} denotes the arc voltage in *Volts*. The *arc current* was varied from 400 A to 500 A. Assuming linear power source characteristic, the *open circuit voltage* and the *short circuit current* for the welding operation are

(*a*) 45 V, 450 A

(*b*) 75 V, 750 A

(*c*) 95 V, 950 A

(*d*) 150 V, 1500 A

2008

6. In arc welding of a butt joint, the welding speed is to be selected such that highest cooling rate is achieved. Melting efficiency and heat transfer efficiency are 0.5 and 0.7, respectively. The area of the weld cross section is 5 mm^2 and the unit energy required to melt the metal is 10 J/mm^3. If the welding power is 2 kW, the welding speed in mm/s is closest to

(*a*) 4

(*b*) 14

(*c*) 24

(*d*) 34

2007

7. Which one of the following is a solid state joining process ?

(*a*) Gas tungsten arc welding

(*b*) Resistance spot welding

(*c*) Friction welding

(*d*) Submerged arc welding

8. A direct current welding machine with a linear power source characteristic provides open circuit voltage of 80 V and short circuit current of 800 A. During welding with the machine, the measured arc current is 500 A corresponding to an arc length of 5.0 mm and the measured arc current is 460 A corresponding to an arc length of 7.0 mm. The linear voltage (E) – arc length (L) characteristic of the welding arc can be given as (where E is in volt and L is in mm)

(*a*) E = 20 + 2L

(*b*) E = 20 + 8L

(*c*) E = 80 + 2L

(*d*) E = 80 + 8L

9. Two metallic sheets, each of 2.0 mm thickness, are welded in a lap joint configuration by resistance spot welding at a welding current of 10 kA and welding time of 10 milli second. A spherical fusion zone extending up to the full thickness of each sheet is formed. The properties of the metallic sheets are given as :

Ambient temperature = 293 K

Melting temperature = 1793 K

Density = 7000 kg/m^3

Latent heat of fusion = 300 kJ/kg

Specific heat = 800 J/kg K

Assume :

(*i*) contact resistance along sheet-sheet interface is 500 micro-ohm and along electrode-sheet interface is zero;

(*ii*) no conductive heat loss through the bulk sheet materials; and

(*iii*) complete weld fusion zone is at the melting temperature.

The melting efficiency (in %) of the process is

(*a*) 50.37

(*b*) 60.37

(*c*) 70.37

(*d*) 80.37

2006

10. In an arc welding process, the voltage and current are 25 V and 300 A respectively. The arc heat transfer efficiency is 0.85 and weldign speed is 8 mm/sec. The net heat input (in J/mm) is

(*a*) 64

(*b*) 797

(*c*) 1103

(*d*) 79700

NUMERICAL TYPE QUESTION

2016

1. The voltage-length characteristic of a direct current arc in an arc welding process is V = (100 + 40*l*), where *l* is the length of the arc in mm and V is arc voltage in volts. During a welding operation, the arc length varies between 1 and 2 mm and the welding current is in the range 200-250 A. Assuming a linear power source, the short circuit current is ______ A.

ANSWERS

MCQ Type Questions

1. (*d*) **2.** (*b*) **3.** (*d*) **4.** (*c*) **5.** (*c*) **6.** (*b*) **7.** (*c*) **8.** (*a*) **9.** (*c*) **10.** (*b*)

Numerical Type Question

1. (425)

EXPLANATIONS

MCQ TYPE QUESTIONS

1. Fusion welding for melting the parent material, this implies the temperature should be greater than the melting temperature. Brazing and soldering are non-fusion welding operations. Temperature in Brazing is above 427°C where in soldering it is below 427°C.

2. In submerged arc welding, the arc is completely submerged inside the granular flux powder and forming as blanket.

4.
$$P = 0.707 \text{ sl. } \tau_{allowable}$$
$$P = 0.707 \times 10 \times 30 \times 94$$
$$= 19937.4\text{N}$$
$$P = 19.934 \text{ kM}$$

5.
$$V_{arc} = 20 + 5l$$
$$l = 5 \text{ mm},$$
$$V_{arch} = 45 \text{ V}$$
$$l = 7 \text{ mm},$$
$$V_{arch} = 55\text{V}$$
$$V = V_0 - \frac{V_0}{I_S} I$$

(where V_0 is open circuit voltage and I_S is short circuit current)

$$45 = V_0 - \frac{V_0}{I_S} \times 500$$

$$55 = V_0 - \frac{V_0}{I_S} \times 400$$

Upon solving (1) and (2) we get

$$V_0 = 95 \text{ Volts}$$
$$I_S = 950 \text{ A}$$

6. Welding power applied = Heating power needed

$$\therefore \quad P \times \eta_m \times \eta_{H.T.} = E_U \times (A) \times f$$
$$\Rightarrow \; 2 \times 10^3 \times 0.5 \times 0.7 = 10 \times 5 \times f$$
$$\Rightarrow \quad f = 14 \text{ mm/sec}$$

7. Friction welding is clearly solid state welding.

8. From power – source characteristics,

$$V = 80 - \frac{80}{100} I$$

$$= 80 - \frac{1}{10} I \qquad ...(i)$$

Arc voltage characteristics

$$V = a + bL \qquad ...(ii)$$

Equating (*i*) and (*ii*), we get

$$\therefore \quad a + bL = 80 - \frac{I}{100}$$

Here, L = 5, I = 500

$$\therefore \quad a + 5b = 30 \qquad ...(iii)$$

But at L = 7, I = 460

$$\therefore \quad a + 7b = 34$$

From (*iii*) and (*iv*), we get

$$b = 2 \; a = 20$$

$$\therefore \quad \text{Voltage charge} = 20 + 2L$$

9. Heat generated due to current,

$$H_1 = I_2RT$$
$$= (10 \times 10^3)^2 \times 500 \times 10^{-6} \times 10 \times 10^{-3}$$
$$= 500 \text{ J}$$

Heat required for melting,

$$H_2 = 300 \times 10^3 \times 7000 \times \text{ volume}$$
$$+ 800 \times (1793 - 293) \times 7000 \times \text{volume}$$

Since fusion zone is spherical is shape

$$\therefore \quad \text{volume} = \frac{4}{3}\pi R^3$$

where, r = t = thickness of sheet = 2mm

Melting efficiency = 70.35%

10. Power generated = VI

$$\therefore \quad \text{Heat input} = \frac{25 \times 300}{\text{Welding speed}} \times 0.85$$

$$= 797 \text{ J/mm}$$

NUMERICAL TYPE QUESTION

1. Given $V = 100 + 40\,l$

$l_1 = 1\text{mm} \Rightarrow V_1 = 140 \text{ V} \qquad I_1 = 250 \text{ A}$

$l_2 = 2 \text{ mm} \Rightarrow V_2 = 180 \text{ V} \qquad I_2 = 200 \text{ A}$

$$\frac{V}{OCV} + \frac{I}{SSC} = 1$$

$$\frac{140}{OCV} + \frac{250}{SSC} = 1$$

$$\frac{180}{OCV} + \frac{200}{SSC} = 1$$

On solving we get

SSC = 424.6 A

13 CHAPTER Machining and Machine Tool Operations

MCQ TYPE QUESTIONS

2016

1. The non-traditional machining process that essentially requires vacuum is
   (a) electron beam machining
   (b) electro chemical machining
   (c) electro chemical discharge machining
   (d) electro discharge machining

2. The tool life equation for HSS tool is $VT^{0.14} f^{0.7} d^{0.4}$ = Constant. The tool life (T) of 30 min is obtained using the following cutting conditions:

   $V = 45$ m/min, $f = 0.35$ mm, $d = 2.0$ mm

   If speed (V), feed (f) and depth of cut (d) are increased individually by 25%, the tool life (in min) is
   (a) 0.15
   (b) 1.06
   (c) 22.50
   (d) 30.0

3. In an ultrasonic machining (USM) process, the material removal rate (MRR) is plotted as a function of the feed force of the USM tool. With increasing feed force, the MRR exhibits the following behavior :
   (a) increases linearly
   (b) decreases linearly
   (c) does not change
   (d) first increases and then decreases

4. In a wire-cut EDM process the necessary conditions that have to be met for making a successful cut are that
   (a) wire and sample are electrically non-conducting
   (b) wire and sample are electrically conducting
   (c) wire is electrically conducting and sample is electrically non-conducting
   (d) sample is electrically conducting and wire is electrically non-conducting

5. Internal gears are manufactured by
   (a) hobbing
   (b) shaping with pinion cutter
   (c) shaping with rack cutter
   (d) milling

6. For an orthogonal cutting operation, tool material is HSS, rake angle is 22°, chip thickness is 0.8 mm, speed is 48 m/min and feed is 0.4 mm/rev. The shear plane angle (in degrees) is
   (a) 19.24 (b) 29.70
   (c) 56.00 (d) 68.75

2012

7. A CNC vertical milling machine has to cut a straight slot of 10 *mm* width and 2 *mm* depth by a cutter of 10 *mm* diameter between points (0, 0) and (100, 100) on the *XY* plane (dimensions in *mm*). The feed rate used for milling is 50 *mm/min*. Milling time for the slot (in *seconds*) is
   (a) 120 (b) 170
   (c) 180 (d) 240

8. Details pertaining to an orthogonal metal cutting process are given below.

Chip thickness ratio	0.4
Undeformed thickness	0.6 *mm*
Rake angle	+10°
Cutting speed	2.5 *m/s*
Mean thickness of primary shear zone	25 *microns*

   The shear strain rate in s^{-1} during the process is
   (a) 0.1781×10^5 (b) 0.7754×10^5
   (c) 1.0104×10^5 (d) 4.397×10^5

9. In a single pass drilling operation, a through hole of 15 mm diameter is to be drilled in a steel plate of 50 mm thickness. Drill spindle speed is 500 rpm, feed is 0.2 *mm/rev* and drill point angle is 118°. Assuming 2 mm clearance at approach and exit, the total drill time (in *seconds*) is
   (a) 35.1 (b) 32.4
   (c) 31.2 (d) 30.1

10. Calculate the punch size in *mm*, for a circular blanking operation for which details are given below.

Size of the blank	25 *mm*
Thickness of the sheet	2 *mm*
Radial clearance between punch and die	0.06 *mm*
Die allowance	0.05 *mm*

(*a*) 24.83 (*b*) 24.89

(*c*) 25.01 (*d*) 25.17

2011

11. Match the following non-traditional machining processes with the corresponding material removal mechanisms :

Machining process

P. Chemical machining

Q. Electro-chemical machining

R. Electro-discharge machining

S. Ultrasonic machining

Mechanism of material removal

1. Erosion
2. Corrosive reaction
3. Ion displacement
4. Fusion and vaporization

(*a*) P-2, Q-3, R-4, S-1 (*b*) P-2, Q-4, R-3, S-1

(*c*) P-3, Q-2, R-4, S-1 (*d*) P-2, Q-3, R-1, S-4

12. A single-point cutting tool with 12° rake angle is used to machine a steel work-piece. The depth of cut, i.e., uncut thickness is 0.81 mm. The chip thickness under orthogonal machining condition is 1.8 mm. The shear angle is approximately

(*a*) 22° (*b*) 26°

(*c*) 56° (*d*) 76°

2010

13. For tool A, Taylor's tool life exponent (n) is 0.45 and constant (K) is 90. Similarly for tool B, $n = 0.3$ and K = 60. The cutting speed (in m/min) above which tool A will have a higher tool life than tool B is

(*a*) 26.7 (*b*) 42.5

(*c*) 80.7 (*d*) 142.9

14. A taper hole is inspected using a CMM, with a probe of 2 mm diameter. At a height, Z = 10 mm from the bottom, 5 points are touched and a diameter of circle (not compensated for probe size) is obtained as 20 mm. Similarly, a 40 mm diameter is obtained at a height Z = 40 mm. The smaller diameter (in mm) of hole at Z = 0 is

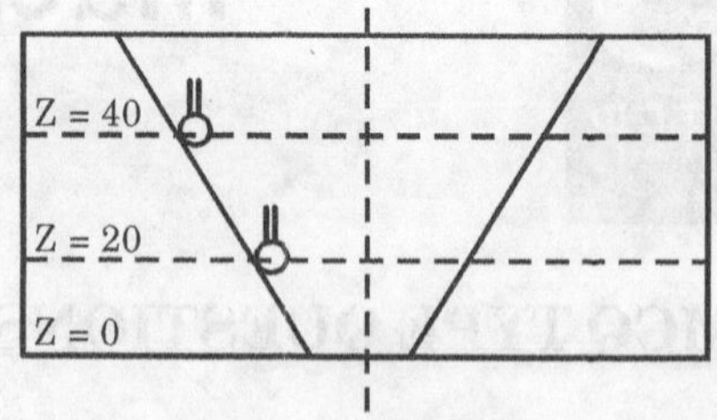

(*a*) 13.334 (*b*) 15.334

(*c*) 15.442 (*d*) 15.542

2009

15. Which of the following is the correct data structure for soild models ?

(*a*) solid part → faces → edges → vertices

(*b*) solid part → edges → faces → vertices

(*c*) vertices → edges → faces → solid parts

(*d*) vertices → faces → edges → solid parts

16. Minimum shear strain in orthogonal turning with a cutting tool of zero rake angle is

(*a*) 0.0 (*b*) 0.5

(*c*) 1.0 (*d*) 2.0

17. Electrochemical machining is performed to remove material from an iron surface of 20 mm × 20 mm under the following conditions:

Inter electrode gap = 0.2 mm

Supply voltage (DC) =12 V

Specific resistance of electrolyte = 2 Ω cm

Atomic weight of Iron = 55.85

Valency of Iron = 2

Farady's constant = 96540 Coulombs

The matrial removal rate (in g/s) is

(*a*) 0.3471 (*b*) 3.471

(*c*) 34.71 (*d*) 347.1

18. Match the following:

NC code	**Definition**
P. M05	1. Absolute coordinate system
Q. G01	2. Dwell
R. G04	3. Spindle stop
S. G09	4. Linear interpolatio

Codes :

	P	**Q**	**R**	**S**
(*a*)	2	3	4	1
(*b*)	3	4	1	2
(*c*)	3	4	2	1
(*d*)	4	3	2	1

2008

19. Internal gear cutting operation can be performed by

(a) milling

(b) shaping with rack cutter

(c) shaping with pinion cutter

(d) hobbing

20. One tooth of a gear having 4 module and 32 teeth is shown in the figure. Assume that the gear tooth and the corresponding tooth space make equal intercepts on the pitch circumference. The dimensions 'a' and 'b', respectively, are closest to

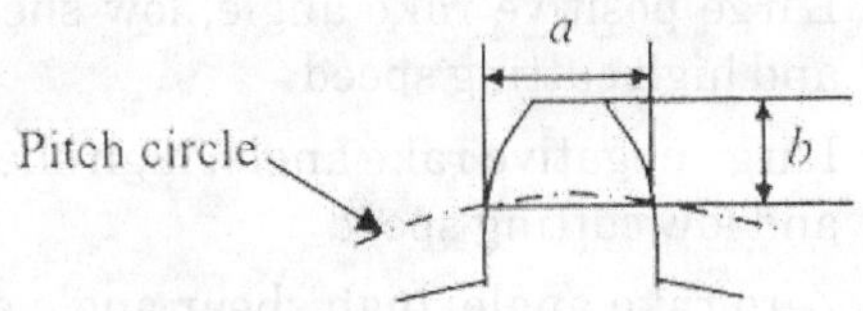

(a) 6.08 mm, 4 mm

(b) 6.48 mm, 4.2 mm

(c) 6.28 mm, 4.3 mm

(d) 6.28 mm, 4.1 mm

21. The figure shows an incomplete schematic of a conventional lathe to be used for cutting threads with, different pitches. The speed gear box U_r is shown and the feed gear box U_s is to be placed. P, Q, R and S denote locations and have no other significance. Changes in U_r should NOT affect the pitch of the thread being cut and changes in U_s should NOT affect the cutting speed.

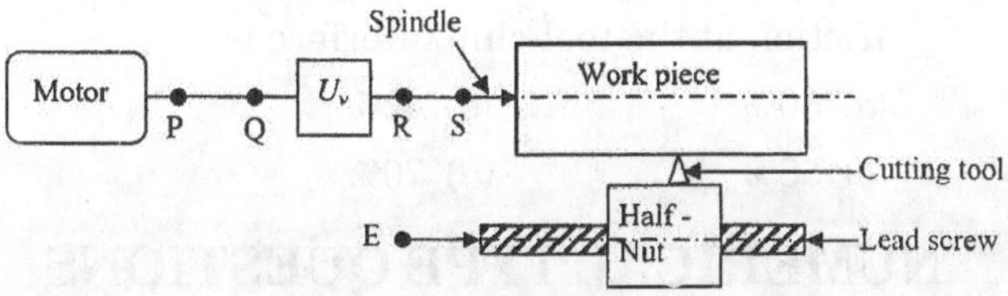

The correct connections and the correct placement of U_s are given by

(a) Q and E are connected. U_s is placed between P and Q.

(b) S and E are connected. U_s is placed between R and S.

(c) Q and E are connected. U_s is placed between Q and E.

(d) S and E are connected. U_s is placed between S and E.

22. A displacement sensor (a dial indicator) measures the lateral displacement of a mandrel mounted on the taper hole inside a drill spindle. The mandrel axis is an extension of the drill spindle taper hole axis and the protruding portion of the mandrel surface is perfectly cylindrical. measurements are taken with the sensor placed at two positions P and Q as shown in the figure. The readings are recorded as R_x = maximum deflection minus minimum deflection, corresponding to sensor position at X, over one rotation.

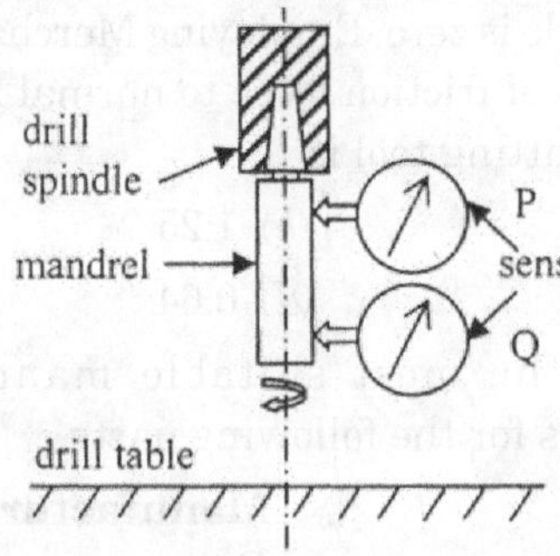

If $R_P = R_Q > 0$, which one of the following would be consistent with the observation?

(a) The drill spindle rotational axis is coincident with the drill spindle taper hole axis

(b) The drill spindle rotational axis intersects the drill spindle taper hole axis at point P

(c) The drill spindle rotational axis is parallel to the drill spindle taper hole axis

(d) The drill spindle rotational axis intersects the drill spindle taper hole axis at point Q

2007

23. In orthogonal turning of a low carbon steel bar of diameter 150 mm with uncoated carbide tool, the cutting velocity is 90 m/min. The feed is 0.24 mm/rev and the depth of cut is 2 mm. The chip thickness obtained is 0.48 mm. If the orthogonal rake angle is zero and the principal cutting edge angle is 90°, the shear angle in degree is

(a) 20.56 (b) 26.56

(c) 30.56 (d) 36.56

24. In electro discharge machining (EDM), if the thermal conductivity of tool is high and the specific heat of work piece is low, then the tool wear rate and material removal rate are expected to be respectively

(a) high and high (b) low and low

(c) high and low (d) low and high

25. In orthogonal turning of medium carbon steel, the specific machining energy is 2.0 J/mm^3. The cutting velocity, feed and depth of cut are 120 m/min, 0.2 mm/rev and 2 mm respectively. The main cutting force in N is

(*a*) 40 (*b*) 80

(*c*) 400 (*d*) 800

26. In orthogonal turning of low carbon steel pipe with principal cutting edge angle of 90 , the main cutting force is 1000 N and the feed force is 800 N. The shear angle is 25. and orthogonal rake angle is zero. Employing Merchant's theory, the ratio of friction force to normal force acting on the cutting tool is

(*a*) 1.56 (*b*) 1.25

(*c*) 0.80 (*d*) 0.64

27. Match the most suitable manufacturing processes for the following parts.

Parts	Manufacturing Processes
P. Computer chip	**1.** Electrochemical Machining
Q. Metal forming dies and molds	**2.** Ultrasonic Machining
R. Turbine blade	**3.** Electrodischarge Machining
S. Glass	**4.** Photochemical Machining

Codes :

	P	Q	R	S
(*a*)	4	3	1	2
(*b*)	4	3	2	1
(*c*)	3	1	4	2
(*d*)	1	2	4	3

2006

28. Match the items in columns I and II.

Column I	Column II
P. Wrinkling	**1.** Yield point elongation
Q. Orange peel	**2.** Anisotropy
R. Stretcher strains	**3.** Large grain size
S. Earing	**4.** Insufficient blank holding force
	5. Fine grain size
	6. Excessive blank holding force

Codes :

	P	Q	R	S
(*a*)	6	3	1	2
(*b*)	4	5	6	1
(*c*)	2	5	3	4
(*d*)	4	3	1	2

29. If each abrasive grain is viewed as a cutting tool, then which of the following represents the cutting parameters in common grinding operations?

(*a*) Large negative rake angle, low shear angle and high cutting speed

(*b*) Large positive rake angle, low shear angle and high cutting speed

(*c*) Large negative rake angle, high shear angle and low cutting speed

(*d*) Zero rake angle, high shear angle and high cutting speed

30. Arrange the processes in the increasing order of their maximum material removal rate.

Electrochemical Machining (ECM)

Ultrasonic Machining (USM)

Electron Beam Machining (EBM)

Laser Beam Machining (LBM) and

Electric Discharge Machining (EDM)

(*a*) USM, LBM, EBM, EDM, ECM

(*b*) EBM, LBM, USM, ECM, EDM

(*c*) LBM, EBM, USM, ECM, EDM

(*d*) LBM, EBM, USM, EDM, ECM

31. The percentage of total energy dissipated due to friction at the tool-chip interface is

(*a*) 30% (*b*) 42%

(*c*) 58% (*d*) 70%

NUMERICAL TYPE QUESTIONS

2012

1. In an orthogonal cutting process the tool used has rake angle of zero degree. The measured cutting force and thrust force are 500 N and 250 N, respectively. The coefficient of friction between the tool and the chip is ________

2. The following data is applicable for a turning operation. The length of job is 900 mm, diameter of job is 200 mm, feed rate is 0.25 mm/rev and optimum cutting speed is 300 m/min. The machining time (in min) is ________

3. For a certain job, the cost of metal cutting is Rs. 18C/V and the cost of tooling is Rs. 270C/(TV), where C is a constant, V is the cutting speed in m/min and T is the tool life in minutes. The Taylor's tool life equation is $VT^{0.25} = 150$. The cutting speed (in m/min) for the minimum total cost is _______.

4. The surface irregularities of electrodes used in an electrochemical machining (ECM) process are 3 μm and 6 μm as shown in the figure. If the work-piece is of pure iron and 12V DC is applied between the electrodes, the largest feed rate is _____ mm/min.

Conductivity of the electrolyte	0.02 ohm^{-1} mm^{-1}
Over-potential voltage	1.5 V
Density of iron	7860 kg/m^3
Atomic weight of iron	55.85 gm

Assume the iron to be dissolved as Fe^{+2} and the Faraday constant to be 96500 Coulomb.

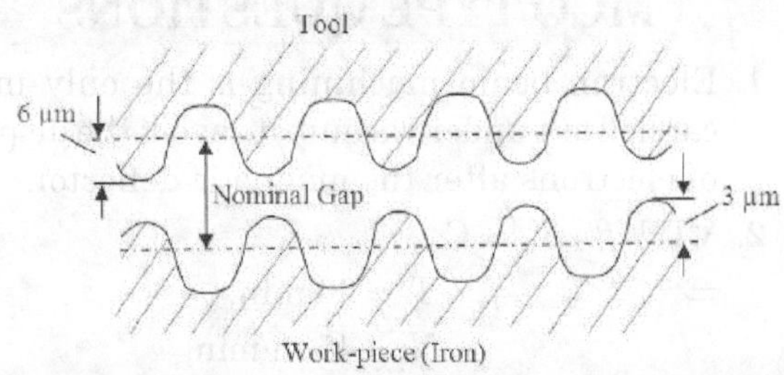

5. In a sheet metal of 2 mm thickness a hole of 10 mm diameter needs to be punched. The yield strength in tension of the sheet material is 100 MPa and its ultimate shear strength is 80 MPa. The force required to punch the hole (in kN) is _________

6. In a single point turning operation with cemented carbide tool and steel work piece, it is found that the Taylor's exponent is 0.25. If the cutting speed is reduced by 50% then the tool life changes by _____ times.

ANSWERS

MCQ Type Questions

1. (*a*)	**2.** (*b*)	**3.** (*d*)	**4.** (*b*)	**5.** (*b*)	**6.** (*b*)	**7.** (*b*)	**8.** (*c*)	**9.** (*a*)	**10.** (*a*)
11. (*a*)	**12.** (*b*)	**13.** (*a*)	**14.** (*a*)	**15.** (*a*)	**16.** (*d*)	**17.** (*a*)	**18.** (*c*)	**19.** (*c*)	**20.** (*d*)
21. (*d*)	**22.** (*c*)	**23.** (*b*)	**24.** (*a*)	**25.** (*d*)	**26.** (*c*)	**27.** (*a*)	**28.** (*d*)	**29.** (*a*)	**30.** (*c*)
31. (*c*)									

Numerical Type Questions

1. 0.5 **2.** (7.53) **3.** (93.2) **4.** (51.542) **5.** (5.024) **6.** (16)

EXPLANATIONS

MCQ TYPE QUESTIONS

1. Electron beam machining is the only method carried out under vacuum, to avoid the dispersion of electrons after the magnetic deflector.

2. $VT^{0.14} f^{0.7} d^{0.4} = C$

$\Rightarrow \quad T_1 = 30$ min,

$V_1 = 45$ m/min,

$f_1 = 0.35$ min,

$d_1 = 2.0$ mm,

$\Rightarrow \quad C = V_1(T_1)^{0.14}(f_1)^{0.7}(d_1)^{0.4}$

$C = 45(30)^{0.14}(0.35)^{0.7}(2)^{0.4}$

$C = 45.8425$

$V_2(T_2)^{0.14}(f_2)^{0.7}(d_2)^{0.4} = 45.8425$

$(1.25 \times 45)(T_2)^{0.14}(1.25 \times 0.35)^{0.7}(1.25 \times 2)^{0.4} = 45.8425$

$\Rightarrow \quad T_2 = 1.06$ min

3. With increase of feed force, the material removal rate MRR is first increases and then decreases.

4. In EDM or wirecut EDM, the work and tool must be electrically conductive otherwise the current passage will not takes place.

5. Internal gears are manufactured by shaping with pinion cutter

6. $\alpha = 22°$

Chip thickness $(t_2) = 0.8$ mm

feed $= 0.4$ mm/rev

speed $= 48$ m/mm

For orthogonal cutting

feed = thickness of uncut chip (t_1)

$$r = \frac{t_2}{t_1} = \frac{0.8}{0.4} = 2$$

$$\tan\theta = \frac{\cos\alpha}{r - \sin\alpha} = \frac{\cos 22}{2 - \sin 22} = 0.57$$

$$\theta = \tan^{-1}(0.57)$$

$$\theta = 29.7$$

8. $r = 0.4$

$t_1 = 0.6$ mm

$\alpha = 10°$

$V_C = 2.5$ m/s

$t_m = 25$ m

$$\tan\phi = \frac{r\cos\alpha}{1 - r\sin\alpha} = 0.4233$$

$\Rightarrow \quad \phi = 22.9°$

$$\text{Shear strain rate} = \frac{V_C \cos\alpha}{\cos(\phi - \alpha) \times t_m}$$

$$= \frac{2.5 \times \cos 10}{\cos 12.9 \times 25 \times 10^{-6}}$$

$$= 1.010^4 \times 10^5/\text{s}$$

9. $$T_c = \frac{(L_h + A + O + C)}{Nf}$$

$L_h = 50$ mm; $A = O = 2$ mm;

$$C = \frac{D}{2}\cot\left(\frac{\beta}{2}\right)$$

$= 7.5 \times \cot(59) = 4.5$ mm

$N = 500$ rpm; $f = 0.2$ mm/rev

$T_c = 0.585$ min or 35.1 Seconds

10. Diameter of punch

= Diameter of Blank – 2 × radial clearance – die allowance

= 25 – 2 × 0.06 – 0.05 = 24.83 mm

12. Relation between shear angle (ϕ), chip thickness ratio (r) and rake angle (α) is given by

$$\tan\phi = \frac{r\cos\alpha}{1 - r\sin\alpha}$$

where, $r = \dfrac{0.81}{1.8} = 0.45$

$$\tan\phi = \frac{0.45\cos 12}{1 - 0.45\sin 12}$$

$\Rightarrow \quad \phi = 26°$

13. Taylor's tool life equation is

VT^n = constant

$\therefore \quad k_1 = v_1 T_1^{n_1}$

and $\quad k_2 = v_2 T_2^{n_2}$

Given conditions for both tools :

Tool A : Constant, $k_1 = 90$;

Exponential constant, $n_1 = 0.45$

Tool B : Constant, $k_2 = 60$;

Exponential constant, $x_2 = 0.3$

$$\therefore \quad \left(\frac{k_1}{k_2}\right) = \left(\frac{v_1}{v_2}\right)\frac{T_1^{n_1}}{T_2^{n_2}}$$

$$\Rightarrow \quad \left(\frac{90}{60}\right) = \left(\frac{v_1}{v_2}\right)\frac{T_1^{0.45}}{T_2^{0.3}}$$

At point of intersection

$v_1 = v_2$, and $T_1 = T_2 = T$

$$\therefore \quad \frac{3}{2} = \frac{T^{0.45}}{T^{0.3}} = T^{0.15}$$

$$\Rightarrow \quad T = (3/2)^{\frac{1}{0.15}} = 14.92 \text{ min}$$

$$\therefore \quad 90 = v_1 (14.92)^{0.45}$$

$$\Rightarrow \quad v_1 = 26.7 \text{ m/min}$$

Above v_1 = 26.7 m/min, tool A will have a higher tool life them too B.

16. Shear strain, $\tau = \dfrac{\cos\alpha}{\sin\phi\cos(\phi-\alpha)}$

where, $\alpha = 0$ = rake angle

ϕ = shear angle

$$\therefore \quad \tau = \frac{1}{\sin\phi\cos\phi}$$

$$\Rightarrow \quad t = \sin\phi\cos\phi$$

$$\therefore \quad \frac{dt}{d\phi} = \cos^2\phi - \sin^2\phi$$

$$\Rightarrow \quad \phi = 45°$$

$$\therefore \quad \tau_{min} = \frac{1}{\sin\frac{\pi}{4}\cos\frac{\pi}{4}} = \frac{1}{\frac{1}{\sqrt{2}}\times\frac{1}{\sqrt{2}}} = 2$$

17. We know, $R = \dfrac{\rho l}{A}$

where, ρ = resistivity

= special resistance of electrolyte

= 2Ω

R = resistance of electrolyte between electrodes

l = interelectrode gap

= 0.2mm

A = electrodes cross-sectional area

= 20 × 20 mm²

$$\therefore \quad R = (2\times 10)\times\frac{0.2}{20\times 20} = 0.01\Omega$$

$$\therefore \quad \text{Current, } I = \frac{V}{R} = \frac{12}{0.01} = 1200 \text{ A}$$

Faraday of electricity, $E = \dfrac{M}{v}$

where, E = equivalent weight

M = molecular weight

v = valency

$$\text{M.R.R (gm/sec)} = \frac{I}{F}\times\frac{M}{v}$$

$$= \frac{1200}{96540}\times\frac{55.85}{2} = 0.3471$$

19. Internal gear cutting is performed by shaping with a pinion cutter.

20. $a \simeq \dfrac{\pi D}{2Z}$

But $D = mZ$

$$\therefore \quad a \simeq \frac{3.14\times m}{2} = 6.28$$

$$\Rightarrow \quad b = \text{addendum} + PR$$

Addendum = module for standard addendum

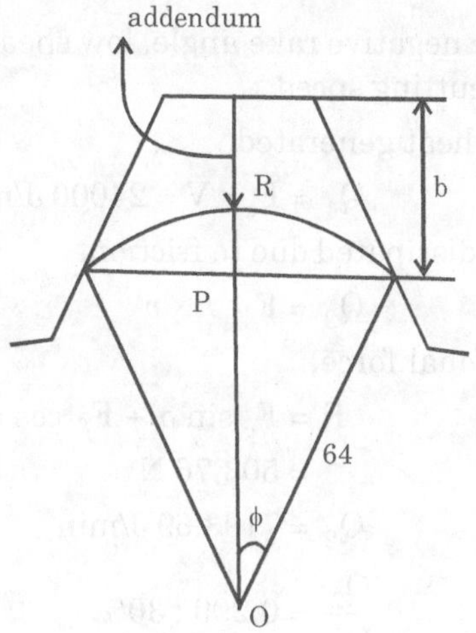

$$\text{Now} \quad \phi = \frac{360°}{32\times 4} = 2.8°$$

$$\therefore \quad OP = 64\cos\phi = 63.9$$

$$\Rightarrow \quad b = m + PR$$

$$= 4 + (64 - 63.9)$$

$$= 4.1 \text{ mm}$$

21. Taking engineering of a conventional lathe, U_v is the back gear arrangement and U_s is the feed gear box; (*a*) and (*b*) are then completely wrong.

Now if (c) is true, then if gear ratio of U_v is changed, the relative velocity of spindle with respect to lead source in changed and hence pitch is changed. Hence (d) is correct

22. Equal deflection in both sensors means eccentricity parallel to the axis.

24. Material removal rate

$\propto$ Thermal conductivity of tool

$\therefore$ Tool wear $\propto$ material removal rate

25. Let F_C = cutting force

V = cutting velocity

Specific machining energy = 2.0 J/mm³

$$\therefore \quad F_C \times V = 2\times \text{mRR}$$

$$\Rightarrow \quad F_C = 2\times .2\times 10^{-3}\times 2\times 10^3$$

$$= 800 \text{ N}$$

26. Here, ϕ = shear angle = 25°

λ = friction angle

α = rake angle = 0°

From Merchant's theory,

$$2\phi + \lambda - \alpha = 90°$$

$$\therefore \quad \lambda = 90° - 50° = 40°$$

$$\therefore \quad \mu = \frac{F}{N} = \frac{\text{Friction force}}{\text{Normal force}} = \tan\lambda = .83$$

29. Large negative rake angle, low shear angle and high cutting speed.

31. Total heat generated,

$$Q_1 = F_c \times V = 24000 \text{ J/min}$$

Heat dissipated due to friction,

$$Q_2 = F \times r \times v$$

Frictional force,

$$F = F_c \sin\alpha + F_f \cos\alpha = 503.76 \text{ N}$$

$$Q_2 = 7193.69 \text{ J/min.}$$

$$\therefore \quad \frac{Q_2}{Q_2} = 0.299 \text{ ; } 30\%$$

NUMERICAL TYPE QUESTIONS

2016

1. Because the rake angle is zero,

F = Friction force = F_c = Cutting force = 500N,

N = normal to friction force = F_t = thrust force = 250 N,

Coefficient of friction = F/N = 250/500 = 0.5

2. L = 900 mm

d = 200 mm

f = 0.25 mm/rev

v = 300 m/min

t = ?

$$v = \pi \text{ DN m/min}$$

$$N = \frac{v}{\pi D} = \frac{300}{\pi \times 0.2} = 478 \text{ rpm}$$

$$t = \frac{L}{f.N} = \frac{900}{0.25 \times 478} \approx 7.539 \text{ min}$$

3. $L_m = \frac{18C}{V}$, $C_g = \frac{270C}{TV}$, $VT^{0.25} = 150$

$$V_{opt} = V = C\left[\frac{n\,L_m}{1 - n\,C_g}\right]^n$$

$$= 150\left[\frac{0.25}{1-0.25} \times \frac{\frac{18C}{V}}{\frac{270C}{TV}}\right]^{0.25}$$

$$= 150\left[\frac{0.25}{1-0.25} \times \frac{18}{270}T\right]^{0.25}$$

$$= 150\left[\frac{1}{3} \times \frac{18}{270} \times \frac{150^4}{V^4}\right]^{0.25}$$

$$= 150(57.914) \times \left(\frac{1}{V^4}\right)^{0.25}$$

$$V = 150 \times 57.914 \times \frac{1}{V}$$

$$V^2 = 150 \times 57.914$$

$$V = 93.2 \text{ m/min}$$

4. $R = \frac{\rho L}{\text{Area}} = \frac{\frac{1}{0.02} \times 0.009}{\text{Area}} = \frac{50 \times 0.009}{\text{Area}}$

$$I = \frac{V}{R} = \frac{(12 - 1.5) \times \text{Area}}{50 \times 0.009} = 23.333 \times \text{Area}$$

$$L = 3 + 6 = 9 \text{ μm} = 0.009$$

$$MRR = \frac{AI}{\rho ZF} = \frac{55.85 \times 23.333 \times \text{Area}}{7860 \times 10^{-6} \times 2 \times 96500} = 0.98189 \times \text{Area}$$

$$\frac{MRR}{\text{Area}} = 0.8590 \text{ mm/sec} = 0.8590 \times 60 \text{ mm/min} = 51.542 \text{ mm/min}$$

5. $F = \pi dt \times Ts$

$$= \pi \times 10 \times 2 \times 80 = 5026.5 \text{ N} = 5.026 \text{ kN}$$

6. $V_2 = 0.5V_1$,

$$\frac{V_1}{V_2} = \frac{V_1}{0.5V_1} = \frac{1}{0.5} = 2$$

$$V_1 T_1^n = V_2 T_2^n$$

$$\left(\frac{T_2}{T_1}\right)^n = \left(\frac{V_1}{V_2}\right)$$

$$\frac{T_2}{T_1} = \left(\frac{V_1}{V_2}\right)^{\frac{1}{n}} = \left(\frac{V_1}{V_2}\right)^{\frac{1}{0.5}} = \left(\frac{V_1}{V_2}\right)^4 = (2)^4 = 16$$

14 CHAPTER Metrology and Inspection

MCQ TYPE QUESTIONS

2016

1. Match the following:

P. Feeler gauge	I. Radius of an object
Q. Fillet gauge	II. Diameter within limits by comparison
R. Snap gauge	III. Clearance or gap between components
S. Cylindrical plug gauge	IV. Inside diameter of straight hole

(*a*) P–III, Q–I, R–II, S–IV
(*b*) P–III, Q–II, R–I, S–IV
(*c*) P–IV, Q–II, R–I, S–III
(*d*) P–IV, Q–I, R–II, S–III

2012

2. In an interchangeable assembly, shafts of size $25.000^{+0.040}_{-0.010}$ *mm* mate with holes of size $25.000^{+0.030}_{-0.020}$ *mm*. The maximum interference (in *microns*) in the assembly is

(*a*) 40 (*b*) 30
(*c*) 20 (*d*) 10

2011

3. A hole is of dimension $\phi 9^{+0.015}_{+0}$ mm. The corresponding shaft is of dimension $\phi 9^{+0.010}_{+0.001}$ mm. The resulting assembly has

(*a*) loose running fit (*b*) close running fit
(*c*) transition fit (*d*) interference fit

2010

4. A shaft has a dimension, $\phi 35^{-0.009}_{-0.025}$. The respective values of fundamental deviation and tolerance are

(*a*) –0.025, ±0.008 (*b*) –0.025, ±0.016
(*c*) –0.009, ±0.008 (*d*) –0.009, ±0.016

2009

5. What are the upper and lower limits of the shaft represented by 60 f_8 ?

Use the following data:

Diameter 60 lies in the diameter step of 50-80 mm.

Fundamental tolerance unit, i, in μm

$$= 0.45\, D^{1/3} + 0.001\, D$$

where D is the representative size in mm;

Tolerance value for IT8 = 25i,

Fundamental deviation for 'f' shaft = – 5.5$D^{0.41}$

(*a*) Lower limit = 59.924 mm,
Upper limit = 59.970 mm
(*b*) Lower limit = 59.954 mm,
Uppr limit = 60.000 mm
(*c*) Lower limit = 59.970 mm,
Upper limit = 60.016 mm
(*d*) Lower limit = 60.00 mm,
Upper limit = 60.046 mm

2007

6. A hole is specified as $40^{0.050}_{0.000}$ mm. The mating shaft has a clearance fit with minimum clearance of 0.01 mm.The tolerance on the shaft is 0.04 mm. The maximum clearance in mm between the hole and the shaft is

(*a*) 0.04 (*b*) 0.05
(*c*) 0.10 (*d*) 0.11

2006

7. A ring gage is used to measure

(*a*) outside diameter but not roundness
(*b*) roundness but not outside diameter
(*c*) both outside diameter and roundness
(*d*) only external threads

NUMERICAL TYPE QUESTIONS

2016

1. Two optically flat plates of glass are kept at a small angle θ as shown in the figure. Monochromatic light is incident vertically.

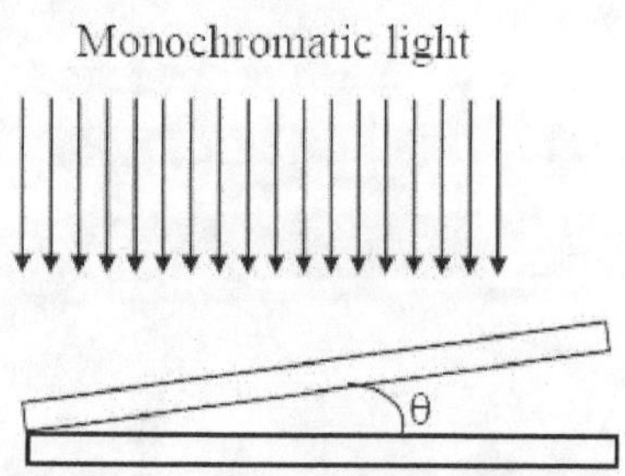

If the wavelength of light used to get a fringe spacing of 1 mm is 450 nm, the wavelength of light (in nm) to get a fringe spacing of 1.5 mm is ______.

ANSWERS

MCQ Type Questions

1. (*a*) **2.** (*c*) **3.** (*c*) **4.** (*d*) **5.** (*a*) **6.** (*c*) **7.** (*c*)

Numerical Type Questions

1. (675)

EXPLANATIONS

MCQ TYPE QUESTIONS

1. Feeler gauge is used for checking the clearance or gap between the parts, radius is checked by fillet gauge, limit of diameter of shaft is checked by snap gauge and plug gauge is used for checking the diameter of hole.

2. Maximum interference

= Maximum size of shaft – Minimum size of hole

= 25.040 – 25.020 = 20 μm

4. Given : $\phi\ 35^{-0.009}_{-0.025}$

Tolerance = Upper deviation – Lower deviation

= 0.025 – 0.009 = 0.016

Fundamental deviation for basic shaft is lower deviation = – 0.009.

5.

$$D = \sqrt{50 \times 80} = 63.25 \text{ mm}$$

$$i = 0.45(D)^{1/3} + .001\, D\ \mu\text{m}$$

$$= 1.856 \times 10^{-6}\text{ m}$$

$$\text{Tolerance} = 25i = 46.4 \times 10^{-6}\text{ m}$$

$$\text{Fundamental deviation} = -(5.5)\, D^{0.41}$$

$$= -30.113 \times 10^{-6}$$

$$= 0.030 \text{ mm (absolute)}$$

High limit = Basic size – Fundametal deviation

= 60 – 0.030 = 59.97

Lower limit = High limit –Tolerance

= 59.77 – 0.0464

= 59.924mm

6. Minimum clearance

= Minimum hole size – Maximum shaft size

∴ Maximum shaft size = 40.00 – .01 = 39.99 mm

Tolerance on shaft = .04 mm

Minimum shaft size = 39.99 – .04 = 39.95

Maximum clearance

= Maximum hole size – Minimum shaft size

= 40.05 – 39.95 = .10

7. Ring gauge is used to measure both outside diameter and roundness.

NUMERICAL TYPE QUESTIONS

1. Frindge space (d_1) = 1mm

$$\lambda_1 = 450 \text{ mm}$$

$$d_2 = 1.5 \text{ mm},\ \lambda_2 = ?$$

$$\tan\theta = \frac{\lambda_1}{2d_1} = \frac{\lambda_2}{2d_2}$$

$$\lambda_2 = 450 \times \left(\frac{1.5}{1}\right)$$

$$\lambda_2 = 675 \text{ mm}$$

■■

15 CHAPTER

Computer Integrated Manufacturing

MCQ TYPE QUESTIONS

2016

1. The figure below represents a triangle PQR with initial coordinates of the vertices as P(1, 3), Q(4, 5) and R(5, 3.5). The triangle is rotated in the X-Y plane about the vertex P by angle θ in clockwise direction. If sinθ = 0.6 and cosθ = 0.8, the new coordinates of the vertex Q are

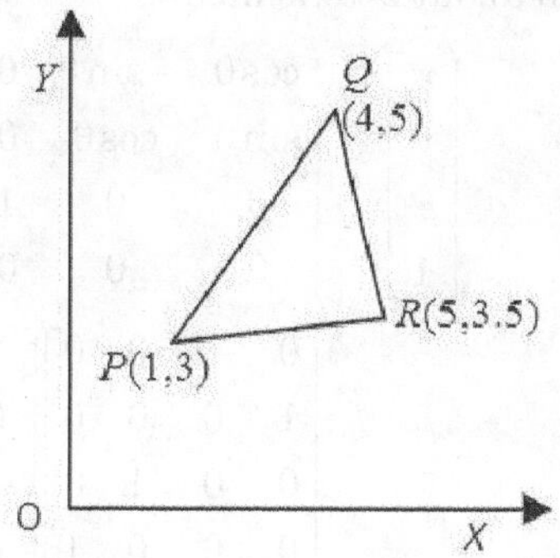

(*a*) (4.6, 2.8)
(*b*) (3.2, 4.6)
(*c*) (7.9, 5.5)
(*d*) (5.5, 7.9)

2. Match the following part programming codes with their respective functions

Part Programming Codes	Functions
P. G01	I. Spindle stop
Q. G03	II. Spindle rotation, clockwise
R. M03	III. Circular interpolation, anticlockwise
S. M05	IV. Linear interpolation

(*a*) P - II, Q - I, R - IV, S - III
(*b*) P - IV, Q - II, R - III, S - I
(*c*) P - IV, Q - III, R - II, S - I
(*d*) P - III, Q - IV, R - II, S - I

3. A point P (1, 3, – 5) is translated by $2\hat{i} + 3\hat{j} - 4\hat{k}$ and then rotated counter clockwise by 90° about the *z*-axis. The new position of the point is

(*a*) (– 6, 3, – 9) (*b*) (– 6, – 3, – 9)
(*c*) (6, 3, – 9) (*d*) (6, 3, 9)

2010

4. In a CNC program block, N002 G02 G91 X40 Z40..., G02 and G91 refer to

(*a*) circular interpolation in counterclockwise direction and incremental dimension
(*b*) circular interpolation in counterclockwise direction and absolute dimension
(*c*) circular interpolation in clockwise direction and incremental dimension
(*d*) circular interpolation in clockwise direction and absolute dimension

2007

5. Which type of motor is **NOT** used in axis or spindle drives of CNC machine tools ?

(*a*) Induction motor
(*b*) DC servo motor
(*c*) Stepper motor
(*d*) Linear servo motor

2006

6. NC contouring is an example of

(*a*) continuous path positioning
(*b*) point-to-point positioning
(*c*) absolute positioning
(*d*) incremental positioning

ANSWERS

MCQ Type Questions

1. (*a*) **2.** (*c*) **3.** (*a*) **4.** (*b*) **5.** (*a*) **6.** (*a*)

EXPLANATIONS

MCQ TYPE QUESTIONS

1.

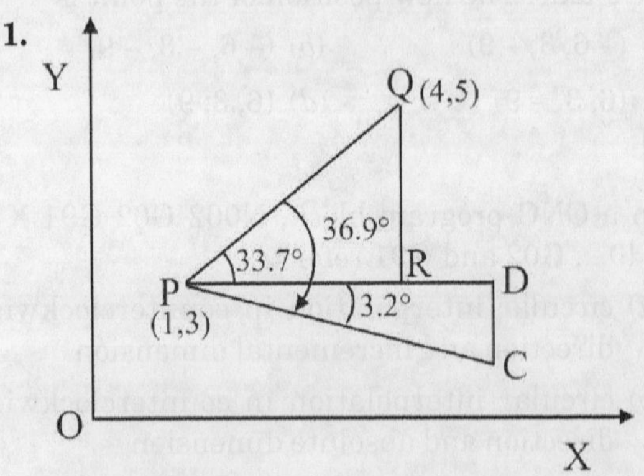

$PQ = \sqrt{2^2 + 3^2} = 3.6055 = PC$

$PD = PC \times \cos 3.2 = 3.6$

X co-ordinate of point C = 1 + 3.6 = 4.6

DC = 3.6 sin 3.2 = 0.2

Y co-ordinate of point C = 3.0 – 0.2 = 2.8

2. G01 is used for linear interpolation , G03 is used for circular interpolation counter clockwise, M03 for spindle rotation clockwise and M05 for spindle top.

3. "P" after translation = (1 + 2, 3 + 3, –5 –4)
= (3, 6, –9)

Rotation about z- axis means

$$\begin{bmatrix} x' \\ y' \\ z' \\ 1 \end{bmatrix} = \begin{bmatrix} \cos\theta & -\sin\theta & 0 & 0 \\ \sin\theta & \cos\theta & 0 & 0 \\ 0 & 0 & 1 & 0 \\ 0 & 0 & 0 & 1 \end{bmatrix} \begin{bmatrix} x \\ y \\ z \\ 1 \end{bmatrix}$$

$$= \begin{bmatrix} 0 & -1 & 0 & 0 \\ 1 & 0 & 0 & 0 \\ 0 & 0 & 1 & 0 \\ 0 & 0 & 0 & 1 \end{bmatrix} \begin{bmatrix} 3 \\ 6 \\ -9 \\ 1 \end{bmatrix}$$

$$= \begin{bmatrix} 0-6+0+0 \\ 3+0+0+0 \\ 0+0-9+0 \\ 0+0+0+1 \end{bmatrix} \begin{bmatrix} -6 \\ 3 \\ -9 \\ 1 \end{bmatrix}$$

Final point = [–6, 3, –9]

4. N002 → circular interpolation in clockwise direction

G91 → Incremental dimension

6. NC contouring is an example of continuous path positioning.

■■

16 CHAPTER

Production, Planning and Control

MCQ TYPE QUESTIONS

2015

1. The function of interpolator in a CNC machine controller is to

(*a*) control spindle speed

(*b*) coordinate feed rates of axes

(*c*) control tool rapid approach speed

(*d*) perform Miscellaneous (M) functions (tool change, coolant control etc.)

2. Holes of diameter $25.0^{+0.040}_{+0.020}$ mm are assembled interchangeably with the pins of diameter $25.0^{+0.005}_{-0.008}$ mm. The minimum clearance in the assembly will be

(*a*) 0.048 mm (*b*) 0.015 mm

(*c*) 0.005 mm (*d*) 0.008 mm

3. Match the following products with preferred manufacturing processes:

	Product	Process
P	Rails	(1) Blow molding
Q	Engine crankshaft	(2) Extrusion
R	Aluminium channels	(3) Forging
S	PET water bottles	(4) Rolling

(*a*) P – 4, Q – 3, R – 1, S – 2

(*b*) P – 4, Q – 3, R – 2, S – 1

(*c*) P – 2, Q – 4, R – 3, S – 1

(*d*) P – 3, Q – 4, R – 2, S – 1

4. In a linear arc welding process, the heat input per unit length is inversely proportional to

(*a*) welding current

(*b*) welding voltage

(*c*) welding speed

(*d*) duty cycle of the power source

5. The probability of obtaining at least two "SIX" in throwing a fair dice 4 time is

(*a*) 425/432 (*b*) 19/144

(*c*) 13/144 (*d*) 125/432

6. The solidification time of a casting is proportional to $\left(\frac{V}{A}\right)^2$, where V is the volume of the casting and A is the total casting surface area losing heat. Two cubes of same material and size are cast using sand casting process. The top face of one of the cubes is completely insulated. The ratio of the solidification time for the cube with top face insulated to that of the other cube is

(*a*) $\frac{25}{36}$ (*b*) $\frac{36}{25}$

(c) 1 (*d*) $\frac{6}{5}$

7. In the assembly shown below, the part dimensions are:

$L_1 = 22.0^{\pm 0.01}$ mm, $L_2 = L_3 = 10.0^{\pm 0.005}$ mm,

Assuming the normal distribution of part dimensions, the dimension L_4 in mm for assembly condition would be:

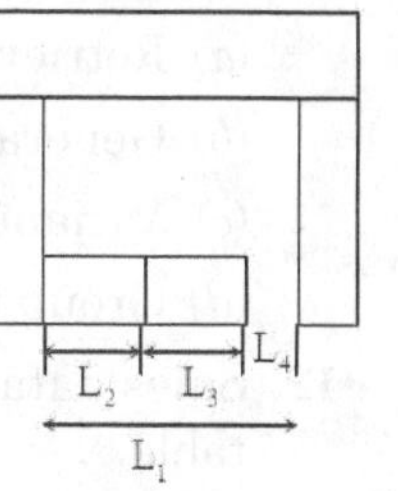

(*a*) $2.0^{\pm 0.008}$

(*b*) $2.0^{\pm 0.012}$

(*c*) $2.0^{\pm 0.016}$

(*d*) $2.0^{\pm 0.020}$

8. A mobile phone has a small motor with an eccentric mass used for vibrator mode. The location of the eccentric mass on motor with respect to center of gravity (CG) of the mobile and the rest of the dimensions of the mobile phone are shown. The mobile is kept on a flat horizontal surface.

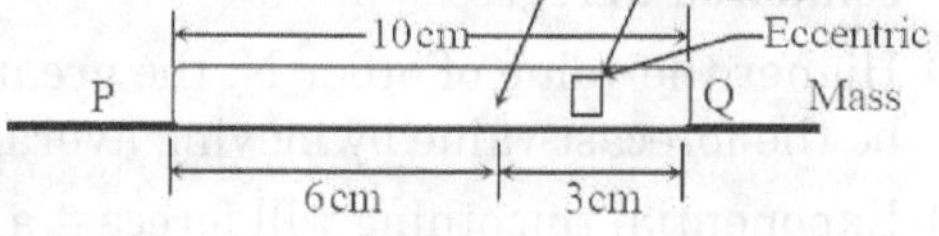

Given in addition that the eccentric mass = 2 grams, eccentricty = 2.19 mm, mass of the mobile = 90 grams, g = 9.81 m/s². Uniform speed of the motor in RPM for which the mobile will get just lifted off the ground at the end Q is approximately

(*a*) 3000 (*b*) 3500

(*c*) 4000 (*d*) 4500

9. Following data refers to the activities of a project, where, node 1 refers to the start and node 5 refers to the end of the project

Activity	Duration(days)
1-2	2
2-3	1
4-3	3
1-4	3
2-5	3
3-5	2
4-5	4

The critical path (CP) in the network is

(*a*) 1-2-3-5 (*b*) 1-4-3-5

(*c*) 1-2-3-4-5 (*d*) 1-4-5

10. The primary mechanism of material removal in electrochemical machining (ECM) is

(*a*) Chemical corrosion (*b*) etching

(*c*) ionic dissolution (*d*) spark erosion

11. During the development of a product an entirely new process plan is made based on design logic, examination of geometry and tolerance information. This type of process planning is known as

(*a*) Retrieval

(*b*) Generative

(*c*) Variant

(*d*) Group technology based

12. Sales data of a product is given in the following table:

Month	January	February	March	April	May
No. of unit Sold	10	11	16	19	25

Regarding forecast for the month of June, which one of the following statements is **TRUE** ?

(*a*) Moving average will forecast a higher value compared to regression

(*b*) Higher the value of order N, the greater will be the forecast value by moving average.

(*c*) Exponential smoothing will forecast a higher value compared to regression.

(*d*) Regression will forecast a higher value compared to moving average

13. At least one eigen value of a singular matrix is

(*a*) positive

(*b*) zero

(*c*) Negative

(*d*) imaginary

14. Which one of the following statement is True?

(*a*) the 'GO' gage represents lower control limit in hole

(*b*) The 'NO' gage controls the lower limit of a shaft

(*c*) The 'GO' gage controls the lower limit of a hole

(*d*) The 'NO GO' gage controls the lower limit of a hole

15. The chance of a student passing in an exam is 20%. The chance of a student passing in a exam and getting above 90% marks in it is 5%. Given that student passes the examination the probability that the student gets above 90% marks is

(*a*) 1/18 (*b*) 1/4

(*c*) 2/9 (*d*) 5/18

16. In a two-stage wire drawing operation, the fractional reduction (ratio of change in cross-sectional area to initial cross-sectional area) in the first stage is 0.4. The fractional reduction in the second stage is 0.3. The overall fractional reduction is

(*a*) 0.24 (*b*) 0.58

(*c*) 0.60 (*d*) 1.00

17. During a TIG welding process, the are current and are voltage were 50 A and 60 V, respectively, when the welding speed was 150 mm/min. In another process, the TIG welding is carried out at a welding speed of 120 mm/min at the same arc voltage and heat input to the material so that weld quality remains the same. The welding current (in A) for this process is

(*a*) 40.00 (*b*) 44.72

(*c*) 55.90 (*d*) 62.65

18. A resistance-capacitance relaxation circuit is used in an electrical discharge machining process. The discharge voltage is 100 V. At a spark cycle time of μs, the average power input required is 1 kW. The capacitance (in μF) in the circuit is.

(*a*) 2.5 (*b*) 5.0

(*c*) 7.5 (*d*) 10.0

19. Which two of the following joining processes are autogenous?

i. Diffusion welding

ii. Electroslag welding

iii. Tungsten inert gas welding

iv. Friction welding

(*a*) *i* and *iv* (*b*) *ii* and *iii*

(*c*) *ii* and *iv* (*d*) *i* and *iii*

20. In a machining operation, if the generatrix and directix both are straight lines, the surface obtained

(*a*) cylindrical (*b*) helical

(*c*) plane (*d*) surface of revolution

21. In full mould (cavity-less) casting process, the pattern is made of

(*a*) expanded polystyrene

(*b*) wax

(*c*) epoxy

(*d*) plaster of Paris

22. Which of the following statements are TRUE for damped vibrations?

P. For a system having critical damping, the value of damping ratio is unity and system does not undergo a vibratory motion.

Q. Logarithmic decrement method is used to determine the amount do damping in a physical system.

R. In case of damping due to dry friction between moving surfaces resisting force of constant magnitude acts opposite to the relative motion.

S. For the case of viscous damping, drag force is directly proportional to the square of relative velocity.

(*a*) P and Q only (*b*) P and S only

(*c*) P, Q and R only (*d*) Q and S only

23. In the notation (*a*/*b*/*c*) : (*d*/*e*/*f*) for summarizing the characteristics of queueing situation, the letters '*b*' and '*d*' stand respectively for

(*a*) service time distribution and queue discipline

(*b*) number of servers and size of calling source

(*c*) number of servers and queue discipline

(*d*) service time distribution and maximum number allowed in system

24. A drill is positioned at point P and its has to proceed to point Q. The coordinates of point Q in the incremental system of defining position of a point in CNC part program will be

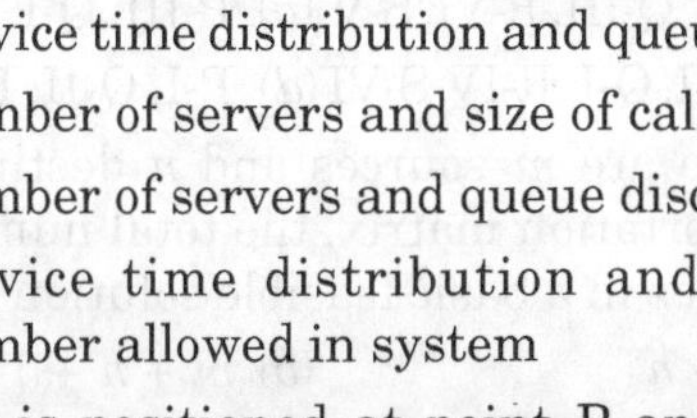

(*a*) (3, 12) (*b*) (5, 7)

(*c*) (7, 12) (*d*) (4, 7)

25. In a CNC milling operation, the tool has to machine the circular arc from point (20, 20) to (10, 10) at sequence number 5 of the CNC part program. If the center of the asrc is at (20, 10) and the machine has incremental mode of defining position coordinates, the correct tool path command is

(*a*) N 05 G 90 G01 X–10 Y–10 R10

(*b*) N 05 G91 G03 X–10 Y–10 R10

(*c*) N 05 G90 G03 X20 Y20 R10

(*d*) N 05 G91 G02 X20 Y20 R10

26. Orthogonal turning of a mild steel tube with a tool of rake angle 10° carried out at a feed of 0.14 mm/rev. If the thickness of the chip produced is 0.28 mm, the values of shear angle and shear strain will be respectively

(*a*) 28°20′ and 2.19 (*b*) 22°20′ and 3.53

(*c*) 24°30′ and 3.53 (*d*) 37°20′ and 5.19

27. A shaft of length 90 mm has a tapered portion of length 55 mm. The diameter of the taper is 80 mm at one end and 65 mm at the other. If the taper is made by tailstock set over method, the taper angle and the set over respectively are

(*a*) 15°32′ and 12.16 mm

(*b*) 15°32′ and 15.66 mm

(*c*) 11°22′ and 10.26 mm

(*d*) 10°32′ and 14.46 mm

2014

28. For machining a rectangular island represented by coordinates P(0,0), Q(100,0) R(100,50) and S(0,50) On a casting using CNC milling machine, an end mill with a diameter of 16 mm is used. The trajectory of the cutter centre to machine the island PQRS is

(*a*) (–8, –8), (108, –8), (108, 58), (–8, 58), (–8,–8)

(*b*) (8, 8), (94, 8),(94, 44), (8, 44), (8, 8)

(*c*) (–8, 8), (94, 0), (94, 44), (8, 44),(–8, 8)

(*d*) (0, 0), (100, 0), (100, 50), (50, 0), (0, 0)

29. Which one of the following instruments is widely used to check and calibrate geometric features of machine tools during their assembly?

(*a*) Ultasonic probe

(*b*) Coordinate Measuring Machine (CMM)

(*c*) Laser interferometer

(*d*) Vernier calipers

30. The majoe difficulty during welding of a aluminium is due to its

(*a*) high tendency of oxidation

(*b*) high thermal conductivity

(*c*) low melting point

(*d*) low density

31. The main cutting force acting on a tool during the turning (othogonal cutting) operation of a metal is 400 N. The turning was performed using 2 mm depth of cut and 0.1 mm/rev feed rate. The specific cutting pressure (in N/mm^2) is

(*a*) 1000 (*b*) 2000
(*c*) 3000 (*d*) 4000

32. The process of reheating the martensitic steel to reduce its brittleness without any significant loss in its hardness is

(*a*) normalising (*b*) annealing
(*c*) quenching (*d*) tempering

33. In solid-state welding, the contamination layers between the surface to be welded are removed by

(*a*) alcohol (*b*) plastic deformation
(*c*) water jet (*d*) sand blasting

34. For the given assembly: 25 H7/g8, match Group A with Group B

Group A	Group B
P. H	I. Shaft Type
Q. IT8	II. Hole Type
R. IT7	III. Hole Tolerance Grade
S. g	IV. Shaft Tolerance Grade

Codes

	P	Q	R	S
(*a*)	I	III	IV	II
(*b*)	I	IV	III	II
(*c*)	II	III	IV	I
(*d*)	II	IV	III	I

35. Jobs arrive at a facility at an average rate of 5 in an 8 hour shift. The arrival of the jobs follows Poisson distribution. The average service time of a job on the facility is 40 minutes. The service time follows exponential distribution. Idle time (in hours) at the facility per shift will be

(*a*) $\frac{5}{7}$ (*b*) $\frac{14}{3}$
(*c*) $\frac{7}{5}$ (*d*) $\frac{10}{3}$

36. The flatness of a machine bed can be measured using

(*a*) Vernier calipers
(*b*) Auto collimator
(*c*) Height gauge
(*d*) Tool maker's microscope

37. A robot arm PQ with end coordinates P(0,0) and Q(2,5) rotates counter clockwise about P in the XY plane by 90°. The new coordinate pair of the end point Q is

(*a*) (–2, 5) (*b*) (–5, 2)
(*c*) (–5, –2) (*d*) (2, –5)

38. Match the Machine Tools (Group A) with the probable Operations (Group B):

Group A	Group B
(P) Centre Lathe	(1) Slotting
(Q) Milling	(2) Counter-boring
(R) Grinding	(3) Knurling
(S) Drilling	(4) Dressing

(*a*) P-1, Q-2, R-4, S-3 (*b*) P-2, Q-1, R-4, S-3
(*c*) P-3, Q-1, R-4, S-2 (*d*) P-3, Q-4, R-2, S-1

39. The following four unconventional machining processes are available in a shop floor. The most appropriate one to drill a hole of square cross section of 6 mm × 6 mm and 25 mm deep is

(*a*) Abrasive Jet Machining
(*b*) Plasma Arc Machining
(*c*) Laser Beam Machining
(*d*) Electro Discharge Machining

40. With respect to metal working, match Group A with Group B:

Group A	Group B
(P) Defect in extrusion	(I) Alligatoring
(Q) Defect in rolling	(II) Scab
(R) Product of skew rolling	(III) Fish tail
(S) Product of rolling through cluster mill	(IV) Seamless tube
	(V) Thin sheet with tight to tolerance
	(VI) Semi-finished balls of ball bearing

(*a*) P-II, Q-III, R-VI, S-V (*b*) P-III, Q-I, R-VI, S-V
(*c*) P-III, Q-I, R-IV, S-VI (*d*) P-I, Q-II, R-V, S-VI

41. If there are m sources and n destinations in a transportation matrix, the total number of basic variables in a basic feasible solution is

(*a*) $m + n$ (*b*) $m + n + 1$
(*c*) $m + n - 1$ (*d*) m

42. A component can be produced by any of the four processes I, II, III and IV. The fixed cost and the variable cost for each of the processes are listed below. The most economical process for producing a batch of 100 pieces is

Process	Fixed cost (in ₹)	Variable cost per piece (in ₹)
I	20	3
II	50	1
III	40	2
IV	10	4

(*a*) I (*b*) II
(*c*) III (*d*) IV

43. For the CNC part programming, match Group A with Group B:

Group A	Group B
(P) Circular Interpolation, counter clock wise	(I) G02
(Q) Dwell	(II) G03
(R) Circular interpolation, clockwise	(III) G04
(S) Point to point countering	(IV) G00

(*a*) P-II, Q-III, R-I, S-IV (*b*) P-I, Q-III, R-II, S-IV
(*c*) P-I, Q-IV, R-II, S-III (*d*) P-II, Q-I, R-III, S-IV

44. A hole of 20 mm diameter is to be drilled in a steel block of 40 mm thickness. The drilling is performed at rotational speed of 400 rpm and feed rate of 0.1 mm/rev. The required approach and over run of the drill together is equal to the radius of drill. The drilling time (in minute) is

(*a*) 1.00 (*b*) 1.25
(*c*) 1.50 (*d*) 1.75

45. A rectangular hole of size 100 mm × 50 mm is to be made on a 5 mm thick sheet of steel having ultimate tensile strength and shear strength of 500 MPa and 300 MPa, respectively. The hole is made by punching process. Neglecting the effect of clearance, the punching force (in kN) is

(*a*) 300 (*b*) 450
(*c*) 600 (*d*) 750

46. A minimal spanning tree in network flow models involves

(*a*) All the nodes with cycle/loop allowed
(*b*) All the nodes with cycle/loop not allowed
(*c*) Shortest path between start and end nodes
(*d*) All the nodes with directed arcs

47. Match the casting defects (Group A) with the probable causes (Group B):

Group A	Group B
P. Hot tears	1. Improper fusion of two streams of liquid metal
Q. Shrinkage	2. Low permeability of the sand mould
R. Blow holes	3. Volumetric contraction both in liquid and solid stage
S. Cold Shut	4. Differential cooling rate

Codes

	P	Q	R	S
(*a*)	1	3	2	4
(*b*)	4	3	2	1
(*c*)	3	4	2	1
(*d*)	1	2	4	3

48. Cutting tool is much harder than the workpiece. Yet the tool wears out during the tool-work interaction, because

(*a*) extra hardness is imparted to the workpiece due to coolant used
(*b*) oxide layers on the workpiece surface impart extra hardness to it
(*c*) extra hardness is imparted to the workpiece due to severe rate of strain
(*d*) vibration is induced in the machine tool

49. The hot tearing in a metal casting is due to

(*a*) high fluidity
(*b*) high melt temperature
(*c*) wide range of solidification temperature
(*d*) low coefficient of thermal expansion

50. Consider the given project network, where numbers along various activities represent the normal time. The free float on activity 4-6 and the project duration, respectively, are

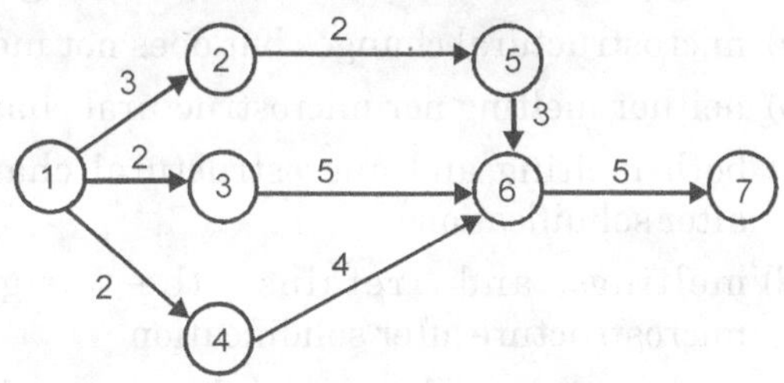

(*a*) 2, 13 (*b*) 0, 13
(*c*) –2, 13 (*d*) 2, 12

51. A manufacturer can produce 12000 bearings per day. The manufacturer received an order of 8000 bearings per day from a customer. The cost of holding a bearing in stock is ₹ 0.20 per month. Setup cost per production run is ₹ 500. Assuming 300 working days in a year, the frequency of production run should be

(*a*) 4.5 days (*b*) 4.5 months
(*c*) 6.8 days (*d*) 6.8 months

52. A cylindrical blind riser with diameter d and height h, is placed on the top of the mold cavity of a closed type sand mold as shown in the figure. If the riser is of constant volume, then the rate of solidification in the riser is the least when the ratio $h : d$ is

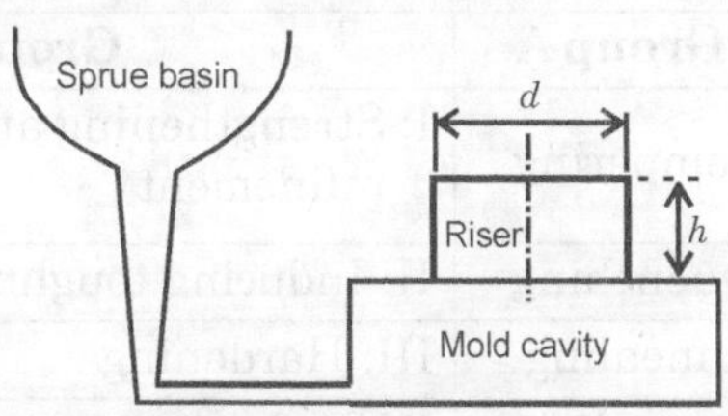

(*a*) 1:2 (*b*) 2:1
(*c*) 1:4 (*d*) 4:1

53. Which pair of following statements is correct for orthogonal cutting using a single-point cutting tool?

P. Reduction in friction angle increases cutting force

Q. Reduction in friction angle decreases cutting force

R. Reduction in friction angle increases chip thickness

S. Reduction in friction angle decreases chip thickness

(*a*) P and R (*b*) P and S

(*c*) Q and R (*d*) Q and S

54. The total number of decision variables in the objective function of an assignment problem of size $n \times n$ (n jobs and n machines) is

(*a*) n^2 (*b*) $2n$

(*c*) $2n - 1$ (*d*) n

55. Within the Heat Affected Zone (HAZ) in a fusion welding process, the work material undergoes

(*a*) microstructural changes but does not melt

(*b*) neither melting nor microstructural changes

(*c*) both melting and microstructural changes after solidification

(*d*) melting and retains the original microstructure after solidification

56. The principle of material removal in Electrochemical machining is

(*a*) Fick's law (*b*) Faraday's laws

(*c*) Kirchhoff's laws (*d*) Ohm's law

57. Better surface finish is obtained with a large rake angle because

(*a*) the area of shear plane decreases resulting in the decrease in shear force and cutting force

(*b*) the tool becomes thinner and the cutting force is reduced

(*c*) less heat is accumulated in the cutting zone

(*d*) the friction between the chip and the tool is less

58. Match the heat treatment processes (Group A) and their associated effects on properties (Group B) of medium carbon steel:

Group A	Group B
P: Tempering	I: Strengthening and grain refinement
Q: Quenching	II: Inducing toughness
R: Annealing	III: Hardening
S: Normalizing	IV: Softening

Codes:

	P	Q	R	S
(*a*)	III	IV	II	I
(*b*)	II	III	IV	I
(*c*)	III	II	IV	I
(*d*)	II	III	I	IV

59. In a rolling process, the maximum possible draft, defined as the difference between the initial and the final thickness of the metal sheet, mainly depends on which pair of the following parameters

P : Strain

Q : Strength of the work material

R : Roll diameter

S : Roll velocity

T : Coefficient of friction between roll and work

(*a*) Q, S (*b*) R, T

(*c*) S, T (*d*) P, R

60. The precedence relations and duration (in days) of activities of a project network are given in the table. The total float (in days) of activities *e* and *f*, respectively, are

Activity	Predecessors	Duration (days)
a	-	2
b	-	4
c	*a*	2
d	*b*	3
e	*c*	2
f	*c*	4
g	*d,e*	5

(*a*) 0 and 4 (*b*) 1 and 4

(*c*) 2 and 3 (*d*) 3 and 1

61. A GO-No GO plug gauge is to be designed for measuring a hole of nominal diameter 25 mm with a hole tolerance of ± 0.015 mm. Considering 10% of work tolerance to be the gauge tolerance and no wear condition, the dimension (in mm) of the GO plug gauge as per the unilateral tolerance system is

(*a*) $24.985^{+0.003}_{-0.003}$ (*b*) $25.015^{+0.000}_{-0.006}$

(*c*) $24.985^{+0.003}_{-0.003}$ (*d*) $24.985^{+0.003}_{-0.000}$

2011

62. The word Kanban is most appropriately associated with

(*a*) economic order quantity

(*b*) just-in-time production

(*c*) capacity planning

(*d*) product design

63. Which one of the following is **NOT** a decision taken during the aggregate production planning stage?

(*a*) Scheduling of machines
(*b*) Amount of labour to be committed
(*c*) Rate at which production should happen
(*d*) Inventory to be carried forward

2010

64. The demand and forecast for February are 12000 and 10275, respectively. Using single exponential smoothening method (smoothening coefficient = 0.25), forecast for the month of March is

(*a*) 431 (*b*) 9587
(*c*) 10706 (*d*) 11000

65. Vehicle manufacturing assembly line is an example of

(*a*) product layout (*b*) process layout
(*c*) manual layout (*d*) fixed layout

2009

66. Which of the following forecasting methods takes a fraction of forecast error into account for the next period forecast ?

(*a*) simple average method
(*b*) moving average method
(*c*) weighted moving average method
(*d*) exponential smoothening method

67. Six jobs arrived in a sequence as given below:

Jobs	Processing Time (days)
I	4
II	9
III	5
IV	10
V	6
VI	8

Average flow time (in days) for the above jobs using Shortest Processing time rule is

(*a*) 20.83 (*b*) 23.16
(*c*) 125.00 (*d*) 139.00

2008

68. In an M/M/1 queuing system, the number of arrivals in an interval of length T is a Poisson random variable (i.e. the probability of there being n arrivals in an interval of length T is $\frac{e^{-lT}(lT)^n}{n!}$). The probability density function f(t) of the inter-arrival time is given by

(*a*) $\lambda^2\left(e^{-\lambda^2 t}\right)$ (*b*) $\frac{e^{-\lambda^2 t}}{\lambda^2}$
(*c*) $\lambda e^{-\lambda t}$ (*d*) $\frac{e^{-\lambda t}}{\lambda}$

69. The product structure of an assembly P is shown in the figure.

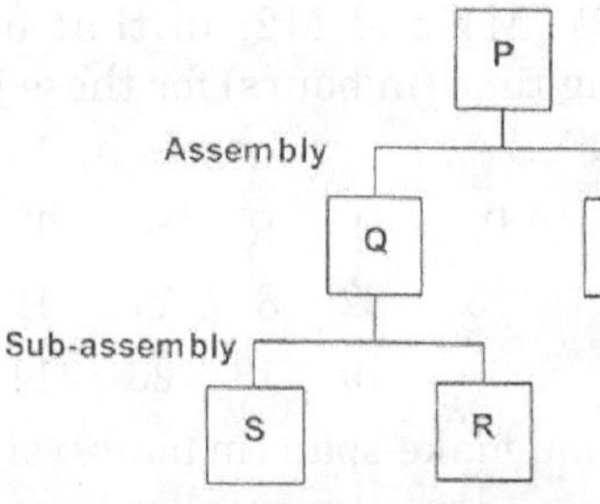

Estimated demand for end product P is as follows

Week	1	2	3	4	5	6
Demand	1000	1000	1000	1000	1200	1200

ignore lead times for assembly and sub-assembly. Production capacity (per week) for component R is the bottleneck operation. Starting with zero inventory, the smallest capacity that will ensure a feasible production plan up to week 6 is

(*a*) 1000 (*b*) 1200
(*c*) 2200 (*d*) 2400

2007

70. Capacities of production of an item over 3 consecutive months in regular time are 100, 100 and 80 and in overtime are 20, 20 and 40. The demands over those 3 months are 90, 130 and 110. The cost of production in regular time and overtime are respectively Rs. 20 per item and Rs. 24 per item. Inventory carrying cost is Rs. 2 per item per month. The levels of starting and final inventory are nil. Backorder is not permitted. For minimum cost of plan, the level of planned production in overtime in the third month is

(*a*) 40 (*b*) 30
(*c*) 20 (*d*) 0

71. In a machine shop, pins of 15 mm diameter are produced at a rate of 1000 per month and the same is consumed at a rate of 500 per month. The production and consumption continue simultaneously till the maximum inventory is reached. Then inventory is allowed to reduced to zero due to consumption. The lot size of production is 1000. If backlog is not allowed, the maximum inventory level is

(*a*) 400 (*b*) 500
(*c*) 600 (*d*) 700

2006

72. In an MRP system, component demand is

(*a*) forecasted
(*b*) established by the master production schedule
(*c*) calculated by the MRP system from the master production schedule
(*d*) ignored

73. A manufacturing shop processes sheet metal jobs, wherein each job must pass through two machines (M1 and M2, in that order). The processing time (in hours) for these jobs is

Machine	Jobs					
	P	Q	R	S	T	U
M1	15	32	8	27	11	16
M2	6	19	13	20	14	7

The optimal make-span (in hours) of the shop is

(*a*) 120 (*b*) 115

(*c*) 109 (*d*) 79

74. The table gives details of an assembly line.

Work station	I	II	III	IV	V	VI
Total task time at the work-station (in minutes)	7	9	7	10	9	6

What is the line efficiency of the assembly line?

(*a*) 70% (*b*) 75%

(*c*) 80% (*d*) 85%

NUMERICAL TYPE QUESTIONS

2016

1. The annual demand for an item is 10,000 units. The unit cost is Rs. 100 and inventory carrying charges are 14.4% of the unit cost per annum. The cost of one procurement is Rs. 2000. The time between two consecutive orders to meet the above demand is _____ month(s).

2015

2. Under certain cutting conditions, doubling the cutting speed reduces the tool life to $\left(\frac{1}{16}\right)^{th}$ of the original. Taylor's tool life index(n) for this tool-workpiece combination will be _______

3. A horizontal plate has been joined to a vertical post using four rivets arranged as shown in figure. The magnitude of the load on the worst loaded rivet (in N) is

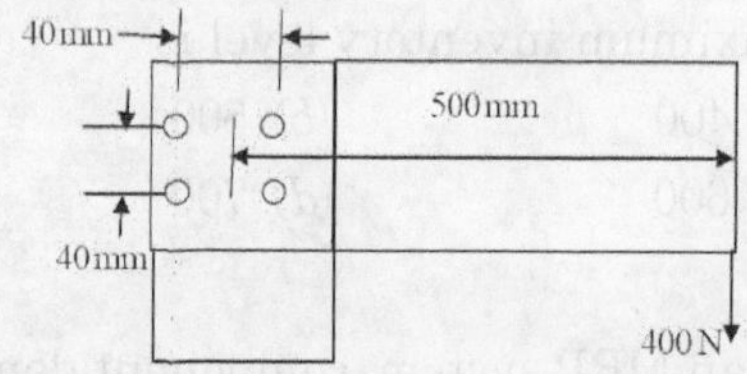

4. For a canteen, the actual demand for disposable cups was 500 units in January and 600 units in February. The forecast for the month of January was 400 units. The forecast for the month of March considering smoothing coefficient as 0.75 is _______.

5. In a slab rolling operation, the maximum thickness reduction (Δh_{max}) is given by (Δh_{max}) μ^2R, where R is the radius of the roll and is the coefficient of friction between the roll and the sheet. If μ = 0.1, the maximum angle subtended by the deformation zone at the centre of the roll (bite angle in degree) is ___.

6. A DC welding power source has a linear voltage current (V-I) characteristic with open circuit voltage of 80 V and a short circuit current of 300 A. For maximum arc power, the current (in Amperes) should be set as _____.

7. A precision instrument package (m = 1 kg) needs to be mounted on a surface vibrating at 60 Hz. It is desired that only 5% of the base surface vibration amplitude be transmitted to the instrument. Assuming that the isolation is designed with its natural frequency significantly lesser than 60 Hz, so that the effect of damping may be ignored. The stiffness (in N/m) of the required mounting pad is _____.

8. An orthogonal turning operation is carried out under the following conditions: rake angle = 5°, spindle rotational speed = 400 rpm; axial feed = 0.4 m/min and radial depth of cut = 5 mm. The chip thickness t_c, is found to be 3 mm. The shear angle (in degrees) in this turning process is _______.

9. The atomic packing factor for a material with body centered cubic structure is _______

10. There vendors were asked to supply a very high precision component. The respective probabilities of their meeting the strict design specifications are 0.8, 0.7 and 0.5. Each vendor supplies one component. The probability that out of total three components supplied by the vendors, at least one will meet the design specifications is _______.

11. A single point cutting tool with 0^0 rake angle is used in an orthogonal machining process. At a cutting speed of 180 m/min, the thrust force is 490N. If the coefficient of friction between the tool and the chip is 0.7, then the power consumption (in kW) for the machining operation is _____.

12. A manufacturer has the following data regarding a product:

Fixed cost per month = ₹ 50000

Variable cost per unit = ₹ 200

Selling price per unit = ₹ 300

Production capacity = 1500 units per month.

If the production is carried out at 80% of the rated capacity, that the monthly profit (in ₹) is

13. A cube and a sphere made of cat iron (each of volume 1000cm^3) were cast under identical conditions. The time taken for solidifying the cube was 4s. The solidification time (in s) for the sphere is _________

14. The values of function $f(x)$ at 5 discrete points are given below:

x	0	0.1	0.2	0.3	0.4
f(x)	0	10	40	90	160

Using Trapezoidal role with step size of 0.1, the value of $\int_0^{0.4} f(x)\,dx$ is ______.

15. A project consists of 7 activities. The network along with the time durations (in days) for various activities is shown in the figure.

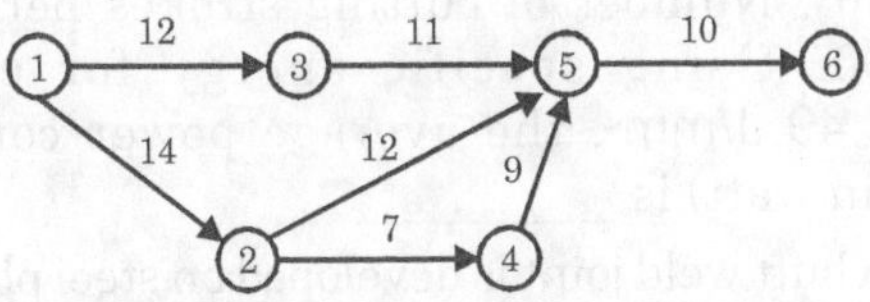

The minimum time for completion of the project is ____.

16. Annual demand of a product is 50000 units and the ordering cost is Rs. 7000 per order considering the basic economic order quantity model, the economic order quantity is 10000 units. When the annual inventory cost is minimized, the annual inventory holding cost (in Rs.) is ______.

17. Ratio of solidification time of a cylindrical casting (height = radius) to the cubic casting of side two times the height of cylindrical casting is _____.

18. The annual requirement of rivets at a ship manufacturing company is 2000 kg. The rivets are supplied in units of 1 kg costing Rs. 25 each. If the costs Rs. 100 to place an order and the annual cost of carrying one unit is 9% of its purchase cost, the cycle length of the order (in days) will be ____

19. The dimensions of a cylindrical side riser (height = diameter) for a 25 cm × 15 cm × 5 cm steel casting are to be determined. For the tabulated shape factor values given below, diameter of the riser (in cm) is ______.

Shape Factor	2	4	6	8	10	12
Riser Volume/Casting Volume	1.0	0.70	0.55	0.50	0.40	0.35

20. In a rolling operation using rolls of diameter 500 mm if a 25 mm thick plate cannot be reduced to less than 20 mm in one pass, the coefficient of friction between the roll and the plate is ________

2014

21. If the Taylor's tool life exponent n is 0.2, and the tool changing time is 1.5 min, then the tool life (in min) for maximum production rate is ______

22. A project has four activities P, Q, R and S as shown below.

Activity	Normal duration (days)	Predec-essor	Cost slop (₹/day)
P	3	-	500
Q	7	P	100
R	4	P	400
S	5	R	200

The normal cost of the project is ₹ 10,000 and the overhead cost is ₹ 200 per day. If the project duration has to be crashed down to 9 days, the total cost (in Rupees) of the project is ________

23. Consider the following data with reference to elementary deterministic economic order quantity model

Annual demand of an item	100000
Unit price of the item (in ₹)	10
Inventory carrying cost per unit per year (in ₹)	1.5
Unit order cost (in ₹)	30

The total number of economic orders per year to meet the annual demand is ________

24. A mild steel plate has to be rolled in one pass such that the final plate thickness is 2/3rd of the initial thickness, with the entrance speed of 10 m/min and roll diameter of 500 mm. If the plate widens by 2% during rolling, the exit velocity (in m/min) is ________

25. The actual sales of a product in different months of a particular year are given below:

September	October	November	December	January	February
180	280	250	190	240	?

The forecast of the sales, using the 4-month moving average method, for the month of February is ________

26. A straight turning operation is carried out using a single point cutting tool on an AISI 1020 steel rod. The feed is 0.2 mm/rev and the depth of cut is 0.5 mm. The tool has a side cutting edge angle of 60°. The uncut chip thickness (in mm) is ____

27. A drum brake is shown in the figure. The drum is rotating in anticlockwise direction. The coefficient of friction between drum and shoe is 0.2. The dimensions shown in the figure are in mm. The braking torque (in N.m.) for the brake shoe is ____

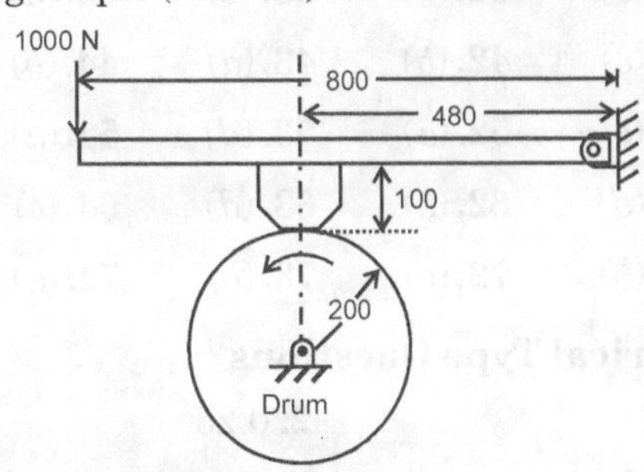

28. Consider an objective function $Z(x_1, x_2) = 3x_1 + 9x_2$ and the constraints

$x_1 + x_2 \le 8,$

$x_1 + 2x_2 \le 4,$

$x_1 \ge 0\ ,\ x_2 \ge 0.$

The maximum value of the objective function is ______

29. For spot welding of two steel sheets (base metal) each of 3 mm thickness, welding current of 10000 A is applied for 0.2s. The heat dissipated to the base metal is 1000 J. Assuming that the heat required for melting 1 mm^3 volume of steel is 20 J and interfacial contact resistance between sheets is 0.0002 Ω, the volume (in mm^3) of weld nugget is ______

30. Demand during lead time with associated probabilities is shown below:

Demand	50	70	75	80	85
Probability	0.15	0.14	0.21	0.20	0.30

Expected demand during lead time is ______

31. Two separate slab milling operations, 1 and 2, are performed with identical milling cutters. The depth of cut in operation 2 is twice that in operation 1. The other cutting parameters are identical.

The ratio of maximum uncut chip thicknesses in operations 1 and 2 is ______

32. At a work station, 5 jobs arrive every minute. The mean time spent on each job in the work station is 1/8 minute. The mean steady state number of jobs in the system is ______

33. A cylindrical riser of 6cm diameter and 6cm height has to be designed for a sand casting mould for producing a steel rectangular plate casting of 7cm × 10cm × 2 cm dimensions having the total solidification time of 1.36 minute. The total solidification time (in minute) of the riser is ______

34. A cast iron block of 200 mm length is being shaped in a shaping machine with a depth of cut of 4 mm, feed of 0.25 mm/stroke and the tool principal cutting edge angle of 30°. Number of cutting strokes per minute is 60. Using specific energy for cutting as 1.49 J/mm^3, the average power consumption (in watt) is ______

35. A butt weld joint is developed on steel plates having yield and ultimate tensile strength of 500 MPa and 700 MPa, respectively. The thickness of the plates is 8 mm and width is 20 mm. Improper selection of welding parameters caused an undercut of 3 mm depth along the weld.

The maximum transverse tensile load (in kN) carrying capacity of the developed weld joint is ______

ANSWERS

MCQ Type Questions

1. (b)	**2.** (b)	**3.** (b)	**4.** (c)	**5.** (b)	**6.** (b)	**7.** (b)	**8.** (b)	**9.** (b)	**10.** (c)
11. (b)	**12.** (d)	**13.** (b)	**14.** (c)	**15.** (b)	**16.** (b)	**17.** (a)	**18.** (b)	**19.** (a)	**20.** (c)
21. (a)	**22.** (c)	**23.** (a)	**24.** (d)	**25.** (b)	**26.** (a)	**27.** (a)	**28.** (a)	**29.** (c)	**30.** (a)
31. (b)	**32.** (a)	**33.** (b)	**34.** (d)	**35.** (b)	**36.** (b)	**37.** (b)	**38.** (c)	**39.** (d)	**40.** (b)
41. (c)	**42.** (b)	**43.** (a)	**44.** (b)	**45.** (b)	**46.** (b)	**47.** (b)	**48.** (c)	**49.** (c)	**50.** (a)
51. (c)	**52.** (a)	**53.** (d)	**54.** (a)	**55.** (a)	**56.** (b)	**57.** (a)	**58.** (b)	**59.** (b)	**60.** (b)
61. (d)	**62.** (b)	**63.** (d)	**64.** (c)	**65.** (c)	**66.** (d)	**67.** (a)	**68.** (c)	**69.** (c)	**70.** (b)
71. (b)	**72.** (c)	**73.** (b)	**74.** (a)						

Numerical Type Questions

1. 2	**2.** 0.25	**3.** 1839.83	**4.** 568.75	**5.** 5.71
6. 150	**7.** 6767.6	**8.** 18.5-19.0	**9.** (0.64)	**10.** (0.97)
11. (2.1)	**12.** (70,000)	**13.** (6.15)	**14.** (22)	**15.** (40)
16. (35000)	**17.** (0.5–0.6)	**18.** 76.94	**19.** (10.5 to 10.7)	**20.** 0.1414
21. (5.9 to 6.1)	**22.** (12490 to 12510)	**23.** (49 to 51)	**24.** (14.6 to 14.8)	**25.** (239 to 241)
26. (0.08 to 0.12)	**27.** (63 to 65)	**28.** (17 to 19)	**29.** (140 to 160)	**30.** (74 to 75)
31. (0.70 to 0.72)	**32.** (1.62 to 1.70)	**33.** (2.5 to 4.5)	**34.** (295 to 305)	**35.** (68 to 72)

EXPLANATIONS

MCQ TYPE QUESTIONS

2. Minimum clearance

$\Rightarrow$ minimum m hole – maxium shaft

$$= 25 + .020 - 25 + .005$$
$$= 0.015 \text{ mm} = 2$$

5. $n = 4; p = \frac{1}{6}$

$$\Rightarrow \quad q = 1 - \frac{1}{6} = \frac{5}{6}$$

$$p(x \geq 2) = 1 - p(x < 2)$$
$$= 1 - [p(x = 0) + p(x = 1)]$$
$$= 1 - \left[4C_0\left(\frac{1}{6}\right)^0\left(\frac{5}{6}\right)^4 + 4C_1\left(\frac{1}{6}\right)^1\left(\frac{5}{6}\right)^3\right] = \frac{19}{144}$$

6. $t_{min} = k\left(V/A\right)^2$

$$t_1 = k\left(\frac{93}{502}\right)^2$$
$$t_2 = k\left(\frac{93}{5}\right)$$
$$\frac{t_1}{t_2} = \frac{36}{25}$$

7. Since all dimensions have bilateral tolerances

$$L_4 = L_1 - L_2 - L_3$$
$$= 22 - 10 - 10$$
$$L_4 = 2 \text{ mm}$$
$$\text{Tolerance} = 0.01 + 0.0015 + 0.0005$$
$$= 0.012$$
$$\therefore \quad L_4 = 2.0^{\pm 0.02} \text{ mm}$$

Tolerance is calculated assuming L_4 to be sink and tolerance of sink will be cumulative sum of all tolerances.

8. When lifted from ground at Q

Reaction = 0

$\therefore$ taking moments about 'p' and equating to 0

$$.09 \times .06 = mr\,\omega^2 \times .09$$
$$9.01 \times .09 \times .06 = .002 \times 2.19 \times 10^{-3} \times \omega^2 \times .09$$
$$\Rightarrow \quad \omega = 366.50 \text{ rad/s} = 3500 \text{ rpm}$$

9.

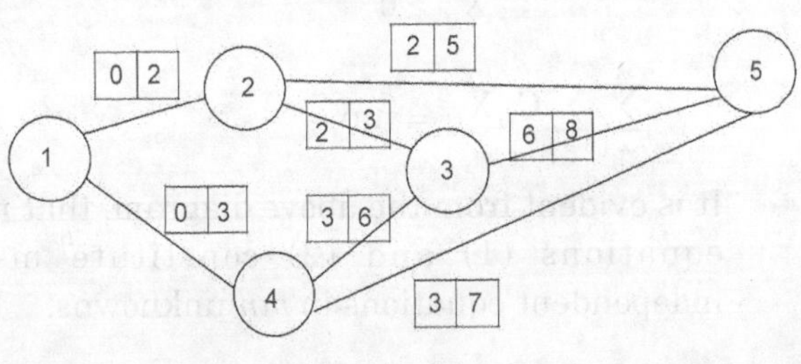

Critical path – 1 – 4 – 3 – 5

Time taken = 8 days

14. GO gauge represents lower limit in hole

15. Given that the student is passing exam, i.e., if only 20 students (out of 100) are considered, of the 5 students get more than 90%

$\therefore$ 5 out of 20 is the probability

or 1/4 is the answer.

Alternative Method :

Let the student pass the examination be A and student pass the examination and got about 90% marks be B

Now P(A) = 20% and P(A∩B) = 5%

$$\therefore \quad P\left(\frac{B}{A}\right) = \frac{P(A \cap B)}{P(A)} = \frac{5\%}{20\%} = \frac{1}{4}$$

17. Total heat input = VIt

Time t is inversely proportional to weld speed (S)

$$\therefore \quad \frac{V_2 I_2}{S_2} = \frac{V_1 I_1}{S_1}$$
$$\therefore \quad I_2 = \frac{120}{150} \times 50$$
$$I_2 = 40 \text{ A}$$

18. Voltage = 100V

Power = 1Kw

$$\text{Power} = \frac{V^2}{R}$$
$$\Rightarrow \quad R = \frac{100 \times 100}{1000} = 10\ \Omega$$

$$\text{Now} \quad \tau = \frac{1}{2} RC$$
$$\Rightarrow C = \frac{\tau}{R} = \frac{25}{10} \times 2 = 5\ \mu F$$

19. Diffusion welding and friction welding are autogenous welding process as they do not require any filler material.

20. The surface obtained is plane.

21. The pattern is made of expanded polystrene.

23. *b* : Service time distribution (usually represented by 'm')

d : Queuing discipline (usually represented by 'GD')

25. For incremental coordinates (G91) and coordinates of final point are (– 10, – 10). The tool moves CCW (counter clockwise), So, G03.

26. $$r = \frac{0.14}{0.28} = 0.5$$

$$\tan\phi = \frac{r\cos\alpha}{1 - r\sin\alpha}$$

$$\therefore \quad \phi = 28.3345^\circ$$

or $28^\circ\ 20'$

Shear strain, $Y = \cot\phi + \tan(\phi - \alpha) = 2.1859 \approx 2.19$

27. Rate of taper,

$$T = \frac{80 - 65}{55} = 0.27$$

$$\text{Set over} = \frac{T \times L}{2}$$

$$= \frac{0.27 \times 90}{2} = 12.15$$

Taper angle = $\tan^{-1}(0.27) = 15.10$

28. End mill centre ≡ (0,0)

Since Radius of end mill is 8 mm

∴ call point 'p' ≡ – 8, – 8

Call point 'Q' ≡ (100 + 8, – 8 + 0) ≡ (108, – 8)

→ *x* direction

Call point 'R' ≡ (108 + 0, 50 + 8) ≡ (108, 58)

→ *y* direction

Call point 'S' ≡ (108 –100 – (2 × 8) , 58 – 0)

≡ (–8, 58)

→ *x* direction

Call point 'P' ≡ (-8-0, 50-50-8) ≡ (-8, -8)

→ –*y* direction

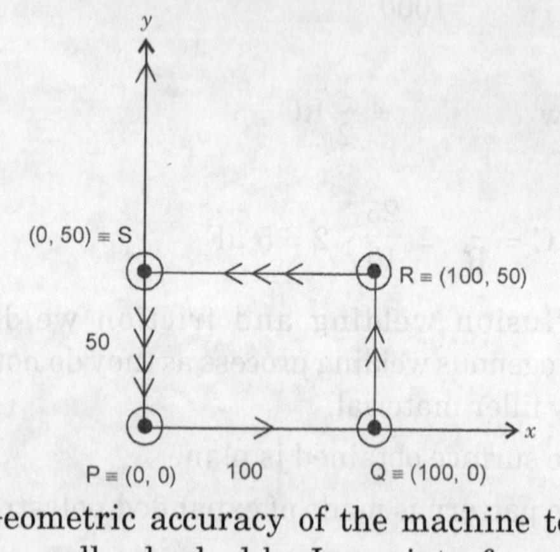

29. Geometric accuracy of the machine tools are generally checked by Laser interferometer, as it is very cheap and easy to handle.

31. specific cutting energy

$$= \frac{F_c}{b \times t_1} = \frac{400}{2 \times 0.1} = 2000 \text{ N/mm}^2.$$

34. H7 is for hole where 7 indicates its tolerance grade g8 is for shaft where 8 indicates its tolerance grade

35. Arrival Rate = 5 jobs in 8 hrs

Service time = 40 min/Job

∴ Total service time = 40 × 5 = 200 min

$$= \frac{200}{60} = \frac{10}{3} \text{ hrs}$$

$$\therefore \quad \text{Idle Time} \,/\, \text{shift} = 8 - \frac{10}{3} = \frac{24 - 10}{3} = \frac{14}{3} \text{ hrs.}$$

36. Flatness of machine bed can be measured from Auto collimator.

37.

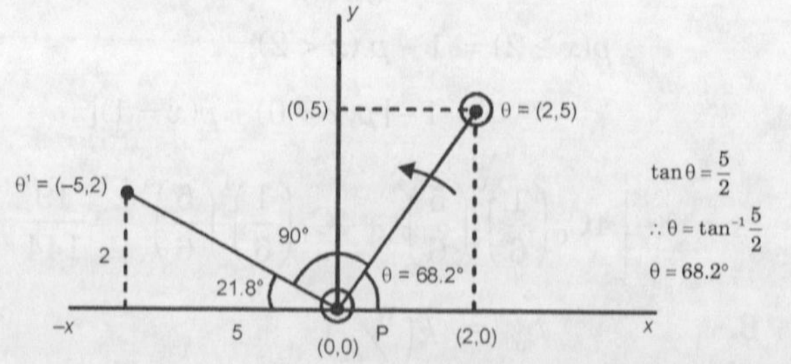

$$\tan 21.8^\circ = \frac{2}{5}$$

$$\therefore \quad \theta' = (-5, 2)$$

41.

ORIGIN ↓	DESTINATION 1	2	3	...	n	TOTAL
1	x_{11}	x_{12}	----			a_1
2	x_{21}	x_{22}	----			a_2
⋮						----
m	X_{m_1}	X_{m_2}			X_{m_n}	a_m
TOTAL	b_1	b_2	------		b_n	$\sum_{i=1}^{m} a_i = \sum_{J=1}^{n} b_J$

$$\sum_{J=1}^{n} X_{iJ} = a_i, \ (i = 1, 2, ...m) \qquad ...(1)$$

$$\sum_{L=1}^{m} X_{iJ} = b_J, \ (J = 1, 2, ...n) \qquad ...(2)$$

$$X_{iJ} \geq 0 \qquad ...(3)$$

$$\sum_{i=1}^{m} \sum_{J=1}^{m} C_{iJ} X_{iJ} = mn \qquad ...(4)$$

It is evident from the above diagram, that m + n equations (1) and (2) constitute m+n–1 independent equations in *mn* unknowns.

42. Process — Cost of production for 100

Process	Cost of production for 100	
I	$F.c + \theta \times v_c = 20 + 100 \times 3$	$= 320$
II	$50 + 100 \times 1$	$= 150$
III	$40 + 100 \times 2$	$= 240$
IV	$10 + 100 \times 4$	$= 410$

Hence process (B) is most economical.

43. G00 – Point to point movement
G02 – Circular interpolation, cw
G02 – Circular interpolation, ccw
G04 – Dwell

44.
$$T = \frac{L}{f_N}$$
$$L = t + Ap_1$$
$$Ap_1 = 0.5\text{ D (holes diameter)}$$
$$= 10 \text{ mm}$$
$$t = 40 \text{ mm}$$
$$T = \frac{50}{0.1 \times 400} = 1.25 \text{ min.}$$

45. $F = \tau_0 \times As = 300 \times 1500 = 450$ kN.

For Rectangular hole,
$$A_s = 2(a + b)t = 2(100 + 50) \times 5 = 1500 \text{ mm}^2$$

46. A path forms a loop or cycle if it connects a node itself. A spanning tree links all the nodes of network with no loops allowed.

50.

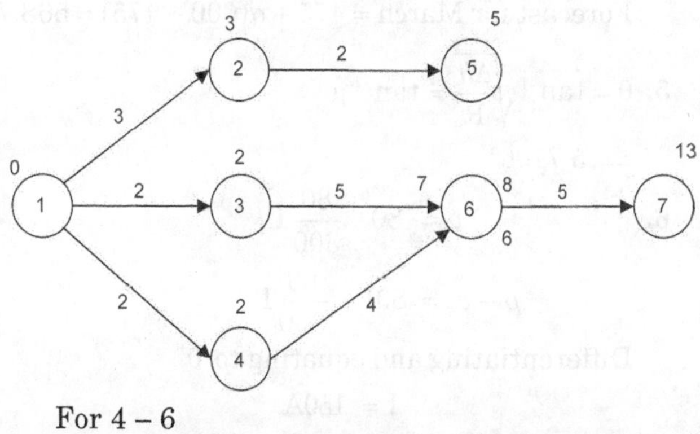

For 4 – 6

$$F \cdot F = (E_5 - E_i) - T_{i5} = (8 - 2) - 4 = 2$$

$\therefore$ 2, 13.

52.
$$v = \frac{\pi}{4} D^2 h$$
$$h = \frac{4v}{\pi D^2} \quad ...(1)$$
$$As = \pi Dh + \frac{\pi}{4} D^2 = \pi D \frac{4v}{\pi D^2} + \frac{\pi}{4} D^2 = \frac{4v}{D} + \frac{\pi}{4} D^2$$

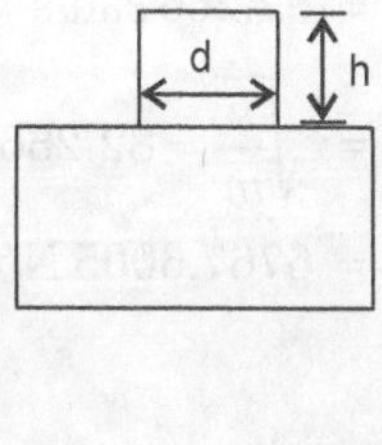

For min A_s, $\frac{dA_s}{dD} = 0$
$$\frac{-4v}{D^2} + \frac{\pi}{4} 2D = 0 \Rightarrow \frac{4v}{D^2} = \frac{d\pi}{2}$$
$$v = \frac{\pi D^3}{8} = \frac{\pi}{4} D^2 h$$
$$D = 2h \Rightarrow \frac{h}{D} = \frac{1}{2}$$
$$h : D = 1 : 2$$

54. A, $n \times n$ assignment problem, if it is solved as a LPP it will have n^2 variables.

60.

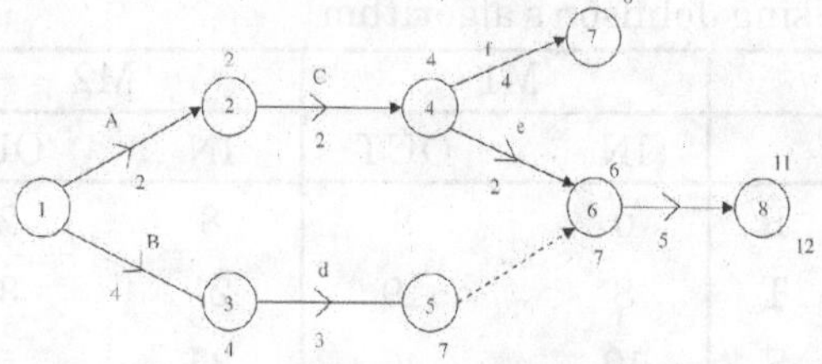

$$(TF)_e = (L_j - E_i) - T_{ij} = (7 - 4) - 2 = 1$$
$$(TF)_f = (8 - 0) - 4 = 4$$

61.

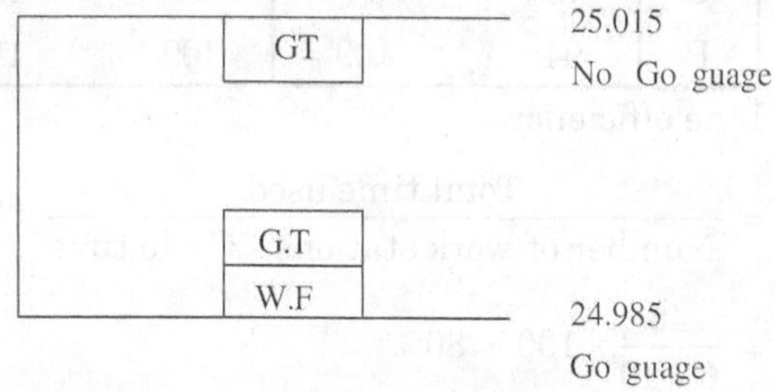

25 ± 0.015, G_O-Gauge, U.L = 24.985

L.L = 24.985, $24.985^{+0.003}_{+0.000}$

64. Given : Demand, $D_{t-1} = 12{,}000$

Forecast, $F_{t-1} = 10275$

$\alpha = 0.25$

$$\therefore \quad F_t = F_{t-1} + \alpha (D_{t-1} - F_{t-1})$$
$$= 10275 + 0.25\,(12000 - 10275) = 10706.25$$

67.

Jobs using shortest possible time rule	Processing time (days)	Cumulative time (days)
I	4	4
III	5	9
V	6	15
VI	8	23
II	9	32
IV	10	42

$$\text{Average time flow} = \frac{4 + 9 + 15 + 23 + 32 + 12}{6}$$
$$= 20.83$$

68. For any Poisson distribution, the probability density function f(t) is given by $\lambda e^{-\lambda t}$.

69. M.R.P. for R is given by

Week	1	2	3	4	5	6
Demand	2000	2000	2000	2000	2400	2400
Inventory Level I = P	P- 2000	2P- 4000	3P- 6000	4P- 8000	5P- 10400	6P- 12800

If $6P - 12800 \geq 0$ then the capacity is feasible

$\therefore$ Minimization, $6P = 12800$

$\Rightarrow$ $P \cong 2200$

72. In a MRP system, component demand is calculated by the MRP system from the Master Production Schedule.

73. Using Johnson's algorithm

	M1		M2	
	IN	OUT	IN	OUT
R	0	8	8	21
T	8	19	21	35
S	19	46	35	55
Q	46	78	78	97
U	78	94	97	104
P	94	109	109	115

74. Line efficiency

$$= \frac{\text{Total time used}}{\text{Number of work stations} \times \text{Cycle time}} \times 100$$

$$= \frac{48}{6 \times 10} \times 100 = 80\%$$

NUMERICAL TYPE QUESTIONS

1. Annual demand (D) = 10000 units

Unit cost (C_u) = Rs. 100

Carrying cost (C_c) = 14.4% of unit cost

Ordering cost (C_o) = Rs. 2000

Cycle time (T) = ?

$$T = \frac{1}{N} = \frac{EOQ}{D}$$

$$EOQ = \sqrt{\frac{2QC_o}{C_c}}$$

$$= \sqrt{\frac{2 \times 10000 \times 2000}{100 \times 0.144}}$$

= 1666.66 units

$T = 0.1666 \times 12$

= 2 months

2. $VT^n = C$

$$V_1 T_1^n = 2V_1 \times \left(\frac{T_1}{16}\right)^2 \text{ on solving we get}$$

$n = 0.25$

3. Shear load on all rivets $\Rightarrow \frac{400}{4} = 100N$

Secondary shear load, due to bending moment

$$\Rightarrow \frac{Pe\, r_1}{r_1^2 + r_1^2 + r_3^2 + r_4^2}$$

$$\Rightarrow \frac{400 \times .5 \times .02\sqrt{2}}{(.02\sqrt{2})^2 \times 4}$$

$\Rightarrow$ 1767.766953 N

P = 1839.837 N.

4. Forecast for Feb $= 400 + \alpha\,(500 - 400)$

$= 400 + 0.75 \times 100 = 475$

Forecast for March $= 475 + \alpha(600 - 475) = 568.75$

5. $\theta = \tan^{-1}\sqrt{\frac{\Delta h}{R}} = \tan^{-1}\mu$

$\Rightarrow 5.7106^{\circ}$

6. $$v = 80 - \frac{80}{300}I$$

$$p - v_7 = 80I - \frac{80}{300}I^2$$

Differentiating and equating to '0'

I = 150A

7. $\omega = 2\pi N = 2 \times \pi \times 60 = 376.99$ rad/second

$$.05 = \frac{1}{1 - \left(\frac{\omega}{\omega_n}\right)^2}$$

$$\Rightarrow -20 = 1 - \left(\frac{\omega}{\omega_n}\right)^2 \Rightarrow \left(\frac{\omega}{\omega_n}\right)^2 = 21$$

$$\omega_2 = 21\omega_n^2$$

$$\omega_n = 82.266 \text{ rad/s}$$

$$\omega_n = \sqrt{\frac{k}{m}} = 82.266$$

1 = 6767.6005 N/m

10. Probability (at least one will meet specification)
= 1 – probability (none will meet specification)
= 1 – (1 – 0.8) × (1 – 0.7) × (1 – 0.5)
= 1 – 0.2 × 0.3 × 0.5 = 1 – 0.03 = 0.97

11. $$\mu = \frac{F}{N} = \frac{F_C \sin\alpha + F_T \cos\alpha}{F_C \cos\alpha + F_T \sin\alpha}$$

Given : $\alpha = 0°$ (rake angle)

$$\therefore \quad \mu = \frac{F_T}{F_C}$$

$$0.7 = \frac{490}{F_C}$$

$$\therefore \quad F_C = 700 \text{ N}$$

Power consumption,

$$P = F_C \times V_C = 700 \times \frac{180}{60} \times \frac{1}{1000} (\text{kW})$$

$$P = 2.1 \text{ kW}$$

12. Profit per month = $0.8 \times 1500 \times (300 - 200) - 50000$
= ₹ (120000 – 50000) = ₹ 70000

13. $$t_s = k\left(\frac{V}{A}\right)^2$$

$$t_{sphere} = t_{cube} \frac{A_{cube}^2}{A_{sphere}^2} \qquad ...(i)$$

$$\frac{4}{3}\pi r^3 = 1000 \text{ cm}^3$$

$$\therefore \quad r = 6.2035 \text{ cm}$$

$$A_{sphere} = 4\pi r^2 = 483.597 \text{ cm}^2$$

$$a^3 = 1000 \text{ cm}^3$$

$$\therefore \quad a = 10 \text{ cm}$$

$$A_{cube} = 6a^2 = 600 \text{ cm}^2$$

From equation (*i*),

$$\therefore \quad t_{sphere} = 4 \times \frac{(600)^2}{(483.597)^2}$$

$$= 6.157 \text{ sec}$$

14.

x	0	0.1	0.2	0.3	0.4
$y = f(x)$	0	10	40	90	160

$y_0 \quad y_1 \quad y_2 \quad y_3 \quad y_4$

$$\int_0^{0.4} f(x)dx = \int_0^{0.4} y\,dx$$

$$= \frac{h}{2}[(y_0 + y_4) + 2(y_1 + y_2 + y_3)]$$

$$= \frac{0.1}{2}[(0 + 160) + 2(10 + 40 + 90)] = 22$$

15. Time taken for 3 paths are as follows
Path 1 = 12 + 11 + 10 = 33
Path 2 = 14 + 12 + 10 = 36
Path 3 = 14 + 7 + 9 + 10 = 40
∴ Path 3 is longest i.e. path 3 is critical path
∴ Project duration = 40 days

16. At optimum total inventory cost (TIC), annual inventory hold is cost is equal to annual inventory ordering cost
= Number of orders × ordering cost per order

$$= \frac{50000}{10000} \times 7000$$

= 5 × 7000 = 35000

18. $$\text{EOQ} = \sqrt{\frac{2DC_0}{C_h}} = \sqrt{\frac{2 \times 2000 \times 100}{0.09 \times 25}} = 421.637$$

$$\text{length of cycle} = \frac{365}{\text{No. of orders}} = 76.948 \text{ days}$$

$$\text{No. of orders} = \frac{2000}{\text{EOQ}}$$

20. $$\mu = \sqrt{\frac{(\Delta h)_{max}}{R}}$$

$$(\Delta h)_{max} = 25 - 20 = 5\text{mm}$$

$$R = 250\text{mm}$$

$$\therefore \quad \mu = \sqrt{\frac{5}{250}} = 0.1414$$

21. $$T_{opt} = \left[\frac{1-n}{n} \times T_C\right] = \frac{1-0.2}{0.2} \times 1.5 = 6 \text{ min.}$$

22.

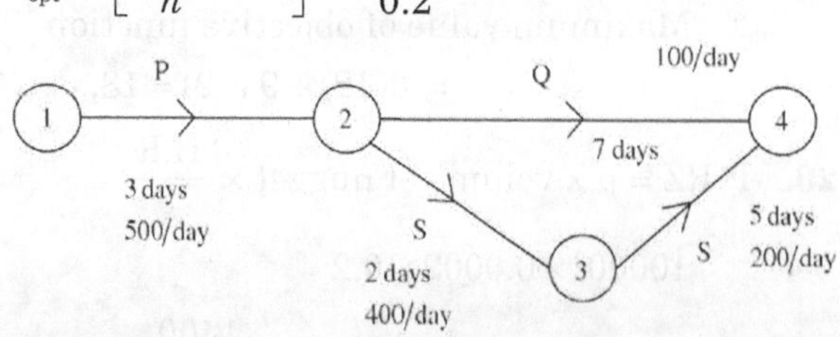

Heuristic model to solve the crashing problem:

Step I – Find the normal time
Step II – Determine critical path ≡ P-R-S = 12days.
Step III – Crash the lowest cost slope from the critical path

crash activity '*s*' by 2 days
Now critical path = P – Q (10 days)
∴ Total cost of the project
= 10000 + 200 × 10days + crashing cost
= 10,000 + 2000 + 200 × 2
= 10,000 + 2000 + 400 = 12,400.

23. Given D = 100000/year

unit cost = 10

$C_c = 1.5$

$C_o = 30$/order

$$\therefore \quad EOQ = \sqrt{\frac{2 \times D \times C_o}{C_c}}$$

$$= \sqrt{\frac{2 \times 100000 \times 30}{1.5}} - 2000$$

$$\therefore \text{ No. of order's/year} = \frac{\text{Demand}}{\text{EOQ}} = \frac{100000}{2000} = 50.$$

25. Number of periods = 4, then the past 4 months average sales is fore cast for next 4 months.

$$\text{So, } \frac{280 + 250 + 190 + 240}{4} = 240.$$

26.
$$t_1 = f \cos \theta$$
$$= 0.2 \,.\, \cos 60$$
$$t_1 = 0.1 \text{ mm}$$

where t_1 = uncut chip thickness

27. Taking moments about pin joint (ΣM)

$$\Rightarrow \quad 1000 \times 0.8 = R_N \times 0.48 + \mu R_N \times 0.1$$

$$\Rightarrow \quad R_N = \frac{800}{0.5} = 1600$$

$$T_B = \mu\, R_N \times r$$

$$= 0.2 \times 1600 \times \frac{200}{1000} = 64 \text{ N} - m.$$

28. Roots are (12, – 4) and (12, – 2)

∴ Maximum value of objective function

$$= 3(12) + 9(-2) = 18.$$

29. $I^2 RZ = p \times \text{volume of nugget} \times \frac{H.R}{g}$

$$10000^2 \times 0.0002 \times 0.2$$

$$= 1 \times \text{volume of nugget} \times \frac{1400}{1000}$$

Volume of nugget = 2857.1 J.mm³

1 mm³ volume of steel is 20J

$$\text{Volume of nugget} = \frac{2857.1}{20} \cdot \frac{\text{J.mm}^3}{\text{J}}$$

$$= 142.8 \text{ mm}^3.$$

30. Probability

x_i	f_i	$x_i f_i$
50	0.15	7.5
70	0.14	9.8
75	0.21	15.75
80	0.2	16
85	0.3	25.5

Expected demand during lead time = 74.55.

31. Ratio of 1 and 2 is = $\frac{\sqrt{d_1}}{\sqrt{d_2}}$

$$\therefore \quad d_2 = 2d_1$$

$$= \frac{\sqrt{d_1}}{\sqrt{2d_2}} = \sqrt{\frac{1}{2}} = 0.707.$$

34. Specific cutting energy = $\frac{F_C}{b \times t_1}$ = 1.49 J/mm³

$$= \frac{F_C}{4 \times 0.25}$$

F_C = 1.49 J/mm. stroke

$$F_C = 1.49 \text{ J/mm} \times \frac{60}{\text{min}}$$

F_C = 89.4 J/mm.min

Power = F_C.1 = 89.4 × 200 mm J/mm.min

= 17880 J/min

Power = 298 J/S or (W).

35. t = 8 mm

b = 20 mm

d = 3 mm

maximum transverse tensile load

= yield strength × $(b - d)t$

= 500 N/mm² × (20 – 3) × 5 mm²

= 68,000 N = 68 kN.

17
CHAPTER

Inventory Control

MCQ TYPE QUESTIONS

2016

1. The part of a gating system which regulates the rate of pouring of molten metal is

(*a*) pouring basin (*b*) runner
(*c*) choke (*d*) ingate

2. A cylindrical job with diameter of 200 mm and height of 100 mm is to be cast using modulus method of riser design. Assume that the bottom surface of cylindrical riser does not contribute as cooling surface. If the diameter of the riser is equal to its height, then the height of the riser (in mm) is

(*a*) 150 (*b*) 200
(*c*) 100 (*d*) 125

2010

3. Little's law is a relationship between

(*a*) stock level and lead time in an inventory system
(*b*) waiting time and length of the queue in a queuing system
(*c*) number of machines and job due dates in a scheduling problem
(*d*) uncertainty in the activity time and project completion time

4. Annual demand for window frames is 10000. Each frame costs Rs. 200 and ordering cost is Rs. 300 per order. Inventory holding cost is Rs. 40 per frame per year. The supplier is willing to offer 2% discount if the order quantity is 1000 or more, and 4% if order quantity is 2000 or more. If total cost is to be minimized, the retailer should

(*a*) order 200 frames every time
(*b*) accept 2% discount
(*c*) accept 4% discount
(*d*) order Economic Order Quantity

2009

5. A company uses 2555 units of an item annually. Delivery lead time is 8 days. The reorder point (in number of units) to achieve optimum inventory is

(*a*) 7 (*b*) 8
(*c*) 56 (*d*) 60

2008

6. A set of 5 jobs is to be processed on a single machine. The processing time (in days) is given in the table below. The holding cost for each job is Rs. K per day.

Job	Processing time
P	5
Q	2
R	3
S	2
T	1

A schedule that minimizes the total inventory cost is

(*a*) T-S-Q-R-P (*b*) P-R-S-Q-T
(*c*) T-R-S-Q-P (*d*) P-Q-R-S-T

2007

7. The maximum level of inventory of an item is 100 and it is achieved with infinite replenishment rate. The inventory becomes zero over one and half month due to consumption at a uniform rate. This cycle continues throughout the year. Ordering cost is Rs. 100 per order and inventory carrying cost is Rs. 10 per item per month. Annual cost (in Rs.) of the plan, neglecting material cost, is

(*a*) 800 (*b*) 2800
(*c*) 4800 (*d*) 6800

8. The net requirements of an item over 5 consecutive weeks are 50 – 0 – 15 – 20 – 20. The inventory carrying cost and ordering cost are Re. 1 per item per week and Rs. 100 per order respectively. Starting inventory is zero. Use "Least Unit Cost Technique" for developing the plan. The cost of the plan (in Rs.) is

(*a*) 200 (*b*) 250
(*c*) 255 (*d*) 260

2006

9. *Consider the following data for an item.*

Annual demand : 2500 units per year
Ordering cost : Rs. 100 per order
Inventory holding rate : 25% of unit price

Price quoted by a supplier

Order quantity (units)	Unit price (Rs.)
< 500	10
≥ 500	9

The optimum order quantity (in units) is

(*a*) 447 (*b*) 471
(*c*) 500 (*d*) ≥ 600

10. A firm is required to procure three items (P, Q, and R). The prices quoted for these items (in Rs.) by suppliers S1, S2 and S3 are given in table. The management policy requires that each item has to be supplied by only one supplier and one supplier supply only one item. The minimum total cost (in Rs.) of procurement to the firm is

Item	Suppliers		
	S1	S2	S3
P	110	120	130
Q	115	140	140
R	125	145	165

(*a*) 350 (*b*) 360
(*c*) 385 (*d*) 395

11. A stockist wishes to optimize the number of perishable items he needs to stock in any month in his store. The demand distribution for this perishable item is

Demand (in units)	2	3	4	5
Probability	0.10	0.35	0.35	0.20

The stockist pays Rs. 70 for each item and he sells each at Rs. 90. If the stock is left unsold in any month, he can sell the item at Rs. 50 each. There is no penalty for unfulfilled demand. To maximize the expected profit, the optimal stock level is

(*a*) 5 units (*b*) 4 units
(*c*) 3 units (*d*) 2 units

ANSWERS

MCQ Type Questions

1. (*a*) **2.** (*a*) **3.** (*d*) **4.** (*d*) **5.** (*c*) **6.** (*a*) **7.** (*c*) **8.** (*b*) **9.** (*a*) **10.** (*c*) **11.** (*c*)

EXPLANATIONS

MCQ TYPE QUESTIONS

1. Rate of pouring of molten metal depends pouring basin.

2. According to modulus method

$$M_R = 1.2\, M_C$$

$$\left[\frac{V}{As}\right]_R = 1.2\left[\frac{V}{As}\right]_C$$

If diameter of riser = height of riser for top riser

D = H

$\Rightarrow$ D = 6 M_C

$$D = 6\times\frac{\frac{\pi}{4}\times(200)^2\times 100}{2\times\frac{\pi}{4}\times(200)^2+\pi\times 200\times 100}$$

$$D = H = \frac{6\times 200\times 100}{400+400} = 150\text{ mm}$$

4. EOQ

The amount of orders that minimises total variable cost required to order and hold inventory.

It gives maximum economy in purchasing any material and contributes towards maintaining the material at optimum level at minimum cost.

5.

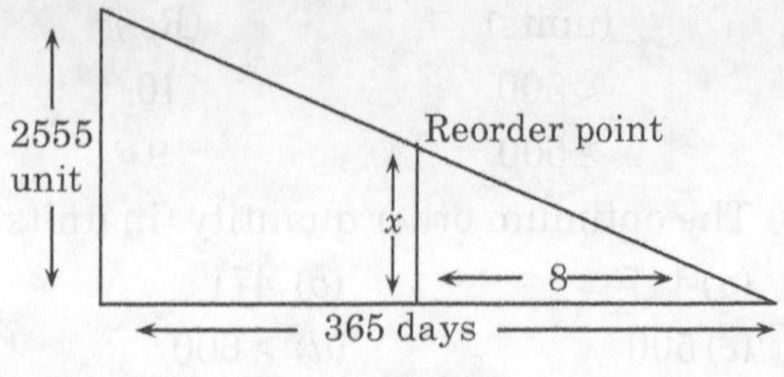

$$\therefore \quad \frac{2555}{365} = \frac{x}{8}$$

$$\Rightarrow \quad x = 56\text{ units}$$

Alternately

Reorder point

= safety stock

$+$ (Delivery lead time $\times$ $\frac{\text{Annual demand}}{365}$)

$$= 0 + \left(8\times\frac{2555}{365}\right) = 56$$

6. To minimizes total inventory cost, i.e. holding cost, the job with the least processing time should be done first.

9. Case I : Let EOQ is less than 500

$$\therefore \quad EOQ = \sqrt{\frac{2Q\times C_0}{C_c}} = 447.21$$

Case II : Let EOQ is greater than 500.

$$\therefore \quad EOQ = \sqrt{\frac{2Q\times C_0}{C_c}} = 471.40$$

which is against the assumption.

$$\therefore \quad EOQ = 447.21$$

■■

18
CHAPTER

Operation Research

MCQ TYPE QUESTIONS

2016

1. In PERT chart, the activity time distribution is
(a) Normal (b) Binomial
(c) Poisson (d) Beta

2. The demand for a two-wheeler was 900 units and 1030 units in April 2015 and May 2015, respectively. The forecast for the month of April 2015 was 850 units. Considering a smoothing constant of 0.6, the forecast for the month of June 2015 is
(a) 850 units
(b) 927 units
(c) 965 units
(d) 970 units

2010

3. Simplex method of solving linear programming problem uses
(a) all the points in the feasible region
(b) only the corner points of the feasible region
(c) intermediate points within the infeasible region
(d) only the interior points in the feasible region

4. The project activities, precedence relationships and durations are described in the table. The critical path of the project is

Activity	Precedence	Duration (in days)
P	-	3
Q	-	4
R	P	5
S	Q	5
T	R, S	7
U	R, S	5
V	T	2
W	U	10

(a) P-R-T-V (b) Q-S-T-V
(c) P-R-U-W (d) Q-S-U-W

2009

5. The expected time (t_e) of a PERT activity in terms of optimistic time (t_o), pessimistic time (t_p) and most likely time (t_l) is given by
(a) $t_e = \dfrac{t_o + 4t_l + t_p}{6}$ (b) $t_e = \dfrac{t_o + 4t_p + t_l}{6}$
(c) $t_e = \dfrac{t_o + 4t_l + t_p}{3}$ (d) $t_e = \dfrac{t_o + 4t_p + t_l}{3}$

2008

6. For the standard transportation linear programme with m sources and n destinations and total supply equaling total demand, an optimal solution (lowest cost) with the smallest number of non-zero x_{ij} values (amounts from source i to destination j) is desired. The best upper bound for this number is
(a) mn (b) 2 (m + n)
(c) m + n (d) m + n – 1

7. For the network below, the objective is to find length of the shortest path from node P to node G. Let d_{ij} be the length of directed are from node i to node j.

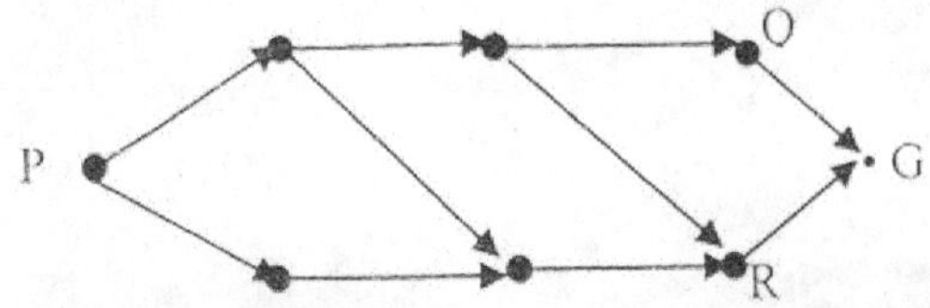

Let s_j be the length of the shortest path from P to node j. Which of the following equations can be used to find s_G?
(a) $s_G = \text{Min}\{s_Q, s_R\}$
(b) $s_G = \text{Min}\{s_Q - d_{QG}, s_R - d_{RG}\}$
(c) $s_G = \text{Min}\{s_Q + d_{QG}, s_R + d_{RG}\}$
(d) $s_G = \text{Min}\{d_{QG}, d_{RG}\}$

NUMERICAL TYPE QUESTIONS

2016

1. In a single-channel queuing model, the customer arrival rate is 12 per hour and the serving rate is 24 per hour. The expected time that a customer is in queue is ______ minutes.

2. A food processing company uses 25,000 kg of corn flour every year. The quantity-discount price of corn flour is provided in the table below :

Quantity (kg)	Unit price (Rs/kg)
1-749	70
750-1499	65
1500 and above	60

The order processing charges are Rs. 500/order. The handling plus carry-over charge on an annual basis is 20% of the purchase price of the corn flour per kg. The optimal order quantity (in kg) is ____.

3. A project consists of 14 activities, A to N. The duration of these activities (in days) are shown in brackets on the network diagram. The latest finish time (in days) for node 10 is ________

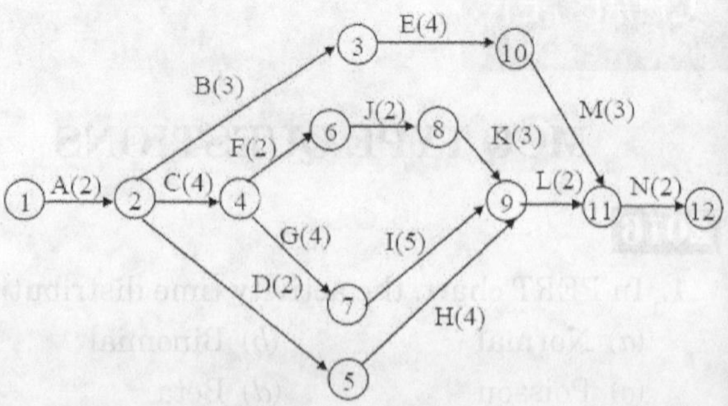

4. A firm uses a turning center, a milling center and a grinding machine to produce two parts. The table below provides the machining time required for each part and the maximum machining time available on each machine. The profit per unit on parts I and II are Rs. 40 and Rs. 100, respectively. The maximum profit per week of the firm is Rs. ________

Type of machine	Machining time required for the machine part (minutes)		Maximum machining time available per week (minutes)
	I	II	
Turning Center	12	6	6000
Milling Center	4	10	4000
Grinding Machine	2	3	1800

ANSWERS

MCQ Type Questions

1. (*d*) **2.** (*d*) **3.** (*d*) **4.** (*d*) **5.** (*a*) **6.** (*d*) **7.** (*c*)

Numerical Type Questions

1. (2.5) **2.** (1500) **3.** (14) **4.** (40,000)

EXPLANATIONS

MCQ TYPE QUESTIONS

2. $F_{MAY} = F_{APRIL} + 2(D_{APRIL} - F_{APRIL})$

$= 850 + 0.6(900 - 850)$

$= 850 + 30 = 880$ Units

$F_{JUNE} = F_{MAY} + 2(D_{MAY} - F_{MAY})$

$= 880 + 0.60(1030 - 880)$

$= 880 + 90 = 970$ Units

4. Time network is

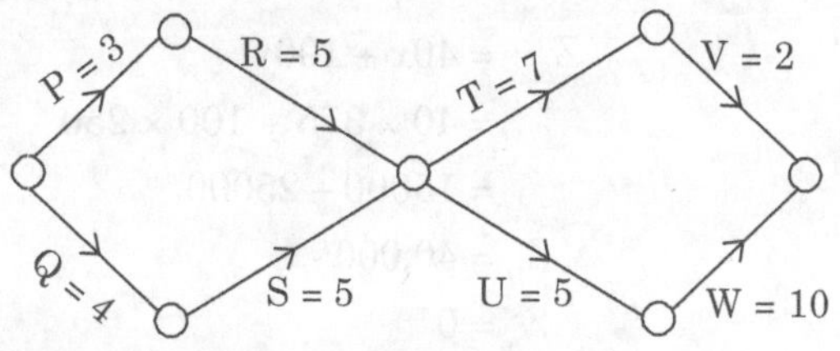

Path QSUW $= Q + S + U + W$

$= 4 + 5 + 5 + 10 = 24$ days

Path QSTV $= Q + S + T + V$

$= 4 + 5 + 7 + 2 = 18$ days

Path PRTV $= P + R + T + V$

$= 3 + 5 + 7 + 2 = 17$ days

Path PRUW $= P + R + U + W$

$= 3 + 5 + 5 + 10 = 23$ days

Hence critical path is QSUW.

6. In such an L.P.P, m × n variables are there and m + n equations/constraints are there (satisfying the demand-supply requirements). But one constraint is removed as total supply equals total demand. The best upper bound x_{ij} values is (m + n – 1).

7. The only way to reach G is by reading Q and then QG or reaching R and then RG.

$\therefore\ S_G = \text{Min}\{S_Q + d_{QG}, S_R + d_{RG}\}$

NUMERICAL TYPE QUESTIONS

1. $\lambda = 12$ /hr, $\mu = 24$ /hr

$$\rho = \frac{\lambda}{\mu} = \frac{12}{24} = 0.5$$

Expected time that the customer in the queue

$$W_q = \frac{L_q}{\lambda} = \frac{L_s - \rho}{\lambda}$$

$$= \frac{\frac{\rho}{1-\rho} - \rho}{\lambda} = \frac{\frac{0.5}{1-0.5} - 0.5}{12}$$

$$= \frac{0.5}{12}\text{hr} = \frac{0.5}{12} \times 60 \text{ min} = 2.5 \text{ min}$$

2. $D = 25000$ kg, C_O = Rs 500/order

$C_h = 20\%$ of C_w

Qty.(kg)	C_w(Rs/kg)	C_h (Rs/years)
$1 \le Q_1 \le 750$	70	$0.2 \times 70 = 14$
$750 \le Q_1 < 1500$	65	$0.2 \times 65 = 13$
$Q_3 \ge 1500$	60	$0.2 \times 60 = 12$

$$Q = \sqrt{\frac{2DC_o}{C_w}}$$

First we check for least unit price

$$Q_3' = \sqrt{\frac{2 \times 25000 \times 500}{12}} = 1443.37$$

$Q_3' = 1443.37 < 1500$

Now check for second maximum unit price

$$Q_2' = \sqrt{\frac{2 \times 25000 \times 500}{13}} = 1386.75$$

$750 < 1386.75 < 1500$

$\therefore$ Total cost for Q_2' unit

$$T_C = 25000 \times 65 + \frac{25000}{1386.75} \times 500 + \frac{1386.75}{2} \times 13$$

$= 1643027.755$

Total cost for 1500 unit

$$T_C = 25000 \times 60 + \frac{25000}{1500} \times 500 + \frac{1500}{2} \times 12$$

$= 1517333.33$

$T_C(1500) < T_C(Q_2')$

So optimum order quantity is 1500

3.

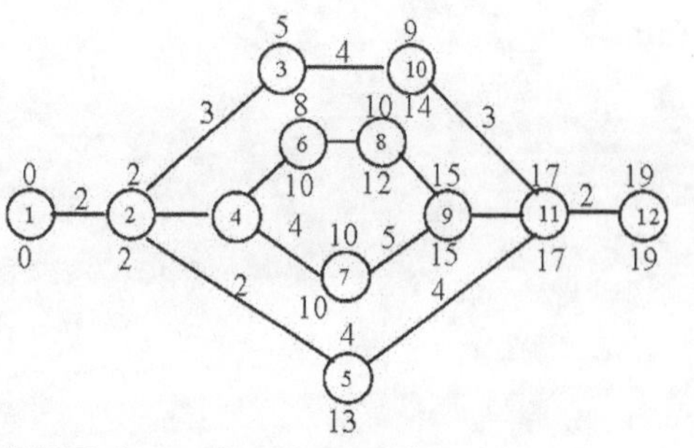

(LFT) for node 10 = 14 days

4.

	M/C I	M/C II	
T C	12	6	6000
M C	4	10	4000
G M	2	3	1800
Profit/unit	40 x	100 y	

$$Z_{max} = 40x + 100y$$

S.t

$12x + 6y \le 6000,\ 4x + 10y \le 4000$

$2x + 3y \le 1800\ x,\ y \ge 0$

$$\frac{x}{500} + \frac{y}{1000} \le 1 \quad \frac{x}{1000} + \frac{y}{400} \le 1$$

$$\frac{x}{900} + \frac{y}{600} \le 1$$

$$12x + 6y = 6000$$

$$4x + 10y = 4000$$

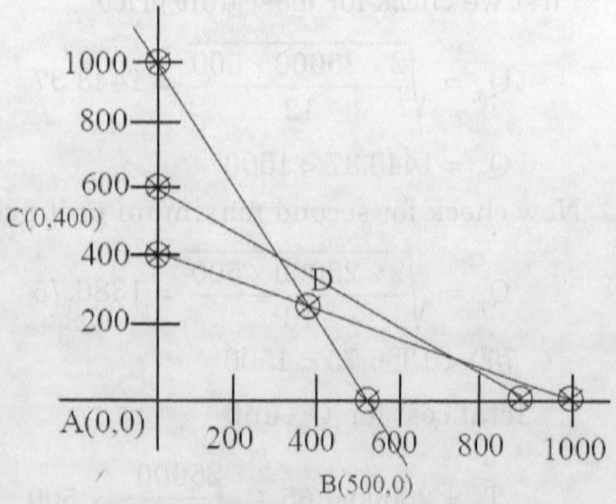

$$12x + 6y = 6000$$
$$12x + 30y = 12000$$
$$-24y = -6000$$
$$y = \frac{6000}{24} = 250$$
$$4x + 10y = 4000$$
$$4x + 250 \times 10 = 4000$$
$$4x = 1500$$
$$x = \frac{1500}{4} = 375$$
$$Z_{max} = 40x + 100y$$
$$= 40 \times 375 + 100 \times 250$$
$$= 15000 + 25000$$
$$Z_D = 40{,}000/-$$
$$Z_A = 0$$
$$Z_B = 40 \times 500 + 100 \times 0$$
$$= 20{,}000$$
$$Z_C = 40 \times 0 + 100 \times 400$$
$$= 40{,}000$$

Optional at (D) & (C).

SOLVED PAPER - 2017

INSTRUCTIONS

1. Total of 65 questions carrying 100 marks, out of which 10 questions carrying a total of 15 marks are in General Aptitude (GA)
2. The Engineering Mathematics will carry around **15% of the total marks,** the General Aptitude section will carry **15% of the total marks** and the **remaining 70% of the total marks.**
3. **Types of Questions**
   (*a*) **Multiple Choice Questions (MCQ)** carrying 1 or 2 marks each in all papers and sections. These questions are objective in nature, and each will have a choice of four options, out of which the candidate has to mark the correct answer(s).
   (*b*) **Numerical Answer Questions** of 1 or 2 marks each in all papers and sections. For these questions the answer is a real number, to be entered by the candidate using the virtual keypad. No choices will be shown for these type of questions.
4. For **1-mark** multiple-choice questions, **1/3 marks** will be deducted for a wrong answer. Likewise, for **2-marks** multiple-choice questions, **2/3** marks will be deducted for a wrong answer. There is no negative marking for numerical answer type questions.

Chapter-Wise Analysis

#	Subject	Marks		
		1	2	Total
1	General Aptitude	5	5	15
2	Engineering Mathematics	5	4	13
3	Engineering Mechanics	-	2	4
4	Strength of Materials	5	4	13
5	Theory of Machines	1	2	5
6	Vibrations	-	1	2
7	Design of Machine Elements	-	1	2
8	Fluid Mechanics	4	3	10
9	Heat Transfer	1	1	3
10	Thermodynamics	5	4	13

#	Subject	Marks		
		1	2	Total
11	Engineering Materials	-	2	4
12	Metal Casting	-	-	-
13	Forming Processes	2	2	6
14	Metal Joining Processes	1	2	5
15	Machining and Machine Tool Operations	1	2	5
16	Metrology and Inspection	-	-	-
17	Computer Integrated Manufacturing	-	-	-
18	Production, Planning and Control	-	-	-
19	Inventory Control	-	-	-
20	Operation Research	-	-	-

TECHNICAL SECTION

(Q.1 – 25) : Carry One Mark Each

1. In a metal forming operation when the material has just started yielding, the principal stresses are

$\sigma_1 = +180$ MPa, $\sigma_2 = -100$ MPa, $\sigma_3 = 0$

Following von Mises' criterion, the yield stress is ______ MPa.

2. A heat pump absorbs 10 KW of heat from outside environment at 250 K while absorbing 15 kW of work. It delivers the heat to a room that must be kept warm at 300 K. The coefficient of Performance (COP) of the heat pump is______

3. Saturated steam at 100°C condenses on the outside of tube. Cold fluid enters the tube at 20°C and exits at 50°C. The value of the log mean Temperature Difference (LMTD) is______°C.

4. The damping ratio for a viscously damped spring mass system, governed by the relationship

$m\frac{d^2x}{dt^2} + c\frac{dx}{dt} + kx = F(t)$, is given by

(*a*) $\sqrt{\frac{c}{mk}}$

(*b*) $\frac{c}{2\sqrt{km}}$

(*c*) $\frac{c}{\sqrt{km}}$

(*d*) $\sqrt{\frac{c}{2mk}}$

5. In the engineering stress-strain curve for mild steel, the Ultimate Tensile Strength (UTS) refers to

(*a*) Yield stress

(*b*) Proportional limit

(*c*) Maximum stress

(*d*) Fracture stress

6. Which one of the following is NOT a rotating machine?

(*a*) Centrifugal pump

(*b*) Gear pump

(*c*) Jet pump

(*d*) Vane pump

7. In an arc welding process, welding speed is doubled. Assuming all other process parameters to be constant, the cross sectional area of the weld bead will

(*a*) Increase by 25% (*b*) Increase by 50%

(*c*) Reduce by 25% (*d*) Reduce by 50%

8. The following figure shows the velocity-time plot for a particle traveling along a straight line. The distance covered by the particle from $t = 0$ to $t = 5s$ is ________m.

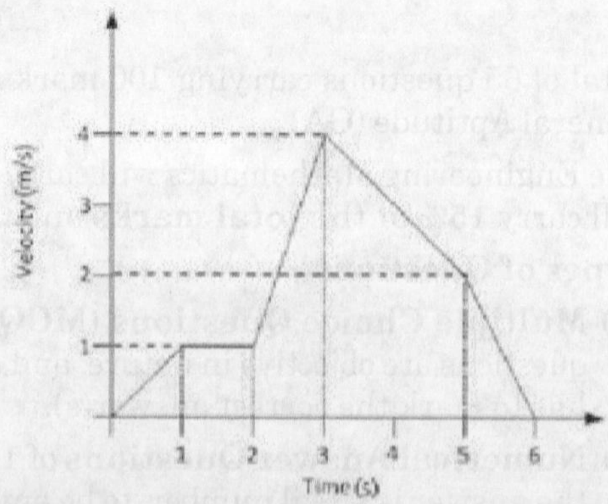

9. A motor driving a solid circular steel shaft transmits 40 kW of power at 500 rpm. If the diameter of the shaft is 40 mm, the maximum shear stress in the shaft is ______MPa.

10. The value of $\lim_{x\to 0} \frac{x^3 - \sin(x)}{x}$ is

(*a*) 0 (*b*) 3

(*c*) 1 (*d*) –1

11. The differential equation $\frac{d^2y}{dx^2} + 16y = 0$ for $ky(x)$ with the two boundary conditions $\left.\frac{dy}{dx}\right|_{x=0} = 1$ and $\left.\frac{dy}{dx}\right|_{x=\frac{\pi}{2}} = -1$ has

(a) No solution

(*b*) Exactly two solutions

(*c*) Exactly one solution

(*d*) Infinitely many solutions

12. Match the processes with their characteristics

Process	Characteristics
P. Electrical Discharge Machining	1. No residual stress
Q. Ultrasonic Machining	2. Machining of electrically conductive materials
R. Chemical Machining	3. Machining of glass
S. Ion Beam Machining	4. Nano-machining

(*a*) P-2, Q-3, R-1, S-4 (*b*) P-3, Q-2, R-1, S-4

(*c*) P-3, Q-2, R-4, S-1 (*d*) P-2, Q-4, R-3, S-1

13. Consider the following partial differential equation for u(x,y) with the constant c>1:

$$\frac{\partial u}{\partial y} + c\frac{\partial u}{\partial x} = 0$$

Solution of this equation is

(*a*) $u(x, y) = f(x + cy)$

(*b*) $u(x, y) = f(x - cy)$

(*c*) $u(x, y) = f(cx + y)$

(*d*) $u(x, y) = f(cx - y)$

14. For steady flow of a viscous incompressible fluid through a circular pipe of constant diameter, the average velocity in the fully developed region is constant. Which one of the following statements about the average velocity in the developing region is TRUE?

(*a*) It increases until the flow is fully developed

(*b*) It is constant and is equal to the average velocity in the fully developed region.

(*c*) It decreases until the flow is fully developed.

(*d*) It is constant but is always lower than the average velocity in the fully developed region.

15. The Poisson's ratio for a perfectly incompressible linear elastic material is

(*a*) 1 (*b*) 0.5

(*c*) 0 (*d*) Infinity

16. Cylindrical pins of diameter $15^{\pm 0.020}$ mm are being produced on a machine. Statistical quality control tests show a mean of 14.995 mm and standard deviation of 0.004 mm. The process capability index C_p is

(*a*) 0.833 (*b*) 1.667

(*c*) 3.333 (*d*) 3.750

17. Consider a beam with circular cross-section of diameter d. The ratio of the second moment of area about the neutral axis to the section modulus of the area is

(*a*) $\frac{d}{2}$ (*b*) $\frac{\pi d}{2}$

(*c*) d (*d*) πd

18. Metric thread of 0.8 mm pitch is to be cut on a lathe. Pitch of the lead screw is 1.5 mm. If the spindle rotates at 1500 rpm, the speed of rotation of the lead screw (rpm) will be________

19. A particle of unit mass is moving on a plane. Its trajectory, in polar coordinates, is given by $r(t) = t^2, \theta(t) = t$ where t is time. The kinetic energy of the particle at time $t = 2$ is

(*a*) 4 (*b*) 12

(*c*) 16 (*d*) 24

20. Consider the two-dimensional velocity field given by $\vec{V} = (5 + a_1 x + b_1 y)\hat{i} + (4 + a_2 x + b_2 y)\hat{j}$, where a_1, b_1, a_2 and b_2 are constants. Which one of the following conditions needs to be satisfied for the flow to be incompressible?

(*a*) $a_1 + b_1 = 0$ (*b*) $a_1 + b_2 = 0$

(*c*) $a_2 + b_2 = 0$ (*d*) $a_2 + b_1 = 0$

21. A six-face fair dice is rolled a large number of times. The mean value of the outcomes is ____

22. Water (density = 1000 kg/m^3) at ambient temperature flows through a horizontal pipe of uniform cross-section at the rate of 1 kg/s. If the pressure drop across the pipe is 100 kPa, the minimum power required to pump the water across the pipe, in watts, is______

23. The molar specific heat at constant volume of an ideal gas is equal to 2.5 times the universal gas constant (8.314 J/mol.K). When the temperature increases by 100 K, the change in molar specific enthalpy is ____ J/mol.

24. Consider the schematic of a riveted lap joint subjected to tensile load F, as shown below. Let d be the diameter of the rivets, and s_f be the maximum permissible tensile stress in the plates. What should be the minimum value for the thickness of the plates to guard against tensile failure of the plates? Assume the plates to be identical.

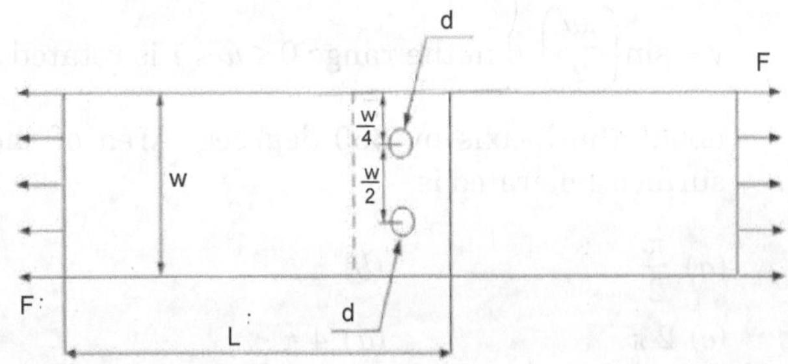

(*a*) $\frac{f}{S_f(W - 2d)}$ (*b*) $\frac{F}{S_f W}$

(*c*) $\frac{f}{S_f(W - d)}$ (*d*) $\frac{2F}{S_f W}$

25. The product of eigenvalues of the matrix P is

$$P = \begin{bmatrix} 2 & 0 & 1 \\ 4 & -3 & 3 \\ 0 & 2 & -1 \end{bmatrix}$$

(*a*) – 6 (*b*) 2

(*c*) 6 (*d*) – 2

(Q.26 – 55) : Carry Two Marks Each

26. Assume that the surface roughness profile is triangular as shown schematically in the figure. If the peak to valley height is 20μm , the central line average surface roughness R_o (in μm) is

(*a*) 5 (*b*) 6.67
(*c*) 10 (*d*) 20

27. Consider steady flow of an incompressible fluid through two long and straight pipes of diameters d_1 and d_2 arranged in series. Both pipes are of equal length and the flow is turbulent in both pipes. The friction factor for turbulent flow though pipes is of the form, $f = K(Re)^{-n}$, where K and *n* are known positive constants and Re is the Reynolds number. Neglecting minor losses, the ratio of the frictional pressure drop in pipe 1 to that in pipe 2, $\left(\frac{\Delta P_1}{\Delta P_2}\right)$, is given by

(*a*) $\left(\frac{d_2}{d_1}\right)^{(5-n)}$ (*b*) $\left(\frac{d_2}{d_1}\right)^{5}$
(*c*) $\left(\frac{d_2}{d_1}\right)^{(3-n)}$ (*d*) $\left(\frac{d_2}{d_1}\right)^{(5+n)}$

28. A parametric curve defined by $x = \cos\left(\frac{\pi u}{2}\right)$, $y = \sin\left(\frac{\pi u}{2}\right)$ d in the range $0 \le u \le 1$ is rotated about the X-axis by 360 degrees. Area of the surface generated is

(*a*) $\frac{\pi}{2}$ (*b*) π
(*c*) 2 π (*d*) 4 π

29. P(0, 3), Q(0.5, 4) and R(1, 5) are three points on the curve defined by *f*(*x*). Numerical integration is carried out using both Trapezoidal rule and Simpson's rule within limits $x = 0$ and $x = 1$ for the

(*a*) 0 (*b*) 0.25
(*c*) 0.5 (*d*) 1

30. Heat is generated uniformly in a long solid cylindrical rod (diameter =10 mm) at the rate of 4×10^7 W/m^3. The thermal conductivity of the rod material is 25 W/m.K. Under steady state conditions, the temperature difference between the centre and the surface of the rod is _____ °C.

31. A sprue in a sand mould has a top diameter of 20 mm and height of 200 mm. The velocity of the molten metal at the entry of the sprue is 0.5 m/s. Assume acceleration due to gravity as 9.8 m/s^2 and neglect all losses. If the mould is well ventilated, the velocity (upto 3 decimal points accuracy) of the molten metal at the bottom of the sprue is _____ m/s

32. In an epicyclic gear train, shown in the figure, the outer ring gear is fixed, while the sum gear rotates counterclockwise at 100 rpm. Let the number of teeth on the sum, planet and outer gears to be 50, 25, and 100, respectively. The ratio of magnitudes of angular velocity of the planet gear to the angular velocity of the carrier arm is _____.

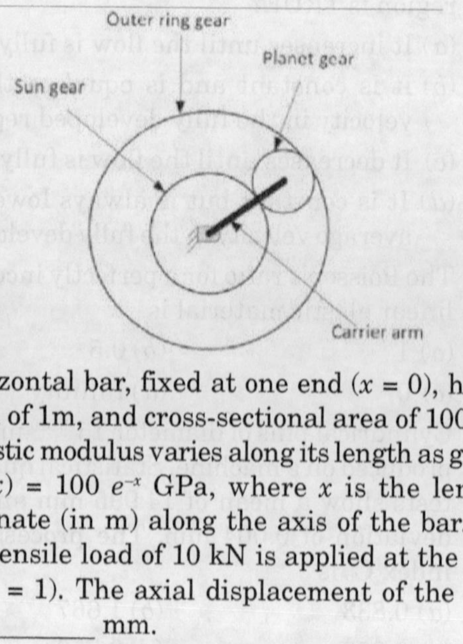

33. A horizontal bar, fixed at one end (x = 0), has a length of 1m, and cross-sectional area of 100 m^2. Its elastic modulus varies along its length as given by $E(x) = 100\ e^{-x}$ GPa, where x is the length coordinate (in m) along the axis of the bar. An axial tensile load of 10 kN is applied at the free end (x = 1). The axial displacement of the free end is _____ mm.

34. One kg of an ideal gas (gas constant, R = 400 J/kg. K; specific heat at constant volume, c_v = 1000 J/kg – K) at 1 bar, and 300 K is contained in a sealed rigid cylinder. During an adiabatic process, 100 kJ of work is done on the system by a stirrer. The increase in entropy of the system is _____ J/K

35. An initially stress-free massless elastic beam of length L and circular cross-section with diameter *d* (*d*<<L) is held fixed between two walls as shown. The beam material has Young's modulus E and coefficient of thermal expansion *α*.

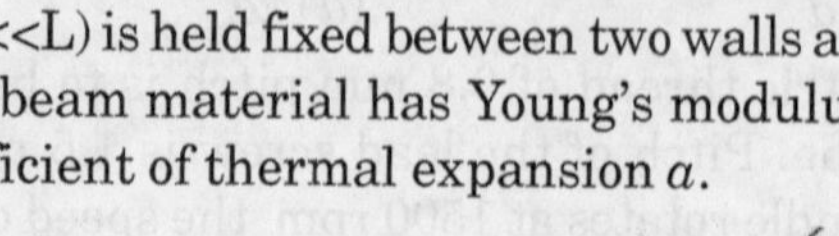

If the beam is slowly and uniformly heated, the temperature rise required to cause the beam to buckle is proportional to

(*a*) d (*b*) d^2
(*c*) d^3 (*d*) d^4

36. Two disks A and B with identical mass (m) and radius (R) are initially at rest. They roll down from the top of identical inclined planes without shipping. Disk A has all of its mass concentrated at the rim, while Disk B has its mass uniformly distributed. At the bottom of the plane, the ratio of velocity of the center of disk A to the velocity of the center of disk B is

(a) $\sqrt{\frac{3}{4}}$ (b) $\sqrt{\frac{3}{2}}$

(c) 1 (d) $\sqrt{2}$

37. A block of length 200 mm is machined by a slab milling cutter 34 mm in diameter. The depth of cut and table feed are set at 2 mm and 18 mm minute, respectively. Considering the approach and the over travel of the cutter to be same, the minimum estimated machining time per pass is ________ minutes.

38. For an inline slider-crank mechanism, the lengths of the crank and connecting rod are 3 m and 4 m. respectively. At the instant when the connecting rod is perpendicular to the crank, if the velocity of the slider is 1m/s, the magnitude of angular velocity (upto 3 decimal points accuracy) of the crank is ________ radian/s.

39. A thin uniform rigid bar of length L and mass M is hinged at point O, located at a distance of $\frac{L}{3}$ from one of its ends. The bar is further supported using springs, each of stiffness k, located at the two ends. A particle of mass $m = \frac{M}{4}$ is fixed at one end of the bar, as shown in the figure. For small rotations of the bar about O, the natural frequency of the system is

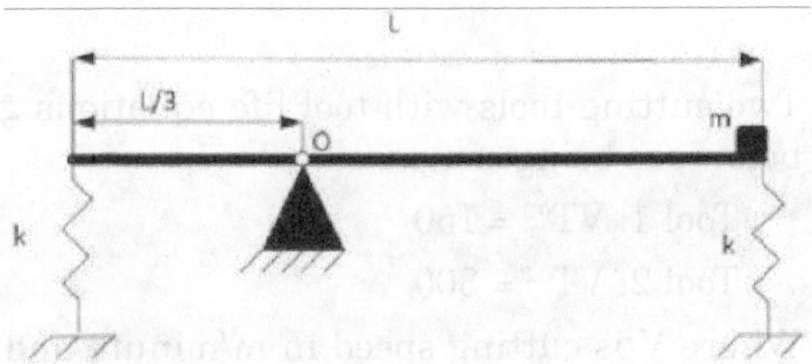

(a) $\sqrt{\frac{5k}{M}}$ (b) $\sqrt{\frac{5k}{2M}}$

(c) $\sqrt{\frac{3k}{2M}}$ (d) $\sqrt{\frac{3k}{M}}$

40. Moist air is treated as an ideal gas mixture of water vapor and dry air (molecular weight of air = 28.84 and molecular weight of water = 18). At a location, the total pressure is 100 kPa, the temperature is 30°C and the relative humidity is 55%. Given that the saturation pressure of water at 30°C is 4246 Pa, the mass of water vapor per kg of dry air is ________ grams.

41. Air contains 78% N_2 and 21% O_2 on a molar basis. Methane (CH_4) is burned with 50% excess air than required stoichiometrically. Assuming complete combustion of methane, the molar percentage of N_2 in the products is ________

42. Two models, P and Q, of a product earn profits of Rs. 100 and Rs. 80 per piece, respectively. Production times for P and Q are 5 hours and 3 hours, respectively, while the total production time available is 150 hours. For a total batch size of 40, to maximize profit, the number of units of P to be produced is ________

43. The velocity profile inside the boundary layer for flow over a flat plate is given as $\frac{u}{U_\infty} = \sin\left(\frac{\pi}{2}\frac{y}{\delta}\right)$, where U_∞ is the free stream velocity and δ is the local boundary layer thickness. If δ^* is the local displacement thickness, the value of $\frac{\delta^*}{\delta}$ is

(a) $\frac{2}{\pi}$ (b) $1-\frac{2}{\pi}$

(c) $1+\frac{2}{\pi}$ (d) 0

44. The pressure ratio across a gas turbine (for air, specific heat at constant pressure, c_p = 1040 J/kg.K and ratio of specific heats, γ = 1.4) is 10. If the inlet temperature to the turbine is 1200 K and the isentropic efficiency is 0.9, the gas temperature at trubine exit is ______ K.

45. A rectangular region in a solid is in a state of plane strain. The (x, y) coordinates of corners of the undeformed rectangle are given by P(0, 0), Q(4, 0), R(4, 3), S(0, 3). The rectangle is subjected to uniform strains, ε_{xx} = 0.001, ε_{yy} = 0.002, γ_{xy} = 0.003. The deformed length of the elongated diagonal, upto three decimal places is ______ units.

46. Two black surfaces. AB and BC, of lengths 5 m and 6 m, respectively, are oriented as shown. Both surfaces extend infinitely into the third dimension. Given that view factor F_{12} = 0.5, T_1 = 800 K, T_2 = 600 K, $T_{surrounding}$ = 300 K and stefan Boltzmann constant, $\sigma = 5.67 \times 10^{-8}$ W/(m^{2}K^4). The heat transfer rate from surface 2 to the surrounding environment is ________ kW.

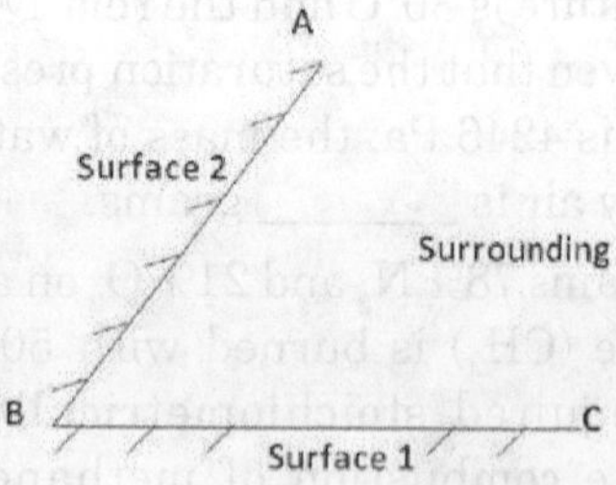

47. Consider the matrix $P = \begin{bmatrix} \frac{1}{\sqrt{2}} & 0 & \frac{1}{\sqrt{2}} \\ 0 & 1 & 0 \\ \frac{-1}{\sqrt{2}} & 0 & \frac{1}{\sqrt{2}} \end{bmatrix}$

Which one of the following statements about P is INCORRECT?

(*a*) Determinant of P is equal to 1.

(*b*) P is orthogonal

(*c*) Inverse of P is equal to its transpose.

(*d*) All eigenvalues of P are real numbers.

48. A 10 mm deep cylindrical cup with diameter of 15 mm is drawn from a circular blank. Neglecting the variation in the sheet thickness, the diameter (upto 2 decimal points accuracy) of the blank is ________ mm.

49. For the vector $\vec{V} = 2yz\hat{i} + 3xz\hat{j} + 4xy\hat{k}$,

the value of $\nabla.(\nabla \times \vec{V})$ is ______

50. A machine element has an ultimate strength (σ_u) of 600 N/mm², and endurance limit (σ_{en}) of 250 N/mm². The fatigue curve for the element on a log-log plot is shown below. If the element is to be designed for a finite life of 10000 cycles, the maximum amplitude of a completely reversed operating stress is ______ N/mm²

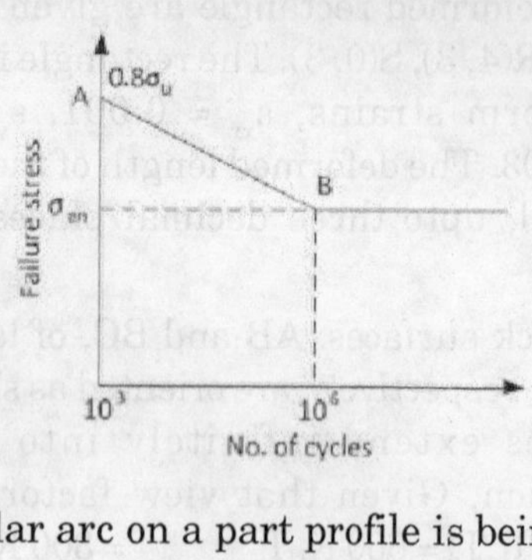

51. Circular arc on a part profile is being machined on a vertical CNC milling machine. CNC part program using meric units with absolute dimensions is listed below :

```
N60 G01 X 30 Y 55 Z-5 F50
N70 G02 X 50 Y 35 R 20
N80 G01 Z 5
```

The coordinates of the centre of the circular arc are :

(*a*) (30, 55) (*b*) (50, 55)

(*c*) (50, 35) (*d*) (30, 35)

52. Following data refers to the job (P, Q, R, S) which have arrived at a machine for scheduling. The shortest possible average flow time is ___ days.

Job	Processing Time (days)
P	15
Q	9
R	22
S	12

53. A point mass of 100 kg is dropped onto a massless elastic bar (cross-sectional area = 100 mm². Length =1m, Young's modulus = 100 (GPa) from a height H of 10 mm as shown (Figure is not to scale). If g =10 m/s², the maximum compression of the elastic bar is ________ mm.

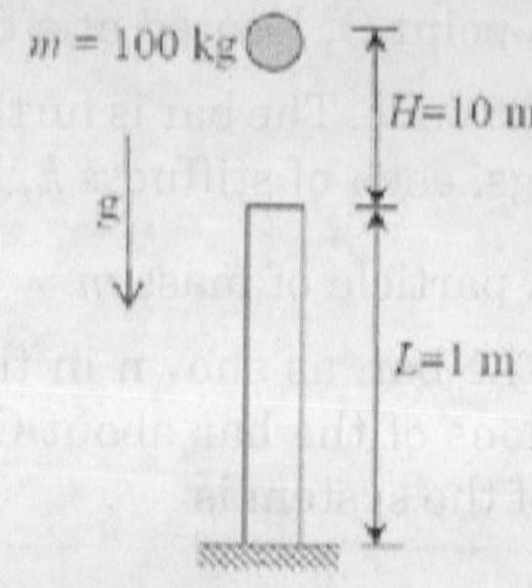

54. Two cutting tools with tool life equations given below are being compared:

Tool 1: $VT^{0.1} = 150$

Tool 2: $VT^{0.3} = 300$

Where V is cutting speed in m/minute and T is tool life in minutes. The breakeven cutting speed beyond which Tool 2 will have a higher tool life is__________ m/minute.

55. For a steady flow, the velocity field is $\vec{V} = (-x^2 + 3y)\hat{i} + (2xy)\hat{j}$. The magnitude of the acceleration of a particle at (1, –1) is

(*a*) 2 (*b*) 1

(*c*) $2\sqrt{5}$ (*d*) 0

GENERAL APTITUDE (GA)

(Q.1 – 5) : Carry One Mark Each

1. He was one of my best_________ and I felt his loss_______

(*a*) Friend, keenly (*b*) Friends, keen

(*c*) Friend, keener (*d*) Friends, keenly

2. A right-angled cone (with base radius 5 cm and height 12 cm), as shown in the figure below , is rolled on the ground keeping the point P fixed until the point Q (at the base of the cone, as shown) touches the ground again.

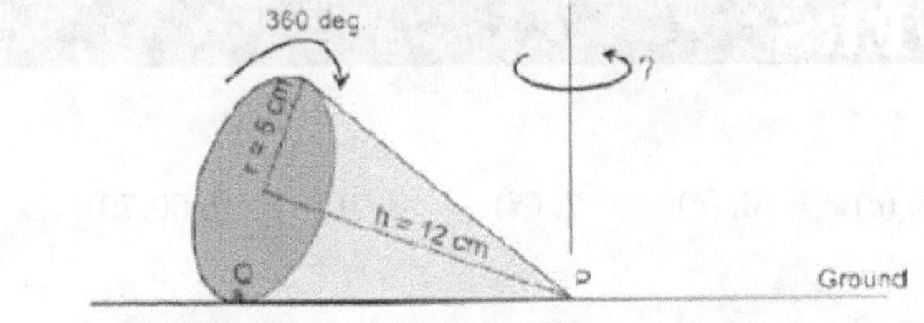

(*a*) $\frac{5\pi}{12}$ (*b*) $\frac{5\pi}{24}$

(*c*) $\frac{24\pi}{5}$ (*d*) $\frac{10\pi}{13}$

3. In a company with 100 employees, 45 earn ₹ 20,000 per month, 25 earn ₹ 30,000, 20 earn ₹ 40,000, 8 earn ₹ 60,000 and 2 earn ₹ 150,000. The median of the salaries is______

(*a*) ₹ 20,000 (*b*) ₹ 30,000

(*c*) ₹ 32,300 (*d*) ₹ 40,000

4. P, Q and R talk about S's car collection. P states that S has at least 3 cars. Q belives that S has less than 3 cars R indicates that to his knowledge, S has at least one car. Only one of P,Q and R is right. The number of cars owned by S is

(*a*) 0

(*b*) 1

(*c*) 3

(*d*) Cannot be determined

5. As the two speakers became increasingly agitated, the debate became_________.

(*a*) Lukewarm (*b*) Poetic

(*c*) Forgiving (*d*) Heated

(Q.6 – 10) : Carry Two Marks Each

6. The growth of bacteria (lactobacillus) in milk leads to curd formation. A minimum bacterial population density of 0.8 (in suitable units) is needed to form curd. In the graph below, the population density of lactobacillus in 1 litre of milk is plotted as a function of time, at two different temperature, 25°C and 37°C.

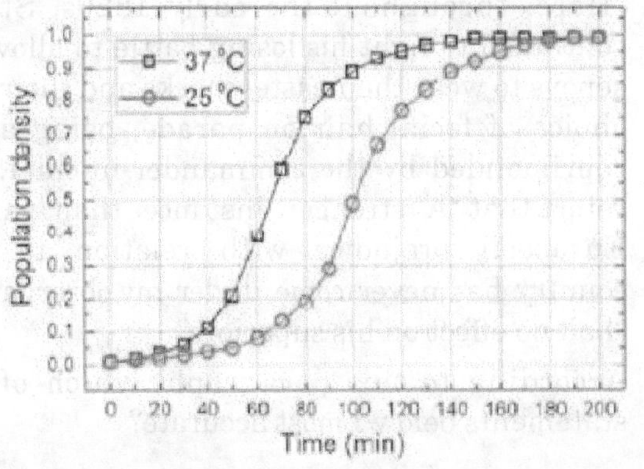

Consider the following statements based on the data shown above:

i. The growth in bacterial population stops earlier at 37°C as compared to 25°C

ii. The time taken for curd formation at 25°C is twice the time taken at 37°C Which one of the following options is correct?

(*a*) Only i (*b*) Only ii

(*c*) Both i and ii (*d*) Neither i nor ii

7. What is the sum of the missing digits in the subtraction problem below?

$$\begin{array}{r} 5\,\_\,\_\,\_\,\_ \\ -\,4\,8\,\_\,8\,9 \\ \hline 1\,1\,1\,1 \end{array}$$

(*a*) 8

(*b*) 10

(*c*) 11

(*d*) Cannot be determined

8. Let S_1 be the plane figure consisting of the points (x, y) given by the inequalities $|x - 1| \le 2$ and $|y + 2| \le 3$. Let S_2 be the plane figure given by the inequalities $x - y \ge -2$, $y \ge 1$, and $x \le 3$ Let S be the union of S_1 and S_2. The area of S is

(*a*) 26 (*b*) 28

(*c*) 32 (*d*) 34

9. Two very famous sportsmen Mark and Steve happened to be brothers, and played for country K. Mark teased James, an opponent from country E, "There is no way you are good enough to play for your country." James replied, "Maybe not, but at least I am the best player in my own family."

Which one of the following can be inferred from this conversation?

(*a*) Mark was known to play better than James

(*b*) Steve was known to play better than Mark

(*c*) James and Steve were good friends

(*d*) James played better than Steve

10. "Here, throughout the early 1820s, Stuart continued to fight his losing battle to allow his sepoys to wear their caste- marks and their own choice of facial hair on parade, being again reprimanded by the commander-in-chief. His retort that "A stronger instance than this of European prejudice with relation to this country has never come under my observations "had no effect on his superiors"

According to this paragraph, which of the statements below I most accurate?

(*a*) Start's commander-in-chief was moved by this demonstration of his prejudice

(*b*) The Europeans were accommodating of the sepoys' desire to wear their caste-marks

(*c*) Stuart's losing battle' refers to his inability to succeed in enabling sepoys to wear cast-marks

(*d*) The commander-in-chief was exempt from the European prejudice that dictated how the sepoys were to dress.

ANSWERS

Technical Section

1. 245.76 **2.** 1.67 **3.** 63.83 **4.** (*b*) **5.** (*c*) **6.** (*c*) **7.** (*d*) **8.** 10 **9.** 60.79

10. (*d*) **11.** (*a*) **12.** (*a*) **13.** (*b*) **14.** (*b*) **15.** (*b*) **16.** (*b*) **17.** (*a*) **18.** 800 rpm

19. (*c*) **20.** (*b*) **21.** 3.5 **22.** 100 **23.** 2909.9 **24.** (*a*) **25.** (*b*)

26. (*c*) **27.** (*a*) **28.** (*c*) **29.** (*a*) **30.** 10 or –10 **31.** 2.043 **32.** 3

33. 1.716 **34.** 287.68 **35.** (*b*) **36.** (*a*) **37.** 12 **38.** 0.267 **39.** (*b*)

40. 14.2 **41.** 73.82 **42.** 15 **43.** (*b*) **44.** 679.4 kelvin **45.** 5.015

46. 13.778 **47.** (*d*) **48.** 28.722 **49.** 0 **50.** 386.19MPa **51.** (*b*) **52.** 31

53. 1.517 **54.** 106.6 **55.** (*c*)

General Aptitude (GA)

1. (*d*) **2.** (*d*) **3.** (*b*) **4.** (*a*) **5.** (*d*) **6.** (*a*) **7.** (8) **8.** (*c*) **9.** (*b*) **10.** (*c*)

EXPLANATIONS

TECHNICAL SECTION

1. According to Von – Mises- theory

$$\sigma_{yt} = \sqrt{\frac{1}{2}(\sigma_1-\sigma_2)^2+(\sigma_2-\sigma_3)^2+(\sigma_3-\sigma_1)^2}$$

$$= \sqrt{\frac{1}{2}(280^2+100^2+180^2)}$$

$= 245.76 \times 10^6$ Pascal

$= 245.76$ MPa

2.

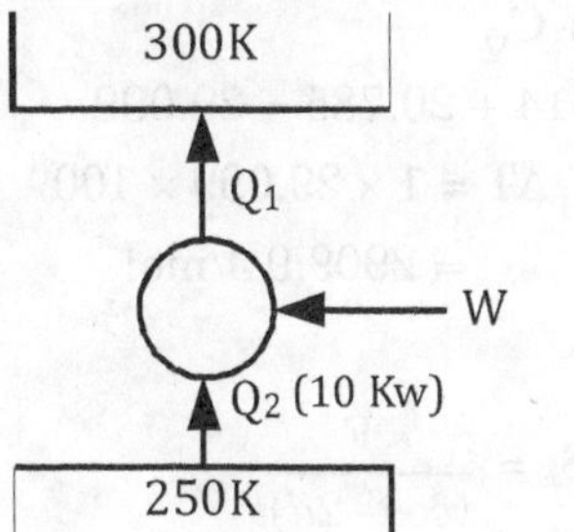

By question, given:

$Q_2 = 10$kW ; W = 15kW

$Q_1 = Q_2 + W = 25$kW

$$\text{COP} = \frac{Q_1}{W} = 1.67$$

3.

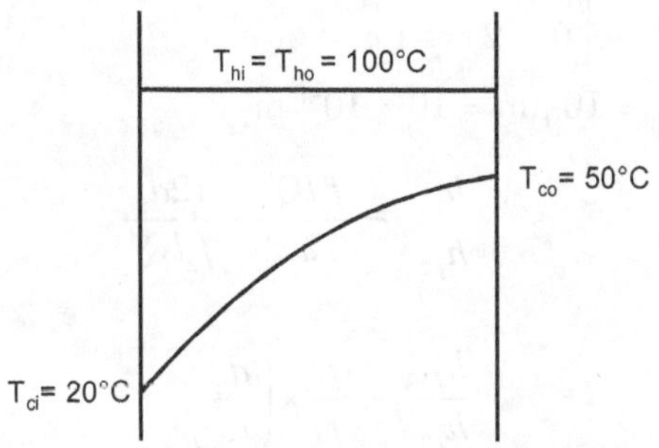

Here

$T_{ci} \rightarrow$ cold fluid which enters the tube (tempt.)

$T_{co} \rightarrow$ cold fluid which exits at tube (tempt.)

$\theta_1 = T_{hi} - T_{ci} = 100 - 20 = 80°C$

$\theta_2 = T_{ho} - T_{co} = 100 - 50 = 50°C$

$$\theta_{im} = \frac{\theta_1-\theta_2}{ln\frac{\theta_1}{\theta_2}} = \frac{80°-50°}{ln\left(\frac{80}{50}\right)} = 63.83°C$$

7. For a weld, volume remains constant.

If welding speed is doubled then original cross sectional area would become half if volume remains constant.

Volume = Area × Velocity

∴ Area reduces by 50%.

8. Distance = Velocity × Time

Distance from $t = 0$ to $t = 5$s

= Area under the curve from $t = 0$ to $t = 5$

$$= \frac{1}{2}\times1\times1+1\times1+\frac{1}{2}\times1\times3+1\times1+\frac{1}{2}\times2\times2+2\times2$$

$$= \frac{1}{2}+1+\frac{3}{2}+\frac{1}{1}+2+4 = 4+2+1+1+2 = 10$$

9. We know that, Power is given by

$$P = \frac{2\pi NT}{60}$$

$$40\times10^3 = \frac{2\pi\times500\times T}{60}$$

T = 763.94 Nm

$$T = \frac{\pi}{16}d^3\tau$$

$$763.94\times10^3 = \frac{\pi}{16}\times40^3\times\tau$$

$\tau = 60.79 \times 10^6$ pascal

$\tau = 60.79$ MPa

10. From L – hospital rule

$$\lim_{x\to0}\frac{x^3-\sin(x)}{x} = \frac{3x^2-\cos x}{1}$$

$$= \frac{3\times0^2-\cos0}{1} = \frac{0-1}{1} = -1$$

11. $(D^2 + 16)y = 0$

A.E. $\rightarrow D^2 + 16 = 0$

$\Rightarrow D = \pm 4i$

$y = C_1\cos 4x + C_2 \sin 4x$

$$\frac{dy}{dx} = -4C_1\sin 4x + 4C_2\cos 4x$$

$$\left.\frac{dy}{dx}\right|_{x=0} = -4C_1\times0 + 4C_2\times1 = 4C_2$$

$4C_2 = 1$

$$C_2 = \frac{1}{4}$$

Now,

$$\left.\frac{dy}{dx}\right|_{x=\pi/2} = -4C_1\times\sin 4/2 + 4C_2\times\cos 4/2$$

$$\left.\frac{dy}{dx}\right|_{x=\pi/2} = 0 + 4C_2 = 4\left(\frac{1}{4}\right) = 1$$

But $\left.\frac{dy}{dx}\right|_{x=\pi/2} = -1$ (given)

Hence no solution.

16. Process capability index $C_P = \frac{USL - LSL}{6\sigma}$

USL = Upper std. limit = 15 + 0.02 = 15.02 mm
LSL = Lower std. limit = 15 − 0.02 = 14.98 mm
σ = std. deviation = 0.004 mm

$$C_P = \frac{15.02 - 14.98}{6 \times 0.004} = 1.67$$

17. Second moment of area or MOI of circular beam = I

Section modulus $Z = \frac{I}{y}$

$$\frac{MOI}{Z} = \frac{I}{I_y} = y = \frac{d}{2}$$

18. 800 rpm (Rotation per minute)

19. $r = t^2$

$\theta = t$

$x = r\cos\theta = t^2 \cos t$

$y = r\sin\theta = t^2 \sin t$

$$u = \frac{dx}{dt} = 2t\cos t - t^2 \sin t$$

at $t = 2$

$u = 4\cos 2 - 4\sin 2 = -5.301$

$$v = \frac{dy}{dt} = 2t\sin t + t^2\cos t$$

at $t = 2$

$v = 4\sin 2 + 4\cos 2 = 1.972$

$$V = \sqrt{u^2 + v^2} = 5.656$$

$$K.E. = \frac{1}{2}mv^2$$

$$= \frac{1}{2} \times 1 \times 5.656^2 = 15.99$$

20. For continuous and incompressible flow

$u_x + u_y = 0$

$a_1 + b_2 = 0$

21.

x.	1	2	3	4	5	6
p(x)	1/6	1/6	1/6	1/6	1/6	1/6

We know that

$\mu = \sum x.p(x)$

$= (1 + 2 + 3 + 4 + 5 + 6) \times \frac{1}{6}$

$= 21 \times \frac{1}{6} = 3.5$

22. $W = \dot{m}v(p_1 - p_2)$

$W = 1 \times \frac{1}{1000} \times 100 \times 10^3 = 100$ Watt

23. $C_V = 2.5\,R$ (R = 8.314)

$C_V = 2.5 \times 8.314 = 20.785$ J/molK

$C_P = R + C_V$

$= 8.314 + 20.785 = 29.099$

$\Delta H = nC_P\Delta T = 1 \times 29.099 \times 100$

$= 2909.9$ J/mol

24. Tensile stress

$$S_f = \frac{F}{(\cos - 2d)t}$$

$$t = \frac{F}{s_f(\omega - 2d)}$$

25. Product of eigen value = |p|

$$= \begin{vmatrix} 2 & 0 & 1 \\ 4 & -3 & 3 \\ 0 & 2 & -1 \end{vmatrix} = 2$$

26. $R_0 = 10\ \mu m = 10 \times 10^{-9}$ m

27.

$$\frac{h_{f1}}{h_{f2}} = \frac{f_1 l Q^2}{12 d_1^5} \times \frac{12 d_2^5}{f_2 l Q^2}$$

$$\frac{h_{f1}}{h_{f2}} = \frac{f_1}{f_2} \times \left(\frac{d_2}{d_1}\right)^5$$

$$f \propto R^{-n}$$

$$f \propto \left(\frac{\rho V d}{\mu}\right)^{-n}$$

$$f \propto \left(\frac{\rho Q d}{\pi/4\, d^2 \mu}\right)^{-n}$$

$$f \propto d^n$$

$$\frac{h_{f1}}{h_{f2}} = \left(\frac{d_1}{d_2}\right)^n \times \left(\frac{d_2}{d_1}\right)^5$$

$$\frac{h_{f1}}{h_{f2}} = \left(\frac{d_2}{d_1}\right)^{5-n}$$

29. $y_0 = 3, y_1 = 4, y_2 = 5$

$$h = 0.5$$

From Trapezoidal rule, we know that

$$I_T = \frac{h}{2}[(y_0 + y_2) + 2y_1]$$

$$= \frac{0.5}{2}[(3+5) + 2 \times 4] = 4$$

Also by Simpson's $\frac{1}{3}rd$ rule

$$I_S = \frac{h}{3}[(y_0 + y_2) + 4y_1]$$

$$= \frac{0.5}{3}[(3+5) + 4 \times 4] = 4$$

$$I_T - I_S = 4 - 4 = 0$$

30. $$\frac{d}{dr}\left(r\frac{dT}{dr}\right) + \frac{q_g r}{k} = 0$$

$$\frac{d}{dr}\left(r\frac{dT}{dr}\right) + \frac{4 \times 10^7}{25} r = 0$$

$$\frac{d}{dr}\left(r\frac{dT}{dr}\right) + 16 \times 10^5 r = 0$$

$$r\frac{dT}{dr} + 8 \times 10^5 r^2 = C_1$$

at $r = 0$, $\frac{dT}{dr} = 0$

$C_1 = 0$

$\therefore$ $$r\frac{dT}{dr} + 8 \times 10^5 r^2 == 0$$

$$\frac{dT}{dr} + 8 \times 10^5 r = 0$$

$$T + 4 \times 10^5 r^2 = C_2$$

$$T = C_2 - 4 \times 10^5 r^2$$

$$T_{r=R} - (T)_{r=0} = (C_2 - 4 \times 10^5 r^2) - C_2$$

$$(T)_{r=R} - (T)_{r=0} = -4 \times 10^5 r^2 = -4 \times 10^5 (0.005)^2$$

$$(T)_{r=R} - (T)_{r=0} = -10°C$$

31.

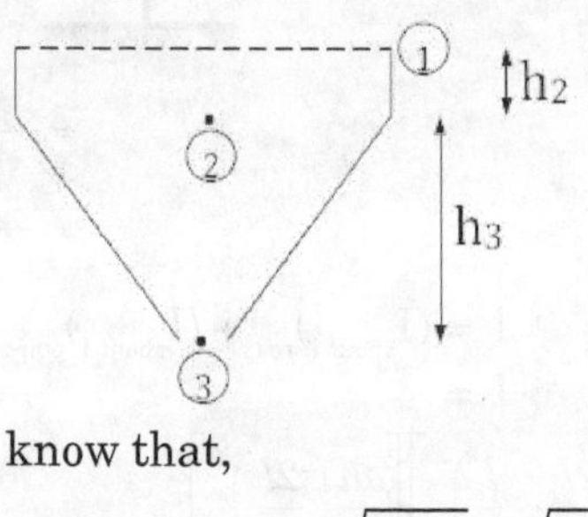

We know that,

$$V_2 = \sqrt{2gh_2} = \sqrt{2 \times 9.81 \times h_2} = 0.5$$

$$h_2 = 0.01274 \text{ m}$$

$$h_3 = h_2 + 0.2 = 0.2127 \text{ m}$$

$$V_3 = \sqrt{2gh_3} = 2.043 \text{ m/sec}$$

32.

	Arm	S	P	R
Arm → Fix S → +1	0	1	–2	–0.5
×x	0	x	$-2x$	$0.5x$
Add y	Y	$x + y$	$-2x + y$	$0.5x + y$

$$N_S = 100 \Rightarrow x + y = 100 \quad ...(i)$$

$$N_R = 0 \Rightarrow 0.5x + y = 0 \quad ...(ii)$$

By solving eqn (*i*) and (*ii*), we get

$$x = \frac{200}{3},$$

$$y = \frac{100}{3}$$

$$\frac{N_P}{N_{arm}} = \frac{-2x + y}{y}$$

$$= \frac{-2\left(\frac{200}{3}\right) + \frac{100}{3}}{\frac{100}{3}} = -3$$

33. $$\Delta l = \int_0^l \frac{Pl}{AE} dx = \frac{Pl}{A}\int_0^l \frac{1}{100e^{-x} \times 10^3} dx$$

$$= \frac{10 \times 10^3 \times 1 \times 10^3}{100} \times \frac{1}{100 \times 10^3} \times \left[e^x\right]_0^1$$

$$= 1.718 \text{ mm}$$

34. Here,

$$m = 1 \text{ kg}$$

$$R = 400 \text{ J/kgK}$$

$$C_V = 1000 \text{ J/kgK}$$

$$P_1 = 1 \text{ bar}$$

$$T_1 = 300 \text{ K}$$

$$W = -100 \text{ kJ} = -100 \times 10^3 \text{ J}$$

$$Q = 0$$

$$0 = -100 + mC_V(T_2 - T_1)$$

$$100 \times 10^3 = 1 \times 1000\,(T_2 - 300)$$

$$T_2 = 400 \text{ K}$$

$$V_1 = V_2$$

$$\Rightarrow \quad \ln\frac{V_2}{V_1} = 0$$

$$S_2 - S_1 = mC_V \ln\frac{T_2}{T_1}$$

$$S_2 - S_1 = 1 \times 1000 \times \ln\left(\frac{400}{300}\right)$$

$$S_2 - S_1 = 287.68 \text{ J/K}$$

35. By Buckling load,

$$P = \frac{4\pi^2 EI}{l^2}$$

Force due to change in temperature,

$$\sigma = \frac{P}{A} = l\alpha\Delta T$$

$$P = Al\alpha\Delta T$$

$$\frac{4\pi^2 EI}{l^2} = Al\alpha\Delta T$$

$$\frac{I}{A} \propto \Delta T$$

$$\Delta T \propto \frac{d^4}{d^2}$$

$$\Delta T \propto d^2$$

36. Disc A $\rightarrow$ I$= mR^2 + mR^2 = 2mR^2$

Disc B $\rightarrow$ I$= \frac{mR^2}{2} + mR^2 = \frac{3}{2}mR^2$

For A $\rightarrow mgx = \frac{1}{2}(2mR^2).\left(\frac{V^2}{R^2}\right)$

$$\sqrt{gx} = V_A$$

For B $\rightarrow mgx = \frac{1}{2}\left(\frac{3}{2}mR^2\right).\left(\frac{V^2}{R^2}\right)$

$$\sqrt{\frac{4}{3}gx} = V_B$$

$$\frac{V_A}{V_B} = \sqrt{\frac{3}{4}}$$

37.

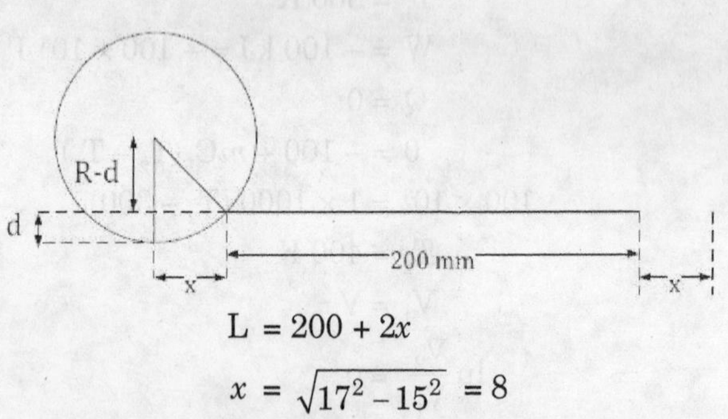

$$L = 200 + 2x$$

$$x = \sqrt{17^2 - 15^2} = 8$$

$\therefore$ $L = 216$ mm

$$\text{Time} = \frac{L}{fN}$$

$$= \frac{216}{18} = 12 \text{ min}$$

38.

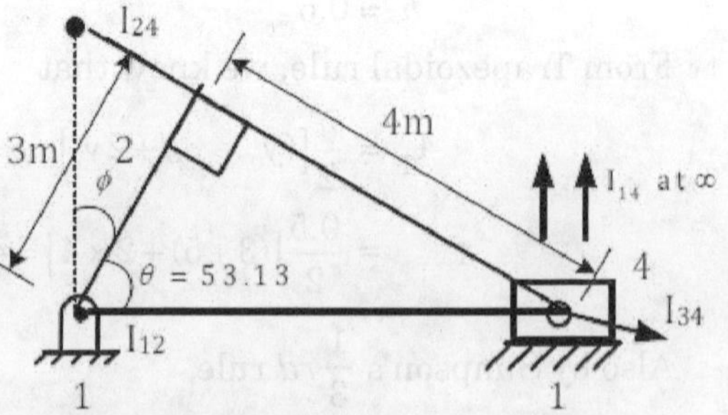

Here, $\tan\theta = \frac{\text{Perpendicular}}{\text{Hypotenuse}}$

$$\tan\theta = \frac{4}{3}$$

$\Rightarrow$ $\theta = 53.13$

$$\phi = 90 - \theta = 36.87$$

$$\cos\phi = \frac{3}{I_{12} - I_{24}}$$

$$\cos 36.87 = \frac{3}{I_{12} - I_{24}}$$

$$I_{12} - I_{24} = 3.75$$

$$V_{I_{24}} = \text{velocity of slider}$$

$$= \omega_2(I_{12} - I_{24})$$

$$I = \omega_2 \times 3.75$$

$$\omega_2 = 0.267 \text{ radian/second}$$

39.

$$x_2 = \frac{2l}{3}\theta$$

$$x_1 = \frac{l}{3}\theta$$

$$I = (I_{about\ I})_{rod} + (I_{about\ I})_{point\ mass}$$

$$I =$$

$$\left[\frac{ml^2}{12} + m\left(\frac{2l}{3} - \frac{l}{2}\right)^2\right]\left[\frac{m}{4}\left(\frac{2l}{3}\right)^2\right]$$

$$I = \frac{ml^2}{9} + \frac{ml^2}{9} = \frac{2ml^2}{9}$$

$$\Sigma M_I = 0$$

$$I\ddot{\theta} + Kx_2 + \frac{2l}{3} + Kx_1 \times \frac{l}{3} = 0$$

$$\frac{2ml^2}{9} + k\left(\frac{5}{9}l^2\right)\theta = 0$$

$$\omega_n = \sqrt{\frac{k \times \frac{5}{9}}{\frac{2m}{9}}} = \sqrt{\frac{5k}{2m}}$$

40.
$$w = 0.622\frac{P_V}{P - P_V}$$

$$\phi = \frac{P_V}{P_S}$$

$$\Rightarrow \quad 0.55 = \frac{P_V}{P_S}$$

$$\Rightarrow \quad P_V = 2235.3$$

$$w = 0.622 \times \frac{2235.3}{(100 \times 10^3) - 2235.3}$$

w = 0.0142 kg/kg of dry air

w = 14.2 gm/kg of dry air

42. Maximize,

$$Z = 100P + 80Q$$

Constraint,

$$5P + 3Q \le 150$$
$$P + Q \le 40$$

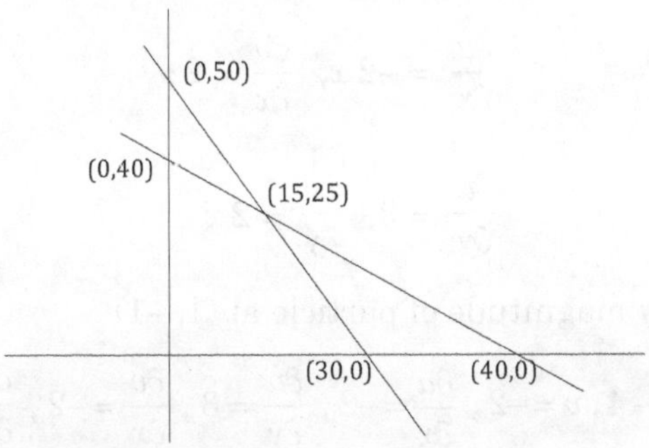

Z (0, 40) = 3200

Z (30, 0) = 3000

Z(15, 25) = 3500 = Z_{max}

$\therefore$ P = 15

43.
$$\frac{\delta^*}{\delta} = \frac{\int_0^\delta \left(1 - \frac{u}{u_\infty}\right)dy}{\delta} = \frac{\int_0^\delta \left(1 - \sin\frac{\pi y}{2\delta}\right)dy}{\delta}$$

By integrating and re-arranging we get,

$$\frac{\delta^*}{\delta} = 1 - \frac{2}{\pi}$$

44.

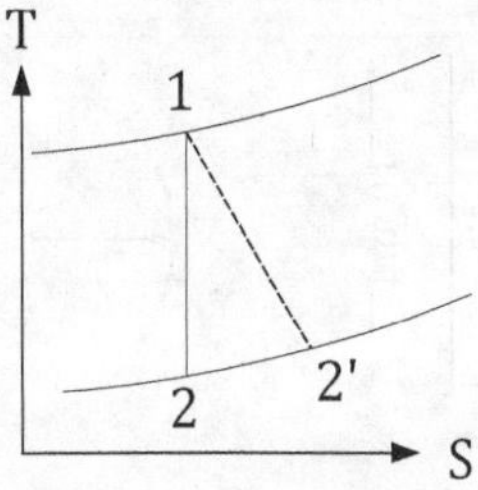

By question

T_1 = 1200 K

$$r_p = 10 = \frac{P_1}{P_2}$$

$\eta_T = 0.9$

$T_2 = ?$

$$\frac{T_2}{T_1} = \left(\frac{P_2}{P_1}\right)^{\frac{\gamma-1}{\gamma}}$$

$$\frac{T_2}{1200} = \left(\frac{1}{10}\right)^{\frac{1.4-1}{1.4}}$$

T_2 = 621.53 K

$$\eta_T = \frac{T_1 - T_{2'}}{T_1 - T_2}$$

$$0.9 = \frac{1200 - T_{2'}}{1200 - 621.53}$$

T_2 = 679.4 K

45.

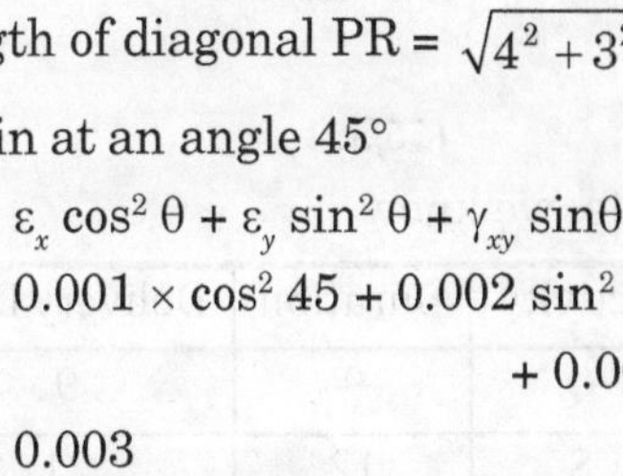

Length of diagonal PR = $\sqrt{4^2 + 3^2} = 5$

Strain at an angle 45°

$\varepsilon = \varepsilon_x \cos^2\theta + \varepsilon_y \sin^2\theta + \gamma_{xy} \sin\theta \cos\theta$

$= 0.001 \times \cos^2 45 + 0.002 \sin^2 45 + 0.003 \sin 45 \cos 45$

$= 0.003$

$$\varepsilon = \frac{\Delta l}{l} = 0.003$$

$\Delta l = 5 \times 0.003 = 0.015$ units

Deformed length = 5 + 0.015 = 5.015 units

47. By finding the eigen values of given matrix, we concludes that all eigen values of P are not real Numbers.

48.

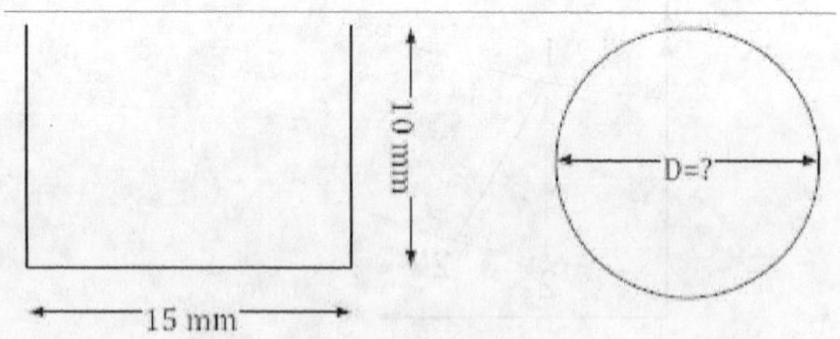

Here

$d = 15$ mm and $h = 10$ mm

Now, by formula

$$D = \sqrt{d^2 + 4dh}$$

$$= \sqrt{(15)^2 + 4 \times 15 \times 10} = 28.72 \text{ mm}$$

49. Given vector $\vec{V} = 2yz\hat{i} + 3xz\hat{j} + 4xy\hat{k}$

$$\text{Now } \nabla \times \vec{V} = \begin{vmatrix} i & j & k \\ \frac{\partial}{\partial x} & \frac{\partial}{\partial y} & \frac{\partial}{\partial z} \\ 2yz & 3xz & 4xy \end{vmatrix} = xi - 2yj + zk$$

$$\nabla.(\nabla \times \vec{V}) = 1 - 2 + 1 = 0$$

50. $$\frac{\log S_f - \log(0.8 \times 600)}{\log 250 - \log(0.6 \times 600)} = \frac{4-3}{6-3}$$

$$\Rightarrow \frac{\log S_f - \log(480)}{\log 250 - \log(360)} = \frac{1}{3}$$

$S_f = 386.19$ MPa

51.

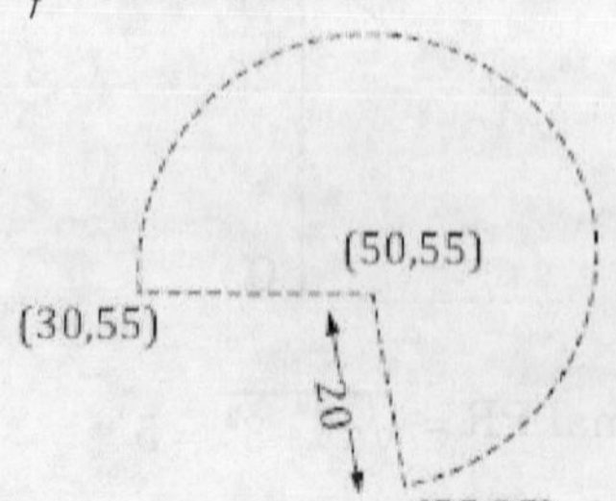

52. By SPT rule, we know

S.No.	Activity	Duration	Delivery Date
1	Q	9	9
2	S	12	21
3	P	15	36
4	R	22	58

$$\text{Average flow time} = \frac{9+21+36+58}{4}$$

$= 31$ days

53. Strain energy stored in bar = potential energy lost by mass

$$\frac{1}{2} \times P \times \Delta l = W(h + \Delta l)$$

$$\frac{1}{2} \times P \times \frac{Pl}{AE} = 100 \times 10 \times 10^3 \left(10 + \frac{Pl}{AE}\right)$$

$$\frac{1}{2} \times P \times \frac{P \times 1}{100 \times 10^{-6} \times 100 \times 10^9}$$

$$= 100 \times 10 \left(10 \times 10^{-3} + \frac{P \times 1}{100 \times 10^{-6} \times 100 \times 10^9}\right)$$

$$P = 15177.45N$$

$$\Delta l = \frac{Pl}{AE} = \frac{15177.45 \times 1}{100 \times 10^{-6} \times 100 \times 10^9}$$

$$= 1.517 \times 10^{-3} \text{ m}$$

$$= 1.517 mm$$

54. $VT^{0.1} = 150$

$VT^{0.3} = 300$

For breakeven cutting speed,

$$3\ln 150 - 3\ln V - \ln 300 - \ln V$$

$$V = 106.66 m/\min$$

55. $$\vec{V} = (-x^2 + 3y)i + (2xy)j$$

$$u = -x^2 + 3y, v = 2xy$$

$$\frac{\partial u}{\partial x} = -2x, \quad \frac{\partial v}{\partial x} = 2y$$

$$\frac{\partial u}{\partial y} = 3, \quad \frac{\partial v}{\partial y} = 2x$$

Now magnitude of particle at (1, –1)

$$u = -4, v = -2, \frac{\partial u}{\partial x} = -2, \frac{\partial u}{\partial y} = 3, \frac{\partial v}{\partial x} = -2, \frac{\partial v}{\partial y} = 2$$

$$a_x = u\frac{\partial u}{\partial x} + v\frac{\partial u}{\partial y} = (-4) \times (-2) + (-2) \times 3$$

$$a_x = 8 - 6 = 2$$

$$a_y = u\frac{\partial v}{\partial x} + v\frac{\partial v}{\partial y}$$

$$= [(-4) \times (-2) + (-2) \times 2]$$

$$a_y = 8 - 4 = 4$$

$$a = \sqrt{a_x^2 + a_y^2}$$

$$= \sqrt{2^2 + 4^2} = 2\sqrt{5}$$

GENERAL APTITUDE (GA)

3.

Salary	No. of employees	Cumulative
2000	45	45
30000	25	70
40000	20	90
60000	8	98
150000	2	100

Median = $\frac{100+1}{2}$ = 50.5

From cumulative column,

Median salary = 30000

7.

$$\begin{array}{r} 50000 \\ 48889 \\ \hline 1111 \end{array}$$

$sum = 0 + 0 + 0 + 0 + 8 = 8$

8.

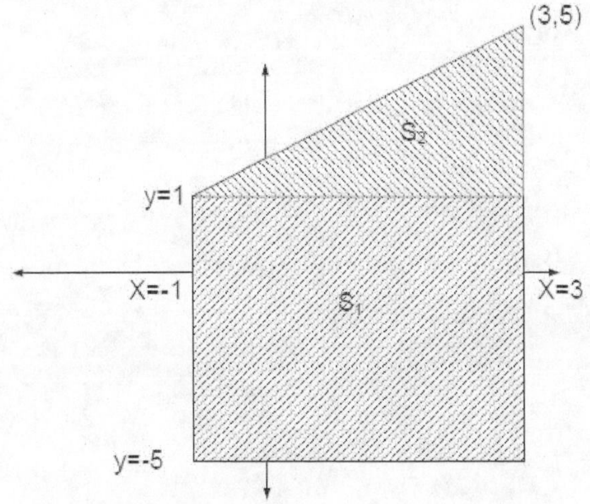

$$S = S_1 + S_2 = 6 \times 4 + \frac{1}{2} \times 4 \times 4$$

$$= 24 + 8 = 32$$

SOLVED PAPER - 2018

INSTRUCTIONS

1. Total of 65 questions carrying 100 marks, out of which 10 questions carrying a total of 15 marks are in General Aptitude (GA)
2. The Engineering Mathematics will carry around **15% of the total marks,** the General Aptitude section will carry **15% of the total marks** and the **remaining 70% of the total marks.**
3. **Types of Questions**
   (*a*) Multiple Choice Questions (MCQ) carrying 1 or 2 marks each in all papers and sections. These questions are objective in nature, and each will have a choice of four options, out of which the candidate has to mark the correct answer(s).
   (*b*) Numerical Answer Questions of 1 or 2 marks each in all papers and sections. For these questions the answer is a real number, to be entered by the candidate using the virtual keypad. No choices will be shown for these type of questions.
4. For **1-mark** multiple-choice questions, **1/3 marks** will be deducted for a wrong answer. Likewise, for **2-marks** multiple-choice questions, **2/3** marks will be deducted for a wrong answer. There is no negative marking for numerical answer type questions.

Chapter-Wise Analysis

#	Chapters	Marks		
		1	2	Total
1	General Aptitude	5	5	15
2	Engineering Mathematics	5	5	15
3	Engineering Mechanics	1	3	7
4	Strength of Materials	3	4	11
5	Theory of Machines	3	4	11

#	Chapters	Marks		
		1	2	Total
6	Design of Machine Elements	1	1	3
7	Fluid Mechanics	3	4	11
8	Heat Transfer	0	1	2
9	Thermodynamics, Refrigeration & Air-Conditioning	2	3	8
10	Materials, Manufacturing & Industrial Engg.	7	5	17

GENERAL APTITUDE

1. Going by the ______ that many hands make light work, the school ______ involved all the students in the task."

(*a*) principle, principal

(*b*) principal, principle

(*c*) principle, principle

(*d*) principal, principal

2. "Her ____ should not be confused with miserliness; she is ever willing to assist those in need."

The word that best fills the blank in the above sentence is

(*a*) cleanliness (*b*) punctuality

(*c*) frugality (*d*) greatness

3. Seven machines take 7 minutes to make 7 identical toys. At the same rate, how many minutes would it take for 100 machines to make 100 toys?

(*a*) 1 (*b*) 7

(*c*) 100 (*d*) 700

4. A rectangle becomes a square when its length and breadth are reduced by 10m and 5 m, respectively, during this process, the rectangle loses 650 m^2 of area. What is the area of the original rectangle in square meters?

(*a*) 1125 (*b*) 2250

(*c*) 2924 (*d*) 4500

5. A number consists of two digits. The sum of the digits is 9. If 45 is subtracted from the number, its digits are interchanged. What is the number?

(*a*) 63 (*b*) 72

(*c*) 81 (*d*) 90

6. For integers a, b and c, what would be the minimum and maximum values respectively of a + b + c if log |a| + log |b| + log |c| = 0

(*a*) –3 and 3

(*b*) –1 and 1

(*c*) –1 and 3

(*d*) 1 and 3

7. Given that a and b are integers and $a + a^2b^3$ is odd, which one of the following statements is correct?

(*a*) a and b are both odd

(*b*) a and b are both even

(*c*) a is even and b is odd

(*d*) a is odd and b is even

8. From the time the front of train enters a platform, it takes 25 seconds for the back of the train to leave the platform, while travelling at a constant speed of 54 km/h. At the same speed, it takes 14 seconds to pass a man running at 9 km/h in the same direction as the train. What is the length of the train and that of the platform in meters, respectively?

(*a*) 210 and 140 (*b*) 162.5 and 187.5

(*c*) 245 and 130 (*d*) 175 and 200

9. Which of the following functions describe the graph shown in the below figure.

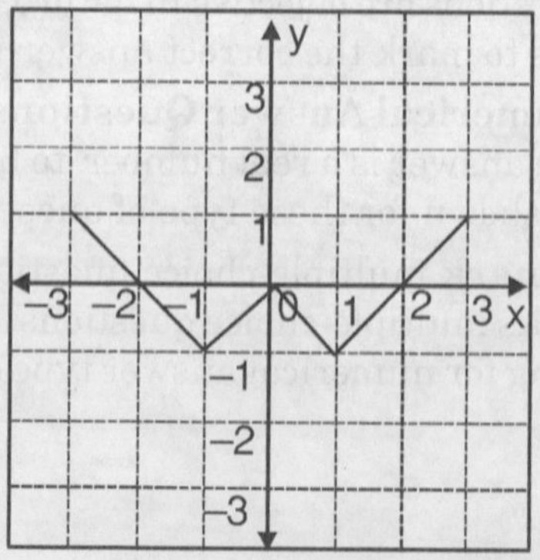

(*a*) $y = ||x| + 1| - 2$ (*b*) $y = ||x| - 1| - 1$

(*c*) $y = ||x| + 1| - 1$ (*d*) $y = ||x - 1| - 1|$

10. Consider the following three statements:

(i) Some roses are red

(ii) All red flowers fade quickly

(iii) Some roses fade quickly

Which of the following statements can be logically inferred from the above statements?

(*a*) If (i) is true and (ii) is false, then (iii) is false.

(*b*) If (i) is true and (ii) is false, then (iii) is true.

(*c*) If (i) and (ii) are true, then (iii) is true.

(*d*) If (i) and (ii) are false, then (iii) is false.

MECHANICAL ENGINEERING

1. Four red balls, four green balls and four blue balls are put in a box. Three balls are pulled out of the box of random one after another without replacement. The probability that all the three balls are red is

(*a*) 1/72 (*b*) 1/55

(*c*) 1/36 (*d*) 1/27

2. The rank of the matrix $\begin{bmatrix} -4 & 1 & -1 \\ -1 & -1 & -1 \\ 7 & -3 & 1 \end{bmatrix}$ is

(*a*) 1 (*b*) 2

(*c*) 3 (*d*) 4

3. According to the Mean Value Theorem, for a continuous function $f(x)$ in the interval $[a, b]$, there exists a value ξ in this interval such that

$$\int_n^b f(x)\, dx =$$

(a) $f(\xi)(b-a)$ (b) $f(b)(\xi-a)$

(c) $f(a)(b-\xi)$ (d) 0

4. $F(z)$ is a function of the complex variable $z = x + iy$ given by

For what value of k will $F(z)$ satisfy the Cauchy-riemann equations?

$F(z) = iz + k\, Re(z) + i\, Im(z)$

(a) 0 (b) 1

(c) –1 (d) y

5. A bar of uniform cross section and weighing 100N is held horizontally using two massless and inextensible strings S_1 and S_2 as shown in the figure.

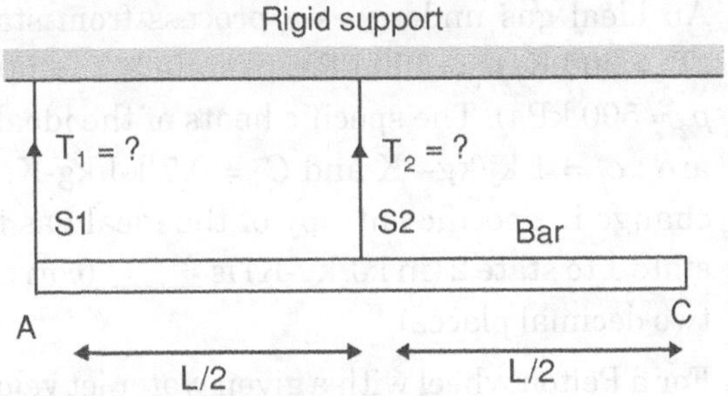

(a) T_1 = 100N and T_2 = 0N

(b) T_1 = 0 N and T_2 = 100 N

(c) T_1 = 75 N and T_2 = 25 N

(d) T_1 = 25 N and T_2 = 75N

6. If σ_1 and σ_3 are the algebraically largest and smallest principal stresses respectively, the value of the maximum shear stress is

(a) $\frac{\sigma_1+\sigma_3}{2}$ (b) $\frac{\sigma_1-\sigma_3}{2}$

(c) $\sqrt{\frac{\sigma_1+\sigma_3}{2}}$ (d) $\sqrt{\frac{\sigma_1-\sigma_3}{2}}$

7. The equation of motion for a spring-mass system excited by a harmonic force is

$$M\ddot{x} + Kx = F\cos(\omega t)$$

where M is the mass, K is the spring stiffness, F is the force amplitude and ω is the angular frequency of excitation. Resonance occurs when ω is equal to

(a) $\sqrt{\frac{M}{K}}$ (b) $\frac{1}{2\pi}\sqrt{\frac{K}{M}}$

(c) $2\pi\sqrt{\frac{K}{M}}$ (d) $\sqrt{\frac{K}{M}}$

8. For an Oldham coupling used between two shafts, which among the following statements are correct?

I. Torsional load is transferred along shaft axis.

II. A velocity ratio of 1 : 2 between shaft is obtained without using gears.

III. Bending load is transferred transverse to shaft axis.

IV. Rotation is transferred along shaft axis.

(a) I and III (b) I and IV

(c) II and III (d) II and IV

9. For a two-dimensional incompressible flow field given by $\vec{u} = A(x\hat{i} - y\hat{j})$, where A > 0, which one of the following statements is FALSE?

A. It satisfies continuity equation

B. It is unidirectional when $x \to 0$ and $y \to \infty$

C. Its streamlines are given by $x = y$.

D. It is irrotational

(a) A (b) B

(c) C (d) D

10. Which one of the following is correct a superheated vapour?

(a) Its pressure is less than the saturation pressure at a given temperature.

(b) Its temperature is less than saturation temperature at a given pressure.

(c) Its volume is less than the volume of the saturated vapour at a given temperature

(d) Its enthalpy is less than enthalpy of the saturated vapour at a given pressure.

11. In a linearly hardening plastic material, the true stress beyond initial yielding

(a) increases linearly with the true strain

(b) decreases linearly with the true strain

(c) first increases linearly and then decreases linearly with the true strain

(d) remain constant

12. The type of weld represented by the shaded region in the figure is

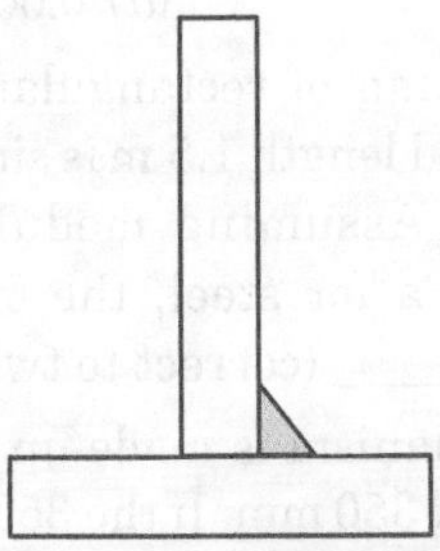

(a) groove (b) spot

(c) fillet (d) plug

13. Using the Taylor's tool life equation with exponent n = 0.5, if the cutting speed is reduced by 50%, the ratio of new tool life to original tool life is

(*a*) 4 (*b*) 2

(*c*) 1 (*d*) 0.5

14. A grinding ratio of 200 implies that the

(*a*) grinding wheel wears 200 times the volume of the material removed.

(*b*) grinding wheel wears 0.005 times the volume of the material removed.

(*c*) aspect ratio of abrasive particles used in the grinding wheel is 200.

(*d*) ratio of volume of abrasive particle to that of grinding wheel is 200.

15. Interpolator in a CNC machine

(*a*) controls spindle speed

(*b*) coordinates axes movements

(*c*) operates tool changer

(*d*) commands canned cycle

16. The time series forecasting method that gives equal weightage to each of the *m* most recent observation is

(*a*) Moving average method

(*b*) Exponential smoothing with linear trend

(*c*) Triple Exponential smoothing

(*d*) Kalman Filter

17. The number of atoms per unit cell and the number of slip systems, respectively, for a face-centered cubic (FCC) crystal are

(*a*) 3, 3

(*b*) 3, 12

(*c*) 4, 12

(*d*) 4, 48

18. A six-faced fair dice is rolled five times. The probability (in%) of obtaining "ONE" at least four times is

(*a*) 33.3 (*b*) 3.33

(*c*) 0.33 (*d*) 0.0033

19. A steel column of rectangular section (15 mm × 10 mm) and length 1.5 m is simply supported at both ends. Assuming modulus of elasticity, E = 200 GPa for steel, the critical axial load (in kN) is ______ (correct to two decimal places)

20. A four mechanism is made up of links of length 100, 200 and 350 mm. If the 350 mm link is fixed, the number of links that can rotate fully is ______.

21. If the wire diameter of a compressive helical spring is increased by 2%, the change in spring stiffness (in %) is ______(correct to two decimal places.)

22. A flat plate of width $L = 1$ m is pushed down with a velocity $U = 0.01$ m/s towards a wall resulting in the drainage of the fluid between the plate and the wall as shown in the figure. Assume two-dimensional incompressible flow and that the plate remains parallel to the wall. The average velocity, u_{avg} of the fluid (in m/s) draining out at the instant shown in the figure is ______(correct to three decimal places).

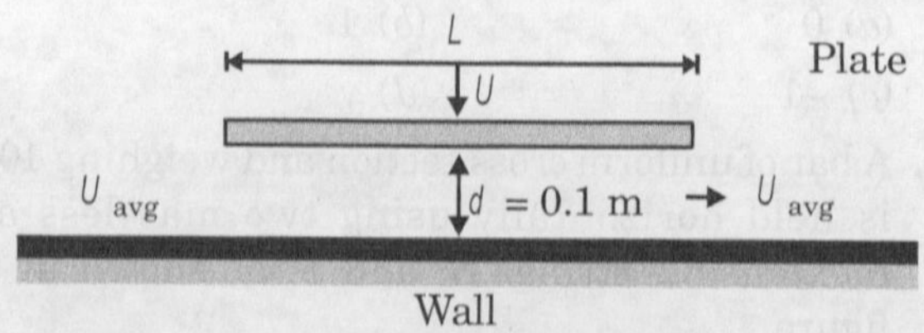

23. An ideal gas undergoes a process from state 1 ($T_1 = 300$ K, $p_1 = 100$ kPa) to state 2($T_2 = 600$ K, $p_2 = 500$ kPa). The specific heats of the ideal gas are : $c_p = 1$ kj/kg– K and $C_v = 0.7$ kJ/kg-K. The change in specific entropy of the ideal gas from state 1 to state 2 (in kJ/kg-K) is ______(correct to two decimal places)

24. For a Pelton wheel with a given water jet velocity, the maximum output power from the Pelton wheel is obtained when the ratio of the bucket speed to the water jet speed is ______(correct to two decimal places).

25. The height (in mm) for a 125 mm sine bar to measure a taper of 27′32′ on a flat work piece is ______ (correct to three decimal places).

26. Let X_1, X_2 be two independent normal random variables with means μ_1, μ_2 and standard deviations σ_1, σ_2 respectively. Consider $Y = X_1 - X_2$; $\mu_1 = \mu_2 = 1$, $\sigma_1 = 1$, $\sigma_2 = 2$, Then,

(*a*) *Y* is normal distributed with mean 0 and variance 1

(*b*) *Y* is normally distributed with mean 0 and variance 5

(*c*) *Y* has mean 0 and variance 5, but is NOT normally distributed

(*d*) *Y* has mean 0 and variance 1, but is NOT normally distributed

27. The value of integral $\oiint_S \vec{r}.\vec{n}\, ds$ over the closed surface S bounding a volume, where $\vec{r} = x\hat{i} + y\hat{i} + z\hat{k}$ is the position vector and $\vec{n}$ is the normal to the surface S, is

(a) V (b) $2V$

(c) $3V$ (d) $4V$

28. A point mass is shot vertically up from ground level with a velocity of 4 m/s at time, t = 0. It loses 20% of its impact velocity after each collision with the ground. Assuming that the acceleration due to gravity is 10 m/s^2 and that air resistance is negligible, the mass stops bouncing and comes to complete rest on the ground after a total time (in seconds) of

(a) 1 (b) 2

(c) 4 (d) ∞

29. The state of stress at a point, for a body in place stress, is shown in the figure below. If the minimum principal stress is 10 kPa, then the normal stress σ_s (in kPa) is

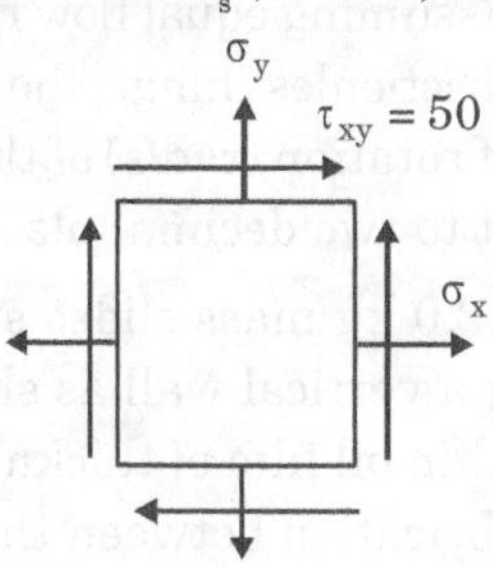

(a) 9.45 (b) 18.88

(c) 37.78 (d) 75.50

30. An epicyclic gear train is shown in the figure below. The number of teeth on the gears A, B and D are 20, 30 and 20, respectively. Gear C has 80 teeth on the inner surface and 100 teeth on the outer surface. If the carrier arm AB is fixed and the sun gear A rotates at 300 rpm in the clockwise direction, then the rpm of D in clockwise direction is

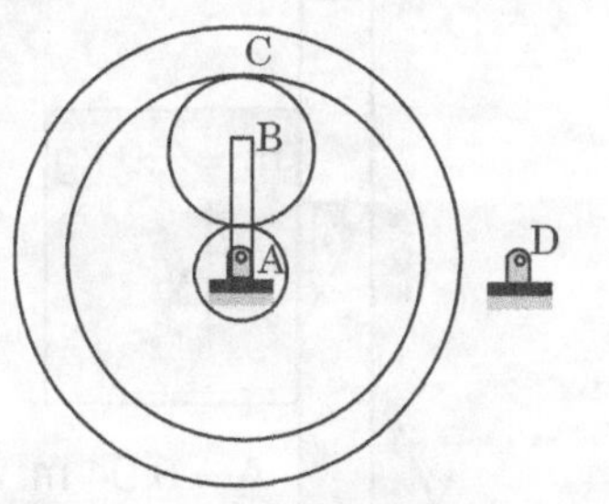

(a) 240 (b) –240

(c) 375 (d) –375

31. A carpenter glues a pair of cylindrical wooden logs by bonding their end faces, at an angle of θ = 30° as shown in the figure.

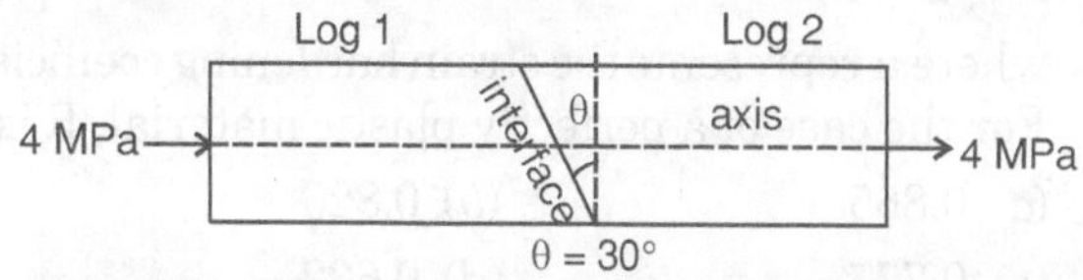

The glue used at the interface fails if

Criterion 1 : the maximum normal stress exceeds 2.5 MPa

Criterion 2 : the maximum shear stress exceeds 1.5 MPa

Assume that the interface fails before the logs fail. When a uniform tensile stress of 4 MPa is applied, the interface

(a) fails only because of criterion 1

(b) fails only because of criterion 2

(c) fails because of both criteria 1 and 2

(d) does not fail

32. A self-aligning ball bearing has a basic dynamic load rating (C_{10}, for 10^6 revolutions) of 35 kN. If the equivalent radial load on the bearing is 45 kN, the expected life (in 10^6 revolutions) is

(a) below 0.5 (b) 0.5 to 0.8

(c) 0.8 to 1.0 (d) above 1.0

33. A tank open at the top with a water level of 1 m, shown in the figure, has a hole at a height of 0.5 m. A free jet leaves horizontally from the smooth hole. The distance X (in m) where the jet strikes the floor is

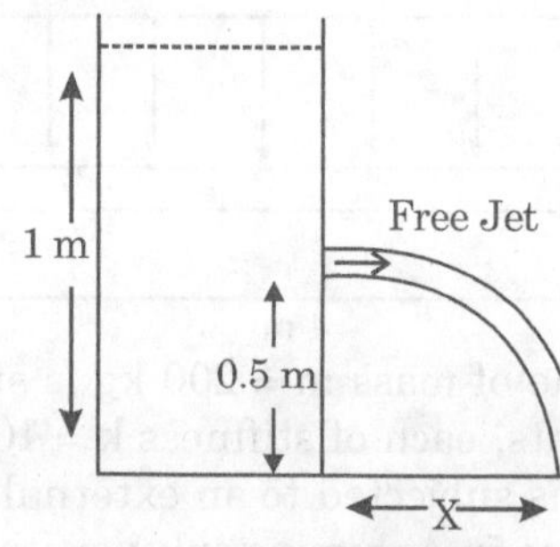

(a) 0.5 (b) 1.0

(c) 2.0 (d) 4.0

34. In a Lagrangian system, the position of fluid particle in a flow is described as $x = x_0 e^{-kt}$ and $y = y_0 e^{kt}$ where t is the time while x_0, y_0, and k are constants. The flow is

(a) unsteady and one-dimensional

(b) steady and two-dimensional

(c) steady and one-dimensional

(d) unsteady and two-dimensional

35. The maximum reduction in cross-sectional area per pass (R) of a cold wire drawing process is

$R = 1 - e^{-(n+1)}$

where n represents the strain hardening coefficient. For the case of a perfectly plastic material, R is

(*a*) 0.865 (*b*) 0.826

(*c*) 0.777 (*d*) 0.632

36. The percentage scrap in a sheet metal blanking operation of a continuous strip of sheet metal as shown in the figure ______(correct to two decimal places).

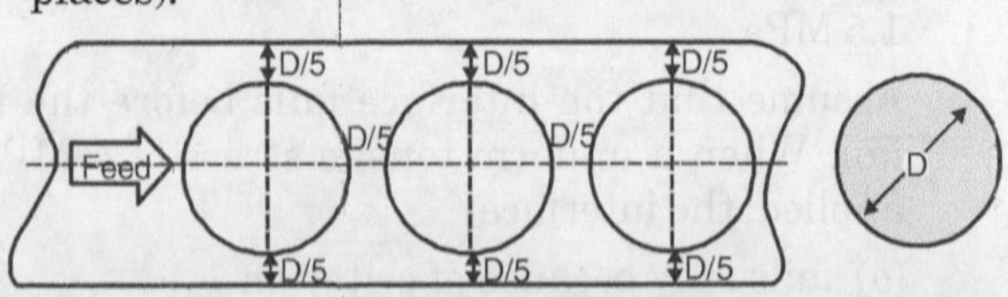

37. An explicit forward Euler method is used to numerically integrate the differential equation $\frac{dy}{dt} = y$ using a time step of 0.1. With the initial condition y(0) = 1, the value of y(1) computed by this method is ____(correct to two decimal places).

38. *F*(*s*) is the Laplace transform of the function

$f(t) = 2t^2 e^{-t}$

F(1) is ______(correct to two decimal places).

39. A simply supported beam of width 100 mm, height 200 mm and length 4 m is carrying a uniformly distributed load of intensity 10 kN/m. The maximum bending stress (in MPa) in the beam is ______(correct to one decimal place)

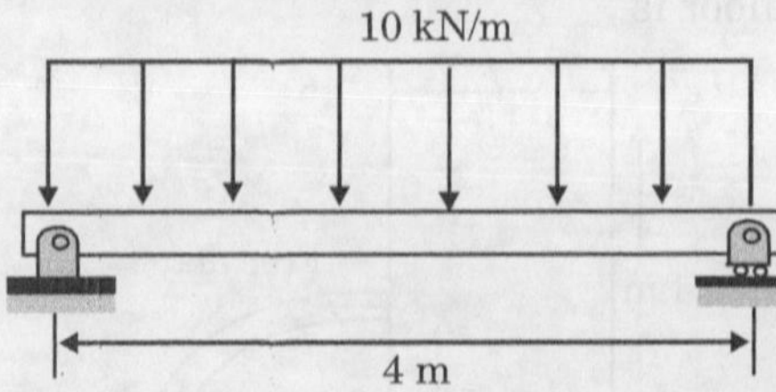

40. A machine of mass m = 200 kg is supported on two mounts, each of stiffness k = 10 kN/m. The machine is subjected to an external force (in N) F(t) = 50 cos 5t. Assuming only vertical translatory motion, the magnitude of the dynamic force (in N) transmitted from each mount to the ground is _____(correct to two decimal places).

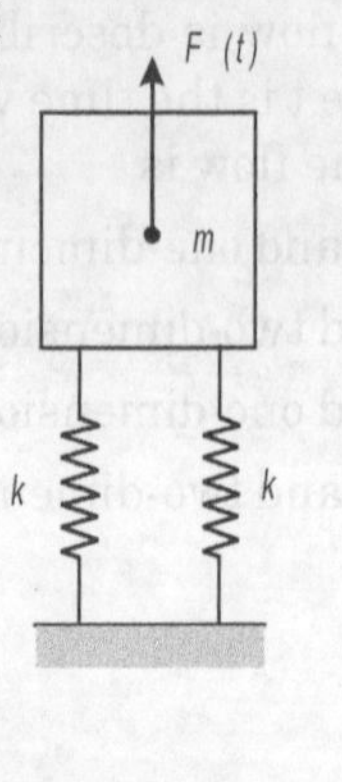

41. A slider crank mechanism is shown in the figure. At some instant, the crank angle is 45° and a force of 40 N is acting towards the left on the slider. The length of the crank is 30 mm and the connecting rod is 70 mm. Ignoring the effect of gravity, friction and inertial forces, the magnitude of the crankshaft torque (in Nm) needed to keep the mechanism in equilibrium is _____(correct to two decimal places).

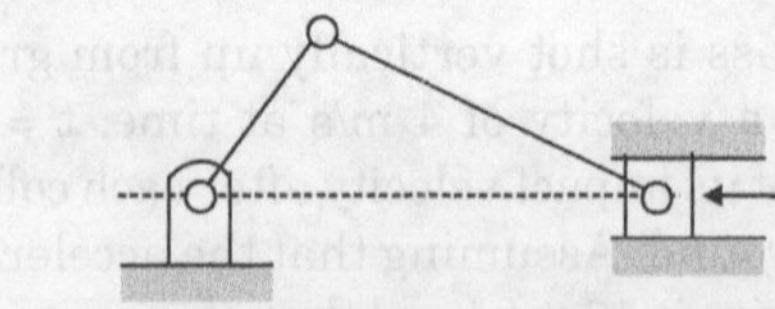

42. A sprinkler shown in the figure rotates about its hinge point in a horizontal plane due to water flow discharged through its two exit nozzles.

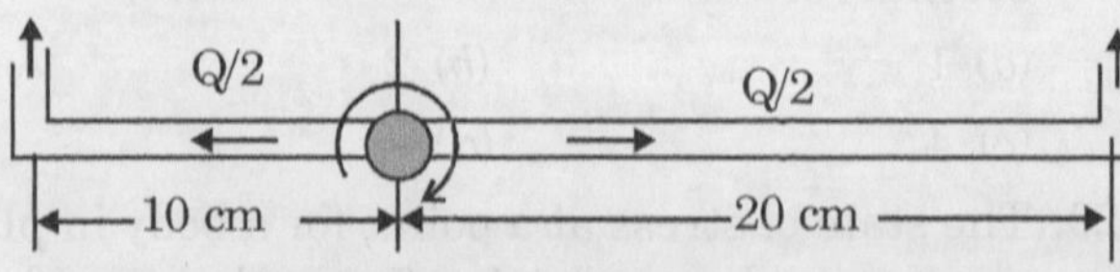

The total flow rate Q through the sprinkler is 1 litre/sec and the cross-sectional area of each exit nozzle is 1 cm^2. Assuming equal flow rate through both arms and a frictionless hinge, the steady state angular speed of rotation (rad/s) of the sprinkler is ______(correct to two decimal places).

43. A solid block of 2.0 kg mass slides steadily at a velocity V along a vertical wall as shown in the figure below. A thin oil film of thickness h = 0.15 mm provides lubrication between the block and the wall. The surface area of the face of the block in contact with the oil film is 0.04 m^2. The velocity distribution within the oil film gap is linear as shown in the figure. Take dynamic viscosity of oil as 7×10^{-3} Pa-s and acceleration due to gravity as 10 m/s^2. Neglect weight of the oil. The terminal velocity V (in m/s) of the 4 block is ______(correct to one decimal place).

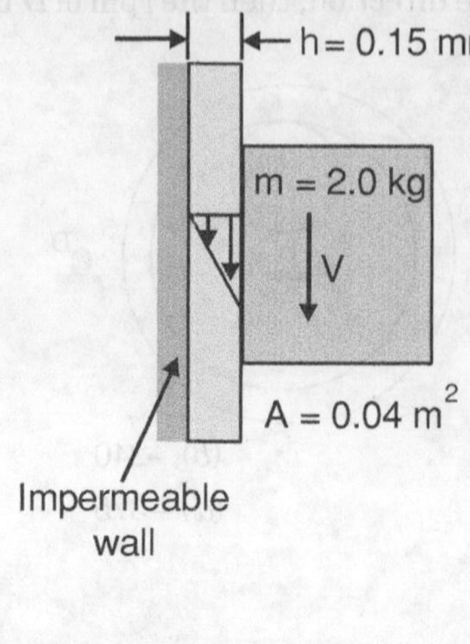

44. A tank of volume 0.05 m^3 contains a mixture of saturated water and saturated steam at 200°C. The mass of the liquid present is 8 kg. The entropy (in kJ/kgK) of the mixture is ______(correct of two decimal places)

Property data for saturated steam and water are:

At 200°C, Psat = 1.5538 MPa

$v_f = 0.001157$ m³/kg,

$v_t = 0.12736$ m³/kg,

$s_{f_g} = 4.1014$ kj/kgK,

$s_f = 2.3309$ kj/kgK

45. Steam flows through a nozzle at mass flow rate of $\dot{m} = 0.1$ kg/s with a heat loss of 5 kW. The enthalpies at inlet and exit are 2500 kJ/kg and 2350 kJ/kg, respectively. Assuming negligible velocity at inlet ($C_1 \approx 0$), the velocity (C_2) of steam (in m/s) at the nozzle exit is ______(correct to two decimal places)

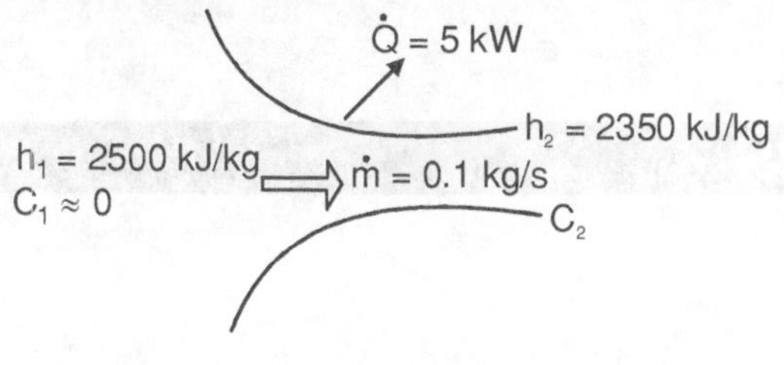

is ______.

46. An engine working on air standard Otto cycle is supplied with air at 0.1 MPa and 35°C. The compression ratio is 8. The heat supplied is 500 kJ/kg. Property data for air : $c_p = 1.005$ kJ/kgK, $c_v = 0.718$ kJ/kgK, R = 0.287 kJ/kgK, the maximum temperature (in K) of the cycle is ______ (correct to one decimal place).

47. A plane slab of thickness L and thermal conductivity k is heated with a fluid on one side (*P*), and the other side (*Q*) is maintained at a constant temperature, *TQ* of 25°C, as shown in the figure. The fluid is at 45°C and the surface heat transfer coefficient, *h*, is 10 W/m²K. The steady state temperature. TP (in °C) of the side which is exposed to the fluid is ______(correct to two decimal places).

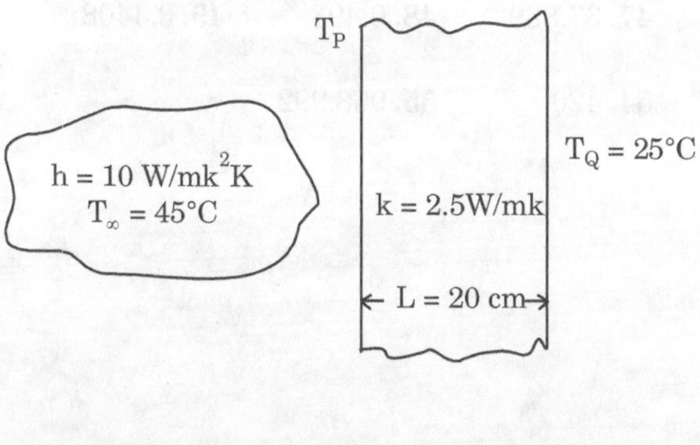

48. The true stress (σ), true strain (ε) diagram of a strain hardening material is shown in figure. First, there is loading up to point A, i.e. up to stress of 500 MPa and strain of 0.5. then from point A, there is unloading up to point B, i.e. to stress of 100 MPa, Given that the Young's modulus E = 200 GPa, the natural straint at point B (ε_B) ______(correct to two decimal places).

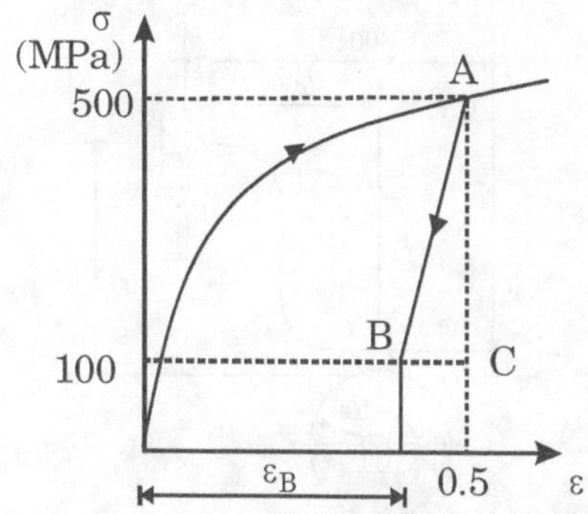

49. An orthogonal cutting operations is being carried out in which uncut thickness is 0.010 mm, cutting speed is 130m/min, rake angle is 15° and width of cut is 6 mm. It is observed that the chip thickness is 0.015 mm, the cutting force is 60 N and the thrust force is 25N. The ratio of friction energy to total energy is ______(correct to two decimal places).

50. A bar is compressed to half of its original length. The magnitude of true strain produced in the deformed bar is ______ (correct to two decimal places).

51. The minimum value of 3x + 5y such that :

$$3x + 5y \le 15$$
$$4x + 9y \le 8$$
$$13x + 2y \le 2$$
$$x \ge 0,\ y \ge 0$$

52. Processing times (including step times) and due dates for six jobs waiting to be processed at a work centre are given in the table. The average tardiness (in days) using shortest processing time rule is ______(correct to two decimal places).

Job	Processing time (days)	Due date (days)
A	3	8
B	7	16
C	4	4
D	9	18
E	5	17
F	13	19

53. The schematic of an external drum rotating clockwise engaging with a short shoe is shown in the figure, the shoe is mounted at point Y on a rigid lever XYZ hinged at point X. A force F = 100 N is applied at the free end of the lever as show. Given that the coefficient of friction between the shoe and the drum is 0.3, the braking torque (in Nm) applied on the drum is ______(correct to two decimal places).

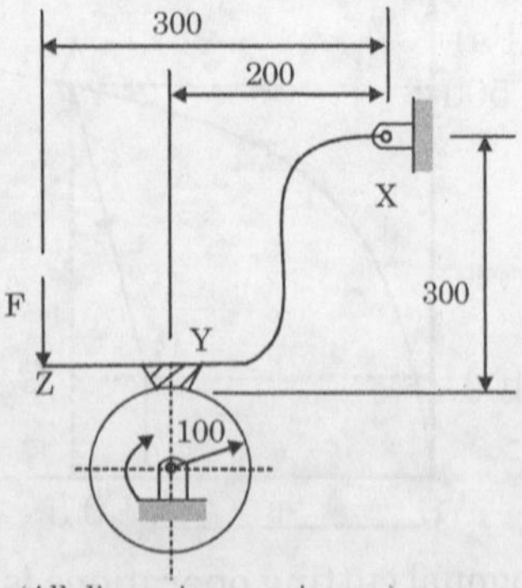

(All dimensions are in mm)

54. Block P of mass 2 kg slides down the surface and has a speed 20 m/s at the lowest point, Q where the local radius of curvature is 2 m as shown in the figure. Assuming $g = 10\ m/s^2$, the normal force (in N) at Q is ______(correct to two decimal places).

55. An electrochemical machining (ECM) is to be used to cut a through hole into a 12 mm thick aluminium plate. The hole has a rectangular cross-section, 10 mm × 30 mm. The ECM operation will be accomplished in 2 minutes, with efficiency of 90%. Assuming specific removal rate for aluminium as $3.44 \times 10^{-2}\ mm^3/(A\ s)$, the current (in A) required is ______(correct to two decimal places).

ANSWERS

General Aptitude (GA)

1. (*a*) **2.** (*c*) **3.** (*b*) **4.** (*b*) **5.** (*b*) **6.** (*a*) **7.** (*d*) **8.** (*d*) **9.** (*b*) **10.** (*c*)

Mechanical Engineering Section

1. (*b*) **2.** (*b*) **3.** (*a*) **4.** (*b*) **5.** (*b*) **6.** (*b*) **7.** (*d*)

8. (*b*) **9.** (*c*) **10.** (*a*) **11.** (*a*) **12.** (*c*) **13.** (*a*) **14.** (*b*)

15. (*b*) **16.** (*a*) **17.** (*c*) **18.** (*c*) **19.** 1.097 **20.** 1 **21.** 8.243

22. 0.05 **23.** 0.21 **24.** 0.50 **25.** 57.782 **26.** (*b*) **27.** (*c*) **28.** (*c*)

29. (*c*) **30.** (*c*) **31.** (*c*) **32.** (*a*) **33.** (*b*) **34.** (*d*) **35.** (*d*)

36. 53.25 **37.** 2.5937 **38.** 0.5 **39.** 30 **40.** 33.333 **41.** 1.12 **42.** 10

43. 10.714 **44.** 2.49 **45.** 447.213 **46.** 1403.97 **47.** 33.889 **48.** 0.498 **49.** 0.4408

50. 0.693 **51.** 0 **52.** 6.33 **53.** 8.18 **54.** 420 **55.** 968.992

EXPLANATIONS

GENERAL APTITUDE

3. $\frac{\text{Machine} \times \text{time}}{\text{Number of toys}} = \frac{M_1 \times t_1}{T_1}$

$$= \frac{M_2 \times t_2}{T_2}$$

$$\therefore \quad \frac{7 \times 7}{7} = \frac{100 \times t_2}{100}$$

$$\therefore \quad \boxed{t_2 = 7 \text{ min}}$$

4. Let the initial length of rectangle is l and breath is b. Now if length is reduced by 10 m and breath is reduced by 5 m then it becomes square.

$\therefore \quad l - 10 = b - 5$

[$\because$ In square length = breadth]

$\therefore \quad l - b = 5$

Given, initial, Area (A) = $l \times b$

Now according to question,

$$(l - 10)(b - 5) = A - 650$$

$\therefore \quad lb - 10b - 5l + 50 = A - 650$

$\therefore \quad A - 10b - 5l = A - 700$

$\therefore \quad 10b + 5l = 700$

$\therefore \quad 10b + 5(b + 5) = 700$

$\therefore \quad 15b + 25 = 700$

$\therefore \quad 15b = 700 - 25$

$\therefore \quad 15b = 675$

$\Rightarrow \quad b = 45$ m

$\therefore \quad l = 45 + 5 = 50$ m

$\therefore$ Area of original rectangle (A) = $(45 \times 50)\text{m}^2$

$= 2250 \text{ m}^2$.

5. Let two digit number be xy.

Now sum of digit is 9

$\therefore \quad x + y = 9 \quad \text{....}(i)$

and when '45' is subtracted from number then digit gets interchanged

$\therefore \quad 10x + y - 45 = 10y + x$

$\therefore \quad x - 5 = y$

$x - y = 5 \quad \text{....}(ii)$

Now solving equation (i) and (ii) we get

$x = 7$ and $y = 2$

Hence the number is 72.

6. $\log |a| + \log |b| + \log |c| = 0$

It is possible only,

when $|a|$, $|b|$ and $|c|$ all are equal to 1.

$\therefore$ a, b, c may be $\pm 1, \pm 1, \pm 1$ respectively.

Now for minimum value of all three will be negative.

$\therefore$ minimum value = -3

and maximum value of all three will be positive.

$\therefore$ maximum value = $+3$

7. From option (a),

Let $\quad a = 1, b = 3$

$\therefore \quad a + a^2b^3 = 1 + 27$

$= 28$ (Even)

$\therefore$ Option (a) is wrong

From option (b)

Let $\quad a = 2, b = 4$

$\therefore \quad a + a^2b^3 = 2 + 4 \times 64$

$= 258$ (Even)

From option (c),

Let $\quad a = 2, b = 1$

From option (d),

for $x = 1$

$$y = ||x - 1| - 1|$$

$$= |-1| = 1$$

$\therefore$ Option (c) is wrong.

Now from option (b),

for $\quad x = 1$

$$y = ||x| - 1| - 1$$

$$= |1 - 1| - 1 = -1$$

$\therefore$ Option (b) is correct.

8. Train speed $(S_T) = 54$ km/h

Time = 25 sec for travelling length of train and length of platform

Man speed $(S_M) = 9$ km/h

Speed of train to man = 45 km/h

Time = 14 sec

So, length of train = time $\times$ speed

$$= 14 \times 45 \times \frac{5}{18}$$

Length of train $(L_T) = (35 \times 5 \text{ m}) = 175$ m

Length of platform (L) + length of train (L_T) = speed $\times$ time

$$= 54 \times \frac{5}{18} \times 25 = 15 \times 25 = 375\text{m}$$

$\therefore$ Length of platform (L) = 375 – 175 = 200 m

9. Here from graph for x = 1, y = – 1

Checking the same from options.

From option (a)

for $x = 1$

$$y = \left\| 1 \right| + 1 \left| - 2\right.$$
$$= 2 - 2 = 0$$

$\therefore$ Option (a) is wrong

From option (c)

for $x = 1$

$$y = \left\| x \right| + 1 \left| - 1\right.$$
$$= |1 + 1| - 1 = 1$$

$\therefore$ Option (c) is wrong.

10.

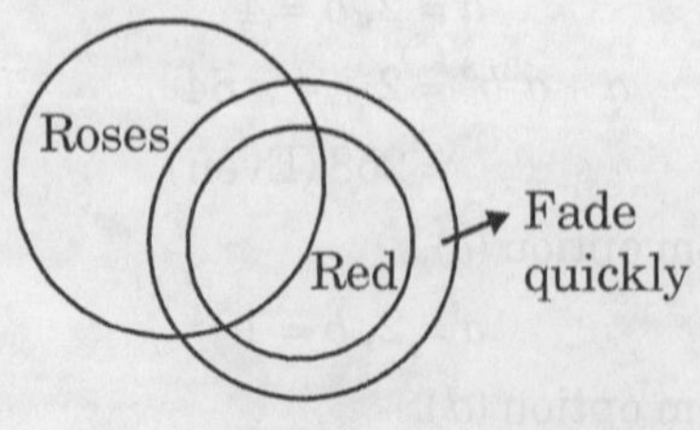

Case-I

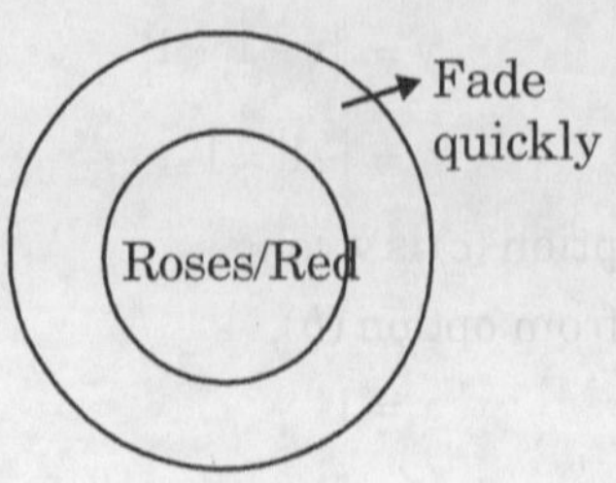

Case-II

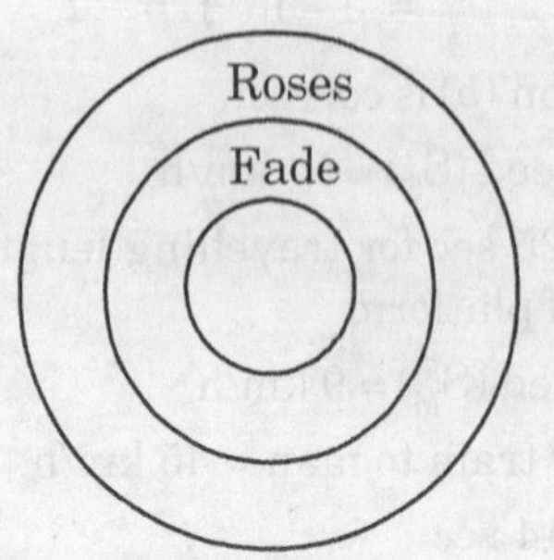

Case-III

MECHANICAL ENGINEERING

1. (4R, 4G, 4B)

P(RRR) (without replacement)

$$= \frac{^4C_1}{^{12}C_1} \times \frac{^3C_1}{^{11}C_1} \times \frac{^2C_1}{^{10}C_1} = \frac{1}{55}$$

2. Let A = $\begin{bmatrix} -4 & 1 & -1 \\ -1 & -1 & -1 \\ 7 & -3 & 1 \end{bmatrix}$

Applying $R_1 \longleftrightarrow R_2$ $\begin{bmatrix} -1 & -1 & -1 \\ -4 & 1 & -1 \\ 7 & -3 & 1 \end{bmatrix}$

Applying $R_2 - 4R_1, R_3 + 7R_1$ $\begin{bmatrix} -1 & -1 & -1 \\ 0 & 5 & 3 \\ 0 & -10 & -6 \end{bmatrix}$

Now applying $R_3 + 2R_2$ $\begin{bmatrix} -1 & -1 & -1 \\ 0 & 5 & 3 \\ 0 & 0 & 0 \end{bmatrix}$

$\therefore$ No. of non zero rows = 2

rank = 2

3. $\int_a^b f(x)\, dx = f(x)(b - a)$

4. $F(z) = iz + k\,Re(z)\ i\,Im(z)$

$$u + iv = i(x + iy) + kx + iy$$
$$u + iv = (kx - y) + i(x + y)$$

Now comparing real and imaginary part we get

$$u = kx - y \text{ and } v = x + y$$

Now $\quad u = kx - y$

Now $\quad \frac{du}{dx} = k \text{ and } \frac{dv}{dx} = -1$

Now $\quad V = x + y$

$\therefore \quad \frac{dv}{dx} = 1$

and $\quad \frac{dv}{dy} = 1$

$\Rightarrow \quad u_x = v_y$

$\therefore \quad K = 1$

5.

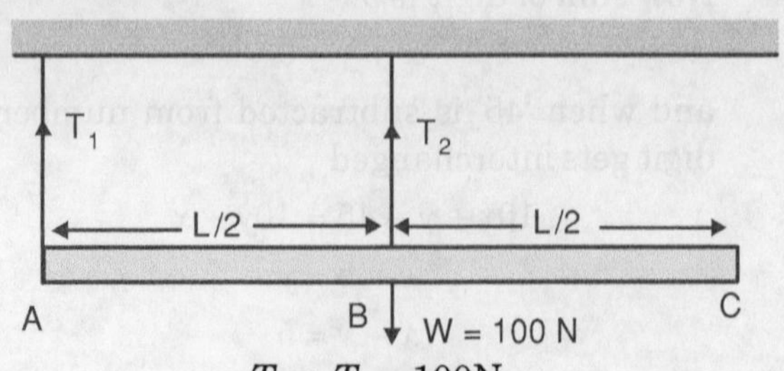

$$T_1 + T_2 = 100\text{N}$$
$$\Sigma M_A = 0$$
$$T_2 . \frac{L}{2} = 100 \times \frac{L}{2}$$
$$T_2 = 100 \text{ N}$$
$$T_1 = 0 \text{ N}$$

6. Maximum shear stress = Radius of Mohr's Circle (for plane stress condition)

$$\tau_{max} = \frac{\sigma_1 - \sigma_2}{2}$$

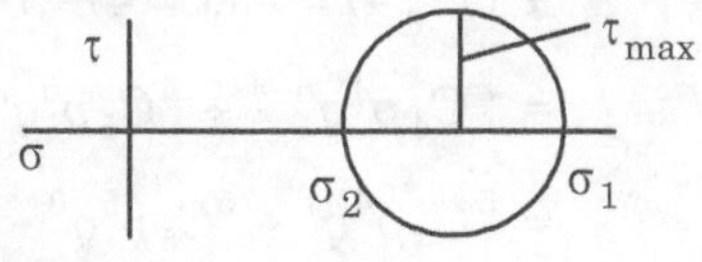

7. The given equation is :

$$Mx + kx = F\cos(\omega t)$$

⇒ it is the equation of frequency under damped forced vibration and for natural frequency of free vibration-

Under root of the co-efficient of x when co-efficient of $\ddot{x}$ is 1.

$$M\ddot{x} + kx = F\cos(\omega t)$$

$$\ddot{x} + \frac{k}{M}x = \frac{F}{M}\cos(\omega t)$$

$$\text{so } \omega_n = \sqrt{\text{co-efficient of } x} = \sqrt{\frac{k}{M}}$$

s for resonance. $\omega = \omega_n = \sqrt{\frac{k}{M}}$

8. The oldham coupling connects parallel shafts having small offset and in this coupling the motion (rotation) and torsion load transfer is along axes of shafts.

9. C is the false statement

2D incompressible flow continuity equation.

$$\frac{\partial u}{\partial x} + \frac{\partial v}{\partial y} = 0$$

$$\frac{\partial (Ax)}{\partial x} + \frac{\partial (-Ay)}{\partial y} = 0$$

A – A = 0 it satisfies continuity equation.

⇒ As $\vec{V} = Ax\,\hat{i} - Ay\hat{j}$

As $y \to \infty$ velocity vector field will not be defined along y axis.

So flow will be along x-axis i.e. 1-D flow.

⇒ Stream line equation for 2D

$$\frac{dx}{u} = \frac{dy}{v}$$

$$\frac{dx}{Ax} = \frac{dy}{-Ay}$$

$$\ln x = -\ln y + \ln c$$

$$\ln xy = \ln c$$

$xy = c \to$ streamline equation

10.

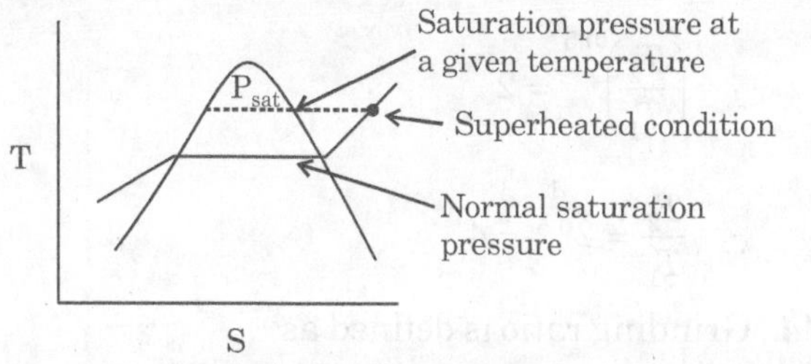

$P < P_{sat.}$ at given T of superheated vapour.

11.

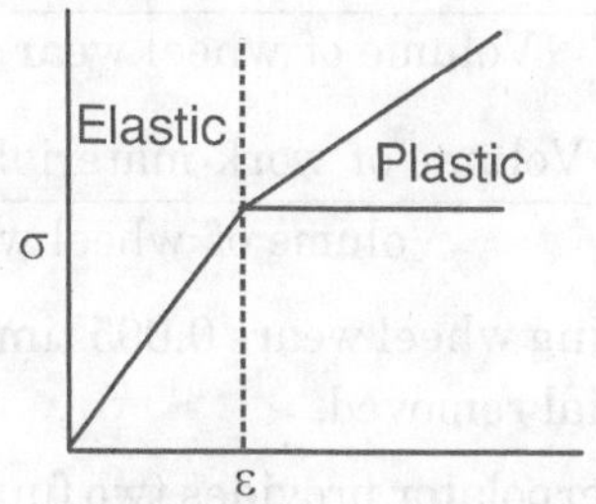

(with Linear hardening)

Fig-1

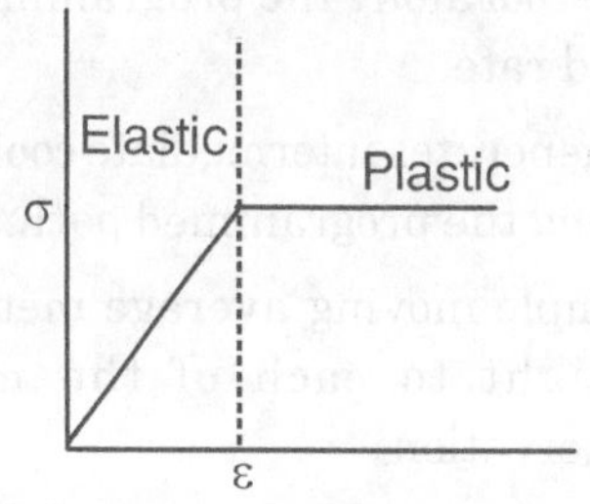

(without hardening)

Fig-2

⇒ increases linearly with true strain.

[as in fig (1)]

12.

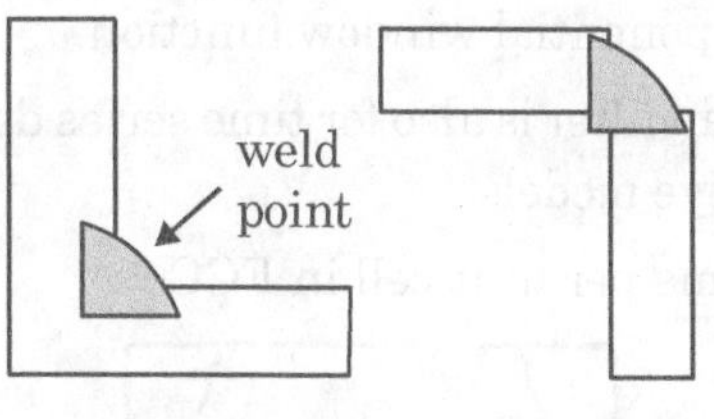

Fillet weld on corner joint

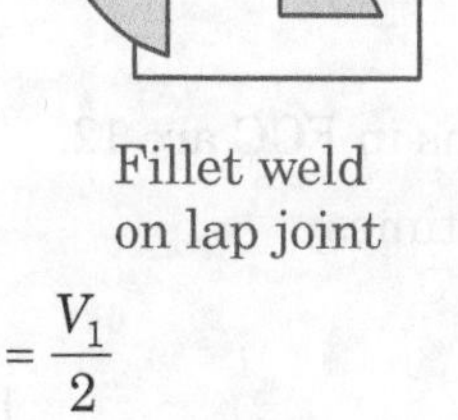

Fillet weld on lap joint

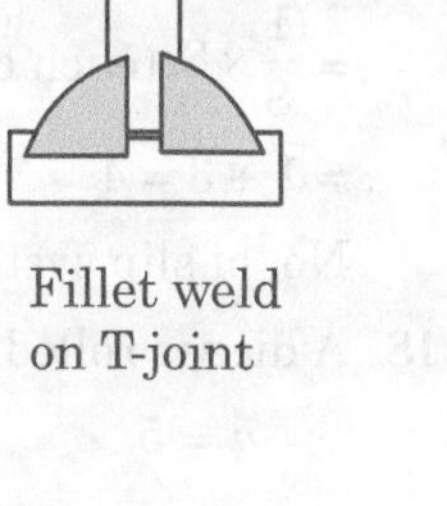

Fillet weld on T-joint

13.

$$V_2 = \frac{V_1}{2}$$

$$V_1 T_1^{0.5} = V_2 T_2^{0.5}$$

$$V_1 T_1^{0.5} = \frac{V_1}{2} T_2^{0.5}$$

$$\therefore \quad \left(\frac{T_2}{T_1}\right)^{0.5} = 2$$

$$\therefore \quad \frac{T_2}{T_1} = 2^{\frac{1}{0.5}} = 4$$

14. Grinding ratio is defined as

$$= \frac{\text{Volume of work material removed } (V_m)}{\text{Volume of wheel wear } (V_w)}$$

$$200 = \frac{\text{Volume of work material removed } (V_m)}{\text{Volume of wheel wear } (V_w)}$$

Grinding wheel wears 0.005 times the volume of material removed.

15. As interpolator provides two functions:

1. It computes individual axis velocities to drive the tool along the programmed path at given feed rate.
2. It generates intermediate coordinate positions along the programmed path.

16. (*a*) Simple moving average method gives equal weight to each of the m most recent observation

(*b*) Weighted moving average method gives more weight to the recent values.

(*c*) Triple exponential smoothing is a rule of thumb technique for smoothing time series data using the exponential window function.

(*d*) Kalman filter is also for time series data using adaptive models.

17. No of atoms per unit cell in FCC ⇒

$$= \frac{1}{8} \times 8 \text{ (each corner)} + \frac{1}{2} \times 6 \text{ (each face)}$$

$$= 1 + 3 = 4$$

No. of slip systems in FCC are 12.

18. A dice is rolled 5 times

$n = 5$

$P = \text{(Probability of getting 1)} = \frac{1}{6}$

$0\%\ q = 1 - p = 1 - \frac{1}{6} = \frac{5}{6}$

Probability of getting 1 at least 4 times is

$$P(x \geq 4) = P(x = 4) + P(x = 5)$$

$$= {}^nC_4 p^4 q^{n-4} + {}^nC_5 p^5 q^{n-5} \quad \left[\because {}^nC_r p^x q^{n-x}\right]$$

$$= {}^5C_4 p^4 q^1 + {}^5C_5 p^5 q^0$$

$$= 5 \times \left(\frac{1}{6}\right)^4 \left(\frac{5}{6}\right)^1 + 1 \times \left(\frac{1}{6}\right)^5 \times \left(\frac{5}{6}\right)^0$$

$$= \frac{25}{(6)^5} + \frac{1}{(6)^5} = \frac{26}{(6)^5} = 0.0033$$

Required probability = 0.33%

19. Steel column supported at both ends = hinged at both ends.

so far hinged condition:-

effective length, Le = L

so buckling load

$$(Pe) = \frac{\pi^2 E I \min}{\{Le(= L)\}^2}$$

$$= \frac{\pi^2 \times 200 \times 10^3 \frac{15 \times 10^3}{12}}{1500^2}$$

$$= 1.097 \text{ kN}$$

20. Here the link, that is fixed is adjacent to the smaller one.

so in this condition we will get on fully rotation (crank)

⇒ if shortest link is fixed

⇒ Double crank (full rotation)

⇒ if opposite to shortest link is fixed

⇒ No Crank.

21. Stiffness of helical spring

$$K = \frac{Gd^4}{64R^3 n}$$

d = spring wire diameter

R = mean coil radius

n = number of turns

$$\therefore \quad K \propto d^4$$

$$\frac{K'}{K} = \left(\frac{d'}{d}\right)^4$$

$$K' = \left(\frac{1.02d}{d}\right)^4 K$$

K' = 1.08243 K

$$\% \text{ increase in stiffness} = \frac{K' - K}{K} \times 100\% = 8.243\%$$

22.

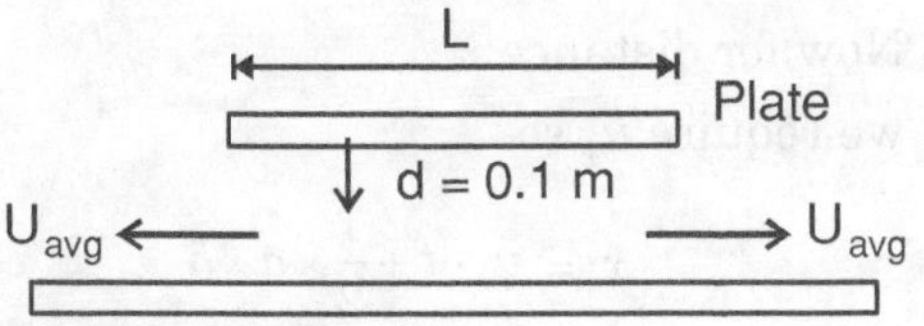

By mass conservation:-

$2(s \times u_{avg} \times db) = (s. u \times Lb)$

$\Rightarrow 2 \times v_{avg} \times 0.1 = 0.01 \times 1$

$\boxed{u_{avg} = 0.05 \text{ m/s}}$

23. Ideal gas

State - 1 : $T_1 = 300$ K, $P_1 = 100$ kPa

State - 2 : $T_2 = 600$ K, $P_2 = 500$ kPa,

$c_p = 1$ kJ/kg-K, $c_p - c_v = R$, $c_v = 0.7$ kJ/kg-K

$\Rightarrow c_p - c_v = 1 - 0.7 = R$

R = 0.3 kJ/kg–K

Change in specific entropy

$$s_2 - s_1 = c_p \ln\frac{T_2}{T_1} - R\ln\frac{P_2}{P_1}$$

$$= 1 \times \ln\frac{600}{300} - 0.3\ln\frac{500}{100} = 0.21 \text{ kj/kg-K}$$

24. In pelton wheel:-

$\dfrac{2(vu - u^2)}{v^2}$ where,

u = bucket speed

v = water jet speed

So, For maximum efficiency:-

$\dfrac{dx}{du} = 0$

$2(v - 2u) = 0$

$v = 2u$

$\boxed{u = \dfrac{v}{2}}$

25. $\theta = 27°32'$

$$= 27 + \left(\frac{32}{60}\right)° = 27.533°$$

In sine bar, $\sin\theta = \dfrac{H}{L}$

$\sin 27.533° = \dfrac{H}{125}$

$H = 57.782427$

$H \approx 57.782$ mm

26.

$Y = X_1 - X_2$ → $E_{(y)} = E(x_1 - x_2) = E(x_1) - E(x_2) = 1 - 1 = 0$

$Y = X_1 - X_2$ → $V_{(y)} = V(x_1 - x_2) = V(x_1) + V(x_2) = 1 + 4 = 5$

Y is normally distributed with mean 0 and variance 5.

27. By Gauss Divergence Theorem

$$\iint_s \vec{r}\cdot\hat{n}ds = \iiint_v \bar{V}\cdot\hat{r}dV = \iiint_v 3\,dV = 3V$$

28.

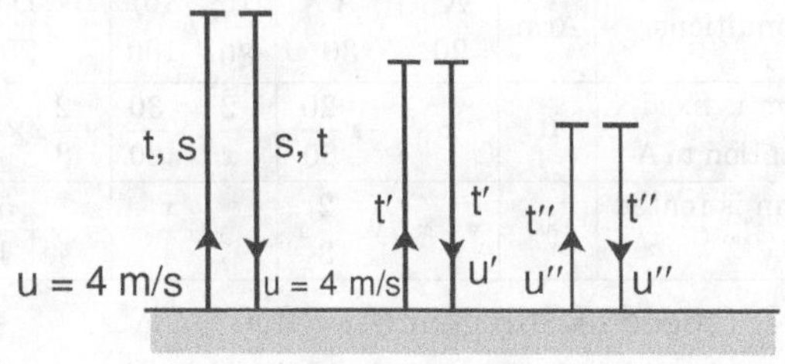

(1) → t = ?

v = u + at

0 = 4 – 10t

$t = \dfrac{4}{10} = 0.4$s

(2) → t′ = ?

u′ = 0.8 × u

= 0.8 × 4 = 3.2 m/s

v′ = u′ + at′

0 = 3.2 – 10t′

$t' = \dfrac{3.2}{10} = 0.32$s

(3) → t″ = ?

u″ = 0.8 u′

= 0.8 × 3.2 = 2.56 m/s

v″ = u″ + at″

0 = 2.56 – 10t″

t″ = 0.256s

So, t, t′, t″ are forming a GP series

So, total time = 2(t + t′ + t″ + ... 0)

= 2[0.4 + 0.32 + 0.256 + ...0]

29. $\sigma_x = 100$ kPa, $\tau_{xy} = 50$ kPa

Minimum principal stress

$$= \frac{\sigma_x + \sigma_y}{2} - \sqrt{\left(\frac{\sigma_x - \sigma_y}{2}\right)^2 + \tau_{xy}^2}$$

$$10 = \frac{100 + \sigma_y}{2} - \sqrt{\left(\frac{100 - \sigma_y}{2}\right)^2 + 50^2}$$

$$\therefore \sqrt{\left(50 - \frac{\sigma y}{2}\right)^2 + 50^2} = 50 + \frac{\sigma_y}{2} - 10 = 40 + \frac{\sigma_y}{2}$$

By squaring

$$2500 + \frac{\sigma_y^2}{4} - 50\sigma_y + 2500 = 1600 + \frac{\sigma_y^2}{4} + 40\sigma_y$$

$\therefore \quad 90\,\sigma_y = 3400$

$\sigma_y = 37.78$ MPa

30.

Conditions	Arm	A 20	A 30	$(i)^C$ / (0) 80 / 100	D 20
1. Arm is fixed x rotation to A	0	x	$-x\frac{20}{30}$	$-\frac{2}{x}x\times\frac{30}{80}$	$+\frac{2}{8}x\times\frac{100}{20}$
2. Arm is roated by y	y	$y+x$	$y-\frac{2}{3}x$	$y-\frac{x}{4}$	$y+\frac{5}{4}x$

N_A (speed of Sun gear, A) = 300

$$x + y = 300$$

Arm (y) is fixed, $y = 0$

Than speed of year,

$$D(N_o) = y + \frac{5}{4}x = 0 + \frac{5}{4}\times 300 = +375 \text{ rpm(C.W.)}$$

31. Normal stress on inclined plane

$T' = \sigma_x \cos^2\theta$

$= 4 \times \cos^2 30 = 3$ MPa

Shear stress on inclined plane $\tau' = \frac{\sigma_x}{2}\sin 2\theta$

$= 2 \times \sin 60° = 1.73$ MPa

Since both the stress exceeds the given limits, answer is option (*c*).

32. Life (in Million Revolution) = $\left(\frac{C}{P_e}\right)^q$

$q \to 3$ = ball bearing

$q \to \frac{10}{3} \Rightarrow$ Roller bearing

Life = $\left(\frac{35}{45}\right)^3$ = 0.4705 **Million Revolution**

C → Dynamic load capacity = 35 KN

P_e equivalent load = 45 KN

33. $u_x = \sqrt{2gh}$

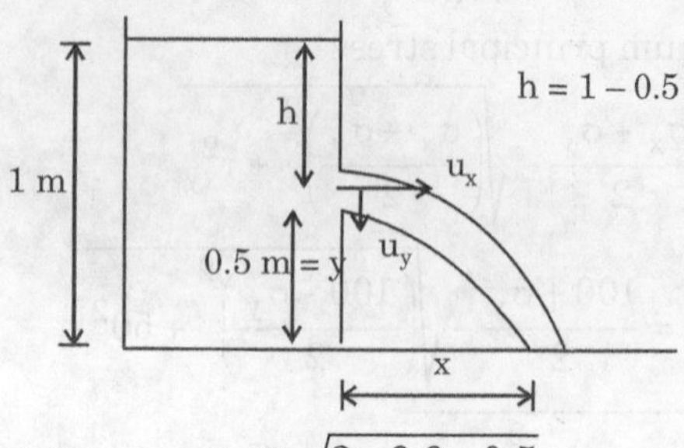

$= \sqrt{2 \times 9.8 \times 0.5}$

$u_x = 3.13$ m/s

Now for distance x

we require u_x so–

$$x = u_x \cdot t + \frac{1}{2}\times a \times b$$

$$x = 3.13\,t$$

($\because$ acceleration (a) = 0 in horizontal direction)

Now for $\quad t \Rightarrow y = u_y t + \frac{1}{2}gt^2$

$\Rightarrow \quad 0.5 = 0 + \frac{1}{2}\times 9.81 \times t^2 \qquad [\because u_y = 0]$

$t = 0.319$ sec.

so distance, $x = 3.13 \times 319 = 1$m

34. x direction scalar of velocity field,

$$u = \frac{dx}{dt}$$

$$u = -kx.e^{-kt}$$

y direction scalar of velocity field

$$v = \frac{dy}{dt}$$

$$v = ky_0e^{kt}$$

$$\vec{V} = u\hat{i} + v\hat{j}$$

$$\vec{V} = -kx_0e^{-kt}\hat{i} + ky_0e^{kt}\hat{j}$$

u & v are non zero scalar t ≥ 0 so it is 2D flow.

2D possible flow field

$$\frac{\partial u}{\partial x} + \frac{\partial v}{\partial y} = 0$$

$$\frac{\partial}{\partial x}\left(-kx_0e^{-kt}\right) + \frac{\partial}{\partial y}\left(ky_0e^{kt}\right) = 0$$

0 + 0 = 0 continuity satisfied.

$$\frac{\partial u}{\partial t} = +k^2x_0e^{-kt}$$

$$\frac{\partial v}{\partial t} = k^2y_0e^{kt}$$

$$\frac{\partial u}{\partial t} \neq 0$$

$$\frac{\partial v}{\partial t} \neq 0$$

So, flow is unsteady.

35. For work hardening

$\sigma = K.\,\varepsilon^n$

where, K = strength Co-efficient

σ = applied stress on material

ε = strain

n = strain hardening co-efficient

For perfectly plastic → n = 0

so $R = 1 - e^{-1(n+1)}$

$= 1 - e^{-(0+1)}$

$= 1 - \frac{1}{e}$

$\boxed{R = 0.632}$

36. For blanking operation the piece obtained through hole is so, Scrap = hatched one area

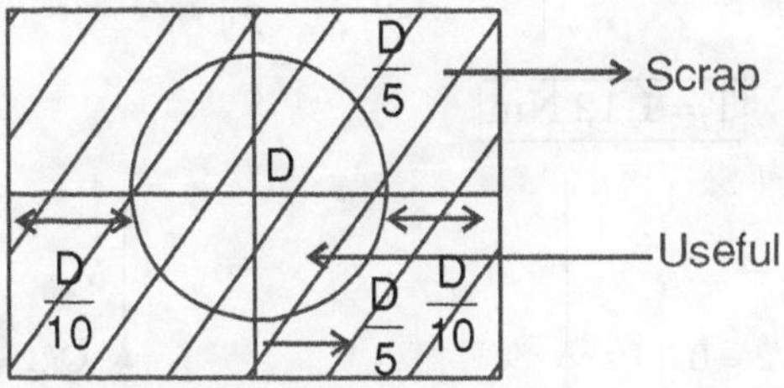

$$\% \text{ scrap} = \left(\frac{\text{scrap area}}{\text{Total Area}}\right) \times 100$$

$$= \left(\frac{\text{Total Area} - \text{hole area}}{\text{Total area}}\right) \times 100$$

$$= \left(1 - \frac{\text{hole area}}{\text{total area}}\right) \times 100$$

$$= \left(1 - \frac{\frac{\pi}{4}D^2}{\left(D + \frac{D}{10} + \frac{D}{10}\right) \times \left(D + \frac{D}{5} + \frac{D}{5}\right)}\right) \times 100$$

= 53.25%

37. General formula

$$y_{n+1} = y_n + h_F(t_n, y_n)$$

For $n = 0$, $y_1 = y_0 + h_F(t_0, y_0)$

$= y_0 + hy_0$

$= 1 + 0.1\,(1)$

$y_1 = 1.1$

For $n = 1$, $y_2 = y_1 + h_F(t_1, y_1)$

$= y_1 + h.y_1$

$= 1.1 + 0.1\,(1.1)$

$y_2 = 1.21$

For $n = 2$, $y_3 = y_2 + hf(t_2, y_2)$

$= y_2 + h.y_2$

$= 1.21 + 0.1 \times 1.21$

$y_3 = 1.331$

For $n = 3$, $y_4 = y_3 + h.f(t_3, y_3)$

$= y_3 + h.y_3$

$= 1.331 + 0.1 \times 1.331$

$y_4 = 1.4641$

For $n = 4$, $y_5 = y_4 + h.f(t_4, y_4)$

$= y_4 + h.y_4$

$= 1.4641 + 0.1 \times (1.4641)$

$= 1.61051$

For $n = 5$, $y_6 = y_5 + h.f(t_5, y_5)$

$= y_5 + h.y_5$

$= 1.61051 + 0.1 \times 1.61051$

$y_6 = 1.771561$

For $n = 6$, $y_7 = y_6 + h.f(t_6, y_6)$

$= y_6 + h \times y_6$

$= 1.771561 + 0.1 \times 1.771561$

$= 1.9487$

For $n = 7$, $y_8 = y_7 + h.f(t_7, y_7)$

$= y_7 + h.y_7$

$= 1.9487 + 0.1 \times (1.9487)$

$y_8 = 2.14357$

For $n = 8$, $y_9 = y_8 + h.f(t_8, y_8)$

$= y_8 + h.y_8$

$= 2.14357 + 0.1 \times 2.14357$

$y_9 = 2.3579$

For $n = 9$, $y_{10} = y_9 + h.f(t_9, y_9)$

$= y_9 + h.y_9$

$= 2.3579 + 0.1 \times (2.3579)$

$y_{10} = 2.5937$

38. Applying Laplace transform

$$L\left(t^2\right) = \frac{L^2}{S^3}$$

$$L\left(2t^2\right) = \frac{4}{S^3}$$

$$F(s) = L\left[e^{-t} \cdot 2t^2\right] = \frac{4}{(S+1)}$$

Putting s = 1 we get

$$F(1) = \frac{4}{(1+1)^3} = \frac{4}{8} = 0.5$$

39. For 0.01 maximum bending moment,

$$M_{max} = \frac{Wl^2}{8}$$

and From bending equation:-

$$\sigma_{max} = \frac{6\,M_{max}}{bd^2} = \frac{6 \times 10 \times (4)^2 \times 1000}{8(0.1) \times (.2)^2}$$

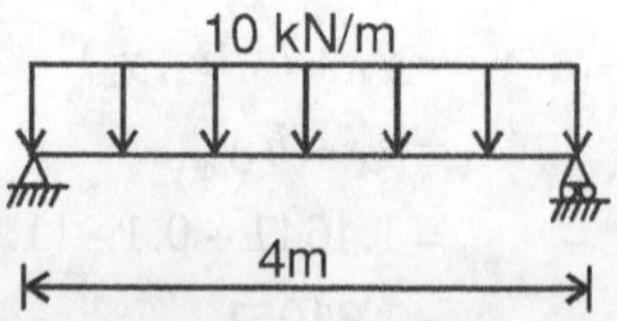

$\sigma_{max} = 30MPa$

40. m = 200 kg, k = 10 kN/m = 10000 N/m

$K_{eq} = K + K = 10000 + 10000 = 20000$ N/m

$F_t = 50 \cos 5t$

$F_0 = 50$ N

$\omega = 5$ rad/s

$$\omega_n = \sqrt{\frac{K_{eq}}{m}} = \sqrt{\frac{20000}{200}} = \sqrt{100} = 10 \text{ rad/s}$$

$$\frac{\omega}{\omega_n} = \frac{5}{10} = \frac{1}{2}$$

No damping $\Rightarrow c = 0 \Rightarrow \xi = 0$

Transmissibility

$$\in = \frac{\sqrt{1+\left(\frac{2\xi\omega}{\omega_n}\right)^2}}{\sqrt{\left\{1-\left(\frac{\omega}{\omega_n}\right)^2\right\}^2+\left\{\frac{2\xi\omega}{\omega_n}\right\}^2}} \qquad (\xi = 0)$$

$$= \frac{1}{\sqrt{\left\{1-\left(\frac{\omega}{\omega_n}\right)^2\right\}^2}} = \frac{1}{\sqrt{\left\{1-\left(\frac{1}{2}\right)^2\right\}^2}}$$

$$= \frac{1}{1-\frac{1}{4}} = \frac{1}{\frac{3}{4}} = \left(\frac{4}{3}\right)$$

$$\in = \frac{F_t}{F_0}$$

$$\frac{4}{3} = \frac{F_t}{50}$$

$$F_t = \frac{200}{3} = 66.666 \text{ N}$$

Force from each mount

$$= \frac{F_t}{2} = \frac{66.666}{2} = 33.333 \text{ N}$$

41.

The Torque required to hold the mechanism in equilibrium is equal & opposite of turning moment.

$$T = r \times F\left\{\sin\theta + \frac{\sin 2\theta}{2\sqrt{x^2-\sin^2\theta}}\right\}$$

$$= 1 \times 2\left\{\frac{1}{\sqrt{2}} + \frac{1}{2\sqrt{\frac{49}{2}-\frac{1}{2}}}\right\}$$

$$\boxed{T = 1.12 \text{ Nm}}$$

42.

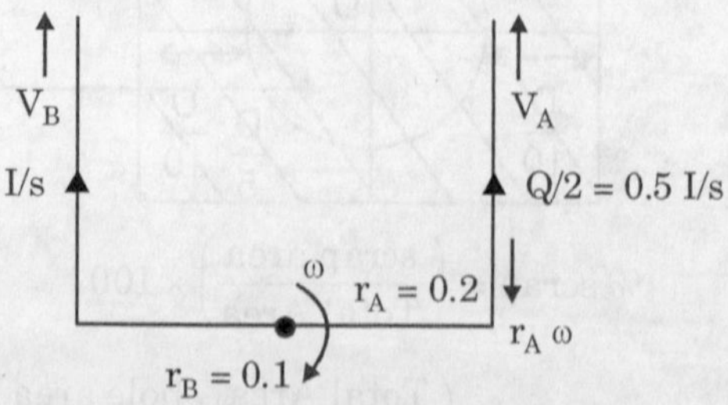

Relative velocities of water with sprinkler

$$V_A = \frac{Q/2}{A}$$

$$= \frac{1 \times 10^{-3}}{2 \times 10^{-4}} = 5 \text{m/s}$$

$V_B = 5$ m/s

Absolute velocity from B side

$$V'_{Abs} - (+r_B\omega) = V_B$$

$$V'_{Abs} = V_B + r_B\omega$$

$$= 5 + 0.1\,\omega$$

Absolute velocity from A side

$$V_{Abs} - (-r_A\omega) = V_A$$

$$V_{Abs} = V_A - r_A\omega$$

$$V_{Abs} = 5 - 0.2\,\omega$$

The external torque to the sprinkler is zero.

So, $\Sigma T = 0$

$$\dot{m}_A V_{Abs} r_A - \dot{m}_B V_{Abs} r_B = 0$$

$$\rho\left(\frac{Q}{2}\right)\{5-0.2\,\omega\}0.2 - \rho\frac{Q}{2}\{5+0.1\,\omega\}0.1 = 0$$

$$1 - 0.04\,\omega - 0.5 - 0.01\,\omega = 0$$

$$0.05\,\omega = 0.5$$

$$\omega = 10 \text{ rad/s}$$

43.

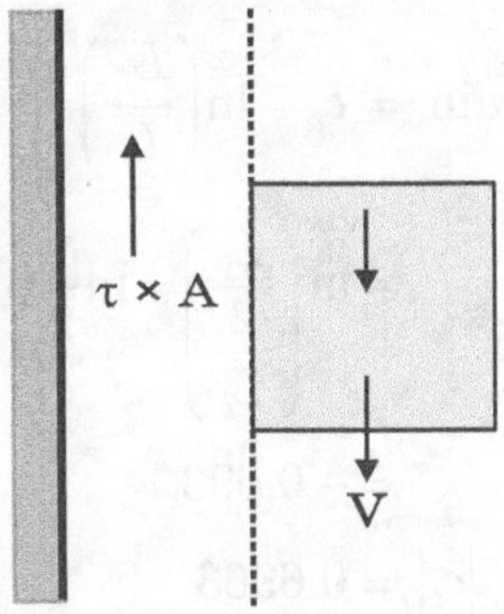

Terminal velocity is a constant velocity i.e. the net acceleration is zero.

So, $\Sigma F_{net} = ma$

$$mg - \tau A = 0$$

$$\tau A = mg$$

$$\mu \frac{V}{h} A = mg$$

$$7 \times 10^{-3} \times \frac{V}{0.15 \times 10^{-3}} \times 0.04 = 2 \times 10$$

$$V = 10.714 \text{ m/s}$$

44. Total volume of tank (V) = 0.05 m^3

Means of liquid (m_2) = 8 kg

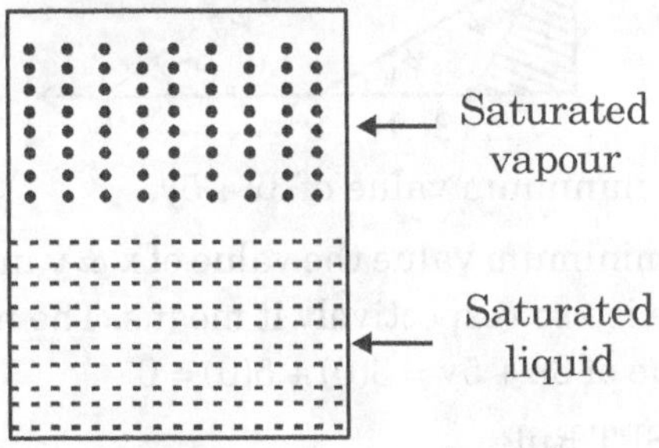

$$x = \frac{m_V}{m_V + m_L} \text{(dryness fraction)}$$

$$V = V_L + V_V \quad \begin{bmatrix} V_L - \text{v of vapour in tank} \\ V_V - \text{v of vapour in tank} \end{bmatrix}$$

$$V_V = m_V V_V$$

$$V_V = V_g @ 200°C$$

$$= 0.12736 \text{ m}^3\text{/kg}$$

$$V_V = m_L V_L$$

$$V_L = V_f @ 200°C$$

$$= 0.001157 \text{ m}^3\text{/kg}$$

$$V_L = 8 \times 0.001157$$

$$= 9.256 \times 10^{-3} \text{ m}^3$$

$$0.05 = 9.256 \times 10^{-3} + V_V$$

$$V_V = 0.0474 \text{ m}^3$$

$$0.04074 = m_V \times 0.12736$$

$$m_V = 0.3198 \text{ kg}$$

$$x = \frac{0.3198}{0.3198 + 8}$$

$$\Rightarrow \quad x = 0.0384$$

Specific entropy of mixture(s)

$$s = s_f + xs_{fg}$$

$$s = 2.3309 + 0.0384 \times 4.1014$$

$$s = 2.4884 \text{ kJ/kg-K}$$

45. $\dot{m} = 0.1$ kg/s, $\dot{Q} = 5$ kW(heat loss)

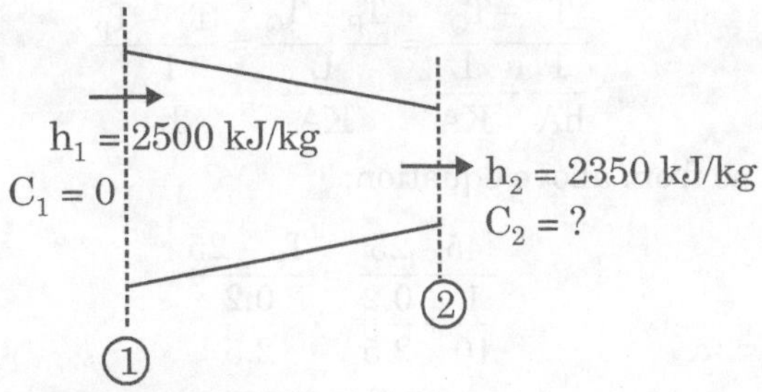

Applying SFEE

$$\dot{m}\left(h_1 + \frac{1}{2}C_1^2 + gz_1\right) + \dot{Q} = \dot{m}\left(h_2 + \frac{1}{2}C_2^2 + gz_2\right) + \dot{W}_{CV}$$

$C_1 = 0$ and $\dot{W}_{CV} = 0$

$z_1 = z_2$ (assume)

$$\dot{m}h_1 + \dot{Q} = \dot{m}h_2 + \dot{m}\frac{1}{2}C_2^2$$

$$\Rightarrow \quad \dot{m} \times \frac{1}{2}C_2^2 = \dot{m}(h_1 - h_2) + \dot{Q}$$

$$\Rightarrow \quad 0.1 \times \frac{1}{2}C_2^2 \times 10^{-3} = 0.1(2500 - 2350) - 5$$

$$C_2 = 447.213 \text{ m/s}$$

46. P1 = 0.1 mpa

T1 = 35°C = 35 + 273 = 308 K

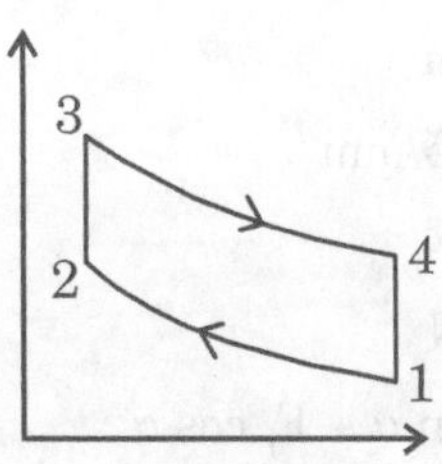

$$V_2 = \frac{V_1}{8}$$

Heat Supplied $Q_s = C_v(T_3 - T_2)$

$$C_v(T_3 - T_2) = 500 \text{ kJ/kg} \quad ...(i)$$

Given, $\left(\frac{V_2}{V_1}\right)^{V-1} = \frac{T_1}{T_2}$

$$\Rightarrow \quad \left(\frac{1}{8}\right)^{14-1} = \frac{308}{T_2}$$

$$\Rightarrow \quad T_2 = 707.60 \text{ K}$$

After putting the value of T_2 is equation(i):-

$0.718\,(T_3 - 707.6) = 500$

$\Rightarrow$ $\boxed{T_3 = 1403.98\text{ K}}$

47. For steady state condition

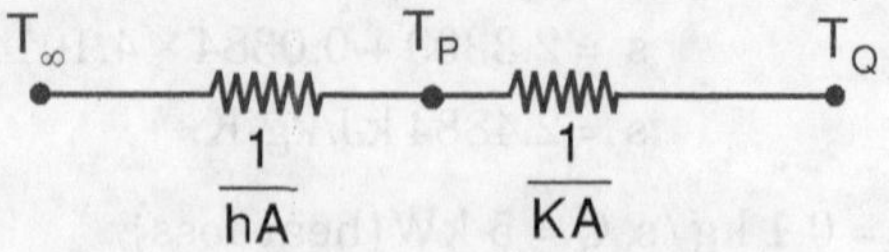

$$\frac{T_\infty - T_Q}{\frac{1}{hA} + \frac{L}{KA}} = \frac{T_P - T_Q}{\frac{L}{KA}} = \frac{T_\infty - T_P}{\frac{1}{h}A}$$

so from above equation:

$$\frac{45-25}{\frac{1}{10} + \frac{0.2}{2.5}} = \frac{T_P - 25}{\frac{0.2}{2.5}}$$

$$\boxed{T_P = 33.889°C}$$

48. We know that

Slope of AB line is E

$$E = \frac{AC}{BC}$$

$$200 \times 10^3\text{ MPa} = \frac{(500-100)\text{MPa}}{BC}$$

$$BC = \frac{400}{200 \times 10^3} = 0.002$$

$$\therefore \quad \epsilon_E = 0.5 - 0.002 = 0.498$$

49. $t = 0.010$ mm

$v = 130$ m/min

$\alpha = 15°$

$b = 6$ mm

$t_c = 0.015$ mm

$F_c = 60$ N

$F_t = 25$ N

$F = F_c \sin\alpha + F_t \cos\alpha$

$= 60 \sin 15 + 25 \cos 15$

$= 39.6773$ N

Ratio of frictional energy to total energy

$$= \frac{F}{F_c}\cdot\frac{V_c}{V} = \frac{F}{F_c}\left(\frac{t}{t_c}\right) \qquad \left[\because \frac{t}{t_c} = \frac{V_c}{V} = r\right]$$

$$= \frac{39.6773}{60} \times \frac{0.010}{0.015} = 0.4408$$

50. True strain $= c_t = \ln\left(\frac{L_f}{L_i}\right)$

$$= \ln\left(\frac{\frac{L_i}{2}}{L_i}\right) = \ln\frac{1}{2}$$

$$= -0.693$$

$$\therefore \quad |c_t| = 0.6963$$

51. $3x + 5y \le 15 \rightarrow \frac{x}{b} + \frac{y}{3} \le 1$

$4x + 9y \le 8 \rightarrow \frac{x}{2} + \frac{y}{(8/9)} \le 1$

$13x + 2y \le 2 \rightarrow \frac{x}{(2/13)} + \frac{y}{1} \le 1$

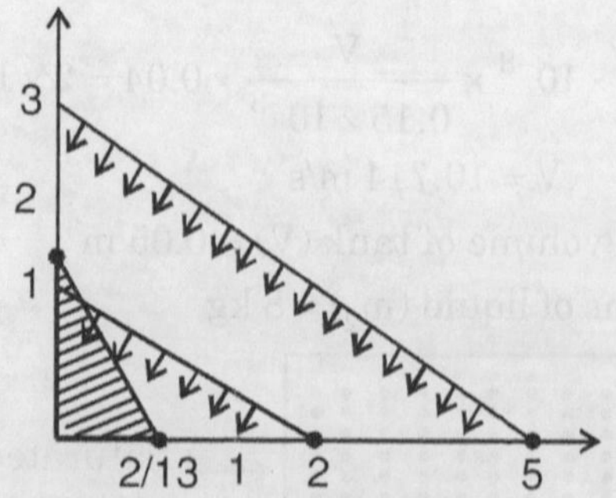

The minimum value of $3x + 5y$.

for minimum value the value of x & y in common area is - so respectively it means. The minimum value of $3x + 5y = 3(0) + 5(0) = 0$

52. By SPT Rule

Job	P.T.	D.D.	Job Flow Time	Tardiness
A	3	8	0 + 3 = 3	0
C	4	4	3 + 4 = 7	3
E	5	17	7 + 5 = 12	0
B	7	16	12 + 7 = 19	3
D	9	18	19 + 9 = 28	10
F	13	19	28 + 13 = 41	22
				38

Total tardiness = 38

Average tardiness per job

$$= \frac{\text{Total tardiness}}{\text{No. of Jobs}}$$

$$= \frac{38}{6} = 6.33\text{ days}$$

53.

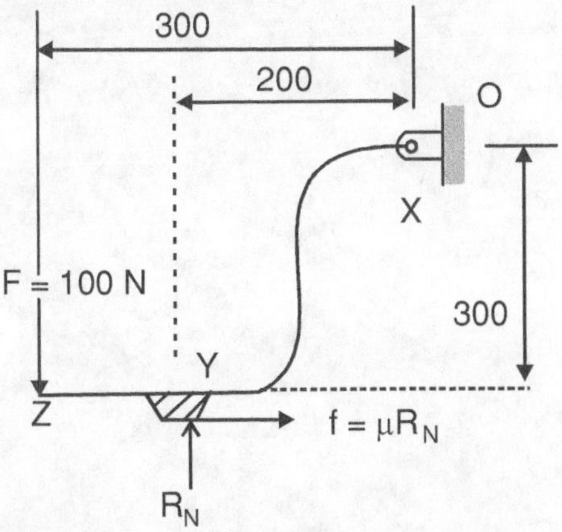

$\Sigma M_0 = 0$

$F \times 300 + f \times 300 = RN \times 200$

$100 \times 300 + \mu R_N \times 300 = R_N \times 200$

$100 \times 300 + 0.3 \times R_N \times 300 = R_N \times 200$

$300 = 1.1\, R_N$

$R_N = 272.72$ N

Braking Torque $= \mu R_N \times R = 0.3 \times 272.72 \times 0.100 = 8.18$ Nm

54.

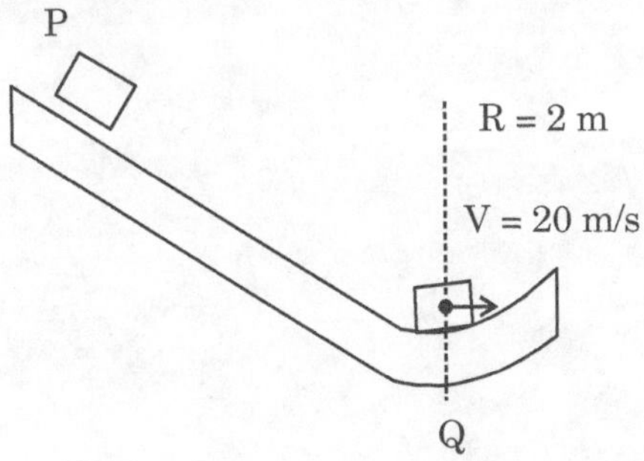

$m = 2$ kg, $g = 10$ m/s^2

FBD of mass P at point Q

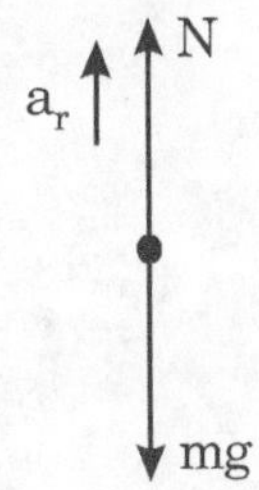

$N - mg = ma_r$ at no slip condition.

$$N = mg + \frac{mv^2}{R} = 2 \times 10 + \frac{2 \times 20 \times 20}{2} = 420 \text{ N}$$

55. $$n \times \text{specific MRR} = \frac{L \times w \times t}{\text{time}} \times \frac{1}{1}$$

$$0.9 \times 3.44 \times 10^{-2} = \frac{10 \times 30 \times 12}{2 \times 60} \times \frac{1}{1}$$

$$\boxed{I = 968.99 \text{ A}}$$

SOLVED PAPER - 2019

INSTRUCTIONS

1. Total of 65 questions carrying 100 marks, out of which 10 questions carrying a total of 15 marks are in General Aptitude (GA)
2. The Engineering Mathematics will carry around **15% of the total marks,** the General Aptitude section will carry **15% of the total marks** and the **remaining 70% of the total marks.**
3. **Types of Questions**
   (*a*) **Multiple Choice Questions (MCQ)** carrying 1 or 2 marks each in all papers and sections. These questions are objective in nature, and each will have a choice of four options, out of which the candidate has to mark the correct answer(s).
   (*b*) **Numerical Answer Questions** of 1 or 2 marks each in all papers and sections. For these questions the answer is a real number, to be entered by the candidate using the virtual keypad. No choices will be shown for these type of questions.
4. For **1-mark** multiple-choice questions, **1/3 marks** will be deducted for a wrong answer. Likewise, for **2-marks** multiple-choice questions, **2/3** marks will be deducted for a wrong answer. There is no negative marking for numerical answer type questions.

Chapter-Wise Analysis

#	Chapters	Marks		
		1	2	Total
1	General Aptitude	5	5	15
2	Engineering Mathematics	4	4	12
3	Engineering Mechanics	2	2	6
4	Strength of Materials	5	5	15
5	Theory of Machines	3	4	11

#	Chapters	Marks		
		1	2	Total
6	Design of Machine Elements	0	0	0
7	Fluid Mechanics	3	2	7
8	Heat Transfer	0	2	4
9	Thermodynamics, Refrigeration & Air-Conditioning	3	4	11
10	Materials, Manufacturing & Industrial Engg.	5	7	19

GENERAL APTITUDE (GA)

Q. 1 to Q. 5 : Carry One Mark Each.

1. John Thomas, an ___ writer, passed away in 2018.

(a) imminent (b) dominant

(c) eminent (d) prominent

2. The minister avoided any mention of the issue of women's reservation in the private sector. He was accused of ____ the issue.

(a) collaring (b) belting

(c) tying (d) skirting

3. The sum and product of two integers are 26 and 165 respectively. The difference between these two integers is _____.

(a) 4 (b) 2

(c) 6 (d) 3

4. _____ I permitted him to leave, I wouldn't have had any problem with him being absent _____ I?

(a) Have, would (b) Have, wouldn't

(c) Had, would (d) Had, wouldn't

5. A worker noticed that the hour hand on the factory clock had moved by 225 degrees during her stay at the factory. For how long did she stay in the factory ?

(a) 8.5 hours

(b) 3.75 hours

(c) 4 hours and 15 minutes

(d) 7.5 hours

Q. 6 to Q. 10 : Carry Two Marks Each.

6. A person divided an amount of Rs. 100,000 into two parts and invested in two different schemes. In one he got 10% profit and in the other he got 12%. If the profit percentages are interchanged with these investments he would have got Rs. 120 less. Find the ratio between his investments in the two schemes.

(a) 9 : 16 (b) 47 : 53

(c) 11 : 14 (d) 37 : 63

7. Under a certain legal system, prisoners are allowed to make one statement. If their statement turns out to be true then they are hanged. If the statement turns out to be false then they are shot. One prisoner made a statement and the judge had no option but to set him free. Which one of the following could be that statement ?

(a) I committed the crime

(b) You committed the crime

(c) I will be shot

(d) I did not commit the crime

8. M and N had four children P, Q, R and S. Of them, only P and R were married. They had children X and Y respectively. If Y is a legitimate child of W, which one of the following statements is necessarily FALSE ?

(a) R is the father of Y

(b) W is the wife of R

(c) M is the grandmother of Y

(d) W is the wife of P

9. Congo was named by Europeans. Congo's dictator Mobuto later changed the name of the country and the river to Zaire with the objective of Africanising names of persons and spaces. However, the name Zaire was a Portuguese alteration of Nzadi o Nzere, a local African term meaning 'River that swallows Rivers'. Zaire was the Portuguese name for the Congo river in the 16th and 17 centuries.

Which one of the following statements can be inferred from the paragraph above ?

(a) The term Nzadi o Nzere was of Portuguese origin

(b) As a dictator Mobuto ordered the Portuguese to alter the name of the river to Zaire

(c) Mobuto's desire to Africanise names was prevented by the Portuguese

(d) Mobuto was not entirely successful in Africanising the name of his country

10. A firm hires employees at five different skill levels P, Q, R, S, T. The shares of employment at these skill levels of total employment in 2010 is given in the pie chart as shown. There were a total of 600 employees in 2010 and the total employment increased by 15% from 2010 to 2016. The total employment at skill levels P, Q and R remained unchanged during this period. If the exmployment at skill level S increased by 40% from 2010 to 2016, how many employees were there at skill level T in 2016 ?

Percentage share of skills in 2010

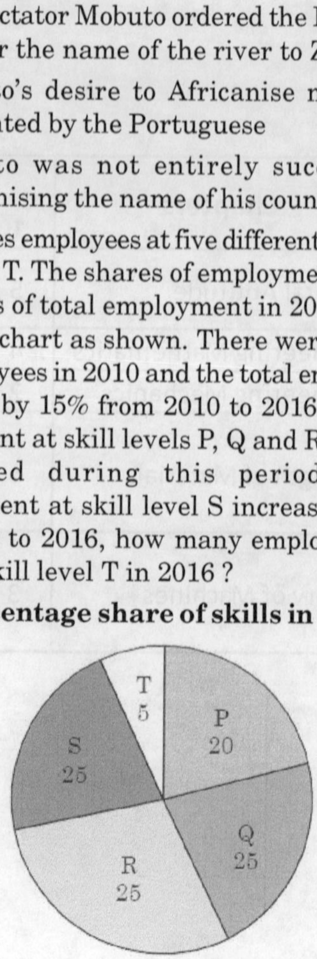

(a) 72 (b) 60

(c) 30 (d) 35

MECHANICAL ENGINEERING

Q. 1 to Q. 25 : Carry One Mark Each.

1. Water flows though a pipe with a velocity given by $\vec{V} = \left(\frac{4}{t} + x + y\right)\hat{j}$ m/s, where $\hat{j}$ is the unit vector in the y direction, t(>0) is in seconds, and x and y are in meters. The magnitude of total acceleration at the point (x, y) = (1, 1) at t = 2 s is ____ m/s^2.

2. A spur gear with 20° full depth teeth is transmitting 20 kW at 200 rad/s. The pitch circle diameter of the gear is 100 mm. The magnitude of the force applied on the gear in the radial direction is

(a) 0.36 kN (b) 2.78 kN
(c) 0.73 kN (d) 1.39 kN

3. The table presents the demand of a product. By simple three-months moving average method, the demand-forecast of the product for the month of September is

Month	Demand
January	450
February	440
March	460
April	510
May	520
June	495
July	475
August	560

(a) 510 (b) 530
(c) 536.67 (d) 490

4. A parabola $x = y^2$ with $0 \le x \le 1$ is shown in the figure. The volume of the solid of rotation obtained by rotating the shaded area by 360° around the x-axis is

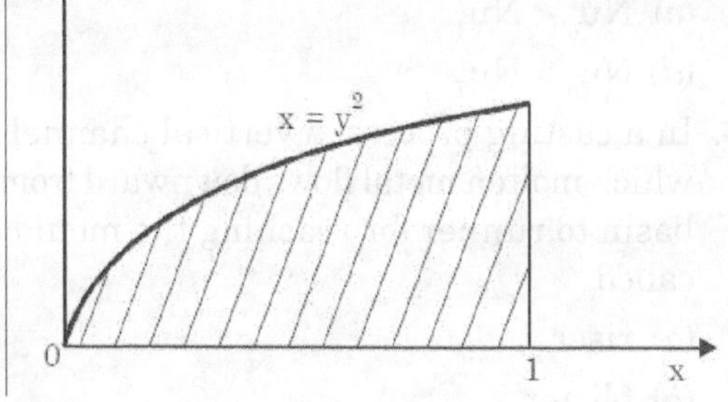

(a) 2π (b) $\frac{\pi}{2}$

(c) $\frac{\pi}{4}$ (d) π

5. As per common design practice, the three types of hydraulic turbines, in descending order of flow rate, are

(a) Pelton, Francis, Kaplan
(b) Francis, Kaplan, Pelton
(c) Pelton, Kaplan, Francis
(d) Kaplan, Francis, Pelton

6. A slender rod of length L, diameter d (L >> d) and thermal conductivity k_1 is joined with another rod of identical dimensions, but of thermal conductivity k_2, to form a composite cylindrical rod of length 2L. The heat transfer in radial direction and contact resistance are negligible. The effective thermal conductivity of the composite rod is

(a) $\frac{2k_1k_2}{k_1 + k_2}$

(b) $\sqrt{k_1k_2}$

(c) $\frac{k_1k_2}{k_1 + k_2}$

(d) $k_1 + k_2$

7. Consider the stress-strain curve for an ideal elastic-plastic strain hardening metal as shown in the figure. The metal was loaded in uniaxial tension starting from O. Upon loading, the stress-strain curve passes through initial yield point at P, and then strain hardens to point Q, where the loading was stopped. From point Q, the specimen was unloaded to point R, where the stress is zero. If the same specimen is reloaded in tension from point R, the value of stress t which the material yields again is _____MPa

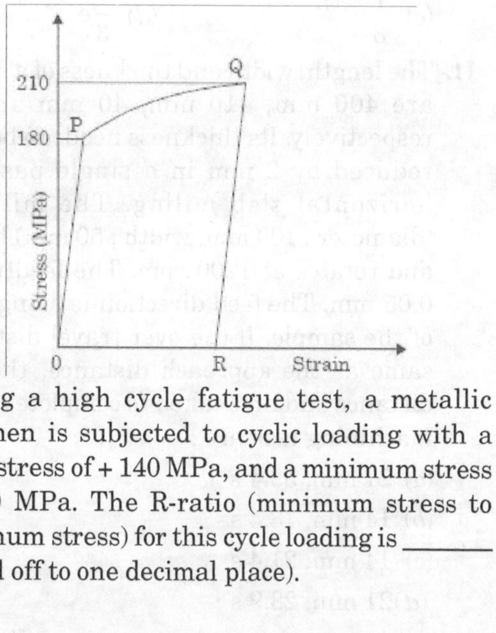

8. During a high cycle fatigue test, a metallic specimen is subjected to cyclic loading with a mean stress of + 140 MPa, and a minimum stress of –70 MPa. The R-ratio (minimum stress to maximum stress) for this cycle loading is ______ (round off to one decimal place).

9. A flat-faced follower is driven using a circular eccentric can rotating at a constant angular velocity ω. At time t = 0, the vertical position of the follower is y(0) = 0, and the system is in the configuration shown below.

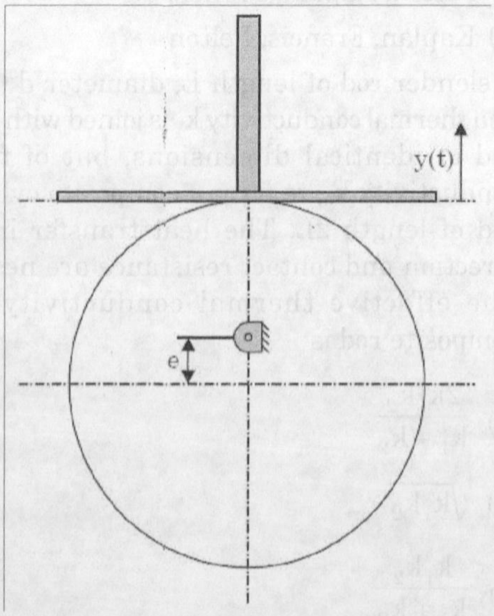

The vertical position of the follower face, y(t) is given by

(*a*) e sin ωt (*b*) e(1 + cos 2ωt)

(c) e sin 2 ωt (*d*) e(1 – cos ωt)

10. For the equation $\frac{dy}{dx} + 7x^2y = 0$, if y(0) = 3/7, then the value of y(1) is

(*a*) $\frac{3}{7}e^{-7/3}$ (*b*) $\frac{3}{7}e^{-3/7}$

(*c*) $\frac{7}{3}e^{-3/7}$ (*d*) $\frac{7}{3}e^{-7/3}$

11. The length, width and thickness of a steel sample are 400 mm, 410 mm, 40 mm and 20 mm, respectively. Its thickness needs to be uniformly reduced by 2 mm in a single pass by using horizontal slab miling. The miling cutter (diameter : 100 mm, width ; 50 mm) has 20 teeth and rotates at 1200 rpm. The feed per tooth is 0.05 mm. The feed direction is along the length of the sample. If the over-travel distance is the same as the approach distance, the approach distance and time taken to complete the required machining task are :

(*a*) 21 mm, 39.4 s

(*b*) 14 mm, 18.4 s

(c) 14 mm, 21.4 s

(*d*)21 mm, 28.9 s

12. A block mass 10 kg rests on a horizontal floor. The acceleration due to gravity is 9.81 m/s². The coefficient of static friction between the floor and the block is 0.2. A horizontal force of 10 N is applied on the block as shown in the figure. The magnitude of force of friction (in N) on the block is ______.

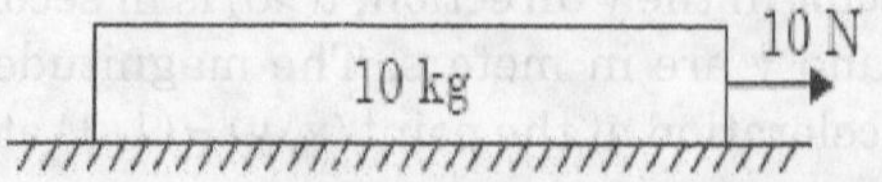

13. Consider an ideal vapor compression refrigeration cycle. If the throttling process is replaced by an isentropic expansion process, keeping all the other processes unchanged, which one of the following statements is true for the modified cycle?

(*a*) Coefficient of performance is higher than that of the original cycle.

(*b*) Coefficient of performance is the same as that of the original cycle.

(*c*) Coefficient of performance is lower than that of the original cycle.

(*d*) Refrigerating effect is lower than that of the original cycle.

14. Consider the matrix :

$$P = \begin{bmatrix} 1 & 1 & 0 \\ 0 & 1 & 1 \\ 0 & 0 & 1 \end{bmatrix}$$

The number of distinct eigenvalues of P is :

(*a*) 3 (*b*) 0

(*c*) 1 (*d*) 2

15. For a hydrodynamically and thermally fully developed laminar flow through a circular pipe of constant cross-section. The Nusselt number at constant wall heat flux (Nu_q) and that at constant wall temperature (Nu_T) are related as

(*a*) $Nu_q = Nu_T$

(*b*) $(Nu_q = Nu_T)^2$

(*c*) $Nu_q < Nu_T$

(*d*) $Nu_q > Nu_T$

16. In a casting process, a vertical channel through which molten metal flows downward from pouring basin to runner for reaching the mould cavity is called

(*a*) riser

(*b*) blister

(*c*) pin hole

(*d*) sprue

17. The position vector $\overline{OP}$ of point P(20, 10) is rotated anti-clockwise in X-Y plane by an angle $\theta = 30°$ a such that point P occupies position Q, as shown in the figure. The coordinates (x, y) of Q are

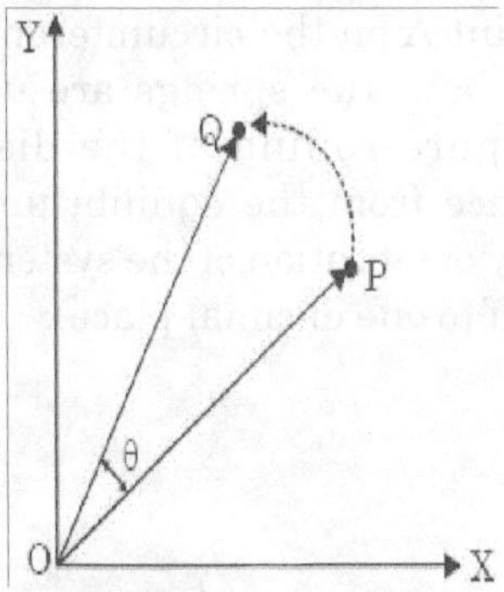

(*a*) (12.32, 18.66) (*b*) (18.66, 12.32)

(*c*) (22.32 ,8.26) (*d*) (13.40, 22.32)

18. A solid cube of side 1 m is kept at a room temperature of 32°C. The coefficient of linear thermal expansion of the cube material is 1×10^{-5}/°C and the bulk modulus is 200 GPa. If the cube is constrained all around and heated uniformly to 42°C, then the magnitude of volumetric (mean) stress (in MPa) induced due to heating is _____.

19. The natural frequencies corresponding to the spring-mass system I and II are ω_I and ω_{II}, respectively. The ratio $\frac{\omega_I}{\omega_{II}}$ is

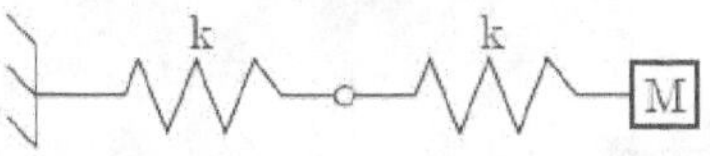

System I

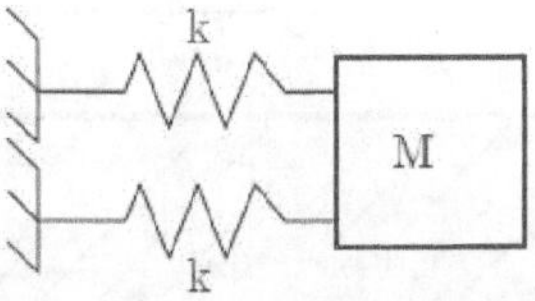

System II

(a) 4 (b) 2

(c) $\frac{1}{4}$ (d) $\frac{1}{2}$

20. The lengths of a large stock of titanium rods follow a normal distribution with a mean (μ) of 440 mm and a standard deviation (σ) of 1 mm. What is the percentage of rods whose lengths lie between 438 mm and 441 mm ?

(a) 68.4% (b) 99.75%

(c) 81.85% (d) 86.64%

21. Air of mass 1 kg. initially at 300 K and 10 bar, is allowed to expand isothermally till it reaches a pressure of 1 bar. Assuming air as an ideal gas with gas constant of 0.287 kJ/kg K, the change in entropy of air (in kJ/kg K, round off to two decimal places) is _______.

22. Which one of the following welding methods provides the highest heat flux(W/mm²)?

(a) Tungsten inert gas welding

(b) Laser beam welding

(c) Oxyacetylene gas welding

(d) Plasma arc welding

23. A cylindrical rod of diameter 10 mm and length 1.0 m fixed at one end. The other end is twisted by angle of 10° by applying a torque. If the maximum shear strain in the rod is $p \times 10^{-3}$, then p is equal to _______ (round off to two decimal places).

24. Evaluation of $\int_2^4 x^3 dx$ using a 2-equal-segment trapezoidal rule gives value of ______.

25. During a non-flow thermodynamic process (1–2) executed by a perfect gas, the heat interaction is equal to the work interaction ($Q_{1.2} = W_{12}$) when the process is

(a) Adiabatic

(b) Polytropic

(c) Isothermal

(d) Isentropic

Q. 26 to Q. 55 : Carry Two Marks Each.

26. A project consists of six activities. The immediate predecessor of each activity and the estimated duration is also provided in the table below:

Activity	Immediate predecessor	Estimated duration (weeks)
P	-	5
Q	-	1
R	Q	2
S	P, R	4
T	P	6
U	S, T	3

If all activities other than S take the estimated amount of time, the maximum duration (in weeks) of the activity S without delaying the completion of the project is ______.

27. Consider an elastic straight beam of length $L = 10\pi m$, with square cross-section of side $a = 5$ mm, and Young's modulus E = 200 GPa. This straight beam was bent in such a way that the two ends meet, to form a circle of mean radius R. Assuming that Euller-Bernoulli beam theory is applicable to this bending problem, the maximum tensile bending stress in the bent beam is _____ MPa.

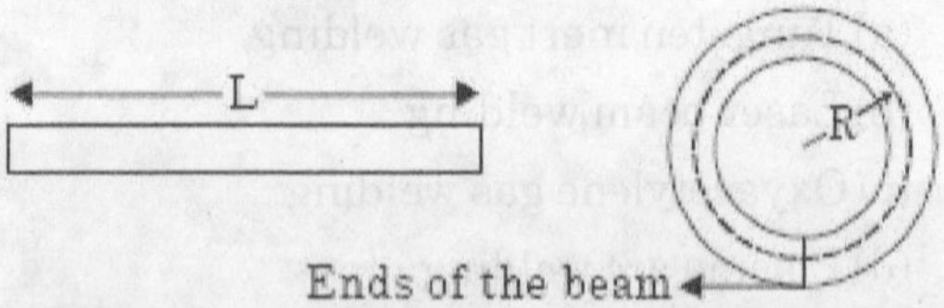

28. In ASA system, the side cutting and end cutting edge angles of a sharp turning tool are 45° and 10°, respectively. The feed during cylindrical turning is 0.1 mm/rev. The center line average surface roughness (in μm, round off to one decimal place) of the generated surface is_____.

29. A car having weight W is moving in the direction as shown in the figure. The center of gravity (CG) of the car is located at height h from the ground, midway between the front and rear wheels. The distance between the front and rear wheels is l. The acceleration of the car is a, and acceleration due to gravity is g. The reaction on the front wheels(R_f) and rear wheels (R_r) are given by:

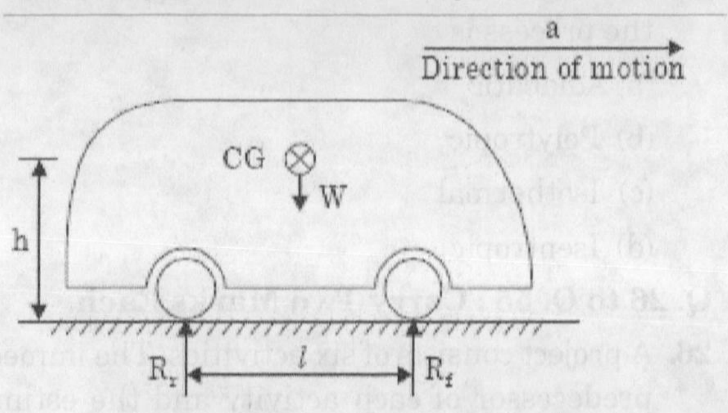

(a) $R_f = \frac{W}{2} - \frac{W}{g}\left(\frac{h}{l}\right)a : R_r = \frac{W}{2} + \frac{W}{g}\left(\frac{h}{l}\right)a$

(b) $R_f = R_r = \frac{W}{2} + \frac{W}{g}\left(\frac{h}{l}\right)a$

(c) $R_f = R_r = \frac{W}{2} - \frac{W}{g}\left(\frac{h}{l}\right)a$

(c) $R_f = \frac{W}{2} + \frac{W}{g}\left(\frac{h}{l}\right)a ; R_r = \frac{W}{2} - \frac{W}{g}\left(\frac{h}{l}\right)a$

30. Taylor's tool life equation is given by $VT^n = C$, where V is in m/min and T is in min. In a turning operation, two tools X and Y are used. For tool X, n = 0.3 and C = 60 and for tool Y, n = 0.6 and C = 90. Both the tools will have the same tool life for the cutting speed (in m/min, round off one decimal place) of ______.

31. A uniform thin disk of mass 1 kg and radius 0.1 m is kept on a surface as shown in the figure. The spring of stiffness k_1 = 400 N/m is connected to the disk center A and antoher spring of stiffness k_2 = 100 N/m is connected at point B just above point A on the circumference of the disk. Initially, both the springs are unstretched. A assume pure rolling of the disk. For small disturbance from the equilibrium, the natural frequency of vibration of the system is _____rad/s (round off to one decimal place).

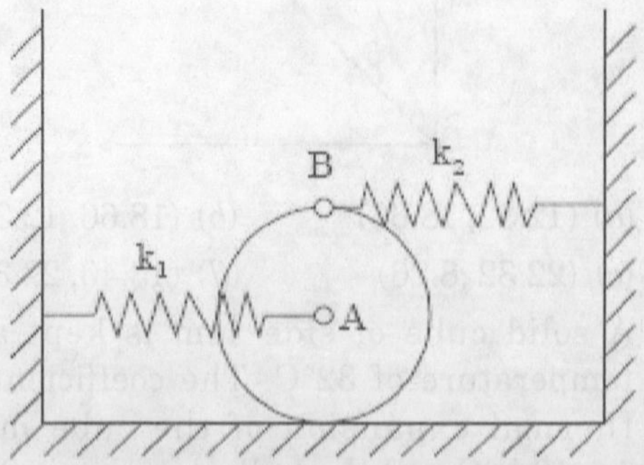

32. A plane-strain compression (forging) of a block is shown in the figure. The strain in the z-direction is zero. The yield strength (S_y) in uniaxial tension/ compression of the material of the block is 300 MPa and its follows the Tresca (maximum shear stress) criterion. Assume that the entire block has started yielding. At a point where σ_x = 40 MPa (compressive) and τ_{xy} = 0, the stress component σ_y is

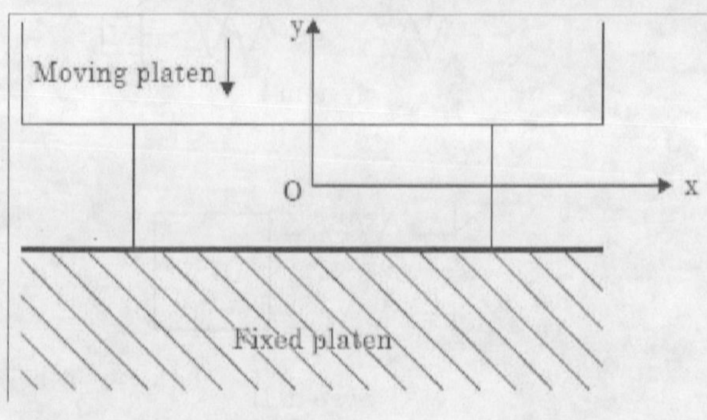

(a) 260 MPa (compressive)

(b) 260 MPa (tensile)

(c) 340 MPa (tensile)

(d) 340 MPa (compressive)

33. A cube of side 100 mm is placed at the bottom of an empty container on one of its faces. The density of the material of the cube is 800 kg/m³. Liquid of density 1000 kg/m³ is now poured into the container. The minimum height to which the liquid needs to be poured into the container for the cube to just lift up is _____ mm.

34. In a UTM experiment, a sample of length 100 mm, was loaded in tension until failure. The failure load was 40 kN. The displacement, measured using the cross-head motion, at failure, was 15 mm. The compliance of the UTM is constant and is given by 5×10^8 m/N. The strain at failure in the sample is _____%.

35. In orthogonal turning of a cylindrical tube of wall thickness 5 mm, the axial and the tangential cutting forces were measured at 1259 N and 1601 N, respectively. The measured chip thickness after machining was found to be 0.3 mm. The rake angel was 10° and the axial feed was 100 mm/min. The rotational speed of the spindle was 1000 rpm. Assuming the material to be perfectly plastic and merchant's first solution, the shear strength of the martial is closest to

(a) 875 MPa

(b) 200 MPa

(c) 722 MPa

(d) 920 MPa

36. The rotor of turbojet engine of an aircraft has a mass 180 kg and polar moment of inertia 10 kg.m^2 about the rotor axis. The rotor rotates at a constant speed of 1100 rad/s in the clockwise direction when viewed from the front of the aircraft. The aircraft while flying at a speed of 800 km per hour takes a turn with a radius of 1.5 km to the left. The gyroscopic moment exerted by the rotor on the aircraft structure and the direction of motion of the nose when the aircraft turns, are

(a) 1629.6 N.m and the nose goes up

(b) 162.9 N.m and the nose goes up

(c) 162.9 N.m and the nose goes down

(d) 1629.6 N.m and the nose goes down

37. A truss is composed of members AB, BC, CD, and AD and BD as shown in the figure. A vertical load of 10 kN is applied at point D. The magnitude of force (in kN) in the member BC is _____.

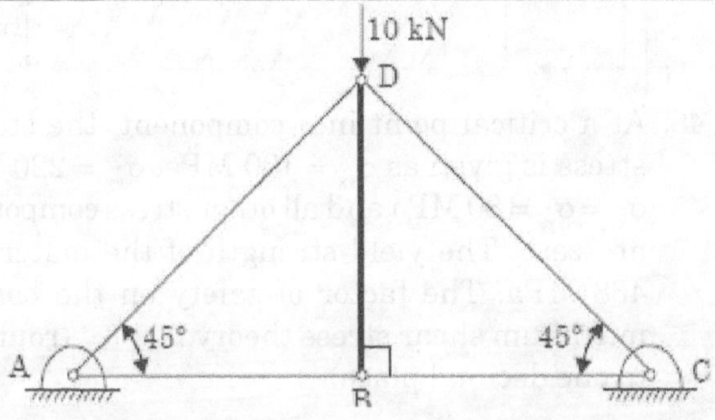

38. In a four bar planar mechanism shown in the figure, AB = 5 cm, AD = 4 cm and DC = 2 cm. In the configuration shown, both AB and DC are perpendicular to AD. The bar AB rotates with an angular velocity of 10 rad/s. The magnitude of angular velocity (in rad/s) of bar DC at this instant is

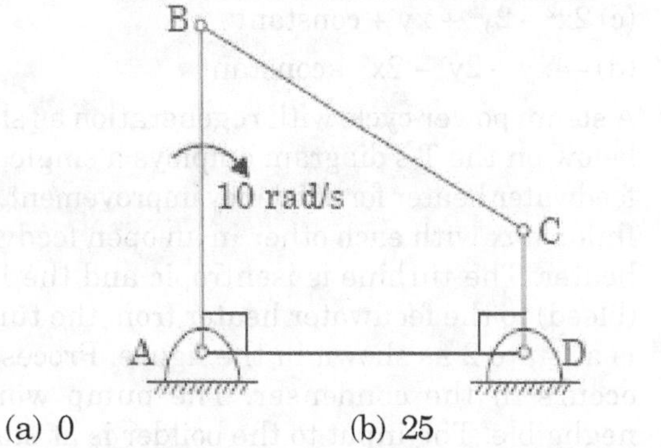

(a) 0 (b) 25

(c) 10 (d) 15

39. The variable x takes a value between 0 and 10 with uniform probability distribution. The variable y takes a value between 0 and 20 with uniform probability distribution. The probability of the sum of variables (x + y) being greater than 20 is

(a) 0.33 (b) 0.25

(c) 0 (d) 0.50

40. The wall of a constant diameter pipe of length 1 m is heated uniformly with flux q" by wrapping a heater coil around it. The flow at the inlet to the pipe is hydrodynamically fully developed. The fluid is incompressible and the flow is assumed to be laminar and steady all through the pipe. The bulk temperature of the fluid is equal to 0°C at the inlet and 50°C at the exit. The wall temperatures are measured at three locations, P, Q and R as shown in the figure. The flow thermally develops after some distance from the inlet. The following measurements are made :

Point	P	Q	R
Wall Temp(°C)	50	80	90

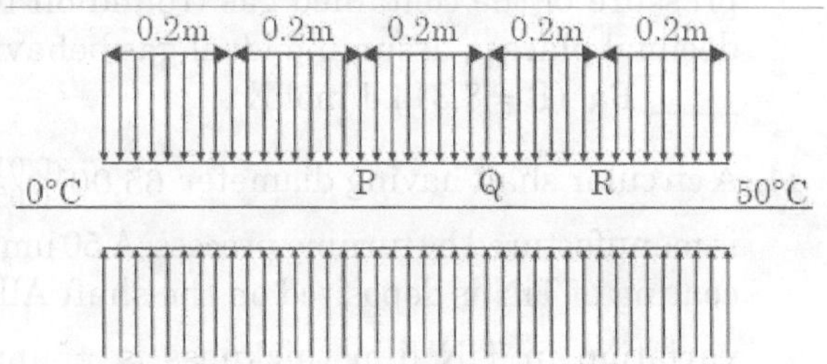

Among the locations P, Q and R, the flow is thermally at :

(a) R only (b) P and Q only

(c) P, Q and R (d) Q and R only

41. A harmonic function is analytic if it satisfies the Laplace equation. If $u(x, y) = 2x^2 - 2y^2 + 4xy$ is a harmonic function, then its conjugate harmonic function $v(x, y)$ is

(a) $4y^2 - 4xy$ + constant

(b) $4xy - 2x^2 + 2y^2$ + constant

(c) $2x^2 - 2y^2 + xy$ + constant

(d) $-4xy + 2y^2 - 2x^2$ + constant

42. A steam power cycle with regeneration as shown below on the T-s diagram employs a single open feedwater heater for efficiency improvement. The fluids mix with each other in an open feedwater heater. The turbine is isentropic and the input (bleed) to the feedwater heater from the turbine is at state 2 as shown in the figure. Process 3-4 occurs in the condenser. The pump work is negligible. The input to the boilder is at state 5. The following information is available from the steam tables:

State	1	2	3	4	5	6
Enthalpy(kJ/kg)	3350	2800	2300	175	700	1000

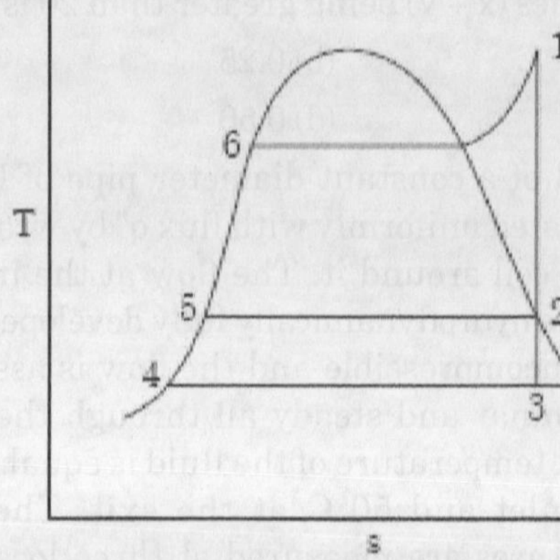

The mass flow rate of steam bled from the turbine as a percentage of the total mass flow rate at the inlet to the turbine at state 1 is ____.

43. If one mole of H_2 gas occupies a rigid container with a capacity of 1000 liters and the temperature is raised from 27°C to 37°C, the change in pressure of the contained gas (round off to two decimal places), assuming ideal gas behavior, is ____ Pa. (R = 8.314 J/mol.K).

44. A circular shaft having diameter $65.00^{+0.01}_{-0.05}$ mm is manufactured by turning process. A 50 μm thick coating of TiN is deposited on the shaft Allowed variation in TiN film thickness is ±5μm. The minimum hole diameter (in mm) to just provide clearance fit is

(a) 65.01 (b) 64.95

(c) 65.12 (d) 65.10

45. The set of equations

$x + y + z = 1$

$a - ay + 3z = 5$

$5x - 3y + az = 6$

has infinite solution, if a =

(a) –4 (b) 3

(c) 4 (d) –3

46. Five jobs (J1, J2, J3, J4 and J5) need to be processed in a factory. Each job can be assigned to any of the five different machines (M1, M2, M3, M4 and M5). The time duration taken (in minutes) by the machines for each of the jobs, are given in the table. However, each job is assigned to a specific machine in such a way that the total processing time is minimum. The total processing time is ____ minutes.

	M1	M2	M3	M4	M5
J1	40	30	50	50	58
J2	26	38	60	26	38
J3	40	34	28	24	30
J4	28	40	40	32	48
J5	28	32	38	22	44

47. Two immiscible, incompressible, viscous fluids having same densities but different viscosities are contained between two infinite horizontal parallel plates. 2 m apart as shown below. The bottom plate is fixed and the upper plate move to the right with a constant velocity of 3 m/s. With the assumptions of Newtonian fluid steady, and fully developed laminar flow with zero pressure gradient in all directions, the momentum equations simplify to $\frac{d^2u}{dy^2} = 0$

If the dynamic viscosity of the lower fluid, μ_2 is twice that of the input fluid, μ_1, then the velocity at the interface (round off to two decimal places) is____m/s.

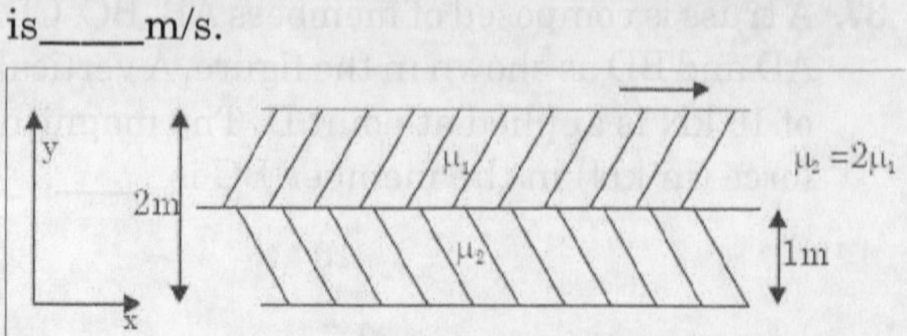

48. At a critical point in a component, the state of stress is given as σ_{xx} = 100 MPa, σ_{yy} = 220 MPa, $\sigma_{xy} = \sigma_{yx}$ = 80 MPa and all other stress components are zero. The yield strength of the material is 468 MPa. The factor of safety on the basis of maximum shear stress theory is____ (round off to one decimal place).

49. Match the following sand mold casting effects with their respective causes.

	Defect		Cause
P	Blow hole	1.	Poor Collapsibility
Q	Misrum	2.	Mold erosion
R	Hot tearing	3.	Poor permeability
S	Wash	4.	Insufficient fluidity

(a) P-3, Q-4, R-1, S-2 (b) P-3, Q-4, R-2, S-1
(c) P-4, Q-3, R-1, S-2 (d) P-2, Q-4, R-1, S-3

50. A gas is heated in a duct as it flows over a resistance heater. Consider a 101 kW electric heating system. The gas enters the heating section of the duct at 100 kPa and 27°C with a volume flow rate of 15 m^3/s. If heat is lost from the gas in the duct to the surroundings at a rate of 51 kW, the exit temperature of the gas is (Assume constant pressure, ideal gas, negligible change in kinetic and potential energies and constant specific heat : C_p = 1 kJ/kg. K; R = 0.5 kJ/kg K).

(a) 37°C (b) 76°C
(c) 53°C (d) 32°C

51. The value of the following definite integral is _____ (round off to there decimal places).

$$\int_1^e (x\ln x)\,dx$$

52. A single block brake with a short shoe and torque capacity of 250 N-m is shown. The cylindrical brake drum rotates anticlockwise at 100 rpm and the coefficient of friction is 0.25. The value of a, in mm (round off to one decimal place), such that the maximum actutating force P is 2000 N, is_____.

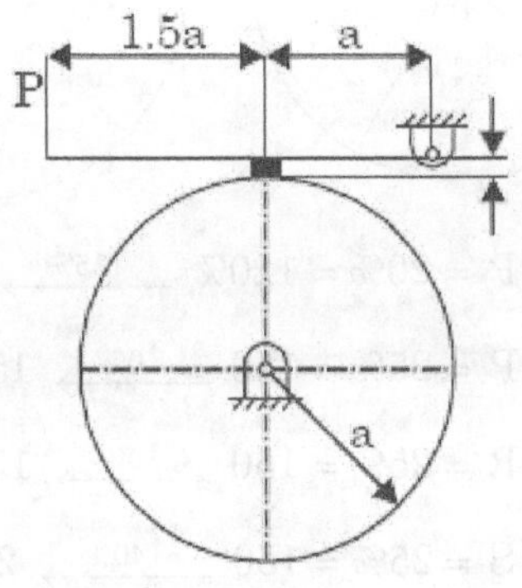

53. Consider a prismatic straight beam of length L = π π m, pinned at the two ends as shown in the figure The beam has a square cross-section of side p = 6 mm. The Young's modulus E = 20 GPa, and the coefficient of thermal expansion α = 3 × 10^{-6} K^{-1}. The minimum temperature rise required to cause Euler buckling of the beam is _____ K.

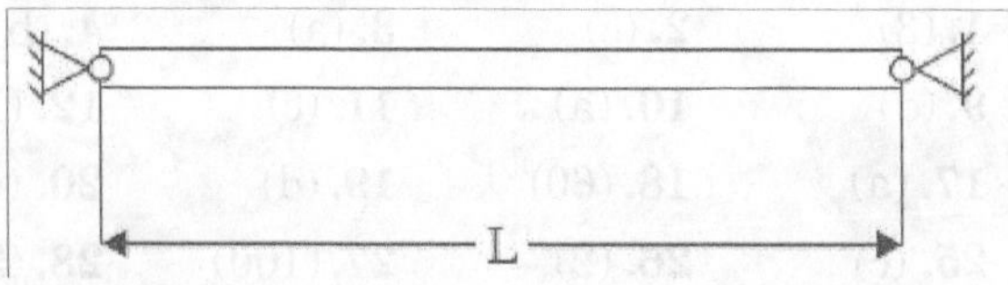

54. A gas turbine with air as the working fluid has an isentropic efficiency of 0.70 when operating at a pressure ratio of 3. Now, the pressure ratio of the turbine is increased to 5, while maintaining the same inlet conditions. Assume air as a perfect gas with specific heat ratio γ = 1.4. If the specific work output remains the same for both the cases, the isentropic efficiency of the turbine at the pressure ratio of 5 is___(round off to two decimal places).

55. Three slabs are joined together as shown in the figure. There is no thermal contact resistance at the interfaces. The center slab experience a nonuniform internal heat generation with an average value equal to 10000 Wm^{-3}, while the left and right slabs have no internal heat generation. All slabs have thickness equal to 1 m and thermal conductivity of each slab is equal to 5 Wm^{-1} K^{-1}. The two extreme faces are exposed to fluid with heat transfer coefficient 100 $Wm^{-2}K^{-1}$ and bulk temperature 30°C as shown. The heat transfer in the slabs is assumed to be one dimensional and steady, and all properties are constant. If the left extreme face temperature T1 is measured to be 100°C, the right extreme faced temperature T_2 is__________°C.

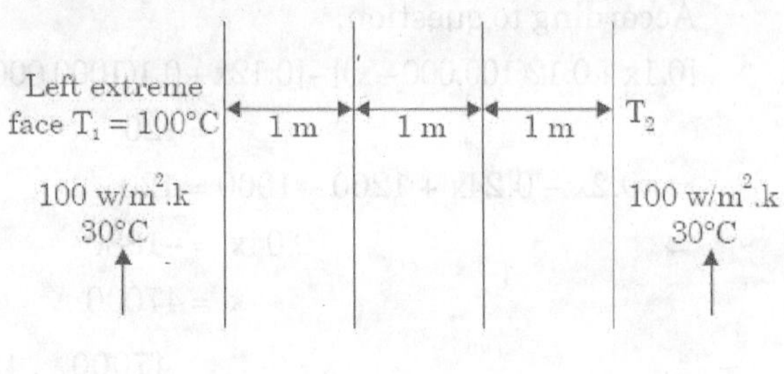

ANSWERS

General Aptitude (GA)

1. (c)	**2.** (d)	**3.** (a)	**4.** (c)	**5.** (d)	**6.** (b)	**7.** (c)	**8.** (d)
9. (a)	**10.** (b)						

Mechanical Engineering

1. (3)	**2.** (c)	**3.** (a)	**4.** (b)	**5.** (d)	**6.** (a)	**7.** (210)	**8.** (–0.2)
9. (d)	**10.** (a)	**11.** (c)	**12.** (10)	**13.** (a)	**14.** (c)	**15.** (d)	**16.** (d)
17. (a)	**18.** (60)	**19.** (d)	**20.** (c)	**21.** (0.66)	**22.** (b)	**23.** (0.87)	**24.** (63)
25. (c)	**26.** (2)	**27.** (100)	**28.** (3.7)	**29.** (a)	**30.** (40)	**31.** (23.1)	**32.** (b)
33. (80)	**34.** (2)	**35.** (c)	**36.** (d)	**37.** (5)	**38.** (b)	**39.** (b)	**40.** (d)
41. (b)	**42.** (20)	**43.** (83.14)	**44.** (c)	**45.** (c)	**46.** (146)	**47.** (1)	**48.** (1.8)
49. (a)	**50.** (d)	**51.** (2.097)	**52.** (212.5)	**53.** (1)	**54.** (0.51)	**55.** (60)	

EXPLANATIONS

GENERAL APTITUDE

3. From prime factorisation of

$$165 = 3 \times 5 \times 11$$

$$165 = 15 \times 11$$

So the two numbers whose sum is 26 and multiplication is 165 is 15 and 11

Now, difference = 15 – 11 = 4

5. Since, we know that

$$360° = 12 \text{ hours}$$

$$\therefore \quad 10 = \frac{12}{360} \text{ hours}$$

$$\therefore \quad 225° = \frac{12 \times 225}{360} \text{ hours}$$

$$= 7.5 \text{ hours}$$

6. Let the first part = x

So second part = 100,000x

According to question,

$$[0.1x + 0.12(100,000 - x)] - [0.12x + 0.1(1000,000 - x)] = 120$$

$$\Rightarrow 0.2x - 0.24x + 1200 - 1000 = 120$$

$$\Rightarrow \quad -0.04x = -1880$$

$$x = 47000$$

$$\frac{x}{100,000 - x} = \frac{47000}{53000} = \frac{47}{53}$$

7. If prisoner make the statement "I will be shot" then only judge have no option and prisoner will be set to free.

8.

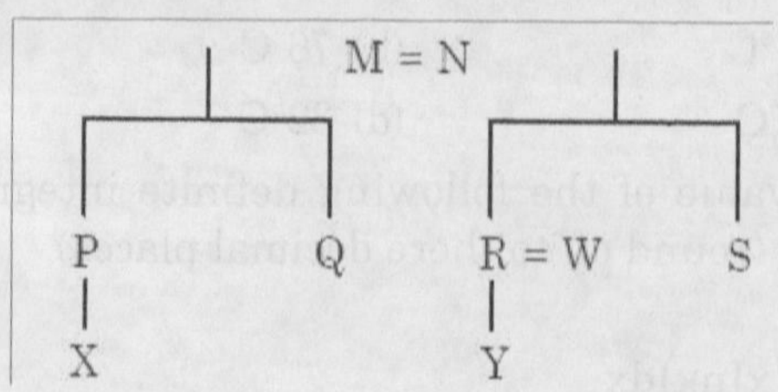

From the above family chart of 'M' and 'N', we can conclude 'W' is the wife of P.

10.

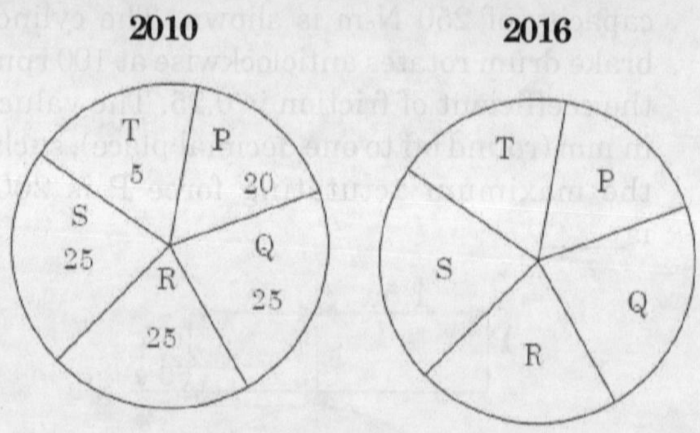

$$P = 20\% = 120\% \xrightarrow{\uparrow 15\%} 1.15 \times 600$$

$$P = 25\% = 150 \xrightarrow{\uparrow 0\%} 150$$

$$R = 25\% = 150 \xrightarrow{\uparrow 0\%} 150$$

$$S = 25\% = 150 \xrightarrow{\uparrow 40\%} 210$$

$$T = 5\% = 30 \longrightarrow ?$$

Now, employees at skill level T in 2016,

$$= 690 - (120 + 150 + 150 + 210)$$

$$= 60$$

MECHANICAL ENGINEERING

1. Given: $\bar{V} = \left(\frac{4}{t} + x + y\right)\hat{j}\ \frac{m}{s}$

$$\bar{V} = u\hat{i} + v\hat{j} + w\hat{k}$$

$$u = 0,\ w = 0$$

$$V = \left(\frac{4}{t} + x + y\right)\hat{j}$$

$$\vec{a} = \vec{a}_x\hat{i} + \vec{a}_y\hat{j} + \vec{a}_z\hat{k}$$

$$\vec{a}_x = u\frac{\partial u}{\partial x} + v\frac{\partial u}{\partial y} + w\frac{\partial u}{\partial z} + \frac{\partial u}{\partial t} = 0$$

$$\vec{a}_z = u\frac{\partial w}{\partial x} + v\frac{\partial w}{\partial y} + w\frac{\partial w}{\partial z} + \frac{\partial w}{\partial t} = 0$$

$$\vec{a}_y = u\frac{\partial v}{\partial x} + v\frac{\partial v}{\partial y} + w\frac{\partial v}{\partial z} + \frac{\partial v}{\partial t}$$

$$= v\frac{\partial v}{\partial y} + \frac{\partial v}{\partial t}$$

$$= \left(\frac{4}{t} + x + y\right) \times 1 + \left(-\frac{4}{t^2}\right)$$

$$\Rightarrow \quad \vec{a}_y = \left(\frac{4}{t} + x + y - \frac{4}{t^2}\right)\hat{j}$$

Now, At (x,y) = 1,1 and t = 2 sec.

Total acceleration is given by,

$$\vec{a} = \vec{a}_y = \left(\frac{4}{2} + 1 + 1 - \frac{4}{4}\right) = 3\frac{m}{s^2}$$

2. Given : $\phi = 20°$

P = 20 kW, w = 200 rad/s

Pitch Circle Diametre = 100 mm

Now, force on radial direction (F_r),

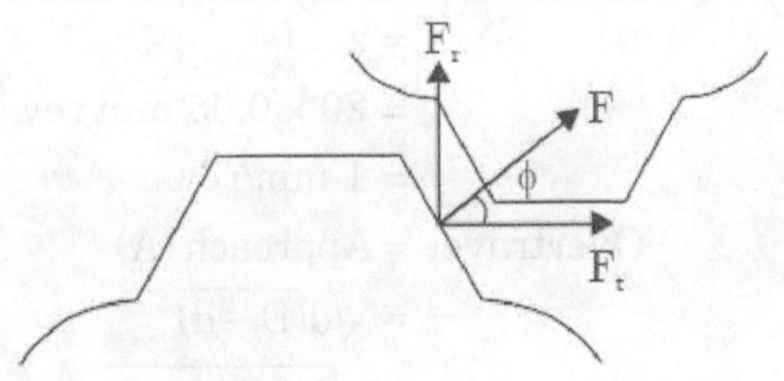

$$F\cos\phi = F_t$$

$$F\sin\phi = F_r$$

Now, $\frac{F_r}{F_t} = \frac{F\sin\phi}{F\cos\phi} = \tan\phi$

$$\Rightarrow \quad F_r = F_t \tan\phi$$

Also, $T = \frac{P}{\omega} = \frac{20 \times 1000}{200} = 100\text{N.m.}$

$$T = F_t \times r$$

$$\Rightarrow \quad F_t = \frac{T}{r} = \frac{100}{\frac{100}{2} \times 10^{-3}} = 2000\text{N}$$

$$F_r = F_t \tan\phi = 2000 \tan 20$$

$$= 727.9\text{ N}$$

$$= 727.9 \times 10^{-3}\text{ kN} = 0.73\text{ kN}$$

3. Three month moving average.

Forecast of the product for the month of September is given by,

$$F = \frac{495 + 475 + 560}{3} = \frac{1530}{3} = 510$$

4.

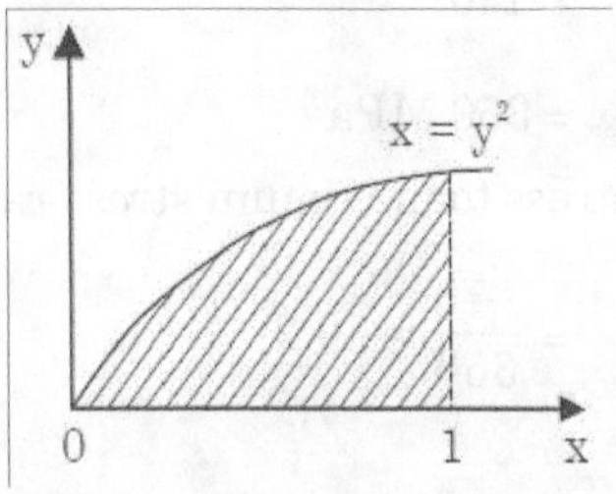

Here, volume of solid of rotation by rotating shaded area by 360° around x-axis is,

$$V = \int_0^1 \pi y^2 dx$$

$$V = \int_0^1 \pi x\, dx$$

$$= \left.\frac{\pi x^2}{2}\right|_0^1 = \frac{\pi}{2}$$

5. In descending order of flow rate.

Kaplan > Francis > Pelton

6.

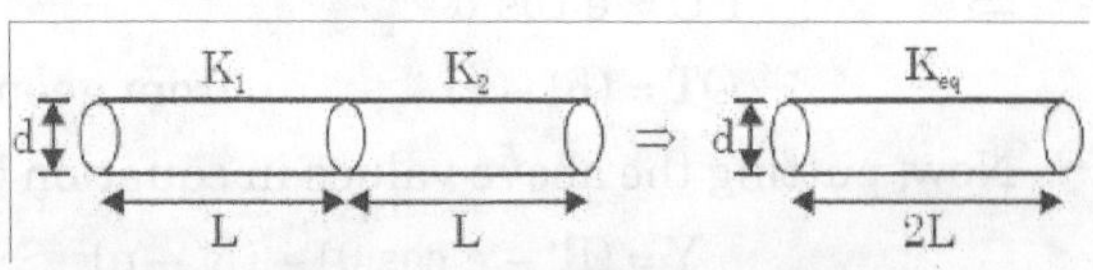

As per question, heat transfer in radial direction and contact resistance are negligible.

In series connection,

$$(R_{th})_{eq} = R_1 + R_2$$

$$\frac{2L}{Ak_{eq}} = \frac{L}{A\,K_1} + \frac{L}{A\,K_2}$$

$$\frac{2}{K_{eq}} = \frac{1}{K_1} + \frac{1}{K_2}$$

$$\Rightarrow \quad K_{eq} = \frac{2K_1K_2}{K_1 + K_2}$$

7. As the point loaded beyond the yield point is unloaded completely and reloaded then because of strain hardening (residual strain remains in body) increase the yield strength.

In this case material with start yielding from point 'R' instead of 'P'.

8. Here, $\sigma_{mean} = +140$ MPa

$\sigma_{min} = -70$ MPa

Now, we know

$$\sigma_{mean} = \frac{\sigma_{max} + \sigma_{min}}{2} = 140$$

$$\frac{\sigma_{max} - 70}{2} = 140$$

$$\Rightarrow \quad \sigma_{max} = 350 \text{ MPa}$$

Now, minimum stress to maximum stress ratio is,

$$\frac{\sigma_{min}}{\sigma_{max}} = \frac{-70}{350} = -0.2$$

9.

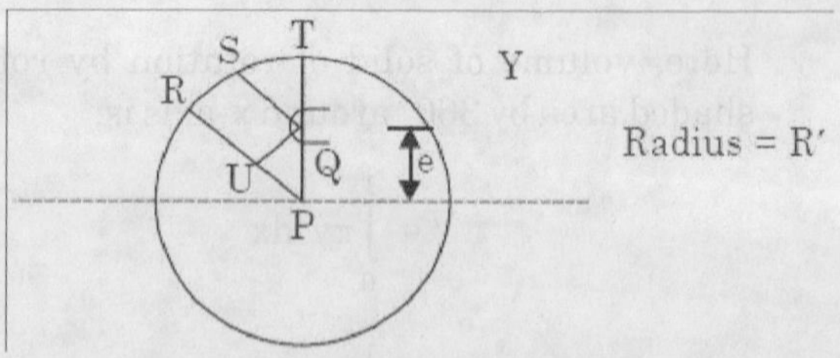

As the follower starts moving, the displacement Y can be shown as marked in the diagram

$$Y = QS - QT$$
$$= RU - QT \quad (\because QS \parallel RU)$$
$$= (PR - PU) - QT \quad ...(i)$$
$$PR = R'$$
$$\Rightarrow \quad PU = e\cos\theta$$
$$QT = (R' - e) \quad \text{from geometry}$$

Now, putting the above values in equation (i)

$$Y = (R' - e\cos\theta) - (R' - e)$$
$$= e - e\cos\theta$$
$$= e(1 - \cos\theta)$$
$$\Rightarrow \quad Y = e(1 - \cos\omega t) \text{ (In time 't', } \theta = \omega t)$$

10. From question, we have

$$\frac{dy}{dx} + 7x^2y = 0,\ y(0) = \frac{3}{7},\ y(1) = ?$$

$$\frac{dy}{dx} = -7x^2 y$$

$$\frac{dy}{y} = -7x^2\,dx$$

$$\ln y = -\frac{7x^3}{3} + C$$

$$y = e^{-\frac{7x^3}{3} + C}$$

$$y(0) = e^C = \frac{3}{7}$$

$$C = \ln\left(\frac{3}{7}\right)$$

$$\ln y = -\frac{7x^3}{3} + \ln\left(\frac{3}{7}\right)$$

$$\ln y - \ln\left(\frac{3}{7}\right) = -\frac{7x^3}{3}$$

$$\ln\left(\frac{7y}{3}\right) = -\frac{7x^3}{3}$$

$$y = \frac{3}{7}e^{-\frac{7x^3}{3}}$$

$$\therefore \quad y(1) = \frac{3}{7}e^{-\frac{7}{3}}$$

11.

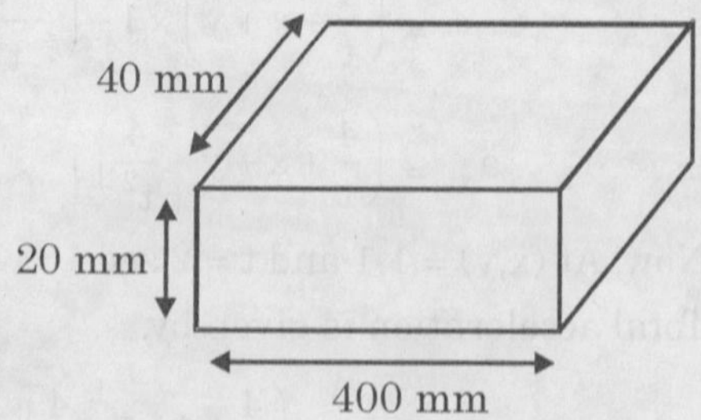

Depth of cut ⇒ (d = 2 mm) Milling cutter,

$D = 100$ mm

$W = 50$ mm

$Z = 20$ tooth,

$N = 1200$ rpm

$f_t = 0.05$ mm/tooth rev

$$f = z \times f_t$$
$$= 20 \times 0.05 \text{ mm/rev}$$
$$= 1 \text{ mm/rev}$$

$$\text{Overtravel} = \text{Approach (A)}$$
$$= \sqrt{d(D-d)}$$
$$= \sqrt{2 \times (100 - 2)}$$
$$= \sqrt{196} = 14 \text{ mm}$$

Total length covered by tool

$$L + 2A = 400 + 2 \times 14 = 428 \text{ mm}$$

Time for one pass,

$$t = \frac{L}{fN} = \frac{428}{1 \times \left(\frac{1200}{60}\right)} = \frac{428}{20}$$

$$= 21.4 \text{ Second}$$

12. 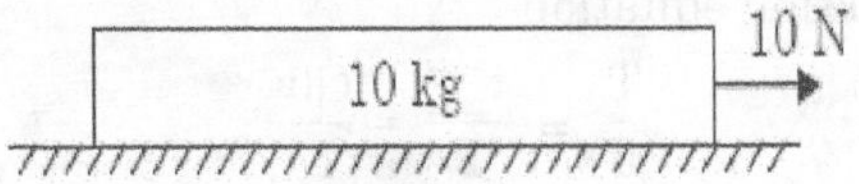

Here, $\mu = 0.2$

$g = 9.81 \text{ m/s}^2$

$M = 10 \text{ kg}$

$(f_s)_{max} = \mu Mg$

$= 0.2 \times 10 \times 9.81$

$= 19.62 \text{ N}$

As, $(F_s)_{max} >$ Force applied

So, magnitude of force of friction on block is,

$(F_s) = 10 \text{ N}$

13. Vapor compression refrigeration cycle :

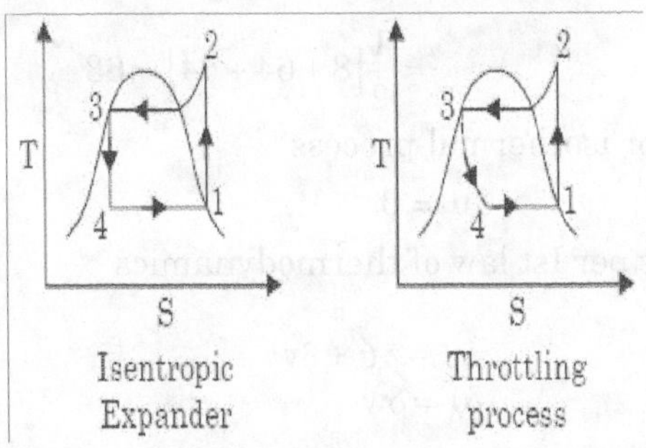

Both cycle required same amount of work input but by use of Isentropic expander, we can get more amount of refrigeration effect, hence

$((COP)_{\text{modified cycle}} > ((COP)_{\text{original cycle}})$

14. $$|P - \lambda I| = \begin{vmatrix} 1-\lambda & 1 & 0 \\ 0 & 1-\lambda & 1 \\ 0 & 0 & 1-\lambda \end{vmatrix}$$

$$(1-\lambda)[(1-\lambda)(1-\lambda) - 0] = 0$$

Eigen values are,

$\lambda = 1,1,1$

So, No. of distinct eigen value = 1

15. $(Nu)_q$ for constant wall heat flux and $(Nu)_T$ at constant wall temperature for a hydrodynamically and thermally fully developed laminar flow through a circular pipe of constant cross-section is 4.36 and 3.66 respectively.

Hence, $(Nu)_q > (Nu)_T$

17. Let Co-ordinates of Q are (x′, y′) and rotation about x-y plane anti clockwise.

$$\begin{bmatrix} x' \\ y' \end{bmatrix} = \begin{bmatrix} \cos\theta & -\sin\theta \\ \sin\theta & \cos\theta \end{bmatrix} \begin{bmatrix} x \\ y \end{bmatrix}$$

$$\begin{bmatrix} x' \\ y' \end{bmatrix} = \begin{bmatrix} \cos 30 & -\sin 30 \\ \sin 30 & \cos 30 \end{bmatrix} \begin{bmatrix} 20 \\ 10 \end{bmatrix}$$

$$= \begin{bmatrix} 0.866 & -0.5 \\ 0.5 & 0.866 \end{bmatrix} \begin{bmatrix} 20 \\ 10 \end{bmatrix}$$

$$\begin{bmatrix} x' \\ y' \end{bmatrix} = \begin{bmatrix} 0.866 \times 20 - 0.5 \times 10 \\ 0.5 \times 20 + 0.866 \times 10 \end{bmatrix}$$

$$= \begin{bmatrix} 12.32 \\ 18.66 \end{bmatrix}$$

$(x', y') = (12.32, 18.66)$

18. 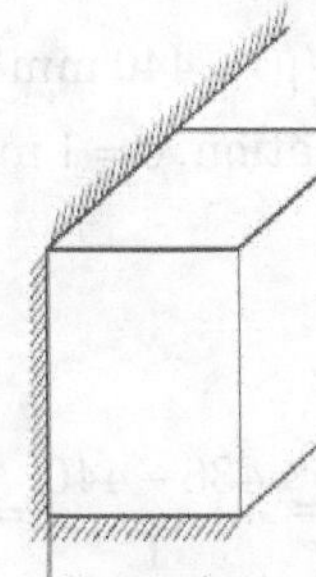

$a = 1 \text{ m}$

$T_a = 32°C$

$\alpha = 1 \times 10^{-5}/°C$

$K = 200 \text{ GPa}$

$\Delta T = 10°C$

$$\epsilon_v = \frac{(1-2\mu)}{E}(3\sigma) = 3\alpha T$$

As $\sigma_1 = \sigma_2 = \sigma_3 = \sigma$

(mean volumetric stress)

Here, $\sigma\dfrac{(1-2\mu)}{E} = \alpha\Delta T,$ $\quad E = 3K(1-2\mu)$

$$\sigma = \frac{E\alpha\Delta T}{1-2\mu} \qquad \left[\frac{E}{1-2\mu} = 3K\right]$$

$\sigma = 3K\alpha\Delta T$

$= 3 \times 200 \times 10^3 \times 1 \times 10^{-5} \times 10$

$\sigma = 60 \text{ MPa}$

19. 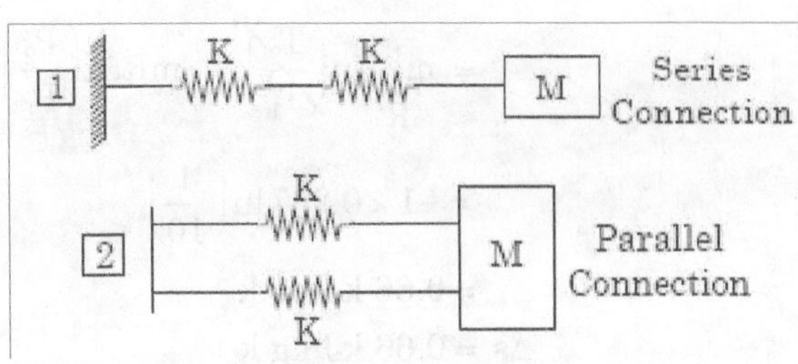

Now, $\dfrac{1}{\{K_{eq}\}} = \dfrac{1}{K_1} + \dfrac{1}{K_2}$ (For series connection)

$$\Rightarrow \left(K_{eq1} = \frac{K_1K_2}{K_1 + K_2}\right) = \frac{K}{2}$$

$(K_{eq})_2 = K_1 + K_2 = 2K$ (For parallel connection)

Natural frequency $\left(\omega = \sqrt{\frac{K}{m}}\right)$

$$\frac{\omega_1}{\omega_2} = \frac{\sqrt{\frac{K}{2m}}}{\sqrt{\frac{2K}{m}}} = \frac{1}{2}$$

20. Given, mean, $(\mu) = 440$ mm

Standard deviation, $\sigma = 1$ mm

$$Z = \frac{X - u}{\sigma}$$

lower limit,

$$Z(x = 438) = \frac{438 - 440}{1} = -2$$

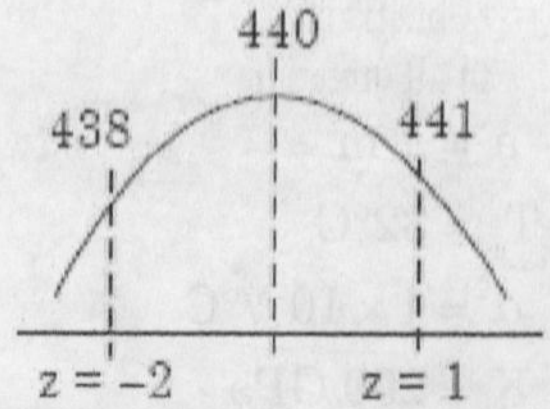

$$Z(x = 441 \text{ mm}) = \frac{441 - 440}{1} = 1$$

Percentage of rods whose lengths lie between 438 mm and 441 mm.

$$= 0.3413 + (0.5 - 0.0228)$$
$$= 0.81854 = 81.854\%$$

21. Air, $m = 1$ kg

$$T = 300K \xrightarrow{\text{Isothermally}} T = 300K, \; P_2 = 1 \text{ bar}$$

$P_1 = 10$ bar

$R = 0.287$ kJ/kg k

Now, $\Delta s = S_2 - S_1$

$$= mc_p \ell n\left(\frac{T_2}{T_1}\right)^{\nearrow 0} - mR\ell n\left(\frac{P_2}{P_1}\right)$$

$$= -1 \times 0.287 \ln\left(\frac{1}{10}\right)$$

$$= 0.66 \text{ kJ/kg k}$$

$\therefore \quad \Delta s = 0.66$ kJ/kg k

22. In laser beam welding, heat flux is highest upto 1 mW/mm²

23. $\theta = 10°$

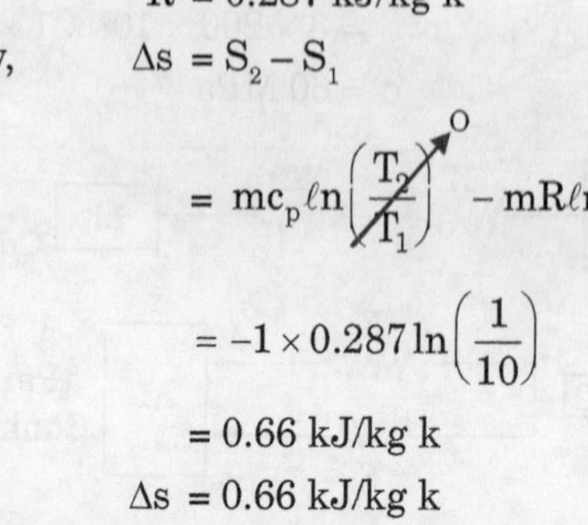

Torsion equation

$$\frac{T}{J} = \frac{\tau_{max}}{R} = \frac{G\theta}{L}$$

$$\frac{\tau_{max}}{G} = \gamma_{max} = \frac{R\theta}{L} = \frac{5 \times 10 \times \frac{\pi}{180}}{1000}$$

$$= 0.87 \times 10^{-3}$$

$$\Rightarrow \quad P \times 10^{-3} = 0.87 \times 10^{-3}$$

$$\therefore \quad P = 0.87$$

24. Trapezoidal Rule, $h = 1$

$$\int y dx = \frac{h}{2}\left[(y_0 + y_1) + 2(y_1 + y_2 + ...)\right]$$

$$= \frac{1}{2}\left[(2^3 + 4^3) + 2(3^3)\right]$$

$$= \frac{1}{2}[8 + 64 + 54] = 63$$

25. For isothermal process

$$\delta u = 0$$

As per Ist law of thermodynamics

$$\delta\theta = \delta u^{\nearrow 0} + \delta w$$
$$\delta\theta = \delta w$$

26.

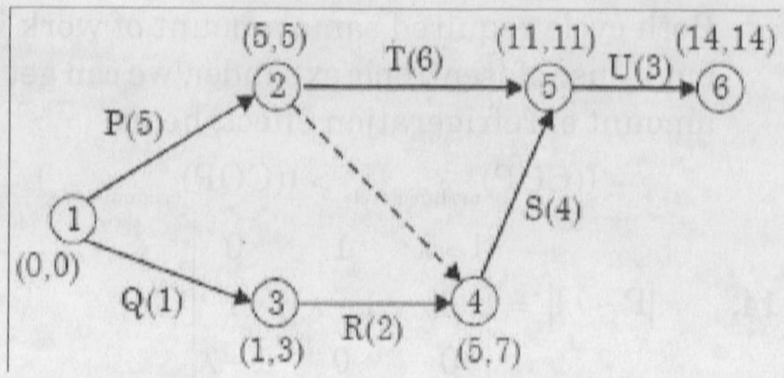

Total float $= L_j - E_i - t_{ij} = 11 - 5 - 4 = 2$

27. By bending equation,

$L = 2\pi R$

$10\pi = 2\pi R$

$R = 5$

$$\frac{\sigma}{y} = \frac{M}{I} = \frac{E}{R}$$

$$\sigma = \frac{Ey}{R}$$

$$= \frac{200 \times 10^3 \times 2.5 \times 10^{-3}}{5} = 100 \text{ Mpa}$$

28. Given, Side cutting edge angles = 45°

End cutting edge angle = 10°

Now, centre line average surface roughness,

$$(R_a) = \frac{f}{4(\tan C_s + \cot C_E)}$$

$$= \frac{0.1}{4(\tan 45 + \cot 10)}$$

$$= 3.7 \times 10^{-3} \text{ mm}$$

$$\boxed{R_a = 3.7 \, \mu m}$$

29. Here, Reaction on front wheels = R_f

Reaction on rear wheels = R_r

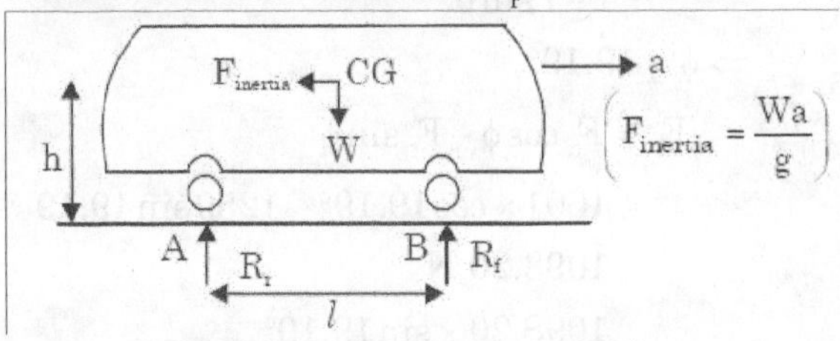

$$\Sigma M_A = 0,\ F_{inertia} \times h + R_f \times l - W \times \frac{l}{2} = 0$$

$$R_f = \frac{\frac{Wl}{2} - h \times \frac{w_a}{g}}{l}$$

$$R_f = \frac{W}{2} - \frac{W}{g}\left(\frac{h}{l}\right)a$$

$$\Sigma f_y = 0 \quad W = R_f + R_r$$

$$R_r = \frac{W}{2} + \frac{W}{g}\left(\frac{h}{l}\right)a$$

30. Taylor's tool life equation,

$$VT^n = c$$

for Tool X, for Tool y

$VT^{0.3} = 60$ $VT^{0.6} = 90$

Velocity when both tools have same tool life.

$$\frac{T^{0.3}}{T^{0.6}} = \frac{60}{90}$$

$$(T)^{0.6-0.6} = \frac{6}{9}$$

$$T^{0.3} = 1.5$$

$$T = (1.5)^{1/0.3} \text{ min}$$

$$V = \frac{60}{(1.5)^{0.3/0.3}} = 40 \text{ m/min}$$

31. Given, Disc mass m = 1 kg, R = 0.1 m

$K_1 = 400$ N/m, $K_2 = 100$ N/m

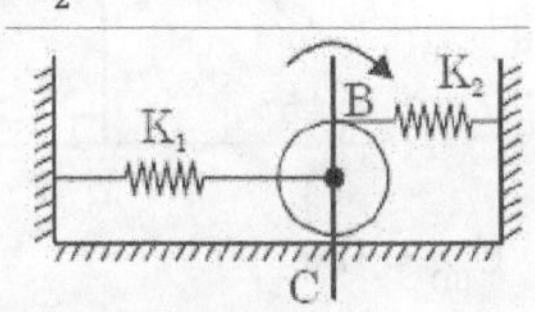

By assuming pure rolling of the disc,

$M_c = 0$ at I.C. at point (C)

$$I_c\ddot{\theta} + K_2(2R)\theta \times 2R + K_1R\theta \times R = 0$$

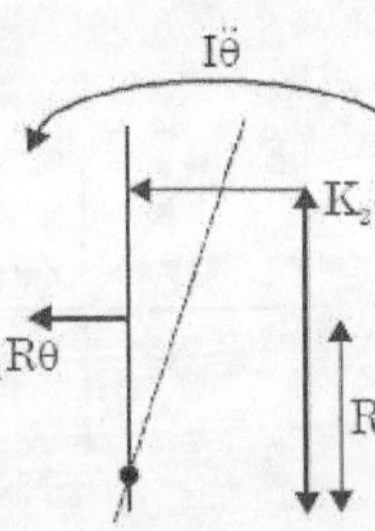

$$I_c = I_{disc} + mR^2$$

$$= \frac{mR^2}{2} + mR^2$$

$$= \frac{3}{2}mR^2 = \frac{3}{2} \times 0.1^2 = 1.5 \times 10^{-2}$$

$$1.5 \times 10^{-2}\ddot{\theta} + 4 \times 100 \times 0.1^2\theta + 400 \times 0.1^2\theta = 0$$

$$1.5 \times 10^{-2}\ddot{\theta} + 800 \times 0.1^2\theta = 0$$

$$\omega_n = \sqrt{\frac{800 \times 0.1^2}{1.5 \times 0.1^2}} = \sqrt{533.3}$$

$$\Rightarrow \omega_n = \sqrt{533.3} \text{ rad/second}$$

$$\omega_n = 23.094 \text{ rad/s}$$

$$\omega_n = 23.1$$

32. σ_x, σ_y & σ_z are principal stresses.

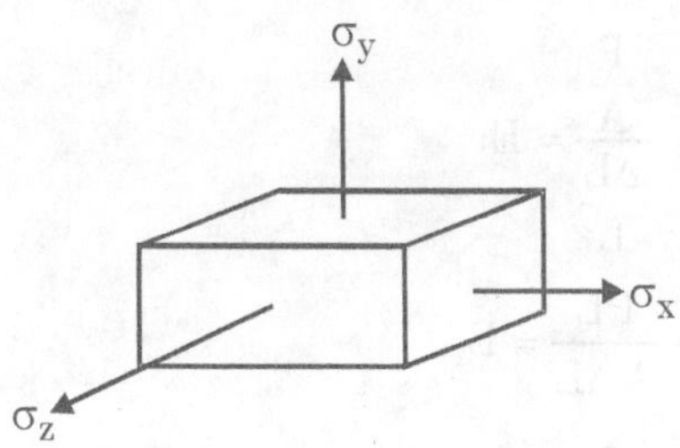

Under plane strain condition, $\epsilon_z = 0$

$$\Rightarrow \frac{\sigma_z}{E} - \frac{\mu\sigma_y}{E} - \frac{\mu\sigma_x}{E} = 0$$

$$\Rightarrow \sigma_z = (\sigma_x + \sigma_y) = \mu(\sigma_x + \sigma_y)$$

but $\sigma_x = -40$

$$\Rightarrow \sigma_z = \mu(-40 + \sigma_y)$$

From Tresca theory,

$$\tau_{max} = \max\left[\frac{|\sigma_x - \sigma_y|}{2}, \frac{|\sigma_y - \sigma_z|}{2}, \frac{|\sigma_z - \sigma_x|}{2}\right]$$

$$= \frac{S_{yt}}{2N}$$

$$\Rightarrow \max\left[\frac{|-40 - \sigma_y|}{2}, \frac{|\sigma_y - \mu(\sigma_y - 40)|}{2}, \frac{|\mu(\sigma_y - 40) - (-40)|}{2}\right]$$

$$\Rightarrow \max\left[|40 + \sigma_y|, |(1-\mu)\sigma_y + 40\mu|, |\mu\sigma_y + 40(1-\mu)|\right]$$

$$= 300$$

But μ is not given

Hence if we neglect σ_z then

max [|40 + σ_y|, |σ_y|, |+40|] = 300

$$\Rightarrow |40 + \sigma_y| = 300$$

$$40 + \sigma_y = 300$$

$$\Rightarrow \sigma_y = 260 \text{ MPa}$$

33. Given, side of cube (a) = 100 m

Density of material of cube = 800 kg/m^3

Poured liquid density = 1000 kg/m^3

Now, Minimum height to which the liquid needs to be poured into the container for the cube to just lift up, should satisfy the condition

$$F_B \geq (mg)_{block}$$

$$\rho_f g V_{fd} \geq (\rho V)_{block} \times g$$

$$1000 \times 0.1 \times 0.1 \times h \geq 800 \times 0.1 \times 0.1 \times 0.1$$

$$h \geq 0.08 \text{ m}$$

$$h \geq 80 \text{ mm}$$

34. UTM constant = $\frac{\Delta L}{P} = C$

As per Hooke's law, $\frac{\sigma}{\in} = E$

or, $\frac{\frac{P}{A}}{\frac{\Delta L}{L}} = E$

$\Rightarrow \frac{PL}{A\,\Delta L} = E$

or, $\frac{\Delta L}{P} = \frac{L}{AE}$ = UTM constant = C

Now, In UTM experiment under tension,

L = 100 mm, ΔL = 15 mm,

$C = 5 \times 10^{-8}$ m/N, P = 40 kN

Also, Strain = $\frac{\Delta L}{L} = \frac{P}{AE} = \frac{P}{L/C}$

$$= \frac{PC}{L} = \frac{40 \times 10^3 \times 5 \times 10^{-8}}{100 \times 10^{-3}} = 2 \times 10^{-2}$$

$\therefore$ Strain in % = $2 \times 10^{-2} \times 100\% = 2\%$

35. For, orthogonal turning, $\lambda = 90°$, F_c = 1601 N chip thickness 't_c' = 0.3, F_t = 1259 N.

N = 1000 rpm, α = 10 m

$$f = 100 \text{ mm/min} = \frac{100}{1000}\frac{\text{mm}}{\text{rev}}$$

$$f = 0.1\frac{\text{mm}}{\text{rev}} = t$$

$$r = \frac{t}{t_c} = \frac{0.1}{0.3} = \frac{1}{3}$$

$$F_a = \tau_s \times \frac{bt}{\sin\phi}$$

$$\Rightarrow \tau_s = \frac{F_s \sin\phi}{6 \times t}$$

$$\tan\phi = \frac{r\cos\alpha}{1 - r\sin\alpha}$$

$$\Rightarrow \phi = 19.19°$$

$$F_s = F_c\cos\phi - F_t\sin\phi$$

$$= 1601 \times \cos 19.19° - 1259 \sin 19.19$$

$$= 1098.20 \text{ N}$$

$$\tau_s = \frac{1098.20 \times \sin 19.19°}{5 \times 0.1}$$

$$= 721.96$$

$$\approx 722 \text{ MPa}$$

36. Given, m = 180 kg, I = 10 kg.m^2, V = 800 km/hr., W = 1100 rad/s, R = 1.5 km = 1500 m

We know that, Gyroscopic couple $G = I\omega\omega_p$

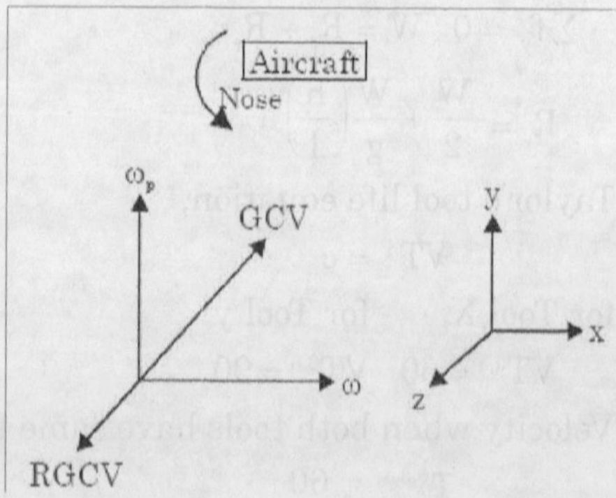

Effect $\Rightarrow$ Nose goes down

$$G = 10 \times 1100 \times \frac{800 \times 5}{1800 \times 18}$$

$$= 1629.63 \text{ Nm}$$

$$G = 1629.63 \text{ N.m}$$

37.

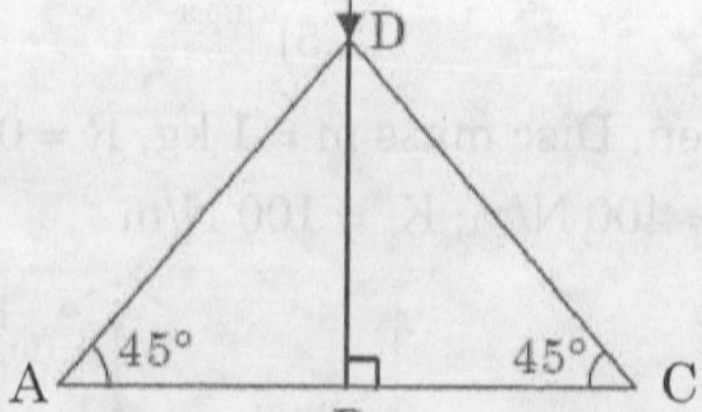

$F_{BD} = 0$

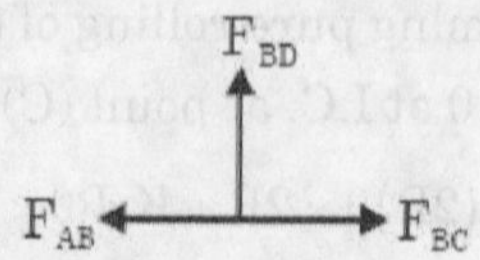

At joint D,

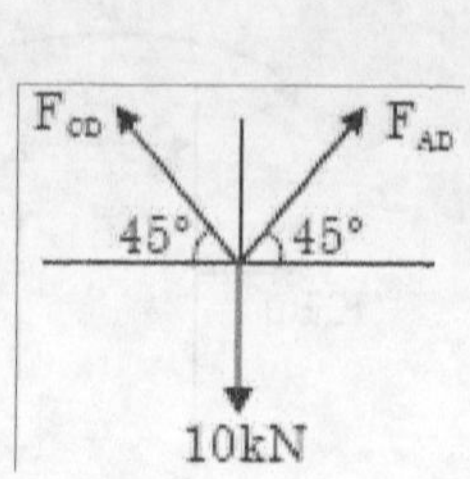

At joint C,

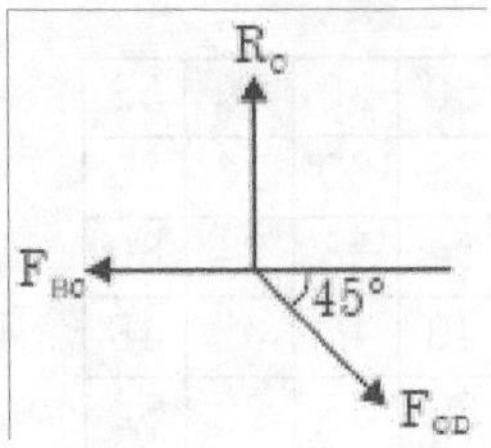

$$\Sigma F_x = 0$$
$$F_{CD} \cos 45 = F_{AD} \cos 45$$
$$F_{CD} = F_{AD}$$
$$\Sigma F_y = 0, 10 = 2F_{CD} \sin 45$$
$$F_{CD} = \frac{10}{\sqrt{2}} kN$$
$$F_{CD} \cos 45 = F_{BC}$$
$$F_{BC} = \frac{10}{\sqrt{2}} \times \frac{1}{\sqrt{2}} = 5 \text{ kN}$$

38.
$$V_C = V_B$$
$$= AB \times \omega AB$$
$$= (0.05 \times 10) \text{ m/s}$$

Now, $\omega_{CD} = \frac{V_C}{CD} = \frac{0.05 \times 10}{0.02} = 25$ rad/s

39.

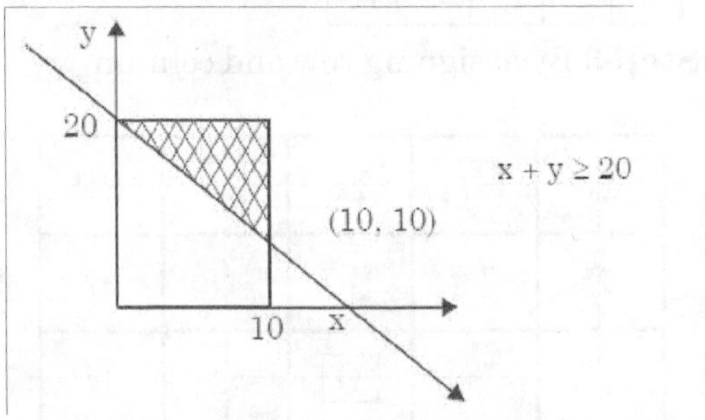

Required Probability (x + y > 20)

$$= \frac{\text{Shaded Area}}{\text{Total Area}}$$
$$= \frac{\frac{1}{2} \times 10 \times 10}{20 \times 10}$$
$$= \frac{50}{200} = \frac{1}{4} = 0.25$$

40. In case of uniform heat flux, bulk mean temperature varies linearly. The difference between bulk mean temperature and wall temperature is constant in thermally developed region so bulk tempeature,

$$T(x) = A + Bx$$
At $x = 0$, $T = 0°C$
$\Rightarrow$ $A = 0°C$
At $x = 1$, $T = 50°C$
$\Rightarrow$ $B = 50°C$
$\therefore$ $T(x) = 50x$

Here,

Location	Wall temp. (T_W)	bulk temp T(x)	$\Delta T = T_W - T(x)$
P, x = 0.4	50°C	20°C	30°C
Q, x = 0.6	80°C	30°C	50°C
R, x = 0.8	90°C	40°C	50°C

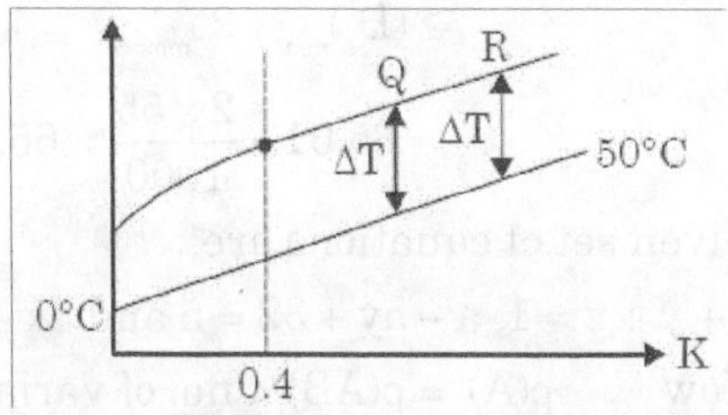

So, Q and R in thermally developed region.

41. $u(x, y) = 2x^2 - 2y^2 + 4xy$

As harmonic function is analytic there fore,

$$u_x = V_y$$
$$u_y = -V_x$$
$$u_x = 4x + 4y$$
$$u_y = -4y + 4x$$
$$\frac{\partial v}{\partial y} = V_y = 4x + 4y$$
$$V = 4xy + 2y^2 + f(x)$$

Now $\frac{\partial v}{\partial y} = 4y + f'(x) = 4y - 4x$

$$f'(x) = -4x$$
$$f(x) = -2x^2 + C$$

So, conjugate harmonic function v(x, y) is,

$$V = 2y^2 - 2x^2 + 4xy + C$$

42. From energy balance equation,

$$\dot{m}h_2 + (1 - \dot{m})h_4 = h_5$$
$$\dot{m}(h_2 + h_4) = h_2 - h_4$$
$$\Rightarrow \dot{m} = \frac{h_5 - h_4}{h_2 - h_4} = \left(\frac{700 - 175}{2800 - 175}\right) \times 100$$
$$= 0.2 \text{ or } 20\%$$

$\therefore$ 20% of mass flow rate of steam bled from the turbine of total mass flow rate.

43.
$$P_1V = n\bar{R}T_1$$
$$P_2V = n\bar{R}T_2$$

Now, Changs in pressure,

$$P_2 - P_1 = n\bar{R}(T_2 - T_1)$$
$$\Rightarrow \Delta P \times 1 = 1 \times 8.314(37 - 27)$$
$$\Rightarrow \Delta P = 83.14 \text{ Pascal}$$

44. Shaft $= 65^{+0.01}_{-0.05}$ mm

Coating thickness 't' = (50 ± 5)μm

To provide clearnace fit, given condition should be satisfied,

$$(D_H)\min \geq ((D)\text{shaft})_{max}$$

$$\geq (D_s)_{max} + 2t_{max}$$

$$= 65.01 + \frac{2 \times 55}{1000} = 65.12\text{mm}$$

45. Given set of equations are,

$x + y + z = 1$, $a - ay + 32 = 5$ and $5x - 3y + az = 6$

Now $\rho(A) = \rho(AB) <$ no. of variables

$|A| = 0$

$$\begin{vmatrix} 1 & 1 & 1 \\ a & -a & 3 \\ 5 & -3 & a \end{vmatrix} = 0$$

$(-a^2 + 9) - (a^2 - 15) + (-3a + 5a) = 0$

$9 - a^2 + 15 - a^2 + 2a = 0$

$2a^2 - 2a - 24 = 0$

$a^2 - a - 12 = 0$

$a^2 - 4a + 3a - 12 = 0$

$(a = 4, -3)$

$\rho(AB) <$ (No. of variable)

$$\begin{vmatrix} 1 & 1 & 1 \\ -a & 3 & 5 \\ -3 & a & 6 \end{vmatrix} = 0$$

$1(18 - 5a) -1(-6a + 15) + 1(-a^2 + 9) = 0$

$18 - 5a - 15 + 6a + 9 - a^2 = 0$

$a^2 - a - 12 = 0$

$a = 3, 4$ or $a = 4$

$$\begin{vmatrix} 1 & 1 & 1 \\ a & -a & 5 \\ 5 & -3 & 6 \end{vmatrix} = 0$$

$(-6a + 15) - (6a - 25) + (-3a + 5a) = 0$

$-6a + 15 - 6a + 25 + 2a = 0$

$10a = 40$

$a = 4$

if $a = 4$

$|AB| = 0$

$$\begin{vmatrix} 1 & 1 & 1 \\ a & 3 & 5 \\ 5 & a & 6 \end{vmatrix} = 0$$

$(18 - 5a) - (6a - 25) + (a^2 - 15) = 0$

$a^2 - 11a + 28 = 0$

$a = 7, 4$ or $\boxed{a = 4}$

46.

	M_1	M_2	M_3	M_4	M_5
J_1	40	30	50	50	58
J_2	26	38	60	26	38
J_3	40	34	28	24	30
J_4	28	40	40	32	48
J_5	28	32	38	22	54

Step-1 Substracting each row with minimum value of each respective row.

10	0	20	20	28
0	12	34	0	12
16	10	4	0	6
0	12	12	4	20
6	10	16	0	22

Step 2: Substracting each column with minimum value of each respectively value.

10	0	16	20	22
0	12	30	0	6
16	10	0	0	0
0	12	8	4	14
6	10	12	0	16

Step-3 By assigning row and column

10	[0]	16	20	22	
~~0~~	12	30	~~0~~	6	✓
16	10	[0]	~~0~~	~~0~~	
[0]	12	8	4	14	✓
6	10	12	[0]	16	✓
✓			✓		

Step-4

16	[0]	16	26	22
~~0~~	6	24	~~0~~	[0]
22	10	[0]	6	~~0~~
[0]	6	2	4	8
6	4	6	[0]	10

Step-5 Minimum time = 28 + 30 + 28 + 22 + 38

= 146 min

47.

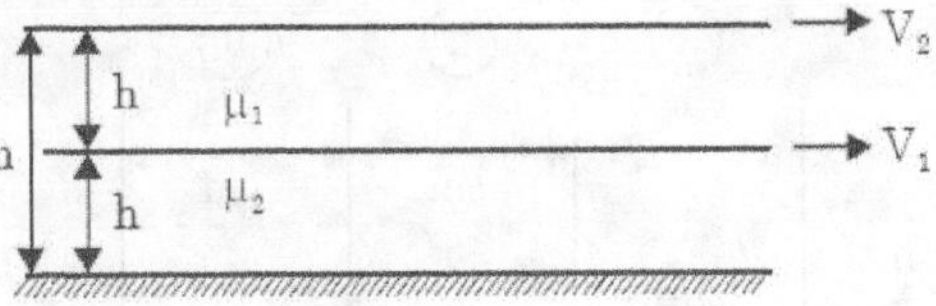

$$\mu_2 = 2\mu_1$$
$$V_2 = 3 \text{ m/s}$$

Shear stress value at intermediate plate will be

$$\tau = \frac{\mu_2 V_1}{h} = \frac{\mu_1 (V_2 - V_1)}{h}$$

Since $\mu_2 = 2\mu_1$

$\therefore$ $2V_1 = V_2 - V_1$

$\Rightarrow$ $V_2 = V_1 + 2V_1$

$V_2 = 3V_1$ $[V_2 = 3 \text{ m/s}]$

$\Rightarrow$ $3 = 3V_1$

$\Rightarrow$ $V_1 = \frac{3}{3} = 1 \text{ m/s}$

48. Given, $\sigma_x = 100$ MPa

$\sigma_y = 220$ MPa

$\sigma_{xy} = 80$ MPa

$$\sigma_{1,2} = \frac{\sigma_x + \sigma_y}{2} \pm \sqrt{\left(\frac{\sigma_x - \sigma_y}{2}\right)^2 + \sigma_{xy}^2}$$

$$= \frac{100 + 200}{2} \pm \sqrt{\left(\frac{100 - 220}{2}\right)^2 + 80^2}$$

$$= 160 \pm \sqrt{55^2 + 80^2}$$

$$\sigma_{1,2} = 160 \pm 97.0824$$

$\Rightarrow$ $\sigma_1 = 257.08$ Mpa, $\sigma_2 = 62.9$ MPa

As per maximum shear stress theory

$$\frac{S_{yt}}{2N} = \text{Max}\left[\left|\frac{\sigma_1 - \sigma_2}{2}\right|, \frac{\sigma_1}{2}, \frac{\sigma_2}{2}\right]$$

$$\frac{S_{yt}}{2N} = \frac{\sigma_1}{2} = \frac{257.08}{2}$$

$$N = \frac{468}{257.08} = 1.8$$

50.

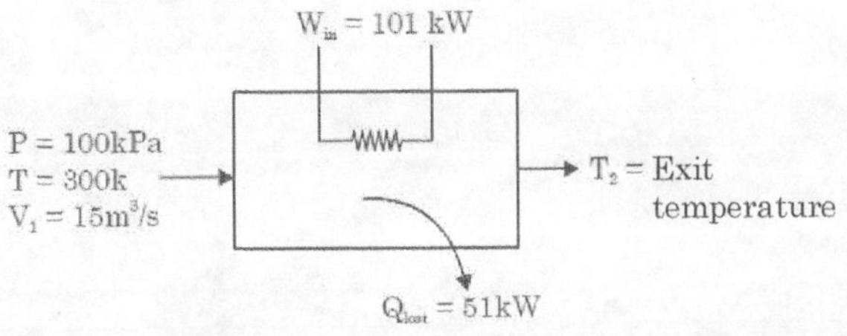

$$C_p = 1 \frac{\text{kJ}}{\text{kgK}}$$

$$R = 0.5 \frac{\text{kJ}}{\text{kgK}}$$

$$\dot{m} = \frac{P_1 \dot{V}_1}{RT_1} = \frac{100 \times 15}{0.5 \times 300} = 10 \frac{\text{kg}}{\text{s}}$$

By applying steady flow energy equation, (SFEE)

$$\dot{m}h_1 + \dot{8} = \dot{m}h_2 + \dot{W}$$

$$10 \times 1 \times 300 + (-51) = 10 \times 1 \times T_2 + (-101)$$

$$T_2 = 305 \text{ K} = 32°\text{C}$$

Hence, Exit temperature

$(T_2) = 32°$C.

51.

$$I = \int_1^e x \, ln \, x dx$$

$$= \frac{ln \, x^2}{2} - \int \left[\frac{1}{x} \times \frac{x^2}{2}\right] dx$$

$$= \frac{x^2 ln \, x}{2} - \frac{x^2}{4}\Bigg|_1^e$$

$$= \left(\frac{e^2}{2} - \frac{e^2}{4}\right) + \frac{1}{4}$$

$$= \frac{e^2 + 1}{4} = 2.097$$

52. $T = 250$ Nm

$F \times r = 250$

$\Rightarrow$ $F = \frac{250}{a}$

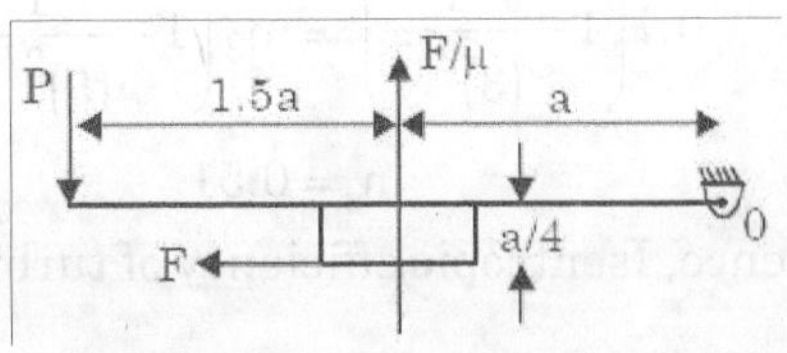

$$\Sigma M_0 = 0$$

$$P \times 2.5a = F \times \frac{a}{4} + \frac{F}{\mu} \times a$$

$$2000 \times 2.5a = \frac{250 \times a}{a \times 0.25} + \frac{250}{a} \times \frac{a}{4}$$

$$a = \frac{1000 + 62.5}{5000}$$

$$= \frac{1062.5}{5000}$$

$$= 212.5 \text{ mm}$$

53. Length of prismatic beam,

$$L = \pi m$$

$$P_{th} = P_{axial}$$

or $$\alpha TEA = \frac{\pi^2 EI}{L^2}$$

$$\Rightarrow 3 \times 10^{-6} \times \Delta T \times 6 \times 6 \times 10^{-6} = \frac{\pi^2 \times \frac{1}{12} \times \left(6 \times 10^{-3}\right)^4}{\pi^2}$$

$$\Delta T = 1\text{ K}$$

54. $$\eta = 0.7, (r_p)_1 = 3$$

$$\eta_2 = ?\ (r_p)_2 = 5$$

$$r = 1.4$$

$$(W_T)_1 = (W_T)_2$$

As Inlet condition same,

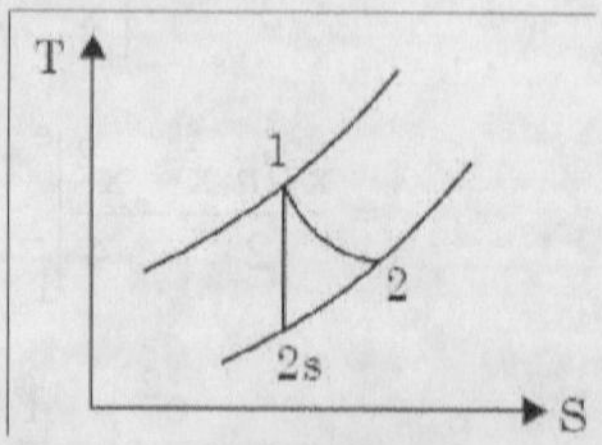

Here, $$\eta_1 = \frac{h_1 - h_2}{h_1 - h_{2s}}$$

$$= \frac{T_1 - T_2}{T_1 - T_{2s}} = \frac{T_1 - T_2}{T_1 - \frac{T_1}{x}}$$

$$\eta_1 \left(T_1 - \frac{T_1}{x}\right) = \eta_2 \left(T_1 - \frac{T_2}{y}\right)$$

$$0.7\left(1 - \frac{1}{(3)^{2/7}}\right) = \eta_2 \left(1 - \frac{1}{(5)^{2/4}}\right)$$

$$\Rightarrow \eta_2 = 0.51$$

Hence, Isentropic efficiency of turbine is 0.51.

55.

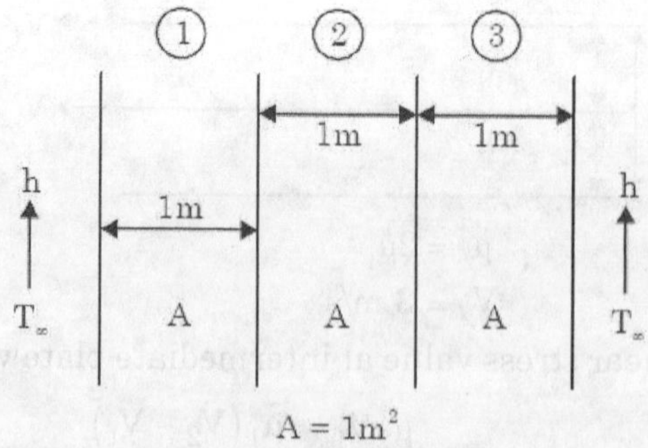

Heat generated in the slab (2) will be convected away from the slabs 1 and 3 i.e.

$$\dot{E}_{gen} = \dot{Q}_{1,\,conv} + \dot{Q}_{3,\,conv}$$

$$\dot{Q}_{1,\,conv} = h \times A \times (T_1 - T_\infty)$$

$$= 100 \times 1 \times (100 - 30)$$

$$= 100 \times 70 = 7000\text{ W}$$

$$\dot{Q}_{3,\,conv} = h \times A \times (T_3 - T_\infty)$$

$$\dot{Q}_{3,\,conv} = 100 \times 1 \times (T_3 - 30)$$

Now $$\dot{E}_{gen} = 10,000 \times A \times \text{width}$$

$$= 10,000 \times 1 \times 1$$

$$\dot{E}_{gen} = 10,000\text{W}$$

$$\Rightarrow 10,000 = 7000 + 100 \times (T_3 - 30)$$

$$\Rightarrow 3000 = 100\,T_3 - 3000$$

$$\Rightarrow 3000 + 3000 = 100\,T_3$$

$$\Rightarrow T_3 = \frac{6000}{100} = 60°\text{C}$$

SOLVED PAPER - 2020

SET 1

INSTRUCTIONS

1. Total of 65 questions carrying 100 marks, out of which 10 questions carrying a total of 15 marks are in General Aptitude (GA)
2. The Engineering Mathematics will carry around **15% of the total marks,** the General Aptitude section will carry **15% of the total marks** and the **remaining 70% of the total marks.**
3. **Types of Questions**
   (*a*) **Multiple Choice Questions (MCQ)** carrying 1 or 2 marks each in all papers and sections. These questions are objective in nature, and each will have a choice of four options, out of which the candidate has to mark the correct answer(s).
   (*b*) **Numerical Answer Questions** of 1 or 2 marks each in all papers and sections. For these questions the answer is a real number, to be entered by the candidate using the virtual keypad. No choices will be shown for these type of questions.
4. For **1-mark** multiple-choice questions, **1/3 marks** will be deducted for a wrong answer. Likewise, for **2-marks** multiple-choice questions, **2/3** marks will be deducted for a wrong answer. There is no negative marking for numerical answer type questions.

Chapter-Wise Analysis

#	Subject	Marks		
		1	2	Total
1	General Aptitude	5	5	15
2	Engineering Mathematics	6	4	14
3	Engineering Mechanics	1	0	1
4	Strength of Materials	1	3	7
5	Theory of Machines	4	3	10
6	Design of Machine Elements	1	2	5
7	Fluid Mechanics	2	3	8
8	Heat Transfer	2	1	4
9	Thermodynamics, IC Engines, Refrigeration & Power Plant Engineering Air-Conditioning	3	5	13
10	Material Manufacturing & Industrial Engineering	5	5	15

GENERAL APTITUDE (GA)

Q. 1 to Q. 5 : Carry One Mark Each.

1. He is known for his unscrupulous ways. He always sheds ________ tears to deceive people.

(a) fox's (b) crocodile's

(c) crocodile (d) fox

2. Jofra Archer, the England fast bowler, is ______ than accurate.

(a) more fast (b) faster

(c) less fast (d) more faster

3. Select the word that fits the analogy:

Build : Building : : Grow :

(a) Grown (b) Grew

(c) Growth (d) Growed

4. I do not think you know the case well enough to have opinions. Having said that, I agree with your other point.

What does the phrase "having said that" mean in the given text?

(a) as opposed to what I have said

(b) despite what I have said

(c) in addition to what I have said

(d) contrary to what I have said

5. Define [x] as the greatest integer less than or equal to x, for each $x \in (-\infty, \infty)$. If $y = [x]$, then area under y for $x \in [1, 4]$ is _______ .

(a) 1 (b) 3

(c) 4 (d) 6

Q. 6 to Q. 10 : Carry Two Marks Each.

6. Crowd funding deals with mobilisation of funds for a project from a large number of people, who would be willing to invest smaller amounts through web-based platforms in the project.

Based on the above paragraph, which of the following is correct about crowd funding?

(a) Funds raised through unwilling contributions on web-based platforms.

(b) Funds raised through large contributions on web-based platforms.

(c) Funds raised through coerced contributions on web-based platforms.

(d) Funds raised through voluntary contributions on web-based platforms.

7. P, Q, R and S are to be uniquely coded using α and β. If P is coded as αα and Q as αβ, then R and S, respectively, can be coded as

(a) βα and αβ (b) ββ and αα

(c) αβ and ββ (d) βα and ββ

8. The sum of the first n terms in the sequence 8, 88, 888, 8888, ... is ________ .

(a) $\frac{81}{80}(10^n - 1) + \frac{9}{8}n$

(b) $\frac{81}{80}(10^n - 1) - \frac{9}{8}n$

(c) $\frac{80}{81}(10^n - 1) + \frac{8}{9}n$

(d) $\frac{80}{81}(10^n - 1) - \frac{8}{9}n$

9. Select the graph that schematically represents BOTH $y = x^m$ and $y = x^{1/m}$ properly in the interval $0 \le x \le 1$, for integer values of m, where $m > 1$.

(a)

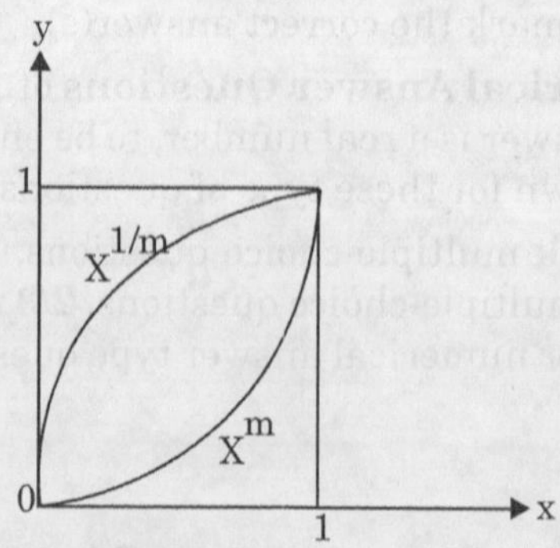

(b)

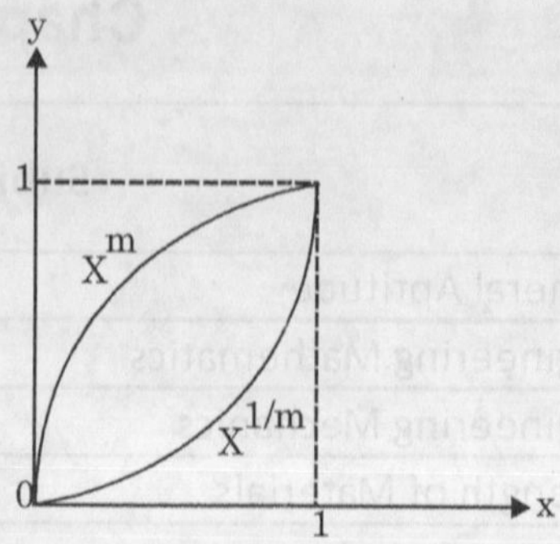

(c)

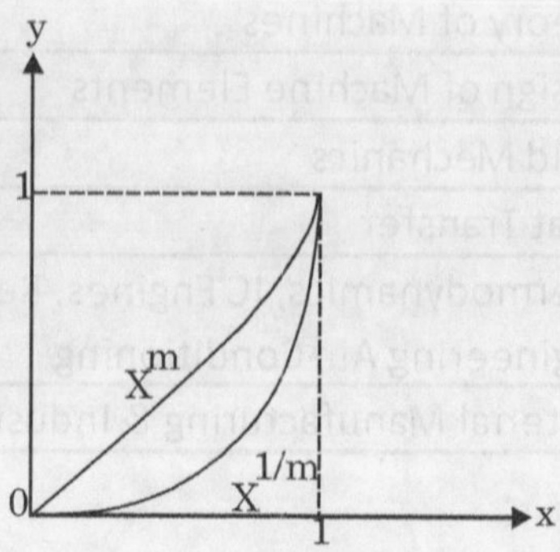

(d)

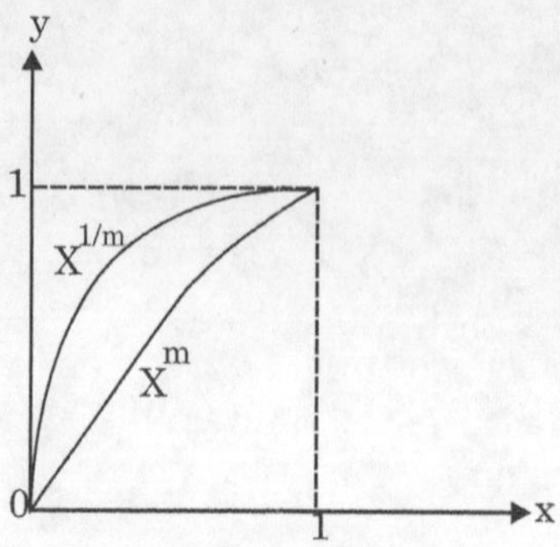

10. The bar graph shows the data of the students who appeared and passed in an examination for four schools P, Q, R and S. The average of success rates (in percentage) of these four schools is ______.

Performance of schools P, Q, R and S

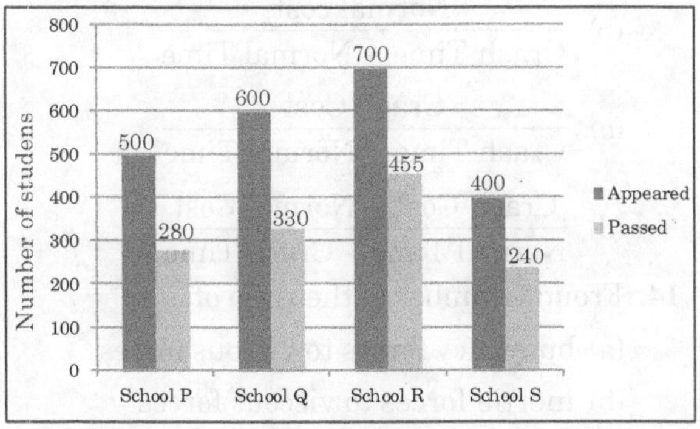

(a) 58.5% (b) 58.8%

(c) 59.0% (d) 59.3%

MECHANICAL ENGINEERING

Q. 1 to Q. 25 : Carry One Mark Each.

1. Multiplication of real valued square matrices of same dimension is

(a) associative

(b) commutative

(c) always positive definite

(d) not always possible to compute

2. The value of $\lim_{x \to 1} \left(\frac{1 - e^{-c(1-x)}}{1 - xe^{-c(1-x)}} \right)$ is

(a) c (b) c + 1

(c) $\frac{c}{c+1}$ (d) $\frac{c+1}{c}$

3. The Laplace transform of a function f(t) is $L(f) = \frac{1}{(s^2 + \omega^2)}$. Then, f(t) is

(a) $f(t) = \frac{1}{\omega^2}(1 - \cos \omega t)$

(b) $f(t) = \frac{1}{\omega} \cos \omega t$

(c) $f(t) = \frac{1}{\omega} \sin \omega t$

(d) $f(t) = \frac{1}{\omega^2}(1 - \sin \omega t)$

4. Which of the following function f(z), of the complex variable z, is NOT analytic at all the points of the complex plane?

(a) $f(z) = z^2$ (b) $f(z) = e^z$

(c) $f(z) = \sin z$ (d) $f(z) = \log z$

5. The members carrying zero force (i.e zero-force members) in the truss shown in the figure, for any load P > 0 with no appreciable deformation of the truss (i.e with no appreciable change in angles between the members) are

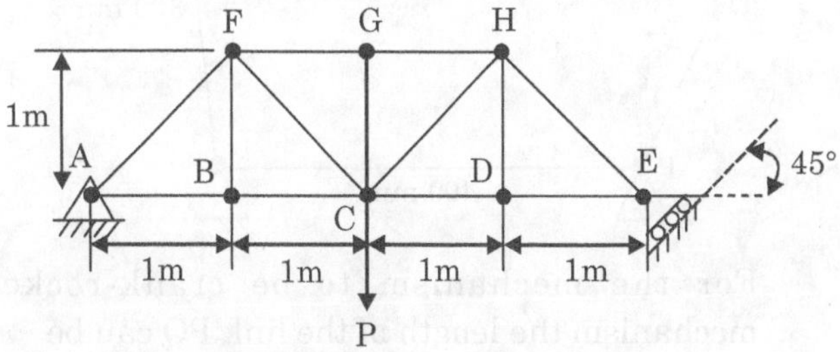

(a) BF and DH only

(b) BF, DH and GC only

(c) BF, DH, GC, CD and DE only

(d) BF, DH, GC, FG and GH only

6. A single-degree-of-freedom oscillator is subjected to harmonic excitation $F(t) = F_0 \cos(\omega t)$ as shown in the figure.

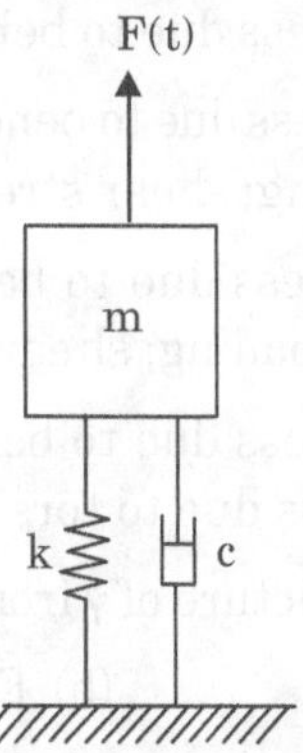

The non-zero value of ω, for which the amplitude of the force transmitted to the ground will be F_0, is

(a) $\sqrt{\frac{k}{2m}}$ (b) $\sqrt{\frac{k}{m}}$

(c) $\sqrt{\frac{2k}{m}}$ (d) $2\sqrt{\frac{k}{m}}$

7. The stress state at a point in a material under plane stress condition is equi-biaxial tension with a magnitude of 10 MPa. If one unit on the σ – τ plane is 1 MPa, the Mohr's circle representation of the state of stress is given by

(a) a circle with a radius equal to principal stress and its center at the origin of the σ – τ plane

(b) a point on the σ axis at a distance of 10 units from the origin

(c) a circle with a radius of 10 units on the σ – τ plane

(d) a point on the τ axis at a distance of 10 units from the origin

8. A four bar mechanism is shown below.

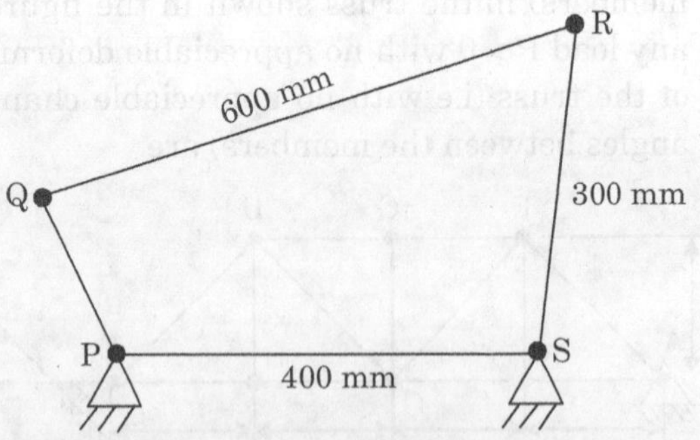

For the mechanism to be crank-rocker mechanism the length of the link PQ can be

(a) 80 mm (b) 200 mm

(c) 300 mm (d) 350 mm

9. A helical gear with 20° pressure angle and 30° helix angle mounted at the midspan of a shaft that is supported between two bearings at the ends. The nature of the stresses induced in the shaft is

(a) normal stress due to bending only

(b) normal stress due to bending in one plane and axial loading; shear stress due to torsion

(c) normal stress due to bending in two planes and axial loading; shear stress due to torsion

(d) normal stress due to bending in two planes; shear stress due to torsion

10. The crstal structure of γ iron (austenite phase) is

(a) BCC (b) FCC

(c) HCP (d) BCT

11. Match the following

Heat treatment process	Effect
P : Tempering	1. Strengthening
Q : Quenching	2. Toughening
R : Annealing	3. Hardening
S : Normalizing	4. Softening

(a) P-2, Q-3, R-4, S-1

(b) P-1, Q-1, R-3, S-2

(c) P-3, Q-3, R-1, S-3

(d) P-4, Q-3, R-2, S-1

12. The base of a brass bracket needs rough grinding. For this purpose,t he most suitable grinding wheel grade specification is

(a) C30Q12V (b) A50G8V

(c) C90J4B (d) A30D12V

13. In the Critical Path Method (CPM), the cost-time slope of an activity is given by

(a) $\dfrac{\text{Crash cost} - \text{Normal cost}}{\text{Crash Time}}$

(b) $\dfrac{\text{Normal cost}}{\text{Crash Time} - \text{Normal Time}}$

(c) $\dfrac{\text{Crash Cost}}{\text{Crash Time} - \text{Normal Time}}$

(d) $\dfrac{\text{Crash Cost} - \text{Normal Cost}}{\text{Normal Time} - \text{Crash Time}}$

14. Froude number is the ratio of

(a) buoyancy forces to viscous forces

(b) inertia forces to viscous forces

(c) buoyancy forces to inertia forces

(d) inertia forces to gravity forces

15. Match the following non-dimensional numbers with the corresponding definitions:

Non-dimensional number	Definition
P : Reynolds number	1. $\dfrac{\text{Buoyancy force}}{\text{Viscous force}}$
Q : Grashof number	2. $\dfrac{\text{Momentum diffusivity}}{\text{Thermal diffusivity}}$
R : Nusselt number	3. $\dfrac{\text{Inertia force}}{\text{Viscous force}}$
S : Prandtl number	4. $\dfrac{\text{Convective heat transfer}}{\text{Conduction heat transfer}}$

(a) P-1, Q-3, R-2, S-4

(b) P-3, Q-1, R-2, S-4

(c) P-4, Q-3, R-1, S-2

(d) P-3, Q-1, R-4, S-2

16. The velocity field of an incompressible flow in a Cartesian system is represented by

$$\vec{V} = 2(x^2 - y^2)\hat{i} + v\hat{j} + 3\hat{k}$$

Which one of the following expressions for v is valid?

(a) $-4xz + 6xy$ (b) $-4xy - 4xz$

(c) $4xz - 6xy$ (d) $4xy + 4xz$

17. For an ideal gas, the value of the Joule-Thomson coefficient is

(a) Positive

(b) Negative

(c) Zero

(d) Indeterminate

18. For an ideal gas, a constant pressure line and a constant volume line intersect at a point, in the Temperature (T) versus specific entropy (s) diagram. Cp is the specific heat at constant pressure and Cv is the specific heat at constant volume. The ratio of the slopes of the constant pressure and constant volume lines at the point of intersection is

(a) $\frac{C_P - C_V}{C_P}$ (b) $\frac{C_P}{C_V}$

(c) $\frac{C_P - C_V}{C_V}$ (d) $\frac{C_V}{C_P}$

19. For three vectors $\vec{A} = 2\hat{j} - 3\hat{k}, \vec{B} = -2\hat{i} + \hat{k}$, and $\vec{C} = 3\hat{i} - \hat{j}$ where $\hat{i}, \hat{j}$ and $\hat{k}$ are unit vectors along the axes of a right-handed rectangular Cartesian coordinate system, the value of $(\vec{A} \cdot (\vec{B} \times \vec{C}) + 6)$ is ________ .

20. A flywheel is attached to an engine to keep its rotational speed between 100 rad/s and 110 rad/s. If the energy fluctuation in the flywheel between these two speeds is 1.05 kJ then the moment of inertia of the flywheel is ________ kg.m^2 (round off to 2 decimal places).

21. A balanced rigid disc mounted on a rigid rotor has four identical point masses, each of 10 grams, attached to four points on the 100mm radius circle shown in the figure.

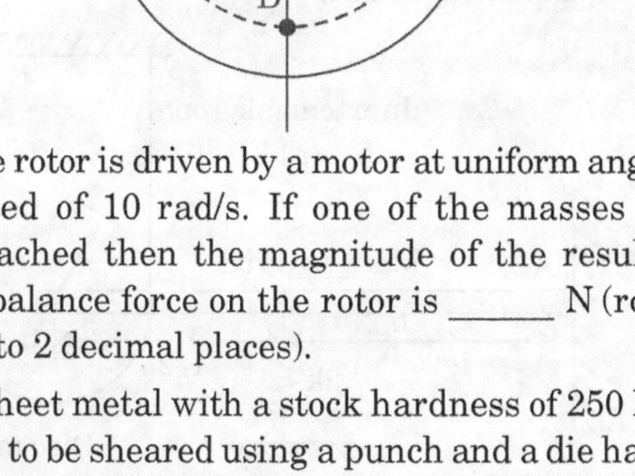

The rotor is driven by a motor at uniform angular speed of 10 rad/s. If one of the masses gets detached then the magnitude of the resultant unbalance force on the rotor is ______ N (round off to 2 decimal places).

22. A sheet metal with a stock hardness of 250 HRC has to be sheared using a punch and a die having a clearance of 1mm between them. If the stock hardness of the sheet metal increases to 400 HRC, the clearance between the punch and the die should be ________ mm.

23. A company is hiring to fill four managerial vacancies. The candidates are five men and three women. If every candidate is equally likely to be chosen then the probability that at least one woman will be selected is ________ (round off to 2 decimal places).

24. The compressor of a gas turbine plant, operating on an ideal intercooled Brayton cycle, accomplishes an overall compression ratio of 6 in a two-stage compression process. Intercooling is used to cool the air coming out from the first stage to the inlet temperature of the first stage, before its entry to the second stage. Air enters the compressor at 300 K and 100 KPa. If the properties of gas are constant, the intercooling pressure for minimum compressor work is ______ kPa (round off to 2 decimal places).

25. In a concentric tube counter flow heat exchanger, hot oil enters at 102°C and leaves at 65°C. Cold water enters at 25 °C and leaves at 42°C. The log mean temperature difference (LMTD) is ______ °C (round off to one decimal place).

Q. 26 to Q. 55 : Carry Two Marks Each.

26. The evaluation of the definite integral $\int_{-1}^{1.4} x|x|\,dx$ by using Simpson's 1/3rd (one third) rule with step size h = 0.6 yields.

(a) 0.914 (b) 1.248

(c) 0.581 (d) 0.592

27. A vector field is defined as

$$\vec{f}(x,y,z) = \frac{x}{\left[x^2+y^2+z^2\right]^{\frac{3}{2}}}\hat{i} + \frac{y}{\left[x^2+y^2+z^2\right]^{\frac{3}{2}}}\hat{j} + \frac{z}{\left[x^2+y^2+z^2\right]^{\frac{3}{2}}}\hat{k}$$

where, $\hat{i}, \hat{j}, \hat{k}$ are unit vectors along the axes of a right handed rectangular– Cartesian coordinate system. The surface integral $\iint \vec{f} \cdot d\vec{S}$ (where $d\vec{S}$ is an elemental surface area vector) evaluated over the inner and outer surfaces of a spherical shell formed by two concentric spheres with origin as the center, and internal and external radii of 1 and 2, respectively is

(a) 0 (b) 2π

(c) 4π (d) 8π

28. Bars of square and circular cross-section with 0.5 m length are made of a material with shear strength of 20 MPa. The square bar cross-section dimension is 4 cm × 4 cm and the cylindrical bar cross-section diameter is 4 cm. The specimens are loaded as shown in the figure.

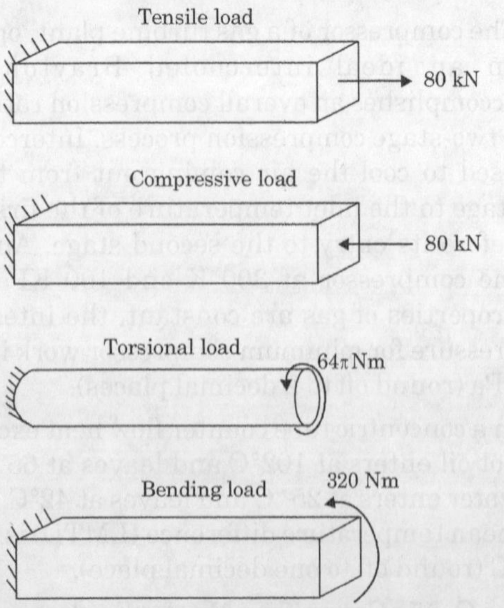

Which specimen(s) will fail due to the applied load as per maximum shear stress theory?

(a) Tensile and compressive load specimens

(b) Torsional load specimen

(c) Bending load specimen

(d) None of the specimens

29. The 2 kg block shown in figure (top view) rests on a smooth horizontal surface and is attached to a massless elastic cord that has a stiffness 5 N/m.

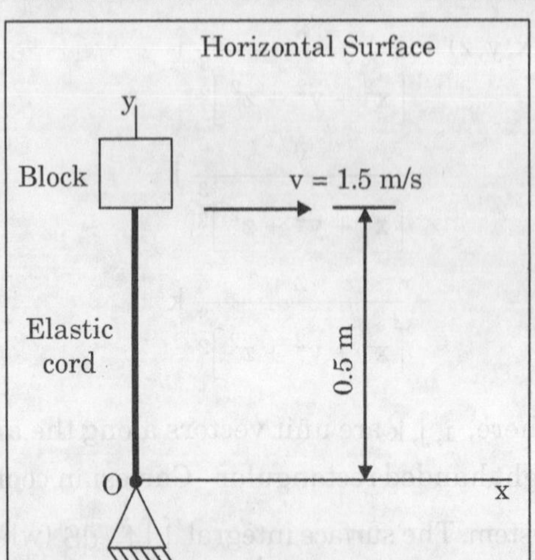

The cord hinged at O is initially unstretched and always remains elastic. The block is given a velocity v of 1.5 m/s perpendicular to the cord. The magnitude of velocity in m/s of the block at the instant the cord is stretched by 0.4 m is

(a) 0.83 (b) 1.07

(c) 1.36 (d) 1.50

30. The truss shown in the figure has four members of length l and flexural rigidity EI, and one member of length $l\sqrt{2}$ and flexural rigidity 4EI. The truss is loaded by a pair of forces of magnitude P, as shown in the figure.

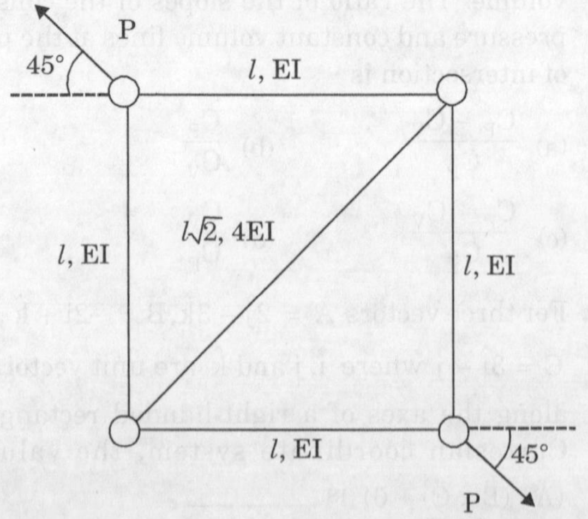

The smallest value of P, at which any of the truss members will buckle is

(a) $\frac{\sqrt{2}\pi^2 EI}{l^2}$ (b) $\frac{\pi^2 EI}{l^2}$

(c) $\frac{2\pi^2 EI}{l^2}$ (d) $\frac{\pi^2 EI}{2l^2}$

31. A rigid mass-less rod of length L is connected to a disc (pulley) of mass m and radius r = L/4 through a friction less revolute joint. The other end of that rod is attached to a wall through a friction less hinge. A spring of stiffness 2k is attached to the rod at its mid span. An inextensible rope passes over half the disc periphery and is securely tied to a spring of stiffness k at point C as shown in the figure. There is no slip between the rope and the pulley. The system is in static equilibrium in the configuration shown in the figure and the rope is always taut.

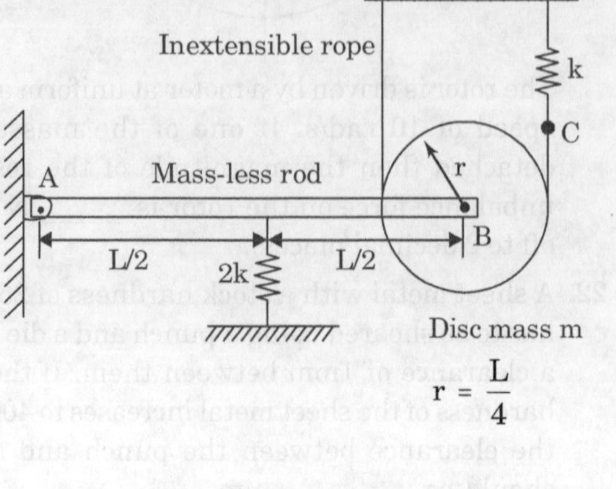

Neglecting the influence of gravity, the natural frequency of the system for small amplitude vibration is

(a) $\sqrt{\frac{3}{2}}\sqrt{\frac{k}{m}}$ (b) $\frac{3}{\sqrt{2}}\sqrt{\frac{k}{m}}$

(c) $\sqrt{3}\sqrt{\frac{k}{m}}$ (d) $\sqrt{\frac{k}{m}}$

32. A strip of thickness 40 mm is to be rolled to a thickness of 20 mm using a two-high mill having rolls of diameter 200 mm. Coefficient of friction and arc length in mm, respectively are

(a) 0.45 and 38.84 (b) 0.39 and 38.84

(c) 0.39 and 44.72 (d) 0.45 and 44.72

33. For an assembly line, the production rate was 4 pieces per hour and the average processing time was 60 minutes. The WIP inventory was calculated. Now, the production rate is kept the same, and the average processing time is brought down by 30 percent. As a result of this change in the processing time, the WIP inventory.

(a) decreases by 25% (b) increases by 25%

(c) decreases by 30% (d) increases by 30%

34. A small metal bead (radius 0.5 mm) initially at 100°C. When placed in a stream of fluid at 20°C, attains a temperature of 28°C in 4.35 seconds. The density and specific heat of the metal are 8500 kg/m^3 and 400 J/kg.K, respectively. If the bead is considered as lumped system, the convective heat transfer coefficient (in W/m^2.K) between the metal bead and the fluid stream is

(a) 283.3 (b) 299.8

(c) 149.9 (d) 449.7

35. Consider two exponentially distributed random variables X and Y, both having a mean of 0.50. Let Z = X + Y and r be the corrrelation coefficient between X and Y. If the variance of Z equals 0, then the value of r is ______ (round off to 2 decimal places)

36. An analytic function of a complex variable

$z = x + iy\left(i = \sqrt{-1}\right)$ is defined as

$f(z) = x^2 - y^2 + i\psi(x, y)$

where $\psi(x, y)$ is a real function. The value of the imaginary part of f(z) at z = (1 + i) is ______ (round off to 2 decimal places)

37. In a disc-type axial clutch, the frictional contact takes place within an annular region with outer and inner diameters 250 mm and 50 mm, respectively. An axial force F_1 is needed to transmit a torque by a new clutch. However, to transmit the same torque, one needs an axial force F_2 when the clutch wears out. If contact pressure remains uniform during operation of a new clutch while the wear is assumed to be uniform for an old clutch, and the coefficient of friction does not change, then the ratio F_1/F_2 is ______ (round off to 2 decimal places)

38. A cam with a translating flat-face follower is desired to have the follower motion

$$y(\theta) = 4\left[2\pi\theta - \theta^2\right], 0 \le \theta \le 2\pi$$

Contact stress considerations dictate that the radius of curvature of the cam profile should not be less than 40 mm anywhere. The minimum permissible base circle radius is ______ mm (round off to one decimal place)

39. A rectangular steel bar of length 500 mm, width 100 mm and thickness 15 mm is cantilevered to a 200 mm steel channel using 4 bolts as shown. For an external load of 10 kN applied at the tip of the steel bar, the resultant shear load on the bolt at B, is ______ kN (round off to one decimal place).

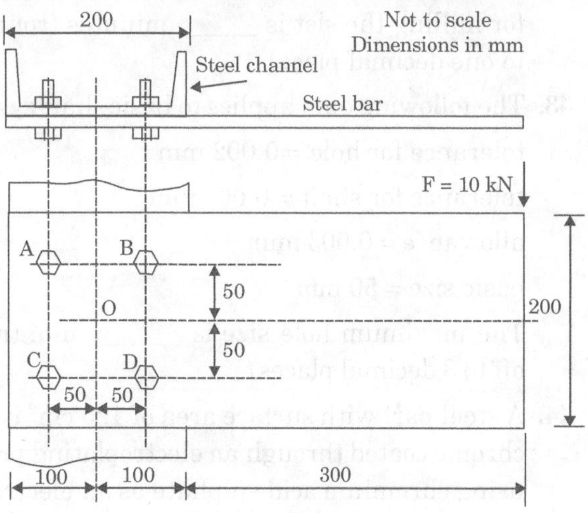

40. The barrier shown between two water tanks of unit width (1m) into the plane of the screen is modeled as a cantilever.

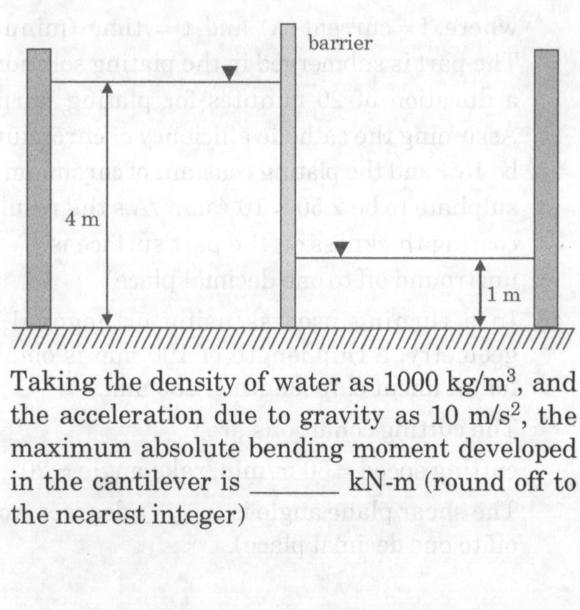

Taking the density of water as 1000 kg/m^3, and the acceleration due to gravity as 10 m/s^2, the maximum absolute bending moment developed in the cantilever is ______ kN-m (round off to the nearest integer)

41. The magnitude of reaction force at joint C of the hinge-beam shown in the figure is ______ kN (round off to 2 decimal places)

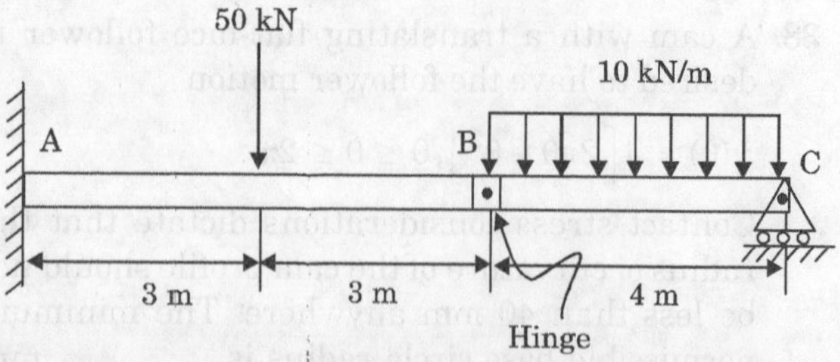

42. A slot of 25 mm × 25 mm is to be milled in a workpiece of 300 mm length using a side and face milling cutter of diameter 100 mm, width 25 mm and having 20 teeth.

For a depth of cut 5 mm, feed per tooth 0.1 mm, cutting speed 35 m/min and approach and over travel distance of 5 mm each, the time required for milling the slot is ______ minutes (round off to one decimal place).

43. The following data applies to basic shaft system :

tolerance for hole = 0.002 mm

tolerance for shaft = 0.001 mm

allowance = 0.003 mm

basic size = 50 mm

The maximum hole size is ______ mm (round off to 3 decimal places)

44. A steel part with surface area of 125 cm^2 is to be chrome coated through an electroplating process using chromium acid sulphate as an electrolyte. An increasing current is applied to the part according to the following current time relation:

$$I = 12 + 0.2t$$

where, I = current (A) and t = time (minutes). The part is submerged in the plating solution for a duration of 20 minutes for plating purpose. Assuming the cathode efficiency of chromium to be 15% and the plating constant of chromium acid sulphate to be 2.50×10^{-2} mm^3/A.s the resulting coating thickness on the part surface is ______ μm (round off to one decimal place).

45. In a turning process using orthogonal tool geometry, a chip length of 100 mm is obtained for an uncut chip length of 250 mm.

The cutting conditions are :

cutting speed = 30 m/min, rake angle = 20°

The shear plane angle is ______ degrees (round off to one decimal place).

46. The thickness of a steel plate with material strength coefficient of 210 MPa has to be reduced from 20 mm to 15 mm in a single pass in a two high rolling mill with a roll radius of 450 mm and rolling velocity of 28 m/min. If the plate has a width of 200 mm and its strain hardening exponent, n is 0.25, the rolling force required for the operation is ______ kN (round off to 2 decimal places).

Note: Average flow stress = Material strength coefficient × $\frac{(\text{True strain})^n}{(1+n)}$

47. Two business owners Shweta and Ashok run their businesses in two different states. Each of them, independent of the other, produces two products A and B, sells them at Rs. 2,000 per kg and Rs. 3,000 per kg, respectively, and uses Linear programming to determine the optimal quantity of A and B to maximize their respective daily revenue. Their constraints are as follows: (i) For each business owner, the production process is such that the daily production of A has to be at least as much as B, and the upper limit for production of B is 10 kg per day, and (ii) the respective state regulations restrict Shveta's production of A to less than 20 kg per day, and Ashok's production of A to less than 15 kg per day. The demand of both A and B in both the states is very high and everything produced is sold.

The absolute value of the difference in daily (optimal) revenue of Shveta and Ashok is ______ thousand Rupees (round off to 2 decimal places).

48. Consider two cases as below.

Case 1: A company buys 1000 pieces per year of a certain part from vendor 'X'. The changeover time is 2 hours and the price is Rs. 10 per piece. The holding cost rate per part is 10% per year.

Case 2: For the same part, another vendor 'Y' offers a design where the changeover time is 6 minutes, with a price of Rs. 5 per piece and a holding cost rate per part of 100% per year. The order size is 800 pieces per year from 'X' and 200 pieces per year from 'Y'.

Assume the cost of downtime as Rs. 200 per hour. The percentage reduction in the annual cost for case 2, as compared to case 1 is ______ (round off to 2 decimal places).

49. Consider steady, viscous, fully developed flow of a fluid through a circular pipe of internal diameter D. We know that the velocity profile forms a paraboloid about the pipe centre line, given by :

$$V = -C\left(r^2 - \frac{D^2}{4}\right) \text{ m/s. Where C is a constant.}$$

The rate of kinetic energy (in J/s) at the control surface A-B, as shown in the figure is proportional to D^n. The value of n is ________

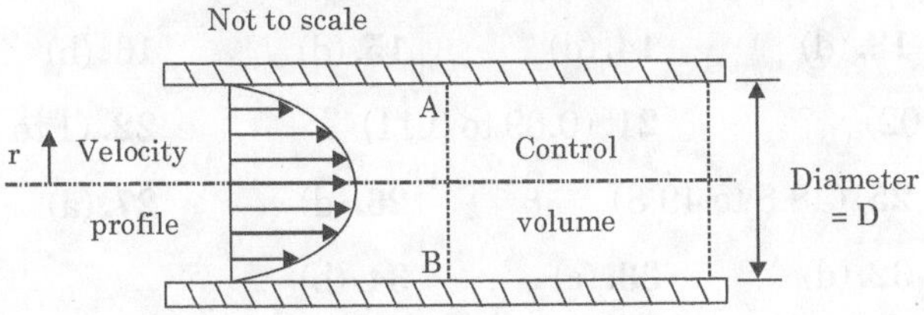

50. Air discharges steadily through a horizontal nozzle and impinges on a stationary vertical plate as shown in figure.

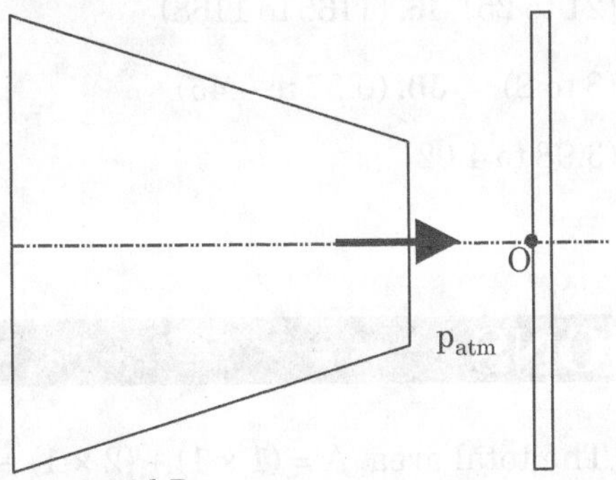

The inlet and outlet areas of the nozzle are 0.1 m^2 and 0.02 m^2, respectively. Take air density as constant and equal to 1.2 kg/m^3. If the inlet gauge pressure of air is 0.36 kPa. the gauge pressure at point O on the plate is ________ kPa (round off to two decimal places).

51. Air (ideal gas) enters a perfectly insulated compressor at a temperature of 310 K. The pressure ratio of the compressor is 6. Specific heat at constant pressure for air is 1005 J/kg.K and ratio of specific heats at constant pressure and constant volume is 1.4. Assume that specific heats of air are constant. If the isentropic efficiency of the compressor is 85 percent, the difference in enthalpies of air between the exit and the inlet of the compressor is________ kJ/ kg (round off to nearest integer)

52. One kg of air, initially at a temperature of 127°C, expands reversibly at a contant pressure until the volume is doubled. If the gas constant of air is 287 J/kg.K, the magnitude of work transfer is ___________ kJ (round off to 2 decimal places).

53. For an ideal Rankine cycle operating between pressures of 30 bar and 0.04 bar, the work output from the turbine is 903 kJ/kg and the work input to the feed pump is 3 kJ/kg. The specific steam consumption is ________ kg/kW.h. (round off to 2 decimal place)

54. For a Kaplan (axial flow) turbine, the outlet blade velocity diagram at a section is shown in figure.

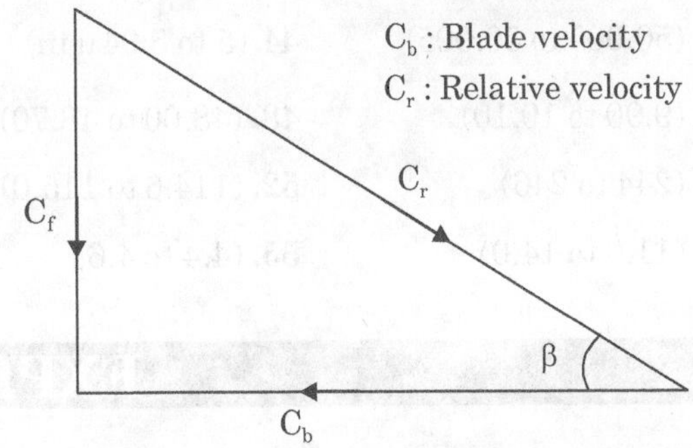

The diameter at this section is 3m. The hub and tip diameters of the blade are 2m and 4m, respectively. The water volume flow rate is 100 m^3/s. The rotational speed of the turbine is 300 rpm. The blade outlet angle β is ________ degrees (round off to one decimal place)

55. The indicated power developed by an engine with compression ratio of 8, is calculated using an air-standard Otto cycle (constant properties). The rate of heat addition is 10 kW. The ratio of specific heats at constant pressure and constant volume is 1.4. The mechanical efficiency of the engine is 80 percent.

The brake power output of the engine is ______ kW (round off to one decimal place)

ANSWERS

General Aptitude (GA)

1. (c) **2.** (a) **3.** (c) **4.** (b) **5.** (d) **6.** (d) **7.** (d) **8.** (d)
9. (a) **10.** (c)

Mechanical Engineering

1. (a) **2.** (c) **3.** (c) **4.** (d) **5.** (b) **6.** (c) **7.** (b) **8.** (a)
9. (c) **10.** (b) **11.** (a) **12.** (a) **13.** (d) **14.** (d) **15.** (d) **16.** (b)
17. (c) **18.** (d) **19.** (6 to 6) **20.** (0.98 to 1.02) **21.** (0.09 to 0.11) **22.** (1 to 1)
23. (0.90 to 0.95) **24.** (243 to 247) **25.** (48.8 to 49.8) **26.** (d) **27.** (a)
28. (a) **29.** (c) **30.** (c) **31.** (c) **32.** (d) **33.** (c) **34.** (b)
35. (–1.00 to –0.98) **36.** (1.99 to 2.01) **37.** (0.85 to 0.89) **38.** (47.9 to 48.1)
39. (15.9 to 16.1) **40.** (105 to 105) **41.** (19.95 to 20.05) **42.** (1.5 to 1.7)
43. (50.005 to 50.005) **44.** (5 to 5.04 μ m) **45.** (21 to 25) **46.** (1166 to 1168)
47. (9.90 to 10.10) **48.** (48.00 to 48.70) **49.** (8 to 8) **50.** (0.37 to 0.45)
51. (244 to 246) **52.** (114.6 to 115.0) **53.** (3.98 to 4.02)
54. (11.0 to 14.0) **55.** (4.4 to 4.6)

EXPLANATIONS

GENERAL APTITUDE

1. Crocodile tears: A false, insincere display of emotion.
2. Jofra Archar, the England fast bowler, is more fast than accurate.
3. Build : Building (Noun)
   so, Grow : Growth (Noun)
4. having said that : despite what one just said.
5. Graph of $y = [x]$ for $x \in [1, 4]$

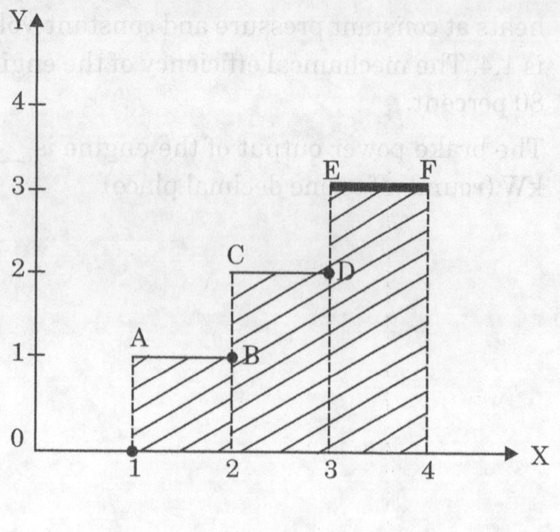

The total area, $A = (1 \times 1) + (2 \times 1) + (3 \times 1)$

$$= 1 + 2 + 3 = 6$$

6. The correct statement about crowd funding is to fund raised through voluntary contributions on web-based platforms.
7.

P,	Q,	R,	S
αα	αβ	[βα]	[ββ]

8. $S = 8 + 88 + 888 + \dots n$

$$= \frac{8}{9}(9 + 99 + 999 + \dots n)$$

$$= \frac{8}{9}\left[(10^1 - 1) + (100^2 - 1) + \dots + (10^n - 1)\right]$$

$$= \frac{8}{9}\left[(10^1 + 100^2 + \dots 100^n) - n\right]$$

$$= \frac{8}{9}\left[\frac{10 \times (10^n - 1)}{10 - 1}\right] - \frac{8}{9}n$$

$$= \frac{80}{81}(10^n - 1) - \frac{8}{9}n$$

9. Graph of $y = x^m$ and $y = x^{1/m}$ for m > 1

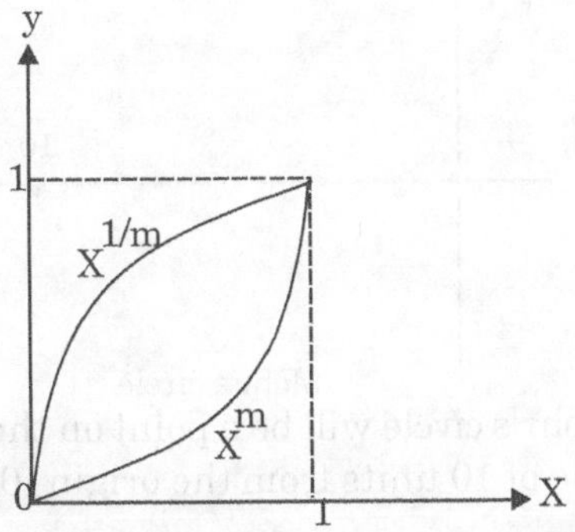

Correct option is (a).

10. Success rate = $\dfrac{\text{Number of candidtes who passed the exam}}{\text{Number of appeared candidates}}$

So, Average success rate,

$$= \frac{\frac{280}{500} + \frac{330}{600} + \frac{455}{700} + \frac{240}{400}}{4} = 0.59 \text{ or } 59\%$$

Correct option is (c).

MECHANICAL ENGINEERING

1. If A, B and C are real valued square matrices of same dimension,

(1) AB ≠ BA

i.e. multiplication of matrices are not commutative.

(2) A (BC) = (AB) C

i.e. multiplication of matrices are associative.

so, the correct option is (a)

2. $\lim\limits_{x \to 1} \left(\dfrac{1 - e^{-c(1-x)}}{1 - xe^{-c(1-x)}} \right)$

At x = 1, $\dfrac{1 - e^{-c(1-x)}}{1 - xe^{-c(1-x)}} = \dfrac{0}{0}$

Apply L-Hospital rule,

$$\lim_{x \to 1} \left[\frac{-ce^{-c(1-x)}}{0 - e^{-c(1-x)} - xe^{-c(1-x)} \times c} \right]$$

At x = 1,

$$\Rightarrow \lim_{x \to 1} \left[\frac{-ce^{-c(1-x)}}{0 - e^{-c(1-x)} - xce^{-c(1-x)}} \right] = \frac{-c \times 1}{-1 - c}$$

$$= \frac{c}{1 + c}$$

3. $L(\sin \omega t) = \dfrac{\omega}{s^2 + \omega^2}$ (Formula to be used)

Given, $L(f) = \dfrac{1}{s^2 + \omega^2} = \dfrac{1}{\omega} \times \dfrac{\omega}{s^2 + \omega^2}$

$L(f) = \dfrac{1}{\omega} L(\sin \omega t)$

$$\boxed{f(t) = \frac{1}{\omega} \sin \omega t}$$

4. Analytic function: A function f(z) is said to be analytic in a given region R, if it is continuous, single-valued and has definite derivartive at every point inside the region R.

f(z) = log z is differentiable and analytic everywhere except at origin.

5. To determine zero force members:

(1) If three members are connected at a joint and there is no external force applied to the joint and two of the members are colinear.

i.e, Joint B, D and G.

(i)

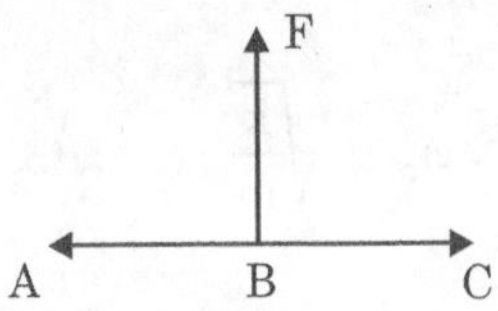

here, $F_{BF} = 0$

(ii)

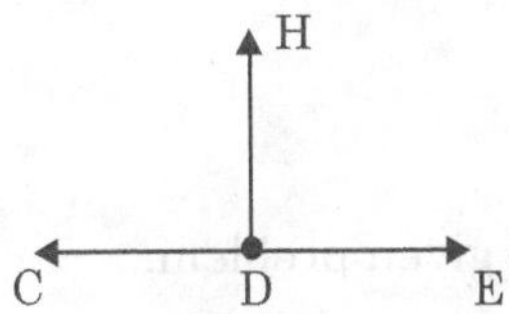

here, $F_{DH} = 0$

(iii)

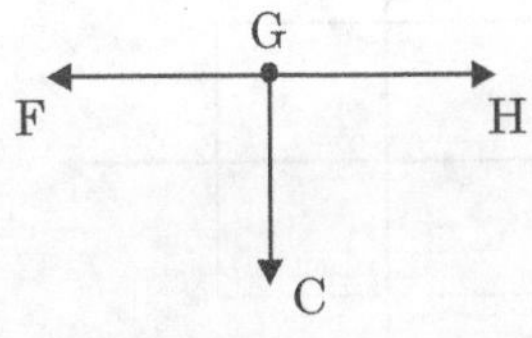

here, $F_{GC} = 0$

so, the members carrying zero force in the truss are BF, DH and GC.

6. Transmissibility ratio,

$$\frac{F_t}{F_o} = \sqrt{\frac{1 + \left(2\xi \frac{\omega}{\omega_n}\right)^2}{\left[1 - \left(\frac{\omega}{\omega_n}\right)^2\right]^2 + \left[2\xi \frac{\omega}{\omega_n}\right]^2}}$$

$\because F_t = F_o$ (Given)

$$\text{so, } 1 = \frac{\sqrt{1 + \left(2\xi \frac{\omega}{\omega_n}\right)^2}}{\sqrt{\left[1 - \left(\frac{\omega}{\omega_n}\right)^2\right]^2 + \left[2\xi \frac{\omega}{\omega_n}\right]^2}}$$

$$\Rightarrow \left[1 - \left(\frac{\omega}{\omega_n}\right)^2\right]^2 + \left(2\xi \frac{\omega}{\omega_n}\right)^2 = \left(2\xi \frac{\omega}{\omega_n}\right)^2 + 1$$

It can be written as, $\left[\left(\frac{\omega}{\omega_n}\right)^2 - 1\right]^2 = 1$

$$\Rightarrow \left(\frac{\omega}{\omega_n}\right)^2 = 2$$

$$\Rightarrow \frac{\omega}{\omega_n} = \sqrt{2}$$

$$\Rightarrow \omega = \sqrt{2}\,\omega_n \qquad \because \omega_n = \sqrt{\frac{k}{m}}$$

$$\text{so, } \omega = \sqrt{2} \times \sqrt{\frac{k}{m}}$$

$$\boxed{\omega = \sqrt{\frac{2k}{m}}}$$

7. According to the given problem:

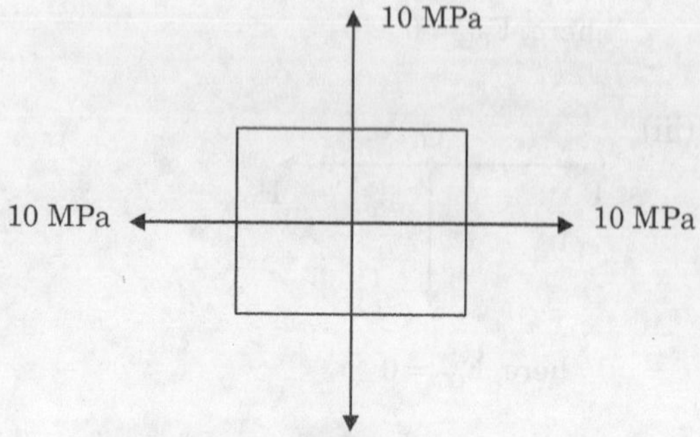

Equi-biaxial tension

Here, $\sigma_x = \sigma_y = 10$ MPa
and $\tau_{xy} = \tau_{yx} = 0$
so, principal stresses,

$$\sigma_{1,2} = \frac{\sigma_x + \sigma_y}{2} \pm \sqrt{\left(\frac{\sigma_x - \sigma_y}{2}\right)^2 + \tau_{xy}^2}$$

$$\sigma_1 = 10 \text{ MPa} \quad \text{and} \quad \sigma_2 = 10 \text{ MPa}$$

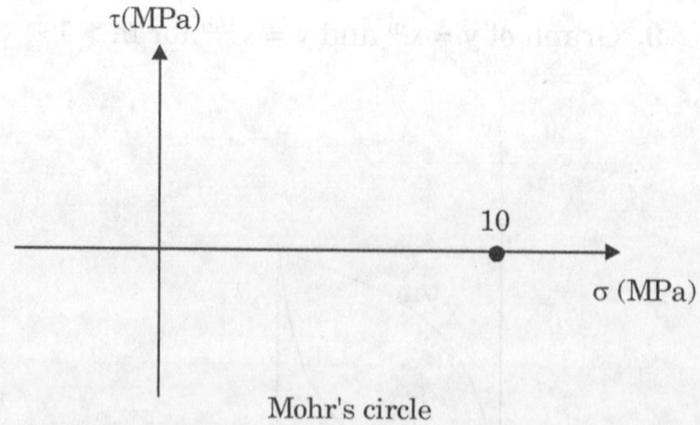

Mohr's circle

i.e. Mohr's circle will be a point on the σ-axis at a distane of 10 units from the origin (0, 0).

8. According to Grashof's Law, (for the continuous relative motion

$$l + S \le P + Q$$

Here, l = length of largest side
S = length of shortest side
P and Q = length of other two sides for, S = 80 mm, the above condition is satisfied.

i.e. 600 + 80 < 300 + 400 to obtain crankrocker mechanism, the adjacent side of shortest link (s = 80 mm) should be fixed.

So, for the mechanism to be a crank-rocker mechanism, the length of the link PQ should be 80 mm.

9.

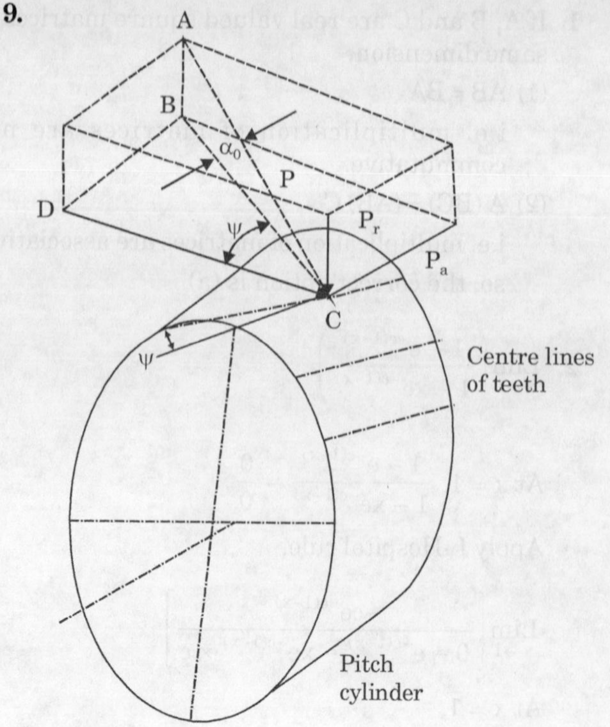

P_t give torsion
P_a give axial loading
P_r give bending

10. The crystal structure of γ-iron (Austenite phase) is FCC i.e. Face Centered Cubic.

11. (i) Tempering: The main effect of tempering is to impart toughness in the metal.
i.e $P \rightarrow 2$

(ii) Quenching : It increases the hardness of the material.
i.e $Q \rightarrow 3$

(iii) Annealing: It is a heat treatment process which 'soften' a metal
i.e $R \rightarrow 4$

(iv) Normalizing: This heat treatment process is used for strengthening and grain refinement.
i.e $S \rightarrow 1$

12. For rough grinding of brass material:

(1) Silicon carbide (SiC) is preferred for grinding of brass material.

(2) For rough cut, open structure is preferred. Range of structure from dense to open varies from 0 to 15 i.e. more the number; more the open structure is.
So, according to that the suitable grinding wheel is,

$$\boxed{C}\,30\,Q\,\boxed{12}\,V$$

13. In critical parth method (CPM), the cost-time slope of an activity.

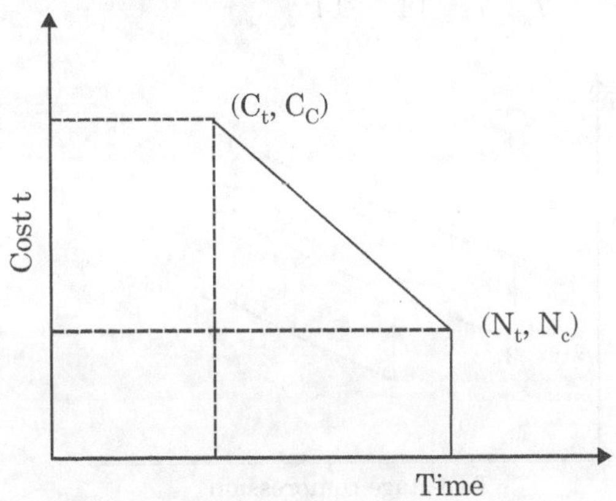

$$= \frac{\text{Crash cost} - \text{Normal cost}}{\text{Normal time} - \text{Crash time}} = \frac{C_c - N_c}{N_t - C_t}$$

14. Froude number, $F_r = \dfrac{\text{Inertia force } (F_i)}{\text{Gravity force } (F_g)}$

15. (i) Reynolds number, $Re = \dfrac{\text{Inertia force}}{\text{Viscous force}}$
i.e. $P \rightarrow 3$

(ii) Grashof Number, $Gr = \dfrac{\text{Buoyancy force}}{\text{Viscous force}}$
i.e. $Q \rightarrow 1$

(iii) Nusselt Number, $Nu = \dfrac{\text{Convective heat transfer}}{\text{Conduction heat transfer}}$
i.e. $R \rightarrow 4$

(iv) Prandtl number, $Pr = \dfrac{\text{Momentum diffusivity}}{\text{Thermal diffusivity}}$
i.e. $S \rightarrow 2$

so, the correct option is (d)

16. Given : $\vec{V} = 2(x^2 - y^2)\hat{i} + v\hat{j} + 3\hat{k}$
$= u\hat{i} + v\hat{j} + w\hat{k}$

For incompressible flow,

$$\Delta \cdot \vec{V} = 0$$

$$\frac{\partial}{\partial x}(u) + \frac{\partial}{\partial y}(v) + \frac{\partial}{\partial z}(w) = 0$$

$$\Rightarrow \frac{\partial}{\partial x}[2(x^2 - y^2)] + \frac{\partial}{\partial y}(v) + \frac{\partial}{\partial z}(3) = 0$$

$$\Rightarrow 4x + \frac{\partial v}{\partial y} + 0 = 0$$

$$\Rightarrow \frac{\partial v}{\partial y} = -4x$$

$$\Rightarrow v = \int -4x\, dy + \text{cons tan t}$$

$$\Rightarrow v = -4xy + \text{constant}$$

17. Joule-Thomson coefficient is defined as,

$$\mu_j = \left(\frac{\partial T}{\partial P}\right)_h$$

Also, $\mu_j = \dfrac{1}{C_P}\left[T\left(\dfrac{\partial V}{\partial T}\right)_P - v\right]$

For ideal gas, $Pv = RT$

$$\Rightarrow \left(\frac{\partial v}{\partial T}\right)_P = \frac{R}{P}$$

$$\Rightarrow \mu_j = \frac{1}{C_P}\left[\frac{T \times R}{P} - v\right]$$

$$\Rightarrow \mu_j = \frac{1}{C_P}\left[\frac{Pv}{P} - v\right]$$

$$\boxed{\mu_j = 0}$$

i.e., for ideal gas, the value of the Joule-Thomson coefficient is zero.

18. For ideal gas,
$Tds = du + Pdv = C_v dT + Pdv$
At constant volume, $(dv = 0)$

$Tds = C_v dT$

$$\Rightarrow \left(\frac{dT}{ds}\right)_v = \frac{T}{C_v} \quad ...(i)$$

Again, for ideal gas.

$$Tds = dh - vdP = C_p dT - vdP$$

For constant pressure,

$$Tds = C_p dT$$

$$\left(\frac{dT}{ds}\right)_P = \frac{T}{C_P} \quad ...(ii)$$

so, the ratio,

$$\frac{\left(\frac{\partial T}{ds}\right)_P}{\left(\frac{dT}{ds}\right)_v} = \frac{T}{C_P} \times \frac{C_v}{T} = \frac{C_v}{C_P}$$

19. Given: $\vec{A} - = 2\hat{j} - 3\hat{k}$

$$\vec{B} = -2\hat{i} + \hat{k}$$

and $\vec{C} = 3\hat{i} - \hat{j}$

$$(\vec{B} \times \vec{C}) = (-2\hat{i} + \hat{k}) \times (3\hat{i} - \hat{j})$$

$$= -2\hat{i} + 3\hat{i} + 2\hat{i} \times \hat{j} + \hat{k} \times 3\hat{i} - \hat{k} \times \hat{j}$$

$[\because \hat{i} \times \hat{i} = 0, \hat{i} \times \hat{j} = \hat{k}, \hat{k} \times \hat{i} = \hat{j} \text{ and } \hat{k} \times \hat{j} = -\hat{i}]$

so, $(\vec{B} \times \vec{C}) = 0 + 2\hat{k} + 3\hat{j} + \hat{i}$

$$= (2\hat{k} + 3\hat{j} + \hat{i})$$

Now, $\vec{A} \cdot (\vec{B} \times \vec{C}) = (2\hat{j} - 3\hat{k}) \cdot (2\hat{k} + 3\hat{j} + \hat{i})$

$$= 4\hat{j} \cdot \hat{k} + 6\hat{j} \cdot \hat{j} + 2\hat{j} \cdot \hat{i} - 6\hat{k} \cdot \hat{k} - 9\hat{k} \cdot \hat{j} - 3\hat{k} \cdot \hat{i}$$

$\because [\hat{j} \cdot \hat{k} = 0, \hat{k} \cdot \hat{k} = 1, \hat{j} \cdot \hat{j} = 1, \hat{j} \cdot \hat{i} = 0, \hat{k} \cdot \hat{j} = 0,$

and $\hat{k} \cdot \hat{i} = 0]$

So $(\vec{A} \cdot (\vec{B} \times \vec{C}) = 0 + 6 + 0 - 6 - 0 - 0$

The value of $(\vec{A} \cdot (\vec{B} \times \vec{C}) + 6)$ is $0 + 6 = 6$

20. Given : Max rotational speed, $\omega_1 = 110$ rad/s

Min rotational speed, $\omega_2 = 100$ rad/s

Energy fluctuation, $\Delta E = 1.05$ kJ

$$\because \Delta E = \frac{1}{2} I\left(\omega_1^2 - \omega^2\right)$$

where, I = Moment of inertia of fly wheel

$$1.05 \times 1000 = \frac{1}{2} \times 1 \times (110^2 - 100^2)$$

$$\boxed{I = 1 \text{ kg} - m^2}$$

21. Mass, m = 10 grams = 10×10^{-3} kg

Radius, r = 100 mm = 100×10^{-3} m

Angular speed, $\omega = 10$ rad/s

Force on each point masses,

$$F = mr\omega^2 = 10 \times 10^{-3} \times 100 \times 10^{-3} \times (10)^2$$
$$= 0.1 \text{ N} = 0.1 \text{ N}$$

22. Clearance in the punch, $C = 0.0032t\sqrt{\tau})$

where, τ is the shear strength of sheet metal.

$\because \tau \propto (BHN) \cdot (HRC)$

We can write, $C \propto \sqrt{HRC}$

$\because$ At, $(HRC)_1 = 250$, $C_1 = 1$ mm

then at $(HRC)_2 = 400$,

$$\text{Clearance, } C_2 = C_1 \times \sqrt{\frac{(HRC)_2}{(HRC)_1}} = 1 \times \sqrt{\frac{400}{250}}$$

$$\boxed{C_2 = 1.26 \text{ mm}}$$

23. Number of Men, M = 5

Number of Women N = 3

We have to hire four candidates.

The required probability (i.e. the Probability that at least one woman will be selected).

$$P = \frac{{}^5C_3 \times {}^3C_1}{{}^8C_4} + \frac{{}^5C_2 + {}^3C_2}{{}^8C_4} + \frac{{}^5C_1 \times {}^3C_3}{{}^8C_4}$$

$$= \frac{3}{7} + \frac{3}{7} + \frac{1}{14} = \frac{13}{14} = 0.93$$

24.

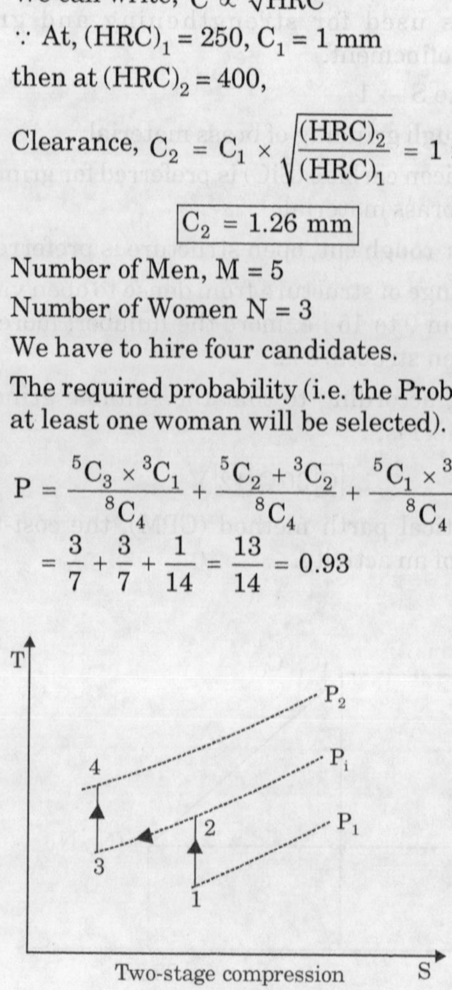

Two-stage compression process

$\because$ There are two (n = 2) at stages.

For minimum work,

$$\text{Overall pressure ratio, } r_o = \left(\frac{P_i}{P_1}\right)^{\frac{1}{2}}$$

$$\Rightarrow \quad 6 = \left(\frac{P_i}{100}\right)^{\frac{1}{2}}$$

Inter cooling pressure, $P_1 = 244.95$ kPa

25. Given :

Temperature of hot oil at the entry, $T_{h_i} = 102°C$

Temperature of hot oil at the exit, $T_{h_e} = 65°C$.

Temperature of cold water at entry, $T_{c_i} = 25°C$

and Temperature of cold water at exit, $T_{c_e} = 42°C$

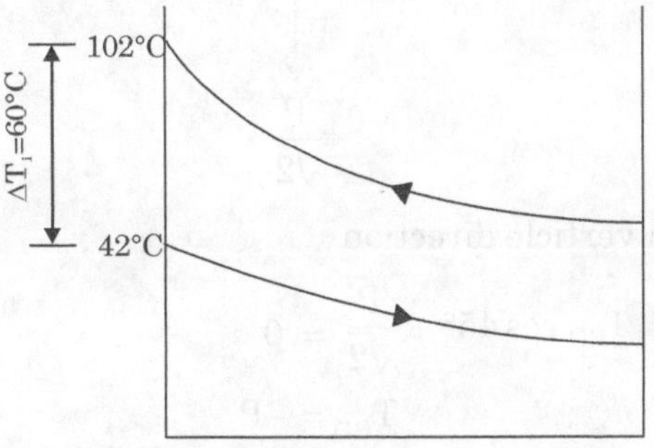

Counter-flow heat exchanger

∵ LMTD is defined as,

$$(\Delta T)_{lm} = \frac{\Delta T_i - \Delta T_e}{\ln\left(\frac{\Delta T_i}{\Delta T_e}\right)} = \frac{60-40}{\ln\left(\frac{60}{40}\right)}$$

$$(\Delta T)_{ln} = 49.3°C$$

26. $\int_{-1}^{1.4} x|x|dx$

$f(x) = -x^2 \quad -1 \le x \le 0$

$f(x) = x^2 \quad 0 \le x < 14$

x	–1	–0.4	0.2	0.8	1.4
f(x)	–1	–0.16	0.04	0.64	1.96
	y_0	y_1	y_2	y_3	y_4

By simpson's 1/3rd rule

$$\int_{-1}^{1} x|x|dx = \frac{h}{3}[(y_0 + y_4) + 2(y_2) + 4(y_1 + y_2)]$$

$$= \frac{0.6}{3}[(-1 + 1.96) + 2 \times 0.04 + 4(-0.16 + 0.64)]$$

$= 0.592$

27. $\vec{f}(x, y, z) = \frac{x}{(x^2+y^2+z^2)^{3/2}}\hat{i} + \frac{y}{(x^2+y^2+z^2)^{3/2}}\hat{j} + \frac{z}{[x^2+y^2+z^2]^{3/2}}\hat{k}$

By gauss-divergence theorem

$$\iint_s \vec{F}\hat{n}ds = \iiint_v \vec{\nabla}\cdot\vec{F}dV \qquad ...(i)$$

So, $\vec{\nabla}\cdot\vec{F} = \frac{\partial}{\partial x}\left(\frac{x}{x^2+y^2+z^2)^{\frac{3}{2}}}\right) + \frac{\partial}{\partial y}\left(\frac{y}{(x^2+y^2+z^2)^{\frac{3}{2}}}\right) + \frac{\partial}{\partial z}\left(\frac{z}{(x^2+y^2+z^2)^{\frac{3}{2}}}\right)$

$$\vec{\nabla}\cdot\vec{F} = \frac{1}{(x^2+y^2+z^2)^{\frac{3}{2}}} - \frac{\frac{3}{2}\times x \times 2x}{(x^2+y^2+z^2)^{\frac{5}{2}}} + \frac{1}{(x^2+y^2+z^2)^{\frac{3}{2}}} - \frac{\frac{3}{2}\times y \times 2y}{(x^2+y^2+z^2)^{\frac{5}{2}}} + \frac{1}{(x^2+y^2+z^2)^{\frac{3}{2}}} - \frac{\frac{3}{2}\times z \times 2z}{(x^2+y^2+z^2)^{\frac{5}{2}}} = 0$$

So by equation (i),

$$\iint_S \vec{F}\hat{n}6dS = \iiint_V \vec{\nabla}\cdot\vec{F}dV = 0$$

28.

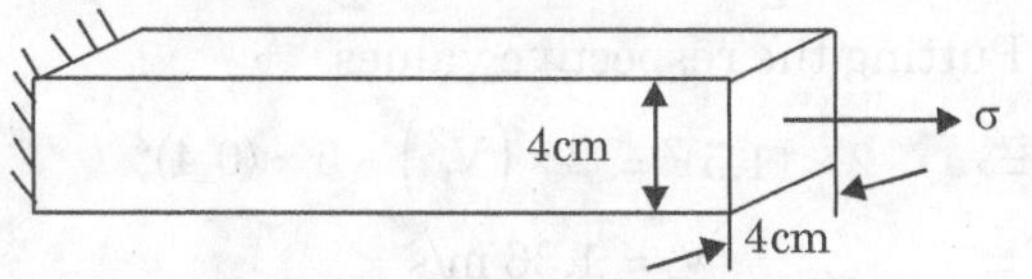

$$\sigma = \frac{P}{A} = \frac{80\times10^3}{40\times40} = 50 \text{ MPa}$$

According to maximum shear stress theory

$$\tau_{max} = \frac{\sigma_1 - \sigma_2}{2} = \frac{50-0}{2}$$

$= 25MPa$

∵ $\tau_{per} < \tau_{max}$

∴ This component will fail

Similarly,

The component, loaded with same load in compression, will also fails.

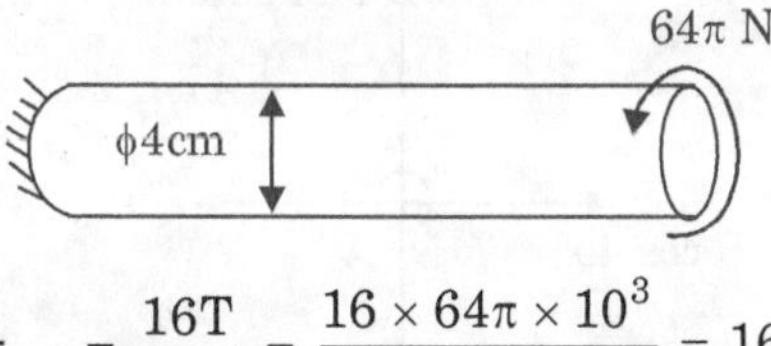

$$\tau_{max} = \frac{16T}{\pi d^3} = \frac{16\times64\pi\times10^3}{\pi\times40^3} = 16 \text{ MPa}$$

∵ $\tau_{max} < \tau_{per}$

∴ Component will not fail.

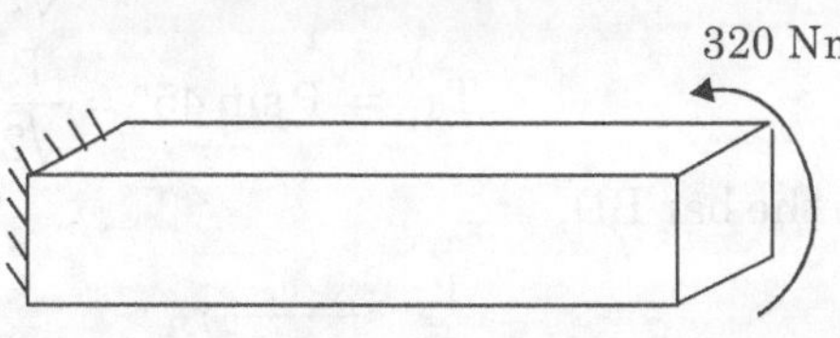

$$\sigma_{max} = \frac{M}{I}\times y = \frac{320\times10^3}{\frac{40\times40^3}{12}}\times 20$$

$= 30 MPa$

According to maximum

$$\tau_{max} = \frac{\sigma_{max} - \sigma_{min}}{2}$$

$$= \frac{30 - 0}{2} = 15 \text{ MPa}$$

$\because \tau_{max} < \tau_{per}$

$\therefore$ component will not fail

29. Given: Mass, m = 2 kg

Stiffness of rod, k = 5 N/m

Initial velocity when the block is unstretched, $V_i = 1.5$ m/s

Now, the cord is stretched by x = 0.4 m, then velocity of the block is V_f.

Using work-energy theorem,

$$\frac{1}{2}mv_i^2 = \frac{1}{2}mv_f^2 + \frac{1}{2}kx^2$$

Putting the respective values,

$$\Rightarrow \quad 2 \times (1.5)^2 = 2 \times \left(V_f^2\right) + 5 \times (0.4)^2$$

$$\Rightarrow \quad V_f = 1.36 \text{ m/s}$$

30.

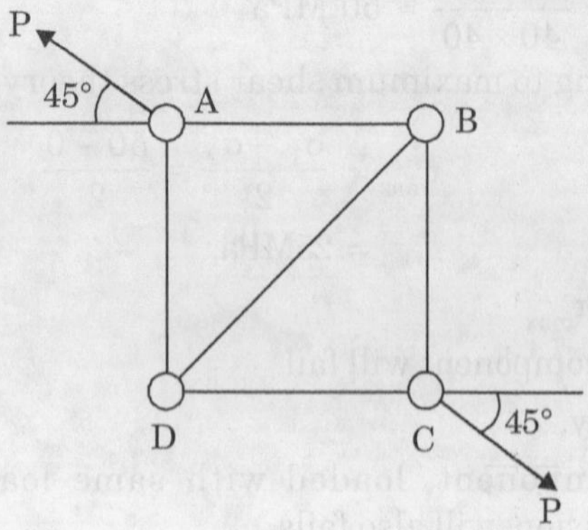

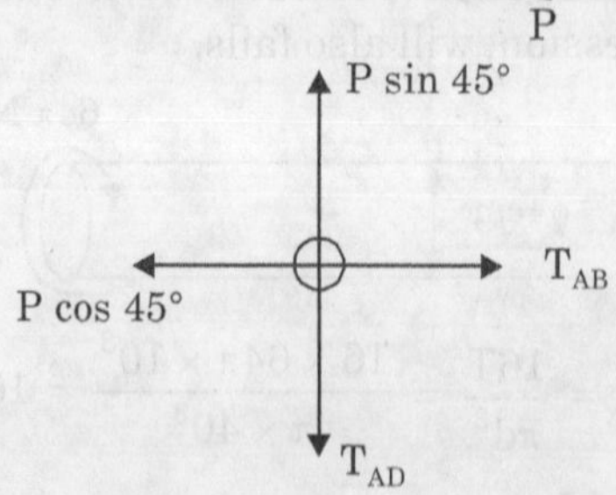

$$T_{AB} = P\cos 45° = \frac{P}{\sqrt{2}}$$

$$T_{AD} = P\sin 45° = \frac{P}{\sqrt{2}}$$

For the bar BD,

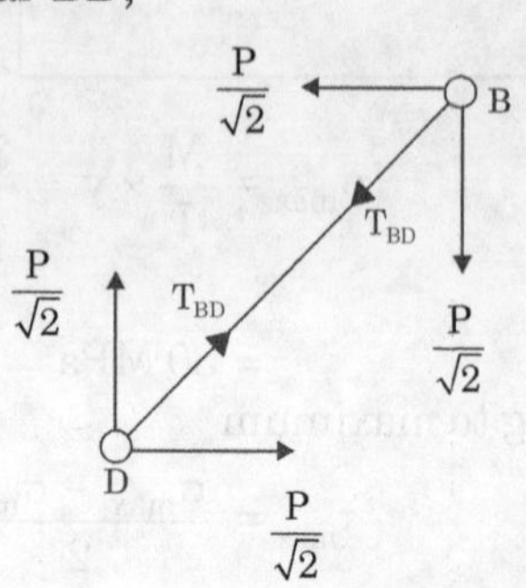

For the joint B,

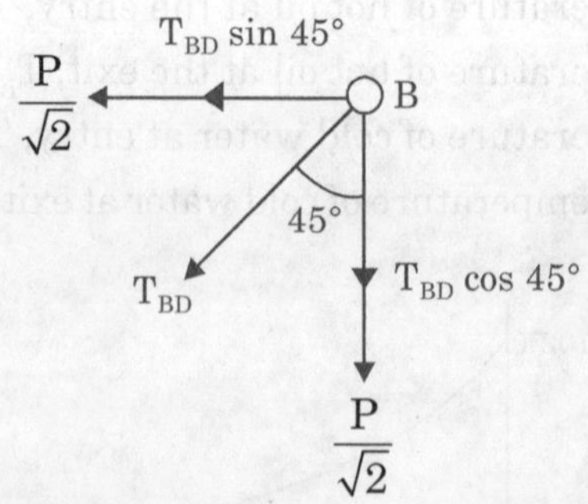

So in verticle direction

$$T_{BD}\cos 45° + \frac{P}{\sqrt{2}} = 0$$

$$T_{BD} = -P$$

so maximum load acting on the bar BD.

so bar BD buckle first

Buckling load,

$$P = \frac{\pi^2 \times 4EI}{\left(\sqrt{2}l\right)^2}$$

[both ends are hinge]

$$P = \frac{2\pi^2 EI}{l^2}$$

31.

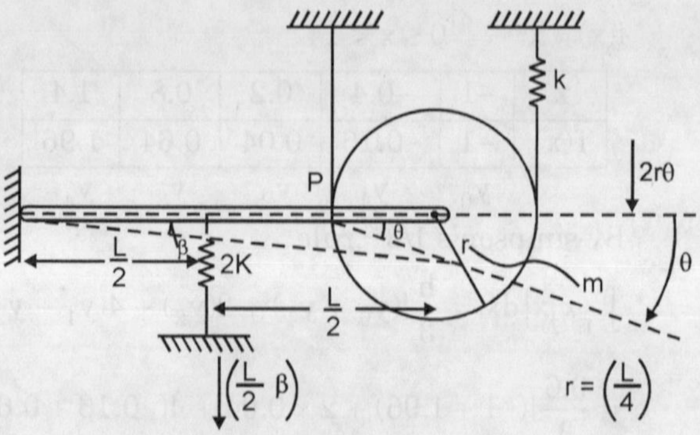

Since the disc is in pull rolling on the string:

$$L\beta = R\theta$$

$$\beta = \frac{\theta}{4} \qquad \because R = \frac{L}{4}$$

Applying Energy method

$$E_{Total} = \frac{1}{2}I_P\dot{\theta}^2 + \frac{1}{2}k(2R\theta)^2 + \frac{1}{2}2k\left(\frac{L\theta}{8}\right)^2$$

$$\frac{dE}{dt} = 0$$

$$\Rightarrow \frac{3}{2}MR^2.\dot{\theta}\ddot{\theta} + 4R^2K\theta\dot{\theta} + \frac{2KL^2\theta\dot{\theta}}{64} = 0$$

$$\Rightarrow \left(\frac{3}{2}\frac{NL^2}{16}\right)\ddot{\theta}+\frac{4L^2}{16}k\theta+\frac{2}{64}kL^2\theta=0$$

$$\Rightarrow \left(\frac{3N}{32}\right)\ddot{\theta}+\left(\frac{9k}{32}\right)\theta=0$$

$$\Rightarrow 3N\ddot{\theta}+9k\theta=0$$

$$\Rightarrow \ddot{\theta}+\left(\frac{3k}{m}\right)\theta=0 \Rightarrow \left[\omega n_1=\sqrt{3}\sqrt{\frac{k}{m}}\right]$$

32. $H_1 = 40$ mm

$H_2 = 20$ mm

$R = 100$ mm

$\Delta H = 40 - 20 = 20$ mm

Coefficient of friction (μ)

$$\mu=\sqrt{\frac{(\Delta H)_{max}}{R}} \qquad (\Delta H=(\Delta H)_{max}]$$

$$=\sqrt{\frac{20}{100}}=0.45$$

Arc length (L) $=\sqrt{R(\Delta H)}$

$$=\sqrt{100\times 20}=44.72 \text{ mm}$$

33. WIP means work in process

In first case $(WIP)_1 = 4\times\frac{60}{60} = 4$ pieces

In second case $(WIP)_2 = 4\times\frac{(60-60\times 0.3)}{60}$

$= 2.8$ piece

So, Increase in (WIP) $=\frac{(WIP)_1-(WIP)_2}{(WIP)_1}\times 100$

$= 30\%$

34. $T_o = 100°C$ $\quad T_\infty = 20°C$

$T = 28°C$ $\quad t = 4.35$ s

$r = 8500$ kg/m^3 $\quad C = 400$ J/kgK

$r = 0.5$ mm $\quad h = ?$

Metal bead is assumed as spherical shape

By lumped system analysis

$$\frac{T-T_\infty}{T_o-T_\infty}=e^{-\frac{hA\tau}{\rho VC}}$$

$$\frac{T-T_\infty}{T_o-T_\infty}=e^{-\frac{h\times 4\pi r^2\tau}{\rho\times V\times\frac{4}{3}\pi r^3}}$$

$$\frac{T-T_\infty}{T_o-T_\infty}=e^{-\frac{3h\tau}{\rho\times V\times r}}=e^{-\frac{3\times 4\times 4.35}{8500\times 400\times 0.5\times 10^{-3}}}$$

$h = 299.95$ W/m^{2}K

35. Mean of exponential distribution $=\frac{1}{\lambda}$

Standard deviation of exponential distribution $=\frac{1}{\lambda}$

For x, $\lambda_1=\frac{1}{0.5}=2$

Standard deviation, $\sigma_x = 0.5$

Variance, $V(X) = (\sigma_x)^2 = 0.25$

For Y, $\lambda_2=\frac{1}{0.5}=2$

Standard deviation, $\sigma_Y = 0.5$

Variance $V(Y) = (\sigma_Y)^2 = 0.25$

$\because V(X+Y) = V(X) + V(Y) + 2COV(X, Y)$

Given : $V(X + Y) = 0$

$\Rightarrow 0 = 0.25 + 0.25 + 2\,COV(X, Y)$

$\Rightarrow COV(X, Y) = -0.25$

Corelation factor,

$$r=\frac{COV(X,Y)}{\sigma_X\sigma_Y}$$

$$=\frac{-0.25}{0.5\times 0.5}=-1$$

36. $f(z) = u + iv$

$u = x^2 - y^2$

$v = \psi(x, y)$ ...(i)

f(z) is an analytic function,

$$\frac{\partial u}{\partial x}=\frac{\partial v}{\partial y}$$

$$\frac{\partial v}{\partial x}=-\frac{\partial u}{\partial y}$$

So, $$dv=\frac{\partial v}{\partial x}dx+\frac{\partial v}{\partial y}dy$$

$$dv=-\frac{\partial u}{\partial y}dx+\frac{\partial u}{\partial x}dy$$

$$dv = 2y\,dx + 2x\,dy$$

$$\int dv = 2\int d(xy)$$

$v = 2xy$ ...(ii)

Given, $z = 1 + i$

So, $x = 1$

$y = 1$

From equation (1) and (2)

So, $v = \psi(x, y)_{(1,1)} = 2$

37. Given:

Inner diameter, $D_i = 50$ mm

or Inner radius, $R_i = \dfrac{50}{2} = 25$ mm

and outer diameter, $D_o = 250$ mm

or outer radius, $R_o = \dfrac{250}{2} = 125$ mm

(1) For uniform pressure theory (for new clutch)

Torque transmitted, $\tau_1 = \mu_1 F_1 R'_m$

where, $R'_m = \dfrac{2}{3}\dfrac{(R_o^3 - R_i^3)}{(R_o^2 - R_i^2)}$

i.e., $\tau_1 = \mu_1 \times F_1 \times \dfrac{2}{3}\left(\dfrac{R_o^3 - R_i^3}{R_o^2 - R_i^2}\right)$

(2) For uniform wear theory (for old clutch),

Torque transmitted, $\tau_2 = \mu_2 F_2 R''_m$

where, $R''_m = \dfrac{R_o + R_i}{2}$

i.e., $\tau_2 = \mu_2 \times F_2 \times \dfrac{(R_o + R_i)}{2}$

$\because$ Torque transmitted are same and coeficient of friction is also same.

so, $F_1 \times \dfrac{2}{3}\left(\dfrac{R_o^3 - R_i^3}{R_o^2 - R_i^2}\right) = F_2 \times \dfrac{(R_o + R_i)}{2}$

$\Rightarrow \dfrac{F_1}{F_2} = \dfrac{3}{4}\left(\dfrac{R_o^2 - R_i^2}{R_o^3 - R_1^3}\right) \times (R_o + R_i)$

Putting the respective values,

$$\frac{F_1}{F_2} = \frac{3}{4}\left[\frac{125^2 - 25^2}{125^3 - 25^3}\right] \times (125 + 25)$$

$$\boxed{\frac{F_1}{F_2} = 0.87}$$

38. Given: $y(\theta) = 4[2p\theta - \theta^2], 0 \le \theta \le 2\pi$

$y'(\theta) = 8\pi - 8\theta$

$y''(\theta) = -8$

The minimum permissible base circle radius,

$(r_0)_{min} = \rho_{min} - y(\theta) + y''(\theta)]_{min}$

Here, $\rho_{min} = 40$ mm

$y(\theta) + y''(\theta) = 4[2\pi\theta - \theta^2] - 8, 0 \le \theta \le 2\pi$

$[y(\theta) \,|\, y''(\theta)]_{min} = 4[2\pi \times 0 - 0^2] - 8$

$= -8$

so, $(r_o)_{min} = 40 - (-8) = 48$ mm

39. 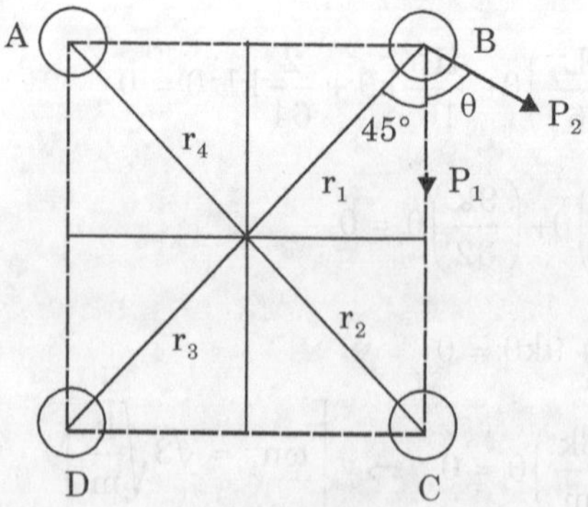

$P = 10$ kN

$r_1 = r_2 = r_3 = r_4$

$= \sqrt{(50)^2 + (50)^2} = 70.71$ mm

Eccentricity, $e = 400$ mm

Primary fore, $P_1 = \dfrac{P}{4} = \dfrac{10}{4} = 2.5$ kN

Secondary force, $P_2 = \dfrac{P \cdot e \cdot r_1}{r_1^2 + r_2^2 + r_3^2 + r_4^2}$

$= \dfrac{10 \times 400 \times 70.71}{4 \times (70.71)^2}$

$= 14.14$ kn

$\theta = 90° - 45° = 45°$

Resultant shear force on bolt B,

$P_B = \sqrt{P_1^2 + P_2^2 + 2P_1P_2 \cos 45°}$

$= \sqrt{(2.5)^2 + (14.14)^2 + 2 \times 2.5 \times 14.14 \cos 45}$

$= 16$ kN

40. 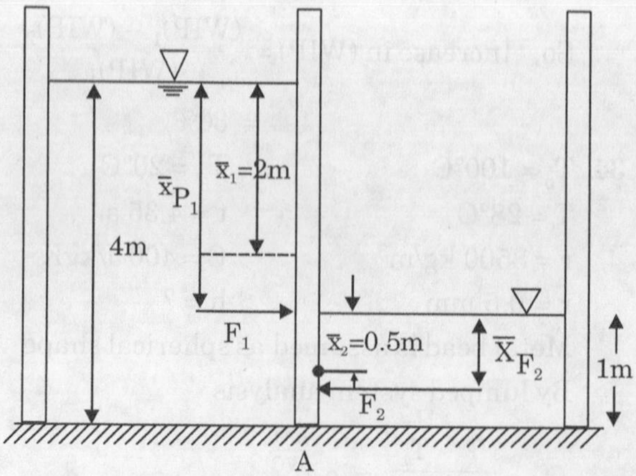

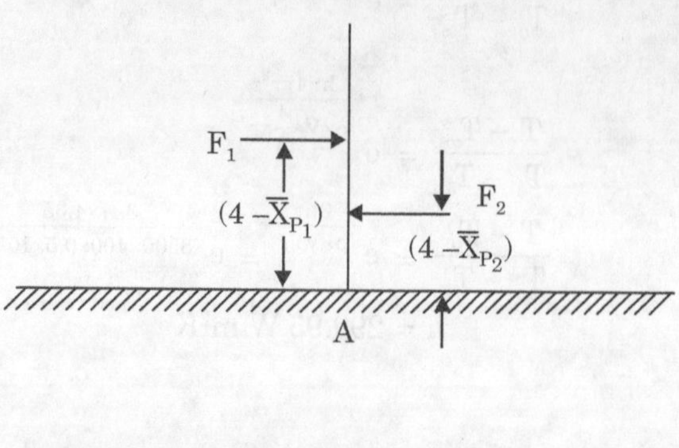

$F_1 = wA\bar{x}_1 = 10 \times (4 \times 1) \times 2 = 80$ kN

$F_2 = wA\bar{x}_2 = 10 \times (1 \times 1) \times 0.50 = 5$ kN

$$\bar{x}_{P_1} = \bar{x}_1 + \frac{1 \times 4^3}{12 \times 1 \times 4 \times 2} = 2 + \frac{2}{3} = \frac{8}{3} \text{ m}$$

$$\bar{x}_{P_2} = \bar{x}_2 + \frac{1 \times 1^3}{12 \times 1 \times 1 \times 0.5} = 0.5 + \frac{1}{6} = \frac{2}{3} \text{ m}$$

The maximum absolute bending moment will be at fixed support i.e. at point A.

$$(BM)_A = F_1 \times (4 - \bar{x}_{P_1}) - F_2 \times (1 - \bar{x}_{P_2})$$

$$= 80 \times \left(4 - \frac{8}{3}\right) - 5 \times \left(1 - \frac{2}{3}\right) = 105 \text{ kN-m}$$

41. At thing point moment is ero

From bar BC

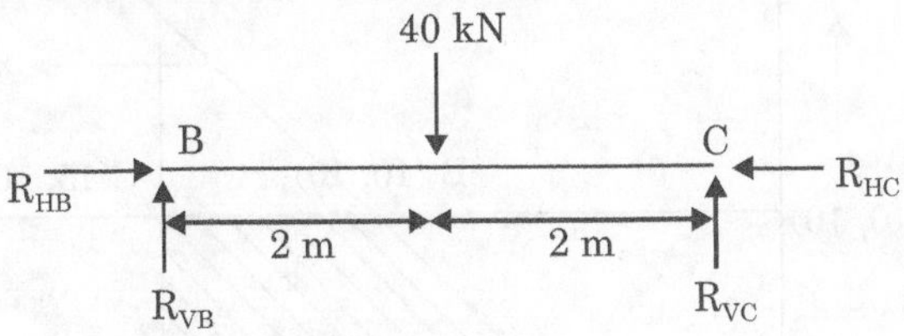

Take moment about B

$R_{VC} \times 4 = 40 \times 2$

$R_{VC} = 20$ kN

There is no force applyinhorizontal direction so

$R_{HB} = R_{HC} = 0$N

Net force on C,

$$F_C = \sqrt{(R_{HC})^2 + (R_{VC})^2}$$

$$= \sqrt{(0)^2 + (20)^2}$$

$F_C = 20$ kN

42. Length of workpiece, $L_w = 300$ mm

Diameter of cutter, $D_c = 100$ mm

Number of teeth, Z = 20

Depth of cut = 5 mm

Feed per tooth, $f_z = 0.10$ mm

Cutting speed, v = 35 m/min.

Approach (A) = over travel (O) = 5 mm

To find, Machining time (t_m)

$$t_m = \frac{\text{length of cut (L)}}{\text{feed rate } (f_r)}$$

$L = L_w + O + A$

$= 300 + 5 + 5 = 310$ mm

$$f_r = f_z \times Z \times N = f_z \times Z \times \frac{v}{\pi D_c}$$

$$= 0.10 \times 20 \times \frac{35 \times 10^3}{\pi \times 100}$$

$= 222.82$ mm/min.

$$t_m = \frac{310}{222.82} = 1.39 \text{ min.}$$

$$\boxed{t_m \simeq 1.4 \text{ min}}$$

43. For basic shaft system,

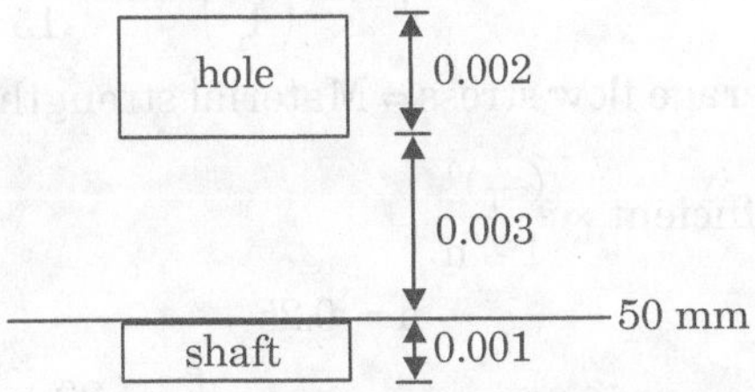

Maximum hole size = 50 + 0.003 +0.002

= 50.005 mm

44. Given: Surface area, A = 125 cm^2

Current-time corelation,

I = 12 + 0.2t, t is in minutes

Duration = 20 minutes

Plating constant = 2.50×10^{-2} mm^3/A.s

Cathode efficiency, $\eta = 0.15$

Let, the coating thickness is dx.

According to the given question,

$A \times dx = \eta \times C \times I\, dt$

$\Rightarrow 125 \times 10^2 \times dx = 0.15 \times 2.50 \times 10^{-2} \times (12 + 0.2t)\, dt$

Integrating both sides,

$$125 \times 10^2 \times \int_0^x dx = (0.15 \times 2.50 \times 10^{-2}) \times \int_0^{20} (12 + 0.2t) \times 60\, dt$$

$$\Rightarrow 125 \times 10^2 \times x = (0.15 \times 2.50 \times 10^{-2}) \times 60 \times \left[12t + \frac{0.2t^2}{2}\right]_0^{20}$$

$\Rightarrow \quad x = 5.04 \times 10^{-3}$ mm

$= 5.04$ μm

45. $l_c = 100$ mm

$l = 250$ mm

V = 30 m/min.

$\alpha = 20°$

Chip thickness ratio (r),

$$r = \frac{l_c}{l} = \frac{100}{250} = 0.4$$

Shear plane angle (ϕ),

$$\tan\phi = \frac{r\cos\alpha}{1 - r\sin\alpha}$$

$$\phi = \tan^{-1}\left(\frac{0.4\cos 20°}{1 - 0.4\sin 20°}\right) = 23.53°$$

46. (Final volume) = (Initial volume)

$$l_f b_f h_f = l_i b_i h_i$$

$$l_f \times 200 \times 15 = l_i \times 200 \times 20$$

[∵ width remains same in rolling]

$$\frac{l_f}{l_i} = \frac{20}{15}$$

Now, true strain, $\epsilon_T = \ln\left(\frac{l_f}{l_i}\right) = \ln\left(\frac{20}{15}\right)$

Average flow stress = Material strength coefficient × $\frac{(\epsilon_T)^n}{1+n}$

n = 0.25 (given)

$$\sigma_f = 210 \times \frac{\left[\ln\left(\frac{20}{15}\right)\right]^{0.25}}{1 + 0.25}$$

$$= 123.04 \text{ MPa}$$

→ Roll contact length,

$$L = \sqrt{R\Delta h}$$

$$= \sqrt{450 \times 5}$$

$$= 47.434 \text{ mm}$$

→ Required force, $F = Lb\sigma_f$

$$= 47.43 \times 200 \times 123.04$$

$$= 1167255.87 \text{ N}$$

$$= 1167.255 \text{ kN}$$

47. Let x_1 = daily production of A

x_2 = daily production of B

Objective function

Maximise Z = 2000 x1 + 3000 x2

For shweta

Constraints $x_1 \geq x_2$

$x_2 \leq 10$

$x_1 \leq 20$

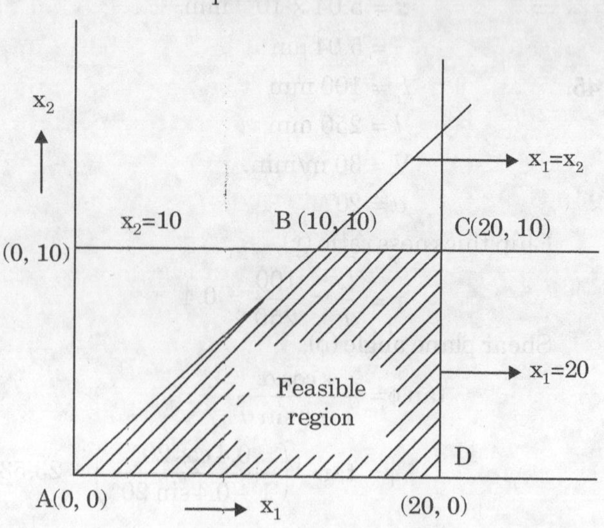

At A Z = 0

At B Z = 2000 × 10 + 3000 × 10 = 50000

At C Z = 2000 × 20 + 3000 × 10 = 70000

At D Z = 2000 × 20 + 3000 × 0 = 40000

Z_{max} = 70000

For Ashok,

Constraints, $x_1 \geq x_2$

$x_2 \leq 10$

$x_1 \leq 15$

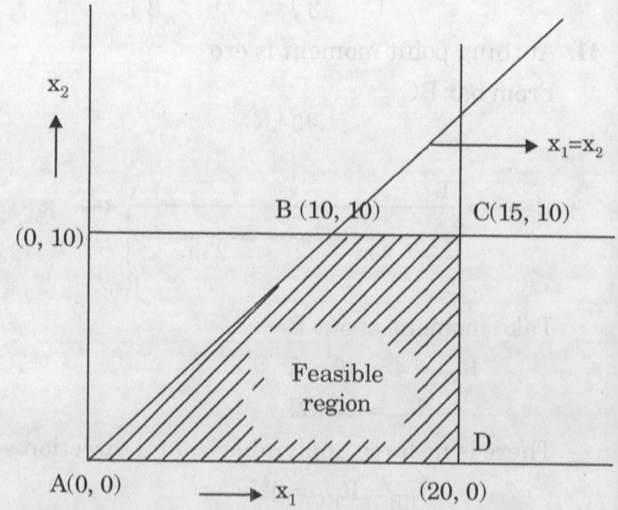

At C,

$$Z_{max} = 2000 \times 15 + 3000 \times 10$$

$$= 60000$$

Difference = 70000 – 60000 = Rs. 10000

48. For vendor x:

Setup cost $C_0 = 2 \times 200 = 400$ Rs.

Carrying cost $C_c = 0.1 \times p = 0.1 \times 10 = 1$ Rs.

p = Rs. 10/price

Assumed demand D = 1000

$$\therefore \text{Total cost} = p \times D + \frac{Q}{2} \times C_c + \frac{D}{Q} \times C_0$$

$$= 10 \times 1000 + \frac{800}{2} \times 1 + \frac{1000}{800} \times 400$$

$$= 10000 + 400 + 500 = 10900 \text{ Rs.}$$

For vendor y

Setup cost $C_0 = \frac{6}{60} \times 200 = 20$

Carrying cost $C_c = 1 \times p = 5$

p = Rs. 5/price

$$\text{Total cost} = p \times D + \frac{Q}{2} + C_c + \frac{D}{Q} \times C_0$$

$$= 5 \times 1000 \frac{200}{2} \times 5 + \frac{1000}{200} \times 20$$

$$= 5000 + 500 + 100 = \text{Rs. } 5600$$

$$\% \text{ reduction in cost} = \frac{10900 - 5600}{10900} \times 100$$

$$= 48.63$$

49.

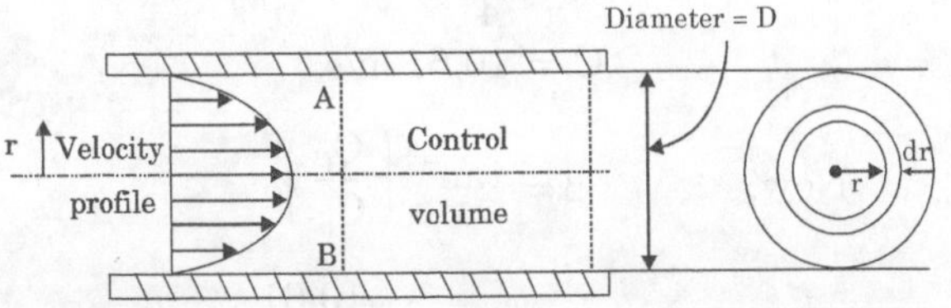

Consider a fluid element of dr thickness at a radius of r from the centre.

Now, kinetic energy rate of this element.

$$d(K\dot{.}E) = \frac{1}{2} d\dot{m}v^2$$

$$= \frac{1}{2}(\rho dA\, v) \cdot v^2$$

$$d(K\dot{.}E) = \frac{1}{2}\rho 2\pi r\, dr\, v^3$$

$$\int d(K\dot{.}E) = \int_0^{D/2} \pi\rho \cdot r dr \left[-C\left(r^2 - \frac{D^2}{4}\right)\right]^3$$

$$K\dot{.}E = -\pi\rho C^3 \int_0^{D/2} r\left(r^2 - \frac{D^2}{4}\right)^3 .dr$$

$$K\dot{.}E = -\pi\rho C^3 \int_0^{D/2} r\left[r^6 - \frac{D^6}{4^3} - 3 \cdot \frac{r^2D^2}{4}\left(r^2 - \frac{D^2}{4}\right)\right] \cdot dr$$

After integration we get,

$$= \text{fn}(D^8)$$

So, $n = 8$

50.

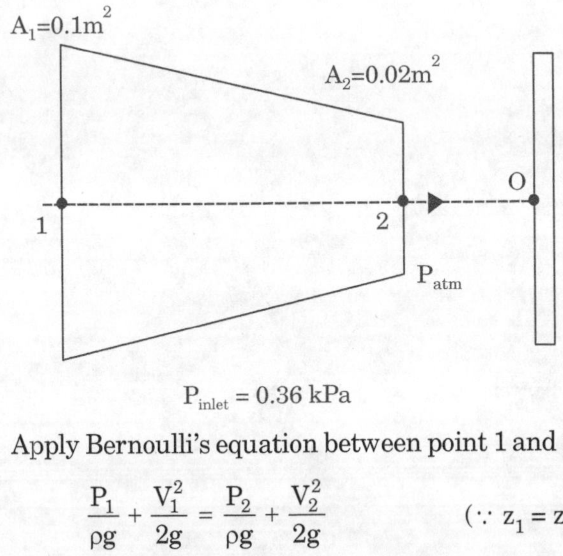

$P_{inlet} = 0.36$ kPa

Apply Bernoulli's equation between point 1 and 2,

$$\frac{P_1}{\rho g} + \frac{V_1^2}{2g} = \frac{P_2}{\rho g} + \frac{V_2^2}{2g} \qquad (\because z_1 = z_2)$$

Here, $P_1 = 0.36$ kPa (gauge)

$P_{atm} = 0$ (gauge)

We get,

$$V_2^2 - V_1^2 = \frac{2P_1}{\rho} = \frac{2 \times 0.36 \times 1000}{1.2} = 600 \qquad \text{...(i)}$$

From continuity equation,

$$A_1V_1 = A_2V_2$$

$$\Rightarrow \quad 0.1 \times V_1 = 0.02 \times V_2$$

$$\Rightarrow \quad \frac{V_2}{V_1} = 5 \qquad \text{...(ii)}$$

Solving both equations (i) and (ii),

We get, $V_2 = 25$ m/s

∵ The pressure betwen 2 and 0 is atmospheric. So, the only pressure at point O will be dynamic pressure due to air which is equal to $\frac{1}{2}\rho V^2$.

so, Gauge pressure at point O on the plate,

$$P_o = \frac{1}{2}\rho V^2 = \frac{1}{2} \times \frac{1.2 \times (25)^2}{1000}$$

$P_o = 0.375$ kPa ≈ 0.38 kPa (Gauge)

51.

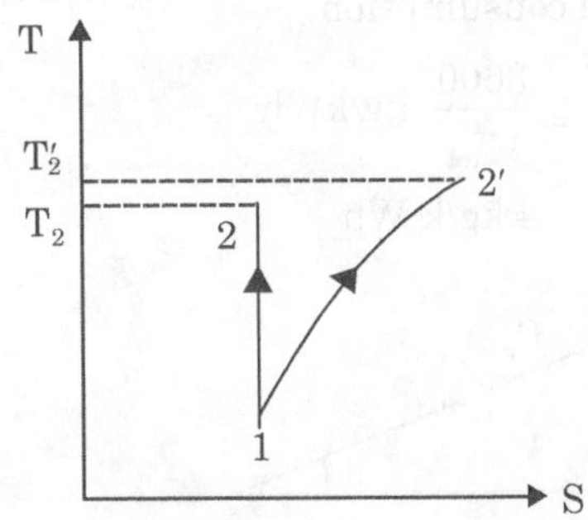

$T_1 = 310$ K

$\frac{P_2}{P_1} = 6$

$C_p = 1005$ J/kgK

$\eta_c = 85\%$

$\gamma = 1.4$

Process (1-2) (Isentropic compression)

$$T_2 = T_1\left(\frac{P_2}{P_1}\right)^{\frac{\gamma-1}{\gamma}}$$

$$T_2 = 310 \times (6)^{\frac{1.4-1}{1.4}} = 517.24 \text{ K}$$

Isentropic efficiency

$$\eta_C = \frac{T_2 - T_1}{T_2' - T_1}$$

$$0.85 = \frac{517.24 - 310}{T_2' - 310}$$

$T'_2 = 553.81 \text{ K}$

Inrease in enthalpy

$$\Delta h = h_2 - h_1 = C_P (T'_2 - T_1)$$
$$= 1.005\,(553.81 - 310)$$
$$= 245.03 \text{ kJ/kg} = 245 \text{ kJ/kg}$$

52. $m = 1$ kg

$R = 287$ J/kgK

$T_1 = 127°C = 127 + 273 = 400$ K

$P_1 = P_2 = P$

$V_1 = V$

$V_2 = 2V$

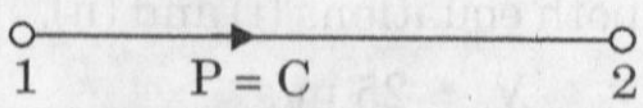

Work done, $w = P\,(V_2 - V_1)$

$$= P\,(2V - V)$$
$$= PV = mRT$$
$$= 1 \times 0.287 \times 400 = 114.80 \text{ kJ}$$

53. Turbine work, $w_T = 903$ kJ/kg

Pump work, $w_P = 3$ kJ/kg

$$w_{net} = w_T - w_P$$
$$= 903 - 3 = 900 \text{ kJ/kg}$$

Specific steam consumption

$$= \frac{3600}{w_{net}} \text{ kg/kWh}$$
$$= 4 \text{ kg/k Wh}$$

54.

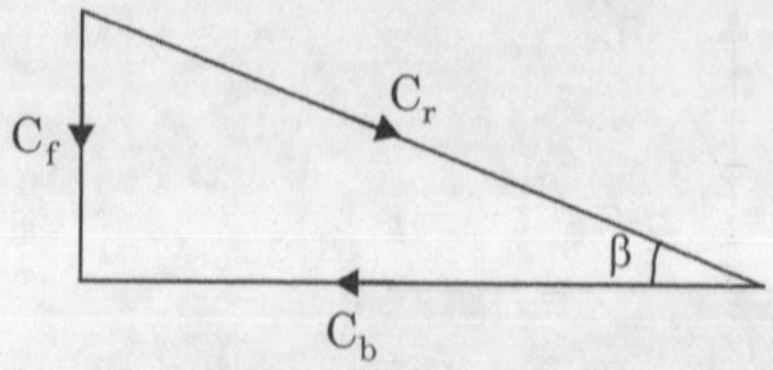

Blade velocity (C_b) at the given section

$$C_b = \frac{\pi DN}{60} = \frac{\pi \times 3 \times 300}{60} = 47.12 \text{ m/s}$$

Discharge (Q) = flow area (A_f) × flow velocity (C_f)

$$100 = \frac{\pi}{4} \times \left(D_t^2 - D_h^2\right) \times C_f$$
$$C_f = 10.61 \text{ m/s}$$

Now, $$\beta = \tan^{-1}\left(\frac{C_f}{C_b}\right)$$
$$= \tan^{-1}\left(\frac{10.61}{47.12}\right) = 12.69°$$

55. Compression ratio, $r = 8$

Het supplied, $\dot{Q}_s = 10$ kW

Mechanical efficiency, $\eta_m = 80\%$

$\because$ Efficiency of otto cycle, $\eta = 1 - \dfrac{1}{r^{\gamma-1}}$

We also know that, $\eta = \dfrac{\text{work output}}{\text{heat supplied}}$

Here, work output is indicated power, I.P. (given in the question)

so, $$\frac{I.P.}{\dot{Q}_s} = 1 - \frac{1}{r^{\gamma-1}}$$

$$\Rightarrow \quad \frac{I.P.}{10} = 1 - \frac{1}{(8)^{1.4-1}}$$

$$\Rightarrow \quad IP. = 5.65 \text{ kW}$$

Now, mechanial efficiency,

$$\Rightarrow \quad 0.8 = \frac{B.P.}{5.65}$$
$$B.P. = 4.5 \text{ kW}$$

SOLVED PAPER - 2020
SET 2

INSTRUCTIONS

1. Total of 65 questions carrying 100 marks, out of which 10 questions carrying a total of 15 marks are in General Aptitude (GA)
2. The Engineering Mathematics will carry around **15% of the total marks,** the General Aptitude section will carry **15% of the total marks** and the **remaining 70% of the total marks.**
3. **Types of Questions**
   (*a*) **Multiple Choice Questions (MCQ)** carrying 1 or 2 marks each in all papers and sections. These questions are objective in nature, and each will have a choice of four options, out of which the candidate has to mark the correct answer(s).
   (*b*) **Numerical Answer Questions** of 1 or 2 marks each in all papers and sections. For these questions the answer is a real number, to be entered by the candidate using the virtual keypad. No choices will be shown for these type of questions.
4. For **1-mark** multiple-choice questions, **1/3 marks** will be deducted for a wrong answer. Likewise, for **2-marks** multiple-choice questions, **2/3** marks will be deducted for a wrong answer. There is no negative marking for numerical answer type questions.

Chapter-Wise Analysis

#	Subject	Marks		
		1	2	Total
1	General Aptitude	5	5	15
2	Engineering Mathematics	5	4	13
3	Engineering Mechanics	1	1	3
4	Strength of Materials	1	3	7
5	Theory of Machines	3	2	7
6	Design of Machine Elements	1	2	5
7	Fluid Mechanics	2	3	8
8	Heat Transfer	1	2	5
9	Thermodynamics, IC Engines, Refrigeration & Power Plant Engineering Air-Conditioning	4	5	14
10	Material Manufacturing & Industrial Engineering	7	8	23

GENERAL APTITUDE

(Q1 – Q5) carry one mark each.

1. While I agree_____ his proposal this time, I do _____ not often agree_____ him.

(a) to, with (b) with, to

(c) with, with (d) to, to

2. The recent measures to improve the output would _____ the level of production to our satisfaction.

(a) increase (b) decrease

(c) speed (d) equalise

3. Select the word that fits the analogy :

White : Whitening : : Light : _____

(a) Lightning (b) Lightening

(c) Lighting (d) Enlightening

4. In one of the greatest innings ever seen in 142 years of Test history, Ben Stokes upped the tempo in a five-and-a-half hour long stay of 219 balls including 11 fours and 8 sixes that saw him finish on a 135 not out as England squared the five-match series.

Based on their connotations in the given passage, which one of the following meanings DOES NOT match?

(a) upped = increased

(b) squared = lost

(c) tempo = enthusiasm

(d) saw = resulted in

5. There are five levels (P, Q, R, S, T), in a linear supply chain before a product reaches customers, as shown in figure.

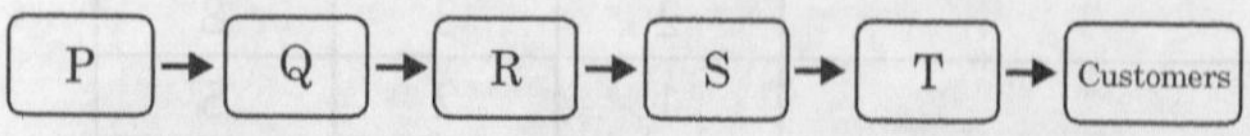

At each of the five levels, the price of the product is increased by 25%. If the product is produced at level P at the cost of Rs. 120 per unit, what is the price paid (in rupees) by the customers?

(a) 187.50 (b) 234.38

(c) 292.96 (d) 366.21

(Q6 – Q10) carry two marks each.

6. Climate change and resilience deal with two aspects - reduction of sources of non-renewable energy resources and reducing vulnerability of climate change aspects. The terms 'mitigation' and 'adaptation' are used to refer to these aspects, respectively.

Which of the following assertions is best supported by the above information?

(a) Mitigation deals with consequences of climate change.

(b) Adaptation deals with causes of climate change.

(c) Mitigation deals with actions taken to reduce the use of fossil fuels.

(d) Adaptation deals with actions taken to combat green-house gas emissions.

7. Find the missing element in the following figure.

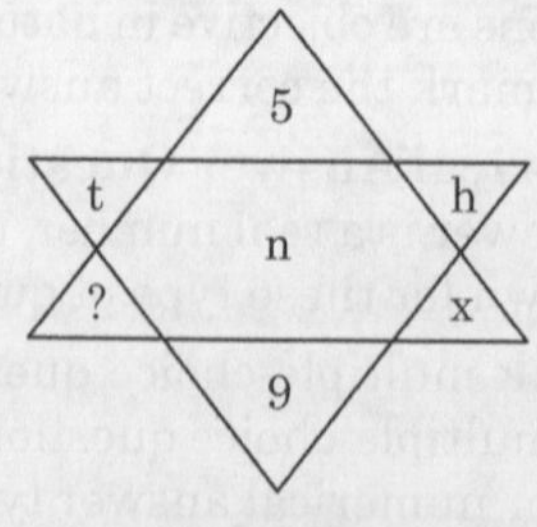

(a) d (b) e

(c) w (d) y

8. It was estimated that 52 men can complete a strip in a newly constructed highway connecting cities P and Q in 10 days. Due to an emergency, 12 men were sent to another project. How many number of days, more than the original estimate, will be required to complete the strip?

(a) 3 days (b) 5 days

(c) 10 days (d) 13 days

9. An engineer measures THREE quantities X, Y and Z in an experiment. She finds that they follow a relationship that is represented in the figure below: (the product of X and Y linearly varies with Z)

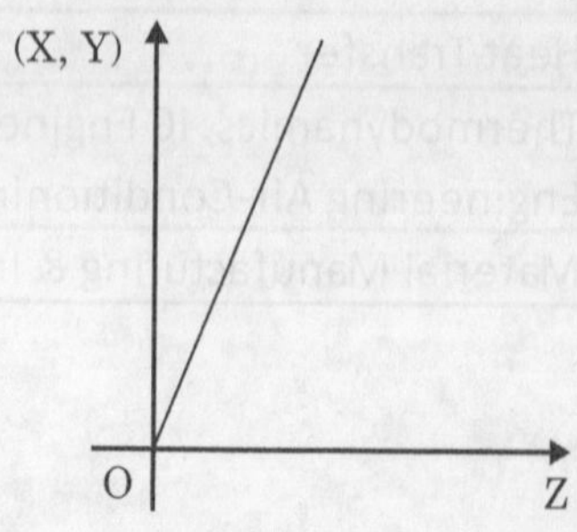

Then, which of the following statements is FALSE?

(a) For fixed Z; X is proportional to Y

(b) For fixed Y; X is proportional to Z

(c) For fixed X; Z is proportional to Y

(d) XY/Z is constant

10. The two pie-charts given below show the data of total students and only girls registered in different streams in a university. If the total number of students registered in the university is 5000, and the total number of the registered girls is 1500; then, the ratio of boys enrolled in Arts to the girls enrolled in Management is ______.

Percentage of students enrolled in different streams in a University

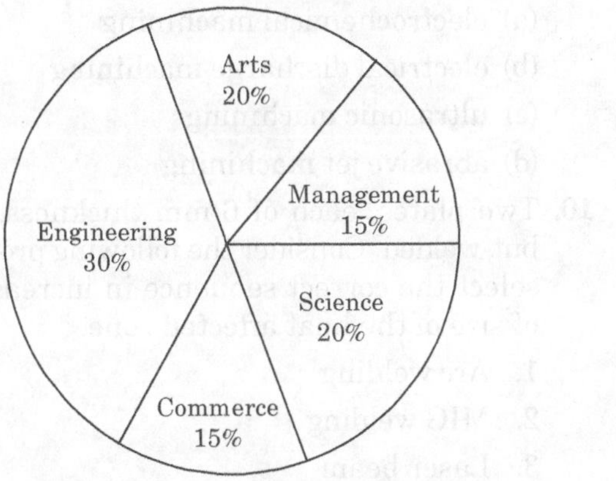

Percentage of girls enrolled in different streams

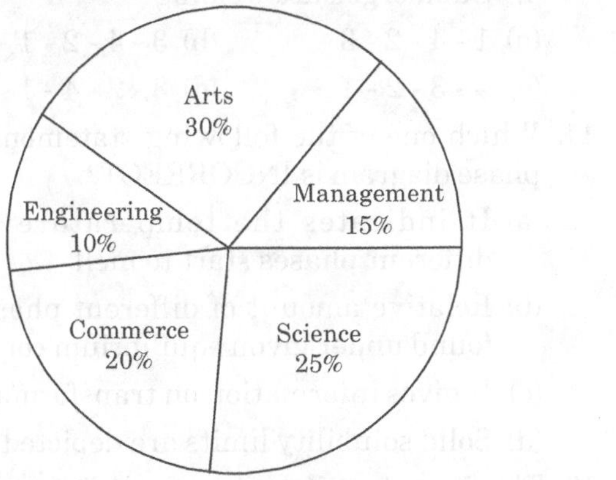

(a) 2 : 1 (b) 9 : 22

(c) 11 : 9 (d) 22 : 9

MECHANICAL ENGINEERING

(Q1 – Q25) carry one mark each.

1. The sum of two normally distributed random variables X and Y is

(a) always normally distributed

(b) normally distributed, only if X and Y are independent

(c) normally distributed, only if X and Y have the same standard deviation

(d) normally distributed, only if X and Y have the same mean

2. A matrix P is decomposed into its symmetric part S and skew symmetric part V. If

$$S = \begin{pmatrix} -4 & 4 & 2 \\ 4 & 3 & 7/2 \\ 2 & 7/2 & 2 \end{pmatrix}, V = \begin{pmatrix} 0 & -2 & 3 \\ 2 & 0 & 7/2 \\ -3 & -7/2 & 0 \end{pmatrix}$$

then matrix P is

(a) $\begin{pmatrix} -4 & 6 & -1 \\ 2 & 3 & 0 \\ 5 & 7 & 2 \end{pmatrix}$ (b) $\begin{pmatrix} -4 & 2 & 5 \\ 6 & 3 & 7 \\ -1 & 0 & 2 \end{pmatrix}$

(c) $\begin{pmatrix} 4 & -6 & 1 \\ -2 & -3 & 0 \\ -5 & -7 & -2 \end{pmatrix}$ (d) $\begin{pmatrix} -2 & 9/2 & -1 \\ -1 & 81/4 & 11 \\ -2 & 45/2 & 73/4 \end{pmatrix}$

3. Let $I = \int_{x=0}^{1} \int_{y=0}^{x^2} xy^2 dy\, dx$. Then, I may also be expressed as

(a) $\int_{y=0}^{1} \int_{x=0}^{\sqrt{y}} xy^2 dx\, dy$ (b) $\int_{y=0}^{1} \int_{x=\sqrt{y}}^{1} yx^2 dx\, dy$

(c) $\int_{y=0}^{1} \int_{x=\sqrt{y}}^{1} xy^2 dx\, dy$ (d) $\int_{y=0}^{1} \int_{x=0}^{\sqrt{y}} yx^2 dy\, dx$

4. The solution of

$$\frac{d^2y}{dt^2} - y = 1,$$

which additionally satisfies $y\Big|_{t=0} = \frac{dy}{dt}\Big|_{t=0} = 0$ in the Laplace s-domain is

(a) $\frac{1}{s(s+1)(s-1)}$ (b) $\frac{1}{s(s+1)}$

(c) $\frac{1}{s(s-1)}$ (d) $\frac{1}{s-1}$

5. An attempt is made to pull a roller of weight W over a curb (step) by applying a horizontal force F as shown in the figure.

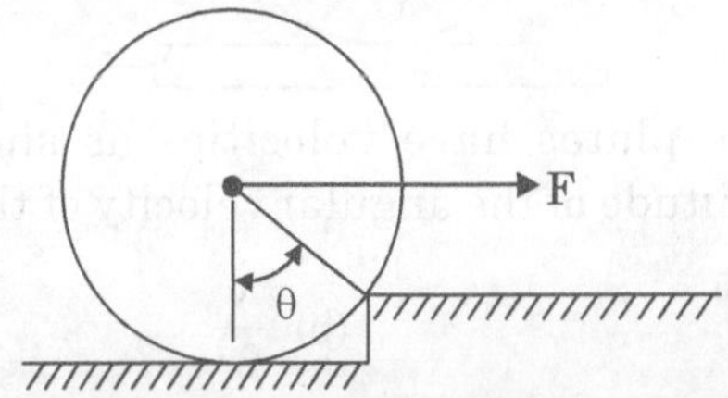

The coefficient of static friction between the roller and the ground (including the edge of the step) is μ. Identify the correct free body diagram (FBD) of the roller when the roller is just about to climb over the step.

(a)

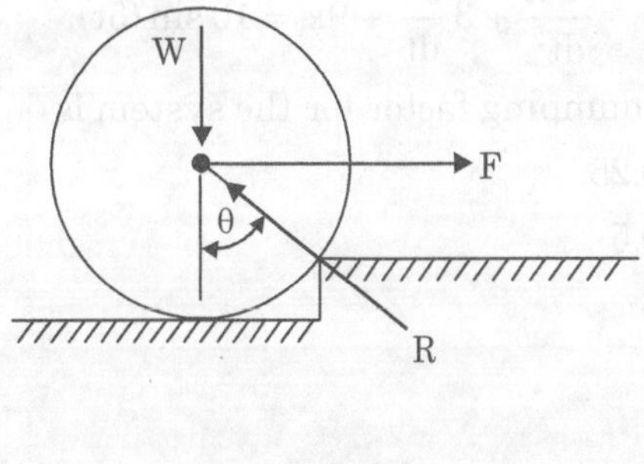

(b) 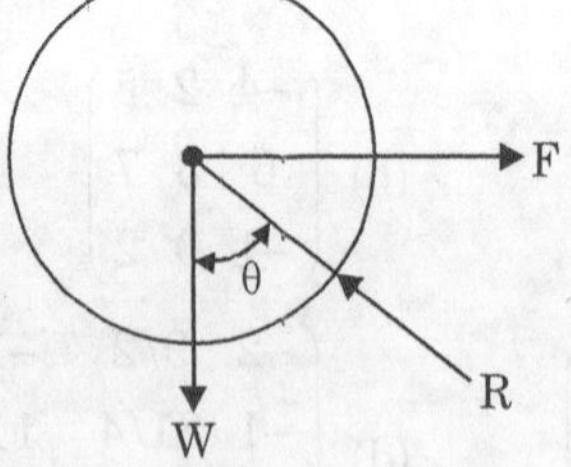

(c) 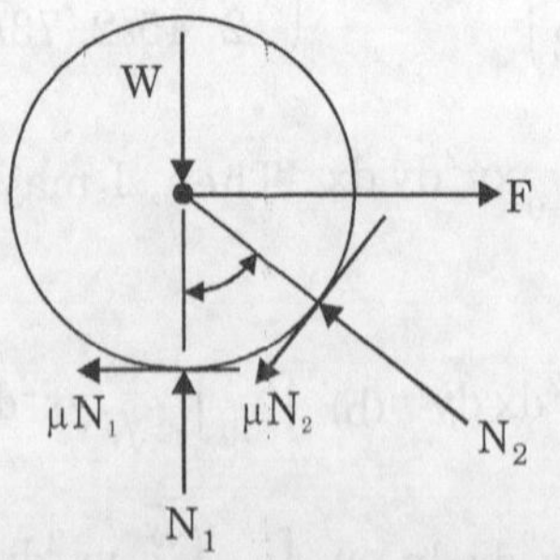

(d) 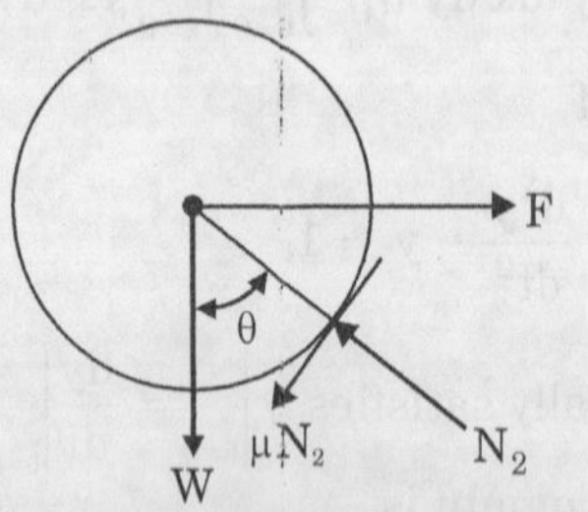

6. A circular disk of radius r is confined to roll without slipping at P and Q as shown in the figure.

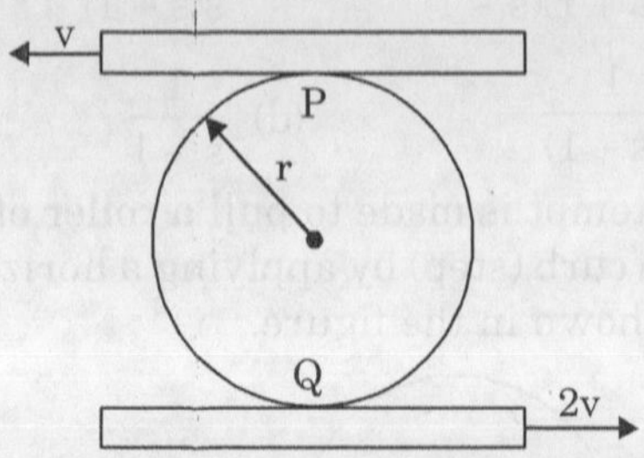

If the plates have velocities as shown, the magnitude of the angular velocity of the disk is

(a) $\frac{v}{r}$ (b) $\frac{v}{2r}$

(c) $\frac{2v}{3r}$ (d) $\frac{3v}{2r}$

7. The equation of motion of a spring-mass-damper system is given by

$$\frac{d^2x}{dt^2} + 3\frac{dx}{dt} + 9x = 10\sin(5t)$$

The damping factor for the system is

(a) 0.25

(b) 0.5

(c) 2

(d) 3

8. The number of qualitatively distinct kinematic inversions possible for a Grashof chain with four revolute pairs is

(a) 1 (b) 2

(c) 3 (d) 4

9. The process, that uses a tapered horn to amplify and focus the mechanical energy for machining of glass, is

(a) electrochemical machining

(b) electrical discharge machining

(c) ultrasonic machining

(d) abrasive jet machining

10. Two plates, each of 6 mm thickness, are to be but-welded. Consider the following processes and select the correct sequence in increasing order of size of the heat affected zone.

1. Arc welding
2. MIG welding
3. Laser beam
4. Submerged arc welding

(a) 1 - 4 - 2 - 3 (b) 3 - 4 - 2 - 1

(c) 4 - 3 - 2 - 1 (d) 3 - 2 - 4 - 1

11. Which one of the following statements about a phase diagram is INCORRECT?

(a) It indicates the temperature at which different phases start to melt

(b) Relative amount of different phases can be found under given equilibrium conditions

(c) It gives information on transformation rates

(d) Solid solubility limits are depicted by it

12. The figure below shows a symbolic representation of the surface texture in a perpendicular lay orientation with indicative values (I through VI) marking the various specifications whose definitions are listed below.

P : Maximum Waviness Height (mm);

Q : Maximum Roughness Height (mm);

R : Minimum Roughness Height (mm);

S : Maximum Waviness Width (mm);

T : Maximum Roughness Width (mm);

U : Roughness Width Cutoff (mm),

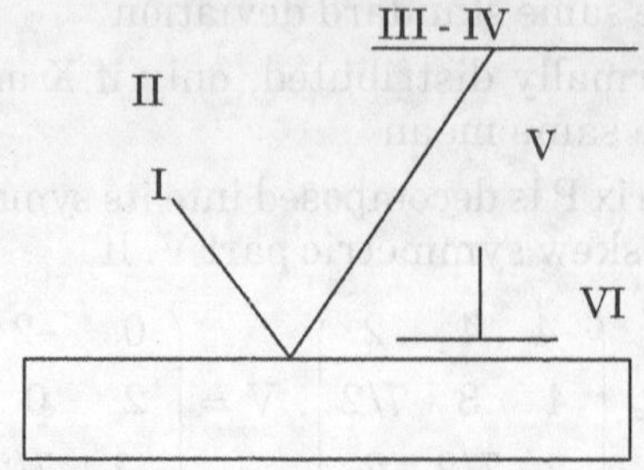

The correct match between the specifications and the symbols (I to VI) s

(a) I-R, II-Q, III-P, IV-S, V-U, VI-T
(b) I-R, II-P, III-U, IV-S, V-T, VI-Q
(c) I-U, II-S, III-Q, IV-T, V-R, VI-P
(d) I-Q, II-U, III-R, IV-T, V-S, VI-P

13. In Materials Requirement Planning, if the inventory holding cost is very high and the setup cost is zero, which one of the following lot sizing approaches should be used?

(a) Economic Order Quantity
(b) Lot-for-Lot
(c) Base Stock Level
(d) Fixed Period Quantity, for 2 periods

14. Which of the following conditions is used to determine the stable equilibrium of all partially submerged floating bodies?

(a) Centre of buoyancy must be above the centre of gravity
(b) Centre of buoyancy must be below the centre of gravity
(c) Metacentre must be at a higher level than the centre of gravity
(d) Metacentre must be at a lower level than the centre of gravity

15. In the space above the mercury column in a barometer tube, the gauge pressure of the vapour is

(a) positive, but more than one atmosphere
(b) negative
(c) zero
(d) positive, but less than one atmosphere

16. A closed vessel contains pure water, in thermal equilibrium with its vapour at 25°C (Stage # 1), as shown.

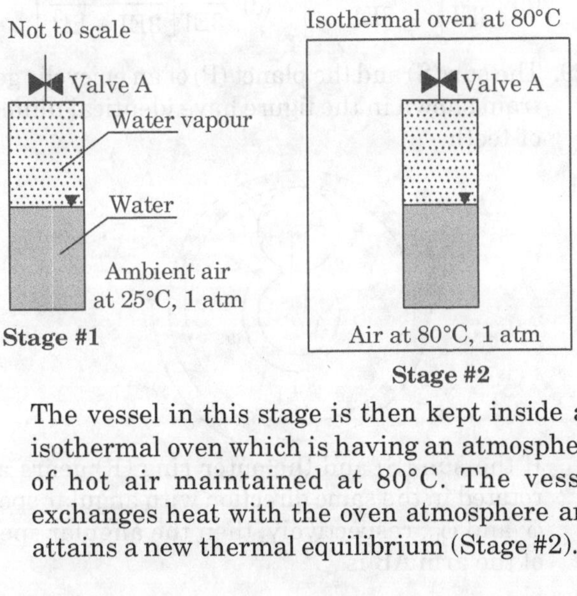

The vessel in this stage is then kept inside an isothermal oven which is having an atmosphere of hot air maintained at 80°C. The vessel exchanges heat with the oven atmosphere and attains a new thermal equilibrium (Stage #2). If the Valve A is now opened inside the oven, what will happen immediately after opening the valve?

(a) Water vapor inside the vessel will come out of the Valve A
(b) Hot air will go inside the vessel through Valve A
(c) Nothing will happen - the vessel will continue to remain in equilibrium
(d) All the vapor inside the vessel will immediately condense

17. For an air-standard Diesel cycle,

(a) heat addition is at constant volume and heat rejection is at constant pressure
(b) heat addition is at constant pressure and heat rejection is at constant pressure
(c) heat addition is at constant pressure and heat rejection is at constant volume
(d) heat addition is at constant volume and heat rejection is at constant volume

18. The values of enthalpies at the stator inlet and rotor outlet of a hydraulic turbomachine stage are h_1 and h_3 respectively. The enthalpy at the stator outlet (or, rotor inlet) is h_2. The condition $(h_2 - h_1) = (h_3 - h_2)$ indicates that the degree of reaction of this stage is

(a) zero (b) 50%
(c) 75% (d) 10%

19. Let I be a 100 dimensional identity matrix and E be the set of its distinct (no value appears more than once in E)real eigenvalues. The number of elements in E is ________.

20. A beam of negligible mass is hinged at support P and has a roller support Q as shown in the figure.

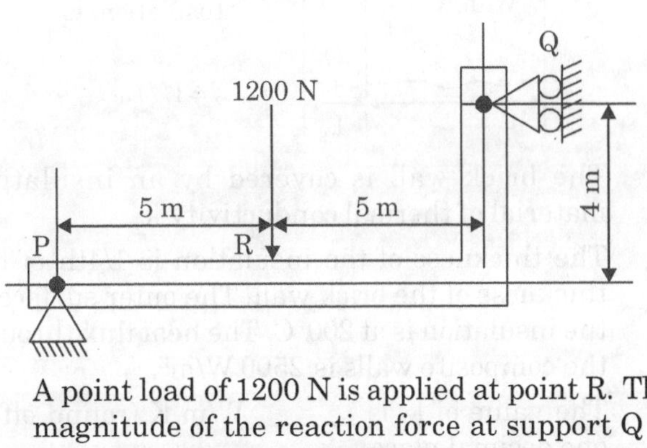

A point load of 1200 N is applied at point R. The magnitude of the reaction force at support Q is ________ N.

21. A machine member is subjected to fluctuating stress $\sigma = \sigma_0 \cos(8\pi t)$. The endurance limit of the material is 350 MPa. If the factor of safety used in the design is 3.5 then the maximum allowable value of σ_0 is ________ MPa (round off to 2 decimal places).

22. A bolt head has to be made at the end of a rod of diameter d = 12 mm by localized forging (upsetting) operation. The length of the unsupported portion of the rod is 40 mm. To avoid buckling of the rod, a closed forging operation has to be performed with a maximum die diameter of ______ mm.

23. Consider the following network of activities, with each activity named A-L, illustrated in the nodes of the network.

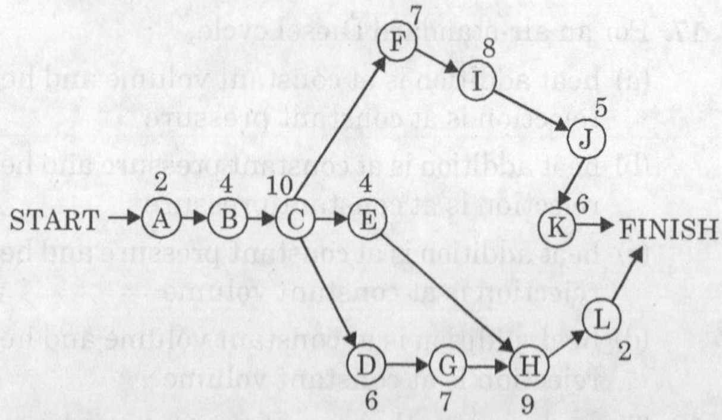

The number of hours required for each activity is shown alongside the nodes. The slack on the activity L, is ______ hours.

24. In a furnace, the inner and outer sides of the brick wall (k_1 = 2.5 W/m.K) are maintained at 1100°C and 700°C, respectively as shown in figure.

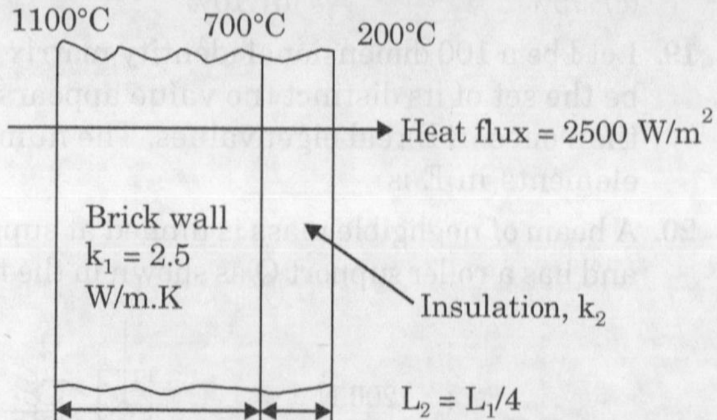

The brick wall is covered by an insulating material of thermal conductivity k_2.

The thickness of the insulation is 1/4th of the thickness of the brick wall. The outer surface of the insulation is at 200°C. The heat flux through the composite walls is 2500 W/m^2.

The value of k_2 is ______ W/m.K (round off to one decimal place).

25. If a reversed Carnot cycle operates between the temperature limits of 27°C and –3°C, then the ratio of the COP of a refrigerator to that of a heat pump (COP of refrigerator / COP of heat pump)based on the cycle is ______ (round off to 2 decimal places).

(Q26 – Q55) carry two marks each.

26. The directional derivative of f (x, y, z) = xyz at point (–1, 1, 3) in the direction of vector $\hat{i} - 2\hat{j} + 2\hat{k}$ is

(a) $3\hat{i} - 3\hat{j} - \hat{k}$ (b) $-\frac{7}{3}$

(c) $\frac{7}{3}$ (d) 7

27. The function f(z) of complex variable z = x + iy, where $i = \sqrt{-1}$, is given as $f(z) = (x^3 - 3xy^2) + i\,v(x, y)$. For this function to be analytic, v(x, y) should be

(a) $(3xy^2 - y^3)$ + constant

(b) $(3x^2y^2 - y^3)$+ constant

(c) $(x^3 - 3x^2y)$ + constant

(d) $(3x^2y - y^3)$ + constant

28. A cantilever of length l, and flexural rigidity EI, stiffened by a spring of stiffness k, is loaded by a transverse force P, as shown.

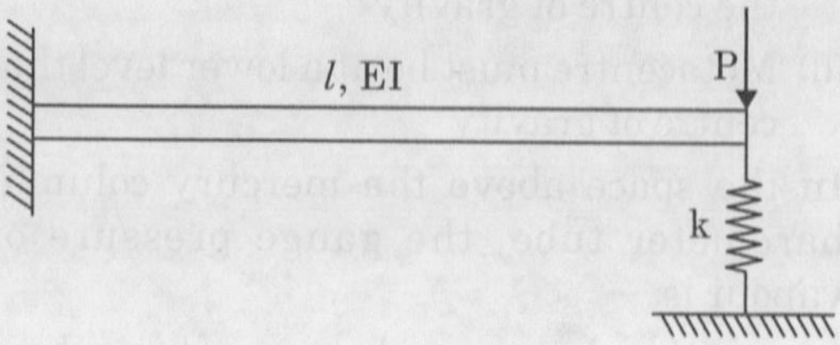

The transverse deflection under the load is

(a) $\frac{Pl^3}{3EI}\left[\frac{3EI}{3EI + 2kl^3}\right]$ (b) $\frac{Pl^3}{3EI}\left[\frac{6EI - kl^3}{6EI}\right]$

(c) $\frac{Pl^3}{3EI}\left[\frac{3EI - kl^3}{3EI}\right]$ (d) $\frac{Pl^3}{3EI}\left[\frac{3EI}{3EI + kl^3}\right]$

29. The sun (S) and the planet (P) of an epicyclic gear train shown in the figure have identical number of teeth.

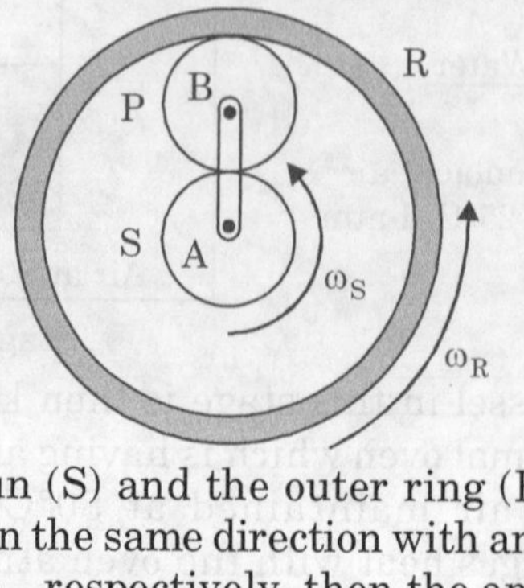

If the sun (S) and the outer ring (R) gears are rotated in the same direction with angular speed ω_s and ω_R, respectively, then the angular speed of the arm AB is

(a) $\frac{3}{4}\omega_R + \frac{1}{4}\omega_S$ (b) $\frac{1}{4}\omega_R + \frac{3}{4}\omega_S$

(c) $\frac{1}{2}\omega_R - \frac{1}{2}\omega_S$ (d) $\frac{3}{4}\omega_R - \frac{4}{4}\omega_S$

30. A thin-walled cylinder of radius r and thickness t is open at both ends, and fits snugly between two rigid walls under ambient conditions, as shown in the figure.

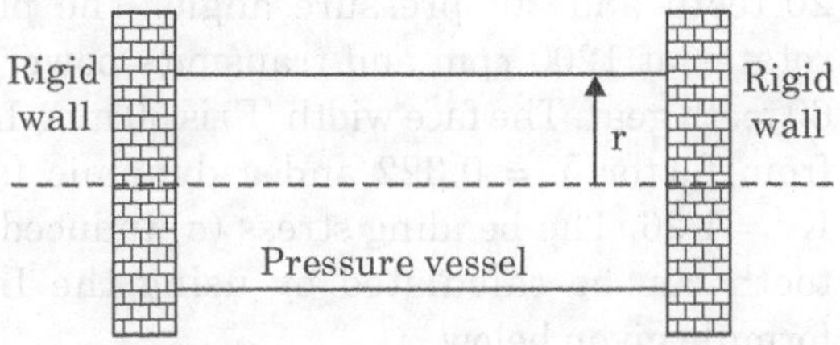

The material of the cylinder has Young's modulus E, Poisson's ratio ν, and coefficient of thermal expansion α. What is the minimum rise in temperature ΔT of the cylinder (assume uniform cylinder temperature with no buckling of the cylinder) required to prevent gas leakage if the cylinder has to store the gas at an internal pressure of p above the atmosphere?

(a) $\Delta T = \frac{3\nu pr}{2\alpha tE}$ (b) $\Delta T = \left(\nu - \frac{1}{4}\right)\frac{pr}{\alpha tE}$

(c) $\Delta T = \frac{\nu pr}{\alpha tE}$ (d) $\Delta T = \left(\nu + \frac{1}{2}\right)\frac{pr}{\alpha tE}$

31. A helical spring has spring constant k. If the wire diameter, spring diameter and the number of coils are all doubled then the spring constant of the new spring becomes

(a) k/2 (b) k

(c) 8 k (d) 16 k

32. Two rollers of diameters D_1 (in mm) and D_2 (in mm) are used to measure the internal taper angle in the V-groove of a machined component. The heights H_1 (in mm) and H_2 (in mm) are measured by using a height gauge after inserting the rollers into the same V-groove as shown in the figure.

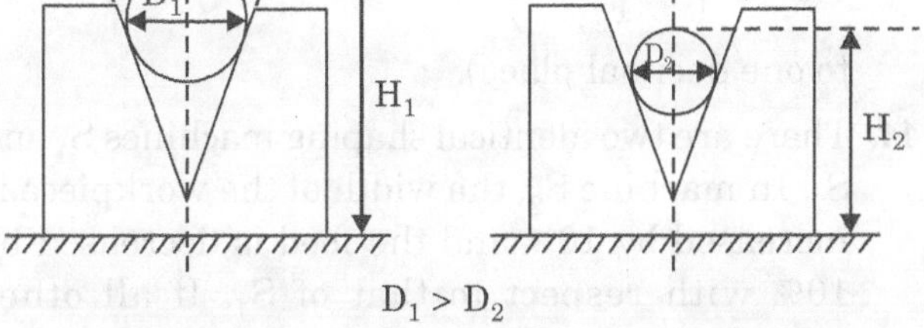

$D_1 > D_2$

Which one of the following is the correct relationship to evaluate the angle α as shown in the figure?

(a) $\sin\alpha = \frac{(D_1 - D_2)}{2(H_1 - H_2) - (D_1 - D_2)}$

(b) $\cos\alpha = \frac{(D_1 - D_2)}{2(H_1 - H_2) - 2(D_1 - D_2)}$

(c) $\operatorname{cosec}\alpha = \frac{(H_1 - H_2) - (D_1 - D_2)}{2(D_1 - D_2)}$

(d) $\sin\alpha = \frac{(H_1 - H_2)}{(D_1 - D_2)}$

33. The forecast for the monthly demand of a product is given in the table below.

Month	Forecast	Actual Sales
1	32.00	30.00
2	31.80	32.00
3	31.82	30.00

The forecast is made by using the exponential smoothing method. The exponential smoothing coefficient used in forecasting the demand is

(a) 0.10 (b) 0.40

(c) 0.50 (d) 1.00

34. One kg of air in a closed system undergoes an irreversible process from an initial state of p_1 = 1 bar (absolute) and T_1 = 27°C, to a final state of p_2 = 3 bar (absolute) and T_2 = 127°C. If the gas constant of air is 287 J/kg.K and the ratio of the specific heats g = 1.4, then the change in the specific entropy (in J/kg.K) of the air in the process is

(a) – 26.3

(b) 28.4

(c) 172.0

(d) indeterminate, as the process is irreversible

35. For the integral $\int_0^{\pi/2} (8 + 4\cos x)\,dx$, the absolute percentage error in numerical evaluation with the Trapezoidal rule, using only the end points, is _______ (round off to one decimal place).

36. A fair coin is tossed 20 times. The probability that 'head' will appear exactly 4 times in the first ten tosses, and 'tail' will appear exactly 4 times in the next ten tosses is _______ (round off to 3 decimal places).

37. A hollow spherical ball of radius 20 cm floats in still water, with half of its volume submerged. Taking the density of water as 1000 kg/m^3, and the acceleration due to gravity as 10 m/s^2, the natural frequency of small oscillations of the ball, normal to the water surface is _____ radians/s (round off to 2 decimal places).

38. Uniaxial compression test data for a solid metal bar of length 1 m is shown in the figure.

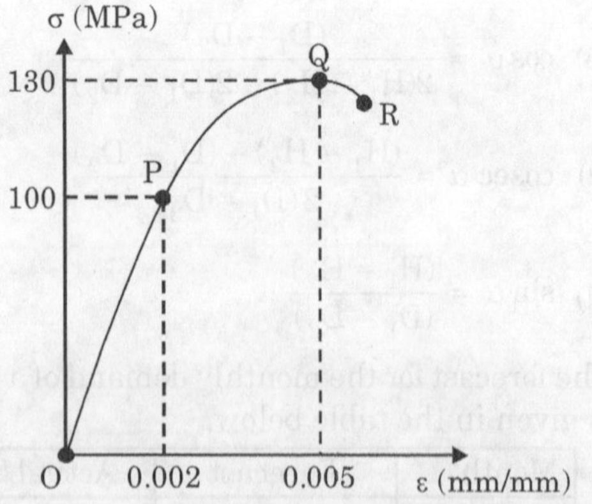

The bar material has a linear elastic response from O to P followed by a nonlinear response. The point P represents the yield point of the material. The rod is pinned at both the ends. The minimum diameter of the bar so that it does not buckle under axial loading before reaching the yield point is ________ mm (round off to one decimal place).

39. The turning moment diagram of a flywheel fitted to a fictitious engine is shown in the figure.

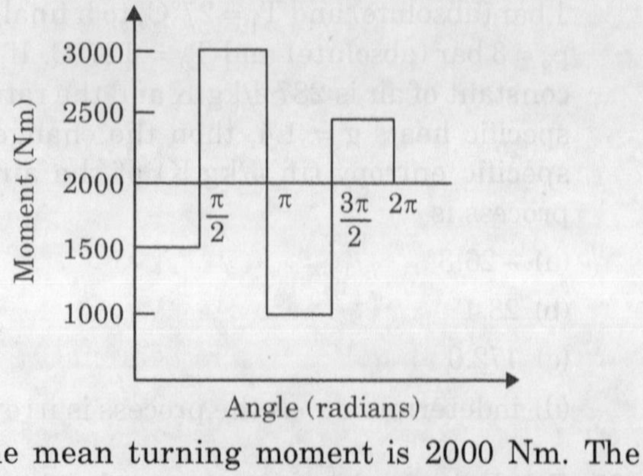

The mean turning moment is 2000 Nm. The average engine speed is 1000 rpm. For fluctuation in the speed to be within ± 2% of the average speed, the mass moment of inertia of the flywheel is ________ kg.m^2.

40. A rigid block of mass m_1 = 10 kg having velocity v_0 = 2 m/s strikes a stationary block of mass m_2 = 30 kg after traveling 1 m along a frictionless horizontal surface as shown in the figure.

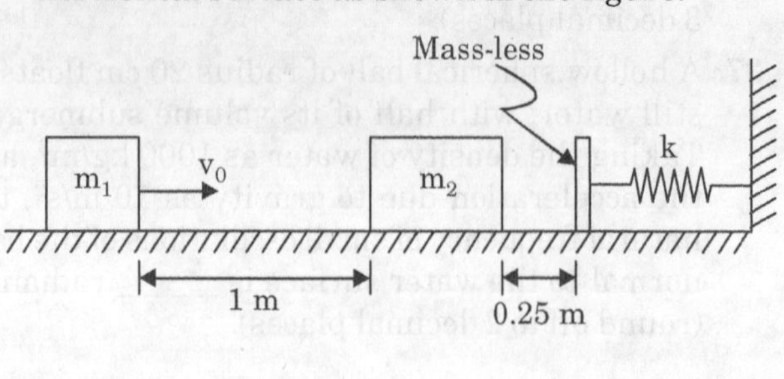

The two masses stick together and jointly move by a distance of 0.25 m further along the same frictionless surface, before they touch the mass-less buffer that is connected to the rigid vertical wall by means of a linear spring having a spring constant k = 10^5 N/m. The maximum deflection of the spring is __________ cm (round off to 2 decimal places).

41. A steel spur pinion has a module (m) of 1.25 mm, 20 teeth and 20° pressure angle. The pinion rotates at 1200 rpm and transmits power to a 60 teeth gear. The face width (F) is 50 mm, Lewis from factor Y = 0.322 and a dynamic factor Kv = 1.26. The bending stress (σ) induced in a tooth can be calculated by using the Lewis formula given below.

If the maximum bending stress experienced by the pinion is 400 MPa, the power transmitted is ________ kW (round off to one decimal place).

Lewis formula : $\sigma = \dfrac{K_v W^t}{FmY}$, where W^t is the tangential load acting on the pinion.

42. A mould cavity of 1200 cm^3 volume has to be filled through a sprue of 10 cm length feeding a horizontal runner. Cross-sectional area at the base of the sprue is 2 cm^2. Consider acceleration due to gravity as 9.81 m/s^2. Neglecting frictional losses due to molten metal flow, the time taken to fill the mould cavity is ____ seconds (round off to 2 decimal places).

43. A cylindrical bar with 200 mm diameter is being turned with a tool having geometry 0° - 9° - 7° - 8° - 15° - 30° - 0.05 inch (Coordinate system, ASA) resulting in a cutting force F_{c1}. If the tool geometry is changed to 0° - 9° - 7° - 8° - 15° - 0° - 0.05 inch (Coordinate system, ASA) and all other parameters remain unchanged, the cutting force changes to F_{c2}. Specific cutting energy (in J/mm^3) is $U_C = U_0 (t_1)^{-0.4}$, where U_0 is the specific energy coefficient, and t_1 is the uncut thickness in mm. The value of percentage change in cutting force F_{c2}, i.e. $\left(\dfrac{F_{c_2} - F_{c_1}}{F_{c_1}}\right) \times 100$, is ________ (round off to one decimal place).

44. There are two identical shaping machines S_1 and S_2. In machine S_2, the width of the workpiece is increased by 10% and the feed is decreased by 10% with respect to that of S_1. If all other conditions remain the same then the ratio of total time per pass in S_1 and S_2 will be ______ (round off to one decimal place).

45. Bars of 250 mm length and 25 mm diameter are to be turned on a lathe with a feed of 0.2 mm/rev. Each regrinding of the tool costs Rs. 20. The time required for each tool change is 1 min. Tool life equation is given as $VT^{0.2} = 24$ (where cutting speed V is in m/min and tool life T is in min). The optimum tool cost per piece for maximum production rate is Rs. ________ (round off to 2 decimal places).

46. A point 'P' on a CNC controlled XY-stage is moved to another point 'Q' using the coordinate system shown in the figure below and rapid positioning command (G00).

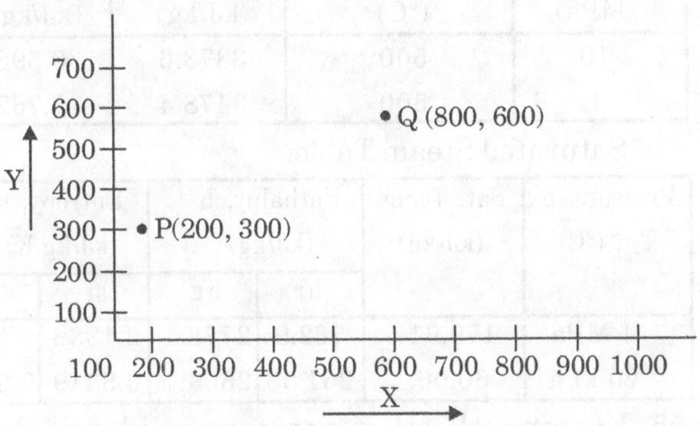

A pair of stepping motors with maximum speed of 800 rpm, controlling both the X and Y motion of the stage, are directly coupled to a pair of lead screws, each with a uniform pitch of 0.5 mm. The time needed to position the point 'P' to the point 'Q' is ________ minutes (round off to 2 decimal places).

47. For a single item inventory system, the demand is continuous, which is 10000 per year. The replenishment is instantaneous and backorders (S units) per cycle are allowed as shown in the figure.

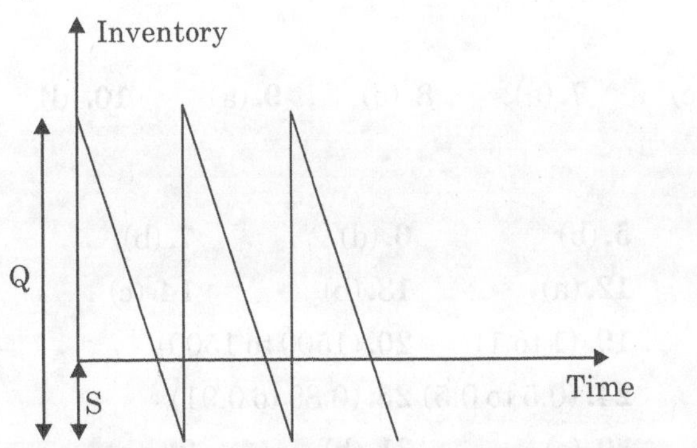

As soon as the quantity (Q units) ordered from the supplier is received, the backordered quantity is issued to the customers. The ordering cost is Rs. 300 per order. The carrying cost is Rs. 4 per unit per year. The cost of backordering is Rs. 25 per unit per year. Based on the total cost minimization criteria, the maximum inventory reached in the system is ________ (round off to nearest integer).

48. Consider a flow through a nozzle, as shown in the figure below.

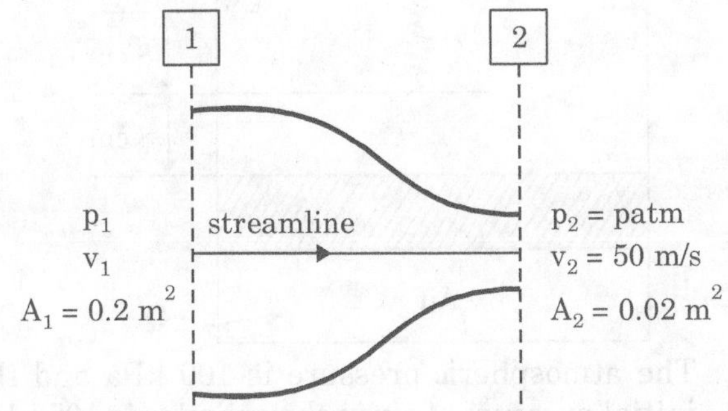

The air flow is steady, incompressible and inviscid. The density of air is 1.23 kg/m^3.

The pressure difference, $(p_1 - p_{atm})$ is ________ kPa (round off to 2 decimal places).

49. Water (density 1000 kg/m^3) flows through an inclined pipe of uniform diameter. The velocity, pressure and elevation at section A are V_A = 3.2 m/s, p_A = 186 kPa and z_A = 24.5 m, respectively, and those at section B are V_B = 3.2 m/s, p_B = 260 kPa and Z_B = 9.1 m, respectively. If acceleration due to gravity is 10 m/s^2 then the head lost due to friction is ________ m (round off to one decimal place).

50. The spectral distribution of radiation from a black body at T_1 = 3000 K has a maximum at wavelength λ_{max}. The body cools down to a temperature T_2. If the wavelength corresponding to the maximum of the spectral distribution at T_2 is 1.2 times of the original wavelength λ_{max}, then the temperature T_2 is ________ K (round off to the nearest integer).

51. Water flows through a tube of 3 cm internal diameter and length 20 m. The outside surface of the tube is heated electrically so that it is subjected to uniform heat flux circumferentially and axially. The mean inlet and exit temperatures of the water are 10°C and 70°C, respectively. The mass flow rate of the water is 720 kg/h. Disregard the thermal resistance of the tube wall. The internal heat transfer coefficient is 1697 W/m^{2}K. Take specific heat C_p of water as 4.179 kJ/kg.K. The inner surface temperature at the exit section of the tube is ________ °C (round off to one decimal place).

52. Air is contained in a frictionless piston-cylinder arrangement as shown in the figure

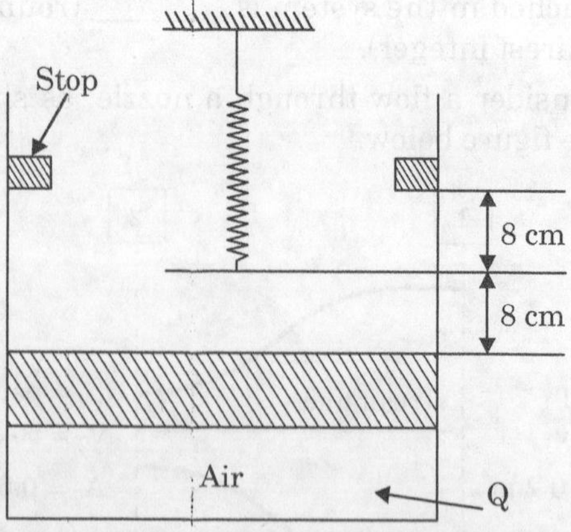

The atmospheric pressure is 100 kPa and the initial pressure of air in the cylinder is 105 kPa. The area of piston is 300 cm^2. Heat is now added and the piston moves slowly from its initial position until it reaches the stops. The spring constant of the linear spring is 12.5 N/mm. Considering the air inside the cylinder as the system, the work interaction is ________ J (round off to the nearest integer).

53. Moist air at 105 kPa, 30°C and 80% relative humidity flows over a cooling coil in an insulated air-conditioning duct. Saturated air exits the duct at 100 kPa and 15°C. The saturation pressures of water at 30°C and 15°C are 4.24 kPa and 1.7 kPa respectively. Molecular weight of water is 18 g/mol and that of air is 28.94 g/ mol. The mass of water condensing out from the duct is g/kg of dry air (round off to the nearest integer).

54. In a steam power plant, superheated steam at 10 MPa and 500°C, is expanded isentropically in a turbine until it becomes a saturated vapour. It is then reheated at constant pressure to 500°C. The steam is next expanded isentropically in another turbine until it reaches the condenser pressure of 20 kPa. Relevant properties of steam are given in the following two tables. The work done by both the turbines together is ________ kJ/kg (round off to the nearest integer).

Superheated Steam Table:

Pressure, p (MPa)	Temperature, T (°C)	Enthapy, h (kJ/kg)	Entropy, s (kJ/kg.K)
10	500	3373.6	6.5965
1	500	3478.4	7.7621

Saturated Steam Table:

Pressure, p T_{sat} (°C)	Sat. Temp. (kJ/kg)	Enthalpy, h (kJ/kg)		Entropy, s (kJ/kg.K)	
		hr	hg	sf	sg
1 MPa	179.91	762.9	2778.1	2.1386	6.5965
20 kPa	60.06	251.38	2609.7	0.8319	7.9085

55. Keeping all other parameters identical, the Compression Ratio (CR) of an air standard diesel cycle is increased from 15 to 21. Take ratio of specific heats = 1.3 and cut-off ratio of the cycle $r_c = 2$.

The difference between the new and the old efficiency values, in percentage,

$(\eta_{new}\,|_{CR=21}) - (\eta_{old}\,|_{Cr=15})$ = % (round off to one decimal place).

ANSWERS

General Aptitude (GA)

1. (a) **2.** (a) **3.** (b) **4.** (b) **5.** (d) **6.** (c) **7.** (a) **8.** (a) **9.** (a) **10.** (d)

Mechanical Engineering

1. (a) **2.** (b) **3.** (c) **4.** (a) **5.** (b) **6.** (d) **7.** (b)
8. (c) **9.** (c) **10.** (d) **11.** (c) **12.** (a) **13.** (b) **14.** (c)
15. (b) **16.** (b) **17.** (c) **18.** (b) **19.** (1 to 1) **20.** (1500 to 1500)
21. (99.98 to 100.02) **22.** (18 to 18) **23.** (2 to 2) **24.** (0.5 to 0.5) **25.** (0.89 to 0.91)
26. (c) **27.** (d) **28.** (d) **29.** (a) **30.** (c) **31.** (b) **32.** (a)
33. (a) **34.** (a) **35.** (5.1 to 5.3) **36.** (0.041 to 0.043) **37.** (8.60 to 8.70)
38. (55.0 to 58.0) **39.** (745 to 755) **40.** (0.98 to 1.02) **41.** (9.8 to 10.8)
42. (4.20 to 4.30) **43.** (– 5.8 to – 5.5) **44.** (0.7 to 0.9) **45.** (26.00 to 28.00)
46. (1.40 to 1.60) **47.** (1130 to 1140) **48.** (1.50 to 1.55) **49.** (7.9 to 8.1)
50. (2499 to 2501) **51.** (85.1 to 86.1) **52.** (543 to 545) **53.** (10 to 10)
54. (1500 to 1525) **55.** (4.6 to 4.9)

EXPLANATIONS

GENERAL APTITUDE

1. "Agree to" is typically used for consent to a request.

   "Agree with" is most often used when two people have the same opinion.

2. The recent measures to improve the output would increase the level of production to our satisfaction.

3. White : Whitening

   Light : Lightening

4. Square means level or draw (in cricket game).

   So, here squared does not mean lost, it means level or draw.

5. Let, product cost at P = x = Rs. 120

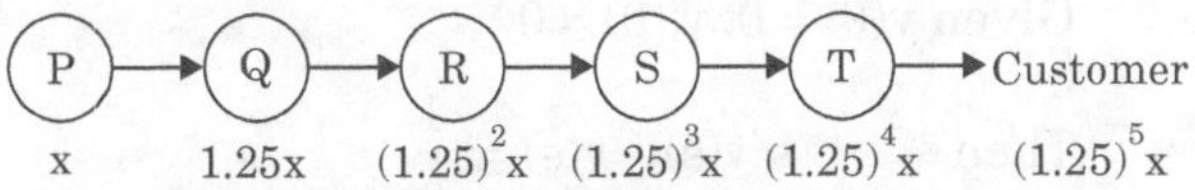

   Price paid by the customer = $(1.25)^5x$

   $= (1.25)^5 \times 120 =$ Rs. 366.21

6. • Mitigation is an intervention to reduce the emissions sources or enhance the sinks of green house gases.

   So, the correct option is (c).

7.

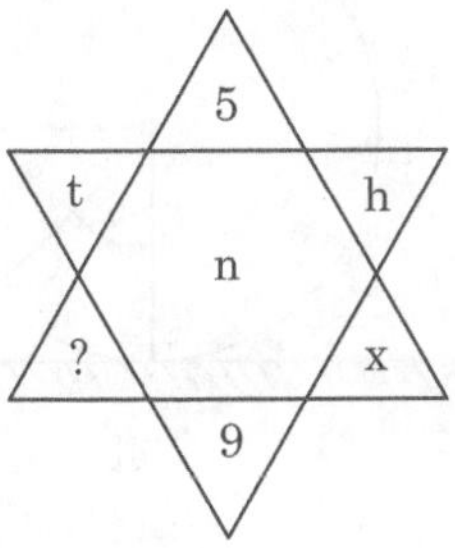

(1)	(2)	(3)	(4)	(5)	(6)	(7)	(8)	(9)
a	b	c	d	e	f	g	h	i
(10)	(11)	(12)	(13)	(14)	(15)	(16)	(17)	(18)
j	k	l	m	n	o	p	q	r
(19)	(20)	(21)	(22)	(23)	(24)	(25)	(26)	
s	t	u	v	w	x	y	z	

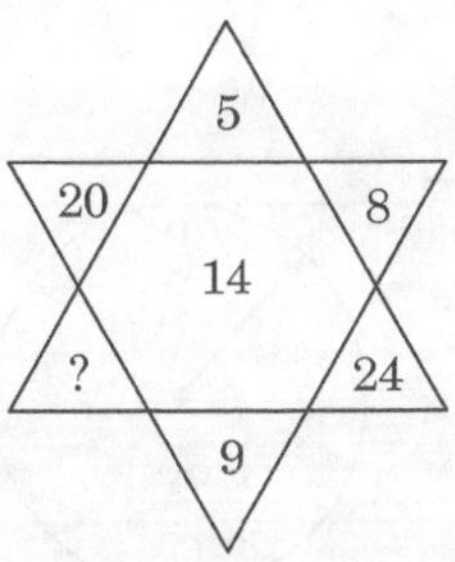

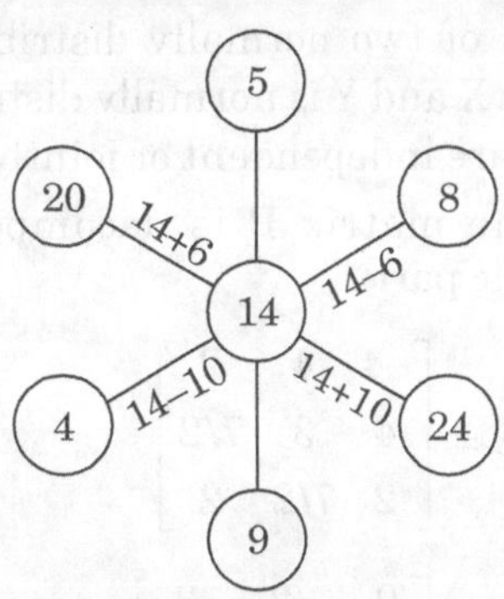

4 represents by d so, the correct option is (a).

8. 52 men can complete a strip in 10 days

   12 men were sent to another project,

   Total men available for work on construction of strip = 52 – 12 = 40

   $52 \times 10 = 40 \times x$

   $x = 13$ days

   Number of days, more than the original is,

   = 13 – 10 = 3 days

9.

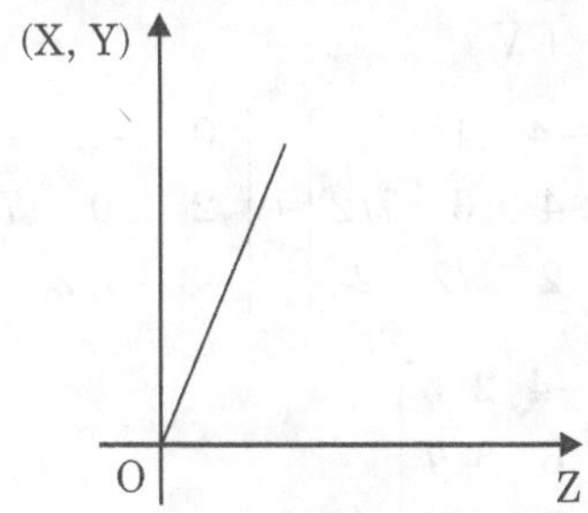

The product of X and Y linearly varies with Z.

XY = aZ + b, where a and b are constants.

For fixed X, Z is proportional to Y.

$\frac{XY}{Z}$ is constant,

For fixed Z, X is inversely proportional to Y.

For fixed, Y, X is proportional to Z.

10. Ratio of boys enrolled in Arts to the girls enrolled in management = $\frac{\text{Boys (Arts)}}{\text{Girls (Management)}}$

    Let, Number of boys in Arts = y

    Number of girls in Management

    $= \frac{15}{100} \times 1500 = 225$

    Number of girls in Arts = $\frac{20}{100} \times 5000 = 1000$

    Number of boys in Arts = 1000 – 450 = 550

    $$\frac{\text{Boys (A)}}{\text{Girls (M)}} = \frac{550}{225} = \frac{110}{45} = \frac{22}{9}$$

MECHANICAL ENGINEERING

1. The sum of two normally distributed random variables X and Y is normally distributed only if X and Y are independent or jointly normal.

2. When the matrix P is decomposed into its symmetric part.

$$S = \begin{bmatrix} -4 & 4 & 2 \\ 4 & 3 & 7/2 \\ 2 & 7/2 & 2 \end{bmatrix}$$

$$V = \begin{bmatrix} 0 & -2 & 3 \\ 2 & 0 & 7/2 \\ -3 & -7/2 & 0 \end{bmatrix}$$

For matrix P its symmetric part is

$$S = \frac{P + P^T}{2}$$

For matrix P its skew symmetric part is

$$V = \frac{P - P^T}{2}$$

$$P = S + V$$

$$= \begin{bmatrix} -4 & 4 & 2 \\ 4 & 3 & 7/2 \\ 2 & 7/2 & 2 \end{bmatrix} + \begin{bmatrix} 0 & -2 & 3 \\ 2 & 0 & 7/2 \\ -3 & -7/2 & 0 \end{bmatrix}$$

$$= \begin{bmatrix} -4 & 2 & 5 \\ 6 & 3 & 7 \\ -1 & 0 & 2 \end{bmatrix}$$

3. Given that

Then by changing the order of integration

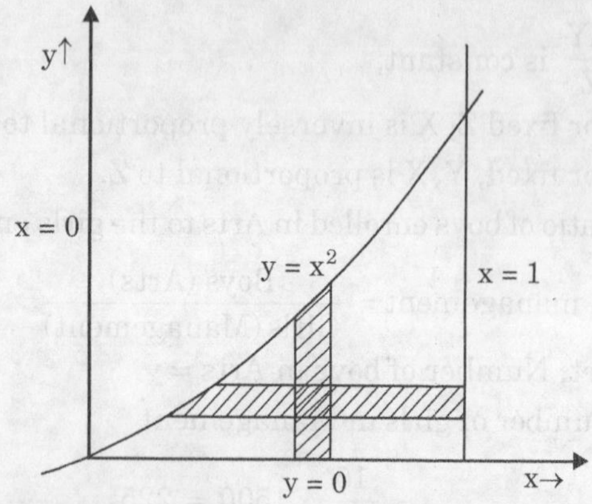

Limit $x = 0$ to $x = 1$

$y = 0$ to $y = x^2$

By changing the order of the integration the strip change from vertical to horizontal.

The new limit, $y = 0$ to $y = 1$

$x = \sqrt{y}$ to $x = 1$

So the integration

$$I = \int_{y=0}^{1} \int_{x=\sqrt{y}}^{1} xy^2 dx\, dy$$

4. Given $\frac{d^2y}{dt^2} - y = 1$

$$y \Big| t = 0 \; \frac{dy}{dt}\Big|_{t=0} = 0$$

$$\frac{d^2y}{dt^2} - y = 1$$

$$y''(t) - y(t) = 1$$

Taking Laplace both side

$$Ly''(t) - Ly(t) = L(1)$$

$$S^2y(s) - sy(0) - y'(0) - y(s) = \frac{1}{s}$$

Given $y(0) = 0$, $y'(0) = 0$

Then $\quad s^2y(s) - y(s) = \frac{1}{s}$

$$y(s)(s^2 - 1) = \frac{1}{s}$$

$$y(s) = \frac{1}{s(s-1)(s+1)}$$

5.

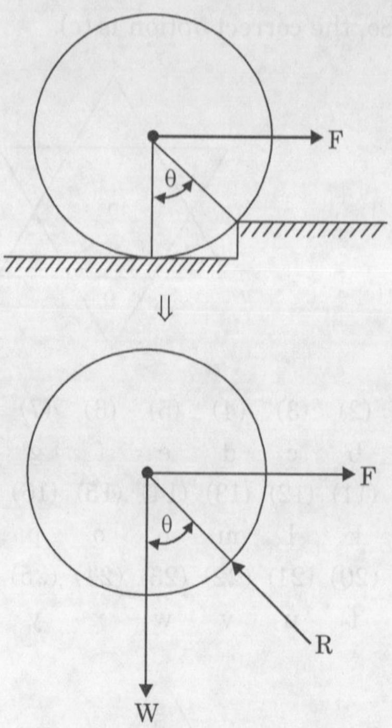

6.

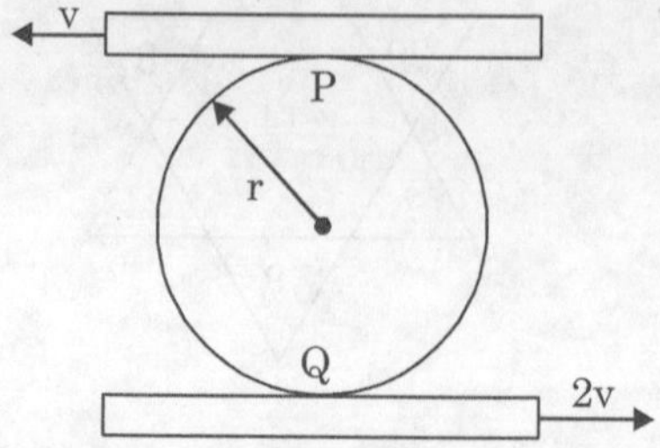

Radius r and roll without slipping at P and Q.

For pure rolling condition P and Q velocity should be zero

Assume, the roller moves with linear velocity with V_0 and rotate with ω angular velocity

So for P $\quad V_0 + V = r\omega$

$$V_0 - r\omega = -V \quad ...(i)$$

For $\quad Q, V_0 + r\omega = 2V \quad ...(ii)$

By solving equation (i) and (ii)

We get $\quad \omega = \frac{3V}{2r}$

7. The equation of motion of spring-mass damper system is given by

$$\frac{d^2x}{dt^2} + \frac{3dx}{dt} + 9x = 10\sin(5t)$$

The general equation of forced vibration system is given by

$$m\ddot{x} + C\dot{x} + Sx = F_0 \sin(\omega t)$$

$$\ddot{x} + \frac{C}{m}\dot{x} + \frac{S}{m}x = \frac{F_0}{m}\sin(\omega t)$$

$$\frac{d^2x}{dt^2} + \frac{C}{m}\frac{dx}{dt} + \frac{S}{m}x = \frac{F_0}{m}\sin(\omega t)$$

By comparing from the general equation, we get

$$\frac{C}{m} = 3, \quad \frac{S}{M} = 9$$

$$C = 3m \quad S = 9m$$

From relation, $\quad C = 2\xi\sqrt{Sm}$

$$3m = 2\xi\sqrt{9m \times m}$$

$$\xi = \frac{1}{2} = 0.5$$

8. Total qualitative inversion when Grashof's law satisfied.

(i) Crank-Crank mechanism when shortest link at the fixed position.

(ii) Rocker-Rocker mechanism when shortest link opposite to the fixed link that is coupler position.

(iii) Crank-rocker mechanism when shortest link at adjacent to the fixed position.

9. Ultrasonic machining is used for machining of glass.

10. Correct sequence is

Laser beam < MIG < SAW < Arc welding

11.
- Phase diagram is a plot on the temperature composition space showing the stability of various phases.
- Phase diagram tells us what will be the melting point of alloy.
- Relative amount of different phases can be found under given equilibrium condition by lever rule.
- Phase diagram tells about solid solubility it does not gives the information on transformation rate.

12. R = minimum Roughness height

Q = maximum Roughness height

P = maximum waviness height

S = maximum waviness width

u = Roughness width cutoff

T = maximum Roughness width

13. In MRP if the inventory holding cost is very high and setup cost is zero then the lot sizing approaches should be used lot-for-lot.

14. For partially submerged floating bodies:
- For stable equilibrium metacentre (M) must be at higher level than the centre of gravity (g)
- For unstable equilibrium metacentre (M) must be at lower level than the centre of gravity (g).
- For neutral the motacentre (M) and the centre of gravity (g) coincide with each other.

15. In the space above the mercury column in a barometer tube the gauge pressure of the vapour is negative.

The absolute pressure is positive but less than one atmosphere.

17. Air standard diesel cycle:

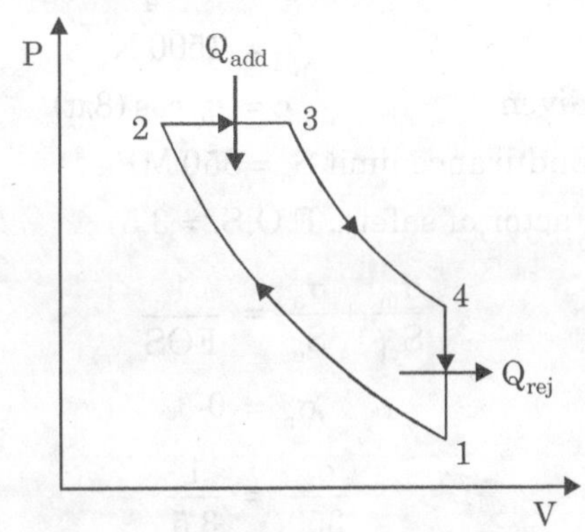

1–2 → Isentropic compression

2–3 → Constant pressure heat addition

3–4 → Reversible adiabatic (Isentropic) expansion

4–1 → Constant volume heat addition

18. Given, $\quad (h_2 - h_1) = (h_3 - h_2)$

$$\text{Degree of reaction} = \frac{\Delta h_{Rotor}}{\Delta h_{Rotor} + \Delta h_{stator}}$$

$$= \frac{(h_3 - h_2)}{(h_3 - h_2) + (h_2 - h_1)}$$

$$= \frac{1}{2} \times 100 = 50\%$$

19. $$I = \begin{bmatrix} 1 & 0 & 0 & - & - & - & 0 \\ 0 & 1 & 0 & - & - & - & 0 \\ 0 & 0 & 1 & - & - & - & 0 \\ - & - & - & - & - & - & 0 \\ - & - & - & - & - & - & 0 \\ - & - & - & - & - & - & 1 \\ - & - & - & - & - & - & 1 \end{bmatrix}_{100\times100}$$

Since, diagonal elements are eigen values of identify matrix.

So, E has only 1 element i.e., 1.

20.

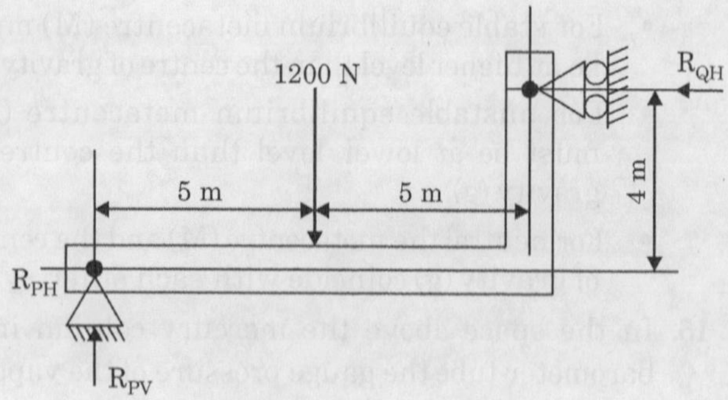

Taking moment about P, we get

$$R_{QH} \times 4 - 1200 \times 5 = 0$$

$$R_{QH} = \frac{1200 \times 5}{4}$$

$$R_{QH} = 1500 \text{ N}$$

21. Given $\sigma = \sigma_0 \cos(8\pi t)$

Endurance limit $S_e = 350$ MPa

Factor of safety, F.O.S. = 3.5

$$\frac{\sigma_m}{S_{yt}} + \frac{\sigma_a}{S_e} = \frac{1}{FOS}$$

$$\sigma_m = 0$$

$$\frac{\sigma_a}{350} = \frac{1}{3.5}$$

$$\sigma_a = \sigma_o$$

$$\sigma_o = 100 \text{ MPa}$$

22. In upsetting forging operation, if the stock is larger than 3 times the diameter then die cavity should not be more than 1.5 times the stock diameter.

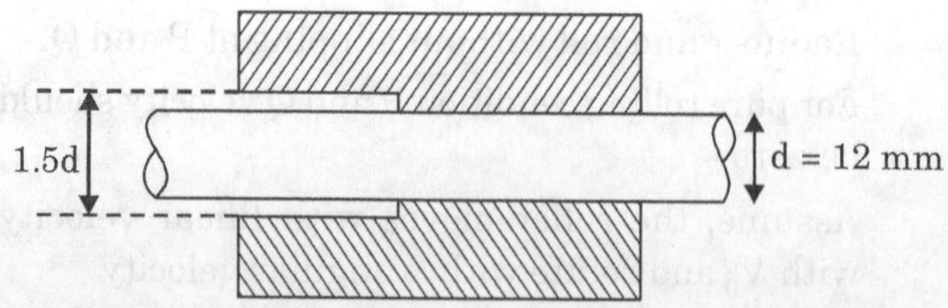

Hence, diameter of die = 1.5 d

$$= 1.5 \times 12$$

$$= 18 \text{ mm}$$

23.

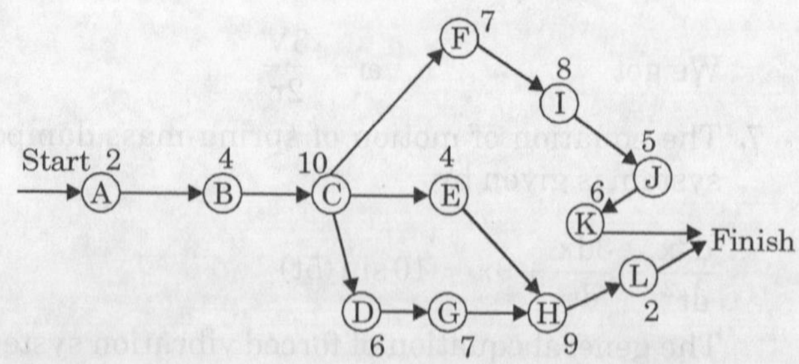

LFT = 42

EFT = 40

ST = LFT − EFT = 42 − 40 = 2

24.

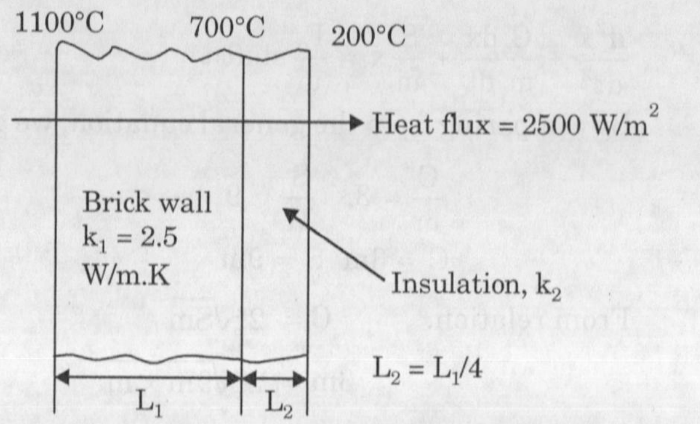

The heat flux $q = \frac{\delta T}{R_{eq.}} = \frac{\Delta T}{R_1}$ both resistance in series so $R = R_1 + R_2$

$$R = \frac{L_1}{K_1 A_1} + \frac{L_2}{K_2 A_2}$$

Then $$q = \frac{\Delta T}{R_{eq.}} = \frac{\Delta T}{R_1}$$

$$2500 = \frac{1100 - 700}{\frac{4}{2.5}}$$

$\Rightarrow$ $$L_1 = 0.4$$

$$L_2 = L_1/4 = 0.1$$

$$2500 = q = \frac{1100 - 200}{\frac{0.4}{2.5} + \frac{0.1}{K_2}}$$

$$K_2 = 0.5 \text{ W/mK}$$

25. Reversed Carnot cycle operate between temperature limit 27°C and – 3°C

COP of the refrigerator, $(COP)_R = \frac{T_L}{T_H - T_L}$

$$T_L = -3 + 273 = 270 \text{ K}$$

$$T_H = 27 + 273 \ 300 \text{ K}$$

$$(COP)_R = \frac{270}{300 - 270} = 9$$

COP of heat pump

$$(COP)_{H,P} = \frac{T_H}{T_H - T_L}$$

$$= \frac{300}{300 - 270} = 10$$

$$\frac{COP_R}{COP_{HP}} = \frac{9}{10} = 0.9$$

26. Directional Derivative $= (\Delta f)_\phi \cdot \hat{a}$

$$f(x) = xyz$$

$$\vec{a} = \hat{i} - 2\hat{j} + 2\hat{k}$$

$$\hat{a} = \frac{\vec{a}}{|a|} = \frac{\hat{i} - 2\hat{j} + 2\hat{k}}{\sqrt{1^2 + (2)^2 + 2^2}}$$

$$= \frac{1}{3}(\hat{i} - 2\hat{j} + 2\hat{k})$$

$$\text{Gradient of (f)} = (\nabla f)_\phi = \left(\frac{\partial}{\partial x}\hat{i} + \frac{\partial}{\partial y}\hat{j} + \frac{\partial}{\partial z}\hat{k}\right) f$$

$$= yz\hat{i} + xz\hat{j} + xy\hat{k}$$

Directional derivative

$$= \left(yz\hat{i} + xz\hat{j} + xy\hat{k}\right) \cdot \frac{1}{3}\left(\hat{i} - 2\hat{j} + 2\hat{k}\right)$$

$$DD = \frac{1}{3}[yz - 2xz + 2xy]$$

DD at (–1, 1, 3) is,

$$DD = \frac{1}{3}[1 \times 3 - 2 \times (-1) \times 3 + 2 \times (-1) \times 1] = \frac{7}{3}$$

27. Given: $u = x^3 - 3xy^2$

For function to be analytic it should follow Cauchy-Riemann's equations

$$u_x = v_y \text{ and } u_y = -v_x \quad \dots (2)$$

Now, ∵ $u_x = v_y$

$$u_x = \frac{\partial u}{\partial x} = 3x^2 - 3y^2$$

$$= v_y = \frac{\partial v}{\partial y}$$

$$\Rightarrow \quad v = \int (3x^2 - 3y^2)\, dy + f(x) + C_1$$

$$\Rightarrow \quad v = 3x^2y - \frac{3y^3}{3} + f(x) + C_1 \quad \dots (3)$$

$$\Rightarrow \quad \frac{\partial v}{\partial x} = 6xy - 0 + f'(x)$$

$$\Rightarrow \quad v_x = 6xy + f'(x) \quad \dots (4)$$

Now, $u_y = \frac{\partial u}{\partial y} = 0 - 6xy \quad \dots (5)$

From (2)

$$u_y = -v_x$$

$$-6xy = -v_x \text{ [From (5)]}$$

$$-6xy = -(6xy + f'(x)) \quad \text{[From (4)]}$$

$$\Rightarrow \quad 6xy = 6xy + f'(x)$$

$$\Rightarrow \quad f'(x) = 0$$

$$\Rightarrow \quad f(x) = \text{Constant} \quad \dots (6)$$

So, from (3)

$$v = 3x^2y - y^3 + \text{Constant} + C_1$$

[From (6)]

$$v = 3x^2y - y^3 + C$$

[Let C = Constant + C_1]

28.

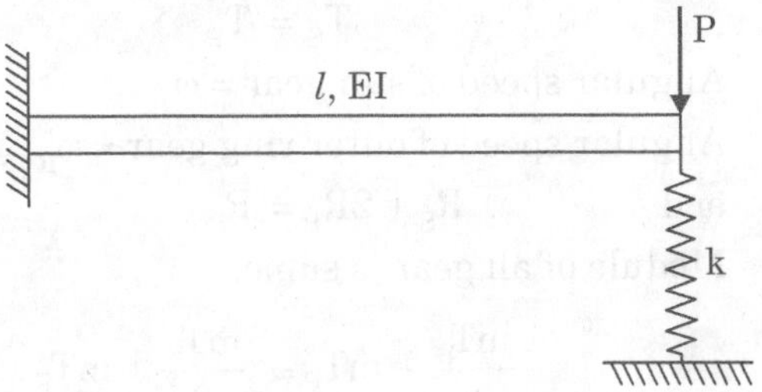

The deflection in spring and cantilever beam is same.

Therefore, stiffness of both (beam and spring) are in parallel combination.

For parallel system,

Equivalent stiffness (K_{eq}) = K + K_{beam} (Stiffness of beam)

$$K_{beam} = \frac{3EI}{l^3}$$

[For point load at cantilever beam]

$$K_{eq} = K + \frac{3EI}{l^3}$$

$$= K + \frac{3EI}{l^3}$$

$$\text{Deflection} = \frac{P}{K_{eq}}$$

$$= \frac{P}{\left(\frac{Kl^3 + 3EI}{l^3}\right)}$$

$$= \frac{Pl^3}{Kl^3 + 3EI}$$

$$= \frac{Pl^3}{3EI}\left(\frac{1}{\frac{Kl^3}{3EI}+1}\right)$$

$$= \frac{Pl^3}{3EI}\left(\frac{3EI}{Kl^3 + 3EI}\right)$$

29.

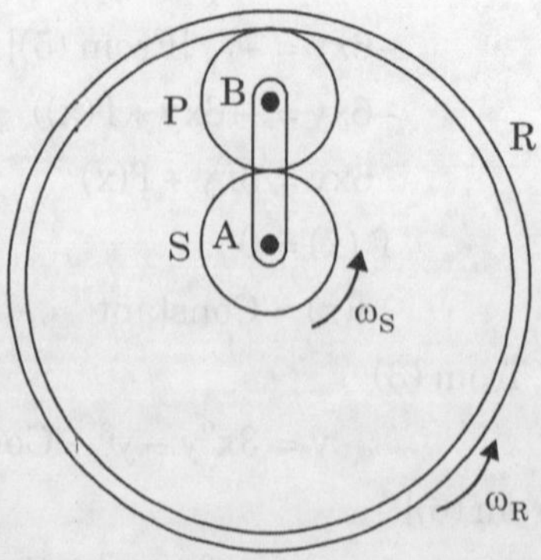

Number of teeth on sun gear (S) = Number of teeth on planet gear (P)

$$T_S = T_P$$

Angular speed of sun gear = ω_s

Angular speed of outer ring gear = ω_R

and $R_S + 2R_P = R_R$

Module of all gear is same,

$$\frac{mT_S}{2} + mT_P = \frac{mT_S}{2} + mT_P$$

$$T_S + 2T_P = T_R \qquad (\because T_S = T_P)$$

$$T_R = 3\,T_P$$

Let, revolution made by (S) = x

Revolution made by arm (a) = y

Line	Action	Rev. of arm	Rev. of S	Rev. of P	Rev. of R
1.	Arm, fixed, S+1 rev.	0	1	$-\frac{T_S}{T_P}$	$\frac{T_S}{T_P} \times \frac{T_P}{T_R}$
2.	Arm fixed,	x	x	$-\frac{T_S}{T_P}x$	$-\frac{T_S}{T_R}x$
3.	Add y	y	x + y	$y - \frac{T_S}{T_P}x$	$y - \frac{T_S}{T_R}x$

Angular speed of (S) = ω_S = x + y

Angular speed of (R) = $\omega_R = y - \frac{T_S}{T_R}x = y - \frac{x}{3}$

$$x + y = \omega_S \qquad ...(i)$$

$$-\frac{x}{3} + y = \omega_R \qquad ...(ii)$$

By subtracting equation (ii) from equation (i)

$$x + \frac{x}{3} = \omega_S - \omega_R$$

$$x = \frac{3}{4}(\omega_S - \omega_R)$$

Putting value of x in equation (i)

$$y = \omega_S - \frac{3}{4}(\omega_S - \omega_R)$$

$\left(y = \frac{\omega_S}{4} + \frac{3}{4}\omega_R\right)$, angular speed of the arm AB.

30. $\alpha\Delta T = \frac{\nu pr}{tE}$

$$\boxed{\Delta T = \frac{\nu pr}{\alpha tE}}$$

31. Since spring constant (K) = $\frac{Gd^4}{8D^3n}$ ...(1)

where,

G = modulus of rigidity of spring matrial

D = Diameter of spring

d = Diameter of wire

n = No. of coils

After doubling the wire diameter, spring diameter and the number of coils,

$$D' = 2D$$

$$d' = 2d$$

$$n' = 2n$$

Putting these values in (1),

$$K' = \frac{G(2d)^4}{8 \times (2D)^3 \times (2n)}$$

$$= \frac{Gd^4 \times 2^4}{8 \times D^3 \times n \times 2^3 \times 2}$$

$$= \frac{Gd^4}{8D^3n} = K$$

32.

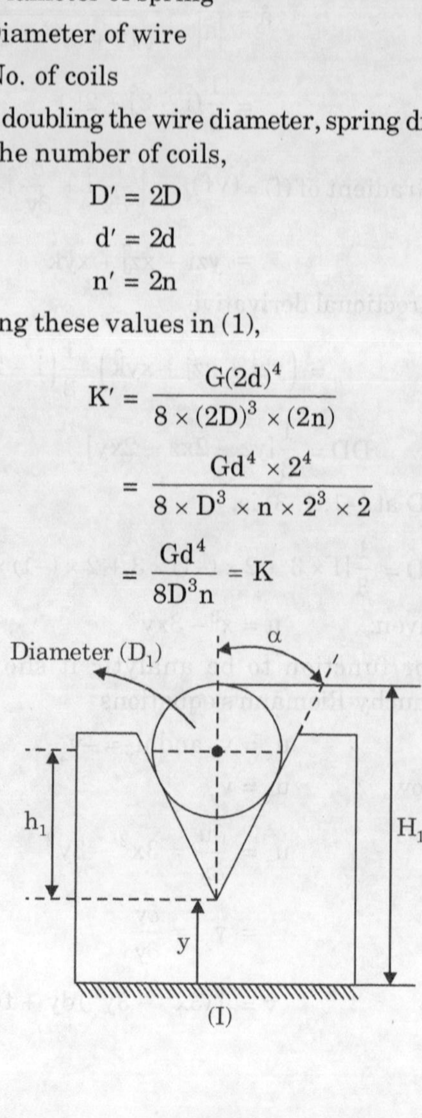

(I)

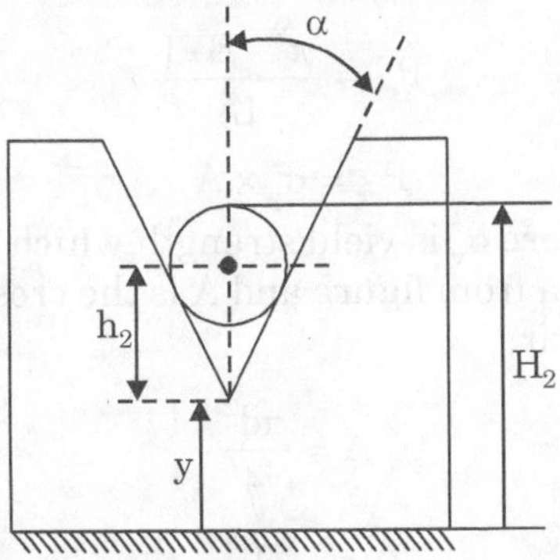

(II)

From Figure (1)

$$\sin\alpha = \frac{\frac{D_1}{2}}{h_1}$$

where $h_1 = H_1 - y - \dfrac{D_1}{2}$

$$\Rightarrow \left(H_1 - y - \frac{D_1}{2}\right)\cdot\sin\alpha = \frac{D_1}{2}$$

Let $\sin\alpha = K$

$$\Rightarrow \left(H_1 - y - \frac{D_1}{2}\right) = \frac{D_1}{2K}$$

$$\Rightarrow \quad -y = \frac{D_1}{2K} + \frac{D_1}{2} - H_1 \quad ...(1)$$

From Fig. (II)

$$\sin\alpha = \frac{\frac{D_2}{2}}{h_2} = \frac{\frac{D_2}{2}}{H_2 - y - \frac{D_2}{2}}$$

$$\Rightarrow K = \frac{\frac{D_2}{2}}{H_2 + \frac{D_1}{2K} + \frac{D_1}{2} - H_1 - \frac{D_2}{2}}$$

$$\Rightarrow K\left(H_2 + \frac{D_1}{2K} + \frac{D_1}{2} - H_1 - \frac{D_2}{2}\right) = \frac{D_2}{2}$$

$$\Rightarrow K\left(H_2 - \frac{D_2}{2} - H_1 + \frac{D_1}{2}\right) = \frac{D_2}{2} - \frac{D_1}{2}$$

$$\Rightarrow K\left(H_2 - \frac{D_2}{2} - H_1 + \frac{D_1}{2}\right) = \frac{D_2}{2} - \frac{D_1}{2}$$

$$\Rightarrow K = \frac{\frac{(D_2 - D_1)}{2}}{H_2 - H_1 - \frac{(D_2 - D_1)}{2}}$$

$$\Rightarrow K = \frac{D_1 - D_2}{2(H_1 - H_2) - (D_1 - D_2)}$$

$$\Rightarrow \sin\alpha = \frac{D_1 - D_2}{2(H_1 - H_2) - (D_1 - D_2)}$$

33.

$$F_t = F_{t-1} + \alpha(\Delta_{t-1} - F_{t-1})$$

$$F_2 = F_1 + \alpha(D_1 - F_1)$$

$$31.8 = 32 + \alpha(30 - 32)$$

$$\alpha = 0.1$$

34. Closed system undergoing irreversible process,

Air, m = 1 kg, R = 0.287 kJ/kgk, $\gamma = 1.4$

Initial state	Final state
P_1 = 1 bar (absolute)	P_2 = 3 bar (absolute)
T_1 = 27°C = 300K	T_2 = 127°C = 400 K

Change in the specific entropy of the air in the process,

$$\Delta S = S_2 - S_1$$

$$= C_p \ln\left(\frac{T_2}{T_1}\right) - R\ln\left(\frac{P_2}{P_1}\right)$$

$$= 1.005\ln\left(\frac{400}{300}\right) - 0.287\ln\left(\frac{3}{1}\right)$$

$$= -0.0262\frac{\text{kJ}}{\text{kgk}} = -26.2\frac{\text{J}}{\text{kgk}}$$

35. Conventional Method (I)

$$I = \int_0^{\pi/2}(8 + 4\cos x)\cdot dx$$

$$= (8x + 4\sin x)_0^{\pi/2}$$

$$= 4\pi + 4 = 16.566$$

Numerical Method using trapezoidal rule,

$$II = \frac{h}{2}(y_0 + y_n)$$

where, $h = \dfrac{\pi}{2} - 0 = \dfrac{\pi}{2}$

$$y_0 = 8 + 4\times\cos(0) = 12$$

$$y_n = 8 + 4\times\cos\left(\frac{\pi}{2}\right) = 8$$

$$\Rightarrow \quad II = \frac{\pi}{4}(12 + 8) = 15.708$$

$$\therefore \quad \%\ \text{error} = \frac{II - I}{I}\times 100$$

$$= \left(\frac{II}{I} - 1\right)\times 100$$

$$= \left(\frac{15.708}{16.566} - 1\right)\times 100$$

$$= -5.18$$

$$= 5.18\ (\text{absolute }\%)$$

36. Fair coin is tossed 20 times,

Probability (Head exactly appear 4 times in the first 10 tosses)

$$P(X) = {}^{10}C_4 \left(\frac{1}{2}\right)^4 \left(\frac{1}{2}\right)^6 = 0.205$$

Probability (Tail exactly appears 4 times in the next 10 tosses)

$$P(Y) = {}^{10}C_4 \left(\frac{1}{2}\right)^4 \left(\frac{1}{2}\right)^6 = 0.205$$

Total probability, P(T) = P(X) × P(Y)

$$= 0.205 \times 0.205 = 0.042$$

37.

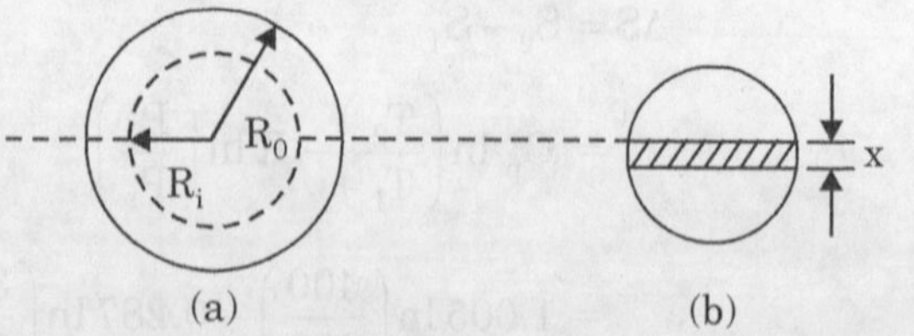

Assuming G = Specific gravity of sphere

In case (a): When hollowsphere is in equilibrium (when half submerged) weight of sphere = upward buoyant force

$$\frac{4}{3} \times \pi\left(R_0^3 - R_i^3\right) \times G\rho_w g = \frac{1}{2} \times \frac{4}{3}\pi R_0^3 \times \rho_{wg}$$

$$\Rightarrow \left(R_0^3 - R_i^3\right) \times G = \frac{1}{2}R_0^3 \quad ...(i)$$

In case (b): When a small downward displaement of 'x' is given

Net upward force = Additional buoyant force due to increase in submergence

$$\Rightarrow F_{net} = m.a_{net} = \left(\pi R_0^2 x\right) \times \rho_w g \quad ...(ii)$$

From (ii)

$$in.a_{net} = \left(\pi R_0^2 x\right) \times \rho_w g$$

$$\frac{4}{3}\pi\left(R_0^3 - R_i^3\right) \times G\rho_w \cdot a_{net} = \left(\pi R_0^2 x\right) \times \rho_w g$$

From (i), putting $\left(R_0^3 - R_i^3\right)G = \frac{R_0^3}{2}$, we get

$$\frac{4}{3}\pi\frac{R_0^3}{2} \times \rho_w \cdot a_{net} = \pi R_0^2 x \times \rho_w g$$

$$\Rightarrow \frac{2}{3}R_o \cdot a_{net} = x.g \quad ...(iii)$$

Comparing with SHM equation

$$a_{net} = \omega^2 x$$

$$\Rightarrow \quad \omega = \sqrt{\frac{3g}{2R_o}} = \sqrt{\frac{3 \times 10}{2 \times 0.2}}$$

$$\Rightarrow \quad \omega = 8.66 \text{ rad/sec}$$

38.

$$P_{cr} = \frac{\pi^2 \cdot E \cdot I}{L_e^2} \quad ...(1)$$

$$P_{cr} = \sigma_y \times A$$

Where σ_y is yield strength which is equal to 100 MPa from figure and A is the cross-section area of bar,

$$A = \frac{\pi d^2}{4}$$

$$\because \quad I = \frac{\pi d^4}{64}$$

$$E = \frac{\sigma_y}{\varepsilon} = \frac{100}{0.002} = 50000 \text{ MPa}$$

From (1), we get

$$100 \times \frac{\pi d^2}{4} = \frac{\pi^2 \times 50000 \times \frac{\pi d^4}{64}}{(1)^2}$$

$$\Rightarrow \frac{100 \times \pi d^2 \times 64}{4 \times \pi^2 \times 50000 \times \pi d^4} = 1$$

$$\Rightarrow \frac{100 \times 16}{\pi^2 \times 50000 \times d^2} = 1$$

$$\Rightarrow d^2 = \frac{16}{\pi^2 \times 500} = 3.2422 \times 10^{-3}$$

$$\Rightarrow \quad d = 0.05694 = 56.94 \text{ mm}$$

39.

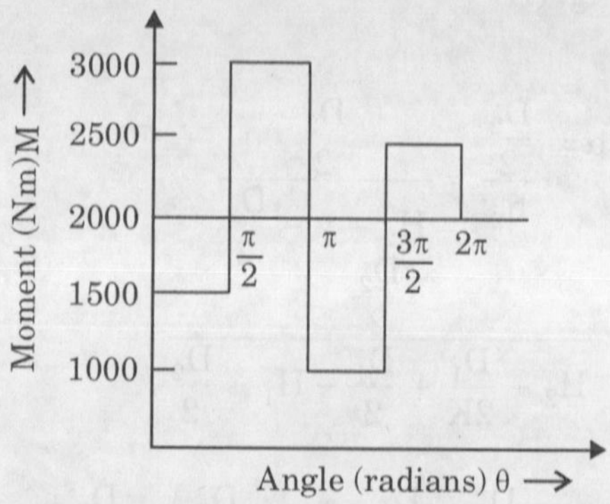

Mean turning moment, T_{mean} = 2000 Nm

Average engine speed, $\bar{N}$ = 1000 rpm

Fluctuation in speed, C_S = 0.04

Let, I = mass moment of inertia of the flywheel (kg-m²)

Maximum fluctuation of energy, ΔE,

$$\Delta E = (T - T_{mean}) \times (\text{Time period})$$

$$= (3000 - 2000) \times \left(\pi - \frac{\pi}{2}\right) = 1570.8 \text{ J}$$

$$\Delta E = I\omega^2 C_s$$

$$1570.8 = I \times \left(\frac{2\pi \times 1000}{60}\right)^2 \times 0.04$$

$$I = 3.58 \text{ kg-m}^2$$

40.

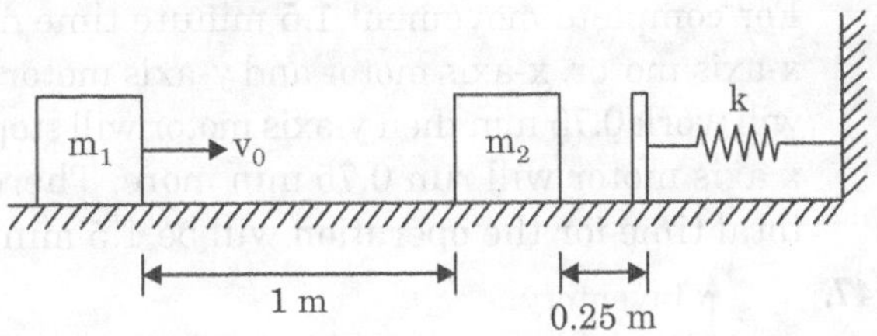

Let velocity of combined masses $(m_1 + m_2)$ just after collision is V m/s.

Since, masses are moving together after collision. So, it is a case of inelastic collision.

$\Rightarrow m_1 v_0 + m_2 \times 0 = (m_1 + m_2) \times V$

$\Rightarrow 10 v_0 = (10 + 30) \times V$

$$\Rightarrow V = \frac{10V_0}{40} = \frac{V_0}{4} = \frac{2}{4} = 0.5 \text{ m/s}$$

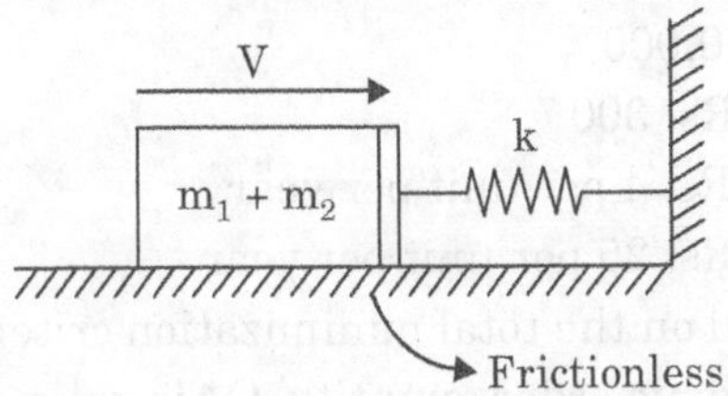

Note : Velocity V_0 will be constant before collision, i.e. during travelling of 1 m (before collision) because the horizontal surface is frictionless.

Similarly, velocity V will also be constant for 0.25 m before collision of combined masses with the spring.

Now, According to conservation of mechanical energy.

Initial K.E. + Initial P.E.

= Final K.E. + Final P.E.

when final velocity of the system (combined masses and spring) is zero. Final K.E will be zero and there will be maximum deflection in spring i.e. A.

$$\text{So, } \frac{1}{2}(m_1 + m_2)V^2 + 0 = 0 + \frac{1}{2}kA^2$$

$$\Rightarrow \frac{1}{2}(10 + 30) \times 0.5^2 + 0 = 0 + \frac{1}{2} + 10^5 \times A^2$$

$$\Rightarrow 40 \times 0.5^2 = 10^5 \times A^2$$

$$\Rightarrow A = 0.01 \text{ m} = 1 \text{ cm}$$

41. Let, W^t = Tangential load acting on the pinion

P = Power transmitted

$$\text{Bending stress } (\sigma) = \frac{k_v W^t}{FmY}$$

$$400 = \frac{1.26 \times W^t}{50 \times 1.25 \times 0.322}$$

$$W^t = 6388.89 \text{ N}$$

Power transmitted (P) = Torque × Speed

$= (T\omega)_{pinion}$

Torque applied on pinion gear

$$= W^t \times r_p = W^t \times \frac{mZ_p}{2}$$

$$T = 6388.89 \times \frac{1.25 \times 20}{2}$$

Torque, (T) = 79861.125 N.mm

= 79.86 N.m

P = Tω = 79.86 × 125.66

= 10035.5 watt = 10 kW

42. Volume = Discharge rate × time

$$\text{time (t)} = \frac{\text{Volume}}{\text{Discharge rate}}$$

$$= \frac{\text{Volume}}{\text{Area} \times \text{Velocity}}$$

$$= \frac{1200}{2 \times \sqrt{2 \times 981 \times 10}}$$

= 4.28 seconds

43. Given diameter, d = 200 mm

Initial geometry (ASA)

0 – 9° – 7° – 15° – 30° – 0.05 inch

After some time

0 – 9° – 7° – 8° – 15° – 0° – 0.05 inch

We know that specific energy consumption

$$u_c = (e) = \frac{F_c}{1000fd}$$

$$f = \left(\frac{t}{\sin\lambda}\right)$$

$$\lambda_1 = 90° - 30° = 60°$$

$$\lambda_2 = 90 - 0 = 90°$$

Specific cutting energy,

$$u_c = u_o(t_1)^{-0.4} = \frac{F_c}{1000fd}$$

$$F_c = 40(f\sin\lambda)^{-0.4} \times 1000 \times fd$$

$$F_c \propto (\sin\lambda)^{-0.4}$$

$$\left(\frac{F_{c2} - F_{cl}}{F_{cl}}\right) \times 100 = \left[\left(\frac{\sin\lambda_2}{\sin\lambda_1}\right)^{-0.4} - 1\right] \times 100$$

$$= \left[\left(\frac{\sin 90}{\sin 60}\right)^{-0.4} - 1\right] \times 100$$

= – 5.59%

44. Two identical shaping machine, S_1 and S_2.

Let width of workpiece for machine $S_1 = b_1$

Feed of workpiece for machine $S_1 = f_1$

For machine S_2, width of workpiece, $b_2 = 1.1\ b_1$

Feed of workpiece, $f_2 = 0.9\ f_1$

Shaper total time of machining,

$$t_m \propto \left(\frac{b(1+R)}{f}\right);\ t_m \ \alpha\ b \ \& \ \tan\alpha \ \frac{1}{f}$$

$$\frac{t_{m_1}}{t_{m_2}} = \frac{b_1}{b_2} \times \frac{f_2}{f_1} = \frac{0.9}{1.1}$$

$$= 0.818 \approx 0.8$$

45. Length of bar (L) = 250 mm

Diameter (D) = 25 mm

Feed (f) = 0.2 mm/rev

Time required for tool change (T_c) = 1 min.

Tool life equation, $VT^{0.2} = 24$

For maximum production rate,

$$T = T_C\left(\frac{1-n}{n}\right) = 1\left(\frac{1-0.2}{0.2}\right)$$

$$= 4 \text{ min}$$

$$V = \frac{24}{(4)^{0.2}} = 18.18 \text{ m/min}$$

Total machining time

$$= \frac{L}{fN} = \frac{L}{f\left(\frac{V}{\pi D}\right)} = \frac{250}{0.2 \times \left(\frac{18.18 \times 1000}{\pi \times 25}\right)}$$

$$= 5.4 \text{ min}$$

Number of tool changes per component.

46.

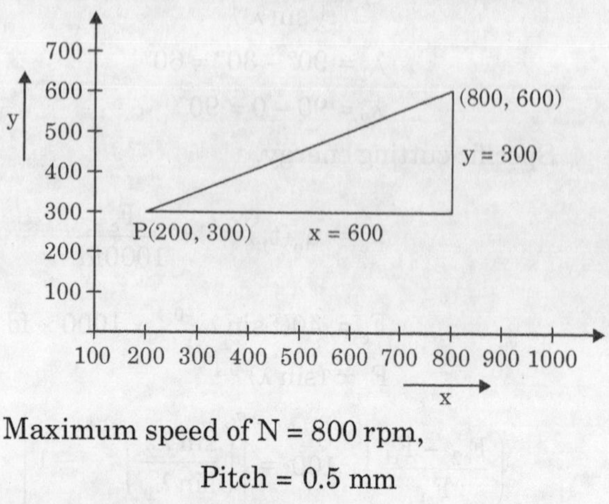

Maximum speed of N = 800 rpm,

Pitch = 0.5 mm

time taken in x direction,

$$t_x = \frac{\Delta x}{P \times N} = \frac{600}{0.5 \times 800} = 1.5 \text{ min}$$

time taken in y direction

$$t_y = \frac{\Delta y}{P \times N} = \frac{300}{0.5 \times 800} = 0.75 \text{ min}$$

For complete movement 1.5 minute time neded x-axis motor. x-axis motor and y-axis motor both will work 0.75 min then y-axis motor will stop and x axis motor will run 0.75 min more. Therefore total time for the operation will be 1.5 min.

47.

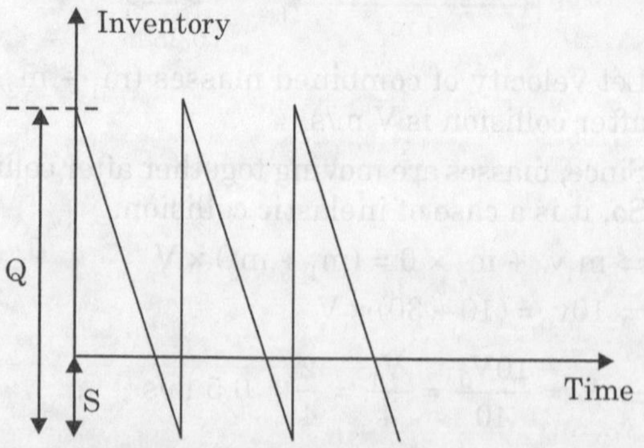

D = 10,000

C_o = Rs. 300

C_h = Rs. 4 per unit per year

C_b = Rs. 25 per unit per year

Based on the total minimization criteria,

Economic order quantity, Q* is given by

$$Q^* = \sqrt{\frac{2DC_o}{C_h}\left(\frac{C_b + C_h}{C_b}\right)}$$

$$Q^* = \sqrt{\frac{2 \times 10000 \times 300}{4}\left(\frac{25+4}{25}\right)} = 1319.09$$

Back order quantity,

$$\frac{Q^*}{S^*} = \frac{C_b + C_h}{C_h}$$

$$S^* = \frac{Q^* C_h}{C_b + C_h} = \frac{1319.09 \times 4}{25 + 4}$$

$$= 181.94$$

Maximum inventory reached in the system

$$Q_m^* = Q^* - S^*$$

$$Q_m^* = 1319.09 - 181.94$$

$$= 1137.15 \approx 1137$$

48. Flow through a nozzle,

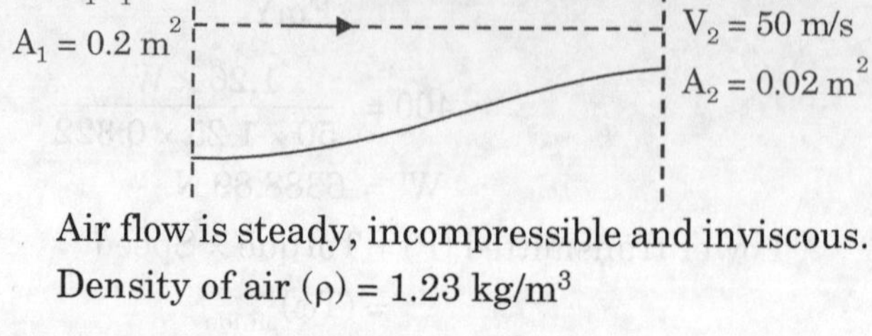

Air flow is steady, incompressible and inviscous.

Density of air (ρ) = 1.23 kg/m^3

Apply Bernoulli equation between (1) and (2)

$$\frac{P_1}{\rho g} + \frac{V_1^2}{2g} + z_1 = \frac{P_2}{\rho g} + \frac{V_2^2}{2g} + z_2$$

$$\frac{P_1 - P_2}{\rho g} = (z_2 - z_1) + \frac{V_2^2 - V_1^2}{2g}$$

$$P_1 - P_{atm} = \rho g (z_2 - z_1) + \frac{\rho}{2}\left(V_2^2 - V_1^2\right)$$

$$= \frac{\rho}{2} \times V_2^2 \left[1 - \left(\frac{V_1}{V_2}\right)^2\right]$$

$$= \frac{\rho}{2} \times V_2^2 \left[1 - \left(\frac{A_2}{A_1}\right)^2\right]$$

$$= \frac{1.23}{2} \times 50^2 \times \left[1 - \left(\frac{0.02}{0.2}\right)^2\right]$$

$= 1522$ Pa $= 1.52$ kPa

49. Density of water $(\rho) = 1000$ kg/m³

$g = 10$ m/s²

Inclined pipe of uniform diameter,

Section A	Section B
$V_A = 3.2$ m/s	$V_B = 3.2$ m/s
$P_A = 186$ kPa	$P_B = 260$ kPa
$Z_A = 24.5$ m	$Z_B = 9.1$ m

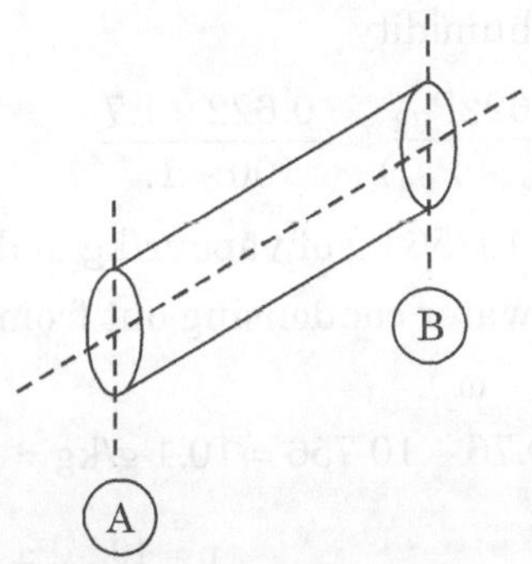

Let, Head lost due to friction $= h_L$

By applying Bernoulli principle between section A and B, we get

Total head at section A,

$$H_A = \frac{P_A}{\rho g} + \frac{V_A^2}{2g} + Z_A$$

$$= \frac{186 \times 1000}{1000 \times 10} + \frac{3.2^2}{2 \times 10} + 24.5$$

$= 43.612$ m

Total head at section B,

$$H_B = \frac{P_B}{\rho g} + \frac{V_B^2}{2g} + z_B$$

$$= \frac{260 \times 1000}{1000} + \frac{3.2^2}{2 \times 10} + 9.1 = 35.612 \text{ m}$$

Head lost due to friction $h_L = H_A - H_B$

$= 43.612 - 35.612 = 8$ m

50. From Wein's displacement Law

$$\lambda_{max} T = b$$

So, $(\lambda_{max})_1 T_1 = b$

and $(\lambda_{max})_2 T_2 = b$ ...(2)

From eqn. (1) divided by eqn. (2), we get

$$\Rightarrow \quad \frac{(\lambda_{max})_1 T_1}{(\lambda_{max})_2 . T_2} = 1$$

Given, $T_1 = 300$ K

and $(\lambda_{max})_2 = 1.2 \times (\lambda_{max})_1$

$$\text{So,} \quad \frac{(\lambda_{max})_1 \times 3000}{1.2 \times (\lambda_{max})_1 \times T_2} = 1$$

$$\Rightarrow \quad T_2 = \frac{3000}{1.2} = 2500 \text{ K}$$

51.

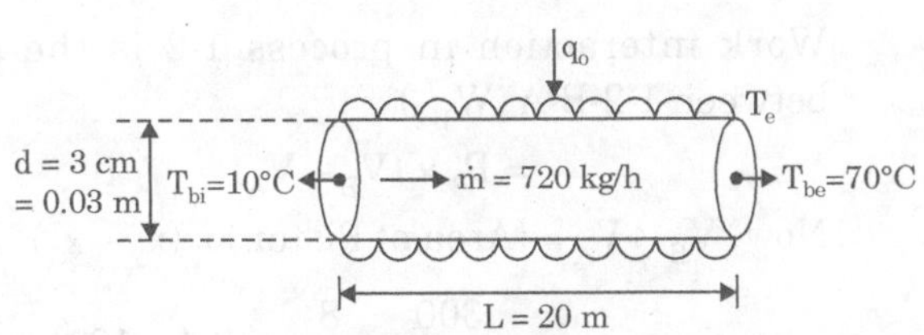

$C_p = 4.179$ kJ/kgK

$h = 1697$ W/m²K

$$m = 720 \times \frac{1}{3600} = 0.2 \text{ Kg/s}$$

$\because$ Heat given to the tube + Heat gain by water in tube.

$$\Rightarrow \quad q \times \pi d \times L = \dot{m} \times C_P \times (T_{be} - T_{bi})$$

$$q = \frac{0.2 \times 4.179 \times (70 - 10)}{\pi \times 0.03 \times 20}$$

$= 26.604$ kW/m²

$= 26604.34$ W/m²

At the end of tube,

$$q = h \times (T_e - T_{be})$$

$$\Rightarrow \quad 26604.34 = 1697 \times (T_e - 70)$$

$$\Rightarrow \quad T_e = 85.67 \approx 85.7°\text{C}$$

52.

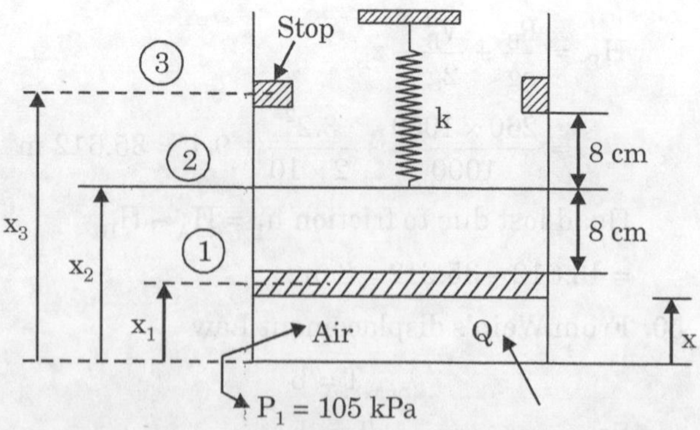

Process (1) – (2) is isobaric (constant pressure)

$\Rightarrow \quad P_1 = P_2$

Process (2) – (3) is a line as shown in P-V diagram

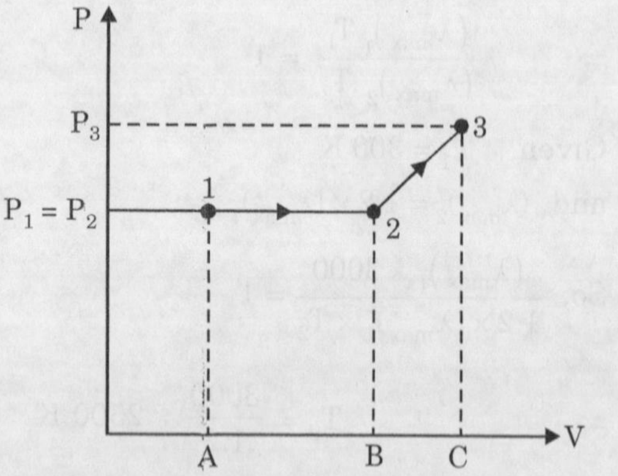

Work interaction in process 1-2 is the area between 1-2-B-A (W_{1-2}).

$$= P_1 \times (V_B - V_A)$$

Now, $V_B - V_A$ = (Area of Piston) × $(x_2 - x_1)$

$$= \frac{300}{10^4} \times \frac{8}{100} = 2.4 \times 10^{-3} \text{ m}^3$$

$W_{1-2} = 105 \times (2.4 \times 10^{-3}) \times 10^3 = 252\text{J}$

Now, W_{2-3} is the area between 2-3-C-B

$$W_{2\text{-}3} = \frac{1}{2}(P_2 + P_3) \times (V_C - V_B)$$

$$= \frac{1}{2}(P_2 + P_3) \times A(x_3 - x_2)$$

where A is the area of piston.

$$\because \quad P_3 = P_2 + \frac{Kx}{A}$$

$$= 105 \times 10^3 + \frac{12.5 \times 10^3 \times 8 \times 10^{-2}}{300 \times 10^{-4}}$$

$$= \frac{415}{3} \times 10^3 \text{Pa}$$

$$\therefore \quad W_{2-3} = \frac{1}{2}\left(105 \times 10^3 + \frac{415}{3} \times 10^3\right) \times 300 \times 10^{-4} (8 \times 10^{-2})$$

$$= 292 \text{ J}$$

$$\therefore \quad W_{Total} = W_{1-2} + W_{2-3}$$

$$= 252 + 292$$

$$= 544 \text{ J}$$

53. Moist air,

State (1), P_1 = 105 kPa, T_1 = 30°C, $(R.H)_1$ = 80%

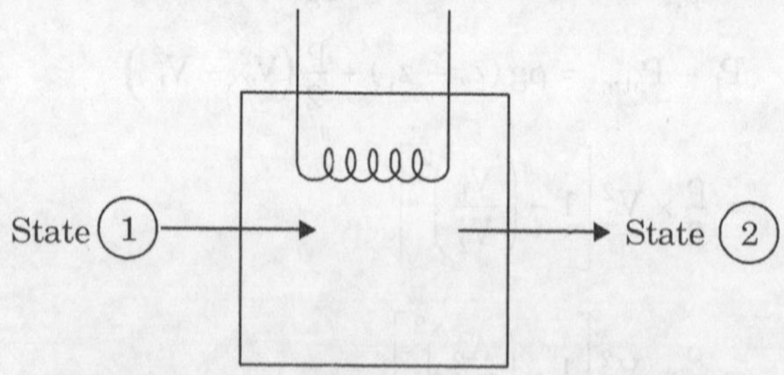

State (2), P_2 = 100 kPa, T_2 = 15°C, (R.H) = 100%

$(P_s)_{@30°C}$ = 4.24 kPa

$(P_s)_{@15°C}$ = 1.7 kPa

$(M.W.)_{water}$ = 18 g/mol, $(M.W)_{air}$ = 28.94 g/mol

Vapour pressure at state (1),

$P_{V1} = (R.H.)_1 \times (P_V)_{S1} = 0.8 \times 4.24 = 3.392$ kPa

Specific humidity,

$$\omega_1 = \frac{0.622(P_{V1})}{(P_1 - P_{V1})} = \frac{0.622 \times 3.392}{105 - 3.392}$$

= 0.02076 k of vapour/kg.d.a.

Vapour pressure at state (2),

$P_{V2} = (P_{VS})_2 = 1.7$ kPa

Specific humidity

$$\omega_2 = \frac{0.622 P_{V2}}{(P_2 - P_{V2})} = \frac{0.622 \times 1.7}{100 - 1.7}$$

= 0.010756 kg of vapour/kg of d.a.

Mass of water condensing out from the duct.

$M_\omega = (\omega_1 - \omega_2)$

= 20.76 – 10.756 = 10.1 g/kg = 10 g/kg

54.

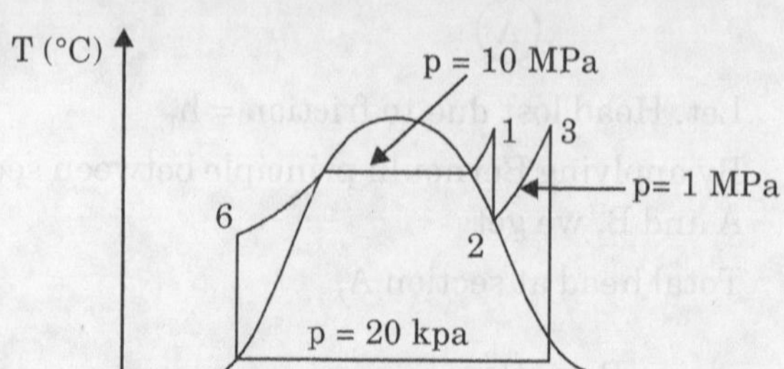

From state (1) to (2), Isentropic expansion

At state (1),

h_1 = 3373.6 kJ/kg, S_1 = 6.5965 kJ/kg-k

At state (2),

$h_2 = h_g = 2778.1$ kJ/kg, $S_2 = 6.5965$ kJ/kg-k

At state (3),

$h_3 = 3478.4$ kJ/kg, $S_4 = 7.7621$ kJ/-kg-k

At state (4),

$h_4 = h_f + xh_{fg}$, $S_3 = S_4$ (Isentropic expansion)

$S_3 = 7.7621 = 0.8319 + x(7.9085 - 0.8319)$

$x = 0.979$

$h_4 = 251.38 + 0.979\,(2609.7 - 251.38)$

$= 2560.91$ kJ/kg

Work done by H.P & L.P. Turbines

$= W_T = W_{1-2} + W_{3-4}$

$= (h_1 - h_2) + (h_3 - h_4)$

$= (3373.6 - 2778.1) + (3478.4 - 2560.91)$

$= 1513$ kJ/kg

55.

$$\eta = 1 - \frac{1}{(r)^{\gamma-1}}\left(\frac{r_c^y - 1}{r_c - 1)\gamma}\right)$$

$$\eta_{old}\big|_{CR=r=15} = 1 - \frac{1}{(15)^{1.3-1}}\left(\frac{2^{1.3} - 1}{(2-1)1.3}\right)$$

$$= 0.5 \text{ or } 50\%$$

$$\eta_{New}\big|_{CR=R=21} = 1 - \frac{1}{(21)^{1.3-1}}\left(\frac{2^{1.3} - 1}{(2-1)1.3}\right)$$

$$= 0.5487 \text{ or } 54.87\%$$

$$\left(\eta_{new}\big|_{CR=21}\right) - \left(\eta_{old}\big|_{CR=15}\right) = 54.87 - 50$$

$$= 4.87 \approx 4.9\%$$

SOLVED PAPER - 2021

SET 1

INSTRUCTIONS

1. Total of 65 questions carrying 100 marks, out of which 10 questions carrying a total of 15 marks are in General Aptitude (GA)
2. The Engineering Mathematics will carry around **15% of the total marks,** the General Aptitude section will carry **15% of the total marks** and the **remaining 70% of the total marks.**
3. **Types of Questions**
   (*a*) **Multiple Choice Questions (MCQ)** carrying 1 or 2 marks each in all papers and sections. These questions are objective in nature, and each will have a choice of four options, out of which the candidate has to mark the correct answer(s).
   (*b*) **Numerical Answer Questions** of 1 or 2 marks each in all papers and sections. For these questions the answer is a real number, to be entered by the candidate using the virtual keypad. No choices will be shown for these type of questions.
   (*c*) **Multiple Select Questions (MSQ)** carrying 1 or 2 marks each in all the papers and sections. These questions are objective in nature, and each will have choice of four answers, out of which ONE or MORE than ONE choice(s) is / are correct.
4. For **1-mark** multiple-choice questions, **1/3 marks** will be deducted for a wrong answer. Likewise, for **2-marks** multiple-choice questions, **2/3** marks will be deducted for a wrong answer. There is no negative marking for numerical answer type questions & multiple selection questions.

Chapter-Wise Analysis

#	Subject	Marks		
		1	2	Total
1	General Aptitude	5	5	15
2	Engineering Mathematics	5	3	11
3	Engineering Mechanics	-	2	4
4	Strength of Materials	4	2	8
5	Theory of Machines	3	3	9

#	Subject	Marks		
		1	2	Total
6	Design of Machine Elements	-	3	6
7	Fluid Mechanics	2	2	6
8	Heat Transfer	2	2	6
9	Thermodynamics, IC Engines, Refrigeration & Power Plant Engineering Air-Conditioning	4	4	12
10	Materials, Manufacturing & Industrial Engineering	5	10	25

GENERAL APTITUDE (GA)

Q. 1 to Q. 5 : Carry One Mark Each.

1. Consider the following sentences:

(i) After his surgery, Raja hardly could walk.

(ii) After his surgery, Raja could barely walk.

(iii) After his surgery, Raja barely could walk.

(iv) After his surgery, Raja could hardly walk.

Which of the above sentences are grammatically CORRECT?

(a) (i) and (ii) (b) (i) and (iii)

(c) (iii) and (iv) (d) (ii) and (iv)

2. Ms. X came out of a building through its front door to find her shadow due to the morning sun falling to her right side with the building to her back. From this, it can be inferred that building is facing ________

(a) North (b) East

(c) West (d) South

3. In the above figure, O is the center of the circle and, M and N lie on the circle.

The area of the right triangle MON is 50 cm^2.

What is the area of the circle in cm^2 ?

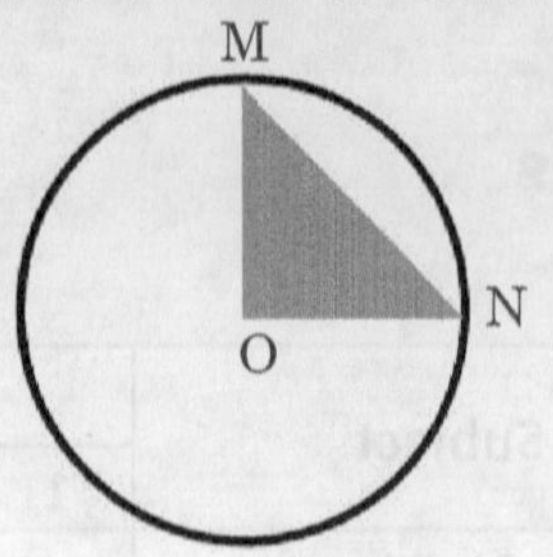

(a) 2π (b) 50π

(c) 75π (d) 100π

4. If $\begin{cases} "\oplus" \text{ means } "-", \\ "\otimes" \text{ means } "\div", \\ "\Delta" \text{ means } "+", \\ "\nabla" \text{ means } "\times", \end{cases}$

then, the value of the expression $\Delta\, 2 \oplus 3\, \Delta\, ((4 \otimes 2) \nabla 4) =$

(a) –1 (b) –0.5

(c) 6 (d) 7

5. "The increased consumption of leafy vegetables in the recent months is a clear indication that the people in the state have begun to lead a healthy lifestyle"

Which of the following can be logically inferred from the information presented in the above statement?

(a) The people in the state did not consume leafy vegetables earlier.

(b) Consumption of leafy vegetables may not be the only indicator of healthy lifestyle.

(c) Leading a healthy lifestyle is related to a diet with leafy vegetables.

(d) The people in the state have increased awareness of health hazards causing by consumption of junk foods.

Q. 6 to Q. 10: Multiple Choice Question (MCQ), Carry Two Marks Each

6. Oxpeckers and rhinos manifest a symbiotic relationship in the wild. The oxpeckers warn the rhinos about approaching poachers, thus possibly saving the lives of the rhinos. Oxpeckers also feed on the parasitic ticks found on rhinos.

In the symbiotic relationship described above, the primary benefits for oxpeckers and rhinos respectively are,

(a) Oxpeckers get a food source, rhinos have no benefit.

(b) Oxpeckers save their habitat from poachers while the rhinos have no benefit.

(c) Oxpeckers get a food source, rhinos may be saved from the poachers.

(d) Oxpeckers save the lives of poachers, rhinos save their own lives.

7. 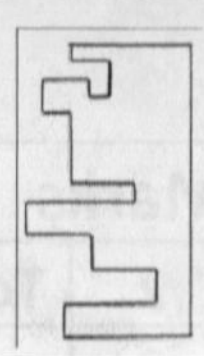

A jigsaw puzzle has 2 pieces. One of the pieces is shown above. Which one of the given options for the missing piece when assembled will form a rectangle? The piece can be moved, rotated or flipped to assemble with the above piece.

(a) 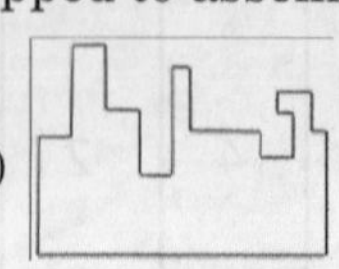 (b) 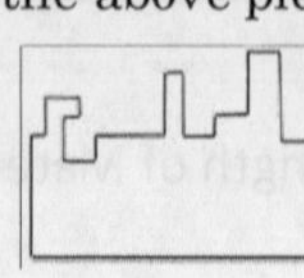

(c) 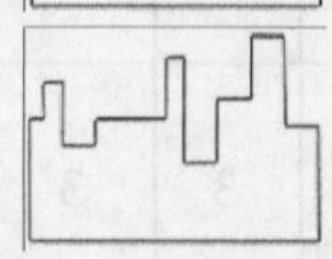 (d) 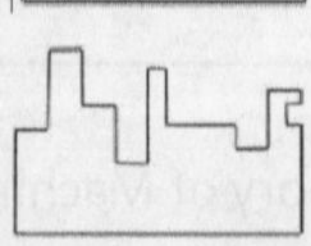

8. The number of hens, ducks and goats in farm P are 65, 91 and 169, respectively. The total number of hens, ducks and goats in a nearby farm Q is 416. The ratio of hens:ducks:goats in farm Q is 5:14:13. All the hens, ducks and goats are sent from farm Q to farm P.

The new ratio of hens:ducks:goats in farm P is

(a) 5:7:13 (b) 5:14:13

(c) 10:21:26 (d) 21:10:26

9.

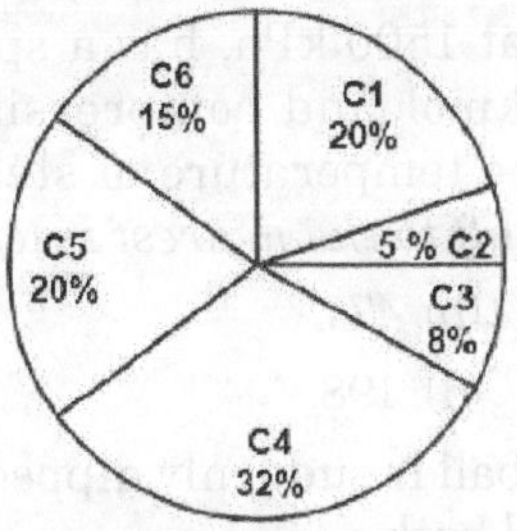

Company	Ratio
C1	3:2
C2	1:4
C3	5:3
C4	2:3
C5	9:1
C6	3:4

The distribution of employees at the rank of executives, across different companies C1, C2, ..., C6 is presented in the chart given above. The ratio of executives with a management degree to those without a management degree in each of these companies is provided in the table above. The total number of executives across all companies is 10,000.

The total number of management degree holders among the executives in companies C2 and C5 together is ______.

(a) 225 (b) 600

(c) 1900 (d) 2500

10. Five persons P, Q, R, S and T are sitting in a row not necessarily in the same order. Q and R are separated by one person, and S should not be seated adjacent to Q.

The number of distinct seating arrangements possible is:

(a) 4 (b) 8

(c) 10 (d) 16

MECHANICAL ENGINEERING

Q.1 to Q.19: Multiple Choice Question (MCQ), Carry One Mark Each

1. If $y(x)$ satisfies the differential equation

$$(\sin x)\frac{dy}{dx} + y\cos x = 1,$$

subject to the condition $y(\pi/2) = \pi/2$, then $y(\pi/6)$ is

(a) 0 (b) $\frac{\pi}{6}$

(c) $\frac{\pi}{3}$ (d) $\frac{\pi}{2}$

2. The value of $\lim_{x\to 0}\left(\frac{1-\cos x}{x^2}\right)$ is

(a) $\frac{1}{4}$ (b) $\frac{1}{3}$

(c) $\frac{1}{2}$ (d) 1

3. The Dirac-delta function $(\delta(t - t_0))$ for $t, t_0 \in \mathbb{R}$, has the following property

$$\int_a^b \phi(t)\delta(t-t_0)\,dt = \int_0^{\phi(t_0)} \begin{matrix} a < t_0 < b \\ \text{otherwise} \end{matrix}$$

The Laplace transform of the Dirac-delta function $\delta(t - a)$ for $a > 0$;

$L(\delta(t - a)) = F(s)$ is

(a) 0 (b) ∞

(c) e^{sa} (d) e^{-sa}

4. The ordinary differential equation $\frac{dy}{dt} = -\pi y$ subject to an initial condition $y(0) = 1$ is solved numerically using the following scheme:

$$\frac{y(t_{n+1}) - y(t_n)}{h} = -xy(t_n)$$

where h is the time step, $t_n = nh$, and $n = 0,1,2,....$ This numerical scheme is stable for all values of h in the interval______.

(a) $0 < h < \frac{2}{\pi}$ (b) $0 < h < 1$

(c) $0 < h < \frac{\pi}{2}$ (d) for all $h > 0$

5. Consider a binomial random variable X. If X_1, X_2, ..., X_n are independent and identically distributed samples from the distribution of X with sum $Y = \sum_{i=1}^{n} X_i$, then the distribution of Y as $n \to \infty$ can be approximated as

(a) Exponential (b) Bernoulli

(c) Binomial (d) Normal

6. The loading and unloading response of a metal is shown in the figure. The elastic and plastic strains corresponding to 200 MPa stress, respectively, are

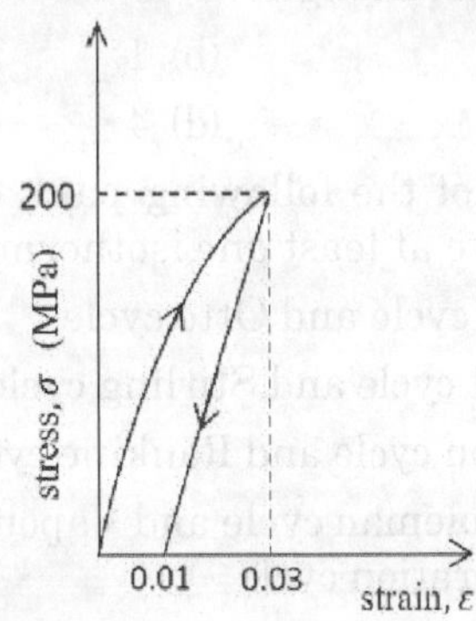

(a) 0.01 and 0.01 (b) 0.02 and 0.01

(c) 0.01 and 0.02 (d) 0.02 and 0.02

7. In a machining operation, if a cutting tool traces the workpiece such that the directrix is perpendicular to the plane of the generatrix as shown in figure, the surface generated is

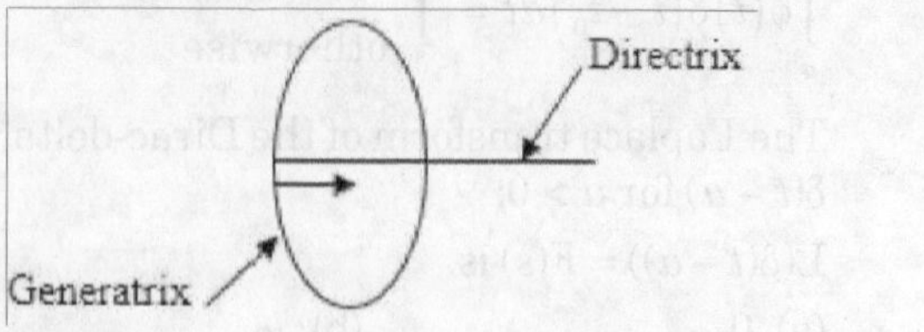

(a) plane (b) cylindrical
(c) spherical (d) a surface of revolution

8. The correct sequence of machining operations to be performed to finish a large diameter through hole is
(a) drilling, boring, reaming
(b) boring, drilling, reaming
(c) drilling, reaming, boring
(d) boring, reaming, drilling

9. In modern CNC machine tools, the backlash has been eliminated by
(a) preloaded ballscrews
(b) rack and pinion
(c) ratchet and pinion
(d) slider crank mechanism

10. Consider the surface roughness profile as shown in the figure.

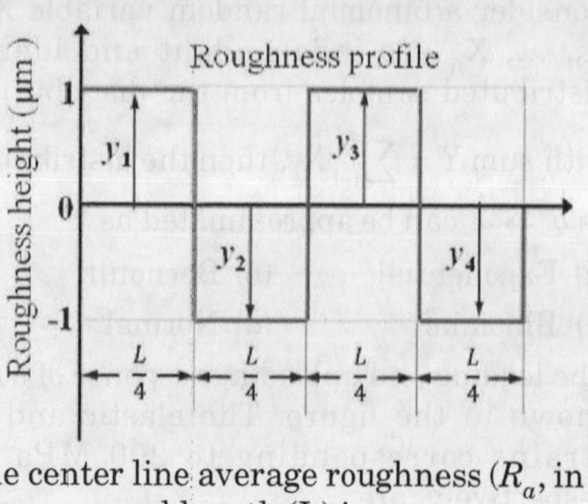

The center line average roughness (R_a, in μm) of the measured length (L) is
(a) 0 (b) 1
(c) 2 (d) 4

11. In which of the following pairs of cycles, both cycles have at least one isothermal process?
(a) Diesel cycle and Otto cycle
(b) Carnot cycle and Stirling cycle
(c) Brayton cycle and Rankine cycle
(d) Bell-Coleman cycle and Vapour compression refrigeration cycle

12. Superheated steam at 1500 kPa, has a specific volume of 2.75 m^3/kmol and compressibility factor (Z) of 0.95. The temperature of steam is ________ °C (*round off to the nearest integer*).
(a) 522 (b) 471
(c) 249 (d) 198

13. A hot steel spherical ball is suddenly dipped into a low temperature oil bath.
Which of the following dimensionless parameters are required to determine instantaneous center temperature of the ball using a Heisler chart?
(a) Biot number and Fourier number
(b) Reynolds number and Prandtl number
(c) Biot number and Froude number
(d) Nusselt number and Grashoff number

14. An infinitely long pin fin, attached to an isothermal hot surface, transfers heat at a steady rate of Q_1 to the ambient air. If the thermal conductivity of the fin material is doubled, while keeping everything else constant, the rate of steady- state heat transfer from the fin becomes Q_2. The ratio Q_2/Q_1 is
(a) $\sqrt{2}$ (b) 2
(c) $\frac{1}{\sqrt{2}}$ (d) $\frac{1}{2}$

15. The relative humidity of ambient air at 300 K is 50% with a partial pressure of water vapour equal to p_v. The saturation pressure of water at 300 K is p_{sat}. The correct relation for the air-water mixture is
(a) $p_v = 0.5\, p_{sat}$ (b) $p_v = p_{sat}$
(c) $p_v = 0.622\, p_{sat}$ (d) $p_v = 2\, p_{sat}$

16. Consider a reciprocating engine with crank radius R and connecting rod of length L. The secondary unbalance force for this case is equivalent to primary unbalance force due to a virtual crank of ________

(a) radius $\frac{L^2}{4R}$ rotating at half the engine speed

(b) radius $\frac{R}{4}$ rotating at half the engine speed

(c) radius $\frac{R^2}{4L}$ rotating at twice the engine speed

(d) radius $\frac{L}{2}$ rotating at twice the engine speed

17. A cantilever beam of length, L, and flexural rigidity, EI, is subjected to an end moment, M as shown in the figure. The deflection of the beam at Let $x = \frac{L}{2}$ is

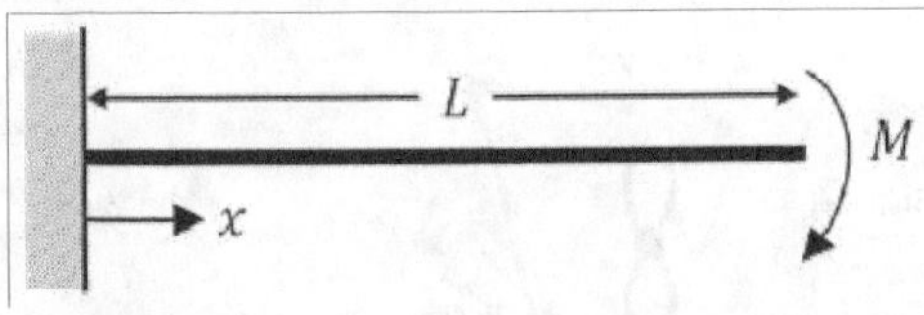

(a) $\frac{ML^2}{2EI}$ (b) $\frac{ML^2}{4EI}$

(c) $\frac{ML^2}{8EI}$ (d) $\frac{ML^2}{16EI}$

18. A prismatic bar PQRST is subjected to axial loads as shown in the figure. The segments having maximum and minimum axial stresses, respectively, are

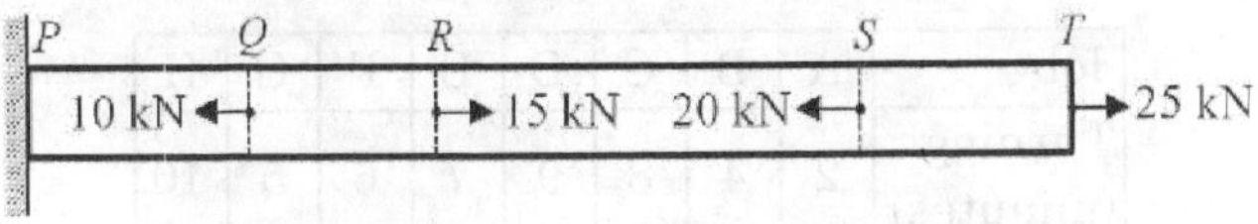

(a) QR and PQ (b) ST and PQ

(c) QR and RS (d) ST and RS

19. Shear stress distribution on the cross-section of the coil wire in a helical compression spring is shown in the figure. This shear stress distribution represents

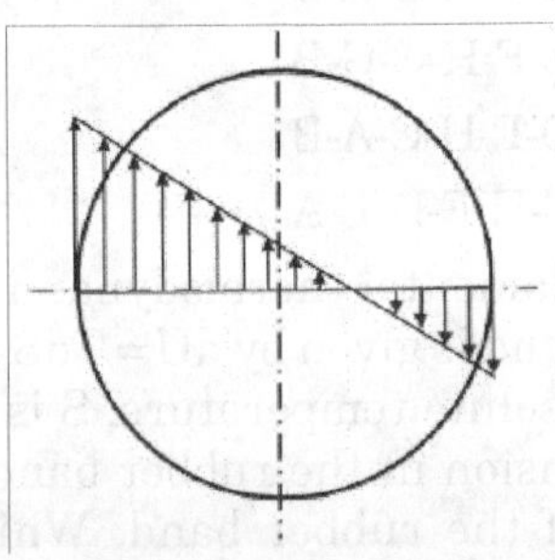

(a) direct shear stress in the coil wire cross-section

(b) torsional shear stress in the coil wire cross-section

(c) combined direct shear and torsional shear stress in the coil wire cross-section

(d) combined direct shear and torsional shear stress along with the effect of stress concentration at inside edge of the coil wire cross-section

Q.20 to Q.25: Numerical Answer Type (NAT), carry ONE mark each

20. Robot Ltd. wishes to maintain enough safety stock during the lead time period between starting a new production run and its completion such that the probability of satisfying the customer demand during the lead time period is 95%. The lead time period is 5 days and daily customer demand can be assumed to follow the Gaussian (normal) distribution with mean 50 units and a standard deviation of 10 units. Using $\phi^{-1}(0.95) = 1.64$, where ϕ represents the cumulative distribution function of the standard normal random variable, the amount of safety stock that must be maintained by Robot Ltd. to achieve this demand fulfillment probability for the lead time period is ________ units (*round off to two decimal places*).

21. A pressure measurement device fitted on the surface of a submarine, located at a depth H below the surface of an ocean, reads an absolute pressure of 4.2 MPa. The density of sea water is 1050 kg/m^3, the atmospheric pressure is 101 kPa, and the acceleration due to gravity is 9.8 m/s^2. The depth H is ________ m

(*round off to the nearest integer*).

22. Consider fully developed, steady state incompressible laminar flow of a viscous fluid between two large parallel horizontal plates. The bottom plate is fixed and the top plate moves with a constant velocity of U = 4 m/s. Separation between the plates is 5 mm. There is no pressure gradient in the direction of flow. The density of fluid is 800 kg/m^3, and the kinematic viscosity is 1.25 × 10^{-4} m^2/s. The average shear stress in the fluid is ________ Pa (*round off to the nearest integer*).

23. A rigid insulated tank is initially evacuated. It is connected through a valve to a supply line that carries air at a constant pressure and temperature of 250 kPa and 400 K respectively. Now the valve is opened and air is allowed to flow into the tank until the pressure inside the tank reaches to 250 kPa at which point the valve is closed. Assume that the air behaves as a perfect gas with constant properties

(c_p = 1.005 kJ/kg.K, c_v = 0.718 kJ/kg.K, R = 0.287 kJ/kg.K). Final temperature of the air inside the tank is ________ K (*round off to one decimal place*).

24. The figure shows an arrangement of a heavy propeller shaft in a ship. The combined polar mass moment of inertia of the propeller and the shaft is 100 kg.m^2. The propeller rotates at ω = 12 rad/s. The waves acting on the ship hull induces a rolling motion as shown in the figure with an angular velocity of 5 rad/s. The gyroscopic moment generated on the shaft due to the motion described is ____________ N.m (*round off to the nearest integer*).

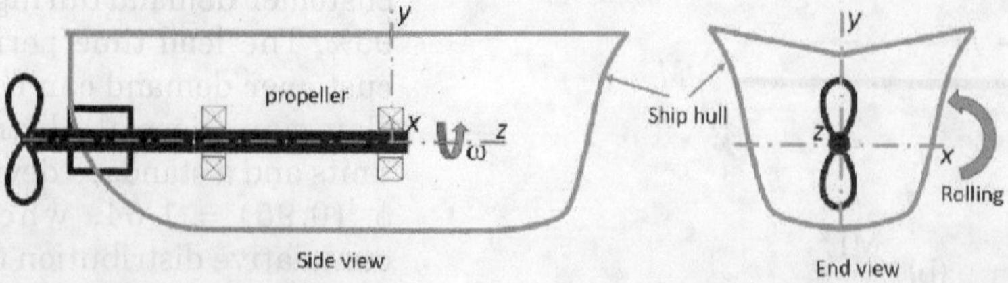

25. Consider a single degree of freedom system comprising a mass M, supported on a spring and a dashpot as shown in the figure.

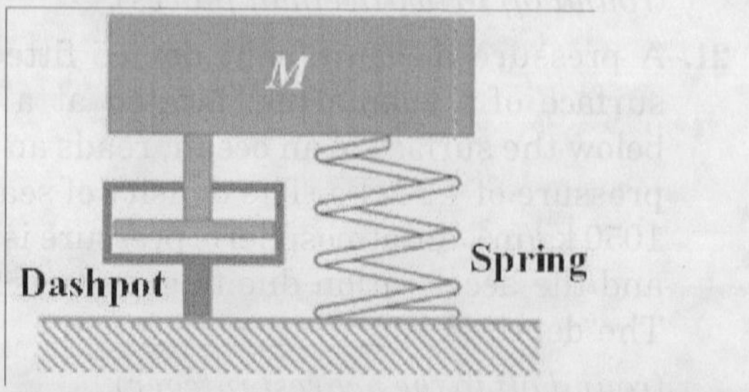

If the amplitude of the free vibration response reduces from 8 mm to 1.5 mm in 3 cycles, the damping ratio of the system is __________ (*round off to three decimal places*).

Q. 26 - Q. 34 Carry Two Mark Each

26. Consider a vector p in 2-dimensional space. Let its direction (counter-clockwise angle with the positive x-axis) be θ. Let p be an eigenvector of a 2×2 matrix A with corresponding eigenvalue λ, $\lambda > 0$. If we denote the magnitude of a vector v by $||v||$, identify the VALID statement regarding p', where $p' = Ap$.

(a) Direction of $p' = \lambda\theta$, $||p'|| = ||p||$

(b) Direction of $p' = \theta$, $||p'|| = \lambda||p||$

(c) Direction of $p' = \lambda\theta$, $||p'|| = \lambda||p||$

(d) Direction of $p' = \theta$, $||p'|| = ||p||$

27. Let C represent the unit circle centered at origin in the complex plane, and complex variable, $z = x + iy$. The value of the contour integral $\int_C \frac{\cosh 3z}{2z} dz$ (where integration is taken counter clockwise) is

(a) 0 (b) 2

(c) πi (d) $2\pi i$

28. A set of jobs A, B, C, D, E, F, G, H arrive at time t = 0 for processing on turning and grinding machines. Each job needs to be processed in sequence - first on the turning machine and second on the grinding machine, and the grinding must occur immediately after turning. The processing times of the jobs are given below.

Job	A	B	C	D	E	F	G	H
Turning (minutes)	2	4	8	9	7	6	5	10
Grinding (minutes)	6	1	3	7	9	5	2	4

If the makespan is to be minimized, then the optimal sequence in which these jobs must be processed on the turning and grinding machines is

(a) A-E-D-F-H-C-G-B

(b) A-D-E-F-H-C-G-B

(c) G-E-D-F-H-C-A-B

(d) B-G-C-H-F-D-E-A

29. The fundamental thermodynamic relation for a rubber band is given by dU = TdS + τdL, where T is the absolute temperature, S is the entropy, τ is the tension in the rubber band, and L is the length of the rubber band. Which one of the following relations is CORRECT:

(a) $\tau = \left(\frac{\partial U}{\partial S}\right)_L$

(b) $\left(\frac{\partial T}{\partial L}\right)_S = \left(\frac{\partial \tau}{\partial S}\right)_L$

(c) $\left(\frac{\partial T}{\partial S}\right)_L = \left(\frac{\partial \tau}{\partial L}\right)_S$

(d) $T = \left(\frac{\partial U}{\partial S}\right)_\tau$

30. Consider a two degree of freedom system as shown in the figure, where PQ is a rigid uniform rod of length, b and mass, m.

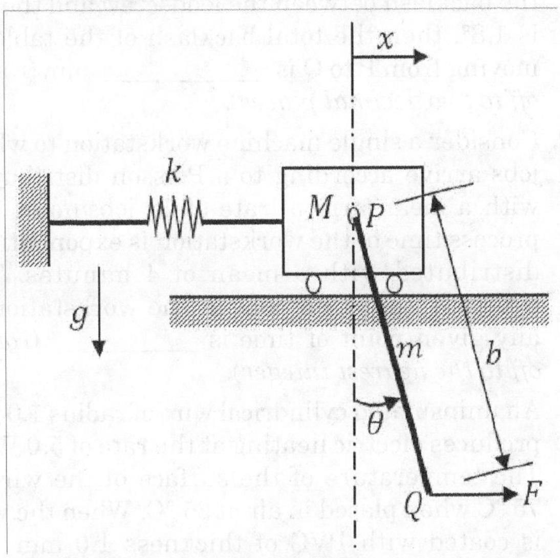

Assume that the spring deflects only horizontally and force F is applied horizontally at Q. For this system, the Lagrangian, L is

(a) $\frac{1}{2}(M+m)\dot{x}^2+\frac{1}{6}mb^2\dot{\theta}^2-\frac{1}{2}kx^2+mg\frac{b}{2}\cos\theta$

(b) $\frac{1}{2}(M+m)\dot{x}^2+\frac{1}{2}mb\dot{\theta}\dot{x}\cos\theta$

$\quad +\frac{1}{6}mb^2\dot{\theta}^2-\frac{1}{2}kx^2+mg\frac{b}{2}\cos\theta$

(c) $\frac{1}{2}M\dot{x}^2+\frac{1}{2}mb\dot{\theta}\dot{x}\cos\theta+\frac{1}{6}mb^2\dot{\theta}^2-\frac{1}{2}kx^2$

(d) $\frac{1}{2}M\dot{x}^2+\frac{1}{2}mb\dot{\theta}\dot{x}\cos\theta+\frac{1}{6}mb^2\dot{\theta}^2-\frac{1}{2}kx^2$

$\quad +mg\frac{b}{2}\cos\theta+Fb\sin\theta$

31. A right solid circular cone standing on its base on a horizontal surface is of height H and base radius R. The cone is made of a material with specific weight w and elastic modulus E. The vertical deflection at the mid-height of the cone due to self-weight is given by

(a) $\frac{wH^2}{8E}$ (b) $\frac{wH^2}{6E}$

(c) $\frac{wRH}{8E}$ (d) $\frac{wRH}{6E}$

32. A tappet valve mechanism in an IC engine comprises a rocker arm ABC that is hinged at B as shown in the figure. The rocker is assumed rigid and it oscillates about the hinge B. The mass moment of inertia of the rocker about B is 10^{-4} kg.m^2. The rocker arm dimensions are a = 3.5 cm and b = 2.5 cm. A pushrod pushes the rocker at location A, when moved vertically by a cam that rotates at N rpm. The pushrod is assumed massless and has a stiffness of 15 N/mm. At the other end C, the rocker pushes a valve against a spring of stiffness 10 N/mm. The valve is assumed massless and rigid.

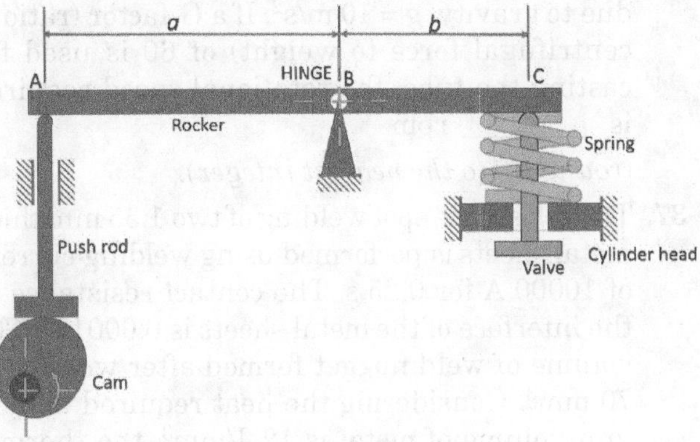

Resonance in the rocker system occurs when the cam shaft runs at a speed of ________ rpm (*round off to the nearest integer*).

(a) 496 (b) 4739

(c) 790 (d) 2369

33. Customers arrive at a shop according to the Poisson distribution with a mean of 10 customers/hour. The manager notes that no customer arrives for the first 3 minutes after the shop opens. The probability that a customer arrives within the next 3 minutes is

(a) 0.39 (b) 0.86

(c) 0.50 (d) 0.61

34. Let f(x) = $x^2 - 2x + 2$ be a continuous function defined on $x \in [1,3]$. The point x at which the tangent of f(x) becomes parallel to the straight line joining f(1) and f(3) is

(a) 0 (b) 1

(c) 2 (d) 3

Q.35 to Q.55: Numerical Answer Type (NAT), carry TWO marks each

35. Activities A, B, C and D form the critical path for a project with a PERT network. The means and variances of the activity duration for each activity are given below. All activity durations follow the Gaussian (normal) distribution, and are independent of each other.

Activity	A	B	C	D
Mean (days)	6	11	8	15
Variance (days^2)	4	9	4	9

The probability that the project will be completed within 40 days is________ *(round off to two decimal places).*

(Note: Probability is a number between 0 and 1).

36. A true centrifugal casting operation needs to be performed horizontally to make copper tube sections with outer diameter of 250 mm and inner diameter of 230 mm. The value of acceleration due to gravity, $g = 10$ m/s^2. If a G-factor (ratio of centrifugal force to weight) of 60 is used for casting the tube, the rotational speed required is ________ rpm

(round off to the nearest integer).

37. The resistance spot welding of two 1.55 mm thick metal sheets is performed using welding current of 10000 A for 0.25 s. The contact resistance at the interface of the metal sheets is 0.0001 Ω. The volume of weld nugget formed after welding is 70 mm^3. Considering the heat required to melt unit volume of metal is 12 J/mm^3, the thermal efficiency of the welding process is ________% *(round off to one decimal place).*

38. An orthogonal cutting operation is performed using a single point cutting tool with a rake angle of 12° on a lathe. During turning, the cutting force and the friction force are 1000 N and 600 N, respectively. If the chip thickness and the uncut chip thickness during turning are 1.5 mm and 0.75 mm, respectively, then the shear force is ________ N *(round off to two decimal places).*

39. In a grinding operation of a metal, specific energy consumption is 15 J/mm^3. If a grinding wheel with a diameter of 200 mm is rotating at 3000 rpm to obtain a material removal rate of 6000 mm^3/min, then the tangential force on the wheel is ________ N *(round off to two decimal places).*

40. A 200 mm wide plate having a thickness of 20 mm is fed through a rolling mill with two rolls. The radius of each roll is 300 mm. The plate thickness is to be reduced to 18 mm in one pass using a roll speed of 50 rpm. The strength coefficient (K) of the work material flow curve is 300 MPa and the strain hardening exponent, n is 0.2. The coefficient of friction between the rolls and the plate is 0.1. If the friction is sufficient to permit the rolling operation then the roll force will be ________ kN *(round off to the nearest integer).*

41. The XY table of a NC machine tool is to move from P(1,1) to Q(51,1); all coordinates are in mm. The pitch of the NC drive leadscrew is 1 mm. If the backlash between the leadscrew and the nut is 1.8°, then the total backlash of the table on moving from P to Q is ________ mm *(round off to two decimal places).*

42. Consider a single machine workstation to which jobs arrive according to a Poisson distribution with a mean arrival rate of 12 jobs/hour. The process time of the workstation is exponentially distributed with a mean of 4 minutes. The expected number of jobs at the workstation at any given point of time is ________ *(round off to the nearest integer).*

43. An uninsulated cylindrical wire of radius 1.0 mm produces electric heating at the rate of 5.0 W/m. The temperature of the surface of the wire is 75 °C when placed in air at 25 °C. When the wire is coated with PVC of thickness 1.0 mm, the temperature of the surface of the wire reduces to 55 °C.

Assume that the heat generation rate from the wire and the convective heat transfer coefficient are same for both uninsulated wire and the coated wire. The thermal conductivity of PVC is ________ W/m.K *(round off to two decimal places).*

44. A solid sphere of radius 10 mm is placed at the centroid of a hollow cubical enclosure of side length 30 mm. The outer surface of the sphere is denoted by 1 and the inner surface of the cube is denoted by 2. The view factor F^{22} for radiation heat transfer is ________ *(rounded off to two decimal places).*

45. Consider a steam power plant operating on an ideal reheat Rankine cycle. The work input to the pump is 20 kJ/kg. The work output from the high pressure turbine is 750 kJ/kg. The work output from the low pressure turbine is 1500 kJ/kg. The thermal efficiency of the cycle is 50 %. The enthalpy of saturated liquid and saturated vapour at condenser pressure are 200 kJ/kg and 2600 kJ/kg, respectively. The quality of steam at the exit of the low pressure turbine is ________ % *(round off to the nearest integer).*

46. In the vicinity of the triple point, the equation of liquid-vapour boundary in the P – T phase diagram for ammonia is In P = 24.38 - 3063/T, where P is pressure (in Pa) and T is temperature (in K). Similarly, the solid-vapour boundary is given by In P = 27.92 – 3754/T. The temperature at the triple point is ________ K *(round off to one decimal place).*

47. A cylindrical jet of water (density = 1000 kg/m^3) impinges at the center of a flat, circular plate and spreads radially outwards, as shown in the figure. The plate is resting on a linear spring with a spring constant k = 1 kN/m. The incoming jet diameter is D = 1 cm.

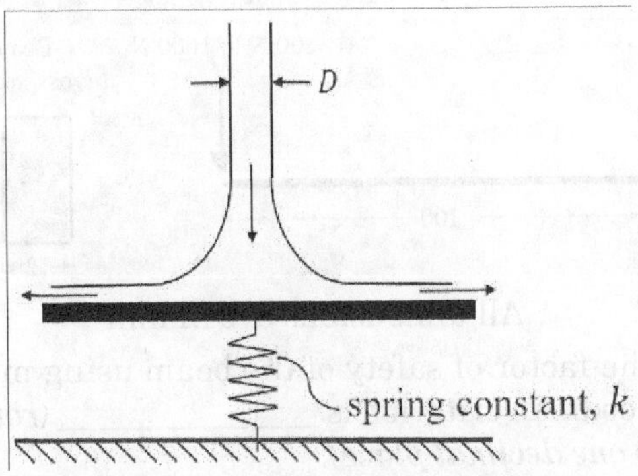

If the spring shows a steady deflection of 1 cm upon impingement of jet, then the velocity of the incoming jet is m/s (*round off to one decimal place*).

48. A single jet Pelton wheel operates at 300 rpm. The mean diameter of the wheel is 2 m. Operating head and dimensions of jet are such that water comes out of the jet with a velocity of 40 m/s and flow rate of 5 m^3/s. The jet is deflected by the bucket at an angle of 165°. Neglecting all losses, the power developed by the Pelton wheel is_______MW (*round off to two decimal places*).

49. An air-conditioning system provides a continuous flow of air to a room using an intake duct and an exit duct, as shown in the figure. To maintain the quality of the indoor air, the intake duct supplies a mixture of fresh air with a cold air stream. The two streams are mixed in an insulated mixing chamber located upstream of the intake duct. Cold air enters the mixing chamber at 5 °C, 105 kPa with a volume flow rate of 1.25 m^3/s during steady state operation. Fresh air enters the mixing chamber at 34 °C and 105 kPa. The mass flow rate of the fresh air is 1.6 times of the cold air stream. Air leaves the room through the exit duct at 24 °C.

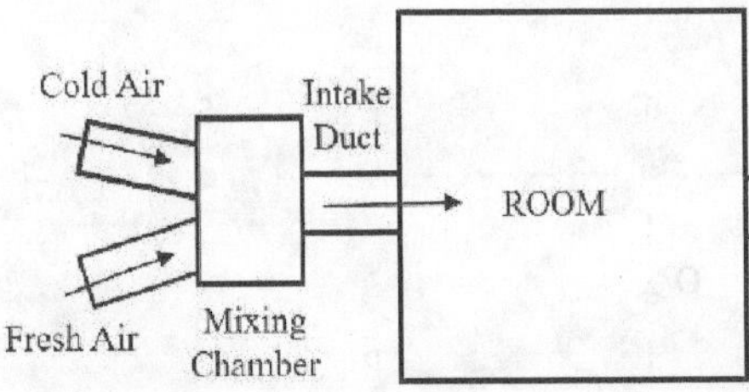

Assuming the air behaves as an ideal gas with c_p = 1.005 kJ/kg.K and R = 0.287 kJ/kg.K, the rate of heat gain by the air from the room is _______kW (*round off to two decimal places*).

50. Two smooth identical spheres each of radius 125 mm and weight 100 N rest in a horizontal channel having vertical walls. The distance between vertical walls of the channel is 400 mm.

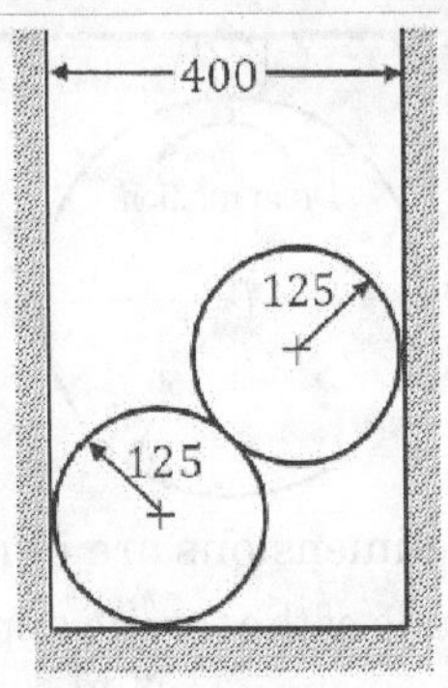

All dimensions are in mm

The reaction at the point of contact between two spheres is ____________ N (*round off to one decimal place*).

51. An overhanging beam PQR is subjected to uniformly distributed load 20 kN/m as shown in the figure.

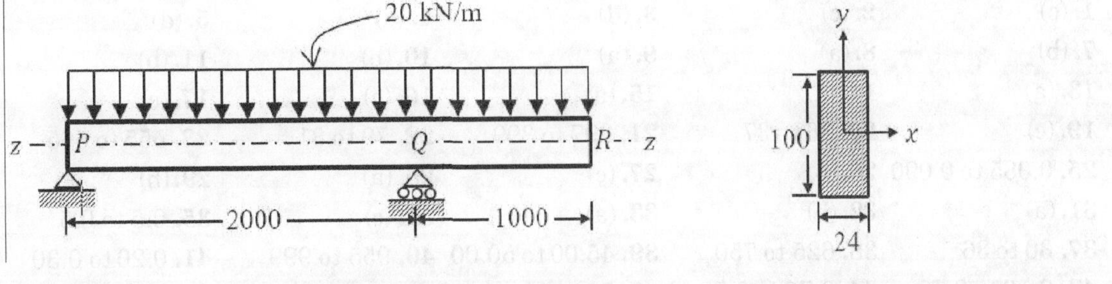

All dimensions are in mm

The maximum bending stress developed in the beam is __________ MPa (*round off to one decimal place*).

52. The Whitworth quick return mechanism is shown in the figure with link lengths as follows: OP = 300 mm, OA = 150 mm, AR = 160 mm, RS = 450 mm.

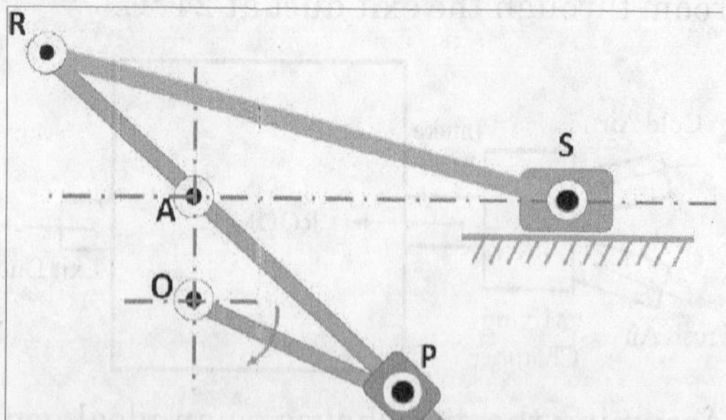

The quick return ratio for the mechanism is ________ *(round off to one decimal place)*.

53. A short shoe drum (radius 260 mm) brake is shown in the figure. A force of 1 kN is applied to the lever. The coefficient of friction is 0.4.

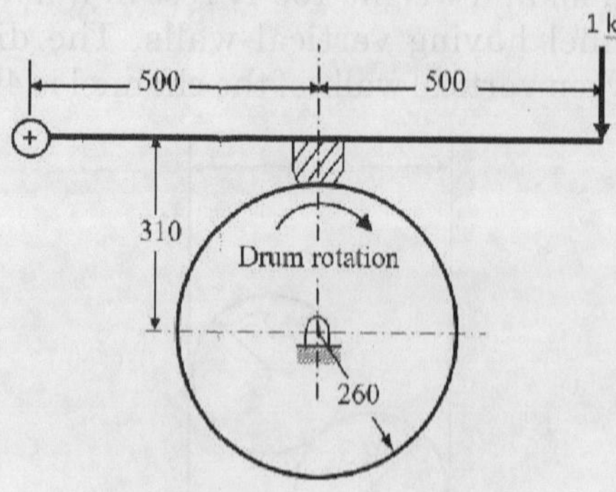

All dimensions are in mm

The magnitude of the torque applied by the brake is ________________ N.m *(round off to one decimal place)*.

54. A machine part in the form of cantilever beam is subjected to fluctuating load as shown in the figure. The load varies from 800 N to 1600 N. The modified endurance, yield and ultimate strengths of the material are 200 MPa, 500 MPa and 600 MPa, respectively.

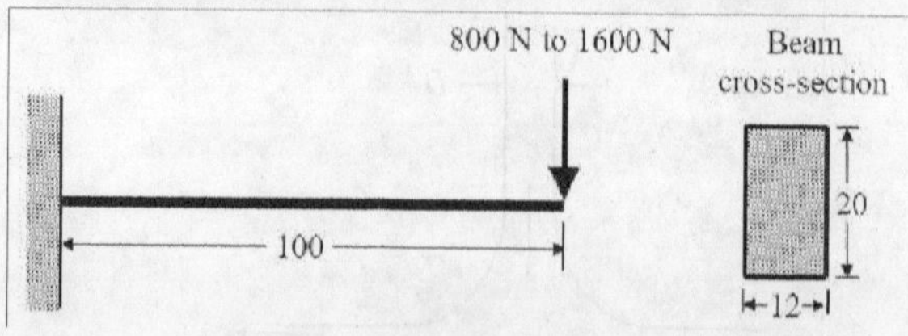

All dimensions are in mm

The factor of safety of the beam using modified Goodman criterion is ________________ *(round off to one decimal place)*.

55. A cantilever beam of rectangular cross-section is welded to a support by means of two fillet welds as shown in figure. A vertical load of 2 kN acts at free end of the beam.

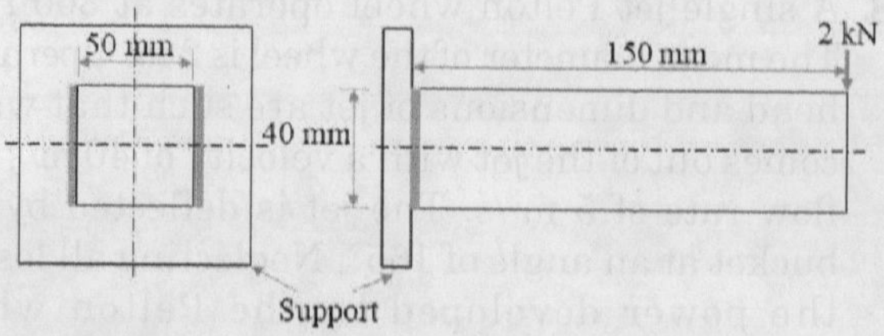

Considering that the allowable shear stress in weld is 60 N/mm^2, the minimum size (leg) of the weld required is ________ mm *(round off to one decimal place)*.

ANSWERS

General Aptitude (GA)

1. (d)	**2.** (d)	**3.** (d)	**4.** (d)	**5.** (c)	**6.** (c)
7. (a)	**8.** (c)	**9.** (c)	**10.** (d)		

Mechanical Engineering

1. (c)	**2.** (c)	**3.** (d)	**4.** (a)	**5.** (d)	**6.** (b)
7. (b)	**8.** (a)	**9.** (a)	**10.** (b)	**11.** (b)	**12.** (c)
13. (a)	**14.** (a)	**15.** (a)	**16.** (c)	**17.** (c)	**18.** (d)
19. (c)	**20.** 36 to 37	**21.** 397 to 399	**22.** 79 to 81	**23.** 555 to 565	**24.** 0 to 0
25. 0.085 to 0.090	**26.** (b)	**27.** (c)	**28.** (a)	**29.** (b)	**30.** (b)
31. (a)	**32.** (b)	**33.** (a)	**34.** (c)	**35.** 0.5 to 0.5	**36.** 660 to 664
37. 30 to 36	**38.** 625 to 750	**39.** 45.00 to 50.00	**40.** 955 to 999	**41.** 0.20 to 0.30	**42.** 4 to 4
43. 0.10 to 0.12	**44.** 0.76 to 0.78	**45.** 92 to 96	**46.** 195.1 to 195.3	**47.** 11.2 to 11.4	
48. 2.50 to 2.80	**49.** 4.90 to 5.10	**50.** 124.0 to 126.0	**51.** 249.0 to 251.0	**52.** 1.9 to 2.1	
53. 199.0 to 201.0	**54.** 1.9 to 2.1	**55.** 6.4 to 6.9			

EXPLANATIONS

GENERAL APTITUDE

1. When you say you could hardly do some thing you are emphasing that it is very difficult for you to do it.

∵ Could + Adverb + Verb (I).

Hence, option (d) is the correct answer.

2. According to the questions:

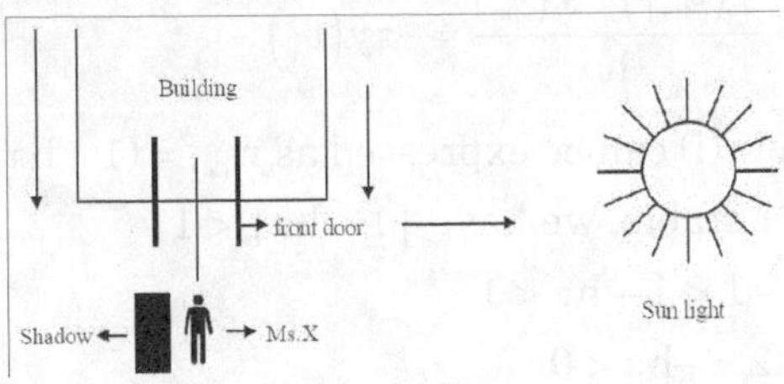

∴ From the above figure, it is clear that building is facing towards **South**.

Hence, option (d) is the correct answer.

3. Given that,

Area of right triangle MON = 50

According to given figure:

$\Rightarrow \frac{1}{2}\times ON \times OM = 50$

[∵ ON = OM = radius of circle = r(say)]

$\Rightarrow \frac{1}{2}\times r \times r = 50$

$\Rightarrow r^2 = 100$

∴ Area of circle with radius r = πr^2

$= \pi(100) = 100\pi$.

Hence option (d) is the correct answer.

4. As given expression =

$\Delta 2 \oplus 3\Delta((4 \otimes 2) \triangledown 4)$

$= +2 - 3 + ((4 \div 2) \times 4)$

$= 2 - 3 + (2 \times 4) = 2 - 3 + 8 = 7$

Hence option (d) is the correct answer.

5. One can clearly infer that eating leafy vegetables undoubtely lead a healthy style.

7. From the options, the figure in option (b) is the missing piece. If the figure is rotated clockwise then it will form a rectangular with the given piece in the picture.

8. For farm P:

Number of hens = 65

Number of ducks = 91

Number of goats = 169

For farm Q:

Number of hens = $416 \times \frac{5}{32} = 65$

Number of ducks = $416 \times \frac{14}{32} = 182$

Number of goats = $416 \times \frac{13}{32} = 169$

∴ Total number of hens = 65 + 65 = 130

Total number of ducks = 91 + 182 = 273

Total number of goats = 169 + 169 = 338

∴ The new ratio of hens: ducks: goats = 130: 273: 338 = 10: 21:26.

9. As given that,

Total number of executives, across all companies = 10,000

∴ Number of executives in $C_2 = 10{,}000 \times \frac{5}{100} = 500$

Number of executives in $C_5 = 10{,}000 \times \frac{20}{100} = 2000$

∴ Number of management degree holders among the executives in $C_2 = 500 \times \frac{1}{5} = 100$

And, Number of management degree holders among the executives in $C_5 = 2000 \times \frac{9}{10} = 1800$

So, the total number of management degree holders among the executives in companies C_5 and C_5 together = 100 + 1800 = 1900.

Hence, option (c) is the correct answer.

10. As given that

Total block is (Q_R)

Total number of ways,

$\frac{1}{Q}\frac{2}{-}\times\frac{1}{R}\times\frac{2}{S}\times\frac{1}{-}\rightarrow 4$ choices

(OR)

$\frac{2}{-}\times\frac{-}{Q}\times\frac{1}{R}\times\frac{S}{-}\rightarrow 2$ choices

$\frac{S\ 2}{-\ -}\times\frac{-}{Q}\times\frac{1}{R}\rightarrow 2$ choices

(OR)

$R\frac{2}{-}\times\frac{-}{Q}\times\frac{1}{S}\rightarrow 2$ choices

(OR)

$\frac{S}{-}\times\frac{R}{-}\times\frac{2}{-}\times\frac{Q}{-}\times\frac{1}{-}\rightarrow$ 2 choices

(OR)

$\frac{2}{S}\times\frac{2}{-}\times\frac{}{R}\times\frac{1}{-}\times\frac{}{Q}\rightarrow$ 4 choices

Hence, total number of distinct seating arrangement possible = 4 + 2 + 2 + 2 + 2 + 4 = 16.

MECHANICAL ENGINEERING

1. As given differential equation,

$$\sin(x)\frac{dy}{dx}+y\cos x=1$$

Differential equation can be written as

$$\frac{dy}{dx}+(\cot x)\,y=\operatorname{cosec} x \quad \text{.... (i)}$$

$$\therefore \text{I.F}=e^{\int \cot x\,dx}=e^{\log(\sin x)}=\sin x$$

$$\therefore \text{solution is } y\times(\sin x)=\int \operatorname{cosec} x\times\sin x\,dx+c$$

$$=\int 1dx+c$$

$$\Rightarrow y\sin x=x+c \quad \text{....(ii)}$$

Using $y\left(\frac{\pi}{2}\right)=\frac{\pi}{2}$, (ii) gives

$$\frac{\pi}{2}\times 1=\frac{\pi}{2}+c\Rightarrow c=0$$

$$\therefore y\sin x=x \quad \text{....(iii)}$$

at $x=\frac{\pi}{6}$, (iii) gives $y\times\frac{1}{2}=\frac{\pi}{6}$

$\Rightarrow y=\frac{\pi}{3}$ is the value of $y\left(\frac{\pi}{6}\right)$

Hence, option (c) is the correct answer.

2. Let, $\ell=\lim_{x\to 0}\left(\frac{1-\cos x}{x^2}\right)$ is $\frac{0}{0}$ form

Then, $\ell=\lim_{x\to 0}\frac{\sin x}{2x}$ (using L – Hospital's rule)

$$\therefore \ell=\frac{1}{2}\lim_{x\to 0}\frac{\sin x}{x}=\frac{1}{2}\times 1=\frac{1}{2}$$

Hence, option (c) is correct answer.

3. As we know that, Dirac Delta function

$$\delta(t-a)=\lim_{\epsilon\to 0}\begin{bmatrix}0, & -\infty<t<a\\ 1/\epsilon, & -a\le t\le a+\epsilon\\ 0, & a+\epsilon<t<\infty\end{bmatrix}$$

$L\{\delta(t-a)\}=e^{-as}\rightarrow$ direct result

$L\delta(t)=1$

4. As we know that,

The Euler's iterative formula (forward) is

$$y_{n+1}=y_n+hf\left(t_n,y_n\right)$$

$$=y_n+h\times(-\pi y_n) \quad \text{...(i)}$$

$$\Rightarrow \frac{y\left(t_{n+1}\right)-y\left(t_n\right)}{h}=-\pi y\left(t_n\right)$$

Now (i) can be expressed as $y_{n+1}=(1-h\pi)\,y_n$

For stable, we have $|1-h\pi|<1$

$\Rightarrow -1<1-h\pi<1$

$\Rightarrow 2<-h\pi<0$

$\Rightarrow 0<h<\frac{2}{\pi}$ $(\because |x|<\ell \Leftrightarrow -\ell<x<\ell)$

Hence, option (a) is the correct answer.

5. Since sum of independent Binomial random variables is also a Binomial random variable and Normal distribution is limiting case of Binomial distribution, if n is indefinitely large i.e., $n\to\infty$

$$\therefore y=\sum_{i=1}^{n}X_i=X_1+X_2+...+X_n$$

And distribution of y as $n\to\infty$ can be approximated by Normal.

Hence option (d) is the correct answer.

6. Let, Loading path is 'OA'

Unloading path is 'AB'

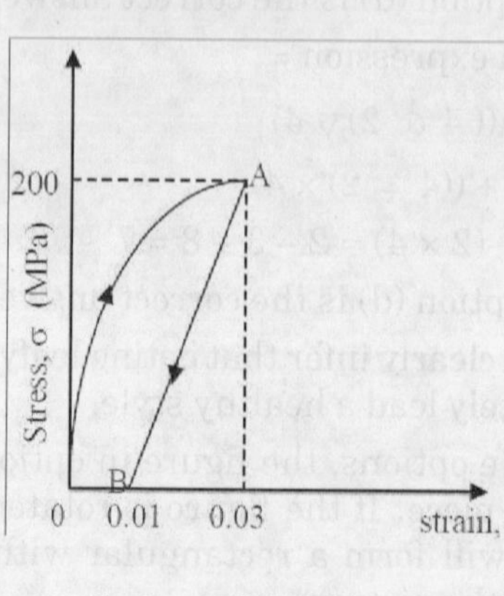

Elastic strain is nothing but the strain which is recovered = 0.03 – 0.01 = 0.02

Plastic strain is nothing but the strain which is retained = 0.01

Hence the correct option is (b)

7. As given figure

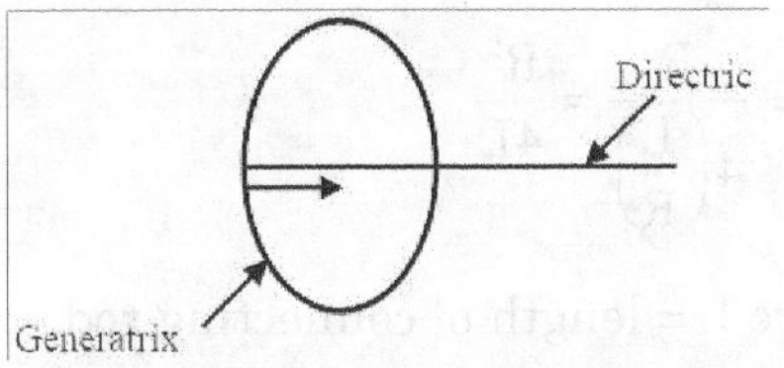

Directrix is perpendicular to generatrix means the surface obtained is cylindrical.

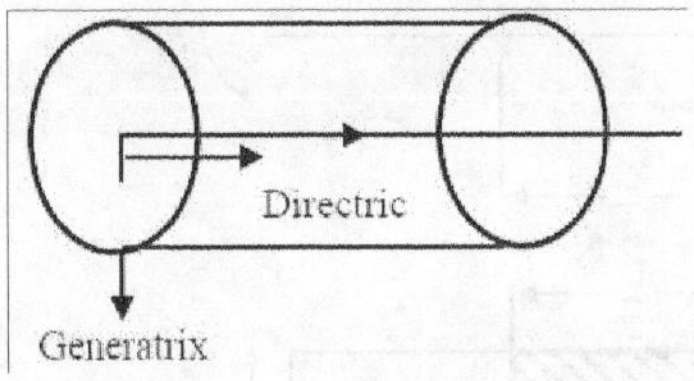

Hence, option (b) is the correct answer.

8. For large diameter through hole machining, the sequence of operation will be drilling, boring and reaming.

Hence option (a) is the correct answer.

9. Total axial displacement is eliminated by using pre-loaded ball screws, which consequently also reduces the backlash. Hence, answer is option (a).

10. The centre line average roughness, (R_a)

$$= \frac{\Sigma A}{L} = \frac{(4 \times 1)L/4}{L} = 1 \text{ mm.}$$

Hence, option (b) is the correct answer.

11. As we know that, Carnot cycle and striling cycle will have one isothermal process.

Carnot cycle

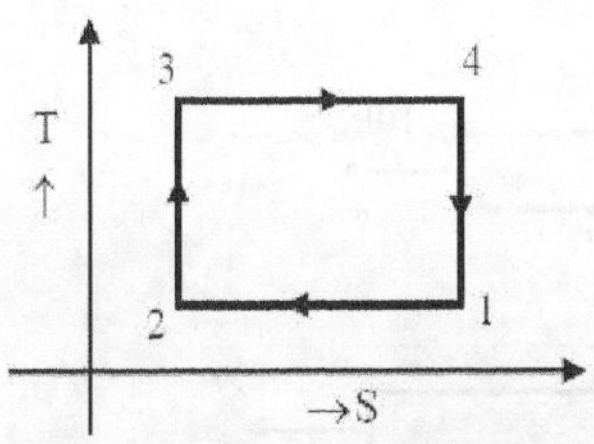

1 – 2 = Isothermal heat rejection process

2 – 3 = Isentropic compression

3 – 4 = Isothermal heat addition process

4 – 1 = Isentropic expansion

Stirling cycle

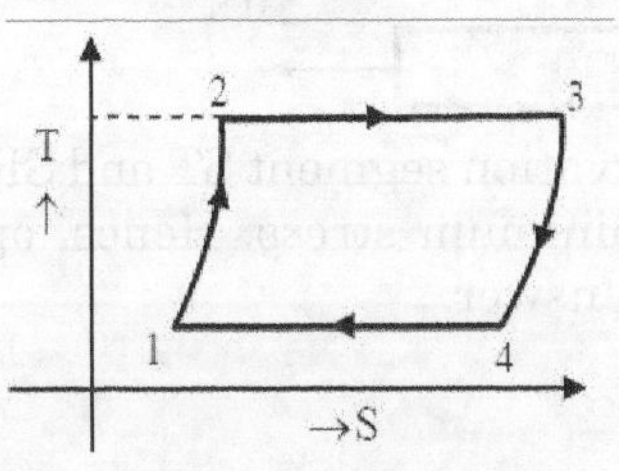

1 – 2 = Constant volume heat addition

2 – 3 = Isothermal compression

3 – 4 = Constant volume heat rejection

4 – 1 = Isothermal expansion

Hence, answer is option (b).

12. As given that,

Specific volume $(\bar{v})$ = 2.75m³/kmol

Compressibility factor (z) = 0.95

Pressure of super heated steam (P) = 1500 kPa

Temperature of steam (T) = ?

As we know that,

$$\text{Compressibility factor (z)} = \frac{P\bar{v}}{RT}$$

$$095 = \frac{500 \times 10^3 \times 2.75}{8.3143 \times T \times 10^3} (\because \bar{R} = 8.3143\text{kJ/kmolK})$$

T = 522.24K = 249.24°C

Hence, so answer is option (c).

13. According to the question:

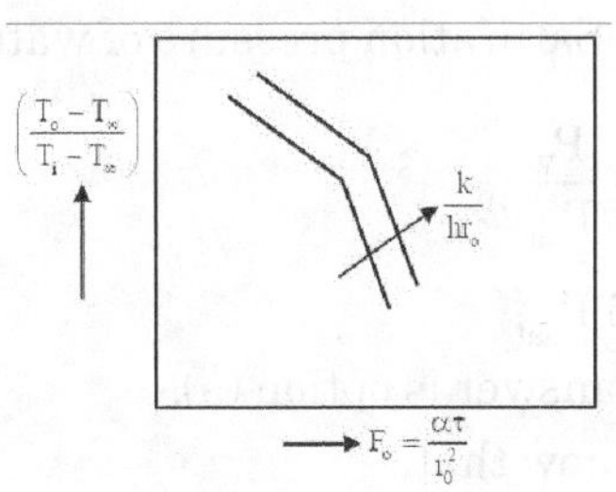

where,

T_0 = Centre line temperature of sphere

T_∞ = Temperature of oil bath

T_i = Initial temperature of sphere

α = Thermal diffusivity of sphere

τ = Time

r_0 = Radius of sphere

k = Thermal conductivity of sphere

h = Convective heat transfer coefficient between Sphere and oil bath

F_o = Fourier number

$$\text{Bi} = \text{Biot number} = \frac{hr_o}{3k} \therefore \frac{k}{hr_o} \propto \frac{1}{Bi}$$

Therefore dimensionless parameters required to determine instantaneous center temperature of ball using Heisler chart are Biot number and Fourier number. Hence option (a) is the correct answer.

14. As we know that,

Heat transfer through a long fin, $(\dot{Q})$ = $\sqrt{hPKA_C}\left(T_S - T_\infty\right)$

From the above we can say that $Q \propto \sqrt{K}$ (keeping everything else is constant)

Given that $K_2 = 2K_1$

$$\frac{\dot{Q}_2}{\dot{Q}_1} = \sqrt{\frac{K_2}{K_1}} = \sqrt{\frac{2K_1}{K_1}} = \sqrt{2}$$

$$\frac{\dot{Q}_2}{\dot{Q}_1} = \sqrt{2}$$

Hence, answer is option (a) is the correct answer.

15. As given that, relative humidity of air at 300 K is $(\phi) = 0.50$

We know that, $\phi = \dfrac{P_V}{P_{sat}}$

where,

P_v = Partial pressure of water vapour

P_{sat} = 6 Saturation pressure of water vapour

$$\Rightarrow 0.5 = \frac{P_V}{P_{sat}}$$

$P_V = 0.5\ P_{sat}$

Hence, answer is option (a).

16. As we know that,

Primary unbalance force in reciprocating engine $= mR\omega^2 \cos\theta$

$\therefore$ Secondary unbalance force in reciprocating engine $= mR\omega^2 \dfrac{\cos 2\theta}{n}$

$\therefore$ Expression for secondary unbalance force can be written as $= mR(2\omega)^2 \dfrac{\cos 2\theta}{4n}$

$$= m\left(\frac{R}{4n}\right)(2\omega)^2 \cos 2\theta$$

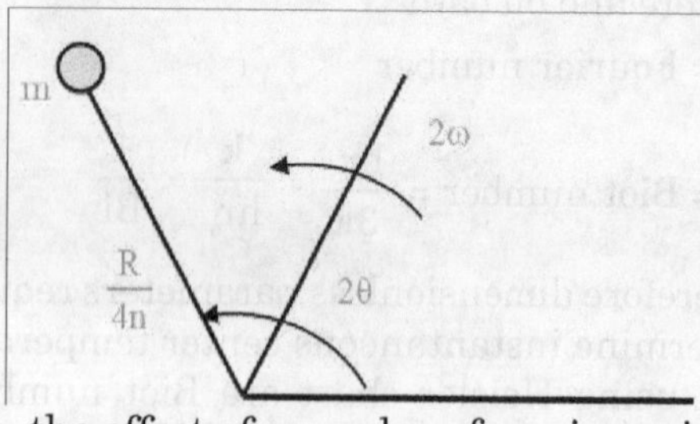

So, the effect of secondary force is equivalent to an imaginary crank of length $\dfrac{R}{4n}$ rotating at a double angular velocity, i.e., twice the engine speed.

$$\frac{R}{4n} = \frac{R}{4\left(\frac{L}{R}\right)} = \frac{R^2}{4L},$$

where L = length of connecting rod.

Hence, answer is option (a).

17.

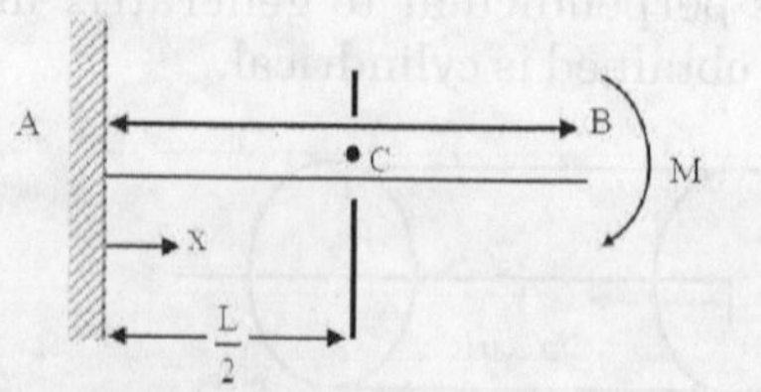

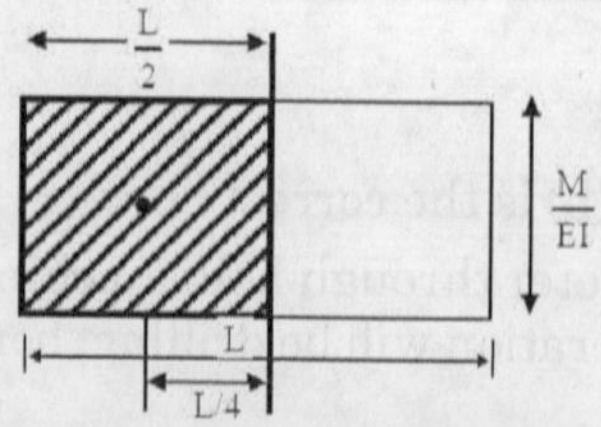

Deflection of the beam at

$x = \dfrac{L}{2}$ (at point 'C')

$= \text{Moment of} \left(\dfrac{M}{EI}\right)$ diagram about 'C' $\left(x = \dfrac{L}{2}\right)$

$$\therefore \delta_C = \left(\frac{M}{EI}\right)\left(\frac{L}{2}\right)\left(\frac{L}{4}\right) = \frac{ML^2}{8EI}\ (\downarrow)$$

Hence answer is option (c).

18. From the given figure,

$R_p = 25 + 15 - 20 - 10 \Rightarrow R_p = 10kN$

For segment PQ

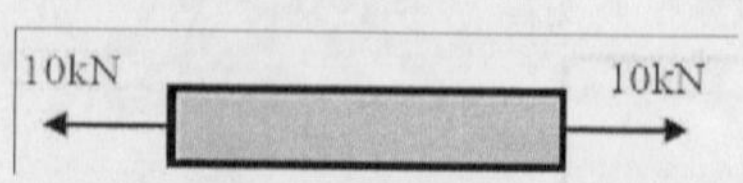

For segment QR

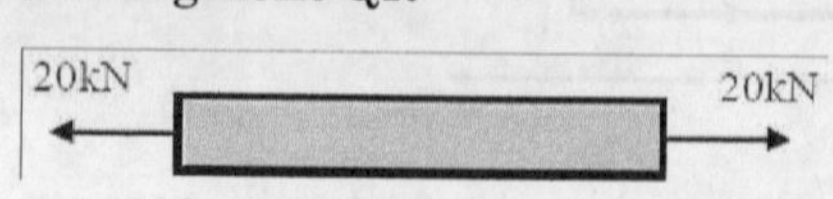

For segment SR

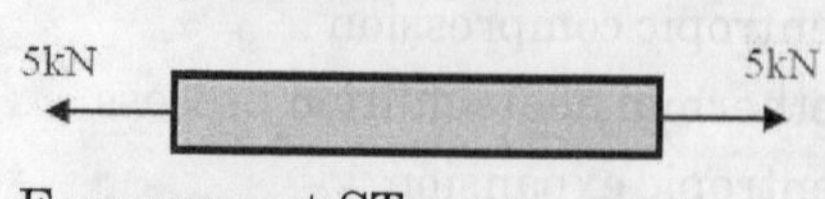

For segment ST

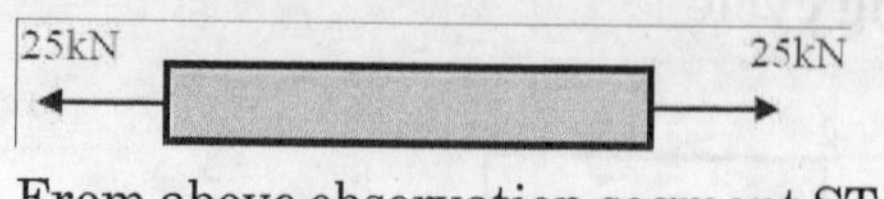

From above observation segment ST and SR will maximum and minimum stress. Hence, option (d) is the correct answer.

19. According to the question, in helical compression spring (or) helical extension spring the spring wire will be subjected to both direct shear stress (τ_1) and torsional shear stress (τ_2).

d = diameter of spring wire

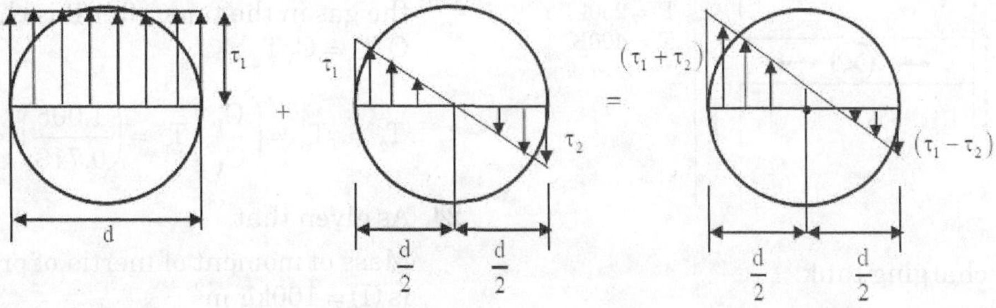

Direct shear stress across spring wire — Torsional shear stress across spring wire — Combined effect of direct and torsional shear stress

Hence, answer is option (c) is the correct answer.

20. As given that,

Probability of satisfaction = 0.95

Lead time (LT) = 5 days

Mean (μ) = 50 units

Standard deviation (σ) = 10 units

Z for 95% satisfaction = 1.64

As we know that,

$$\text{Safety stock} = Z\sqrt{LT}\,\sigma = (1.64)\sqrt{5} \times 10$$

$$= 36.67\text{units}$$

21. As given,

Density of sea water, (ρ) = 1050kg/m^3

P_{atm} = 101kPa

$= 101 \times 10^3\text{Pa} = 101 \times 10^3\ \text{N/m}^2$

$g = 9.8\ \text{m/sec}^2$

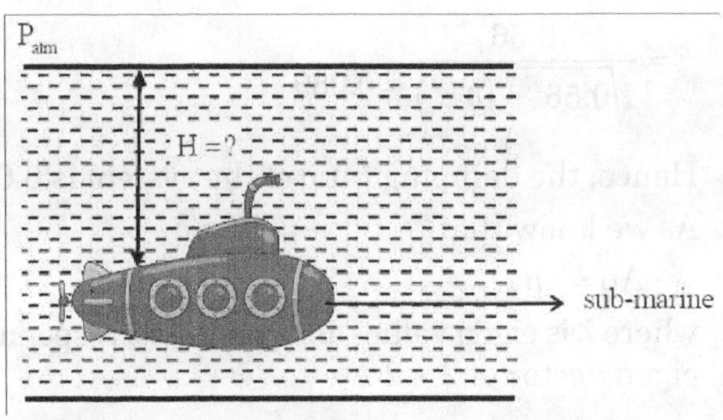

As we know that,

Pressure on the submarine,

$(P_{abs}) = P_{atm} + \rho gH$

$4.2 \times 10^6 = 101 \times 10^3 + (1050 \times 9.8 \times H)$

$H = 398.35\text{m}$

22. As given that

$u = 4\text{m/sec}$

$du = (u - 0) = (4 - 0) = 4\text{m/sec}$

$$dy = 5\text{mm} = \left(\frac{5}{1000}\text{m}\right)$$

$\rho = 800\text{kg/m}^3$

$v = 1.25 \times 10^{-4}\text{m}^2\text{/sec}$

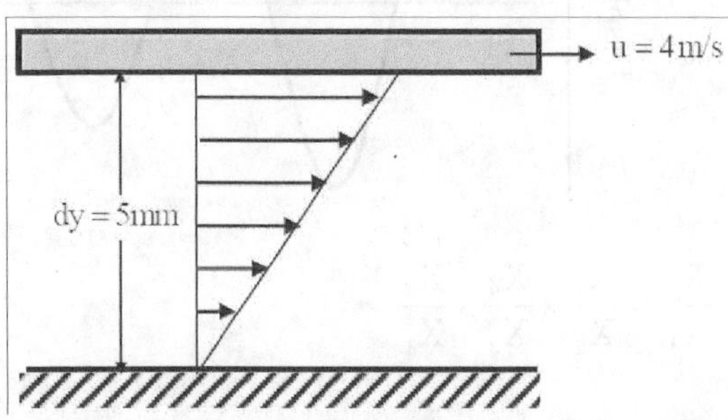

As we know that,

$$\text{Average shear stress in the fluid is } (\tau) = \mu\left(\frac{du}{dy}\right)$$

where,

$$v = \frac{\mu}{\rho} \Rightarrow \mu = (v)(\rho)$$

$\mu = (1.25 \times 10^{-4})(800) = 0.1\ \text{Nsec/m}^2$

$$\Rightarrow \tau = (0.1)\left(\frac{4}{\left(\frac{5}{1000}\right)}\right) = 80\text{N/m}^2$$

23. According to the question:

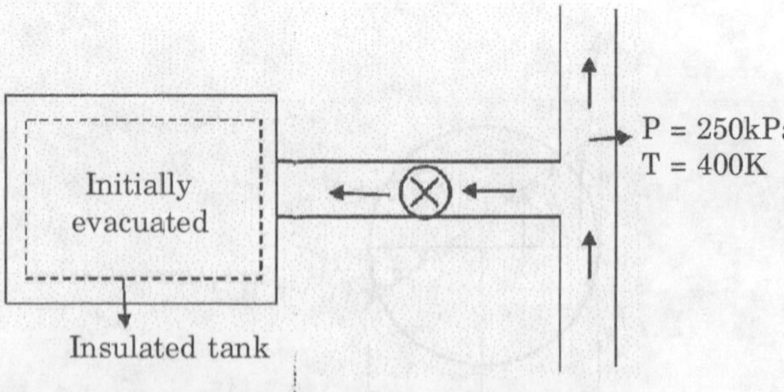

It look like a charging tank.

As we know that,

For charging tank, $m_p h_p = m_2 u_2 - m_1 u_1$

where, the subscript 'p' refers to the constant state of fluid in the pipeline. Since the tank is initially empty, $m_1 = 0$

Since, $m_p h_p = m_2 u_2$

$\therefore\ m_p = m_2$

$\therefore\ h_p = u_2$

Since, the fluid is perfect gas, the temperature of the gas in the tank after it is charged is given by $C_p T_p = C_V T_2$

$$T_2 = \gamma T_p = \left(\frac{C_P}{C_V}\right) T_p = \left(\frac{1.005}{0.718}\right)(400) = 559.89\text{K}$$

24. As given that,

Mass of moment of inertia of propeller and shaft is $(I) = 100\text{kg.m}^2$

Rotational speed of propeller, $(\omega_S) = 12\text{rad / sec}$

Ship is rolling, hence precession angular velocity, $(\omega_p) = 0$

$\therefore$ Gyroscopic couple (moment) generated on the shaft $(C) = I \cdot (\vec{\omega}_s \times \vec{\omega}_p) = 0$ N.m

25. As given that,

$X_1 = 8$ mm; $X_4 = 1.5$ mm

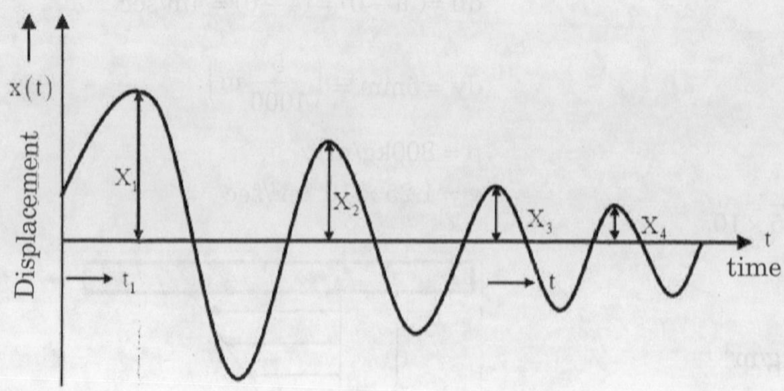

$$\frac{X_1}{X_4} = \frac{X_1}{X_2} \times \frac{X_2}{X_3} \times \frac{X_3}{X_4}$$

$$\frac{X_1}{X_4} = \left(\frac{X_1}{X_2}\right)^3 \left(\because \frac{X_1}{X_2} = \frac{X_2}{X_3} = \frac{X_3}{X_4}\right)$$

$$\frac{X_1}{X_2} = \left(\frac{X_1}{X_4}\right)^{\frac{1}{3}}$$

$$\ell n\left(\frac{X_1}{X_2}\right) = \frac{1}{3} \ell n\left(\frac{X_1}{X_4}\right)$$

Logarithmic decrement, (δ)

$$= \ell n\left(\frac{X_1}{X_2}\right) = \frac{1}{3} \ell n\left(\frac{8}{1.5}\right) = 0.56$$

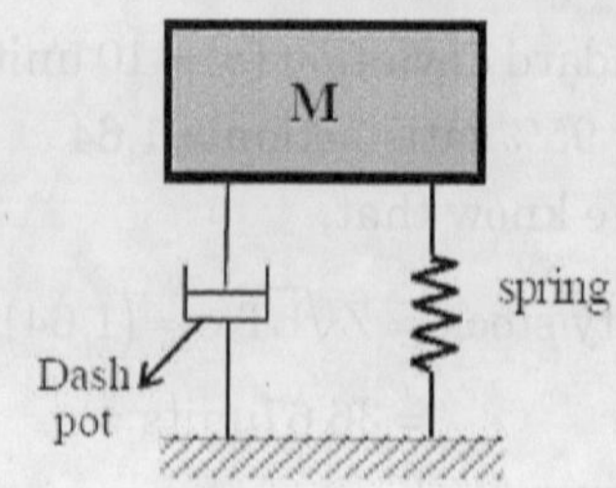

$$\text{Damping factor, } (\xi) = \frac{\delta}{\sqrt{\delta^2 + 4\pi^2}}$$

$$\xi = \frac{0.56}{\sqrt{0.56^2 + (4\pi^2)}} = 0.089$$

Hence, the damping ratio of the system is 0.089.

26. As we know that

$Ap = \lambda p$,

where λ is eigen values and p is the corresponding eigen vector of 2×2 A

$\Rightarrow \|p'\| = |\lambda p'\| = \lambda\ |p\|$

Since p is 2-dimensional vector

$\Rightarrow \lambda$ p and p' are also 2-dimensional vectors. Also p and λp has same direction vectors of p' = θ

Hence, option (b) is the correct answer.

27. Let, $I = \oint \frac{\cosh 3z}{2z}$, where

C : |z| = 1 and 2z = 0 gives z= 0 is a singular point lies inside C.

$$I = \int_C \frac{\cosh 3z}{2z} dz = \int_C \frac{\left(\frac{\cosh 3z}{2}\right)}{z-0} dz$$

$$I = 2\pi i \left(\frac{\cosh 3z}{2}\right)_{at\, z=0}$$

(By using Cauchys integral formula)

$$I = 2\pi i \frac{1}{2} = \pi i$$

28. As given table,

Job	M_1 Turing (Minutes)	M_2 Grinding (Minutes)
A	(2)	6
B	4	(1)
C	8	(3)
D	9	(7)
E	(7)	9
F	6	(5)
G	5	(2)
H	10	(4)

M_1	A	E	D	F	H	C	H	B	M_2

If least time taken job is to perform on machine (M_1) should be performed first and place at near to M_1.

If least time taken job is to perform on machine 2 (M_2) should be performed at last and placed near to M_2.

If tie occurs first place near to machine 1 (M_1)

So the optimal sequence is A E D F H C G B.

Hence, the correct answer is option (a).

29. As given that,

$dU = Tds + \tau dL$

Since all the thermodynamic properties and 'dU' is exact differential like

$dz = Mdx + Ndy$

The condition for exact differential is

$$\left(\frac{\partial M}{dy}\right)_x = \left(\frac{\partial N}{dx}\right)_y$$

Based on above condition $\left(\frac{\partial T}{dL}\right)_S = \left(\frac{\partial \tau}{\partial S}\right)_L$

Hence, option (b) is the correct answer.

30. As we now that,

Lagrangian (L) = T – V

where,

T = kinetic energy of system

V = Potential energy of systems.

T = Linear kinetic energy of 'M' + Linear kinetic energy of 'm' in 'x' and 'y' directions

Linear kinetic energy of 'M = $\frac{1}{2}Mx^2$

Linear kinetic energy of 'm' = Sum of kinetic energy as of small strips of mass 'dm' with length 'dl'

dm = mass of small strip of rod

dl = length of strip of rod

Because force 'F' mass 'M' moves a distance 'x' and rod of mass 'm' moves an angle 'θ'.

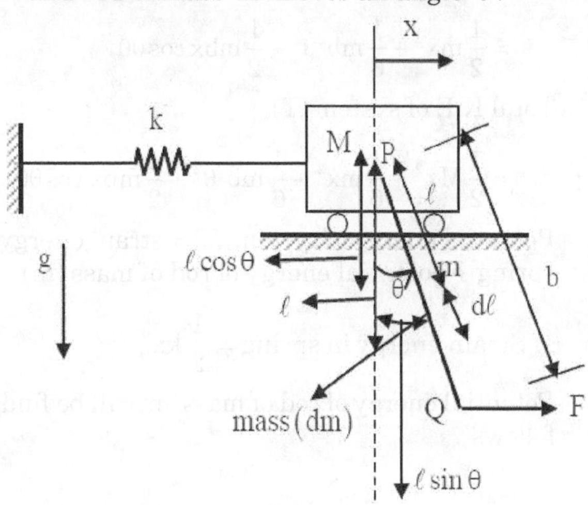

Mass 'm' is over 'b' length,

'dm' is over 'dl' length

$$\therefore \frac{m}{b} = \frac{dm}{dl}$$

$$dm = \left(\frac{m}{b}\right)(dl)$$

Linear displacement of strip 'dm' which is at a distance of 'l' from P' is in x-Direction = $(x + l\sin\theta)$

Linear velocity of strip (dm) in x-direction $= (x + l\cos\theta\,\dot{\theta})$

Linear displacement of strip 'dm' in y-direction $= (\ell - \ell\cos 0)$

Linear velocity of strip 'dm' in y-direction

$= +\ell\sin\theta\,\dot{\theta}$

Resultant velocity of 'dm'

$$= \sqrt{\left(\dot{x} + \ell\cos\theta\dot{\theta}\right)^2 + \left(\ell\sin\theta\dot{\theta}\right)}$$

$\therefore$ Resultant K.E of small strip

$$= \frac{1}{2}(\text{dm})\left(\left(\dot{x} + \ell\cos\theta\dot{\theta}\right)^2 + \left(\ell\sin\theta\dot{\theta}\right)^2\right)$$

$$= \frac{1}{2}\left(\frac{m}{b}\right)(d\ell)\left(\dot{x}^2 + \ell^2\cos\theta\dot{\theta}^2 + 2\dot{x}\ell\cos\theta\dot{\theta} + \ell^2\sin^2\theta\dot{\theta}^2\right)$$

$$= \frac{1}{2}\left(\frac{m}{b}\right)(d\ell)\left(\dot{x}^2 + \ell^2\dot{\theta}^2 + 2\dot{x}\ell\cos\theta\dot{\theta}\right)$$

So, Total K.E of Rod:

$$\int_0^b\left(\frac{1}{2}\right)\left(\frac{m}{b}\right)(d\ell)\left(\dot{x}^2 + \ell^2\dot{\theta}^2 + 2\dot{x}\ell\cos\theta\dot{\theta}\right)$$

$$= \left(\frac{1}{2}\right)\left(\frac{m}{b}\right)\left(\dot{x}^2(b) + \left(\frac{b^3}{3}\right)\dot{\theta}^2(b) + 2\dot{x}\frac{b^2}{2}\cos\theta.\dot{\theta}\right)$$

$$= \frac{1}{2}m\dot{x}^2 + \frac{1}{6}mb^2\dot{\theta}^2 + \frac{1}{2}mb\dot{x}\cos\theta\dot{\theta}$$

Total K.E of system (T)

$$= \frac{1}{2}M\dot{x}^2 + \frac{1}{6}m\dot{x}^2 + \frac{1}{6}mb^2\dot{\theta}^2 + \frac{1}{2}mb\dot{x}\cos\theta\dot{\theta}$$

Potential energy of system (V) = strain energy in spring + potential energy of rod of mass (m)

$\therefore$ Strain energy in spring $= \frac{1}{2}kx^2$

Potential energy of rod of mass 'm' will be find as follows

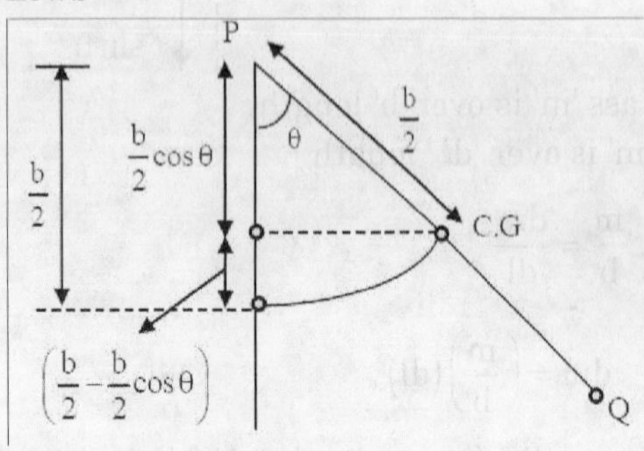

$\therefore$ Potential energy of mass 'm' $= mg\left(\frac{b}{2} - \frac{b}{2}\cos\theta\right)$

But in calculation of langrangian, will take the terms which are function 'θ' hence P.E of mass

$$\text{'m'} = \frac{-mgb}{2}\cos\theta$$

As we know that,

$\therefore$ Total potential energy of system (V)

$$= \frac{1}{2}kx^2 - \frac{mgb}{2}\cos\theta$$

$$L = T + V = \frac{1}{2}(m+M)\dot{x}^2 + \frac{1}{6}mb^2\dot{\theta}^2$$

$$+ \frac{1}{2}mb\dot{x}\cos\theta\dot{\theta} - \frac{1}{2}kx^2 + \frac{mgb}{2}(\cos\theta)$$

$$L = \frac{1}{2}(m+M)\dot{x}^2 + \frac{1}{6}mb^2\dot{\theta}^2$$

$$+ \frac{1}{2}mb\dot{x}\cos\theta\dot{\theta} - \frac{1}{2}kx^2 + \frac{mgb}{2}(\cos\theta)$$

Hence, option (b) is the correct answer.

31. As to the question:

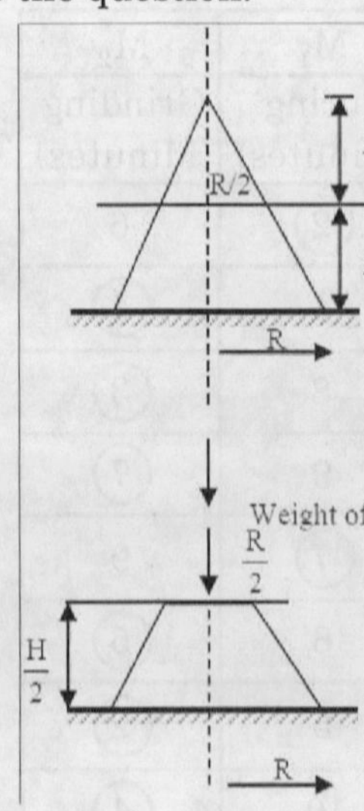

Weight of cone (P), above $\frac{H}{2}$

= volume of Cone above $\frac{H}{2}$ × specific weight of cone

$$P = \pi\left(\frac{R}{2}\right)^2\left(\frac{H}{2}\right)(W)$$

$$\text{Deflection of cone}(\delta) = \frac{4P\left(\frac{H}{2}\right)}{\pi E(4R)\left(\frac{R}{2}\right)} \quad \left(\therefore \delta = \frac{4PL}{\pi E d_1 d_2}\right)$$

$$\delta = \frac{(4)(\pi)\left(\frac{R}{2}\right)^2\left(\frac{H}{2}\right)(W)\left(\frac{H}{2}\right)}{(\pi E)(4R)\left(\frac{R}{2}\right)} = \frac{WH^2}{8E}$$

Hence, option (a) is the correct answer.

32. As given that

Mass moment of inertia of rocker about 'B' is $(I_B) = 10^{-4} kg.m^2$

$a = 3.5cm = 0.035m$, $b = 2.5cm = 0.025m$

Stiffness of push rod $(K_P) = 15N/mm = 15,000\ N/m$

Stiffness of valve spring $(K_V) = 10N/mm = 10,000\ N/m$

As given figure:

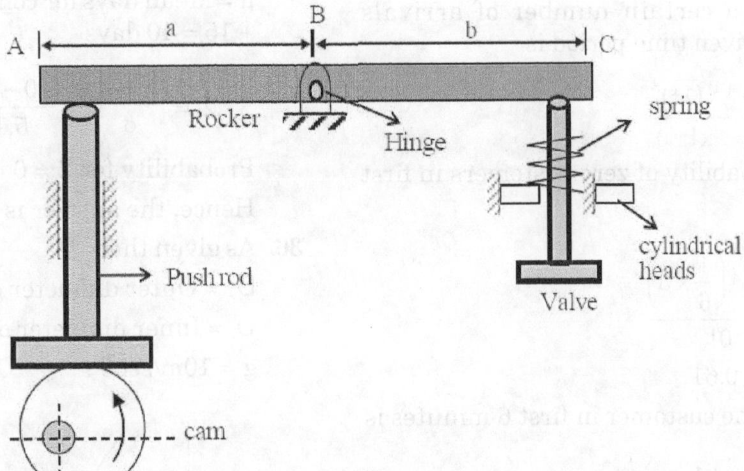

According to the question the mass of push rod and valve is negligible, then the given system will be looking like this

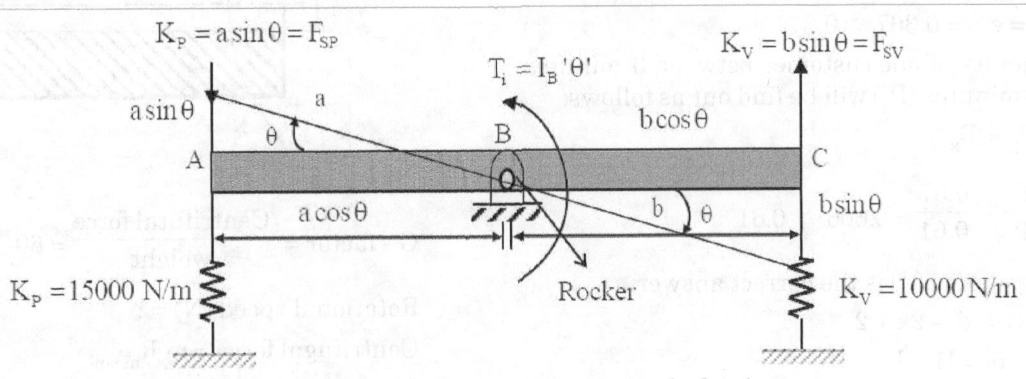

Here, that give some small angular displacement to Rocker in clock wise,

F_{SP} = Spring force of push rod = $K_p a \sin\theta$

F_{SV} = Spring force of valve spring = $K_v b \sin\theta$

$\Sigma M_B = 0$

$I_B\ddot{\theta} + (K_v b \sin\theta)(b\cos\theta) + (K_p a\sin\theta)(a\cos\theta) = 0$

For small angles

$\sin\theta \cong \theta$, $\cos\theta \cong 1$

$I_B\ddot{\theta} + (K_V b^2 + K_p a^2)\theta = 0$

$10^{-4}\ddot{\theta} + (10000 \times 0.025^2 + 15000 \times 0.035^2)\theta = 0$

$10^{-4}\ddot{\theta} + 24.6250 = 0$

$$\therefore \omega_n = \sqrt{\frac{24.625}{10^{-4}}} = 496.23 \text{ rad/sec}$$

$$\frac{2\pi N}{60} = 496.23 \Rightarrow N = 4738.65 \cong 4739 \text{ rpm}$$

Hence, option 'b' is the correct answer.

33. As given that,

Mean arrival rate, $(\lambda) = \dfrac{10 \text{ customers}}{\text{hr}}$

$= \dfrac{10}{60} = \dfrac{1}{6}$ customers/min

Probability of a certain number of arrivals occurring in a given time period is

$$\therefore P(X = x) = \frac{e^{-(\lambda t)}(\lambda t)^x}{x!}$$

Given that probability of zero customers in first 3 minutes is

$$P(X = 0) = \frac{e^{-\left(\frac{1}{6}\times 3\right)}\left(\frac{1}{6}\times 3\right)^0}{0!}$$

$\Rightarrow P_1 = 0.606 \cong 0.61$

Probability of one customer in first 6 minutes is

$$P(X = 1) = \frac{e^{-\left(\frac{1}{6}\times 6\right)}\left(\frac{1}{6}\times 6\right)^1}{1!}$$

$\Rightarrow P_3 = e^{-1} = 0.367 \cong 0.37$

Probability of one customer between 3 minutes and 6 minutes (P_2) will be find out as follows:

$P_1 \times P_2 = P_3$

$$P_2 = \frac{P_3}{P_1} = \frac{0.37}{0.61} = 0.606 \cong 0.61$$

Hence, option (d) is the correct answer.

34. Let $f(x) = x^2 - 2x + 2$

and $[a, b] = [1, 3]$

than $f'((x) = (2x - 2)$

Since the tangent of f(x) is parallel to the straight line joining f(1) and f(3), which is geometrical interpretation of Lagrange's mean value theorem.

$$\therefore f'(x) = \frac{f(b) - f(a)}{b - a}, \text{ where } a = 1, b = 3$$

$$\Rightarrow 2x - 2 = \frac{5-1}{2} = 2$$

$\Rightarrow x = 2 \in [1, 3]$

$\therefore x = 2$

Hence, option (c) is the correct answer.

35. As given table;

Activity	A	B	C	D
Mean (days)	6	11	8	15
Variance (days^2)	4	9	4	9

Sum of variances of all activities in critical path $= 4 + 9 + 4 + 9 = 26$ days

Standard deviation of all activities in critical path $= \sqrt{26} = 5.099 \cong 5.1$

X = No. of days project to be completed = 40

μ = Mean days for completing project = 6 + 11 + 8 + 15 = 40 days

$$\therefore Z = \frac{X - \mu}{\sigma} = \frac{40 - 40}{5.1} = 0$$

Probability for Z = 0 is 50% i.e., 0.5

Hence, the answer is (0.5)

36. As given that

D_o = Outer diameter of casting = 250mm

D_i = Inner diameter of casting = 230mm

g = 10m/sec^2

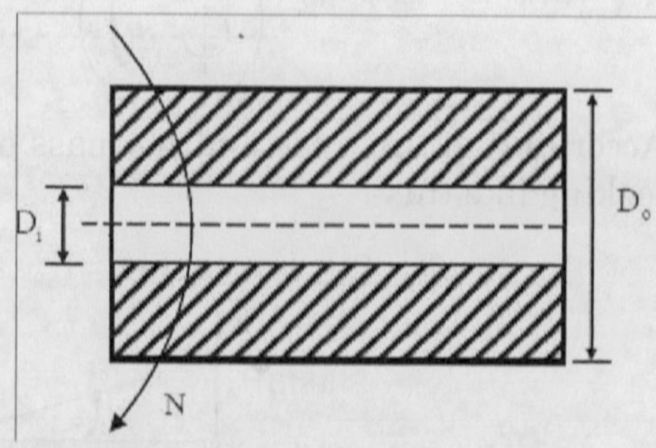

$$\text{G - factor} = \frac{\text{Centrifutal force}}{\text{weight}} = 60$$

Rotational speed (N) = ?

Centrifugal force = m $R_{mean}\omega^2$

As we know that

$$R_{mean} = \left(\frac{D_o + D_i}{4}\right) = \left(\frac{250 + 230}{4}\right) = 120 \text{ mm}$$

$= 0.12$ m

Centrifugal force = (m)(0.12)ω^2

$$\text{G-factor} = \frac{mRm\omega^2}{mg} = \frac{(m)(0.12)\omega^2}{(mg)} = 60$$

$$= \frac{(0.12)\omega^2}{(10)} = 60$$

$\Rightarrow \omega = 70.71$ rad/sec

$$\frac{2\pi N}{60} = 70.71 \Rightarrow N = 675.23 \cong 675 \text{ rpm}$$

37. As given that

Welding current (I) = 10,000A

Time taken for welding, (t) = 0.25 seconds

Contact resistance of metal sheets, (R) = 0.0001Ω

Volume of weld nugget, (V) = 70mm^3

Heat required to melt unit volume, (Q_R) = 12 J / mm^3

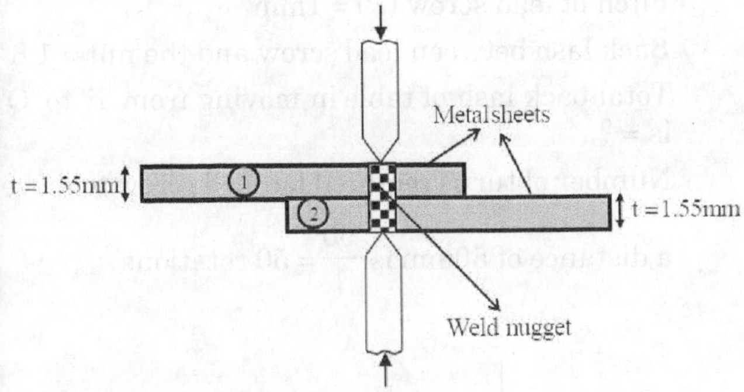

As we know that,

Heat supplied for welding, $Q_s = I^2RT$

$= (10{,}000)^2 (0.0001)(0.25) = 2500$ Joules

Total heat required, $(Q_R) = 12 \times 70 = 840$J

We also know that,

Thermal efficiency of welding process (μ)

$$= \frac{Q_R}{Q_S} = \frac{840}{2500} = 33.6\%$$

38. As given that

Cutting force (F_C) = 1000N

Friction force (F_f) = 600N

Uncut chip thickness (t_1) = 0.35mm

Chip thickness (t_2) = 1.5mm

Rake angle (α) = 120°

Shear force (F_S) = ?

Merchant's circle is as follows

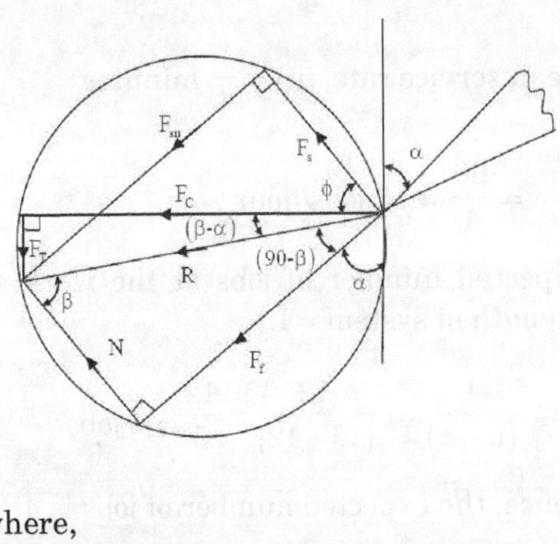

where,

φ = shear angle

$$r = \text{Chip thickness ratio} = \frac{t_1}{t_2} = \frac{0.75}{1.5} = 0.5$$

$$\therefore \tan\phi = \frac{r\cos\alpha}{1 - r\sin\alpha}$$

$$= \frac{r\cos(12)}{1 - r\sin(12)} = \frac{(0.5)\cos(12)}{1 - (0.5)\sin 12}$$

$\phi = 28.63°$

From the above circle

$R\cos(90 - \beta) = F_f = 600$N

$R\sin\beta = 600$N ...(i)

$R\cos(\beta - \alpha) = 1000$N ...(ii)

$$\frac{\sin\beta}{\cos(\beta - \alpha)} = \frac{600}{1000\text{N}}$$

$$\frac{\sin\beta}{\cos(\beta - 12)} = \frac{600}{1000\text{N}}$$

$1000\sin\beta = 600(\cos\beta\cos 12 + \sin\beta\sin 12)$

$1000\sin\beta = 586.89\cos\beta + 124.75\sin\beta$

$= 875.25\sin\beta = 586.89\cos\beta$

$\tan\beta = 0.67 \Rightarrow \beta = 33.82°$

$R\sin\beta = 600$

$$R = \frac{600}{\sin(33.82)} = 1078$$

$R\cos(\phi + \beta - \alpha) = F_s$

$(1078)\cos(28.63 + 33.82 - 12) = F_s$

$F_S = 686.42$N

Hence, the shear force is 686.42N

39. As given that, grinding operations;

Specific energy consumption = 15J/mm^3

Diameter of grinding wheel (D) = 200mm

Rotating speed of grinding wheel (N) = 3000 rpm

Material removal rate (MRR) = 6000 mm^3/min

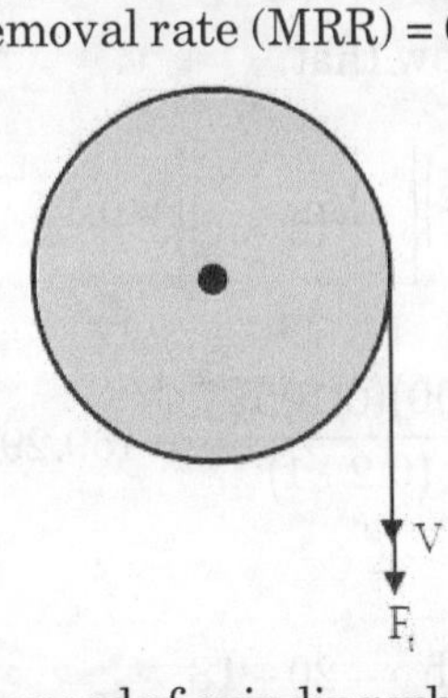

Peripheral speed of grinding wheel, (V)

$$= \frac{\pi DN}{1000}\ (\text{m/min})$$

$$= \frac{\pi \times 200 \times 3000}{1000} = 600\ \pi\text{m/min}$$

Let F_t = Tangential force

We know that, Specific energy consumption

$$= \frac{(F_t \times V)}{MRR} = 15 = \frac{F_t \times 600\pi}{6000} = \frac{150}{\pi} = 47.77\text{N}$$

Hence, the tangential force on the wheel is 47.77N

40. As given that

Width of plate (W) = 200mm

Initial thickness of plate, (h_o) = 20mm

Radius of roller, (R) = 300mm

Final thickness of plate, (h_f) = 18mm

Speed of roller (N) = 50 rpm

Strength coefficient (K) of the work material flow curve is 300MPa

Strain hardening coefficient (n) = 0.2

Coefficient of friction between rolls and plate, (μ) = 0.1

Roll force or roll separating force = ?

As we know that

Roll force per each roller,

$$F|_{Roller} = 1.15\, LW\bar{y}_{flow}\left(1 + \frac{\mu L}{2h_{avg}}\right)$$

where,

L = Length of contact roller with plate = $\sqrt{R\Delta h}$

R = 300mm, $\Delta h = h_o - h_f = 20 - 18 = 2$mm

$L = \sqrt{300 \times 2} = 24.49$mm

W = width of plate = 200mm

$\bar{y}_{flow}$ = Flow stress = $\frac{K\varepsilon_f^n}{n+1}$

K = 300Mpa, n = 0.2

We also know that,

$$\varepsilon_f = \left|\ell n\left(\frac{h_f}{h_o}\right)\right| = \left|\ell n\left(\frac{18}{20}\right)\right| = 0.15$$

$$\therefore \bar{y}_{flow} = \frac{(300)(0.105)^{0.2}}{(0.2+1)} = 159.29\text{MPa}$$

μ = 0.1

$$h_{avg} = \frac{h_o + h_f}{2} = \frac{20+18}{2} = 19\text{ mm}$$

$$F|_{Roller} = (1.15)(24.49)(200)(159.29)\left(1 + \frac{(0.1 \times 24.49)}{2 \times 19}\right)$$

= 955057.1N = 955.05kN

Hence, the roll force is 955.05 kN

41. As given that, distance moved by tool = 51 – 1 = 50 mm

Pitch of lead screw (P) = 1mm

Back lash between lead screw and the nut = 1.8°

Total back lash of table in moving from 'P' to 'Q' is = ?

Number of turns required for lead screw to move a distance of 50mm is $\frac{50}{1}$ = 50 rotations

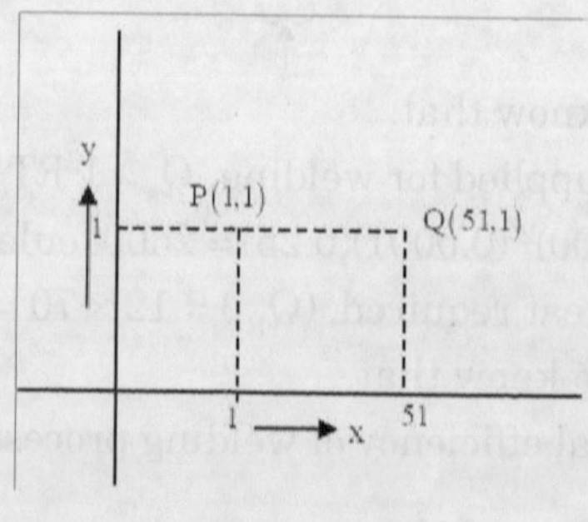

For each turn, the back lash = 1.8°

Total back lash = 50 × 1.8 = 90°

For every 360°, lead screw will move a distance of 1 mm

Back lash in mm is = $\left(\frac{90}{360}\right)$ = 0.25mm

Hence, the total backlash of the table on moving P to Q is 0.25 mm.

42. As given that,

Mean arrival rate, (λ) = 12 jobs/hour

Mean service rate, (μ) = $\frac{1}{4}$ minutes

$$= \frac{60}{4} = 15\text{ jobs/hour}$$

Expected number of jobs at the work station = length of system = L_S

$$L_S = \frac{\lambda}{(\mu - \lambda)} = \frac{12}{(15-12)} = \frac{12}{3} = 4\text{ jobs}$$

Hence, the expected number of jobs is 4.

43. As given that,

Heat rate in wire = 5 W/m

Assume unit length Heat transfer from wire,

Q = 5 watts

h = convective heat air transfer coefficient between wire and air.

As we know that,

$$\dot{Q} = \frac{T_S - T_\infty}{\left(\frac{1}{hA}\right)} \Rightarrow 5 = \frac{\frac{(348-298)}{1}}{(h)\left(2\pi\left(\frac{1}{1000}\right)(1)\right)}$$

So, h = 15.91 W/m^{2}K

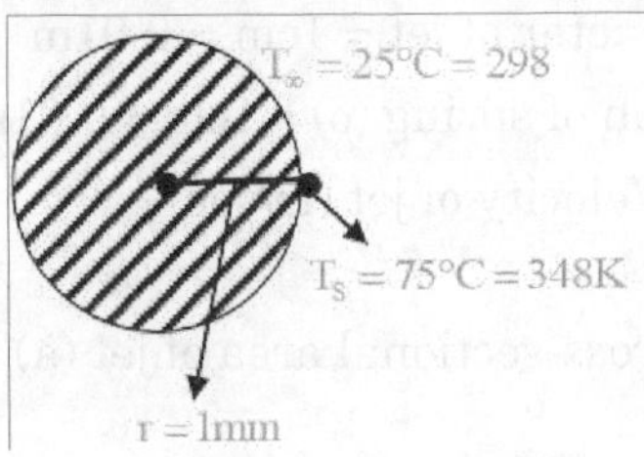

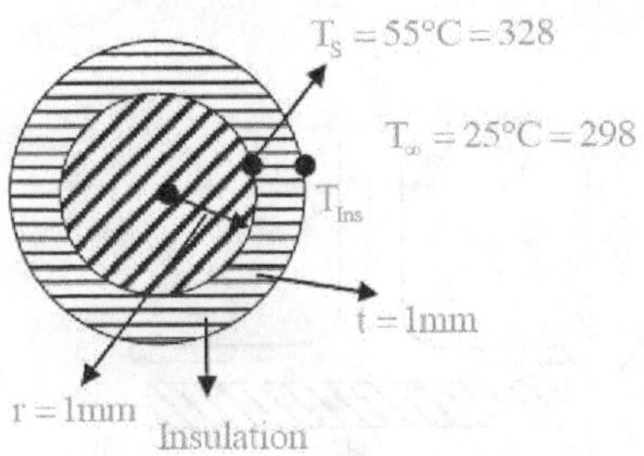

Where, T_{ins} = Temperature at the outer surface of Insulation

$$\therefore \dot{Q} = 5 = \frac{\frac{T_{Ins} - 298}{1}}{(15.91)(2\pi)\left(\frac{2}{1000}\right)(1)}$$

$$5 = \frac{(15.91)(2\pi)(2)(T_{Ins} - 298)}{(1000)}$$

T_{Ins} = 323K

K_{Ins} = Thermal conductivity of PVC insulation

$$\therefore \dot{Q} = \frac{\frac{T_S - T_{Ins}}{\ell n\left(\frac{2}{1}\right)}}{(2\pi) \times K_{Ins} L}$$

$$\Rightarrow 5 = \frac{(328-323)(2\pi(K_{Ins})(1))}{\ell n(2)}$$

K_{Ins} = 0.11 W/mK

Hence, the terminal conductivity of PVC is 0.11 with K.

44. As given that,

Radius of sphere (r) = 10 mm

Side of cube (a) = 30mm

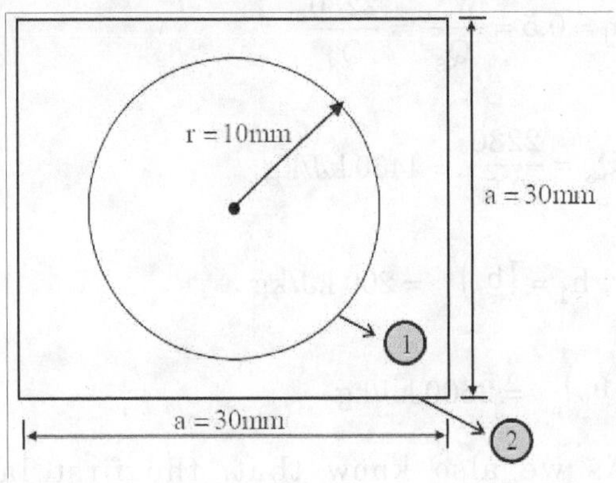

F_{22} = ?

$F_{11} + F_{12} = 1$

$F_{12} = 1$

$F_{21} + F_{22} = 1$

$A_1F_{12} = A_2F_{21}$, $A_1 = 4\pi r^2$, $A_2 = 6a^2$

$$F_{21} = \frac{A_1 \times F_{12}}{A_2} = \frac{4\pi r^2}{6a^2} \times 1 = \frac{4\pi \times 10^2}{6 \times 30^2} = 0.23$$

$F_{22} = 1 - F_{21} = 1 - 0.23 = 0.77$

45. As given that, work input to the compressor (W_P) = 20 kJ/kg

Work output of high pressure turbine (W_T)

= $(h_3 - h_4)$ = 750 kJ/kg

Work output of low pressure turbine $(W_T)_L$

= $(h_5 - h_6)$ = 1500 kJ/kg

As we know that,

$$\text{Thermal efficiency if cycle } (\eta) = \frac{W_{net}}{Q_S} = 0.5$$

Where,

W_{net} = Network output of cycle

= $((W_T)_H + (W_T)_L - W_P)$

So, W_{net} = (1500 + 750 – 20) = 2230 kJ/kg

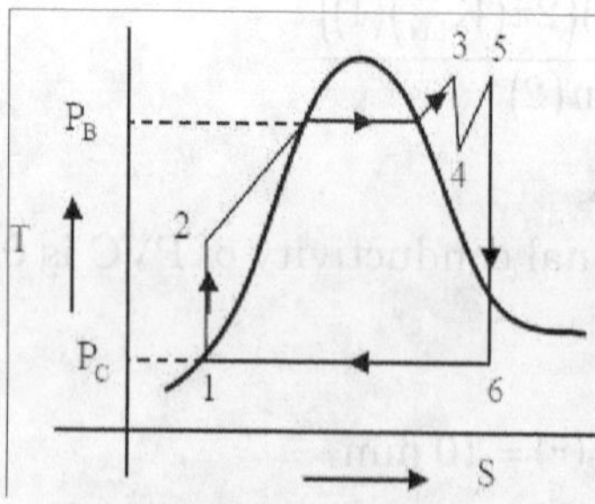

$$\eta = 0.5 = \frac{W_{net}}{Q_S} = \frac{2230}{Q_S}$$

$$Q_S = \frac{2230}{0.5} = 4460 \text{ kJ/kg}$$

$$\therefore h_1 = (h_f)_{P_c} = 200 \text{ kJ/kg}$$

$$(h_g)_{P_C} = 2600 \text{ kJ/kg}$$

As we also know that, the first law of thermodynamics:

$$\oint \delta Q = \oint dw$$

Heat supplied (Q_S) – Heat rejected (Q_R)
= Network done

$4460 - Q_R = 2230$

$Q_R = 4460 - 2230 = 2230$ kJ/kg

$$\therefore Q_R = (h_6 - h_1),\ h_6 = \left((h_f)\right)_{P_C} + x\left((h_g)_{P_C} - (h_f)_{P_C}\right)$$

$$h_f = (h_f)_{P_C}$$

$\therefore 2230 = (200 + x\,(2600 - 200) - (200))$

$x = 0.929 \cong 0.93 = 93\%$

Hence, answer is '93'

46. As we know that, liquid-vapour boundary in the P-T phase diagram for ammonia is

$$\ell nP = 24.38 - \frac{3063}{T}$$

Solid-vapour boundary in the P-T phase diagram for ammonia is

$$\ell nP = 27.92 - \frac{3754}{T}$$

For a triple point in P-T curve same pressure & Temperature will be there for solid-vapour, liquid – vapour phases.

$$\therefore 24.38 - \frac{3063}{T} = 27.92 - \frac{3754}{T}$$

$$\Rightarrow \frac{3754 - 3063}{T} = (27.92 - 24.38)$$

$$T = \frac{3754 - 3063}{27.92 - 24.38} = 195.2\text{K}$$

Hence, the temperature at the triple point is 195.2 K.

47. As given that

$\rho = 1000$ kg/m^2

K = 1 kN/m=1000 N/m

D = Diameter of jet = 1cm = 0.01m

Deflection of spring (δ) = 1cm = 0.01m

Let V= Velocity of jet (m/sec)

$\therefore$ The cross-sectional area of jet (a) = $\frac{\pi}{4} D^2$

$$= \frac{\pi}{4}(0.01^2) = 7.85 \times 10^{-5} \text{ m}^2$$

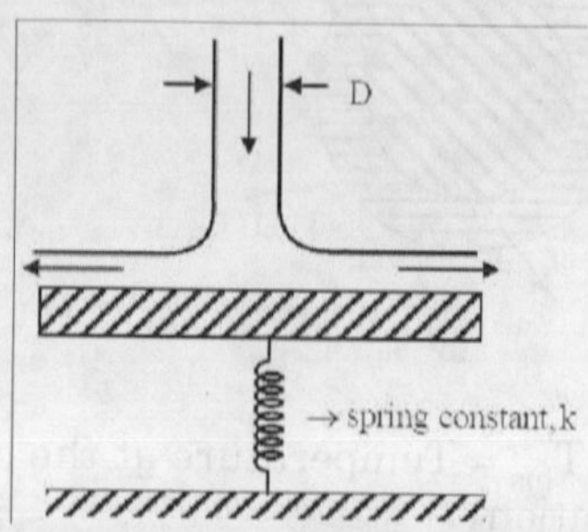

Force exerted by jet on circular plate in the direction of jet, (F) = m(V – 0)

m = mass flow rate of jet of water = ρaV

V = Velocity of jet of water striking the circular plate in the downward direction.

As we know that,

$F = \rho a V^2 = (1000)(7.85 \times 10^{-5})(V^2)$

As we also know that,

Stiffness of spring, (K) = $\frac{F}{\delta}$

$$1000 = \frac{\left(1000 \times 7.85 \times 10^{-5} \times V^2\right)}{(0.01)} = 11.29 \text{m/sec}$$

Hence the velocity of the incoming jet is 11.29 m/sec.

48. As given that,

$N = 300\text{rpm}, D = 2\text{m}$

$V = 40\text{m/sec}, Q = 5\text{m}^3\text{/sec}$

As we know that,

Power developed by

Pelton wheel $(P) = \rho av(v - u) \times u(1 + \cos\phi)$

$\phi = 180 - 165 = 15°$

$$u = \frac{\pi DN}{60} = \frac{\pi \times 2 \times 300}{60} = 10\pi \text{ m/sec}$$

$Q = \rho av = 5\text{m}^3\text{/sec}$

$\therefore P = (1000)(5)(40 - 10\pi)(10\pi)(1 + \cos 15) = 2.65\text{MW}$

Hence, the power developed by the pelton wheel is 2.65 MW.

49. As given figure:

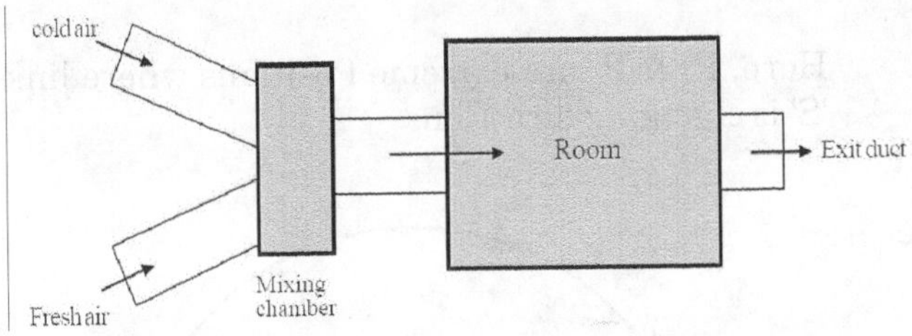

As given that,

Temperature of cold air $(T_c) = 5°C = 273 + 5 = 278K$

Pressure of cold air $(P_C) = 105\text{kPa} = 105 \times 10^3\text{pa}$

Volume flow rate of cold air $(V_c) = 1.25 \text{ m}^3\text{/sec}$

$(C_P)_{air} = 1.005\text{kJ/kgK}; R = 0.287 \text{ kJ/kgK}$

As we know that,

ideal gas mass flow rate of cold air $(\dot{m}_c)$

$$P_C V_C = m_C R T_C$$

$(105 \times 10^3)(1.25) = \dot{m}_c(0.287 \times 10^3)(278)$

$\dot{m}_c = 1.64 \text{ kg/sec}$

Mass flow rate of fresh air $(\dot{m}_F) = 1.6 \times \dot{m}_c$

$= 1.6 \times 1.64 = 2.624\text{kg/sec}$

Temperature of fresh air $(T_F) = 34°C$

$= 34 + 2.73 = 307K$

In mixing chambers,

Heat lost of fresh air = Heat gained by cold air

$$\dot{m}_F C_P (T_F - T_f) = \dot{m}_C C_P (T_f - T_C)$$

$(2.624)(1.005)(307 - T_f) = (1.64)(1.005)(T_f - 278)$

$(307 - T_f) = (0.625)(T_f - 278)$

$307 - T_f = 0.625T_f - 173.75$

$1.625T_f = 480.75$

$T_f = 298.85K$

T_f = Final temperature of mixed air = 295.85K

Final mass flow rate of air from the mixed chamber

$$\dot{m}_f = (\dot{m}_C + \dot{m}_F) = 4.264 \text{ kg / sec}$$

Exit temperature of air, $(T_E) = 24°C = 297K$

Energy balance in room:

Heat gain by air from the room $(\dot{Q}) = \dot{m}_f C_p (T_E - T_f)$

$= (4.264)(1.005)(297 - 295.85) = 4.93\text{kW}$

50. As given figure:

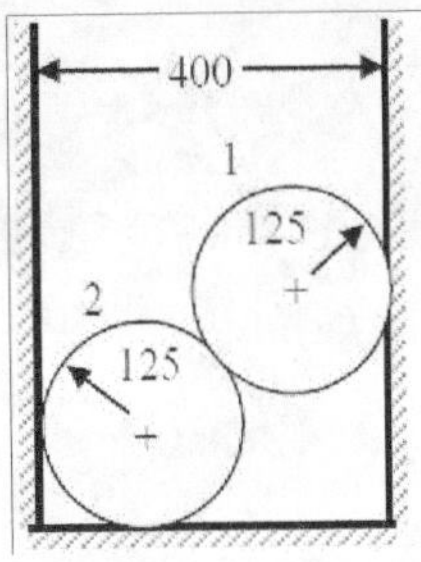

(i)

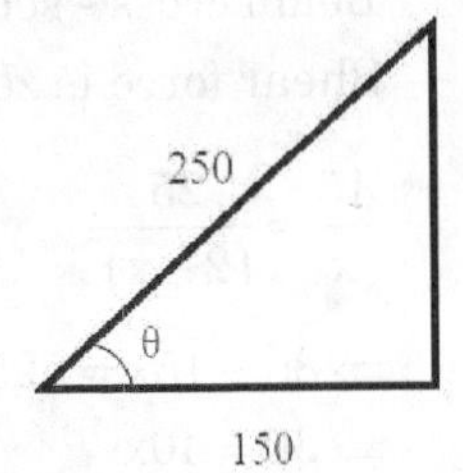

(ii)

From (ii):

$$\theta = \cos^{-1}\left(\frac{150}{250}\right)$$

$\theta = 53.13°$

Considering sphere 1

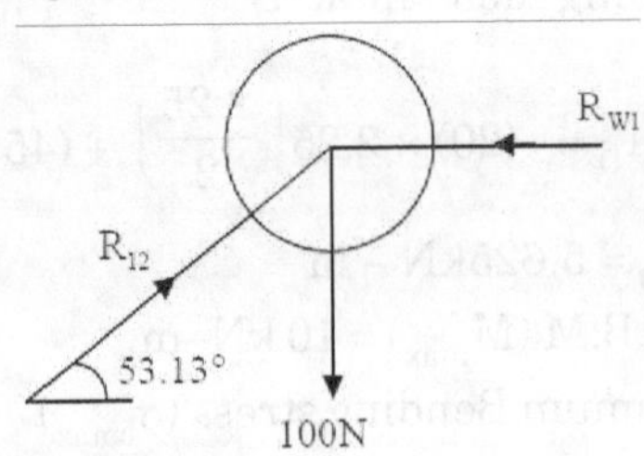

$\Sigma F_y = 0$

$R_{12} \sin 53.13° - 100 = 0$

$$R_{12} = \frac{100}{\sin 53.13} = 125N$$

Hence, the reaction at the point of constant between two sphere is 125 N.

51. As given figure:

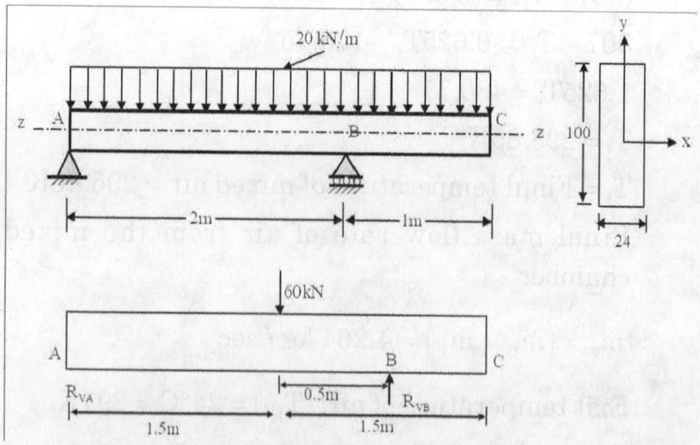

By taking the bending moment at

$$\sum M_A = 0$$

$R_{VB} \times 2 = 60 \times 1.5 = 45kN$

$R_{VA} + R_{VB} = 60$

$R_{VA} = 60 - 45 = 15kN$

Beam cross - section

Shear force is zero at 'D'

$$\frac{15}{x} = \frac{25}{(2-x)}$$

$\Rightarrow 30 - 15x = 525x$

$\Rightarrow 30 = 40x$

$\Rightarrow 40x = 30 \Rightarrow x = 0.75m$

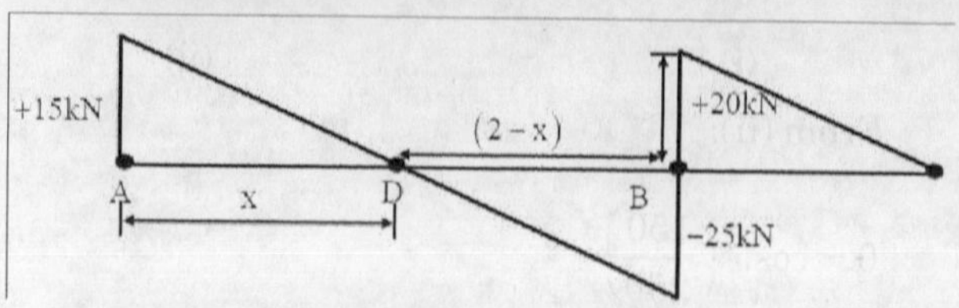

Bending moment at 'B'

$BM_B = 20 \times 1 \times 0.5 = 10kN\text{–}m$

Bending moment at 'D'

$$= BM_D = -(20) \times 2.25\left(\frac{2.25}{2}\right) + (45 \times 1.25)$$

$BM_D = 5.625kN - m$

Max B.M (M_{max}) = 10 kN–m

Maximum Bending stress (σ_{bmax})

$$= \frac{M_{max} \times y_{max}}{I}$$

$$= \frac{10 \times 10^3 \times 10^3 \times \left(\frac{100}{2}\right)}{\left(\frac{24 \times 100^3}{12}\right)} = 250MPa\ (N/mm^2)$$

Hence, the maximum bending stress developed in the beam is 250 mPa.

52. As given that

OA = 150mm, OP = 300mm,

AR = 160mm, RS = 450mm

The link 'S' will change from left to right or from right to left when 'P' crosses the horizontal line.

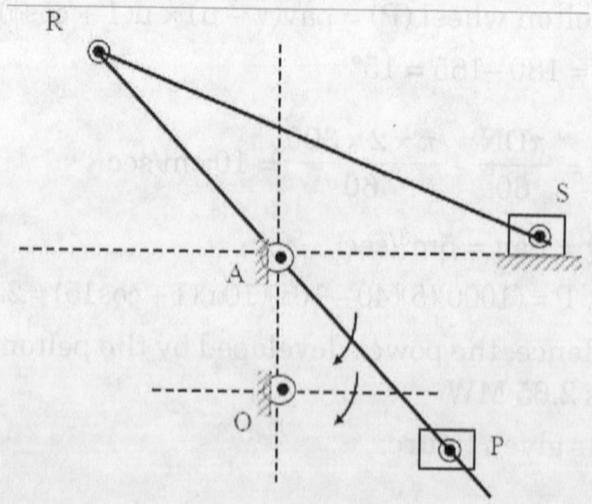

Here, P_1 & P_2 are extreme Positions where link 'S' is changing directions

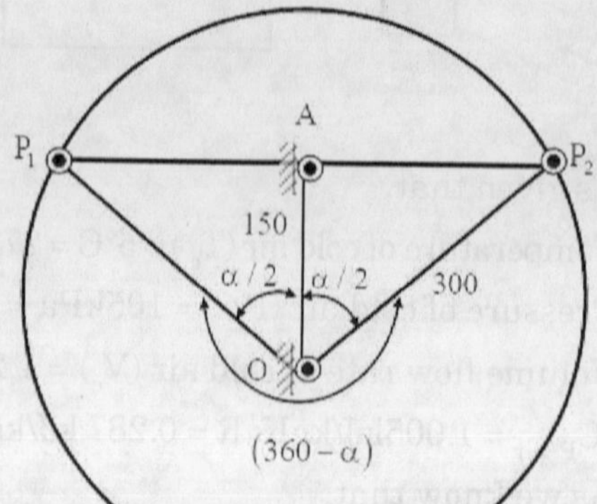

From figure above:

$$\therefore \cos\frac{\alpha}{2} = \frac{150}{300}$$

$$\frac{\alpha}{2} = \cos^{-1}\left(\frac{1}{2}\right) \Rightarrow \frac{\alpha}{2} = 60°$$

$\alpha = 120°$

As we know that,

$$\text{Quick return ratio} = \frac{360-\alpha}{\alpha}$$

$$= \frac{(360-120)}{120} = \frac{240}{120} = 2.0$$

Hence, the quick return ratio for the mechanism is 2.0.

53. As we know that,

Freebody diagram of Brake Drum Braking torque applied $(T_b) = F_f R = (\mu NR)$ ...(i)

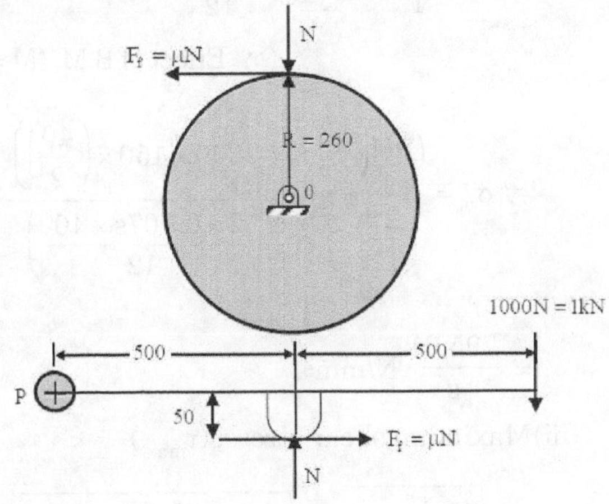

$\therefore$ Free body diagram of lever is $\Sigma M_p = 0$

$\Rightarrow (1000)(1000) - N \times 500 - (\mu N)(50) = 0$

$\Rightarrow (1000)(1000) - 500N - (0.4)(50N) = 0$

$\Rightarrow 520N = (1000)(1000) \Rightarrow N = 1923.1N$

From (i)

$\therefore$ Braking torque, $(T_b) = (\mu N)(R)$

$= (0.4)(1923.1)(260)$

$= 200N.m$

54. As given that,

$S_e' = 200MPa$, $S_{yt} = 500Mpa$

$S_{ut} = 600Mpa$,

$$I = \frac{12 \times 20^3}{12} = 8000mm^4$$

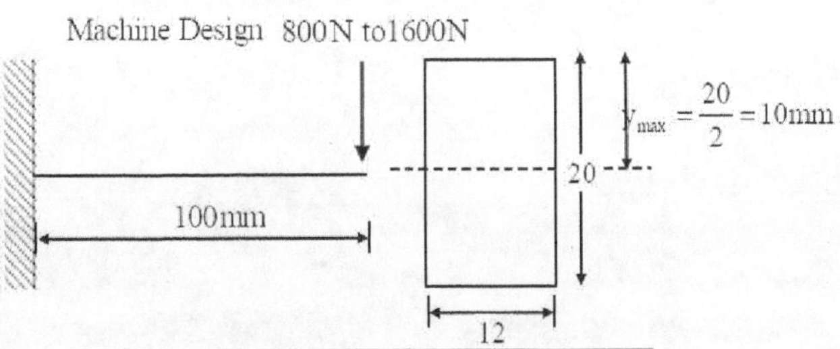

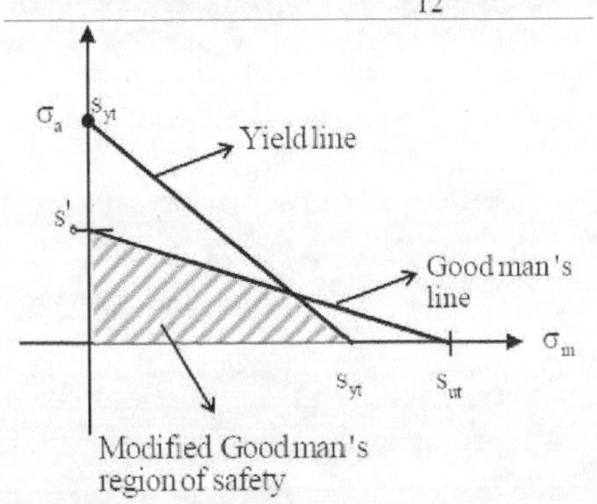

$P_{min} = 800N$, $P_{max} = 1600N$

$M_{min} = P_{min} \times 100$, $M_{max} = P_{max} \times 100$,

$= 800 \times 100$ $= 1600 \times 100$

$= 80{,}000$ Nmm $= 1{,}60{,}000$ Nmm

As we know that,

$$(\sigma_b)_{min} = \frac{M_{min} y_{max}}{I} = \frac{80{,}000 \times 10}{8000} = 100 \text{ N/mm}^2$$

$$= \sigma_{min}$$

$$(\sigma_b)_{max} = \frac{M_{min} y_{max}}{I} = \frac{1{,}60{,}000 \times 10}{8000} = 200 \text{ N/mm}^2$$

$$= \sigma_{max}$$

$$\text{So, } \sigma_m = \frac{\sigma_{max} + \sigma_{min}}{2} = \frac{200 + 100}{2} = 150 \text{ N/mm}^2$$

$$\sigma_a = \frac{\sigma_{max} - \sigma_{min}}{2} = \frac{200 - 100}{2} = 50 \text{ N/mm}^2$$

Now, according to yield line,

$$\frac{\sigma_m}{S_{yt}} + \frac{\sigma_a}{S_{yt}} \le \frac{1}{FOS}$$

$$\left(\frac{150}{500}\right) + \left(\frac{50}{500}\right) \le \frac{1}{FOS} \Rightarrow FOS \ge 2.5$$

According to Good man's line

$$\frac{\sigma_m}{S_{ut}} + \frac{\sigma_a'}{S_e'} \le \frac{1}{FOS}$$

$$\left(\frac{150}{600}\right) + \left(\frac{50}{200}\right) \le \frac{1}{FOS} \Rightarrow FOS \ge 2$$

FOS in beam using modified good man criterion is = min[2.5, 2] = 2

Hence the correct answer is '2'.

55. As given that,

Allowable shear stress $(\tau_{all}) = 60N/mm^2$

After transferring the load to weld, it will be as follows weld will be subjected to direct shear force (P) and a bending moment (M = pe)

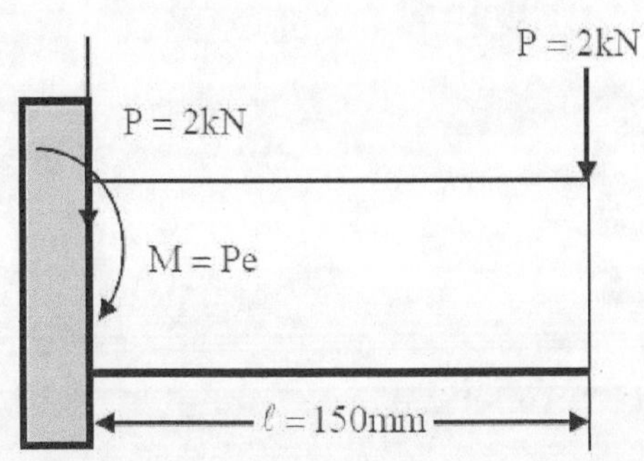

Effect of 'P':

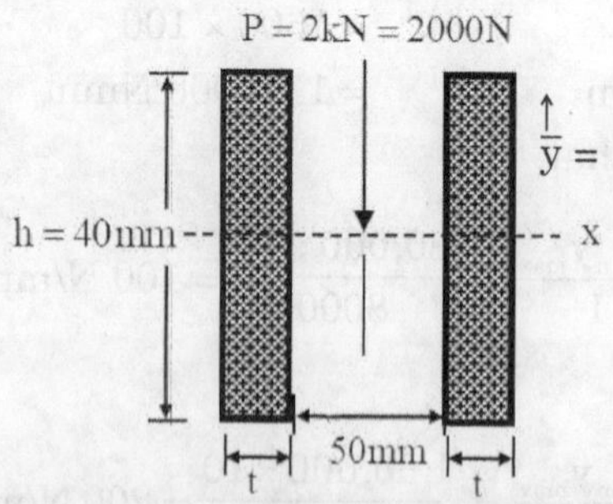

$\bar{y} = \frac{h}{2} = \frac{40}{2} = 20$ mm

It will produce direct shear stress t = throat thickness of weld = 0.707s

where, s = size of weld

As we know that

(i) Direct shear stress, $(\tau) = \frac{P}{2th}$

$$(\tau) = \frac{2000}{2 \times 0.707s \times 40} = \frac{35.36}{S} N/mm^2$$

(ii) B.M. produce bending stress,

$$(\sigma_b) = \frac{M\bar{y}}{I}, \ I = I_{xx} = \frac{2th^3}{12}$$

($\because$ Effect of B.M. $(M = P\ell)$)

$$\therefore \sigma_b = \frac{(P\ell)\left(\frac{h}{2}\right)}{\left(\frac{2th^3}{12}\right)} = \frac{\left(2000 \times 150 \times \left(\frac{40}{2}\right)\right)}{\left(\frac{2 \times 0.707s \times 40^3}{12}\right)}$$

$$= \frac{795.61}{s} N/mm^2$$

(iii) Maximum shear stress, (τ_{max})

$$= \sqrt{\left(\frac{\sigma_b}{2}\right)^2 + \tau^2} = \sqrt{\left(\frac{795.6}{2s}\right)^2 + \left(\frac{35.36}{5}\right)^2}$$

For safe design, $\tau_{max} \le \tau_{all}$

$$= \sqrt{\left(\frac{795.6}{2s}\right)^2 + \left(\frac{35.36}{5}\right)^2} \le 60$$

$s \ge 6.65$mm

Hence, minimum Leg size of weld (s) is 6.65 mm.

SOLVED PAPER - 2021

SET 2

INSTRUCTIONS

1. Total of 65 questions carrying 100 marks, out of which 10 questions carrying a total of 15 marks are in General Aptitude (GA)
2. The Engineering Mathematics will carry around **15% of the total marks,** the General Aptitude section will carry **15% of the total marks** and the **remaining 70% of the total marks.**
3. **Types of Questions**
   (*a*) **Multiple Choice Questions (MCQ)** carrying 1 or 2 marks each in all papers and sections. These questions are objective in nature, and each will have a choice of four options, out of which the candidate has to mark the correct answer(s).
   (*b*) **Numerical Answer Questions** of 1 or 2 marks each in all papers and sections. For these questions the answer is a real number, to be entered by the candidate using the virtual keypad. No choices will be shown for these type of questions.
   (*c*) **Multiple Select Questions (MSQ)** carrying 1 or 2 marks each in all the papers and sections. These questions are objective in nature, and each will have choice of four answers, out of which ONE or MORE than ONE choice(s) is / are correct.
4. For **1-mark** multiple-choice questions, **1/3 marks** will be deducted for a wrong answer. Likewise, for **2-marks** multiple-choice questions, 2/3 marks will be deducted for a wrong answer. There is no negative marking for numerical answer type questions & multiple selection questions.

Chapter-Wise Analysis

#	Subject	Marks		
		1	2	Total
1	General Aptitude	5	5	15
2	Engineering Mathematics	5	4	13
3	Engineering Mechanics	1	2	5
4	Strength of Materials	3	2	7
5	Theory of Machines	2	4	10

#	Subject	Marks		
		1	2	Total
6	Design of Machine Elements	-	1	2
7	Fluid Mechanics	5	3	11
8	Heat Transfer	1	2	5
9	Thermodynamics, IC Engines, Refrigeration & Power Plant Engineering Air-Conditioning	2	3	8
10	Materials, Manufacturing & Industrial Engineering	6	9	24

GENERAL APTITUDE (GA)

Q. 1 to Q. 5 : Carry One Mark Each.

1. Five persons P, Q, R, S and T are to be seated in a row, all facing the same direction, but not necessarily in the same order. P and T cannot be seated at either end of the row. P should not be seated adjacent to S. R is to be seated at the second position from the left end of the row. The number of distinct seating arrangements possible is:

(*a*) 2 (*b*) 3

(*c*) 4 (*d*) 5

2. Consider the following sentences:

(i) The number of candidates who appear for the GATE examination is staggering.

(ii) A number of candidates from my class are appearing for the GATE examination.

(iii) The number of candidates who appear for the GATE examination are staggering.

(iv) A number of candidates from my class is appearing for the GATE examination.

Which of the above sentences are grammatically CORRECT?

(*a*) (i) and (ii) (*b*) (i) and (iii)

(*c*) (ii) and (iii) (*d*) (ii) and (iv)

3. A digital watch X beeps every 30 seconds while watch Y beeps every 32 seconds. They beeped together at 10 AM.

The immediate next time that they will beep together is ________

(*a*) 10.08 AM (*b*) 10.42 AM

(*c*) 11.00 AM (*d*) 10.00 PM

4. If $\oplus \div \odot = 2; \oplus \div \Delta = 3; \odot + \Delta = 5; \Delta \times \otimes = 10,$

Then, the value of $(\otimes - \oplus)^2$, is:

(*a*) 0 (*b*) 1

(*c*) 4 (*d*) 16

5. The front door of Mr. X's house faces East. Mr. X leaves the house, walking 50 m straight from the back door that is situated directly opposite to the front door. He then turns to his right, walks for another 50 m and stops. The direction of the point Mr. X is now located at with respect to the starting point is ________

(*a*) South-East (*b*) North-East

(*c*) West (*d*) North-West

Q. 6 to Q. 10: Multiple Choice Questions (MCQ), Carry Two Marks Each

6. Given below are two statements 1 and 2, and two conclusions I and II.

Statement 1: All entrepreneurs are wealthy.

Statement 2: All wealthy are risk seekers.

Conclusion I: All risk seekers are wealthy.

Conclusion II: Only some entrepreneurs are risk seekers.

Based on the above statements and conclusions, which one of the following options is CORRECT?

(*a*) Only conclusion I is correct

(*b*) Only conclusion II is correct

(*c*) Neither conclusion I nor II is correct

(*d*) Both conclusions I and II are correct

7. A box contains 15 blue balls and 45 black balls. If 2 balls are selected randomly, without replacement, the probability of an outcome in which the first selected is a blue ball and the second selected is a black ball, is ________

(*a*) $\frac{3}{16}$ (*b*) $\frac{45}{236}$

(*c*) $\frac{1}{4}$ (*d*) $\frac{3}{4}$

8.

The ratio of the area of the inscribed circle to the area of the circumscribed circle of an equilateral triangle is ________

(*a*) $\frac{1}{8}$ (*b*) $\frac{1}{6}$

(*c*) $\frac{1}{4}$ (*d*) $\frac{1}{2}$

9. Consider a square sheet of side 1 unit. The sheet is first folded along the main diagonal. This is followed by a fold along its line of symmetry. The resulting folded shape is again folded along its line of symmetry. The area of each face of the final folded shape, in square units, equal to _____

(*a*) $\frac{1}{4}$ (*b*) $\frac{1}{8}$

(*c*) $\frac{1}{16}$ (*d*) $\frac{1}{32}$

10. The world is going through the worst pandemic in the past hundred years. The air travel industry is facing a crisis, as the resulting quarantine requirement for travelers led to weak demand. In relation to the first sentence above, what does the second sentence do?

(*a*) Restates an idea from the first sentence.

(*b*) Second sentence entirely contradicts the first sentence.

(*c*) The two statements are unrelated.

(*d*) States an effect of the first sentence.

MECHANICAL ENGINEERING

Q.1 to Q.19: Multiple Choice Question (MCQ), Carry One Mark Each

1. Consider an $n \times n$ matrix A and a non-zero $n \times l$ vector p. Their product $Ap = \alpha^2 p$, where $\alpha \in R$ and $\alpha \notin \{-1,0,1\}$. Based on the given information, the eigen value of A^2 is:

(a) α (b) α^2

(c) $\sqrt{\alpha}$ (d) α^4

2. If the Laplace transform of a function f(t) is given by $\frac{s+3}{(s+1)(s+2)}$, then f(0) is

(a) 0 (b) $\frac{1}{2}$

(c) 1 (d) $\frac{3}{2}$

3. The mean and variance, respectively, of a binomial distribution for n independent trials with the probability of success as p, are

(a) $\sqrt{np}$, $np(1-2p)$ (b) $\sqrt{np}, \sqrt{np(1-p)}$

(c) np, np (d) $np, np(1-p)$

4. The Cast Iron which possesses all the carbon in the combined form as cementite is known as

(a) Grey Cast Iron

(b) Spheroidal Cast Iron

(c) Malleable Cast Iron

(d) White Cast Iron

5. The size distribution of the powder particles used in Powder Metallurgy process can be determined by

(a) Laser scattering (b) Laser reflection

(c) Laser absorption (d) Laser penetration

6. In a CNC machine tool, the function of an interpolator is to generate

(a) signal for the lubrication pump during machining

(b) error signal for tool radius compensation during machining

(c) NC code from the part drawing during post processing

(d) reference signal prescribing the shape of the part to be machined

7. The machining process that involves ablation is

(a) Abrasive Jet Machining

(b) Chemical Machining

(c) Electrochemical Machining

(d) Laser Beam Machining

8. A PERT network has 9 activities on its critical path. The standard deviation of each activity on the critical path is 3. The standard deviation of the critical path is

(a) 3 (b) 9

(c) 27 (d) 81

9. The allowance provided in between a hole and a shaft is calculated from the difference between

(a) lower limit of the shaft and the upper limit of the hole

(b) upper limit of the shaft and the upper limit of the hole

(c) upper limit of the shaft and the lower limit of the hole

(d) lower limit of the shaft and the lower limit of the hole

10. In forced convective heat transfer, Stanton number (St), Nusselt number (Nu), Reynolds number (Re) and Prandtl number (Pr) are related as

(a) $St = \frac{Nu}{Re\,Pr}$ (b) $St = \frac{Nu\,Pr}{Re}$

(c) $St = Nu\,Pr\,Re$ (d) $St = \frac{Nu\,Re}{Pr}$

11. For a two-dimensional, incompressible flow having velocity components u and v in the x and y directions, respectively, the expression

$$\frac{\partial(u^2)}{\partial x} + \frac{\partial(uv)}{\partial y}$$

can be simplified to

(a) $u\frac{\partial u}{\partial x} + u\frac{\partial v}{\partial y}$ (b) $2u\frac{\partial u}{\partial x} + u\frac{\partial v}{\partial y}$

(c) $2u\frac{\partial u}{\partial x} + v\frac{\partial u}{\partial y}$ (d) $u\frac{\partial u}{\partial x} + v\frac{\partial u}{\partial y}$

12. Which of the following is responsible for eddy viscosity (or turbulent viscosity) in a turbulent boundary layer on a flat plate?

(a) Nikuradse stresses

(b) Reynolds stresses

(c) Boussinesq stresses

(d) Prandtl stresses

13. A two dimensional flow has velocities in x and y directions given by $u = 2xyt$ and $v = -y^2t$, where t denotes time. The equation for streamline passing through $x = 1, y = 1$ is

(a) $x^2y = 1$

(b) $xy^2 = 1$

(c) $x^2y^2 = 1$

(d) $x/y^2 = 1$

14. A plane truss PQRS (PQ = RS, and ∠PQR = 90°) is shown in the figure.

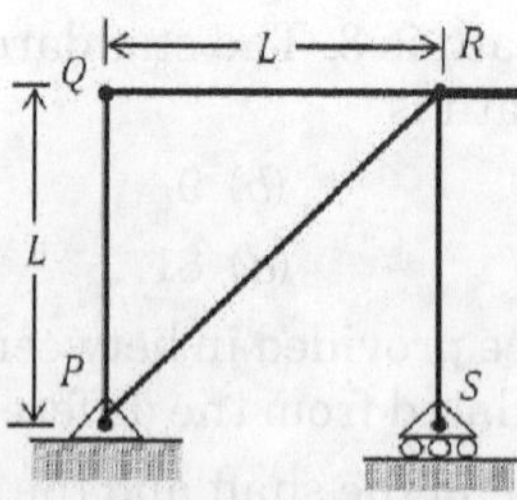

The forces in the members PR and RS, respectively, are

(*a*) F $\sqrt{2}$ (tensile) and F (tensile)

(*b*) F $\sqrt{2}$ (tensile) and F (compressive)

(*c*) F (compressive) and F $\sqrt{2}$ (compressive)

(*d*) F (tensile) and F $\sqrt{2}$ (tensile)

15. Consider the mechanism shown in the figure. There is rolling contact without slip between the disc and ground.

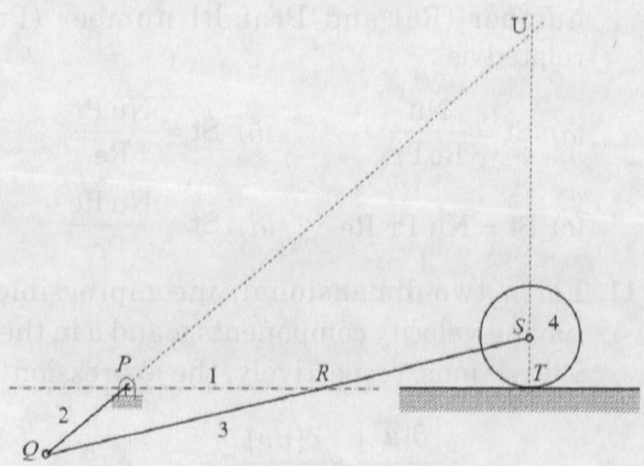

Select the correct statement about instantaneous centers in the mechanism.

(*a*) Only points P, Q, and S are instantaneous centers of mechanism

(*b*) Only points P, Q, S and T are instantaneous centers of mechanism

(*c*) Only points P, Q, R, S, and U are instantaneous centers of mechanism

(*d*) All points P, Q, R, S, T and U are instantaneous centers of mechanism

16. The controlling force curves P, Q and R for a spring controlled governor are shown in the figure, where r_1 and r_2 are any two radii of rotation.

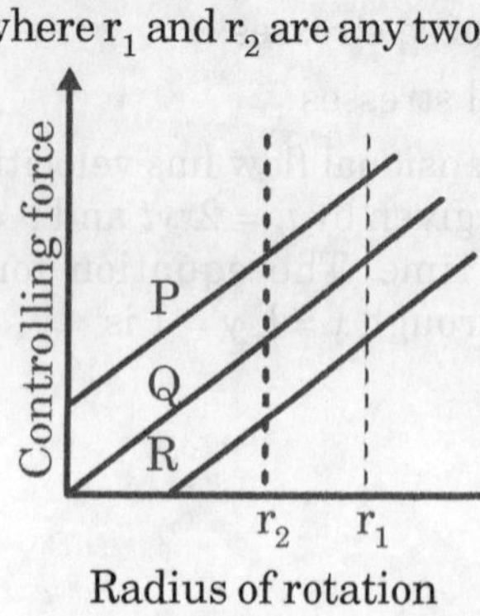

The characteristics shown by the curves are

(*a*) P - Unstable; Q - Stable; R - Isochronous

(*b*) P - Unstable; Q - Isochronous; R - Stable

(*c*) P- Stable; Q - Isochronous; R - Unstable

(*d*) P - Stable; Q - Unstable; R - Isochronous

17. The von Mises stress at a point in a body subjected to forces is proportional to the square root of the

(*a*) total strain energy per unit volume

(*b*) plastic strain energy per unit volume

(*c*) dilatational strain energy per unit volume

(*d*) distortional strain energy per unit volume

18. Value of $\int_{4}^{5.2} \ln x \, dx$ using Simpson's one-third rule with interval size 0.3 is

(*a*) 1.83 (*b*) 1.60

(*c*) 1.51 (*d*) 1.06

19. Value of $(1 + i)^8$, where $i = \sqrt{-1}$, is equal to

(*a*) 4 (*b*) 16

(*c*) 4*i* (*d*) 16*i*

Q.20 to Q.25: Numerical Answer Type (NAT), carry ONE mark each

20. Consider adiabatic flow of air through a duct. At a given point in the duct, velocity of air is 300 m/s, temperature is 330 K and pressure is 180 kPa. Assume that the air behaves as a perfect gas with constant c_p =1.005 kJ/kg.K. The stagnation temperature at this point is________K (*round off to two decimal places*).

21. Consider an ideal vapour compression refrigeration cycle working on R-134a refrigerant. The COP of the cycle is 10 and the refrigeration capacity is 150 kJ/kg. The heat rejected by the refrigerant in the condenser is ________kJ/kg (*round off to the nearest integer*).

22. A rigid tank of volume 50 m³ contains a pure substance as a saturated liquid vapour mixture at 400 kPa. Of the total mass of the mixture, 20% mass is liquid and 80% mass is vapour. Properties at 400 kPa are: Tsat = 143.61 °C, V_f = 0.001084 m³/kg, v_g = 0.46242 m³/kg. The total mass of liquid vapour mixture in the tank is ________kg (*round off to the nearest integer*).

23. An object is moving with a Mach number of 0.6 in an ideal gas environment, which is at a temperature of 350 K. The gas constant is 320 J/kg.K and ratio of specific heats is 1.3. The speed of object is ________m/s (*round off to the nearest integer*).

24. A column with one end fixed and one end free has a critical buckling load of 100 N. For the same column, if the free end is replaced with a pinned end then the critical buckling load will be__________N (*round off to the nearest integer*).

25. A steel cubic block of side 200 mm is subjected to hydrostatic pressure of 250 N/mm^2. The elastic modulus is 2×10^5 N/mm^2 and Poisson ratio is 0.3 for steel. The side of the block is reduced by________mm (*round off to two decimal places*).

Q. 26 to Q.34: Multiple Choice Questions (MCQ), Carry Two Mark Each

26. The value of $\int_0^{\frac{\pi}{2}} \int_0^{\cos\theta} r \sin\theta \, dr \, d\theta$ is

(*a*) 0 (*b*) $\frac{1}{6}$

(*c*) $\frac{4}{3}$ (*d*) π

27. Let the superscript T represent the transpose operation. Consider the function $f(X) = \frac{1}{2}x^T Qx - r^T x$, where x and r are $n \times 1$ vectors and Q is a symmetric $n \times n$ matrix. The stationary point of f(*x*) is

(*a*) $Q^T r$ (*b*) $Q^{-1} r$

(*c*) $\frac{r}{r^T r}$ (*d*) r

28. Consider the following differential equation

$$(1+y)\frac{dy}{dx} = y$$

The solution of the equation that satisfies the condition y(1) = 1 is

(*a*) $2ye^y = e^x + e$ (*b*) $y^2 e^y = e^x$

(*c*) $ye^y = e^x$ (*d*) $(1 + y)e^y = 2e^x$

29. A factory produces m(i = 1, 2, ..., m) products, each of which requires processing on $n(j = 1,2,..,n)$ workstations. Let a_{ij} be the amount of processing time that one unit of the ith product requires on the jth workstation. Let the revenue from selling one unit of the ith product be r_i and h_i be the holding cost per unit per time period for the ith product. The planning horizon consists of T (t = 1,2,...., T) time periods. The minimum demand that must be satisfied in time period t is dit, and the capacity of the jth workstation in time period t is c_{jt}. Consider the aggregate planning formulation below, with decision variables S_{it} (amount of product i sold in time period t), X_{it} (amount of product i manufactured in time period t) and I_{it} (amount of product i held in inventory at the end of time period t.

$$\max \sum_{t=1}^{T} \sum_{i=1}^{m} \left(r_i S_{it} - h_i I_{it}\right)$$

subject to

$S_{it} \ge d_{it} \ \forall \ i,t$

<capacity constraint>

<inventory balance constraint>

$X_{it}, S_{it}, I_{it} \ge 0; I_{i0} = 0$

The capacity constraints and inventory balance constraints for this formulation are

(*a*) $\sum_i^m a_{ij} X_{it} \le C_{jt} \ \forall \ j,t$ and $I_{it} = I_{i,t-1} + X_{it} - S_{it} \ \forall \ i,t$

(*b*) $\sum_i^m a_{ij} X_{it} \le C_{jt} \ \forall \ i,t$ and $I_{it} = I_{i,t-1} + X_{it} - d_{it} \ \forall \ i,t$

(*c*) $\sum_i^m a_{ij} X_{it} \le d_{it} \ \forall \ i,t$ and $I_{it} = I_{i,t-1} + X_{it} - S_{it} \ \forall \ i,t$

(*d*) $\sum_i^m a_{ij} X_{it} \le d_{it} \ \forall \ i,t$ and $I_{it} = I_{i,t-1} + S_{it} - X_{it} \ \forall \ i,t$

30. Ambient pressure, temperature, and relative humidity at a location are 101 kPa, 300 K, and 60%, respectively. The saturation pressure of water at 300 K is 3.6 kPa. The specific humidity of ambient air is__________g/kg of dry air.

(*a*) 21.4 (*b*) 35.1

(*c*) 21.9 (*d*) 13.6

31. A plane frame PQR (fixed at P and free at R) is shown in the figure. Both members (PQ and QR) have length, L, and flexural rigidity, EI. Neglecting the effect of axial stress and transverse shear, the horizontal deflection at free end, R, is

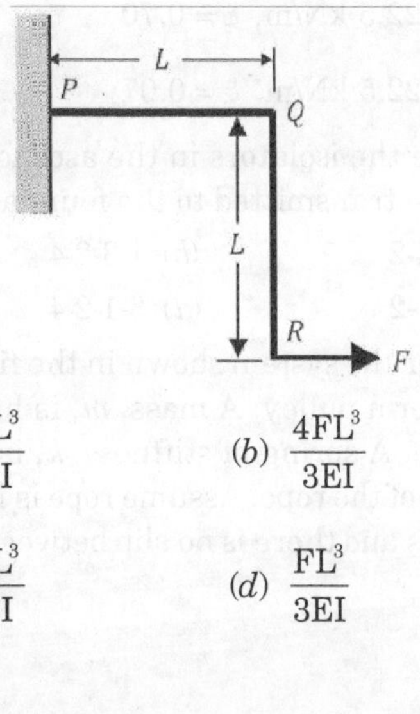

(*a*) $\frac{5FL^3}{3EI}$ (*b*) $\frac{4FL^3}{3EI}$

(*c*) $\frac{2FL^3}{3EI}$ (*d*) $\frac{FL^3}{3EI}$

32. A power transmission mechanism consists of a belt drive and a gear train as shown in the figure.

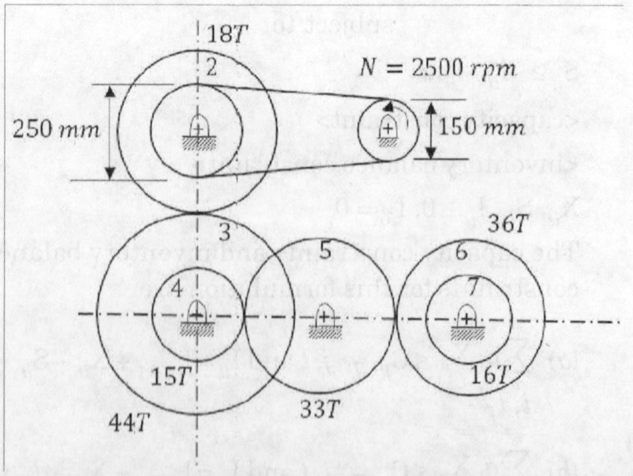

Diameters of pulleys of belt drive and number of teeth (T) on the gears 2 to 7 are indicated in the figure. The speed and direction of rotation of gear 7, respectively, are

(*a*) 255.68 rpm; clockwise

(*b*) 255.68 rpm; anticlockwise

(*c*) 575.28 rpm; clockwise

(*d*) 575.28 rpm; anticlockwise

33. A machine of mass 100 kg is subjected to an external harmonic force with a frequency of 40 rad/s. The designer decides to mount the machine on an isolator to reduce the force transmitted to the foundation. The isolator can be considered as a combination of stiffness (K) and damper (damping factor, ξ) in parallel. The designer has the following four isolators:

1) K = 640 kN/m, $\xi = 0.70$

2) K = 640 kN/m, $\xi = 0.07$

3) K = 22.5 kN/m, $\xi = 0.70$

4) K = 22.5 kN/m, $\xi = 0.07$

Arrange the isolators in the ascending order of the force transmitted to the foundation.

(*a*) 1-3-4-2 (*b*) 1-3-2-4

(*c*) 4-3-1-2 (*d*) 3-1-2-4

34. Consider the system shown in the figure. A rope goes over a pulley. A mass, m, is hanging from the rope. A spring of stiffness, k, is attached at one end of the rope. Assume rope is inextensible, massless and there is no slip between pulley and rope.

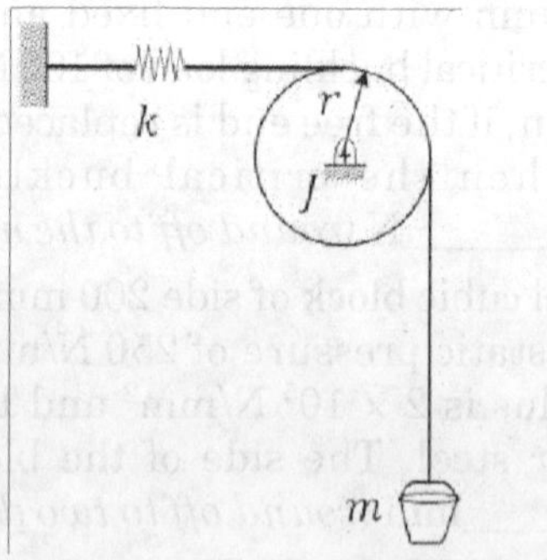

The pulley radius is r and its mass moment of inertia is J. Assume that the mass is vibrating harmonically about its static equilibrium position. The natural frequency of the system is

(*a*) $\sqrt{\dfrac{kr^2}{J - mr^2}}$ (*b*) $\sqrt{\dfrac{kr^2}{J + mr^2}}$

(*c*) $\sqrt{k/m}$ (*d*) $\sqrt{\dfrac{kr^2}{J}}$

Q.35 to Q.55: Numerical Answer Type (NAT), Carry Two Mark Each

35. Find the positive real root of $x^3 - x - 3 = 0$ using Newton-Raphson method. If the starting guess (x_0) is 2, the numerical value of the root after two iterations (x_2) is__________*(round off to two decimal places)*.

36. Daily production capacity of a bearing manufacturing company is 30000 bearings. The daily demand of the bearing is 15000. The holding cost per year of keeping a bearing in the inventory is ₹20. The setup cost for the production of a batch is ₹1800. Assuming 300 working days in a year, the economic batch quantity in number of bearings is________*(in integer)*.

37. A cast product of a particular material has dimensions 75 mm × 125 mm × 20 mm. The total solidification time for the cast product is found to be 2.0 minutes as calculated using Chvorinov's rule having the index, $n = 2$. If under the identical casting conditions, the cast product shape is changed to a cylinder having diameter = 50 mm and height = 50 mm, the total solidification time will be ________ minutes *(round off to two decimal places)*.

38. A spot welding operation performed on two pieces of steel yielded a nugget with a diameter of 5 mm and a thickness of 1 mm. The welding time was 0.1 s. The melting energy for the steel is 20 J/mm^3. Assuming the heat conversion efficiency as 10%, the power required for performing the spot welding operation is________kW *(round off to two decimal places)*.

39. A surface grinding operation has been performed on a Cast Iron plate having dimensions 300 mm (*length*) × 10 mm (*width*) × 50 mm (*height*). The grinding was performed using an alumina wheel having a wheel diameter of 150 mm and wheel width of 12 mm. The grinding velocity used is 40 m/s, table speed is 5 m/min, depth of cut per pass is 50 μm and the number of grinding passes is 20. The average tangential and average normal force for each pass is found to be 40 N and 60 N respectively. The value of the specific grinding energy under the aforesaid grinding conditions is ________ J/mm^3 (*round off to one decimal place*).

40. In a pure orthogonal turning by a zero rake angle single point carbide cutting tool, the shear force has been computed to be 400 N. The cutting velocity, V_c = 100 m/min, depth of cut, t = 2.0 mm, feed, s_0 = 0.1 mm/revolution and chip velocity, V_f = 20 m/min, the shear strength, τ_s of the material will be ________ MPa (*round off to two decimal places*).

41. The thickness, width and length of a metal slab are 50 mm, 250 mm and 3600 mm, respectively. A rolling operation on this slab reduces the thickness by 10% and increases the width by 3%. The length of the rolled slab is ________ mm (*round off to one decimal place*).

42. A 76.2 mm gauge block is used under one end of a 254 mm sine bar with roll diameter of 25.4 mm. The height of gauge blocks required at the other end of the sine bar to measure an angle of 30° is________mm (*round off to two decimal places*).

43. The demand and forecast of an item for five months are given in the table.

Month	Demand	Forecast
April	225	200
May	220	240
June	285	300
July	290	270
August	250	230

The Mean Absolute Percent Error (MAPE) in the forecast is________% (*round off to two decimal places*)

44. A shell and tube heat exchanger is used as a steam condenser. Coolant water enters the tube at 300 K at a rate of 100 kg/s. The overall heat transfer coefficient is 1500 W/m^2.K, and total heat transfer area is 400 m^2. Steam condenses at a saturation temperature of 350 K. Assume that the specific heat of coolant water is 4000 J/kg.K. The temperature of the coolant water coming out of the condenser is________K (*round off to the nearest integer*).

45. Ambient air flows over a heated slab having flat, top surface at y = 0. The local temperature (in Kelvin) profile within the thermal boundary layer is given by $T(y) = 300 + 200 \exp(-5y)$, where y is the distance measured from the slab surface in meter. If the thermal conductivity of air is 1.0 W/m.K and that of the slab is 100 W/m.K, then the magnitude of temperature gradient $|dT/dy|$ within the slab at y = 0 is________K/m (*round off to the nearest integer*).

46. Water flows out from a large tank of cross-sectional area A_t = 1 m^2 through a small rounded orifice of cross-sectional area A_0 = 1 cm^2, located at y = 0. Initially the water level, measured from y = 0, is H = 1 m. The acceleration due to gravity is 9.8 m/s^2.

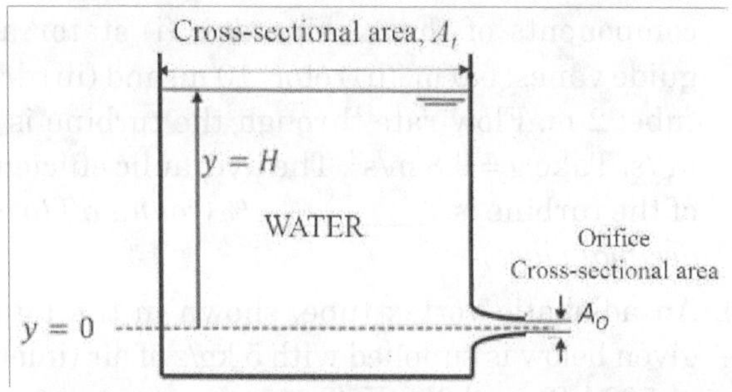

Neglecting any losses, the time taken by water in the tank to reach a level of y = H / 4 is ________seconds (*round off to one decimal place*).

47. Consider the open feed water heater (FWH) shown in the figure given below:

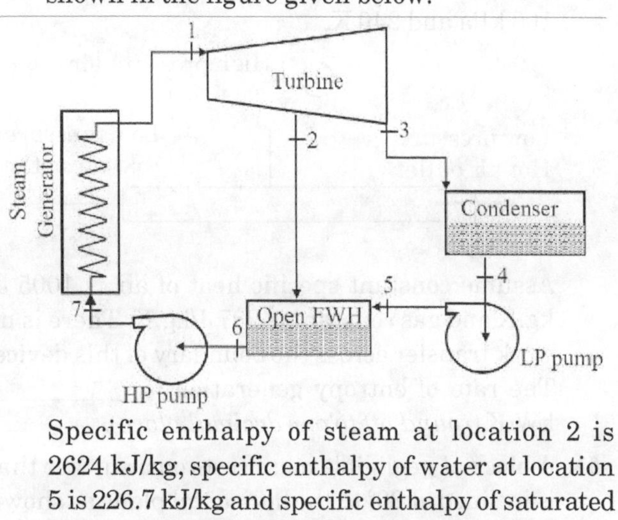

Specific enthalpy of steam at location 2 is 2624 kJ/kg, specific enthalpy of water at location 5 is 226.7 kJ/kg and specific enthalpy of saturated water at location 6 is 708.6 kJ/kg. If the mass flow rate of water entering the open feed water heater (at location 5) is 100 kg/s then the mass flow rate of steam at location 2 will be ________ kg/s (*round off to one decimal place*).

48. A high velocity water jet of cross section area = 0.01 m^2 and velocity = 35 m/s enters a pipe filled with stagnant water. The diameter of the pipe is 0.32 m. This high velocity water jet entrains additional water from the pipe and the total water leaves the pipe with a velocity 6m/s as shown in the figure.

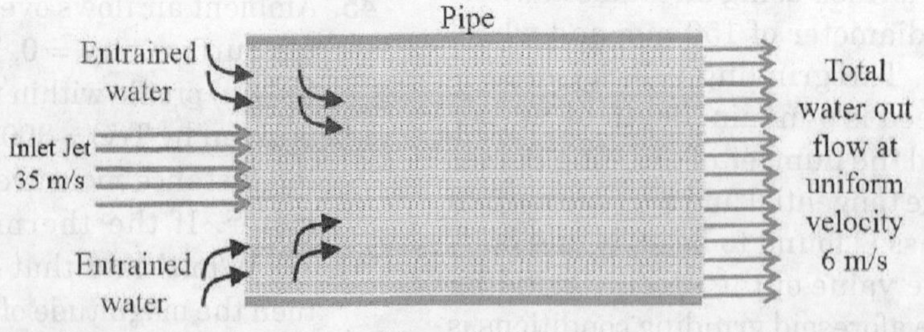

The flow rate of entrained water is ________ litres/s (*round off to two decimal places*).

49. A vertical shaft Francis turbine rotates at 300 rpm. The available head at the inlet to the turbine is 200 m. The tip speed of the rotor is 40 m/s. Water leaves the runner of the turbine without whirl. Velocity at the exit of the draft tube is 3.5 m/s. The head losses in different components of the turbine are: (i) stator and guide vanes: 5.0 m, (ii) rotor: 10 m, and (iii) draft tube: 2 m. Flow rate through the turbine is 20 m^3/s. Take g = 9.8 m/s^2. The hydraulic efficiency of the turbine is ________ % (*round off to one decimal place*).

50. An adiabatic vortex tube, shown in the figure given below is supplied with 5 kg/s of air (inlet 1) at 500 kPa and 300 K. Two separate streams of air are leaving the device from outlets 2 and 3. Hot air leaves the device at a rate of 3 kg/s from outlet 2 at 100 kPa and 340 K, and 2 kg/s of cold air stream is leaving the device from outlet 3 at 100 kPa and 240 K.

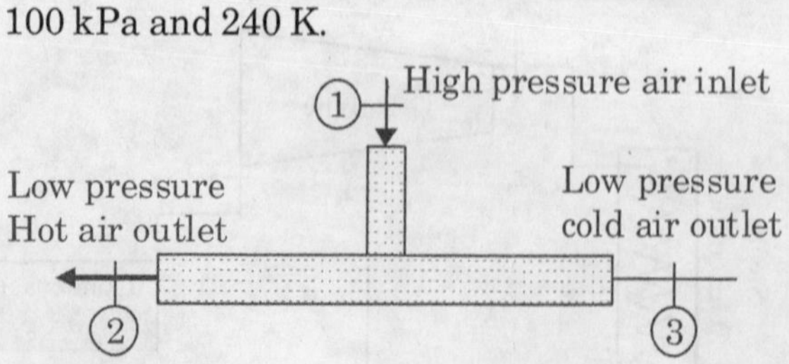

Assume constant specific heat of air is 1005 J/kg.K and gas constant is 287 J/kg.K. There is no work transfer across the boundary of this device. The rate of entropy generation is ________ kW/K (*round off to one decimal place*).

51. A block of negligible mass rests on a surface that is inclined at 30° to the horizontal plane as shown in the figure. When a vertical force of 900 N and a horizontal force of 750 N are applied, the block is just about to slide.

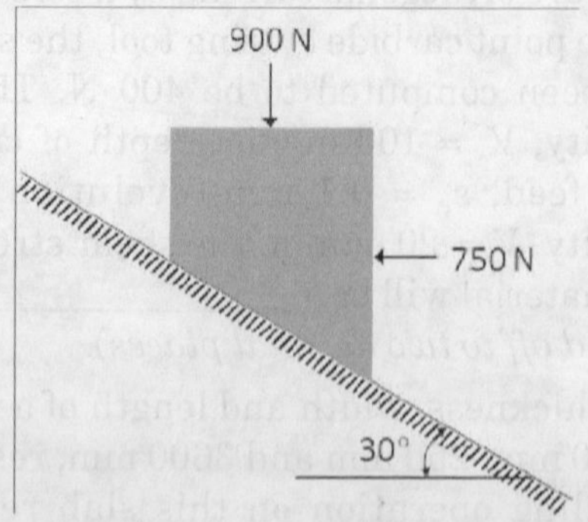

The coefficient of static friction between the block and surface is ________ (*round off to two decimal places*).

52. The wheels and axle system lying on a rough surface is shown in the figure.

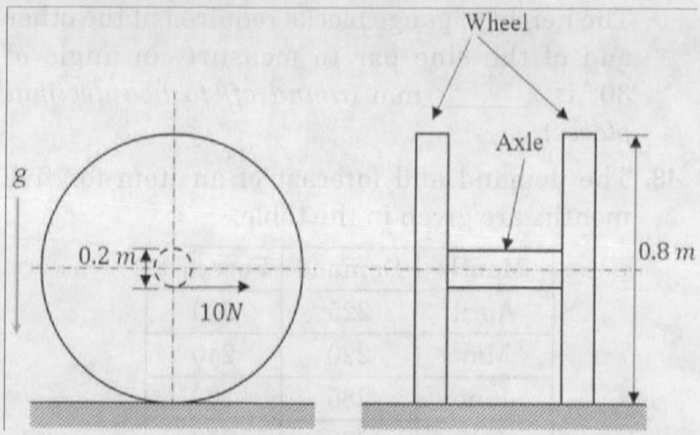

Each wheel has diameter 0.8 m and mass 1 kg. Assume that the mass of the wheel is concentrated at rim and neglect the mass of the spokes. The diameter of axle is 0.2 m and its mass is 1.5 kg. Neglect the moment of inertia of the axle and assume g = 9.8 m/s^2. An effort of 10 N is applied on the axle in the horizontal direction shown at mid span of the axle. Assume that the wheels move on a horizontal surface without slip. The acceleration of the wheel axle system in horizontal direction is ________ m/s^2 (*round off to one decimal place*).

53. A cantilever beam with a uniform flexural rigidity (EI = 200×10^6 N.m^2) is loaded with a concentrated force at its free end. The area of the bending moment diagram corresponding to the full length of the beam is 10000 N.m^2. The magnitude of the slope of the beam at its free end is ______ micro radian (*round off to the nearest integer*).

54. The torque provided by an engine is given by $T(\theta) = 12000 + 2500 \sin(2\theta)$ N.m, where θ is the angle turned by the crank from inner dead center. The mean speed of the engine is 200 rpm and it drives a machine that provides a constant resisting torque. If variation of the speed from the mean speed is not to exceed $\pm 0.5\%$, the minimum mass moment of inertia of the flywheel should be ______ kg.m^2 (*round off to the nearest integer*).

55. The figure shows the relationship between fatigue strength (S) and fatigue life (N) of a material. The fatigue strength of the material for a life of 1000 cycles is 450 MPa, while its fatigue strength for a life of 10^6 cycles is 150 MPa.

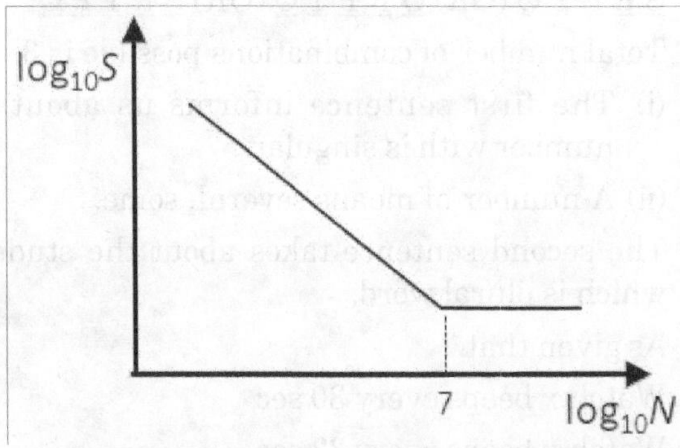

The life of a cylindrical shaft made of this material subjected to an alternating stress of 200 MPa will then be ______ cycles (*round off to the nearest integer*).

ANSWERS

General Aptitude (GA)

1. (*b*) **2.** (*a*) **3.** (*a*) **4.** (*b*) **5.** (*d*) **6.** (*c*)
7. (*b*) **8.** (*c*) **9.** (*b*) **10.** (*d*)

Mechanical Engineering

1. (*d*) **2.** (*c*) **3.** (*d*) **4.** (*d*) **5.** (*a*) **6.** (*d*)
7. (*d*) **8.** (*b*) **9.** (*c*) **10.** (*a*) **11.** (*d*) **12.** (*b*)
13. (*b*) **14.** (*b*) **15.** (*d*) **16.** (*b*) **17.** (*d*) **18.** (*a*)
19. (*b*) **20.** 373.00 to 377.00 **21.** 165 to 165 **22.** 134 to 136 **23.** 228 to 230
24. 800 to 840 **25.** 0.08 to 0.12 **26.** (*b*) **27.** (*b*) **28.** (*c*) **29.** (*a*)
30. (*d*) **31.** (*b*) **32.** (*a*) **33.** (*c*) **34.** (*b*)
35. 1.66 to 1.68 **36.** 40200 to 40300 **37.** 2.60 to 3.00 **38.** 39.00 to 40.00
39. 57 to 58 **40.** 388.00 to 400.00 **41.** 3880.0 to 3886.0
42. 200.00 to 206.00 **43.** 6.00 to 10.00 **44.** 337 to 341 **45.** 10 to 10
46. 2257.0 to 2259.0 **47.** 25.0 to 25.4 **48.** 130.00 to 134.00
49. 90.0 to 92.0 **50.** 2.1 to 2.3 **51.** 0.16 to 0.19 **52.** 1.0 to 1.5 **53.** 48 to 52 **54.** 560 to 580
55. 152000 to 165000

EXPLANATIONS

GENERAL APTITUDE

1. Positive arrangements are

S R P T Q (OR) Q R P T Q (OR) S R T P Q

Total number of combinations possible is 3.

2. (i) The first sentence informs us about the number with is singular.

(ii) A number of means several, some.

The second sentence takes about the students which is plural word.

3. As given that,

Watch x beeps every 30 sec

Watch y beeps every 32 sec

L.C.M of 30, 32 = 480 seconds = 8 minutes

They beeps together every 8 min

∴ i.e., 10 AM, 10.08 AM; 10.16 AM, 10.24 AM, ...

Hence option (a) is the correct answer.

4. (i) $\oplus \div \odot \Rightarrow \frac{\oplus}{\odot} = 2$

(ii) $\oplus \div \Delta = 3 \Rightarrow \frac{\oplus}{\Delta} = 3$

(iii) $\odot + \Delta = 5 \Rightarrow \frac{\oplus}{2} + \frac{\oplus}{3} = 5,\ \oplus = 6$ and $\Delta = \frac{6}{3} = 2.$

(iv) $\Delta \times \otimes = 10 \Rightarrow 2 \times \otimes = 10\ \otimes = 5$

So, $(\otimes - \oplus)^2 = (5-6)^2 = 1$

5. From the figure, it is clear that, the direction of X is North West

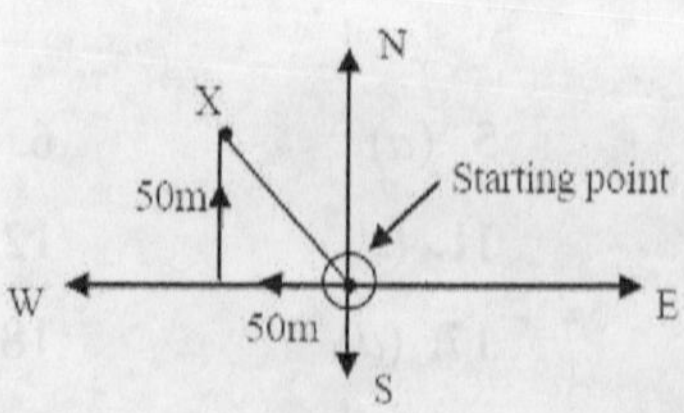

6. E → Entrepreneurs

W → Wealthy

R → Risk seekers

Statement-1

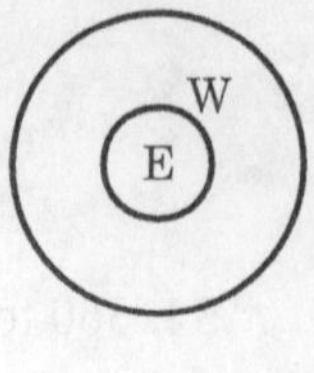

All entrepreneurs are wealthy

Statement-2

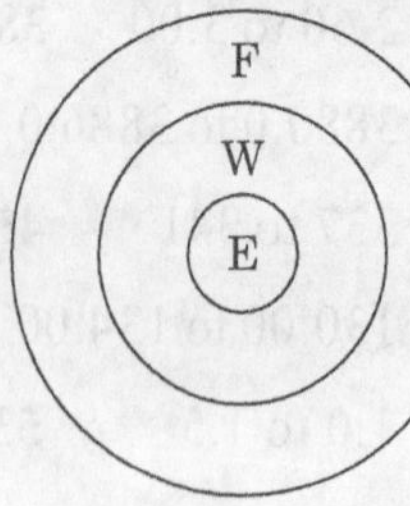

All wealthy are risk seekers

∴ All risk sectors are not wealthy ⇒ Conclusion I is incorrect.

All entrepreneurs are risk seekers ⇒ Conclusion II is incorrect

Hence, neither conclusion I nor II is correct.

7. Box

Total balls = 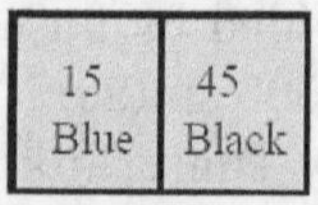

If the two balls are drawn without replacement the required probability of getting first ball blue and second ball black will be

$$\therefore \text{Required probability} = \frac{15}{60} \times \frac{45}{59}$$

$$= \frac{45}{236}.$$

8. ∴ From the figure,

$$\sin 30° = \frac{r}{R} \Rightarrow \frac{1}{2} = \frac{r}{R} \Rightarrow R = 2r$$

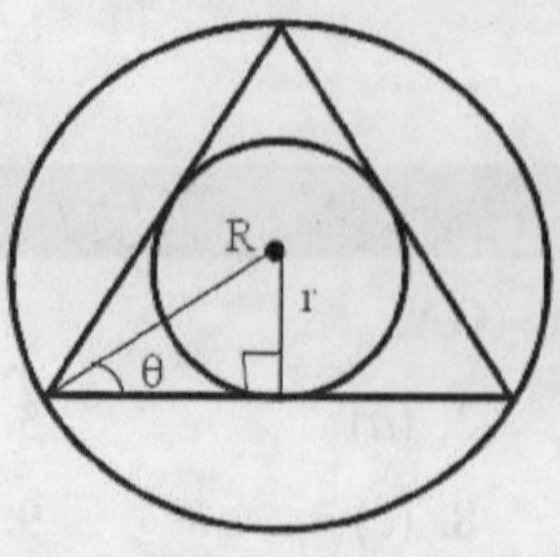

r → radius of inscribed circle

R → radius of circumscribed circle

∴ Area of inscribed circle = πr^2

Area of circumscribed circle = πR^2

$= \pi(4r^2)$

∴ Required ratio = $\pi r^2 : 4\pi r^2$

$= 1 : 4$

$= \frac{1}{4}$

9. ∴ Area of each face of the final folded shape. (Right angle triangle)

$$= \frac{1}{2} \times \frac{1}{2} \times \frac{1}{2}\left(\frac{1}{2} \times \text{Box} \times \text{weight}\right)$$

$$= \frac{1}{8} \text{ square units}$$

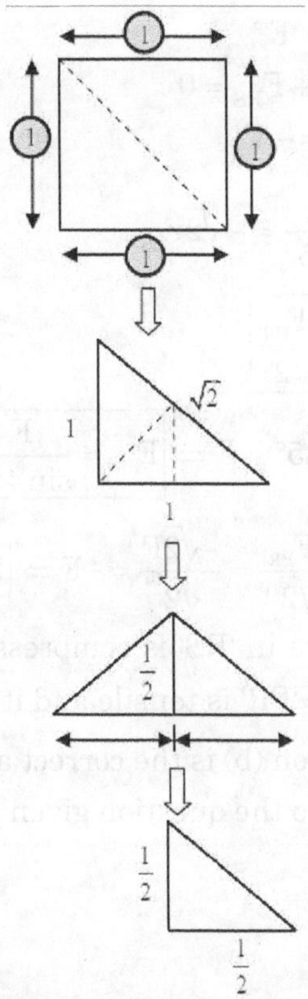

Hence, option (b) is the correct answer.

10. Due to worst pandemic the air travel industry is facing a crisis is nothing but restarting the idea from first sentence.

MECHANICAL ENGINEERING

1. As given that,

$Ap = \alpha^2 p$ ...(i)

Pre-multiply both sides by A, we get

$A(AP) = A(\alpha^2 p)$

$\Rightarrow (AA)p = \alpha^2 (Ap)$

$\Rightarrow A^2 p = \alpha^2(\alpha^2 p) = \alpha^4 p$ ($\because$ From (i))

Hence, α^4 is the eigenvalue of A^2

2. As given that,

$$F(s) = L\{f(t)\} = \frac{s+3}{(s+1)(s+2)}$$

$\therefore f(0) = \lim_{s\to\infty} sF(s)$ (Using initial value theorem)

$$= \lim_{s\to\infty} \frac{s(s+3)}{s(s+1)(s+2)}$$

$$= \lim_{s\to\infty} \frac{\left(1+\frac{3}{s}\right)}{\left(1+\frac{1}{s}\right)\left(1+\frac{2}{s}\right)} = \frac{1}{1\times 1} = 1$$

Hence, option (c) is correct answer.

3. Mean and variance of a binomial distribution are np and npq respectively, where q = 1 – p i.e.., np and np(1 – p)

Hence, option (d) is correct answer.

4. While cast iron has carbon in the form of cementite hence the answer is option (d).

5. Laser diffraction (or) laser scattering used in size distribution of powder particles used in powder metallurgy process. Hence option 'a' is the correct answer.

6. In CNC machine tool, the function of an interpolator is to generate reference signal prescribing the shape of the part to be machined. Hence option (d) is correct.

7. Ablation means removal of material by melting or evaporation.

From the given options, the only laser beam machining method will remove the material by melting and vaporization.

Hence the answer is option (d).

8. Number of activities on the critical path = 9

Standard deviation of each activity in the critical path (σ) = 3

Variance of each activity in the critical path = σ^2 = 9

Sum of variances of all activities in the critical path = $9\sigma^2 = 9 \times 9 = 81$

Standard deviation of critical path = $\sqrt{81} = 9$

Hence answer is option (b).

9. As we know that, Allowance

= Lower limit of hole – Upper limit of shaft

Hence option (c) is the correct answer.

10. As we know that,

$$\text{Stanton Number (st)} \frac{j}{ec_p\mu_m} = \frac{h\Delta T}{ec_p\mu_m\Delta T}$$

$$= \frac{\dfrac{h\Delta T}{k\left(\dfrac{\Delta L}{L}\right)}}{\dfrac{ec_p(\mu_m\Delta_T(\mu))}{k\left(\dfrac{\Delta T}{L}\right)\mu}} = \frac{Nu}{(Pr)\left(\dfrac{e\mu_m L}{\mu}\right)} = \frac{Nu}{Pr.Re}$$

Hence option (a) is the correct answer.

11. As we know that,

For a two-dimensional compressible flow, continuity equation is

$$\frac{\partial u}{\partial^2 x} + \frac{\partial v}{\partial y} = 0$$

$$\Rightarrow \frac{\partial u}{\partial x} = \frac{-\partial v}{\partial y} \Rightarrow \frac{\partial v}{\partial y} = \frac{-\partial u}{\partial x}$$

As given that,

$$\frac{\partial(u^2)}{\partial x}+\frac{\partial(uv)}{\partial y}=2u\frac{\partial u}{\partial x}+\left[\frac{\partial u}{\partial y}(v)+(u)\frac{\partial v}{\partial y}\right]$$

$$=2u\frac{\partial u}{\partial x}+(v)\left(\frac{\partial u}{\partial y}\right)+(u)\left(\frac{\partial v}{\partial y}\right)$$

$$=2u\frac{\partial u}{\partial x}+(V)\left(\frac{\partial u}{\partial y}\right)+(u)\left(\frac{-\partial u}{\partial x}\right)$$

$$=(2u-u)\left(\frac{\partial u}{\partial x}\right)+v\frac{\partial u}{\partial y}+u\frac{\partial u}{\partial x}+v\frac{\partial u}{\partial y}$$

Hence option (d) is the correct answer.

12. Boussinesq stress is responsible for eddy viscosity in turbulent boundary layer. Hence, option (c) is the correct answer.

13. $u = 2xyt$, $v = -y^2t$

Equation for steam line is

$$\frac{dx}{u}=\frac{dy}{v}$$

$$\frac{dx}{2xyt}=\frac{dy}{-y^2t}$$

$$\frac{dx}{2xy}=\frac{-dy}{y^2}\Rightarrow\frac{dx}{2x}=\frac{-dy}{y}$$

Integrating on both sides

$$\frac{1}{2}\log x=-\log y+c$$

at $x = 1$, $y = 1$

$$\frac{1}{2}(\log 1)=-\log(1)+c$$

$c = 0$

$\therefore$ equation of stream line is

$$\frac{1}{2}\log x=-\log y$$

$$\log^{\sqrt{x}}=-\log y\Rightarrow\log^{\sqrt{x}}+\log y=0$$

$$\sqrt{xy}=1$$

$xy^2 = 1$

Hence, option (b) is correct answer.

14. PQ = RS

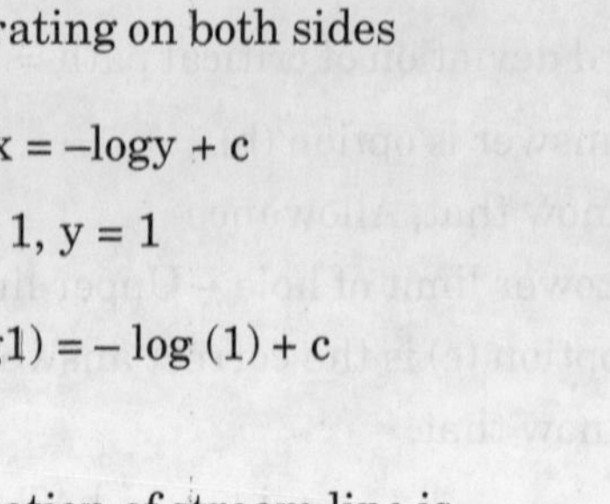

$\angle PQR = 90°$

At junction 'F'

$$F_{PR}\cos 45° + F_{RS} = 0$$

$$F_{PR}\cos 45° = -F_{RS}$$

$$F_{PR}=\frac{-F_{Rs}}{\cos 45°}=-\sqrt{2}F_{RS}$$

So, $\boxed{F_{Rs}=\frac{-F_{PR}}{\sqrt{2}}}$

$$\therefore\quad F_{Pr}\sin 45° = F \Rightarrow \boxed{F_{PR}=\frac{F}{\sin 45°}=\sqrt{2}F}$$

$$F_{RS}=\frac{-F_{PR}}{\sqrt{2}}=\frac{-\sqrt{2}F}{\sqrt{2}}=-F\Rightarrow\boxed{F_{RS}=-F}$$

So, the force in 'RS' is compressive and it is 'F' the force in 'PR' is tensile and it is $\sqrt{2}$ F

Hence, option (b) is the correct answer.

15. According to the question given figure is :

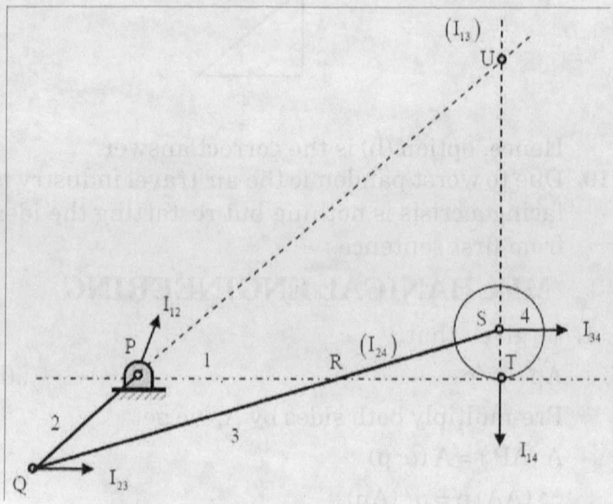

Number of instantaneous centresare

$$\frac{n(n-1)}{2}=\frac{4(4-1)}{2}$$

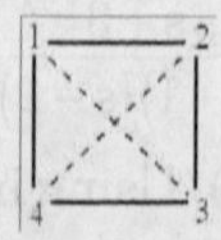

So, P, Q, R, S, T and U are instantaneous centres of mechanism

Hence option (d) is the correct answer.

16. The controlling force (F) (Vs) radius of rotation (r) in a string controlled governor is as follows.

F = ar + b

If b < 0 i.e., F = ar – b → It is a stable governor

'R' is stable governor

If b = 0 i.e., F = ar → It is isochronous governor

'Q' is isochronous governor

If b > 0 i.e., F = ar + b → It is a unstable governor

'P' is unstable governor

Hence, option (b) is the correct answer.

17. According to distortion energy theorem,

For 2-D dimensional case,

$$\sigma_1^2 + \sigma_2^2 - \sigma_1\sigma_2 \le s_{yt}^2$$

$$\sqrt{\sigma_1^2 + \sigma_2^2 - \sigma_1\sigma_2} \le s_{yt}$$

Von mises stress $(\sigma_{von}) = \sqrt{\sigma_1^2 + \sigma_2^2 - \sigma_1\sigma_2}$

Where, $\sigma_1^2 + \sigma_2^2 - \sigma_1\sigma_2$ represent distortion strain energy per unit volume Hence von mises stress is proportional to distortion strain energy per unit volume

Hence answer is option (d).

18. As given that,

f(x) = fnx, h = 0.3; a = 4, b = 5.2

$y_0 = f(4) = \ln 4 = 1.386$

$y_0 = f(a + h) = \ln(4.3) = 1.458$

$y_2 = f(a + 2h) = l.n(4.6) = 1.526$

$y_3 = f(a + 3h) = \ln(4.9) = 1.589$

$y_4 = f(a + 4h) = \ln(5.2) = 1.648$

∴ By Simpson's one-third rule,

$$\int_4^{5.2} \ln x dx = \frac{h}{3}\left[y_0 + y_4 + 4\times(y_1 + y_3) + 2(y_2)\right]$$

$$= \frac{0.3}{3}[(1.386 + 1.648) + 4(1.458 + 1.589) + 2\pi(1.526)]$$

$= 1.827 \approx 1.83$

Hence, option (a) is the correct answer.

19. As we know that,

$(1 + i)^2 = 1^2 + i^2 + 2i = 2i$

$(1 + i)^4 = 2^2 i^2 = -4$

∴ $(1 + i)^8 = (-4)^2 = 16.$

Hence, option (b) is the correct answer.

20. As given that,

Velocity of air (V) = 300 msec

Temperature of air (T) = 330K

Pressure of air (P) = 180 kPa

$C_P = 1.005$ kJ/kg.K

As we know that,

Stagnation temperature $(T_{stag}) = T + \frac{V^2}{2C_p}$

$$T_{stag} = 330 + \left(\frac{300^2}{2\times 1.005\times 10^3}\right) = 374.78 \text{ K}$$

21. As given that, COP = 10

Refrigeration capacity = 150 kJ/kg

Work input to refrigeration (W) = ?

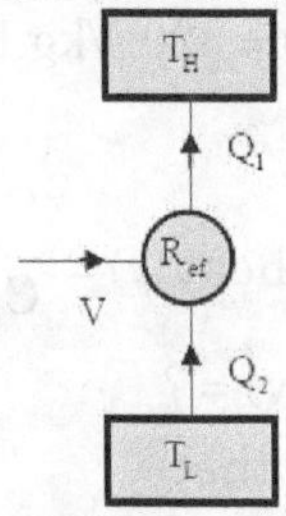

As we know that,

$$COP = \frac{Q_2}{W}$$

$$10 = \frac{150}{W}$$

$$W = = \frac{150}{10} = 15 \text{ kJ/kg}$$

Q_1 = Heat rejected by refrigeration

$W = (Q_1 - Q_2)$

$1.5 = Q_1 - 150$

$Q_1 = 150 + 15 = 165$ kJ/kg

22. As given that,

∴ Volume of tank (V) = 50m³

Saturation temperature at 400 KPa, (T_{sat}) = 143.61° C

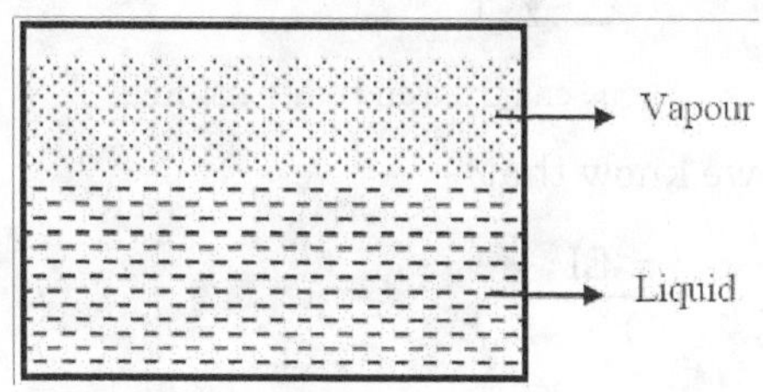

$V_f = 0.001084$ m³/kg

$V_g = 0.46242$ m³/kg

Total mass of liquid-vapour-mixture (m) = ?

$m = m_L + m_v$

Specific volume of vapour $(V_V) = 0.46242$ m³/kg

Mass of vapour $(m_v) = \frac{V}{V_g} = \left(\frac{50}{0.46242}\right)$

= 108.13kg

According to the question.

Mass of vapour (m_v) + mass of liquid (m_L) = Total mass (m) of liquid vapour mixture.

So, 108.13 + m_L = m

1081.13 + (0.2m) = m

108.13 = 0.8m

m = 135.16 kg

23. As given that,

Mach number (M) = 0.6

Temperature of air (T) = 350K

Gas constant (R) = 320 J/kg.K

As we know that

Ratio of specific heat (r) = $\frac{C_p}{C_v}$ = 1.3

Speed of object (v) = ?

As we also know that

$$\text{Mach number (M)} = \frac{\text{velocity of object}(V)}{\text{sonic velocity}(C)} = 0.6$$

$$\text{And } M = \frac{V}{\sqrt{rRT}}$$

$$\text{So, } V = M\sqrt{rRt} = (0.6)\sqrt{1.3 \times 320 \times 350}$$

$$= 228.9 \cong 229 \text{ m/sec}$$

24. According to the question,

Case (i)

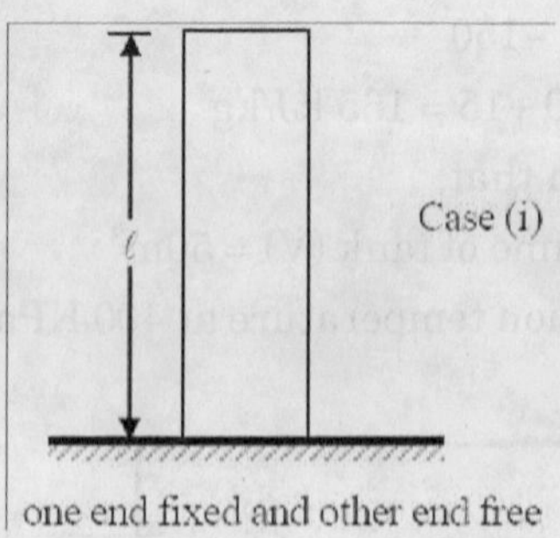

one end fixed and other end free

As we know that

$$(P_{cr})_1 = \frac{\pi^2 EI}{L_e^2}$$

$$= \frac{\pi^2 EI}{(2\ell)^2} = 100N$$

$$\Rightarrow \frac{\pi^2 EI}{\ell^2} = 400N$$

Case (ii)

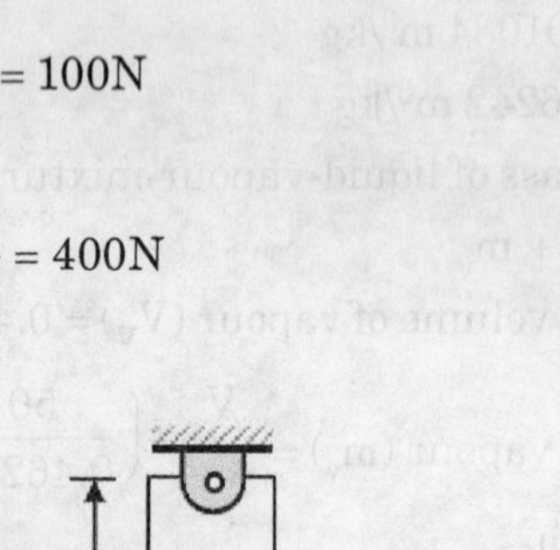

one end fixed and other end pinned

$$(P_{cr})_2 = \frac{\pi^2 EI}{L_e^2}$$

$$= \frac{\pi^2 EI}{\left(\frac{\ell}{\sqrt{2}}\right)^2} = \frac{2\pi^2 EI}{\ell^2}$$

= 2 × 400 = 800N

Hence, answer is 800N

25. As given that,

$E = 2 \times 10^5 N/mm^2$, $\mu = 0.3$

$E = 3k(1 - 2\mu)$

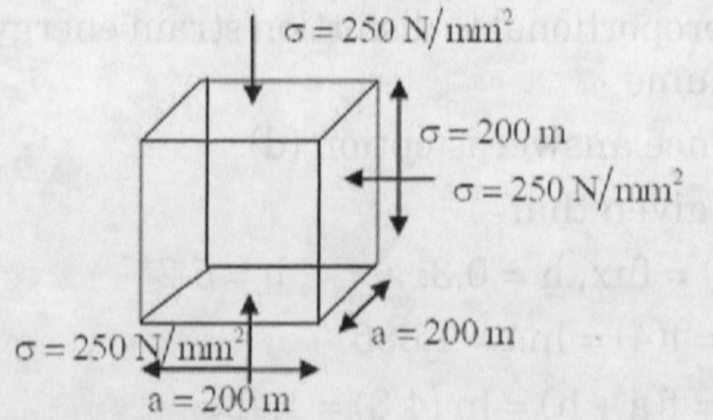

As we know that,

$$\text{Bulk modulus (k)} = \frac{E}{3(1-2\mu)} = \frac{(2\times10^5)}{3(1-2\times0.3)}$$

$k = 166666.67 \text{ N/mm}^2$

Bulk modulus (k)

$$= \frac{\text{volumetric stress(or) hydrostatic stress}(\sigma)}{\text{volumetric strain}(e_v)}$$

$$166666.67 = \frac{250}{e_v}$$

$$e_v = \frac{250}{166666.67} = 1.499 \times 10^{-3} \cong 1.5 \times 10^{-3}$$

$$e_v = \frac{\text{change in volume}(\Delta V)}{\text{original volume(or) initial }(a^3)}$$

$$\Delta V = e_v(a^3) = (1.5 \times 10^{-3})(200)^3 = 12{,}000$$

ΔV = Initial volume – Final volume

$12{,}000 = V_i - V_f$

$V_f = (200)^3 - (12{,}000) = 7988000$

Let a' = New size of cube

$(a')^3 = 7988000$

a' = 199.89

a – a' – = 200 –199.89 = 0.11mm

Hence, the side of the block is reduced by 0.11 mm.

26. As given double integral

$$\Rightarrow \int_{\theta=0}^{\pi/2}\left[\int_{r=0}^{\cos\theta} r\sin\theta dr\right] d\theta$$

$$= \int_{\theta=0}^{\pi/2} \sin\theta\left(\frac{r^2}{2}\right)_0^{\cos\theta} d\theta = \frac{1}{2}\int_0^{\pi/2} \sin\theta\cos^2\theta d\theta$$

$$= \frac{1}{2}\int_0^1 t^2 dt$$

[∵ By putting $\cos\theta = t, \sin\theta\, d\theta$

$= dt$ & $\theta = \frac{\pi}{2} \Rightarrow t = 1$]

$$= \frac{1}{2}\left(\frac{t^3}{3}\right)_0^1 = \frac{1}{6}$$

Hence, option (b) is the correct answer.

27. As we know that

If x is (n × 1) vector = $\begin{bmatrix} x_1 \\ x_2 \\ \vdots \\ x_n \end{bmatrix}$

So, x^T is (1 × n) vector = $[x_1, x_2, \dots x_n]$

As given that,

Take n = 2 and Let Q = $\begin{bmatrix} a & c \\ c & b \end{bmatrix}$ be a symmetric (2 × 2) matrix, x = $\begin{bmatrix} x_1 \\ x_2 \end{bmatrix}$; r = $\begin{bmatrix} r_1 \\ r_2 \end{bmatrix}$

Then $f(x) = \frac{1}{2}\{x, x_2\}\begin{bmatrix} a & c \\ c & b \end{bmatrix}\begin{bmatrix} x_1 \\ x_2 \end{bmatrix} - \begin{bmatrix} r_1 & r_2 \end{bmatrix}\begin{bmatrix} x_1 \\ x_2 \end{bmatrix}$

$$= \frac{1}{2}\left[ax_1^2 + bx_2^2 + 2cx_1x_2\right] - r_1x_1 - r_2x_2$$

Let $f(x) = \phi(x_1, x_2) = \frac{ax_1^2}{2} + \frac{bx_2^2}{2} + cx_1x_2 - r_1x_1 - r_2x_2$

For stationary points, we have $= \frac{\partial\phi}{\partial x_1} = 0$ and $\frac{\partial\phi}{\partial x_2} = 0$

$\Rightarrow ax_1 + cx_2 - r_1 = 0$ ....(i)

And $bx_2 + cx_1 - r_2 = 0$ ....(ii)

Equation (i) and (ii) in matrix form is

$$\begin{bmatrix} a & c \\ c & b \end{bmatrix}\begin{bmatrix} x_1 \\ x_2 \end{bmatrix} = \begin{bmatrix} r_1 \\ r_2 \end{bmatrix}$$

$\Rightarrow Q.x = r$

Multiply by Q^{-1}, we get

Hence, $x = Q^{-1}r$, option (b) is the stationary point.

28. As given that,

The differential equation

$$\Rightarrow \left(\frac{1+y}{y}\right) dy = dx, \text{ which is VSF}$$

Integrating, we get

$$\int\left(\frac{1}{y} + 1\right) dy = \int dx + C$$

$\Rightarrow \ell ny + y = x + C$ ...(1)

Using y(1) = 1, (1) gives 0 + 1 = 1+ C

$\Rightarrow C = 0$

$\therefore\ \ell ny + y = x$

$\Rightarrow y = e^{x-y}$ or $ye^y = e^x$

Hence option (c) is the correct answer.

29. As given that,

j → work station

n → number of work station

i → product

m → number of products.

Capacity constraint is :

$a_{ij}x_{it} \le c_{jt}$

Where, a_{ij} → processing time that one unit of ith product requires on j^{th} work station

x_{it} → Amount of product 'r' manufacture in time period 't'.

C_{ij} → Capacity work station 'j' in time 't'

Inventory balance constraint is :

$$I_{it} = I_{i,t-1} + X_{it} - S_{it}$$

Where, I_{it} → Amount of product 'i' held in inventory at the end of time period 't'

$I_{i,t-1}$ → Amount of product 'i' held in inventory at the time period 't–1'

X_{it} → Amount product 'I' manufacture in time period 't'

$I_{i,t-1}$ → Amount of product 'I' held C_n Inventory at the time period 't-x'

X_{it} → Amount product is manufactured in time period 't'

S_{it} → Amount of product 'i' sold in time period 't'.

Hence option (a) is the correct answer.

30. As given that,

Ambient pressure (P_{atm}) = 101kPa = P

Temperature of air (T) = 300K

Relative humidity (ϕ) = 0.6

Saturation pressure of water at 300K is

(P_{sat}) = 3.6 kPa

Specific humidity of air (W) = ?

As we know that,

$$\phi = \frac{P_V}{P_{sat}} \Rightarrow P_V = (0.6)(3.6)$$

$P_V = 2.16$ kPa

We also know that,

$$W = \frac{0.622P_V}{P - P_V} = \frac{(0.622)(2.16)}{(101 - 2.16)}$$

W = 0.01359 kg of water vapour/kg of dry air

= 13.59 gms of water vapour/kg of dry air

Hence option (d) is the correct answer.

31. According to the question.

Flexural rigidity of members PQ and QR having length 'L' are 'EI'

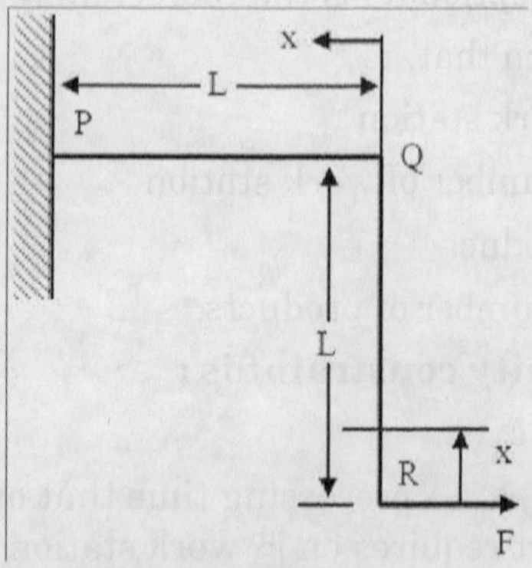

Bending moment at any distance 'x ' from 'R' is $(m_x)_{QR} = Fx$

Bending moment in 'PQ' portion is constant and is equal to $M_{PQ} = FL$

Strain energy stored in frame $(U = U_{QR} + U_{PQ})$

$$\int_0^L \frac{(m_x)_{QR}^2 dx}{2EI} + \int_0^L \frac{(m_{PQ})^2 dx}{2EI}$$

$$U = \int_0^L \frac{(Fx)^2 dx}{2EI} + \int_0^L \frac{(FL)^2 dx}{2EI}$$

$$U = \frac{F^2}{2EI}\left[\frac{x^3}{3}\right]_0^L + \frac{F^2L^2}{2EI}[dx]_0^L$$

$$= \frac{F^2}{2EI}\left[\frac{L^3}{3}\right]_0^L + \frac{F^2L^2}{2EI}[L]$$

$$= \frac{F^2L^3}{6EI} + \frac{F^2L^3}{2EI}$$

$$= \frac{F^2L^3 + 3F^2L^3}{6EI} = \frac{2}{3}\frac{F^2L^3}{EI}$$

Horizontal deflection at 'R' is $(\delta_H)_R$

$$U = \frac{\partial U}{\partial F} = \left(\frac{2}{3}\right)\left(\frac{2FL^3}{EI}\right) = \frac{4FL^3}{3EI}$$

Hence option (b) is the correct answer.

32. As given that,

$d_1 = 150$ mm, $d_2 = 250$ mm

$$\frac{N_2}{N_1} = (2500)\left(\frac{150}{250}\right) = 1500 \text{ rpm (ACW)}$$

$$\frac{N_3}{N_2} = \frac{d_2}{d_3} = \frac{z_1}{z_3}$$

$$\frac{N_3}{1500} = \left(\frac{18}{44}\right) \Rightarrow N_3 = 613.64 \text{ rpm (CW)}$$

$N_3 = N_4$, since they are mounted on same shaft

$$\frac{N_5}{N_4} = \frac{d_4}{d_5} = \frac{z_4}{z_5}$$

$$\frac{N_5}{613.64} = \frac{15}{33}$$

$N_5 = 278.93$ rpm (ACW)

$$\frac{N_6}{N_5} = \frac{d_5}{d_6} = \frac{z_5}{z_6}$$

$$\frac{N_6}{278.93} = \frac{33}{36}$$

$N_6 = 255.68$ rpm (CW)

$N_7 = N_6$, since they are mounted on same shaft

Hence option (a) is the correct answer.

33. As given that,

Mass of machine (m) = 100 kg

Applied frequency (ω) = 40rad/sec

As we know that,

$$\text{Transmissibility (TR)} = \frac{\sqrt{\left(1 + (2gr)^2\right)}}{\sqrt{(1 - r^2)^2 + (2gr)^2}} = \left|\frac{F_T}{F_0}\right|$$

Where, F_T = Force transmitted

F_0 = Applied force

For first Isolators

$\therefore K = 640$ kN/m, $\xi = 0.70$

$$\therefore \omega_n = \sqrt{\frac{k}{m}} = \sqrt{\frac{640 \times 10^3}{100}} = 80 \text{ rad/sec}$$

$$\therefore r = \frac{\omega}{\omega_n} = \frac{40}{80} = 0.5$$

$$\text{So, } \left|\frac{F_T}{F_0}\right| = \frac{\sqrt{\left(1 + (2 \times 0.70 \times 0.5)^2\right)}}{\sqrt{(1 - 0.5^2)^2 + (2 \times 0.7 \times 0.5)^2}}$$

$$\left|\frac{F_T}{F_0}\right| = \frac{1.22}{1.026} = 1.189$$

For second Isolators

$\therefore K = 640$ kN/m, $\xi = 0.07$

$$\therefore \omega_n = \sqrt{\frac{k}{m}} = = 80 \text{ rad/sec}, r = 0.5$$

$$\left|\frac{F_T}{F_0}\right| = \frac{\sqrt{\left(1+(2\times0.07\times0.5)^2\right)}}{\sqrt{(1-0.5^2)^2+(2\times0.07\times0.5)^2}}$$

$$\left|\frac{F_T}{F_0}\right| = \frac{1}{0.753} = 1.32$$

For third Isolators

$\therefore K = 22.5$ kN/m, $\xi = 0.70$

$$\therefore \omega_n = \sqrt{\frac{k}{m}}\sqrt{\frac{225\times10^3}{100}} = 15 \text{ rad/sec}$$

$$\therefore r = \frac{\omega}{\omega_n} = \frac{40}{15} = 2.67$$

$$\left|\frac{F_T}{F_0}\right| = \frac{\sqrt{\left(1+(2\times0.7\times2.67)^2\right)}}{\sqrt{(1-2.67^2)^2+(2\times0.7\times2.67)^2}}$$

$$\left|\frac{T_T}{F_0}\right|\frac{3.87}{4.18} = 0.54$$

For forth Isolators

$\therefore K = 22.5$ kN/m, $\xi = 0.07$

$\therefore \omega_n = 15$ rad/sec, $r = 2.67$

$$\left|\frac{F_T}{F_0}\right| = \frac{\sqrt{\left(1+(2\times0.07\times2.67)^2\right)}}{\sqrt{(1-2.67^2)^2+(2\times0.07\times2.67)^2}}$$

$$\left|\frac{T_T}{F_0}\right| = \frac{1.067}{6.14} = 0.17$$

So ascending order is 4 – 3 – 1 – 2.

Hence, option (c) is the correct answer.

34. As we know that,

A small angular displacement to pulley

F_s = spring force = $kr\sin\theta$

$$F_i = m\frac{d^2}{dt^2}(r\sin\theta)$$

For small angles $r\sin\theta \cong r\theta = (mr)\frac{d^2}{dt^2}(\theta) = mr\theta$

$T_i = J\,\theta$

Where,

J = Mass moment of inertia of purely about '0'

$\Sigma M_0 = \theta$

$J\ddot{\theta} + (F_i)(r\cos\theta) + (F_s)(r\cos\theta) = 0$

$J\ddot{\theta} + (mr\ddot{\theta})(r\cos\theta) + (kr\sin\theta)(r\cos\theta) = 0$

For small angels $\cos\theta \cong 1$, $\sin\theta = \theta$

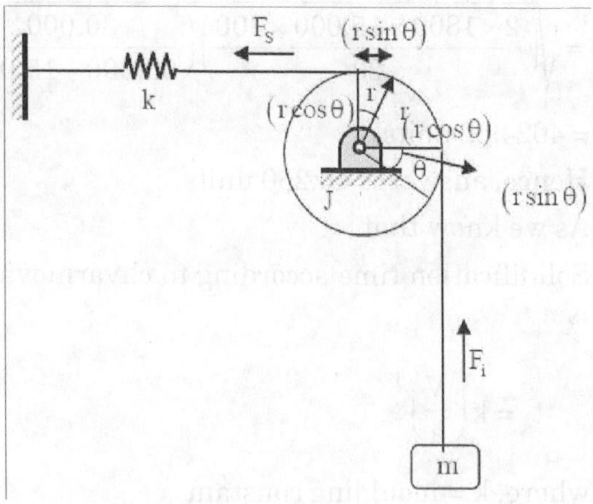

$J\ddot{\theta} + mr^2)\ddot{\theta} + kr^2\theta = 0$

$(J + mr^2)\ddot{\theta} + kr^2\,\theta = 0$

$$\omega_n = \sqrt{\frac{kr^2}{J+mr^2}}$$

Hence option (b) is the correct answer.

35. As given that,

$f(x) = x^3 - x - 3$

$\Rightarrow f'(x) = 3x^2 - 1$ and $x_0 = 2$

As we know that,

Newton-Raphson method :

$$x_1 = x_0 - \frac{f(x_0)}{f'(x_0)} = 2 - \frac{(3)}{11} = 1.73$$

$$x_2 = x_1 - \frac{f(x_1)}{f'(x_1)} = 1.73 - \frac{(0.48)}{7.98} \approx 1.67$$

Hence the numerical value of the root after two iterations (x_2) is 1.67.

36. As given that,

Daily Demand = 15,000 bearings

Annual Demand (R) = 15,000 × 300

C_3 = ordering cost (or) set up cost = 1800 Rs

C_1 = carrying cost = 20 Rs | Bearing | year

K = Production rate = 30, 000, Bearings per day

As we know that,

Economic batch quantity $(E_BQ) = \sqrt{\dfrac{2C_3R}{C_1\left(1-\dfrac{R}{K}\right)}}$

$$= \sqrt{\frac{2C_3R}{C_1}\left(\frac{k}{k-R}\right)}$$

$$= \sqrt{\left(\frac{2\times1800\times15{,}000\times300}{20}\right)\left(\frac{30{,}000}{30{,}000-15{,}000}\right)}$$

= 40249.2 units

Hence, answer is 40,250 units.

37. As we know that :

Solidification time according to chvarinovs rule is :

$$t_s = k\left(\frac{V}{a}\right)^{\eta}$$

where, k = moulding constant

n = 2

For cube :

Volume of the cube (V) = 75 × 125 × 20

= 18750 mm^3

Surface are of cube (A) = 2(75 × 125 + 125 × 20 + 20 × 75) = 26750 mm^2

Solidification time (t_s) for cube is = 2 minutes

$$2 = (k)\left(\frac{187500}{26750}\right)^2$$

k = 0.041

For cylinder :

Volume of cylinder (V)

$$= \frac{\pi}{4}(d^2)(h) = \frac{\pi}{4}(50^2)(50) = 31250\ \text{imm}^3$$

Surface area of cylinder (A)

$$= \left(\pi dl + 2\times\frac{\pi}{4}d^2\right) = \left(\pi(50)(50)+\frac{\pi}{2}(50)^2\right)$$

= 3750 πmm^2

Solidification of time (t_s) for cylindrical is

$$t_s = (k)\left(\frac{V}{A}\right)^n = (0.041)\left(\frac{31250\pi}{3750\pi}\right)^2 = 2.85\ \text{mins}$$

Hence total solidification time will be 2.85 mm.

38. As given that,

Diameter of Nugget (d) = 5mm

Thickness of Nugget (t) = 1mm

Welding time (t) = 0.1 seconds

Melting energy of steel or Heat required for melting = +20 J/mm^3

Heat conversion efficiency (η) = 0.10

Heat supplied (Qs) = ?

As we know that

$$\text{Volume of Nugget (V)} = \left(\frac{\pi}{4}d\right)(t)$$

$$V = \left(\frac{\pi}{5}(5^2)(1)\right) = 6.25\ \pi\ \text{mm}^3$$

Total heat required (Q_R) = (20 × 6.25m) = 125πJ

we also know that,

$$\eta = \frac{Q_R}{Q_S}$$

$$0.10 = \frac{125\pi}{Q_S} \Rightarrow Q_S = \frac{125\pi}{0.1}$$

= 3926.99 ≅ 3927 joules

So, heat supplied per seconds.

$$= \frac{Q_S}{t} = \frac{3927}{0.1} = 39270\ \text{watts} = 39.27\ \text{kW}$$

39. As given that,

Grinding wheel diameter (D) = 150 mm

Width of wheel (b)= 12 mm

Grinding velocity (V) = 40 m/sec

Table speed (V_{Table}) = 5m/min

Depth of cut per pass = 50μm = 50 × 10^{-6}m

Number of grinding passes (n) = 20

Tangential force = 40N = F_t

Normal force = 60N = F_n

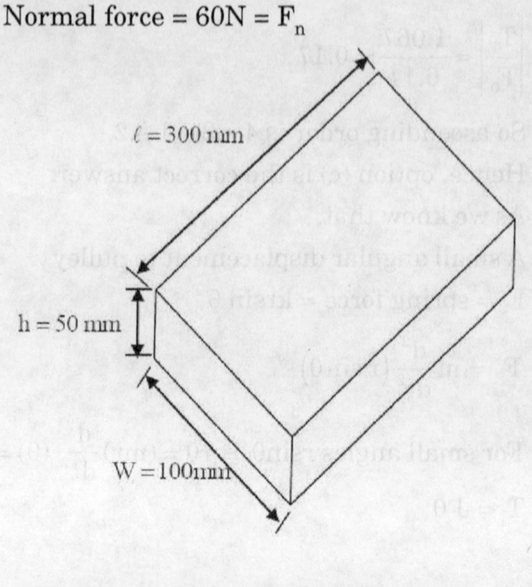

As we know that
Resultant force acting on grinding wheel

$$(F)_{grinding} = \sqrt{F_t^2 + F_n^2}$$

$$F_{grinding} = \sqrt{40^2 + 60^2} = 72.11N$$

In grinding : We also know that,

$$\text{Total grinding force } (F_{grinding}) = u \times \frac{\partial db}{V}$$

Where,
u = specific grinding (J/mm³)
∂ = work piece velocity = 5 mmin = 500 mm/min
V = grinding wheel velocity = 40 m/sec
d = depth of cut = 50×10^{-6}m = 50×10^{-3}mm
b = width of cut = 12 mm
By putting respective values:

$$\text{So, } 72.11 = \frac{(u)\times(5000\text{mm/min})(50\times10^{-3})\text{mm}(12)\text{mm}}{(40\times60)(\text{m/mm})}$$

u = 57.7 J/mm³

40. As given that,
Rake angle (α) = 0 9
Shear force (F_S) = 400N
Cutting velocity (V_C) = 100 min/min
Depth of cut (t) = 2mm
Feed (S_0) = 0.1mm/rev
Chip flow velocity (V_f) = 20m/min
Shear strength of material (τ_s) = ?

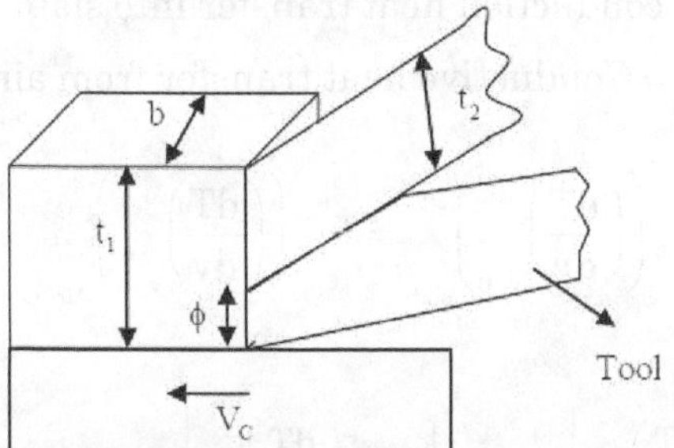

As we know that,

$$\text{Chip thickness ratio (r)} = \frac{t_1}{t_2} = \frac{V_f}{V_c} = \frac{20}{100} = 0.2$$

Shear angle 'ϕ' will be find out as follows,

$$\tan\phi = \frac{r\cos\alpha}{1 - r\sin\alpha} = \frac{(0.2)\cos 0°}{1-(0.2)\sin 0°}$$

$\phi = \tan^{-1}(0.2) = 11.31°$
We also know that,

Shear strength (or) shear stress in the work piece is

$$(\tau_s) = \left(\frac{F_s}{t_1 b}\right)(\sin\phi)$$

$$= \left(\frac{400}{2\times0.1}\right)\sin(1.31)$$

$$= 392.23 \text{ MPa}$$

t_1 = uncut chip thickness = depth of cut = 2mm
b = width of chip = feed = 0.1 mm
Hence, the shear strength, τ_s of the material is 392.23 MPa.

41. As given that,
t_1 = 50 mm
W_1 = 250 mm
ℓ_1 = 3600 min
$t_2 = 0.9 \times 50 = 45$ mm
$W_2 = 1.03 \times 250 = 257.5$ mm
ℓ_2 = ?

According to the question
Volume of slab before rolling (V_1) = Volume of slab after (V_2) Rolling

So, $t_1 w_1 \ell_1 = t_2 w_2 \ell_2$

$(50)(25)(3600) = (45)(257.5)(\ell_2)$

$$\ell_2 = \left[\frac{50\times25\times3600}{45\times257.5}\right]$$

Hence, Length of slab after rolling (ℓ_2) = 3883.5 mm

42. As given that,
Length of sine bar (l) = 254 mm
Diameter of roll (d) = 25.4 mm
Angle measured (θ) = 30°
Height of gauge blocks
On one side (h_2) = 76.2 mm

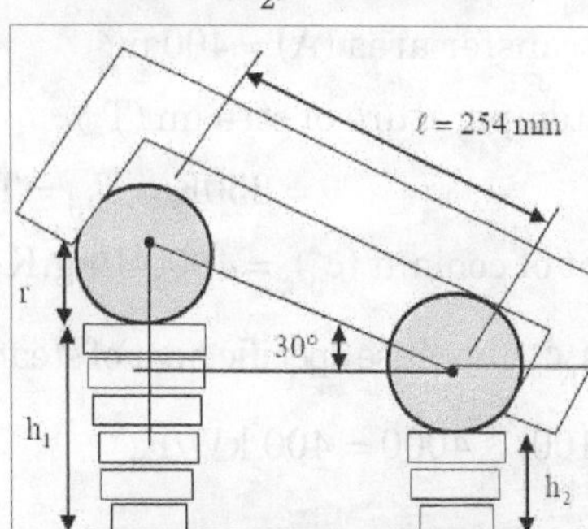

So, height of gauge blocks on another side (h_1) = ?

$$\sin 30° = \frac{(h_1 + r) - (h_2 + r)}{l} \quad (\because \text{ from figure})$$

$$\frac{1}{2} = \frac{(h_1 - h_2)}{254} = \frac{(h_1 - 76.2)}{254}$$

$h_1 = 203.2$ mm

43. According to given table :

Month	Demand(D)	Forecast(F)	Percentage absolute error $\left(\left\|\frac{D-F}{D}\right\|\right)$
April	225	200	$\frac{25}{225} = 11.11\%$
May	220	240	$\frac{20}{220} = 9.09\%$
June	285	300	$\frac{15}{285} = 5.26\%$
July	290	270	$\frac{20}{290} = 6.89\%$
August	250	230	$\frac{20}{250} = 8\%$

Mean absolute percentage :

$$\text{error (MAPE)} = \frac{11.11 + 9.09 + 5.26 + 6.89 + 8}{5}$$

$$= 8.07\%$$

Hence (MAPE) in the forecast is 8%.

44. As given that,

Mass flow rate of coolant (m_c) = 100 kg/sec

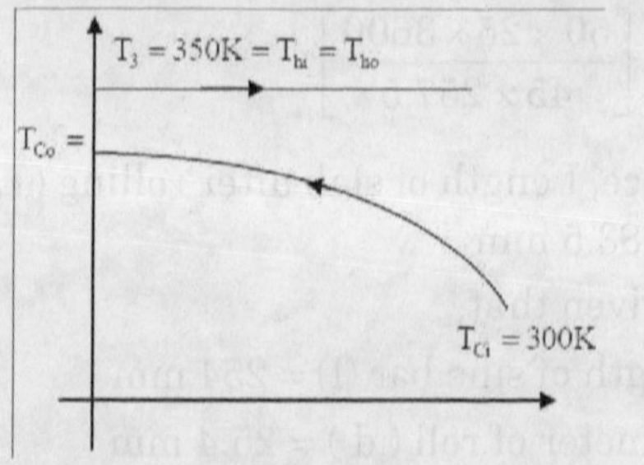

Inlet temperature of coolant water overall heat transfer $T_{ci} = 300$K

Coefficient (U) = 1500 W/m^{2}K

Total heat transfer area (A) = 400 m^2

Saturation temperature of stream (T_s)

$= 350K = T_{hi} = T_{ho}$

Specific heat of coolant $(c_p)_c = 4000$ J/kg.K

$\therefore \; C_{min} = \dot{m}_c c_{pc}$ (because specific heat of steam is ∞)

$= (100) * 4000 = 400$ kW/K,

$$NTU = \frac{UA}{C_{min}} = \frac{(1500)(400)}{(400 \times 10^3)} = 1.5$$

For condensing of boiling effectiveness of heat exchanger is

$\varepsilon = 1 - e^{-NTU} = 1 - e^{-15} = 0.78$

As we know that

$$\varepsilon = \frac{m_c C_{pc}(T_{co} - T_{ci})}{C_{min}(T_{hi} - T_{ci})} \quad (\because \; C_{min} = \dot{m}C_{pc})$$

$$\text{So, } 0.78 = \frac{(T_{co} - 300)}{(350 - 300)}$$

$T_{co} = 339$K

Hence, the temperature of the coolant water coming out of the condenser is 339 K.

45. As given that,

Thermal conductivity of air (k_{air}) = 1 W/mK

Thermal conductivity of slab (k_{slab}) = 100 W/mK

According to the question :

At y = 0

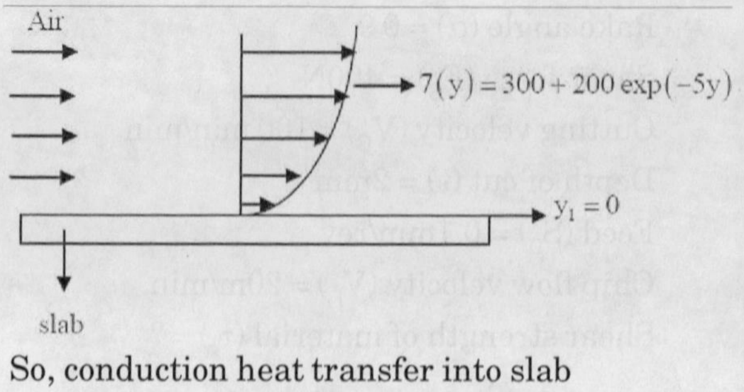

So, conduction heat transfer into slab

= Conductive heat transfer from air to slab

$$k_{slab}\left(\left(\frac{dT}{dy}\right)_{y=0}\right)_{slab} = k_{air}\left(\left(\frac{dT}{dy}\right)_{y=0}\right)_{Air}$$

$$\left(\frac{dT}{dy}\right)_{y=0}\bigg|_{slab} = \left(\frac{k_{air}}{k_{slab}}\right)\left(\frac{dT}{dy}\right)_{y=0}\bigg|_{air}$$

$$= \left(\frac{1}{100}\right)\left(-(5)(200)e^{-5y}\right)_{y=0}$$

$$-\left(\frac{1}{100}\right)(-(5)(200)) = -10 \text{ k/m}$$

Hence, magnitude of temperature gradient within the slab at y = 0 is '10'

46. As given that,

Cross sectional area of tank $(A_t) = 1m^2$

Cross sectional of area orifice $(A_0) = 1cm^2$

$= 1 \times 10^{-4}\ m^2$

Initially Height of water, $(H) = 1m$

$g = 9.8\ m/sec^2$

According to question.

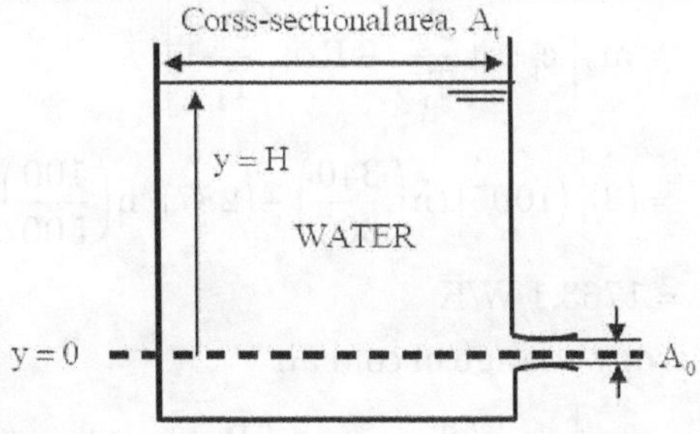

Now, time taken to reach a level of water $y = \frac{H}{4}$ will be find out as follows,

$m_{in} - m_{out} = m_{cu}$

$m_{in} = 0$

$$-\rho A_0\sqrt{2gH} = \frac{d}{dt}(\rho A_t dH)$$

$$-A_0\sqrt{2gH} = \frac{d}{dt}(A_t dH)$$

$$-A_0\sqrt{2gH} = d(A_t dH)$$

By integration on both sides :

$$-A_0\sqrt{2gH}(t-0) = A_t(dH)$$

$$t = \frac{-A}{A_0\sqrt{2g}}\left(\frac{dH}{\sqrt{H}}\right)$$

$$t = \frac{-A_t}{A_0\sqrt{2g}}\int_1^{0.25}\frac{dH}{\sqrt{H}} = \frac{A}{A_0(\sqrt{29})}\int_{0.25}^{1}\frac{dH}{\sqrt{H}}$$

$$= \frac{(1)}{\left(1\times10^{-4}\times\sqrt{2\times9.8}\right)}\left[2\sqrt{H}\right]_{0.25}^{1}$$

$t = 2258$ seconds

47. As given figure :

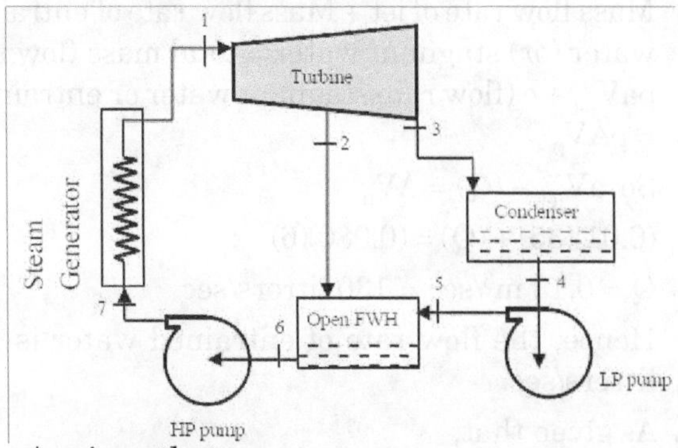

As given that,

$h_2 = 2624$ kJ/kg

$h_5 = 226.7$ kJ/kg

$h_6 = 708.6$ kJ/kg

$m_5 = 100$ kg/sec

$m_2 = ?$

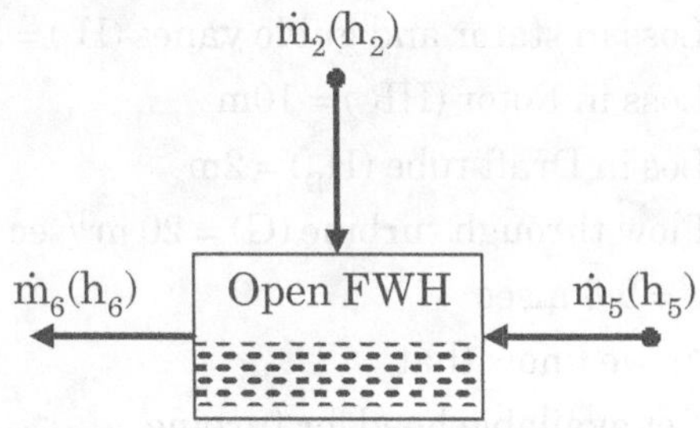

According to above figure

$m_2h_2 + m_5h_5 = m_6h_6$

$m_2(2624) + (100)(226.7) = m_6(708.6)$

Let 'm_5' be the mass of flow rate steam taken at location '2'

Let m_1 be the total mass flow rate of steam entering at location '1'

$m_1 = m_2 + m_5$ and $m_1 = m_6$

$m_6 = m_2 + m_5 = m_2 + 100$

$(m_2)(2624) + (100)(226.7) = (m_2 + 100)(708.6)$

$2624m_2 + 22670 = 708.6h + 70860$

$1915.4m_2 = 48190$

$m_2 = 25.16$ kg/sec

Hence, the mass flow rate of stream at location 2 is 25.16 kg/sec.

48. As given figure :

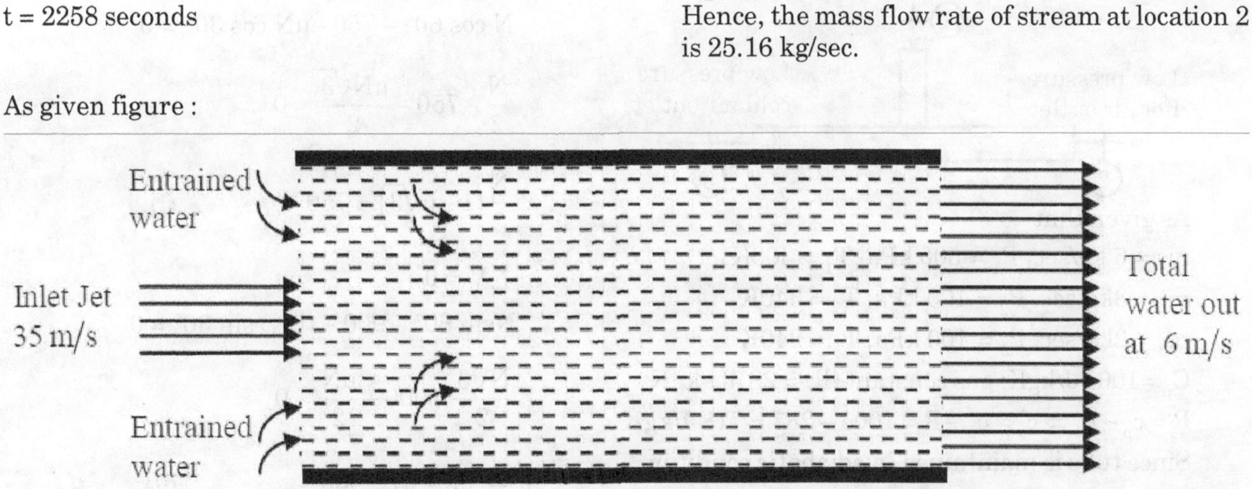

Now, applying conservation of mass of principle,
Mass flow rate of jet + Mass flow rate of entrained water (or) stagnant water = total mass flow rate

$\rho a V_{jet}$ + e (flow rate stagnant water or entrained) $= \rho A V_0$

So, $aV_{jet} + (Q) = AV_0$

$(0.01)(35) + (Q) = (0.080)(6)$

Q = 0.13 m^3/sec = 130 litrers/sec

Hence, the flow rate of entrained water is 130 litrers/sec.

49. As given that,

N = 300 rpm

H = 200 mm

u = 40 msec

Velocity of water at exit of draft tube (V_2) = 3.5 m/sec

Loss in stator and guide vanes (H_s) = 5m

Loss in Rotor (HR) = 10m

Los in Draft tube (H_D) = 2m

Flow through turbine (G) = 20 m^3/sec

g = 9.8 $msec^2$

As we know that

Net available head for turbine

$$(H_{net}) = H - H_S - H_R - H_D - \frac{V_0^2}{2g}$$

$$H_{Net} = 200 - 5 - 10 - 2 - \left(\frac{3.5^2}{2 \times 9.8}\right) = 182.375\text{m}$$

$$\therefore \text{ Hydraulic efficiency } \eta_H = \frac{H_{net}}{H} = \frac{182.375}{200}$$

$$\eta_H = 91.187\%$$

Hence, answer is 91.2

50. As given figure

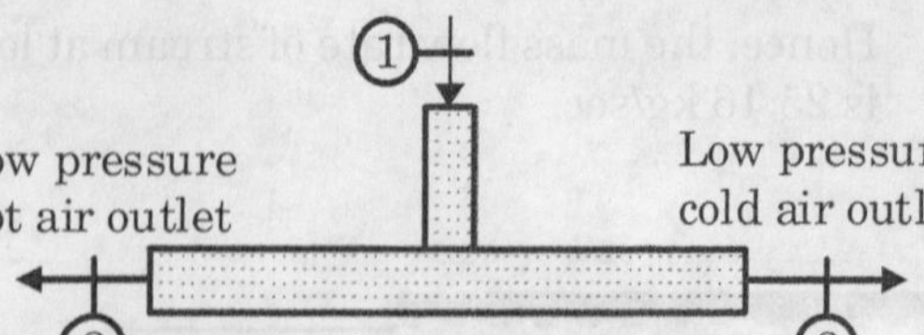

As given that

m_1 = 5 kg/sec, P_1 = 500 kPa, T_1 = 300K

m_2 = 3kg/sec, P_2 = 100 kPa, T_2 = 340K

m_3 = 2kg/sec, P_3 = 100 kPa, T_3 = 240K

C_p =1005 J/kgK; gas constant (R) = 287J / kgK

$R = c_p - c_v \Rightarrow c_v = c_p - R = 1005 - 287 = 718$ J/kgK

Since tube is maintained in adiabatic condition

No entropy transfer from the vortex tube

$\therefore$ Entropy generation = Entropy change – Entropy transfer

$\therefore$ Entropy generation = Entropy change ($\because$ entropy transfer = 0)

As we know that

Entropy change of hot air

$$= m_2\left[c_p \ell n\left(\frac{T_2}{T_1}\right) - R\ell n\left(\frac{P_2}{P_1}\right)\right]$$

$$= (3)\left[(1005)\ell n\left(\frac{340}{300}\right) - (287)\ell n\left(\frac{100}{500}\right)\right]$$

$$= 1763.1 \text{ W/K}$$

Entropy change of cold air

$$= m_3\left[c_p \ell n\left(\frac{T_3}{T_1}\right) - R\ell n\left(\frac{P_3}{P_1}\right)\right]$$

$$= (2)\left[(1005)\ell n\left(\frac{240}{300}\right) - (287)\ell n\left(\frac{100}{500}\right)\right]$$

$$= 475.30 \text{ W/K}$$

$\therefore$ Entropy generation (s_{gen})

= Total entropy change

= 1763.1 + 475.30

= 2238.4 W/K = 2.24 kW/K

51. As given that, free body diagram of block is

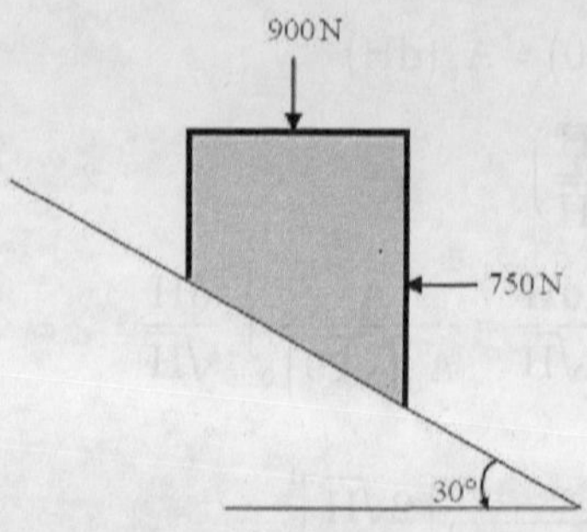

$\Sigma H = 0$

$N \cos 60° - 750 - \mu N \cos 30° = 0$

$$\frac{N}{2} - 750 - \frac{\mu N\sqrt{3}}{2} = 0$$

$$\frac{N}{2}\left(1 - \mu\sqrt{3}\right) = 750 \qquad \text{....(i)}$$

$\Sigma V = 0$

$N \sin 60° - 900 + (\mu N) \sin 30° = 0$

$$\frac{N\sqrt{3}}{2} - 900 + \frac{\mu N}{2} = 0$$

$$\frac{N}{2}\left(\sqrt{3} + \mu\right) = 900 \qquad \text{....(ii)}$$

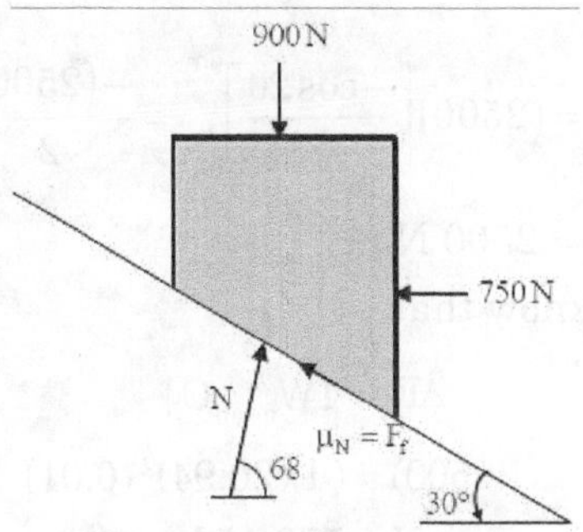

$$\text{Equation (i)/(ii)} = \frac{(1-\mu\sqrt{3})}{(\sqrt{3}+\mu)} = \frac{750}{900}$$

$$(1-\mu\sqrt{3}) = \left(\frac{5}{6}\right)(\sqrt{3}+\mu)$$

$$6-6\sqrt{3}\mu = 5\sqrt{3}+5\mu$$

$$6-5\sqrt{3} = (5+6\sqrt{3})\mu$$

$$\mu = \frac{(6-5\sqrt{3})}{(5+6\sqrt{3})} = -0.17$$

'μ' is negative means our assumed direction of frictional force is not correct.

Hence, answer is $\mu = 0.17$

52. As given figure :

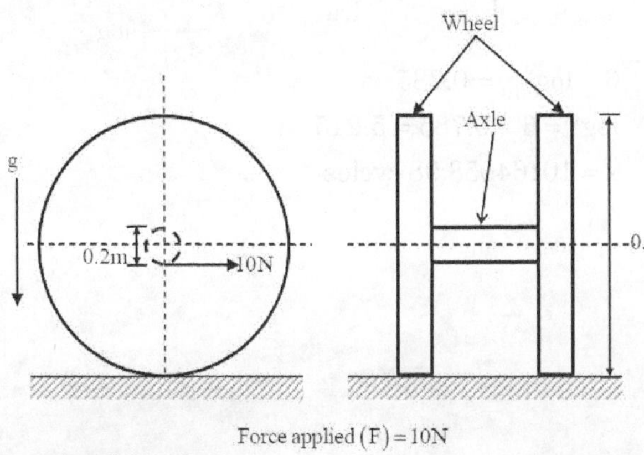

Force applied (F) = 10N

As given that,

Diameter of wheel (D) = 0.8m, Radius of wheel

$(R) = \frac{D}{2} = 0.4\text{m}$

Mess of wheel $(m_w) = 1$ kg

Diameter of axle $(\alpha) = 0.2$m,

Radius of axle $(r) = \frac{d}{2} = \frac{0.2}{2} = 0.1\text{m}$

Mass of axle $(m_a) = 1.5$kg

$g = 9.8$ m/sec^2

Mass moment inertia of wheels and axle is $(I) = 2m_wR^2$ (Neglecting mass moment of inertia of axle)

$I = (2)(1)(0.4)^2 = 0.32\text{kg.m}^2$

Torque applied on wheel axle $(T) = (F)(r)$

$= (10)(0.1) = 1$ N.m

$T = I\alpha$

$$\alpha = \frac{T}{I} = \frac{1}{0.32} \simeq 3.125\ \text{rad/sec}^2$$

Tangential acceleration of wheel axle system $(a_t) = R\alpha$

$a_t = (0.4)(3.125) = 1.25\text{m/sec}^2$

53. As given that,

Area of bending moment diagram between A and B is = 10,000 N.m^2

$E_I = 200 \times 10^6$N.m^2

According to the question:

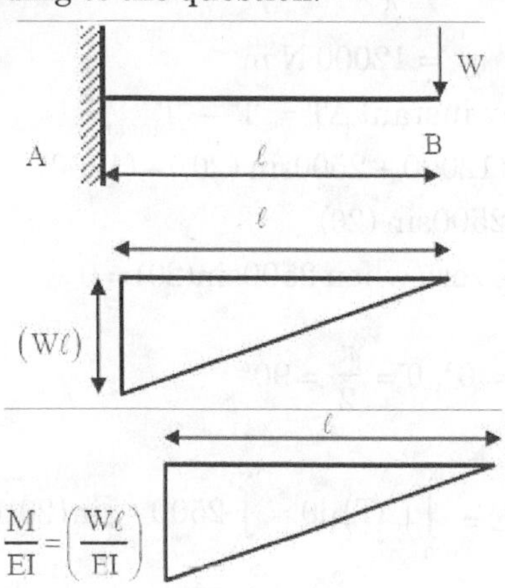

$$\therefore\ 10{,}000 = \frac{1}{2}(W\ell)(\ell)$$

$$10{,}000 = \frac{W\ell^2}{2} \Rightarrow W\ell^2 = 20{,}000\ \text{N.m}^2$$

$\therefore$ Slope at 'B' = Area of $\left(\frac{M}{EI}\right)$ diagram between A and B.

$$= \frac{1}{2}\times\left(\frac{W\ell}{EI}\right)(\ell) = \frac{1}{2}\times\frac{W\ell^2}{EI}$$

$$= \frac{1}{2}\times\frac{20{,}000}{200\times10^6} = 5\times10^{-5}\ \text{radians}$$

= 50 micro radians

54. As given that,

$T(\theta) = 1200 + 2500\sin(2\theta)$

$N_{mean} = 200rpm$

Coefficient of fluctuation of speed $(C_S) = 2 \times \frac{0.5}{100}$

$C_s = 0.01$

$W_{mean} = \frac{2\pi N_{mean}}{60} = \frac{2\pi \times 200}{60} = 20.94$ rad/sec

ΔE = maximum fluctuation in energy

Since T (θ) is a function of 'θ'

$$\text{So, } T_{mean} = \frac{1}{\pi}\int_0^{\pi} T d\theta = \frac{1}{\pi}\int_0^{\pi}(12000 + 2500\sin(2\theta))\,d\theta$$

$$= \frac{1}{\pi}\left[12000(\pi - 0) + \frac{2500}{2}(-\cos 2\theta)_0^{\pi}\right]$$

$$= \frac{1}{\pi}\left[1200\pi + 1250(-\cos 360° + \cos 0°)\right]$$

$= 12000$ N.m

At any instant $\Delta T = T - T_{mean}$

$\Delta T = (12000 + 2500\sin(2\theta)) - (1200)$

$\Delta T = 2500\sin(2\theta)$

'ΔT' is zero when $2500\sin(2\theta) = 0$

$\Rightarrow \theta = 0°, \theta = \frac{\pi}{2} = 90°$

$$\therefore \Delta E = \int_0^{90°}(\Delta T)\,d\theta = \int_0^{90°} 2500 - \sin(2\theta)\,d\theta$$

$$= (2500)\left[\frac{-\cos 2\theta}{2}\right]_0^{90°} = \frac{-(2500)}{2}[-1-1]$$

$= 2500$ N.m

As we know that,

$$\Delta E = IW_{mean}^2 C_S$$

$$(2500) = (I)(20.94)^2(0.01)$$

$$I = 570.15 \text{ kg.m}^2$$

Hence, the minimum mass moment of inertia of flywheel should be 570.15 kg.m².

55. According to question,

⇒ Comparing similar triangle ΔABC and ΔADE

$$\frac{AB}{BC} = \frac{AD}{ED}$$

$$\frac{\log_{10}^{450} - \log_{10}^{150}}{(6-3)} = \frac{\log_{10}^{200} - \log_{10}^{150}}{(6 - \log_{10}^{x})}$$

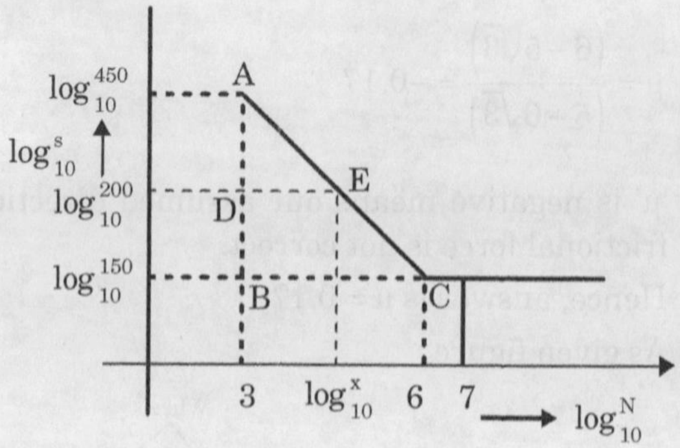

$6 - \log^x{}_{10} = 0.785$

$\log^x = 6 - 0.785 = 5.215$

x = 10164058.98 cycles